AF766079

Units for cross sections

$$\frac{d\sigma}{dt}\left[\frac{\text{mb}}{\text{GeV}^2}\right]=0.3894\frac{d\sigma}{dt}[\text{GeV}^{-4}]=1026.0\frac{d\sigma}{dt}[\text{n.u.}];$$

$$\frac{d\sigma}{d\Omega}\left[\frac{\text{mb}}{\text{sr}}\right]=0.3894\frac{d\sigma}{d\Omega}[\text{GeV}^{-2}]=19.99\frac{d\sigma}{d\Omega}[\text{n.u.}]=\frac{q^2[\text{GeV}^2]}{\pi}\frac{d\sigma}{dt}\left[\frac{\text{mb}}{\text{GeV}^2}\right]$$

$$1\,\text{mb}=10^3\,\mu\text{b}=10^6\,\text{nb}=10^{-27}\,\text{cm}^2.$$

Masses and derived quantities

Pion: $\mu=\mu_\pm=0.13957$ GeV, $\mu_0=0.13496$ GeV.
Nucleon: $m=m_p=0.93828$ GeV, $m_n=0.93957$ GeV.

$$1/\mu=7.1649\,\text{GeV}^{-1},\quad 1/\mu^2=51.336\,\text{GeV}^{-2},\quad m/\mu=6.7227,\quad m^2=45.195\,\mu^2.$$

Compton wavelength: $\lambda\!\!\!^{-}_\pi=\hbar c/\mu c^2=1.4138$ fm, $\lambda\!\!\!^{-2}_\pi=19.989$ mb, $\lambda\!\!\!^{-}_\pi/c=4.716\cdot10^{-24}$ s.

We have always used the proton mass and the charged pion mass in kinematical calculations.

Pion-Nucleon coupling constant and other parameters

The pseudoscalar and pseudovector coupling constants g and f are related by $g^2/4\pi=(2m/\mu)^2f^2$. Taking $f^2=0.079$, we have $g^2/4\pi=14.28$, $g=13.40$, $g^2/m=26.70\,\mu^{-1}=191.3\,\text{GeV}^{-1}$, $g^2/2m^2=1.99\,\mu^{-2}$.

Pion decay constant: $f_\pi=(0.945\pm0.001)\mu=132$ MeV. Axial vector coupling constant: $g_A=1.26\pm0.01$.

Isospin relations

Amplitudes for the reactions $\pi^\pm p\to\pi^\pm p$ and $\pi^- p\to\pi^0 n$

$$A_+=A^{3/2}\qquad\qquad =A^+-A^-=\tfrac{1}{3}(2A^N+A^\Delta),$$

$$A_-=\tfrac{1}{3}(2A^{1/2}+A^{3/2})\qquad =A^++A^-=A^\Delta,$$

$$A_0=-\frac{\sqrt{2}}{3}(A^{1/2}-A^{3/2})=-\sqrt{2}A^-=\frac{\sqrt{2}}{3}(A^N-A^\Delta)=\frac{1}{\sqrt{2}}(A_+-A_-).$$

$I=1/2$ amplitude

$$A^{1/2}=\tfrac{1}{2}(3A_--A_+)=A^++2A^-.$$

Isospin even and odd combinations

$$A^+=\tfrac{1}{2}(A_-+A_+)=\tfrac{1}{3}(A^{1/2}+2A^{3/2});\qquad A^-=\tfrac{1}{2}(A_--A_+)=-\frac{1}{\sqrt{2}}A_0=\tfrac{1}{3}(A^{1/2}-A^{3/2}).$$

Upper indices: 1/2 and 3/2 give the value of I. N and Δ belong to $I_u=1/2$ and 3/2, respectively.

Parameters of meson resonances

	η	ϱ	ω	S*	φ	f	g
Mass [MeV]	548.8	769	782.6	975	1019.61	1273	1691
	±0.6	±3	±0.2	±4	±0.07	±5	±5
Full width [MeV]	$0.83\cdot10^{-3}$	154	9.9	33	4.21	179	200
	$\pm0.12\cdot10^{-3}$	±5	±0.3	±6	±0.13	±20	±20
$I^G(J^P)C_n$	$0^+(0^-)+$	$1^+(1^-)-$	$0^-(1^-)-$	$0^+(0^+)+$	$0^-(1^-)-$	$0^+(2^+)+$	$1^+(3^-)-$

I=isospin, G=G-parity, J=spin, P=space parity, C_n=charge-conjugation parity for the neutral member of the multiplet (Review of Particle Properties: 82 R Roos).

LANDOLT-BÖRNSTEIN

Numerical Data and Functional Relationships in Science and Technology

New Series

Editor in Chief: K.-H. Hellwege

Group I: Nuclear and Particle Physics

Volume 9

Elastic and Charge Exchange Scattering of Elementary Particles

Supplement to Volume I/7 and Extension to High Energies

Subvolume b: Pion Nucleon Scattering

Part 2: Methods and Results of Phenomenological Analyses

G. Höhler

Editor: H. Schopper

Springer-Verlag Berlin · Heidelberg · New York 1983

LANDOLT-BÖRNSTEIN

Zahlenwerte und Funktionen aus Naturwissenschaften und Technik

Neue Serie

Gesamtherausgabe: K.-H. Hellwege

Gruppe I: Kern- und Teilchenphysik

Band 9

Elastische und Ladungsaustausch-Streuung von Elementarteilchen

Ergänzung zu Band I/7 und Erweiterung für hohe Energien

Teilband b: Pion-Nukleon-Streuung

Teil 2: Methoden und Ergebnisse phänomenologischer Analysen

G. Höhler

Herausgeber: H. Schopper

Springer-Verlag Berlin · Heidelberg · New York 1983

CIP-Kurztitelaufnahme der Deutschen Bibliothek. *Zahlenwerte und Funktionen aus Naturwissenschaften und Technik*/Landolt-Börnstein. – Berlin; Heidelberg; New York: Springer. Parallelt.: Numerical data and functional relationships in science and technology.
NE: Landolt, Hans [Begr.]; PT. N.S./Gesamthrsg.: K.-H. Hellwege. Gruppe 1, Kern- und Teilchenphysik. Bd. 9. Elastische und Ladungsaustausch-Streuung von Elementarteilchen: Erg. zu Bd. I/7 u. Erw. für hohe Energien. Teilbd. b. Pion-Nukleon-Streuung/ G. Höhler. Hrsg.: H. Schopper. Teil 2. Methoden und Ergebnisse phänomenologischer Analysen. – 1983.

ISBN 3-540-11282-0 (Berlin, Heidelberg, New York)
ISBN 0-387-11282-0 (New York, Heidelberg, Berlin)

NE: Hellwege, Karl-Heinz [Hrsg.]; Höhler, Gerhard [Mitverf.]; Schopper, Herwig [Hrsg.]

Printed in Germany

Typesetting, printing and bookbinding: Brühlsche Universitätsdruckerei, 6300 Giessen
2163/3020—543210

Vorwort

Band I/9, der Daten über die Elastische und Ladungsaustausch-Streuung von Elementarteilchen enthält, besteht aus zwei Teilbänden a, b, von denen I/9a bereits 1980 erschienen ist.

In Teilband I/9b sind die Daten über die π-N Streuung zusammengestellt. Da dieser Prozess über viele Jahrzehnte das besondere Interesse sowohl der Experimentalphysiker wie der Theoretiker fand, wurde es nötig, den Teilband noch einmal zu unterteilen. Teil I/9b1 (erschienen Ende 1981) enthält die experimentellen Daten. Die π-N Streuung ist einer der wenigen Prozesse, bei denen es gelingt, aus den Experimenten Streuamplituden abzuleiten. Die Zusammenstellung dieser Streuamplituden ist der wesentliche Inhalt dieses (letzten) Teiles I/9b2.

Da die Berechnung der Streuamplituden aus den experimentellen Daten nicht ganz einfach ist und gewisse Annahmen erfordert, hat sich um diesen Problemkreis eine reiche Literatur entwickelt. Im ersten Kapitel wird eine kritische Übersicht über die verschiedenen Methoden der Partialwellenanalyse und der Bestimmung der Streuamplituden gegeben.

Da die Streuamplituden Mittelwerte über die einzelnen Experimente darstellen, sind sie zuverlässiger als die ursprünglichen Daten. Es ist daher für manche Zwecke sinnvoll, die aus den Streuamplituden zurückberechneten meßbaren Größen zu verwenden. Diese Größen sind in einem weiteren Kapitel zusammengestellt. Weiterhin werden Anwendungen der Streuamplituden auf eine Reihe von speziellen Problemen diskutiert.

Sehr nützlich sollte auch die Formelzusammenstellung im Anhang sein, insbesondere bei der Lektüre von Originalarbeiten mit ihren sehr uneinheitlichen Bezeichnungsweisen.

Die kritische und umfassende Zusammenstellung der Ergebnisse und Methoden der π-N Streuung dürfte einmalig sein. Sie wurde nur möglich durch die große Kompetenz und Mühe des Autors, der diesen Problemen einen großen Teil seiner wissenschaftlichen Aktivität gewidmet hat. Alle bei der Fertigstellung des Bandes aufgetretenen technischen und administrativen Schwierigkeiten wurden in der üblichen souveränen Weise durch den Gesamtherausgeber, die Redaktion, besonders Herrn Dr. W. Polzin und Frau D. Dolle, und den Verlag überwunden. Ihnen allen gebührt besonderer Dank. Dieser Bandteil wurde, wie alle Bände des Landolt-Börnstein, ohne finanzielle Unterstützung von anderer Seite veröffentlicht.

Genf, Oktober 1982 **Der Herausgeber**

Preface

Volume I/9 containing data on Elastic and Charge Exchange Scattering of Elementary Particles, consists of two subvolumes a, b, the first of which, I/9a, is already available since 1980.

In subvolume I/9b the data of the π-N scattering have been compiled. Since this process has attracted the special interest of both experimentalists and theorists during many decades, it became necessary to subdivide this subvolume into two parts. Part I/9b1 (published late 1981) contains the experimental data. π-N scattering is one of the few processes for which it is possible to derive scattering amplitudes from the experiments. The compilation of these scattering amplitudes is the essential content of part I/9b2.

Since the derivation of the scattering amplitudes from the experimental data is not simple and requires certain assumptions, a comprehensive literature developed around this field of problems. Hence a critical review of the various methods of partial wave analysis and the determination of scattering amplitudes is presented in the first chapter.

Since the scattering amplitudes represent average values over individual experiments, they are more reliable than the original data. For certain purposes it seems useful therefore to utilize measurable quantities reconstructed from the scattering amplitudes. These quantities are compiled in another chapter. In addition, applications of the scattering amplitudes to a number of special problems are discussed.

The compilation of formulae in the Appendix should turn out to be very useful, in particular facilitating the reading of original papers with their rather non-uniform notation.

The critical and comprehensive review of the results and methods of π-N scattering should be regarded as unique. It became possible only by the remarkable competence and efforts of the author who has dedicated a large part of his scientific activity to these problems. All technical and administrative difficulties which arose during the production of this volume were mastered in the usual expert way by the editor in chief, the editorial office, especially Dr. W. Polzin and Frau D. Dolle, and the publisher. Special thanks are due to all of them. This part of the volume, like all other volumes of Landolt-Börnstein, has been published without financial support from any outside source.

Geneva, October 1982 **The Editor**

Contents

Pion-nucleon scattering

Subvolume b2: Methods and results of phenomenological analyses

Appendix: Collection of pion-nucleon scattering formulas

Subvolume b1: Tables of data

2 Methods and results of phenomenological analyses

2.0 Introduction

Volumes I/7 and I/9b1 contain tables of the experimental data on elastic and charge-exchange pion-nucleon scattering. The present volume deals with the determination of scattering amplitudes from these data and applications within the framework of phenomenological theories.

The methods for partial wave and amplitude analysis are treated in Sect. 2.1 and results for various scattering amplitudes and for measurable quantities reconstructed from them are presented in Sects. 2.2 and 2.3, respectively. These tables give a condensed version of the experimental information.

Scattering amplitudes can be determined from data only if one uses in addition a theoretical input. Besides general theoretical principles (Lorentz invariance, T, C, and P-invariance, unitarity, fixed-t analyticity) one employs Mandelstam analyticity, which has the status of a working hypothesis, and isospin invariance, which is thought to be a good approximation for the strong interaction amplitudes.

Thus one could argue that the theoretical input leads to a model dependence of the result, which is due for instance to the necessary assumptions on the high energy behaviour and on the smoothing procedure in applications of the analyticity constraint. The magnitude of the resulting uncertainty can be estimated by repeating the analysis with reasonable modifications of these assumptions. It turns out to be much less serious than the model dependence in those cases, for which the analysis is based on a parametrization derived from a specific phenomenological model.

For a comparison with theoretical predictions, it is not sufficient to have long tables of four complex-valued amplitudes, which depend on two variables. One needs in addition a description of the amplitudes in certain kinematical regions in terms of a relatively small number of parameters, which have a direct physical interpretation and can be compared with the values calculated from models.

The most important parameters of the pion-nucleon system are those of the poles and zeros of the scattering amplitudes. The pole parameters are listed and discussed in Sects. 2.4.1 and 2.4.2. They are the "experimental" quantities, which are used in tests of various quark models of the nucleon. The properties of the zero trajectories (Sect. 2.4.3) show a remarkable similarity to the prediction from Veneziano-type dual models.

Section 2.4.4 deals with the saturation of isospin bounds. Kinematical regions in which these saturations occur are of interest in connection with the search for small violations of isospin invariance.

Analytic continuations of scattering amplitudes into unphysical regions have to be performed for several reasons, for instance for the determination of $\pi\pi N\bar{N}$ amplitudes, for the test of predictions derived from Current Algebra and PCAC and for the study of zero trajectories. A brief treatment of $\pi\pi$-scattering at low energies is included, because $\pi\pi$-phase shifts are needed as part of the input (Sects. 2.4.5 and 2.4.6). The result is given not only in the form of tables (Tables 2.2.3.2 and 2.4.6.1) but also by the coefficients of the subthreshold expansion (Sect. 2.4.7). The coefficients of the effective range expansion at threshold are in practice also determined by a continuation, because it is not possible to obtain accurate values of the strong interaction amplitudes from πN scattering experiments at very low energies (Sect. 2.4.7).

Three applications of the continued amplitudes are treated in Sect. 2.5:

i) Many authors have proposed models for the πN system at low energies, which incorporate the predictions derived from Current Algebra and PCAC. These models have to be compared with amplitudes in the threshold and subthreshold regions (Sect. 2.5.1). One of the reasons why these models are studied is the possibility to calculate off-shell extrapolations of πN scattering amplitudes, which are needed in nuclear physics.

Another problem of current interest is a test of the prediction for the πN sigma term from quantum chromodynamics.

ii) Important properties of the electromagnetic nucleon form factors can be determined only if one applies the dispersion approach and includes the $I=J=1$ $\pi\pi N\bar{N}$ partial waves of Sect. 2.4.6 as part of the input (Sect. 2.5.2). Since the pion form factor is also needed, we have added a summary on recent experimental results and investigations of its dispersion relation.

iii) Originally, simple one-boson-exchange models were used for the description of two-pion-exchange in nucleon-nucleon scattering. The results for the $\pi\pi N\bar{N}$ partial waves are the basic input for a quantitative treatment (Sect. 2.5.3).

Since the physical range $t>4m^2$ occurs in dispersion analyses of $\pi\pi N\bar{N}$-amplitudes, we have added in Sect. 2.5.4 a short summary of the $\bar{N}N\leftrightarrow\pi\pi$ experiments available in this region, mentioning the first attempts to determine the partial wave amplitudes.

Section 2.6 is devoted to the analysis of πN scattering data at high energies. It can be expected that, in the coming years, a considerable part of the experimental and theoretical activity in pion-nucleon physics will concentrate in this field. Therefore, I have summarized the most popular models. pp-scattering is included because of the close similarity of the diffraction phenomena.

The last section contains a list of references for books and reviews (Sect. 3.1), original papers and recent preprints (Sect. 3.2), and papers on pion-nucleon scattering experiments (Sect. 3.3, see also Table 1.2.1 in Volume I/9b1).

A collection of pion-nucleon scattering formulas has been added in an Appendix, because it is my experience that one can lose considerable time on more or less trivial but tedious calculations, if it is necessary to work with formulas taken from various books, reviews and original papers. The reason is that the authors employ different conventions and notations, which are not always explained in full detail, or at a place where the definition is not easily found. Furthermore, the number of misprints is far from negligible.

To my knowledge, there exists no single book which covers all topics treated in the Appendix. Of course, printing and other errors cannot be avoided, but the present volume has a useful redundancy: it contains many numerical tables and figures, which have been calculated from the formulas of the Appendix. Furthermore, a shorter version of the collection of formulas has been used in our group for many years.

A reader who is interested in the derivation of a formula can find the reference in Sect. A.12.

The present volume has been written for physicists who are familiar with the main experimental facts and at least with the elementary parts of the theoretical formalism, which is described in several excellent textbooks and monographs. It will be useful for experimentalists who are studying possible interpretations of their results or planning new experiments, and also for theorists who want to test phenomenological models or rigorous predictions.

Part of the contents will be of interest to physicists who are working in other fields, where πN or $\pi\pi N\bar{N}$-amplitudes are needed, for instance in nuclear physics or in investigations of pionic atoms or of nucleon form factors.

The chapters on pion-nucleon scattering in the existing textbooks and monographs were written about 10 years ago or even earlier. Therefore, this book will be valuable for graduate students who want to become acquainted with the more recent developments.

Volume I/9b1 includes the results of all experimental investigations on pion-nucleon scattering which have been performed since the publication of Volume I/7. Because of the very large number of phenomenological and theoretical papers in this field, only a small fraction could be taken into account in Volume I/9b2. My choice was to give a fairly complete treatment of the phenomenological analysis of the experimental data, which leads to results for the scattering amplitudes and for various parameters of the pion-nucleon system. Special attention has been paid to applications of dispersion relations as constraints and for analytic continuations.

Since the work done in the fifties and sixties has been described in several excellent books and in many reviews and summer school lectures, I have concentrated on the progress made during the last decade.

It would have been unsatisfactory to present the amplitudes and the parameters without explaining why such a great effort has been made for their determination. Therefore, I have also discussed the present status of various models which have had some success in explaining properties of the pion-nucleon system, and a few applications of πN and $\pi\pi N\bar{N}$-amplitudes in other fields. The choice of topics reflects my own interests and limitations.

One of my aims was to present a "guide to the literature" in the following sense. In Sects. 2 and A.12, I have not only mentioned the publications actually used in writing this book but also many others, which would have to be discussed in a more detailed review. In particular, I have added references of the most recent papers, even if I have not been able to give a comment, because my manuscript had already been submitted. Furthermore, Sect. 3.1 contains a fairly complete list of books and reviews, which are related to the topics treated in this volume.

A well-known difficulty of books of this type is that parts will be outdated after a period which is appreciably smaller than the time interval between two editions. One can extend the space of time during which the book fulfills its purpose and is actually used by physicists working in this field by producing an informal "Supplement", which contains updated versions of the main tables and lists, corrections of misprints and short remarks on errors or omissions, which have been brought to the attention of the author. Updated versions of the tables will be prepared in any case for the continuation of the work of our group and we are willing to distribute copies on request.

In the fifties and sixties, pion-nucleon scattering was one of the main topics at international conferences on high energy physics. Later, many physicists left this field and started to work on other problems which seemed to be more promising.

Nowadays, interest is increasing again because of first attempts to calculate parameters of the pion-nucleon system using methods based on quantum chromodynamics, which is believed to have a good chance to be the correct theory of strong interactions. At present, the best evidence in favour of this theory comes from reactions other than hadron-hadron scattering, e.g. electron-positron interactions. However, as soon as the serious difficulties in deriving quantitative predictions for the interactions between hadrons have been overcome, pion-nucleon scattering amplitudes will play an important role for a critical test. They are known with a higher accuracy than other strong scattering amplitudes and furthermore, they contain the experimental information on the excited states of the nucleon. A calculation of these states is one of the fundamental problems in a theory of strong interactions.

Acknowledgements

Our group has been working on pion-nucleon scattering and related topics for two decades. During this period I have had several generations of coworkers who investigated very actively various problems in this field. Among them are: J. Baacke, E. Borie, K. Dietz, G. Ebel, J. Engels, G. von Gehlen, W. Grein, M. Hutt, H.P. Jakob, F. Kaiser, R. Koch, P. Kroll, E. Pietarinen, I. Sabba-Stefanescu, W. Schmidt, F. Steiner, and R. Strauss. I wish to express my gratitude for many discussions, which helped to clarify the ideas presented in this book.

The treatment of partial wave analysis and of analytic continuations was strongly influenced by the work of E. Pietarinen. When he left Karlsruhe and changed his field, the analysis was continued by R. Koch and the study of related mathematical problems by I. Sabba-Stefanescu.

I am grateful to J. Hamilton for a stimulating exchange of ideas on dispersion methods over a long period and to R.E. Cutkosky and R. Kelly for valuable discussions on partial wave analysis as well as for sending us tables of their amplitudes and amalgamated data on magnetic tapes.

It is a great pleasure to thank all those physicists who sent us their unpublished results and contributed to my understanding of special questions. I can only mention a few: D. Atkinson, D. Chew, T. Ekelöf, J. Fischer, J. Gasser, A. Hendry, P. Koehler, P. Kroll, C.B. Lang, G.C. Oades, K.P. Pretzl, M. Scadron, and B. and F. Schrempp.

The writing of this book would have been much more difficult without the untiring help of my coworkers M. Hutt, F. Kaiser, R. Koch, and I. Sabba-Stefanescu, who have read part of the manuscript, contributed figures and tables and noticed a number of misprints and errors in the collection of formulas. I was very glad to have had the possibility to discuss various questions with my colleagues E. Borie, H. Genz, H. Pilkuhn, and H.M. Staudenmaier. Furthermore, I also wish to thank S. Kunze for her assistance with the preparation of the list of references on a microcomputer as well as J.G. Otto and J. Stahov for supplying some figures and tables.

The permission of the Fachinformationszentrum and the Kernforschungszentrum Karlsruhe to include several figures and tables, which belong to earlier publications, is gratefully acknowledged.

The search for references was simplified by using the High Energy Physics Index compiled at DESY.

I am deeply indebted to the editor, H. Schopper, for supporting my proposal to include a treatment of the phenomenological analysis and for his advice and encouragement during the long time it took to complete this project.

This book is dedicated to my wife Hildegard and I wish to thank her for her patience.

G. H.

2.1 Methods of partial wave and amplitude analysis

The aim of partial wave analysis (which is also called "phase shift analysis") is to determine the partial wave amplitudes from differential cross section and polarization data. This is equivalent to a determination of the invariant or helicity amplitudes at all angles.

At high energies, a partial wave analysis is not possible, because data for the polarization parameter P are available only in small momentum transfer ranges near the forward and backward directions (see Sect. 1.2.2, Vol. I/9b 1). However, one has the advantage that spin-rotation data exist at a few energies in small near-forward t-intervals (Sect. 1.6). If these data are included, one can determine the invariant or helicity amplitudes up to a common phase factor. This type of analysis is called "amplitude analysis" (Sect. 2.1.15).

An improved version of amplitude analysis, which can be carried out in the range $0 \geqq t \gtrsim -0.5\,\text{GeV}^2$, uses constraints from fixed-t analyticity. The advantage is that the invariant amplitudes are determined completely at all energies. The input consists of all data in this strip and one has to make assumptions on the asymptotic behaviour of the amplitudes. At low and intermediate energies, one can use amplitudes reconstructed from phase shifts instead of data. (Sect. 2.1.16).

Although a complete partial wave analysis at high energies (for instance at 10 GeV) is not possible due to the lack of data, one can use the available data and the result of the last mentioned method. The projection integrals are dominated by the contribution from the range $0 \geqq t < -0.5\,\text{GeV}^2$. It is possible to estimate the errors, using bounds derived from the small values of the differential cross section at $t < -0.5\,\text{GeV}^2$ (76 L PIETARINEN-4).

We shall start with some problems which occur in all partial wave analyses (Sects. 2.1.1 to 2.1.3). Then we shall discuss the methods and results of the CERN 72 and SACLAY 73 analyses and subsequent work. The historical development of partial wave analysis is treated for instance in the following books and reviews: 56 R BETHE-DE HOFFMANN (Vol. II), 67 R DONNACHIE, 68 R LOVELACE, 69 R JOHNSOHN-STEINER, 69 R LOVELACE, 72 R LOVELACE, 73 R BRANSDEN-MOORHOUSE, 73 R DONNACHIE. See also 65 L ROPER-1 and 2.

2.1.1 General remarks on partial wave and amplitude analyses

If isospin invariance is assumed to be valid, all elastic and charge-exchange pion-nucleon reactions can be described by 4 invariant amplitudes, for instance by $A^{\pm}(s,t)$, $B^{\pm}(s,t)$, i.e. by 8 real numbers for each pair (s,t) of kinematic variables. If one wants to discuss the relation of these 8 numbers to measurable quantities, it is useful to work with transversity amplitudes $F_{+}(\pm)$, $F_{-}(\pm)$, $F_{0}(\pm)$, Eq. (A.2.4), whose modulus is determined by differential cross section and polarization measurements

$$|F(\pm)|^2 = \frac{d\sigma}{d\Omega}(1 \pm P). \tag{1.1.1}$$

The subscripts $+$, $-$, 0 refer to the three reactions

$$\pi^+ p \to \pi^+ p, \quad \pi^- p \to \pi^- p, \quad \pi^- p \to \pi^0 n,$$

which can be measured with a good accuracy.

From differential cross section data and polarization parameters for all three reactions at a given value of (s,t), one can construct the two isospin triangles of Fig. A.4.2. There remain three types of ambiguities:

i) the relative angle between the two triangles is not determined,

ii) each of the triangles can be reflected at one of its sides,

iii) a common phase factor of all amplitudes remains arbitrary, i.e. the system of two triangles can be rotated by an arbitrary angle.

If one spin-rotation parameter is measured, for instance R_+, the spin-rotation angle β_+ in Fig. A.4.2 is known, except for a two-fold ambiguity, from Eq. (A.4.2).

If two *suitable* spin-rotation parameters are measured, the ambiguities i) and ii) can be resolved completely as described by DEAN et al. (72 L). Their result differs from that obtained in 67 L BILENKI.

However, the common phase factor mentioned in iii) cannot be measured in elastic or charge-exchange pion-nucleon scattering experiments, except for two cases. At $t=0$, the imaginary part of the forward amplitude follows from the total cross section according to the optical theorem, and the real part can be derived from a Coulomb interference experiment (Sect. 2.6.2). The second case is a determination of the phase of the no-flip amplitude G for $\pi^- p$ elastic scattering from the angular dependence of the cusp structure of the differential cross section at the η-production threshold (see Sect. 2.1.13).

In principle, the absolute phase can also be determined by analyzing pion-deuteron scattering data in terms of pion-nucleon scattering amplitudes. However, the accuracy is poor, since one needs a model for the deuteron (75 L ZOVKO). – In 80 L CHAN the phase is constructed by iteration of unitarity bounds.

One should note that the absolute phase of a scattering amplitude plays a role different from that of the phase of a wave function in quantum mechanics. The former corresponds to the *relative phase* of the incident and scattered waves. A general discussion is given by GOLDBERGER (63 L) and by BOWCOCK-BURKHARDT (75 R).

Since spin-rotation data exist only at 6, 16 and near 40 GeV/c (Sect. 1.6), the usual situation in partial wave analysis is that at each value of the kinematic variables (s,t) or $(k, \cos\theta)$, one has to determine 4 complex amplitudes (i.e. 8 real numbers) from an input which consists of at most 6 real numbers ($d\sigma/d\Omega$ and P for all three reactions). The missing two quantities belong to the two angles in i) and iii). In addition one has to resolve the discrete ambiguities ii) mentioned above.

From these remarks it is obvious that all methods for the determination of partial waves or amplitudes require a *theoretical input*.

2.1.2 Elastic phase shift analysis

Several authors studied purely elastic scattering of spinless particles, assuming that data of infinite accuracy are available at all angles. Even in this simplified case, where unitarity is a powerful constraint, the problem of partial wave analysis cannot be solved completely. Aside from a trivial ambiguity, which corresponds to a sign change of all phase shifts, the solution is unique near threshold, where only the S-wave is large and all other phase shifts are small. At higher energies, a non-trivial two-fold ambiguity was shown to exist in a case where only S, P, and D-waves are different from zero (66 L CRICHTON).

Ambiguities can be studied by considering a product representation for the scattering amplitude (69 L and 74 L GERSTEN).

$$F(z) = \prod_i \frac{z - z_i}{1 - z_i} f(z), \tag{1.2.1}$$

where $f(z)$ has no zeros. The modulus of F and therefore the differential cross section is unchanged, if one or several zeros z_i are replaced by their complex conjugates. The problem is to determine the conditions under which the new amplitude is compatible with the unitarity constraint.

The progress in this field until 1975 is reviewed in 74 R MARTIN and 75 R BOWCOCK-BURKHARDT. See also 73 L ATKINSON-1 and 76 L BRUNET.

In 78 L ATKINSON the authors studied the CRICHTON ambiguity in a case which is more realistic than the earlier ones. They considered an infinite number of partial waves which fall off exponentially as the angular momentum tends to infinity, limiting their search for an alternative unitary amplitude to the case that only one of the zeros is complex conjugated. The result is that the ambiguities are much more common than one might have thought on the basis of the earlier examples.

Conclusion: even in the elastic region, a unique determination of the amplitude from data is possible only if one uses further theoretical constraints in addition to unitarity (except near threshold).

2.1.3 Inelastic partial wave analysis — the continuum ambiguity

Above the threshold for the production of a second pion, the situation changes considerably, because the unitarity condition, Eq. (A.5.20), becomes an *inequality constraint*: $\mathrm{Im}\, T_{\ell\pm} \geqq |T_{\ell\pm}|^2$. It states that the vector $T_{\ell\pm}$ in the Argand diagram, Fig. A.3.1, must remain *inside or upon* the unitarity circle.

If one starts from a partial wave solution at a certain energy, one can derive a *continuum* of solutions, which belong exactly to the same data, by multiplying both invariant or transversity amplitudes with the above mentioned common phase factor $\phi(s,t)$.

In general, spin-rotation data do not exist. Then, one even has a *two-dimensional continuum ambiguity*, the second unknown quantity being the spin-rotation angle β of Eq. A.4.2.

The real functions $\phi(s,t)$ and $\beta(s,t)$ are restricted by unitarity and by the "short range nature" of the strong interaction, which is expressed mathematically by the location of the nearest singularity of the scattering amplitude at fixed s in the z-plane.

In partial wave analysis, it is easy to impose unitarity. The second condition has also been taken into account in the CMU–LBL (Sect. 2.1.9) and KH (Sect. 2.1.7) analyses. However, one should note that it leads only to a statement for partial waves of angular momentum $\ell > L$, Eq. (A. 5.39), where the value of L is not known. There is no reason to believe that the exponential decrease starts *immediately* after the ℓ-value of a peripheral resonance.

The continuum of solutions is represented graphically by islands around the original points in the Argand diagrams of the partial waves. The existence of disconnected islands cannot be excluded.

Up to now, we have considered islands belonging to infinitely accurate data. The islands are enlarged by the experimental errors.

The important question is now, how large these islands are in the case of πN scattering. In particular, one would like to know, whether the continuum ambiguity can lead to doubts concerning the existence of resonances.

A careful numerical study of the continuum ambiguity has been performed by Atkinson, Heemskerk, and Swierstra (76 L Atkinson), using the result of an earlier paper (73 L Atkinson-2). The authors started from the SACLAY solution for π^+p scattering and added higher partial waves of reasonable magnitudes, taking into account the unitarity constraint. They found ambiguity patches which, in some cases, are comparable with the size of structures ascribed to weak resonances, although the experimental errors were not yet taken into account, (see also the reviews 73 R and 76 R Atkinson, 73 R Burkhardt, and 75 R Bowcock-Burkhardt, and the papers 75 L Heemskerk and 76 L Kok).

The work of Atkinson et al. has recently been criticized by D. Chew and Ely (81 L Chew) but, in my opinion, it is difficult to understand or to accept the arguments of these authors. Already the title ("Is there a continuum ambiguity for elastic πN amplitudes?") looks strange, since the existence of this ambiguity cannot be doubted. At each kinematical point (s,t) one has at most two real values from data ($d\sigma/d\Omega$ and P) and one derives two complex-valued amplitudes. As mentioned above, there is a two-dimensional continuum ambiguity due to the unknown overall phase and to the not yet measured spin-rotation angle.

The fact that the results of two phase shift analyses lead to a small influence of the high partial waves on the phase of the transversity amplitudes is an empirical observation which, apparently, has been made so far only for one of the 3 reactions at one energy. It has no relation to the *existence* of the continuum ambiguity.

Some of the conclusions are drawn from the exponential decrease of the partial wave expansion for ℓ-values just above the angular momentum of a peripheral resonance. As mentioned above, the exponential decrease can be proven only for "large" ℓ. — Other conclusions are derived from the assumption that the high partial waves are approximately real. Again, this can be wrong. In the inelastic region, the behaviour depends on the sign and magnitude of the left hand cut contributions in the partial wave dispersion relation, which can cancel the contribution from the rescattering integral, leading to a predominantly imaginary partial wave. This is observed in some results of partial wave analyses.

Reduction of the ambiguity by analyticity constraints

In the spinless case, Burkhardt and Martin (75 L) were able to show that a unique solution can be obtained if one incorporates fixed-t analyticity, $s-u$ crossing symmetry, some t-plane analyticity, positivity and the S-wave scattering length.

Pietarinens work (76 L-3) is closely related to this approach, but he considers a situation which is nearer to that in a practical partial wave analysis. He uses special transversity amplitudes $g(\pm;\tilde{\nu})$, Eq. (A.2.17), whose modulus is assumed to be known in the physical region from data in the s- and u-channels. The variable $\tilde{\nu} \equiv \nu/\bar{\nu}$ [see Eq. (A.2.19) for $\bar{\nu}$], is defined in such a way that the line $t=$const intersects the backward directions at $|\tilde{\nu}|=1$. Between the thresholds and $\tilde{\nu}=\pm 1$, it is assumed that Img can be obtained from an analytic continuation of the absorptive parts.

The idealized mathematical problem is now to find all functions $g(\tilde{\nu})$, which have the same modulus at $|\tilde{\nu}|>1$ and the same imaginary parts in the range $-1<\tilde{\nu}<+1$. The solution is constructed in analogy to that for the N/D-problem for partial wave amplitudes. There remains an ambiguity, but it is shown how all solutions can be constructed explicitly, starting from a given solution. See also the discussion in 76 R Atkinson and in 76 L Pietarinen-1.

In recent papers, Sabba-Stefanescu (80 L-2, 81 L, 82 L-1 and 2) investigated the conditions under which a partial wave analysis with analyticity constraints can lead to a unique solution, paying attention in particular to consequences of isospin invariance.

The aim is to determine amplitudes which fulfill the following conditions:

i) All available data ($d\sigma/d\Omega$, P, σ^{tot}) are reproduced within their errors;

ii) isospin invariance is fulfilled;

iii) the unitarity condition is fulfilled for all partial waves;

iv) the invariant amplitudes have the correct analyticity properties in s at fixed t;

v) the amplitudes at fixed s are analytic in $\cos\theta$ in the small Lehmann ellipse.

It follows from condition v) that the area of the isospin triangle is an analytic function of $\cos\theta$ in the small Lehmann ellipse (apart from a cut at $-1 \leqq \cos\theta \leqq +1$). This property is important for a reduction of the fourfold discrete ambiguity mentioned in Sect. 2.1.1. The ambiguity belongs to reflections of the triangles at one of their sides, which can also be described by a change of sign of the area of the triangle. If the transversity amplitudes, and therefore also the signs of the areas of the triangles, are known at $t=0$, they follow at $t<0$ by analytic continuation.

Assuming some properties of the invariant amplitudes (see below), which are fulfilled for instance in the usual diffraction models, SABBA-STEFANESCU shows that the phases of the transversity amplitudes $g(\pm)$, Eq. (A.2.17), are determined completely and in a stable way in an interval $0>t>t_0$ by the data i) and the conditions ii) to v).

The assumptions are:

(a) $|A^+(\nu,t)/D^+(\nu,t)|<$const at all $\nu>\nu_0$ in an interval $0\geqq t>t_0$. The analysis of the spin-rotation data shows that the ratio is even decreasing at high energies (Sect. 2.6.3.8).

(b) $|\mathrm{Re}\,C^+(\nu,t)/\mathrm{Im}\,C^+(\nu,t)|<$const at all $\nu>\nu_0$ in the interval $0\geqq t>t_0$.

(c) In the physical region it does not happen that one of the isospin triangles reduces to a point.

(d) $|g^+(\pm;\nu,t)/\nu|^2 \approx 64\pi m^2(1\pm P^+)d\sigma^+/dt \geqq \mathrm{const}\cdot|\nu|^{\gamma-2}$ at all $\nu>\nu_0$ in an interval $0\geqq t>t_0$; $\gamma>0$.

(e) $C_+(\nu,0)$ does not vanish at symmetrical points $(\pm\nu)$ in the complex ν-plane.

The method can be extended to the interval $0\geqq t>-4m\mu\approx-0.5\,\mathrm{GeV}^2$.

We conclude that the constraints from fixed-t analyticity and from isospin invariance are strong enough to resolve the ambiguities of partial wave analysis. However, there remain of course uncertainties resulting from experimental errors and, frequently more important, from discrepancies between different data sets

2.1.4 Classification of partial wave analyses

The different partial wave analyses can be classified according to the following points of view:

i) Treatment of the high partial waves. In most of the analyses the partial wave expansion is cut off at a certain value of the angular momentum $\ell=\ell_{max}$. The highest partial waves are sometimes not determined from the data but taken from theoretical predictions. Alternatively, one can use a soft cut-off, admitting a rapidly decreasing tail of high partial waves.

ii) Energy dependent and energy independent analyses. In energy-dependent analyses the energy dependence of all partial waves is expressed by parametrizations and the parameters are determined simultaneously by a fit to all data in the energy interval under consideration.

In energy-independent analyses one chooses a sequence of suitable energy bins and shifts the data to the centers of these bins. A partial wave analysis is carried out separately at each energy. It leads to many solutions which describe the data equally well. Finally, one starts with the unique solution at low energies in the elastic region and applies an empirical prescription or a theoretical constraint, which selects one of the solutions at each of the higher energies.

In the sixties and early seventies several authors applied the "shortest path method" which demands that the energy dependence of all partial waves should be as smooth as possible. Improved techniques made it feasible to use instead analyticity constraints. Another useful constraint is to demand the continuity of the zero trajectories or, in a stronger version, that these trajectories have the correct analytic behaviour.

Some determinations of πN amplitudes are not based on new analyses of the data but on fits to amplitudes derived in several earlier partial wave analyses (75 L LANGBEIN, 78 L ROWE). It is difficult to see the usefulness of this procedure, since the old data occur in all analyses and get therefore a larger influence. Furthermore, it does not make sense to assign a weight to different methods. The idea of forming a "world average" is reasonable for data like masses or lifetimes, but not for results of complicated phenomenological analyses which have been carried out in quite different ways.

It would be a major progress to work with two-variable expansions, but up to now there exist only proposals [PRESNAJDER (74 L, 76 L), 76 L BYSTRICKY, 80 L DAUMENS].

Table 2.1.4.1 Survey of partial wave analyses. The list only includes results published in 1972 or later.

	Lab. momentum range [GeV/c]	Mass range [GeV]	Cut-off of p.w. expansion	Method and constraints	Ref.
CERN 72	0···2.12	1.08···2.2	$\ell \leqq 5$	Constraints from forward and partial wave dispersion relations. Shortest path method	72 L Almehed
SACLAY 73	0···2.85	1.08···2.5	$\ell \leqq 6$	Forward dispersion relation. Shortest path method	76 L Ayed
LONDON 73	0.18···0.43	1.15···1.32	$\ell \leqq 3$	Forward dispersion relation for flip and no-flip amplitudes. Theoretical D- and F-waves	73 L Carter-1
KH 78	0···10	1.08···4.4	soft	Fixed-t and fixed c.m. angle dispersion relations	79 R Höhler-Kaiser-Koch-Pietarinen
LENINGRAD 78	0.34···0.83	1.26···1.57	$\ell \leqq 4$	Shortest path method	78 L Abaev
CMU-LBL 79	0.43···2.0	1.32···2.16	soft	Dispersion relations along hyperbolas. Models for singularities in the z-plane. Data amalgamation	79 L Cutkosky-1
VPI 80	0···0.47	1.08···1.35	$\ell \leqq 3$	Energy-dependent empirical parametrization. Forward dispersion relation	80 L Zidell-1
CHEW 80	0.6···2.3	1.43···2.9	$J \leqq \frac{7}{2}, \frac{11}{2}$	Barrelet method	80 L Chew
HENDRY 80	1.6···10	2.0···4.4		1st step: fit to ansatz for the partial waves; 2nd step: small variation of partial waves	80 L Hendry
KH 80	0···0.5	1.08···1.37	$\ell \leqq 3$	Updated version of KH 78	80 L Koch
SIN 81	0.35, 0.41, 0.43		$\ell \leqq 3$	Energy-independent analysis	81 L Alder

2.1.5 The SACLAY and CERN phase shift analyses

The pioneering work on phase shift analysis in the early sixties is described in the reviews mentioned in the introduction to Sect. 2.1. We shall treat only the methods of the SACLAY and CERN analyses which were developed in 1965–1973. Both analyses were very successful in determining the excited states of the nucleon (cf. Table 2.4.1.4).

The *SACLAY method* starts with the preparation of a relatively coherent subset of about 6000 data points, renormalizing the cross sections if necessary for a smooth energy dependence. Legendre expansions for $d\sigma/d\Omega$, Eq. (A.4.21), and for the parameter P were used for this purpose and also for the interpolation to 43 selected energy values. Next, all acceptable phase shift solutions were determined at each energy, truncating the partial wave expansion at $\ell_{max}=6$ (I-waves) above a total c.m. energy $W=1900$ MeV and at smaller ℓ-values at lower energies. Finally, a unique partial wave solution was derived, starting from the unique low energy solution and applying a shortest path method.

An evaluation of the forward dispersion relation for the amplitudes $C_{\pm}(s,0)$ was taken as part of the input. The dispersion relation for $B_{\pm}(s,0)$ was evaluated in order to test the compatibility with the phase shift solution and to determine the coupling constant f^2, using the method of 64 L Hueper. The result $f^2=0.074\pm0.001$ is at variance with earlier and also with subsequent determinations based on other phase shifts (Sect. 2.4.2.1, see also Fig. 4.2.1). There are also some discrepancies with fixed-t dispersion relations above 1 GeV/c (74 L Hecht).

Details of the SACLAY analysis can be found in 70 L and 76 L Ayed. One should note that the results of 76 L Ayed are based on the data available in 1973.

The *CERN analysis* used a considerably larger data set (almost 12000 data points at 84 energies in the latest version). The method starts with a random search, which leads to a large number of solutions at each energy (30 or more in the high energy part of the range). Next, a shortest path method is used in order to link the solutions at different energies.

The phase shifts that emerge from this linking process are still too bumpy to read off resonance parameters. They are smoothed by imposing constraints from partial wave dispersion relations. Unfortunately, the errors of the different partial waves are strongly correlated. As a consequence, satisfactory smooth fits to the separate partial waves lead to totally unacceptable results, if one calculates the experimental quantities.

One could hope that the situation improves, if the smoothed partial waves are used as the starting values for a new phase shift analysis, but it turned out that this is not true. Therefore, the CERN group presented two solutions

i) "CERN TH", which fulfills the partial wave dispersion relations, but does not give a satisfactory fit to the data,

ii) "CERN EXP", which was obtained from a fit to the experimental data and additional "theoretical data" calculated from the "CERN TH" solution. The errors of the latter "data" were enlarged by a suitable factor, such that a good fit to the experimental data was possible.

The discrepancy between the solutions "CERN TH" and "CERN EXP" was a technical problem and had nothing to do with difficulties to fulfill the analyticity constraint. This is evident from the fact that the KH 80 solution not only gives a good fit to the data, but also is compatible with the partial wave dispersion relations (80 L Hutt).

One should notice that Almehed and Lovelace (72 L) have not recalculated the left hand cuts from their solution. Instead, they used an evaluation made in 1963 from the phase shifts available at that time (64 L Donnachie).

Hecht, Jakob, and Kroll (74 L) found appreciable systematic discrepancies of the CERN 72 solution with fixed-t analyticity.

The results of the CERN and SACLAY groups for the nucleon resonances show a remarkable agreement, except for some details. However, there are large discrepancies, if one calculates the invariant amplitudes or predictions for spin-rotation parameters (73 L Höhler-2).

The work of the CERN group is described in the following papers and reviews: 67 R Donnachie, 67 R Lovelace, 68 L Donnachie, 69 R Lovelace, 72 L Almehed, 72 R Lovelace.

2.1.6 The analysis LONDON 73 ("CBC")

Using their new accurate data (71 Bugg, 71 Carter, 73 Bussey), Carter et al. (73 L-1) performed an energy-independent partial wave analysis in the range $186 \leqq k \leqq 427$ MeV/c ($94.5 \leqq T_\pi \leqq 310$ MeV), taking into account S, P, D, and F-waves. For the D 33, D 35, and F-waves, the authors inserted theoretical results derived from partial wave dispersion relations (65 L Donnachie-1, 71 L Nielsen). The larger phase shifts D 13 and D 15 were determined from data at 310 MeV, assuming a scattering length approximation.

S- and P-wave phase shifts are determined by minimizing χ^2. For π^-p scattering above resonance, a strong correlation develops between P 33, S 11, and P 11 phase shifts. It is broken by adding as part of the input the real part of the forward flip amplitude ReH as calculated from the dispersion relation.

Close to resonance, the phase shift solution becomes ill-defined, because the resonant amplitude is imaginary and all others are almost real.

Because of the high accuracy, it was necessary to treat the Coulomb corrections more carefully than in earlier analyses (73 L Bugg). The evidence for different P 33 phase shifts in π^+p and π^-p scattering will be treated in Sect. 2.1.17.5.

Nowadays, the LONDON 73 solution is outdated because, due to experiments at meson factories, the experimental information has been significantly improved, in particular for the polarization parameters. The phase shift solutions of the SIN group can be considered as updated versions of the LONDON 73 solution.

2.1.7 The Karlsruhe–Helsinki analyses KH 78 and KH 80

The main difference between the Karlsruhe–Helsinki analyses and all others is that fixed-t analyticity is used as a constraint. One-dimensional dispersion relations along lines t=const in the Mandelstam diagram (Fig. A.1.5) have two advantages over all other dispersion relations:

i) due to $s-u$ crossing symmetry, there are no unknown left hand cuts,

ii) experimental data are available in the very large momentum range up to 200 GeV/c for t-values in the range $0 \geqq t \gtrsim -1\,\mathrm{GeV}^2$. A disadvantage is that for $t<0$ the dispersion integral has to be calculated in the unphysical region, between threshold ($s=(m+\mu)^2$) and the point where the line t=const meets the backward hyperbola (Fig. A.6.6). However, the uncertainty is small as long as the partial wave expansion for the imaginary parts of the invariant amplitudes A and B converges: $t \geqq -0.5\,\mathrm{GeV}^2$ (Sect. A.6.2.4). Somewhat beyond this range, a truncated partial wave expansion still is a reasonable approximation.

A systematic treatment of fixed-t dispersion relations at larger negative values of t is not yet available, but one could use the same discrepancy methods, which have been applied in other cases where the amplitude is not known along parts of the cut.

It is a consequence of the fixed-t constraint that one has to include near-forward data at *all* energies. The partial wave analysis has to be performed simultaneously with an amplitude analysis of the near-forward scattering data at high momenta up to 200 GeV/c.

One should note that, according to Sect. 2.1.3, the fixed-t constraint is helpful for the resolution of the ambiguities.

In the KH analysis, the basic input is the forward amplitude as derived from total cross sections and forward dispersion relations for the amplitude $C^{\pm}$ (Sect. 2.2.1). The analysis consists of three stages, which are iterated until the amplitudes fit the data and are compatible with the dispersion relations at fixed-t and fixed c.m. angle.

The three stages of the analysis can be described briefly as follows:

i) Fixed-t analysis

The analysis would be too complicated if one would insist on working with dispersion integrals. It is possible only by the use of PIETARINENs expansion of the invariant amplitudes in terms of functions which have the correct analytic properties (Sect. A.6.3.4). The coefficients are determined from fits to the data, using the fixed-s solution as a constraint.

Other attempts to apply the fixed-t analyticity constraint at $t \neq 0$: 69 L BAACKE, 72 L ENGELS, 72 L MCCLURE, 72 L NIELSEN-1, 73 L and 74 L ARGYRES, 73 L DEVENISH, 74 L HECHT, 74 L HÖHLER, 75 R CONTOGOURIS, 73 R HÖHLER.

ii) Fixed-θ analysis

Since the fixed-t constraint can be applied only in a strip $0 \geqq t \gtrsim -1\,\mathrm{GeV}^2$, it is useful to introduce a second analyticity constraint, which covers the remaining parts of the angular range at intermediate and high energies. Fixed-c.m. angle analyticity is a suitable choice. Again, an expansion method is used and the coefficients are fitted to the data and to the fixed-s solution.

iii) Fixed-s analysis

This stage is the usual "phase shift analysis" insofar as partial waves are fitted to the data and to forward amplitudes calculated from total cross sections and dispersion relations, taking into account the unitarity constraint. However, in the KH-method, the partial waves are fitted simultaneously to the fixed-t and fixed-θ amplitudes.

The three analyses are iterated until the results for the amplitudes agree up to about 3%. In the following we give some details. The flow diagram of Fig. 1.7.1 shows the interrelations between the different procedures.

Treatment of the data

At low momenta ($k<0.5\,\mathrm{GeV/c}$), the electromagnetic effects were treated as proposed by TROMBORG, WALDENSTRØM and ØVERBØ (77 L). At higher momenta, the one-photon exchange correction was applied (including form factors) to the no-flip and flip amplitudes, taking into account the Coulomb phase.

For the 3 different analyses, the data were shifted to the selected values of s, t, and θ, using the previous solution of the iteration for the calculation of the correction. Some data were omitted, because the required momentum shift was too large.

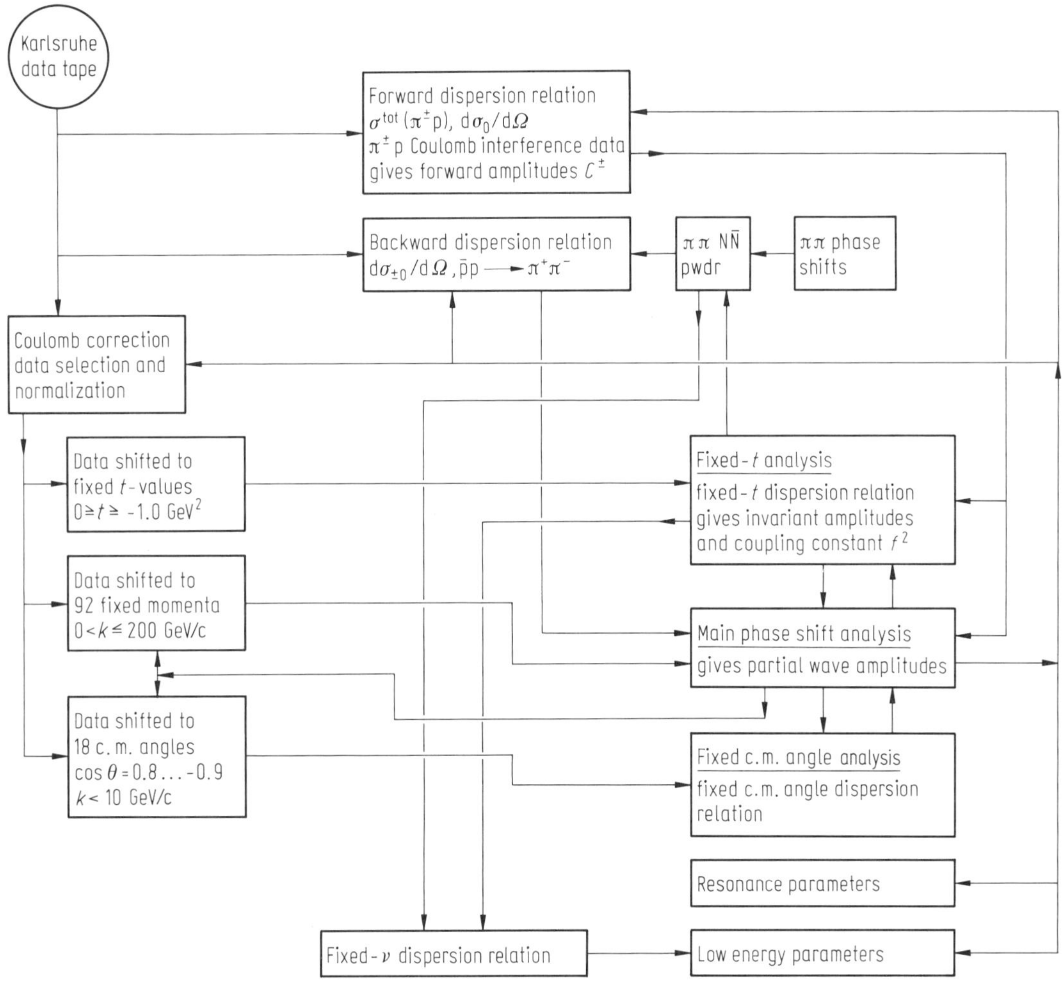

Tests of the result:
i) Compatibility with the s-channel partial wave dispersion relations
ii) Compatibility with dispersion relations along hyperbolas in the Mandelstam plane (for instance at fixed lab. scattering angle and at $u=0$)
iii) Reasonable behaviour of zero trajectories of invariant and transversity amplitudes

Fig. 1.7.1. Flow diagram of the KH 78 partial wave analysis.

As usual in phase shift analysis, it was necessary to correct the normalization of part of the cross sections in order to secure a smooth extrapolation to the input for the forward amplitude and to avoid problems with the analyticity, unitarity and isospin constraints. For a first group of accurate data a renormalization was not necessary. For a second group, it was sufficient to change the normalization by less than 3%. But for about 30 data sets (listed in Table 1.1 in 79 R Höhler-Kaiser-Koch-Pietarinen) a much larger normalization correction had to be admitted, even if the authors estimated a smaller systematic error.

For the experimental input, the Karlsruhe data base was used, which agrees with the tables in the first part of this volume, except for earlier data included in Vol. I/7 and new data, which were added since 1978. An earlier version of the data base is described in 77 L Augenstein.

Fixed-*t*-analysis

The fixed-t analysis was carried out at 40 t-values in the range $0 > t > -1.0\,\mathrm{GeV}^2$, paying attention to the overlap of the s- and u-channel cuts at $t < -4m\mu$. For $t > -4m\mu$ the expansions are written as in Eq. (A.6.42), treating the nucleon pole term separately and assuming a factor in front of the sum, which describes the expected asymptotic behaviour. Details of this factor are unimportant at low and intermediate energies.

The expansion coefficients are determined by minimizing χ^2 which consists of three parts

$$\chi^2 = \chi^2_{\text{data}} + \chi^2_{\text{p.w.}} + \sum_{n=0}^{N} (n+1)^3 \{\lambda_1 (c_n^+)^2 + \lambda_2 (c_n^-)^2 + \lambda_3 (b_n^+)^2 + \lambda_4 (b_n^-)^2\}. \tag{1.7.1}$$

c_n^+ etc. are the coefficients belonging to the invariant amplitudes $C^\pm$, $B^\pm$, and the λ_i are weight factors chosen according to the convergence test function method (76 L PIETARINEN-1). $\chi^2_{\text{p.w.}}$ belongs to the deviation from the fixed-s partial wave solution. The third term is a "penalty" which suppresses large values of the higher coefficients of the expansion.

Analyticity constraints can be used in practice only if one makes an assumption on the degree of smoothing. The aim is, of course, to smooth the statistical fluctuations of the data without distorting small structures which are physically relevant.

It is obvious that this cannot be achieved in all cases, because a narrow resonance cannot be distinguished from a fluctuation, if the data are not accurate enough and the experimental points are not spaced densely enough along the energy axis.

Except for the η-production threshold (Sect. 2.1.13), no narrow structure has been detected up to now in πN scattering, including the high energy region (80 BARREL). Therefore, it is reasonable to work with the convergence test function method, which is flexible enough to describe even details of the shapes of the known resonances.

Arguments based on probability theory for the choice of the factor $(n+1)^3$ and of the λ_i are given in 72 L PIETARINEN-3. Earlier work in this field is due to CUTKOSKY (69 L), PRESNAJDER and PISUT (71 L), and ROSS (71 L), but PIETARINEN comes to a different conclusion.

It is important to note that the result for the amplitudes is practically independent of the number N of terms in the expansion, as long as N is large enough. We have made several tests, comparing the expansion with an evaluation of the dispersion integral for the forward amplitude and with a Breit-Wigner type amplitude ($\Gamma = 100$ MeV), where the analytic function is explicitly known. It was found that for $N \approx 50$ the remaining discrepancies are of the order of a few per cent.

PIETARINENs expansion is not much different from CUTKOSKYs expansion in terms of reproducing kernels (73 L CUTKOSKY). The reason why PIETARINEN prefers to work with Taylor expansions is that it is much easier to perform the numerical calculations (73 L PIETARINEN). The minimization program is an adaptation of the conjugate gradient minimizer of FLETCHER and REEVES (64 L).

Fixed-θ analysis

The fixed c.m. angle dispersion relation is usually considered in the q^2-plane. It has a rather complicated left hand cut structure (Sect. A.6.8). Furthermore, at high energies, there must be large cancellations between the different contributions, because the modulus of the invariant amplitudes is rapidly decreasing. An evaluation of this dispersion relation is discussed in 78 L KOCH and 77 L HÖHLER-1.

In the KH analysis, the left hand cut contributions were expressed by an expansion, because the evaluation of the nearby part was not yet available. Since the constraint is not needed in the low energy region, the contribution of the dispersion integral in this range was calculated from the well known low energy phase shifts. A second expansion was used at intermediate and high energies up to 10 GeV/c. The calculation was made at 18 angles $\cos\theta = -0.9 \cdots 0.8$. The backward dispersion relation was treated in a separate analysis (78 L PIETARINEN-2), including the knowledge of the nearby left hand cut.

The expansions were made for combinations of invariant amplitudes, Eq. (A.2.17), which have simple analytic properties and the same relations to data, Eq. (A.2.18), as BARRELETs transversity amplitudes.

Again, the convergence test function method was used for the determination of the coefficients, adding to χ^2_{data} the χ^2 belonging to the fixed-s amplitudes.

Fixed-s analysis (Phase shift analysis)

At 92 selected momenta (of which 6 lie at 6···200 GeV/c), the coefficients of truncated partial wave expansions were fitted to the data and to the forward, fixed-t, fixed-θ and backward amplitudes. Since the parameters are $\text{Re}\, T_{\ell\pm}$ and $\text{Im}\, T_{\ell\pm}$, it is necessary to enforce unitarity by adding a suitable term to the χ^2.

The truncation was made at about 2 times $\ell_{\max} \approx qR$, where $R \approx 1$ fermi. For instance partial waves up to L 1, 17 and L 3, 17 were admitted in the neighborhood of 1 GeV/c. However, the exact number is unimportant, because the effective cut-off comes from the contribution

$$\lambda \sum_{\ell\pm, I} \{|\text{Re}\, T^I_{\ell\pm}|^2 R_1^{2\ell} + |\text{Im}\, T^I_{\ell\pm}|^2 R_2^{2\ell}\} \tag{1.7.2}$$

to χ^2. R_1 and R_2 are determined from the location of the nearest singularity z_i in the $z=\cos\theta$ plane (Table A.6.2)

$$R_i = z_i + \sqrt{z_i^2 - 1}\,, \tag{1.7.3}$$

which is different for the real and imaginary parts. The soft cut-off by the expression Eq. (1.7.2) takes into account the "short range nature" of the strong interaction forces. — Of course, the suppression of the high partial waves depends not only on the location of the nearest singularity but also on its strength, which is not yet well known.

Pietarinen (78 L-2) also discusses the conformal mapping expansion of the CMU-LBL group (Sect. 2.1.9). He notes that problems can arise from the fact that for the description of a resonance in a single partial wave one needs a coherent superposition of structures in the conformally mapped expansion. — The analyticity constraint in z at fixed s is much weaker than the fixed-t constraint, where one can calculate the real part if the imaginary part is given at all energies.

Tests of the KH-78 partial wave solution

Although the KH 78 analysis already uses stronger dispersion constraints than all other methods, it is still of interest to test the compatibility with further dispersion relations.

The usefulness of fixed lab. angle ("interior") dispersion relations for this purpose is considerably improved, if the parameter θ_π (Sect. A.6.7.2) is chosen in the range $95° < \theta_\pi < 180°$, because only then the t-channel partial wave expansion converges and one can evaluate the nearby part of the left hand cut. A study of this dispersion relation shows small discrepancies but, in general, the compatibility is satisfactory. This is not trivial, because the fixed-t constraint cannot be applied in the backward hemisphere at intermediate energies, and the fixed-θ constraint is weaker (76 L Kaiser, 77 L Borie, 78 L Kaiser).

A second test has been made by Hutt and Koch, who used the solution KH 78 (replacing the low energy part by the updated version KH 80) for an evaluation of the *partial wave dispersion relation* (Sect. A.7.1). They concluded that the compatibility is satisfactory (80 L and 81 L Hutt).

A third test follows from the calculation of the *zero trajectories* of invariant and transversity amplitudes. As discussed in Sect. 2.4.3, these trajectories have analytic properties from which one can conclude that their projections onto the physical s, t-plane must show a smooth energy dependence except for structures, which occur at "intersections" of two trajectories. However, these structures must fulfill certain conditions (78 L Sabba-Stefanescu).

The determination of the zero trajectories of invariant and transversity amplitudes from the KH 78 partial wave solution shows that the result is satisfactory, except for the partial waves at a few energies, where the zeros do not lie on a smooth curve connecting the zeros at lower and higher energies. This question will be investigated together with the preparation of an updated solution (see Sect. 2.4.3 for a further discussion of zero trajectories).

The solution KH 80

Koch and Pietarinen (80 L-1) have performed an energy independent partial wave analysis in the range from threshold up to 500 MeV/c, taking into account the new data and constraints from fixed-t amplitudes, which were improved by iteration in the course of the analysis. For the electromagnetic corrections, the authors used the formalism of Tromborg et al. (77 L), except for a few simplifications. Results are given for partial waves up to F 17 and F 37 (see Figs. 2.2.22···2.2.33). Further results of this work are discussed in Sects. 2.1.17, 2.4.2, and 2.4.7.

Preliminary results in the high energy region

After the completion of the KH 78 analysis, the experimental information on pion-nucleon scattering above 2 GeV/c has improved considerably (see Sect. 2.6). Preliminary results for updated partial waves are available below 10 GeV/c. (79 R Höhler, 80 L Höhler-2, 80 L Koch-2, 80 R Koch). However, we have not yet completed the determination of an updated fixed-t solution. This is more difficult than at intermediate energies, because one cannot work with the expansion alone. The first step is an amplitude analysis with dispersion constraints of the new data (Sect. 2.1.16), the aim being to find parametrizations for the different amplitudes at high energies, which are compatible with the old and new data. Presumably, this will lead to changes of the factors in front of the sums in the expansion discussed in Sect. A.6.3.4.

Further details on the KH 78 and KH 80 analyses can be found in papers by Pietarinen (72 L-2, 72 L-3, 73 L, 76 L-1, 76 L-2, 76 L-3, 76 L-4, 78 L-1, 78 L-2), by Koch (78 L, 80 L-1, 80 L-2, 80 R) and in our "Handbook" (79 R Höhler-Kaiser-Koch-Pietarinen).

Starting from the KH solution, HIROSHIGE et al. (81 L, 82 L) determined new phase shifts in an energy range where new data became available (80 FUKUSH, 80 TERADA).

A detailed study of the uniqueness problem has recently been performed by SABBA-STEFANESCU (80 L-2, 81 L, 82 L-1 and 2).

2.1.8 The analysis LENINGRAD 78 and the SIN analyses

Including their own new data, the Leningrad group performed an energy-independent phase shift analysis at 12 momenta in the range $350 < k < 830$ MeV/c ($230 < T_\pi < 700$ MeV), admitting S, P, D, and F-waves up to 450 MeV/c and also G-waves at higher momenta (78 L ABAEV). Starting from a random search, they found a unique best fit up to 520 MeV/c and selected a unique continuation up to 830 MeV/c by demanding that all partial waves have a smooth momentum dependence in their respective Argand diagrams.

At $k = 685$ MeV/c, the Leningrad data (79 GORDEE) agree well with the differential cross sections of the new experiment 79 SARMA at Rutherford Laboratory.

A comparison with other recent results is shown in Figs. 2.2.22···2.2.33. The authors claim to have found additional evidence for a splitting of the P 11 resonance N(1470), which was already noticed by the SACLAY group (76 L AYED).

However, the evidence for the splitting of this resonance is a small effect, which is comparable with irregularities in other partial waves and with discrepancies between recent phase shift analyses.

For a critical analysis of a possible splitting of the P 11 resonance, it is necessary to improve two points of the work of ABAEV et al.:

i) If the $I = 1/2$ forward amplitude is reconstructed from the Leningrad partial waves, one finds a large discrepancy with the prediction from the forward dispersion relation in the energy region of the P 11 resonance. It would be useful to include the predicted forward amplitudes as part of the input, as it is done in other partial wave analyses (The discrepancy is shown in Fig. 3 of the paper 81 L STAHOV).

ii) The determination of the unique solution by choosing a smooth continuation of all partial waves is questionable if one passes the η-production threshold, where the partial wave S 11 has a pronounced cusp (Sect. 2.1.13). This occurs at $W = 1488$ MeV, within the width of the N(1470) P 11 resonance.

Furthermore, it is possible that the "splitting" of the P 11 resonance is only another aspect of a difficulty noted in the KH 78 and CMU-LBL 79 analyses: in this energy region, it is not possible to find a good fit to all data and to the dispersion constraints. In both solutions, the fit to part of the charge-exchange data 76 BROWN and 78 BROWN is poor.

The analyses of the SIN groups

The SIN analyses were carried out at fixed energies. The authors started from phase shifts which were obtained or used as part of the input in the analysis 73 L CARTER-1. They adjusted some of these phase shifts in order to fit the new polarization and total cross section data of the SIN groups.

In 75 AMSLER and 76 AMSLER the authors modified the S 31, P 31, and P 33 phase shifts at $T_\pi = 166$ and 194.3 MeV. At 236.3 MeV, they modified in addition the D 33 and D 35 phase shifts in order to improve the fit to their P_+-data.

In 78 ALDER and 79 ALDER they used their P_--data and the total cross section data of 78 PEDRON in order to redetermine the S 11, P 11, P 13, and P 33 phase shifts at $T_\pi = 98$ MeV and in addition also the D 13 and D 15 phase shifts at 238, 291.5, and 310 MeV.

The P_--data deviate considerably from the prediction derived from the phase shifts of 73 L CARTER-1. The main correction occurs in the S 11-wave. The results are shown in Figs. 2.2.22···2.2.33.

The most recent SIN phase shift analysis was performed at three energies, where accurate differential cross section data of BUSSEY et al. (72) are available and phase shifts were determined in the analysis LONDON 73 (73 L CARTER-1): $T_\pi = 237.9$, 291.6, and 310.0 MeV. The additional input comes mainly from polarization experiments of the SIN group (P_+: 75 AMSLER, 77 DUBAL; P_-: 78 ALDER, 82 ALDER; P_0: 82 ALDER).

The new SIN phase shift analysis (82 L ALDER) can be considered as an updated version of the analysis LONDON 73. In some cases, the changes are considerably larger than expected from the errors given by 73 L CARTER-1. It is very useful to have careful single-energy analyses of this type in addition to analyses which cover a large energy range and use dispersion constraints.

2.1.9 The Carnegie-Mellon-Berkeley analysis (CMU-LBL79)

The CMU-LBL analysis covers the momentum range in which data of high precision are available at all angles and for all 3 reactions. The range of the version CMU-LBL 79 (79 L CUTKOSKY-1) is 0.43···2.0 GeV/c. Further results up to 2.5 GeV/c were reported in 80 L CUTKOSKY. The main aim is to determine the parameters of the nucleon resonances.

The analysis consists of 4 stages

i) Data amalgamation. As a preparation for single-energy partial wave analyses, smoothed sets of "amalgamated data" at 35 momenta are constructed from the original data and error estimates are calculated.

ii) Partial wave analysis. At each momentum, the amalgamated data are partial wave analysed, taking into account theoretical information on singularities in the $z=\cos\theta$-plane. There are many solutions due to discrete and continuum ambiguities.

iii) Analyticity constraint for the energy dependence. The final solution at all momenta is chosen by demanding that the invariant amplitudes constructed from the partial waves are analytic along certain hyperbolas in the Mandelstam plane. The fits at fixed s and along the hyperbolas are iterated, until the partial wave amplitudes are consistent with the amalgamated data and with the analyticity constraint for the s-dependence. The result is a unique solution.

iv) Determination of resonance parameters. The parameters are determined by fitting the partial waves to suitable parametrizations. The results will be treated in Sect. 2.4.1.

In the following we shall give some details. A complete description can be found in the original papers: stage i): 79 L KELLY, stages ii) and iii): 79 L CUTKOSKY-1, stage iv): 79 L CUTKOSKY-2. Subsequent developments are reported in 80 L-1, 80 L-2 and 80 R CUTKOSKY, 80 R KELLY. Earlier versions: 76 L CUTKOSKY, 76 L HODGKINSON. Papers on related mathematical problems: CUTKOSKY (68 L, 69 L, 70 L, 71 R, 73 L).

Data amalgamation

The momentum and c.m. angle ranges are divided into bins which have widths of about 50 MeV/c and 3°, respectively. Starting from the original data, the authors construct for each data type smooth momentum and angle-dependent "interpolation surfaces" which cover about three adjacent momentum bins. These surfaces are used in order to shift the data of the central bin to the nearest multiple of $\theta=3°$ and to the central momentum k_i. The result is a set of shifted data at 35 momenta k_i and 61 c.m. angles θ_i.

These data are statistically combined, taking into account all kinds of errors and the effects of thresholds. An error matrix is calculated and the statistical errors are separated from collective errors, which belong to angle-dependent normalization errors and errors of the momentum calibration.

The "amalgamated data" and their error matrix will be made available to interested users.

Partial wave analysis

The real and imaginary parts of each invariant amplitude are separately parametrized as a sum of a fixed "Born term" and a polynomial term, whose coefficients are adjustable parameters to be determined in the fits. The ansatz reads (omitting the fixed value of s in the argument) for instance for the invariant amplitude B

$$B(z)=B_{\mathrm{Born}}(z)+R(z)P(Z), \tag{1.9.1}$$

where $z=\cos\theta$. $Z(z)$ is the CUTKOSKY-DEO variable (see below). The name "Born term" is used for a contribution which follows from a Regge model for N, Δ, ϱ and f-exchanges. Absorptive corrections are taken into account. In the imaginary part, a diffractive Pomeron term is included and also a two-pion exchange term as proposed in 73 L ALCOCK-1. [Earlier papers: Nucl. Phys. B **31** (1971) 443 and B **41** (1972) 141.]

The variable $Z(z)$ is chosen to map the singularities of $B(z)$ (except those which are taken into account in the Born term) onto the periphery of a unifocal ellipse. The physical region is mapped onto $-1<Z<+1$, and the expansion in Z is formally convergent in the entire physical-sheet analyticity domain of $B(z)$.

Within the physical region, the polynomial $P(Z)$ converges more rapidly than the usual partial wave expansion. One can expect that this stabilizes the fit against the effects of statistical fluctuations of the data. [ACE=accelerated convergence expansion; CUTKOSKY and DEO (68 L and 70 L)]. An earlier version of the ACE method was criticized by GRISS and FOX (73 L), because it led (in K^+p scattering) to much larger high partial waves than other methods. The ACE method takes into account only the location of the cut. But the effect of the cut also depends of course strongly on the magnitude of its discontinuity. In order to diminish this difficulty, the CMU-LBL group has included two-pion exchange in the Born term as mentioned above.

The fixed modulating factor $R(z)$ is chosen to be proportional to $z^{\alpha(s)}$ as suggested by an s-channel Regge behaviour for $z\to\infty$ (Sect. A.9.2.2).

A direct fit of the parametrization (1.9.1) to the data does not ensure that the resulting partial waves are unitary. The authors developed a method which allowed them to vary the "low" partial waves freely, subject only to the unitarity constraint. The "high" partial waves are restricted by the ansatz (1.9.1) and unitarity is enforced by adding a penalty function to χ^2.

The expansion $P(Z)$ is cut off in a smooth way by another penalty function (the "truncation function"). It suppresses results which have large values of high order coefficients.

The fits lead to a large number of local minima of the total χ^2, which are analysed in a multi-dimensional space of search parameters (low partial waves). In this space the minima tend to lie in clusters. The size and shape of a cluster is related to the continuum ambiguity, whereas the existence of separated clusters is a consequence of the discrete ambiguities.

If the inverse error matrix is used as a metric, one can choose the clusters such that they are statistically distinguishable.

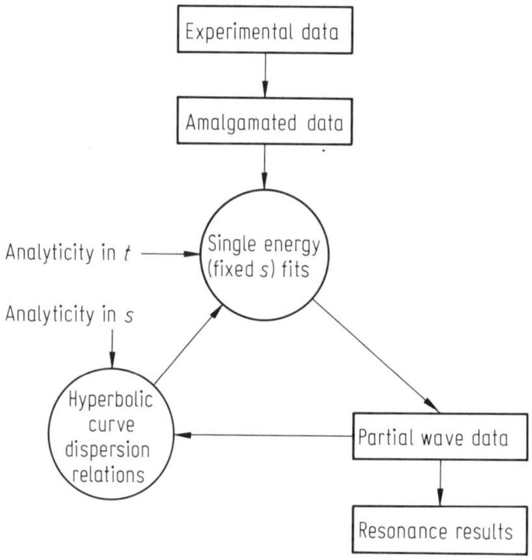

Fig. 1.9.1. Flow diagram of the CMU-LBL 79 partial wave analysis.

Analyticity constraints for the energy dependence

The authors impose the constraint that the invariant amplitudes are analytic along 5 crossing symmetric hyperbolas in the Mandelstam plane, which pass through the s- and u-channel thresholds and are asymptotic at large s to fixed c.m. scattering angles in the range $31°\cdots180°$. In addition, an evaluation of the forward dispersion relations for the C-amplitudes is used as part of the input (77 L Engelmann).

The unique final solution is determined by an iterative procedure (Fig. 1.9.1), in which the predictions from the dispersion relations along hyperbolas are used as part of the input in fixed-s analyses. It turns out that 5 hyperbolas are sufficient for a resolution of all discrete ambiguities.

The result of the CMU-LBL analysis is given in terms of the usual partial waves and by the error correlation matrix. The partial waves are listed in Sect. 2.2.2 (see also Figs. 2.2.1$\cdots$2.2.21).

Comparison of the CMU-LBL and KH analyses and comments

i) Aims of the analyses

The CMU-LBL analysis is directed towards a comprehensive knowledge of the nucleon resonance parameters in the momentum range where accurate data for all 3 reactions are available. The group spent a great effort in calculating the error matrix of the partial waves and in applying an elaborate multichannel resonance formalism for the determination of the resonance parameters.

The KH analysis also has the aim to determine partial waves and resonance parameters, but other objects of equal importance were preferred to detailed studies of the error propagation and of the multichannel character of the resonances.

One of them is the application of dispersion constraints for the s-dependence, which are much stronger than those used in the CMU-LBL analysis. Some of them are part of the KH-analysis and others were tested in subsequent calculations (Fig. 1.7.1). It can be expected that, in general, the strong dispersion constraints will pull the amplitudes towards their correct values. Furthermore, the remaining discrepancies between the amplitudes and different dispersion constraints give an estimate of the uncertainties. These can be larger than the errors calculated from the CMU-LBL error matrix, since "accurate" data can have additional unknown systematic errors.

Another test is to study the zero trajectories which are rather sensitive to errors of partial waves. It is, therefore, necessary to check whether they show the behaviour expected from two-variable analyticity (Sect. 2.4.3).

Finally, the different dispersion relations employed in the KH-method make it possible to use an *additional experimental input*: πN data at high energies in fixed-t dispersion relations and ππ-phase shifts in dispersion relations which require $\pi\pi N\bar{N}$ amplitudes.

ii) The fixed-s analysis

The KH method uses the simple partial wave expansion, whereas the CMU-LBL method prefers the ansatz (1.9.1), which requires a considerably greater effort. The question is now whether the effort is rewarded by a significantly better reliability. Furthermore, one would like to know to what extent the CMU-LBL method is "model-dependent" due to the use of the "Born term".

One can have doubts concerning the application of Regge models at rather low energies since, even at high energies, a generally accepted success of absorptive Regge models in πN scattering does not exist (Sect. 2.6).

As stated in 79 L Cutkosky-1, the most critical parts of the Born term are the nucleon term and the two-pion exchange term. The contribution of the latter term to the imaginary parts of the high partial waves as given by Alcock and Cottingham (73 L-1) is more reliable than any value that could be extracted from the data. Errors of the other terms can be compensated by the ACE expansion.

In my opinion, it is necessary to study in detail the question, how much one can trust the high partial waves of both analyses. An evaluation of the forward dispersion relations for $\partial C/\partial t$ and $\partial^2 C/\partial t^2$ by Koch has shown that both solutions are not yet satisfactory. See also the discrepancies in Figs. 6.2.8 and 6.2.10, which are due to a poor accuracy of some data at small t. The discussion of the high partial waves will be continued in Sect. 2.1.14.

iii) Comparison between the CMU-LBL and KH partial wave amplitudes

Argand diagrams for the partial waves of both analyses are shown in Figs. 2.2.1···2.2.21. In general, the results agree within one or two error bars. This is remarkable, because the two methods differ considerably.They work with different data bases, the interpolations and renormalizations are made in a different way, the single-energy analyses are based on the ansatz (1.9.1) in one case and on a simple partial wave expansion in the other case and finally, different dispersion relations are used for the continuation in s and the resolution of the discrete ambiguities.

It will be interesting to study the origin of several larger discrepancies in the partial waves which will be mentioned in Sect. 2.4.1. One reason is probably that the two solutions behave differently in regions where experimental discrepancies between data sets exist which are appreciably larger than the error estimates.

iv) Quality of the fit to the data and desirable experiments

The authors of both analyses came to the conclusion that it is difficult to reconcile the charge-exchange data of the Rutherford group (76 Brown and 78 Brown) with the elastic data and the dispersion constraints. Although the Rutherford experiment has led to a great improvement of our knowledge of the charge-exchange reaction, it is desirable to perform a second experiment in this range.

As far as the elastic differential cross sections are concerned, the CMU-LBL analysis included the preliminary data set 76 Bardsl of the Bristol-Southampton-Rutherford group, which was not available to us at the time of the KH 78 analysis. Nevertheless, the agreement with the predictions from our partial waves is satisfactory in general. Unfortunately, the final result of this important large data set has not yet been distributed or published, although the preliminary data were presented already at the Oxford Conference in 1976.

The experimental situation in near-forward and near-backward scattering is not yet satisfactory due to discrepancies between different data sets and to the lack of Coulomb interference data. Furthermore, there are still fairly large discrepancies between the most accurate total cross section data below 1 GeV/c (see Sect. 2.2.1). As a consequence, there are discrepancies for instance between the predictions for the slope of the cross section at

$t=0$ (Figs. 6.2.8 and 6.2.9) and for the backward amplitudes (Fig. 10 in 79 L CUTKOSKY-1), which are related to discrepancies between high partial waves.

Another shortcoming is the relatively poor information on the π^-p polarization. Although new results are available from several groups (SIN, Leningrad, KEK), there are still large gaps in the data. One can only hope that further measurements will be made before the last accelerator in this energy range is closed down.

Usually, it is pointed out that spin-rotation data would be of great importance and this is certainly true in principle. However, since these experiments require a great effort and since one cannot expect a high accuracy, it is necessary to choose carefully a suitable kinematic region. It should lie outside the range where the amplitudes are dominated by a resonance.

2.1.10 New energy-dependent analyses

The analysis VPI 80 (ZIDELL 80-1)

Data base. The data base consists of 1572 data points in the range 0···350 MeV kinetic energy of the pion ($k \leqq 470$ MeV/c). It is difficult to understand why some data of the pioneers in this field (1952) were included. Some of the recent data sets are missing (77 JENEFS, 78 PEDRON, 79 AULD, 79 ALDER).

Pruning. The authors eliminated all data points contributing a χ^2 greater than 6 for π^+p and 5 for π^-p-scattering. It is remarkable that 10 data points of the accurate experiment 73 BUSSEY belong to this category and 4 data points of the experiment 71 CARTER.

Method: an energy dependent parametrization is assumed for S, P, D and F-waves

$$\tan\delta_{0+} = \varrho_0 \Big/ \sum_1^3 c_n T_\pi^{n-1}; \quad \text{S 31},$$

$$\tan\delta_{\ell\pm} = \varrho_\ell \sum_1^N c_n T_\pi^{n-1}; \quad \text{N}=1: \text{D 35 and all F-waves}; \ \text{N}=3: \text{S 11, P 13}, \tag{1.10.1}$$

$$\text{N}=2: \text{P 31, D 13, D 33, D 15}; \ \text{N}=6: \text{P 11}.$$

$T_\pi = \omega - \mu = (s - m^2 - \mu^2)/2m - \mu$ is the kinetic energy of the pion in the lab. system, which is linear in s. The expression

$$\varrho_\ell = [W^2 - (m+\mu)^2]^{\ell + 1/2} \tag{1.10.2}$$

differs from the usual phase space factor $q^{2\ell+1}$, but the threshold behaviour is the same. The coefficients c_n depend on isospin I and on J and ℓ.

For the resonant P 33-wave the authors use LICHTENBERGs parametrization, which has the advantage that the parameters determine directly the location of the pole of the S-matrix. In order to fulfil the unitarity condition, the background is described by a real phase δ_B which is added to the resonant phase δ_R, i.e. the S-matrix element S_{33} is the product of the resonance and background S-matrix elements (73 L LICHTENBERG)

$$\delta(\text{P 33}) = \delta_R + \delta_B; \quad S_{33} \equiv \exp(2i\delta(\text{P 33})) = S_R S_B. \tag{1.10.3}$$

Denoting the corresponding K- and T-matrix elements by K_R, K_B and T_R, T_B one obtains for the total matrix elements

$$K_{33} = \tan\delta(\text{P 33}) = -i\frac{S_{33}-1}{S_{33}+1} = \frac{K_R + K_B}{1 - K_R K_B}; \quad T_{33} = T_R S_B + T_B. \tag{1.10.4}$$

If a pole term ansatz is made for K_R

$$K_R = \frac{\Gamma/2}{M-W}; \quad T_R = \frac{K_R}{1 - iK_R} = \frac{\Gamma/2}{M - W - i\Gamma/2}; \quad S_R = 1 + 2iT_R = \frac{M - W + i\Gamma/2}{M - W - i\Gamma/2}, \tag{1.10.5}$$

it is seen that T_{33} and S_{33} have a complex pole at

$$W = M - i\Gamma/2. \tag{1.10.6}$$

In order to ensure the correct threshold behaviour, K_B is parametrized as follows

$$K_B = K_R\left(\frac{\varrho_1(W)}{\varrho_1(M)} - 1\right) + \frac{\varrho_1(W)}{\varrho_1(M)}(c_1 + c_2 T_\pi). \tag{1.10.7}$$

Electromagnetic corrections. The authors present "nuclear phase shifts". They eliminated only the effects due to single-photon exchange between pointlike pions and protons. Their discussion of the difference between the π^+p and π^-p Δ-exchange pole positions will be treated in Sect. 2.1.17.

Comments

i) The authors regard their parametrization as the *most economic description* of the πN scattering data that has been made up to date. However, the aim of partial wave analysis is a different one: one wants to find a good approximation to the true partial waves. There is no reason why a parametrization of the true partial waves should be "economic" in the sense of ZIDELL et al.

As discussed above, even data of infinite accuracy can be fitted equally well by a variety of parametrizations due to the ambiguities. An approximation to the true partial waves can be obtained only by imposing theoretical constraints.

ZIDELL et al. assumed theoretical constraints for the threshold behaviour and by using the prediction for the forward amplitude as part of the input. It is necessary to employ in addition the theoretical information on partial wave dispersion relations (Sect. A.7.1) in order to check, whether the energy dependence of the nonresonant partial waves can be described correctly by simple effective range type formulas.

Figure 1.10.1 shows that a more complicated energy dependence can be expected because, in the *s*-plane, the point $T_\pi = 350$ MeV is *twice* as far from threshold as two important left hand cut singularities: the short nucleon cut and the front of the circle.

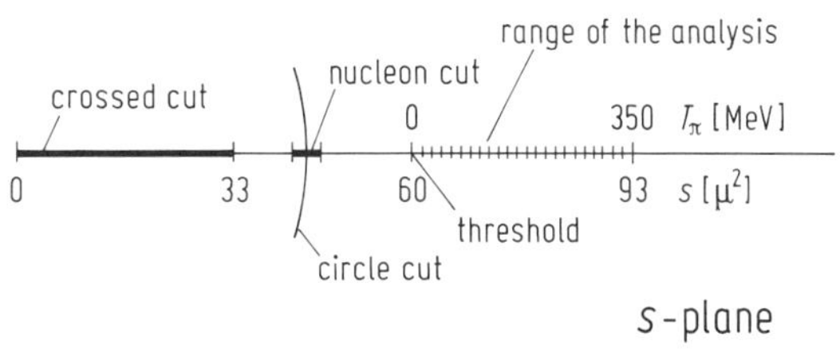

Fig. 1.10.1. Singularities of πN partial waves in the *s*-plane. ZIDELL et al. have used simple parametrizations, which ignore the nearby left hand cut singularities.

The question is whether one can find partial waves which do not only fit the data but also are compatible with the partial wave dispersion relations (Sect. A.7.1) and partial wave relations (Sect. A.7.2).

The most recent determination of such partial waves is due to KOCH and PIETARINEN (80 L-1), HUTT and KOCH (80 L) have shown the compatibility with the partial wave dispersion relation and, for $\text{Re} f_{0+}^-$, KOCH has also shown the compatibility with the partial wave relation, which is based on an independent numerical calculation, although it is essentially equivalent from a theoretical point of view.

Figures 2.2.22···2.2.33 show that for some partial waves (for instance for $\text{Re} f_{0+}^-$ and for $\text{Re} f_{1-}^{1/2}$) the results of KOCH and PIETARINEN differ strongly from those of ZIDELL et al., in particular at very low energies, where data do not exist (pp. 66···71).

The discrepancy of the amplitudes $B_\pm$ as reconstructed from the phase shifts of ZIDELL et al. with the forward dispersion relation (Fig. 5 in 80 R KOCH) is related to the discrepancy in $\text{Re} f_{1-}^{1/2}$.

ii) An important application of pion-nucleon amplitudes is the continuation into the unphysical region between the *s*- and *u*-channel thresholds, where one can test predictions from Current Algebra and PCAC (Sect. 2.5.1). For this purpose, simple parametrizations of the partial waves as those of ZIDELL et al. are not applicable, because the center of the interesting region is the point $\nu = 0$, $t = 2\mu^2$, which belongs to $s = m^2 = 45.2\mu^2$, and this point lies practically at the nearest singularities of the partial wave amplitudes (Fig. 1.10.1).

The analysis of ROWE et al.

The "energy-dependent phase shift analysis" of ROWE et al. (78 L) is not an analysis of original data. Instead, the authors fitted an energy-dependent parametrization to the results of the phase shift analyses 72 L ALMEHED, 73 L CARTER-1, 76 BERTIN, 76 AMSLER, including also several earlier analyses listed in 72 R HERNDON and scattering lengths listed in 76 R NAGELS.

The following parametrization was used in the range $0 < T_\pi < 400$ MeV for the S-waves, P-waves and D 13-, D 15-waves

$$\tan\delta_{\ell\pm} = q^{2\ell+1}(b + cq^2 + dq^4) - \left(\frac{q}{q_r}\right)^{2\ell+1} \frac{x_r \Gamma_r W_r}{W_r^2 - W^2}. \tag{1.10.8}$$

The parameters of the lowest resonance in each partial wave: W_r (mass), Γ_r (width), x_r (elasticity) were taken from the table of the Particle Data Group (1976).

It is difficult to see the usefulness of a "world average" over old and new phase shifts. For example, the experimental information on elastic scattering in the low energy region was considerably improved by the data of BUSSEY et al. (73), which were used in the analysis 73 L CARTER-1. If one takes an average over these phase shifts together with earlier ones and uses in addition an empirical energy-dependent parametrization, one has certainly not produced a more reliable solution but a deterioration.

Conclusion on energy-dependent parametrizations in partial wave analysis

In the low energy range, simple energy dependent parametrizations were of interest at a time, when the experimental information was poor and detailed investigations of the partial wave dispersion relations did not yet exist.

Nowadays, it is much better to perform either energy-independent analyses as those of the SIN group or energy-dependent analyses based on flexible dispersion constraints.

2.1.11 BARRELETs method

The basic idea is to represent the transversity amplitudes (Sect. A.2.1.2) at fixed energy by the following ansatz, which exhibits the zeros in the complex $z=\cos\theta$-plane (72 L and 76 L BARRELET, see also 69 L and 74 L GERSTEN)

$$F(\pm;z)=F(1)\prod_{i=1}^{N}\frac{z-z_i^{\pm}}{1-z_i^{\pm}}R(\pm;z); \qquad R(\pm;1)=1. \tag{1.11.1}$$

$F(1)$ is the forward amplitude. Its real part is taken from an evaluation of the forward dispersion relation and its imaginary part from the optical theorem, inserting the experimental total cross sections.

The $z_i^{\pm}$ are those zeros which lie near the physical range in the z-plane and can therefore be determined from structures in the data for $\Sigma(\pm)\equiv(1\pm P)\mathrm{d}\sigma/\mathrm{d}\Omega$. The choice of N will be discussed later.

$R(\pm;z)$ is the remainder. It is a factor which produces the correct phases of $F(\pm;z)$. The relative phase $\arg F(+;z)-\arg F(-;z)$ can be determined from spin-rotation measurements which, however, have not been made below 6 GeV/c. Therefore, R contains two undetermined real parameters which depend on z: there is a *twofold continuum ambiguity*.

i) R contains the effects of those zeros which are not included in the product, because they lie at a larger distance from the physical region.

ii) Finally, R contains the effects of branch cuts and poles in the z-plane, which are known to exist but cannot be described by a polynomial.

To begin with, we ignore the remainder. The amplitudes can now be described in two equivalent ways:

i) by their zeros $z_i^{\pm}$. $F(+;z)$ and $F(-;z)$ have together 2N complex zeros. This gives, including $F(1)$, 2N+1 *complex parameters.*

ii) by a truncated partial wave expansion, Eq. (A.3.1), where the highest ℓ-value is $\ell_{max}=\mathrm{N}$ and the highest J-value is $J_{max}=\ell_{max}+1/2$. Again, one has $2\mathrm{N}+1=2\ell_{max}+1$ complex parameters: the partial waves $f_{\ell\pm}$.

The z-dependence of the data can be continued into the complex plane if we write

$$\Sigma(\pm;z)\equiv\frac{\mathrm{d}\sigma}{\mathrm{d}\Omega}(1\pm P)=F(\pm;z)F^*(\pm;z^*). \tag{1.11.2}$$

If Eq. (1.11.1) is substituted and R is ignored, one obtains a polynomial of degree 2N.

The data can also be described in two equivalent ways:

i) by the 4N+1 coefficients of suitable fits to polynomials in z, for instance Legendre fits, Eq. (A.4.21). [2N+1 coefficients for $\mathrm{d}\sigma/\mathrm{d}\Omega$ and 2N for P.] A detailed discussion of other fits can be found in 72 L BARRELET.

ii) By the $2\cdot 2\mathrm{N}$ roots of the polynomials and $|F(1)|$. Since the polynomials are real functions, the roots occur in pairs z_i, z_i^*, i.e. only 2N of the complex roots are independent and we have again 4N+1 *real parameters.*

It is seen that the descriptions by Legendre coefficients of the data, by the zeros of the amplitudes and by partial wave amplitudes are equivalent, if the expansions are truncated as described above. The only difference is that the amplitudes have an additional real parameter, $\arg F(1)$, which cannot be determined from the Σ-data.

Equation (1.11.2) shows that the zeros of the amplitude can be derived from the zeros of the data but, unfortunately, there is a $2^{2\mathrm{N}}$ fold ambiguity because, for each pair of zeros z_i, z_i^* of Σ, one has the choice whether z_i or z_i^* belongs to the amplitude F.

In some applications it is preferable to work with a single analytic function $F(w)$ instead of $F(\pm;z)$ (see Sect. A.2.1.2). Now, the amplitude is represented in terms of its zeros in the w-plane [$w=e^{i\theta}$; $z=(w+w^{-1})/2$].

$$F(w)=F(1)w^{-N}\prod_{i=1}^{2N}\frac{w-w_i}{1-w_1}=\sum_{J,\varepsilon}R_{J,\varepsilon}(w)f_{J,\varepsilon}. \qquad (1.11.3)$$

The partial wave expansion of $F(w)$ is given in Sect. A.3.2. $f_{J,\varepsilon}=T_{J,\varepsilon}/q$ is another notation for $f_{\ell\pm}$. The 2N zeros w_i follow from the $z_i^{\pm}$, $i=1\cdots N$. Again, we have $2N+1$ complex parameters: w_i and $F(1)$, which can be expressed by $2N+1$ partial wave amplitudes $f_{\ell\pm}$.

The inversion formula for the partial wave expansion (1.11.3) reads

$$f_{J,\varepsilon}=\frac{1}{2(2J+1)}\int_0^{2\pi}F(w)R^*_{J,\varepsilon}|\sin\theta|\,d\theta. \qquad (1.11.4)$$

For the evaluation of the integral it is useful to have the formula

$$\int_0^{2\pi}w^n|\sin\theta|\,d\theta=\begin{cases}0, & \text{if n is odd,}\\ 4/(1-n^2), & \text{if n is even.}\end{cases} \qquad (1.11.5)$$

If one writes the expression for $\Sigma(w)$, one should notice that $z_i\to z_i^*$ belongs to $w_i\to 1/w_i^*$, because the first operation is carried out within the same sheet of the z-plane (see Fig. A.2.1)

$$\Sigma(w)=F(w)F^*(1/w^*)=|F(1)|^2\prod_{i=1}^{2N}\frac{w-w_i}{1-w_i}\,\frac{w^{-1}-w_i^*}{1-w_i^*}=\Sigma^*(1/w^*). \qquad (1.11.6)$$

Energy independent phase shift analysis

Now we summarize the first step of a simplified version of BARRELETs method and compare it with the conventional method in phase shift analysis. The data for $\Sigma(+;z)$ and $\Sigma(-;z)$ (including the prediction for $\Sigma(1)$) are fitted by polynomials, whose degree 2N is chosen such that the highest coefficients are still significant. Then the roots of these polynomials are determined and transversity amplitudes are constructed from the expression Eq. (1.11.1) with $R\equiv 1$. Finally, partial waves are derived using the projection formula Eq. (A.3.5), or the equivalent expression Eq. (1.11.4) if one works in the w-plane. There is a 2^{2N}-fold ambiguity.

The simplified version of BARRELETs method is equivalent to a conventional phase shift analysis, in which the partial wave expansion is truncated at $\ell=N$. However, the use of the roots as intermediary parameters has a clear advantage, since one does not have to perform a random search and it is easy to survey the 2^{2N} ambiguities. The error analysis also is more transparent.

On the other hand, it is a disadvantage that the unitarity constraint is not automatically fulfilled. It can happen that the result for a partial wave lies outside the unitarity circle (see Fig. 2 in 80 L CHEW). Furthermore, it is known that in special cases the location of some zeros can have an unexpected sensitivity with respect to tiny variations of one of the coefficients of the polynomial (72 R CONTE-DEBOER, p. 73).

Continuation of zeros and of partial wave amplitudes in energy

As discussed in Sect. 2.4.3 the zeros of the amplitudes lie on two-dimensional surfaces in a complex s, t-space. These "zero trajectories" are expected to vary smoothly except in the vicinity of their known singularities. At present we are interested only in real s, i.e. in the intersection of the surfaces with $\mathrm{Im}\,s=0$, which occurs along a curve in a 3-dimensional s, $\mathrm{Re}\,t$, $\mathrm{Im}\,t$ (or s, $\mathrm{Re}\,z$, $\mathrm{Im}\,z$)-space. One considers usually the projection of these curves onto the complex z-plane or w-plane.

The constraint which demands a smooth energy dependence of all zeros in the w-plane can be used in a similar way as the "shortest path method" in conventional phase shift analysis. Again, BARRELETs method has an advantage, since one can use the additional information that the zeros can pass a resonance surface [$s=(M-i\Gamma/2)^2$] only at certain points ("Legendre zeros" and intersections with other resonance surfaces, see Sect. 2.4.3). Furthermore, this method is more transparent, since one sees immediately under which conditions the continuation becomes difficult: if a zero passes or touches the physical s, t-plane ("critical points", $P=\pm 1$).

Shortcomings and possible improvements

The simplified version of BARRELETs method has clearly an advantage over the SACLAY method (76 L AYED), except for possible problems with the unitarity constraint. However, two points have to be improved, if one wants to obtain results which can compete with the work described in Sects. 2.1.7 and 2.1.9.

i) It is not justified to ignore the contribution of the remainder R in Eq. (1.11.1). The twofold continuum ambiguity certainly exists and, from the simplified BARRELET method alone, one cannot decide how much the lower partial waves are affected, if R is neglected. Since R cannot be determined from the experimental data, one has to add theoretical constraints, of which some are discussed in Sect. 2.1.14. Others are mentioned in BARRELETs paper (72 L).

ii) In BARRELETs method, analyticity constraints for zero trajectories are taken into account only in a semi-quantitative way, ignoring problems which arise when zero trajectories "intersect" each other. This occurs frequently, since the trajectories have a tendency to run along lines t = const and u = const. It is an improvement to take into account *quantitative* analyticity constraints for zero trajectories, as it is done in the work of SABBA-STEFANESCU (78 L, 80 L-1 and 2, 81 L, 82 L-1 and 2).

Remarks on distant zeros

One can distinguish two types of distant zeros:

i) Zeros of the exact amplitude $F(w)$. Some properties of these distant zeros are well-known. There exist many intersections of pole surfaces ("double poles") and some Legendre zeros outside the physical region. Each of these double poles and Legendre zeros lies on a zero trajectory. Some of these trajectories approach the physical region as the energy increases and become "nearby zeros".

It is possible that these distant zeros approximately agree with distant zeros of the truncated Legendre expansion, sometimes even outside the region of convergence. They are "unstable", if determined from data, but their energy dependence can be relatively smooth, if they are determined from a partial wave analysis which uses analyticity constraints.

ii) Zeros of the truncated expansion, which are *not* related to zeros of the exact amplitude. These zeros occur, because the expansion has to simulate the effects of branch cuts and poles in the z-plane. As an example, we mention the zeros belonging to a pole

$$\frac{1}{1-z/z_0} \approx 1 + \frac{z}{z_0} + \left(\frac{z}{z_0}\right)^2 + \ldots + \left(\frac{z}{z_0}\right)^{\mathrm{N}} = \frac{1-(z/z_0)^{\mathrm{N}+1}}{1-z/z_0}. \tag{1.11.7}$$

The exact pole term has no zeros, but the truncated expansion, which is a good approximation for physical values of z, has N zeros at the roots of $(z/z_0)^{\mathrm{N}+1} - 1 = 0$, which lie on the circle $|z| = |z_0|$. Such a "crown" of zeros is found, if zeros are calculated from KH 78 or CMU-LBL 79 partial waves at high energies (see for instance Fig. 1 in 81 L CHEW).

If the distant zeros of the "crown" are neglected, as it is done in the simplified BARRELET method, one simulates somehow the effects of the nucleon pole and of the ϱ-pole by a *shift of the nearby zeros.* It is well-known that the contribution of the nucleon pole is important for near-backward scattering and that the ϱ-pole gives the main background in the isospin odd amplitude.

D. CHEW'S work

D. CHEW and coworkers applied the simplified BARRELET method to the analysis of the π^+p elastic scattering data. In the most recent papers (79 L CHEW-1, 80 L CHEW) the transversity amplitudes are approximated by a product ansatz, Eq. (1.11.3) with 6 zeros in the momentum range $0.6 < k < 1.2$ GeV/c (highest partial wave F 37) and with 10 zeros in the range $1.2 < k < 2.3$ GeV/c (highest partial wave H 3, 11).

An important part of D. CHEWs work is a careful investigation of the behaviour of the zero trajectories near the critical points, which is necessary for the resolution of the discrete ambiguities. This was also done for charge-exchange scattering (78 L CHEW).

Some of the results for the π^+p partial waves differ strongly from those of all earlier authors. The origin of the discrepancies is not known. There are several possible reasons:

i) The data input. D. CHEW used only a selection of a small number of experiments, whereas all accurate data were taken into account by other authors. In particular, the preliminary data set 76 BARDSL has a large weight although the analysis has not yet been completed.

ii) The treatment of the data. D. CHEW developed a method which is expected to identify systematic errors larger than the statistical ones. This leads to a "severe selection of scattering data" (79 L CHEW-1, 78 L CHEW). Other authors think that the removal of doubtful data points should be made only by the authors of an experiment and that a phase shift analyst should use all published points.

iii) The neglection of the remainder in Eq. (1.11.1).

iv) The use of a Breit-Wigner formula which is simpler than that of other authors (cf. Sect. 2.4.1.2).

The most serious problem is

v) The violation of unitarity (Figs. 2 and 3 in 80 L CHEW). The method is of interest only if a procedure is found which enforces the unitarity constraint.

In 80 L CHEW, Figs. 1···3, the Breit-Wigner fits to the points in the Argand diagrams are so poor that the evidence for new resonances and the determination of the parameters of the known resonances cannot be accepted.

Finally, D. CHEW claims that her final result satisfies fixed-t dispersion relations, because the partial waves are fitted to Breit-Wigner parametrizations and background terms. One should note that all other authors in this field consider *a much stronger constraint*, if they test the "compatibility with the fixed-t dispersion relation".

One has to calculate the invariant amplitudes from the partial waves and to add an assumption on the high energy behaviour, which is compatible with the data up to 200 GeV/c. Then one has the choice either to evaluate the dispersion integral or to employ an expansion method. It would be of interest to calculate the invariant amplitudes from D. CHEWs parametrizations of the partial waves and to compare the Argand diagrams with those shown in our Handbook (79 R HÖHLER-KAISER-KOCH-PIETARINEN).

Further details can be found in BARRELETs papers (72 L, 76 L) and in papers by D. Chew and coworkers (76 L, 78 L, 79 L-1, 79 L-2, 80 L, 81 L). See also 76 L BAKER. The present status has been reviewed recently by SABBA-STEFANESCU (81 L), who treats in particular the application of quantitative methods for the resolution of the ambiguities. He also gives a critical discussion of the meaning of errors calculated with BARRELETs method (see also SABBA-STEFANESCU 78 L, 80 L-2, 82 L-1).

Conclusion

I agree with the main point of BARRELETs work: a detailed study of zero trajectories is of great importance for partial wave analysis, in particular for the resolution of the discrete ambiguities. It is of interest to apply quantitative constraints for zero trajectories, which were derived from local two-variable analyticity by SABBA-STEFANESCU.

As far as the *simplified* BARRELET method is concerned, it might be useful for partial wave analyses in other reactions, where the data are not yet very accurate and the theoretical constraints are weaker. In pion-nucleon scattering, however, one nowadays can obtain information on the higher partial waves, if the generally accepted theoretical constraints are applied in a quantitative way (Sect. 2.1.14). There is no reason to be satisfied with the neglection of these partial waves and with semi-quantitative applications of the theoretical constraints.

2.1.12 The analysis HENDRY 80

The analysis covers data at 14 momenta in the range 1.6···10 GeV/c for all 3 reactions. It is done in two stages (HENDRY 78 L, 80 L, and 81 L).

Stage 1. The method is an energy independent analysis which differs from the usual ones insofar as the adjustable parameters are not the partial waves, but the coefficients of a parametrization, which is based on the impact parameter picture. As an example we give the ansatz for the imaginary part of the partial wave $f_{\ell+}$

$$\mathrm{Im}\, f_{\ell+}(s) = y_1 \mathrm{e}^{-y_2 B^2} + (y_3 + (-1)^{\ell+1} y_4)\mathrm{e}^{-y_5(B_+ - y_6)^2}, \quad (1.12.1)$$

where $B=(J-1/2)/q$, $B_+=(J+1/2)/q$, and y_i are the adjustable parameters. The first term describes diffraction scattering and the second term peripheral scattering. The advantage of this ansatz is that the parameters have a slow momentum dependence, since part of the variation is taken into account by $B(s)$ and $B_+(s)$.

A similar ansatz is made for $\mathrm{Re}\, f_{\ell+}$, $\mathrm{Re}\, f_{(\ell+1)-}$ and $\mathrm{Im}\, f_{(\ell+1)-}$. At each momentum 48 parameters are determined from the data.

Stage 2. In order to improve the solution, the individual partial waves are allowed to vary freely in the vicinity of the values determined in the first stage.

At momenta below 2.75 GeV/c, it is necessary to remove the P 33, P 13, and P 11 partial waves from the sum and to let them vary freely. The results will be discussed in Sect. 2.4.1.

Comments

i) It would be useful to take into account all available data, including for instance the wide-angle data of the Argonne group (78 JENKINS) and the spin-rotation data at 6 GeV/c.

ii) In the high energy region, the isospin $I=1/2$ and 3/2 amplitudes consist of two contributions which have a quite different physical origin: the dominant diffraction scattering and the small charge-exchange scattering, which is related to ϱ-exchange. Instead of representing the latter as the difference of two large quantities, it would be preferable to parametrize the isospin even and odd combinations separately.

iii) Several quantities are not determined by the experimental data and therefore depend strongly on the theoretical input, for instance the real/imaginary part ratios or the ratio of the flip/nonflip amplitudes, which can be determined only from spin-rotation data [Eq. (A.4.1)]. The ansatz for the imaginary parts, which is motivated by geometrical ideas and by duality, is known to have some success. However, there are no good arguments for using the same ansatz also for the *real parts* (see 72 L HARARI).

iv) A crucial test for HENDRYs method is to check, whether the amplitudes are compatible with fixed-t dispersion relations in the near-forward region and with interior dispersion relations in the near-backward region.

v) The comparison with the KH 78 solution is simplified, if one uses the impact parameter transforms calculated in 80 L HÖHLER-2. Our aim is similar to that of HENDRY insofar as we want to find a simple parametrization for the imaginary parts of the amplitudes at high energies. However, we start from a much more flexible ansatz and include the analyticity constraints.

2.1.13 π^-p scattering near the η-production threshold

The reaction $\pi^- p \to \eta n$ has its threshold at the total c.m. energy $W = 1488$ MeV, corresponding to $k = 687$ MeV/c, $q = 433$ MeV/c, $T_\pi = 562$ MeV. The η-width is so small ($\Gamma = 0.85$ keV) that η can be treated as a stable particle for our purpose.

The η-production threshold lies just below the S 11 resonance N(1535), which has a width of about 150 MeV and two main decay modes: ηn (≈55%) and πN (40%). We shall neglect the other decay modes and apply a two-channel formalism (48 L WIGNER, 59 L BAZ, 61 L ROSS, 62 R DALITZ, 65 L KATO, 77 L BHANDARI, 79 L SARMA).

For the S 11 partial wave, the T-matrix can be expressed in terms of a real symmetric matrix M and a diagonal matrix Q as follows

$$\mathsf{T} = \frac{1}{\mathsf{M} - i\mathsf{Q}}; \quad \mathsf{T} = \begin{pmatrix} T_{\pi\pi} & T_{\pi\eta} \\ T_{\eta\pi} & T_{\eta\eta} \end{pmatrix}; \quad \mathsf{M} = \begin{pmatrix} \mathrm{M}_{\pi\pi} & \mathrm{M}_{\pi\eta} \\ \mathrm{M}_{\pi\eta} & \mathrm{M}_{\eta\eta} \end{pmatrix}; \quad \mathsf{Q} = \begin{pmatrix} q & 0 \\ 0 & q_\eta \end{pmatrix}, \tag{1.13.1}$$

where $T_{\pi\pi} = f_{0+}^{1/2}$ is the S 11 πN partial wave and $T_{\pi\eta}$ and $T_{\eta\eta}$ are the S 11 partial waves of the reactions $\pi^- p \to \eta n$ and $\eta n \to \eta n$. Since we are only interested in a scattering length approximation, we retain only terms of first order in the ηn c.m. momentum q_η and take the πN c.m. momentum q as well as M and $T_{\pi\eta}$ as constants, evaluated at the η-threshold.

From Eq. (1.13.1) one obtains

$$T_{\pi\pi}(q_\eta) = f_{0+}^{1/2} = (f_{0+}^{1/2})_{\mathrm{th}} + i q_\eta T_{\pi\eta}^2 . \tag{1.13.2}$$

The unitarity relation leads to

$$\begin{aligned} \mathrm{Im}\, T_{\pi\pi} &= q|T_{\pi\pi}|^2 + \frac{1}{4q}[1 - (\eta_{0+}^{1/2})^2] = q|(f_{0+}^{1/2})_{\mathrm{th}} + i q_\eta T_{\pi\eta}^2|^2 + q_\eta |T_{\pi\eta}|^2 \\ &= q|(f_{0+}^{1/2})_{\mathrm{th}}|^2 + q_\eta \,\mathrm{Re}(T_{\pi\eta}^2). \end{aligned} \tag{1.13.3}$$

The modulus of $T_{\pi\eta}$ can be obtained from data for the total cross section for η-production near threshold

$$\sigma(\pi^- p \to \eta n) = \tfrac{2}{3} 4\pi (q_\eta / q)|T_{\pi\eta}|^2 = \tfrac{2}{3} \frac{\pi}{q^2} [1 - (\eta_{0+}^{1/2})^2]. \tag{1.13.4}$$

The experiment 73 L BINNIE leads to the result

$$|T_{\pi\eta}|^2 = (1.10 \pm 0.09)\,\mathrm{mb}. \tag{1.13.5}$$

In order to derive an expression for the shape of the π^-p elastic cross section at the η-threshold, we insert Eq. (1.13.2) into the expression for the differential cross section, Eq. (A.4.13) and keep all quantities constant except q_η

$$\begin{aligned} \frac{d\sigma_-}{d\Omega} - \left(\frac{d\sigma_-}{d\Omega}\right)_{\mathrm{th}} &= \tfrac{4}{3} \mathrm{Re}\{i q_\eta T_{\pi\eta}^2 G_-^*\} = -\tfrac{4}{3} |T_{\pi\eta}|^2 |G_-| \sin(2\phi_\eta - \phi_G) q_\eta \quad \text{if} \quad q_\eta^2 > 0; \\ &= -\tfrac{4}{3} |T_{\pi\eta}|^2 |G_-| \cos(2\phi_\eta - \phi_G) |q_\eta| \quad \text{if} \quad q_\eta^2 < 0. \end{aligned} \tag{1.13.6}$$

ϕ_η and ϕ_G are the phases of $T_{\pi\eta}$ and of G_-, respectively. Below threshold, q_η has to be replaced by $i|q_\eta|$.

Equation (1.13.6) describes a cusp effect in the elastic π^-p cross section. The best graphical representation is to plot the cross section at fixed angle as a function of $q_\eta^2/|q_\eta|$. The shape of the cusp depends on the values of the phases. In the present case the cross section decreases on both sides of the tip of the cusp (Fig. 1.13.1 and 79 L Sarma).

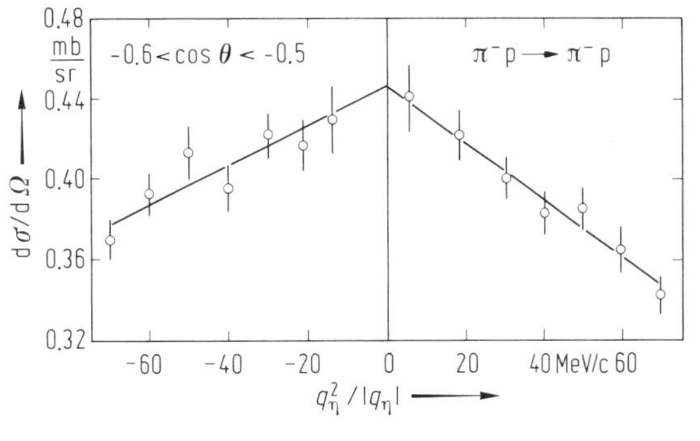

Fig. 1.13.1. The η-threshold cusp in π^-p elastic scattering (79 L Sarma).

The discontinuity of the slope of $d\sigma/d\Omega$ at the cusp can be measured as a function of the scattering angle in π^-p elastic scattering. Since the phase ϕ_η is a constant, one can measure in this case the *absolute phase* of the no-flip amplitude G_- at the energy of the η-threshold, since the phase of the forward amplitude is known from the optical theorem and Coulomb interference data or from an evaluation of the forward dispersion relation.

Up to now, the cusp has been observed only in two experiments at Rutherford laboratory: 75 Debenh and 79 Sarma. Sarma et al. give values for the absolute phase ϕ_G and for G_- as determined from their data and Eq. (1.13.6). The result agrees as far as it can be expected with the prediction from phase shift analyses, some deviations being probably due to the approximate analysis of Sarma et al. They obtain the value $\phi_\eta = 19°$ for the phase of the production amplitude.

The η-threshold leads to a characteristic cusp phenomenon also in the Argand diagram of the S 11 πN amplitude $T_{0+}^{1/2}$, which can be calculated from Eq. (1.13.2), since the modulus and the phase of $T_{\pi\eta}$ are known. Below threshold, the Argand vector lies in our approximation on the circle. Above threshold, it moves rapidly towards the center because $\eta_{0+}^{1/2}$ is rapidly decreasing (77 L Bhandari). In the Argand diagram of the forward amplitude the cusp is not seen, because it is practically not visible in the total cross section data.

Isospin violation in πN scattering from $\pi^\circ - \eta$ mixing has recently been investigated by Cutkosky (79 L). A significant effect is predicted only near and somewhat above the η-threshold (650···900 MeV/c).

The ω-production threshold

The situation is similar at the ω-production threshold, which lies at $W = 1722$ MeV, $k = 1092$ MeV/c, but the analysis is more complicated, because the width is now much larger ($\Gamma = 10.1$ MeV). Again, a resonance lies close to the threshold: P 11 N(1710).

ω-production and π^-p elastic scattering in this region have recently been investigated by a Rutherford group (79 Karami). The ω-production is close to isotropic. One expects that the S 11 and P 11 πN partial waves will be affected, but the interpretation is not yet clear.

2.1.14 The tail of high partial waves

Phase shift analysis is an efficient method for the determination of amplitudes, because the higher terms are suppressed by the centrifugal barrier and by the short range nature of the strong interaction forces. The number of partial waves required increases with energy and so does the complexity of the angular structures of the data.

The problem is the following: at a given energy, one can determine from the data only a certain finite number of parameters, which depends on the angular structures and on the experimental errors.

Now there are 3 possible attitudes:

i) One assumes a sharp cut-off in the partial wave expansion, neglecting all higher partial waves with $\ell > \ell_{max}$. ℓ_{max} is determined by the condition that an attempt to determine partial waves with $\ell > \ell_{max}$ leads to results which do not differ significantly from zero.

ii) One admits a tail of higher partial waves, which is restricted in such a way that the short range nature of the strong interaction is taken into account. The methods used in the KH 78 and CMU-LBL 79 partial wave analyses are described in Sects. 2.1.7 and 2.1.9, respectively. Both groups do not claim that the result for a single high partial wave should be taken seriously, but some collective properties of the tail are relevant.

iii) One uses theoretical predictions for those partial waves which cannot be determined from data, but give nevertheless non-negligible contributions to the amplitude. The simplest example is the approximation of the long range force in NN-scattering by one-pion exchange. In πN scattering, it is necessary to apply more sophisticated methods based on partial wave dispersion relations, (67 R HAMILTON-1, KOCH 82 L-2) or on approximations for the double spectral function (ALCOCK 73 L-1 and 2, 75 L).

Low energies

It is known from the history of phase shift analysis that method i) is not satisfactory in the low energy region. LOVELACE (69 R) noticed that in the range below $T_\pi = 200$ MeV the S- and P-waves can be well determined from the data, but it can be shown analytically that the S 31-wave is unstable against small contributions from D-waves. — At $T_\pi = 310$ MeV, the S, P, and D-waves can be determined, but now there is an instability with respect to small contributions from F-waves.

Because of this problem, CARTER et al. (73 L-1) used method iii) in their phase shift analysis, which covers the range $T_\pi = 94...310$ MeV. They assumed theoretical predictions for D- and F-waves from partial wave dispersion relations.

It is true that HAMILTON and coworkers (67 R-1) had to use phase shifts in order to calculate these predictions. But the point is that the predictions for D- and F-waves at low energies are not sensitive to the input for these partial waves at low energies and in the unphysical region.

In my opinion, the best method for new phase shift analyses in the low energy region is to proceed in the same way as CARTER et al. (73 L-1), using an updated prediction, and to add a constraint from the KH fixed-t solution.

The most recent prediction for the high partial waves at low energies has been made by KOCH and OTTO (82 L KOCH-2), who started from the evaluation of the partial wave dispersion relations by HUTT and KOCH (80 L). The result is a smooth interpolation of the real part of the reduced partial wave, Eq. (A.7.3), between threshold, where it is compatible with the scattering length derived from the fixed-s dispersion relation (78 L KOCH), and the range where the phase shift can be well determined from data. This method is much more reliable than an effective range expansion, because the nearby parts of the left hand cuts are taken into account.

The fixed-t constraint can be applied to the amplitudes $\mathrm{Re}\,\tilde{C}$ and $\mathrm{Re}\,\tilde{B}$ ($\tilde{B} \equiv B - B_N$, etc.) in the range where the partial wave expansion converges (Sect. A.6.2.4) by demanding that the fixed-t amplitudes of the Karlsruhe-Helsinki analysis agree approximately with the amplitudes reconstructed from phase shifts. The unphysical range beyond the backward direction is of special interest, because the relative contribution of the higher partial waves is larger than in the physical region.

Intermediate energies

Theoretical predictions for high partial waves at intermediate energies can be derived, if one has some knowledge of the double spectral functions A_{st} and B_{st} in the neighborhood of the physical region (Fig. A.6.5). The longest range force follows from those unitarity diagrams where two pions are exchanged in the t-channel (Fig. A.5.1b). Because of G-parity conservation at the pion vertices, these diagrams have inelastic intermediate states in the s-channel.

Quantitative estimates of the double spectral functions have been made by ALCOCK and COTTINGHAM (ALCOCK 73 L-1 and 2, 75 L). They used s-channel partial wave expansions for the extrapolations of Im A and Im B into the unphysical region $t > 4\mu^2$ and resonance saturation models for the $\pi\pi$ scattering amplitudes.

The prediction of ALCOCK and COTTINGHAM was included as part of the input in the CMU-LBL partial wave analysis, assuming a rather large error. It would be of interest, to test an updated version of the ALCOCK-COTTINGHAM method by comparing the small high partial waves with those obtained by KOCH and OTTO from partial wave dispersion relations.

Since the most important resonances were successfully determined in the CERN and SACLAY partial wave analyses, which used a sharp cut-off at $\ell = 5$ or 6, it is clear that the effect of the tail of high partial waves is relatively small. Nevertheless it cannot be ignored, if one wants to improve the accuracy of the resonance parameters or of the information on the weaker resonances.

As to the data, the higher partial waves contribute characteristic effects mainly in the near-forward and near-backward directions, where their Legendre functions have large and rapid variations. The most sensitive experimental quantity is probably the Coulomb interference structure, which depends on $\mathrm{Re}\,C$ near $t = 0$. In the expression for the slope $\partial \mathrm{Re}\,C/\partial t$ at $t = 0$, the higher partial waves are enhanced by a factor which is approximately ℓ^3 [see Eq. (A.3.3)]. Unfortunately, Coulomb interference experiments below 2 GeV/c have been performed up to now only at a few energies.

In π^-p charge-exchange scattering experiments, contributions of high partial waves can be detected by performing a Legendre expansion, Eq. (A.4.21). In our analysis of the data 76 BROWN, we noticed a peak of the coefficient $A_{10}(k)$ at $k=1$ GeV/c, which indicates either an unexpected large interference of a $J=15/2$ background term with the resonances or a systematic experimental error (see Fig. 6.5.7 and 79 L HÖHLER-1).

In order to test the tail of high partial waves in the KH 78 and CMU-LBL 79 solutions, one can check the compatibility with the dispersion relation for $\partial C^+/\partial t$ at $t=0$, inserting the amplitude as reconstructed from phase shifts. In this relation, there is no problem with the high energy behaviour, because the slope of $\mathrm{Im}\, C^+$ is known up to 345 GeV/c (82 BURQ).

The result is disappointing for both solutions. The discrepancy with the KH 78 solution (Fig. 4.7.1) is considerably reduced, if the amplitude is calculated with the corrected high partial waves of KOCH and OTTO (82 L-2). The present status is discussed in 81 L KOCH-2.

Another possibility to improve the determination of high partial waves is to use a suitable set of one-dimensional dispersion relations as a constraint in partial wave analysis. In the range $4\mu^2 < t < 50\mu^2$, the curve should pass the region where the absorptive parts can be calculated from a convergent t-channel partial wave expansion. Then, it should enter the physical region at $\theta=0°$ or $180°$ and remain there up to infinity. In this way, one can exploit the experimental information on $\pi\pi$ phase shifts for πN partial wave analysis (J. STAHOV, Karlsruhe, work in progress).

High energies

Above 2 GeV/c, the shape of the near-forward differential cross section changes only slowly with momentum, because the highest resonances are superimposed on a large background which is mainly due to diffraction, except for π^-p charge-exchange scattering where the background is ascribed to reggeized ϱ-exchange.

The tail of high partial waves can easily be calculated, if one assumes that the diffraction peak of $\mathrm{Im}\, G$ has an exponential decrease

$$\mathrm{Im}\, G(s,t) = \frac{q\sigma}{4\pi} e^{Bt/2}; \quad \mathrm{Im}\, \hat{T}_\ell = \frac{q^2\sigma}{4\pi} \frac{\sqrt{\pi}}{\sqrt{2y}} e^{-y} I_{\ell+1/2}(y); \quad y = Bq^2, \tag{1.14.1}$$

where $I_{\ell+1/2}$ is the Bessel function with imaginary argument and $\hat{T}_\ell$ is defined in Eq. (A.3.17).

If one considers the impact parameter transform instead of the partial wave expansion (Sect. A.3.3), $\mathrm{Im}\, \hat{T}_\ell$ is replaced by the profile function $\Gamma(s,b)$, which decreases in the present case as $\exp(-b^2/2B)$, where $b=(\ell+1/2)/q$ [except for complications discussed in connection with Eq. (A.3.30).]

We shall see in Sect. 2.6.3 to what extent the diffraction peak can be approximated by an exponential. In any case, a determination of the high partial waves in the 3⋯10 GeV/c range can be used together with the theoretical predictions at lower momenta for an estimate of these partial waves in the 1⋯3 GeV/c region.

The situation can be described as follows (69 L LOVELACE): "No energy region can be securely pacified until outposts have been established in the region above. The top end of any phase shift analysis will be relatively uncertain. This is just the old imperialist situation: we are continually forced to expand our territory in order to protect the boundaries".

2.1.15 Amplitude analyses without dispersion constraints

First we discuss the determination of the transversity amplitudes, if the following experimental data are given at a certain energy in a range of t-values: the differential cross sections and the polarization parameters for all three reactions $\pi^\pm p \to \pi^\pm p$ and $\pi^- p \to \pi^0 n$ and one of the spin-rotation parameters, for instance R_-. At present, these data are available only at 6 and near 40 GeV/c.

From the differential cross section and polarization data one obtains the modulus of the transversity amplitudes $F(\pm)$

$$|F(\pm)|^2 = \Sigma(\pm) \equiv \frac{d\sigma}{d\Omega}(1 \pm P) \tag{1.15.1}$$

for all three reactions. Therefore, one can construct at each value of s and t the two isospin triangles of Fig. A.4.2, but the relative orientation is not yet fixed.

If in addition the spin-rotation parameter R_- is known, one can determine the spin-rotation angle β_- from Eq. (A.4.2) up to a twofold ambiguity. According to Eq. (A.4.16), β_- is the angle between the transversity amplitudes $F(-)$ and $F(+)$ for elastic π^-p scattering, i.e. β_- determines the relative angle between the two isospin triangles.

The situation is now that the two triangles are determined at each s, t value where data are available, except for two ambiguities:

i) A *continuum ambiguity* due to the fact that all transversity amplitudes can be rotated by a common angle $\phi(s,t)$. Only at $t=0$ this angle is determined by the optical theorem, which gives Im F, and Coulomb interference data or an evaluation of the dispersion relation, which give Re F.

ii) An 8-fold *discrete ambiguity*.*). Assuming an arbitrary direction for $F(+)$ in π^-p elastic scattering because of i), the value of R_- gives two possible directions for $F(-)$ in π^-p elastic scattering. In each case, the isospin triangles can be drawn either to the right or to the left of the π^-p amplitudes.

It was shown in 72 L DEAN that the discrete ambiguity ii) can be resolved completely, if one knows *one* suitably chosen additional spin-rotation parameter. These authors criticize the earlier results of BILENKII et al. (67 L).

At fixed s, the amplitudes $F(\pm;z)$ and, according to Eq. (1.11.2), $\Sigma(\pm;z)$ are analytic functions of z. This is also true for the area of the isospin triangles, except for an additional cut, cf. (Eqs. (A.10.3) and (A.10.4). Therefore, it is of interest to study the z-dependence of the area of the isospin triangles, starting with the known value at $t=0$, where both triangles coincide. The spin-rotation parameters at $t=0$ are $R=0$, $A=1$ and the recoil angle is $\theta_N=90°$.

The analyticity in z leads to a unique continuation for increasing negative values of t, even across the singular points which belong to the case that

i) the area of one of the isospin triangles vanishes (and one of the isospin bounds is saturated, see Sect. A.10),

ii) the spin-rotation parameter A has a zero, which is equivalent to the statement that the spin-rotation angle has the unique value $\beta=\theta+\theta_N$.

In practice, this continuation might lead to difficulties due to data which are not closely spaced or have an insufficient accuracy. Therefore, it is useful to check the continuation beyond the singular points if further spin-rotation data are available.

The above description looks simple, but if one wants to have exact formulas, giving for instance the invariant amplitudes, except for a common phase factor, in terms of the above mentioned experimental quantities, rather complicated expressions are obtained. We refer to work in progress by HUTT and KOCH, who have listed these formulas and studied applications to πN data at high energies.

The determination of amplitudes from data has been investigated by several authors, frequently by using approximations or numerical methods. We mention here the work of GIFFON et al. (71 L, 72 L-1, 72 L-2). Further references will be given below.

Approximations in amplitude analysis at high energies and $t \gtrsim -0.5\,\text{GeV}^2$

Since complete data sets are available only in tiny kinematical regions, it is important to discuss the question to what extent information on scattering amplitudes can be derived from the existing incomplete data sets. The hope is that for a suitable choice of the four independent amplitudes one can find a hierarchy such that it is possible to neglect some of the amplitudes in a first approximation.

From the total cross section data and the optical theorem, it is clear that $\text{Im}\,C^+ \gg \text{Im}\,C^-$ at $t=0$. Coulomb interference data and evaluations of the forward dispersion relation (Sect. 2.2.1) show that $|\text{Re}\,C^+| \ll \text{Im}\,C^+$ at $t=0$, whereas $\text{Re}\,C^-$ is almost equal to $\text{Im}\,C^-$.

The approximately exponential decrease of the diffraction peak in elastic scattering suggests that the dominance of $\text{Im}\,C^+$ remains true up to at least $t=-0.5\,\text{GeV}^2$, whereas the peak in the charge-exchange differential cross section at $t=-0.035\,\text{GeV}^2$ indicates a strong contribution from the amplitude B^-.

These results show clearly that high energy scattering should be described in terms of isospin even and odd amplitudes. Diffraction is a property of the isospin even amplitude which belongs to elastic π^0p scattering. In practice, it is observed in $\pi^\pm$p elastic scattering. One can remove the small isospin odd contribution in the latter case by forming the isospin even combination as in Eq. (A.10.11).

At very small $|t|$, diffraction can be described equally well by each of the different "no-flip" amplitudes C^+, D^+, F^+_{++}, G^+. If $|t|$ becomes somewhat larger, one expects an increasing contribution of the "flip" amplitudes. The high energy behaviour of the modulus of the flip/no-flip ratio will be different for the different choices. The best choice is a pair of amplitudes, for which this ratio is already small just above the resonance region and decreases at higher energies.

*) There are two types of discrete ambiguities. In the present case data for *all 3 reactions* and one spin-rotation parameter are assumed to be known. The 2^{2N} fold ambiguity in Sect. 2.1.11 occurs in the amplitude of a *single reaction*, when only $d\sigma/d\Omega$ and P are known and a truncated partial wave expansion is used.

Unfortunately, there exists no well-founded theoretical prediction for the high energy behaviour of the isospin even amplitudes or for the flip/no-flip ratio. At fixed t and $s\to\infty$, A^+ can tend to infinity as well as to a finite constant or to zero. In the latter case, the dispersion relation for A^+ does not require a subtraction and the subtraction function for the amplitude C^+ can be calculated from $\mathrm{Im}\,A^+(s,t)$. Asymptotic s-channel helicity conservation ($F^+_{+-}/F^+_{++}\to 0$) is a consequence (70 L Höhler-1).

Experimental information on the flip/no-flip ratio can be derived from polarization and spin-rotation data. The first step is to determine the angle ϕ_{RA} of the complex vector $(-R+\mathrm{i}A)$, the spin-rotation angle β and the parameter T [Eqs. (A.4.2) and (A.4.13)]

$$\cos\phi_{RA}=\frac{-R}{\sqrt{1-P^2}};\quad \sin\phi_{RA}=\frac{A}{\sqrt{1-P^2}};\quad \beta=\theta+\theta_{\mathrm{N}}-\phi_{RA};\quad \cos\theta=1+t/2q^2;$$
$$\sin\theta_{\mathrm{N}}=\left[\frac{1+t/4q^2}{1-t/4m^2}\right]^{1/2};\quad T=\sqrt{1-P^2}\sin(\phi_{RA}-\theta_{\mathrm{N}})=\sqrt{1-P^2}\sin(\theta-\beta). \tag{1.15.2}$$

Next, the flip/no-flip ratio κ is obtained from

$$\frac{4\kappa^2}{(1+\kappa^2)^2}=1-S^2=T^2+P^2=\sin^2(\theta-\beta)+P^2\cos^2(\theta-\beta);\quad \kappa=\frac{|F_{+-}|}{|F_{++}|}. \tag{1.15.3}$$

The components of F_{+-} parallel and perpendicular to F_{++} follow from

$$2|F_{++}|(F_{+-})_{\perp}=P\frac{\mathrm{d}\sigma}{\mathrm{d}t};\quad 2|F_{++}|(F_{+-})_{\parallel}=T\frac{\mathrm{d}\sigma}{\mathrm{d}t}. \tag{1.15.4}$$

An evaluation of these equations for the isospin even amplitudes at 6 GeV/c and 40 GeV/c (where spin-rotation data exist) shows that the flip contribution $|F_{+-}|^2$ to the differential cross section is negligible at least up to $t=-0.5\,\mathrm{GeV}^2$. For instance, at $t=-0.3\,\mathrm{GeV}^2$, one finds $\kappa^2=0.006$ at 6 GeV/c and a comparable or smaller value at 40 GeV/c. Even at 3 GeV/c the ratio ($\kappa^2\approx 0.02$) is smaller than the experimental errors of $\mathrm{d}\sigma/\mathrm{d}t$. We conclude that $|F^+_{++}|$ is well-determined by the differential cross section data alone. Our statement that the πN scattering data favor asymptotic s-channel helicity conservation was one of the arguments of Gilman et al. (70 L), when they proposed that asymptotic helicity conservation could be true for a more general class of reactions. Part of our discussion was repeated in the subsequent paper of Pfeffer et al. (70 L).

The relative smallness of $\mathrm{d}\sigma_0/\mathrm{d}t$ in comparison with $\mathrm{d}\sigma_{\pm}/\mathrm{d}t$ remains valid for all t-values of our range. The ratio amounts to 1/50···1/100 at 6 GeV/c and decreases approximately as $1/k$ for increasing values of the lab. momentum. Therefore, $\mathrm{d}\sigma_0/\mathrm{d}t$ can be neglected, if $\mathrm{d}\sigma^+/\mathrm{d}t$ is calculated from Eq. (A.10.11) (see Fig. 6.6.2).

Another consequence of the smallness of the isospin odd amplitudes is that the difference $\mathrm{d}\sigma_-/\mathrm{d}t-\mathrm{d}\sigma_+/\mathrm{d}t$ becomes small in comparison with $\mathrm{d}\sigma_{\pm}/\mathrm{d}t$. Even in the most accurate elastic scattering data above 10.4 GeV/c, the difference is comparable with the experimental errors or smaller (80 L Höhler-1, Fig. 2).

Taking into account the last three remarks, one obtains the following approximation

$$\tfrac{1}{2}\left(\frac{\mathrm{d}\sigma_-}{\mathrm{d}t}+\frac{\mathrm{d}\sigma_+}{\mathrm{d}t}\right)\approx\frac{\mathrm{d}\sigma^+}{\mathrm{d}t}\approx|F^+_{++}|^2;\quad P^+\approx\tfrac{1}{2}(P_++P_-);$$
$$(F^+_{+-})_{\perp}\approx\tfrac{1}{2}P^+\sqrt{\mathrm{d}\sigma^+/\mathrm{d}t}\approx\tfrac{1}{2}P^+|F^+_{++}|;\quad (F^+_{+-})_{\parallel}\approx\tfrac{1}{2}T^+\sqrt{\mathrm{d}\sigma^+/\mathrm{d}t}\approx\tfrac{1}{2}T^+|F^+_{++}|. \tag{1.15.5}$$

If one wants to apply analyticity constraints, it is preferable to work with invariant amplitudes. It turns out that the choice (C^+,A^+) for the pair of amplitudes is better than (C^+,B^+), because the error of neglecting A^+ in the differential cross section Eq. (A.4.17), is smaller than that of neglecting B^+. For instance at 6 GeV/c and $t=-0.3\,\mathrm{GeV}^2$, the error is 2.5% in the first case and 7% in the second.

One could say that the amplitudes at high energies are nearer to asymptotic s-channel helicity conservation ($|F^+_{+-}/F^+_{++}|\to 0$) than to t-channel helicity conservation [$|G^+_{+-}/G^+_{++}|\to 0$, see Eq. (A.2.39)], since a decrease of the first ratio is related to a decrease of $|A^+/C^+|$:

$$\frac{A^+}{C^+}\approx\left[-\frac{m}{k}+\frac{2m}{\sqrt{-t}}\frac{F^+_{+-}}{F^+_{++}}\right](1-t/4m^2)\quad\text{for large } s \text{ and fixed } t. \tag{1.15.6}$$

If A^+ is neglected, $|C^+|$ follows from the differential cross section data

$$\frac{\mathrm{d}\sigma^+}{\mathrm{d}t}\approx\frac{1}{16\pi}(1-t/4m^2)^2|C^+/k|^2. \tag{1.15.7}$$

The small quantity $A_\perp^+$ can be determined from the polarization data [see Eq. (A.4.19)]

$$\left|\frac{A_\perp^+}{C^+}\right| \approx m(1-t/4m^2)P^+/\sqrt{-t}\,. \tag{1.15.8}$$

The formulas for $A_\perp^+$ and for $(F_{+-}^+)_\perp$, Eq. (1.15.5), follow from each other, if we insert the kinematic approximations

$$A_\perp^+ \approx \frac{8\sqrt{\pi}mk}{\sqrt{-t}}(F_{+-})_\perp\,; \qquad |C^+| \approx \frac{4\sqrt{\pi}k}{1-t/4m^2}|F_{++}^+|\,. \tag{1.15.9}$$

They are valid if

$$|F_{+-}^+/F_{++}^+| \ll \frac{4m^2k}{\sqrt{-t}(s+m^2)}, \tag{1.15.10}$$

which is fulfilled in our case.

$A_\parallel^+$ can be determined from

$$A_\parallel^+ \approx 4\sqrt{\pi}m\sqrt{\mathrm{d}\sigma^+/\mathrm{d}t}\{T^+k/\sqrt{-t}-1\}, \tag{1.15.11}$$

Unfortunately, the uncertainties of the spin-rotation data are so large that the small quantity $A_\parallel^+$ is comparable with its errors. An average over 7 points in the range $-0.22 \leqq t \leqq -0.44\,\mathrm{GeV}^2$ at 6 GeV/c gives $A_\parallel^+/|C^+| = 0.16 \pm 0.06$ (74 L Höhler).

If one compares this result with the earlier one for $(F_{+-}^+)_\parallel$

$$\frac{(F_{+-}^+)_\parallel}{|F_{++}^+|} \approx \frac{\sqrt{-t}}{2k}\left\{1+\frac{1}{m(1-t/4m^2)}\frac{A_\parallel^+}{|C^+/k|}\right\}, \tag{1.15.12}$$

it is seen that the two terms in the bracket have the same sign and comparable magnitudes.

We conclude that the high energy behaviour of A^+ should be studied in addition to that of F_{+-}^+.

Amplitude analyses at 6 GeV/c and $t > -0.6\,\mathrm{GeV}^2$

Amplitude analyses were possible as soon as spin-rotation data were available (72 Delesq). Since a common phase factor of all amplitudes cannot be determined, the result consists of $|F_{++}^+|$ and of the parallel and perpendicular components of all other amplitudes with respect to the dominant amplitude F_{++}^+.

The first investigations are outdated, because the accurate $\pi^\pm$p elastic scattering data 74 Ambats were not yet available (71 L Halzen, 72 L Barger, 72 L Cozzika, 72 L Kelly, 73 L Johnson).

In 74 Ambats the authors shifted the existing data to 6.0 GeV/c and to 7 t-values in the range $0 > t > -0.55$ GeV/c and determined amplitudes, which fit the data and fulfill the exact relations for the measurable quantities. They started at $t=0$ and discussed carefully the ambiguities which occur at t-values where isospin bounds are saturated ($t=-0.04$ and $-0.4\,\mathrm{GeV}^2$).

The analysis 74 L Miller of the 6 GeV/c data uses a parametrization of the s-channel helicity amplitudes which is similar to that of Kelly (72 L). The amplitudes are represented by a form factor times an expansion in terms of a variable Ω obtained from a conformal mapping, which carries the left and right hand cuts onto an ellipse in the cosθ-plane. The parametrization is written for the components of the helicity amplitudes with respect to the dominant amplitude F_{++}^+, divided by the appropriate half-angle factors, for instance ($q_-^2 = \mu^2 - t/4$)

$$(F_{+-}^+)_\parallel/\sqrt{1-z} = c\mathrm{e}^{-a(q_- - \mu)}\left\{1+\sum_{n=1}^{6} b_n(\Omega-1)^n\right\}. \tag{1.15.13}$$

The data in the range $0 \geqq t > -0.65\,\mathrm{GeV}^2$ are described by 56 real parameters.

There is an indication that the parametrization is not flexible enough, because the authors were not able to fit the charge-exchange differential cross section near its peak at $t \approx -0.04\,\mathrm{GeV}^2$.

The results of the two amplitude analyses approximately agree with each other.

It is difficult to see the advantage of the t-dependent parametrization, since it introduces a model-dependence. Furthermore, simplicity is expected for the real and imaginary parts of the amplitudes rather than for the components with respect to F_{++}^+. One should note that ReF_{++}^+/ImF_{++}^+ is far from being negligible according to the dispersion methods (Sect. 2.1.16).

Amplitude analysis at 40 GeV/c

Using all available data in the neighborhood of 40 GeV/c, PIERRARD et al. (76 L) performed an amplitude analysis at 12 t-values in the range $0 > t > -0.6\,\mathrm{GeV}^2$. The experimental input is less complete than at 6 GeV/c, because P_0-data were missing and the R_+ data are much poorer. Therefore, he obtained two solutions which are outdated nowadays, because the Serpuchov group has measured P_0 (80 AVVAKU) and remeasured $d\sigma_0/dt$ in collaboration with a CERN group (79 APEL and 81 BINON).

The results of amplitude analyses will be discussed in Sects. 2.6.3 and 2.6.5. If one compares tables of different authors, it is of interest to know that the flip amplitude is sometimes defined with a minus sign with respect to our convention.

2.1.16 Amplitude analyses with dispersion constraints

If the amplitudes are determined from a complete set of data, a common s- and t-dependent phase factor of all amplitudes remains unknown except at $t=0$, where the phase can be derived from total cross sections and Coulomb interference experiments. In the general case, analyticity constraints offer the only possibility to determine the absolute phases of all amplitudes.

Dispersion relations are a powerful tool only if data are available in a large kinematical region. For near-forward scattering, fixed-t dispersion relations have a clear advantage over all other choices, because in a strip $0 \geqq t \gtrsim -1\,\mathrm{GeV}^2$ experimental results exist up to 200 GeV/c.

Problems can arise only from the unknown high energy behaviour above 200 GeV/c and from the unphysical part of the cut between threshold and the point where the line $t=\text{const}$ enters the physical region at $\theta=180°$ (Fig. A.6.3).

For our purpose, the effect of the theoretical restrictions on the asymptotic behaviour (Sects. A.5.3 and A.5.4) is weak, because they are valid in the high energy limit, whereas the uncertainty of the prediction comes mainly from that of the absorptive parts in the range 200···2000 GeV/c. It will be seen that up to about 50 GeV/c the uncertainties of the real parts are small.

The problems connected with the unphysical part of the cut are sometimes overemphasized. For $t \geqq -0.6\,\mathrm{GeV}^2$ the absorptive parts can be calculated from the partial wave expansion. In fact, it turns out that a truncated partial wave expansion still gives an approximation up to about $t=-1\,\mathrm{GeV}^2$. One reason for this behaviour is seen, if the expansions of the "elastic" and "inelastic" parts of $\mathrm{Im}\,A$ and $\mathrm{Im}\,B$ [in the sense of Eq. (A.5.19)] are discussed separately. The dominant "elastic" part has a larger ellipse of convergence which is determined by the boundary T_{II} in Fig. A.6.5. It should be noted that possible errors of the input along the unphysical part of the cut are reduced by the fact that one knows not only the imaginary part but also the real part in the physical region as far as partial waves are available.

Of course, the applicability of the dispersion relations could also be endangered by insufficient data above 3 GeV/c. This is indeed a serious problem, but it turns out that one can at least evaluate rather accurately the dispersion relation for the dominant amplitude C^+. The result for its phase can then be used in order to determine the phases of all other amplitudes from the analyses at 6 and 40 GeV/c.

As mentioned in Sect. 2.1.15, the contribution of the amplitude A^+ in Eq. (A.4.17) to the differential cross section is negligible. Therefore, the modulus $|C^+|$ can be obtained from the differential cross section [Eq. (1.15.7)].

The problem is now to construct a fixed-t amplitude, whose modulus is known at high energies and whose real and imaginary parts are known from partial wave analysis at intermediate and low energies. In 74 L HÖHLER, we used a dispersion relation for this purpose and gave results for $\arg C^+(s,t)$ in a range $0 \geqq t \gtrsim -0.4\,\mathrm{GeV}^2$.

One could think that it would be preferable to employ a logarithmic dispersion relation, where the phase is expressed by a dispersion integral over the modulus. However, this method led to difficulties which were not successfully solved at that time (72 L MCCLURE).

A much more ambitious program was started in 1972 by PIETARINEN. It is described in detail in Sect. 2.1.7. In particular, he obtained a result for $\arg C^+$ at intermediate energies. In the high energy region, the flexibility of the expansion can lead to problems. Therefore, it is useful to determine in a separate analysis an ansatz for the asymptotic behaviour of the amplitudes and to write it as a factor in front of the sum of the expansion as in Eq. (A.6.42).

HECHT, KROLL, and JAKOB (74 L) applied fixed-t dispersion relations in order to determine all 4 invariant amplitudes in the range 1.5···14 GeV/c, $0 > t > -0.4\,\mathrm{GeV}^2$, starting with a careful data amalgamation and using CERN 72 partial waves below 1.5 GeV/c. They found appreciable discrepancies between their result and the invariant amplitudes reconstructed from the CERN partial waves.

The work of HECHT et al. is an improvement over earlier determinations of amplitudes, which used finite energy or continuous moment sum rules and a model-dependent Regge pole ansatz [BARGER and PHILLIPS (69 L)]. Although the parameters of the Regge poles in the BARGER-PHILLIPS model should not be taken seriously (for instance for the ϱ'), their amplitudes were remarkably successful, fitting even some data which were measured several years later (74 L BARGER).

Some discrepancies between the results of HECHT, KROLL, and JAKOB and PIETARINEN are discussed in 74 L HECHT and 76 L PIETARINEN-1.

PIERRARD (77 L) extended the method of HECHT et al. to 100 GeV/c. Up to 10 GeV/c, he used the amplitudes of these authors. At higher momenta, he parametrized the imaginary parts by a sum of power laws and calculated the real parts from the dispersion relations.

The result is satisfactory only up to $t=-0.2\,\mathrm{GeV}^2$. At larger negative t-values the author was not able to fit the R_--data. This problem cannot be serious. It must be possible to adjust the amplitude A^+, since the spin-rotation parameters give the *only* information on $A_{||}^+$ as mentioned in Sect. 2.1.15.

FERRARI (78 L) used the results of HECHT, KROLL, and JAKOB (74 L) in order to determine the lower partial waves by projection. His result is outdated, because PIETARINEN (76 L-4) performed a more accurate determination.

Fixed-t analyticity was also used as a constraint in the analyses of 69 L BAACKE-2, ARGYRES et al. (73 L, 74 L), 73 L DEVENISH, CONTOGOURIS (75 R) and 77 L SVEC.

The results of the KH 78 analysis for amplitudes at high energies are preliminary and not yet quite satisfactory. An updated and improved version will be presented soon.

2.1.17 Electromagnetic corrections

2.1.17.1 Introduction

In analyses of accurate pion-nucleon scattering data at and below the first resonance, it is important to treat carefully a number of corrections which are related to electromagnetic interactions and other isospin-breaking effects. A classification given by TROMBORG et al. (78 L) distinguishes corrections of the phase shifts and of the absorption parameters.

There are the following phase shift corrections, which include an $I=1/2$ and 3/2 mixing term:

i) Corrections due to Coulomb scattering of the external particles. This term includes the exchange of a photon between a pion and a nucleon in πN intermediate states.

ii) Corrections due to the $\pi^\pm-\pi^0$ and n − p mass differences. The pion mass difference is an electromagnetic effect, whereas the nucleon mass difference is partially due to the u − d quark mass difference (see the remarks in Sect. 2.4.4.1).

iii) Corrections due to the difference between the π^-pn, π^0pp, and π^0nn coupling constants.

iv) Corrections following from the coupling to the γn-channel in π^-p-scattering.

v) Other corrections, which are classified as "short range contributions" in the dispersion approach. Some effects of the u − d quark mass difference belong to this category. Others are included in ii).

The "inelasticity corrections" of the absorption parameters, which also include an isospin $I=1/2$ and 3/2 mixing term, can be calculated from the unitarity relation. In π^+p scattering this correction is only due to bremsstrahlung and turns out to be negligible. The bremsstrahlung correction is also negligible in π^-p scattering, but it is important to take into account the effect of the γn-channel.

2.1.17.2 Models for the electromagnetic corrections

The theoretical methods developed until 1975 are treated in the reviews by HAMILTON (75 R) and by RASCHE and WOOLCOCK (76 R). There are two different approaches which are based either on potential models or on the application of dispersion relations.

i) Potential models

The early work in this field is discussed in the above mentioned reviews. The first complete treatment of the coupled (π^-p), (π^0n)-system was given in papers by OADES and RASCHE (76 R RASCHE-WOOLCOCK). This work was continued by ZIMMERMANN (74 L, 75 L), who developed a charge-independent potential model, taking into account mass difference effects. In the second paper, he analyzed the π^+p and π^-p scattering data simultaneously and obtained a good fit to the data of BUGG et al. (73 BUSSEY, 71 CARTER, 71 BUGG). The result is given as a set of charge-independent phase shifts.

Another potential model was investigated earlier by BUGG (73 L-1) and applied to the analysis of the data of his group (73 L CARTER-1). These authors give different P 33 phase shifts for π^+p and π^-p scattering. Some problems of their method were discussed by RASCHE and WOOLCOCK (76 R).

The investigations of RASCHE and WOOLCOCK were continued in subsequent papers (76 L-1, 76 L-2, 77 L-1, 77 L-2), which deal with the analysis of $\pi^\pm$p scattering experiments at very low energies, with the effects of radiative capture and with extended unitarity in hadronic multichannel scattering with Coulomb interactions present. Further work: 74 L PILKUHN, 80 L KRIESCHE and 81 L POEPPING.

It is obvious that the potential models have several unsatisfactory features, of which some are discussed in 76 R RASCHE-WOOLCOCK and others in 76 L MYHRER. Presumably, it will be difficult to develop a version which meets with unanimous approval. Therefore, other authors preferred a dispersion method.

ii) Dispersion methods

The early work on dispersion methods is treated in the above mentioned reviews. Since 1973 the NORDITA group (HAMILTON, TROMBORG, ØVERBØ, and WALDENSTRØM) spent a great effort in improving and extending the formalism, which is nowadays considered as a standard method by most phase shift analysts. It was used for instance in the CMU-LBL and KH analyses and also in the recent analysis of the SIN group (82 L ALDER).

The following description is based on 77 L and 78 L TROMBORG. (Earlier work: 73 L HAMILTON, 74 L and 76 L TROMBORG.)

The "nuclear amplitude" F_N is obtained from the $\pi^\pm$p scattering data by subtracting the pure Coulomb scattering and the Coulomb-nuclear interference term. F_N differs from the purely "hadronic amplitude" F_H, because it still contains some electromagnetic contributions. It is the aim to determine the correction

$$\delta F = F_N - F_H, \tag{1.17.1}$$

using as few assumptions as possible about the strong interaction.

Sometimes F_H is defined as the scattering amplitude in the limit of vanishing electromagnetic interaction, but this statement is empty as long as a complete theory including both interactions does not exist.

It is assumed that

i) F_H fulfills isospin invariance and unitarity in the space of purely hadronic states (ignoring effects of the u − d quark mass difference), and that

ii) F_H has the usual analyticity and crossing properties.

A difficulty of this method is that one has to work with specific values for the masses and coupling constants, but there is no unique prescription how these values have to be chosen.

The important observation is now that one can derive a dispersion relation for δF, Eq. (1.17.1) from the analyticity and unitarity properties of F_N and F_H [DASHEN and FRAUTSCHI (64 L) and (65 L)]. However, the form of this relation is not unique. The final version of the NORDITA method is based on the paper 76 L TROMBORG, suitably extended to include the multichannel case (78 L TROMBORG).

Aside from the fact that the difference between the three πNN coupling constants is not known, there is another uncertainty, since it is not possible to evaluate or even estimate the contribution of the distant parts of the left hand cut of the dispersion relation. As a consequence, one cannot expect that the output of the analysis fulfills exactly charge-independence.

Although it is not possible to test charge-independence, there is a constraint: the electromagnetic contribution belonging to the distant part of the cut (the "short range contribution") must be slowly variable in energy in the physical region.

Since the formalism is very complicated, the authors prepared a "user's manual" for the application of their method (77 L TROMBORG), which includes tables of S- and P-wave Coulomb phase shifts, phase shift corrections to $\pi^\pm$p scattering and inelasticity corrections to π^-p scattering.

They suggested to follow PIETARINENs proposal for the analysis: using an earlier version of the phase shifts, one calculates the differential cross sections and polarizations not only directly, but also after having taken into account the electromagnetic corrections. The difference between the two results is then added to the experimental data, in order to obtain "hadronic" data, which can be phase shift analyzed without considering any electromagnetic effects.

One can ask for a comparison between the dispersion method and other methods. In 76 L TROMBORG the authors discuss the relation of their method to that of WEST and YENNIE (68 L). In 78 L TROMBORG they point out that their correction to the differential cross sections differs from that of SOGARD (74 L) and BORIE (78 L-1). Furthermore, they mention that a detailed comparison with the work of ZIMMERMANN (74 L, 75 L) is not possible.

Final remark

It would be of interest to study in detail how the dispersion model can accommodate the isospin breaking effects of the u − d quark mass difference.

2.1.17.3 Radiative corrections

The early work on radiative corrections was discussed by LOCHER (67 L) and by WEST and YENNIE (68 L), whose results have been used in the analysis of the experimental data at high energies. In a more recent paper, SOGARD (74 L) found unexpected large effects from radiative corrections to π^-p elastic backward scattering. The correction to the backward cross section exceeds 30% in the 10···20 GeV/c region. It can be understood classically as being due to the rapid change of the direction of the moving charge. The effects is much smaller for π^+p backward scattering. The slope of the π^-p backward peak is not much affected. SOGARD also studied radiative corrections to near-forward elastic scattering and to π^-p charge-exchange scattering. Starting from SOGARDs work, E. BORIE (78 L-1) calculated tables of radiative corrections which show their magnitude under different kinematic conditions. One should notice that radiative corrections can be calculated only for a particular experimental situation. They are different if the elastically scattered particle or the recoil particle are detected.

The problem of radiative corrections to backward scattering and to π^-p-charge-exchange scattering was also treated by GINZBURG, KOTKIN, and SERBO (78 L), who obtained simple formulas. They gave some critical comments on SOGARDs work.

Further investigations in this field are desirable, since reliable radiative corrections are needed not only for the analysis of high energy backward scattering data but also for charge-exchange forward scattering as mentioned in Sect. 2.2.1.2.

The investigation of radiative pion-proton scattering ($\pi^\pm p\gamma$ in the final state) is of interest because one obtains information on pion-nucleon off-shell amplitudes and on electromagnetic properties of the nucleon resonances. We mention only a few recent papers: 77 L HO-KIM, 77 L and 78 L LIOU, 79 L PICCIOTTO, 80 L SMITH, from which earlier references can be traced.

2.1.17.4 Electromagnetic corrections to polarization data

At earlier times, electromagnetic corrections to polarization data were ignored by experimentalists. Following a suggestion by BOURRELY and SOFFER (77 L, earlier work: 73 L VANZHA) BORIE and JAKOB (78 L-2) investigated the corrections in detail and found that they are comparable with the errors or larger at and above 40 GeV/c in the range $t > -0.4\,\mathrm{GeV}^2$. For more accurate data, these corrections should even be considered at lower momenta (Fig. 1.17.1, next page).

In the low energy region, the electromagnetic effects should lead to a structure (Fig. 3.2.3a, p. 171), which has not yet been observed in experimental data. Polarization data in this range would be an interesting new input in phase shift analysis.

See also 80 R BOURRELY-LEADER-SOFFER and, for $\dot{\mathrm{N}}\mathrm{N}$ scattering, 78 L BUTTIMORE.

2.1.17.5 The $\Delta^{++}-\Delta^0$ mass difference

In the energy range of the Δ(1232)-resonance, the phase shift analysis of CARTER et al. (73 L-1) led to a difference between the nuclear P 33 phase shifts for π^+p- and π^-p-scattering, which cannot be explained by the usually considered electromagnetic effects. The discrepancy amounts to two standard deviations.

In the analysis of their accurate total cross section data, PEDRONI et al. (78) also found some evidence for a violation of charge-independence. CARTER et al. and PEDRONI et al. suggested an interpretation in terms of different isobar masses and widths of the Δ^{++} and Δ^0 states:

$$M_0 - M_{++} = (2.7 \pm 0.3)\,\mathrm{MeV}; \quad \Gamma_0 - \Gamma_{++} = (6.6 \pm 1.0)\,\mathrm{MeV}; \text{ 78 PEDRON.} \qquad (1.17.2)$$

The lower indices refer to the charge of the isobar. The authors mentioned that the total uncertainty is larger than the errors in Eq. (1.17.2).

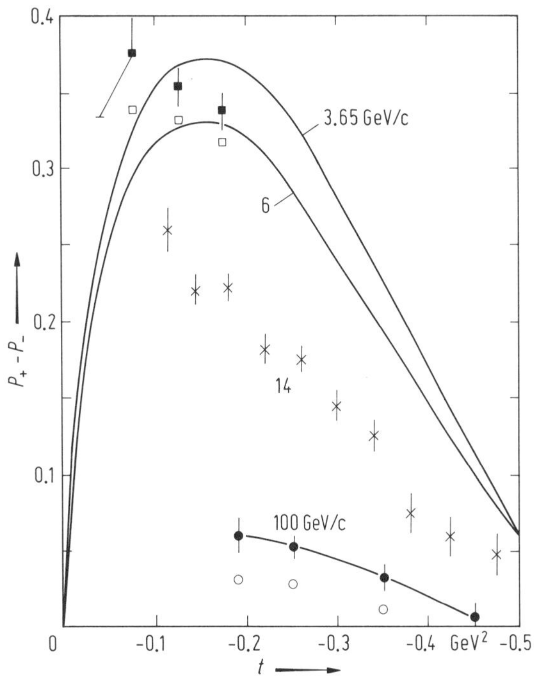

Fig. 1.17.1. The magnitude of the electromagnetic correction to $P_+ - P_-$. Data points: ■ 70 Borghi (6 GeV/c); × 72 Borghi (14 GeV/c); ● 80 Kline (100 GeV/c). The open symbols below the data points show the corrected values. Solid lines: reconstruction of $P_+ - P_-$ from the KH 78 partial wave solution. The discrepancy at 6 GeV/c and small t is related to the violation of the isospin bound in Fig. 4.4.4. The shape of the t-dependence of $P_+ - P_-$ will be discussed in Sect. 2.6.5.3.

Koch and Pietarinen (80 L-1), who used the formalism of Tromborg et al. (77 L) for the electromagnetic corrections, found no clear evidence for isospin-breaking effects in the differential cross section and polarization data. A reanalysis of the data set 78 Pedroni confirmed the result for the mass splitting, but a considerably smaller splitting of the widths (2 ± 1.8) MeV was found.

The authors of the new SIN phase shift analysis (82 L Alder) had an additional input from their P_- and P_0 polarization data. They found a difference between the P 33 phase shifts for π^-p- and π^+p-scattering after having applied the electromagnetic corrections (77 L Tromborg), but no indication for isospin violations in the S-waves.

Zidell, Arndt, and Roper (80-1, Table VI) give the following result for the isobar pole parameters

$$\begin{aligned} M_{++} &= (1210.7 \pm 0.08)\,\text{MeV}; \quad & M_0 &= (1210.3 \pm 0.36)\,\text{MeV}; \\ \Gamma_{++} &= (99.21 \pm 0.23)\,\text{MeV}; \quad & \Gamma_0 &= (108.0 \pm 0.52)\,\text{MeV}. \end{aligned} \tag{1.17.3}$$

In a "world average", which was used by some authors for coupling constants, scattering lengths etc., this result would dominate strongly due to the extremely small errors. However, one should notice that $\Delta M_{++}/M_{++} = 7 \cdot 10^{-5}$ corresponds to $\Delta k/k = 4 \cdot 10^{-4}$ or 0.04% for the accuracy of the pion lab. momentum. This is much better than the accuracy of the beam momentum of the most accurate experiment.

Furthermore, the discussion of the method of Zidell et al. in Sect. 2.1.10 shows that the true uncertainties are much larger. The result derived from the new SIN analysis (82 L Alder) is, in my opinion, at present the most reliable one.

There exist a number of theoretical papers on the $\Delta^{++} - \Delta^0$ mass splitting. Thirring (65 L) investigated the electromagnetic properties of hadrons in a static SU(6) model. His mass formula depends on 3 parameters which are determined from well-known mass differences. For our case, his calculation gives

$$M(\Delta^0) - M(\Delta^{++}) = 2(m_n - m_p) - \{M(\Sigma^+) + M(\Sigma^-) - 2M(\Sigma^0)\} = (0.8 \pm 0.2)\,\text{MeV}. \tag{1.17.4}$$

Rubinstein et al. (67 L) started from a mass formula which contains a contribution from the quark masses and spin-dependent two-particle terms. They found 9 relations between the 18 masses of the octet and decuplet ground states. For the $\Delta^0 - \Delta^{++}$ mass difference the result agrees with that of Thirring.

Thirrings formula also follows as an approximation in the model of de Rujula, Georgi, and Glashow (75 L), who studied hadron masses in a gauge theory. In this model, the quarks are bound by *long-range confining forces*, which depend only on the spatial separation, but not on the quark masses and spins. In addition, they assume *short range forces*, depending on quark masses and spins. This "Coulomb-like" force has the form of a two-body BREIT-FERMI interaction between charged particles. It corresponds to the exchange of gluons and photons. Since specific assumptions on the confining potential are not made, the authors obtain only relations between different masses. They use first order mass formulas and neglect terms of order $\alpha(m_d - m_u)$.

Further work in this field was done by BALL and GOBLE (75 L), by CELMASTER (77 L) [who also predicted a resonance N(1335) S 11, which is not seen in phase shift analysis], by LICHTENBERG (75 L, 76 L) and by DESHPANDE et al. (77 L), who estimate the mass difference in the MIT bag model.

Final remark

In my opinion, one should not simply draw conclusions by comparing numerical values for the mass splitting obtained in phase shift analysis and in the study of quark models. It is necessary to discuss carefully the theoretical meaning of these quantities. For instance in πN scattering, one can define a resonance mass in different ways (see Sect. 2.4.1), since the resonance pole lies in the complex domain. But in quark models, the isobar is treated as a stable particle.

In a realistic model, one would have to include the decay. The effect would be a shift of the Δ^0 and Δ^{++}-poles in the s-plane from the real axis by about 50 MeV in the negative imaginary direction. Such a shift is usually accompanied by a shift in the direction of the real s-axis. The question is, whether the small distance of the poles, as calculated from Eq. (1.17.4), remains valid for the locations of the complex poles. The literature on the closely related problem of the neutron-proton mass difference is given in the review 82 R GASSER-LEUTWYLER. See also 80 L WEST and 82 L FUCHS.

2.2 Results for pion-nucleon scattering amplitudes

2.2.1 πN forward amplitudes from total cross sections and dispersion relations

2.2.1.1 Introduction

A first step in (almost) all partial wave and amplitude analyses is the determination of the forward amplitudes $C^{\pm}$ or $C_{\pm}$. The imaginary parts follow from total cross section data and the optical theorem, Eq. (A.5.5)

$$\operatorname{Im} C_{\pm}(\omega) = k\sigma(\pi^{\pm}p) \equiv k\sigma_{\pm}\,. \tag{2.1.1}$$

The real parts can be obtained from the analysis of charge-exchange forward scattering data, of Coulomb interference data (Sect. 2.6.2) or by an evaluation of the forward dispersion relations (Sect. A.6.4.).

There are three problems in the last mentioned case:

i) One has to make an assumption on the behaviour of the total cross sections above the experimental region, i.e. above 370 GeV/c.

ii) One has to determine the subtraction constant of the dispersion relation for C^+. This constant is given either by the subthreshold coefficient c_{00}^+ or by the S-wave scattering length a^+. The method for its determination is discussed in Sect. 2.4.7.

iii) One has to determine the coupling constant f^2. The methods are described in Sect. 2.4.2.1.

The evaluation of the dispersion relation can either be made by calculating the principal value integral (80 R HÖHLER-KAISER) or by applying an expansion method (72 L PIETARINEN-3).

In the following, we shall discuss some points which are of interest for the determination of the real parts.

2.2.1.2 The input for the total cross section

Although the total cross section data have a very good statistical accuracy, there are several relatively large systematic discrepancies. Figures 2.1.1 and 2.1.2 show the situation below 0.7 GeV/c, where the data sets 72 DAVIDS ("Arizona data") and 68 CARTER ("Rutherford data") differ considerably from each other (up to 8%). Part of the discrepancy seems to be due to an error in the determination of the momentum of the incoming beam.

These figures also show another problem in the low energy region: a deviation from charge independence due to different masses and widths of the Δ^{++} and Δ^0 isobars. This effect was noticed in the data sets 71 CARTER and 78 PEDRON.

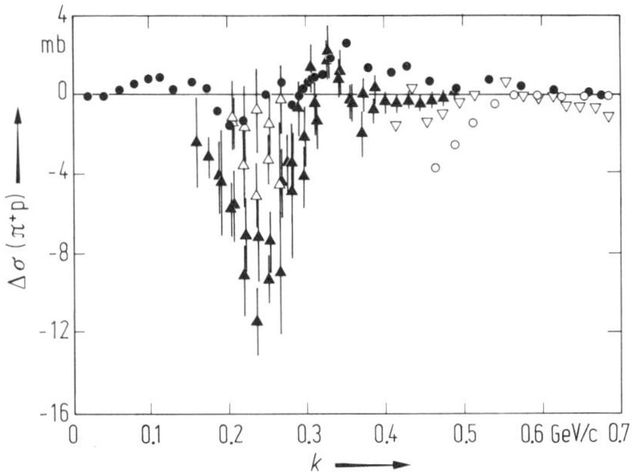

Fig. 2.1.1. Discrepancies between π^+p total cross sections below 0.7 GeV/c. The plot shows the data minus our interpolation in Table 2.2.1.1. ▲ 78 PEDRON (systematic error 0.6%, not corrected for Coulomb distortion); △ the same data, but corrected for Coulomb distortion (80 L KOCH-1). The latter points are only shown if the correction is large. ○ 68 CARTER; ▽ 72 DAVIDS. ● total cross sections reconstructed from the KH 80 partial wave analysis (Sect. 2.1.7, 80 L KOCH-1), in which charge-independence is assumed after the usual Coulomb corrections have been taken into account.

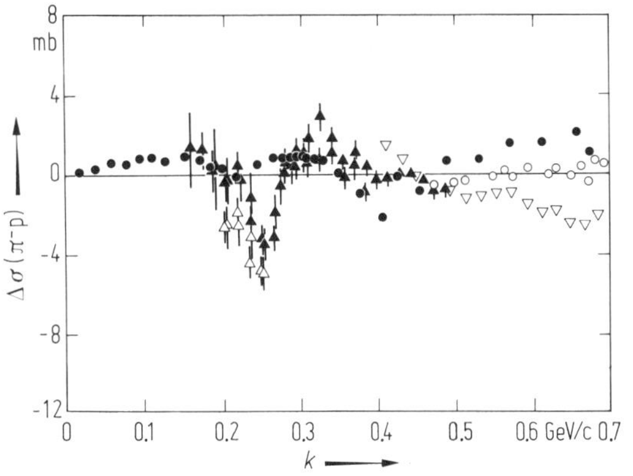

Fig. 2.1.2. Discrepancies between π^-p total cross sections below 0.7 GeV/c. Notation: see Fig. 2.1.1.

Figure 2.1.3 shows the isospin 1/2 total cross section, calculated from the usual formula after having removed the electromagnetic effects including the Coulomb distortion correction according to the formalism of TROMBORG et al. (77 L). The remaining structure indicates that isospin violating effects are comparable with the magnitude of $\sigma^{1/2}$ as calculated from the dominant S11-contribution.

Figures 2.1.4···2.1.7 show the total cross sections in the high energy region. There is a discrepancy between the data sets 66 CITRON and 67 FOLEY at 8 GeV/c. Part of it has been corrected by GIACOMELLI (70 R). For π^-p-scattering the situation will become much better as soon as the data 80 BARREL are available (2···14 GeV/c).

Another discrepancy is shown in Fig. 2.1.7 around 30···50 GeV/c. The trend of the Serpuchov data led to many speculations, but nowadays it is clear that the correct behaviour is given by the Fermilab data sets 76 CARROL and 79 CARROL.

The first step in a dispersion analysis is to assume a smooth interpolation and a parametrization for the high energy behaviour up to infinity. A number of possibilities is listed in Sect. A.6.4.4.

Table 2.2.1.1 is based on the following assumptions

$$\sigma^+(k)=\sigma_0+\sigma_1\ln^2(k/k_1)+bk^{\alpha-1};\quad k\geqq 10\,\text{GeV/c};\quad \text{fit I}$$
$$\sigma_0=22.26\,\text{mb};\quad \sigma_1=0.4183\,\text{mb};\quad k_1=37.83\,\text{GeV/c};\quad b=8.229\,\text{mb GeV}^{1-\alpha};\quad \alpha=0.534; \tag{2.1.2}$$

$$\sigma^-(k)=c/\sqrt{k};\quad c=3.30\,\text{mb GeV}^{1/2}. \tag{2.1.3}$$

The large number of decimals is not significant, but it is needed in order to reproduce our table.

Figure 2.1.4 shows in addition the following fits:

$$\sigma^+=b_1k^{\alpha_1-1}+b_2k^{\alpha_2-1};\quad \text{fit II};\quad b_1=8.30\,\text{mb GeV}^{1-\alpha_1};$$
$$b_2=27.99\,\text{mb GeV}^{1-\alpha_2};\quad \alpha_1=1.15;\quad \alpha_2=0.702; \tag{2.1.2a}$$

$$\sigma^+=\sigma_0+\sigma_1\ln(k/k_1)+bk^{\alpha-1};\quad \text{fit III};\quad \sigma_0=26.14\,\text{mb};\quad \sigma_1=3.30\,\text{mb};$$
$$b=38.85\,\text{mb GeV}^{1-\alpha};\quad k_1=4516\,\text{GeV/c};\quad \alpha=0.708. \tag{2.1.2b}$$

A fit of the type II will be used in Sect. 2.6.4.9.

The analytic parametrizations of the forward amplitude belonging to these and other fits are discussed in 80 R HÖHLER-KAISER. The real parts of the forward amplitude, which are derived from these fits, will be treated in Sect. 2.6.2.

Figure 2.1.6 shows that there is a large momentum range in which the total cross section σ^+ decreases as $1/k$.

The simple law Eq. (2.1.3) was chosen although the best fit to the total cross section difference gives a power law with $\alpha_\varrho = 0.54 \pm 0.03$ (79 CARROL). The reason is that the decrease of the charge-exchange forward cross section can be well fitted with a power law and $\alpha_\varrho = 0.48 \pm 0.01$ (76 BARNES). According to 78 L GINZBURG, there are radiative corrections to the charge-exchange forward cross section, which cannot be accurately evaluated, but the tendency is to increase the value of α_ϱ. If one takes Figs. 2.1.7 and 2.1.8 together, it is seen that the simple choice $\alpha_\varrho = 0.50$ is not yet excluded. Table 80/2 in 79 R HÖHLER-KAISER is based on $\alpha_\varrho = 0.534$. — The decrease of the difference of the pp and $\bar{p}$p total cross sections at very high energies is also compatible with a power law and $\alpha = 0.50$ (82 L BLOCK).

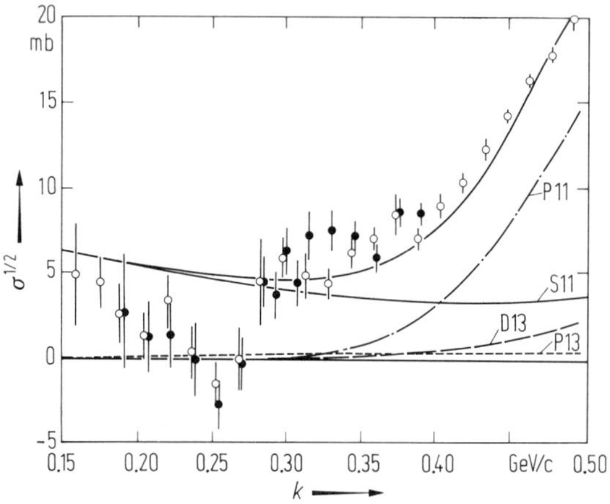

Fig. 2.1.3. Isospin $I = 1/2$ combination of the total cross sections: $\sigma^{1/2} = (3\sigma_- - \sigma_+)/2$. The electromagnetic effects including the Coulomb distortion have been removed. The structure is probably due to a charge-dependence of the parameters of $\Delta(1232)$. The upper solid line shows $\sigma^{1/2}$ as reconstructed from the KH 80 partial wave amplitudes. The lower lines show the contributions of the S-, P-, and D-waves. It is seen that the S 11 contribution is dominant at and below the momentum belonging to the $\Delta(1232)$ resonance ($k = 0.30$ GeV/c). The P 11-resonance N(1470) occurs near $k = 0.56$ GeV/c. Data: 78 PEDRON; ⚲ 20 cm target, ● 10 cm target.

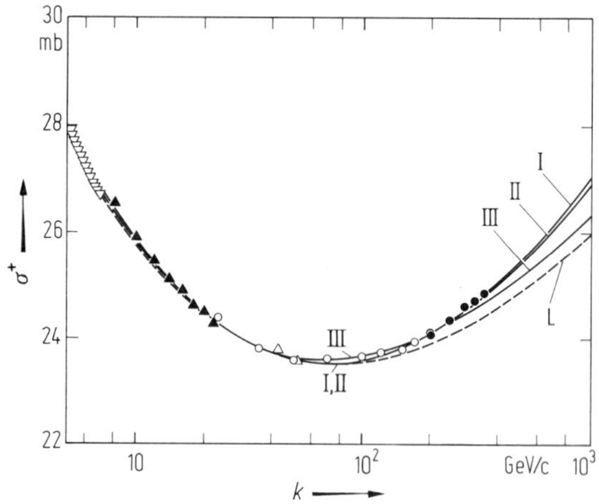

Fig. 2.1.4. The isospin even combination $\sigma^+ = (\sigma_+ + \sigma_-)/2$ of the total cross sections at high energies. Data: ▽ 66 CITRON (including GIACOMELLIs correction (70 R)), ▲ 67 FOLEY, ○ 76 CARROL, ● 79 CARROL, △ 76 APOKIN. The error bars are not visible in our scale. — The solid lines are fits to the high energy behaviour of σ^+ which are explained in Eqs. (2.1.2), (2.1.2a), and (2.1.2b). The dashed line (L) belongs to LIPKINs fit: Eq. (6.4.42), which does not include the data 79 CARROL.

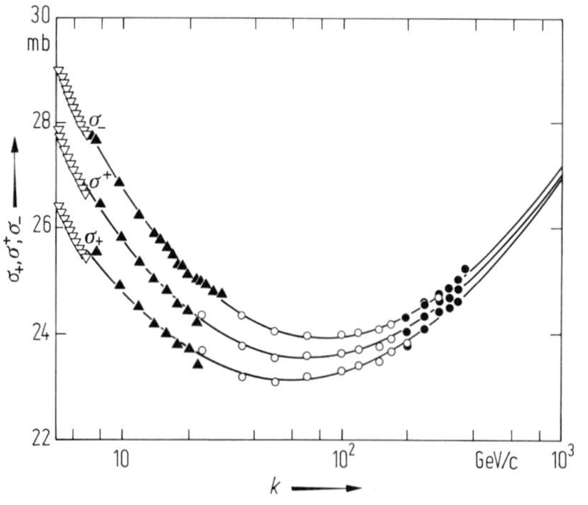

Fig. 2.1.5. Total $\pi^\pm$p cross sections above 5 GeV/c. The solid lines belong to the fits of Eqs. (2.1.2) and (2.1.3). Notation: see Fig. 2.1.4.

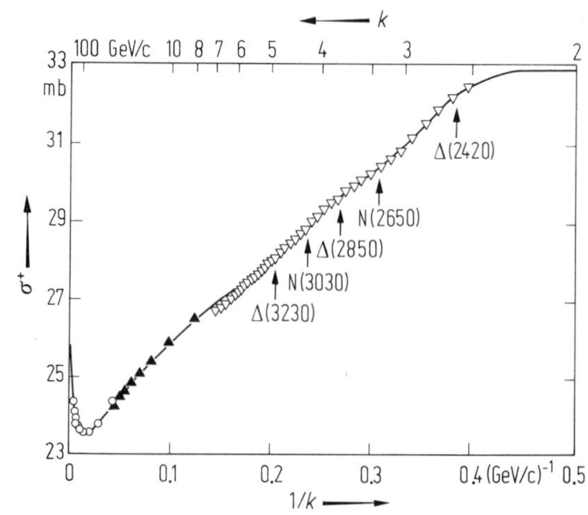

Fig. 2.1.6. The decrease of σ^+ between 2 and 100 GeV/c. We have plotted σ^+ vs. $1/k$ in order to show that the decrease is almost linear in $1/k$ in a large momentum range. It is difficult to see the small resonance structures. Notation: see Fig. 2.1.4.

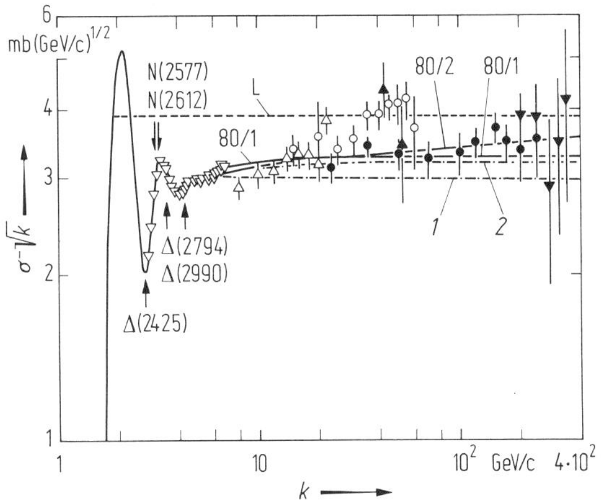

Fig. 2.1.7. Plot of $\sqrt{k}\sigma^-$ vs k. $2\sigma^-$ is the difference between the $\pi^{\pm}p$ total cross sections. Data: ▽ 66 CITRON [including GIACOMELLIs corretion (70 R)]; △ 67 FOLEY; ○ 73 DENISO; ▲ 76 APOKIN; ● 76 CARROL; ▼ 79 CARROL. — The solid curves denoted by "80/1" and "80/2" belong to fits in which a power law $\sigma^- = ck^{\alpha-1}$ is assumed for the asymptotic behaviour and $\alpha=0.5$ and 0.534, respectively. Tables of forward amplitudes are given in both cases in 80 R HÖHLER. Table 2.2.1.1 agrees with Table 80/1, i.e. it follows with the choice $\alpha=0.5$. — The dot-dashed lines denoted by "1" and "2" belong to fits which are constrained by the superconvergent sum rule, Eq. (6.5.50), and have the high energy ansatz of Tables 80/1 and 80/2, respectively. $\alpha_{\varrho'}$ has been chosen <-1. It is seen that the sum rule is compatible with the data. — The masses used in the names of the resonances in Figs. 2.1.6 and 2.1.7 are slightly different from those in the tables of Sect. 2.4.1.

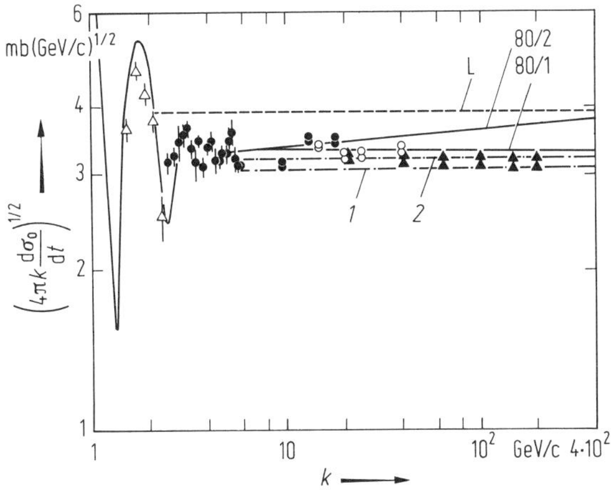

Fig. 2.1.8. High energy behaviour of the charge-exchange forward cross section. The plot shows $(4\pi k\, d\sigma_0/dt)^{1/2}$, because this quantity is equal to $\sigma^- \sqrt{k}/[\sqrt{2}\cos(\alpha\pi/2)]$ if $\sigma^- = ck^{\alpha-1}$ in the asymptotic region. With our choice $\alpha=1/2$ the quantities plotted in Figs. 2.1.7 and 2.1.8 should be equal at high energies, if this assumption is correct. Data: △ 73 NELSON, ● a fit to the data 68 YVERT and 65 STIRLI, ▲ 76 BARNES, ○ 79 APEL. The data points of 79 APEL and 76 BARNES have been plotted twice. An estimate for the radiative correction is included in the higher points. The lines "80/1" and "80/2" belong to power laws for σ^- with $\alpha=0.5$ and 0.534, respectively. The dot-dashed lines belong to the dot-dashed lines in Fig. 2.1.7. "L" is the fit used in LIPKINs work (78 L).

2.2.1.3 Evaluation of the dispersion relation

The evaluation consists of two parts:

i) The principal value integral from $k=0$ to $k=k_h$, where the parametrization starts, has to be calculated on a computer.

ii) The remaining part of the integral from $k=k_h$ to infinity is calculated analytically. We have expanded the denominator of the dispersion integral as a geometrical series, calculating the integral over each term separately. Depending on the k-range, one expands either the integral from 0 to k_h or the integral from k_h to infinity, using the fact that for our choice of the parametrizations the integral from 0 to infinity can be expressed by a known function. Details of the evaluation are given in 79 R HÖHLER-KAISER and HÖHLER, JAKOB, and KAISER (75 L-1).

Do-it-yourself-method

It is of interest to try a variety of different high energy parametrizations in order to see the consequences for the energy dependence of the real parts. For calculations of this type, it is not necessary to repeat the numerical evaluation of the principal value integral on the computer.

We choose an analytic parametrization $C^+_{\rm par}$ for the high energy behaviour such that $\operatorname{Im} C^+_{\rm par} = k\sigma^+_{\rm par}$ is a fit to the total cross section data above $k=k_h$. For simplicity, we assume that $C^+_{\rm par}$ fulfills the dispersion relation

$$\operatorname{Re} C^+_{\rm par}(\omega) = k^2 \frac{2}{\pi} \fint_0^\infty \frac{\sigma^+_{\rm par}(k')}{k'^2 - k^2}\, dk', \tag{2.1.4}$$

although it is not important that the high energy ansatz has the correct location of the threshold branch point.

Now, the dispersion relation, Eq. (2.1.4), is subtracted from the dispersion relation Eq. (A.6.46) for the C^+-amplitude

$$\mathrm{Re}\,C^+(\omega) = \mathrm{Re}\,C^+_{\mathrm{par}}(\omega) + C^+_{\mathrm{N}}(\omega) - C^+_{\mathrm{N}}(\mu) + C^+(\mu) + k^2 \frac{2}{\pi} \mathop{⨍}_0^{k_\mathrm{h}} \frac{\sigma^+ - \sigma^+_{\mathrm{par}}}{k'^2 - k^2} \mathrm{d}k'. \tag{2.1.5}$$

We consider $\mathrm{Re}\,C^+(\omega)$ for momentum values k somewhat higher than k_h and expand the integral

$$\mathrm{Re}\,C^+(\omega) = \mathrm{Re}\,C^+_{\mathrm{par}}(\omega) + C^+(\mu) + 4\pi f^2/m + (c_0^{\mathrm{par}} - c_0) + (c_2^{\mathrm{par}} - c_2)(k_\mathrm{h}/k)^2 + \cdots. \tag{2.1.6}$$

Small terms in the expansion of C^+_N are neglected.

The coefficients

$$c_{2n} = \frac{2}{\pi} \int_0^{k_\mathrm{h}} \left(\frac{k}{k_\mathrm{h}}\right)^{2n} \sigma^+(k)\mathrm{d}k \tag{2.1.7}$$

are calculated only once from the fit to the total cross section data below $k = k_\mathrm{h}$. Our evaluation gives for $k_\mathrm{h} = 10\,\mathrm{GeV/c}$

$$\bar{c}_0 \equiv c_0 - C^+(\mu) - 4\pi f^2/m = 490.7\,\mathrm{GeV}^{-1}; \quad c_2 = 146.4\,\mathrm{GeV}^{-1}. \tag{2.1.8}$$

The coefficients c_0^{par} and c_2^{par} follow from the same integral, Eq. (2.1.7), by inserting σ^+_{par} instead of σ^+. It is remarkable that σ^+_{par} has to be used in a range far away from that where it was fitted to the data.

For some parametrizations, $c_2^{\mathrm{par}} - c_2$ and $c_0^{\mathrm{par}} - \bar{c}_0$ are small and for others, which fit the high energy data equally well, these quantities can be large. The correction $c_0^{\mathrm{par}} - c_0$ is small, if the high energy parametrization also fits *in the average* the resonance structures at lower energies.

The "do-it-yourself method" can now be summarized as follows: First, one has to choose a parametrization for the high energy amplitude, whose coefficients are fitted to the total cross section data. In general, one takes a linear combination of the following expressions

$$C^+_{\mathrm{par}} = (-\mathrm{i}k)^\alpha \{\ln(k/k_1) - \mathrm{i}\pi/2\}^\beta, \tag{2.1.9}$$

which includes Regge poles ($\beta = 0$), Regge dipoles ($\beta = 1$) and special Regge cuts ($\beta < -1$). (65 R van Hove, 72 L Barger-5). Then, one has to calculate

$$c_0^{\mathrm{par}} = \frac{2}{\pi} \int_0^{k_\mathrm{h}} \sigma^+_{\mathrm{par}}(k)\mathrm{d}k. \tag{2.1.10}$$

The exact evaluation of the dispersion integral at high energies is given by

$$\mathrm{Re}\,C^+(\omega) = \mathrm{Re}\,C^+_{\mathrm{par}}(\omega) + (c_0^{\mathrm{par}} - \bar{c}_0) + O[(k_\mathrm{h}/k)^{-2}], \tag{2.1.11}$$

where the value of $\bar{c}_0$ can be taken from Eq. (2.1.8).

We conclude that at high energies it is sufficient to correct $\mathrm{Re}\,C^+_{\mathrm{par}}$ by a constant. The isospin odd amplitude can be treated in an analogous way.

In the case of our parametrization, Eqs. (2.1.2) and (2.1.3), the real parts are given by

$$\begin{aligned} \mathrm{Re}\,C^+_{\mathrm{par}}(\omega) &= \pi\sigma_1 k \ln(k/k_1) - b\cot(\alpha\pi/2)k^\alpha; \\ \mathrm{Re}\,C^-_{\mathrm{par}}(\omega) &= c\sqrt{k}, \end{aligned} \tag{2.1.12}$$

and the first terms of the expansion for $k > k_\mathrm{h}$ read (in GeV^{-1}-units)

$$\mathrm{Re}\,C^+ = \mathrm{Re}\,C^+_{\mathrm{par}} + 3.28; \quad \mathrm{Re}\,C^- = \mathrm{Re}\,C^-_{\mathrm{par}} + 14.0/k. \tag{2.1.13}$$

We have treated this method in some detail, because the results are needed for the discussion of two other methods.

Analytic parametrization methods

Several authors fitted parametrizations with the correct analyticity and crossing properties to the total cross section data and sometimes also to the real part data (for instance 70 L Barger-2, 72 L Barger-5, 73 L Bourrely). The ansatz is usually a combination of expressions of the type Eq. (2.1.9) and a similar expression for $C^-(\omega)$

$$C^-(\omega) = \mathrm{i}(-\mathrm{i}\omega)^\alpha(\ln\omega - \mathrm{i}\pi/2)^\beta. \tag{2.1.14}$$

It can happen that the results are at variance with an exact evaluation of the dispersion relation at intermediate energies. In this case $\mathrm{Re}C^{+}_{\mathrm{par}}$ in Eq. (2.1.11) is not yet large enough in comparison with the constant term. A further discussion is given in my reviews (HÖHLER 73 R, 76 R, and HÖHLER-KAISER 80 R).

In principle, it could happen that the correction term is non-negligible even in the high energy limit. This occurs if $\mathrm{Re}\,C^{+}(\omega)$ tends to a constant as ω goes to infinity, or even tends to zero. However, this is possible only if the total cross section goes to a constant (or to zero) and the secondary term decreases rapidly enough (62 L LEHMANN), and this is far from the present experimental situation.

Derivative analyticity relation methods

Several authors claimed that the analyticity properties, which are embodied in dispersion relations, can *equally well* be implemented by writing "derivative analyticity relations" (Sect. A.6.4.5; 74 L BRONZAN, 75 L SIDHU, 75 L SUKHATME, 74 R JACKSON). This method gives the correct result in the case of the ansatz Eq. (2.1.9), which was known much earlier. However, EICHMANN and DRONKERS (74 L) noticed that the method fails in the region where resonance poles exist. This was confirmed by BUJAK and DUMBRAIS (76 L).

SUKHATME et al. (75 L) treated resonance contributions separately, but then the method becomes more complicated than PIETARINENs expansion or the evaluation of the principal value integral.

At momenta above the resonance region, the derivative analyticity method takes into account only the "local" contribution ("local" in energy). This corresponds essentially to the approximation of using only $\mathrm{Re}\,C^{+}_{\mathrm{par}}$ and neglecting the "non-local" contribution from the dispersion integral at low and intermediate energies, which corresponds to the constant term in Eq. (2.1.11).

From the discussion of the "do-it-yourself method" it is clear that the "derivative analyticity method" can give either good or bad results, depending on the choice of the parametrization. The results are good, if the chosen parametrization *happens to fit the resonance structures in the average.* This is not known, unless one includes *non-local information.* Since the main point of the derivative analyticity method is to ignore non-local information, it is *unreliable*, except at very high energies. But then one can as well use a simple analytic parametrization of the types of Eqs. (2.1.9) and (2.1.14).

A rigorous mathematical investigation of derivative analyticity relations has been performed by J. FISCHER and KOLAR (78 L). The authors showed that under certain assumptions, which are stronger than the general principles of axiomatic field theory, derivative analyticity relations exist, but they contain only first derivatives. The authors pointed out that there is no theoretical basis for applications of such relations at a finite energy. See also 81 R FISCHER.

2.2.1.4 Results for forward amplitudes

The results for the forward amplitudes are given in Tables 2.2.1.1 and 2.2.1.2. In the low energy region, the table is based on charge independence and should be corrected for charge-dependent effects as discussed above. Some Argand diagrams for forward amplitudes are shown in Fig. 2.1.9···2.1.15. The results for the ratio of the real and imaginary parts will be discussed in Sect. 2.6.2.

Other determinations of forward amplitudes are discussed in the review 80 R HÖHLER-KAISER. The table of CARTER (73 L-3) has problems with charge-exchange forward cross sections at intermediate and high energies. The table of HENDRICK and LAUTRUP (75 L) shows structures of the real parts at high energies which are due to their unsatisfactory method. The improved version of ENGELMANN and HENDRICK (77 L) essentially agrees with earlier versions of our table. However, I think that it is better to parametrize the large sum and the small difference of the total cross sections at high energies than the total cross sections for $\pi^{\pm}p$ scattering, since the latter method led to a non-smooth behaviour of the difference in the 5···20 GeV/c region. — At high energies, the error channel is misleading, since it is based on assumptions on the behaviour of the total cross sections above the experimental region which can be wrong.

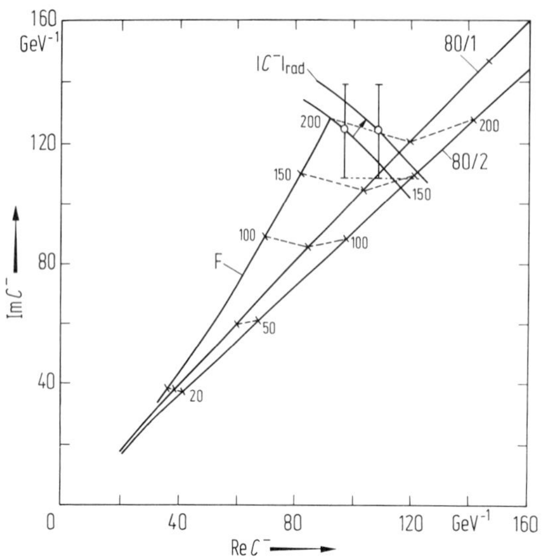

Fig. 2.1.9. Argand diagram of the charge-exchange forward scattering amplitude C^-. The curves "80/1" and "80/2" (which are straight lines above 20 GeV/c) belong to $\alpha=0.50$ and $\alpha=0.534$, respectively. The curve "F" has been constructed from $|C^-|$ as following from the fit of BARNES et al. (76) to their charge-exchange data and $\mathrm{Im}\,C^-=k\sigma^-$ as following from the fit of CARROLL et al. (79) to their total cross section data. The numbers are pion lab. momenta in GeV/c. — At 200 GeV/c, the data point is shown without (on the left) and with the radiative correction, whose magnitude is, however, not well known. See Fig. 2.1.8. — The discrepancy between the curve "F" and the expectation from the Regge pole model has led to speculations about a non-vanishing additional real term $c^-\omega$ in the forward dispersion relation for C^-, Eq. (A.6.48) (see Sect. A.6.3.2). These are discussed in Sect. 8.2 of the review 80 R HÖHLER-KAISER. The conclusion is that the effect is comparable with the errors.

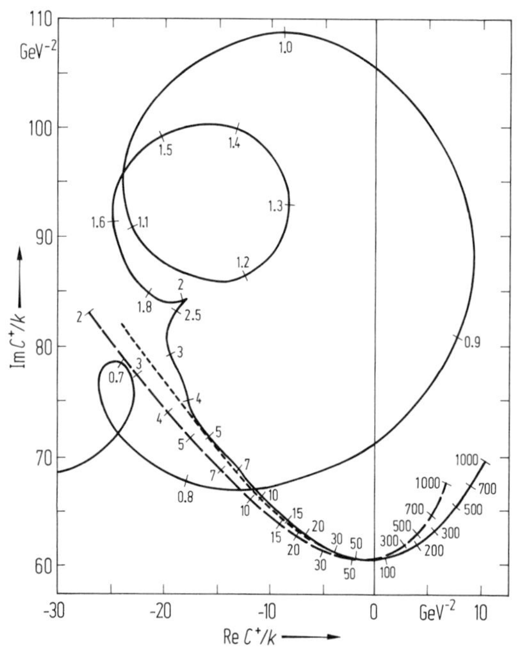

Fig. 2.1.10. Argand diagram of the forward amplitude C^+/k. Solid line: from Table 2.2.1.1. The numbers give the pion lab. momenta k in GeV/c. The dashed lines belong to analytic parametrizations. The expressions for σ^+ are fitted to the data above 10 GeV/c. Short-dashed line: fit I, long-dashed line: fit III in Fig. 2.1.4.

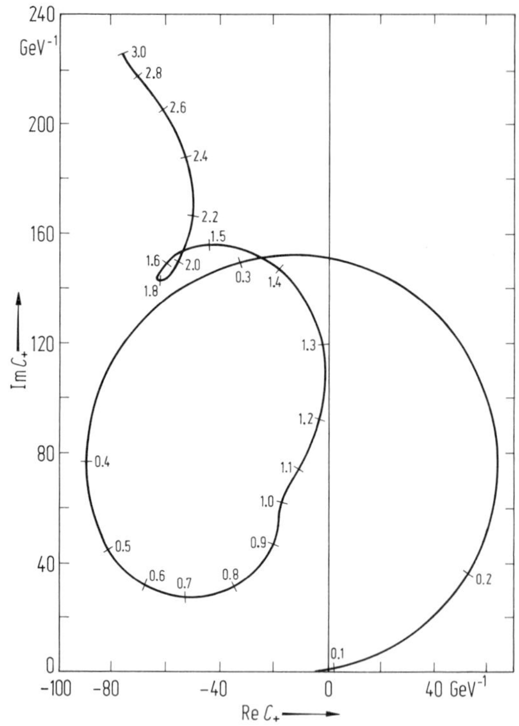

Fig. 2.1.11. Argand diagram for the elastic π^+p forward amplitude. The numbers give the lab. momentum in GeV/c.

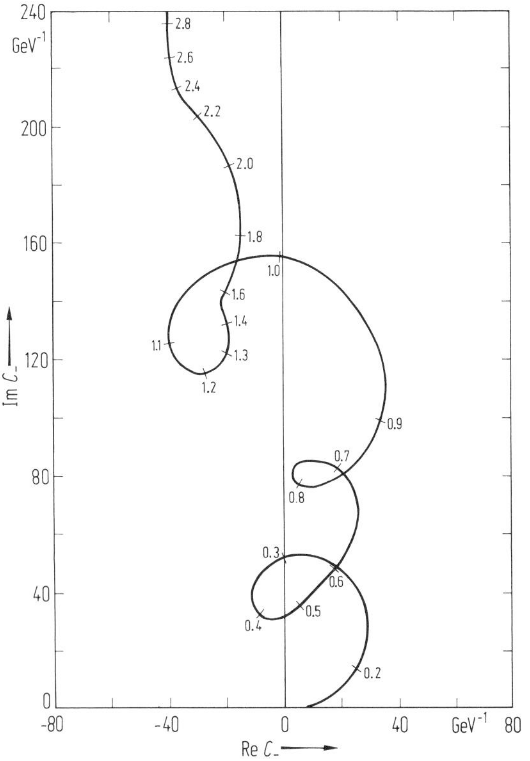

Fig. 2.1.12. Argand diagram of the elastic π^-p forward amplitude.

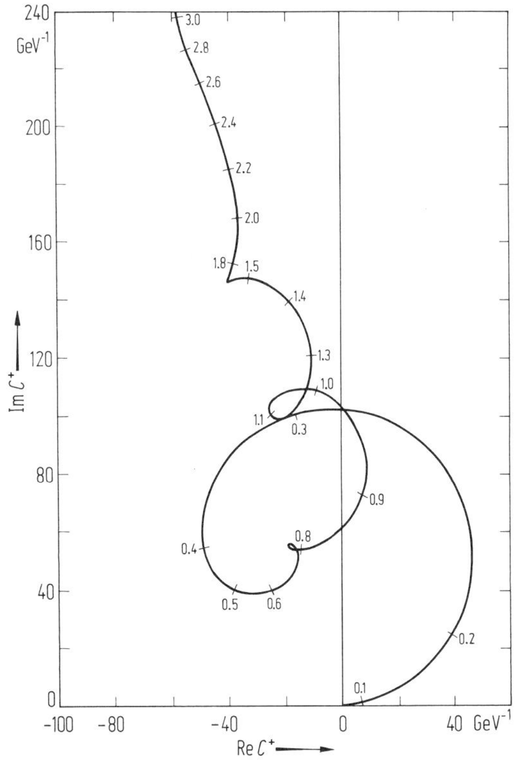

Fig. 2.1.13. Argand diagram of the isospin even forward amplitude.

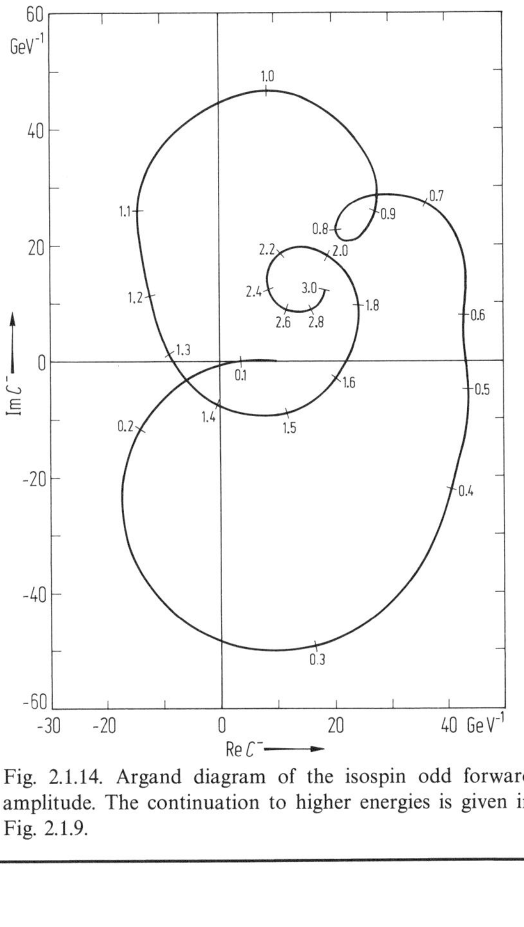

Fig. 2.1.14. Argand diagram of the isospin odd forward amplitude. The continuation to higher energies is given in Fig. 2.1.9.

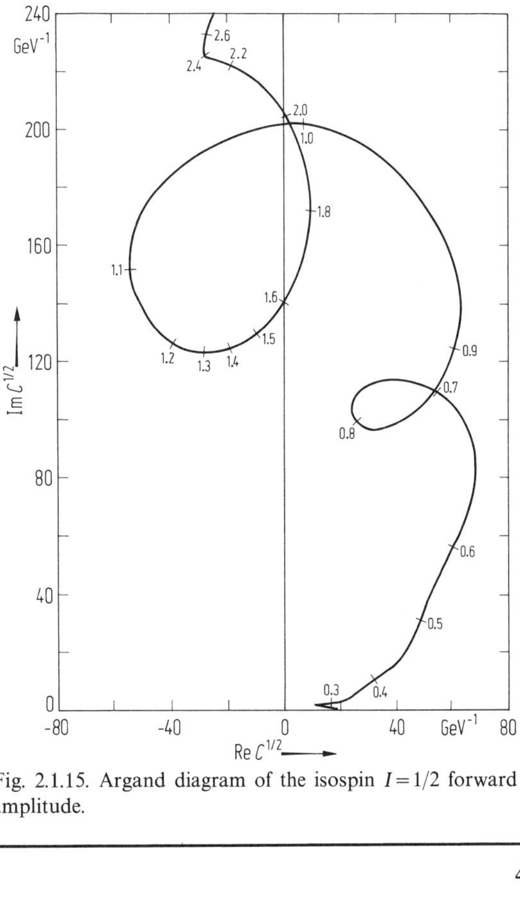

Fig. 2.1.15. Argand diagram of the isospin $I=1/2$ forward amplitude.

Table 2.2.1.1. Pion-nucleon forward amplitudes and cross sections. "EVEN" and "ODD" refer to the isospin even and odd combinations, "CHEX" to the reaction $\pi^- p \to \pi^0 n$ and "DS/DO" = $d\sigma/d\Omega$, "DS/DT" = $d\sigma/dt$. Above 1.9 GeV/c, the list gives "RHO" = Re C/Im C instead of $d\sigma/d\Omega$. It is based on the high energy ansatz of Eqs. (2.1.2) and (2.1.3). See the text and 80 R Höhler-Kaiser for further details.

K GEV/C	REAC-TION	RE C 1/GEV	IM C 1/GEV	SIGMA TOT MB	DS/DO MB/SR	DS/DT MB/GEV**2
0.001	PI+P	-10.53	0.01	2.60	0.21	
	PI-P	8.55	0.02	5.90	0.14	
	EVEN	-0.99	0.01	4.25	0.00	7598.6
	ODD	9.54	0.00	1.65	CHEX: 0.34	
0.010	PI+P	-10.40	0.07	2.65	0.20	8379.6
	PI-P	8.55	0.15	5.94	0.14	5664.0
	EVEN	-0.93	0.11	4.29	0.00	67.3
	ODD	9.47	0.04	1.64	CHEX: 0.34	
0.020	PI+P	-10.01	0.14	2.70	0.19	1940.0
	PI-P	8.55	0.30	5.90	0.14	1417.1
	EVEN	-0.73	0.22	4.30	0.00	11.2
	ODD	9.28	0.08	1.60	CHEX: 0.32	3334.6
0.040	PI+P	-8.44	0.30	2.90	0.13	345.7
	PI-P	8.58	0.60	5.80	0.14	358.1
	EVEN	0.07	0.45	4.35	0.00	1.0
	ODD	8.51	0.15	1.45	CHEX: 0.27	701.9
0.060	PI+P	-5.81	0.49	3.20	0.06	73.1
	PI-P	8.75	0.86	5.60	0.14	166.5
	EVEN	1.47	0.68	4.40	0.00	5.7
	ODD	7.28	0.18	1.20	CHEX: 0.19	228.3
0.080	PI+P	-2.04	0.87	4.25	0.01	6.0
	PI-P	9.21	1.16	5.65	0.16	104.3
	EVEN	3.58	1.02	4.95	0.03	16.8
	ODD	5.63	0.14	0.70	CHEX: 0.11	76.7
0.100	PI+P	3.03	1.52	5.90	0.02	8.9
	PI-P	10.11	1.52	5.90	0.19	81.0
	EVEN	6.57	1.52	5.90	0.08	35.2
	ODD	3.54	0.00	0.00	CHEX: 0.04	19.5
0.120	PI+P	9.77	2.97	9.64	0.18	56.1
	PI-P	11.67	2.12	6.88	0.25	75.7
	EVEN	10.72	2.55	8.26	0.21	65.3
	ODD	0.95	-0.43	-1.38	CHEX: 0.00	1.2
0.140	PI+P	18.17	6.03	16.78	0.63	144.9
	PI-P	13.92	3.23	8.99	0.35	80.7
	EVEN	16.05	4.63	12.89	0.48	110.3
	ODD	-2.12	-1.40	-3.89	CHEX: 0.02	5.1
0.160	PI+P	28.23	11.43	27.83	1.55	280.8
	PI-P	16.92	5.13	12.48	0.52	94.5
	EVEN	22.57	8.28	20.15	0.97	175.0
	ODD	-5.66	-3.15	-7.68	CHEX: 0.14	25.4
0.180	PI+P	39.93	20.62	44.60	3.30	483.0
	PI-P	20.64	8.27	17.90	0.81	118.2
	EVEN	30.29	14.45	31.25	1.84	269.2
	ODD	-9.65	-6.17	-13.35	CHEX: 0.43	62.7
0.200	PI+P	52.33	35.92	69.94	6.45	780.3
	PI-P	24.80	13.45	26.18	1.27	154.1
	EVEN	38.57	24.68	48.06	3.35	406.1
	ODD	-13.77	-11.24	-21.88	CHEX: 1.01	122.3
0.220	PI+P	61.98	60.23	106.61	11.68	1195.5
	PI-P	28.25	21.62	38.27	1.98	202.6
	EVEN	45.11	40.92	72.44	5.80	593.8
	ODD	-16.86	-19.30	-34.17	CHEX: 2.05	210.3
0.240	PI+P	61.32	93.81	152.21	19.20	1689.3
	PI-P	28.51	32.87	53.34	2.89	254.6
	EVEN	44.91	63.34	102.77	9.21	810.9
	ODD	-16.40	-30.47	-49.44	CHEX: 3.66	322.1
0.260	PI+P	42.06	128.83	192.95	27.44	2104.7
	PI-P	22.81	44.63	66.85	3.75	288.0
	EVEN	32.44	86.73	129.90	12.81	982.6
	ODD	-9.62	-42.10	-63.05	CHEX: 5.57	427.4
0.280	PI+P	5.70	150.08	208.72	32.94	2228.8
	PI-P	11.70	51.80	72.04	4.12	278.7
	EVEN	8.70	100.94	140.38	14.99	1014.2
	ODD	3.00	-49.14	-68.34	CHEX: 7.08	479.0

K GEV/C	REAC- TION	RE C 1/GEV	IM C 1/GEV	SIGMA TOT MB	DS/DO MB/SR	DS/DT MB/GEV**2
0.300	PI+P	-32.60	149.64	194.24	33.49	2019.0
	PI-P	0.17	51.78	67.21	3.83	230.8
	EVEN	-16.21	100.71	130.73	14.86	895.7
	ODD	16.38	-48.93	-63.51	CHEX: 7.60	458.4
0.320	PI+P	-60.55	135.85	165.32	30.89	1673.6
	PI-P	-7.24	47.31	57.57	3.20	173.3
	EVEN	-33.89	91.58	111.45	13.32	721.4
	ODD	26.65	-44.27	-53.87	CHEX: 7.46	404.0
0.340	PI+P	-76.93	117.94	135.07	27.08	1328.7
	PI-P	-10.50	42.30	48.45	2.60	127.3
	EVEN	-43.71	80.12	91.76	11.38	558.2
	ODD	33.21	-37.82	-43.31	CHEX: 6.92	339.5
0.360	PI+P	-85.05	101.55	109.85	23.45	1048.9
	PI-P	-11.21	38.03	41.14	2.10	94.0
	EVEN	-48.13	69.79	75.49	9.61	429.6
	ODD	36.92	-31.76	-34.36	CHEX: 6.34	283.6
0.380	PI+P	-88.52	87.78	89.95	20.33	833.7
	PI-P	-10.32	34.78	35.64	1.72	70.6
	EVEN	-49.42	61.28	62.80	8.11	332.5
	ODD	39.10	-26.50	-27.15	CHEX: 5.84	239.4
0.400	PI+P	-89.30	76.70	74.67	17.75	671.0
	PI-P	-8.45	32.55	31.68	1.45	54.7
	EVEN	-48.87	54.62	53.18	6.88	260.1
	ODD	40.42	-22.08	-21.49	CHEX: 5.44	205.4
0.420	PI+P	-88.73	67.79	62.85	15.65	547.6
	PI-P	-5.84	31.19	28.92	1.26	44.2
	EVEN	-47.28	49.49	45.89	5.88	205.8
	ODD	41.45	-18.30	-16.97	CHEX: 5.15	180.3
0.440	PI+P	-87.40	60.50	53.54	13.89	452.1
	PI-P	-2.62	30.93	27.37	1.18	38.6
	EVEN	-45.01	45.71	40.46	5.06	164.7
	ODD	42.39	-14.78	-13.08	CHEX: 4.96	161.3
0.460	PI+P	-85.60	54.47	46.11	12.40	376.8
	PI-P	0.51	31.93	27.03	1.23	37.3
	EVEN	-42.54	43.20	36.57	4.43	134.6
	ODD	43.05	-11.27	-9.54	CHEX: 4.77	145.0
0.480	PI+P	-83.53	49.43	40.10	11.13	316.7
	PI-P	3.24	33.51	27.19	1.34	38.1
	EVEN	-40.14	41.47	33.64	3.94	112.0
	ODD	43.38	-7.96	-6.46	CHEX: 4.60	130.8

K GEV/C	REAC- TION	RE C 1/GEV	IM C 1/GEV	SIGMA TOT MB	DS/DO MB/SR	DS/DT MB/GEV**2
0.500	PI+P	-81.28	45.10	35.13	10.01	267.8
	PI-P	5.66	35.54	27.68	1.50	40.1
	EVEN	-37.81	40.32	31.40	3.54	94.7
	ODD	43.47	-4.78	-3.73	CHEX: 4.43	118.5
0.520	PI+P	-78.78	41.32	30.95	9.00	226.7
	PI-P	7.71	37.67	28.21	1.68	42.4
	EVEN	-35.54	39.50	29.58	3.21	80.9
	ODD	43.25	-1.83	-1.37	CHEX: 4.26	107.4
0.540	PI+P	-75.93	38.17	27.53	8.06	191.9
	PI-P	9.89	39.64	28.58	1.86	44.3
	EVEN	-33.02	38.90	28.05	2.90	69.2
	ODD	42.91	0.73	0.53	CHEX: 4.11	97.9
0.560	PI+P	-72.97	35.83	24.92	7.24	163.3
	PI-P	12.48	42.00	29.21	2.10	47.4
	EVEN	-30.25	38.92	27.06	2.66	60.0
	ODD	42.72	3.08	2.14	CHEX: 4.02	90.7
0.580	PI+P	-70.31	34.03	22.85	6.56	140.5
	PI-P	15.05	44.92	30.16	2.41	51.7
	EVEN	-27.63	39.47	26.50	2.50	53.5
	ODD	42.68	5.44	3.65	CHEX: 3.98	85.3
0.600	PI+P	-67.68	32.23	20.92	5.94	120.9
	PI-P	17.83	48.10	31.22	2.78	56.6
	EVEN	-24.93	40.17	26.07	2.36	48.1
	ODD	42.76	7.94	5.15	CHEX: 3.99	81.4
0.620	PI+P	-64.77	30.79	19.34	5.34	103.6
	PI-P	21.02	52.30	32.85	3.30	64.0
	EVEN	-21.88	41.54	26.09	2.29	44.4
	ODD	42.89	10.75	6.75	CHEX: 4.06	78.8
0.640	PI+P	-62.00	29.71	18.08	4.82	89.4
	PI-P	23.91	58.00	35.29	4.01	74.4
	EVEN	-19.04	43.86	26.68	2.33	43.2
	ODD	42.96	14.15	8.61	CHEX: 4.17	77.4
0.660	PI+P	-58.84	28.54	16.84	4.29	76.1
	PI-P	25.83	65.41	38.59	4.96	88.0
	EVEN	-16.51	46.97	27.72	2.49	44.1
	ODD	42.33	18.44	10.88	CHEX: 4.27	75.8
0.680	PI+P	-55.65	28.27	16.19	3.84	65.3
	PI-P	24.72	74.62	42.73	6.09	103.5
	EVEN	-15.46	51.44	29.46	2.84	48.3
	ODD	40.19	23.17	13.27	CHEX: 4.24	72.1

K GEV/C	REAC-TION	RE C 1/GEV	IM C 1/GEV	SIGMA TOT MB		DS/DO MB/SR	DS/DT MB/GEV**2
0.700	PI+P	-52.60	27.73	15.43		3.43	55.9
	PI-P	18.98	82.12	45.68		6.89	112.3
	EVEN	-16.81	54.92	30.55		3.20	52.2
	ODD	35.79	27.19	15.13	CHEX:	3.92	63.9
0.720	PI+P	-49.05	27.64	14.95		3.02	47.4
	PI-P	12.22	84.61	45.76		6.97	109.2
	EVEN	-18.42	56.13	30.36		3.33	52.1
	ODD	30.63	28.49	15.41	CHEX:	3.34	52.3
0.740	PI+P	-45.52	28.00	14.73		2.68	40.4
	PI-P	7.12	84.57	44.50		6.76	101.9
	EVEN	-19.20	56.29	29.62		3.32	50.0
	ODD	26.32	28.29	14.89	CHEX:	2.80	42.2
0.760	PI+P	-41.82	28.63	14.67		2.37	34.5
	PI-P	3.86	82.42	42.23		6.29	91.3
	EVEN	-18.98	55.52	28.45		3.18	46.2
	ODD	22.84	26.89	13.78	CHEX:	2.30	33.4
0.780	PI+P	-38.02	29.89	14.92		2.13	29.8
	PI-P	3.41	79.36	39.62		5.74	80.3
	EVEN	-17.30	54.63	27.27		2.99	41.8
	ODD	20.71	24.74	12.35	CHEX:	1.89	26.5
0.800	PI+P	-34.37	31.60	15.38		1.95	26.4
	PI-P	5.83	76.71	37.34		5.30	71.6
	EVEN	-14.27	54.16	26.36		2.81	38.0
	ODD	20.10	22.56	10.98	CHEX:	1.64	22.1
0.820	PI+P	-30.71	33.76	16.03		1.84	24.0
	PI-P	10.81	75.90	36.04		5.19	67.7
	EVEN	-9.95	54.83	26.04		2.74	35.8
	ODD	20.76	21.07	10.01	CHEX:	1.54	20.2
0.840	PI+P	-27.38	36.64	16.99		1.82	23.0
	PI-P	16.47	77.93	36.13		5.52	69.7
	EVEN	-5.45	57.29	26.56		2.88	36.4
	ODD	21.93	20.65	9.57	CHEX:	1.58	19.9
0.860	PI+P	-24.45	39.75	18.00		1.87	22.8
	PI-P	22.70	81.49	36.90		6.13	75.0
	EVEN	-0.88	60.62	27.45		3.15	38.5
	ODD	23.57	20.87	9.45	CHEX:	1.70	20.8
0.880	PI+P	-21.94	43.39	19.20		2.00	23.6
	PI-P	29.08	88.54	39.18		7.33	86.9
	EVEN	3.57	65.96	29.19		3.68	43.7
	ODD	25.51	22.58	9.99	CHEX:	1.96	23.2
0.900	PI+P	-20.05	47.15	20.40		2.18	25.1
	PI-P	33.99	98.46	42.60		9.03	103.8
	EVEN	6.97	72.80	31.50		4.45	51.2
	ODD	27.02	25.65	11.10	CHEX:	2.31	26.6
0.920	PI+P	-18.80	51.03	21.60		2.43	27.1
	PI-P	35.86	111.70	47.28		11.29	126.0
	EVEN	8.53	81.37	34.44		5.49	61.3
	ODD	27.33	30.34	12.84	CHEX:	2.74	30.5
0.940	PI+P	-18.26	54.55	22.60		2.68	29.0
	PI-P	32.49	125.72	52.08		13.64	147.8
	EVEN	7.12	90.14	37.34		6.61	71.7
	ODD	25.38	35.58	14.74	CHEX:	3.09	33.5
0.960	PI+P	-17.96	57.44	23.30		2.89	30.4
	PI-P	25.36	137.81	55.90		15.67	165.1
	EVEN	3.70	97.63	39.60		7.62	80.2
	ODD	21.66	40.18	16.30	CHEX:	3.33	35.0
0.980	PI+P	-17.51	59.90	23.80		3.07	31.4
	PI-P	14.84	148.68	59.08		17.58	180.1
	EVEN	-1.33	104.29	41.44		8.56	87.7
	ODD	16.18	44.39	17.64	CHEX:	3.52	36.0
1.000	PI+P	-16.79	62.15	24.20		3.22	32.1
	PI-P	-0.68	155.31	60.48		18.74	186.9
	EVEN	-8.73	108.73	42.34		9.24	92.2
	ODD	8.05	46.58	18.14	CHEX:	3.47	34.6
1.020	PI+P	-15.86	64.44	24.60		3.38	32.8
	PI-P	-15.59	153.49	58.60		18.25	177.2
	EVEN	-15.72	108.97	41.60		9.29	90.3
	ODD	0.14	44.53	17.00	CHEX:	3.040	29.53
1.040	PI+P	-14.82	66.77	25.00		3.54	33.5
	PI-P	-26.43	148.92	55.76		17.31	163.8
	EVEN	-20.63	107.84	40.38		9.12	86.4
	ODD	-5.81	41.08	15.38	CHEX:	2.605	24.65
1.060	PI+P	-13.67	69.14	25.40		3.71	34.2
	PI-P	-34.73	141.93	52.14		15.95	147.2
	EVEN	-24.20	105.54	38.77		8.76	80.8
	ODD	-10.53	36.39	13.37	CHEX:	2.145	19.79
1.080	PI+P	-12.42	71.58	25.81		3.89	35.1
	PI-P	-39.12	133.54	48.15		14.28	128.6
	EVEN	-25.77	102.56	36.98		8.25	74.3
	ODD	-13.35	30.98	11.17	CHEX:	1.679	15.12

K GEV/C	REAC-TION	RE C 1/GEV	IM C 1/GEV	SIGMA TOT MB		DS/DO MB/SR	DS/DT MB/GEV**2
1.100	PI+P	-10.88	74.07	26.22		4.08	35.9
	PI-P	-39.64	125.82	44.54		12.67	111.4
	EVEN	-25.26	99.94	35.38		7.74	68.0
	ODD	-14.38	25.88	9.16	CHEX:	1.277	11.22
1.120	PI+P	-9.12	77.02	26.78		4.33	37.2
	PI-P	-37.41	120.28	41.82		11.41	98.0
	EVEN	-23.26	98.65	34.30		7.39	63.4
	ODD	-14.15	21.63	7.52	CHEX:	0.961	8.25
1.140	PI+P	-7.46	80.36	27.45		4.63	38.8
	PI-P	-34.61	117.54	40.15		10.67	89.5
	EVEN	-21.03	98.95	33.80		7.27	61.0
	ODD	-13.57	18.59	6.35	CHEX:	0.753	6.32
1.160	PI+P	-5.88	83.94	28.18		4.97	40.8
	PI-P	-32.41	115.88	38.90		10.16	83.4
	EVEN	-19.14	99.91	33.54		7.27	59.6
	ODD	-13.27	15.97	5.36	CHEX:	0.605	4.96
1.180	PI+P	-4.31	87.88	29.00		5.37	43.1
	PI-P	-29.74	114.66	37.84		9.73	78.1
	EVEN	-17.03	101.27	33.42		7.32	58.7
	ODD	-12.72	13.39	4.42	CHEX:	0.473	3.80
1.200	PI+P	-2.86	92.26	29.94		5.84	45.8
	PI-P	-27.11	115.04	37.33		9.58	75.1
	EVEN	-14.98	103.65	33.63		7.52	59.0
	ODD	-12.12	11.39	3.70	CHEX:	0.379	2.98
1.220	PI+P	-1.77	97.12	31.00		6.39	49.1
	PI-P	-25.27	115.79	36.96		9.52	73.1
	EVEN	-13.52	106.46	33.98		7.80	59.9
	ODD	-11.75	9.34	2.98	CHEX:	0.305	2.34
1.240	PI+P	-1.12	102.22	32.10		7.00	52.6
	PI-P	-23.49	116.74	36.66		9.50	71.4
	EVEN	-12.30	109.48	34.38		8.13	61.1
	ODD	-11.19	7.26	2.28	CHEX:	0.238	1.79
1.260	PI+P	-0.72	107.55	33.24		7.66	56.5
	PI-P	-21.93	118.20	36.53		9.57	70.5
	EVEN	-11.32	112.88	34.88		8.52	62.8
	ODD	-10.60	5.32	1.64	CHEX:	0.186	1.37
1.280	PI+P	-0.80	113.40	34.50		8.42	60.8
	PI-P	-20.63	119.85	36.46		9.68	69.9
	EVEN	-10.71	116.62	35.48		8.98	64.9
	ODD	-9.92	3.22	0.98	CHEX:	0.142	1.03

K GEV/C	REAC-TION	RE C 1/GEV	IM C 1/GEV	SIGMA TOT MB		DS/DO MB/SR	DS/DT MB/GEV**2
1.300	PI+P	-1.70	119.51	35.80		9.25	65.5
	PI-P	-19.62	121.85	36.50		9.86	69.8
	EVEN	-10.66	120.68	36.15		9.50	67.3
	ODD	-8.96	1.17	0.35	CHEX:	0.106	0.75
1.320	PI+P	-3.36	125.59	37.05		10.11	70.2
	PI-P	-19.00	123.90	36.55		10.06	69.9
	EVEN	-11.18	124.74	36.80		10.04	69.7
	ODD	-7.82	-0.85	-0.25	CHEX:	0.079	0.55
1.340	PI+P	-5.85	131.62	38.25		10.99	74.9
	PI-P	-18.61	125.98	36.61		10.27	70.0
	EVEN	-12.23	128.80	37.43		10.60	72.2
	ODD	-6.38	-2.82	-0.82	CHEX:	0.062	0.42
1.360	PI+P	-9.17	137.25	39.30		11.86	79.3
	PI-P	-18.49	128.03	36.66		10.48	70.1
	EVEN	-13.83	132.64	37.98		11.14	74.5
	ODD	-4.66	-4.61	-1.32	CHEX:	0.054	0.36
1.380	PI+P	-13.08	142.46	40.20		12.69	83.3
	PI-P	-18.43	130.06	36.70		10.69	70.2
	EVEN	-15.75	136.26	38.45		11.66	76.5
	ODD	-2.68	-6.20	-1.75	CHEX:	0.057	0.37
1.400	PI+P	-17.97	147.26	40.96		13.50	87.0
	PI-P	-18.92	132.30	36.80		10.95	70.6
	EVEN	-18.44	139.78	38.88		12.19	78.6
	ODD	-0.48	-7.48	-2.08	CHEX:	0.069	0.44
1.420	PI+P	-23.44	150.64	41.31		14.10	89.3
	PI-P	-19.66	133.72	36.67		11.09	70.2
	EVEN	-21.55	142.18	38.99		12.55	79.5
	ODD	1.89	-8.46	-2.32	CHEX:	0.091	0.58
1.440	PI+P	-28.75	153.13	41.41		14.58	90.7
	PI-P	-20.04	135.08	36.53		11.20	69.7
	EVEN	-24.40	144.11	38.97		12.83	79.8
	ODD	4.36	-9.02	-2.44	CHEX:	0.121	0.75
1.460	PI+P	-34.01	154.88	41.31		14.95	91.4
	PI-P	-20.52	136.36	36.37		11.30	69.1
	EVEN	-27.27	145.62	38.84		13.05	79.8
	ODD	6.74	-9.26	-2.47	CHEX:	0.156	0.95
1.480	PI+P	-39.25	155.94	41.03		15.22	91.5
	PI-P	-20.87	137.39	36.15		11.36	68.3
	EVEN	-30.06	146.67	38.59		13.19	79.3
	ODD	9.19	-9.27	-2.44	CHEX:	0.201	1.21

K GEV/C	REAC- TION	RE C 1/GEV	IM C 1/GEV	SIGMA TOT MB		DS/DO MB/SR	DS/DT MB/GEV**2
1.500	PI+P	-44.29	156.04	40.51		15.33	90.6
	PI-P	-21.12	138.40	35.93		11.42	67.5
	EVEN	-32.71	147.22	38.22		13.25	78.3
	ODD	11.58	-8.82	-2.29	CHEX:	0.247	1.46
1.520	PI+P	-48.75	155.39	39.81		15.30	88.9
	PI-P	-21.08	139.23	35.67		11.44	66.5
	EVEN	-34.92	147.31	37.74		13.22	76.9
	ODD	13.84	-8.08	-2.07	CHEX:	0.296	1.72
1.540	PI+P	-52.61	154.23	39.00		15.17	86.7
	PI-P	-21.09	140.39	35.50		11.51	65.8
	EVEN	-36.85	147.31	37.25		13.17	75.3
	ODD	15.76	-6.92	-1.75	CHEX:	0.338	1.94
1.560	PI+P	-55.72	152.63	38.10		14.93	84.0
	PI-P	-20.96	141.01	35.20		11.49	64.7
	EVEN	-38.34	146.82	36.65		13.02	73.3
	ODD	17.38	-5.81	-1.45	CHEX:	0.380	2.14
1.580	PI+P	-58.15	150.94	37.20		14.65	81.2
	PI-P	-20.35	142.01	35.00		11.53	63.9
	EVEN	-39.25	146.47	36.10		12.88	71.4
	ODD	18.90	-4.46	-1.10	CHEX:	0.423	2.34
1.600	PI+P	-59.74	149.07	36.28		14.31	78.0
	PI-P	-19.73	143.15	34.84		11.58	63.2
	EVEN	-39.74	146.11	35.56		12.72	69.4
	ODD	20.00	-2.96	-0.72	CHEX:	0.454	2.47
1.650	PI+P	-61.83	146.61	34.60		13.72	72.0
	PI-P	-18.00	146.86	34.66		11.86	62.3
	EVEN	-39.91	146.73	34.63		12.53	65.8
	ODD	21.92	0.13	0.03	CHEX:	0.520	2.73
1.700	PI+P	-63.22	144.55	33.11		13.18	66.7
	PI-P	-16.48	151.27	34.65		12.26	62.1
	EVEN	-39.85	147.91	33.88		12.42	62.9
	ODD	23.37	3.36	0.77	CHEX:	0.590	2.99
1.750	PI+P	-63.12	143.36	31.90		12.70	62.1
	PI-P	-15.11	156.39	34.80		12.78	62.4
	EVEN	-39.11	149.87	33.35		12.42	60.7
	ODD	24.01	6.52	1.45	CHEX:	0.641	3.13
1.800	PI+P	-62.47	143.06	30.95		12.34	58.3
	PI-P	-14.38	162.20	35.09		13.43	63.4
	EVEN	-38.42	152.63	33.02		12.54	59.2
	ODD	24.04	9.57	2.07	CHEX:	0.678	3.20

K GEV/C	REAC- TION	RE C 1/GEV	IM C 1/GEV	SIGMA TOT MB		DS/DO MB/SR	DS/DT MB/GEV**2
1.850	PI+P	-61.04	143.24	30.15		12.01	54.9
	PI-P	-14.24	168.13	35.39		14.11	64.4
	EVEN	-37.64	155.68	32.77		12.71	58.1
	ODD	23.40	12.45	2.62	CHEX:	0.696	3.18
1.900	PI+P	-58.94	144.67	29.65		11.84	52.4
	PI-P	-14.75	174.43	35.75		14.87	65.8
	EVEN	-36.84	159.55	32.70		13.01	57.5
	ODD	22.10	14.88	3.05	CHEX:	0.689	3.05

K GEV/C	REAC- TION	RE C 1/GEV	IM C 1/GEV	SIGMA TOT MB	RHO		DS/DT MB/GEV**2
1.950	PI+P	-57.1	147.2	29.40	-0.388		50.8
	PI-P	-16.1	180.6	36.06	-0.089		67.0
	EVEN	-36.6	163.9	32.73	-0.223		57.5
	ODD	20.5	16.7	3.33	1.228	CHEX:	2.84
2.000	PI+P	-55.6	150.2	29.24	-0.370		49.7
	PI-P	-18.2	186.3	36.28	-0.097		67.9
	EVEN	-36.9	168.3	32.76	-0.219		57.5
	ODD	18.7	18.1	3.52	1.036	CHEX:	2.63
2.050	PI+P	-54.2	153.5	29.15	-0.353		48.8
	PI-P	-20.6	191.6	36.39	-0.108		68.4
	EVEN	-37.4	172.5	32.77	-0.217		57.4
	ODD	16.8	19.1	3.62	0.880	CHEX:	2.38
2.100	PI+P	-52.7	157.2	29.15	-0.335		48.3
	PI-P	-23.4	196.2	36.39	-0.119		68.6
	EVEN	-38.1	176.7	32.77	-0.215		57.4
	ODD	14.6	19.5	3.62	0.749	CHEX:	2.09
2.150	PI+P	-51.3	161.6	29.26	-0.317		48.1
	PI-P	-26.4	200.2	36.26	-0.132		68.3
	EVEN	-38.8	180.9	32.76	-0.215		57.4
	ODD	12.4	19.3	3.50	0.643	CHEX:	1.77
2.200	PI+P	-50.2	166.7	29.50	-0.301		48.5
	PI-P	-29.1	203.4	36.00	-0.143		67.6
	EVEN	-39.7	185.0	32.75	-0.215		57.3
	ODD	10.5	18.4	3.25	0.575	CHEX:	1.44
2.250	PI+P	-50.0	172.2	29.80	-0.290		49.2
	PI-P	-31.5	206.3	35.70	-0.153		66.6
	EVEN	-40.8	189.2	32.75	-0.215		57.3
	ODD	9.2	17.0	2.95	0.541	CHEX:	1.15
2.300	PI+P	-50.5	177.8	30.10	-0.284		50.0
	PI-P	-33.6	208.7	35.34	-0.161		65.5
	EVEN	-42.1	193.3	32.72	-0.218		57.3
	ODD	8.4	15.5	2.62	0.544	CHEX:	0.91

K GEV/C	REAC-TION	RE C 1/GEV	IM C 1/GEV	SIGMA TOT MB	RHO		DS/DT MB/GEV**2
2.350	PI+P	-51.6	183.2	30.35	-0.282		50.8
	PI-P	-35.2	210.9	34.95	-0.167		64.1
	EVEN	-43.4	197.0	32.65	-0.220		57.1
	ODD	8.2	13.9	2.30	0.590	CHEX:	0.73
2.400	PI+P	-53.1	188.2	30.54	-0.282		51.4
	PI-P	-36.3	213.1	34.58	-0.170		62.9
	EVEN	-44.7	200.7	32.56	-0.223		56.8
	ODD	8.4	12.4	2.02	0.676	CHEX:	0.607
2.450	PI+P	-54.9	193.0	30.68	-0.284		52.0
	PI-P	-37.0	215.6	34.26	-0.171		61.7
	EVEN	-45.9	204.3	32.47	-0.225		56.6
	ODD	9.0	11.3	1.79	0.796	CHEX:	0.535
2.500	PI+P	-56.9	197.6	30.78	-0.288		52.4
	PI-P	-37.6	218.3	34.00	-0.172		60.8
	EVEN	-47.2	207.9	32.39	-0.227		56.4
	ODD	9.7	10.3	1.61	0.936	CHEX:	0.497
2.550	PI+P	-59.2	201.9	30.83	-0.293		52.7
	PI-P	-38.2	221.0	33.75	-0.173		59.9
	EVEN	-48.7	211.4	32.29	-0.230		56.1
	ODD	10.5	9.6	1.46	1.100	CHEX:	0.481
2.600	PI+P	-61.7	205.7	30.81	-0.300		52.9
	PI-P	-38.7	223.7	33.51	-0.173		59.1
	EVEN	-50.2	214.7	32.16	-0.234		55.7
	ODD	11.5	9.0	1.35	1.274	CHEX:	0.488
2.650	PI+P	-64.1	209.2	30.74	-0.307		52.8
	PI-P	-39.1	226.5	33.28	-0.172		58.3
	EVEN	-51.6	217.8	32.01	-0.237		55.3
	ODD	12.5	8.6	1.27	1.450	CHEX:	0.512
2.700	PI+P	-66.6	212.3	30.62	-0.314		52.6
	PI-P	-39.2	229.4	33.08	-0.171		57.5
	EVEN	-52.9	220.8	31.85	-0.240		54.8
	ODD	13.7	8.5	1.23	1.603	CHEX:	0.552
2.750	PI+P	-68.9	215.0	30.45	-0.320		52.2
	PI-P	-39.3	232.4	32.91	-0.169		56.9
	EVEN	-54.1	223.7	31.68	-0.242		54.3
	ODD	14.8	8.7	1.23	1.702	CHEX:	0.603
2.800	PI+P	-71.0	217.4	30.24	-0.326		51.7
	PI-P	-39.6	235.6	32.76	-0.168		56.4
	EVEN	-55.3	226.5	31.50	-0.244		53.7
	ODD	15.7	9.1	1.26	1.733	CHEX:	0.650

K GEV/C	REAC-TION	RE C 1/GEV	IM C 1/GEV	SIGMA TOT MB	RHO		DS/DT MB/GEV**2
2.850	PI+P	-72.8	219.7	30.02	-0.332		51.1
	PI-P	-39.3	238.5	32.59	-0.165		55.7
	EVEN	-56.1	229.1	31.31	-0.245		53.1
	ODD	16.8	9.4	1.28	1.783	CHEX:	0.705
2.900	PI+P	-74.4	221.7	29.77	-0.336		50.4
	PI-P	-39.2	242.6	32.57	-0.162		55.6
	EVEN	-56.8	232.1	31.17	-0.245		52.6
	ODD	17.6	10.4	1.40	1.691	CHEX:	0.773
2.950	PI+P	-75.6	223.6	29.52	-0.338		49.6
	PI-P	-39.6	246.2	32.50	-0.161		55.4
	EVEN	-57.6	234.9	31.01	-0.245		52.1
	ODD	18.0	11.3	1.49	1.594	CHEX:	0.803
3.000	PI+P	-76.5	225.6	29.28	-0.339		48.8
	PI-P	-40.0	249.9	32.44	-0.160		55.1
	EVEN	-58.2	237.7	30.86	-0.245		51.6
	ODD	18.3	12.2	1.58	1.500	CHEX:	0.829
3.100	PI+P	-77.4	230.1	28.91	-0.336		47.5
	PI-P	-41.2	257.5	32.35	-0.160		54.8
	EVEN	-59.3	243.8	30.63	-0.243		50.8
	ODD	18.1	13.7	1.72	1.325	CHEX:	0.833
3.200	PI+P	-78.1	235.3	28.63	-0.332		46.5
	PI-P	-42.9	264.7	32.21	-0.162		54.4
	EVEN	-60.5	250.0	30.42	-0.242		50.0
	ODD	17.6	14.7	1.79	1.196	CHEX:	0.796
3.300	PI+P	-78.5	241.0	28.44	-0.326		45.7
	PI-P	-44.6	271.4	32.02	-0.164		53.8
	EVEN	-61.6	256.2	30.23	-0.240		49.4
	ODD	17.0	15.2	1.79	1.118	CHEX:	0.737
3.400	PI+P	-79.1	247.4	28.33	-0.320		45.2
	PI-P	-46.2	277.9	31.83	-0.166		53.2
	EVEN	-62.6	262.6	30.08	-0.239		48.9
	ODD	16.5	15.3	1.75	1.077	CHEX:	0.676
3.500	PI+P	-80.1	254.0	28.26	-0.315		44.9
	PI-P	-47.9	284.2	31.62	-0.169		52.5
	EVEN	-64.0	269.1	29.94	-0.238		48.4
	ODD	16.1	15.1	1.68	1.066	CHEX:	0.616
3.600	PI+P	-81.5	260.6	28.19	-0.313		44.6
	PI-P	-49.3	290.0	31.37	-0.170		51.7
	EVEN	-65.4	275.3	29.78	-0.238		47.9
	ODD	16.1	14.7	1.59	1.095	CHEX:	0.568

K GEV/C	REAC-TION	RE C 1/GEV	IM C 1/GEV	SIGMA TOT MB	RHO		DS/DT MB/GEV**2
3.700	PI+P	-83.1	267.2	28.12	-0.311		44.3
	PI-P	-50.3	296.1	31.16	-0.170		51.0
	EVEN	-66.7	281.6	29.64	-0.237		47.4
	ODD	16.4	14.4	1.52	1.135	CHEX:	0.540
3.800	PI+P	-85.0	273.5	28.03	-0.311		44.0
	PI-P	-51.3	302.0	30.95	-0.170		50.4
	EVEN	-68.2	287.8	29.49	-0.237		46.9
	ODD	16.8	14.2	1.46	1.181	CHEX:	0.522
3.900	PI+P	-87.0	279.6	27.92	-0.311		43.7
	PI-P	-52.2	308.1	30.76	-0.169		49.7
	EVEN	-69.6	293.8	29.34	-0.237		46.4
	ODD	17.4	14.2	1.42	1.225	CHEX:	0.515
4.000	PI+P	-88.9	285.3	27.78	-0.312		43.2
	PI-P	-52.9	314.1	30.58	-0.168		49.1
	EVEN	-70.9	299.7	29.18	-0.237		45.9
	ODD	18.0	14.4	1.40	1.249	CHEX:	0.515
4.100	PI+P	-90.7	290.8	27.62	-0.312		42.8
	PI-P	-53.6	320.3	30.42	-0.168		48.6
	EVEN	-72.2	305.5	29.02	-0.236		45.4
	ODD	18.5	14.7	1.40	1.259	CHEX:	0.516
4.200	PI+P	-92.2	296.1	27.46	-0.311		42.2
	PI-P	-54.3	326.4	30.26	-0.166		48.1
	EVEN	-73.3	311.3	28.86	-0.235		44.9
	ODD	19.0	15.1	1.40	1.253	CHEX:	0.517
4.300	PI+P	-93.4	301.4	27.30	-0.310		41.7
	PI-P	-55.0	332.6	30.12	-0.165		47.6
	EVEN	-74.2	317.0	28.71	-0.234		44.4
	ODD	19.2	15.6	1.41	1.233	CHEX:	0.513
4.400	PI+P	-94.4	306.9	27.16	-0.308		41.3
	PI-P	-55.7	338.7	29.98	-0.164		47.2
	EVEN	-75.0	322.8	28.57	-0.232		44.0
	ODD	19.4	15.9	1.41	1.216	CHEX:	0.503
4.500	PI+P	-95.3	312.5	27.04	-0.305		40.8
	PI-P	-56.3	344.8	29.84	-0.163		46.7
	EVEN	-75.8	328.7	28.44	-0.231		43.5
	ODD	19.5	16.2	1.40	1.207	CHEX:	0.493
4.600	PI+P	-96.2	318.1	26.93	-0.302		40.4
	PI-P	-56.8	351.0	29.71	-0.162		46.3
	EVEN	-76.5	334.5	28.32	-0.229		43.1
	ODD	19.7	16.4	1.39	1.200	CHEX:	0.483

K GEV/C	REAC-TION	RE C 1/GEV	IM C 1/GEV	SIGMA TOT MB	RHO		DS/DT MB/GEV**2
4.700	PI+P	-97.1	323.8	26.83	-0.300		40.1
	PI-P	-57.2	357.2	29.59	-0.160		45.9
	EVEN	-77.1	340.5	28.21	-0.227		42.7
	ODD	19.9	16.7	1.38	1.195	CHEX:	0.473
4.800	PI+P	-97.9	329.6	26.74	-0.297		39.7
	PI-P	-57.6	363.4	29.48	-0.159		45.5
	EVEN	-77.7	346.5	28.11	-0.224		42.4
	ODD	20.1	16.9	1.37	1.189	CHEX:	0.465
4.900	PI+P	-98.7	335.4	26.66	-0.294		39.4
	PI-P	-58.1	369.7	29.38	-0.157		45.2
	EVEN	-78.4	352.6	28.02	-0.222		42.1
	ODD	20.3	17.1	1.36	1.184	CHEX:	0.456
5.000	PI+P	-99.6	341.2	26.58	-0.292		39.2
	PI-P	-58.6	376.0	29.28	-0.156		44.9
	EVEN	-79.1	358.6	27.93	-0.220		41.8
	ODD	20.5	17.4	1.35	1.180	CHEX:	0.448
5.200	PI+P	-101.2	353.0	26.43	-0.287		38.6
	PI-P	-59.4	388.7	29.11	-0.153		44.3
	EVEN	-80.3	370.8	27.77	-0.216		41.2
	ODD	20.9	17.8	1.34	1.171	CHEX:	0.432
5.400	PI+P	-102.8	364.8	26.31	-0.282		38.2
	PI-P	-60.3	401.3	28.94	-0.150		43.8
	EVEN	-81.5	383.1	27.62	-0.213		40.8
	ODD	21.2	18.3	1.32	1.163	CHEX:	0.417
5.600	PI+P	-104.4	376.7	26.19	-0.277		37.7
	PI-P	-61.2	414.0	28.79	-0.148		43.3
	EVEN	-82.8	395.4	27.49	-0.209		40.3
	ODD	21.6	18.7	1.30	1.155	CHEX:	0.403
5.800	PI+P	-106.0	388.5	26.09	-0.273		37.4
	PI-P	-62.1	426.7	28.65	-0.146		42.8
	EVEN	-84.0	407.6	27.37	-0.206		39.9
	ODD	21.9	19.1	1.28	1.148	CHEX:	0.389
6.000	PI+P	-107.5	400.5	25.99	-0.269		37.0
	PI-P	-63.0	439.5	28.52	-0.143		42.4
	EVEN	-85.3	420.0	27.26	-0.203		39.5
	ODD	22.3	19.5	1.27	1.142	CHEX:	0.377
6.200	PI+P	-109.1	412.4	25.90	-0.265		36.7
	PI-P	-64.0	452.2	28.40	-0.141		42.0
	EVEN	-86.5	432.3	27.15	-0.200		39.2
	ODD	22.6	19.9	1.25	1.136	CHEX:	0.365

K GEV/C	REAC-TION	RE C 1/GEV	IM C 1/GEV	SIGMA TOT MB	RHO	DS/DT MB/GEV**2
6.400	PI+P	-110.7	424.3	25.81	-0.261	36.4
	PI-P	-64.9	464.8	28.28	-0.140	41.7
	EVEN	-87.8	444.5	27.05	-0.198	38.8
	ODD	22.9	20.3	1.23	1.130	CHEX: 0.354
6.600	PI+P	-112.3	436.1	25.73	-0.257	36.1
	PI-P	-65.8	477.5	28.17	-0.138	41.3
	EVEN	-89.0	456.8	26.95	-0.195	38.5
	ODD	23.2	20.7	1.22	1.125	CHEX: 0.344
6.800	PI+P	-113.8	448.0	25.66	-0.254	35.8
	PI-P	-66.7	490.1	28.06	-0.136	41.0
	EVEN	-90.3	469.0	26.86	-0.192	38.2
	ODD	23.5	21.0	1.20	1.120	CHEX: 0.334
7.000	PI+P	-115.3	459.9	25.58	-0.251	35.5
	PI-P	-67.6	502.6	27.96	-0.135	40.7
	EVEN	-91.5	481.2	26.77	-0.190	37.9
	ODD	23.8	21.4	1.19	1.115	CHEX: 0.324
7.200	PI+P	-116.8	471.7	25.51	-0.248	35.3
	PI-P	-68.5	515.2	27.86	-0.133	40.4
	EVEN	-92.6	493.5	26.69	-0.188	37.7
	ODD	24.1	21.7	1.18	1.111	CHEX: 0.315
7.400	PI+P	-118.2	483.6	25.45	-0.244	35.1
	PI-P	-69.3	527.7	27.77	-0.131	40.1
	EVEN	-93.8	505.7	26.61	-0.185	37.4
	ODD	24.4	22.1	1.16	1.107	CHEX: 0.307
7.600	PI+P	-119.6	495.4	25.38	-0.242	34.8
	PI-P	-70.2	540.3	27.68	-0.130	39.8
	EVEN	-94.9	517.8	26.53	-0.183	37.2
	ODD	24.7	22.4	1.15	1.103	CHEX: 0.299
7.800	PI+P	-121.1	507.2	25.32	-0.239	34.6
	PI-P	-71.0	552.7	27.60	-0.128	39.5
	EVEN	-96.0	530.0	26.46	-0.181	36.9
	ODD	25.0	22.8	1.14	1.100	CHEX: 0.291
8.000	PI+P	-122.4	519.0	25.27	-0.236	34.4
	PI-P	-71.8	565.2	27.51	-0.127	39.3
	EVEN	-97.1	542.1	26.39	-0.179	36.7
	ODD	25.3	23.1	1.12	1.096	CHEX: 0.284
9.000	PI+P	-129.0	578.0	25.01	-0.223	33.5
	PI-P	-75.6	627.3	27.14	-0.121	38.2
	EVEN	-102.3	602.6	26.08	-0.170	35.7
	ODD	26.7	24.7	1.07	1.082	CHEX: 0.252

K GEV/C	REAC-TION	RE C 1/GEV	IM C 1/GEV	SIGMA TOT MB	RHO	DS/DT MB/GEV**2
10.0	PI+P	-135.0	636.7	24.79	-0.212	32.8
	PI-P	-79.1	689.0	26.83	-0.115	37.3
	EVEN	-107.0	662.9	25.81	-0.161	34.9
	ODD	28.0	26.1	1.02	1.071	CHEX: 0.227
11.0	PI+P	-140.5	695.3	24.61	-0.202	32.2
	PI-P	-82.1	750.3	26.56	-0.109	36.5
	EVEN	-111.3	722.8	25.59	-0.154	34.2
	ODD	29.2	27.5	0.97	1.062	CHEX: 0.206
12.0	PI+P	-145.5	753.7	24.46	-0.193	31.7
	PI-P	-84.7	811.3	26.33	-0.104	35.8
	EVEN	-115.1	782.5	25.39	-0.147	33.7
	ODD	30.4	28.8	0.94	1.055	CHEX: 0.189
13.0	PI+P	-150.1	812.0	24.32	-0.185	31.3
	PI-P	-87.0	872.2	26.13	-0.100	35.2
	EVEN	-118.5	842.1	25.22	-0.141	33.2
	ODD	31.6	30.1	0.90	1.049	CHEX: 0.174
14.0	PI+P	-154.3	870.3	24.21	-0.177	30.9
	PI-P	-89.0	932.8	25.95	-0.095	34.7
	EVEN	-121.6	901.5	25.08	-0.135	32.7
	ODD	32.7	31.3	0.87	1.045	CHEX: 0.162
15.0	PI+P	-158.1	928.5	24.10	-0.170	30.5
	PI-P	-90.7	993.3	25.79	-0.091	34.3
	EVEN	-124.4	960.9	24.94	-0.129	32.3
	ODD	33.7	32.4	0.84	1.040	CHEX: 0.151
16.0	PI+P	-161.6	986.6	24.01	-0.164	30.2
	PI-P	-92.1	1053.6	25.64	-0.087	33.9
	EVEN	-126.8	1020.1	24.83	-0.124	32.0
	ODD	34.8	33.5	0.82	1.037	CHEX: 0.141
17.0	PI+P	-164.7	1044.7	23.93	-0.158	30.0
	PI-P	-93.2	1113.9	25.51	-0.084	33.5
	EVEN	-129.0	1079.3	24.72	-0.119	31.7
	ODD	35.8	34.6	0.79	1.034	CHEX: 0.133
18.0	PI+P	-167.6	1102.8	23.86	-0.152	29.7
	PI-P	-94.1	1174.0	25.40	-0.080	33.2
	EVEN	-130.8	1138.4	24.63	-0.115	31.4
	ODD	36.7	35.6	0.77	1.031	CHEX: 0.125
19.0	PI+P	-170.2	1160.8	23.79	-0.147	29.5
	PI-P	-94.8	1234.1	25.29	-0.077	32.9
	EVEN	-132.5	1197.5	24.54	-0.111	31.1
	ODD	37.7	36.6	0.75	1.029	CHEX: 0.119

K GEV/C	REACTION	RE C 1/GEV	IM C 1/GEV	SIGMA TOT MB	RHO	DS/DT MB/GEV**2
20.0	PI+P	-172.5	1218.9	23.73	-0.142	29.3
	PI-P	-95.2	1294.1	25.20	-0.074	32.6
	EVEN	-133.9	1256.5	24.46	-0.107	30.9
	ODD	38.6	37.6	0.73	1.027	CHEX: 0.113
22.0	PI+P	-176.4	1335.0	23.63	-0.132	29.0
	PI-P	-95.6	1414.0	25.03	-0.068	32.1
	EVEN	-136.0	1374.5	24.33	-0.099	30.5
	ODD	40.4	39.5	0.70	1.024	CHEX: 0.102
24.0	PI+P	-179.4	1451.1	23.55	-0.124	28.8
	PI-P	-95.1	1533.7	24.88	-0.062	31.8
	EVEN	-137.3	1492.4	24.22	-0.092	30.2
	ODD	42.2	41.3	0.67	1.021	CHEX: 0.094
26.0	PI+P	-181.7	1567.4	23.47	-0.116	28.5
	PI-P	-94.0	1653.4	24.76	-0.057	31.4
	EVEN	-137.8	1610.4	24.12	-0.086	29.9
	ODD	43.8	43.0	0.64	1.019	CHEX: 0.086
28.0	PI+P	-183.1	1683.7	23.42	-0.109	28.3
	PI-P	-92.3	1773.0	24.66	-0.052	31.1
	EVEN	-137.7	1728.3	24.04	-0.080	29.7
	ODD	45.4	44.7	0.62	1.017	CHEX:0.0801
30.0	PI+P	-183.9	1800.1	23.37	-0.102	28.2
	PI-P	-90.0	1892.6	24.57	-0.048	30.9
	EVEN	-137.0	1846.4	23.97	-0.074	29.5
	ODD	47.0	46.2	0.60	1.015	CHEX:0.0748
35.0	PI+P	-183.3	2091.8	23.27	-0.088	27.9
	PI-P	-82.1	2191.8	24.39	-0.037	30.4
	EVEN	-132.7	2141.8	23.83	-0.062	29.1
	ODD	50.6	50.0	0.56	1.012	CHEX:0.0640
40.0	PI+P	-179.4	2384.4	23.21	-0.075	27.7
	PI-P	-71.3	2491.4	24.25	-0.029	30.1
	EVEN	-125.4	2437.9	23.73	-0.051	28.9
	ODD	54.0	53.5	0.52	1.010	CHEX:0.0559
45.0	PI+P	-172.6	2678.1	23.17	-0.064	27.6
	PI-P	-58.1	2791.6	24.16	-0.021	29.8
	EVEN	-115.4	2734.8	23.67	-0.042	28.7
	ODD	57.2	56.7	0.49	1.009	CHEX:0.0497
50.0	PI+P	-163.3	2972.8	23.15	-0.055	27.5
	PI-P	-42.8	3092.5	24.08	-0.014	29.6
	EVEN	-103.0	3032.7	23.62	-0.034	28.5
	ODD	60.3	59.8	0.47	1.007	CHEX:0.0447

K GEV/C	REACTION	RE C 1/GEV	IM C 1/GEV	SIGMA TOT MB	RHO	DS/DT MB/GEV**2
55.0	PI+P	-151.8	3268.7	23.14	-0.046	27.4
	PI-P	-25.5	3394.2	24.03	-0.008	29.5
	EVEN	-88.7	3331.5	23.59	-0.027	28.4
	ODD	63.2	62.8	0.44	1.006	CHEX:0.0406
60.0	PI+P	-138.3	3565.6	23.14	-0.039	27.4
	PI-P	-6.5	3696.8	23.99	-0.002	29.4
	EVEN	-72.4	3631.2	23.57	-0.020	28.4
	ODD	65.9	65.6	0.43	1.006	CHEX:0.0372
65.0	PI+P	-123.0	3863.7	23.15	-0.032	27.4
	PI-P	14.2	4000.2	23.96	0.004	29.3
	EVEN	-54.4	3932.0	23.56	-0.014	28.4
	ODD	68.6	68.3	0.41	1.005	CHEX:0.0343
70.0	PI+P	-106.1	4162.8	23.16	-0.025	27.4
	PI-P	36.2	4304.5	23.95	0.008	29.3
	EVEN	-34.9	4233.7	23.55	-0.008	28.3
	ODD	71.2	70.8	0.39	1.005	CHEX:0.0319
80.0	PI+P	-67.7	4764.3	23.19	-0.014	27.5
	PI-P	84.4	4915.8	23.93	0.017	29.3
	EVEN	8.3	4840.1	23.56	0.002	28.4
	ODD	76.0	75.7	0.37	1.004	CHEX:0.0279
90.0	PI+P	-24.0	5369.9	23.23	-0.004	27.6
	PI-P	137.2	5530.6	23.93	0.025	29.3
	EVEN	56.6	5450.3	23.58	0.010	28.4
	ODD	80.6	80.3	0.35	1.003	CHEX:0.0248
100.0	PI+P	24.3	5979.5	23.28	0.004	27.7
	PI-P	194.2	6148.9	23.94	0.032	29.3
	EVEN	109.2	6064.2	23.61	0.018	28.5
	ODD	84.9	84.7	0.33	1.003	CHEX:0.0223
110.0	PI+P	76.7	6593.0	23.34	0.012	27.8
	PI-P	254.9	6770.7	23.97	0.038	29.4
	EVEN	165.8	6681.8	23.65	0.025	28.6
	ODD	89.1	88.8	0.31	1.002	CHEX:0.0203
120.0	PI+P	132.9	7210.1	23.40	0.018	28.0
	PI-P	318.9	7395.7	24.00	0.043	29.5
	EVEN	225.9	7302.9	23.70	0.031	28.7
	ODD	93.0	92.8	0.30	1.002	CHEX:0.0186
130.0	PI+P	192.5	7830.8	23.46	0.025	28.1
	PI-P	386.1	8024.0	24.04	0.048	29.6
	EVEN	289.3	7927.4	23.75	0.036	28.8
	ODD	96.8	96.6	0.29	1.002	CHEX:0.0171

K GEV/C	REAC-TION	RE C 1/GEV	IM C 1/GEV	SIGMA TOT MB	RHO	DS/DT MB/GEV**2
140.0	PI+P	255.3	8454.9	23.52	0.030	28.3
	PI-P	456.1	8655.4	24.07	0.053	29.7
	EVEN	355.7	8555.1	23.80	0.042	29.0
	ODD	100.4	100.2	0.28	1.002	CHEX:0.0159
150.0	PI+P	320.8	9082.3	23.58	0.035	28.4
	PI-P	528.7	9289.8	24.12	0.057	29.8
	EVEN	424.8	9186.0	23.85	0.046	29.1
	ODD	103.9	103.8	0.27	1.002	CHEX:0.0149
160.0	PI+P	389.1	9712.8	23.64	0.040	28.6
	PI-P	603.7	9927.2	24.16	0.061	29.9
	EVEN	496.4	9820.0	23.90	0.051	29.3
	ODD	107.3	107.2	0.26	1.001	CHEX:0.0139
170.0	PI+P	459.9	10346.5	23.70	0.044	28.8
	PI-P	681.1	10567.4	24.21	0.064	30.1
	EVEN	570.5	10457.0	23.95	0.055	29.4
	ODD	110.6	110.5	0.25	1.001	CHEX:0.0131
180.0	PI+P	532.9	10983.1	23.76	0.049	28.9
	PI-P	760.6	11210.5	24.25	0.068	30.2
	EVEN	646.8	11096.8	24.01	0.058	29.5
	ODD	113.8	113.7	0.25	1.001	CHEX:0.0124
200.0	PI+P	685.6	12264.9	23.88	0.056	29.2
	PI-P	925.5	12504.6	24.35	0.074	30.4
	EVEN	805.5	12384.7	24.11	0.065	29.8
	ODD	120.0	119.8	0.23	1.001	CHEX:0.0111
220.0	PI+P	846.1	13557.6	24.00	0.062	29.5
	PI-P	1097.7	13808.9	24.44	0.079	30.7
	EVEN	971.9	13683.3	24.22	0.071	30.1
	ODD	125.8	125.7	0.22	1.001	CHEX:0.0101
240.0	PI+P	1013.7	14860.5	24.11	0.068	29.8
	PI-P	1276.4	15123.1	24.54	0.084	31.0
	EVEN	1145.1	14991.8	24.32	0.076	30.4
	ODD	131.4	131.3	0.21	1.001	CHEX:0.0093
260.0	PI+P	1187.8	16173.2	24.22	0.073	30.1
	PI-P	1461.2	16446.5	24.63	0.089	31.2
	EVEN	1324.5	16309.9	24.43	0.081	30.7
	ODD	136.7	136.6	0.20	1.001	CHEX:0.0086
280.0	PI+P	1367.8	17495.2	24.33	0.078	30.4
	PI-P	1651.6	17778.8	24.73	0.093	31.5
	EVEN	1509.7	17637.0	24.53	0.086	31.0
	ODD	141.9	141.8	0.20	1.001	CHEX:0.0080

K GEV/C	REAC-TION	RE C 1/GEV	IM C 1/GEV	SIGMA TOT MB	RHO	DS/DT MB/GEV**2
300.0	PI+P	1553.3	18826.0	24.44	0.083	30.7
	PI-P	1847.0	19119.6	24.82	0.097	31.8
	EVEN	1700.2	18972.8	24.63	0.090	31.2
	ODD	146.9	146.8	0.19	1.001	CHEX:0.0074
350.0	PI+P	2038.7	22189.4	24.69	0.092	31.4
	PI-P	2355.9	22506.4	25.04	0.105	32.4
	EVEN	2197.3	22347.9	24.86	0.098	31.9
	ODD	158.6	158.5	0.18	1.000	CHEX:0.0064
400.0	PI+P	2551.3	25600.7	24.92	0.100	32.0
	PI-P	2890.4	25939.6	25.25	0.111	33.0
	EVEN	2720.9	25770.1	25.09	0.106	32.5
	ODD	169.5	169.5	0.16	1.000	CHEX:0.0056
450.0	PI+P	3087.6	29055.7	25.14	0.106	32.7
	PI-P	3447.2	29415.3	25.45	0.117	33.6
	EVEN	3267.4	29235.5	25.30	0.112	33.1
	ODD	179.8	179.8	0.16	1.000	CHEX:0.0049
500.0	PI+P	3644.7	32551.2	25.35	0.112	33.2
	PI-P	4023.8	32930.2	25.65	0.122	34.1
	EVEN	3834.2	32740.7	25.50	0.117	33.7
	ODD	189.5	189.5	0.15	1.000	CHEX:0.0045
600.0	PI+P	4813.2	39652.3	25.73	0.121	34.3
	PI-P	5228.4	40067.4	26.00	0.130	35.1
	EVEN	5020.8	39859.8	25.87	0.126	34.7
	ODD	207.6	207.6	0.13	1.000	CHEX:0.0037
700.0	PI+P	6043.7	46885.4	26.08	0.129	35.3
	PI-P	6492.1	47333.8	26.33	0.137	36.1
	EVEN	6267.9	47109.6	26.21	0.133	35.7
	ODD	224.2	224.2	0.12	1.000	CHEX:0.0032
800.0	PI+P	7326.9	54236.6	26.40	0.135	36.3
	PI-P	7806.3	54716.0	26.63	0.143	37.0
	EVEN	7566.6	54476.3	26.52	0.139	36.6
	ODD	239.7	239.7	0.12	1.000	CHEX:0.0028
900.0	PI+P	8655.9	61694.8	26.69	0.140	37.1
	PI-P	9164.5	62203.3	26.91	0.147	37.8
	EVEN	8910.2	61949.1	26.80	0.144	37.5
	ODD	254.3	254.2	0.11	1.000	CHEX:0.0025
1000.0	PI+P	10025.6	69251.0	26.97	0.145	37.9
	PI-P	10561.6	69787.0	27.18	0.151	38.6
	EVEN	10293.6	69519.0	27.07	0.148	38.3
	ODD	268.0	268.0	0.10	1.000	CHEX:0.0022

Table 2.2.1.2. Analytic continuation of Table 2.2.1.1 into the unphysical region. $\tilde{C}$ denotes the forward amplitude minus the contribution of the pseudoscalar nucleon Born term. See Sect. 2.4.7 for applications.

ω^2	$\tilde{C}^-/\omega$	$\tilde{C}^+ - \frac{g^2}{m}$
μ^2	μ^{-2}	μ^{-1}
0.00	−0.463	−1.468
0.10	−0.480	−1.353
0.20	−0.498	−1.234
0.30	−0.517	−1.110
0.40	−0.536	−0.980
0.50	−0.557	−0.844
0.60	−0.579	−0.702
0.70	−0.602	−0.551
0.80	−0.626	−0.389
0.81	−0.629	−0.373
0.82	−0.631	−0.356
0.83	−0.634	−0.339
0.84	−0.636	−0.321
0.85	−0.639	−0.304
0.86	−0.641	−0.286
0.87	−0.644	−0.269
0.88	−0.646	−0.251
0.89	−0.649	−0.232
0.90	−0.651	−0.214
0.91	−0.654	−0.195
0.92	−0.656	−0.176
0.93	−0.658	−0.157
0.94	−0.660	−0.137
0.95	−0.663	−0.117
0.96	−0.665	−0.097
0.97	−0.667	−0.075
0.98	−0.668	−0.052
0.99	−0.669	−0.027
1.00	−0.665	0.010

2.2.1.5 Remarks on the literature on pion-nucleon forward amplitudes

There exists a very large number of papers on the pion-nucleon forward amplitude, since the theory of functions supplies many theorems which are applicable in this case. A selection of these papers is discussed in the review 80 R Höhler-Kaiser. In the following, I shall make only a few remarks.

i) Alternative versions of the ordinary dispersion relation

A great effort was spent in deriving alternative versions, claiming that they have advantages in comparison with simple-minded evaluations of the ordinary dispersion relations as used in the first part of this section (for example: dispersion relations for functions of the amplitudes, conformal mapping methods, finite contour dispersion relations, weighted dispersion relations). Up to now, I have not seen a demonstration that these claims are justified. It is true that Pietarinens expansion (Sect. A.6.3.4) is much more suitable than a dispersion relation in the case that the real part is known with an accuracy comparable with that of the imaginary part, but for $t=0$ it is only an equally practical alternative.

ii) Rigorous results for the asymptotic behaviour (Sects. A.5.3 and A.5.4)

Some of the results are of great theoretical interest but, in general, the theorems on the asymptotic behaviour have no direct consequences for the dispersion analysis of data and for models in the energy range which is accessible to experiments. Unfortunately, the dispersion relation is not a "crystal ball which can view the road to infinity" (69 R LINDENBAUM). Reasonable changes of the assumptions on the total cross sections above 5000 GeV/c have only a very small influence on the real parts below 200 GeV/c.

There is no objection against using a power law as Eq. (2.1.2a) up to the ISR range in pp-scattering, since a transition to another law, which satisfies the FROISSART bound, can occur at very much higher energies (78 L COLLINS-1, see Sect. 2.6.4.9).

iii) Microcausality

In earlier discussions of dispersion relations, the possibility to test the assumption of microcausality played an important role. The question is whether the dispersion relation is compatible with the total cross section data, the data for the real part and with a "reasonable" ansatz for the high energy behaviour of the total cross sections above the experimental region. Up to now, all discrepancies have disappeared when the experiments were repeated. – It was pointed out by CREUTZ and JAFFE (70 L) that, in this situation, one cannot derive a bound for a "fundamental length" by an application of acausal models. The only argument for a bound is a dimensional one: since there is no discrepancy up to 200 GeV/c, it is unlikely that the fundamental length is larger than $\hbar c/200\,\text{GeV/c} \approx 10^{-16}$ cm. The pp-scattering data measured at the ISR give a value which is 10 times smaller (see also 71 R OEHME-1).

iv) Sum rules

There exists a large number of papers on sum rules [the review of FERRO-FONTAN, QUEEN, and VIOLINI (72 R) lists more than 600 papers]. Most of them are applicable to pion-nucleon scattering.

The dispersion sum rules follow from the same analyticity properties from which the forward dispersion relation is derived and from additional assumptions on the high energy behaviour. In general, the authors use superpositions of the expressions Eqs. (2.1.9) and (2.1.14). It is the aim to find relations between the parameters of the high energy expansion and integrals over the forward amplitudes at low and intermediate energies.

Many of the results are outdated and some of the proposed methods cannot be accepted. In 80 R HÖHLER-KAISER, Sect. 9, we have discussed two of these sum rules which are of interest in connection with our parametrizations (see also Sect. 2.6.5.6).

2.2.2 πN partial wave amplitudes

In this section we present results of recent partial wave analyses. The KH 78 solution (Sect. 2.1.7) is given in Table 2.2.2.1 and Argand diagrams are shown in Figs. 2.2.1···2.2.21. Table 2.2.2.2 belongs to an updated version of this solution, which includes new experimental results, in particular the large data set 80 JENKIN. The data set 80 TERADA was not yet available.

The latter solution is preliminary, because the fixed-t solution has not yet been adjusted to the new experimental data up to 200 GeV/c and the iterations have not yet been performed. We hope that the final result will be available at the end of 1982.

Table 2.2.2.3 is the updated low-energy solution published by KOCH and PIETARINEN (80 L-1). It is compared with other recent solutions in Figs. 2.2.22···2.2.33 (see Sects. 2.1.8 and 2.1.10).

Table 2.2.2.4 gives the most recent version of the CMU–LBL solution (80 L CUTKOSKY-1). The Argand plots in Figs. 2.2.1···2.2.21 show the earlier solution of 1979 (Sect. 2.1.9), because the new result was not yet available. It is seen that the agreement between the KH 78 and CMU–LBL 79 solutions is reasonable, except for a few discrepancies. Some of them are probably due to the different data input. For instance, we have not used the preliminary large data set 76 BARDSL, of which the final result is not yet available.

The diagonal errors of the CMU–LBL partial waves are shown in the Argand diagrams (Figs. 2.2.1···2.2.21). They have not been listed, because they are of limited interest due to strong error correlations. If someone wants to work with the error matrix, he should ask R.E. Cutkosky or R. Kelly for a copy of the magnetic tape.

The number of decimal places in the tables is larger than required by the accuracy, but it can happen that it is not large enough to reproduce exactly some of the values in other tables.

Table 2.2.2.5 gives the result of the most recent partial wave analysis of the SIN group (82 L ALDER), which includes their new polarization data. Some of these data have not yet been available to KOCH and PIETARINEN (80 L-1).

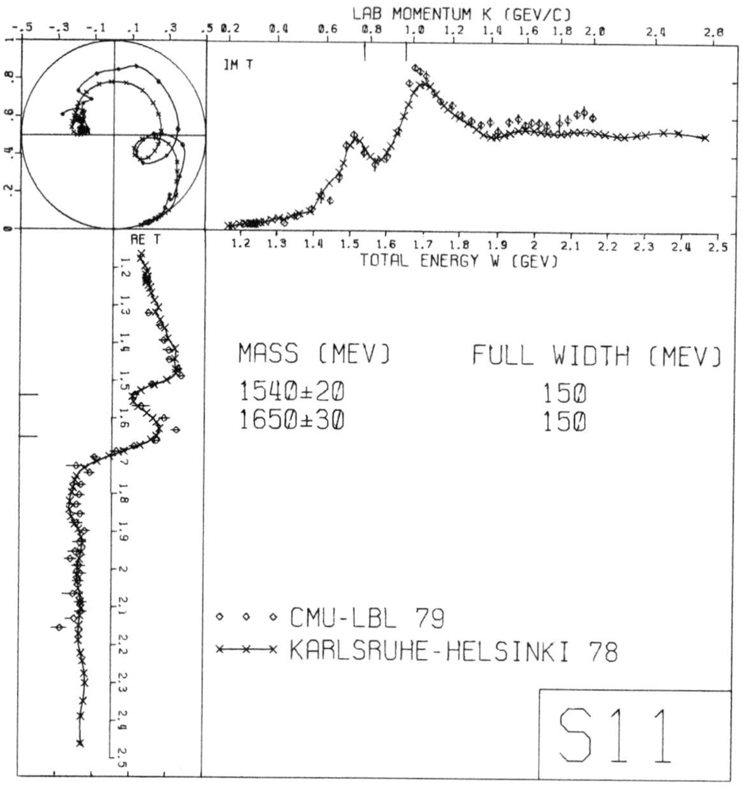

Figs. 2.2.1···2.2.21. Argand diagrams of πN partial waves. The plots show the results of the CMU–LBL 79 and KH 78 analyses. The parameters of the most important nucleon resonances are listed and the locations are marked on the horizontal scale for Im T and on the vertical scale for Re T (Values of the Particle Data Group 80 R Bricman).

Fig. 2.2.1.

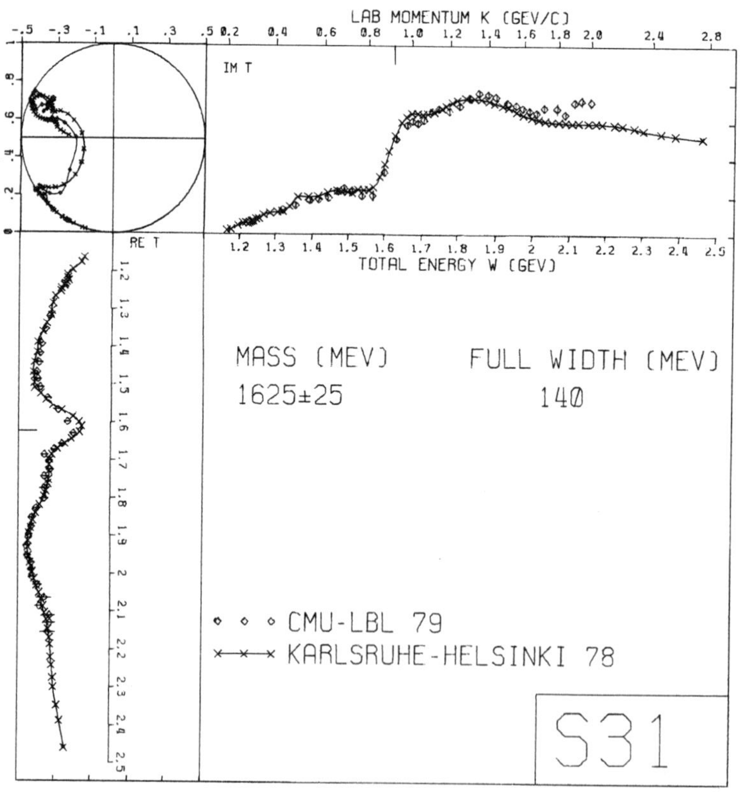

Fig. 2.2.2.

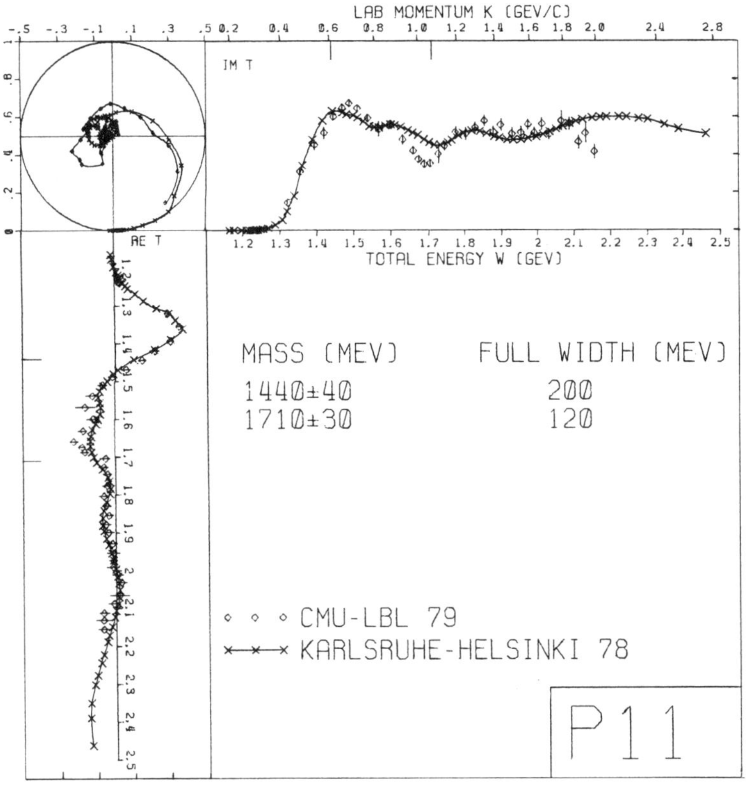

Fig. 2.2.3.

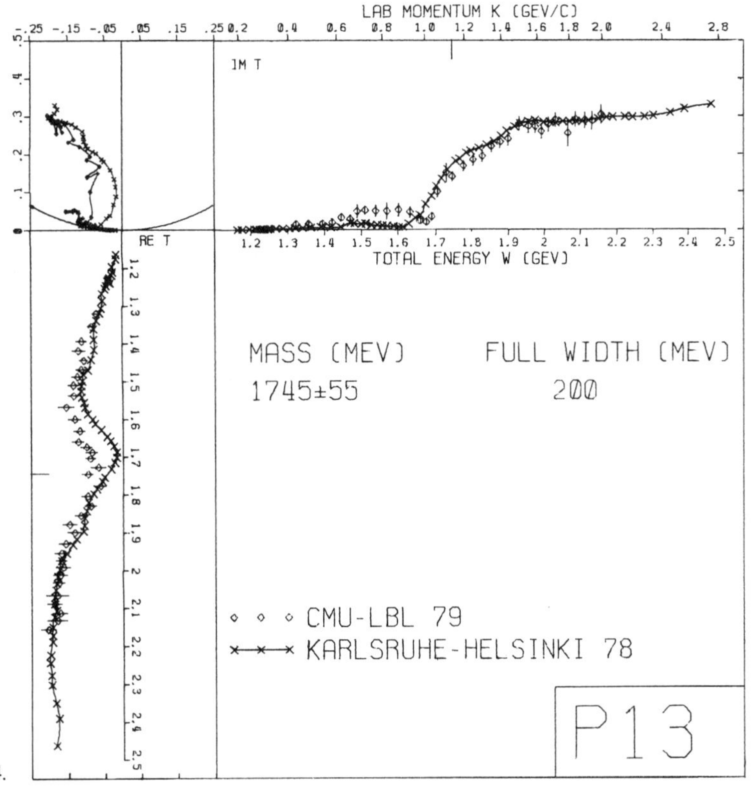

Fig. 2.2.4.

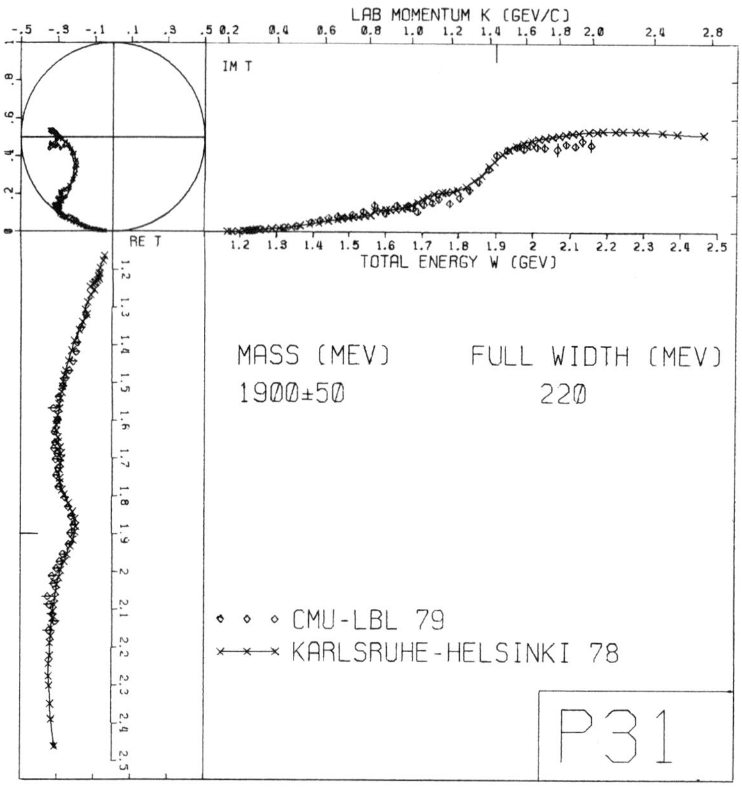

Fig. 2.2.5.

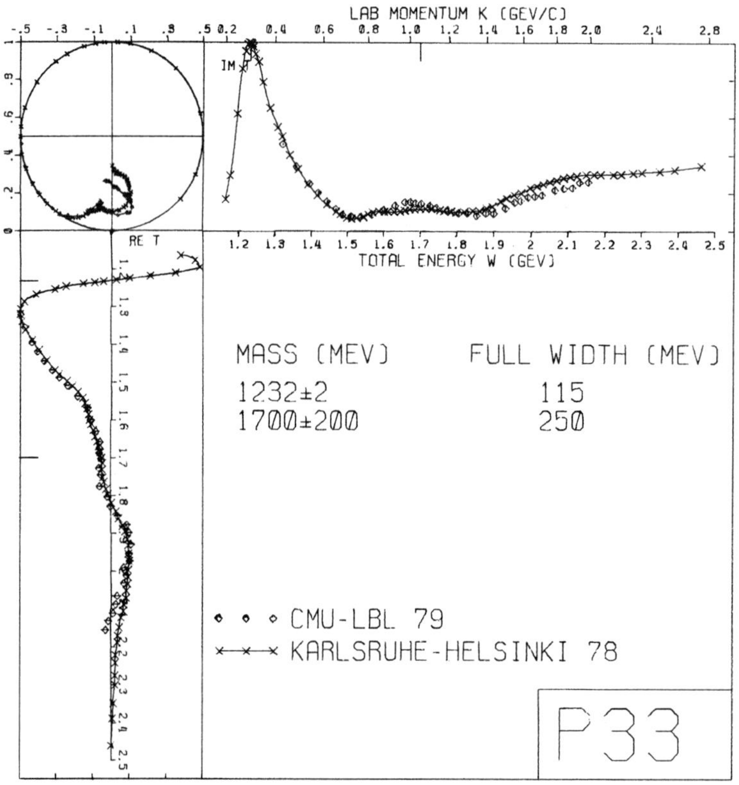

Fig. 2.2.6.

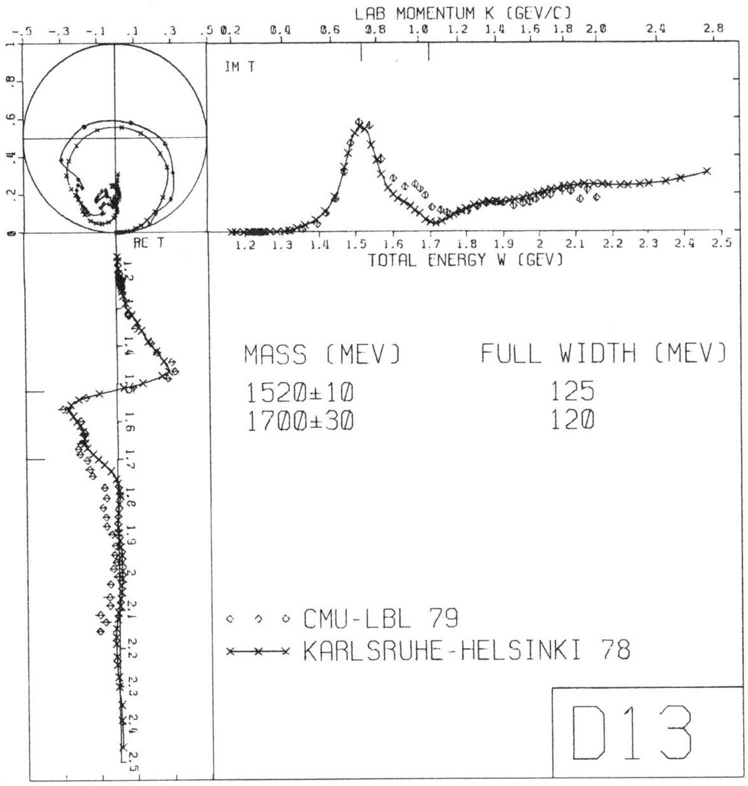

Fig. 2.2.7.

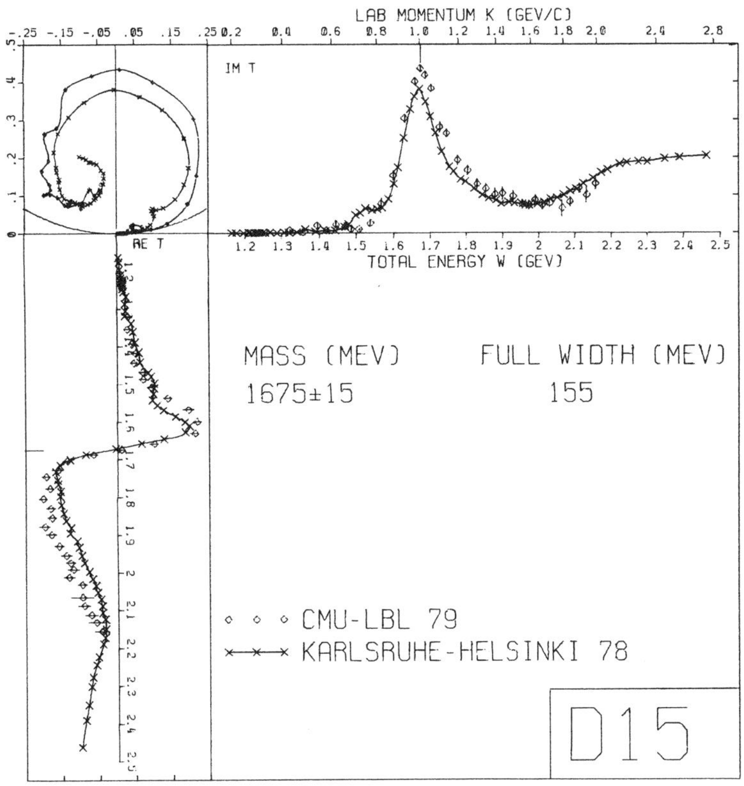

Fig. 2.2.8.

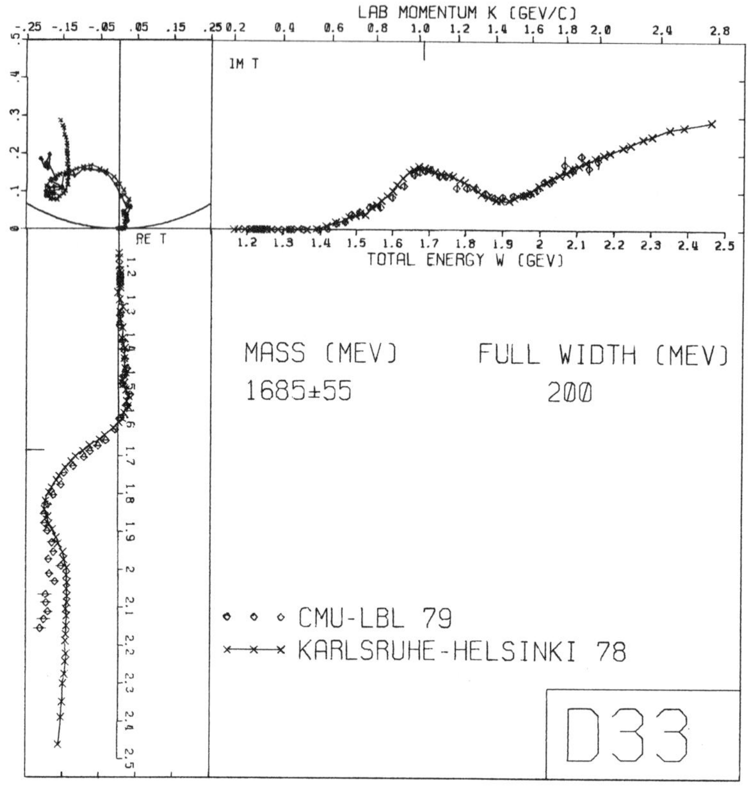

Fig. 2.2.9.

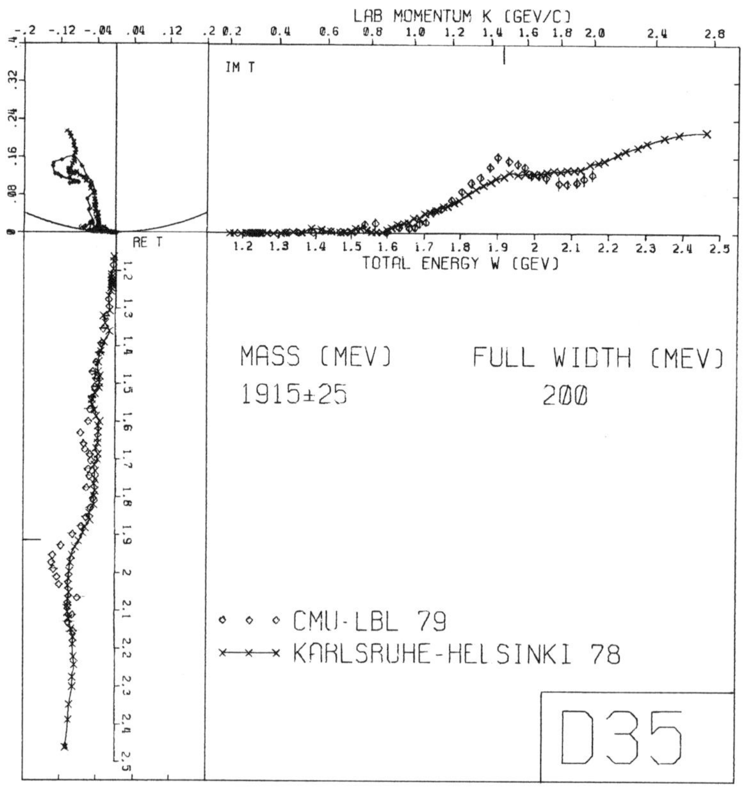

Fig. 2.2.10.

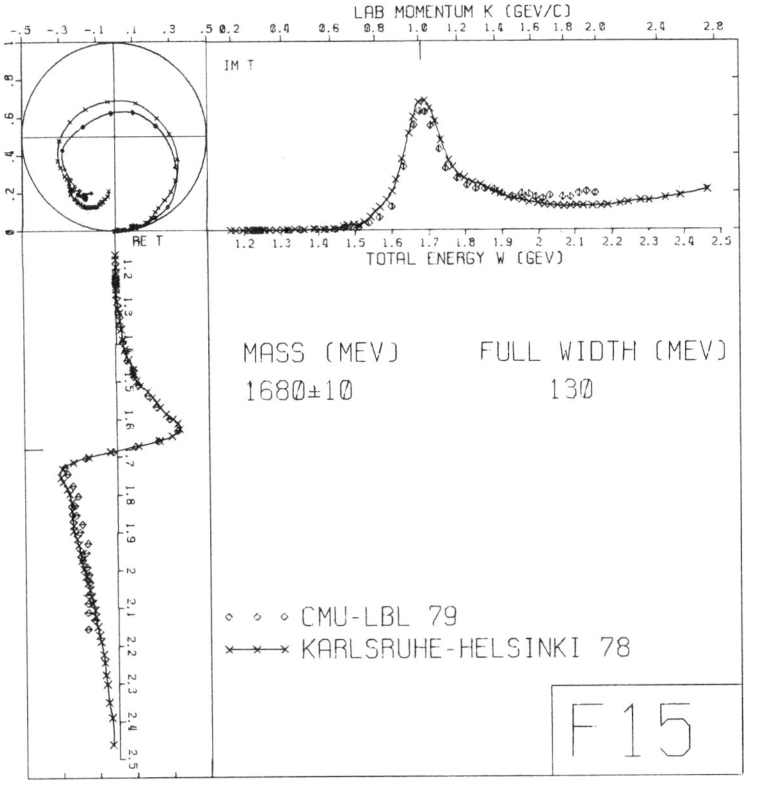

Fig. 2.2.11.

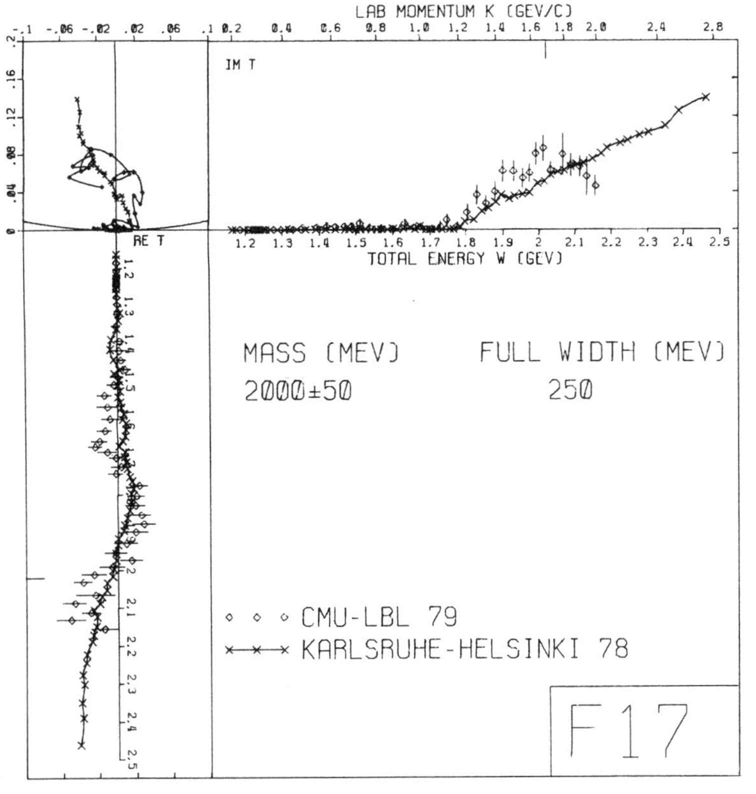

Fig. 2.2.12.

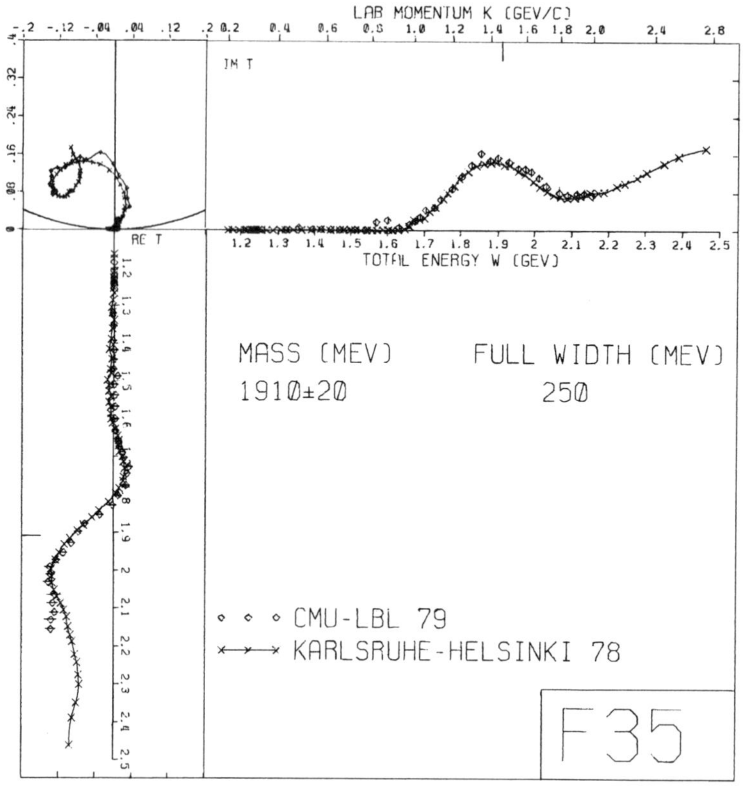

Fig. 2.2.13.

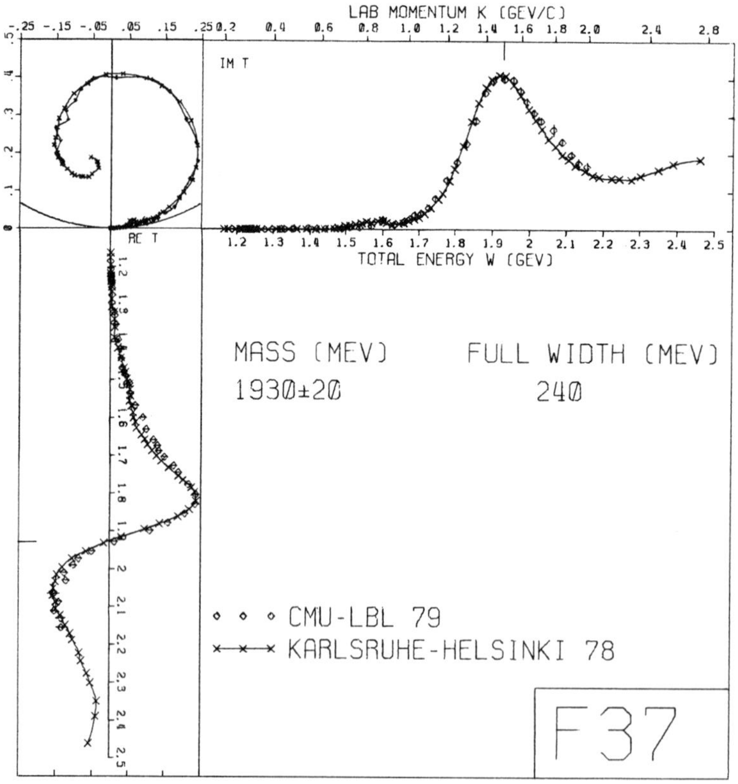

Fig. 2.2.14.

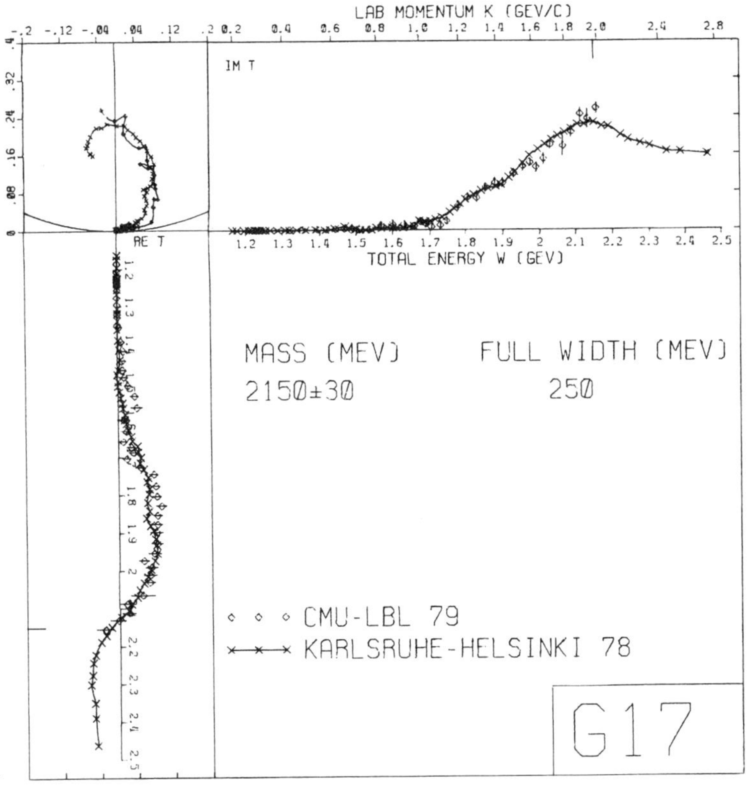

Fig. 2.2.15.

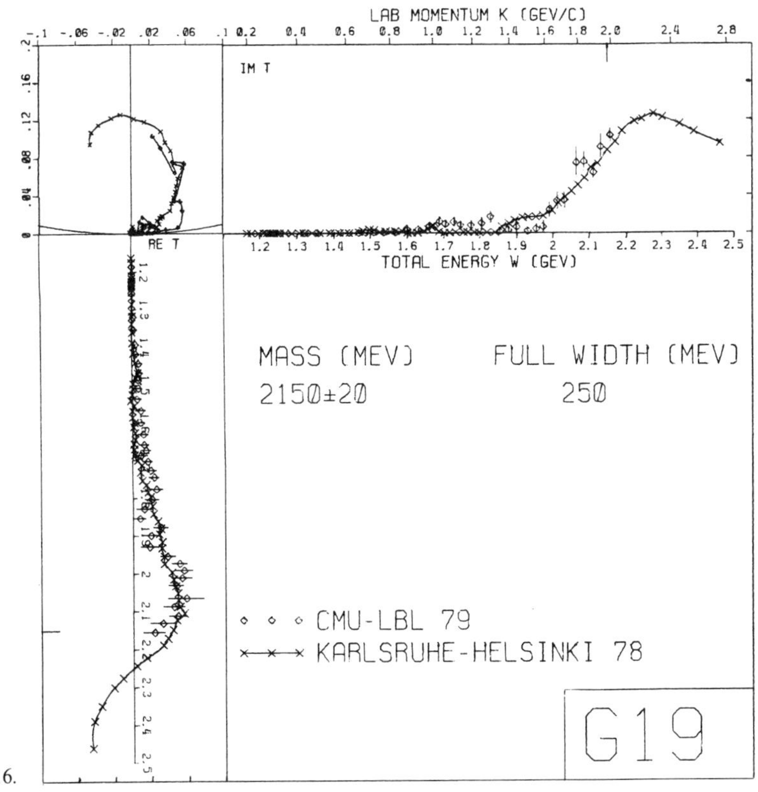

Fig. 2.2.16.

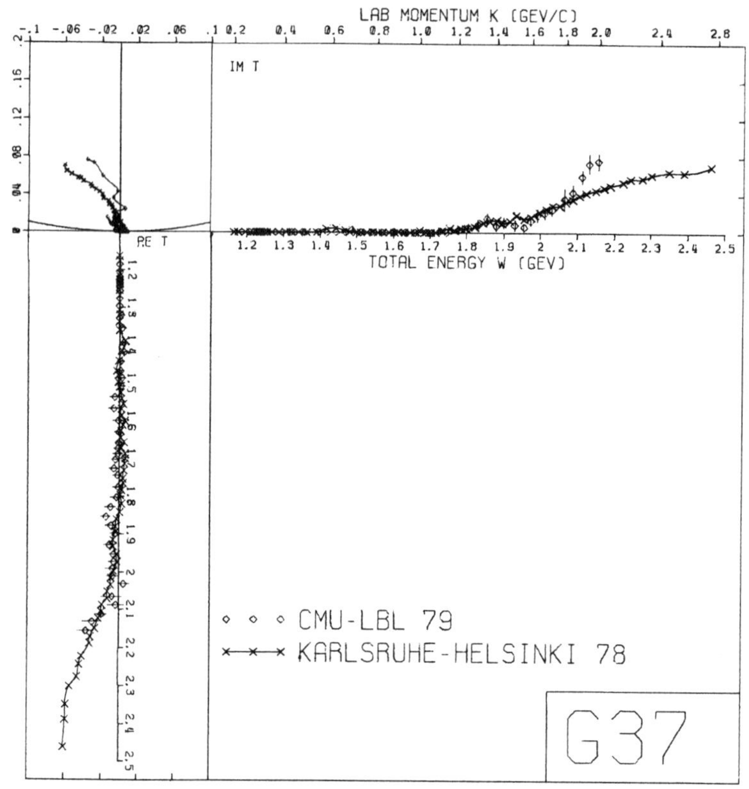

Fig. 2.2.17.

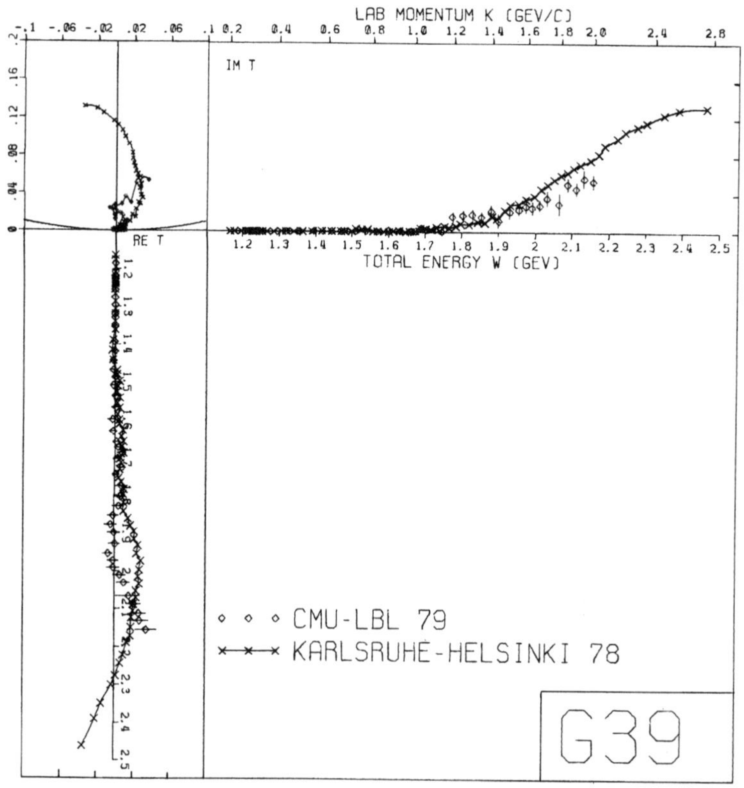

Fig. 2.2.18.

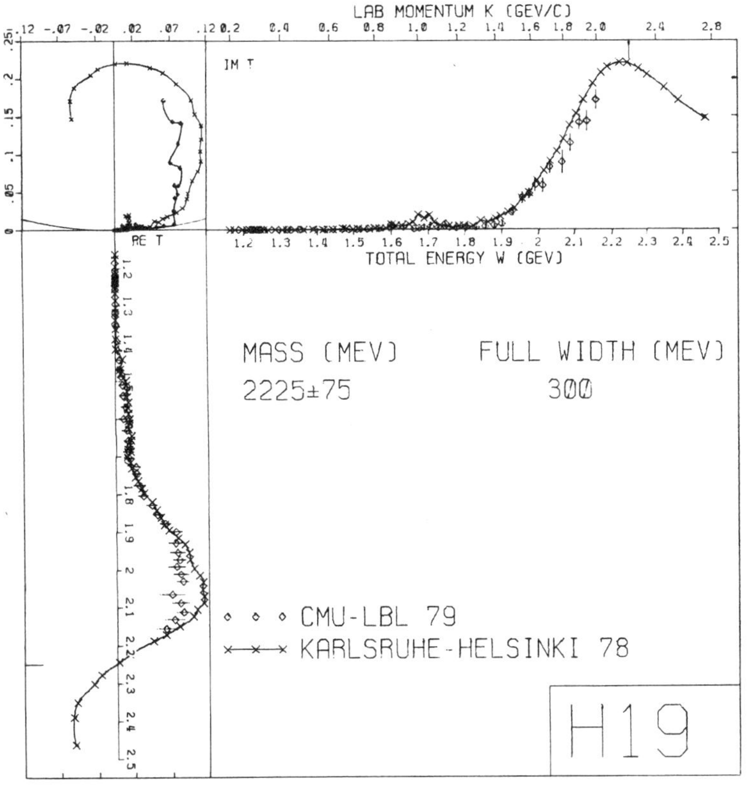

Fig. 2.2.19.

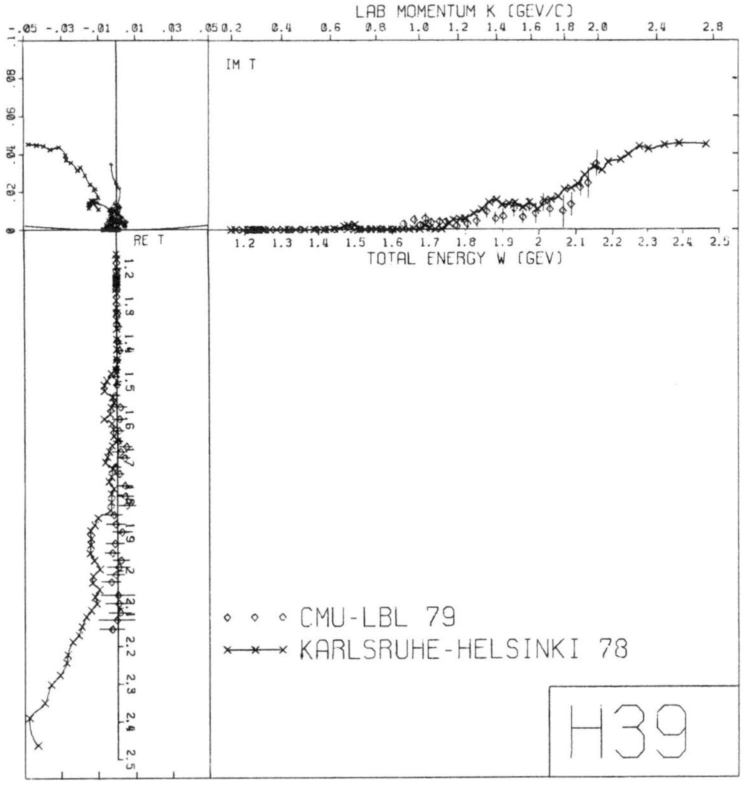

Fig. 2.2.20.

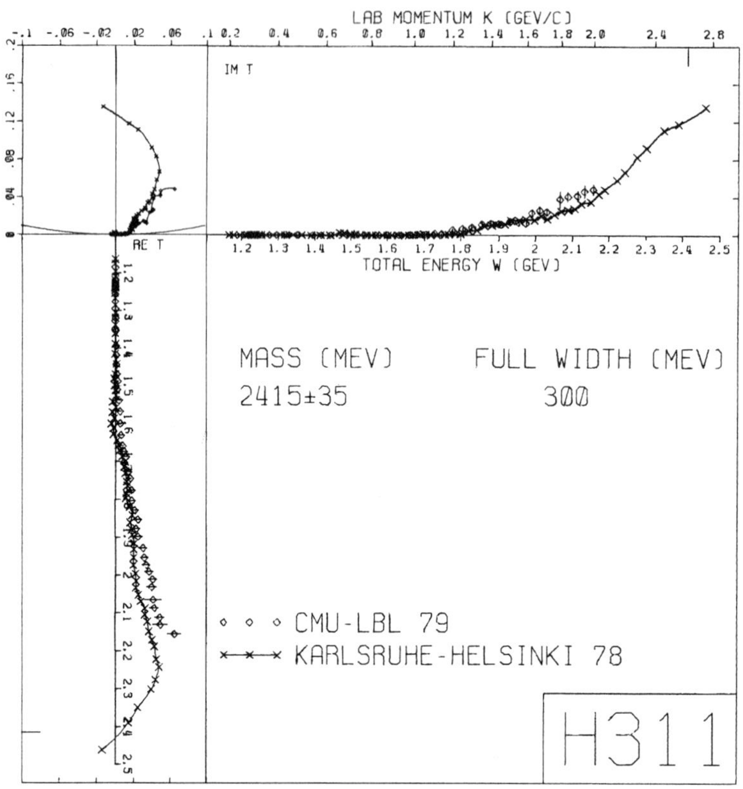

Fig. 2.2.21.

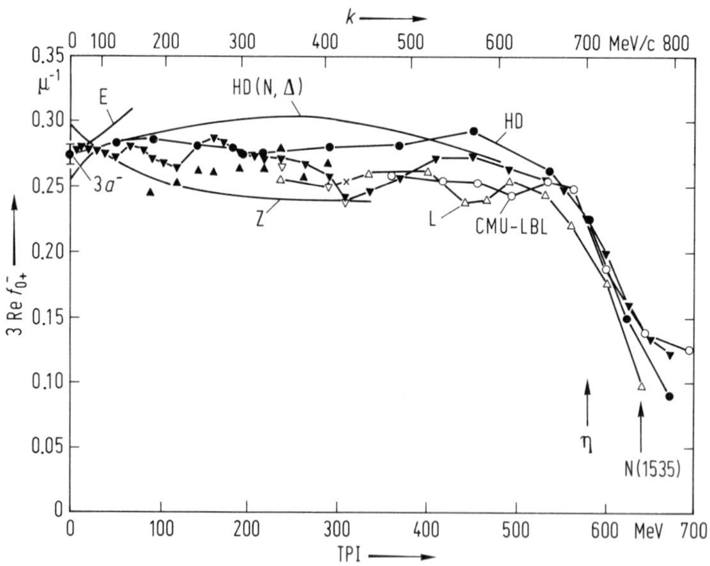

Fig. 2.2.22.

Figs. 2.2.22···2.2.33. The real parts of the lowest πN partial waves below $k = 800$ MeV/c. -▼- 80 L Koch-1 ("KH"); -○- 79 L Cutkosky ("CMU–LBL"); -△- 78 L Abaev (Leningrad, "L"); ▲ 73 L Carter (only a few typical error bars are shown); ▽ 78 Alder, 79 Alder (SIN); Z: 80 L Zidell; + 76 Bertin; * 79 Auld. — Error bars at threshold show the values of the scattering lengths (80 L Koch-1). — In the plots for f_{0+}^{-} and for f_{0+}^{+} the short lines denoted by "E" shows the first two terms of the effective range expansion. In the first case the validity of this approximation has been overestimated by several authors. — In the plot for f_{0+}^{-} the curve "HD" is the prediction from the projection of the fixed-t dispersion relation [Eq. (A.7.11)]. HD (N, Δ) belongs to an approximation where only the nucleon pole term and the P 33 contribution to the dispersion integral are taken into account (60 L Höhler). Data are not available below 0.079 GeV/c. The first three points of 80 L Koch-1 are extrapolations.

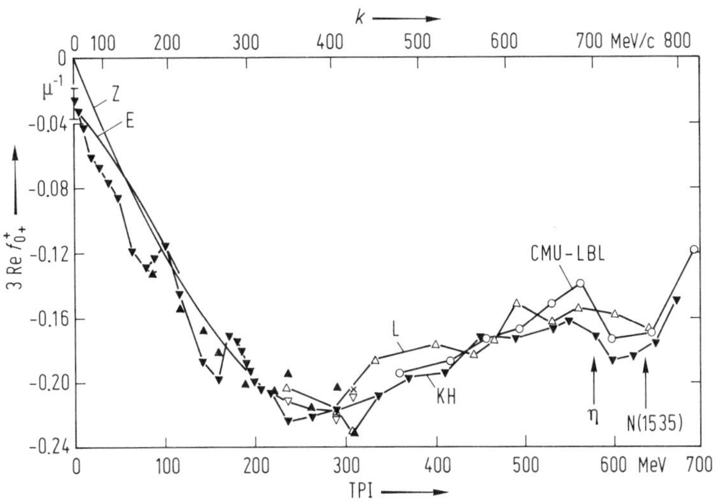

Fig. 2.2.23.

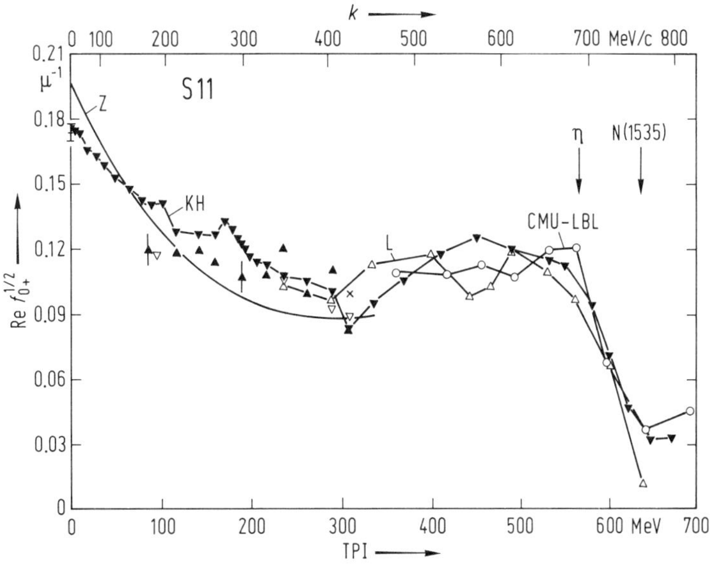

Fig. 2.2.24.

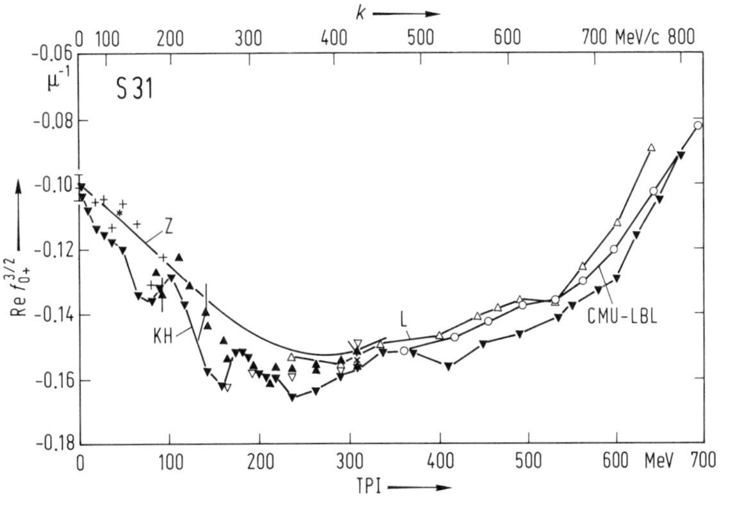

Fig. 2.2.25.

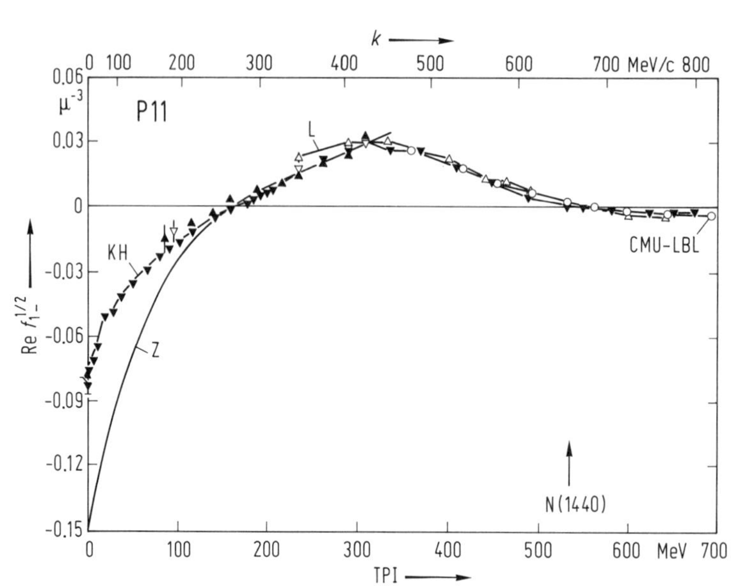

Fig. 2.2.26.

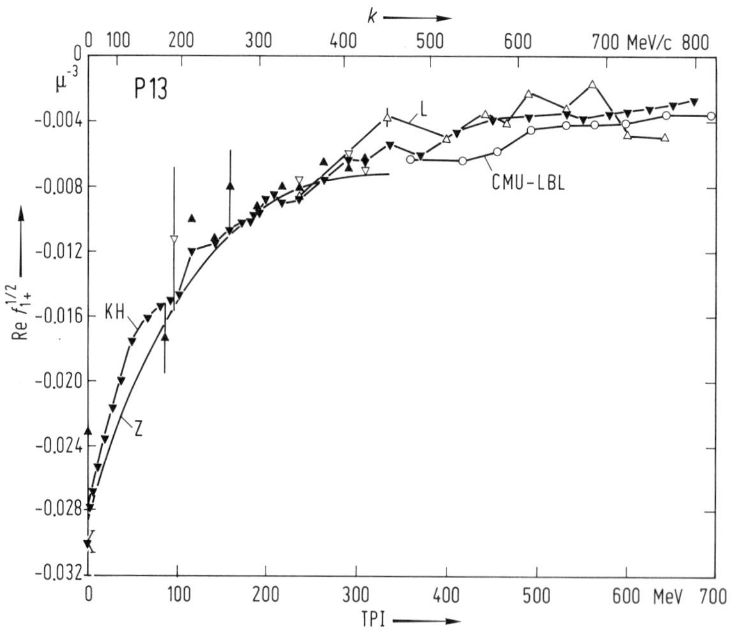

Fig. 2.2.27.

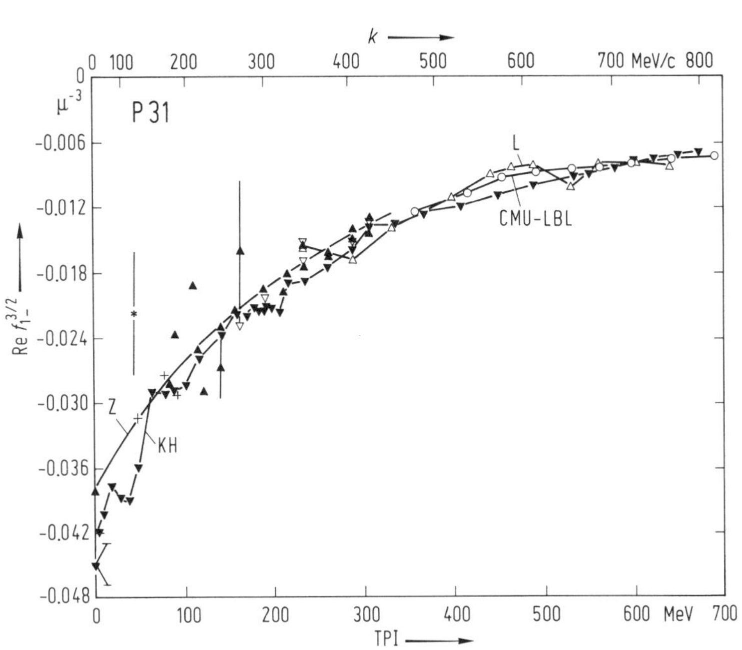

Fig. 2.2.28.

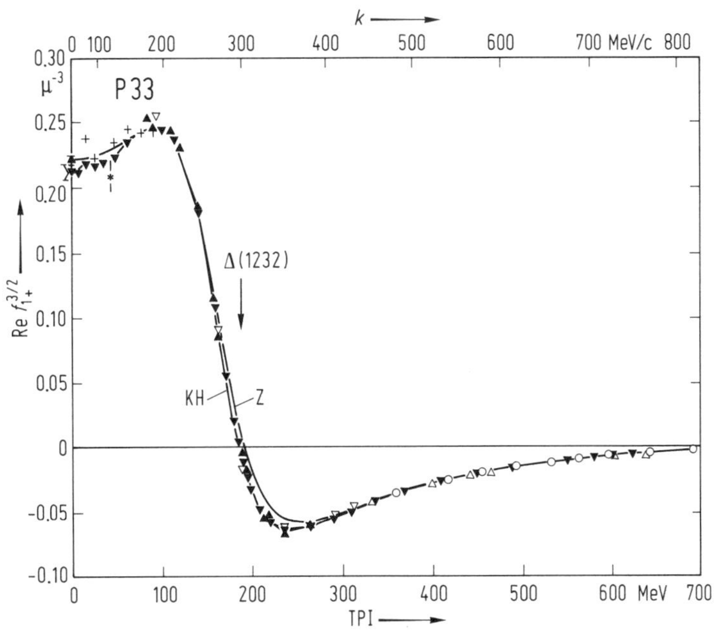

Fig. 2.2.29.

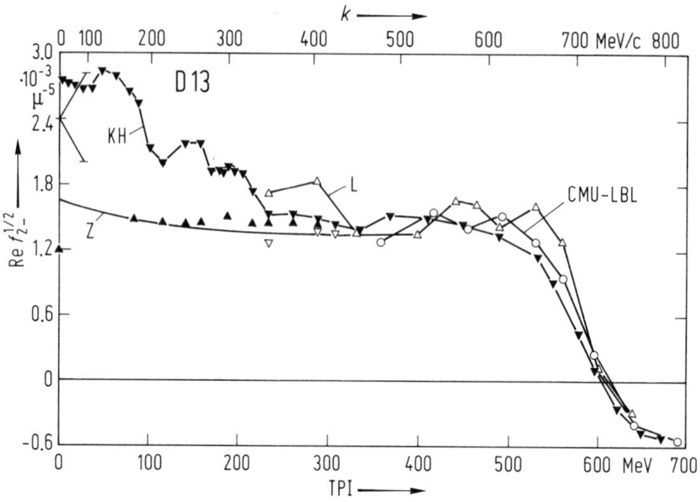

Fig. 2.2.30.

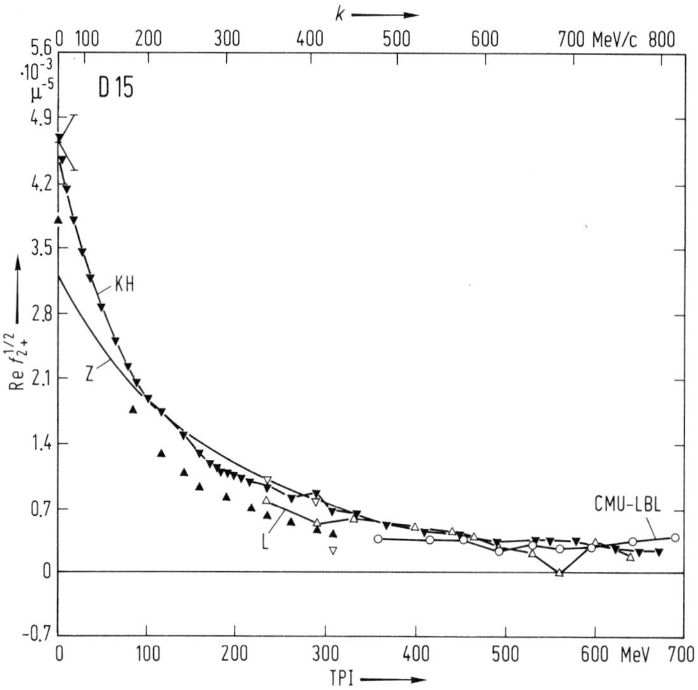

Fig. 2.2.31.

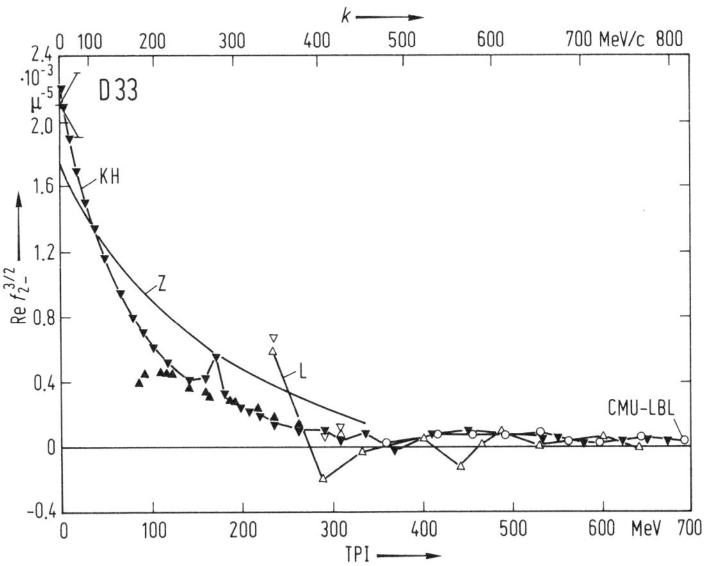

Fig. 2.2.32.

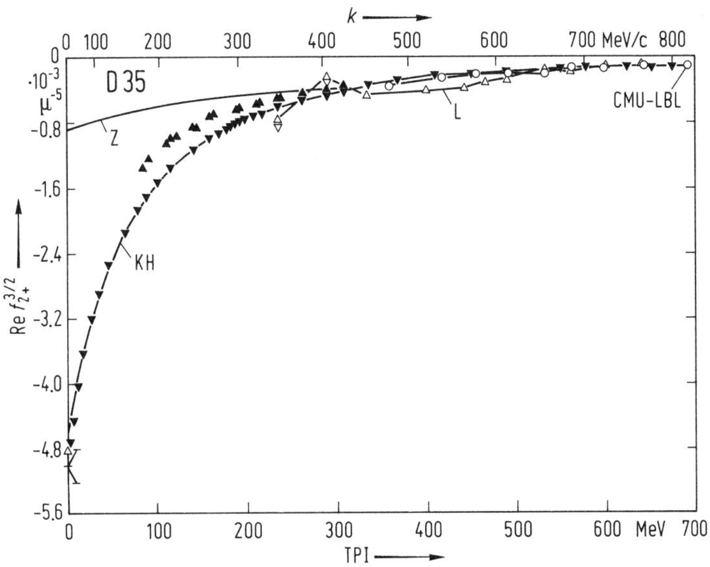

Fig. 2.2.33.

Table 2.2.2.1. Partial wave amplitudes of the solution KH 78 (Sect. 2.1.7). The kinematical variables listed in the first line are given for each energy in the line denoted by "KIN". Argand diagrams are shown in Figs. 2.2.1···2.2.21.

KH 78	K GEV/C		W GEV		S GEV**2		Q**2 (GEV/C)**2		ω_q GEV		E GEV	
	S1		P1		P3		D3		D5		F5	
KIN:	.020		1.079		1.165		.0003		.1407		.9385	
I=												
1/2	.022	.000	-.000	.000	-.000	.000	.000	.000	.000	.000	.000	.000
3/2	-.012	.000	-.000	.000	.000	.000	.000	.000	.000	.000	.000	.000
KIN:	.040		1.083		1.172		.0012		.1438		.9389	
I=												
1/2	.043	.002	-.001	.000	-.001	.000	.000	.000	.000	.000	.000	.000
3/2	-.024	.001	-.001	.000	.003	.000	.000	.000	.000	.000	.000	.000
KIN:	.060		1.089		1.185		.0027		.1489		.9397	
I=												
1/2	.064	.004	-.004	.000	-.002	.000	.000	.000	.000	.000	.000	.000
3/2	-.037	.001	-.002	.000	.011	.000	.000	.000	-.000	.000	.000	.000
KIN:	.079		1.096		1.201		.0046		.1551		.9407	
I=												
1/2	.088	.008	-.006	.000	-.003	.000	.000	.000	.000	.000	.000	.000
3/2	-.051	.003	-.004	.000	.026	.001	.000	.000	-.000	.000	.000	.000
KIN:	.097		1.104		1.219		.0068		.1621		.9419	
I=												
1/2	.106	.011	-.010	.000	-.005	.000	.000	.000	.000	.000	.000	.000
3/2	-.064	.004	-.008	.000	.046	.002	.000	.000	-.000	.000	.000	.000
KIN:	.112		1.112		1.236		.0089		.1686		.9431	
I=												
1/2	.118	.014	-.013	.000	-.007	.000	.000	.000	.000	.000	.000	.000
3/2	-.076	.006	-.011	.000	.070	.005	.000	.000	-.000	.000	.000	.000
KIN:	.130		1.122		1.258		.0118		.1770		.9446	
I=												
1/2	.129	.017	-.017	.000	-.009	.000	.001	.000	.001	.000	.000	.000
3/2	-.091	.008	-.015	.000	.109	.012	.000	.000	-.001	.000	.000	.000
KIN:	.153		1.135		1.289		.0160		.1884		.9468	
I=												
1/2	.143	.021	-.021	.000	-.013	.000	.001	.000	.002	.000	.000	.000
3/2	-.115	.013	-.021	.001	.177	.032	.001	.000	-.001	.000	.000	.000
KIN:	.172		1.147		1.316		.0198		.1982		.9488	
I=												
1/2	.157	.025	-.023	.001	-.016	.000	.002	.000	.002	.000	.000	.000
3/2	-.135	.019	-.027	.001	.249	.066	.001	.000	-.002	.000	.000	.000

KH 78	K GEV/C		W GEV		S GEV**2		Q**2 (GEV/C)**2		$\|\omega_q$ GEV		E GEV	
	S1		P1		P3		D3		D5		F5	
KIN:	.185		1.155		1.335		.0226		.2051		.9503	
I=												
1/2	.156	.025	-.022	.000	-.017	.000	.003	.000	.003	.000	.000	.000
3/2	-.139	.020	-.034	.001	.304	.103	.001	.000	-.002	.000	.000	.000
KIN:	.200		1.165		1.358		.0259		.2131		.9520	
I=												
1/2	.154	.024	-.019	.000	-.017	.000	.005	.000	.005	.000	.000	.000
	-.000	.000										
3/2	-.147	.022	-.043	.002	.375	.169	.001	.000	-.003	.000	.000	.000
	.001	.000										
KIN:	.218		1.177		1.386		.0302		.2229		.9543	
I=												
1/2	.150	.023	-.016	.000	-.018	.000	.007	.000	.007	.000	.001	.000
	-.000	.000										
3/2	-.164	.028	-.050	.003	.457	.298	.001	.000	-.004	.000	.000	.000
	.001	.000										
KIN:	.247		1.197		1.432		.0375		.2387		.9581	
I=												
1/2	.176	.032	-.004	.000	-.028	.001	.012	.000	.009	.000	.001	.000
	-.000	.000										
3/2	-.213	.048	-.065	.004	.485	.622	.002	.000	-.005	.000	.000	.000
	.001	.000										
KIN:	.267		1.211		1.466		.0429		.2498		.9609	
I=												
1/2	.185	.035	.011	.000	-.027	.001	.015	.000	.011	.000	.001	.000
	-.000	.000										
3/2	-.240	.061	-.072	.005	.348	.859	.004	.000	-.008	.000	.000	.000
	.001	.000										
KIN:	.280		1.219		1.487		.0464		.2567		.9627	
I=												
1/2	.179	.033	.020	.000	-.029	.001	.017	.000	.012	.000	.002	.000
	-.000	.000										
3/2	-.240	.061	-.075	.006	.213	.952	.005	.000	-.009	.000	.000	.000
	.001	.000										
KIN:	.290		1.226		1.504		.0492		.2621		.9642	
I=												
1/2	.188	.037	.025	.001	-.036	.001	.019	.000	.012	.000	.002	.000
	-.000	.000										
3/2	-.234	.058	-.072	.005	.100	.990	.004	.000	-.007	.000	-.000	.000
	.002	.000										
KIN:	.295		1.230		1.512		.0507		.2649		.9649	
I=												
1/2	.183	.035	.026	.001	-.038	.001	.022	.000	.014	.000	.003	.000
	-.000	.000										
3/2	-.244	.064	-.081	.007	.030	.999	.004	.000	-.007	.000	-.000	.000
	.002	.000										

KH 78	K GEV/C	W GEV	S GEV**2	Q**2 (GEV/C)**2	ω_q GEV	E GEV
	S1 F7	P1 G7	P3 G9	D3 H9	D5 H11	F5 I11
KIN:	.301	1.234	1.522	.0523	.2679	.9658
I=						
1/2	.183 .035	.031 .001	-.041 .002	.025 .001	.016 .000	.003 .000
	-.000 .000					
3/2	-.252 .068	-.092 .009	-.042 .998	.004 .000	-.007 .000	-.000 .000
	.003 .000					
KIN:	.305	1.237	1.529	.0536	.2703	.9664
I=						
1/2	.183 .035	.037 .001	-.039 .002	.025 .001	.016 .000	.003 .000
	-.000 .000					
3/2	-.256 .070	-.096 .009	-.092 .991	.004 .000	-.007 .000	-.000 .000
	.003 .000					
KIN:	.310	1.240	1.538	.0550	.2730	.9672
I=						
1/2	.182 .034	.047 .002	-.033 .001	.027 .001	.017 .000	.004 .000
	-.000 .000					
3/2	-.258 .072	-.105 .011	-.154 .976	.004 .000	-.007 .000	-.000 .000
	.003 .000					
KIN:	.320	1.247	1.555	.0580	.2783	.9687
I=						
1/2	.195 .039	.058 .003	-.045 .002	.027 .001	.016 .000	.004 .000
	-.000 .000					
3/2	-.273 .081	-.119 .014	-.246 .935	.004 .000	-.009 .000	-.000 .000
	.004 .000					
KIN:	.331	1.255	1.574	.0613	.2841	.9704
I=						
1/2	.202 .042	.073 .005	-.048 .002	.029 .001	.019 .000	.004 .000
	-.000 .000					
3/2	-.277 .084	-.098 .010	-.305 .896	.005 .000	-.010 .000	-.000 .000
	.004 .000					
KIN:	.351	1.268	1.608	.0673	.2946	.9735
I=						
1/2	.210 .046	.113 .013	-.056 .003	.038 .001	.026 .001	.005 .000
	.000 .000					
3/2	-.306 .104	-.119 .014	-.409 .787	-.002 .000	-.014 .000	.001 .000
	.006 .000					
KIN:	.378	1.287	1.657	.0761	.3091	.9780
I=						
1/2	.229 .055	.158 .029	-.054 .003	.049 .002	.022 .001	.008 .000
	.000 .000					
3/2	-.318 .114	-.134 .018	-.476 .653	.004 .000	-.014 .000	-.001 .000
	.005 .000					

KH 78	K GEV/C		W GEV		S GEV**2		Q**2 (GEV/C)**2		ω_q GEV		E GEV	
	S1 F7		P1 G7		P3 G9		D3 H9		D5 H11		F5 I11	
KIN:	.408		1.307		1.709		.0857		.3244		.9829	
I= 1/2	.248	.066	.228	.055	-.059	.003	.065	.004	.024	.001	.011	.000
	.001	.000										
3/2	-.325	.120	-.149	.023	-.497	.552	.012	.000	-.012	.000	-.005	.000
	.004	.000										
KIN:	.427		1.320		1.743		.0921		.3340		.9862	
I= 1/2	.235	.059	.300	.100	-.063	.004	.082	.007	.022	.001	.020	.000
	.003	.000										
3/2	-.326	.121	-.139	.020	-.500	.501	.004	.000	-.026	.001	-.004	.000
	.012	.000										
KIN:	.456		1.340		1.795		.1020		.3485		.9912	
I= 1/2	.260	.076	.331	.183	-.072	.005	.110	.023	.040	.002	.023	.003
	.002	.000										
3/2	-.351	.150	-.162	.028	-.490	.404	.005	.002	-.022	.002	.001	.000
	.012	.000										
KIN:	.490		1.362		1.856		.1139		.3652		.9971	
I= 1/2	.285	.089	.371	.344	-.079	.006	.136	.038	.045	.010	.030	.001
	.000	.000										
3/2	-.370	.200	-.181	.035	-.468	.331	.011	.000	-.013	.001	-.003	.000
	.013	.000										
KIN:	.532		1.390		1.932		.1290		.3853		1.0047	
I= 1/2	.303	.103	.306	.480	-.079	.007	.169	.065	.051	.003	.035	.005
	-.006	.000	.001	.001	-.000	.000						
3/2	-.399	.199	-.205	.044	-.432	.249	.011	.002	-.025	.011	-.006	.000
	.012	.000	.006	.000	-.002	.000						
KIN:	.573		1.417		2.007		.1441		.4044		1.0122	
I= 1/2	.343	.182	.217	.581	-.078	.006	.215	.111	.061	.006	.047	.002
	-.007	.000	.003	.000	.001	.000						
3/2	-.403	.204	-.220	.051	-.393	.191	.017	.010	-.032	.008	-.009	.001
	.021	.000	.003	.004	-.003	.001						
KIN:	.614		1.443		2.082		.1595		.4230		1.0197	
I= 1/2	.342	.253	.104	.629	-.086	.008	.257	.186	.064	.006	.068	.005
	-.004	.000	.005	.003	.003	.000	.010	.000	.001	.000	.005	.000
	.002	.000										
3/2	-.416	.223	-.237	.060	-.350	.143	.017	.020	-.036	.004	-.004	.000
	.030	.001	-.000	.005	-.002	.000	.000	.000	.001	.000	-.002	.000
	-.001	.000										

KH 78	K GEV/C		W GEV		S GEV**2		Q**2 (GEV/C)**2		ω_q GEV		E GEV	
	S1 F7 I13		P1 G7 K13		P3 G9 K15		D3 H9 L15		D5 H11 L17		F5 I11 M17	
KIN:	.658		1.470		2.162		.1763		.4425		1.0280	
I=												
1/2	.347	.310	.018	.625	-.095	.019	.282	.346	.086	.016	.093	.021
	.003	.000	.002	.008	.006	.001	.006	.001	.006	.000	.005	.005
	.004	.002										
3/2	-.422	.232	-.252	.068	-.305	.106	.019	.026	-.037	.001	-.007	.001
	.038	.002	-.003	.002	.002	.000	-.001	.002	-.000	.002	-.003	.000
	-.003	.000										
KIN:	.675		1.481		2.193		.1829		.4499		1.0312	
I=												
1/2	.346	.358	-.008	.614	-.108	.018	.246	.419	.094	.023	.097	.027
	-.003	.000	-.002	.005	.003	.002	.007	.001	.008	.000	.004	.006
	.004	.002										
3/2	-.419	.227	-.259	.072	-.286	.090	.021	.034	-.034	.003	-.006	.001
	.042	.006	-.001	.000	.002	.000	-.003	.003	-.001	.002	-.002	.000
	-.003	.000										
KIN:	.705		1.500		2.249		.1946		.4627		1.0368	
I=												
1/2	.298	.454	-.042	.610	-.110	.016	.141	.525	.103	.047	.111	.030
	.002	.000	.001	.004	.001	.004	.014	.000	.007	.000	.003	.008
	.004	.002	-.001	.000	-.000	.000						
3/2	-.417	.224	-.266	.076	-.240	.072	.013	.038	-.034	.004	-.013	.001
	.046	.010	-.001	.000	.006	.000	-.005	.003	-.001	.002	.001	.000
	-.004	.001	-.001	.000	-.002	.000						
KIN:	.725		1.512		2.285		.2025		.4711		1.0406	
I=												
1/2	.231	.485	-.064	.596	-.113	.017	.038	.557	.102	.054	.131	.027
	.001	.000	.000	.000	.001	.003	.015	.001	.008	.000	.001	.010
	.002	.004	.001	.000	.001	.000						
3/2	-.415	.221	-.270	.079	-.213	.068	.014	.043	-.036	.010	-.010	.001
	.053	.011	.000	.000	.004	.003	-.007	.000	.000	.001	.001	.000
	-.002	.001	-.001	.000	-.004	.000						
KIN:	.750		1.527		2.332		.2124		.4816		1.0454	
I=												
1/2	.157	.479	-.081	.577	-.113	.013	-.097	.544	.100	.066	.172	.043
	.003	.000	.005	.000	.000	.001	.015	.000	.007	.000	.000	.007
	.005	.002	.001	.001	.000	.000						
3/2	-.380	.238	-.273	.081	-.181	.069	.021	.041	-.044	.011	-.010	.000
	.054	.014	.002	.000	.003	.002	-.007	.000	.001	.000	.001	.000
	-.003	.000	-.003	.000	-.003	.000						
KIN:	.777		1.543		2.381		.2232		.4926		1.0505	
I=												
1/2	.111	.432	-.088	.552	-.112	.013	-.210	.463	.097	.061	.200	.070
	.001	.000	.007	.000	-.002	.002	.016	.001	.009	.000	.007	.009
	.008	.003	-.001	.002	.000	.000						
3/2	-.351	.235	-.279	.085	-.155	.076	.029	.060	-.051	.006	-.005	.000
	.056	.017	.003	.000	.006	.003	-.002	.000	-.003	.000	-.004	.000
	-.002	.000	-.002	.000	-.001	.000						

KH 78	K GEV/C		W GEV		S GEV**2		Q**2 (GEV/C)**2		ω_q GEV		E GEV	
	S1 F7 I13		P1 G7 K13		P3 G9 K15		D3 H9 L15		D5 H11 L17		F5 I11 M17	
KIN:	.800		1.557		2.424		.2325		.5019		1.0549	
I=												
1/2	.115	.394	-.082	.545	-.106	.012	-.253	.381	.110	.063	.220	.100
	.004	.000	.010	.005	.002	.001	.017	.002	.003	.000	.006	.012
	.009	.005	.001	.000	.003	.000						
3/2	-.313	.237	-.288	.091	-.144	.092	.024	.066	-.050	.003	-.011	.000
	.054	.021	.005	.000	.003	.000	-.002	.000	.001	.000	-.002	.000
	-.003	.001	-.000	.000	.002	.000						
KIN:	.822		1.570		2.465		.2414		.5107		1.0591	
I=												
1/2	.142	.373	-.077	.544	-.103	.011	-.265	.304	.128	.067	.238	.122
	.004	.000	.011	.003	.000	.002	.016	.002	.005	.000	.008	.018
	.005	.006	.002	.000	.005	.000	-.001	.000	.002	.000		
3/2	-.266	.253	-.286	.110	-.132	.101	.027	.078	-.045	.002	-.007	.000
	.058	.019	.002	.000	.006	.001	-.003	.000	-.003	.000	-.001	.000
	.003	.000	.002	.000	.001	.000	-.003	.000	-.004	.000		
KIN:	.851		1.587		2.518		.2532		.5222		1.0647	
I=												
1/2	.190	.382	-.081	.554	-.095	.009	-.242	.232	.160	.090	.271	.162
	.007	.000	.015	.006	.000	.002	.018	.002	.004	.000	.001	.019
	.003	.007	.003	.000	.005	.001	-.002	.000	-.000	.000		
3/2	-.206	.308	-.294	.116	-.123	.106	.016	.090	-.042	.002	-.006	.001
	.064	.022	.003	.000	.003	.000	-.004	.000	-.001	.001	-.000	.000
	.001	.000	-.002	.000	-.000	.000	.001	.000	-.001	.000		
KIN:	.875		1.601		2.563		.2630		.5315		1.0693	
I=												
1/2	.224	.417	-.092	.561	-.082	.007	-.220	.199	.185	.130	.304	.207
	.006	.001	.015	.004	.003	.000	.019	.007	.005	.001	.001	.016
	.002	.006	.003	.000	.005	.002	-.002	.000	-.002	.001		
3/2	-.171	.372	-.300	.119	-.120	.103	.012	.103	-.034	.007	-.007	.001
	.065	.023	.007	.000	.005	.000	-.007	.000	-.004	.000	-.002	.000
	.002	.000	.002	.000	.003	.000	-.001	.001	-.002	.000		
KIN:	.895		1.612		2.600		.2713		.5392		1.0732	
I=												
1/2	.243	.454	-.105	.557	-.075	.007	-.203	.178	.197	.174	.331	.268
	.010	.000	.019	.005	.001	.000	.019	.005	.003	.002	-.002	.014
	.002	.003	.004	.000	.003	.004	-.003	.000	-.000	.000		
3/2	-.158	.441	-.303	.120	-.111	.104	.002	.119	-.038	.012	-.002	.000
	.070	.018	.004	.001	.005	.001	-.003	.000	-.002	.000	.001	.000
	.003	.000	-.002	.000	.000	.000	-.001	.000	-.001	.000		
KIN:	.923		1.628		2.652		.2829		.5499		1.0785	
I=												
1/2	.258	.522	-.121	.547	-.058	.016	-.191	.158	.186	.252	.343	.378
	.009	.001	.022	.004	.003	.000	.018	.005	.001	.003	-.003	.014
	.001	.001	.005	.001	.003	.005	.001	.001	.003	.000		
3/2	-.170	.526	-.303	.122	-.100	.103	-.013	.139	-.036	.018	.003	.000
	.074	.015	.002	.000	.009	.001	-.002	.000	-.002	.000	-.001	.000
	.001	.000	.001	.000	.001	.000	-.002	.000	-.000	.000		

KH 78	K GEV/C		W GEV		S GEV**2		Q**2 (GEV/C)**2		ω_q GEV		E GEV	
	S1 F7 I13		P1 G7 K13		P3 G9 K15		D3 H9 L15		D5 H11 L17		F5 I11 M17	
KIN:	.954		1.646		2.709		.2958		.5615		1.0845	
I=												
1/2	.243	.610	-.131	.530	-.042	.033	-.182	.136	.128	.328	.299	.515
	.009	.002	.029	.008	.003	.003	.022	.007	-.001	.000	-.005	.013
	.001	.001	.003	.003	.003	.005	.004	.001	.004	.002		
3/2	-.214	.593	-.294	.128	-.089	.107	-.036	.154	-.038	.022	.010	.003
	.087	.017	.002	.000	.008	.000	-.002	.000	.002	.000	.000	.001
	.002	.000	.001	.000	-.000	.000	-.003	.000	-.000	.000		
KIN:	.975		1.658		2.748		.3045		.5692		1.0886	
I=												
1/2	.213	.669	-.133	.518	-.032	.039	-.179	.114	.068	.362	.233	.599
	.006	.000	.031	.009	.002	.004	.021	.010	.001	.000	-.003	.009
	.002	.000	-.002	.003	.005	.008	.005	.007	.000	.003		
3/2	-.250	.624	-.289	.131	-.077	.110	-.050	.159	-.037	.022	.011	.009
	.096	.018	.006	.000	.010	.000	.000	.000	.004	.000	-.003	.000
	-.001	.001	.001	.001	-.002	.000	-.004	.000	.003	.000		
KIN:	1.000		1.672		2.795		.3150		.5784		1.0934	
I=												
1/2	.154	.735	-.133	.504	-.022	.068	-.166	.097	-.004	.381	.119	.678
	.002	.003	.043	.020	.002	.008	.020	.020	.005	.004	-.006	.012
	.000	.000	-.001	.000	.001	.002	.001	.003	.000	.003		
3/2	-.290	.641	-.283	.145	-.070	.115	-.076	.169	-.042	.032	.012	.022
	.104	.024	.003	.002	.009	.001	-.003	.000	.004	.001	-.003	.000
	-.001	.000	.002	.000	.002	.000	.001	.000	.004	.000		
KIN:	1.030		1.688		2.850		.3277		.5892		1.0991	
I=												
1/2	.066	.774	-.128	.482	-.015	.089	-.137	.064	-.085	.348	-.033	.686
	.009	.000	.044	.018	.003	.001	.020	.016	.006	.000	-.004	.005
	.003	.000	-.003	.002	-.002	.005	-.001	.008	-.004	.005		
3/2	-.320	.639	-.276	.159	-.060	.116	-.095	.164	-.037	.032	.013	.023
	.117	.028	.007	.000	.010	.004	-.004	.002	.008	.000	-.003	.000
	-.003	.001	.001	.000	.002	.001	-.000	.000	.003	.000		
KIN:	1.055		1.702		2.897		.3383		.5981		1.1039	
I=												
1/2	-.013	.780	-.113	.466	-.016	.115	-.109	.052	-.128	.309	-.151	.648
	.008	.000	.049	.020	.006	.000	.016	.019	.007	.000	-.005	.000
	.004	.000	.001	.000	-.004	.000	-.003	.008	-.005	.005		
3/2	-.333	.638	-.277	.176	-.055	.120	-.114	.160	-.038	.042	.019	.026
	.130	.032	.007	.000	.007	.003	-.005	.001	.011	.001	-.001	.000
	-.001	.001	-.000	.000	.000	.000	.001	.001	.004	.001		
KIN:	1.080		1.716		2.943		.3489		.6069		1.1087	
I=												
1/2	-.080	.767	-.098	.453	-.020	.134	-.076	.047	-.157	.266	-.235	.581
	.009	.001	.047	.023	.010	.000	.020	.010	.007	.000	-.002	.000
	.006	.000	-.000	.000	-.004	.000	-.004	.005	-.006	.001		
3/2	-.336	.640	-.280	.189	-.049	.118	-.126	.154	-.044	.044	.022	.040
	.143	.044	.006	.000	.007	.003	-.006	.000	.009	.000	-.001	.000
	-.001	.004	-.000	.000	.002	.000	.003	.000	.004	.000		

KH 78	K GEV/C		W GEV		S GEV**2		Q**2 (GEV/C)**2		ω_q GEV		E GEV	
	S1 F7 I13		P1 G7 K13		P3 G9 K15		D3 H9 L15		D5 H11 L17		F5 I11 M17	
KIN:	1.113		1.733		3.005		.3629		.6184		1.1151	
I=												
1/2	-.147	.726	-.070	.451	-.031	.160	-.040	.056	-.168	.217	-.295	.480
	.011	.000	.055	.027	.009	.000	.022	.006	.004	.000	.002	.000
	.006	.001	.001	.000	-.002	.000	-.003	.002	-.004	.000		
3/2	-.334	.648	-.284	.202	-.049	.117	-.142	.150	-.043	.049	.024	.048
	.163	.058	.006	.002	.006	.005	-.003	.000	.011	.000	.001	.000
	.002	.004	-.000	.000	.002	.000	-.000	.000	.003	.001		
KIN:	1.154		1.755		3.081		.3805		.6325		1.1229	
I=												
1/2	-.188	.669	-.040	.466	-.049	.182	-.010	.079	-.162	.176	-.304	.373
	.014	.000	.064	.039	.011	.001	.026	.005	.001	.000	.005	.001
	.003	.002	.005	.000	.001	.000	-.003	.000	-.002	.000		
3/2	-.335	.668	-.284	.210	-.044	.112	-.159	.146	-.044	.057	.022	.067
	.190	.081	.004	.005	.005	.006	-.003	.003	.012	.000	.003	.000
	.001	.003	.000	.000	.003	.001	.002	.000	.001	.000		
KIN:	1.174		1.766		3.119		.3891		.6392		1.1267	
I=												
1/2	-.195	.647	-.032	.478	-.055	.188	-.001	.092	-.161	.164	-.291	.337
	.016	.000	.062	.046	.015	.000	.030	.003	.001	.000	.008	.000
	-.000	.000	.007	.005	-.002	.003	-.005	.001	.001	.000		
3/2	-.340	.680	-.283	.211	-.038	.109	-.166	.142	-.046	.058	.021	.079
	.203	.097	.005	.002	.007	.005	-.004	.005	.013	.001	.006	.000
	.000	.002	-.001	.000	.003	.000	.003	.000	-.002	.002		
KIN:	1.210		1.785		3.186		.4046		.6512		1.1336	
I=												
1/2	-.206	.618	-.027	.501	-.071	.204	.007	.110	-.154	.143	-.269	.292
	.018	.003	.067	.055	.017	.000	.035	.005	-.001	.000	.007	.000
	-.000	.000	.007	.009	-.003	.000	-.004	.003	-.001	.001		
3/2	-.350	.700	-.272	.218	-.027	.105	-.179	.133	-.044	.065	.013	.095
	.225	.130	.003	.005	.010	.007	-.002	.006	.012	.002	.001	.000
	-.000	.002	-.000	.001	.002	.000	.002	.001	-.002	.000		
KIN:	1.235		1.798		3.232		.4154		.6595		1.1384	
I=												
1/2	-.212	.602	-.031	.514	-.080	.209	.010	.122	-.155	.137	-.257	.278
	.015	.009	.066	.065	.020	.000	.038	.006	.003	.000	.011	.000
	.004	.000	.005	.012	-.003	.000	-.005	.004	.001	.000		
3/2	-.362	.709	-.260	.223	-.015	.100	-.187	.127	-.046	.070	.005	.110
	.234	.161	.002	.005	.009	.009	-.003	.006	.012	.001	.001	.000
	-.000	.001	-.001	.001	.003	.000	.002	.000	-.000	.000		
KIN:	1.280		1.821		3.316		.4350		.6741		1.1469	
I=												
1/2	-.222	.572	-.047	.528	-.093	.217	.004	.142	-.153	.118	-.244	.254
	.015	.011	.062	.075	.022	.000	.049	.004	.006	.000	.015	.001
	.005	.000	-.003	.015	.000	.000	-.003	.003	.006	.000		
3/2	-.385	.720	-.237	.243	.002	.105	-.196	.114	-.046	.083	-.011	.125
	.238	.222	.003	.006	.010	.008	-.003	.009	.015	.005	.003	.000
	-.001	.001	-.005	.000	.005	.000	.005	.000	-.002	.000		

KH 78	K GEV/C		W GEV		S GEV**2		Q**2 (GEV/C)**2		ω_q GEV		E GEV	
	S1 F7 I13		P1 G7 K13		P3 G9 K15		D3 H9 L15		D5 H11 L17		F5 I11 M17	
KIN:	1.324		1.843		3.398		.4542		.6882		1.1552	
I=												
1/2	-.225	.543	-.064	.524	-.099	.224	.000	.152	-.146	.101	-.244	.241
	.013	.019	.067	.084	.023	.001	.055	.012	.009	.000	.015	.001
	.010	.000	-.003	.010	.004	.001	-.002	.001	.007	.001		
3/2	-.407	.720	-.214	.277	.027	.108	-.198	.097	-.054	.093	-.031	.138
	.219	.287	.001	.011	.010	.011	-.004	.011	.018	.005	.002	.001
	.000	.001	-.004	.003	.005	.000	.006	.000	-.001	.000		
KIN:	1.360		1.862		3.465		.4699		.6996		1.1620	
I=												
1/2	-.215	.516	-.069	.514	-.105	.235	-.003	.159	-.140	.092	-.240	.223
	.011	.023	.060	.088	.028	.007	.060	.010	.014	.000	.016	.004
	.009	.001	-.002	.009	.001	.001	.001	.000	.009	.003		
3/2	-.421	.716	-.202	.305	.042	.111	-.190	.089	-.054	.102	-.045	.141
	.189	.336	-.002	.013	.015	.010	-.010	.014	.017	.010	.002	.002
	-.002	.000	-.003	.002	.005	.000	.005	.004	-.002	.000		
KIN:	1.400		1.882		3.540		.4874		.7120		1.1695	
I=												
1/2	-.195	.503	-.071	.497	-.106	.249	-.003	.160	-.126	.087	-.236	.205
	.008	.029	.069	.092	.031	.010	.065	.015	.012	.001	.016	.006
	.011	.004	-.003	.005	.001	.003	.001	.000	.002	.004		
3/2	-.434	.708	-.198	.349	.066	.124	-.186	.079	-.064	.110	-.066	.144
	.138	.377	-.003	.014	.018	.017	-.012	.016	.019	.011	-.000	.004
	.001	.000	-.003	.002	.004	.000	.005	.002	-.001	.000		
KIN:	1.430		1.896		3.596		.5006		.7212		1.1752	
I=												
1/2	-.177	.503	-.064	.490	-.108	.262	.000	.156	-.129	.078	-.235	.195
	.007	.037	.075	.096	.029	.014	.071	.018	.014	.001	.016	.007
	.011	.007	-.001	.006	-.001	.002	.002	.000	.003	.002	.001	.000
	.000	.001										
3/2	-.442	.697	-.200	.376	.078	.134	-.177	.079	-.069	.117	-.079	.145
	.096	.398	-.004	.010	.022	.015	-.014	.013	.019	.012	-.001	.004
	.003	.000	-.002	.002	.003	.000	.004	.006	-.001	.000	.001	.001
	-.000	.002										
KIN:	1.473		1.917		3.677		.5196		.7342		1.1832	
I=												
1/2	-.161	.512	-.049	.480	-.127	.273	.005	.164	-.110	.082	-.216	.174
	.001	.033	.079	.111	.032	.016	.084	.023	.010	.003	.019	.006
	.010	.006	-.001	.003	-.000	.004	.001	.000	-.001	.001	.000	.001
	-.000	.000										
3/2	-.447	.684	-.212	.414	.091	.157	-.166	.078	-.078	.121	-.096	.138
	.032	.409	-.006	.013	.022	.023	-.014	.014	.021	.012	-.005	.004
	.003	.001	-.004	.004	.005	.000	.002	.003	.003	.001	.001	.002
	-.002	.000										

KH 78	K GEV/C		W GEV		S GEV**2		Q**2 (GEV/C)**2		ω_q GEV		E GEV	
	S1 F7 I13		P1 G7 K13		P3 G9 K15		D3 H9 L15		D5 H11 L17		F5 I11 M17	
KIN:	1.505		1.933		3.736		.5337		.7438		1.1892	
I=												
1/2	-.156	.523	-.038	.478	-.137	.281	.009	.168	-.105	.080	-.209	.168
	.000	.036	.081	.122	.032	.017	.092	.029	.010	.004	.023	.006
	.009	.007	.001	.001	.001	.003	.003	.000	-.000	.000	-.001	.000
	.000	.000										
3/2	-.448	.673	-.227	.437	.094	.176	-.161	.086	-.085	.129	-.107	.138
	-.015	.407	-.006	.018	.027	.029	-.014	.014	.020	.016	-.005	.003
	.003	.000	-.007	.001	.004	.001	.002	.001	.004	.000	.003	.000
	.000	.000										
KIN:	1.550		1.955		3.820		.5537		.7571		1.1975	
I=												
1/2	-.161	.536	-.023	.481	-.153	.286	.017	.177	-.098	.077	-.200	.159
	-.002	.038	.083	.141	.034	.017	.098	.041	.006	.004	.024	.010
	.008	.006	.004	.001	-.000	.004	.002	.000	-.001	.000	.000	.000
	.003	.000										
3/2	-.444	.652	-.244	.460	.099	.194	-.148	.093	-.093	.125	-.119	.126
	-.065	.383	-.002	.015	.025	.030	-.015	.012	.020	.016	-.004	.003
	.004	.000	-.004	.002	.007	.000	.001	.001	.005	.000	.000	.001
	-.003	.000										
KIN:	1.590		1.974		3.895		.5714		.7687		1.2049	
I=												
1/2	-.170	.539	-.011	.488	-.168	.290	.021	.189	-.091	.073	-.192	.150
	-.002	.039	.077	.158	.034	.019	.099	.049	.006	.003	.018	.010
	.006	.005	.007	.007	-.002	.005	.008	.000	-.000	.003	-.000	.000
	.005	.000										
3/2	-.437	.636	-.259	.473	.103	.204	-.145	.101	-.097	.127	-.126	.118
	-.103	.356	-.002	.017	.029	.035	-.012	.014	.020	.017	-.003	.004
	.004	.000	-.005	.000	.007	.000	.002	.001	.006	.000	.001	.000
	-.001	.001										
KIN:	1.640		1.997		3.989		.5937		.7830		1.2141	
I=												
1/2	-.180	.535	.001	.496	-.173	.285	.022	.201	-.078	.077	-.182	.143
	-.004	.049	.069	.171	.042	.024	.106	.065	.005	.002	.022	.016
	.006	.003	.008	.005	-.002	.005	.005	.000	-.000	.002	.003	.000
	.005	.000										
3/2	-.424	.621	-.278	.488	.096	.228	-.137	.117	-.099	.126	-.136	.104
	-.129	.319	-.006	.022	.028	.038	-.009	.011	.022	.017	-.004	.002
	.000	.002	-.004	.000	.008	.000	.002	.000	.006	.001	-.003	.000
	-.003	.001										
KIN:	1.680		2.016		4.063		.6115		.7944		1.2214	
I=												
1/2	-.183	.529	.011	.508	-.179	.285	.020	.211	-.069	.081	-.171	.135
	-.006	.051	.062	.181	.044	.032	.113	.077	.008	.005	.027	.014
	.008	.005	.006	.008	-.002	.003	.003	.001	-.001	.001	.003	.000
	.002	.001										
3/2	-.412	.612	-.288	.496	.094	.240	-.137	.127	-.100	.127	-.135	.092
	-.144	.291	-.009	.025	.027	.045	-.013	.015	.023	.020	-.003	.006
	.004	.002	-.003	.000	.008	.000	.002	.000	.005	.001	-.002	.000
	-.003	.000										

KH 78	K GEV/C		W GEV		S GEV**2		Q**2 (GEV/C)**2		ω_q GEV		E GEV	
	S1 F7 I13		P1 G7 K13		P3 G9 K15		D3 H9 L15		D5 H11 L17		F5 I11 M17	
KIN:	1.720		2.034		4.138		.6294		.8055		1.2287	
I=												
1/2	-.179	.524	.017	.521	-.180	.283	.016	.220	-.060	.090	-.163	.129
	-.012	.058	.053	.192	.046	.038	.118	.091	.005	.003	.023	.016
	.008	.003	.006	.007	-.003	.004	.007	.000	.000	.001	.004	.000
	.004	.000										
3/2	-.400	.604	-.296	.501	.092	.251	-.136	.138	-.102	.130	-.134	.083
	-.148	.267	-.009	.028	.028	.050	-.013	.016	.023	.017	-.004	.003
	-.000	.001	-.001	.001	.010	.000	.002	.000	.005	.003	-.004	.001
	-.002	.000										
KIN:	1.760		2.053		4.213		.6473		.8166		1.2360	
I=												
1/2	-.173	.520	.021	.535	-.184	.287	.015	.227	-.055	.093	-.151	.127
	-.012	.061	.045	.200	.049	.044	.117	.104	.005	.004	.029	.016
	.008	.006	.008	.010	-.005	.002	.006	.001	.000	.001	.003	.001
	.005	.000										
3/2	-.385	.598	-.304	.508	.086	.263	-.137	.148	-.099	.133	-.130	.075
	-.153	.240	-.012	.029	.024	.056	-.010	.017	.025	.022	-.004	.007
	.005	.000	-.002	.000	.011	.001	.001	.000	.003	.001	-.002	.000
	-.003	.001										
KIN:	1.800		2.071		4.288		.6652		.8275		1.2432	
I=												
1/2	-.166	.520	.017	.550	-.187	.286	.010	.236	-.046	.101	-.142	.125
	-.017	.063	.039	.207	.049	.051	.118	.120	.003	.004	.031	.019
	.010	.006	.010	.009	-.005	.002	.004	.003	-.001	.000	.004	.000
	.005	.000										
3/2	-.374	.597	-.308	.514	.083	.274	-.136	.153	-.102	.131	-.123	.071
	-.156	.221	-.013	.033	.024	.061	-.012	.021	.028	.025	-.003	.008
	.003	.002	-.002	.000	.013	.001	.001	.000	.005	.000	-.001	.000
	-.002	.000										
KIN:	1.840		2.089		4.363		.6832		.8383		1.2504	
I=												
1/2	-.164	.524	.013	.564	-.188	.287	.006	.242	-.043	.111	-.134	.127
	-.019	.067	.027	.215	.051	.058	.118	.138	.004	.002	.035	.023
	.008	.006	.011	.009	-.003	.001	.005	.002	.002	.001	.003	.001
	.003	.000										
3/2	-.364	.594	-.312	.519	.072	.284	-.136	.163	-.102	.135	-.115	.069
	-.147	.199	-.018	.035	.022	.063	-.011	.022	.032	.027	-.004	.007
	.002	.001	-.001	.002	.010	.001	.003	.000	.005	.001	-.004	.000
	-.003	.001										
KIN:	1.880		2.107		4.438		.7012		.8489		1.2576	
I=												
1/2	-.161	.531	.001	.576	-.182	.286	-.004	.244	-.042	.116	-.123	.128
	-.026	.069	.020	.224	.056	.070	.109	.153	.008	.000	.037	.029
	.002	.004	.013	.012	-.004	.001	.003	.000	.004	.004	.004	.001
	.003	.000										
3/2	-.354	.593	-.321	.526	.070	.290	-.139	.172	-.102	.134	-.108	.070
	-.146	.187	-.019	.039	.021	.068	-.014	.024	.032	.028	-.005	.010
	.001	.000	.000	.000	.013	.001	.003	.000	.010	.001	-.001	.001
	-.004	.002										

KH 78	K GEV/C		W GEV		S GEV**2		Q**2 (GEV/C)**2		ω_q GEV		E GEV	
	S1		P1		P3		D3		D5		F5	
	F7		G7		G9		H9		H11		I11	
	I13		K13		K15		L15		L17		M17	
KIN:	1.920		2.124		4.512		.7192		.8595		1.2648	
I=												
1/2	-.165	.532	-.006	.586	-.188	.289	-.008	.249	-.035	.131	-.119	.127
	-.022	.071	.004	.224	.048	.074	.104	.172	.007	.000	.038	.033
	.008	.007	.013	.010	-.002	.001	.005	.002	.002	.002	.005	.000
	.003	.000										
3/2	-.345	.594	-.322	.528	.058	.295	-.137	.180	-.099	.136	-.103	.072
	-.135	.171	-.022	.042	.019	.072	-.017	.029	.035	.034	-.004	.009
	.002	.001	-.001	.001	.012	.000	.003	.000	.006	.001	-.003	.000
	-.002	.001										
KIN:	1.980		2.151		4.625		.7463		.8751		1.2754	
I=												
1/2	-.171	.531	-.022	.596	-.192	.293	-.015	.249	-.033	.146	-.110	.127
	-.023	.075	-.017	.228	.043	.088	.085	.193	.009	.004	.037	.039
	.007	.006	.015	.011	.000	.001	.003	.003	.003	.003	.006	.000
	.002	.000										
3/2	-.335	.593	-.328	.534	.046	.302	-.138	.192	-.095	.146	-.100	.081
	-.121	.157	-.026	.045	.019	.076	-.019	.033	.037	.035	-.004	.011
	.004	.001	.001	.000	.014	.000	.004	.002	.007	.002	-.002	.000
	-.001	.002										
KIN:	2.030		2.172		4.718		.7689		.8879		1.2843	
I=												
1/2	-.173	.525	-.038	.599	-.193	.297	-.017	.247	-.035	.161	-.101	.129
	-.026	.081	-.030	.220	.037	.097	.066	.208	.011	.003	.037	.043
	.006	.010	.016	.010	.004	.002	.004	.003	.001	.002	.006	.000
	.002	.000										
3/2	-.329	.592	-.332	.535	.037	.302	-.139	.200	-.091	.151	-.094	.081
	-.108	.146	-.031	.048	.018	.083	-.021	.031	.040	.044	-.005	.012
	.005	.000	-.001	.004	.013	.001	.001	.002	.005	.001	-.003	.000
	-.002	.001										
KIN:	2.070		2.189		4.793		.7870		.8981		1.2913	
I=												
1/2	-.174	.520	-.047	.600	-.193	.298	-.018	.245	-.041	.169	-.093	.130
	-.028	.087	-.041	.219	.033	.108	.049	.215	.011	.007	.037	.045
	.006	.007	.019	.011	.003	.001	.002	.001	.003	.005	.008	.000
	.003	.000										
3/2	-.324	.591	-.336	.538	.036	.302	-.141	.206	-.091	.155	-.090	.083
	-.101	.141	-.032	.050	.014	.092	-.024	.036	.042	.048	-.007	.015
	.008	.000	-.001	.001	.016	.000	.003	.001	.005	.000	-.002	.001
	-.001	.002										
KIN:	2.150		2.223		4.943		.8233		.9180		1.3053	
I=												
1/2	-.162	.509	-.069	.601	-.197	.299	-.016	.245	-.053	.184	-.079	.140
	-.034	.092	-.054	.202	.015	.119	.018	.221	.012	.013	.035	.049
	.006	.012	.024	.011	.006	.002	.007	.000	.003	.003	.009	.000
	.003	.000										
3/2	-.320	.585	-.338	.538	.027	.303	-.139	.220	-.089	.168	-.084	.094
	-.083	.138	-.041	.054	.010	.100	-.027	.037	.045	.059	-.006	.013
	.010	.000	-.001	.002	.018	.000	-.001	.002	.006	.003	-.001	.001
	-.001	.001										

KH 78	K GEV/C		W GEV		S GEV**2		Q**2 (GEV/C)**2		ω_q GEV		E GEV	
	S1 F7 I13		P1 G7 K13		P3 G9 K15		D3 H9 L15		D5 H11 L17		F5 I11 M17	
KIN:	2.200		2.244		5.037		.8460		.9303		1.3139	
I=												
1/2	-.152	.508	-.080	.601	-.200	.298	-.013	.244	-.059	.187	-.074	.146
	-.035	.095	-.059	.192	.004	.122	.003	.220	.010	.017	.037	.050
	.004	.015	.027	.011	.006	.005	.007	.001	.003	.005	.009	.001
	.003	.001										
3/2	-.315	.580	-.341	.538	.025	.305	-.140	.227	-.087	.175	-.079	.101
	-.078	.137	-.043	.057	.007	.107	-.028	.040	.048	.067	-.010	.014
	.011	.000	-.000	.005	.018	.001	-.001	.000	.007	.004	-.000	.001
	-.000	.001										
KIN:	2.280		2.277		5.187		.8824		.9497		1.3277	
I=												
1/2	-.141	.515	-.101	.593	-.196	.299	-.006	.247	-.069	.190	-.069	.155
	-.039	.101	-.060	.184	-.011	.126	-.021	.213	.007	.019	.040	.055
	.005	.012	.034	.016	.005	.009	.012	.000	-.000	.008	.011	.000
	.006	.001										
3/2	-.309	.568	-.342	.538	.023	.310	-.143	.242	-.092	.183	-.076	.113
	-.061	.136	-.046	.058	.002	.112	-.031	.044	.044	.083	-.012	.014
	.011	.004	-.000	.003	.021	.000	-.003	.001	.008	.003	-.000	.002
	.001	.004										
KIN:	2.340		2.302		5.299		.9098		.9640		1.3380	
I=												
1/2	-.139	.523	-.114	.587	-.195	.302	-.004	.253	-.073	.190	-.063	.155
	-.037	.103	-.063	.179	-.021	.123	-.031	.205	.003	.024	.044	.062
	.003	.013	.036	.019	.003	.012	.011	.001	.000	.006	.013	.000
	.004	.002										
3/2	-.306	.556	-.340	.535	.024	.313	-.147	.250	-.092	.192	-.075	.124
	-.052	.145	-.053	.061	-.003	.116	-.036	.043	.039	.093	-.008	.016
	.017	.004	-.001	.007	.023	.006	-.003	.001	.004	.004	.000	.001
	.001	.002										
KIN:	2.460		2.350		5.524		.9645		.9920		1.3583	
I=												
1/2	-.146	.533	-.135	.560	-.184	.309	.010	.261	-.081	.197	-.053	.170
	-.039	.110	-.055	.167	-.035	.116	-.054	.188	.004	.027	.046	.074
	.002	.014	.045	.031	.001	.010	.023	.001	-.006	.006	.014	.000
	.004	.002										
3/2	-.289	.540	-.334	.530	.014	.319	-.149	.268	-.098	.203	-.080	.143
	-.034	.160	-.058	.065	-.014	.125	-.039	.045	.025	.111	-.013	.019
	.019	.006	-.002	.003	.026	.011	-.003	.001	.004	.000	-.001	.004
	.001	.004										
KIN:	2.560		2.390		5.711		1.0103		1.0148		1.3750	
I=												
1/2	-.159	.529	-.139	.537	-.176	.319	.012	.277	-.088	.201	-.038	.183
	-.038	.125	-.054	.165	-.043	.108	-.058	.171	.006	.032	.043	.085
	.002	.013	.049	.037	-.002	.009	.031	.011	-.002	.003	.014	.000
	-.002	.004										
3/2	-.272	.529	-.328	.524	.009	.327	-.152	.274	-.100	.210	-.090	.158
	-.037	.176	-.058	.064	-.021	.130	-.047	.046	.015	.118	-.008	.017
	.018	.013	-.007	.004	.027	.014	-.005	.001	-.000	.000	-.003	.007
	.004	.001										

KH 78	K GEV/C		W GEV		S GEV**2		Q**2 (GEV/C)**2		ω_q GEV		E GEV	
	S1 F7 I13		P1 G7 K13		P3 G9 K15		D3 H9 L15		D5 H11 L17		F5 I11 M17	
KIN:	2.750		2.463		6.067		1.0974		1.0568		1.4063	
I= 1/2	-.157	.510	-.128	.511	-.182	.331	.016	.311	-.099	.205	-.032	.213
	-.041	.139	-.049	.160	-.044	.095	-.057	.147	-.004	.040	.038	.105
	.007	.018	.053	.058	-.002	.007	.031	.015	-.001	.000	.021	.003
	.001	.002	.006	.005	.003	.001						
3/2	-.243	.514	-.310	.518	.002	.346	-.159	.288	-.107	.215	-.095	.174
	-.056	.188	-.061	.071	-.035	.132	-.043	.045	-.013	.136	-.013	.027
	.025	.020	-.007	.007	.023	.019	-.003	.003	.004	.000	-.004	.004
	.006	.002	-.007	.001	.002	.000						
KIN:	3.000		2.557		6.536		1.2124		1.1099		1.4466	
I= 1/2	-.152	.515	-.123	.502	-.170	.332	-.012	.346	-.105	.204	-.042	.235
	-.049	.153	-.050	.168	-.041	.097	-.043	.138	-.009	.049	.022	.127
	.011	.024	.045	.078	-.003	.010	.033	.027	-.001	.000	.023	.011
	.003	.001	.011	.007	.006	.002	.003	.000	-.002	.002		
3/2	-.212	.496	-.291	.515	-.030	.366	-.153	.309	-.112	.221	-.108	.194
	-.067	.186	-.064	.090	-.046	.128	-.046	.050	-.031	.124	-.019	.034
	.022	.039	-.003	.011	.022	.026	-.005	.004	.005	.000	-.001	.003
	.005	.004	-.005	.000	.007	.002	-.005	.002	.005	.002		
KIN:	3.400		2.699		7.286		1.3969		1.1901		1.5091	
I= 1/2	-.174	.495	-.133	.463	-.161	.330	-.048	.373	-.081	.234	-.053	.238
	-.068	.169	-.041	.183	-.028	.118	-.036	.137	-.015	.062	.001	.134
	-.001	.039	.014	.105	.006	.023	.030	.041	.004	.000	.032	.023
	-.000	.000	.015	.008	.000	.002	.008	.002	.001	.006		
3/2	-.180	.487	-.267	.495	-.067	.369	-.154	.326	-.103	.226	-.122	.235
	-.064	.194	-.077	.107	-.041	.130	-.050	.064	-.033	.109	-.027	.043
	.012	.067	-.005	.019	.024	.038	-.008	.009	.012	.008	-.001	.003
	.008	.002	-.006	.000	.003	.002	-.005	.006	.005	.009		
KIN:	3.650		2.785		7.754		1.5126		1.2378		1.5469	
I= 1/2	-.169	.462	-.119	.453	-.146	.341	-.069	.369	-.096	.260	-.049	.246
	-.070	.173	-.044	.192	-.035	.131	-.029	.139	-.017	.070	-.005	.133
	.008	.043	.008	.098	.003	.026	.033	.049	.001	.003	.029	.026
	.001	.004	.017	.011	.004	.000	.010	.006	.000	.002	.003	.000
	.002	.000	.005	.003	-.002	.006						
3/2	-.180	.478	-.246	.483	-.076	.360	-.151	.338	-.094	.242	-.134	.242
	-.073	.204	-.076	.124	-.042	.138	-.056	.074	-.033	.119	-.034	.045
	.006	.073	-.010	.020	.016	.048	-.015	.013	.014	.013	-.001	.005
	.009	.009	-.004	.002	.008	.000	-.003	.005	.008	.005	-.001	.001
	-.005	.000	-.001	.000	.003	.002						
KIN:	4.000		2.900		8.411		1.6748		1.3016		1.5985	
I= 1/2	-.158	.449	-.117	.457	-.139	.364	-.082	.357	-.121	.250	-.059	.265
	-.069	.186	-.042	.189	-.045	.132	-.024	.150	-.021	.086	-.002	.130
	-.001	.058	-.008	.105	.005	.037	.030	.063	-.003	.013	.026	.033
	.000	.001	.030	.021	.004	.002	.010	.006	.003	.002	.011	.000
	-.000	.002	.004	.000	.001	.002						
3/2	-.183	.445	-.218	.476	-.083	.372	-.155	.346	-.099	.260	-.137	.249
	-.076	.202	-.083	.145	-.049	.146	-.055	.085	-.030	.127	-.039	.054
	-.009	.086	-.008	.024	.011	.059	-.018	.017	.013	.021	-.001	.002
	.013	.014	-.001	.002	.008	.000	-.006	.003	.006	.007	-.000	.000
	.003	.000	-.007	.007	.002	.003						

Table 2.2.2.2. Partial wave amplitudes of the preliminary updated solution KH 80 (Sect. 2.1.7).

KH 80	K GEV/C		W GEV		S GEV**2		Q**2 (GEV/C)**2		ω_q GEV		E GEV	
	S1 F7 I13		P1 G7 K13		P3 G9 K15		D3 H9 L15		D5 H11 L17		F5 I11 M17	
KIN:	2.075		2.191		4.803		.7893		.8993		1.2922	
I=												
1/2	-.174	.517	-.046	.581	-.182	.295	-.023	.248	-.039	.174	-.095	.144
	-.033	.092	-.031	.222	.035	.107	.049	.203	.015	.002	.034	.046
	-.000	.008	.023	.014	.003	.004	-.001	.001	.011	.002	.011	.000
	-.000	.000										
3/2	-.322	.571	-.337	.529	.046	.294	-.145	.211	-.093	.166	-.085	.087
	-.105	.138	-.030	.045	.009	.093	-.032	.033	.042	.043	-.012	.017
	.006	.002	-.001	.005	.018	.000	.006	.001	.008	.001	-.001	.001
	-.001	.004										
KIN:	2.140		2.219		4.924		.8188		.9156		1.3035	
I=												
1/2	-.160	.506	-.067	.591	-.200	.300	-.014	.247	-.048	.191	-.085	.153
	-.025	.095	-.053	.200	.017	.110	.024	.213	.009	.011	.036	.050
	.001	.016	.024	.012	.011	.000	.008	.001	.007	.004	.010	.001
	.002	.001										
3/2	-.324	.564	-.327	.524	.033	.294	-.129	.214	-.102	.189	-.090	.103
	-.081	.135	-.042	.052	.015	.095	-.027	.033	.040	.050	-.011	.015
	.004	.002	-.001	.000	.019	.001	-.003	.000	.013	.002	.002	.005
	-.006	.005										
KIN:	2.210		2.248		5.055		.8506		.9328		1.3157	
I=												
1/2	-.159	.497	-.061	.591	-.197	.308	-.031	.246	-.072	.188	-.069	.157
	-.016	.104	-.057	.187	-.010	.119	-.006	.217	.015	.009	.046	.053
	.002	.024	.025	.013	.002	.004	.006	.000	.017	.003	.014	.000
	-.002	.000										
3/2	-.317	.548	-.318	.514	.016	.303	-.143	.234	-.100	.200	-.089	.122
	-.059	.143	-.046	.058	.007	.097	-.025	.028	.036	.060	-.017	.013
	.013	.004	.002	.001	.024	.001	.003	.000	.006	.005	-.004	.006
	-.003	.003										
KIN:	2.260		2.269		5.149		.8733		.9449		1.3243	
I=												
1/2	-.154	.496	-.068	.590	-.191	.297	-.017	.235	-.072	.195	-.077	.159
	-.030	.101	-.059	.189	-.004	.129	-.003	.221	.002	.016	.030	.050
	.003	.021	.030	.016	.001	.007	.014	.001	.010	.006	.013	.002
	.005	.000										
3/2	-.313	.554	-.328	.521	.018	.308	-.147	.238	-.098	.201	-.086	.122
	-.061	.141	-.045	.062	-.003	.104	-.031	.036	.044	.070	-.019	.011
	.007	.010	-.003	.009	.026	.004	.008	.001	.007	.001	-.003	.000
	.001	.002										
KIN:	2.300		2.286		5.224		.8915		.9545		1.3311	
I=												
1/2	-.137	.503	-.097	.574	-.198	.301	-.009	.248	-.070	.203	-.077	.168
	-.038	.107	-.058	.185	-.016	.114	-.018	.199	.007	.019	.039	.055
	.004	.021	.032	.022	.010	.011	.020	.000	-.003	.005	.011	.000
	.008	.001										
3/2	-.317	.557	-.343	.539	.037	.303	-.138	.244	-.099	.193	-.079	.120
	-.056	.133	-.044	.053	-.006	.118	-.032	.047	.044	.080	-.020	.016
	.009	.006	.002	.000	.019	.004	-.002	.002	.012	.001	-.003	.002
	-.002	.005										

KH80	K GEV/C		W GEV		S GEV**2		Q**2 (GEV/C)**2		ω_q GEV		E GEV	
	S1 F7 I13		P1 G7 K13		P3 G9 K15		D3 H9 L15		D5 H11 L17		F5 I11 M17	
KIN:	2.360		2.310		5.336		.9189		.9687		1.3414	
I=												
1/2	-.139	.509	-.107	.576	-.225	.302	.000	.263	-.077	.204	-.071	.170
	-.017	.120	-.068	.179	-.020	.107	-.027	.187	-.013	.013	.042	.060
	-.004	.018	.039	.023	.013	.018	.012	.002	.009	.006	.019	.000
	.001	.000										
3/2	-.303	.539	-.320	.516	.008	.322	-.152	.259	-.100	.217	-.080	.136
	-.046	.137	-.067	.063	-.003	.114	-.031	.046	.030	.089	-.012	.007
	.015	.006	-.001	.005	.028	.008	-.000	.002	.006	.005	.002	.005
	-.001	.002										
KIN:	2.440		2.342		5.486		.9554		.9874		1.3549	
I=												
1/2	-.143	.510	-.123	.548	-.175	.311	-.000	.247	-.092	.203	-.060	.190
	-.031	.129	-.051	.171	-.041	.100	-.055	.179	.007	.022	.052	.077
	.002	.025	.039	.028	-.000	.009	.021	.000	.000	.007	.020	.001
	.004	.001										
3/2	-.292	.525	-.327	.518	.016	.315	-.155	.265	-.105	.212	-.083	.151
	-.033	.165	-.061	.067	-.009	.118	-.039	.040	.019	.104	-.015	.014
	.021	.009	-.002	.004	.028	.006	-.001	.001	-.001	.004	-.001	.006
	.003	.003										
KIN:	2.500		2.366		5.599		.9828		1.0012		1.3650	
I=												
1/2	-.140	.516	-.125	.544	-.186	.310	.002	.259	-.084	.207	-.060	.186
	-.034	.122	-.049	.168	-.041	.107	-.047	.181	.005	.023	.045	.076
	.003	.021	.041	.034	-.003	.013	.028	.003	-.005	.003	.019	.001
	.006	.000										
3/2	-.294	.532	-.329	.522	.015	.315	-.153	.277	-.104	.214	-.081	.148
	-.032	.166	-.061	.064	-.018	.125	-.038	.040	.016	.114	-.016	.018
	.021	.007	-.005	.006	.030	.016	-.003	.003	.003	.000	-.002	.005
	-.000	.002										
KIN:	2.560		2.390		5.711		1.0103		1.0148		1.3750	
I=												
1/2	-.142	.507	-.121	.539	-.189	.320	-.005	.268	-.091	.213	-.059	.200
	-.026	.128	-.047	.165	-.036	.096	-.041	.158	-.001	.023	.047	.074
	.001	.020	.044	.042	-.003	.011	.028	.015	.000	.007	.021	.004
	.001	.002										
3/2	-.285	.527	-.320	.514	.013	.327	-.156	.273	-.110	.223	-.089	.155
	-.036	.171	-.063	.065	-.021	.122	-.037	.042	.014	.123	-.016	.024
	.021	.012	-.010	.006	.028	.013	.004	.000	-.005	.001	-.009	.007
	.004	.002										
KIN:	2.630		2.417		5.842		1.0423		1.0305		1.3866	
I=												
1/2	-.149	.505	-.119	.533	-.194	.315	.002	.278	-.094	.216	-.046	.200
	-.023	.138	-.054	.170	-.038	.093	-.046	.159	-.008	.028	.046	.085
	-.002	.015	.047	.043	.003	.013	.031	.014	-.001	.004	.020	.002
	-.002	.000	.004	.001	.002	.000						
3/2	-.265	.521	-.316	.507	.005	.332	-.156	.280	-.108	.226	-.095	.164
	-.041	.176	-.056	.057	-.023	.128	-.045	.046	.009	.122	-.012	.016
	.020	.022	-.007	.013	.028	.020	.001	.002	-.007	.000	-.004	.006
	.004	.000	-.007	.003	.004	.001						
KIN:	2.725		2.454		6.020		1.0859		1.0514		1.4023	
I=												
1/2	-.155	.514	-.124	.525	-.176	.329	.006	.281	-.104	.214	-.042	.205
	-.036	.127	-.045	.165	-.036	.095	-.041	.159	-.006	.032	.037	.096
	.004	.023	.047	.055	-.002	.013	.034	.017	-.004	.002	.018	.006
	.002	.001	.007	.001	.002	.000						
3/2	-.267	.513	-.317	.511	.003	.332	-.159	.277	-.106	.222	-.100	.174
	-.044	.186	-.049	.067	-.034	.129	-.041	.045	-.007	.132	-.008	.023
	.017	.024	-.013	.010	.031	.023	-.002	.005	-.003	.000	-.006	.005
	.006	.000	-.009	.002	.003	.000						

KH 80	K GEV/C		W GEV		S GEV**2		Q**2 (GEV/C)**2		ω_q GEV		E GEV	
	S1 F7 I13		P1 G7 K13		P3 G9 K15		D3 H9 L15		D5 H11 L17		F5 I11 M17	
KIN:	2.780		2.475		6.123		1.1112		1.0633		1.4112	
I=												
1/2	-.153	.509	-.121	.519	-.183	.330	-.006	.293	-.097	.213	-.043	.213
	-.036	.132	-.048	.170	-.035	.099	-.038	.164	-.009	.042	.036	.091
	.003	.025	.049	.054	-.007	.004	.028	.021	-.001	.000	.023	.009
	.003	.000	.010	.002	.006	.001						
3/2	-.266	.509	-.309	.503	-.007	.339	-.159	.283	-.116	.234	-.101	.184
	-.054	.192	-.047	.069	-.031	.125	-.041	.046	-.007	.129	-.015	.024
	.016	.024	-.015	.016	.024	.025	.001	.003	-.005	.000	-.002	.000
	.009	.001	-.009	.004	.008	.000						
KIN:	2.880		2.512		6.311		1.1571		1.0847		1.4274	
I=												
1/2	-.149	.504	-.132	.506	-.174	.339	.000	.319	-.099	.212	-.040	.222
	-.043	.130	-.046	.162	-.045	.092	-.039	.146	-.006	.049	.036	.107
	.008	.021	.040	.060	.003	.004	.029	.032	.001	.001	.028	.012
	-.003	.001	.014	.005	.001	.000	-.002	.001	.004	.007		
3/2	-.230	.501	-.298	.507	-.024	.355	-.164	.296	-.105	.228	-.104	.189
	-.063	.184	-.057	.076	-.037	.122	-.029	.047	-.018	.125	-.015	.019
	.015	.031	-.013	.006	.025	.029	-.002	.011	.001	.005	-.005	.007
	.006	.002	-.005	.000	.002	.000	-.005	.001	.005	.001		
KIN:	2.965		2.544		6.470		1.1962		1.1026		1.4411	
I=												
1/2	-.159	.501	-.119	.508	-.176	.336	-.006	.321	-.105	.212	-.036	.228
	-.035	.142	-.054	.172	-.036	.090	-.042	.140	-.019	.046	.031	.115
	.010	.028	.046	.071	.002	.017	.031	.032	-.001	.000	.025	.012
	.002	.000	.011	.006	.007	.001	.003	.000	-.003	.000		
3/2	-.233	.503	-.292	.510	-.028	.354	-.160	.298	-.115	.231	-.102	.192
	-.064	.184	-.054	.087	-.038	.126	-.038	.052	-.020	.126	-.018	.035
	.015	.036	-.008	.013	.019	.029	-.002	.003	.005	.000	-.001	.001
	.006	.001	-.009	.000	.005	.001	-.005	.000	.002	.004		
KIN:	3.090		2.589		6.705		1.2538		1.1284		1.4609	
I=												
1/2	-.173	.493	-.133	.461	-.161	.327	-.044	.370	-.093	.226	-.037	.242
	-.051	.154	-.044	.169	-.040	.096	-.039	.135	-.014	.056	.024	.130
	.012	.023	.034	.086	-.000	.010	.027	.031	.006	.002	.031	.017
	.000	.004	.020	.006	-.002	.001	-.002	.001	.005	.003		
3/2	-.210	.491	-.283	.511	-.040	.367	-.150	.315	-.121	.229	-.114	.207
	-.060	.189	-.068	.097	-.041	.116	-.040	.046	-.033	.117	-.015	.032
	.016	.038	-.004	.008	.021	.033	-.010	.005	.003	.001	-.001	.003
	.006	.004	-.008	.006	.011	.005	-.002	.001	.002	.006		
KIN:	3.250		2.647		7.004		1.3276		1.1607		1.4859	
I=												
1/2	-.144	.496	-.123	.516	-.175	.340	-.018	.304	-.104	.237	-.054	.243
	-.034	.138	-.055	.178	-.023	.091	-.024	.146	-.017	.054	.012	.119
	-.002	.043	.032	.080	.003	.008	.026	.043	.001	.009	.012	.038
	.004	.000	.029	.009	.001	.008	.007	.001	.003	.010		
3/2	-.236	.494	-.296	.490	-.037	.357	-.164	.303	-.111	.251	-.118	.224
	-.077	.175	-.070	.108	-.020	.106	-.031	.038	-.020	.113	-.012	.035
	.002	.054	-.022	.024	.030	.035	.009	.005	.006	.009	.002	.002
	.004	.010	-.021	.008	.004	.001	-.004	.007	.001	.008		
KIN:	3.400		2.699		7.286		1.3969		1.1901		1.5091	
I=												
1/2	-.142	.491	-.126	.494	-.174	.344	-.016	.334	-.093	.231	-.060	.247
	-.044	.157	-.051	.178	-.030	.104	-.041	.142	-.014	.062	.013	.123
	-.005	.049	.018	.097	.010	.014	.032	.051	-.001	.003	.022	.024
	-.000	.001	.016	.009	.001	.001	.012	.002	.001	.008		
3/2	-.227	.490	-.279	.497	-.032	.354	-.162	.318	-.132	.244	-.122	.228
	-.051	.183	-.068	.113	-.023	.115	-.036	.054	-.036	.116	-.027	.031
	.011	.068	-.017	.022	.019	.039	-.000	.009	.010	.009	-.000	.010
	.012	.000	-.010	.005	.003	.002	-.005	.000	.001	.011		

KH 80	K GEV/C		W GEV		S GEV**2		Q**2 (GEV/C)**2		ω_q GEV		E GEV	
	S1 F7 I13		P1 G7 K13		P3 G9 K15		D3 H9 L15		D5 H11 L17		F5 I11 M17	
KIN:	3.500		2.734		7.473		1.4432		1.2094		1.5243	
I=												
1/2	-.140	.483	-.116	.480	-.188	.345	-.028	.334	-.094	.240	-.052	.250
	-.055	.168	-.050	.191	-.021	.109	-.037	.150	-.011	.083	.002	.122
	-.006	.049	.014	.101	.004	.011	.027	.052	.001	.004	.024	.028
	.000	.000	.023	.006	-.001	.009	.012	.002	.003	.000		
3/2	-.215	.475	-.254	.479	-.067	.357	-.170	.317	-.105	.254	-.126	.226
	-.060	.184	-.070	.114	-.040	.131	-.040	.072	-.021	.103	-.033	.043
	-.001	.079	-.018	.026	.017	.053	-.009	.010	.012	.005	.003	.006
	.013	.009	.000	.000	.009	.000	-.007	.003	.002	.005		
KIN:	3.640		2.781		7.736		1.5079		1.2359		1.5454	
I=												
1/2	-.169	.489	-.129	.456	-.162	.336	-.053	.369	-.087	.247	-.057	.244
	-.072	.174	-.038	.188	-.030	.124	-.032	.137	-.015	.075	-.006	.128
	-.003	.051	.005	.107	.003	.029	.030	.056	.006	.002	.033	.033
	-.000	.000	.018	.006	-.001	.002	.009	.000	.003	.000	-.000	.002
	.001	.000	.005	.001	-.003	.007						
3/2	-.176	.473	-.253	.485	-.077	.370	-.155	.326	-.109	.241	-.133	.248
	-.059	.201	-.080	.121	-.040	.133	-.048	.074	-.033	.118	-.032	.044
	.004	.077	-.012	.019	.014	.047	-.013	.013	.016	.015	-.000	.007
	.015	.004	-.005	.001	.004	.000	-.003	.002	.006	.005	-.002	.001
	.001	.000	.002	.001	-.001	.002						
KIN:	3.750		2.818		7.942		1.5589		1.2563		1.5618	
I=												
1/2	-.168	.462	-.116	.450	-.151	.340	-.066	.368	-.099	.260	-.050	.247
	-.073	.172	-.044	.195	-.035	.133	-.028	.139	-.022	.079	-.009	.135
	.003	.052	.002	.105	-.001	.031	.028	.054	.006	.001	.031	.032
	.002	.003	.020	.011	.005	.000	.015	.006	-.005	.002	-.005	.000
	.005	.000	.009	.001	-.004	.006						
3/2	-.181	.471	-.239	.482	-.077	.365	-.153	.338	-.102	.244	-.136	.255
	-.064	.207	-.079	.128	-.042	.134	-.050	.074	-.033	.118	-.034	.042
	-.002	.083	-.013	.015	.013	.051	-.013	.019	.018	.013	-.002	.006
	.010	.009	-.003	.002	.004	.000	-.005	.002	.007	.007	-.002	.002
	-.002	.000	.002	.002	.004	.004						
KIN:	3.900		2.868		8.223		1.6284		1.2837		1.5839	
I=												
1/2	-.167	.462	-.117	.450	-.147	.344	-.069	.368	-.102	.265	-.053	.252
	-.072	.176	-.044	.199	-.033	.137	-.028	.141	-.025	.082	-.005	.135
	.001	.058	-.001	.107	-.005	.038	.027	.062	.005	.006	.031	.036
	.000	.006	.020	.009	.002	.000	.015	.005	.002	.002	.002	.000
	.004	.000	.010	.005	-.003	.004						
3/2	-.178	.471	-.235	.476	-.083	.363	-.149	.338	-.102	.252	-.143	.254
	-.067	.209	-.077	.131	-.039	.135	-.051	.077	-.039	.124	-.035	.046
	-.002	.083	-.014	.023	.010	.062	-.016	.022	.015	.021	-.001	.011
	.014	.015	-.007	.003	.003	.000	-.003	.001	.007	.003	-.003	.001
	.002	.000	.001	.000	.005	.001						
KIN:	4.000		2.900		8.411		1.6748		1.3016		1.5985	
I=												
1/2	-.166	.461	-.117	.450	-.151	.347	-.066	.366	-.103	.262	-.054	.258
	-.076	.178	-.041	.195	-.036	.134	-.028	.144	-.025	.086	-.010	.128
	-.004	.059	-.001	.102	.001	.038	.025	.068	.005	.008	.033	.038
	-.001	.005	.024	.018	.004	.001	.016	.009	-.001	.004	.005	.002
	.001	.000	.005	.001	-.003	.005						
3/2	-.174	.459	-.229	.470	-.095	.364	-.147	.336	-.102	.258	-.146	.257
	-.066	.205	-.076	.145	-.043	.150	-.056	.087	-.037	.121	-.035	.048
	-.006	.087	-.012	.027	.007	.061	-.019	.012	.015	.021	.002	.005
	.014	.014	-.005	.000	.009	.000	-.001	.002	.007	.005	-.008	.004
	.002	.001	.001	.004	.002	.004						

KH 80	K GEV/C		W GEV		S GEV**2		Q**2 (GEV/C)**2		ω_q GEV		E GEV	
	S1 F7 I13		P1 G7 K13		P3 G9 K15		D3 H9 L15		D5 H11 L17		F5 I11 M17	
KIN:	4.200		2.964		8.786		1.7676		1.3368		1.6273	
I=												
1/2	-.158	.449	-.115	.454	-.140	.364	-.079	.357	-.123	.254	-.059	.272
	-.070	.186	-.042	.195	-.046	.133	-.026	.149	-.024	.091	-.001	.124
	-.001	.064	-.010	.106	.001	.042	.027	.070	-.002	.012	.026	.041
	.003	.004	.031	.026	.000	.009	.015	.008	.006	.000	.014	.006
	-.002	.001	.003	.003	-.002	.006						
3/2	-.180	.438	-.213	.471	-.088	.372	-.154	.347	-.099	.262	-.143	.264
	-.074	.207	-.082	.157	-.050	.144	-.050	.088	-.035	.126	-.038	.049
	-.010	.086	-.009	.024	.010	.067	-.019	.022	.012	.033	.002	.008
	.012	.017	-.007	.002	.007	.003	-.004	.005	.003	.003	-.002	.000
	.005	.000	-.006	.005	.003	.005						
KIN:	4.400		3.027		9.161		1.8606		1.3711		1.6556	
I=												
1/2	-.156	.445	-.115	.454	-.141	.367	-.080	.360	-.125	.260	-.057	.272
	-.072	.192	-.041	.195	-.045	.129	-.024	.154	-.023	.091	-.002	.124
	-.005	.063	-.008	.106	.003	.043	.029	.071	-.004	.021	.024	.048
	-.000	.008	.032	.026	.004	.005	.016	.011	-.001	.004	.011	.010
	.001	.003	.005	.005	-.001	.003	.001	.004	-.003	.006		
3/2	-.176	.427	-.202	.453	-.101	.368	-.152	.342	-.108	.272	-.142	.266
	-.069	.219	-.085	.172	-.052	.150	-.058	.099	-.032	.127	-.038	.059
	-.011	.080	-.008	.022	.007	.063	-.014	.019	.011	.030	.002	.002
	.011	.019	-.004	.000	.007	.004	-.006	.004	.004	.013	-.005	.000
	.004	.002	-.006	.010	.005	.007	-.001	.006	.001	.003		
KIN:	4.600		3.088		9.536		1.9535		1.4046		1.6834	
I=												
1/2	-.155	.443	-.116	.453	-.140	.368	-.082	.362	-.127	.265	-.059	.277
	-.074	.194	-.043	.200	-.046	.130	-.029	.156	-.027	.092	-.002	.128
	-.009	.071	-.009	.104	-.000	.049	.025	.073	-.005	.023	.017	.053
	-.002	.011	.033	.032	.002	.002	.016	.018	.003	.005	.018	.010
	-.002	.006	.008	.005	.004	.003	.005	.006	-.003	.007		
3/2	-.177	.423	-.199	.450	-.098	.364	-.146	.339	-.110	.275	-.143	.275
	-.070	.218	-.085	.182	-.047	.157	-.058	.111	-.036	.130	-.042	.070
	-.009	.092	-.007	.030	.004	.070	-.014	.021	.008	.036	.000	.007
	.012	.026	.001	.000	.007	.001	-.009	.002	.005	.011	-.003	.000
	.004	.000	-.005	.008	.005	.005	-.006	.003	.000	.005		
KIN:	4.840		3.160		9.986		2.0652		1.4438		1.7163	
I=												
1/2	-.155	.387	-.135	.412	-.139	.354	-.105	.366	-.099	.258	-.061	.295
	-.083	.211	-.036	.219	-.044	.150	-.015	.171	-.040	.109	-.012	.121
	-.015	.079	-.010	.107	-.015	.054	.002	.078	-.004	.025	.016	.065
	.001	.010	.026	.043	.001	.010	.027	.021	.003	.000	.024	.010
	.003	.006	.010	.002	.002	.000	.008	.001	.002	.005	.004	.002
	-.000	.000										
3/2	-.128	.381	-.169	.427	-.119	.335	-.139	.354	-.103	.269	-.130	.286
	-.071	.222	-.107	.190	-.058	.176	-.063	.117	-.038	.136	-.051	.070
	-.021	.105	-.014	.043	-.001	.075	-.022	.018	.002	.034	.002	.017
	.010	.026	-.003	.003	.012	.006	-.004	.004	.009	.012	.001	.004
	.009	.002	-.008	.003	.004	.004	.003	.001	.002	.004	-.001	.000
	.001	.001										
KIN:	5.000		3.207		10.287		2.1397		1.4694		1.7378	
I=												
1/2	-.155	.388	-.135	.413	-.139	.354	-.104	.368	-.103	.261	-.062	.296
	-.082	.212	-.037	.222	-.047	.150	-.020	.172	-.038	.110	-.013	.122
	-.018	.081	-.011	.107	-.016	.057	.002	.081	-.005	.030	.017	.067
	.001	.012	.025	.050	.002	.009	.028	.023	.001	.004	.024	.014
	.004	.004	.011	.002	.005	.003	.008	.002	.001	.003	.003	.000
	-.001	.001	.003	.001	-.001	.000						
3/2	-.128	.382	-.171	.431	-.119	.339	-.139	.358	-.105	.270	-.135	.293
	-.068	.223	-.109	.197	-.061	.178	-.064	.119	-.038	.136	-.050	.073
	-.022	.106	-.015	.048	-.006	.082	-.021	.020	-.001	.042	-.003	.022
	.010	.030	-.006	.008	.009	.010	-.006	.003	.009	.013	-.001	.005
	.011	.002	-.005	.000	.004	.004	.004	.000	.003	.000	-.002	.000
	.002	.001	.001	.000	.001	.000						

KH80	K GEV/C		W GEV		S GEV**2		Q**2 (GEV/C)**2		ω_q GEV		E GEV	
	S1 F7 I13		P1 G7 K13		P3 G9 K15		D3 H9 L15		D5 H11 L17		F5 I11 M17	
KIN:	5.150		3.251		10.568		2.2096		1.4930		1.7578	
I=												
1/2	-.154	.389	-.135	.411	-.140	.354	-.103	.367	-.102	.261	-.063	.296
	-.085	.211	-.037	.223	-.044	.153	-.018	.171	-.040	.110	-.013	.122
	-.016	.083	-.012	.109	-.018	.058	.000	.082	-.004	.033	.016	.069
	-.000	.013	.025	.050	.004	.012	.029	.025	.001	.003	.024	.016
	.003	.005	.014	.006	.004	.003	.007	.004	.001	.005	.006	.005
	-.003	.002	-.001	.000	-.001	.001						
3/2	-.127	.380	-.170	.427	-.121	.338	-.135	.352	-.102	.271	-.134	.290
	-.070	.223	-.104	.198	-.059	.178	-.065	.124	-.037	.133	-.050	.075
	-.023	.104	-.015	.047	-.003	.080	-.020	.020	-.001	.043	-.003	.020
	.011	.031	-.005	.007	.009	.012	-.007	.006	.009	.015	.001	.006
	.009	.004	-.006	.004	.003	.007	.002	.002	.002	.002	-.002	.002
	.004	.002	.001	.001	.000	.001						
KIN:	5.400		3.322		11.037		2.3261		1.5315		1.7907	
I=												
1/2	-.154	.388	-.134	.411	-.138	.356	-.104	.367	-.105	.261	-.066	.297
	-.083	.214	-.039	.228	-.047	.157	-.022	.177	-.036	.115	-.012	.127
	-.020	.086	-.011	.110	-.019	.063	.003	.083	-.004	.035	.013	.070
	.001	.021	.024	.053	.001	.011	.026	.032	-.002	.002	.025	.020
	.002	.007	.014	.008	.006	.006	.012	.008	-.003	.004	.001	.003
	.004	.001	.006	.002	-.003	.003						
3/2	-.129	.380	-.168	.420	-.122	.341	-.136	.350	-.102	.276	-.127	.290
	-.068	.223	-.102	.206	-.055	.176	-.068	.135	-.039	.140	-.052	.082
	-.024	.109	-.020	.050	-.009	.083	-.020	.023	-.003	.051	-.004	.020
	.009	.037	-.004	.006	.010	.011	-.001	.011	.011	.015	-.004	.005
	.008	.005	-.005	.007	.002	.008	-.001	.002	.002	.004	-.002	.001
	.000	.001	-.000	.004	.004	.003						
KIN:	5.650		3.392		11.506		2.4426		1.5691		1.8229	
I=												
1/2	-.138	.362	-.144	.382	-.122	.320	-.113	.365	-.100	.268	-.079	.304
	-.076	.224	-.045	.257	-.061	.181	-.019	.190	-.040	.133	-.013	.142
	-.032	.096	-.001	.104	-.019	.064	-.004	.085	-.007	.039	.006	.067
	-.009	.021	.012	.056	.005	.006	.021	.042	.001	.006	.027	.024
	.004	.002	.024	.009	.002	.005	.014	.000	.008	.002	.011	.000
	-.004	.003	-.000	.001	.001	.002	-.000	.002	-.001	.003	.002	.004
	.000	.000										
3/2	-.122	.372	-.160	.391	-.108	.318	-.151	.349	-.084	.291	-.119	.291
	-.072	.235	-.103	.220	-.059	.189	-.068	.161	-.045	.157	-.046	.092
	-.030	.108	-.027	.064	-.012	.093	-.020	.025	.002	.055	-.010	.020
	.011	.039	-.004	.009	.008	.017	-.002	.009	.012	.009	-.001	.004
	.006	.004	-.000	.003	.003	.003	-.003	.001	.004	.005	-.003	.002
	.002	.005	.000	.001	.003	.002	-.002	.000	.002	.001	.000	.001
	-.002	.000										
KIN:	6.000		3.487		12.163		2.6059		1.6203		1.8672	
I=												
1/2	-.137	.361	-.143	.383	-.122	.319	-.111	.365	-.101	.269	-.077	.304
	-.077	.226	-.048	.259	-.062	.182	-.020	.192	-.041	.134	-.014	.148
	-.034	.101	-.001	.108	-.018	.069	-.004	.087	-.007	.044	.006	.071
	-.008	.028	.011	.059	.001	.014	.016	.046	-.002	.011	.024	.029
	.002	.003	.024	.019	-.001	.005	.016	.004	.007	.002	.011	.002
	.000	.001	.004	.001	.005	.003	.003	.002	-.001	.001	.002	.001
	-.000	.002										
3/2	-.121	.372	-.156	.390	-.107	.316	-.151	.348	-.087	.288	-.117	.290
	-.073	.234	-.102	.221	-.056	.191	-.064	.165	-.043	.158	-.049	.098
	-.031	.113	-.029	.072	-.015	.098	-.020	.034	-.004	.062	-.012	.027
	.006	.050	-.007	.015	.008	.027	-.003	.013	.012	.015	-.003	.008
	.007	.008	-.002	.005	.006	.004	-.002	.001	.004	.003	-.001	.000
	.002	.003	-.002	.000	.001	.001	-.003	.000	.001	.001	.001	.000
	.002	.001										

Table 2.2.2.3. Results of the KH 80 partial wave analysis (80 L Koch and Pietarinen). The hadronic phase shifts are given in degrees. The values at $k<0.079$ GeV/c are extrapolations, since data in this range do not exist.

k (GeV/c)	S11	P11	P13	D13	D15	F15	F17
0.020	1.26	−0.01	−0.00	0.000	0.000	0.000	−0.000
0.040	2.50	−0.06	−0.02	0.000	0.000	0.000	−0.000
0.060	3.69	−0.19	−0.07	0.001	0.002	0.000	−0.000
0.079	4.62	−0.33	−0.15	0.004	0.006	0.000	−0.000
0.097	5.55	−0.57	−0.25	0.011	0.014	0.000	−0.000
0.112	6.19	−0.74	−0.35	0.022	0.026	0.000	−0.001
0.130	6.90	−0.96	−0.48	0.047	0.047	0.000	−0.002
0.153	7.79	−1.23	−0.69	0.098	0.087	0.004	−0.003
0.172	8.35	−1.33	−0.90	0.158	0.132	0.007	−0.005
0.185	8.78	−1.40	−1.08	0.212	0.170	0.017	−0.006
0.200	9.48	−1.47	−1.29	0.251	0.219	0.025	−0.007
0.218	9.30	−1.32	−1.33	0.343	0.298	0.031	−0.027
0.247	10.30	−0.72	−1.77	0.645	0.441	0.083	−0.004
0.267	11.00	−0.25	−1.99	0.899	0.532	0.122	−0.000
0.280	12.12	0.14	−2.16	0.967	0.596	0.112	−0.032
0.290	12.09	0.49	−2.35	1.126	0.663	0.155	−0.037
0.295	11.96	0.83	−2.37	1.201	0.696	0.188	−0.026
0.301	11.85	1.05	−2.45	1.332	0.751	0.228	−0.026
0.305	11.79	1.33	−2.45	1.410	0.784	0.249	−0.029
0.310	11.63	1.78	−2.40	1.484	0.819	0.263	−0.031
0.320	11.61	2.43	−2.49	1.675	0.909	0.325	−0.010
0.331	11.86	3.37	−2.89	1.758	0.994	0.410	0.093
0.351	11.78	5.97	−3.25	1.941	1.218	0.473	0.170
0.378	12.33	9.08	−3.36	2.640	1.411	0.633	0.281
0.408	12.55	13.52	−3.39	3.475	1.994	0.470	−0.132
0.427	11.13	18.96	−3.89	4.031	1.910	0.710	0.099
0.456	13.03	22.84	−3.71	5.146	2.247	0.945	0.031
0.490	15.50	33.53	−4.98	7.566	2.520	1.485	0.076

k (GeV/c)	S31	P31	P33	D33	D35	F35	F37
0.020	−0.72	−0.00	0.02	0.000	−0.000	0.000	0.000
0.040	−1.48	−0.04	0.18	0.000	−0.000	0.000	0.000
0.060	−2.30	−0.12	0.62	0.001	−0.002	0.000	0.000
0.079	−3.15	−0.25	1.42	0.003	−0.006	0.000	0.000
0.097	−3.91	−0.46	2.55	0.006	−0.013	0.000	0.001
0.112	−4.58	−0.69	3.90	0.011	−0.023	0.000	0.001
0.130	−5.36	−0.97	6.07	0.019	−0.042	0.000	0.004
0.153	−7.00	−1.23	10.23	0.034	−0.075	−0.002	0.005
0.172	−7.95	−1.71	14.83	0.048	−0.111	−0.000	0.006
0.185	−8.20	−2.06	18.69	0.060	−0.140	0.006	0.016
0.200	−8.58	−2.50	24.28	0.073	−0.179	0.012	0.027
0.218	−9.93	−2.87	32.97	0.091	−0.231	0.016	0.041
0.247	−12.86	−3.66	51.78	0.122	−0.329	−0.023	0.043
0.267	−14.34	−4.08	67.51	0.180	−0.410	−0.039	0.067
0.280	−13.91	−4.66	77.78	0.282	−0.471	−0.006	0.095
0.290	−14.35	−4.90	85.19	0.202	−0.505	−0.062	0.084
0.295	−14.71	−5.20	88.83	0.191	−0.526	−0.039	0.113
0.301	−15.21	−5.41	92.63	0.186	−0.549	−0.046	0.122
0.305	−15.55	−5.55	95.42	0.189	−0.569	−0.060	0.120

(continued)

Table 2.2.2.3 (continued)

k (GeV/c)	S31	P31	P33	D33	D35	F35	F37
0.310	−15.95	− 5.81	98.55	0.195	−0.591	−0.050	0.144
0.320	−16.58	− 6.44	104.42	0.210	−0.638	0.037	0.241
0.331	−17.15	− 6.16	109.17	0.221	−0.691	−0.086	0.147
0.351	−18.89	− 7.01	117.28	0.204	−0.788	0.227	0.422
0.378	−19.96	− 7.88	126.07	0.205	−0.921	−0.213	0.113
0.408	−20.75	− 8.54	132.39	0.263	−1.111	−0.241	0.237
0.427	−21.43	− 8.26	135.55	0.149	−1.198	−0.102	0.409
0.456	−22.38	− 9.64	141.06	0.353	−1.302	−0.199	0.129
0.490	−24.89	−10.48	145.64	−0.010	−1.375	−0.410	0.538

Table 2.2.2.4. Partial wave amplitudes of the updated CMU-LBL solution (Sect. 2.1.9, 80 L Cutkosky).

CMU-LBL 1980	K GEV/C	W GEV	S GEV**2	Q**2 (GEV/C)**2	ω_q GEV	E GEV
	S1	P1	P3	D3	D5	F5
	F7	G7	G9	H9	H11	I11
	I13	K13	K15	L15	L17	M17
KIN:	.429	1.322	1.747	.0928	.3351	.9865
I=						
1/2	.1899 .0376	.2838 .1361	−.0695 .0146	.0659 .0096	.0325 .0069	.0129 .0003
	.0011 .0012	−.0002 .0001	.0010 .0001	−.0003 .0000	.0001 .0000	−.0002 .0000
	.0000 .0000	−.0001 .0000				
3/2	−.3370 .1311	−.1476 .0228	−.4978 .4637	.0049 .0009	−.0212 .0011	−.0022 .0002
	.0100 .0009	.0011 .0002	−.0004 .0001	.0002 .0000	.0001 .0000	.0001 .0000
	−.0000 .0000	.0000 .0000				
KIN:	.480	1.356	1.838	.1104	.3603	.9954
I=						
1/2	.2329 .0672	.3491 .2910	−.0875 .0229	.1003 .0149	.0351 .0016	.0233 .0011
	.0018 .0018	.0020 .0002	.0017 .0001	.0001 .0000	.0002 .0000	−.0001 .0000
	.0001 .0000	−.0001 .0000				
3/2	−.3567 .1541	−.1687 .0305	−.4748 .3446	.0030 .0000	−.0259 .0007	−.0014 .0052
	.0148 .0021	.0035 .0001	−.0002 .0002	.0009 .0001	.0005 .0000	.0003 .0000
	.0001 .0000	.0001 .0000				
KIN:	.539	1.395	1.945	.1315	.3886	1.0059
I=						
1/2	.2548 .0983	.2936 .4714	−.1098 .0194	.1778 .0498	.0480 .0142	.0402 .0027
	.0041 .0028	.0084 .0007	.0035 .0001	.0019 .0001	.0008 .0000	.0004 .0000
	.0003 .0000	.0001 .0000				
3/2	−.3808 .1788	−.1885 .0540	−.4343 .2525	.0120 .0002	−.0317 .0011	−.0007 .0028
	.0208 .0032	.0036 .0000	−.0009 .0003	.0001 .0001	.0005 .0000	−.0001 .0000
	.0000 .0000	−.0001 .0000				
KIN:	.578	1.420	2.016	.1459	.4067	1.0131
I=						
1/2	.3075 .1859	.2182 .5176	−.1200 .0197	.2178 .0993	.0521 .0056	.0513 .0045
	.0020 .0039	.0129 .0011	.0034 .0002	.0032 .0001	.0007 .0000	.0007 .0000
	.0003 .0000	.0002 .0000				
3/2	−.3877 .1865	−.1961 .0623	−.4027 .2044	.0139 .0019	−.0335 .0038	.0010 .0013
	.0279 .0020	.0055 .0003	−.0003 .0003	.0014 .0001	.0011 .0000	.0005 .0000
	.0002 .0000	.0002 .0000				

CMU-LBL 1980	K GEV/C		W GEV		S GEV**2		Q**2 (GEV/C)**2		ω_q GEV		E GEV	
	S1 F7 I13 M19		P1 G7 K13 N19		P3 G9 K15 N21		D3 H9 L15 021		D5 H11 L17 023		F5 I11 M17 F23	
KIN:	.617		1.445		2.087		.1606		.4244		1.0203	
I=												
1/2	.3141	.1517	.1458	.6096	-.0987	.0336	.2940	.1765	.0527	.0181	.0545	.0053
	.0010	.0033	.0106	.0018	.0058	.0002	.0047	.0002	.0018	.0000	.0017	.0001
	.0009	.0000	.0006	.0000	.0003	.0000	.0003	.0000	.0001	.0000	.0001	.0000
3/2	-.3925	.1947	-.2137	.0742	-.3644	.1588	.0185	.0155	-.0428	.0039	.0006	.0020
	.0296	.0017	.0008	.0002	-.0016	.0004	.0000	.0001	.0011	.0000	.0001	.0000
	.0001	.0000	.0000	.0000								
KIN:	.656		1.469		2.159		.1755		.4416		1.0276	
I=												
1/2	.3593	.2817	.0624	.6475	-.1136	.0289	.3133	.3140	.0708	.0169	.0905	.0111
	.0086	.0034	.0100	.0027	.0075	.0001	.0048	.0003	.0021	.0001	.0016	.0001
	.0010	.0000	.0006	.0000	.0004	.0000	.0003	.0000	.0002	.0000	.0001	.0000
	.0001	.0000	.0000	.0000	.0000	.0000						
3/2	-.4053	.2302	-.2351	.0839	-.3221	.1181	.0241	.0213	-.0496	.0028	-.0005	.0007
	.0322	.0030	.0013	.0004	-.0020	.0005	-.0001	.0000	.0015	.0000	.0001	.0000
	.0002	.0000	.0001	.0000	.0002	.0000	.0000	.0000	.0001	.0000	.0000	.0000
	.0000	.0000	.0000	.0000	.0000	.0000						
KIN:	.687		1.488		2.215		.1876		.4550		1.0334	
I=												
1/2	.3747	.4511	-.0122	.6718	-.1237	.0494	.2736	.4691	.0725	.0101	.0942	.0135
	-.0002	.0035	.0133	.0003	.0064	.0004	.0073	.0002	.0020	.0001	.0027	.0001
	.0011	.0000	.0011	.0000	.0004	.0000	.0005	.0000	.0002	.0000	.0002	.0000
	.0001	.0000	.0001	.0000	.0000	.0000						
3/2	-.4013	.2464	-.2536	.0785	-.2832	.0887	.0156	.0408	-.0412	.0020	.0083	.0001
	.0401	.0021	.0030	.0026	.0001	.0005	-.0005	.0002	.0026	.0000	-.0002	.0000
	.0005	.0000	-.0001	.0000	.0003	.0000	-.0001	.0000	.0001	.0000	-.0000	.0000
	.0000	.0000	-.0000	.0000	.0000	.0000						
KIN:	.723		1.511		2.282		.2017		.4703		1.0402	
I=												
1/2	.2365	.5073	-.0703	.6406	-.1325	.0603	.0869	.5833	.0959	.0129	.1140	.0137
	-.0075	.0039	.0221	.0005	.0060	.0002	.0093	.0003	.0020	.0001	.0033	.0000
	.0013	.0000	.0014	.0000	.0006	.0000	.0006	.0000	.0003	.0000	.0003	.0000
	.0001	.0000	.0001	.0000	.0001	.0000						
3/2	-.3805	.2330	-.2697	.0936	-.2394	.0833	.0126	.0451	-.0449	.0027	.0009	.0014
	.0523	.0046	.0024	.0002	-.0018	.0006	.0001	.0001	.0020	.0000	.0000	.0000
	.0000	.0000	-.0001	.0000	.0001	.0000	-.0000	.0000	.0000	.0000	-.0000	.0000
	.0000	.0000	-.0000	.0000	.0000	.0000						
KIN:	.770		1.539		2.368		.2204		.4898		1.0492	
I=												
1/2	.1242	.4151	-.1202	.5942	-.1343	.0504	-.1711	.5611	.1388	.0262	.1723	.0394
	-.0133	.0018	.0361	.0017	.0054	.0003	.0097	.0008	.0020	.0002	.0033	.0001
	.0015	.0000	.0014	.0000	.0007	.0000	.0006	.0000	.0004	.0000	.0003	.0000
	.0002	.0000	.0001	.0000	.0001	.0000						
3/2	-.3442	.2038	-.2914	.1081	-.1845	.0778	.0315	.0596	-.0501	.0196	.0034	.0007
	.0557	.0047	-.0050	.0005	.0005	.0006	-.0007	.0001	.0040	.0000	.0001	.0001
	.0006	.0001	.0000	.0000	.0004	.0000	.0000	.0000	.0001	.0000	-.0000	.0000
	.0000	.0000	-.0000	.0000	.0000	.0000						
KIN:	.820		1.569		2.461		.2406		.5100		1.0588	
I=												
1/2	.1587	.3527	-.1597	.5370	-.1544	.0494	-.2956	.3850	.1947	.0772	.2186	.0682
	-.0101	.0010	.0411	.0097	.0093	.0006	.0133	.0012	.0043	.0003	.0058	.0002
	.0029	.0000	.0028	.0000	.0014	.0000	.0014	.0000	.0008	.0000	.0007	.0000
	.0004	.0000	.0003	.0000	.0002	.0000						
3/2	-.2869	.2064	-.3208	.1428	-.1301	.0890	.0235	.0602	-.0546	.0214	.0034	.0177
	.0701	.0095	-.0057	.0001	.0014	.0002	.0020	.0001	.0049	.0001	.0011	.0001
	.0005	.0001	.0004	.0000	.0003	.0000	.0001	.0000	.0000	.0000	.0000	.0000
	.0000	.0000	.0000	.0000	-.0000	.0000						

CMU-LBL 1980	K GEV/C	W GEV	S GEV**2	Q**2 (GEV/C)**2	ω_q GEV	E GEV
	S1 F7 I13 M19	P1 G7 K13 N19	P3 G9 K15 N21	D3 H9 L15 O21	D5 H11 L17 O23	F5 I11 M17 P23
KIN:	.874	1.600	2.561	.2626	.5311	1.0691
I=						
1/2	.2863 .3921	-.1189 .5616	-.1314 .0532	-.2005 .2805	.2184 .1527	.2860 .1244
	-.0078 .0006	.0125 .0132	.0090 .0043	.0103 .0062	.0028 .0013	.0044 .0012
	.0021 .0003	.0021 .0003	.0009 .0001	.0010 .0001	.0005 .0000	.0005 .0000
	.0002 .0000	.0002 .0000	.0001 .0000			
3/2	-.2379 .3301	-.2969 .0984	-.1148 .1139	.0060 .0864	-.0577 .0038	.0036 .0218
	.0926 .0258	-.0010 .0012	-.0024 .0017	.0013 .0000	.0051 .0001	.0012 .0001
	.0004 .0001	.0006 .0001	.0004 .0000	.0003 .0000	.0001 .0000	.0001 .0000
	.0001 .0000	.0001 .0000	.0000 .0000			
KIN:	.928	1.631	2.661	.2849	.5517	1.0795
I=						
1/2	.3493 .5326	-.1759 .4815	-.1179 .0472	-.1805 .2368	.2113 .3040	.3329 .3369
	-.0132 .0063	.0179 .0116	.0121 .0035	.0148 .0032	.0028 .0006	.0062 .0005
	.0021 .0000	.0028 .0000	.0008 .0000	.0013 .0000	.0005 .0000	.0006 .0000
	.0002 .0000	.0003 .0000	.0001 .0000			
3/2	-.2044 .5064	-.3137 .1422	-.0816 .1351	-.0089 .1169	-.0742 .0126	.0042 .0067
	.1001 .0133	-.0001 .0002	-.0018 .0007	.0015 .0030	.0065 .0001	.0021 .0009
	.0005 .0001	.0012 .0003	.0006 .0000	.0007 .0001	.0002 .0000	.0003 .0000
	.0001 .0000	.0002 .0000	.0001 .0000			
KIN:	.978	1.659	2.754	.3058	.5703	1.0891
I=						
1/2	.2422 .7840	-.2225 .4217	-.1217 .0239	-.2025 .2502	.1013 .4006	.2261 .5570
	-.0191 .0018	.0116 .0088	.0127 .0069	.0163 .0009	.0045 .0021	.0082 .0004
	.0037 .0004	.0042 .0000	.0018 .0001	.0021 .0000	.0010 .0000	.0011 .0000
	.0005 .0000	.0006 .0000	.0003 .0000			
3/2	-.2666 .5785	-.3112 .1382	-.0636 .1540	-.0344 .1465	-.0680 .0111	.0109 .0124
	.1205 .0215	-.0003 .0000	.0024 .0004	.0008 .0052	.0078 .0001	.0021 .0010
	-.0001 .0001	.0010 .0003	.0002 .0000	.0005 .0001	-.0001 .0000	.0002 .0000
	-.0000 .0000	.0001 .0000	-.0000 .0000			
KIN:	1.005	1.675	2.804	.3171	.5802	1.0943
I=						
1/2	.1193 .8656	-.1801 .3766	-.0994 .0207	-.2120 .2244	.0115 .4351	.1003 .6305
	-.0237 .0027	.0282 .0183	.0149 .0065	.0175 .0017	.0043 .0020	.0074 .0006
	.0035 .0003	.0036 .0000	.0016 .0001	.0017 .0000	.0009 .0000	.0009 .0000
	.0005 .0000	.0004 .0000	.0003 .0000			
3/2	-.3076 .5998	-.3147 .1298	-.0683 .1566	-.0550 .1541	-.0648 .0132	.0124 .0199
	.1301 .0277	-.0011 .0002	.0040 .0010	.0050 .0021	.0096 .0003	.0027 .0007
	.0004 .0001	.0011 .0002	.0005 .0000	.0005 .0001	.0000 .0000	.0002 .0000
	.0000 .0000	.0001 .0000	-.0000 .0000			
KIN:	1.030	1.688	2.850	.3277	.5892	1.0991
I=						
1/2	.0232 .8462	-.1653 .3509	-.0841 .0347	-.2079 .1898	-.0655 .4180	-.0193 .6272
	-.0111 .0001	.0323 .0137	.0095 .0122	.0176 .0029	.0017 .0032	.0060 .0007
	.0025 .0005	.0028 .0000	.0012 .0001	.0013 .0000	.0007 .0000	.0007 .0000
	.0004 .0000	.0003 .0000	.0002 .0000			
3/2	-.3597 .5894	-.2983 .1071	-.0682 .1496	-.0757 .1570	-.0539 .0202	.0182 .0281
	.1348 .0392	-.0011 .0000	.0079 .0038	.0017 .0059	.0115 .0010	.0010 .0017
	.0008 .0002	.0004 .0004	.0007 .0000	.0003 .0001	.0001 .0000	.0001 .0000
	.0001 .0000	.0001 .0000	.0000 .0000			
KIN:	1.060	1.705	2.906	.3404	.5999	1.1049
I=						
1/2	-.0932 .8229	-.0546 .3544	-.0873 .1003	-.1674 .1282	-.1359 .3827	-.1667 .5550
	-.0016 .0001	.0192 .0067	.0167 .0095	.0176 .0045	.0050 .0027	.0079 .0010
	.0044 .0003	.0043 .0001	.0021 .0001	.0022 .0001	.0012 .0000	.0011 .0000
	.0006 .0000	.0006 .0000	.0003 .0000			
3/2	-.3352 .6063	-.3056 .1463	-.0549 .1474	-.0924 .1597	-.0507 .0238	.0225 .0429
	.1516 .0430	-.0039 .0001	.0023 .0025	.0040 .0040	.0101 .0009	.0017 .0012
	-.0002 .0002	.0005 .0003	.0003 .0000	.0002 .0001	-.0001 .0000	.0001 .0000
	-.0000 .0000	.0000 .0000	-.0000 .0000			

CMU-LBL 1980	K GEV/C	W GEV	S GEV**2	Q**2 (GEV/C)**2	ω_q GEV	E GEV
	S1	P1	P3	D3	D5	F5
	F7	G7	G9	H9	H11	I11
	I13	K13	K15	L15	L17	M17
	M19	N19	N21	O21	O23	P23
KIN:	1.103	1.728	2.986	.3587	.6149	1.1131
I=						
1/2	-.1905 .7326	-.0616 .4049	-.0660 .1502	-.1553 .1111	-.1613 .2805	-.2812 .4298
	.0041 .0029	.0486 .0102	.0196 .0120	.0274 .0027	.0016 .0024	.0098 .0006
	.0022 .0002	.0045 .0000	.0007 .0001	.0019 .0001	.0005 .0000	.0009 .0000
	.0002 .0000	.0004 .0000	.0001 .0000	.0002 .0000	.0001 .0000	.0001 .0000
	.0000 .0000	.0001 .0000				
3/2	-.3302 .6525	-.2930 .1556	-.0709 .1355	-.1199 .1427	-.0583 .0478	.0333 .0480
	.1760 .0571	-.0057 .0020	.0079 .0026	-.0012 .0041	.0145 .0017	.0009 .0010
	.0012 .0003	.0004 .0002	.0012 .0000	.0003 .0000	.0003 .0000	.0002 .0000
	.0002 .0000	.0001 .0000	.0001 .0000	.0001 .0000	.0000 .0000	.0000 .0000
	.0000 .0000	.0000 .0000				
KIN:	1.137	1.746	3.050	.3732	.6267	1.1197
I=						
1/2	-.1187 .6882	-.0452 .4461	-.0958 .1396	-.1425 .0990	-.1946 .2643	-.2672 .3278
	-.0017 .0102	.0765 .0194	.0233 .0085	.0287 .0076	.0022 .0016	.0091 .0009
	.0029-.0001	.0043 .0004	.0011 .0000	.0019 .0002	.0007 .0000	.0009 .0001
	.0004 .0000	.0005 .0000	.0002 .0000	.0003 .0000	.0001 .0000	.0001 .0000
	.0001 .0000	.0001 .0000				
3/2	-.3629 .6662	-.3032 .1765	-.0600 .1247	-.1461 .1434	-.0553 .0509	.0290 .0649
	.1895 .0820	-.0018 .0003	.0013 .0007	.0012 .0043	.0161 .0016	.0029 .0008
	.0024 .0004	.0021 .0001	.0023 .0000	.0014 .0000	.0009 .0000	.0009 .0000
	.0006 .0000	.0005 .0000	.0003 .0000	.0003 .0000	.0002 .0000	.0002 .0000
	.0001 .0000	.0001 .0000				
KIN:	1.196	1.778	3.160	.3986	.6466	1.1309
I=						
1/2	-.1653 .6662	-.0470 .5189	-.0609 .1691	-.0768 .1012	-.1842 .1933	-.2372 .2741
	.0234 .0036	.0804 .0518	.0256 .0080	.0320 .0020	-.0013 .0017	.0080 .0003
	.0006 .0000	.0028 .0000	-.0005 .0001	.0006 .0001	-.0002 .0000	.0001 .0000
	-.0002 .0000	-.0000 .0000	-.0001 .0000	-.0001 .0000	-.0001 .0000	-.0000 .0000
	-.0000 .0000	-.0000 .0000				
3/2	-.3624 .6530	-.2914 .1532	-.0632 .1144	-.1538 .1111	-.0616 .0705	.0270 .0879
	.2168 .1321	-.0005 .0027	.0075 .0162	.0042 .0020	.0175 .0054	.0033 .0012
	.0014 .0008	.0018 .0008	.0019 .0000	.0012 .0003	.0006 .0000	.0007 .0001
	.0004 .0000	.0004 .0000	.0002 .0000	.0003 .0000	.0001 .0000	.0001 .0000
	.0001 .0000	.0001 .0000				
KIN:	1.247	1.804	3.255	.4206	.6634	1.1406
I=						
1/2	-.1739 .6149	-.0627 .5104	-.0967 .1858	-.0653 .1074	-.2028 .1665	-.2105 .2364
	.0202 .0182	.0829 .0686	.0210 .0104	.0370 .0034	-.0011 .0017	.0143 .0003
	.0023 .0000	.0069 .0001	.0007 .0001	.0029 .0001	.0006 .0000	.0014 .0001
	.0003 .0000	.0007 .0000	.0002 .0000	.0003 .0000	.0001 .0000	.0002 .0000
	.0001 .0000	.0001 .0000				
3/2	-.3612 .6813	-.2564 .1829	-.0226 .0963	-.1752 .1113	-.0471 .0890	.0097 .1149
	.2373 .1800	-.0024 .0057	.0061 .0178	.0045 .0031	.0179 .0068	.0020 .0010
	.0003 .0010	.0009 .0001	.0014 .0000	.0007 .0000	.0002 .0000	.0004 .0000
	.0002 .0000	.0003 .0000	.0001 .0000	.0001 .0000	.0001 .0000	.0001 .0000
	.0000 .0000	.0001 .0000				
KIN:	1.297	1.830	3.348	.4424	.6796	1.1501
I=						
1/2	-.1883 .5862	-.0491 .5447	-.0877 .1954	-.0832 .1324	-.1820 .1298	-.2265 .2188
	.0192 .0368	.0926 .0677	.0124 .0173	.0487 .0028	-.0010 .0014	.0194 .0010
	.0035 .0000	.0095 .0003	.0013 .0001	.0040 .0001	.0010 .0000	.0019 .0001
	.0005 .0000	.0009 .0000	.0003 .0000	.0005 .0000	.0002 .0000	.0003 .0000
	.0001 .0000	.0001 .0000				
3/2	-.3995 .7241	-.2374 .2248	-.0050 .0962	-.1889 .1085	-.0527 .1076	-.0030 .1387
	.2352 .2285	-.0092 .0088	.0043 .0187	.0056 .0044	.0214 .0085	.0035 .0008
	.0014 .0011	.0020 .0001	.0024 .0000	.0014 .0000	.0006 .0000	.0008 .0000
	.0005 .0000	.0005 .0000	.0002 .0000	.0003 .0000	.0001 .0000	.0002 .0000
	.0001 .0000	.0001 .0000				

CMU-LBL 1980	K GEV/C	W GEV	S GEV**2	Q**2 (GEV/C)**2	ω_q GEV	E GEV
	S1	P1	P3	D3	D5	F5
	F7	G7	G9	H9	H11	I11
	I13	K13	K15	L15	L17	M17
	M19	N19	N21	O21	O23	P23
KIN:	1.347	1.855	3.441	.4642	.6955	1.1596
I=						
1/2	-.1676 .5663	-.0344 .5799	-.1151 .2213	-.0684 .1491	-.1785 .1168	-.2177 .2174
	.0259 .0276	.0850 .0907	.0079 .0018	.0562 .0041	-.0032 .0004	.0197 .0008
	.0036 .0000	.0099 .0005	.0015 .0002	.0042 .0002	.0013 .0001	.0021 .0001
	.0007 .0000	.0011 .0001	.0005 .0000	.0006 .0000	.0003 .0000	.0003 .0000
	.0002 .0000	.0002 .0000	.0001 .0000	.0001 .0000		
3/2	-.4292 .7496	-.2233 .2667	.0353 .0823	-.1985 .0973	-.0627 .1191	-.0296 .1633
	.2072 .2895	-.0142 .0154	-.0015 .0161	-.0018 .0096	.0248 .0118	.0039 .0004
	.0031 .0014	.0034 .0001	.0040 .0000	.0025 .0000	.0014 .0000	.0016 .0000
	.0010 .0000	.0010 .0000	.0005 .0000	.0006 .0000	.0003 .0000	.0003 .0000
	.0002 .0000	.0002 .0000	.0001 .0000	.0001 .0000		
KIN:	1.395	1.879	3.531	.4852	.7104	1.1686
I=						
1/2	-.1875 .5813	-.0528 .5210	-.1458 .2332	-.0656 .1540	-.1978 .1018	-.1873 .2144
	.0284 .0401	.0838 .0999	.0305 .0032	.0663 .0056	.0023 .0014	.0222 .0006
	.0053 .0000	.0103 .0007	.0017 .0002	.0040 .0002	.0012 .0000	.0018 .0001
	.0006 .0000	.0009 .0001	.0004 .0000	.0004 .0000	.0002 .0000	.0002 .0000
	.0001 .0000	.0001 .0000	.0001 .0000	.0001 .0000		
3/2	-.4312 .7423	-.2108 .3352	.0853 .0952	-.1986 .0861	-.0731 .1407	-.0617 .1478
	.1614 .3625	-.0086 .0074	-.0041 .0228	-.0007 .0057	.0228 .0125	.0010 .0028
	.0009 .0016	.0014 .0004	.0032 .0000	.0014 .0001	.0009 .0001	.0010 .0001
	.0007 .0000	.0006 .0000	.0004 .0000	.0004 .0000	.0002 .0000	.0002 .0000
	.0001 .0000	.0001 .0000	.0001 .0000	.0001 .0000		
KIN:	1.437	1.900	3.609	.5037	.7233	1.1765
I=						
1/2	-.1445 .5340	-.0384 .5565	-.1324 .2407	-.0373 .1511	-.1804 .1045	-.1980 .1945
	.0196 .0623	.0820 .0998	.0201 .0066	.0803 .0075	-.0002 .0002	.0222 .0047
	.0048 .0000	.0101 .0021	.0014 .0002	.0037 .0006	.0012 .0000	.0017 .0002
	.0006 .0000	.0008 .0001	.0004 .0000	.0004 .0001	.0002 .0000	.0003 .0000
	.0001 .0000	.0001 .0000	.0001 .0000	.0001 .0000		
3/2	-.4435 .7298	-.2254 .4083	.0965 .0945	-.1906 .0877	-.0920 .1617	-.0755 .1539
	.1115 .3932	-.0036 .0108	-.0005 .0121	.0029 .0070	.0251 .0120	-.0003 .0010
	.0003 .0014	-.0002 .0001	.0030 .0000	.0002 .0001	.0004 .0001	.0001 .0001
	.0004 .0000	.0001 .0000	.0001 .0000	.0001 .0000	.0001 .0000	.0001 .0000
	.0001 .0000	.0000 .0000	.0000 .0000	.0000 .0000		
KIN:	1.498	1.930	3.723	.5306	.7417	1.1879
I=						
1/2	-.1673 .5828	-.0182 .5102	-.1565 .2807	-.0152 .1347	-.1591 .0963	-.1604 .1774
	.0092 .0618	.0832 .1196	.0187 .0021	.0797 .0251	-.0007 .0009	.0235 .0055
	.0056 .0001	.0114 .0030	.0017 .0002	.0043 .0008	.0015 .0000	.0021 .0003
	.0008 .0000	.0011 .0002	.0005 .0000	.0006 .0001	.0003 .0000	.0003 .0000
	.0002 .0000	.0002 .0000	.0001 .0000	.0001 .0000		
3/2	-.4560 .6981	-.2324 .4366	.1096 .1227	-.1784 .0910	-.1179 .1536	-.0922 .1454
	.0133 .3987	-.0106 .0085	.0003 .0220	-.0011 .0111	.0300 .0145	-.0013 .0026
	.0019 .0015	.0001 .0009	.0046 .0000	.0006 .0004	.0011 .0001	.0004 .0002
	.0009 .0000	.0003 .0001	.0004 .0000	.0002 .0000	.0003 .0000	.0001 .0000
	.0001 .0000	.0001 .0000	.0001 .0000	.0001 .0000		
KIN:	1.550	1.955	3.820	.5537	.7571	1.1975
I=						
1/2	-.1933 .5993	-.0128 .5169	-.1671 .2737	-.0194 .1472	-.1391 .0757	-.1684 .1935
	-.0019 .0545	.0774 .1348	.0377 .0045	.0828 .0427	.0013 .0010	.0297 .0069
	.0066 .0004	.0147 .0038	.0018 .0003	.0059 .0010	.0016 .0000	.0029 .0004
	.0008 .0000	.0015 .0002	.0006 .0000	.0008 .0001	.0003 .0000	.0004 .0001
	.0002 .0000	.0003 .0000	.0001 .0000	.0002 .0000		
3/2	-.4475 .6858	-.2668 .4521	.1005 .1535	-.1742 .0936	-.1360 .1472	-.1095 .1313
	-.0493 .3972	-.0053 .0063	-.0068 .0241	-.0026 .0064	.0316 .0137	.0020 .0022
	.0027 .0014	.0026 .0007	.0060 .0000	.0021 .0005	.0017 .0001	.0013 .0003
	.0013 .0000	.0008 .0001	.0006 .0000	.0005 .0000	.0004 .0000	.0003 .0000
	.0002 .0000	.0002 .0000	.0001 .0000	.0001 .0000		

CMU-LBL 1980	K GEV/C	W GEV	S GEV**2	Q**2 (GEV/C)**2	ω_q GEV	E GEV
	S1 F7 I13 M19	P1 G7 K13 N19	P3 G9 K15 N21	D3 H9 L15 O21	D5 H11 L17 O23	F5 I11 M17 P23
KIN:	1.590	1.974	3.895	.5714	.7687	1.2049
I=						
1/2	-.2194 .5653	-.0129 .5592	-.1675 .2733	-.0158 .1530	-.1275 .0738	-.1871 .1904
	.0149 .0599	.0540 .1429	.0507 .0067	.0849 .0476	.0042 .0013	.0319 .0125
	.0085 .0007	.0166 .0060	.0028 .0004	.0072 .0017	.0023 .0000	.0037 .0006
	.0013 .0000	.0020 .0003	.0008 .0000	.0011 .0001	.0005 .0000	.0007 .0001
	.0003 .0000	.0004 .0000	.0002 .0000	.0002 .0000		
3/2	-.4300 .6704	-.2821 .4413	.1019 .1623	-.1880 .0974	-.1379 .1406	-.1248 .1312
	-.0845 .3687	-.0053 .0128	-.0011 .0277	.0022 .0123	.0336 .0126	.0010 .0005
	.0031 .0010	.0020 .0012	.0068 .0001	.0017 .0008	.0020 .0001	.0012 .0004
	.0016 .0000	.0008 .0002	.0008 .0000	.0005 .0001	.0005 .0000	.0003 .0000
	.0003 .0000	.0002 .0000	.0002 .0000	.0001 .0000		
KIN:	1.627	1.991	3.964	.5879	.7793	1.2117
I=						
1/2	-.1735 .5799	-.0155 .5152	-.1630 .2600	-.0271 .1697	-.1197 .0866	-.1645 .1739
	-.0056 .0804	.0698 .1329	.0554 .0247	.0810 .0594	.0062 .0011	.0320 .0133
	.0090 .0011	.0164 .0064	.0026 .0005	.0068 .0018	.0021 .0000	.0034 .0007
	.0011 .0000	.0017 .0003	.0007 .0000	.0010 .0001	.0004 .0000	.0005 .0001
	.0003 .0000	.0003 .0000	.0002 .0000	.0002 .0000		
3/2	-.4292 .6575	-.3010 .4598	.0728 .1823	-.1529 .1073	-.1332 .1241	-.1396 .1244
	-.1005 .3392	-.0064 .0189	-.0010 .0261	.0008 .0085	.0367 .0236	-.0007 .0065
	.0035 .0025	.0014 .0006	.0073 .0005	.0014 .0004	.0019 .0002	.0009 .0002
	.0015 .0000	.0007 .0001	.0007 .0000	.0004 .0000	.0005 .0000	.0003 .0000
	.0003 .0000	.0002 .0000	.0002 .0000	.0001 .0000		
KIN:	1.671	2.012	4.047	.6075	.7918	1.2198
I=						
1/2	-.1668 .5772	.0094 .5634	-.1707 .2785	-.0075 .1898	-.1317 .0732	-.1557 .1657
	-.0252 .0861	.0681 .1505	.0530 .0341	.0869 .0584	.0112 .0020	.0312 .0114
	.0125 .0016	.0172 .0073	.0044 .0006	.0076 .0020	.0033 .0000	.0041 .0008
	.0018 .0000	.0022 .0004	.0012 .0000	.0013 .0002	.0007 .0000	.0007 .0001
	.0004 .0000	.0004 .0001	.0003 .0000	.0003 .0000		
3/2	-.4235 .6464	-.3254 .4527	.0844 .1868	-.1852 .1292	-.1264 .1234	-.1370 .1124
	-.1245 .3110	-.0070 .0211	.0060 .0287	-.0007 .0141	.0411 .0266	.0007 .0042
	.0045 .0027	.0020 .0007	.0082 .0008	.0014 .0006	.0020 .0002	.0008 .0003
	.0016 .0000	.0005 .0001	.0007 .0000	.0003 .0001	.0004 .0000	.0002 .0000
	.0002 .0000	.0001 .0000	.0001 .0000	.0001 .0000		
KIN:	1.715	2.032	4.129	.6272	.8041	1.2278
I=						
1/2	-.1751 .5634	.0304 .5087	-.1746 .2901	-.0477 .1916	-.0967 .0789	-.1521 .1816
	-.0373 .0630	.0661 .1804	.0469 .0346	.0896 .0824	.0091 .0079	.0370 .0190
	.0121 .0027	.0201 .0091	.0039 .0008	.0090 .0025	.0032 .0000	.0048 .0009
	.0017 .0000	.0025 .0004	.0011 .0000	.0014 .0002	.0007 .0000	.0008 .0001
	.0004 .0000	.0005 .0001	.0003 .0000	.0003 .0000		
3/2	-.3918 .6728	-.3155 .4504	.0726 .1921	-.1705 .1323	-.1204 .1185	-.1425 .0947
	-.1193 .2888	.0055 .0246	.0106 .0363	-.0032 .0108	.0394 .0248	-.0006 .0071
	.0025 .0025	.0006 .0003	.0075 .0010	.0002 .0005	.0012 .0003	-.0001 .0002
	.0011 .0000	-.0000 .0001	.0003 .0000	-.0000 .0000	.0002 .0000	-.0000 .0000
	.0001 .0000	-.0000 .0000	.0001 .0000	-.0000 .0000		
KIN:	1.790	2.066	4.269	.6608	.8248	1.2414
I=						
1/2	-.2049 .5838	.0151 .5645	-.1837 .2591	-.0616 .2179	-.1002 .0665	-.1606 .1735
	-.0250 .0787	.0484 .1793	.0583 .0723	.0746 .0910	.0140 .0110	.0351 .0279
	.0158 .0041	.0217 .0129	.0058 .0011	.0106 .0037	.0046 .0001	.0061 .0015
	.0026 .0001	.0033 .0007	.0017 .0000	.0020 .0003	.0011 .0000	.0012 .0002
	.0007 .0000	.0007 .0001	.0004 .0000	.0005 .0001		
3/2	-.3536 .6706	-.3501 .4301	.0306 .2203	-.2015 .1637	-.0891 .1053	-.1339 .0754
	-.1570 .2624	-.0096 .0371	.0174 .0284	.0011 .0123	.0442 .0398	-.0032 .0043
	.0053 .0041	.0024 .0011	.0108 .0020	.0021 .0009	.0028 .0004	.0014 .0004
	.0023 .0000	.0011 .0002	.0011 .0000	.0007 .0001	.0007 .0000	.0005 .0000
	.0004 .0000	.0003 .0000	.0003 .0000	.0002 .0000		

CMU-LBL 1980	K GEV/C	W GEV	S GEV**2	Q**2 (GEV/C)**2	ω_q GEV	E GEV
	S1	P1	P3	D3	D5	F5
	F7	G7	G9	H9	H11	I11
	I13	K13	K15	L15	L17	M17
	M19	N19	N21	O21	O23	P23
KIN:	1.840	2.089	4.363	.6832	.8383	1.2504
I=						
1/2	-.1628 .5998	-.0275 .5384	-.1899 .2830	-.0684 .2138	-.0905 .0834	-.1607 .1789
	-.0475 .0716	.0259 .2122	.0414 .0714	.0869 .1105	.0136 .0074	.0440 .0288
	.0177 .0043	.0269 .0140	.0070 .0011	.0138 .0041	.0057 .0000	.0080 .0017
	.0033 .0001	.0044 .0008	.0022 .0000	.0027 .0004	.0014 .0000	.0016 .0002
	.0009 .0000	.0010 .0001	.0006 .0000	.0006 .0001		
3/2	-.3770 .6530	-.3230 .4631	.0191 .2450	-.1795 .1385	-.1208 .1039	-.1462 .0611
	-.1437 .2203	-.0119 .0512	.0305 .0511	.0059 .0201	.0439 .0463	.0038 .0102
	.0038 .0048	.0043 .0016	.0103 .0026	.0021 .0011	.0019 .0005	.0011 .0004
	.0018 .0000	.0007 .0002	.0006 .0000	.0004 .0001	.0004 .0000	.0003 .0000
	.0002 .0000	.0002 .0000	.0001 .0000	.0001 .0000		
KIN:	1.895	2.113	4.466	.7080	.8529	1.2603
I=						
1/2	-.1644 .6147	-.0550 .5144	-.1775 .2884	-.0954 .1886	-.0750 .0991	-.1594 .1870
	-.0311 .0703	.0148 .2412	.0444 .0671	.0862 .1355	.0030 .0079	.0442 .0357
	.0123 .0051	.0259 .0160	.0041 .0012	.0133 .0048	.0044 .0000	.0078 .0020
	.0026 .0001	.0043 .0009	.0019 .0000	.0026 .0005	.0013 .0000	.0016 .0003
	.0009 .0000	.0010 .0001	.0006 .0000	.0007 .0001		
3/2	-.3457 .6798	-.3251 .4476	.0148 .2397	-.1926 .1886	-.0965 .1087	-.1278 .0713
	-.1611 .1931	-.0213 .0598	.0301 .0485	.0025 .0220	.0484 .0438	-.0053 .0066
	.0068 .0023	.0041 .0015	.0130 .0028	.0028 .0013	.0030 .0005	.0017 .0005
	.0026 .0000	.0012 .0002	.0011 .0000	.0008 .0001	.0007 .0000	.0005 .0001
	.0004 .0000	.0003 .0000	.0002 .0000	.0002 .0000		
KIN:	1.940	2.133	4.550	.7283	.8647	1.2683
I=						
1/2	-.1944 .6227	-.0739 .5167	-.1804 .3073	-.0845 .1963	-.0522 .1135	-.1498 .1805
	-.0389 .0664	-.0141 .2383	.0365 .0944	.0837 .1440	.0156 .0096	.0436 .0343
	.0166 .0049	.0266 .0181	.0053 .0013	.0136 .0057	.0048 .0000	.0079 .0025
	.0026 .0001	.0041 .0012	.0018 .0000	.0024 .0006	.0011 .0000	.0014 .0003
	.0008 .0000	.0009 .0002	.0005 .0000	.0006 .0001		
3/2	-.3439 .6872	-.3302 .4787	-.0077 .2583	-.2121 .1686	-.0975 .1156	-.1337 .0765
	-.1425 .1774	-.0300 .0670	.0239 .0574	-.0060 .0231	.0531 .0460	-.0031 .0084
	.0097 .0027	.0061 .0008	.0152 .0035	.0036 .0011	.0039 .0005	.0022 .0003
	.0032 .0000	.0015 .0002	.0014 .0001	.0010 .0001	.0009 .0000	.0006 .0001
	.0005 .0000	.0004 .0000	.0003 .0000	.0003 .0000		
KIN:	1.995	2.157	4.653	.7531	.8790	1.2781
I=						
1/2	-.2371 .6277	-.0374 .4605	-.2034 .3103	-.1118 .2064	-.0477 .1263	-.1716 .1774
	-.0241 .0589	-.0348 .2337	.0135 .1156	.0722 .1559	.0209 .0225	.0430 .0314
	.0204 .0078	.0280 .0206	.0069 .0018	.0148 .0065	.0060 .0001	.0087 .0029
	.0032 .0002	.0046 .0014	.0022 .0000	.0027 .0007	.0014 .0000	.0016 .0004
	.0009 .0000	.0010 .0002	.0006 .0000	.0006 .0001		
3/2	-.3518 .6918	-.3533 .4498	-.0307 .2673	-.2072 .1692	-.0991 .1361	-.1301 .0625
	-.1420 .1723	-.0281 .0679	.0304 .0624	.0009 .0265	.0604 .0540	-.0048 .0099
	.0124 .0039	.0044 .0018	.0174 .0046	.0029 .0014	.0048 .0006	.0020 .0004
	.0040 .0000	.0016 .0002	.0018 .0000	.0011 .0001	.0013 .0000	.0008 .0001
	.0007 .0000	.0005 .0000	.0005 .0000	.0004 .0000		
KIN:	2.055	2.183	4.765	.7802	.8943	1.2887
I=						
1/2	-.2750 .6003	-.0612 .4191	-.2075 .3273	-.0833 .1865	-.0705 .1391	-.1385 .1867
	-.0481 .0568	-.0130 .2374	.0158 .1207	.0636 .1713	.0200 .0259	.0400 .0363
	.0178 .0095	.0297 .0230	.0049 .0020	.0160 .0075	.0049 .0002	.0095 .0035
	.0024 .0002	.0049 .0016	.0017 .0000	.0028 .0008	.0010 .0000	.0016 .0005
	.0007 .0000	.0010 .0003	.0004 .0000	.0006 .0001		
3/2	-.3557 .6839	-.3270 .4482	-.0482 .2726	-.1978 .1663	-.0978 .1289	-.1222 .0626
	-.1434 .1574	-.0358 .0755	.0252 .0714	.0020 .0365	.0581 .0622	-.0024 .0135
	.0125 .0049	.0073 .0014	.0181 .0057	.0039 .0015	.0047 .0007	.0024 .0003
	.0041 .0000	.0017 .0002	.0017 .0000	.0011 .0001	.0012 .0000	.0007 .0001
	.0007 .0000	.0005 .0001	.0004 .0000	.0003 .0000		

CMU-LBL 1980	K GEV/C	W GEV	S GEV**2	Q**2 (GEV/C)**2	ω_q GEV	E GEV
	S1	P1	P3	D3	D5	F5
	F7	G7	G9	H9	H11	I11
	I13	K13	K15	L15	L17	M17
	M19	N19	N21	O21	O23	P23
KIN:	2.100	2.202	4.849	.8006	.9056	1.2965
I=						
1/2	-.2309 .5834	-.0669 .4470	-.2130 .3309	-.0483 .2193	-.0695 .1197	-.1306 .2076
	-.0356 .0505	-.0497 .2267	-.0041 .1163	.0450 .1910	.0121 .0151	.0442 .0455
	.0204 .0109	.0300 .0268	.0069 .0024	.0167 .0090	.0066 .0004	.0101 .0042
	.0034 .0003	.0052 .0019	.0024 .0001	.0031 .0010	.0015 .0000	.0018 .0005
	.0010 .0000	.0011 .0003	.0006 .0000	.0007 .0002		
3/2	-.3556 .6992	-.2860 .4424	-.0571 .2625	-.1961 .1611	-.0798 .1316	-.1178 .0794
	-.1249 .1399	-.0459 .0887	.0341 .0843	.0044 .0348	.0560 .0678	-.0043 .0150
	.0108 .0057	.0033 .0009	.0171 .0068	.0008 .0014	.0036 .0009	.0004 .0002
	.0035 .0001	.0004 .0002	.0012 .0001	.0003 .0001	.0008 .0000	.0002 .0001
	.0004 .0000	.0001 .0001	.0003 .0000	.0001 .0000		
KIN:	2.150	2.223	4.943	.8233	.9180	1.3053
I=						
1/2	-.2478 .5803	-.0449 .4257	-.2105 .3232	-.0738 .2059	-.0725 .1424	-.1598 .2198
	-.0209 .0439	-.0567 .2201	.0028 .1252	.0250 .1921	.0124 .0243	.0401 .0470
	.0201 .0124	.0300 .0284	.0067 .0028	.0175 .0098	.0069 .0006	.0109 .0047
	.0035 .0004	.0058 .0021	.0026 .0001	.0034 .0011	.0016 .0001	.0021 .0006
	.0011 .0000	.0013 .0003	.0007 .0000	.0008 .0002		
3/2	-.3783 .6894	-.2891 .4382	-.0420 .2417	-.1792 .1972	-.0927 .1574	-.1119 .0668
	-.1217 .1239	-.0560 .0796	.0144 .0908	-.0017 .0456	.0504 .0762	-.0026 .0157
	.0134 .0067	.0075 .0040	.0198 .0082	.0034 .0023	.0051 .0011	.0021 .0005
	.0047 .0001	.0014 .0003	.0019 .0001	.0009 .0002	.0013 .0000	.0006 .0001
	.0007 .0000	.0004 .0001	.0004 .0000	.0003 .0000		
KIN:	2.205	2.246	5.046	.8483	.9315	1.3148
I=						
1/2	-.2326 .5215	-.0024 .3847	-.2272 .3105	-.0531 .2105	-.0636 .1416	-.1576 .2204
	-.0105 .0468	-.0877 .2003	-.0155 .1210	.0141 .1896	.0028 .0231	.0419 .0503
	.0202 .0185	.0312 .0349	.0071 .0048	.0187 .0126	.0075 .0015	.0118 .0061
	.0039 .0008	.0062 .0027	.0028 .0002	.0037 .0014	.0018 .0001	.0022 .0008
	.0012 .0001	.0014 .0004	.0008 .0000	.0009 .0003		
3/2	-.3878 .6482	-.2750 .4678	-.0364 .2485	-.1918 .1955	-.1093 .1567	-.1116 .0734
	-.1014 .1226	-.0582 .0897	.0240 .1032	.0007 .0489	.0534 .0777	.0006 .0209
	.0137 .0077	.0066 .0041	.0195 .0095	.0017 .0027	.0043 .0012	.0007 .0006
	.0043 .0002	.0005 .0004	.0014 .0001	.0002 .0003	.0010 .0000	.0001 .0001
	.0004 .0000	.0001 .0001	.0003 .0000	.0001 .0001		
KIN:	2.267	2.272	5.162	.8765	.9466	1.3255
I=						
1/2	-.2733 .4972	-.0426 .3527	-.2154 .3116	-.0624 .1648	-.0815 .1574	-.1155 .1896
	-.0375 .0337	-.0700 .2241	-.0260 .1380	.0027 .1883	.0191 .0377	.0436 .0548
	.0243 .0182	.0349 .0383	.0072 .0048	.0201 .0142	.0070 .0016	.0122 .0071
	.0030 .0008	.0061 .0031	.0021 .0002	.0034 .0016	.0012 .0001	.0019 .0009
	.0008 .0001	.0011 .0005	.0005 .0000	.0007 .0003		
3/2	-.3385 .6155	-.3050 .4572	-.0653 .2913	-.1867 .2076	-.0808 .1669	-.1095 .0895
	-.0901 .1325	-.0613 .0815	-.0010 .0971	-.0145 .0410	.0577 .0884	-.0119 .0209
	.0164 .0111	.0113 .0041	.0213 .0116	.0047 .0027	.0050 .0016	.0026 .0005
	.0050 .0004	.0016 .0004	.0016 .0001	.0008 .0003	.0011 .0000	.0005 .0001
	.0005 .0000	.0003 .0001	.0003 .0000	.0002 .0001		
KIN:	2.340	2.302	5.299	.9098	.9640	1.3380
I=						
1/2	-.2201 .5240	-.0356 .4180	-.2075 .3385	-.0673 .2166	-.0709 .1571	-.1564 .2256
	-.0124 .0620	-.0739 .2141	-.0266 .1188	-.0214 .1945	.0209 .0260	.0324 .0589
	.0185 .0191	.0286 .0394	.0040 .0045	.0177 .0148	.0058 .0016	.0114 .0076
	.0023 .0007	.0058 .0032	.0020 .0002	.0034 .0017	.0012 .0001	.0019 .0010
	.0009 .0001	.0012 .0005	.0006 .0000	.0007 .0003		
3/2	-.3616 .6419	-.2681 .4491	-.0364 .2534	-.1752 .2231	-.0858 .1779	-.1097 .0859
	-.0828 .1108	-.0659 .0770	-.0015 .1124	-.0105 .0514	.0447 .0919	-.0107 .0185
	.0182 .0096	.0102 .0063	.0231 .0133	.0038 .0032	.0059 .0017	.0022 .0006
	.0059 .0006	.0013 .0005	.0020 .0002	.0007 .0003	.0014 .0000	.0004 .0002
	.0006 .0000	.0002 .0001	.0004 .0000	.0001 .0001		

CMU-LBL 1980	K GEV/C	W GEV	S GEV**2	Q**2 (GEV/C)**2	ω_q GEV	E GEV
	S1	P1	P3	D3	D5	F5
	F7	G7	G9	H9	H11	I11
	I13	K13	K15	L15	L17	M17
	M19	N19	N21	O21	O23	P23
KIN:	2.420	2.334	5.449	.9463	.9827	1.3515
I= 1/2	-.2197 .4649	.0128 .3962	-.2218 .3324	-.0597 .2470	-.0850 .1518	-.1624 .2478
	-.0013 .0321	-.1095 .2055	-.0439 .1068	-.0421 .1522	.0154 .0460	.0354 .0520
	.0259 .0266	.0331 .0393	.0087 .0060	.0216 .0154	.0091 .0024	.0141 .0083
	.0041 .0010	.0074 .0034	.0032 .0003	.0045 .0018	.0019 .0002	.0026 .0011
	.0014 .0001	.0016 .0006	.0009 .0001	.0010 .0004		
3/2	-.3341 .5777	-.2609 .4707	-.0599 .2895	-.2019 .2145	-.1236 .1761	-.1248 .1109
	-.0652 .1445	-.0697 .0814	-.0010 .1232	-.0171 .0703	.0433 .1021	-.0139 .0211
	.0199 .0176	.0092 .0055	.0245 .0166	.0036 .0037	.0068 .0024	.0026 .0007
	.0070 .0010	.0017 .0006	.0025 .0002	.0010 .0003	.0018 .0000	.0007 .0002
	.0009 .0000	.0004 .0001	.0006 .0000	.0003 .0001		
KIN:	2.500	2.366	5.599	.9828	1.0012	1.3650
I= 1/2	-.2683 .5108	-.0733 .3715	-.2367 .3734	-.0641 .1902	-.1037 .1576	-.1223 .2408
	-.0189 .0494	-.0912 .2069	-.0548 .0931	-.0439 .1559	.0181 .0334	.0408 .0806
	.0273 .0232	.0370 .0475	.0091 .0047	.0240 .0179	.0095 .0021	.0155 .0097
	.0040 .0008	.0082 .0039	.0032 .0001	.0049 .0021	.0019 .0001	.0028 .0012
	.0013 .0001	.0017 .0007	.0008 .0000	.0011 .0004		
3/2	-.4058 .5707	-.2399 .4655	-.0634 .3267	-.2094 .2209	-.1288 .1664	-.1361 .1095
	-.0719 .1320	-.0680 .0755	-.0157 .1284	-.0219 .0700	.0363 .1209	.0028 .0153
	.0189 .0225	.0108 .0078	.0245 .0196	.0040 .0045	.0073 .0030	.0035 .0009
	.0081 .0014	.0024 .0007	.0031 .0003	.0016 .0004	.0024 .0000	.0012 .0002
	.0013 .0000	.0008 .0001	.0009 .0000	.0006 .0001		

Table 2.2.2.5 Hadronic phase shifts in degrees as obtained in the most recent analysis of the SIN group (82 L Alder). δ(P33) is slightly different for π^+p and π^-p scattering. The results of the analysis LONDON 73 (Sect. 2.1.6) were used as starting values.

T_π [MeV]	237.9	291.6	310.0
k [MeV/c]	350.7	408.0	427.4
δ(S11)	10.98 ± 0.49	10.95 ± 0.32	12.94 ± 0.78
δ(S31)	− 18.05 ± 0.32	− 20.62 ± 0.24	− 21.01 ± 0.32
δ(P11)	6.01 ± 0.40	13.27 ± 0.39	17.92 ± 0.86
δ(P13)	− 2.92 ± 0.36	− 3.81 ± 0.26	− 2.39 ± 0.81
δ(P31)	− 6.22 ± 0.49	− 8.26 ± 0.30	− 8.25 ± 0.37
δ(P33)π^+p	118.19 ± 0.27	133.06 ± 0.20	135.86 ± 0.21
δ(P33)π^-p	116.58 ± 0.48	131.11 ± 0.39	134.58 ± 0.99
δ(D13)	1.83 ± 0.28	3.08 ± 0.23	4.70 ± 0.62
δ(D15)	1.37 ± 0.36	1.39 ± 0.20	2.64 ± 0.42
δ(D33)	0.21 ± 0.30	0.47 ± 0.20	− 0.03 ± 0.24
δ(D35)	− 0.74 ± 0.33	− 0.39 ± 0.19	− 1.28 ± 0.26
δ(F15)	0.43 ± 0.20	0.54 ± 0.14	0.87 ± 0.33
δ(F17)	0.0	0.0	0.0
δ(F35)	0.0	0.0	0.0
δ(F37)	0.30	0.53	0.58
η(P11)	0.9952 ± 0.0025	0.9646 ± 0.0037	0.9544 ± 0.0049

2.2.3 Invariant amplitudes

The result of the KH 78 "fixed-t analysis" (Sect. 2.1.7) is listed in Tables 2.2.3.1 and 2.2.3.2. These amplitudes have the correct analytic properties in s at fixed t and the measurable quantities reconstructed from them fit the data.

Table 2.2.3.1 includes only s-values above threshold, i.e. the pion laboratory momentum k is real and positive. At $t<0$ the low-energy part of the table lies in the unphysical region beyond $\theta=180°$ (Fig. A.3.4). The k-values belonging to backward scattering are marked by arrows.

Due to the u-channel nucleon pole term, the real parts of the amplitudes B and C show a pole-like behaviour at the intersection of the lines $u=m^2$ and t=const in the Mandelstam diagram, Fig. A.3.4, i.e. at $s=-t+m^2+2\mu^2$, for instance at $t=-0.3$ GeV^2 and $k\approx 0.08$ GeV/c. Another less pronounced pole-like behaviour of the real parts of all amplitudes occurs at the location of the s-channel $\Delta(1232)$ resonance ($k=0.300$ GeV/c).

As discussed in Sects 2.4.7 and 2.5.1, it is also of great interest to know the continuation of the invariant amplitudes into the unphysical region below threshold, where they assume real values (if $t>-4m\mu$). It is convenient to subtract the well-known and rapidly varying contributions of the nucleon pole terms and to divide the crossing-odd amplitudes by ν. The amplitudes listed in Table 2.2.3.2 agree with those considered in the subthreshold expansion, Eq. (4.7.2), except for the difference between $\tilde{B}^-$, $\tilde{C}^-$ and $\bar{B}^-$ and $\bar{C}^-$ and a slightly different value of $\bar{A}^+=\bar{C}^+$ at $\nu=0$. This table includes ν^2-values somewhat beyond threshold ($\nu_{th}=\mu+t/4m$) and therefore, it shows the magnitude of the threshold cusps. It is also of interest, if one wants to search for the continuations of the zero trajectories into the unphysical region (if one adds to pole term contributions).

In order to present invariant amplitudes outside the strip considered in Table 2.2.3.1, we have added a table of amplitudes at fixed c.m. scattering angles as calculated from the KH 78 partial wave solution (Table 2.2.3.3).

Figure 2.3.1 shows an Argand diagram of the amplitude A^+ at $t=0$ as following from the fixed-t solution. Due to the low accuracy of the spin-rotation data, the information on the high energy behaviour of this amplitude is very poor. It is not even known, whether it goes in the asymptotic limit to zero, to infinity or to a finite value (Sects. 2.1.15 and 2.6.3.8).

Argand diagrams of invariant forward amplitudes are shown in Figs. 2.1.9···2.1.15. Many other Argand diagrams of invariant amplitudes at fixed values of t or of the c.m. scattering angle can be found in our "Handbook" (79 R Höhler-Kaiser-Koch-Pietarinen).

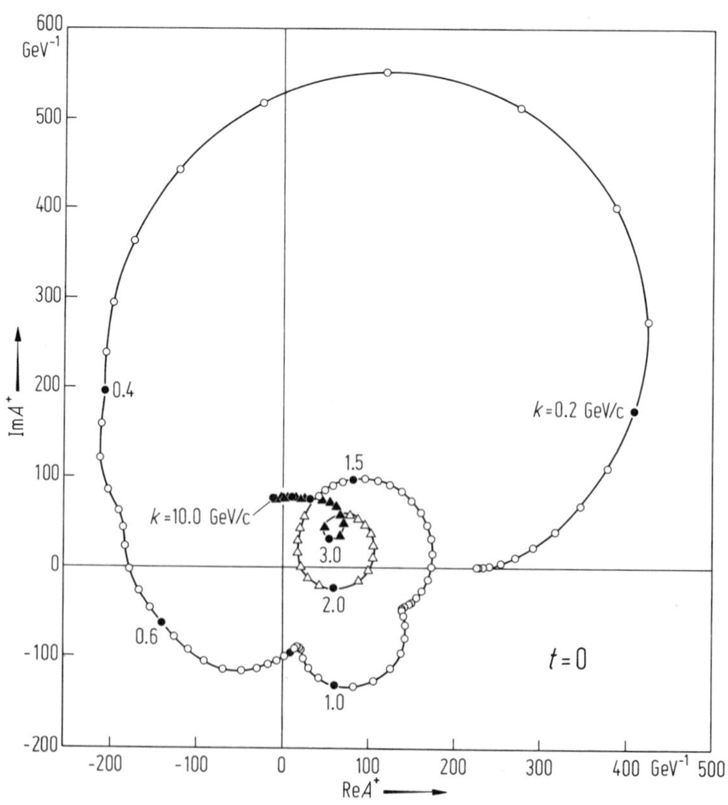

Fig. 2.3.1. Argand diagram of the invariant amplitude A^+ at $t=0$ as following from the KH 78 solution. The high energy behaviour of A^+ is not yet known (see Sect. 2.6.3.8).

Table 2.2.3.1. Result of the KH 78 fixed-t analysis above threshold (Sect. 2.1.7). The arrows mark the k-values in the first column which belong to $\theta=180°$. A sequence of stars indicates that the number is too large for the given format. The uncertainties are discussed in Sect. 2.4.5.

INVARIANT FIXED-T AMPLITUDES — ISOSPIN EVEN AND ODD:+- — T=0.00 GEV**2

K GEV/C	RE A+ 1/GEV	IM A+ 1/GEV	RE A- 1/GEV	IM A- 1/GEV	RE B+ 1/GEV**2	IM B+ 1/GEV**2	RE B- 1/GEV**2	IM B- 1/GEV**2	RE C+ 1/GEV	IM C+ 1/GEV	RE C- 1/GEV	IM C- 1/GEV
.00	227.3	-.0	-75.4	.0	-1639.	0.	608.1	-.0	-1.44	.00	9.51	-.00
.02	228.8	.1	-77.0	.4	-1631.	1.	612.0	-2.4	-1.26	.25	9.30	.07
.04	233.5	.6	-80.3	.2	-1611.	-2.	611.8	.0	-.32	.33	8.51	.16
.06	241.4	2.0	-87.5	1.0	-1583.	-8.	623.8	-5.4	1.01	.83	7.29	.18
.08	254.0	4.6	-98.6	-1.8	-1561.	-23.	647.9	12.5	2.94	.90	5.62	.17
.10	271.0	11.7	-109.4	-6.7	-1542.	-59.	658.0	39.0	6.16	1.53	3.54	-.02
.12	291.0	23.0	-123.2	-12.0	-1527.	-111.	674.7	63.2	9.83	2.67	.98	-.40
.14	316.4	40.8	-139.9	-21.4	-1524.	-184.	697.0	101.1	15.12	4.29	-2.12	-1.40
.16	345.0	69.5	-158.5	-35.0	-1523.	-290.	720.2	150.0	21.63	8.02	-5.57	-3.11
.18	375.9	111.4	-179.8	-56.0	-1523.	-429.	746.9	219.4	29.01	13.71	-9.67	-6.03
.20	407.1	176.3	-199.8	-89.1	-1515.	-626.	762.6	319.4	37.57	23.52	-13.79	-11.17
.22	421.1	275.8	-211.9	-137.3	-1447.	-908.	748.4	453.5	44.06	39.34	-16.90	-19.12
.24	384.2	403.0	-201.0	-200.6	-1223.	-1233.	663.9	613.2	44.65	60.74	-16.64	-30.36
.26	273.5	513.3	-151.4	-256.9	-812.	-1453.	480.0	728.0	33.78	84.38	-9.75	-42.08
.28	118.1	553.7	-79.6	-275.6	-343.	-1451.	263.5	724.4	10.67	99.69	2.79	-48.95
.30	-23.3	519.7	-18.3	-258.0	28.	-1267.	103.8	631.9	-14.19	100.57	16.08	-48.92
.32	-119.4	446.2	21.6	-223.1	249.	-1013.	13.5	512.5	-32.43	92.44	26.26	-44.23
.34	-172.1	366.0	39.8	-183.1	350.	-775.	-19.1	394.9	-43.51	81.13	32.78	-38.01
.36	-195.6	295.2	40.6	-149.3	381.	-584.	-10.6	304.1	-48.35	69.88	36.52	-31.91
.38	-202.6	238.8	34.8	-124.2	379.	-439.	9.0	240.5	-49.21	61.12	38.47	-26.83
.40	-204.3	196.1	23.8	-103.4	367.	-334.	38.1	190.1	-48.61	54.59	39.96	-22.86
.42	-208.0	159.5	7.6	-91.4	364.	-249.	76.3	163.6	-47.00	49.43	41.41	-19.02
.44	-209.2	120.6	-4.7	-88.2	356.	-162.	101.5	157.9	-44.73	45.78	42.17	-15.34
.46	-200.2	84.7	-12.8	-84.3	328.	-86.	115.7	149.5	-42.43	43.17	42.80	-12.41
.48	-187.3	60.6	-24.4	-80.9	295.	-39.	136.8	143.2	-39.87	41.26	44.00	-9.34
.50	-181.3	43.4	-35.3	-84.9	277.	-6.	154.1	153.0	-37.27	40.32	44.69	-5.43
.52	-180.3	22.4	-38.2	-90.1	270.	32.	153.0	163.9	-35.17	39.85	44.15	-1.82
.54	-175.4	-3.3	-39.4	-88.3	255.	76.	148.5	159.6	-33.04	39.19	43.43	.74
.56	-164.3	-27.5	-47.8	-84.7	232.	115.	157.8	152.0	-30.26	38.85	43.26	3.02
.58	-150.5	-47.2	-60.2	-88.7	206.	145.	173.2	158.0	-27.32	39.43	43.17	5.55
.60	-136.6	-63.8	-67.7	-99.8	182.	169.	179.7	175.0	-24.74	40.48	42.99	8.02
.62	-122.7	-79.3	-67.9	-111.3	158.	190.	174.8	191.9	-22.03	41.65	43.17	10.60
.64	-107.0	-94.3	-64.1	-119.7	135.	211.	164.2	204.2	-18.91	43.68	43.44	14.11
.66	-88.0	-107.4	-58.9	-125.0	106.	229.	150.6	213.0	-16.33	47.19	42.68	18.66
.68	-66.1	-115.5	-53.1	-128.0	73.	240.	134.3	217.8	-15.55	51.43	40.12	23.25
.70	-44.8	-117.2	-47.4	-128.9	39.	241.	116.9	218.0	-16.61	54.81	36.02	26.70

INVARIANT FIXED-T AMPLITUDES ISOSPIN EVEN AND ODD:+- T=0.00 GEV**2

K GEV/C	RE A+ 1/GEV	IM A+ 1/GEV	RE A- 1/GEV	IM A- 1/GEV	RE B+ 1/GEV**2	IM B+ 1/GEV**2	RE B- 1/GEV**2	IM B- 1/GEV**2	RE C+ 1/GEV	IM C+ 1/GEV	RE C- 1/GEV	IM C- 1/GEV
.72	-27.1	-114.1	-42.1	-128.2	12.	232.	100.0	213.6	-18.29	56.34	31.27	28.43
.74	-14.0	-109.4	-36.9	-126.1	-7.	220.	84.5	205.3	-19.27	56.27	26.72	28.42
.76	-3.9	-105.0	-32.0	-122.0	-19.	208.	71.3	192.8	-18.88	55.43	23.11	26.99
.78	4.5	-100.9	-28.5	-115.4	-27.	196.	62.4	176.8	-17.10	54.62	20.97	24.76
.80	11.3	-96.7	-28.1	-107.0	-31.	186.	59.7	159.5	-14.11	54.34	20.39	22.57
.82	16.0	-92.8	-31.4	-98.8	-31.	178.	63.0	144.1	-10.13	55.00	21.01	21.09
.84	19.0	-90.1	-37.7	-92.4	-29.	173.	70.4	132.7	-5.47	57.00	22.31	20.59
.86	21.1	-88.8	-45.6	-88.5	-25.	171.	79.8	125.8	-.64	60.64	23.93	21.10
.88	22.5	-88.8	-54.3	-87.3	-21.	174.	89.8	123.4	3.74	66.04	25.64	22.72
.90	23.2	-90.5	-63.2	-88.7	-18.	180.	99.1	125.6	6.94	73.06	27.07	25.69
.92	23.6	-94.9	-71.4	-93.1	-16.	189.	106.2	132.4	8.29	81.27	27.48	30.08
.94	25.5	-102.9	-77.5	-100.4	-19.	203.	108.9	142.9	7.28	89.88	26.00	35.42
.96	31.5	-114.1	-80.1	-109.4	-29.	218.	105.4	154.8	3.79	97.85	22.08	40.69
.98	43.8	-125.8	-78.5	-118.5	-46.	232.	95.4	164.8	-1.76	104.14	15.89	44.56
1.00	62.6	-134.0	-73.0	-125.5	-70.	240.	80.6	169.9	-8.46	107.97	8.33	46.03
1.02	85.5	-135.5	-65.1	-128.6	-98.	238.	64.0	168.4	-15.10	109.12	.73	44.77
1.04	108.3	-128.7	-56.9	-127.2	-123.	226.	48.8	160.5	-20.57	108.00	-5.72	41.21
1.06	127.1	-115.0	-50.6	-121.7	-141.	206.	37.7	147.8	-24.10	105.47	-10.31	36.27
1.08	139.3	-97.4	-47.7	-113.5	-151.	184.	31.9	132.7	-25.51	102.56	-12.94	30.96
1.10	144.5	-79.9	-48.9	-104.5	-153.	162.	31.5	117.7	-25.08	100.16	-13.97	26.04
1.12	144.6	-65.5	-54.2	-96.5	-149.	146.	35.6	104.9	-23.41	98.82	-13.98	21.91
1.14	142.3	-55.5	-62.6	-91.0	-142.	134.	42.7	95.4	-21.15	98.70	-13.51	18.59
1.16	140.4	-49.6	-72.6	-89.1	-136.	128.	51.0	89.9	-18.81	99.67	-12.94	15.90
1.18	140.6	-46.5	-82.6	-90.9	-132.	124.	59.0	88.0	-16.71	101.46	-12.43	13.59
1.20	143.8	-43.9	-91.1	-96.0	-131.	122.	65.5	89.0	-14.94	103.80	-11.99	11.47
1.22	149.6	-40.3	-96.9	-103.4	-133.	120.	69.5	91.8	-13.48	106.52	-11.56	9.40
1.24	156.9	-34.2	-99.7	-111.5	-136.	115.	71.0	95.3	-12.30	109.55	-11.07	7.33
1.26	164.2	-25.2	-99.4	-119.2	-139.	109.	70.2	98.2	-11.39	112.87	-10.45	5.26
1.28	170.5	-13.4	-96.8	-125.3	-141.	101.	67.7	99.8	-10.79	116.51	-9.68	3.19
1.30	174.5	.6	-93.0	-129.1	-142.	92.	64.5	99.6	-10.61	120.43	-8.72	1.15
1.32	175.8	15.9	-89.0	-130.5	-141.	82.	61.4	97.7	-10.97	124.54	-7.55	-.85
1.34	173.9	31.8	-85.8	-130.0	-138.	72.	59.2	94.5	-11.94	128.70	-6.12	-2.77
1.36	169.0	47.2	-84.1	-128.3	-134.	63.	58.3	90.5	-13.54	132.72	-4.44	-4.57
1.38	161.2	61.5	-84.1	-126.1	-128.	54.	58.8	86.5	-15.72	136.41	-2.49	-6.16
1.40	150.9	74.0	-85.5	-124.3	-120.	47.	60.6	83.0	-18.36	139.62	-.30	-7.50
1.42	138.8	84.2	-88.1	-123.4	-112.	41.	63.2	80.5	-21.29	142.24	2.08	-8.50
1.44	125.3	91.8	-91.2	-123.7	-103.	36.	66.2	79.2	-24.34	144.24	4.58	-9.13
1.46	111.2	96.6	-94.2	-125.3	-94.	33.	69.1	79.1	-27.33	145.64	7.11	-9.36
1.48	97.1	98.7	-96.5	-127.9	-86.	32.	71.4	79.9	-30.12	146.51	9.57	-9.21

INVARIANT FIXED-T AMPLITUDES ISOSPIN EVEN AND ODD:+- T=0.00 GEV**2

K GEV/C	RE A+ 1/GEV	IM A+ 1/GEV	RE A- 1/GEV	IM A- 1/GEV	RE B+ 1/GEV**2	IM B+ 1/GEV**2	RE B- 1/GEV**2	IM B- 1/GEV**2	RE C+ 1/GEV	IM C+ 1/GEV	RE C- 1/GEV	IM C- 1/GEV
1.50	83.6	98.4	-98.0	-131.1	-77.	32.	73.0	81.3	-32.61	146.95	11.90	-8.72
1.52	71.2	95.9	-98.5	-134.5	-69.	34.	73.7	82.9	-34.73	147.08	14.03	-7.93
1.54	60.2	91.7	-98.0	-137.7	-63.	36.	73.7	84.6	-36.45	147.01	15.92	-6.92
1.56	50.9	86.2	-96.7	-140.4	-57.	39.	73.0	85.9	-37.80	146.85	17.56	-5.76
1.58	43.1	80.0	-95.0	-142.3	-52.	42.	71.9	86.9	-38.79	146.68	18.97	-4.50
1.60	37.0	73.5	-93.2	-143.4	-47.6	45.5	70.6	87.3	-39.5	146.6	20.15	-3.19
1.65	27.0	57.3	-89.6	-143.7	-40.5	54.0	67.6	86.8	-40.1	146.7	22.33	.10
1.70	21.9	42.9	-88.7	-142.5	-36.2	61.4	65.9	85.5	-39.8	147.8	23.68	3.32
1.75	19.4	29.6	-89.7	-142.5	-33.3	68.4	65.0	84.9	-39.0	149.7	24.38	6.48
1.80	19.3	15.9	-90.4	-144.3	-31.8	75.6	63.6	85.2	-38.1	152.4	24.43	9.58
1.85	23.0	1.9	-89.3	-146.6	-32.5	82.9	61.0	85.8	-37.3	155.7	23.79	12.47
1.90	31.5	-11.1	-86.3	-147.8	-35.8	89.6	57.1	85.5	-36.7	159.6	22.54	14.96
1.95	44.6	-20.5	-82.4	-146.7	-41.5	94.3	52.8	83.7	-36.5	163.8	20.82	16.92
2.00	60.6	-24.6	-79.1	-142.8	-48.5	96.1	48.8	80.3	-36.6	168.1	18.84	18.27
2.05	77.1	-22.5	-77.7	-136.9	-55.5	94.9	46.0	75.9	-37.0	172.4	16.76	19.02
2.10	91.4	-14.7	-78.9	-130.0	-61.3	91.0	44.5	70.9	-37.6	176.8	14.73	19.21
2.15	101.7	-2.8	-82.9	-123.3	-65.0	85.3	44.4	66.0	-38.5	181.0	12.85	18.89
2.20	107.1	11.2	-89.1	-117.7	-66.5	78.9	45.5	61.6	-39.5	185.2	11.23	18.14
2.25	107.6	25.2	-97.0	-113.9	-65.7	72.8	47.4	58.1	-40.6	189.2	9.95	17.05
2.30	104.1	37.6	-105.8	-112.1	-63.3	67.5	49.9	55.5	-41.8	193.2	9.06	15.70
2.35	97.6	47.5	-114.9	-112.5	-59.7	63.5	52.5	53.8	-43.0	197.1	8.62	14.21
2.40	89.4	54.5	-123.5	-114.8	-55.7	60.9	54.9	53.1	-44.4	200.9	8.62	12.72
2.45	80.7	58.4	-131.2	-118.8	-51.5	59.5	57.1	53.0	-45.8	204.5	9.04	11.33
2.50	72.3	59.6	-137.7	-123.9	-47.7	59.2	58.9	53.5	-47.2	207.9	9.81	10.16
2.75	49.5	45.0	-150.5	-154.0	-37.4	64.8	60.3	59.1	-53.4	223.5	15.52	8.84
3.00	55.6	32.0	-144.0	-170.0	-37.7	68.7	54.0	60.7	-57.6	238.3	18.26	12.20
3.25	67.5	35.3	-140.7	-171.2	-39.5	67.1	48.7	57.1	-60.9	253.6	17.76	14.53
3.50	72.3	47.9	-145.9	-169.8	-39.0	63.1	46.6	52.7	-64.2	268.9	17.14	14.89
3.75	68.8	60.6	-155.0	-171.7	-36.2	59.5	46.0	49.7	-67.2	284.0	17.48	14.70
4.00	61.0	69.2	-163.9	-176.5	-32.7	57.4	45.5	47.8	-69.9	299.0	18.37	14.81
4.25	52.5	73.5	-171.3	-182.4	-29.3	56.6	44.8	46.5	-72.2	314.0	19.30	15.29
4.50	45.2	75.2	-177.7	-188.3	-26.5	56.4	43.9	45.4	-74.3	329.1	20.06	15.95
4.75	39.5	75.6	-183.4	-193.9	-24.3	56.5	42.9	44.3	-76.2	344.2	20.66	16.61
5.00	35.0	75.7	-188.9	-199.2	-22.6	56.7	42.0	43.3	-78.0	359.4	21.17	17.20
5.50	28.0	76.2	-199.3	-209.3	-19.9	57.0	40.2	41.4	-81.4	389.9	22.12	18.23
6.00	22.1	76.8	-209.0	-219.1	-17.8	57.3	38.7	39.7	-84.5	420.4	23.08	19.13

INVARIANT FIXED-T AMPLITUDES ISOSPIN EVEN AND ODD:+- T=-.10 GEV**2

K GEV/C	RE A+ 1/GEV	IM A+	RE A- 1/GEV	IM A-	RE B+ 1/GEV**2	IM B+	RE B- 1/GEV**2	IM B-	RE C+ 1/GEV	IM C+	RE C- 1/GEV	IM C-
.00	170.6	-.0	-47.4	.0	-2165.	0.	1080.4	-.0	-67.12	.00	71.24	-.00
.02	171.7	.1	-48.3	.2	-2143.	1.	1066.9	-.5	-66.67	.22	70.30	.11
.04	175.0	.6	-51.0	.0	-2085.	-4.	1029.8	2.1	-65.44	.17	67.69	.29
.06	180.5	1.8	-55.5	-.2	-2006.	-16.	980.6	7.1	-63.85	-.11	63.97	.71
.08	188.5	4.0	-62.0	-1.0	-1920.	-39.	932.4	17.9	-62.14	-1.07	59.66	1.35
.10	199.2	7.8	-70.4	-2.7	-1839.	-72.	891.0	35.0	-60.25	-2.41	55.27	2.19
.12	214.0	13.5	-81.1	-5.7	-1780.	-116.	863.4	59.2	-58.59	-4.31	51.03	3.32
.14	234.3	24.7	-94.2	-11.6	-1748.	-191.	848.6	98.5	-56.53	-7.09	46.96	4.77
.16	257.2	45.4	-108.6	-20.8	-1722.	-308.	839.0	151.6	-53.75	-10.19	42.88	6.54
.18	280.9	75.9	-125.5	-34.7	-1693.	-459.	838.3	221.2	-50.15	-13.81	38.50	8.53
→ .20	305.5	122.2	-143.2	-58.3	-1657.	-662.	835.9	325.2	-44.62	-17.68	33.41	10.40
.22	319.3	194.0	-154.6	-94.9	-1564.	-943.	800.5	467.3	-36.41	-20.44	27.49	11.40
.24	299.2	287.7	-148.9	-142.0	-1331.	-1260.	696.4	626.7	-25.70	-19.83	21.09	10.94
.26	225.9	376.5	-116.8	-186.2	-923.	-1495.	503.8	744.9	-15.08	-13.71	14.73	8.26
.28	110.8	419.3	-65.0	-207.1	-429.	-1517.	270.1	754.7	-8.64	-2.94	10.21	2.94
.30	-1.2	401.1	-15.1	-198.6	-24.	-1327.	82.4	661.0	-8.38	8.47	9.29	-3.07
.32	-77.9	349.2	17.4	-172.8	209.	-1060.	-20.6	527.8	-12.32	16.67	10.93	-7.33
.34	-123.4	291.9	32.0	-144.6	321.	-816.	-56.4	406.9	-17.11	21.34	13.29	-9.69
.36	-147.1	236.9	35.7	-120.6	359.	-610.	-57.2	314.4	-21.59	23.73	15.72	-10.72
.38	-153.7	192.1	33.3	-100.8	350.	-456.	-42.2	245.1	-25.09	24.24	17.83	-10.70
.40	-156.9	160.8	25.7	-85.9	337.	-354.	-16.3	196.1	-26.95	24.30	19.38	-10.18
.42	-164.0	131.1	16.3	-77.7	335.	-263.	10.7	168.5	-28.47	24.80	20.66	-9.54
.44	-165.1	97.9	9.0	-73.1	319.	-173.	30.4	152.2	-29.92	24.62	21.85	-8.70
.46	-157.6	70.7	1.4	-69.3	289.	-105.	48.7	139.7	-30.05	24.31	22.92	-7.60
.48	-151.0	51.2	-7.2	-69.6	263.	-56.	67.5	137.5	-30.02	25.33	23.80	-6.29
.50	-146.3	31.6	-12.0	-74.0	240.	-11.	76.1	144.5	-31.36	26.14	24.49	-4.84
.52	-137.8	12.6	-11.6	-76.9	212.	26.	73.8	148.1	-32.23	25.26	25.09	-3.24
.54	-127.7	-1.0	-11.3	-76.0	187.	50.	71.3	144.6	-31.17	24.70	25.50	-1.31
.56	-121.0	-12.0	-13.5	-74.8	170.	71.	72.7	141.4	-30.14	25.94	25.38	.89
.58	-115.4	-25.0	-16.0	-75.1	153.	94.	73.3	140.7	-30.66	27.25	24.63	2.90
.60	-107.0	-38.1	-18.0	-75.4	132.	114.	72.6	139.2	-31.31	27.15	23.61	4.40
.62	-98.3	-48.5	-21.2	-76.0	114.	127.	74.2	137.6	-30.71	26.63	22.76	5.52
.64	-91.1	-59.6	-24.9	-79.0	101.	142.	77.1	140.3	-29.38	26.93	22.22	6.68
.66	-81.1	-73.9	-26.3	-84.8	84.	161.	76.3	147.7	-28.20	27.88	21.75	8.26
.68	-64.1	-86.6	-23.9	-90.7	57.	178.	68.9	155.6	-27.17	29.13	20.83	10.37
.70	-42.9	-91.2	-18.9	-94.3	25.	183.	56.6	160.1	-26.36	30.84	18.88	12.61
.72	-25.1	-87.3	-12.8	-95.5	-2.	175.	41.7	159.5	-26.38	32.84	15.82	14.13
.74	-14.3	-80.7	-6.5	-94.1	-18.	163.	26.6	153.4	-27.35	34.13	12.30	14.18
.76	-8.1	-76.2	-.6	-90.0	-28.	152.	13.8	141.6	-28.30	34.00	9.38	12.77

INVARIANT FIXED-T AMPLITUDES ISOSPIN EVEN AND ODD:+- T=-.10 GEV**2

K GEV/C	RE A+ 1/GEV	IM A+ 1/GEV	RE A- 1/GEV	IM A- 1/GEV	RE B+ 1/GEV**2	IM B+ 1/GEV**2	RE B- 1/GEV**2	IM B- 1/GEV**2	RE C+ 1/GEV	IM C+ 1/GEV	RE C- 1/GEV	IM C- 1/GEV
.78	-2.4	-74.0	3.1	-83.1	-34.	144.	6.3	125.8	-28.00	32.87	7.74	10.61
.80	3.9	-71.6	3.3	-75.1	-39.	136.	5.3	109.6	-26.10	31.99	7.35	8.61
.82	9.2	-67.8	.3	-68.1	-41.	128.	9.5	96.2	-23.24	32.21	7.69	7.22
.84	12.1	-63.7	-4.7	-63.3	-40.	121.	16.2	87.0	-20.22	33.66	8.30	6.41
.86	12.7	-60.9	-10.0	-60.7	-37.	118.	23.1	81.3	-17.43	36.05	9.02	6.04
.88	12.3	-60.4	-15.1	-59.5	-32.	119.	29.8	78.0	-15.01	39.26	9.87	6.09
.90	12.0	-62.0	-20.3	-59.4	-29.	122.	36.1	76.9	-13.22	43.28	10.75	6.72
.92	12.5	-65.5	-25.3	-60.7	-28.	129.	41.7	78.1	-12.47	47.89	11.34	7.98
.94	14.6	-70.9	-29.7	-63.7	-31.	137.	45.7	81.7	-13.02	52.55	11.29	9.72
.96	19.2	-77.4	-32.3	-68.3	-37.	146.	46.6	87.0	-14.71	56.61	10.41	11.52
.98	27.3	-83.8	-32.4	-73.3	-47.	153.	43.9	92.1	-17.03	59.69	8.79	12.97
1.00	38.6	-88.1	-29.8	-77.3	-61.	157.	38.2	95.3	-19.40	61.82	6.73	13.79
1.02	51.8	-88.7	-25.5	-79.0	-75.	156.	30.8	95.3	-21.54	63.30	4.59	13.92
1.04	64.7	-85.1	-20.7	-77.9	-89.	150.	23.5	91.8	-23.41	64.40	2.64	13.48
1.06	75.5	-78.3	-17.1	-74.0	-99.	142.	17.9	85.6	-25.11	65.23	1.02	12.69
1.08	83.1	-70.1	-15.5	-68.4	-106.	131.	14.8	77.6	-26.66	65.71	-.24	11.73
1.10	88.1	-62.1	-16.6	-62.1	-110.	122.	14.6	69.2	-27.94	65.78	-1.22	10.74
1.12	91.4	-55.4	-20.3	-56.2	-112.	113.	17.1	61.6	-28.69	65.53	-1.97	9.78
1.14	94.4	-49.8	-26.1	-51.7	-113.	105.	21.5	55.6	-28.76	65.21	-2.58	8.88
1.16	97.8	-44.8	-33.2	-49.3	-113.	99.	27.1	51.7	-28.17	65.18	-3.11	8.04
1.18	101.6	-39.6	-40.7	-49.2	-114.	93.	32.9	50.0	-27.14	65.73	-3.59	7.24
1.20	105.6	-33.6	-47.9	-51.2	-115.	88.	38.1	50.2	-26.04	66.96	-4.05	6.46
1.22	109.3	-26.8	-53.9	-55.1	-115.	82.	42.3	52.0	-25.18	68.79	-4.52	5.66
1.24	112.1	-19.3	-58.4	-60.1	-115.	76.	45.0	54.6	-24.79	70.97	-4.98	4.79
1.26	114.1	-11.4	-61.1	-65.5	-115.	70.	46.2	57.4	-24.90	73.22	-5.40	3.83
1.28	115.2	-3.3	-62.3	-70.8	-115.	64.	46.2	60.0	-25.41	75.30	-5.73	2.76
1.30	115.6	4.8	-62.3	-75.4	-114.	58.	45.2	61.8	-26.12	77.11	-5.93	1.58
1.32	115.4	13.2	-61.5	-79.2	-112.	52.	43.9	62.9	-26.85	78.66	-5.95	.34
1.34	114.5	21.8	-60.3	-82.1	-111.	45.	42.5	63.2	-27.51	80.07	-5.76	-.93
1.36	112.5	30.9	-59.2	-84.2	-108.	39.	41.3	62.9	-28.05	81.44	-5.35	-2.15
1.38	109.1	40.3	-58.3	-85.8	-104.	32.	40.5	62.4	-28.54	82.88	-4.74	-3.27
1.40	104.0	49.5	-57.7	-87.1	-99.	26.	40.1	61.7	-29.06	84.43	-3.96	-4.26
1.42	97.1	58.0	-57.4	-88.4	-93.	21.	39.9	61.2	-29.72	86.05	-3.06	-5.07
1.44	88.4	65.2	-57.2	-89.9	-86.	16.	39.9	61.0	-30.57	87.68	-2.08	-5.70
1.46	78.4	70.6	-56.9	-91.5	-79.	13.	39.9	61.0	-31.62	89.21	-1.07	-6.15
1.48	67.7	73.8	-56.4	-93.2	-71.	12.	39.7	61.1	-32.82	90.56	-.06	-6.43
1.50	56.8	74.7	-55.6	-94.9	-63.	12.	39.3	61.4	-34.11	91.68	.91	-6.56
1.52	46.6	73.5	-54.5	-96.5	-56.	13.	38.7	61.7	-35.39	92.55	1.84	-6.56
1.54	37.3	70.4	-53.2	-97.7	-50.	15.	37.8	61.8	-36.58	93.21	2.71	-6.45

INVARIANT FIXED-T AMPLITUDES ISOSPIN EVEN AND ODD:+- T=-.10 GEV**2

K GEV/C	RE A+ 1/GEV	IM A+ 1/GEV	RE A- 1/GEV	IM A- 1/GEV	RE B+ 1/GEV**2	IM B+ 1/GEV**2	RE B- 1/GEV**2	IM B- 1/GEV**2	RE C+ 1/GEV	IM C+ 1/GEV	RE C- 1/GEV	IM C- 1/GEV
1.56	29.6	66.0	-51.7	-98.6	-45.	18.	36.9	61.7	-37.64	93.70	3.52	-6.26
1.58	23.3	60.9	-50.1	-99.1	-41.	22.	35.9	61.4	-38.54	94.08	4.26	-5.99
1.60	18.6	55.4	-48.6	-99.2	-37.7	25.4	34.9	60.9	-39.3	94.4	4.92	-5.66
1.65	11.9	42.7	-46.0	-98.3	-33.1	33.2	33.0	59.1	-40.6	95.3	6.31	-4.65
1.70	8.8	32.7	-45.2	-96.9	-30.8	39.0	32.2	57.3	-41.5	96.4	7.31	-3.46
1.75	6.4	23.8	-45.5	-96.4	-28.9	43.8	31.8	56.0	-42.1	97.4	7.96	-2.18
1.80	5.1	14.0	-45.7	-96.8	-27.4	48.8	31.2	55.5	-42.4	98.3	8.30	-.86
1.85	6.7	2.8	-45.1	-97.6	-27.3	54.3	30.0	55.2	-41.9	99.3	8.35	.47
1.90	12.7	-7.8	-43.5	-97.8	-29.3	59.6	28.2	54.5	-40.8	101.0	8.08	1.77
1.95	22.8	-15.5	-41.5	-96.8	-33.3	63.5	26.1	53.2	-39.7	103.5	7.48	2.98
2.00	35.0	-18.5	-40.1	-94.3	-38.5	65.1	24.2	51.1	-38.9	106.8	6.56	3.98
2.05	47.0	-16.5	-39.8	-90.8	-43.6	64.4	22.9	48.4	-38.9	110.5	5.38	4.67
2.10	56.8	-10.5	-41.1	-87.0	-47.7	61.6	22.3	45.5	-39.6	114.1	4.07	4.96
2.15	63.2	-2.2	-43.9	-83.4	-50.2	57.7	22.6	42.6	-40.7	117.2	2.76	4.84
2.20	66.2	6.7	-48.1	-80.8	-51.1	53.4	23.5	40.2	-42.0	119.8	1.61	4.32
2.25	66.4	15.1	-53.1	-79.4	-50.5	49.3	24.8	38.3	-43.0	122.0	.72	3.52
2.30	64.7	22.2	-58.4	-79.4	-48.9	46.0	26.4	37.0	-43.6	124.0	.16	2.53
2.35	61.8	28.1	-63.5	-80.8	-46.7	43.3	28.0	36.4	-44.0	126.2	-.06	1.50
2.40	58.2	32.7	-68.1	-83.3	-44.3	41.5	29.5	36.3	-44.1	128.6	.05	.53
2.45	54.3	36.2	-71.8	-86.7	-41.8	40.3	30.6	36.6	-44.3	131.3	.40	-.30
2.50	50.2	38.9	-74.6	-90.6	-39.4	39.6	31.4	37.2	-44.7	134.3	.94	-.93
2.75	28.8	40.0	-75.8	-108.6	-30.0	40.9	30.0	40.5	-50.9	148.4	3.81	-1.22
3.00	20.3	23.8	-69.1	-114.8	-26.6	45.8	25.5	39.8	-56.7	156.4	4.59	.28
3.25	32.0	12.9	-66.7	-113.8	-28.7	48.5	22.6	36.6	-58.0	165.0	4.21	1.05
3.50	46.0	19.2	-69.1	-112.5	-31.2	46.5	21.6	33.6	-59.6	176.3	3.89	1.19
3.75	49.3	33.5	-72.9	-113.2	-31.0	42.5	21.2	31.6	-63.0	187.4	3.82	1.18
4.00	43.4	44.8	-76.6	-115.1	-28.4	39.4	20.8	30.1	-66.6	197.2	3.88	1.22
4.25	34.5	49.9	-79.8	-117.3	-25.3	38.1	20.4	28.9	-69.3	206.4	3.95	1.31
4.50	26.9	50.5	-82.7	-119.5	-22.6	38.0	19.9	27.8	-71.6	215.9	4.02	1.42
4.75	21.9	49.1	-85.5	-121.6	-20.9	38.4	19.5	26.8	-74.0	225.5	4.07	1.54
5.00	18.9	47.6	-88.3	-123.8	-19.7	38.7	19.1	25.9	-76.7	235.0	4.13	1.65
5.50	15.6	45.9	-93.6	-128.1	-18.3	38.8	18.4	24.4	-82.0	252.4	4.22	1.88
6.00	13.2	45.4	-98.5	-132.5	-17.0	38.3	17.7	23.2	-85.3	268.2	4.30	2.08

INVARIANT FIXED-T AMPLITUDES ISOSPIN EVEN AND ODD:+- T=-.20 GEV**2

K GEV/C	RE A+ 1/GEV	IM A+ 1/GEV	RE A- 1/GEV	IM A- 1/GEV	RE B+ 1/GEV**2	IM B+ 1/GEV**2	RE B- 1/GEV**2	IM B- 1/GEV**2	RE C+ 1/GEV	IM C+ 1/GEV	RE C- 1/GEV	IM C- 1/GEV
.00	116.6	-.0	-25.7	.0	-5142.	0.	4037.1	-.0	-303.25	.00	303.95	-.00
.02	117.0	.5	-26.2	.2	-4892.	-3.	3795.2	-.7	-289.08	.29	288.76	.11
.04	118.4	.7	-27.8	.2	-4305.	-9.	3228.9	2.1	-256.03	-.13	253.01	.39
.06	121.2	1.1	-30.5	.5	-3663.	-22.	2615.8	6.8	-220.64	-.99	213.62	1.13
.08	125.6	1.5	-34.8	.5	-3124.	-45.	2109.9	18.7	-192.40	-3.13	179.98	2.36
.10	132.9	2.6	-40.3	-.1	-2721.	-84.	1735.8	37.7	-171.97	-6.84	154.24	4.16
.12	142.8	6.6	-47.6	-1.0	-2436.	-146.	1477.4	63.6	-158.58	-11.48	135.27	6.82
.14	154.1	13.3	-56.6	-4.1	-2231.	-235.	1302.1	107.2	-150.78	-18.88	121.36	10.55
.16	168.4	23.1	-66.3	-8.7	-2081.	-349.	1176.9	162.3	-144.74	-29.39	110.77	15.74
.18	186.0	39.8	-79.2	-15.8	-1973.	-494.	1096.8	231.3	-139.75	-41.86	101.89	22.36
.20	204.6	67.0	-93.4	-30.4	-1880.	-701.	1038.7	339.6	-134.51	-59.35	93.90	30.87
.22	219.4	109.8	-104.1	-53.4	-1741.	-978.	958.6	488.2	-122.03	-82.07	83.86	42.39
.24	217.7	171.2	-106.0	-85.2	-1484.	-1292.	817.4	653.9	-97.35	-103.10	67.52	53.59
.26	180.5	239.7	-89.1	-119.6	-1059.	-1542.	591.2	777.2	-61.85	-113.17	46.14	58.26
.28	105.7	283.0	-55.1	-139.3	-531.	-1580.	326.8	787.1	-24.62	-105.17	25.13	54.03
→ .30	25.2	280.9	-20.3	-137.9	-91.	-1384.	114.3	687.8	1.38	-82.50	9.71	42.73
.32	-32.6	252.3	3.7	-123.5	160.	-1105.	-4.6	546.7	12.23	-57.03	2.40	29.54
.34	-71.5	217.4	15.7	-106.4	285.	-852.	-50.6	422.1	13.23	-36.10	.65	19.07
.36	-94.7	178.6	20.9	-91.6	331.	-633.	-62.8	327.4	9.50	-20.88	1.08	11.49
.38	-101.7	146.3	21.5	-78.0	319.	-472.	-56.5	253.0	4.37	-10.68	2.74	6.18
.40	-107.4	125.6	17.0	-67.8	305.	-369.	-35.8	202.0	-.42	-3.70	4.47	3.00
.42	-116.7	103.3	12.1	-63.0	304.	-277.	-17.5	173.1	-4.75	1.32	5.62	.73
.44	-119.2	77.8	8.5	-59.2	287.	-189.	-4.4	150.1	-8.42	4.89	6.80	-1.23
.46	-116.2	57.8	2.8	-56.8	259.	-124.	13.3	133.9	-11.56	7.84	8.19	-2.65
.48	-114.5	40.4	-2.0	-59.3	236.	-72.	27.6	132.1	-14.87	10.07	9.70	-3.49
.50	-109.7	21.7	-1.7	-62.5	209.	-25.	29.2	133.8	-17.78	10.83	11.17	-3.53
.52	-100.0	7.6	-.5	-62.2	176.	7.	27.1	129.0	-19.14	10.93	11.99	-2.98
.54	-92.2	-.3	-1.4	-61.7	152.	25.	28.6	123.8	-19.47	11.82	12.27	-2.65
.56	-88.4	-8.4	-1.5	-63.0	138.	44.	29.0	122.1	-20.09	13.48	12.82	-2.48
.58	-83.5	-18.4	-.1	-63.2	121.	65.	26.4	119.7	-21.25	15.02	13.44	-1.65
.60	-76.6	-26.6	.1	-61.7	101.	80.	24.7	114.9	-22.59	16.20	13.23	-.48
.62	-71.3	-32.6	-1.6	-61.3	86.	90.	25.6	111.2	-24.06	16.98	12.46	-.02
.64	-67.9	-40.4	-3.2	-63.1	75.	101.	27.0	110.6	-25.41	17.04	12.23	-.16
.66	-62.4	-51.3	-3.1	-66.2	62.	115.	26.8	112.8	-25.95	16.56	12.70	.15
.68	-52.0	-61.8	-1.0	-69.5	44.	129.	23.1	116.9	-25.52	16.45	12.95	1.36
.70	-38.2	-68.2	3.2	-72.2	21.	137.	14.6	120.2	-24.98	17.28	12.29	2.85
.72	-24.5	-69.7	9.2	-72.9	-1.	137.	2.6	119.3	-25.20	18.49	10.91	3.87
.74	-13.1	-68.1	15.7	-70.5	-20.	132.	-9.6	112.9	-26.02	19.05	9.34	4.21
.76	-3.8	-65.4	20.7	-65.5	-33.	123.	-18.8	102.2	-26.46	18.68	7.92	4.05

INVARIANT FIXED-T AMPLITUDES ISOSPIN EVEN AND ODD:+- T=-.20 GEV**2

K GEV/C	RE A+ 1/GEV	IM A+ 1/GEV	RE A- 1/GEV	IM A- 1/GEV	RE B+ 1/GEV**2	IM B+ 1/GEV**2	RE B- 1/GEV**2	IM B- 1/GEV**2	RE C+ 1/GEV	IM C+ 1/GEV	RE C- 1/GEV	IM C- 1/GEV
.78	4.1	-61.9	23.5	-59.2	-43.	114.	-23.9	89.7	-25.86	18.08	6.78	3.57
.80	10.6	-57.6	24.1	-52.6	-49.	105.	-25.3	77.3	-24.45	18.07	5.94	2.95
.82	15.3	-52.6	22.7	-46.4	-51.7	97.0	-23.6	66.1	-22.84	18.89	5.36	2.32
.84	17.8	-47.9	19.9	-41.2	-52.0	90.3	-19.8	56.8	-21.47	20.30	4.91	1.73
.86	19.0	-44.5	16.2	-37.5	-50.9	86.0	-15.1	49.8	-20.38	22.03	4.53	1.10
.88	19.7	-42.4	12.4	-35.0	-49.5	83.8	-10.2	44.6	-19.53	24.02	4.31	.35
.90	20.5	-41.2	8.8	-33.4	-48.7	83.1	-5.4	40.5	-19.02	26.29	4.41	-.46
.92	21.3	-40.4	5.2	-32.3	-48.5	83.2	-.4	37.5	-19.01	28.69	4.90	-1.13
.94	21.9	-40.3	1.6	-31.9	-48.8	83.9	4.9	35.9	-19.53	30.94	5.68	-1.40
.96	22.7	-41.3	-1.9	-32.5	-49.7	85.4	9.6	36.2	-20.39	32.79	6.47	-1.18
.98	24.6	-43.2	-4.4	-34.1	-51.8	87.4	12.9	37.9	-21.28	34.23	7.02	-.54
1.00	28.2	-45.3	-5.5	-36.0	-55.4	89.3	14.0	40.1	-22.01	35.44	7.18	.30
1.02	33.2	-46.5	-5.3	-37.1	-60.4	90.0	13.3	41.4	-22.62	36.65	6.96	1.10
1.04	38.8	-46.1	-4.6	-37.0	-65.9	89.1	11.7	41.1	-23.28	37.93	6.48	1.73
1.06	44.1	-44.2	-4.2	-35.5	-71.0	86.7	10.5	39.2	-24.15	39.16	5.89	2.13
1.08	48.5	-41.5	-4.8	-33.3	-75.2	83.3	10.3	36.4	-25.21	40.16	5.32	2.35
1.10	52.1	-38.7	-6.6	-30.9	-78.5	79.6	11.4	33.5	-26.29	40.78	4.83	2.48
1.12	55.4	-36.1	-9.3	-29.0	-81.2	75.8	13.5	31.1	-27.17	41.04	4.41	2.58
1.14	58.9	-33.5	-12.6	-27.8	-83.6	72.0	16.0	29.4	-27.69	41.13	4.01	2.73
1.16	62.7	-30.5	-16.3	-27.2	-85.8	68.1	18.8	28.5	-27.84	41.27	3.58	2.92
1.18	66.5	-26.7	-20.1	-27.3	-87.8	63.6	21.5	28.3	-27.74	41.62	3.06	3.12
1.20	69.9	-22.1	-24.0	-28.1	-89.2	58.8	24.2	28.7	-27.59	42.24	2.43	3.29
1.22	72.4	-16.8	-28.0	-29.6	-89.9	53.9	26.7	29.7	-27.54	43.05	1.70	3.35
1.24	74.0	-11.4	-31.7	-32.1	-89.9	49.0	28.9	31.3	-27.64	43.91	.93	3.27
1.26	74.9	-6.2	-34.9	-35.4	-89.5	44.3	30.5	33.4	-27.86	44.68	.16	3.02
1.28	75.5	-1.2	-37.1	-39.4	-88.7	39.9	31.3	36.0	-28.08	45.32	-.55	2.64
1.30	76.1	3.7	-38.3	-43.6	-87.9	35.5	31.3	38.6	-28.18	45.84	-1.15	2.14
1.32	76.6	9.0	-38.4	-47.8	-86.8	31.0	30.5	40.9	-28.10	46.36	-1.64	1.59
1.34	76.7	14.9	-37.6	-51.4	-85.4	26.2	29.0	42.8	-27.82	46.99	-2.01	1.01
1.36	76.1	21.6	-36.2	-54.3	-83.2	21.1	27.3	44.1	-27.43	47.84	-2.30	.43
1.38	74.1	28.7	-34.6	-56.6	-80.1	16.0	25.4	44.7	-27.06	48.95	-2.51	-.13
1.40	70.5	35.9	-33.0	-58.1	-76.0	11.3	23.7	44.9	-26.82	50.30	-2.68	-.67
1.42	65.1	42.3	-31.6	-59.3	-70.8	7.3	22.2	44.7	-26.82	51.81	-2.80	-1.22
1.44	58.3	47.5	-30.4	-60.2	-64.8	4.4	20.9	44.3	-27.10	53.37	-2.87	-1.78
1.46	50.5	51.0	-29.4	-61.1	-58.5	2.9	19.8	43.9	-27.65	54.85	-2.88	-2.35
1.48	42.4	52.4	-28.5	-62.0	-52.2	2.8	18.9	43.5	-28.41	56.16	-2.82	-2.93
1.50	34.7	52.0	-27.4	-62.8	-46.5	3.8	18.0	43.2	-29.30	57.25	-2.67	-3.49
1.52	27.9	49.9	-26.3	-63.6	-41.7	5.8	17.1	42.7	-30.21	58.10	-2.43	-4.02
1.54	22.3	46.8	-25.0	-64.2	-37.8	8.4	16.2	42.3	-31.07	58.75	-2.11	-4.49

INVARIANT FIXED-T AMPLITUDES ISOSPIN EVEN AND ODD:+- T=-.20 GEV**2

K GEV/C	RE A+ 1/GEV	IM A+ 1/GEV	RE A- 1/GEV	IM A- 1/GEV	RE B+ 1/GEV**2	IM B+ 1/GEV**2	RE B- 1/GEV**2	IM B- 1/GEV**2	RE C+ 1/GEV	IM C+ 1/GEV	RE C- 1/GEV	IM C- 1/GEV
1.56	18.1	43.2	-23.6	-64.5	-34.8	11.2	15.3	41.7	-31.81	59.24	-1.71	-4.87
1.58	15.1	39.5	-22.2	-64.6	-32.8	13.9	14.4	40.9	-32.42	59.64	-1.28	-5.16
1.60	13.2	36.1	-20.9	-64.3	-31.3	16.3	13.7	40.1	-32.9	60.0	-.83	-5.35
1.65	10.5	29.9	-18.9	-62.6	-29.1	20.6	12.6	37.7	-33.6	61.1	.19	-5.43
1.70	7.8	25.8	-18.5	-60.9	-26.9	23.5	12.4	35.6	-34.2	62.5	.90	-5.16
1.75	4.2	20.7	-19.2	-60.2	-24.3	26.8	12.7	34.4	-35.0	63.9	1.30	-4.82
1.80	2.1	13.2	-19.5	-60.6	-22.8	31.3	12.7	33.8	-35.8	65.1	1.53	-4.54
1.85	3.5	4.8	-18.9	-61.1	-23.3	35.9	12.1	33.3	-36.2	66.1	1.71	-4.30
1.90	8.8	-1.9	-17.8	-61.0	-25.7	39.4	11.2	32.5	-36.4	67.1	1.87	-4.03
1.95	16.0	-5.2	-16.7	-60.1	-29.0	40.9	10.4	31.3	-36.2	68.4	1.97	-3.70
2.00	23.0	-4.9	-16.3	-58.4	-32.0	40.6	9.9	29.8	-36.0	70.0	1.95	-3.33
2.05	28.1	-2.3	-16.7	-56.7	-33.9	39.2	9.8	28.3	-36.0	71.9	1.79	-2.98
2.10	31.1	1.0	-17.8	-55.1	-34.7	37.5	10.0	27.0	-36.3	73.8	1.54	-2.72
2.15	32.7	3.9	-19.4	-54.0	-34.9	36.1	10.4	25.9	-36.7	75.7	1.23	-2.56
2.20	33.8	6.0	-21.2	-53.3	-34.8	35.1	10.9	25.0	-37.1	77.4	.92	-2.50
2.25	35.2	7.7	-23.1	-53.0	-34.9	34.3	11.4	24.3	-37.5	79.1	.63	-2.51
2.30	37.1	9.8	-25.1	-53.1	-35.2	33.3	12.0	23.7	-37.8	80.8	.39	-2.57
2.35	39.1	12.8	-27.1	-53.6	-35.4	32.0	12.5	23.4	-38.0	82.5	.17	-2.65
2.40	40.7	16.9	-29.1	-54.5	-35.5	30.3	13.1	23.3	-38.3	84.3	-.02	-2.75
2.45	41.0	22.1	-30.9	-55.8	-35.0	28.3	13.5	23.3	-38.6	86.3	-.20	-2.86
2.50	39.6	27.6	-32.4	-57.6	-33.9	26.2	13.8	23.5	-39.1	88.3	-.37	-3.00
2.75	13.4	38.8	-32.5	-67.7	-22.3	22.9	12.4	24.9	-43.4	97.3	-.73	-4.04
3.00	1.0	16.5	-26.3	-69.7	-17.2	31.0	9.4	23.3	-47.0	103.1	-.05	-4.68
3.25	17.2	4.4	-25.5	-66.2	-21.7	34.7	8.6	20.4	-48.5	109.5	.55	-4.33
3.50	30.7	14.7	-28.7	-65.4	-24.8	31.3	8.9	18.9	-50.3	117.0	.44	-3.79
3.75	28.6	29.0	-31.0	-67.2	-23.3	27.2	8.9	18.2	-52.9	124.1	.02	-3.60
4.00	18.8	35.3	-31.5	-69.0	-19.9	25.5	8.4	17.5	-55.4	130.8	-.32	-3.71
4.25	10.2	34.1	-31.8	-69.5	-17.1	26.0	7.9	16.5	-57.8	137.3	-.49	-3.90
4.50	6.2	30.0	-32.7	-69.5	-15.8	27.0	7.6	15.6	-60.3	143.8	-.55	-4.04
4.75	5.6	26.7	-34.2	-69.7	-15.4	27.7	7.6	14.8	-63.1	149.9	-.58	-4.12
5.00	6.3	25.4	-36.0	-70.4	-15.4	27.8	7.6	14.2	-65.9	155.4	-.61	-4.17
5.50	6.6	26.8	-39.0	-73.0	-15.0	26.7	7.4	13.3	-70.6	164.5	-.72	-4.25
6.00	3.8	28.5	-40.6	-75.5	-13.6	25.6	7.1	12.6	-72.5	172.5	-.81	-4.34

INVARIANT FIXED-T AMPLITUDES ISOSPIN EVEN AND ODD:+- T=-.30 GEV**2

K GEV/C	RE A+ 1/GEV	IM A+ 1/GEV	RE A- 1/GEV	IM A- 1/GEV	RE B+ 1/GEV**2	IM B+ 1/GEV**2	RE B- 1/GEV**2	IM B- 1/GEV**2	RE C+ 1/GEV	IM C+ 1/GEV	RE C- 1/GEV	IM C- 1/GEV
.00	58.4	.0	-9.4	-.0	2239.	0.	-3298.0	-.0	181.44	.00	-190.62	-.00
.02	58.4	.2	-9.4	-.1	2388.	-2.	-3443.0	3.9	192.75	.15	-203.12	.09
.04	58.2	.2	-9.8	.4	2925.	-6.	-3964.0	1.4	234.04	-.20	-248.17	.45
.06	58.2	-.7	-11.2	.9	4303.	-19.	-5312.4	7.1	343.67	-1.89	-363.62	1.34
.08	59.3	-2.2	-12.7	1.2	9219.	-43.	*******	25.7	746.99	-5.41	-773.85	3.08
.10	66.0	-3.5	-18.8	2.3	-69186.	-85.	68219.4	42.6	*******	-10.73	5750.20	5.89
.12	67.4	-4.5	-20.3	3.1	-7946.	-149.	7005.8	70.2	-695.13	-18.79	651.95	9.87
.14	73.5	-5.0	-24.9	2.5	-4536.	-241.	3614.9	122.9	-418.71	-31.09	367.39	15.80
.16	82.5	-3.9	-30.0	3.0	-3359.	-360.	2449.3	176.9	-327.37	-47.74	268.79	24.55
.18	94.2	.6	-39.6	2.3	-2784.	-516.	1900.7	246.2	-285.04	-69.70	219.35	35.85
.20	108.1	9.8	-50.4	-3.9	-2433.	-730.	1586.3	363.0	-259.55	-100.48	189.26	50.95
.22	124.1	26.5	-60.4	-13.9	-2142.	-1010.	1348.5	515.5	-232.35	-141.57	164.01	71.90
.24	138.1	57.5	-68.9	-31.5	-1789.	-1332.	1105.6	688.7	-187.85	-185.23	132.50	93.99
.26	134.2	103.8	-65.6	-55.0	-1291.	-1593.	798.4	817.9	-121.72	-212.12	92.66	107.14
.28	100.5	145.0	-50.3	-71.7	-703.	-1635.	475.2	820.2	-50.35	-205.96	51.65	104.38
.30	52.7	159.6	-32.1	-78.3	-216.	-1435.	226.1	716.4	2.85	-172.25	20.20	87.42
.32	13.0	153.2	-15.4	-76.0	70.	-1144.	74.5	573.8	30.28	-130.47	3.06	66.28
.34	-16.3	139.3	-5.5	-68.6	209.	-880.	3.3	441.5	39.07	-93.86	-4.57	48.40
.36	-37.6	120.5	-.2	-62.2	268.	-661.	-25.1	341.3	37.98	-65.92	-7.30	34.07
.38	-48.9	101.7	3.0	-54.9	273.	-492.	-32.7	261.8	32.75	-45.76	-6.75	23.48
.40	-56.3	88.2	.9	-48.7	260.	-378.	-20.3	205.6	26.19	-31.39	-5.58	16.42
.42	-65.3	75.2	-1.7	-48.1	254.	-289.	-8.5	176.0	19.70	-21.40	-4.56	10.73
.44	-71.8	59.2	-1.4	-47.2	245.	-208.	-4.3	151.8	14.29	-14.15	-2.88	6.18
.46	-74.0	43.8	-3.7	-45.8	226.	-141.	6.8	132.1	9.45	-8.07	-1.21	2.97
.48	-74.7	29.7	-5.7	-49.4	204.	-85.	16.1	128.3	4.34	-3.40	.48	.23
.50	-72.5	15.1	-2.0	-52.8	178.	-39.	11.8	126.8	-.32	-.65	2.77	-1.46
.52	-66.3	3.3	1.7	-51.2	148.	-5.	6.5	117.2	-3.70	1.03	4.41	-1.63
.54	-59.9	-4.0	1.8	-49.9	122.	15.	6.8	109.2	-6.18	2.57	4.81	-1.82
.56	-55.4	-9.9	3.0	-50.9	103.	31.	5.2	105.4	-8.25	4.08	5.39	-2.59
.58	-50.9	-15.8	5.8	-50.1	86.	45.	1.4	99.6	-10.08	5.65	6.41	-2.70
.60	-46.1	-20.5	6.5	-48.1	69.	56.	.1	93.1	-12.15	7.27	6.60	-2.11
.62	-42.5	-24.0	6.2	-47.9	54.	63.	-.4	89.2	-14.67	8.29	5.97	-2.19
.64	-40.0	-28.5	7.2	-48.5	44.	69.	-2.5	85.7	-16.96	8.23	5.86	-3.07
.66	-36.9	-34.4	8.6	-48.2	34.	77.	-3.6	81.3	-18.26	7.56	6.57	-3.62
.68	-32.1	-40.8	9.0	-48.3	24.	85.	-3.1	79.3	-18.63	7.10	7.23	-3.38
.70	-25.1	-46.9	10.3	-50.0	11.	92.	-5.0	80.6	-18.66	7.13	7.37	-2.93
.72	-16.0	-51.2	14.2	-51.7	-5.	97.	-11.5	81.4	-18.68	7.39	7.30	-2.68
.74	-5.7	-52.3	19.8	-50.7	-21.	97.	-20.2	77.9	-18.64	7.70	7.29	-2.41
.76	3.4	-50.0	24.6	-47.0	-34.	91.	-27.3	70.6	-18.45	8.08	7.15	-1.95

INVARIANT FIXED-T AMPLITUDES ISOSPIN EVEN AND ODD:+- T=-.30 GEV**2

K GEV/C	RE A+ 1/GEV	IM A+ 1/GEV	RE A- 1/GEV	IM A- 1/GEV	RE B+ 1/GEV**2	IM B+ 1/GEV**2	RE B- 1/GEV**2	IM B- 1/GEV**2	RE C+ 1/GEV	IM C+ 1/GEV	RE C- 1/GEV	IM C- 1/GEV
.78	9.8	-46.1	27.4	-42.0	-43.	83.	-31.6	61.8	-18.11	8.58	6.68	-1.48
.80	14.2	-42.4	28.5	-37.0	-47.	77.	-33.4	52.9	-17.66	9.27	5.96	-1.24
.82	17.8	-39.5	28.5	-32.2	-50.5	71.8	-33.6	44.7	-17.17	10.21	5.19	-1.23
.84	21.4	-36.5	27.5	-28.0	-53.7	67.4	-32.5	37.4	-16.78	11.44	4.38	-1.40
.86	24.4	-32.8	26.1	-24.4	-56.3	62.7	-31.0	31.0	-16.65	12.87	3.48	-1.81
.88	26.2	-28.9	24.5	-21.4	-57.6	57.8	-29.3	25.1	-16.86	14.29	2.59	-2.66
.90	26.8	-25.4	22.8	-18.6	-57.6	53.5	-27.2	19.0	-17.30	15.54	2.04	-4.02
.92	26.9	-22.7	20.9	-15.7	-57.1	50.1	-23.9	12.9	-17.83	16.56	2.16	-5.63
.94	26.8	-20.5	18.4	-13.0	-56.3	47.2	-19.1	7.4	-18.31	17.39	3.05	-7.09
.96	26.5	-18.6	15.4	-10.9	-55.1	44.7	-13.3	3.5	-18.70	18.12	4.49	-8.01
.98	25.7	-17.3	12.1	-9.5	-53.3	43.0	-7.1	1.6	-19.00	18.80	6.13	-8.23
1.00	24.8	-16.9	8.9	-8.8	-51.4	42.4	-1.5	1.2	-19.20	19.46	7.64	-7.83
1.02	24.5	-17.3	6.0	-8.7	-50.0	42.8	3.2	1.9	-19.30	20.18	8.81	-7.00
1.04	25.0	-18.2	3.3	-8.8	-49.6	43.9	7.0	3.2	-19.33	21.03	9.61	-5.95
1.06	26.4	-18.8	1.0	-9.3	-50.3	44.8	10.0	4.8	-19.41	22.07	10.09	-4.84
1.08	28.3	-18.8	-1.1	-9.8	-51.6	45.2	12.2	6.6	-19.66	23.25	10.31	-3.71
1.10	30.2	-18.2	-2.8	-10.4	-53.1	45.0	13.8	8.2	-20.18	24.47	10.31	-2.61
1.12	31.7	-17.4	-4.2	-10.8	-54.5	44.4	14.9	9.6	-20.98	25.56	10.12	-1.54
1.14	33.1	-16.6	-5.8	-11.0	-56.0	43.7	15.7	10.6	-21.98	26.41	9.73	-.55
1.16	34.7	-16.0	-7.6	-11.0	-57.6	42.8	16.7	11.3	-23.04	26.94	9.18	.35
1.18	36.7	-15.4	-9.8	-11.2	-59.5	41.7	17.9	12.0	-24.04	27.17	8.50	1.10
1.20	39.1	-14.3	-12.4	-11.9	-61.6	39.9	19.3	13.1	-24.87	27.17	7.73	1.70
1.22	41.7	-12.5	-14.9	-13.2	-63.5	37.4	20.7	14.5	-25.50	27.01	6.92	2.15
1.24	44.2	-10.0	-17.2	-15.3	-65.1	34.1	21.7	16.5	-25.92	26.77	6.10	2.46
1.26	46.1	-6.8	-18.9	-17.8	-66.0	30.4	22.1	18.7	-26.13	26.50	5.31	2.66
1.28	47.4	-3.2	-19.9	-20.6	-66.1	26.4	22.0	21.0	-26.15	26.24	4.54	2.78
1.30	48.0	.6	-20.2	-23.2	-65.4	22.5	21.2	23.0	-25.96	26.02	3.79	2.82
1.32	48.1	4.3	-20.0	-25.6	-64.1	18.8	20.0	24.7	-25.58	25.88	3.06	2.80
1.34	47.7	7.9	-19.5	-27.5	-62.3	15.4	18.7	25.9	-25.04	25.90	2.35	2.72
1.36	46.9	11.5	-18.9	-29.1	-60.1	12.3	17.3	26.7	-24.36	26.12	1.65	2.57
1.38	45.7	15.0	-18.2	-30.5	-57.6	9.6	15.9	27.3	-23.64	26.57	.98	2.35
1.40	44.0	18.5	-17.6	-31.8	-54.7	7.1	14.7	27.7	-22.94	27.26	.35	2.06
1.42	41.6	21.8	-16.9	-32.9	-51.5	5.1	13.4	27.9	-22.34	28.16	-.24	1.71
1.44	38.6	24.7	-16.1	-34.1	-48.1	3.6	12.2	28.1	-21.90	29.21	-.76	1.29
1.46	35.1	27.0	-15.2	-35.1	-44.4	2.6	11.0	28.1	-21.66	30.35	-1.22	.83
1.48	31.2	28.6	-14.1	-35.9	-40.8	2.3	9.7	27.9	-21.61	31.51	-1.59	.34
1.50	27.3	29.2	-13.0	-36.4	-37.3	2.6	8.4	27.6	-21.72	32.61	-1.89	-.17
1.52	23.4	29.0	-11.8	-36.7	-34.1	3.5	7.3	27.1	-21.95	33.64	-2.12	-.68
1.54	20.0	28.1	-10.7	-36.8	-31.3	4.7	6.3	26.3	-22.26	34.56	-2.28	-1.18

INVARIANT FIXED-T AMPLITUDES ISOSPIN EVEN AND ODD:+- T=-.30 GEV**2

K GEV/C	RE A+ 1/GEV	IM A+	RE A- 1/GEV	IM A-	RE B+ 1/GEV**2	IM B+	RE B- 1/GEV**2	IM B-	RE C+ 1/GEV	IM C+	RE C- 1/GEV	IM C-
1.56	17.2	26.8	-9.8	-36.6	-29.1	6.3	5.4	25.5	-22.59	35.38	-2.37	-1.67
1.58	15.0	25.1	-9.1	-36.2	-27.3	8.0	4.8	24.6	-22.92	36.13	-2.41	-2.12
1.60	13.3	23.4	-8.6	-35.8	-26.0	9.6	4.40	23.68	-23.24	36.83	-2.40	-2.54
1.65	11.0	19.8	-8.1	-35.1	-24.1	12.9	4.04	21.79	-23.96	38.47	-2.25	-3.45
1.70	9.5	17.5	-8.0	-35.0	-22.8	15.1	4.01	20.62	-24.73	40.08	-1.99	-4.14
1.75	7.5	15.3	-7.5	-35.3	-21.5	17.0	3.77	19.85	-25.66	41.53	-1.70	-4.68
1.80	5.7	12.2	-6.6	-35.4	-20.4	19.1	3.29	19.03	-26.64	42.68	-1.40	-5.10
1.85	5.2	8.3	-5.7	-34.8	-20.0	21.5	2.84	17.96	-27.45	43.53	-1.09	-5.45
1.90	6.5	4.6	-5.2	-33.9	-20.5	23.6	2.67	16.77	-27.98	44.27	-.75	-5.72
1.95	9.0	2.1	-5.2	-33.0	-21.6	24.9	2.81	15.66	-28.28	45.08	-.38	-5.91
2.00	11.9	1.1	-5.6	-32.2	-22.7	25.3	3.15	14.77	-28.49	46.04	.00	-5.99
2.05	14.2	1.2	-6.1	-31.6	-23.6	25.2	3.54	14.11	-28.75	47.09	.36	-5.97
2.10	15.9	1.8	-6.6	-31.2	-24.1	24.8	3.90	13.60	-29.07	48.12	.67	-5.84
2.15	17.1	2.3	-7.2	-30.9	-24.3	24.5	4.24	13.20	-29.39	49.07	.90	-5.64
2.20	18.2	2.5	-7.9	-30.6	-24.4	24.2	4.57	12.86	-29.63	49.92	1.03	-5.40
2.25	19.7	2.9	-8.8	-30.4	-24.7	23.9	4.93	12.60	-29.69	50.75	1.08	-5.15
2.30	21.6	3.6	-9.8	-30.5	-25.0	23.4	5.30	12.46	-29.59	51.67	1.03	-4.92
2.35	23.6	5.1	-10.9	-30.8	-25.3	22.7	5.64	12.46	-29.37	52.76	.93	-4.72
2.40	25.3	7.6	-11.9	-31.5	-25.4	21.7	5.91	12.59	-29.14	54.09	.77	-4.58
2.45	26.1	10.7	-12.6	-32.5	-25.2	20.5	6.05	12.82	-29.00	55.62	.59	-4.49
2.50	26.0	14.2	-13.1	-33.7	-24.6	19.3	6.04	13.10	-29.05	57.30	.41	-4.46
2.75	12.2	23.4	-10.7	-38.2	-18.0	16.9	4.24	13.54	-32.16	65.00	-.25	-4.85
3.00	3.7	13.5	-7.6	-37.1	-14.5	20.8	2.78	11.79	-35.40	69.39	-.15	-5.34
3.25	9.2	7.8	-8.1	-35.1	-15.8	22.6	2.80	10.21	-36.90	73.96	.13	-5.30
3.50	12.8	11.4	-9.4	-34.8	-16.4	21.5	3.01	9.45	-39.02	79.20	.06	-4.98
3.75	10.2	14.9	-10.3	-35.2	-15.3	20.3	2.96	8.96	-41.74	83.48	-.31	-4.84
4.00	6.5	14.3	-10.8	-35.4	-13.9	20.1	2.81	8.43	-43.79	86.77	-.68	-4.97
4.25	5.4	11.7	-11.4	-35.6	-13.1	20.4	2.73	7.89	-44.84	90.03	-.89	-5.24
4.50	6.5	9.8	-12.0	-35.8	-12.7	20.6	2.72	7.42	-45.41	93.85	-.95	-5.51
4.75	8.4	9.9	-12.7	-36.0	-12.7	20.5	2.74	7.04	-46.09	98.14	-.91	-5.70
5.00	9.6	11.4	-13.4	-36.4	-12.5	20.1	2.76	6.74	-47.09	102.56	-.85	-5.81
5.50	8.5	15.1	-14.4	-37.2	-11.6	19.1	2.74	6.26	-49.60	110.71	-.77	-5.89
6.00	4.8	16.5	-15.4	-37.9	-10.3	18.6	2.68	5.86	-51.53	117.90	-.75	-5.91

INVARIANT FIXED-T AMPLITUDES ISOSPIN EVEN AND ODD:+- T=-.40 GEV**2

K GEV/C	RE A+ 1/GEV	IM A+ 1/GEV	RE A- 1/GEV	IM A- 1/GEV	RE B+ 1/GEV**2	IM B+ 1/GEV**2	RE B- 1/GEV**2	IM B- 1/GEV**2	RE C+ 1/GEV	IM C+ 1/GEV	RE C- 1/GEV	IM C- 1/GEV
.00	-5.2	.0	1.7	-.0	318.	0.	-1289.5	-.0	4.24	.00	-36.54	-.00
.02	-5.8	.4	1.8	.3	334.	-4.	-1298.6	-2.6	4.57	.31	-38.34	.17
.04	-7.6	-.0	2.1	.6	386.	-10.	-1330.5	2.0	5.78	-.39	-44.03	.62
.06	-9.9	-1.9	2.8	1.2	480.	-22.	-1406.0	13.7	9.67	-2.86	-54.45	1.77
.08	-12.0	-5.3	3.6	2.8	638.	-47.	-1550.0	26.5	19.09	-7.60	-72.00	4.10
.10	-13.8	-10.3	3.5	5.3	913.	-90.	-1807.6	44.7	39.56	-15.59	-102.25	7.94
.12	-14.4	-17.6	2.4	7.7	1441.	-157.	-2316.9	83.6	85.87	-28.58	-158.85	13.50
.14	-11.7	-26.5	1.5	10.3	2661.	-251.	-3533.0	138.6	206.09	-47.08	-287.61	21.60
.16	-5.7	-34.6	-.5	14.4	7406.	-378.	-8274.3	198.9	697.60	-70.55	-786.23	33.31
.18	3.9	-42.0	-7.0	18.2	-23963.	-547.	23112.4	283.9	*******	-101.52	2508.53	49.13
.20	15.1	-50.0	-12.1	21.1	-5836.	-766.	5008.7	403.2	-704.55	-144.54	605.54	70.79
.22	35.0	-54.9	-21.3	24.1	-3662.	-1050.	2881.5	553.3	-471.29	-200.09	377.09	100.55
.24	61.1	-48.8	-35.2	21.2	-2656.	-1378.	1981.0	730.5	-346.90	-260.45	269.09	133.42
.26	84.4	-26.5	-46.2	10.0	-1846.	-1638.	1353.7	857.9	-228.09	-303.84	182.99	155.26
.28	92.3	7.6	-50.4	-4.3	-1088.	-1682.	852.0	854.0	-109.16	-303.87	107.38	153.87
.30	80.8	39.2	-47.9	-18.4	-490.	-1488.	492.9	744.9	-17.81	-260.62	51.40	131.60
.32	59.1	58.0	-40.1	-27.5	-122.	-1195.	269.0	596.2	32.43	-202.21	18.44	102.41
.34	36.5	65.1	-33.2	-31.5	72.	-918.	148.6	459.3	53.31	-150.11	1.59	76.12
.36	17.2	64.3	-26.5	-34.1	162.	-689.	83.9	354.5	57.75	-108.62	-5.49	54.94
.38	3.0	59.4	-20.9	-32.3	187.	-516.	51.5	269.9	53.00	-78.67	-7.10	39.98
.40	-8.0	54.5	-21.0	-30.7	191.	-398.	46.4	212.8	46.30	-58.95	-7.75	29.87
.42	-18.8	48.6	-20.4	-33.9	197.	-308.	40.4	183.4	40.81	-44.28	-8.16	21.44
.44	-28.1	39.5	-16.7	-34.9	198.	-225.	29.5	154.8	35.03	-32.08	-7.31	14.45
→ .46	-33.9	28.7	-16.7	-34.7	186.	-154.	31.9	130.8	28.65	-22.94	-6.00	9.21
.48	-36.5	17.7	-16.3	-39.4	168.	-97.	33.5	125.4	22.93	-16.48	-4.44	4.91
.50	-36.0	7.0	-10.3	-42.9	146.	-50.	21.6	120.6	18.09	-11.43	-2.32	1.82
.52	-32.2	-1.9	-5.5	-41.3	118.	-14.	12.9	106.9	13.64	-7.36	-.55	.12
.54	-27.1	-7.4	-4.1	-41.1	90.	8.	11.4	98.5	9.54	-4.32	.55	-1.20
.56	-23.3	-10.4	-.5	-42.8	69.	20.	5.2	95.5	6.03	-2.08	1.72	-2.41
.58	-20.5	-13.4	4.8	-41.6	53.	30.	-4.3	87.9	3.06	-.22	2.90	-2.95
.60	-16.9	-16.4	7.7	-38.6	38.	39.	-9.1	77.7	.25	1.42	3.50	-3.04
.62	-13.0	-17.7	9.3	-36.8	22.	43.	-11.9	70.4	-2.58	2.57	3.61	-3.32
.64	-10.7	-17.5	11.6	-34.8	11.	42.	-15.9	62.9	-5.14	2.99	3.75	-3.86
.66	-10.4	-18.2	12.8	-31.6	6.	41.	-17.2	53.2	-7.08	2.91	4.04	-4.42
.68	-10.2	-20.9	11.7	-29.2	3.	45.	-13.7	46.0	-8.39	2.69	4.47	-4.89
.70	-8.4	-24.9	10.5	-29.5	-2.	50.	-10.0	44.7	-9.26	2.49	5.02	-5.19
.72	-4.4	-28.7	11.5	-31.2	-9.	55.	-10.6	46.2	-9.72	2.30	5.58	-5.23
.74	1.2	-30.9	14.7	-31.8	-19.	57.	-14.9	46.1	-9.74	2.29	6.00	-5.04
.76	7.0	-30.9	18.1	-30.5	-28.	56.	-19.8	43.0	-9.48	2.66	6.25	-4.72

INVARIANT FIXED-T AMPLITUDES ISOSPIN EVEN AND ODD:+- T=-.40 GEV**2

K GEV/C	RE A+ 1/GEV	IM A+ 1/GEV	RE A- 1/GEV	IM A- 1/GEV	RE B+ 1/GEV**2	IM B+ 1/GEV**2	RE B- 1/GEV**2	IM B- 1/GEV**2	RE C+ 1/GEV	IM C+ 1/GEV	RE C- 1/GEV	IM C- 1/GEV
.78	11.7	-29.1	20.7	-28.0	-34.	53.	-23.3	38.4	-9.27	3.44	6.34	-4.31
.80	14.8	-27.1	22.3	-25.1	-38.	50.	-25.5	33.7	-9.28	4.41	6.18	-3.79
.82	17.4	-25.7	23.4	-22.5	-41.2	47.8	-27.3	29.4	-9.45	5.37	5.64	-3.32
.84	20.5	-24.6	24.3	-20.0	-45.1	46.2	-29.2	25.2	-9.64	6.32	4.76	-3.14
.86	24.3	-22.5	25.4	-17.5	-49.8	43.5	-31.5	20.5	-9.90	7.36	3.73	-3.45
.88	27.6	-18.8	26.5	-14.4	-53.9	38.8	-33.6	14.4	-10.38	8.47	2.80	-4.27
.90	29.4	-14.2	27.1	-10.6	-56.2	32.8	-34.5	7.0	-11.17	9.43	2.20	-5.55
.92	29.6	-9.7	26.8	-6.3	-56.3	26.7	-33.4	-1.2	-12.11	10.00	2.12	-7.16
.94	28.7	-6.0	25.4	-2.0	-54.9	21.2	-29.9	-9.0	-12.90	10.11	2.75	-8.85
.96	27.3	-2.9	23.0	1.9	-52.3	16.5	-24.4	-15.7	-13.24	9.97	4.11	-10.28
.98	25.5	-.2	19.9	5.2	-48.6	12.7	-17.5	-20.5	-13.05	9.92	6.06	-11.11
1.00	23.1	2.0	16.2	7.8	-44.0	10.2	-9.9	-23.3	-12.52	10.28	8.23	-11.09
1.02	20.2	3.4	12.0	9.5	-38.8	9.4	-2.2	-23.9	-11.97	11.17	10.20	-10.25
1.04	17.1	3.5	7.5	10.2	-34.0	10.6	4.9	-22.5	-11.70	12.49	11.66	-8.83
1.06	14.4	2.6	3.2	9.6	-30.5	13.3	10.7	-19.4	-11.91	14.01	12.49	-7.17
1.08	12.7	.8	-.4	8.1	-28.6	16.6	14.9	-15.5	-12.59	15.47	12.76	-5.55
1.10	11.8	-1.1	-3.0	6.0	-28.3	19.8	17.4	-11.3	-13.64	16.65	12.65	-4.13
1.12	11.8	-2.9	-4.7	3.9	-29.0	22.2	18.5	-7.5	-14.88	17.48	12.33	-2.96
1.14	12.1	-4.5	-5.7	2.1	-30.2	24.0	18.9	-4.4	-16.16	17.95	11.92	-1.99
1.16	12.8	-5.8	-6.5	.8	-31.7	25.1	18.9	-2.0	-17.38	18.10	11.48	-1.15
1.18	13.8	-7.0	-7.4	-.3	-33.2	25.7	18.9	-.1	-18.48	18.01	11.02	-.40
1.20	15.1	-8.0	-8.3	-1.4	-34.9	26.0	19.0	1.7	-19.43	17.71	10.55	.28
1.22	16.8	-8.7	-9.2	-2.6	-36.7	25.8	19.1	3.5	-20.21	17.24	10.04	.92
1.24	18.8	-9.0	-10.0	-4.0	-38.6	25.0	19.0	5.4	-20.78	16.62	9.49	1.52
1.26	20.9	-8.7	-10.5	-5.5	-40.3	23.6	18.6	7.2	-21.10	15.89	8.90	2.08
1.28	23.0	-7.8	-10.7	-6.9	-41.6	21.6	17.8	9.0	-21.15	15.13	8.26	2.59
1.30	24.8	-6.4	-10.6	-8.3	-42.3	19.3	16.9	10.5	-20.90	14.40	7.57	3.06
1.32	26.2	-4.6	-10.4	-9.4	-42.5	16.8	15.7	11.7	-20.38	13.79	6.83	3.46
1.34	27.2	-2.6	-10.1	-10.3	-42.0	14.3	14.5	12.7	-19.64	13.38	6.04	3.78
1.36	27.7	-.4	-9.9	-11.1	-41.1	12.1	13.3	13.4	-18.76	13.22	5.21	4.01
1.38	27.9	1.7	-9.6	-11.9	-39.7	10.1	12.2	13.9	-17.81	13.32	4.36	4.14
1.40	27.7	3.8	-9.4	-12.6	-38.1	8.4	11.1	14.4	-16.88	13.68	3.51	4.16
1.42	27.1	5.8	-9.1	-13.3	-36.4	7.1	9.9	14.7	-16.03	14.25	2.68	4.10
1.44	26.2	7.7	-8.7	-14.0	-34.5	6.1	8.8	14.9	-15.31	14.99	1.88	3.94
1.46	25.0	9.2	-8.2	-14.6	-32.5	5.4	7.7	15.0	-14.73	15.84	1.12	3.71
1.48	23.6	10.5	-7.7	-15.1	-30.6	5.1	6.5	14.9	-14.30	16.76	.41	3.40
1.50	22.1	11.4	-7.1	-15.4	-28.7	5.0	5.5	14.7	-14.00	17.69	-.25	3.04
1.52	20.5	12.0	-6.6	-15.6	-26.9	5.2	4.5	14.3	-13.82	18.62	-.85	2.62
1.54	19.1	12.2	-6.1	-15.6	-25.4	5.7	3.6	13.8	-13.72	19.54	-1.39	2.15

INVARIANT FIXED-T AMPLITUDES — ISOSPIN EVEN AND ODD:+- — T=-.40 GEV**2

K GEV/C	RE A+ 1/GEV	IM A+ 1/GEV	RE A- 1/GEV	IM A- 1/GEV	RE B+ 1/GEV**2	IM B+ 1/GEV**2	RE B- 1/GEV**2	IM B- 1/GEV**2	RE C+ 1/GEV	IM C+ 1/GEV	RE C- 1/GEV	IM C- 1/GEV
1.56	17.8	12.3	-5.7	-15.6	-24.1	6.2	2.9	13.2	-13.70	20.43	-1.87	1.64
1.58	16.8	12.1	-5.5	-15.6	-23.0	6.9	2.4	12.5	-13.74	21.30	-2.27	1.10
1.60	15.9	12.0	-5.4	-15.5	-22.1	7.6	2.06	11.94	-13.84	22.14	-2.61	.54
1.65	14.2	11.6	-5.3	-15.8	-20.4	9.0	1.57	10.74	-14.29	24.15	-3.11	-.87
1.70	12.6	11.5	-5.0	-16.6	-19.2	10.1	1.23	10.02	-15.04	25.95	-3.22	-2.17
1.75	10.8	11.0	-4.1	-17.2	-18.1	11.1	.69	9.45	-15.99	27.46	-3.06	-3.25
1.80	9.1	9.9	-2.8	-17.4	-17.1	12.3	.03	8.68	-17.00	28.63	-2.77	-4.12
1.85	8.2	8.3	-1.7	-16.8	-16.7	13.5	-.47	7.61	-17.95	29.51	-2.43	-4.84
1.90	8.1	6.8	-1.1	-15.8	-16.6	14.5	-.57	6.40	-18.75	30.20	-2.02	-5.46
1.95	8.6	5.8	-1.1	-14.8	-16.9	15.1	-.25	5.30	-19.42	30.76	-1.54	-5.98
2.00	9.0	5.3	-1.6	-14.1	-17.0	15.2	.36	4.51	-19.94	31.26	-.98	-6.37
2.05	9.2	5.1	-2.2	-13.7	-16.9	15.2	1.05	4.08	-20.34	31.71	-.36	-6.61
2.10	9.1	4.9	-2.7	-13.7	-16.6	15.2	1.68	3.94	-20.61	32.14	.28	-6.68
2.15	9.0	4.2	-3.1	-13.9	-16.2	15.4	2.16	3.99	-20.74	32.58	.87	-6.56
2.20	9.3	3.4	-3.3	-14.1	-16.0	15.8	2.48	4.13	-20.74	33.08	1.36	-6.29
2.25	10.2	2.6	-3.5	-14.2	-16.0	16.1	2.69	4.28	-20.63	33.69	1.72	-5.92
2.30	11.5	2.2	-3.6	-14.2	-16.2	16.3	2.82	4.43	-20.47	34.43	1.94	-5.48
2.35	13.1	2.4	-3.8	-14.3	-16.5	16.3	2.90	4.58	-20.31	35.32	2.00	-5.04
2.40	14.6	3.4	-4.1	-14.4	-16.9	16.0	2.94	4.73	-20.21	36.34	1.94	-4.62
2.45	15.6	5.0	-4.4	-14.6	-17.0	15.4	2.95	4.89	-20.22	37.45	1.78	-4.27
2.50	16.0	6.9	-4.7	-14.9	-16.9	14.7	2.91	5.07	-20.35	38.61	1.53	-4.00
2.75	9.4	13.0	-4.3	-17.3	-13.4	12.9	1.88	5.63	-22.40	43.71	.16	-3.90
3.00	4.4	8.3	-2.1	-17.4	-11.1	14.9	.73	4.88	-24.52	47.12	-.21	-4.68
3.25	6.6	5.1	-1.6	-15.8	-11.5	16.0	.58	3.84	-25.85	50.34	.03	-4.98
3.50	8.5	6.6	-2.5	-15.1	-11.7	15.5	.84	3.35	-27.16	53.79	.09	-4.90
3.75	7.4	8.4	-3.2	-15.3	-11.0	14.9	.95	3.20	-28.67	57.07	-.07	-4.85
4.00	5.5	8.5	-3.3	-15.7	-10.2	14.7	.89	3.07	-30.18	60.03	-.25	-4.96
4.25	4.2	7.4	-3.3	-15.8	-9.6	14.9	.80	2.87	-31.53	62.74	-.33	-5.13
4.50	4.0	6.1	-3.4	-15.8	-9.3	15.0	.76	2.66	-32.68	65.31	-.34	-5.27
4.75	4.4	5.3	-3.5	-15.6	-9.1	15.0	.77	2.46	-33.66	67.80	-.32	-5.37
5.00	5.1	4.9	-3.8	-15.6	-9.0	14.9	.80	2.31	-34.46	70.25	-.30	-5.44
5.50	6.2	5.4	-4.4	-15.8	-8.6	14.4	.86	2.11	-35.57	75.20	-.28	-5.56
6.00	6.4	6.5	-4.9	-16.2	-8.1	14.0	.88	1.98	-36.22	80.45	-.20	-5.68

INVARIANT FIXED-T AMPLITUDES ISOSPIN EVEN AND ODD:+- T=-.50 GEV**2

K GEV/C	RE A+ 1/GEV	IM A+	RE A- 1/GEV	IM A-	RE B+ 1/GEV**2	IM B+	RE B- 1/GEV**2	IM B-	RE C+ 1/GEV	IM C+	RE C- 1/GEV	IM C-
.00	-84.0	.1	3.2	-.0	8.	0.	-803.6	-.0	-83.96	.07	-1.23	-.03
.02	-84.8	.1	4.3	.2	8.	-0.	-803.4	.1	-84.79	.09	-1.18	.17
.04	-86.4	-.6	6.3	.7	12.	-4.	-803.6	2.4	-86.32	-.69	-2.11	.70
.06	-89.8	-3.3	10.1	2.1	19.	-16.	-808.1	9.0	-89.52	-3.60	-3.15	2.24
.08	-93.1	-8.8	13.9	4.8	36.	-39.	-823.6	21.6	-92.24	-9.71	-6.10	5.28
.10	-96.8	-16.3	18.1	8.6	68.	-76.	-857.2	40.1	-94.45	-18.91	-10.77	9.91
.12	-100.3	-26.9	22.0	13.9	124.	-133.	-920.5	68.8	-94.80	-32.85	-19.03	16.93
.14	-103.1	-41.8	25.5	21.2	226.	-220.	-1035.3	112.2	-90.34	-54.18	-32.90	27.50
.16	-102.1	-60.3	26.9	30.9	428.	-348.	-1249.1	175.8	-72.46	-84.43	-59.62	43.03
.18	-97.8	-81.4	26.0	41.7	829.	-515.	-1660.7	258.9	-29.18	-124.04	-111.50	63.16
.20	-87.6	-108.0	22.0	55.1	1727.	-753.	-2564.7	377.3	79.75	-180.99	-226.59	91.70
.22	-63.9	-137.3	11.1	70.1	4711.	-1092.	-5514.3	545.9	461.36	-259.08	-603.70	130.96
.24	-14.9	-159.1	-16.3	81.6	-127603.	-1474.	*******	735.2	*******	-345.52	*******	174.60
.26	34.4	-157.1	-38.4	80.8	-5741.	-1749.	5297.5	872.0	-779.34	-404.95	712.55	204.40
.28	80.6	-125.1	-61.6	64.7	-2621.	-1768.	2457.5	882.0	-331.67	-403.22	324.94	203.44
.30	100.8	-80.1	-72.4	42.3	-1308.	-1546.	1373.7	770.3	-125.59	-347.72	165.32	175.65
.32	99.7	-40.2	-73.1	22.0	-628.	-1234.	826.5	615.7	-19.04	-273.48	83.13	138.37
.34	86.4	-11.2	-67.9	6.6	-271.	-947.	528.0	475.6	30.89	-205.49	40.42	104.21
.36	69.1	5.7	-60.8	-3.2	-84.	-714.	358.0	363.2	50.56	-152.38	18.43	77.20
.38	53.7	14.8	-54.5	-9.1	7.	-537.	260.2	278.3	55.35	-112.98	7.44	57.04
.40	39.2	19.8	-50.2	-13.1	53.	-414.	204.5	220.5	52.58	-85.58	1.77	43.01
.42	25.9	20.0	-47.2	-18.2	88.	-322.	164.8	186.4	49.77	-67.25	-2.53	32.32
.44	15.6	16.8	-42.0	-22.8	110.	-234.	130.7	157.2	47.14	-50.54	-4.46	22.42
.46	7.8	11.6	-38.6	-25.8	106.	-165.	113.7	135.3	40.07	-38.76	-3.94	15.39
.48	4.1	6.4	-34.4	-30.0	103.	-109.	94.8	125.5	37.23	-28.63	-3.93	10.30
.50	1.5	1.0	-28.3	-33.2	87.	-63.	75.1	115.0	30.95	-20.37	-2.94	5.63
.52	.9	-7.8	-24.1	-37.6	73.	-26.	62.2	111.6	26.80	-17.02	-2.06	1.99
.54	6.9	-9.6	-17.0	-39.6	41.	-5.	47.4	104.7	22.21	-11.27	.60	-.66
.56	9.9	-11.3	-5.1	-38.3	24.	5.	18.8	96.0	19.09	-9.28	2.20	-1.02
.58	10.9	-8.9	-5.4	-34.1	13.	7.	18.6	79.1	16.14	-5.87	2.15	-2.02
.60	13.4	-9.9	4.5	-34.9	1.	15.	-2.0	75.6	13.81	-3.47	3.64	-2.99
.62	14.8	-6.3	7.6	-26.8	-9.	13.	-8.7	54.3	10.72	-.50	3:77	-2.93
.64	14.0	-5.4	10.1	-24.7	-15.	13.	-12.5	45.5	7.22	.58	4.37	-3.88
.66	14.1	-2.6	12.6	-17.1	-19.	8.	-15.9	28.5	5.07	1.21	5.09	-3.55
.68	9.9	-.9	8.4	-12.4	-14.	6.	-7.4	17.8	2.86	2.06	4.74	-3.62
.70	6.7	-3.2	5.9	-12.1	-12.	11.	-2.0	15.6	.61	2.12	4.89	-4.16
.72	5.5	-6.2	4.2	-12.1	-12.	15.	1.9	14.6	-.83	1.47	5.15	-4.41
.74	6.0	-9.9	3.4	-14.0	-13.	20.	3.8	17.1	-1.16	.93	5.49	-4.70
.76	8.9	-12.3	5.1	-15.4	-18.	24.	1.7	19.3	-1.01	1.16	6.05	-4.63

INVARIANT FIXED-T AMPLITUDES ISOSPIN EVEN AND ODD:+- T=-.50 GEV**2

K GEV/C	RE A+ 1/GEV	IM A+ 1/GEV	RE A- 1/GEV	IM A- 1/GEV	RE B+ 1/GEV**2	IM B+ 1/GEV**2	RE B- 1/GEV**2	IM B- 1/GEV**2	RE C+ 1/GEV	IM C+ 1/GEV	RE C- 1/GEV	IM C- 1/GEV
.78	11.4	-12.5	6.9	-15.5	-22.	25.	-1.1	19.7	-1.13	1.88	6.26	-4.18
.80	12.8	-13.0	8.7	-15.7	-24.	26.	-4.4	20.0	-1.41	2.52	6.12	-3.81
.82	15.1	-14.0	11.2	-15.5	-27.3	28.2	-8.7	19.6	-1.53	3.20	5.83	-3.49
.84	18.5	-14.0	13.7	-14.7	-32.1	28.9	-13.5	18.1	-1.67	4.19	5.23	-3.28
.86	21.9	-12.8	16.7	-13.5	-37.2	28.0	-19.0	15.5	-2.12	5.29	4.42	-3.49
.88	25.4	-10.8	20.4	-11.1	-42.5	25.6	-25.1	10.4	-2.81	6.24	3.76	-4.17
.90	28.8	-7.0	23.4	-6.7	-47.5	20.5	-29.3	2.2	-3.58	6.99	3.39	-5.13
.92	30.5	-1.4	24.5	-1.3	-50.0	12.8	-30.2	-7.2	-4.36	7.55	3.39	-6.29
.94	29.7	4.5	24.0	3.8	-48.6	4.8	-28.1	-15.9	-5.07	7.91	3.90	-7.54
.96	26.9	9.2	22.7	8.6	-44.4	-1.4	-24.2	-23.5	-5.58	8.15	5.02	-8.56
.98	23.4	12.6	20.3	13.4	-39.0	-5.6	-18.4	-29.9	-5.87	8.40	6.52	-9.01
1.00	19.5	15.0	16.3	17.6	-33.2	-8.1	-10.9	-34.5	-6.01	8.82	8.01	-8.84
1.02	15.0	16.4	11.1	20.2	-26.9	-8.9	-2.4	-36.3	-6.15	9.44	9.24	-8.24
1.04	10.3	16.4	5.8	20.9	-20.8	-7.7	5.5	-35.3	-6.40	10.23	10.20	-7.42
1.06	6.1	15.0	1.2	20.1	-15.7	-4.7	11.9	-32.4	-6.83	11.12	10.91	-6.48
1.08	2.8	12.7	-2.5	18.5	-12.3	-.8	16.6	-28.5	-7.49	12.01	11.40	-5.43
1.10	.6	10.1	-5.5	16.4	-10.5	3.2	20.0	-24.3	-8.40	12.81	11.61	-4.33
1.12	-1.0	7.4	-7.8	14.0	-9.8	6.9	22.2	-19.9	-9.51	13.41	11.56	-3.31
1.14	-2.1	4.7	-9.3	11.3	-9.7	10.1	23.2	-15.5	-10.73	13.72	11.33	-2.45
1.16	-2.8	1.8	-9.9	8.7	-10.1	13.2	23.1	-11.5	-11.94	13.73	11.04	-1.78
1.18	-2.6	-1.3	-9.7	6.5	-11.3	16.0	22.2	-8.3	-13.06	13.50	10.79	-1.22
1.20	-1.7	-4.1	-9.2	4.8	-13.1	18.2	21.1	-5.9	-14.04	13.05	10.61	-.70
1.22	-.1	-6.4	-8.6	3.7	-15.4	19.6	19.9	-4.1	-14.87	12.44	10.46	-.15
1.24	1.9	-7.9	-8.2	2.9	-17.8	20.1	18.9	-2.5	-15.53	11.68	10.29	.45
1.26	3.8	-8.8	-7.9	2.3	-19.9	19.8	18.0	-1.2	-15.98	10.78	10.05	1.10
1.28	5.6	-9.3	-7.5	1.6	-21.5	18.9	17.1	.2	-16.17	9.78	9.73	1.77
1.30	7.1	-9.6	-7.1	1.1	-22.5	17.9	16.0	1.3	-16.05	8.74	9.33	2.43
1.32	8.6	-9.8	-6.6	.7	-23.2	16.8	14.8	2.3	-15.61	7.77	8.83	3.06
1.34	10.1	-9.9	-6.2	.5	-23.5	15.9	13.6	3.0	-14.88	6.97	8.24	3.66
1.36	11.8	-9.8	-5.8	.5	-23.8	15.0	12.4	3.5	-13.92	6.41	7.57	4.19
1.38	13.4	-9.3	-5.7	.5	-23.9	14.1	11.4	3.8	-12.84	6.13	6.83	4.65
1.40	14.9	-8.5	-5.6	.5	-23.9	13.1	10.5	4.0	-11.72	6.15	6.02	5.02
1.42	16.3	-7.4	-5.7	.5	-23.7	12.2	9.6	4.2	-10.64	6.43	5.16	5.29
1.44	17.2	-6.0	-5.8	.4	-23.4	11.2	8.8	4.4	-9.65	6.92	4.28	5.45
1.46	17.9	-4.5	-5.9	.2	-22.8	10.4	8.0	4.6	-8.79	7.59	3.39	5.51
1.48	18.2	-3.1	-5.9	-.0	-22.1	9.7	7.1	4.6	-8.07	8.37	2.51	5.45
1.50	18.2	-1.8	-5.9	-.2	-21.3	9.2	6.3	4.6	-7.49	9.24	1.65	5.29
1.52	18.0	-.6	-6.0	-.3	-20.5	8.8	5.6	4.4	-7.05	10.16	.83	5.03
1.54	17.7	.4	-6.0	-.5	-19.8	8.6	4.9	4.1	-6.74	11.10	.06	4.67

INVARIANT FIXED-T AMPLITUDES ISOSPIN EVEN AND ODD:+- T=-.50 GEV**2

K GEV/C	RE A+ 1/GEV	IM A+ 1/GEV	RE A- 1/GEV	IM A- 1/GEV	RE B+ 1/GEV**2	IM B+ 1/GEV**2	RE B- 1/GEV**2	IM B- 1/GEV**2	RE C+ 1/GEV	IM C+ 1/GEV	RE C- 1/GEV	IM C- 1/GEV
1.56	17.3	1.3	-6.2	-.6	-19.0	8.6	4.4	3.9	-6.55	12.04	-.65	4.24
1.58	16.9	2.0	-6.3	-.8	-18.4	8.6	4.0	3.6	-6.47	12.97	-1.27	3.73
1.60	16.51	2.71	-6.51	-1.11	-17.83	8.65	3.65	3.32	-6.49	13.87	-1.81	3.17
1.65	15.34	4.13	-6.76	-2.28	-16.64	8.86	3.01	2.95	-6.85	15.94	-2.75	1.65
1.70	13.97	5.13	-6.28	-3.69	-15.61	9.14	2.27	2.79	-7.53	17.71	-3.16	.16
1.75	12.59	5.60	-5.07	-4.69	-14.76	9.58	1.33	2.51	-8.38	19.21	-3.18	-1.13
1.80	11.44	5.67	-3.58	-4.93	-14.21	10.10	.39	1.86	-9.37	20.46	-3.00	-2.20
1.85	10.62	5.63	-2.33	-4.44	-13.97	10.48	-.26	.89	-10.44	21.43	-2.72	-3.09
1.90	9.95	5.68	-1.61	-3.55	-13.85	10.59	-.47	-.21	-11.54	22.11	-2.34	-3.87
1.95	9.19	5.84	-1.48	-2.62	-13.63	10.44	-.23	-1.20	-12.55	22.48	-1.85	-4.54
2.00	8.20	5.88	-1.78	-1.92	-13.16	10.22	.32	-1.92	-13.37	22.63	-1.25	-5.07
2.05	7.07	5.58	-2.29	-1.58	-12.49	10.14	1.01	-2.28	-13.94	22.64	-.58	-5.41
2.10	6.05	4.81	-2.78	-1.59	-11.78	10.33	1.68	-2.30	-14.28	22.64	.12	-5.56
2.15	5.40	3.65	-3.11	-1.83	-11.20	10.75	2.19	-2.09	-14.43	22.68	.78	-5.52
2.20	5.30	2.34	-3.22	-2.13	-10.89	11.28	2.52	-1.75	-14.44	22.80	1.35	-5.31
2.25	5.78	1.16	-3.15	-2.38	-10.85	11.77	2.67	-1.40	-14.36	23.02	1.81	-4.98
2.30	6.71	.36	-3.00	-2.50	-11.00	12.09	2.69	-1.08	-14.21	23.34	2.12	-4.56
2.35	7.84	.09	-2.87	-2.49	-11.24	12.17	2.65	-.84	-14.02	23.77	2.29	-4.12
2.40	8.92	.36	-2.84	-2.40	-11.44	12.04	2.60	-.64	-13.82	24.30	2.32	-3.68
2.45	9.74	1.05	-2.93	-2.31	-11.50	11.75	2.54	-.48	-13.63	24.93	2.24	-3.28
2.50	10.17	2.01	-3.12	-2.30	-11.39	11.39	2.50	-.31	-13.49	25.66	2.06	-2.95
2.75	7.44	5.19	-3.54	-3.77	-9.37	10.79	1.87	.56	-14.05	29.96	.75	-2.48
3.00	5.03	3.18	-1.74	-4.51	-8.46	11.91	.75	.56	-16.22	33.10	.13	-3.10
3.25	5.79	1.76	-.81	-3.35	-8.63	12.14	.36	-.08	-17.78	34.93	.17	-3.56
3.50	6.30	2.14	-1.23	-2.62	-8.42	11.76	.52	-.37	-18.53	36.85	.30	-3.72
3.75	5.78	2.52	-1.61	-2.71	-7.89	11.56	.62	-.33	-19.24	39.15	.36	-3.76
4.00	5.09	2.28	-1.55	-2.90	-7.46	11.56	.57	-.26	-20.19	41.45	.39	-3.80
4.25	4.74	1.74	-1.36	-2.85	-7.20	11.58	.49	-.27	-21.22	43.51	.41	-3.82
4.50	4.70	1.26	-1.30	-2.63	-7.02	11.53	.45	-.31	-22.15	45.37	.43	-3.83
4.75	4.77	.95	-1.40	-2.43	-6.84	11.42	.46	-.35	-22.91	47.13	.44	-3.83
5.00	4.83	.77	-1.59	-2.32	-6.65	11.29	.48	-.35	-23.53	48.89	.45	-3.83
5.50	4.84	.53	-1.94	-2.33	-6.25	11.06	.51	-.32	-24.53	52.51	.45	-3.82
6.00	4.78	.28	-2.17	-2.44	-5.88	10.89	.51	-.26	-25.42	56.23	.44	-3.80

INVARIANT FIXED-T AMPLITUDES — ISOSPIN EVEN AND ODD:+- — T=-.60 GEV**2

K GEV/C	RE A+ (1/GEV)	IM A+ (1/GEV)	RE A- (1/GEV)	IM A- (1/GEV)	RE B+ (1/GEV**2)	IM B+ (1/GEV**2)	RE B- (1/GEV**2)	IM B- (1/GEV**2)	RE C+ (1/GEV)	IM C+ (1/GEV)	RE C- (1/GEV)	IM C- (1/GEV)
.00	-211.3	-31.9	-21.3	-16.5	103.	118.	-452.7	61.0	-213.11	-33.96	-13.47	-17.52
.02	-212.1	-30.2	-19.5	-15.3	96.	110.	-450.7	57.6	-213.60	-31.98	-12.26	-16.27
.04	-212.9	-26.3	-15.6	-12.2	75.	87.	-445.6	50.2	-213.86	-27.41	-10.03	-12.87
.06	-214.9	-22.7	-8.1	-6.6	40.	48.	-440.0	43.5	-215.19	-23.06	-5.07	-6.93
.08	-215.4	-21.9	1.0	.7	-5.	-6.	-437.8	41.6	-215.36	-21.92	.59	.77
.10	-213.9	-24.2	12.4	10.0	-60.	-70.	-442.8	46.6	-214.50	-24.93	7.88	10.50
.12	-211.6	-36.9	24.9	18.8	-119.	-137.	-456.8	71.2	-214.02	-39.75	15.41	20.26
.14	-213.4	-58.3	35.8	29.3	-158.	-228.	-474.7	116.2	-218.47	-65.62	20.44	33.05
.16	-212.9	-86.2	43.3	43.8	-165.	-360.	-513.6	182.0	-220.26	-102.40	20.33	51.98
.18	-210.1	-119.7	47.9	61.0	-126.	-533.	-588.7	268.1	-217.41	-150.69	13.72	76.55
.20	-201.1	-164.5	48.3	83.6	-20.	-780.	-721.8	390.6	-202.56	-220.45	-3.50	111.59
.22	-172.8	-219.5	38.5	111.4	265.	-1130.	-989.2	565.2	-150.03	-316.71	-46.62	160.05
.24	-112.8	-270.7	11.6	137.8	925.	-1527.	-1533.7	761.3	-19.70	-424.31	-142.74	214.41
.26	-23.2	-290.4	-31.2	147.9	2270.	-1811.	-2648.4	902.7	239.14	-499.66	-337.18	252.15
.28	64.7	-260.9	-73.1	132.9	5608.	-1831.	-5706.1	913.2	797.74	-500.21	-818.98	252.32
.30	126.2	-200.0	-106.1	102.7	-134351.	-1602.	*******	798.2	*******	-434.11	*******	219.29
.32	138.3	-137.0	-108.3	70.8	-4289.	-1279.	4548.2	638.0	-555.22	-343.85	627.19	173.95
.34	133.5	-86.4	-105.5	44.7	-1715.	-981.	2027.3	492.7	-170.68	-260.49	254.22	132.12
.36	118.4	-51.8	-97.9	26.2	-862.	-741.	1184.8	376.0	-48.25	-195.01	131.12	98.87
.38	101.5	-29.3	-89.4	13.6	-475.	-558.	785.8	287.6	2.07	-146.19	75.02	73.79
.40	84.6	-15.0	-82.4	5.0	-273.	-431.	567.6	227.5	22.96	-112.12	45.53	56.25
.42	69.1	-7.9	-77.1	-3.3	-144.	-337.	430.0	191.6	34.36	-89.26	26.81	42.94
.44	57.2	-6.1	-69.9	-11.6	-62.	-243.	332.3	162.3	41.26	-68.62	15.80	30.23
.46	49.1	-6.7	-62.7	-17.1	-33.	-170.	270.6	140.1	40.06	-53.41	11.45	21.35
.48	45.7	-5.7	-54.9	-19.9	-12.	-119.	214.9	124.2	42.27	-40.38	7.53	16.22
.50	41.2	-7.4	-50.6	-25.1	-7.	-75.	177.7	114.7	38.90	-30.45	3.99	10.12
.52	41.0	-8.1	-44.7	-33.0	-10.	-43.	147.8	113.4	37.68	-21.90	3.07	3.65
.54	43.8	-5.7	-31.9	-39.2	-30.	-24.	112.3	112.1	33.58	-13.89	6.31	-1.07
.56	41.5	-.1	-15.4	-34.0	-35.	-24.	68.7	93.1	28.93	-8.77	9.15	-.85
.58	36.8	1.8	-13.3	-28.4	-34.	-21.	59.9	76.8	24.21	-5.99	9.07	.26
→ .60	36.4	1.6	-3.8	-25.3	-35.	-16.	30.5	65.5	22.57	-4.86	8.10	.24
.62	36.5	5.0	-.9	-18.6	-38.	-14.	20.9	43.8	21.17	-.91	7.59	-.81
.64	35.0	8.6	4.4	-13.5	-42.	-17.	8.8	29.3	17.15	1.41	8.15	-1.12
.66	32.4	13.6	4.0	-2.8	-39.	-26.	9.0	4.6	15.21	2.06	7.91	-.74
.68	26.1	17.6	-2.0	.4	-28.	-29.	20.8	-2.6	13.16	4.22	7.51	-.82
.70	18.7	17.9	-3.4	1.3	-19.	-27.	22.2	-4.1	9.79	4.93	7.10	-.67
.72	12.3	15.1	-6.9	3.9	-9.	-22.	26.8	-9.8	7.77	4.18	6.22	-.89
.74	8.5	9.8	-12.1	1.9	-3.	-12.	35.4	-7.2	6.79	3.82	5.82	-1.79
.76	8.4	5.0	-13.4	-3.1	-5.	-3.	37.5	1.5	5.98	3.58	6.24	-2.39

INVARIANT FIXED-T AMPLITUDES ISOSPIN EVEN AND ODD:+- T=-.60 GEV**2

K GEV/C	RE A+ 1/GEV	IM A+ 1/GEV	RE A- 1/GEV	IM A- 1/GEV	RE B+ 1/GEV**2	IM B+ 1/GEV**2	RE B- 1/GEV**2	IM B- 1/GEV**2	RE C+ 1/GEV	IM C+ 1/GEV	RE C- 1/GEV	IM C- 1/GEV
.78	9.4	2.3	-11.7	-7.2	-7.	2.	33.8	9.2	5.65	3.25	6.63	-2.23
.80	10.2	-.1	-8.6	-10.0	-8.	6.	27.2	14.5	5.85	3.24	6.52	-1.88
.82	12.0	-2.4	-4.8	-11.5	-9.9	10.8	19.1	17.0	6.28	3.78	6.15	-1.74
.84	14.9	-3.7	-.9	-12.0	-14.2	14.5	11.3	17.3	6.56	4.90	5.75	-1.77
.86	18.1	-3.6	3.3	-11.9	-19.3	16.3	3.2	16.4	6.32	6.34	5.29	-1.93
.88	21.3	-2.6	8.4	-10.6	-25.2	16.3	-5.6	13.2	5.59	7.60	4.86	-2.36
.90	24.8	-.0	13.3	-7.1	-31.3	13.3	-13.4	6.3	4.73	8.53	4.69	-3.02
.92	27.4	4.7	16.6	-1.7	-35.7	7.0	-17.8	-2.9	3.93	9.31	4.87	-3.66
.94	27.7	10.7	17.7	4.1	-36.4	-1.0	-18.5	-12.3	3.10	10.07	5.27	-4.14
.96	25.2	16.4	17.3	9.7	-33.3	-8.3	-16.6	-20.5	2.18	10.71	5.80	-4.45
.98	21.1	20.6	15.6	14.9	-28.0	-13.4	-12.9	-27.5	1.26	11.17	6.45	-4.57
1.00	16.2	23.3	12.5	19.5	-21.7	-16.2	-7.4	-32.9	.46	11.54	7.14	-4.38
1.02	11.1	24.7	8.1	22.9	-15.4	-17.1	-.6	-36.1	-.30	11.94	7.63	-3.91
1.04	5.8	24.8	2.9	24.5	-9.1	-16.4	6.4	-36.7	-1.12	12.36	7.79	-3.37
1.06	.7	23.5	-1.9	24.2	-3.5	-14.0	12.4	-35.0	-2.02	12.70	7.73	-3.03
1.08	-3.7	21.0	-5.8	22.5	.9	-10.2	17.0	-32.0	-2.95	12.91	7.72	-2.91
1.10	-7.0	17.7	-8.4	20.2	3.9	-5.7	20.1	-28.5	-3.86	13.02	7.91	-2.86
1.12	-9.1	14.1	-10.0	17.9	5.3	-1.2	22.0	-24.8	-4.74	13.07	8.24	-2.66
1.14	-10.3	10.6	-10.9	15.7	5.5	2.9	23.0	-21.2	-5.65	13.06	8.53	-2.25
1.16	-10.9	7.5	-11.3	13.7	5.0	6.3	23.2	-17.9	-6.60	12.93	8.67	-1.76
1.18	-11.2	4.6	-11.3	11.9	4.2	9.1	22.7	-15.0	-7.55	12.64	8.66	-1.33
1.20	-11.2	1.9	-10.9	10.3	3.1	11.5	21.8	-12.6	-8.44	12.16	8.65	-1.01
1.22	-10.9	-.7	-10.2	9.0	1.8	13.5	20.7	-10.7	-9.23	11.53	8.72	-.74
1.24	-10.3	-3.3	-9.3	8.2	.4	15.1	19.6	-9.3	-9.88	10.77	8.91	-.39
1.26	-9.3	-5.7	-8.4	7.8	-1.1	16.4	18.5	-8.1	-10.38	9.88	9.12	.11
1.28	-8.1	-7.9	-7.6	7.7	-2.7	17.4	17.5	-7.1	-10.69	8.90	9.24	.78
1.30	-6.6	-9.9	-7.1	7.7	-4.3	18.0	16.6	-6.2	-10.78	7.83	9.21	1.55
1.32	-4.8	-11.6	-6.6	7.8	-5.8	18.4	15.7	-5.5	-10.59	6.73	8.99	2.32
1.34	-2.8	-13.0	-6.3	7.9	-7.2	18.4	14.7	-4.8	-10.12	5.69	8.64	3.04
1.36	-.6	-14.1	-6.0	8.1	-8.5	18.3	13.7	-4.2	-9.37	4.79	8.20	3.69
1.38	1.7	-14.8	-5.7	8.3	-9.6	18.0	12.8	-3.9	-8.39	4.11	7.70	4.27
1.40	4.1	-15.0	-5.6	8.7	-10.7	17.6	11.9	-3.6	-7.28	3.71	7.14	4.83
1.42	6.4	-14.8	-5.5	9.1	-11.6	17.0	11.1	-3.5	-6.12	3.60	6.51	5.36
1.44	8.6	-14.1	-5.7	9.6	-12.3	16.3	10.5	-3.4	-4.98	3.75	5.77	5.83
1.46	10.4	-13.1	-6.1	10.0	-12.9	15.5	9.9	-3.3	-3.94	4.13	4.93	6.22
1.48	12.0	-11.9	-6.6	10.2	-13.2	14.6	9.4	-3.3	-3.01	4.68	4.00	6.47
1.50	13.1	-10.5	-7.3	10.3	-13.4	13.8	9.0	-3.2	-2.23	5.36	3.02	6.55
1.52	14.0	-9.1	-8.0	10.1	-13.3	13.1	8.6	-3.1	-1.58	6.13	2.04	6.45
1.54	14.6	-7.8	-8.6	9.8	-13.2	12.4	8.2	-3.0	-1.07	6.96	1.13	6.19

INVARIANT FIXED-T AMPLITUDES ISOSPIN EVEN AND ODD:+- T=-.60 GEV**2

K GEV/C	RE A+ 1/GEV	IM A+ 1/GEV	RE A- 1/GEV	IM A- 1/GEV	RE B+ 1/GEV**2	IM B+ 1/GEV**2	RE B- 1/GEV**2	IM B- 1/GEV**2	RE C+ 1/GEV	IM C+ 1/GEV	RE C- 1/GEV	IM C- 1/GEV
1.56	14.9	-6.5	-9.1	9.3	-13.0	11.9	7.8	-2.9	-.69	7.83	.32	5.79
1.58	15.1	-5.3	-9.5	8.6	-12.7	11.5	7.5	-2.8	-.43	8.70	-.37	5.29
1.60	15.09	-4.25	-9.69	7.98	-12.44	11.19	7.09	-2.61	-.28	9.58	-.93	4.75
1.65	14.82	-1.99	-9.69	6.30	-11.86	10.67	6.11	-2.29	-.34	11.64	-1.88	3.38
1.70	14.28	-.19	-9.08	4.84	-11.43	10.33	5.05	-2.06	-.82	13.45	-2.40	2.12
1.75	13.55	1.27	-8.00	3.74	-11.08	10.05	3.91	-2.05	-1.55	14.98	-2.67	.95
1.80	12.72	2.48	-6.61	3.17	-10.80	9.81	2.79	-2.37	-2.46	16.27	-2.69	-.16
1.85	11.79	3.51	-5.20	3.26	-10.58	9.53	1.88	-3.03	-3.54	17.31	-2.47	-1.13
1.90	10.72	4.38	-4.14	3.92	-10.36	9.16	1.37	-3.92	-4.73	18.04	-2.10	-1.92
1.95	9.45	5.04	-3.66	4.83	-10.02	8.71	1.31	-4.80	-5.91	18.41	-1.65	-2.54
2.00	8.00	5.37	-3.72	5.61	-9.48	8.30	1.63	-5.47	-6.94	18.46	-1.16	-3.02
2.05	6.51	5.25	-4.08	6.04	-8.79	8.06	2.14	-5.82	-7.73	18.30	-.62	-3.39
2.10	5.16	4.65	-4.46	6.09	-8.07	8.07	2.66	-5.86	-8.25	18.06	-.03	-3.64
2.15	4.16	3.67	-4.66	5.92	-7.46	8.32	3.08	-5.67	-8.55	17.85	.59	-3.73
2.20	3.62	2.51	-4.64	5.72	-7.05	8.69	3.35	-5.36	-8.69	17.70	1.21	-3.65
2.25	3.56	1.37	-4.48	5.61	-6.87	9.08	3.48	-5.03	-8.73	17.62	1.74	-3.38
2.30	3.88	.43	-4.31	5.67	-6.86	9.37	3.52	-4.72	-8.69	17.60	2.15	-2.99
2.35	4.45	-.20	-4.24	5.84	-6.95	9.52	3.53	-4.46	-8.58	17.64	2.38	-2.52
2.40	5.09	-.50	-4.31	6.04	-7.03	9.51	3.52	-4.21	-8.40	17.75	2.45	-2.04
2.45	5.68	-.50	-4.53	6.17	-7.06	9.41	3.52	-3.97	-8.16	17.94	2.37	-1.61
2.50	6.12	-.30	-4.83	6.17	-6.99	9.26	3.50	-3.71	-7.88	18.24	2.18	-1.26
2.75	6.03	1.14	-5.40	4.39	-5.95	9.11	2.87	-2.35	-7.16	21.34	.96	-.82
3.00	4.95	1.10	-3.61	3.47	-5.65	9.61	1.65	-1.94	-8.77	24.46	.39	-1.24
3.25	4.31	.63	-2.55	4.30	-5.63	9.53	1.06	-2.25	-10.57	25.81	.26	-1.65
3.50	3.83	.01	-2.67	4.95	-5.32	9.38	1.05	-2.42	-11.35	26.80	.34	-1.96
3.75	3.66	-.80	-2.95	4.96	-5.04	9.45	1.13	-2.30	-11.83	28.22	.51	-2.09
4.00	3.87	-1.53	-2.99	4.74	-4.98	9.54	1.11	-2.08	-12.49	29.78	.66	-2.10
4.25	4.26	-1.97	-2.87	4.57	-5.02	9.48	1.03	-1.89	-13.28	31.18	.75	-2.04
4.50	4.65	-2.17	-2.72	4.49	-5.03	9.32	.94	-1.74	-14.00	32.40	.78	-1.98
4.75	4.93	-2.21	-2.60	4.45	-4.97	9.11	.87	-1.63	-14.56	33.54	.79	-1.93
5.00	5.09	-2.21	-2.52	4.42	-4.85	8.92	.80	-1.53	-14.98	34.71	.79	-1.89
5.50	5.23	-2.22	-2.40	4.34	-4.58	8.64	.69	-1.35	-15.68	37.21	.77	-1.83
6.00	5.25	-2.29	-2.28	4.25	-4.34	8.44	.61	-1.21	-16.39	39.84	.75	-1.78

Table 2.2.3.2. Result of the KH 78 fixed-t analysis below and near threshold. Natural units: $\hbar=c=\mu=1$. The notation is explained in the text.

FIXED-T AMPLITUDES AT UNPHYSICAL AND LOW ENERGIES T=0.00 GEV**2
NU=(S-U)/4M AND AMPLITUDES IN NATURAL UNITS ISOSPIN EVEN:+, ISOSPIN ODD:-

NU**2	RE $\bar{A}$+	RE A-/NU	RE $\tilde{B}$+/NU	RE $\bar{B}$-	RE $\bar{C}$+	RE $\tilde{C}$-/NU
0.00	-1.51	-8.83	-3.54	8.37	-1.51	-0.464
0.10	-1.03	-8.96	-3.64	8.48	-1.39	-0.481
0.20	-0.53	-9.09	-3.75	8.60	-1.28	-0.499
0.30	0.01	-9.24	-3.87	8.72	-1.15	-0.518
0.40	0.57	-9.39	-4.00	8.85	-1.03	-0.538
0.50	1.18	-9.55	-4.14	8.99	-0.89	-0.558
0.60	1.82	-9.73	-4.29	9.15	-0.75	-0.580
0.70	2.52	-9.91	-4.46	9.31	-0.60	-0.603
0.80	3.27	-10.11	-4.64	9.49	-0.44	-0.628
0.90	4.09	-10.33	-4.85	9.68	-0.27	-0.652
1.00	5.05	-10.50	-5.09	9.83	-0.04	-0.669
1.20	7.19	-11.32	-5.75	10.56	0.29	-0.757
1.40	9.71	-12.16	-6.44	11.30	0.69	-0.859
1.60	12.21	-12.64	-6.91	11.68	1.15	-0.962
1.80	14.73	-13.23	-7.29	12.15	1.61	-1.081
2.00	17.41	-13.78	-7.62	12.58	2.17	-1.200
2.20	20.05	-14.27	-7.86	12.96	2.76	-1.310
2.40	22.56	-14.77	-8.01	13.34	3.33	-1.423
2.60	25.03	-15.25	-8.12	13.71	3.92	-1.541
2.80	27.45	-15.65	-8.18	14.00	4.54	-1.649
3.00	29.65	-15.92	-8.17	14.18	5.15	-1.736
3.20	31.32	-16.03	-8.01	14.24	5.68	-1.799
3.40	32.11	-15.95	-7.66	14.12	6.07	-1.833
3.60	31.65	-15.61	-7.04	13.78	6.32	-1.826
3.80	29.63	-14.91	-6.12	13.15	6.38	-1.764
4.00	25.92	-13.82	-4.93	12.18	6.21	-1.635
4.20	20.59	-12.36	-3.53	10.92	5.77	-1.439
4.40	13.92	-10.62	-2.02	9.43	5.05	-1.184
4.60	6.37	-8.74	-0.50	7.85	4.07	-0.889
4.80	-1.56	-6.87	0.93	6.30	2.91	-0.572
5.00	-9.41	-5.13	2.21	4.88	1.65	-0.254

FIXED-T AMPLITUDES AT UNPHYSICAL AND LOW ENERGIES T=-.10 GEV**2
NU=(S-U)/4M AND AMPLITUDES IN NATURAL UNITS ISOSPIN EVEN:+, ISOSPIN ODD:-

NU**2	RE $\bar{A}$+	RE A-/NU	RE $\tilde{B}$+/NU	RE $\bar{B}$-	RE $\bar{C}$+	RE $\tilde{C}$-/NU
0.00	-6.88	-7.08	-4.78	7.58	-6.88	0.290
0.10	-6.40	-7.22	-4.95	7.74	-6.88	0.306
0.20	-5.89	-7.36	-5.14	7.90	-6.89	0.324
0.30	-5.34	-7.52	-5.36	8.09	-6.90	0.345
0.40	-4.74	-7.68	-5.60	8.29	-6.92	0.371
0.50	-4.08	-7.87	-5.88	8.51	-6.94	0.405
0.60	-3.35	-8.07	-6.21	8.77	-6.97	0.452
0.70	-2.47	-8.33	-6.72	9.11	-7.04	0.531
0.80	-1.55	-8.61	-7.21	9.45	-7.16	0.579
0.90	-0.62	-8.93	-7.58	9.80	-7.26	0.597
1.00	0.34	-9.23	-7.86	10.11	-7.31	0.603
1.20	2.36	-9.82	-8.35	10.71	-7.38	0.594
1.40	4.74	-10.44	-8.91	11.32	-7.39	0.562
1.60	7.24	-10.99	-9.34	11.82	-7.29	0.509
1.80	9.56	-11.47	-9.54	12.25	-7.14	0.434
2.00	11.72	-11.97	-9.59	12.66	-6.93	0.338
2.20	13.83	-12.45	-9.56	13.03	-6.63	0.221
2.40	15.76	-12.82	-9.43	13.27	-6.23	0.083
2.60	17.23	-12.98	-9.09	13.28	-5.75	-0.067
2.80	17.87	-12.89	-8.46	13.03	-5.17	-0.221

FIXED-T AMPLITUDES AT UNPHYSICAL AND LOW ENERGIES T=-.10 GEV**2
NU=(S-U)/4M AND AMPLITUDES IN NATURAL UNITS ISOSPIN EVEN:+, ISOSPIN ODD:-

NU**2	RE Ã+	RE A-/NU	RE B̃+/NU	RE B̃-	RE C̃+	RE C̃-/NU
3.00	17.36	-12.47	-7.50	12.44	-4.53	-0.372
3.20	15.49	-11.71	-6.21	11.51	-3.85	-0.518
3.40	12.16	-10.61	-4.64	10.23	-3.19	-0.655
3.60	7.49	-9.22	-2.88	8.68	-2.59	-0.778
3.80	1.75	-7.65	-1.04	6.96	-2.10	-0.877
4.00	-4.62	-6.02	0.74	5.22	-1.73	-0.943
4.20	-11.16	-4.44	2.37	3.56	-1.50	-0.971
4.40	-17.44	-3.00	3.75	2.09	-1.41	-0.962
4.60	-23.13	-1.75	4.85	0.85	-1.44	-0.921
4.80	-28.07	-0.72	5.68	-0.15	-1.55	-0.859
5.00	-32.21	0.10	6.27	-0.91	-1.74	-0.783

FIXED-T AMPLITUDES AT UNPHYSICAL AND LOW ENERGIES T=-.20 GEV**2
NU=(S-U)/4M AND AMPLITUDES IN NATURAL UNITS ISOSPIN EVEN:+, ISOSPIN ODD:-

NU**2	RE Ã+	RE A-/NU	RE B̃+/NU	RE B̃-	RE C̃+	RE C̃-/NU
0.00	-12.40	-5.30	-6.58	7.45	-12.40	1.748
0.10	-11.97	-5.42	-6.92	7.68	-12.63	1.850
0.20	-11.51	-5.55	-7.32	7.94	-12.90	1.971
0.30	-10.99	-5.68	-7.82	8.25	-13.21	2.126
0.40	-10.37	-5.81	-8.61	8.65	-13.63	2.376
0.50	-9.78	-6.01	-9.27	9.11	-14.17	2.603
0.60	-9.13	-6.31	-9.79	9.56	-14.69	2.732
0.70	-8.32	-6.57	-10.22	9.93	-15.09	2.829
0.80	-7.43	-6.84	-10.58	10.30	-15.44	2.903
0.90	-6.58	-7.15	-10.84	10.68	-15.80	2.955
1.00	-5.76	-7.45	-11.00	11.03	-16.17	2.990
1.20	-4.07	-7.91	-11.10	11.55	-16.68	3.016
1.40	-2.24	-8.38	-11.08	11.99	-16.92	2.961
1.60	-0.39	-8.95	-11.03	12.47	-17.09	2.847
1.80	1.34	-9.44	-10.83	12.83	-17.10	2.701
2.00	2.85	-9.72	-10.35	12.89	-16.74	2.481
2.20	3.91	-9.79	-9.50	12.59	-15.85	2.119
2.40	4.26	-9.62	-8.24	11.86	-14.45	1.595
2.60	3.58	-9.15	-6.59	10.67	-12.62	0.950
2.80	1.66	-8.32	-4.58	9.06	-10.48	0.260
3.00	-1.52	-7.18	-2.34	7.15	-8.17	-0.412
3.20	-5.73	-5.86	-0.05	5.11	-5.88	-1.018
3.40	-10.55	-4.50	2.10	3.14	-3.79	-1.534
3.60	-15.48	-3.24	3.95	1.37	-2.01	-1.944
3.80	-20.07	-2.14	5.41	-0.11	-0.63	-2.241
4.00	-24.06	-1.23	6.45	-1.27	0.36	-2.430
4.20	-27.41	-0.50	7.15	-2.14	1.02	-2.524
4.40	-30.17	0.05	7.58	-2.74	1.40	-2.543
4.60	-32.48	0.45	7.82	-3.13	1.58	-2.509
4.80	-34.46	0.73	7.94	-3.36	1.61	-2.443
5.00	-36.17	0.92	7.97	-3.47	1.54	-2.360

FIXED-T AMPLITUDES AT UNPHYSICAL AND LOW ENERGIES T=-.30 GEV**2
NU=(S-U)/4M AND AMPLITUDES IN NATURAL UNITS ISOSPIN EVEN:+, ISOSPIN ODD:-

NU**2	RE Ã+	RE A-/NU	RE B̃+/NU	RE B̃-	RE C̃+	RE C̃-/NU
0.00	-18.86	-3.10	-9.04	7.91	-18.86	4.189
0.10	-18.68	-3.12	-9.78	8.31	-19.59	4.543
0.20	-18.54	-2.99	-10.97	8.82	-20.56	5.134
0.30	-18.46	-3.15	-12.08	9.57	-21.80	5.662
0.40	-18.08	-3.24	-12.76	9.97	-22.78	5.951

FIXED-T AMPLITUDES AT UNPHYSICAL AND LOW ENERGIES T=-.30 GEV**2
NU=(S-U)/4M AND AMPLITUDES IN NATURAL UNITS ISOSPIN EVEN:+, ISOSPIN ODD:-

NU**2	RE $\bar{A}$+	RE A-/NU	RE $\tilde{B}$+/NU	RE $\tilde{B}$-	RE $\bar{C}$+	RE $\tilde{C}$-/NU
0.50	-17.62	-3.54	-13.19	10.49	-23.69	6.126
0.60	-17.11	-3.88	-13.44	11.02	-24.54	6.271
0.70	-16.53	-4.08	-13.54	11.36	-25.26	6.390
0.80	-15.89	-4.20	-13.53	11.56	-25.86	6.450
0.90	-15.19	-4.41	-13.47	11.77	-26.36	6.439
1.00	-14.46	-4.74	-13.38	12.06	-26.79	6.375
1.20	-12.96	-5.50	-13.09	12.67	-27.44	6.179
1.40	-11.45	-6.03	-12.50	12.93	-27.58	5.885
1.60	-9.93	-6.39	-11.48	12.71	-26.86	5.318
1.80	-8.48	-6.70	-9.95	12.00	-24.99	4.360
2.00	-7.43	-6.79	-7.85	10.70	-21.91	3.063
2.20	-7.23	-6.50	-5.23	8.76	-17.83	1.577
2.40	-8.14	-5.84	-2.29	6.42	-13.22	0.069
2.60	-10.15	-5.00	0.65	3.99	-8.60	-1.324
2.80	-12.92	-4.14	3.29	1.78	-4.43	-2.500
3.00	-15.97	-3.33	5.42	-0.08	-0.99	-3.402
3.20	-18.92	-2.59	6.97	-1.55	1.64	-4.021
3.40	-21.53	-1.94	7.99	-2.66	3.51	-4.389
3.60	-23.74	-1.39	8.59	-3.45	4.76	-4.565
3.80	-25.62	-0.95	8.89	-3.97	5.52	-4.609
4.00	-27.25	-0.63	9.00	-4.28	5.91	-4.574
4.20	-28.69	-0.41	8.97	-4.43	6.05	-4.492
4.40	-29.95	-0.25	8.87	-4.48	6.00	-4.381
4.60	-31.03	-0.13	8.69	-4.47	5.82	-4.247
4.80	-31.91	-0.02	8.47	-4.43	5.56	-4.096
5.00	-32.59	0.08	8.21	-4.35	5.26	-3.934

FIXED-T AMPLITUDES AT UNPHYSICAL AND LOW ENERGIES T=-.40 GEV**2
NU=(S-U)/4M AND AMPLITUDES IN NATURAL UNITS ISOSPIN EVEN:+, ISOSPIN ODD:-

NU**2	RE $\bar{A}$+	RE A-/NU	RE $\tilde{B}$+/NU	RE $\tilde{B}$-	RE $\bar{C}$+	RE $\tilde{C}$-/NU
0.00	-27.07	0.44	-13.40	9.04	-27.07	8.558
0.10	-28.02	1.16	-15.81	10.04	-29.44	10.171
0.20	-28.57	1.10	-16.81	10.77	-31.59	10.771
0.30	-28.70	0.60	-17.08	11.53	-33.30	10.953
0.40	-28.42	0.33	-17.06	11.96	-34.55	11.069
0.50	-27.88	0.09	-16.91	12.25	-35.47	11.094
0.60	-27.27	-0.27	-16.65	12.57	-36.24	11.010
0.70	-26.65	-0.71	-16.29	12.87	-36.89	10.847
0.80	-25.99	-1.12	-15.84	13.09	-37.36	10.635
0.90	-25.22	-1.46	-15.29	13.17	-37.58	10.364
1.00	-24.30	-1.79	-14.64	13.12	-37.45	9.995
1.20	-22.03	-2.61	-12.92	12.67	-35.95	8.773
1.40	-19.45	-3.57	-10.41	11.52	-32.54	6.776
1.60	-17.01	-4.34	-7.09	9.47	-27.20	4.166
1.80	-15.09	-4.74	-3.27	6.75	-20.37	1.320
2.00	-14.01	-4.84	0.57	3.88	-12.98	-1.363
2.20	-13.83	-4.75	3.95	1.29	-6.02	-3.592
2.40	-14.43	-4.49	6.59	-0.82	-0.23	-5.235
2.60	-15.54	-4.12	8.42	-2.44	4.11	-6.306
2.80	-16.90	-3.68	9.53	-3.60	7.08	-6.910
3.00	-18.33	-3.26	10.13	-4.37	8.95	-7.181
3.20	-19.71	-2.90	10.36	-4.83	10.05	-7.233
3.40	-21.02	-2.60	10.36	-5.06	10.62	-7.149
3.60	-22.22	-2.35	10.22	-5.15	10.81	-6.975
3.80	-23.30	-2.10	9.97	-5.17	10.73	-6.740

FIXED-T AMPLITUDES AT UNPHYSICAL AND LOW ENERGIES T=-.40 GEV**2
NU=(S-U)/4M AND AMPLITUDES IN NATURAL UNITS ISOSPIN EVEN:+, ISOSPIN ODD:-

NU**2	RE $\bar{A}$+	RE A-/NU	RE $\tilde{B}$+/NU	RE $\tilde{B}$-	RE $\bar{C}$+	RE $\tilde{C}$-/NU
4.00	-24.26	-1.86	9.66	-5.13	10.44	-6.468
4.20	-25.08	-1.64	9.30	-5.06	10.01	-6.178
4.40	-25.78	-1.46	8.92	-4.93	9.49	-5.889
4.60	-26.38	-1.35	8.55	-4.75	8.93	-5.616
4.80	-26.92	-1.30	8.19	-4.53	8.39	-5.367
5.00	-27.42	-1.29	7.86	-4.30	7.88	-5.143

FIXED-T AMPLITUDES AT UNPHYSICAL AND LOW ENERGIES T=-.50 GEV**2
NU=(S-U)/4M AND AMPLITUDES IN NATURAL UNITS ISOSPIN EVEN:+, ISOSPIN ODD:-

NU**2	RE $\bar{A}$+	RE A-/NU	RE $\tilde{B}$+/NU	RE $\tilde{B}$-	RE $\bar{C}$+	RE $\tilde{C}$-/NU
0.00	-38.35	9.70	-21.55	10.29	-38.35	18.714
0.10	-40.44	8.81	-22.24	11.45	-42.38	18.836
0.20	-41.05	7.83	-22.05	12.22	-44.92	18.527
0.30	-41.00	6.83	-21.60	12.79	-46.68	18.034
0.40	-40.62	5.85	-20.97	13.20	-47.97	17.406
0.50	-40.08	4.99	-20.40	13.54	-49.02	16.845
0.60	-39.23	4.11	-19.76	13.78	-49.61	16.176
0.70	-37.94	3.12	-18.80	13.82	-49.46	15.225
0.80	-36.20	2.02	-17.37	13.58	-48.37	13.909
0.90	-34.05	0.82	-15.43	13.00	-46.21	12.203
1.00	-31.55	-0.45	-13.04	12.09	-42.97	10.134
1.20	-25.92	-3.00	-7.34	9.43	-33.63	5.249
1.40	-20.48	-5.11	-1.38	6.08	-22.17	0.216
1.60	-16.31	-6.45	3.83	2.72	-10.95	-4.061
1.80	-13.78	-7.05	7.70	-0.13	-1.64	-7.158
2.00	-12.63	-7.14	10.19	-2.24	5.21	-9.101
2.20	-12.43	-6.94	11.55	-3.65	9.82	-10.139
2.40	-12.81	-6.58	12.13	-4.53	12.69	-10.550
2.60	-13.57	-6.14	12.24	-5.05	14.30	-10.563
2.80	-14.55	-5.68	12.06	-5.32	15.03	-10.340
3.00	-15.61	-5.24	11.73	-5.42	15.20	-9.989
3.20	-16.65	-4.84	11.30	-5.41	15.01	-9.574
3.40	-17.60	-4.48	10.82	-5.32	14.61	-9.134
3.60	-18.47	-4.16	10.32	-5.17	14.07	-8.689
3.80	-19.26	-3.89	9.82	-4.99	13.43	-8.253
4.00	-20.02	-3.66	9.35	-4.78	12.73	-7.839
4.20	-20.77	-3.47	8.91	-4.55	12.02	-7.458
4.40	-21.48	-3.32	8.52	-4.34	11.35	-7.117
4.60	-22.15	-3.18	8.17	-4.15	10.77	-6.814
4.80	-22.77	-3.05	7.86	-3.99	10.28	-6.542
5.00	-23.32	-2.90	7.58	-3.87	9.87	-6.289

FIXED-T AMPLITUDES AT UNPHYSICAL AND LOW ENERGIES T=-.60 GEV**2
NU=(S-U)/4M AND AMPLITUDES IN NATURAL UNITS ISOSPIN EVEN:+, ISOSPIN ODD:-

NU**2	RE $\bar{A}$+	RE A-/NU	RE $\tilde{B}$+/NU	RE $\tilde{B}$-	RE $\bar{C}$+	RE $\tilde{C}$-/NU
0.00	-56.75	19.18	-31.20	13.35	-56.75	30.594
0.10	-56.51	17.49	-29.48	13.95	-59.03	29.412
0.20	-56.18	14.53	-27.05	14.41	-60.81	26.843
0.30	-55.56	12.47	-25.41	14.75	-62.07	25.075
0.40	-54.07	10.37	-23.60	14.85	-62.14	23.057
0.50	-51.46	7.95	-21.02	14.49	-60.44	20.333
0.60	-47.72	5.21	-17.58	13.60	-56.73	16.832
0.70	-43.00	2.24	-13.49	12.22	-51.06	12.683
0.80	-37.61	-0.76	-9.02	10.45	-43.78	8.167
0.90	-31.99	-3.55	-4.51	8.41	-35.46	3.630

FIXED-T AMPLITUDES AT UNPHYSICAL AND LOW ENERGIES T=-.60 GEV**2
NU=(S-U)/4M AND AMPLITUDES IN NATURAL UNITS ISOSPIN EVEN:+, ISOSPIN ODD:-

NU**2	RE $\tilde{A}$+	RE A-/NU	RE $\tilde{B}$+/NU	RE $\tilde{B}$-	RE $\tilde{C}$+	RE $\tilde{C}$-/NU
1.00	-26.58	-5.96	-0.24	6.26	-26.79	-0.618
1.20	-17.63	-9.34	6.77	2.19	-10.69	-7.471
1.40	-11.84	-10.98	11.31	-0.98	1.69	-11.816
1.60	-8.75	-11.43	13.71	-3.10	9.99	-14.079
1.80	-7.50	-11.21	14.65	-4.36	15.03	-14.941
2.00	-7.41	-10.65	14.75	-5.06	17.79	-14.972
2.20	-8.01	-9.94	14.41	-5.41	19.07	-14.561
2.40	-8.96	-9.21	13.84	-5.54	19.42	-13.940
2.60	-10.03	-8.51	13.17	-5.53	19.23	-13.235
2.80	-11.10	-7.88	12.47	-5.41	18.74	-12.508
3.00	-12.13	-7.31	11.77	-5.24	18.04	-11.790
3.20	-13.13	-6.82	11.10	-5.02	17.21	-11.105
3.40	-14.10	-6.39	10.48	-4.78	16.33	-10.475
3.60	-15.03	-6.03	9.92	-4.54	15.49	-9.914
3.80	-15.90	-5.73	9.44	-4.33	14.75	-9.425
4.00	-16.69	-5.45	9.02	-4.15	14.15	-8.996
4.20	-17.39	-5.17	8.65	-4.01	13.66	-8.604
4.40	-18.01	-4.89	8.30	-3.90	13.20	-8.226
4.60	-18.54	-4.61	7.95	-3.79	12.71	-7.850
4.80	-18.98	-4.34	7.60	-3.67	12.17	-7.473
5.00	-19.34	-4.09	7.23	-3.54	11.57	-7.107

For Table 2.2.3.3, see next page.

2.2.4 *s*-channel helicity amplitudes

In the high energy region, most of the authors preferred to work with *s*-channel helicity amplitudes. These are a natural choice in various models, and the simplicity of the connection with the measurable quantities, Eq. (A.4.13), is an advantage in amplitude analysis. We have listed in Table 2.2.4.1 the isospin even and odd combinations of the amplitudes at fixed values of *t*, because this simplifies the discussion of the dominant mechanisms: diffraction and reggeized ϱ-exchange. Furthermore, the isospin even and charge-exchange differential cross sections are given in the last two columns.

The amplitudes have been calculated from the preliminary updated solution KH 80 (Table 2.2.2.2). Beyond $t=-0.5\,\mathrm{GeV}^2$ the decomposition into no-flip and flip contributions is rather uncertain, because spin-rotation data are not available and the constraints from suitable one-dimensional dispersion relations have not yet been studied.

If one compares our table with those of earlier authors, one should note that there exist different conventions, of which some are listed in the following table.

Table 2.2.4.2. Various notations for the πN *s*-channel helicity amplitudes.

Our notation 72 L Barger-4 76 L Pierrard 78 L Morrow	71 L Halzen 72 L Cozzika 74 L Ambats	73 L Johnson 74 L Miller	70 R Martin-Spearman 74 R Hamilton-Petersen
F_{++}	F_{++}	H_{++}	$f_{++}\sqrt{\pi}/q$
F_{+-}	$-F_{+-}$	$-H_{+-}$	$f_{+-}\sqrt{\pi}/q$

We have denoted the isospin even and odd combinations of all amplitudes (Table A.2.1) by upper indices + and −, respectively. Upper indices $I_t=0$ and 1 are introduced by Eq. (A.2.78). Earlier authors frequently used upper indices 0 and 1 for the isospin even and odd combinations. – Instead of our units ($\sqrt{\mathrm{mb}}/\mathrm{GeV}$) one could take GeV^{-2}.

For Table 2.2.4.1, see p. 145.

Table 2.2.3.3. Invariant amplitudes at fixed c.m. scattering angle as calculated from the KH 78 partial wave solution.

INVARIANT FIXED C.M.ANGLE AMPLITUDES — ISOSPIN EVEN AND ODD:+- — THETA= 0 DEGREES

K GEV/C	RE A+ 1/GEV	IM A+ 1/GEV	RE A- 1/GEV	IM A- 1/GEV	RE B+ 1/GEV**2	IM B+ 1/GEV**2	RE B- 1/GEV**2	IM B- 1/GEV**2	RE C+ 1/GEV	IM C+ 1/GEV	RE C- 1/GEV	IM C- 1/GEV
.020	229.4	.2	-77.9	.1	-1629.	0.	617.0	.0	-.35	.21	9.09	.09
.040	231.0	1.0	-79.9	-.1	-1587.	-4.	607.8	2.0	.52	.43	8.33	.18
.060	240.4	2.6	-86.2	-.9	-1569.	-13.	614.1	7.5	2.05	.69	7.10	.24
.079	261.7	6.1	-101.7	-2.5	-1600.	-31.	669.8	17.4	5.03	1.14	5.77	.29
.097	271.9	11.3	-109.0	-5.0	-1554.	-56.	664.4	30.4	7.63	1.69	3.96	.18
.112	287.2	18.1	-119.3	-8.4	-1547.	-88.	678.6	46.3	10.35	2.42	2.14	-.08
.130	305.9	30.5	-131.9	-14.5	-1527.	-140.	688.3	72.5	14.55	3.79	-.56	-.70
.153	337.9	57.8	-153.2	-28.0	-1528.	-244.	718.2	124.3	21.44	7.19	-4.45	-2.29
.172	364.5	92.7	-171.7	-45.3	-1518.	-364.	738.7	183.9	28.33	12.00	-8.01	-4.57
.185	386.3	124.3	-186.0	-61.1	-1522.	-465.	755.9	233.8	33.65	16.48	-10.80	-6.89
.200	409.7	175.5	-199.9	-86.7	-1516.	-620.	762.7	310.8	39.80	24.23	-13.84	-10.89
.218	424.2	263.3	-210.5	-130.7	-1461.	-868.	748.0	434.5	46.07	38.53	-16.90	-18.17
.247	355.7	443.8	-190.3	-220.4	-1103.	-1309.	620.6	654.0	42.62	72.45	-14.21	-34.82
.267	213.3	538.4	-123.2	-267.8	-625.	-1474.	394.4	737.1	24.76	94.12	-4.26	-45.59
.280	115.1	553.6	-78.1	-275.6	-334.	-1448.	259.0	724.5	10.43	100.54	2.98	-48.89
.290	47.0	544.8	-48.8	-271.1	-146.	-1376.	177.0	688.8	-.04	101.80	8.19	-49.38
.295	12.1	535.2	-34.1	-266.4	-60.	-1328.	141.2	664.5	-7.55	101.86	12.01	-49.47
.301	-22.6	519.0	-19.4	-258.3	23.	-1262.	107.1	631.8	-15.01	100.83	16.04	-48.94
.305	-47.6	503.9	-8.3	-251.0	83.	-1206.	80.7	604.6	-19.61	99.37	18.75	-48.21
.310	-75.0	482.9	4.1	-241.0	147.	-1135.	54.0	570.5	-25.21	96.98	22.46	-47.04
.320	-117.2	440.9	20.0	-219.8	236.	-999.	20.0	502.8	-34.62	92.03	26.95	-44.31
.331	-150.4	403.7	34.5	-201.6	313.	-883.	-14.2	446.0	-37.94	86.56	29.39	-41.40
.351	-179.3	325.1	40.9	-163.7	351.	-661.	-13.3	339.8	-46.74	75.44	35.85	-35.41
.378	-200.2	239.7	36.6	-124.6	374.	-441.	6.7	239.6	-49.43	62.00	39.34	-27.97
.408	-202.1	178.7	20.5	-98.7	360.	-293.	48.5	177.5	-46.93	52.50	41.38	-22.11
.427	-209.0	147.7	1.1	-91.5	366.	-222.	93.3	162.0	-44.41	48.04	42.97	-18.70
.456	-196.5	92.5	-9.4	-83.2	328.	-103.	112.4	148.2	-39.94	43.55	44.21	-12.50
.490	-177.8	53.1	-30.0	-81.1	281.	-21.	145.7	146.8	-34.70	42.54	44.25	-6.27
.532	-178.1	11.4	-28.6	-86.9	263.	52.	130.6	157.2	-33.23	39.98	43.22	-.43
.573	-149.6	-28.0	-42.4	-80.0	206.	116.	144.4	144.6	-28.15	40.62	42.81	5.27
.614	-134.6	-63.0	-72.1	-86.1	178.	167.	182.4	153.9	-22.24	42.37	42.70	10.76
.658	-88.2	-102.8	-61.0	-116.3	106.	225.	153.8	201.8	-17.14	48.35	42.44	19.45
.675	-69.1	-110.8	-54.4	-122.6	76.	235.	135.6	210.8	-16.86	51.11	39.06	22.75
.705	-40.6	-115.1	-47.9	-129.8	32.	238.	114.5	220.1	-17.62	56.17	34.36	28.38
.725	-23.7	-106.8	-40.4	-127.6	7.	222.	94.4	212.6	-18.78	57.45	29.26	29.37

INVARIANT FIXED C.M.ANGLE AMPLITUDES ISOSPIN EVEN AND ODD:+- THETA= 0 DEGREES

K GEV/C	RE A+ 1/GEV	IM A+ 1/GEV	RE A- 1/GEV	IM A- 1/GEV	RE B+ 1/GEV**2	IM B+ 1/GEV**2	RE B- 1/GEV**2	IM B- 1/GEV**2	RE C+ 1/GEV	IM C+ 1/GEV	RE C- 1/GEV	IM C- 1/GEV
.750	-6.8	-101.3	-36.0	-125.3	-16.	207.	79.3	201.6	-19.31	56.97	24.50	28.48
.777	2.4	-100.2	-29.7	-115.3	-25.	198.	63.6	178.3	-17.70	55.75	20.48	25.41
.800	8.3	-95.8	-30.0	-110.3	-28.	186.	62.0	164.3	-14.18	55.37	20.28	23.15
.822	11.5	-94.7	-28.6	-101.4	-26.	181.	59.1	147.1	-9.76	55.90	20.64	21.26
.851	18.6	-89.0	-40.2	-95.4	-25.	172.	72.9	134.8	-3.14	59.41	22.63	20.87
.875	17.1	-88.8	-50.6	-90.8	-16.6	173.8	85.5	127.3	2.46	65.18	25.17	22.07
.895	20.0	-90.7	-60.7	-91.5	-15.4	179.0	96.5	128.2	6.07	71.39	26.68	24.67
.923	22.9	-102.9	-71.7	-100.4	-15.5	198.7	106.2	140.7	8.44	82.58	27.41	30.98
.954	33.8	-119.3	-79.8	-112.4	-29.3	222.6	107.2	156.7	5.61	95.36	23.61	38.68
.975	48.4	-132.3	-78.5	-120.8	-48.6	238.8	97.8	166.5	.54	102.94	17.83	43.28
1.00	77.3	-147.3	-71.1	-136.0	-83.7	256.6	79.6	180.7	-7.23	111.85	9.26	46.53
1.03	104.0	-136.5	-61.4	-134.4	-115.8	237.6	58.5	171.4	-16.41	110.55	-.57	43.79
1.06	130.9	-121.3	-51.7	-131.4	-143.1	215.3	41.4	160.0	-21.38	107.78	-7.63	38.88
1.08	144.0	-100.4	-43.5	-118.2	-154.2	187.5	29.2	138.0	-23.95	103.82	-11.75	32.06
1.11	154.7	-69.0	-49.7	-99.8	-158.6	151.2	32.0	110.5	-23.24	100.58	-13.79	24.16
1.15	143.0	-52.3	-64.0	-82.0	-139.4	131.3	43.8	85.0	-19.11	100.33	-13.05	16.77
1.17	133.3	-44.6	-69.5	-84.0	-127.4	123.7	48.1	83.5	-17.25	101.66	-12.58	14.64
1.21	140.6	-46.1	-91.2	-93.4	-127.5	124.7	65.3	85.8	-14.62	105.81	-11.70	11.11
1.24	159.7	-37.6	-94.6	-103.5	-139.0	118.9	67.6	91.0	-13.16	110.16	-10.57	9.66
1.28	173.4	-8.5	-87.4	-120.1	-143.3	98.1	60.7	96.7	-11.13	117.91	-9.21	4.49
1.32	179.1	19.5	-79.3	-124.4	-143.0	80.3	54.5	93.1	-11.33	126.48	-6.80	-.44
1.36	168.7	41.9	-76.8	-120.9	-133.9	67.0	53.5	85.1	-14.39	133.44	-3.60	-4.47
1.40	149.4	75.9	-84.0	-119.2	-119.7	45.4	59.9	79.5	-19.05	139.69	.20	-7.34
1.43	128.5	86.5	-88.9	-119.4	-104.8	39.8	64.2	77.0	-22.03	143.68	3.28	-8.69
1.47	102.3	96.0	-99.2	-123.1	-88.1	34.3	72.9	76.7	-28.07	146.74	8.70	-9.69
1.51	84.8	111.6	-103.2	-133.1	-76.4	24.8	76.7	81.6	-30.66	149.04	12.66	-9.78
1.55	54.6	90.4	-99.1	-137.0	-57.4	36.8	74.2	83.9	-34.84	147.63	16.44	-6.35
1.59	41.9	80.3	-96.7	-140.8	-49.3	42.3	71.8	85.8	-36.75	147.88	17.84	-3.90
1.64	28.4	63.4	-92.9	-144.4	-41.5	50.8	69.6	87.6	-40.02	146.97	21.65	-.30
1.68	21.3	45.5	-92.1	-141.5	-36.6	61.3	68.4	85.0	-40.36	148.88	23.15	1.76
1.72	19.4	36.6	-93.1	-141.7	-34.4	65.0	67.5	84.7	-39.99	148.79	23.45	4.39
1.76	15.7	25.2	-92.9	-143.4	-31.0	70.9	66.1	85.1	-39.03	150.46	23.77	6.88
1.80	17.0	11.3	-93.3	-144.5	-30.2	78.3	64.6	85.0	-37.60	152.60	23.38	8.87
1.84	19.8	.2	-92.2	-147.5	-30.9	84.0	62.8	86.3	-37.27	155.15	23.62	11.78
1.88	25.6	-12.2	-88.8	-147.7	-32.6	90.7	58.8	85.6	-35.80	158.76	22.17	13.72
1.92	35.1	-22.0	-86.4	-148.5	-37.2	95.3	56.0	85.3	-36.44	161.54	21.49	15.64
1.98	53.7	-28.6	-81.0	-146.1	-45.1	98.8	49.8	82.5	-35.80	167.60	17.73	17.64
2.03	72.8	-31.0	-80.4	-139.2	-54.5	99.1	47.6	77.7	-38.08	170.71	16.47	18.92
2.07	87.8	-23.1	-80.1	-133.6	-60.5	95.1	45.4	73.7	-37.65	174.21	13.99	19.23

INVARIANT FIXED C.M.ANGLE AMPLITUDES ISOSPIN EVEN AND ODD:+- THETA= 0 DEGREES

K GEV/C	RE A+ 1/GEV	IM A+ 1/GEV	RE A- 1/GEV	IM A- 1/GEV	RE B+ 1/GEV**2	IM B+ 1/GEV**2	RE B- 1/GEV**2	IM B- 1/GEV**2	RE C+ 1/GEV	IM C+ 1/GEV	RE C- 1/GEV	IM C- 1/GEV
2.15	104.5	-4.5	-86.2	-124.3	-66.4	85.6	45.2	66.4	-38.63	179.96	11.17	18.66
2.20	107.2	11.9	-91.4	-118.4	-66.5	78.3	45.5	61.9	-39.29	184.46	8.86	17.93
2.28	106.4	33.5	-105.5	-116.1	-64.2	69.0	49.5	57.8	-40.34	191.07	7.64	16.02
2.34	97.8	44.7	-114.8	-116.8	-59.6	64.5	51.7	55.9	-41.94	195.99	6.42	14.21
2.46	79.4	57.4	-133.1	-127.5	-50.6	59.6	57.3	56.1	-45.32	204.18	8.04	10.82
2.56	65.7	56.7	-143.9	-137.8	-44.8	60.2	60.1	57.5	-49.10	211.09	10.29	9.70
2.75	54.8	47.1	-148.5	-161.2	-39.4	63.9	59.5	61.8	-53.60	223.05	15.20	9.07
3.00	59.0	29.0	-143.2	-175.4	-39.1	69.8	53.4	62.5	-58.39	238.74	17.26	12.13

INVARIANT FIXED C.M.ANGLE AMPLITUDES ISOSPIN EVEN AND ODD:+- THETA= 30 DEGREES

K GEV/C	RE A+ 1/GEV	IM A+ 1/GEV	RE A- 1/GEV	IM A- 1/GEV	RE B+ 1/GEV**2	IM B+ 1/GEV**2	RE B- 1/GEV**2	IM B- 1/GEV**2	RE C+ 1/GEV	IM C+ 1/GEV	RE C- 1/GEV	IM C- 1/GEV
.020	229.3	.2	-77.9	.1	-1629.	0.	617.0	.0	-.39	.21	9.13	.09
.040	230.7	1.0	-79.7	-.1	-1587.	-4.	607.9	2.0	.34	.43	8.48	.18
.060	239.9	2.6	-86.0	-.9	-1571.	-13.	615.5	7.5	1.64	.69	7.42	.24
.079	260.8	6.1	-101.2	-2.5	-1603.	-31.	671.9	17.4	4.23	1.11	6.34	.30
.097	270.5	11.2	-108.2	-4.9	-1558.	-57.	666.7	30.4	6.43	1.62	4.76	.22
.112	285.2	17.9	-118.3	-8.3	-1551.	-88.	681.4	46.3	8.71	2.28	3.18	-.01
.130	303.1	30.1	-130.4	-14.3	-1533.	-140.	691.4	72.6	12.24	3.49	.81	-.55
.153	333.8	56.8	-150.9	-27.6	-1535.	-245.	721.0	124.5	18.05	6.48	-2.56	-1.94
.172	359.4	91.0	-168.7	-44.4	-1526.	-365.	741.8	184.3	23.91	10.70	-5.63	-3.93
.185	379.9	121.7	-182.3	-59.8	-1530.	-466.	757.9	234.4	28.46	14.56	-8.16	-5.94
.200	402.2	171.6	-195.8	-84.7	-1525.	-622.	765.2	311.7	33.72	21.27	-10.93	-9.42
.218	415.1	256.8	-205.6	-127.4	-1469.	-871.	749.2	436.0	38.93	33.68	-13.68	-15.76
.247	347.4	431.6	-184.7	-214.3	-1112.	-1314.	618.9	656.8	35.58	63.20	-11.27	-30.21
.267	209.8	522.6	-120.0	-259.9	-638.	-1481.	394.4	740.7	20.07	82.05	-2.72	-39.57
.280	114.4	536.7	-76.2	-267.1	-346.	-1456.	258.2	728.3	7.62	87.58	3.43	-42.44
.290	47.2	527.7	-46.9	-262.5	-153.	-1384.	173.2	692.7	-1.50	88.65	7.97	-42.84
.295	13.0	518.1	-32.3	-257.8	-66.	-1335.	135.4	668.3	-8.23	88.72	11.22	-42.94
.301	-20.9	502.2	-17.9	-249.9	19.	-1269.	99.7	635.5	-14.81	87.85	14.66	-42.50
.305	-45.1	487.4	-7.0	-242.8	79.	-1214.	72.6	608.2	-18.85	86.59	16.99	-41.87
.310	-71.6	466.9	5.1	-233.0	143.	-1143.	44.8	574.0	-23.77	84.52	20.14	-40.87
.320	-112.2	425.9	20.3	-212.3	234.	-1006.	11.1	506.0	-31.91	80.29	24.09	-38.51
.331	-144.4	389.7	34.1	-194.5	310.	-890.	-22.6	449.0	-34.78	75.56	26.12	-35.99
.351	-173.4	313.3	40.4	-157.8	352.	-667.	-23.7	342.2	-42.77	66.02	31.56	-30.86
.378	-192.0	230.4	36.8	-119.8	372.	-445.	-6.0	241.3	-45.07	54.41	34.41	-24.41
.408	-193.4	171.4	21.9	-94.8	356.	-296.	33.0	178.6	-43.06	46.18	35.82	-19.34

INVARIANT FIXED C.M.ANGLE AMPLITUDES ISOSPIN EVEN AND ODD:+- THETA= 30 DEGREES

K GEV/C	RE A+ 1/GEV	IM A+ 1/GEV	RE A- 1/GEV	IM A- 1/GEV	RE B+ 1/GEV**2	IM B+ 1/GEV**2	RE B- 1/GEV**2	IM B- 1/GEV**2	RE C+ 1/GEV	IM C+ 1/GEV	RE C- 1/GEV	IM C- 1/GEV
.427	-200.9	141.3	5.3	-87.8	363.	-225.	71.5	162.4	-41.49	42.23	36.72	-16.44
.456	-187.1	88.8	-5.0	-79.3	319.	-109.	91.4	145.9	-38.53	38.12	37.58	-11.31
.490	-169.4	49.7	-22.7	-78.1	271.	-25.	120.9	144.6	-34.51	37.32	37.37	-6.22
.532	-159.0	10.0	-23.3	-82.0	235.	46.	111.1	150.6	-33.40	34.44	36.24	-1.30
.573	-135.0	-24.6	-31.9	-77.1	184.	102.	116.5	141.2	-29.60	34.13	34.86	3.81
.614	-118.0	-54.2	-45.4	-81.1	148.	145.	126.2	145.6	-27.65	34.60	31.75	7.88
.658	-85.6	-86.4	-45.6	-99.3	94.	189.	115.7	172.1	-24.17	36.49	29.78	12.76
.675	-70.8	-95.7	-41.4	-104.0	71.	200.	103.1	178.4	-23.62	37.76	27.39	15.02
.705	-35.8	-101.7	-34.9	-108.0	17.	205.	83.8	182.9	-23.63	40.34	23.29	18.99
.725	-18.3	-96.3	-27.1	-106.9	-9.	193.	64.7	177.2	-24.61	41.28	19.01	19.41
.750	-8.9	-88.9	-17.4	-103.0	-22.	177.	43.0	165.5	-24.99	41.47	14.22	18.74
.777	-4.8	-85.1	-9.6	-91.6	-27.	164.	26.0	141.0	-25.21	39.38	10.21	15.62
.800	1.4	-79.6	-8.1	-86.2	-33.	151.	23.2	127.2	-24.48	38.34	10.04	13.20
.822	8.6	-75.3	-11.7	-75.9	-36.	141.	26.3	107.9	-20.29	38.14	9.39	10.67
.851	12.8	-70.0	-18.5	-69.9	-34.	133.	36.0	95.6	-15.41	39.91	11.31	9.31
.875	14.3	-67.3	-23.7	-67.1	-30.7	130.3	42.8	90.2	-11.82	43.52	12.72	9.57
.895	11.8	-70.7	-29.2	-66.4	-25.1	136.4	49.5	88.2	-10.00	47.81	13.80	10.24
.923	13.3	-74.7	-35.5	-68.2	-25.6	144.4	55.3	90.3	-9.63	54.37	13.96	12.54
.954	19.2	-83.0	-40.0	-74.9	-33.9	156.2	56.6	98.1	-12.03	61.14	12.23	15.52
.975	26.2	-89.4	-37.8	-78.7	-43.1	162.8	50.4	101.3	-14.42	63.83	9.70	16.75
1.00	40.8	-97.3	-35.0	-84.8	-62.8	172.1	44.1	106.8	-19.71	68.61	7.56	18.13
1.03	62.8	-91.6	-25.8	-83.7	-86.5	160.3	29.4	101.8	-22.96	67.31	3.37	17.17
1.06	77.3	-81.5	-17.5	-78.9	-101.8	146.4	18.1	93.3	-25.89	66.99	.81	15.75
1.08	87.2	-70.0	-14.5	-68.8	-111.1	131.1	13.0	79.7	-27.94	65.93	-1.09	13.83
1.11	92.1	-53.9	-18.7	-58.7	-113.6	111.1	15.0	65.4	-29.08	64.54	-2.66	11.00
1.15	96.6	-41.5	-31.0	-50.7	-113.6	95.3	25.2	53.2	-28.67	63.66	-3.24	8.06
1.17	99.9	-37.7	-37.8	-49.2	-113.9	90.2	30.9	50.0	-27.80	63.39	-3.18	6.94
1.21	105.0	-25.5	-48.3	-51.3	-113.8	77.8	38.8	49.0	-26.30	64.17	-3.55	5.26
1.24	106.5	-17.3	-53.2	-54.5	-113.0	70.4	41.6	49.9	-26.40	65.46	-4.21	4.17
1.28	107.6	-.2	-58.2	-63.8	-110.7	55.6	43.6	54.1	-26.99	67.50	-5.13	1.95
1.32	105.4	16.0	-57.0	-72.8	-106.4	42.5	41.1	57.9	-28.19	69.37	-5.41	-.02
1.36	102.3	32.7	-53.1	-76.5	-101.4	29.3	37.0	58.2	-28.22	70.38	-5.53	-1.56
1.40	91.6	48.2	-48.5	-78.5	-91.1	18.1	33.6	56.8	-28.91	72.15	-4.00	-3.40
1.43	80.4	58.1	-46.6	-79.0	-81.7	11.0	32.1	55.4	-29.94	72.90	-3.28	-4.24
1.47	59.3	65.0	-43.5	-79.8	-65.5	6.7	30.2	54.0	-31.52	74.35	-1.57	-4.92
1.51	44.4	63.6	-41.3	-79.9	-55.1	8.9	28.6	52.5	-33.63	76.13	-.75	-5.62
1.55	25.3	55.9	-36.9	-80.6	-41.8	13.2	25.6	51.6	-35.62	75.13	.38	-5.47
1.59	15.7	45.3	-32.0	-79.2	-35.4	19.2	22.2	49.5	-37.10	73.95	1.10	-5.45
1.64	10.6	36.1	-28.5	-76.4	-31.5	24.2	19.7	46.3	-37.66	73.25	1.67	-5.30

INVARIANT FIXED C.M.ANGLE AMPLITUDES ISOSPIN EVEN AND ODD:+- THETA= 30 DEGREES

K	RE A+	IM A+	RE A-	IM A-	RE B+	IM B+	RE B-	IM B-	RE C+	IM C+	RE C-	IM C-
GEV/C	1/GEV		1/GEV		1/GEV**2		1/GEV**2		1/GEV		1/GEV	
1.68	7.9	30.7	-27.2	-73.3	-29.3	26.3	18.7	43.5	-38.08	71.97	2.04	-4.95
1.72	5.7	24.6	-25.1	-71.0	-27.2	29.4	17.1	41.4	-37.87	71.80	2.33	-4.63
1.76	4.6	19.7	-24.7	-68.8	-25.9	31.3	16.6	39.3	-37.82	70.95	2.58	-4.41
1.80	2.4	14.3	-23.5	-67.6	-24.1	33.5	15.6	38.0	-37.85	70.38	2.70	-4.06
1.84	4.1	7.5	-22.7	-66.4	-24.4	36.7	14.7	36.6	-37.57	70.19	2.39	-3.84
1.88	5.6	1.4	-21.2	-65.2	-24.9	39.2	13.6	35.2	-37.73	69.79	2.59	-3.85
1.92	10.2	-3.0	-19.1	-62.7	-26.6	40.8	11.9	33.3	-37.06	69.42	2.07	-3.59
1.98	18.2	-6.8	-16.8	-58.6	-30.3	41.7	10.4	30.0	-37.08	69.36	2.11	-3.77
2.03	23.5	-4.1	-16.4	-55.9	-31.8	39.0	9.6	28.0	-35.97	68.84	1.55	-3.57
2.07	26.3	-2.4	-16.1	-53.4	-32.7	37.4	9.3	26.2	-36.08	68.71	1.58	-3.52
2.15	29.3	2.2	-16.5	-48.7	-33.0	33.7	8.9	22.8	-35.77	68.72	1.01	-3.68
2.20	29.5	3.9	-17.2	-46.3	-32.2	31.9	9.0	21.0	-35.38	68.14	.95	-3.92
2.28	29.3	6.4	-17.8	-44.1	-30.8	29.4	8.8	19.2	-34.87	67.65	.51	-4.00
2.34	31.3	9.2	-19.2	-43.0	-30.7	27.2	9.2	18.2	-34.29	67.18	.47	-4.18
2.46	29.9	15.1	-18.9	-41.5	-28.3	23.4	8.5	16.6	-33.29	67.25	-.04	-4.47
2.56	26.4	21.8	-17.9	-41.1	-25.5	19.6	7.5	15.8	-32.64	67.22	-.45	-4.56
2.75	10.0	23.2	-12.6	-38.9	-17.3	17.2	4.8	13.8	-32.83	65.61	-.70	-4.68
3.00	2.7	11.5	-6.2	-31.6	-13.1	19.2	2.1	9.9	-32.32	62.71	-.59	-5.27

INVARIANT FIXED C.M.ANGLE AMPLITUDES ISOSPIN EVEN AND ODD:+- THETA= 60 DEGREES

K	RE A+	IM A+	RE A-	IM A-	RE B+	IM B+	RE B-	IM B-	RE C+	IM C+	RE C-	IM C-
GEV/C	1/GEV		1/GEV		1/GEV**2		1/GEV**2		1/GEV		1/GEV	
.020	229.1	.2	-77.7	.1	-1629.	0.	617.1	.0	-.51	.21	9.23	.09
.040	229.8	.9	-79.2	-.1	-1588.	-4.	608.1	2.0	-.13	.43	8.89	.18
.060	238.7	2.6	-85.3	-.9	-1576.	-13.	619.2	7.5	.53	.67	8.31	.25
.079	258.3	6.0	-99.8	-2.4	-1612.	-31.	677.7	17.4	2.03	1.04	7.89	.34
.097	266.6	10.9	-106.1	-4.8	-1569.	-57.	673.5	30.5	3.14	1.42	6.98	.31
.112	279.8	17.4	-115.4	-8.0	-1564.	-88.	689.5	46.5	4.24	1.88	6.06	.18
.130	295.7	29.0	-126.3	-13.8	-1549.	-141.	700.4	72.9	5.99	2.66	4.62	-.16
.153	322.8	54.2	-144.7	-26.3	-1554.	-246.	729.9	125.1	8.87	4.55	2.69	-.99
.172	345.4	86.2	-160.7	-42.0	-1551.	-368.	751.6	185.4	11.97	7.13	.97	-2.17
.185	362.8	114.7	-172.5	-56.3	-1553.	-470.	765.3	236.0	14.51	9.34	-.80	-3.35
.200	382.3	160.8	-184.8	-79.4	-1551.	-627.	773.9	314.3	17.38	13.21	-2.75	-5.42
.218	391.2	239.2	-192.4	-118.6	-1493.	-880.	755.3	440.1	19.88	20.45	-4.57	-9.17
.247	325.5	398.3	-170.4	-197.6	-1139.	-1330.	618.5	664.5	17.05	38.01	-2.84	-17.66
.267	200.6	479.4	-111.9	-238.3	-673.	-1501.	397.6	750.6	7.88	49.17	2.04	-23.18
.280	112.7	490.5	-71.6	-244.0	-378.	-1477.	259.1	738.8	.60	52.31	5.29	-24.87

INVARIANT FIXED C.M.ANGLE AMPLITUDES ISOSPIN EVEN AND ODD:+- THETA= 60 DEGREES

K GEV/C	RE A+ 1/GEV	IM A+	RE A- 1/GEV	IM A-	RE B+ 1/GEV**2	IM B+	RE B- 1/GEV**2	IM B-	RE C+ 1/GEV	IM C+	RE C- 1/GEV	IM C-
.290	48.6	481.0	-42.9	-239.1	-175.	-1406.	166.8	703.2	-4.60	52.86	7.91	-25.02
.295	16.8	471.6	-28.7	-234.5	-83.	-1357.	124.4	678.7	-8.96	52.98	9.69	-25.17
.301	-14.6	456.3	-15.1	-227.0	5.	-1291.	85.1	645.7	-13.03	52.54	11.57	-24.96
.305	-36.5	442.4	-5.0	-220.2	66.	-1235.	56.6	618.2	-15.46	51.84	12.90	-24.62
.310	-60.4	423.2	6.2	-211.0	131.	-1163.	26.1	583.6	-18.40	50.65	14.58	-24.05
.320	-96.6	385.0	19.3	-191.8	224.	-1025.	-7.0	515.0	-23.13	48.39	17.06	-22.72
.331	-126.2	351.4	31.4	-175.4	301.	-907.	-39.5	457.3	-24.78	45.69	18.08	-21.26
.351	-154.3	281.0	37.0	-141.6	352.	-682.	-45.6	348.9	-29.98	40.47	20.89	-18.45
.378	-167.2	205.2	35.0	-106.9	362.	-457.	-33.6	245.9	-31.36	33.87	22.41	-14.72
.408	-166.9	151.5	23.0	-84.1	342.	-307.	-.6	181.6	-30.54	29.12	22.79	-11.77
.427	-174.0	123.9	12.2	-77.9	348.	-235.	24.3	163.6	-30.02	26.62	22.29	-10.20
.456	-158.2	78.1	2.5	-69.1	293.	-124.	45.3	140.5	-29.90	24.08	22.34	-7.67
.490	-142.9	40.6	-8.7	-69.9	246.	-36.	66.2	138.8	-28.64	23.97	21.99	-5.48
.532	-117.6	6.3	-10.2	-69.7	180.	30.	61.9	134.7	-27.81	21.19	20.64	-2.72
.573	-99.5	-18.4	-9.3	-67.7	138.	73.	53.0	128.2	-26.58	20.32	18.82	.22
.614	-82.2	-37.2	-7.7	-66.5	98.	101.	43.5	121.2	-27.29	19.81	16.71	1.63
.658	-67.8	-55.7	-9.0	-69.5	68.	124.	39.4	119.9	-27.60	18.40	14.47	1.94
.675	-57.9	-64.1	-6.3	-71.1	51.	134.	33.5	120.8	-27.11	17.53	14.08	2.49
.705	-34.7	-70.8	3.4	-72.7	14.	139.	13.8	120.6	-26.18	17.19	12.15	3.55
.725	-22.1	-71.1	10.4	-70.4	-7.	136.	.6	115.2	-26.52	16.80	10.74	4.18
.750	-7.8	-64.7	18.6	-65.9	-27.	122.	-15.3	103.7	-25.76	16.28	8.33	3.19
.777	4.3	-59.2	25.2	-55.4	-42.	108.	-26.1	84.1	-24.71	15.05	7.32	2.34
.800	13.2	-50.2	26.5	-47.4	-51.	92.	-29.0	69.1	-22.53	14.53	6.03	1.27
.822	16.0	-45.4	27.4	-39.2	-51.	84.	-30.9	55.4	-20.49	15.05	5.20	.63
.851	20.9	-38.3	24.5	-31.7	-54.	73.	-27.5	42.2	-19.33	15.99	4.12	-.43
.875	23.5	-33.4	22.4	-26.4	-55.0	66.3	-25.1	32.7	-18.33	16.87	3.34	-1.54
.895	25.9	-29.6	20.6	-22.9	-57.1	61.3	-23.2	25.3	-18.30	17.83	2.62	-3.25
.923	26.8	-24.6	18.6	-17.7	-57.1	54.4	-20.1	15.3	-18.56	18.58	2.68	-5.48
.954	26.1	-19.1	16.1	-12.2	-54.9	46.0	-14.9	5.5	-18.72	18.48	3.99	-7.66
.975	26.1	-16.7	13.2	-9.6	-54.0	42.0	-8.8	1.0	-18.84	18.17	5.93	-8.72
1.00	24.7	-13.3	10.7	-5.4	-50.1	36.0	-3.8	-4.2	-17.86	17.22	7.44	-8.97
1.03	22.6	-11.3	6.2	-3.1	-45.8	33.4	3.9	-5.7	-17.27	17.83	9.59	-8.06
1.06	21.3	-9.2	2.2	-1.8	-42.7	31.1	9.7	-5.9	-16.67	18.46	10.82	-7.03
1.08	19.2	-9.1	-.7	-1.1	-39.1	31.0	13.5	-4.8	-16.32	18.95	11.51	-5.43
1.11	19.4	-8.5	-3.8	-.6	-38.5	30.2	16.6	-3.0	-16.35	19.50	11.63	-3.43
1.15	17.9	-7.9	-6.4	-.6	-36.8	28.5	18.3	-.9	-17.35	19.40	11.11	-1.47
1.17	16.0	-8.1	-7.2	-1.2	-35.2	27.8	18.5	.6	-18.16	18.94	10.72	-.65
1.21	16.0	-9.1	-8.9	-2.2	-35.6	27.0	19.0	2.8	-19.47	17.77	10.11	.56
1.24	16.0	-9.8	-9.8	-2.7	-34.9	25.3	19.4	4.0	-19.38	15.82	9.82	1.38

INVARIANT FIXED C.M.ANGLE AMPLITUDES ISOSPIN EVEN AND ODD:+- THETA= 60 DEGREES

K GEV/C	RE A+ 1/GEV	IM A+	RE A- 1/GEV	IM A-	RE B+ 1/GEV**2	IM B+	RE B- 1/GEV**2	IM B-	RE C+ 1/GEV	IM C+	RE C- 1/GEV	IM C-
1.28	17.3	-9.6	-8.6	-4.4	-34.8	22.1	16.9	6.6	-19.02	13.46	8.99	2.53
1.32	18.0	-8.5	-7.0	-3.7	-32.5	17.6	13.9	6.6	-16.89	10.42	7.95	3.41
1.36	15.9	-8.6	-6.6	-3.0	-28.4	15.5	12.5	6.5	-15.25	8.38	7.10	4.14
1.40	16.2	-7.2	-5.7	-1.1	-25.3	12.8	10.2	5.3	-12.17	7.25	5.70	4.84
1.43	15.6	-7.8	-5.5	.3	-22.5	12.7	9.0	4.4	-10.03	6.63	4.86	5.33
1.47	16.6	-5.9	-5.6	2.0	-20.1	11.1	7.5	3.0	-6.98	7.12	3.15	5.42
1.51	16.7	-5.0	-5.8	2.8	-18.6	10.2	6.9	2.2	-5.47	7.20	2.43	5.47
1.55	16.7	-4.6	-7.2	4.9	-16.2	10.9	6.4	.1	-3.02	8.71	.53	5.00
1.59	15.7	-3.8	-9.2	5.3	-14.3	10.9	6.8	-.6	-2.08	9.69	-.74	4.58
1.64	15.4	-3.0	-9.7	6.0	-12.8	11.1	6.3	-2.1	-.82	11.12	-1.71	3.41
1.68	15.0	-3.2	-9.4	5.6	-11.4	11.9	5.7	-2.3	.27	12.21	-2.01	2.55
1.72	14.6	-2.1	-9.5	6.0	-10.8	11.1	5.5	-3.0	.32	12.58	-2.22	1.96
1.76	14.4	-2.0	-10.1	6.0	-10.0	11.7	5.6	-3.5	.98	13.72	-2.53	1.25
1.80	13.9	-1.6	-9.2	6.2	-9.3	11.8	5.0	-4.0	1.14	14.60	-2.35	.71
1.84	12.3	-.6	-8.4	7.1	-8.2	11.3	4.6	-4.9	.91	15.13	-2.00	.31
1.88	12.2	-.3	-8.6	7.9	-8.2	10.9	4.7	-5.5	.61	15.18	-1.99	.19
1.92	10.3	.4	-7.9	8.8	-6.9	10.8	4.4	-6.1	.32	15.86	-1.55	.06
1.98	8.2	1.3	-8.2	9.6	-5.8	9.8	4.7	-6.5	-.23	15.76	-1.32	.05
2.03	6.6	.4	-8.8	11.0	-4.9	10.3	5.1	-7.3	-.70	15.93	-1.10	-.01
2.07	5.3	-.1	-9.5	11.3	-4.1	10.1	5.5	-7.5	-.94	15.29	-1.08	-.18
2.15	3.0	-2.1	-10.7	11.4	-2.7	10.4	6.3	-7.1	-1.27	14.25	-.81	.18
2.20	2.1	-3.1	-10.8	11.1	-2.2	10.6	6.4	-6.8	-1.41	13.82	-.59	.33
2.28	1.3	-5.4	-11.7	11.4	-1.8	11.2	7.0	-6.5	-1.63	12.88	-.26	.66
2.34	1.4	-6.4	-11.6	11.2	-1.6	11.3	6.7	-6.0	-1.23	12.50	-.46	1.24
2.46	.6	-8.8	-13.3	11.3	-1.1	11.4	7.2	-5.2	-1.21	10.92	-.83	2.19
2.56	.9	-10.5	-14.3	10.2	-.9	11.3	7.2	-4.4	-.69	9.66	-1.37	2.24
2.75	2.35	-12.38	-14.87	7.87	-.96	11.39	6.51	-3.15	.55	8.99	-2.66	1.96
3.00	2.07	-11.76	-11.77	4.17	-.59	10.37	4.56	-1.39	.90	8.91	-2.67	1.40

INVARIANT FIXED C.M.ANGLE AMPLITUDES ISOSPIN EVEN AND ODD:+- THETA= 90 DEGREES

K GEV/C	RE A+ 1/GEV	IM A+	RE A- 1/GEV	IM A-	RE B+ 1/GEV**2	IM B+	RE B- 1/GEV**2	IM B-	RE C+ 1/GEV	IM C+	RE C- 1/GEV	IM C-
.020	228.8	.2	-77.5	.1	-1630.	0.	617.2	.0	-.68	.21	9.38	.09
.040	228.7	.9	-78.5	-.1	-1588.	-4.	608.5	2.0	-.79	.43	9.44	.18
.060	236.9	2.6	-84.3	-.9	-1583.	-13.	624.3	7.5	-.99	.65	9.53	.26
.079	254.9	5.8	-98.0	-2.4	-1624.	-31.	685.6	17.5	-.97	.94	10.05	.38
.097	261.4	10.6	-103.2	-4.6	-1585.	-57.	683.6	30.6	-1.33	1.16	10.07	.43

INVARIANT FIXED C.M.ANGLE AMPLITUDES ISOSPIN EVEN AND ODD:+- THETA= 90 DEGREES

K GEV/C	RE A+ 1/GEV	IM A+ 1/GEV	RE A- 1/GEV	IM A- 1/GEV	RE B+ 1/GEV**2	IM B+ 1/GEV**2	RE B- 1/GEV**2	IM B- 1/GEV**2	RE C+ 1/GEV	IM C+ 1/GEV	RE C- 1/GEV	IM C- 1/GEV
.112	272.6	16.7	-111.5	-7.7	-1583.	-88.	701.4	46.6	-1.83	1.34	10.09	.43
.130	285.6	27.4	-120.9	-13.0	-1572.	-141.	714.2	73.2	-2.48	1.54	9.96	.39
.153	308.0	50.6	-136.5	-24.5	-1583.	-248.	745.1	126.0	-3.50	1.91	10.10	.30
.172	326.7	79.6	-150.0	-38.8	-1585.	-371.	768.1	187.0	-4.11	2.26	10.26	.23
.185	340.3	105.1	-159.6	-51.5	-1588.	-474.	780.2	238.3	-4.15	2.22	9.66	.17
.200	355.9	146.0	-170.3	-72.0	-1589.	-634.	790.9	317.7	-4.47	2.24	9.03	.04
.218	359.9	215.1	-175.3	-106.6	-1530.	-891.	770.6	445.7	-5.36	2.45	8.62	-.20
.247	297.5	352.8	-152.4	-174.9	-1179.	-1351.	627.3	675.0	-6.99	3.76	9.63	-.58
.267	188.5	420.3	-101.8	-208.7	-722.	-1529.	410.1	764.1	-7.73	4.52	9.81	-.91
.280	110.8	427.4	-66.1	-212.4	-423.	-1506.	268.5	753.2	-7.82	4.42	9.27	-.99
.290	52.3	417.2	-39.2	-207.2	-207.	-1436.	168.3	717.7	-7.29	4.29	9.15	-.82
.295	24.4	408.0	-26.2	-202.7	-112.	-1387.	121.3	693.0	-8.06	4.50	9.06	-1.02
.301	-2.7	393.8	-14.2	-195.6	-19.	-1320.	78.6	659.6	-8.44	4.66	9.00	-1.13
.305	-21.6	380.9	-5.5	-189.4	44.	-1263.	48.6	631.9	-8.60	4.72	8.99	-1.17
.310	-41.7	363.5	4.1	-181.0	109.	-1191.	15.8	596.8	-8.71	4.74	8.83	-1.22
.320	-71.7	329.1	14.4	-163.7	204.	-1051.	-16.8	527.2	-8.91	5.18	9.27	-1.28
.331	-97.9	299.1	24.5	-149.1	282.	-931.	-48.4	468.7	-8.86	5.25	9.17	-1.26
.351	-122.8	236.9	28.5	-119.4	344.	-702.	-59.4	358.2	-9.56	5.97	8.94	-1.59
.378	-128.9	170.9	27.9	-89.3	343.	-474.	-53.8	252.4	-9.71	6.22	9.19	-1.53
.408	-126.0	124.3	18.6	-69.7	316.	-321.	-25.8	185.8	-9.95	6.27	9.10	-1.39
.427	-129.2	100.3	11.7	-64.6	316.	-248.	-10.2	165.8	-9.25	5.91	7.80	-1.56
.456	-114.3	62.8	3.8	-56.1	260.	-141.	9.1	135.5	-10.65	6.49	7.49	-1.99
.490	-101.3	29.0	-1.9	-58.8	212.	-51.	22.7	132.1	-11.74	7.67	7.69	-3.13
.532	-75.4	1.1	.4	-55.2	142.	12.	14.8	116.9	-11.74	6.58	7.06	-2.84
.573	-55.8	-14.7	5.9	-52.6	93.	45.	1.6	105.8	-11.65	6.87	6.68	-2.47
.614	-37.2	-23.3	8.9	-47.0	50.	61.	-6.4	88.8	-12.07	6.97	5.71	-2.62
.658	-22.8	-25.5	12.1	-39.6	20.	58.	-14.1	66.9	-12.23	4.94	4.65	-4.39
.675	-17.6	-27.4	12.4	-36.0	11.	58.	-14.1	58.6	-11.91	3.72	4.81	-4.62
.705	-9.8	-29.4	11.3	-32.4	-1.	58.	-11.0	49.3	-10.54	2.59	5.18	-5.08
.725	-3.0	-29.0	12.4	-30.7	-12.	55.	-11.8	45.1	-9.77	1.84	5.77	-5.20
.750	5.0	-25.9	14.7	-27.4	-22.	49.	-14.9	38.0	-7.46	2.24	6.04	-5.43
.777	10.5	-21.6	15.6	-22.5	-26.	40.	-16.1	30.0	-5.26	2.38	6.01	-4.62
.800	13.7	-18.2	15.1	-19.4	-28.	34.	-14.9	25.3	-3.59	2.72	6.03	-3.95
.822	15.6	-16.2	15.4	-16.6	-30.	31.	-15.0	21.2	-3.15	3.32	6.11	-3.49
.851	19.6	-12.4	15.8	-13.9	-32.	27.	-17.3	16.5	-.91	5.02	4.85	-3.36
.875	23.3	-9.6	17.7	-11.5	-36.5	24.5	-20.7	12.1	-.40	6.29	4.23	-3.67
.895	25.7	-3.8	19.0	-8.0	-38.7	17.2	-23.0	5.8	.21	7.51	3.85	-4.17
.923	28.8	4.5	20.2	-1.6	-40.8	6.2	-23.6	-5.1	1.30	8.67	4.30	-4.99
.954	26.2	14.1	18.4	6.9	-35.6	-5.9	-18.5	-17.2	1.62	10.00	5.63	-5.05

INVARIANT FIXED C.M.ANGLE AMPLITUDES ISOSPIN EVEN AND ODD:+- THETA= 90 DEGREES

K GEV/C	RE A+ 1/GEV	IM A+ 1/GEV	RE A- 1/GEV	IM A- 1/GEV	RE B+ 1/GEV**2	IM B+ 1/GEV**2	RE B- 1/GEV**2	IM B- 1/GEV**2	RE C+ 1/GEV	IM C+ 1/GEV	RE C- 1/GEV	IM C- 1/GEV
.975	22.5	20.5	15.1	11.6	-29.0	-12.9	-13.0	-22.8	2.21	11.39	5.97	-4.42
1.00	16.7	26.2	10.8	17.0	-19.8	-19.6	-6.1	-28.7	2.63	12.17	6.47	-3.49
1.03	7.7	27.1	2.7	20.6	-7.8	-18.4	4.7	-30.6	2.02	13.72	6.12	-1.73
1.06	1.6	26.1	-3.3	20.6	-.5	-15.8	11.7	-28.5	1.30	14.36	5.44	-.58
1.08	-4.3	23.3	-8.7	19.2	6.8	-11.6	17.0	-25.5	.89	14.56	4.18	.00
1.11	-11.0	18.2	-12.5	15.8	13.9	-4.4	20.5	-20.5	-.36	14.81	3.23	.03
1.15	-15.4	11.8	-13.9	11.2	18.2	3.7	20.5	-14.7	-1.00	14.68	2.33	-.38
1.17	-15.9	9.7	-13.1	9.5	18.7	6.4	19.1	-12.7	-1.01	14.84	2.18	-.61
1.21	-18.1	6.2	-11.6	8.2	20.9	10.9	16.4	-11.1	-1.06	15.04	1.83	-.84
1.24	-18.9	3.4	-10.3	7.9	21.3	14.2	14.4	-10.8	-1.33	15.07	1.69	-1.01
1.28	-21.2	-1.9	-9.1	8.8	22.9	19.5	12.8	-11.6	-1.73	14.56	1.73	-1.04
1.32	-20.2	-8.2	-9.0	9.9	21.1	26.0	12.3	-12.7	-1.92	14.32	1.68	-1.10
1.36	-17.5	-12.9	-9.0	11.5	17.5	30.3	12.5	-14.3	-2.05	13.82	1.99	-1.09
1.40	-13.4	-17.4	-10.6	12.7	12.8	34.2	13.9	-15.3	-1.87	13.31	1.82	-.97
1.43	-10.2	-19.1	-11.9	13.6	9.5	35.0	15.3	-15.4	-1.61	12.82	2.08	-.48
1.47	-6.5	-20.9	-14.3	13.9	5.4	36.0	17.2	-15.1	-1.45	12.54	1.63	-.14
1.51	-3.7	-20.6	-16.0	13.6	2.8	35.1	18.4	-14.1	-1.07	12.39	1.33	.28
1.55	-.6	-19.9	-17.9	12.3	-.5	33.3	19.6	-12.0	-1.06	12.03	.94	.80
1.59	2.3	-18.9	-19.0	11.3	-2.8	31.5	20.1	-10.3	-.44	11.88	.56	1.27
1.64	3.2	-18.1	-19.9	8.5	-3.8	29.6	20.0	-7.3	-.63	11.33	-.05	1.23
1.68	4.2	-17.2	-20.9	7.3	-5.1	28.1	20.1	-5.9	-.99	11.10	-.51	1.35
1.72	5.4	-17.1	-20.9	5.3	-5.6	27.4	19.6	-4.0	-.37	10.97	-.75	1.18
1.76	5.5	-15.9	-20.1	3.9	-6.1	25.5	18.6	-2.5	-.87	10.60	-.79	1.28
1.80	6.6	-15.5	-20.0	2.3	-6.8	24.7	17.8	-1.2	-.57	10.51	-1.29	1.08
1.84	7.2	-14.6	-19.5	.6	-7.2	23.4	17.0	.4	-.48	10.42	-1.37	1.02
1.88	9.1	-14.3	-18.0	-.2	-8.1	23.0	15.4	1.3	.36	10.54	-1.37	1.19
1.92	8.2	-13.5	-17.9	-1.9	-8.1	21.7	14.7	2.4	-.69	10.25	-1.80	.74
1.98	8.2	-12.9	-16.4	-3.0	-8.3	20.5	12.8	3.4	-1.03	9.85	-2.09	.80
2.03	8.5	-12.2	-14.7	-4.5	-8.8	19.1	11.2	4.2	-1.49	9.38	-2.05	.30
2.07	9.2	-12.0	-13.6	-4.6	-9.4	18.6	10.1	4.4	-1.54	9.30	-2.12	.42
2.15	9.5	-10.7	-10.9	-5.3	-9.8	16.4	7.8	4.4	-2.02	8.49	-1.73	-.16
2.20	10.3	-9.6	-10.1	-5.6	-10.4	15.0	7.1	4.5	-2.05	8.20	-1.74	-.27
2.28	10.7	-7.6	-8.1	-5.4	-10.4	12.5	5.2	4.4	-1.93	7.57	-1.78	-.01
2.34	9.6	-6.6	-7.0	-5.2	-10.1	11.1	4.5	4.2	-2.82	6.98	-1.47	-.09
2.46	8.5	-4.8	-4.0	-3.8	-9.0	8.3	2.1	2.9	-2.82	5.67	-1.32	-.11
2.56	8.1	-3.8	-2.8	-2.6	-8.5	6.7	1.4	1.9	-2.79	4.82	-1.02	-.15
2.75	6.34	-1.19	-2.62	-1.76	-6.42	3.13	1.02	.74	-2.24	2.99	-1.26	-.77
3.00	2.98	-.72	-1.24	-1.03	-3.63	1.49	.33	.43	-2.08	1.37	-.77	-.43

INVARIANT FIXED C.M.ANGLE AMPLITUDES ISOSPIN EVEN AND ODD:+- THETA= 120 DEGREES

K GEV/C	RE A+ 1/GEV	IM A+ 1/GEV	RE A- 1/GEV	IM A- 1/GEV	RE B+ 1/GEV**2	IM B+ 1/GEV**2	RE B- 1/GEV**2	IM B- 1/GEV**2	RE C+ 1/GEV	IM C+ 1/GEV	RE C- 1/GEV	IM C- 1/GEV
.020	228.5	.2	-77.4	.1	-1630.	0.	617.3	.0	-.84	.21	9.52	.09
.040	227.5	.9	-77.7	-.1	-1589.	-4.	608.9	2.0	-1.44	.42	10.00	.18
.060	235.2	2.5	-83.3	-.9	-1590.	-13.	629.3	7.5	-2.51	.62	10.76	.27
.079	251.5	5.7	-96.1	-2.3	-1636.	-31.	693.5	17.5	-3.97	.84	12.23	.43
.097	256.2	10.2	-100.5	-4.5	-1601.	-57.	694.9	30.6	-5.80	.89	13.22	.55
.112	265.5	15.9	-107.6	-7.3	-1603.	-89.	714.5	46.8	-7.89	.81	14.22	.69
.130	275.7	25.9	-115.5	-12.3	-1597.	-142.	729.7	73.6	-10.91	.41	15.47	.93
.153	293.6	47.1	-128.6	-22.7	-1614.	-249.	763.6	126.8	-15.79	-.71	17.80	1.59
.172	308.1	73.0	-139.5	-35.5	-1622.	-374.	788.2	188.6	-20.04	-2.58	19.92	2.62
.185	318.5	95.5	-147.3	-46.7	-1627.	-479.	800.7	240.5	-22.59	-4.87	20.64	3.69
.200	330.2	131.3	-156.2	-64.7	-1632.	-641.	813.9	321.2	-26.06	-8.69	21.56	5.48
.218	329.8	191.0	-159.0	-94.6	-1573.	-902.	794.1	451.3	-30.15	-15.46	22.77	8.74
.247	270.8	307.3	-135.8	-152.2	-1224.	-1373.	647.2	685.5	-30.28	-30.29	23.31	16.41
.267	176.4	361.3	-92.4	-179.2	-773.	-1556.	432.3	777.6	-22.76	-39.84	19.06	21.25
.280	108.7	364.3	-61.4	-180.9	-468.	-1536.	287.3	767.5	-15.51	-43.13	14.87	22.75
.290	57.2	353.4	-37.3	-175.3	-245.	-1465.	181.6	732.1	-9.10	-43.91	11.96	23.23
.295	33.8	344.5	-26.0	-170.9	-146.	-1416.	132.3	707.3	-6.08	-43.61	10.20	22.98
.301	11.3	331.2	-16.0	-164.2	-51.	-1349.	87.9	673.6	-2.68	-42.83	8.40	22.56
.305	-4.4	319.5	-8.9	-158.6	14.	-1292.	57.1	645.6	-.54	-41.99	7.08	22.13
.310	-20.6	303.8	-1.5	-151.0	81.	-1219.	23.7	610.1	2.17	-40.77	5.23	21.48
.320	-44.3	273.2	6.3	-135.7	176.	-1077.	-9.0	539.6	6.35	-37.63	3.65	20.02
.331	-67.4	246.9	14.6	-122.9	256.	-955.	-40.5	480.1	8.15	-34.79	2.61	18.62
.351	-87.8	193.0	16.9	-97.3	325.	-722.	-54.0	367.5	11.66	-28.10	.38	15.20
.378	-87.8	136.6	16.3	-71.7	314.	-490.	-53.3	259.0	13.22	-21.02	-.81	11.62
.408	-81.6	97.2	7.8	-55.3	280.	-335.	-26.7	190.2	12.94	-16.11	-1.22	9.01
.427	-80.6	76.7	1.2	-51.5	267.	-261.	-9.7	168.4	12.54	-14.22	-2.16	7.17
.456	-70.2	46.6	-3.4	-44.2	224.	-155.	4.8	133.0	11.44	-9.83	-1.65	4.21
.490	-58.0	17.9	-7.0	-47.7	178.	-65.	16.4	126.7	9.75	-6.73	-.75	.60
.532	-35.2	-4.4	-2.4	-42.8	107.	-2.	8.1	102.4	8.04	-5.04	.85	-1.54
.573	-10.8	-11.6	2.6	-37.7	46.	21.	-.7	84.1	8.44	-2.96	2.32	-2.13
.614	9.5	-9.7	9.8	-28.9	-1.	21.	-13.0	59.7	8.79	-.56	4.06	-2.56
.658	20.9	1.8	10.1	-12.7	-27.	-1.	-9.4	22.0	8.62	1.30	5.77	-2.56
.675	19.8	7.8	6.2	-5.9	-23.	-12.	-.0	8.1	8.87	2.39	6.20	-2.06
.705	15.8	13.5	-1.4	1.7	-16.	-20.	17.0	-5.9	7.79	3.95	6.79	-1.18
.725	11.0	15.5	-9.0	5.3	-7.	-23.	31.3	-11.9	7.59	4.30	6.35	-.57

INVARIANT FIXED C.M.ANGLE AMPLITUDES ISOSPIN EVEN AND ODD:+- THETA= 120 DEGREES

K GEV/C	RE A+ 1/GEV	IM A+	RE A- 1/GEV	IM A-	RE B+ 1/GEV**2	IM B+	RE B- 1/GEV**2	IM B-	RE C+ 1/GEV	IM C+	RE C- 1/GEV	IM C-
.750	7.3	13.6	-20.0	3.7	1.	-17.	52.5	-8.4	8.03	4.99	6.36	-.58
.777	6.9	12.3	-29.2	-2.0	3.	-14.	69.8	3.7	8.28	4.97	6.69	-.14
.800	4.1	8.7	-30.8	-8.0	9.	-7.	71.3	15.7	8.79	5.15	6.51	.25
.822	4.6	4.2	-33.5	-12.5	12.	2.	74.3	25.2	10.75	5.30	5.96	.88
.851	5.6	.7	-32.2	-18.8	13.	12.	69.5	36.4	12.94	7.26	5.57	.94
.875	6.2	-.8	-29.9	-22.7	13.9	19.0	64.7	43.0	13.89	9.65	5.83	1.06
.895	8.4	-2.1	-29.0	-26.3	12.4	25.0	62.1	49.7	15.33	11.91	5.82	1.47
.923	9.2	-2.9	-25.0	-30.4	9.3	32.5	54.5	57.9	14.54	15.59	6.11	2.58
.954	10.3	-3.4	-17.9	-33.3	4.8	39.6	40.0	64.7	13.13	19.57	5.37	4.33
.975	11.1	-3.0	-12.2	-33.8	-.7	41.8	29.0	66.6	10.71	21.68	4.88	5.41
1.00	11.2	-3.5	-9.3	-32.6	-6.5	45.4	20.1	65.9	7.28	23.60	2.72	6.80
1.03	12.2	-1.4	-2.8	-29.5	-13.8	41.4	5.1	60.5	3.86	23.78	.28	7.26
1.06	11.2	-1.3	-1.3	-25.0	-15.3	39.3	-2.4	51.5	1.83	22.92	-2.78	6.76
1.08	14.2	.3	-.0	-22.2	-21.9	35.8	-7.6	43.9	.49	22.65	-4.76	5.22
1.11	13.7	1.6	-.4	-19.0	-21.6	30.6	-10.5	34.3	-.03	21.06	-7.13	2.84
1.15	12.7	3.3	-1.8	-17.2	-19.3	25.7	-10.2	25.8	.17	19.92	-8.36	-.48
1.17	13.5	3.5	-2.7	-17.3	-19.7	25.1	-8.8	23.0	.56	19.92	-8.42	-2.25
1.21	12.4	4.1	-2.8	-16.7	-17.8	24.2	-7.8	18.5	.60	20.21	-7.98	-4.43
1.24	11.8	4.7	-2.3	-16.8	-16.8	24.5	-7.6	16.3	.48	21.19	-7.41	-5.80
1.28	11.0	5.1	-1.0	-16.3	-16.7	25.2	-7.6	12.3	-.45	22.44	-6.18	-7.81
1.32	9.4	5.5	-.2	-14.2	-17.4	26.0	-6.8	7.3	-2.78	23.67	-4.93	-9.04
1.36	7.4	6.0	.6	-13.2	-17.5	25.5	-5.9	4.4	-4.93	24.07	-3.57	-10.06
1.40	5.9	6.2	1.3	-11.2	-19.0	24.6	-4.5	1.1	-7.71	23.93	-1.90	-10.37
1.43	4.6	6.6	1.7	-10.1	-19.6	23.8	-3.3	-.5	-9.68	23.88	-.70	-10.42
1.47	2.8	5.1	2.0	-8.8	-20.6	22.5	-1.3	-1.9	-12.45	21.75	1.05	-10.18
1.51	2.4	4.3	1.9	-7.7	-21.7	21.2	-.1	-2.4	-13.76	20.07	1.84	-9.49
1.55	1.0	3.1	2.0	-6.7	-22.0	19.0	1.3	-2.5	-15.66	17.49	3.00	-8.60
1.59	.5	1.8	1.7	-6.4	-22.6	17.3	2.4	-2.2	-16.82	14.98	3.52	-8.12
1.64	.2	.9	1.4	-5.0	-22.6	14.2	3.3	-2.3	-17.41	11.93	4.00	-6.85
1.68	-.1	-.2	1.2	-4.6	-21.3	12.3	3.5	-2.1	-16.87	9.43	3.98	-6.27
1.72	.1	-.9	1.3	-4.2	-21.3	10.2	3.4	-1.8	-16.80	7.21	4.07	-5.66
1.76	.6	-1.6	1.1	-3.7	-20.0	8.6	3.6	-1.7	-15.44	5.29	3.99	-5.11
1.80	.3	-1.7	1.0	-3.2	-18.6	6.8	3.7	-1.8	-14.77	3.83	4.08	-4.72
1.84	.6	-2.5	.9	-2.5	-17.4	5.9	3.4	-2.0	-13.67	2.33	3.67	-4.15
1.88	-.2	-2.7	.6	-2.6	-15.6	5.1	3.5	-1.8	-13.23	1.51	3.49	-4.04
1.92	.8	-2.8	.7	-1.7	-14.6	4.0	3.2	-2.2	-11.42	.50	3.41	-3.61
1.98	.9	-3.0	.3	-1.6	-12.7	3.0	3.5	-2.3	-9.89	-.43	3.27	-3.58
2.03	1.0	-3.5	.2	-.4	-11.3	2.6	3.3	-3.1	-8.63	-1.24	3.05	-3.01
2.07	1.2	-3.6	-.1	-.6	-10.2	2.2	3.6	-2.9	-7.59	-1.69	3.06	-3.04

INVARIANT FIXED C.M.ANGLE AMPLITUDES ISOSPIN EVEN AND ODD:+- THETA= 120 DEGREES

K GEV/C	RE A+ 1/GEV	IM A+ 1/GEV	RE A- 1/GEV	IM A- 1/GEV	RE B+ 1/GEV**2	IM B+ 1/GEV**2	RE B- 1/GEV**2	IM B- 1/GEV**2	RE C+ 1/GEV	IM C+ 1/GEV	RE C- 1/GEV	IM C- 1/GEV
2.15	1.5	-4.1	-.3	.4	-8.2	2.3	3.8	-3.2	-5.68	-2.08	3.03	-2.46
2.20	1.6	-4.7	-.3	1.0	-7.1	2.6	3.9	-3.4	-4.75	-2.39	3.16	-1.99
2.28	2.3	-4.6	-.4	1.0	-6.2	2.5	3.9	-3.0	-3.33	-2.26	3.19	-1.73
2.34	3.4	-5.0	-.8	1.5	-5.8	2.8	4.2	-3.0	-1.86	-2.41	2.96	-1.25
2.46	4.8	-4.2	-1.4	1.1	-5.8	3.0	4.2	-2.0	-.58	-1.43	2.44	-.77
2.56	4.9	-2.9	-1.8	1.3	-5.2	2.5	4.0	-1.8	.02	-.54	1.94	-.39
2.75	6.09	-2.85	-2.25	1.15	-5.34	2.92	3.89	-1.05	.91	-.02	1.53	.13
3.00	5.05	-1.23	-1.17	.68	-4.28	1.78	2.28	-.18	.77	.55	1.11	.51

INVARIANT FIXED C.M.ANGLE AMPLITUDES ISOSPIN EVEN AND ODD:+- THETA= 150 DEGREES

K GEV/C	RE A+ 1/GEV	IM A+ 1/GEV	RE A- 1/GEV	IM A- 1/GEV	RE B+ 1/GEV**2	IM B+ 1/GEV**2	RE B- 1/GEV**2	IM B- 1/GEV**2	RE C+ 1/GEV	IM C+ 1/GEV	RE C- 1/GEV	IM C- 1/GEV
.020	228.3	.2	-77.2	.1	-1630.	0.	617.4	.0	-.96	.21	9.63	.09
.040	226.6	.9	-77.2	-.1	-1589.	-4.	609.1	2.0	-1.92	.42	10.40	.18
.060	233.9	2.5	-82.6	-.8	-1596.	-13.	633.1	7.5	-3.62	.61	11.68	.28
.079	249.0	5.6	-94.6	-2.3	-1645.	-31.	699.4	17.5	-6.17	.77	13.85	.46
.097	252.4	10.0	-98.5	-4.3	-1613.	-57.	703.9	30.7	-9.07	.70	15.57	.65
.112	260.3	15.4	-104.8	-7.0	-1618.	-89.	724.8	46.9	-12.32	.42	17.31	.87
.130	268.5	24.8	-111.6	-11.7	-1615.	-143.	742.2	73.9	-17.09	-.41	19.62	1.33
.153	283.1	44.5	-122.9	-21.4	-1638.	-251.	779.4	127.5	-24.79	-2.63	23.64	2.53
.172	294.5	68.2	-132.0	-33.1	-1650.	-376.	805.2	189.7	-31.70	-6.12	27.23	4.37
.185	302.7	88.5	-138.5	-43.2	-1658.	-482.	819.4	242.1	-36.10	-10.04	29.04	6.25
.200	311.6	120.5	-145.9	-59.3	-1665.	-646.	834.6	323.7	-41.89	-16.66	31.26	9.45
.218	308.1	173.4	-147.3	-85.7	-1607.	-910.	816.5	455.5	-48.28	-28.52	33.79	15.26
.247	251.5	274.0	-124.4	-135.5	-1261.	-1388.	668.9	693.2	-47.30	-55.09	34.14	28.80
.267	167.2	318.1	-85.8	-157.6	-812.	-1576.	454.7	787.5	-33.77	-72.13	26.78	37.39
.280	106.7	318.1	-58.1	-157.7	-502.	-1557.	307.2	778.0	-21.05	-77.72	20.00	40.05
.290	60.9	306.7	-36.6	-151.9	-275.	-1487.	199.2	742.7	-10.44	-78.98	15.06	40.75
.295	41.1	298.0	-26.9	-147.6	-175.	-1438.	149.5	717.8	-4.64	-78.60	12.21	40.46
.301	22.2	285.5	-18.5	-141.3	-78.	-1370.	104.9	683.9	1.47	-77.36	9.28	39.81
.305	8.7	274.6	-12.8	-136.0	-13.	-1313.	74.0	655.7	5.27	-75.95	6.99	39.10
.310	-4.8	260.1	-7.1	-129.1	55.	-1239.	41.2	619.8	9.95	-73.84	4.00	38.02
.320	-23.8	232.3	-1.3	-115.1	149.	-1096.	8.0	548.7	17.21	-68.73	.94	35.52
.331	-44.8	208.7	6.0	-103.7	233.	-973.	-23.7	488.6	20.36	-63.85	-.68	33.09
.351	-61.9	160.8	7.3	-81.1	303.	-737.	-37.4	374.4	26.08	-52.79	-3.50	27.43
.378	-57.5	111.5	5.7	-58.8	287.	-502.	-39.6	263.9	29.53	-40.71	-6.31	21.22
.408	-48.2	77.3	-3.7	-44.8	247.	-345.	-11.8	193.6	30.05	-32.22	-7.43	16.64

INVARIANT FIXED C.M.ANGLE AMPLITUDES ISOSPIN EVEN AND ODD:+- THETA= 150 DEGREES

K	RE A+	IM A+	RE A-	IM A-	RE B+	IM B+	RE B-	IM B-	RE C+	IM C+	RE C-	IM C-
GEV/C	1/GEV		1/GEV		1/GEV**2		1/GEV**2		1/GEV		1/GEV	
.427	-45.7	59.4	-11.8	-42.0	220.	-270.	13.4	170.6	26.14	-28.66	-7.41	13.61
.456	-40.6	34.0	-12.9	-36.1	196.	-163.	22.3	132.9	25.92	-21.26	-5.36	8.92
.490	-28.2	10.1	-16.9	-39.6	150.	-75.	35.8	123.6	24.97	-16.27	-4.22	4.12
.532	.4	-8.5	-15.2	-34.6	61.	-9.	39.7	93.1	23.02	-11.92	-.50	-.02
.573	25.5	-6.0	-10.0	-28.8	-7.	-4.	34.7	72.7	22.68	-7.34	3.45	-.66
.614	37.4	5.4	-4.0	-17.0	-40.	-19.	26.8	41.3	21.27	-2.35	6.77	-.37
.658	42.5	24.3	-4.1	5.6	-53.	-50.	32.3	-8.5	20.31	3.19	9.45	2.02
.675	38.8	31.9	-12.3	13.7	-45.	-61.	52.5	-23.1	19.52	6.12	10.00	3.82
.705	26.8	42.3	-29.6	24.7	-28.	-73.	88.4	-40.2	14.52	10.48	8.94	7.15
.725	14.7	44.4	-46.7	28.3	-11.	-74.	122.1	-45.7	9.98	11.86	7.33	8.08
.750	-.6	39.7	-70.1	22.7	13.	-61.	164.0	-32.6	5.30	12.17	3.73	8.06
.777	-17.3	30.8	-83.2	6.3	42.	-46.	182.2	-1.2	2.02	9.57	.42	5.75
.800	-21.7	20.0	-92.6	-9.2	52.	-29.	196.6	28.0	2.43	6.24	-.96	3.87
.822	-26.2	8.2	-91.2	-21.8	63.	-8.	190.7	50.1	3.57	4.26	-.97	1.88
.851	-26.6	-2.6	-91.3	-37.2	66.	14.	190.7	77.3	5.10	4.41	.54	.05
.875	-28.3	-12.0	-91.9	-49.8	73.2	38.7	189.5	100.7	7.49	6.90	.63	-.66
.895	-30.9	-21.0	-89.2	-61.9	75.5	62.7	183.8	122.5	6.45	10.00	1.59	-1.33
.923	-32.9	-36.4	-82.1	-80.8	75.7	99.9	166.8	157.2	5.12	13.68	1.63	-1.91
.954	-23.5	-53.5	-67.8	-101.0	47.9	135.6	136.8	193.5	.91	15.70	1.97	-2.26
.975	-11.5	-66.3	-54.0	-112.4	18.6	160.1	108.3	213.3	-1.97	16.26	1.87	-2.42
1.00	4.9	-74.4	-29.6	-117.5	-19.1	172.5	61.6	219.1	-5.10	15.65	2.61	-3.07
1.03	28.1	-75.3	-5.1	-112.4	-67.4	168.3	14.5	205.1	-7.55	13.80	2.61	-3.69
1.06	44.4	-69.0	12.5	-102.0	-100.4	151.5	-16.4	182.5	-9.39	12.17	3.71	-4.21
1.08	53.6	-59.1	23.1	-86.6	-118.5	128.7	-34.8	152.1	-10.59	10.60	4.24	-4.15
1.11	60.9	-43.3	30.9	-68.5	-131.0	95.8	-47.8	116.7	-11.01	9.35	4.65	-4.42
1.15	62.2	-28.1	30.5	-49.8	-132.5	65.6	-44.6	80.8	-11.79	8.58	5.55	-4.64
1.17	60.4	-22.3	28.0	-42.8	-129.9	54.1	-39.2	68.6	-12.72	8.22	5.96	-4.17
1.21	56.3	-14.6	24.4	-35.1	-122.5	38.0	-31.0	53.8	-13.62	7.11	6.65	-4.38
1.24	54.8	-10.9	21.6	-31.1	-120.9	29.5	-25.1	46.8	-14.86	6.11	7.12	-4.12
1.28	50.3	-2.3	17.9	-26.5	-114.0	10.4	-17.6	39.3	-16.40	3.82	7.53	-3.46
1.32	47.6	3.1	17.0	-23.7	-108.8	-4.0	-14.5	35.3	-16.95	.78	8.38	-2.72
1.36	42.3	8.9	15.6	-21.7	-97.3	-17.3	-12.1	32.5	-16.17	-1.46	8.34	-2.17
1.40	36.6	10.8	15.5	-19.3	-85.3	-24.5	-11.0	29.6	-15.23	-4.10	8.80	-1.30
1.43	34.3	11.7	16.7	-17.7	-78.9	-28.5	-13.2	27.3	-14.08	-5.77	8.53	-.96
1.47	27.3	14.1	14.1	-14.2	-62.7	-33.3	-9.2	23.6	-11.61	-6.57	8.42	.40
1.51	22.7	14.7	13.8	-12.4	-52.3	-34.6	-9.3	20.8	-10.08	-6.98	7.98	.62
1.55	19.8	14.6	12.1	-8.9	-43.8	-34.2	-7.3	15.9	-7.95	-7.05	7.44	1.19
1.59	14.6	13.6	11.2	-5.1	-32.9	-31.6	-6.6	10.0	-6.49	-6.57	6.95	1.34
1.64	11.8	13.4	8.3	-3.6	-25.1	-30.1	-3.5	8.0	-4.43	-6.16	6.07	1.54

INVARIANT FIXED C.M.ANGLE AMPLITUDES ISOSPIN EVEN AND ODD:+- THETA= 150 DEGREES

K GEV/C	RE A+ 1/GEV	IM A+ 1/GEV	RE A- 1/GEV	IM A- 1/GEV	RE B+ 1/GEV**2	IM B+ 1/GEV**2	RE B- 1/GEV**2	IM B- 1/GEV**2	RE C+ 1/GEV	IM C+ 1/GEV	RE C- 1/GEV	IM C- 1/GEV
1.68	8.7	12.4	7.0	-2.9	-19.2	-27.1	-1.9	6.7	-3.88	-5.38	5.73	1.54
1.72	5.6	11.0	4.1	-1.9	-12.9	-23.6	1.7	5.2	-2.90	-4.54	5.20	1.53
1.76	3.1	9.0	2.8	-1.6	-8.1	-19.4	2.8	4.2	-2.31	-3.89	4.70	1.16
1.80	2.0	7.0	1.1	-1.8	-5.4	-15.7	5.4	4.5	-1.69	-3.49	4.70	1.27
1.84	.1	5.0	.3	-2.1	-2.7	-11.4	6.4	4.7	-1.73	-2.74	4.61	1.07
1.88	.5	3.4	-.8	-2.5	-2.8	-8.7	7.5	5.3	-1.42	-2.46	4.34	1.14
1.92	-.5	.8	-1.5	-3.4	-.9	-3.7	8.2	6.4	-1.13	-1.80	4.11	1.02
1.98	.2	-2.9	-2.0	-3.4	-1.5	2.3	8.4	6.3	-.81	-1.29	3.82	1.01
2.03	.7	-4.3	-2.0	-5.6	-1.7	5.4	8.2	8.7	-.49	-.50	3.71	.52
2.07	1.6	-5.4	-1.9	-5.6	-3.4	7.5	7.7	8.5	-.75	-.06	3.54	.41
2.15	4.4	-7.4	-1.2	-7.5	-6.9	11.1	6.5	10.5	-.54	.52	3.38	.00
2.20	5.4	-7.7	-.2	-8.1	-8.0	12.0	4.8	10.9	-.34	.89	3.24	-.27
2.28	7.9	-7.7	.8	-7.5	-11.7	12.3	3.4	10.0	-.59	1.28	3.33	-.25
2.34	8.0	-5.4	2.2	-7.1	-12.6	9.7	1.6	9.2	-1.22	1.70	3.38	-.32
2.46	9.7	-2.7	3.5	-6.3	-15.3	6.6	-.3	7.8	-1.67	2.16	3.28	-.44
2.56	9.6	-1.3	3.3	-4.8	-15.6	3.7	-1.1	5.6	-2.16	1.45	2.55	-.53
2.75	5.69	1.69	4.07	-3.33	-10.98	-1.40	-1.74	3.76	-2.75	.61	2.73	-.43
3.00	3.76	2.03	2.05	-.91	-6.88	-3.15	-.51	1.32	-1.65	-.45	1.65	.13

INVARIANT FIXED C.M.ANGLE AMPLITUDES ISOSPIN EVEN AND ODD:+- THETA= 180 DEGREES

K GEV/C	RE A+ 1/GEV	IM A+ 1/GEV	RE A- 1/GEV	IM A- 1/GEV	RE B+ 1/GEV**2	IM B+ 1/GEV**2	RE B- 1/GEV**2	IM B- 1/GEV**2	RE C+ 1/GEV	IM C+ 1/GEV	RE C- 1/GEV	IM C- 1/GEV
.020	228.2	.2	-77.2	.1	-1630.	0.	617.4	.0	-1.00	.21	9.67	.09
.040	226.3	.9	-77.0	-.1	-1589.	-4.	609.2	2.0	-2.09	.42	10.55	.18
.060	233.5	2.5	-82.3	-.8	-1598.	-13.	634.4	7.5	-4.03	.60	12.01	.28
.079	248.0	5.6	-94.1	-2.2	-1648.	-31.	701.5	17.5	-6.97	.74	14.45	.48
.097	251.0	9.9	-97.8	-4.3	-1618.	-57.	707.3	30.7	-10.27	.63	16.44	.68
.112	258.4	15.2	-103.8	-6.9	-1624.	-89.	728.7	47.0	-13.95	.27	18.46	.94
.130	265.8	24.4	-110.1	-11.5	-1622.	-143.	747.0	74.0	-19.36	-.71	21.17	1.47
.153	279.3	43.5	-120.8	-21.0	-1647.	-251.	785.6	127.7	-28.10	-3.34	25.82	2.87
.172	289.5	66.5	-129.2	-32.3	-1660.	-377.	811.9	190.1	-35.97	-7.42	29.96	5.01
.185	297.0	85.9	-135.3	-42.0	-1669.	-483.	827.0	242.7	-41.07	-11.93	32.20	7.19
.200	304.8	116.6	-142.2	-57.3	-1678.	-648.	843.0	324.7	-47.72	-19.57	34.92	10.90
.218	300.2	166.9	-143.0	-82.5	-1621.	-914.	825.9	457.0	-54.96	-33.29	37.98	17.64
.247	244.4	261.8	-120.3	-129.4	-1275.	-1394.	678.4	696.0	-53.60	-64.15	38.28	33.33
.267	163.6	302.3	-83.4	-149.7	-826.	-1584.	464.2	791.1	-37.87	-83.92	29.82	43.28
.280	105.8	301.2	-56.9	-149.3	-515.	-1565.	315.8	781.9	-23.12	-90.35	22.09	46.36

INVARIANT FIXED C.M.ANGLE AMPLITUDES ISOSPIN EVEN AND ODD:+- THETA= 180 DEGREES

K GEV/C	RE A+ 1/GEV	IM A+ 1/GEV	RE A- 1/GEV	IM A- 1/GEV	RE B+ 1/GEV**2	IM B+ 1/GEV**2	RE B- 1/GEV**2	IM B- 1/GEV**2	RE C+ 1/GEV	IM C+ 1/GEV	RE C- 1/GEV	IM C- 1/GEV
.290	62.2	289.6	-36.4	-143.3	-287.	-1495.	207.2	746.6	-11.02	-91.77	16.42	47.14
.295	43.7	281.0	-27.4	-139.0	-186.	-1446.	157.7	721.7	-4.22	-91.36	13.21	46.84
.301	26.2	268.7	-19.6	-132.9	-90.	-1378.	113.3	687.7	2.85	-89.95	9.89	46.11
.305	13.5	258.1	-14.4	-127.7	-24.	-1320.	82.5	659.4	7.25	-88.33	7.25	45.29
.310	1.0	244.2	-9.4	-121.0	44.	-1246.	50.1	623.4	12.62	-85.90	3.85	44.05
.320	-16.4	217.3	-4.3	-107.6	139.	-1103.	16.7	552.0	20.98	-80.06	.26	41.18
.331	-36.5	194.7	2.6	-96.7	223.	-980.	-15.3	491.6	24.63	-74.44	-1.57	38.38
.351	-52.5	149.1	3.7	-75.1	294.	-742.	-28.6	376.9	30.87	-61.77	-4.38	31.90
.378	-46.5	102.4	1.4	-54.1	276.	-507.	-31.8	265.8	35.21	-47.87	-7.97	24.73
.408	-35.9	70.0	-8.6	-40.9	233.	-349.	-3.0	194.9	36.23	-38.07	-9.53	19.44
.427	-33.6	53.1	-17.5	-38.5	201.	-274.	26.8	171.5	30.18	-33.89	-9.01	15.98
.456	-30.9	29.3	-17.1	-33.3	186.	-166.	33.1	133.2	30.35	-25.38	-6.17	10.65
.490	-18.1	7.3	-21.6	-36.6	140.	-78.	48.1	122.7	29.93	-19.58	-5.05	5.56
.532	15.7	-9.9	-22.5	-31.7	36.	-11.	60.9	89.9	28.53	-14.09	-.59	.62
.573	39.7	-2.5	-16.3	-26.3	-33.	-15.	56.5	70.0	27.30	-8.01	4.88	-.09
.614	44.3	13.2	-10.2	-13.1	-55.	-38.	51.2	35.7	22.94	-1.62	9.69	.74
.658	47.6	36.8	-10.4	15.5	-66.	-78.	56.5	-23.9	21.06	5.21	12.46	5.85
.675	44.3	43.0	-22.0	25.2	-59.	-84.	83.0	-40.6	20.10	8.63	11.94	8.56
.705	33.0	55.0	-48.7	39.1	-44.	-96.	136.0	-61.4	14.76	14.69	8.21	13.40
.725	14.8	59.8	-70.3	41.7	-15.	-98.	174.7	-64.6	8.30	18.03	3.90	14.25
.750	-3.0	52.0	-106.2	32.8	10.	-82.	242.8	-45.1	1.21	16.65	-1.20	13.27
.777	-30.8	38.9	-107.0	11.0	60.	-59.	234.3	-3.0	-4.38	12.98	-3.96	9.73
.800	-33.7	27.6	-126.6	-5.3	61.	-43.	272.1	28.2	-6.32	8.22	-5.14	7.28
.822	-48.7	15.7	-114.3	-18.5	94.	-22.	239.8	51.7	-6.21	5.97	-5.82	4.90
.851	-38.8	3.6	-127.9	-36.4	75.	3.	262.8	83.6	-4.06	4.78	-6.91	2.06
.875	-47.1	-11.6	-126.9	-57.4	94.3	37.3	256.9	121.1	-3.11	5.84	-7.08	-.93
.895	-56.5	-23.8	-129.1	-70.1	112.4	67.6	260.1	140.8	-3.53	8.04	-6.42	-3.69
.923	-73.2	-43.9	-115.7	-94.6	139.4	113.2	232.6	183.5	-6.50	10.21	-4.39	-6.79
.954	-57.8	-72.6	-92.3	-127.5	95.4	170.6	186.2	243.7	-11.45	10.32	-1.82	-9.10
.975	-40.9	-88.7	-54.3	-138.7	54.1	199.2	110.2	261.1	-14.36	9.01	-.19	-10.60
1.00	-9.0	-106.7	-37.0	-153.9	-15.6	227.9	79.3	285.7	-16.72	6.45	2.34	-12.12
1.03	30.0	-107.5	-3.0	-145.6	-96.0	220.8	17.8	262.7	-18.35	3.60	5.93	-13.48
1.06	49.8	-104.5	16.3	-136.2	-133.4	210.9	-14.2	241.0	-18.05	2.69	9.11	-13.68
1.08	70.5	-89.6	27.6	-111.6	-172.6	176.1	-30.5	193.2	-18.17	.89	11.94	-12.33
1.11	77.4	-69.1	43.7	-87.9	-185.4	134.3	-54.2	148.7	-19.07	.85	15.51	-10.55
1.15	86.5	-46.4	38.9	-62.9	-204.5	88.0	-39.6	105.6	-21.53	.06	17.92	-7.08
1.17	91.5	-43.0	23.2	-50.0	-215.0	77.4	-9.3	83.0	-23.00	-1.80	18.22	-5.83
1.21	74.8	-33.5	23.9	-43.2	-186.6	53.3	-8.3	73.9	-25.76	-4.80	19.44	-3.38
1.24	70.9	-28.8	19.9	-36.5	-179.9	38.8	.5	64.0	-26.93	-7.70	20.20	-1.76

INVARIANT FIXED C.M.ANGLE AMPLITUDES ISOSPIN EVEN AND ODD:+- THETA= 180 DEGREES

K	RE A+	IM A+	RE A-	IM A-	RE B+	IM B+	RE B-	IM B-	RE C+	IM C+	RE C-	IM C-
GEV/C	1/GEV		1/GEV		1/GEV**2		1/GEV**2		1/GEV		1/GEV	
1.28	85.5	-19.4	10.8	-31.4	-204.1	10.2	18.4	59.8	-27.06	-13.77	20.94	1.59
1.32	82.6	-15.7	9.4	-25.3	-194.8	-9.0	20.8	53.4	-26.27	-20.76	21.01	4.54
1.36	75.5	-.9	6.4	-24.9	-174.0	-41.9	23.3	57.6	-22.81	-24.50	19.59	7.68
1.40	67.7	2.6	11.8	-19.6	-149.8	-53.5	10.6	52.0	-17.87	-27.98	17.91	10.12
1.43	56.1	9.3	13.4	-21.1	-121.2	-69.5	4.3	56.0	-13.76	-30.72	15.86	11.11
1.47	42.0	4.1	14.2	-9.0	-86.0	-59.3	-1.5	37.4	-8.07	-30.46	13.34	12.72
1.51	39.0	12.0	16.2	-7.6	-72.7	-71.2	-8.8	35.5	-3.70	-29.79	11.05	13.22
1.55	33.6	11.1	12.3	-2.0	-54.0	-65.8	-7.1	25.3	1.61	-27.93	8.08	13.00
1.59	31.6	11.9	9.9	-4.1	-44.5	-61.4	-6.8	27.0	4.90	-24.88	5.87	11.99
1.64	23.7	13.0	.3	.8	-27.4	-55.1	6.3	15.6	7.15	-20.32	4.14	10.24
1.68	17.3	10.3	-2.9	-1.9	-14.5	-44.8	9.7	17.7	8.39	-17.07	3.06	8.96
1.72	16.1	5.4	-9.4	.2	-13.5	-31.5	20.5	12.1	7.80	-13.96	3.18	7.69
1.76	8.1	6.5	-8.2	-7.2	-.9	-29.7	18.6	22.6	7.56	-11.88	3.25	6.82
1.80	4.5	5.5	-13.1	-6.0	3.9	-24.6	26.2	19.8	6.96	-9.86	3.25	6.34
1.84	5.2	-.7	-15.8	-7.7	1.7	-11.7	30.5	21.6	6.26	-8.06	3.35	5.84
1.88	-3.6	-3.6	-12.6	-12.9	14.2	-4.8	25.1	29.1	5.33	-6.61	3.30	5.54
1.92	4.4	-7.1	-17.1	-12.4	-.4	2.6	31.6	27.7	4.20	-5.45	3.05	5.22
1.98	1.5	-9.6	-18.3	-15.1	1.9	9.0	32.7	30.4	2.76	-3.79	2.80	4.47
2.03	5.2	-12.1	-15.0	-17.9	-6.3	14.1	27.4	33.5	1.15	-2.94	2.77	3.81
2.07	6.2	-14.6	-16.8	-20.6	-9.6	18.5	30.1	36.5	-.08	-2.46	2.92	3.23
2.15	6.2	-20.2	-9.7	-22.2	-13.8	26.9	20.2	37.3	-2.85	-2.44	3.63	2.34
2.20	5.6	-22.1	-4.9	-21.7	-14.9	28.8	13.8	35.8	-4.33	-2.96	4.24	2.05
2.28	10.5	-20.0	.7	-21.4	-24.7	23.1	6.8	35.2	-6.08	-4.57	5.20	2.20
2.34	12.0	-12.8	-.3	-17.8	-27.8	10.0	8.8	30.2	-6.78	-6.04	5.72	2.62
2.46	20.7	-5.1	7.6	-19.5	-39.9	-5.6	-2.4	34.1	-6.61	-8.95	6.00	3.84
2.56	28.9	-2.1	.2	-7.0	-49.3	-11.8	7.4	16.7	-5.22	-10.31	5.32	4.57
2.75	6.54	10.13	4.33	-6.89	-12.19	-30.05	-.55	16.92	-2.05	-11.06	3.94	5.04
3.00	1.12	9.18	-.28	-1.56	-1.33	-25.23	4.12	8.20	.16	-8.99	2.68	4.34

Table 2.2.4.1. s-channel helicity amplitudes as calculated from the preliminary updated solution KH 80. The last two columns give the isospin even and charge-exchange cross sections. Re $F++$ = ReF_{++}, IM $F+-$ = Im F_{+-}, DS/DT = $d\sigma/dt$, SQRT (MB)/GEV = $\sqrt{\text{mb}}$/GeV.

S-CHANNEL HELICITY AMPLITUDES

ISOSPIN	EVEN		ODD		EVEN		ODD		EVEN	CHEX
K	RE F++	IM F++	RE F++	IM F++	RE F+-	IM F+-	RE F+-	IM F+-	DS/DT	DS/DT
GEV/C	UNITS: SQRT(MB)/GEV								MB/GEV**2	
					T =-0.05 GEV**2					
2.03	-1.873	6.030	0.533	0.562	0.182	0.230	-0.265	-0.559	39.95	1.966
2.30	-1.777	5.971	0.208	0.423	0.282	0.422	-0.352	-0.420	39.07	1.045
2.50	-1.756	5.946	0.222	0.260	0.178	0.466	-0.419	-0.444	38.69	0.980
2.78	-1.766	5.837	0.320	0.194	0.064	0.399	-0.387	-0.493	37.36	1.065
3.09	-1.736	5.629	0.336	0.267	0.058	0.288	-0.316	-0.479	34.78	1.025
3.25	-1.633	5.497	0.313	0.256	0.086	0.260	-0.296	-0.441	32.96	0.892
3.40	-1.616	5.469	0.292	0.255	0.107	0.254	-0.287	-0.422	32.60	0.822
3.50	-1.634	5.516	0.295	0.261	0.115	0.262	-0.285	-0.408	33.18	0.807
3.64	-1.628	5.472	0.283	0.237	0.120	0.273	-0.287	-0.387	32.68	0.737
3.75	-1.624	5.430	0.273	0.236	0.113	0.281	-0.287	-0.379	32.22	0.713
3.90	-1.599	5.421	0.278	0.227	0.099	0.288	-0.288	-0.368	32.04	0.694
4.00	-1.579	5.359	0.278	0.215	0.091	0.289	-0.289	-0.363	31.31	0.677
4.20	-1.539	5.288	0.267	0.207	0.070	0.283	-0.286	-0.354	30.42	0.643
4.40	-1.485	5.165	0.270	0.208	0.057	0.269	-0.283	-0.345	28.96	0.630
4.60	-1.450	5.164	0.239	0.182	0.044	0.260	-0.278	-0.338	28.84	0.563
4.84	-1.485	5.147	0.235	0.201	0.034	0.243	-0.273	-0.329	28.76	0.556
5.00	-1.485	5.143	0.244	0.179	0.032	0.234	-0.271	-0.322	28.71	0.537
5.15	-1.426	5.043	0.235	0.191	0.029	0.224	-0.267	-0.316	27.51	0.526
5.40	-1.381	4.999	0.225	0.188	0.023	0.214	-0.262	-0.309	26.95	0.500
5.65	-1.356	5.026	0.188	0.174	0.020	0.206	-0.258	-0.302	27.14	0.447
6.00	-1.314	4.984	0.186	0.160	0.014	0.195	-0.251	-0.293	26.60	0.418
					T =-0.10 GEV**2					
2.03	-1.947	4.918	0.364	0.352	0.157	0.279	-0.262	-0.645	28.08	1.483
2.30	-1.818	4.839	0.125	0.275	0.272	0.465	-0.352	-0.494	27.02	0.919
2.50	-1.754	4.838	0.139	0.172	0.196	0.543	-0.422	-0.521	26.82	0.998
2.78	-1.727	4.771	0.193	0.125	0.027	0.490	-0.378	-0.561	25.99	1.023
3.09	-1.693	4.592	0.205	0.167	0.013	0.326	-0.303	-0.536	24.06	0.898
3.25	-1.580	4.419	0.188	0.147	0.064	0.287	-0.287	-0.502	22.11	0.782
3.40	-1.562	4.451	0.189	0.161	0.099	0.285	-0.281	-0.474	22.35	0.730
3.50	-1.596	4.501	0.201	0.172	0.108	0.301	-0.279	-0.459	22.90	0.717
3.64	-1.583	4.480	0.188	0.145	0.121	0.310	-0.280	-0.440	22.69	0.657
3.75	-1.578	4.441	0.177	0.152	0.105	0.320	-0.276	-0.431	22.33	0.632
3.90	-1.560	4.433	0.178	0.145	0.088	0.337	-0.276	-0.415	22.21	0.603
4.00	-1.538	4.381	0.183	0.130	0.080	0.336	-0.279	-0.412	21.68	0.595
4.20	-1.486	4.312	0.165	0.124	0.052	0.326	-0.275	-0.398	20.92	0.553
4.40	-1.434	4.188	0.177	0.122	0.036	0.305	-0.269	-0.387	19.70	0.537
4.60	-1.403	4.202	0.141	0.088	0.024	0.290	-0.266	-0.376	19.71	0.480
4.84	-1.459	4.196	0.141	0.117	0.019	0.269	-0.263	-0.363	19.81	0.470
5.00	-1.459	4.197	0.149	0.093	0.018	0.263	-0.260	-0.358	19.82	0.454
5.15	-1.399	4.080	0.143	0.107	0.015	0.250	-0.257	-0.350	18.67	0.442
5.40	-1.345	4.036	0.131	0.105	0.010	0.240	-0.252	-0.341	18.15	0.416
5.65	-1.331	4.099	0.098	0.095	0.007	0.230	-0.248	-0.332	18.63	0.381
6.00	-1.287	4.047	0.097	0.085	-0.001	0.217	-0.241	-0.322	18.09	0.357
					T =-0.20 GEV**2					
2.03	-1.762	3.276	0.150	0.050	0.055	0.318	-0.162	-0.586	13.94	0.788
2.30	-1.632	3.181	0.061	0.057	0.157	0.403	-0.221	-0.462	12.97	0.539
2.50	-1.555	3.190	0.065	0.039	0.162	0.510	-0.270	-0.482	12.88	0.623
2.78	-1.478	3.174	0.060	0.030	-0.032	0.508	-0.228	-0.494	12.52	0.601
3.09	-1.450	3.111	0.053	0.020	-0.063	0.311	-0.171	-0.456	11.88	0.481
3.25	-1.381	2.944	0.071	0.006	0.005	0.264	-0.166	-0.434	10.65	0.442
3.40	-1.355	2.991	0.077	0.026	0.050	0.270	-0.168	-0.405	10.86	0.398
3.50	-1.380	3.005	0.090	0.045	0.060	0.283	-0.168	-0.391	11.02	0.382
3.64	-1.367	3.033	0.074	0.022	0.074	0.297	-0.170	-0.379	11.16	0.357
3.75	-1.345	3.014	0.059	0.029	0.054	0.306	-0.165	-0.372	10.99	0.340
3.90	-1.328	2.976	0.060	0.038	0.040	0.324	-0.164	-0.357	10.73	0.319
4.00	-1.326	2.968	0.065	0.011	0.030	0.321	-0.166	-0.355	10.67	0.315
4.20	-1.285	2.910	0.046	0.014	0.003	0.307	-0.161	-0.339	10.22	0.286
4.40	-1.251	2.859	0.061	0.001	-0.012	0.279	-0.157	-0.326	9.82	0.269
4.60	-1.226	2.857	0.040	-0.021	-0.015	0.259	-0.155	-0.313	9.74	0.249
4.84	-1.248	2.830	0.045	0.004	-0.012	0.240	-0.156	-0.301	9.63	0.234
5.00	-1.239	2.818	0.041	-0.008	-0.010	0.236	-0.154	-0.297	9.53	0.228
5.15	-1.203	2.736	0.037	0.003	-0.012	0.224	-0.153	-0.288	8.98	0.216
5.40	-1.156	2.702	0.031	-0.000	-0.013	0.218	-0.149	-0.280	8.69	0.204
5.65	-1.143	2.785	0.015	-0.001	-0.015	0.208	-0.148	-0.271	9.11	0.191
6.00	-1.095	2.701	0.014	-0.000	-0.023	0.196	-0.145	-0.260	8.54	0.178

S-CHANNEL HELICITY AMPLITUDES

ISOSPIN	EVEN		ODD		EVEN		ODD		EVEN	CHEX
K	RE F++	IM F++	RE F++	IM F++	RE F+-	IM F+-	RE F+-	IM F+-	DS/DT	DS/DT
GEV/C	UNITS: SQRT(MB)/GEV								MB/GEV**2	
					T =-0.30 GEV**2					
2.03	-1.379	2.194	0.035	-0.125	-0.014	0.333	-0.070	-0.424	6.83	0.402
2.30	-1.280	2.107	0.063	-0.074	0.056	0.311	-0.094	-0.342	6.18	0.270
2.50	-1.228	2.108	0.058	-0.045	0.108	0.400	-0.116	-0.350	6.13	0.283
2.78	-1.146	2.115	0.025	-0.025	-0.038	0.434	-0.091	-0.347	5.98	0.259
3.09	-1.127	2.162	-0.003	-0.066	-0.079	0.269	-0.059	-0.309	6.03	0.207
3.25	-1.128	2.057	0.050	-0.064	-0.025	0.228	-0.062	-0.290	5.56	0.189
3.40	-1.089	2.060	0.040	-0.052	0.013	0.238	-0.065	-0.275	5.49	0.168
3.50	-1.087	2.028	0.043	-0.029	0.024	0.232	-0.067	-0.261	5.35	0.151
3.64	-1.077	2.079	0.025	-0.047	0.024	0.256	-0.073	-0.252	5.55	0.143
3.75	-1.047	2.083	0.011	-0.046	0.013	0.259	-0.076	-0.249	5.50	0.140
3.90	-1.027	2.019	0.014	-0.025	0.008	0.264	-0.071	-0.243	5.20	0.130
4.00	-1.044	2.049	0.015	-0.054	-0.005	0.261	-0.070	-0.237	5.36	0.128
4.20	-1.030	2.011	0.002	-0.045	-0.022	0.244	-0.065	-0.226	5.17	0.115
4.40	-1.022	2.042	0.012	-0.065	-0.032	0.218	-0.066	-0.215	5.26	0.110
4.60	-1.000	2.010	0.013	-0.064	-0.028	0.199	-0.063	-0.203	5.08	0.099
4.84	-0.960	1.949	0.016	-0.053	-0.022	0.189	-0.065	-0.197	4.76	0.092
5.00	-0.943	1.922	0.004	-0.051	-0.020	0.182	-0.066	-0.192	4.62	0.088
5.15	-0.936	1.897	-0.001	-0.048	-0.021	0.175	-0.065	-0.185	4.51	0.082
5.40	-0.911	1.878	0.002	-0.050	-0.018	0.174	-0.064	-0.180	4.39	0.078
5.65	-0.887	1.933	-0.000	-0.046	-0.019	0.166	-0.064	-0.174	4.55	0.073
6.00	-0.844	1.836	-0.001	-0.035	-0.024	0.156	-0.064	-0.165	4.11	0.065
					T =-0.40 GEV**2					
2.03	-0.973	1.495	-0.024	-0.205	-0.035	0.328	-0.027	-0.244	3.29	0.206
2.30	-0.908	1.426	0.086	-0.143	-0.001	0.233	-0.024	-0.199	2.91	0.137
2.50	-0.883	1.420	0.077	-0.092	0.060	0.282	-0.025	-0.198	2.88	0.109
2.78	-0.829	1.431	0.033	-0.053	-0.012	0.329	-0.018	-0.197	2.84	0.086
3.09	-0.814	1.546	-0.007	-0.105	-0.052	0.222	-0.006	-0.167	3.10	0.078
3.25	-0.870	1.510	0.062	-0.091	-0.024	0.195	-0.011	-0.148	3.08	0.068
3.40	-0.821	1.465	0.037	-0.088	-0.000	0.201	-0.011	-0.147	2.86	0.062
3.50	-0.795	1.400	0.029	-0.067	0.009	0.178	-0.014	-0.135	2.62	0.048
3.64	-0.790	1.447	0.010	-0.083	-0.006	0.205	-0.022	-0.127	2.76	0.048
3.75	-0.764	1.466	0.003	-0.087	-0.005	0.202	-0.032	-0.127	2.77	0.049
3.90	-0.745	1.398	0.005	-0.061	-0.004	0.194	-0.025	-0.132	2.55	0.043
4.00	-0.767	1.444	0.002	-0.081	-0.018	0.189	-0.021	-0.120	2.71	0.043
4.20	-0.777	1.429	-0.002	-0.070	-0.024	0.172	-0.017	-0.116	2.68	0.037
4.40	-0.791	1.506	0.001	-0.091	-0.026	0.154	-0.022	-0.109	2.92	0.041
4.60	-0.772	1.454	0.016	-0.071	-0.019	0.136	-0.019	-0.099	2.73	0.031
4.84	-0.683	1.370	0.015	-0.073	-0.015	0.138	-0.019	-0.100	2.36	0.032
5.00	-0.667	1.339	0.001	-0.063	-0.014	0.126	-0.021	-0.094	2.26	0.026
5.15	-0.679	1.360	-0.002	-0.066	-0.014	0.124	-0.019	-0.091	2.33	0.026
5.40	-0.675	1.351	0.002	-0.066	-0.009	0.125	-0.021	-0.089	2.30	0.026
5.65	-0.638	1.359	0.009	-0.061	-0.009	0.122	-0.021	-0.087	2.27	0.024
6.00	-0.611	1.274	0.007	-0.043	-0.014	0.112	-0.021	-0.081	2.01	0.018
					T =-0.50 GEV**2					
2.03	-0.628	1.064	-0.042	-0.220	-0.020	0.303	-0.030	-0.082	1.62	0.115
2.30	-0.589	1.012	0.111	-0.170	-0.022	0.173	-0.013	-0.064	1.40	0.091
2.50	-0.580	1.000	0.100	-0.113	0.027	0.177	-0.001	-0.058	1.37	0.052
2.78	-0.568	1.005	0.056	-0.063	0.020	0.222	-0.003	-0.073	1.38	0.025
3.09	-0.550	1.139	0.012	-0.111	-0.009	0.169	-0.002	-0.052	1.63	0.030
3.25	-0.638	1.159	0.076	-0.092	-0.004	0.157	-0.005	-0.035	1.77	0.031
3.40	-0.585	1.084	0.046	-0.098	0.003	0.157	-0.001	-0.043	1.54	0.027
3.50	-0.545	1.004	0.029	-0.081	0.007	0.128	-0.005	-0.034	1.32	0.017
3.64	-0.542	1.031	0.012	-0.096	-0.013	0.147	-0.011	-0.028	1.38	0.021
3.75	-0.526	1.054	0.012	-0.103	-0.008	0.139	-0.023	-0.028	1.41	0.024
3.90	-0.516	1.003	0.013	-0.081	-0.002	0.126	-0.015	-0.042	1.29	0.017
4.00	-0.529	1.042	0.007	-0.084	-0.014	0.117	-0.012	-0.026	1.38	0.016
4.20	-0.554	1.049	0.009	-0.076	-0.014	0.103	-0.008	-0.028	1.42	0.013
4.40	-0.584	1.130	0.009	-0.090	-0.008	0.094	-0.015	-0.027	1.63	0.018
4.60	-0.569	1.073	0.026	-0.061	-0.000	0.076	-0.011	-0.020	1.48	0.010
4.84	-0.455	0.983	0.020	-0.070	0.000	0.088	-0.010	-0.025	1.18	0.012
5.00	-0.447	0.961	0.010	-0.058	-0.000	0.074	-0.009	-0.018	1.13	0.008
5.15	-0.467	1.003	0.011	-0.066	-0.001	0.074	-0.008	-0.019	1.23	0.010
5.40	-0.477	0.999	0.011	-0.063	0.005	0.076	-0.009	-0.020	1.23	0.009
5.65	-0.432	0.960	0.022	-0.060	0.005	0.077	-0.010	-0.020	1.12	0.009
6.00	-0.423	0.906	0.017	-0.039	-0.002	0.067	-0.008	-0.019	1.00	0.005

S-CHANNEL HELICITY AMPLITUDES

ISOSPIN	EVEN		ODD		EVEN		ODD		EVEN	CHEX
K	RE F++	IM F++	RE F++	IM F++	RE F+-	IM F+-	RE F+-	IM F+-	DS/DT	DS/DT
GEV/C	UNITS: SQRT(MB)/GEV								MB/GEV**2	
				T =-0.75 GEV**2						
2.03	-0.144	0.690	0.038	-0.117	0.081	0.187	-0.142	0.175	0.54	0.132
2.30	-0.153	0.642	0.144	-0.146	-0.004	0.079	-0.120	0.157	0.44	0.162
2.50	-0.135	0.596	0.131	-0.099	0.001	0.010	-0.108	0.167	0.37	0.133
2.78	-0.189	0.567	0.093	-0.053	0.042	0.018	-0.089	0.102	0.36	0.060
3.09	-0.149	0.630	0.058	-0.065	0.062	0.024	-0.089	0.101	0.42	0.052
3.25	-0.232	0.695	0.068	-0.055	0.053	0.046	-0.081	0.095	0.54	0.046
3.40	-0.195	0.635	0.060	-0.066	0.035	0.019	-0.068	0.098	0.44	0.044
3.50	-0.172	0.565	0.039	-0.067	0.017	0.027	-0.075	0.091	0.35	0.040
3.64	-0.159	0.551	0.027	-0.069	0.021	0.002	-0.066	0.094	0.33	0.037
3.75	-0.160	0.550	0.036	-0.082	0.017	-0.011	-0.066	0.095	0.33	0.042
3.90	-0.179	0.560	0.030	-0.084	0.020	-0.003	-0.055	0.079	0.35	0.034
4.00	-0.160	0.536	0.029	-0.047	0.018	-0.025	-0.063	0.089	0.31	0.030
4.20	-0.186	0.576	0.032	-0.054	0.016	-0.034	-0.057	0.082	0.37	0.028
4.40	-0.220	0.567	0.029	-0.037	0.028	-0.018	-0.057	0.068	0.37	0.020
4.60	-0.220	0.540	0.030	-0.016	0.031	-0.040	-0.055	0.072	0.34	0.019
4.84	-0.134	0.471	0.011	-0.025	0.031	-0.032	-0.052	0.066	0.24	0.016
5.00	-0.150	0.492	0.021	-0.025	0.033	-0.037	-0.045	0.069	0.27	0.016
5.15	-0.159	0.516	0.027	-0.033	0.026	-0.035	-0.043	0.062	0.29	0.015
5.40	-0.171	0.512	0.013	-0.026	0.030	-0.040	-0.040	0.058	0.29	0.012
5.65	-0.134	0.414	0.018	-0.025	0.028	-0.037	-0.037	0.056	0.19	0.011
6.00	-0.155	0.441	0.009	-0.017	0.019	-0.042	-0.033	0.047	0.22	0.007
				T =-1.00 GEV**2						
2.03	-0.036	0.747	0.194	0.020	0.176	0.088	-0.258	0.209	0.60	0.297
2.30	-0.103	0.681	0.147	-0.087	0.030	0.037	-0.226	0.188	0.48	0.232
2.50	-0.075	0.564	0.128	-0.056	0.017	-0.042	-0.226	0.204	0.33	0.224
2.78	-0.077	0.494	0.088	-0.031	0.010	-0.063	-0.182	0.134	0.25	0.119
3.09	-0.021	0.471	0.051	-0.018	0.063	-0.076	-0.154	0.109	0.23	0.077
3.25	-0.061	0.502	0.027	-0.018	0.065	-0.038	-0.142	0.085	0.26	0.057
3.40	-0.057	0.504	0.052	-0.023	0.049	-0.091	-0.130	0.113	0.27	0.066
3.50	-0.076	0.453	0.040	-0.036	0.012	-0.033	-0.139	0.086	0.21	0.059
3.64	-0.040	0.444	0.026	-0.016	0.046	-0.081	-0.121	0.097	0.21	0.050
3.75	-0.031	0.413	0.029	-0.037	0.037	-0.097	-0.109	0.101	0.18	0.048
3.90	-0.059	0.445	0.020	-0.051	0.036	-0.065	-0.097	0.094	0.21	0.043
4.00	-0.041	0.379	0.027	-0.008	0.033	-0.084	-0.106	0.086	0.15	0.039
4.20	-0.044	0.413	0.018	-0.026	0.023	-0.094	-0.099	0.086	0.18	0.036
4.40	-0.053	0.328	0.015	0.016	0.022	-0.060	-0.088	0.071	0.11	0.026
4.60	-0.068	0.328	0.005	0.018	0.020	-0.087	-0.084	0.069	0.12	0.024
4.84	-0.046	0.274	-0.022	0.013	0.030	-0.119	-0.080	0.069	0.09	0.024
5.00	-0.068	0.317	-0.004	0.004	0.038	-0.112	-0.077	0.065	0.12	0.021
5.15	-0.063	0.303	-0.001	0.004	0.027	-0.106	-0.073	0.060	0.11	0.018
5.40	-0.060	0.297	-0.013	0.007	0.030	-0.115	-0.067	0.058	0.11	0.016
5.65	-0.049	0.210	-0.021	0.008	0.034	-0.124	-0.056	0.050	0.06	0.012
6.00	-0.070	0.262	-0.022	-0.000	0.028	-0.114	-0.056	0.040	0.09	0.010
				T =-1.50 GEV**2						
2.03	-0.193	0.737	0.265	0.103	0.131	0.007	-0.314	-0.038	0.60	0.362
2.30	-0.276	0.721	0.122	0.039	0.120	0.036	-0.227	-0.015	0.61	0.136
2.50	-0.287	0.564	0.087	0.020	0.045	-0.010	-0.173	0.023	0.40	0.077
2.78	-0.157	0.432	0.066	0.007	0.058	-0.028	-0.199	0.059	0.21	0.095
3.09	-0.081	0.372	-0.002	-0.000	0.062	-0.061	-0.118	-0.009	0.15	0.028
3.25	-0.079	0.347	0.036	0.005	0.031	0.004	-0.131	0.029	0.13	0.039
3.40	-0.095	0.378	0.040	0.013	0.022	-0.043	-0.119	0.017	0.15	0.033
3.50	-0.112	0.353	0.038	0.006	0.008	-0.022	-0.120	-0.007	0.14	0.032
3.64	-0.072	0.346	0.002	0.020	0.028	-0.054	-0.105	0.010	0.13	0.023
3.75	-0.060	0.326	0.002	0.007	0.038	-0.069	-0.102	0.028	0.12	0.022
3.90	-0.046	0.322	0.004	0.012	0.044	-0.061	-0.099	0.029	0.11	0.022
4.00	-0.065	0.299	0.003	0.010	0.020	-0.044	-0.086	0.012	0.10	0.015
4.20	-0.035	0.286	-0.010	0.008	0.038	-0.052	-0.085	0.019	0.09	0.016
4.40	-0.012	0.243	-0.015	0.031	-0.000	-0.029	-0.073	0.027	0.06	0.014
4.60	-0.019	0.240	-0.013	0.032	-0.006	-0.028	-0.062	0.017	0.06	0.011
4.84	-0.039	0.162	-0.053	0.016	0.006	-0.130	-0.036	0.017	0.04	0.009
5.00	-0.039	0.170	-0.047	0.015	0.014	-0.133	-0.051	0.017	0.05	0.011
5.15	-0.042	0.160	-0.053	0.027	0.013	-0.114	-0.050	0.016	0.04	0.012
5.40	-0.046	0.158	-0.036	0.022	0.007	-0.111	-0.044	0.016	0.04	0.008
5.65	-0.054	0.091	-0.049	0.002	0.061	-0.130	-0.028	-0.013	0.03	0.007
6.00	-0.057	0.101	-0.029	-0.001	0.052	-0.111	-0.031	-0.007	0.03	0.004

2.3 Measurable quantities reconstructed from partial waves

If one wants to apply πN scattering cross sections for another purpose than partial wave analysis, it is useful to have a smooth interpolation in energy and scattering angle of the existing data.

A first possibility is to employ an empirical "amalgamation" of the data, which has been performed by the CMU–LBL group. It is described in detail in the paper 79 L Kelly. The authors offered to distribute copies of the magnetic tape on request.

The second possibility is to reconstruct the measurable quantities from a partial wave solution. This has the advantage that the smoothing is constrained by the general theoretical principles mentioned in Sects 2.0, 2.1.7, and 2.1.9. Furthermore, one can calculate complete tables, which include angular ranges where data do not yet exist and quantities which have not yet been measured.

In all cases, the tables give quantities calculated from the strong interaction amplitudes, using the proton and charged pion masses in the kinematical calculations and assuming the validity of isospin invariance.

An estimate of the errors follows from a comparison with tables calculated from other partial wave solutions and with tables of the experimental data.

A smoothed table of the $\pi^{\pm}p$ total cross sections is included in Table 2.2.1.1.

2.3.1 Differential cross sections

$d\sigma/d\Omega$ for $\pi^{\pm}p$ elastic and $\pi^{-}p$ charge-exchange scattering as calculated from the Karlsruhe-Helsinki partial wave solution is given in Table 2.3.1.1. Figures 3.1.1···3.1.3 show three-dimensional plots of these cross sections.

$d\sigma/dt$-values at fixed t and momenta above 2 GeV/c are listed in the last two columns of Table 2.2.4.1 and forward cross sections in Table 2.2.1.1.

The electromagnetic contributions to $d\sigma/dt$ are discussed in Sects. 2.1.17 and 2.6.2. Since interesting measurements of differential cross sections in the Coulomb interference region are planned at meson factories, we show in Fig. 3.1.4 the prediction for the shape of the angular dependence and in Table 2.3.1.2 the magnitude of the different terms. These calculations have been made using the partial waves of Koch and Pietarinen (80 L-1) and the formalism employed by these authors.

Legendre expansions of differential cross sections have been discussed in many papers on scattering experiments and they have played an important role in the SACLAY partial wave analysis (Sect. 2.1.5). Therefore, we have included a list of the dimensionless Legendre coefficients defined in Eq. (A.4.21) as calculated from the KH 78 partial wave solution (Table 2.3.1.3).

Formulas which express Legendre coefficients in terms of partial waves have been given for the general case in 79 Höhler-1 and 78 L Oyanagi. Tables for the coefficients used in the SACLAY partial wave analysis can be found in 76 L Ayed and Tables which include higher coefficients in 79 L von Schlippe. Equation (A.4.22) is the simplest example.

The most recent discussions of Legendre coefficients for charge-exchange scattering are those in 76 Brown, 79 L Höhler-1 and 79 Minowa (see for instance Fig. 6.5.8).

For Table 2.3.1.1, see p. 154.

For Table 2.3.1.2, see p. 151.

For Table 2.3.1.3, see p. 164.

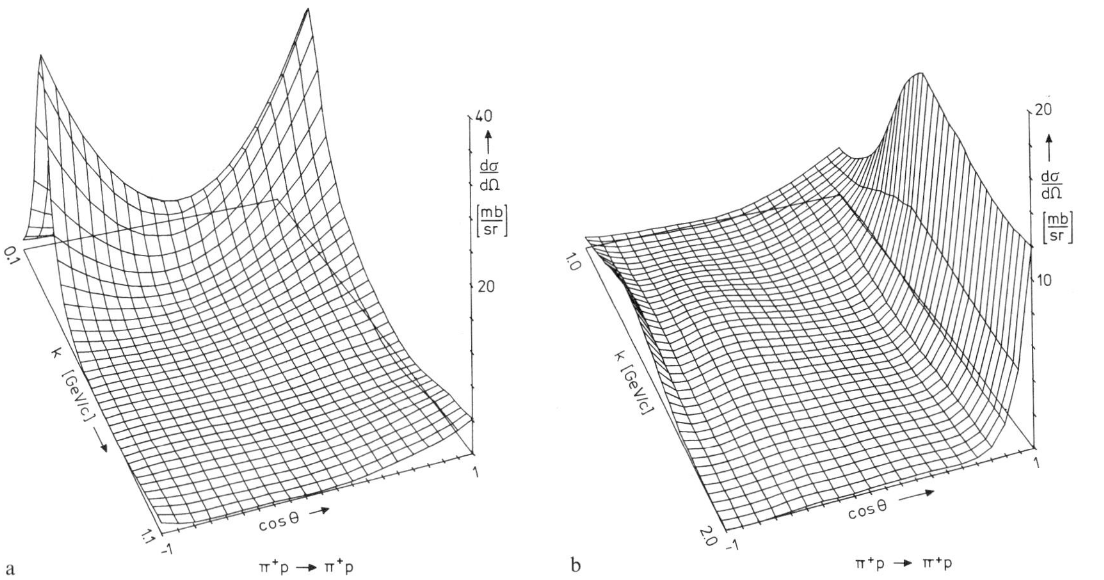

Fig. 3.1.1a, b. Plots of $d\sigma/d\Omega$ for π^+p elastic scattering in the momentum range 0.1···2.0 GeV/c. In this figure and in the two following ones the scales of $d\sigma/d\Omega$ in the two parts of the momentum range are different.

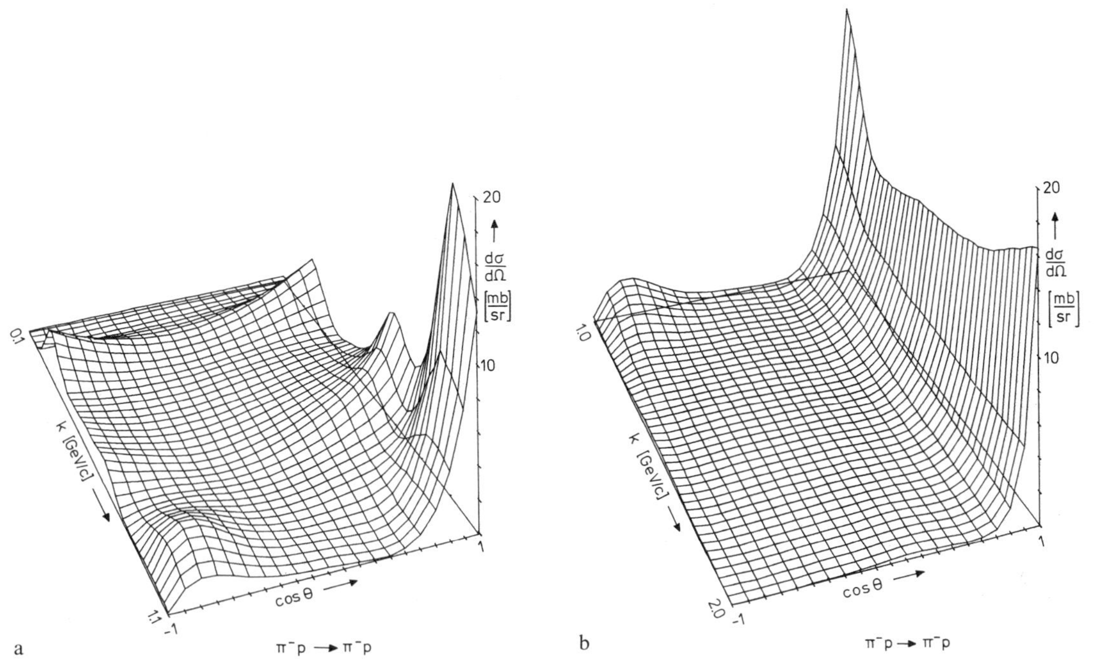

Fig. 3.1.2a, b. Plots of $d\sigma/d\Omega$ for π^-p elastic scattering in the momentum range 0.1···2.0 GeV/c.

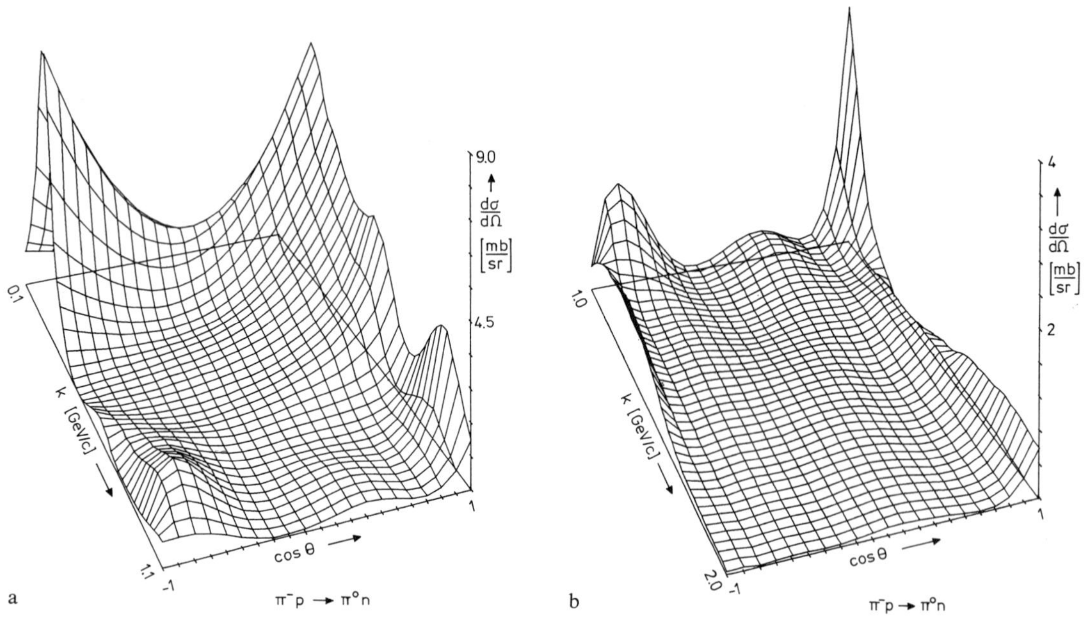

Fig. 3.1.3a, b. Plots of dσ/dΩ for $\pi^- p$ charge-exchange scattering in the momentum range 0.1⋯2.0 GeV/c.

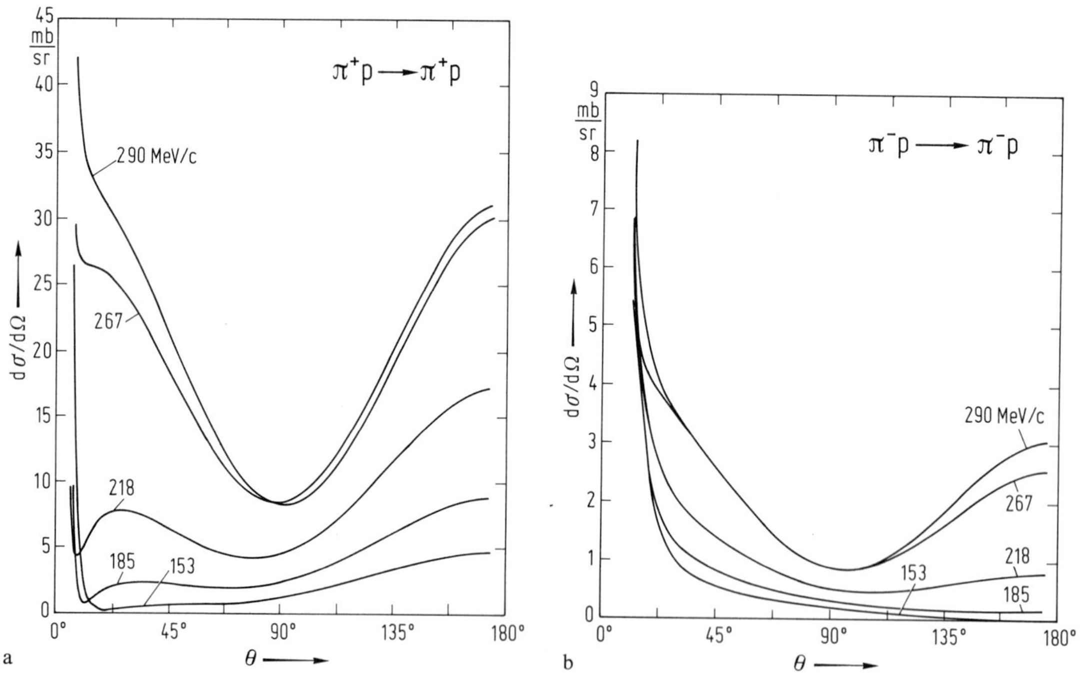

Fig. 3.1.4a, b. Shape of the elastic $\pi^{\pm} p$ cross sections between 153 and 290 MeV/c. The Coulomb and the Coulomb interference terms are included.

Table 2.3.1.2. Decomposition of the differential cross sections in the presence of Coulomb effects. These tables are based on the work of KOCH and PIETARINEN (80 L-1) (cf. Sect. 2.1.17.2 for the definition of nuclear (NUCL) and hadronic (HADR) contributions, INT: interference term). TPI $= T_\pi$, DS/DO $= d\sigma/d\Omega$. Lab. angle in degrees.

K= 79.0 MEV/C; TPI= 20.8 MEV — PI+P->PI+P

COS THETA	-1000*T GEV**2	THETA LAB.	DS/DO[MB/SR] EXP =	NUCL +	INT +	COUL	HADR
0.990	0.092	6.9	189.249	0.013	2.583	186.653	0.013
0.980	0.183	9.8	47.981	0.015	1.355	46.611	0.015
0.970	0.275	12.0	21.653	0.017	0.944	20.692	0.017
0.960	0.366	13.9	12.383	0.019	0.737	11.626	0.020
0.950	0.458	15.6	8.067	0.021	0.613	7.432	0.022
0.940	0.549	17.1	5.709	0.023	0.530	5.156	0.024
0.930	0.641	18.5	4.280	0.026	0.471	3.783	0.026
0.900	0.915	22.2	2.243	0.032	0.364	1.848	0.033
0.800	1.830	31.8	0.749	0.055	0.237	0.457	0.058
0.700	2.745	39.4	0.475	0.080	0.195	0.201	0.084
0.600	3.660	46.2	0.391	0.106	0.173	0.112	0.112
0.500	4.576	52.4	0.365	0.135	0.159	0.071	0.143
0.400	5.491	58.2	0.364	0.165	0.150	0.048	0.176
0.300	6.406	63.9	0.376	0.198	0.143	0.035	0.211
0.200	7.321	69.5	0.396	0.232	0.137	0.027	0.248
0.100	8.236	75.0	0.422	0.268	0.133	0.021	0.287
0.000	9.151	80.6	0.452	0.306	0.129	0.017	0.328
-0.100	10.066	86.2	0.486	0.346	0.126	0.014	0.371
-0.200	10.981	92.0	0.523	0.388	0.123	0.011	0.417
-0.300	11.897	98.0	0.562	0.432	0.121	0.010	0.464

K=112.0 MEV/C; TPI= 39.4 MEV — PI+P->PI+P

COS THETA	-1000*T GEV**2	THETA LAB.	DS/DO[MB/SR] EXP =	NUCL +	INT +	COUL	HADR
0.995	0.089	4.8	228.693	0.083	-8.186	236.796	0.097
0.990	0.179	6.8	55.208	0.083	-4.013	59.138	0.097
0.980	0.357	9.6	12.905	0.084	-1.932	14.754	0.098
0.970	0.536	11.8	5.388	0.085	-1.241	6.543	0.099
0.960	0.715	13.7	2.864	0.086	-0.895	3.673	0.099
0.950	0.894	15.3	1.744	0.087	-0.689	2.346	0.100
0.940	1.072	16.8	1.163	0.088	-0.551	1.626	0.101
0.930	1.251	18.2	0.829	0.089	-0.453	1.192	0.103
0.900	1.787	21.9	0.398	0.093	-0.276	0.580	0.106
0.800	3.575	31.3	0.182	0.112	-0.072	0.142	0.124
0.700	5.362	38.9	0.194	0.138	-0.006	0.062	0.150
0.600	7.150	45.6	0.231	0.171	0.026	0.034	0.184
0.500	8.937	51.7	0.278	0.212	0.045	0.021	0.226
0.400	10.725	57.5	0.330	0.259	0.057	0.014	0.276
0.300	12.512	63.2	0.389	0.314	0.065	0.010	0.333
0.200	14.299	68.7	0.454	0.377	0.070	0.008	0.399
0.100	16.087	74.2	0.526	0.446	0.074	0.006	0.473
0.000	17.874	79.8	0.605	0.523	0.077	0.005	0.554
-0.100	19.662	85.4	0.690	0.607	0.079	0.004	0.644
-0.200	21.449	91.2	0.782	0.699	0.080	0.003	0.742

K= 79.0 MEV/C; TPI= 20.8 MEV — PI-P->PI-P

COS THETA	-1000*T GEV**2	THETA LAB.	DS/DO[MB/SR] EXP =	NUCL +	INT +	COUL	HADR
0.990	0.092	6.9	197.903	0.172	11.078	186.653	0.161
0.980	0.183	9.8	52.316	0.171	5.534	46.611	0.160
0.970	0.275	12.0	24.546	0.171	3.683	20.692	0.160
0.960	0.366	13.9	14.553	0.170	2.756	11.626	0.160
0.950	0.458	15.6	9.802	0.170	2.200	7.432	0.159
0.940	0.549	17.1	7.154	0.169	1.829	5.156	0.159
0.930	0.641	18.5	5.516	0.169	1.564	3.783	0.158
0.900	0.915	22.2	3.101	0.168	1.086	1.848	0.157
0.800	1.830	31.8	1.148	0.163	0.528	0.457	0.153
0.700	2.745	39.4	0.702	0.159	0.342	0.201	0.149
0.600	3.660	46.2	0.515	0.154	0.249	0.112	0.145
0.500	4.576	52.4	0.414	0.150	0.194	0.071	0.141
0.400	5.491	58.2	0.350	0.145	0.157	0.048	0.137
0.300	6.406	63.9	0.306	0.140	0.131	0.035	0.132
0.200	7.321	69.5	0.273	0.136	0.111	0.027	0.128
0.100	8.236	75.0	0.247	0.131	0.096	0.021	0.123
0.000	9.151	80.6	0.226	0.126	0.084	0.017	0.119
-0.100	10.066	86.2	0.208	0.121	0.074	0.014	0.114
-0.200	10.981	92.0	0.193	0.116	0.066	0.011	0.109
-0.300	11.897	98.0	0.179	0.110	0.059	0.010	0.105

K=112.0 MEV/C; TPI= 39.4 MEV — PI-P->PI-P

COS THETA	-1000*T GEV**2	THETA LAB.	DS/DO[MB/SR] EXP =	NUCL +	INT +	COUL	HADR
0.995	0.089	4.8	251.545	0.240	14.509	236.796	0.224
0.990	0.179	6.8	66.626	0.239	7.249	59.138	0.223
0.980	0.357	9.6	18.602	0.238	3.610	14.754	0.222
0.970	0.536	11.8	9.176	0.237	2.396	6.543	0.221
0.960	0.715	13.7	5.697	0.236	1.787	3.673	0.220
0.950	0.894	15.3	4.003	0.235	1.422	2.346	0.219
0.940	1.072	16.8	3.038	0.234	1.179	1.626	0.219
0.930	1.251	18.2	2.430	0.233	1.005	1.192	0.218
0.900	1.787	21.9	1.502	0.230	0.691	0.580	0.215
0.800	3.575	31.3	0.688	0.220	0.326	0.142	0.206
0.700	5.362	38.9	0.476	0.210	0.204	0.062	0.197
0.600	7.150	45.6	0.378	0.200	0.144	0.034	0.188
0.500	8.937	51.7	0.319	0.190	0.108	0.021	0.179
0.400	10.725	57.5	0.278	0.180	0.084	0.014	0.170
0.300	12.512	63.2	0.247	0.170	0.067	0.010	0.161
0.200	14.299	68.7	0.222	0.159	0.055	0.008	0.151
0.100	16.087	74.2	0.201	0.149	0.046	0.006	0.142
0.000	17.874	79.8	0.182	0.139	0.038	0.005	0.132
-0.100	19.662	85.4	0.164	0.128	0.032	0.004	0.123
-0.200	21.449	91.2	0.148	0.118	0.027	0.003	0.113

K=153.0 MEV/C; TPI= 67.5 MEV PI+P->PI+P

COS THETA	-1000*T GEV**2	THETA LAB.	DS/DO[MB/SR] EXP =	NUCL +	INT +	COUL	HADR
0.995	0.160	4.7	77.067	1.064	-18.887	94.890	1.172
0.990	0.320	6.7	15.413	1.057	-9.325	23.681	1.164
0.980	0.640	9.4	2.387	1.043	-4.556	5.900	1.148
0.970	0.960	11.6	0.672	1.029	-2.969	2.613	1.133
0.960	1.280	13.4	0.302	1.015	-2.177	1.465	1.118
0.950	1.599	15.0	0.233	1.002	-1.703	0.934	1.103
0.940	1.919	16.5	0.249	0.989	-1.387	0.646	1.089
0.930	2.239	17.8	0.288	0.977	-1.161	0.473	1.075
0.900	3.199	21.4	0.414	0.941	-0.757	0.229	1.035
0.800	6.398	30.7	0.611	0.843	-0.288	0.055	0.925
0.700	9.597	38.2	0.667	0.778	-0.135	0.024	0.850
0.600	12.796	44.7	0.698	0.746	-0.061	0.013	0.810
0.500	15.995	50.8	0.736	0.746	-0.018	0.008	0.806

K=200.0 MEV/C; TPI=104.3 MEV PI+P->PI+P

COS THETA	-1000*T GEV**2	THETA LAB.	DS/DO[MB/SR] EXP =	NUCL +	INT +	COUL	HADR
0.995	0.259	4.6	24.394	6.091	-29.139	47.442	6.570
0.990	0.519	6.5	3.494	6.047	-14.382	11.829	6.523
0.980	1.038	9.2	1.870	5.960	-7.031	2.941	6.429
0.970	1.556	11.3	2.585	5.874	-4.589	1.300	6.336
0.960	2.075	13.1	3.146	5.789	-3.371	0.728	6.245
0.950	2.594	14.6	3.527	5.706	-2.642	0.463	6.155
0.940	3.113	16.1	3.787	5.624	-2.156	0.320	6.066
0.930	3.632	17.4	3.966	5.543	-1.810	0.234	5.978
0.900	5.188	20.9	4.231	5.307	-1.189	0.113	5.724
0.800	10.376	30.0	4.155	4.600	-0.472	0.027	4.959
0.700	15.564	37.3	3.785	4.014	-0.240	0.011	4.322
0.600	20.753	43.7	3.424	3.546	-0.128	0.006	3.812
0.500	25.941	49.7	3.135	3.195	-0.064	0.004	3.428

K=247.0 MEV/C; TPI=144.1 MEV PI+P->PI+P

COS THETA	-1000*T GEV**2	THETA LAB.	DS/DO[MB/SR] EXP =	NUCL +	INT +	COUL	HADR
0.995	0.375	4.4	24.723	20.889	-25.253	29.087	21.966
0.990	0.750	6.3	15.608	20.739	-12.375	7.245	21.808
0.980	1.500	9.0	16.238	20.439	-6.000	1.798	21.493
0.970	2.250	11.0	17.044	20.144	-3.893	0.793	21.182
0.960	3.000	12.7	17.448	19.851	-2.846	0.443	20.875
0.950	3.750	14.3	17.623	19.562	-2.220	0.281	20.571
0.940	4.500	15.6	17.665	19.276	-1.805	0.194	20.270
0.930	5.250	16.9	17.626	18.993	-1.509	0.141	19.973
0.900	7.500	20.3	17.253	18.165	-0.980	0.068	19.101
0.800	15.000	29.2	15.260	15.620	-0.376	0.016	16.421
0.700	22.501	36.3	13.229	13.405	-0.183	0.006	14.087
0.600	30.001	42.6	11.435	11.523	-0.091	0.003	12.100

K=153.0 MEV/C; TPI= 67.5 MEV PI-P->PI-P

COS THETA	-1000*T GEV**2	THETA LAB.	DS/DO[MB/SR] EXP =	NUCL +	INT +	COUL	HADR
0.995	0.160	4.7	108.115	0.491	12.733	94.890	0.463
0.990	0.320	6.7	30.520	0.489	6.350	23.681	0.461
0.980	0.640	9.4	9.534	0.485	3.150	5.900	0.457
0.970	0.960	11.6	5.174	0.480	2.081	2.613	0.453
0.960	1.280	13.4	3.486	0.476	1.546	1.465	0.449
0.950	1.599	15.0	2.630	0.472	1.224	0.934	0.446
0.940	1.919	16.5	2.124	0.468	1.010	0.646	0.442
0.930	2.239	17.8	1.794	0.464	0.857	0.473	0.438
0.900	3.199	21.4	1.263	0.452	0.581	0.229	0.427
0.800	6.398	30.7	0.729	0.414	0.260	0.055	0.391
0.700	9.597	38.2	0.557	0.379	0.154	0.024	0.359
0.600	12.796	44.7	0.461	0.346	0.102	0.013	0.328
0.500	15.995	50.8	0.396	0.316	0.072	0.008	0.301

K=200.0 MEV/C; TPI=104.3 MEV PI-P->PI-P

COS THETA	-1000*T GEV**2	THETA LAB.	DS/DO[MB/SR] EXP =	NUCL +	INT +	COUL	HADR
0.995	0.259	4.6	62.296	1.353	13.501	47.442	1.300
0.990	0.519	6.5	19.903	1.344	6.730	11.829	1.292
0.980	1.038	9.2	7.600	1.328	3.331	2.941	1.276
0.970	1.556	11.3	4.806	1.311	2.194	1.300	1.260
0.960	2.075	13.1	3.647	1.295	1.625	0.728	1.245
0.950	2.594	14.6	3.025	1.278	1.283	0.463	1.229
0.940	3.113	16.1	2.637	1.262	1.055	0.320	1.214
0.930	3.632	17.4	2.372	1.247	0.892	0.234	1.198
0.900	5.188	20.9	1.912	1.200	0.599	0.113	1.154
0.800	10.376	30.0	1.342	1.057	0.258	0.027	1.017
0.700	15.564	37.3	1.088	0.931	0.146	0.011	0.896
0.600	20.753	43.7	0.918	0.820	0.092	0.006	0.790
0.500	25.941	49.7	0.787	0.724	0.060	0.004	0.698

K=247.0 MEV/C; TPI=182.3 MEV PI-P->PI-P

COS THETA	-1000*T GEV**2	THETA LAB.	DS/DO[MB/SR] EXP =	NUCL +	INT +	COUL	HADR
0.995	0.375	4.4	42.767	3.265	10.415	29.087	3.213
0.990	0.750	6.3	15.695	3.240	5.209	7.245	3.188
0.980	1.500	9.0	7.572	3.192	2.582	1.798	3.140
0.970	2.250	11.0	5.636	3.143	1.700	0.793	3.092
0.960	3.000	12.7	4.795	3.095	1.257	0.443	3.045
0.950	3.750	14.3	4.320	3.048	0.991	0.281	2.998
0.940	4.500	15.6	4.008	3.002	0.813	0.194	2.952
0.930	5.250	16.9	3.782	2.955	0.685	0.141	2.907
0.900	7.500	20.3	3.345	2.821	0.456	0.068	2.774
0.800	15.000	29.2	2.613	2.408	0.189	0.016	2.366
0.700	22.501	36.3	2.156	2.047	0.103	0.006	2.012
0.600	30.001	42.6	1.801	1.737	0.061	0.003	1.706

K=290.0 MEV/C; TPI=182.3 MEV PI+P->PI+P

COS THETA	-1000*T GEV**2	THETA LAB.	DS/DO[MB/SR] EXP =	NUCL +	INT +	COUL	HADR
0.995	0.492	4.3	55.568	33.099	1.806	20.662	33.499
0.990	0.985	6.2	39.060	32.858	1.061	5.142	33.256
0.980	1.969	8.7	34.266	32.380	0.613	1.273	32.771
0.970	2.954	10.7	32.909	31.906	0.442	0.561	32.292
0.960	3.939	12.4	32.100	31.438	0.350	0.312	31.817
0.950	4.923	13.9	31.464	30.973	0.292	0.198	31.348
0.940	5.908	15.3	30.902	30.514	0.252	0.136	30.882
0.930	6.893	16.5	30.380	30.059	0.222	0.099	30.422
0.900	9.847	19.8	28.936	28.723	0.166	0.047	29.069

K=331.0 MEV/C; TPI=219.6 MEV PI+P->PI+P

COS THETA	-1000*T GEV**2	THETA LAB.	DS/DO[MB/SR] EXP =	NUCL +	INT +	COUL	HADR
0.995	0.613	4.2	64.164	29.126	19.179	15.859	29.152
0.990	1.226	6.0	42.475	28.914	9.619	3.943	28.939
0.980	2.451	8.5	34.252	28.492	4.786	0.975	28.517
0.970	3.677	10.5	31.663	28.074	3.160	0.428	28.099
0.960	4.903	12.1	30.242	27.661	2.343	0.238	27.685
0.950	6.128	13.6	29.252	27.251	1.851	0.151	27.274
0.940	7.354	14.9	28.470	26.844	1.522	0.104	26.868
0.930	8.579	16.2	27.804	26.442	1.287	0.075	26.465
0.900	12.256	19.4	26.156	25.258	0.863	0.036	25.280

K=427.0 MEV/C; TPI=309.7 MEV PI+P->PI+P

COS THETA	-1000*T GEV**2	THETA LAB.	DS/DO[MB/SR] EXP =	NUCL +	INT +	COUL	HADR
0.995	0.921	4.0	43.922	15.016	19.014	9.892	15.168
0.990	1.842	5.7	26.814	14.907	9.454	2.453	15.058
0.980	3.684	8.1	19.948	14.690	4.655	0.603	14.839
0.970	5.526	10.0	17.789	14.475	3.050	0.264	14.622
0.960	7.368	11.5	16.654	14.262	2.246	0.146	14.407
0.950	9.210	12.9	15.906	14.050	1.764	0.092	14.193
0.940	11.052	14.2	15.346	13.841	1.443	0.063	13.981
0.930	12.894	15.4	14.891	13.632	1.213	0.045	13.770

K=532.0 MEV/C; TPI=410.4 MEV PI+P->PI+P

COS THETA	-1000*T GEV**2	THETA LAB.	DS/DO[MB/SR] EXP =	NUCL +	INT +	COUL	HADR
0.995	1.290	3.8	27.993	8.198	12.993	6.802	8.358
0.990	2.579	5.4	16.276	8.148	6.446	1.682	8.307
0.980	5.159	7.7	11.623	8.048	3.164	0.411	8.205
0.970	7.738	9.5	10.194	7.947	2.068	0.179	8.102
0.960	10.318	11.0	9.465	7.847	1.520	0.098	8.000
0.950	12.897	12.3	8.999	7.746	1.191	0.062	7.897
0.940	15.476	13.5	8.660	7.646	0.972	0.042	7.794
0.930	18.056	14.6	8.391	7.545	0.816	0.030	7.691

K=290.0 MEV/C; TPI=182.3 MEV PI-P->PI-P

COS THETA	-1000*T GEV**2	THETA LAB.	DS/DO[MB/SR] EXP =	NUCL +	INT +	COUL	HADR
0.995	0.492	4.3	26.039	4.108	1.269	20.662	4.096
0.990	0.985	6.2	9.897	4.078	0.678	5.142	4.066
0.980	1.969	8.7	5.648	4.019	0.356	1.273	4.007
0.970	2.954	10.7	4.762	3.960	0.241	0.561	3.948
0.960	3.939	12.4	4.396	3.901	0.182	0.312	3.889
0.950	4.923	13.9	4.187	3.844	0.145	0.198	3.832
0.940	5.908	15.3	4.043	3.786	0.120	0.136	3.774
0.930	6.893	16.5	3.931	3.730	0.102	0.099	3.718
0.900	9.847	19.8	3.680	3.563	0.069	0.047	3.551

K=331.0 MEV/C; TPI=219.6 MEV PI-P->PI-P

COS THETA	-1000*T GEV**2	THETA LAB.	DS/DO[MB/SR] EXP =	NUCL +	INT +	COUL	HADR
0.995	0.613	4.2	15.521	2.918	-3.255	15.859	2.911
0.990	1.226	6.0	5.261	2.900	-1.582	3.943	2.894
0.980	2.451	8.5	3.077	2.866	-0.764	0.975	2.859
0.970	3.677	10.5	2.764	2.831	-0.496	0.428	2.825
0.960	4.903	12.1	2.672	2.797	-0.364	0.238	2.791
0.950	6.128	13.6	2.629	2.763	-0.285	0.151	2.757
0.940	7.354	14.9	2.600	2.729	-0.233	0.104	2.723
0.930	8.579	16.2	2.574	2.695	-0.196	0.075	2.690
0.900	12.256	19.4	2.501	2.595	-0.130	0.036	2.590

K=427.0 MEV/C; TPI=309.7 MEV PI-P->PI-P

COS THETA	-1000*T GEV**2	THETA LAB.	DS/DO[MB/SR] EXP =	NUCL +	INT +	COUL	HADR
0.995	0.921	4.0	9.806	1.237	-1.323	9.892	1.202
0.990	1.842	5.7	3.051	1.237	-0.639	2.453	1.203
0.980	3.684	8.1	1.535	1.238	-0.307	0.603	1.204
0.970	5.526	10.0	1.304	1.238	-0.198	0.264	1.205
0.960	7.368	11.5	1.240	1.239	-0.145	0.146	1.206
0.950	9.210	12.9	1.217	1.239	-0.113	0.092	1.206
0.940	11.052	14.2	1.209	1.239	-0.092	0.063	1.207
0.930	12.894	15.4	1.206	1.239	-0.078	0.045	1.207

K=532.0 MEV/C; TPI=410.4 MEV PI-P->PI-P

COS THETA	-1000*T GEV**2	THETA LAB.	DS/DO[MB/SR] EXP =	NUCL +	INT +	COUL	HADR
0.995	1.290	3.8	10.097	1.876	1.419	6.802	1.862
0.990	2.579	5.4	4.268	1.870	0.715	1.682	1.856
0.980	5.159	7.7	2.625	1.860	0.354	0.411	1.846
0.970	7.738	9.5	2.259	1.849	0.231	0.179	1.835
0.960	10.318	11.0	2.106	1.838	0.169	0.098	1.823
0.950	12.897	12.3	2.020	1.827	0.131	0.062	1.812
0.940	15.476	13.5	1.964	1.816	0.106	0.042	1.800
0.930	18.056	14.6	1.923	1.805	0.088	0.030	1.789

Table 2.3.1.1. $d\sigma/d\Omega$ as reconstructed from the KH partial wave solution. $d\sigma/dt$ in mb/GeV^2 follows by multiplication with "DOM/DT"$=\pi/q^2$, which is listed below the k-value. Notation: $0.92-02=0.92\cdot10^{-2}$, etc. "THETA" is the c.m. scattering angle θ. KH 80 was used below 0.5 GeV/c and KH 78 at higher momenta.

K[MEV/C] DOM/DT		COS THETA 1.00	.90	.80	.70	.60	.50	.40	.30	DSIG/DOM[MB/SR] .20	.10
20.0	PI+P	.18	.18	.19	.19	.19	.19	.19	.20	.20	.20
10388.2	PI-P	.14	.14	.14	.14	.14	.14	.14	.14	.14	.14
	CHEX	.32	.32	.33	.33	.33	.33	.33	.34	.34	.34
40.0	PI+P	.13	.14	.15	.16	.16	.17	.18	.19	.20	.21
2614.6	PI-P	.14	.14	.14	.14	.14	.14	.14	.14	.14	.14
	CHEX	.27	.28	.29	.29	.30	.31	.32	.32	.33	.34
60.0	PI+P	.06	.08	.10	.11	.13	.15	.17	.19	.21	.23
1174.6	PI-P	.15	.15	.15	.15	.14	.14	.14	.14	.14	.13
	CHEX	.20	.22	.23	.24	.26	.27	.29	.30	.32	.34
79.0	PI+P	.01	.03	.06	.08	.11	.14	.18	.21	.25	.29
686.6	PI-P	.16	.16	.15	.15	.15	.14	.14	.13	.13	.12
	CHEX	.12	.14	.16	.18	.20	.23	.25	.27	.30	.33
97.0	PI+P	.01	.03	.06	.09	.12	.16	.21	.25	.30	.36
462.2	PI-P	.19	.18	.18	.17	.17	.16	.15	.15	.14	.13
	CHEX	.06	.08	.10	.13	.16	.19	.22	.25	.29	.33
112.0	PI+P	.09	.10	.12	.15	.18	.22	.27	.33	.40	.47
351.5	PI-P	.22	.22	.21	.20	.19	.18	.17	.16	.15	.14
	CHEX	.02	.04	.06	.09	.12	.16	.20	.24	.28	.33
130.0	PI+P	.36	.33	.32	.32	.34	.38	.43	.49	.57	.67
265.6	PI-P	.30	.28	.27	.25	.23	.22	.21	.19	.18	.16
	CHEX	.00	.02	.04	.07	.10	.14	.18	.23	.29	.35
153.0	PI+P	1.17	1.02	.92	.84	.80	.80	.83	.90	1.00	1.14
196.4	PI-P	.47	.43	.39	.36	.33	.30	.28	.25	.23	.21
	CHEX	.08	.08	.09	.10	.13	.17	.22	.28	.35	.44
172.0	PI+P	2.52	2.19	1.92	1.72	1.57	1.48	1.46	1.49	1.59	1.74
158.7	PI-P	.69	.62	.56	.50	.45	.41	.37	.33	.30	.27
	CHEX	.29	.25	.23	.23	.24	.27	.31	.37	.45	.54
185.0	PI+P	4.02	3.48	3.03	2.67	2.39	2.20	2.09	2.07	2.13	2.27
139.2	PI-P	.92	.82	.73	.65	.58	.52	.46	.41	.37	.34
	CHEX	.55	.48	.42	.39	.38	.38	.41	.47	.54	.64
200.0	PI+P	6.59	5.70	4.93	4.30	3.79	3.41	3.16	3.02	3.01	3.13
121.1	PI-P	1.31	1.15	1.02	.90	.79	.70	.62	.55	.49	.44
	CHEX	1.02	.88	.77	.68	.63	.60	.61	.64	.71	.82
218.0	PI+P	11.13	9.60	8.28	7.15	6.22	5.49	4.96	4.62	4.48	4.53
104.0	PI-P	1.88	1.64	1.42	1.23	1.07	.93	.81	.71	.63	.57
	CHEX	1.95	1.69	1.46	1.28	1.13	1.04	.98	.98	1.02	1.11
247.0	PI+P	22.03	19.02	16.35	14.02	12.05	10.42	9.14	8.20	7.62	7.39
83.8	PI-P	3.24	2.77	2.37	2.01	1.70	1.45	1.23	1.06	.93	.83
	CHEX	4.27	3.72	3.23	2.80	2.44	2.16	1.95	1.81	1.76	1.79
267.0	PI+P	29.90	25.79	22.12	18.88	16.09	13.75	11.84	10.38	9.37	8.80
73.3	PI-P	4.02	3.44	2.92	2.47	2.08	1.75	1.48	1.26	1.08	.96
	CHEX	6.15	5.34	4.61	3.97	3.42	2.96	2.60	2.34	2.18	2.12
280.0	PI+P	32.79	28.25	24.18	20.58	17.46	14.82	12.65	10.95	9.72	8.97
67.7	PI-P	4.22	3.62	3.09	2.62	2.21	1.85	1.56	1.32	1.13	1.00
	CHEX	6.95	6.01	5.16	4.42	3.79	3.25	2.82	2.50	2.28	2.16
290.0	PI+P	33.64	28.98	24.79	21.07	17.83	15.06	12.77	10.95	9.61	8.74
63.8	PI-P	4.12	3.55	3.03	2.57	2.16	1.82	1.52	1.28	1.10	.97
	CHEX	7.34	6.33	5.43	4.63	3.94	3.36	2.88	2.52	2.26	2.10
295.0	PI+P	33.74	29.06	24.86	21.12	17.85	15.05	12.72	10.86	9.47	8.55
62.0	PI-P	4.03	3.47	2.97	2.52	2.12	1.78	1.49	1.26	1.07	.95
	CHEX	7.48	6.45	5.52	4.70	3.99	3.39	2.89	2.51	2.23	2.05

K[MEV/C] DOM/DT	COS THETA .00	-.10	-.20	-.30	-.40	-.50	-.60	-.70	-.80	-.90	DSIG/DOM[MB/SR] -1.00
20.0	.20	.21	.21	.21	.21	.22	.22	.22	.22	.23	.23
10388.2	.14	.14	.14	.14	.14	.14	.14	.14	.14	.14	.14
	.34	.34	.35	.35	.35	.35	.35	.36	.36	.36	.36
40.0	.22	.23	.24	.25	.26	.27	.28	.29	.30	.31	.32
2614.6	.14	.14	.14	.13	.13	.13	.13	.13	.13	.13	.13
	.35	.35	.36	.37	.38	.39	.39	.40	.41	.42	.43
60.0	.26	.28	.30	.33	.35	.38	.41	.43	.46	.49	.52
1174.6	.13	.13	.13	.13	.12	.12	.12	.11	.11	.11	.11
	.35	.37	.39	.41	.43	.44	.46	.48	.50	.52	.55
79.0	.33	.37	.42	.46	.51	.57	.62	.68	.73	.79	.86
686.6	.12	.11	.11	.10	.10	.09	.09	.08	.08	.07	.07
	.36	.39	.42	.45	.48	.51	.55	.58	.62	.66	.70
97.0	.42	.49	.56	.63	.71	.80	.88	.98	1.07	1.18	1.28
462.2	.13	.12	.11	.10	.10	.09	.08	.07	.06	.06	.05
	.37	.41	.46	.50	.55	.61	.66	.72	.78	.84	.90
112.0	.55	.64	.74	.85	.96	1.08	1.21	1.35	1.50	1.65	1.82
351.5	.13	.12	.11	.10	.09	.08	.07	.06	.05	.04	.02
	.38	.44	.50	.56	.63	.70	.77	.85	.94	1.02	1.11
130.0	.78	.90	1.05	1.20	1.38	1.56	1.77	1.99	2.22	2.47	2.74
265.6	.15	.14	.12	.11	.10	.08	.07	.05	.04	.02	.00
	.42	.49	.57	.66	.75	.85	.95	1.07	1.19	1.32	1.45
153.0	1.31	1.52	1.77	2.05	2.37	2.72	3.11	3.54	4.00	4.50	5.03
196.4	.19	.17	.15	.14	.12	.10	.09	.07	.05	.03	.02
	.53	.64	.76	.90	1.04	1.21	1.38	1.57	1.78	2.00	2.24
172.0	1.96	2.23	2.57	2.97	3.43	3.95	4.53	5.17	5.88	6.65	7.48
158.7	.25	.22	.20	.19	.17	.15	.14	.13	.11	.10	.09
	.66	.80	.95	1.13	1.32	1.54	1.79	2.05	2.34	2.66	3.00
185.0	2.50	2.81	3.21	3.69	4.26	4.91	5.65	6.48	7.40	8.41	9.51
139.2	.31	.28	.26	.24	.22	.21	.20	.19	.18	.18	.17
	.77	.92	1.10	1.31	1.54	1.81	2.10	2.43	2.79	3.18	3.61
200.0	3.36	3.72	4.20	4.81	5.55	6.41	7.40	8.52	9.78	11.17	12.70
121.1	.40	.37	.34	.32	.31	.30	.30	.30	.31	.32	.33
	.96	1.13	1.35	1.60	1.89	2.22	2.60	3.02	3.48	3.99	4.54
218.0	4.78	5.22	5.85	6.68	7.71	8.94	10.38	12.02	13.87	15.93	18.21
104.0	.53	.50	.48	.48	.49	.51	.55	.59	.65	.71	.79
	1.25	1.45	1.70	2.00	2.36	2.78	3.27	3.81	4.43	5.11	5.86
247.0	7.51	7.98	8.80	9.98	11.51	13.40	15.64	18.24	21.19	24.50	28.17
83.8	.77	.74	.75	.78	.85	.94	1.07	1.22	1.39	1.60	1.83
	1.91	2.11	2.40	2.78	3.26	3.83	4.50	5.27	6.13	7.10	8.16
267.0	8.68	9.00	9.77	10.98	12.65	14.76	17.31	20.31	23.76	27.66	32.00
73.3	.89	.87	.89	.95	1.06	1.21	1.40	1.64	1.92	2.24	2.59
	2.16	2.31	2.57	2.94	3.42	4.01	4.71	5.53	6.46	7.51	8.67
280.0	8.69	8.88	9.54	10.68	12.29	14.37	16.93	19.96	23.46	27.43	31.88
67.7	.92	.89	.92	.99	1.12	1.30	1.53	1.81	2.14	2.52	2.94
	2.16	2.27	2.49	2.82	3.26	3.81	4.49	5.27	6.18	7.20	8.34
290.0	8.35	8.43	8.99	10.02	11.52	13.50	15.95	18.87	22.26	26.12	30.44
63.8	.89	.87	.90	.99	1.13	1.32	1.57	1.87	2.23	2.64	3.11
	2.06	2.12	2.30	2.58	2.98	3.48	4.10	4.83	5.67	6.62	7.68
295.0	8.10	8.12	8.61	9.57	11.00	12.90	15.27	18.10	21.40	25.16	29.39
62.0	.87	.85	.88	.97	1.12	1.31	1.57	1.88	2.24	2.66	3.14
	1.99	2.03	2.18	2.44	2.80	3.27	3.86	4.54	5.34	6.25	7.26

K[MEV/C] DOM/DT		COS THETA 1.00	.90	.80	.70	.60	.50	.40	.30	DSIG/DOM[MB/SR] .20	.10
301.0	PI+P	33.58	28.92	24.73	20.99	17.72	14.91	12.56	10.67	9.24	8.28
60.1	PI-P	3.90	3.36	2.88	2.45	2.07	1.73	1.45	1.22	1.04	.92
	CHEX	7.57	6.51	5.57	4.73	4.00	3.38	2.87	2.47	2.17	1.98
305.0	PI+P	33.25	28.64	24.48	20.77	17.51	14.71	12.36	10.46	9.02	8.03
58.7	PI-P	3.77	3.26	2.80	2.38	2.01	1.69	1.42	1.19	1.01	.89
	CHEX	7.60	6.53	5.57	4.72	3.99	3.36	2.84	2.43	2.12	1.92
310.0	PI+P	32.74	28.21	24.10	20.44	17.22	14.44	12.09	10.19	8.73	7.71
57.1	PI-P	3.61	3.13	2.70	2.30	1.95	1.64	1.37	1.15	.98	.86
	CHEX	7.61	6.53	5.56	4.70	3.96	3.33	2.80	2.38	2.06	1.84
320.0	PI+P	31.22	26.93	23.03	19.53	16.42	13.72	11.43	9.54	8.07	7.01
54.2	PI-P	3.27	2.86	2.48	2.13	1.81	1.53	1.28	1.08	.91	.79
	CHEX	7.49	6.41	5.44	4.58	3.84	3.20	2.67	2.23	1.90	1.67
331.0	PI+P	29.31	25.23	21.54	18.22	15.29	12.75	10.58	8.80	7.39	6.36
51.3	PI-P	2.93	2.59	2.27	1.96	1.69	1.43	1.21	1.02	.87	.76
	CHEX	7.24	6.14	5.17	4.33	3.60	2.98	2.47	2.05	1.73	1.50
351.0	PI+P	25.54	22.19	19.08	16.22	13.63	11.32	9.32	7.63	6.26	5.22
46.7	PI-P	2.32	2.10	1.89	1.68	1.47	1.28	1.10	.93	.80	.69
	CHEX	6.71	5.69	4.79	3.99	3.30	2.70	2.20	1.79	1.46	1.21
378.0	PI+P	21.46	18.38	15.59	13.09	10.88	8.94	7.28	5.88	4.74	3.85
41.3	PI-P	1.67	1.57	1.45	1.33	1.20	1.06	.93	.80	.69	.60
	CHEX	6.24	5.14	4.21	3.42	2.76	2.21	1.76	1.39	1.11	.89
408.0	PI+P	17.47	14.96	12.68	10.64	8.82	7.22	5.83	4.65	3.68	2.91
36.7	PI-P	1.18	1.17	1.13	1.07	1.00	.92	.84	.75	.67	.59
	CHEX	5.64	4.65	3.79	3.06	2.44	1.92	1.49	1.14	.87	.66
427.0	PI+P	15.26	13.13	11.18	9.41	7.81	6.39	5.16	4.10	3.23	2.53
34.1	PI-P	1.20	1.20	1.19	1.16	1.10	1.03	.95	.87	.78	.69
	CHEX	5.14	4.20	3.40	2.72	2.15	1.68	1.29	.99	.75	.57
456.0	PI+P	12.75	10.79	9.05	7.51	6.17	5.00	3.99	3.13	2.42	1.85
30.8	PI-P	1.13	1.10	1.06	1.01	.94	.87	.79	.71	.63	.55
	CHEX	4.78	3.85	3.07	2.43	1.91	1.49	1.15	.89	.68	.53
490.0	PI+P	10.68	9.10	7.68	6.41	5.27	4.28	3.41	2.66	2.03	1.52
27.6	PI-P	1.47	1.42	1.36	1.29	1.21	1.12	1.01	.90	.79	.68
	CHEX	4.61	3.69	2.93	2.31	1.81	1.42	1.11	.87	.69	.55
532.0	PI+P	8.40	7.38	6.36	5.38	4.45	3.61	2.86	2.20	1.65	1.19
24.4	PI-P	1.87	1.75	1.62	1.48	1.33	1.18	1.03	.88	.74	.60
	CHEX	4.20	3.42	2.76	2.21	1.75	1.38	1.08	.85	.68	.54
573.0	PI+P	6.80	5.87	5.02	4.24	3.53	2.88	2.31	1.79	1.35	.97
21.8	PI-P	2.51	2.23	1.98	1.76	1.55	1.35	1.15	.96	.78	.62
	CHEX	4.03	3.18	2.50	1.96	1.53	1.20	.95	.76	.61	.50
614.0	PI+P	5.44	4.70	4.10	3.55	3.03	2.51	2.02	1.56	1.15	.81
19.7	PI-P	3.38	2.77	2.38	2.06	1.78	1.51	1.26	1.03	.82	.63
	CHEX	4.04	2.87	2.15	1.66	1.31	1.04	.82	.65	.52	.42
658.0	PI+P	4.40	3.71	3.27	2.91	2.55	2.16	1.76	1.36	.99	.68
17.8	PI-P	5.26	3.75	2.97	2.47	2.08	1.72	1.39	1.08	.81	.57
	CHEX	4.38	2.92	2.11	1.60	1.23	.94	.73	.56	.43	.34
675.0	PI+P	3.89	3.28	2.91	2.63	2.34	2.01	1.65	1.28	.93	.63
17.2	PI-P	5.88	4.18	3.27	2.67	2.21	1.80	1.42	1.09	.80	.55
	CHEX	4.04	2.74	2.02	1.56	1.22	.95	.73	.56	.43	.33
705.0	PI+P	3.35	2.75	2.46	2.23	1.98	1.69	1.37	1.06	.77	.52
16.1	PI-P	7.17	4.98	3.78	3.01	2.42	1.92	1.49	1.11	.79	.52
	CHEX	3.83	2.58	1.88	1.43	1.12	.88	.70	.54	.42	.32

K[MEV/C] DOM/DT	COS THETA .00	-.10	-.20	-.30	-.40	-.50	-.60	-.70	-.80	DSIG/DOM[MB/SR] -.90	-1.00
301.0	7.77	7.73	8.15	9.03	10.36	12.16	14.41	17.12	20.29	23.91	27.98
60.1	.84	.82	.86	.95	1.09	1.29	1.55	1.86	2.23	2.66	3.14
	1.90	1.92	2.04	2.27	2.60	3.04	3.58	4.22	4.97	5.82	6.78
305.0	7.49	7.40	7.76	8.58	9.84	11.55	13.71	16.32	19.37	22.86	26.79
58.7	.82	.80	.83	.92	1.06	1.26	1.52	1.83	2.20	2.63	3.11
	1.82	1.82	1.93	2.14	2.44	2.85	3.36	3.97	4.68	5.48	6.39
310.0	7.14	7.00	7.30	8.04	9.22	10.83	12.87	15.35	18.26	21.60	25.36
57.1	.79	.76	.80	.88	1.02	1.22	1.47	1.78	2.15	2.57	3.05
	1.73	1.71	1.80	1.98	2.26	2.63	3.10	3.67	4.33	5.08	5.93
320.0	6.37	6.15	6.34	6.94	7.96	9.38	11.20	13.43	16.04	19.05	22.44
54.2	.72	.70	.73	.81	.95	1.14	1.38	1.68	2.03	2.44	2.90
	1.53	1.49	1.53	1.67	1.90	2.22	2.62	3.11	3.68	4.34	5.08
331.0	5.71	5.44	5.54	6.02	6.86	8.08	9.66	11.61	13.93	16.60	19.63
51.3	.69	.66	.68	.76	.88	1.06	1.29	1.57	1.89	2.27	2.70
	1.36	1.30	1.32	1.43	1.61	1.87	2.21	2.62	3.11	3.67	4.31
351.0	4.51	4.13	4.08	4.36	4.96	5.86	7.07	8.56	10.33	12.35	14.61
46.7	.62	.59	.60	.66	.76	.90	1.09	1.32	1.58	1.87	2.19
	1.05	.96	.94	.98	1.10	1.28	1.52	1.81	2.17	2.58	3.04
378.0	3.21	2.83	2.68	2.78	3.12	3.70	4.52	5.58	6.88	8.42	10.19
41.3	.53	.49	.49	.52	.58	.69	.83	1.02	1.24	1.49	1.77
	.74	.65	.60	.61	.66	.76	.90	1.09	1.32	1.59	1.91
408.0	2.34	1.97	1.79	1.80	2.00	2.39	2.96	3.71	4.64	5.75	7.03
36.7	.53	.48	.46	.45	.48	.54	.64	.77	.96	1.18	1.47
	.52	.42	.37	.37	.40	.46	.56	.68	.82	.99	1.18
427.0	2.00	1.65	1.46	1.44	1.57	1.86	2.30	2.87	3.58	4.41	5.35
34.1	.61	.55	.50	.47	.47	.50	.56	.65	.77	.92	1.09
	.45	.36	.32	.31	.33	.37	.43	.51	.61	.72	.83
456.0	1.40	1.08	.89	.81	.86	1.03	1.32	1.74	2.29	2.98	3.81
30.8	.48	.42	.37	.35	.36	.39	.45	.55	.69	.86	1.08
	.41	.33	.27	.24	.23	.23	.25	.27	.32	.37	.44
490.0	1.11	.81	.60	.50	.49	.57	.74	1.00	1.35	1.78	2.28
27.6	.58	.49	.43	.40	.39	.42	.48	.58	.71	.86	1.04
	.45	.37	.31	.27	.23	.20	.18	.16	.15	.13	.12
532.0	.83	.55	.37	.25	.21	.24	.34	.50	.73	1.02	1.37
24.4	.49	.40	.33	.30	.30	.34	.42	.55	.73	.96	1.24
	.44	.36	.30	.25	.21	.17	.14	.11	.07	.04	.00
573.0	.66	.41	.24	.13	.09	.11	.18	.30	.44	.59	.71
21.8	.48	.37	.29	.24	.25	.30	.40	.56	.77	1.05	1.39
	.42	.35	.30	.25	.20	.16	.13	.10	.08	.07	.06
614.0	.53	.32	.17	.08	.04	.05	.10	.18	.26	.29	.22
19.7	.46	.33	.23	.19	.21	.29	.42	.60	.82	1.05	1.31
	.35	.29	.25	.21	.17	.14	.12	.11	.11	.15	.23
658.0	.44	.25	.13	.06	.03	.04	.07	.13	.17	.18	.09
17.8	.37	.23	.13	.11	.15	.27	.44	.67	.93	1.20	1.50
	.28	.22	.18	.14	.12	.11	.11	.14	.18	.27	.46
675.0	.39	.22	.11	.05	.03	.04	.08	.12	.16	.16	.08
17.2	.35	.19	.10	.08	.13	.27	.47	.72	1.00	1.28	1.58
	.25	.19	.14	.11	.09	.10	.13	.17	.24	.34	.52
705.0	.33	.20	.11	.06	.05	.05	.08	.10	.13	.12	.05
16.1	.31	.16	.07	.06	.12	.26	.46	.69	.93	1.19	1.55
	.23	.17	.12	.10	.09	.12	.17	.24	.34	.45	.58

K[MEV/C] DOM/DT		COS THETA 1.00	.90	.80	.70	.60	.50	.40	.30	DSIG/DOM[MB/SR] .20	.10
725.	PI+P	2.94	2.42	2.22	2.05	1.84	1.56	1.26	.96	.69	.48
15.515	PI-P	7.26	5.05	3.86	3.10	2.50	1.98	1.51	1.11	.78	.51
	CHEX	3.26	2.15	1.60	1.25	1.00	.81	.65	.51	.41	.32
750.	PI+P	2.54	2.06	1.87	1.74	1.57	1.33	1.06	.79	.57	.39
14.791	PI-P	6.82	4.90	3.77	3.00	2.38	1.85	1.40	1.02	.70	.45
	CHEX	2.63	1.69	1.23	.97	.78	.65	.54	.45	.37	.30
777.	PI+P	2.17	1.78	1.62	1.52	1.37	1.17	.93	.69	.47	.31
14.075	PI-P	6.01	4.18	3.25	2.60	2.07	1.61	1.21	.87	.60	.37
	CHEX	1.94	1.14	.81	.65	.56	.50	.44	.38	.31	.25
800.	PI+P	1.99	1.77	1.57	1.38	1.18	.97	.76	.56	.39	.25
13.514	PI-P	5.55	3.69	2.80	2.20	1.74	1.36	1.03	.75	.51	.32
	CHEX	1.70	.94	.64	.49	.42	.37	.33	.30	.25	.21
822.	PI+P	1.87	1.56	1.44	1.29	1.08	.85	.65	.49	.36	.24
13.016	PI-P	5.35	3.24	2.37	1.87	1.51	1.20	.92	.68	.47	.30
	CHEX	1.55	.73	.47	.36	.31	.28	.26	.24	.21	.18
851.	PI+P	1.85	1.59	1.39	1.20	1.01	.81	.62	.44	.30	.19
12.408	PI-P	5.88	3.18	2.12	1.62	1.31	1.07	.84	.63	.44	.29
	CHEX	1.63	.76	.41	.26	.21	.19	.19	.18	.17	.14
875.	PI+P	2.01	1.69	1.44	1.22	1.00	.80	.62	.46	.33	.21
11.944	PI-P	7.09	3.63	2.23	1.58	1.21	.97	.78	.61	.45	.31
	CHEX	1.90	.87	.42	.22	.15	.14	.15	.16	.16	.14
895.	PI+P	2.18	1.87	1.62	1.37	1.12	.88	.66	.48	.34	.23
11.581	PI-P	8.60	4.32	2.49	1.63	1.18	.91	.73	.57	.43	.30
	CHEX	2.20	.99	.45	.21	.12	.12	.14	.17	.17	.15
923.	PI+P	2.47	2.16	1.87	1.58	1.29	1.02	.78	.57	.40	.28
11.107	PI-P	11.61	5.65	3.02	1.74	1.11	.80	.63	.50	.39	.29
	CHEX	2.80	1.19	.51	.21	.11	.12	.17	.21	.22	.19
954.	PI+P	2.83	2.47	2.16	1.84	1.51	1.18	.89	.65	.46	.33
10.622	PI-P	15.08	7.30	3.72	1.90	1.01	.61	.45	.37	.31	.26
	CHEX	3.29	1.37	.57	.22	.12	.15	.23	.29	.30	.25
975.	PI+P	3.05	2.58	2.26	1.98	1.66	1.31	.98	.70	.48	.34
10.316	PI-P	17.15	8.05	4.09	2.04	.97	.48	.32	.28	.27	.26
	CHEX	3.46	1.36	.57	.23	.14	.20	.30	.37	.35	.28
1000.	PI+P	3.52	3.07	2.53	2.04	1.64	1.30	.99	.72	.49	.34
9.973	PI-P	19.48	9.48	4.56	2.04	.83	.34	.19	.19	.23	.27
	CHEX	3.50	1.46	.56	.21	.15	.23	.34	.39	.38	.30
1030.	PI+P	3.59	2.99	2.50	2.09	1.72	1.36	1.03	.73	.50	.34
9.588	PI-P	18.36	9.02	4.61	2.19	.86	.26	.09	.11	.20	.27
	CHEX	2.92	1.19	.49	.20	.16	.27	.39	.44	.41	.31
1055.	PI+P	3.70	3.08	2.58	2.15	1.76	1.38	1.03	.72	.48	.32
9.288	PI-P	16.75	8.68	4.56	2.17	.84	.23	.06	.09	.19	.28
	CHEX	2.35	.99	.40	.18	.18	.29	.39	.43	.38	.29
1080.	PI+P	3.91	3.19	2.59	2.10	1.70	1.34	1.01	.70	.45	.29
9.005	PI-P	14.56	8.06	4.34	2.11	.85	.25	.07	.09	.19	.28
	CHEX	1.72	.75	.31	.14	.17	.29	.39	.42	.36	.26
1113.	PI+P	4.28	3.31	2.66	2.15	1.70	1.30	.95	.65	.43	.29
8.656	PI-P	12.23	7.13	3.99	2.02	.87	.30	.10	.11	.19	.27
	CHEX	1.12	.52	.22	.13	.17	.27	.34	.36	.31	.23
1154.	PI+P	4.95	3.54	2.73	2.13	1.65	1.24	.90	.61	.40	.28
8.256	PI-P	10.39	6.22	3.60	1.91	.89	.36	.15	.13	.19	.26
	CHEX	.64	.35	.18	.13	.17	.24	.29	.29	.25	.18

COS THETA — DSIG/DOM[MB/SR]

K[MEV/C] DOM/DT	.00	-.10	-.20	-.30	-.40	-.50	-.60	-.70	-.80	-.90	-1.00
725.	.31	.19	.11	.06	.05	.06	.08	.11	.13	.12	.04
15.515	.30	.15	.07	.06	.13	.26	.45	.65	.84	1.04	1.39
	.24	.18	.13	.10	.10	.13	.20	.30	.41	.50	.51
750.	.26	.16	.10	.07	.07	.09	.12	.15	.17	.16	.02
14.791	.26	.14	.08	.10	.19	.32	.47	.62	.74	.86	1.03
	.23	.18	.15	.14	.15	.20	.28	.39	.49	.54	.41
777.	.19	.11	.08	.08	.10	.14	.18	.21	.20	.15	.01
14.075	.21	.12	.10	.15	.25	.39	.51	.59	.63	.65	.67
	.19	.15	.13	.15	.21	.31	.43	.52	.57	.50	.25
800.	.15	.09	.07	.07	.10	.14	.18	.21	.22	.18	.00
13.514	.19	.13	.14	.21	.32	.44	.54	.61	.62	.55	.42
	.16	.14	.14	.17	.24	.34	.46	.58	.65	.59	.18
822.	.14	.08	.05	.07	.11	.16	.20	.23	.22	.16	.00
13.016	.18	.14	.17	.27	.41	.54	.65	.71	.68	.52	.29
	.15	.12	.12	.15	.24	.39	.57	.71	.73	.56	.13
851.	.12	.08	.07	.10	.15	.20	.25	.26	.25	.19	.02
12.408	.19	.17	.22	.35	.51	.69	.83	.91	.86	.60	.19
	.11	.09	.10	.15	.25	.41	.59	.76	.82	.66	.12
875.	.13	.10	.11	.14	.19	.24	.28	.30	.30	.25	.07
11.944	.22	.20	.27	.41	.61	.82	1.01	1.15	1.13	.79	.14
	.11	.08	.08	.13	.24	.43	.66	.88	.96	.77	.11
895.	.16	.13	.14	.18	.24	.30	.34	.36	.34	.29	.16
11.581	.22	.21	.30	.48	.72	.99	1.25	1.41	1.38	.98	.13
	.11	.08	.08	.13	.26	.47	.73	.97	1.08	.87	.12
923.	.21	.18	.19	.23	.29	.34	.38	.41	.42	.38	.32
11.107	.24	.26	.37	.56	.84	1.18	1.55	1.81	1.80	1.31	.14
	.14	.09	.07	.12	.27	.52	.83	1.13	1.26	1.00	.14
954.	.25	.22	.22	.26	.31	.36	.41	.44	.47	.47	.50
10.622	.25	.30	.43	.65	.97	1.39	1.83	2.16	2.16	1.59	.19
	.17	.10	.07	.11	.26	.52	.88	1.24	1.43	1.18	.18
975.	.26	.24	.24	.27	.31	.36	.41	.46	.51	.52	.62
10.316	.28	.35	.48	.70	1.02	1.46	1.97	2.41	2.46	1.78	.23
	.18	.10	.06	.11	.25	.51	.90	1.32	1.54	1.24	.24
1000.	.25	.23	.24	.27	.30	.34	.39	.43	.49	.58	.75
9.973	.32	.39	.50	.71	1.05	1.53	2.07	2.49	2.53	1.88	.25
	.20	.10	.05	.08	.23	.50	.86	1.23	1.44	1.25	.32
1030.	.25	.22	.23	.25	.28	.31	.34	.38	.45	.59	.92
9.588	.34	.42	.53	.71	1.00	1.42	1.92	2.34	2.41	1.82	.26
	.19	.10	.05	.08	.20	.43	.74	1.07	1.26	1.13	.45
1055.	.24	.21	.21	.23	.26	.29	.31	.35	.44	.66	1.04
9.288	.35	.41	.49	.63	.88	1.25	1.70	2.09	2.18	1.69	.21
	.18	.10	.05	.07	.16	.35	.61	.86	1.03	1.02	.56
1080.	.22	.21	.23	.27	.31	.33	.34	.34	.41	.68	1.11
9.005	.35	.41	.49	.61	.82	1.12	1.48	1.80	1.88	1.45	.17
	.16	.09	.05	.06	.14	.29	.49	.70	.82	.81	.61
1113.	.22	.22	.26	.31	.35	.37	.36	.34	.40	.69	1.35
8.656	.34	.39	.43	.51	.65	.87	1.14	1.40	1.50	1.17	.11
	.15	.08	.06	.07	.13	.25	.39	.51	.58	.63	.72
1154.	.23	.26	.32	.40	.45	.46	.43	.38	.43	.78	1.62
8.256	.31	.34	.37	.40	.48	.62	.81	1.01	1.10	.88	.06
	.12	.07	.05	.08	.14	.22	.30	.34	.37	.46	.75

K[MEV/C] DOM/DT		COS THETA 1.00	.90	.80	.70	.60	.50	.40	.30	DSIG/DOM[MB/SR] .20	.10
1174.	PI+P	5.29	3.63	2.79	2.15	1.63	1.20	.86	.59	.39	.28
8.074	PI-P	10.03	5.86	3.46	1.90	.92	.38	.16	.13	.19	.25
	CHEX	.52	.32	.18	.13	.17	.23	.27	.28	.24	.17
1210.	PI+P	6.12	3.91	2.82	2.11	1.59	1.16	.82	.54	.36	.26
7.764	PI-P	9.79	5.52	3.27	1.85	.94	.41	.18	.14	.18	.25
	CHEX	.35	.32	.20	.15	.17	.21	.25	.24	.20	.14
1235.	PI+P	6.79	4.21	2.84	2.01	1.47	1.07	.76	.51	.34	.26
7.562	PI-P	10.02	5.52	3.19	1.78	.89	.38	.16	.12	.17	.24
	CHEX	.28	.33	.22	.16	.17	.21	.24	.23	.19	.13
1280.	PI+P	8.42	4.76	2.90	1.88	1.30	.92	.64	.43	.30	.26
7.223	PI-P	10.08	5.45	3.09	1.69	.84	.36	.15	.11	.16	.24
	CHEX	.14	.37	.27	.18	.16	.19	.21	.20	.17	.12
1324.	PI+P	10.30	5.31	2.88	1.66	1.03	.70	.49	.34	.27	.28
6.918	PI-P	10.36	5.44	2.95	1.52	.71	.28	.11	.09	.14	.22
	CHEX	.06	.39	.32	.19	.14	.15	.17	.18	.16	.12
1360.	PI+P	11.99	5.64	2.82	1.48	.84	.55	.39	.30	.27	.31
6.686	PI-P	10.62	5.30	2.77	1.37	.61	.23	.09	.09	.14	.21
	CHEX	.04	.40	.32	.19	.13	.14	.16	.17	.16	.13
1400.	PI+P	13.48	6.14	2.80	1.28	.62	.37	.28	.25	.27	.35
6.445	PI-P	10.96	5.12	2.58	1.25	.53	.19	.08	.08	.14	.20
	CHEX	.07	.38	.31	.17	.11	.11	.14	.16	.16	.14
1430.	PI+P	14.40	6.29	2.82	1.20	.50	.28	.23	.23	.26	.35
6.275	PI-P	11.21	5.02	2.47	1.16	.48	.17	.08	.09	.14	.20
	CHEX	.10	.39	.30	.16	.09	.10	.13	.16	.17	.16
1473.	PI+P	15.25	6.58	2.74	1.03	.36	.17	.17	.22	.29	.39
6.046	PI-P	11.31	4.97	2.38	1.09	.44	.16	.09	.10	.16	.21
	CHEX	.20	.40	.29	.14	.06	.07	.11	.16	.18	.17
1505.	PI+P	15.75	6.95	2.709	.923	.294	.137	.146	.207	.288	.391
5.886	PI-P	11.46	5.10	2.365	1.036	.405	.151	.093	.121	.174	.222
	CHEX	.30	.42	.273	.125	.055	.062	.108	.153	.177	.179
1550.	PI+P	14.97	6.64	2.632	.875	.242	.105	.142	.221	.304	.392
5.674	PI-P	11.53	5.02	2.277	.986	.389	.158	.111	.142	.191	.230
	CHEX	.35	.43	.257	.103	.035	.045	.098	.154	.186	.190
1590.	PI+P	14.50	6.43	2.502	.815	.225	.103	.148	.233	.315	.393
5.498	PI-P	11.75	4.94	2.191	.935	.376	.161	.118	.150	.200	.236
	CHEX	.37	.41	.236	.092	.031	.042	.091	.143	.177	.187
1640.	PI+P	13.88	6.16	2.451	.828	.247	.123	.158	.227	.296	.361
5.292	PI-P	11.89	4.88	2.102	.878	.351	.158	.128	.167	.216	.245
	CHEX	.51	.40	.206	.069	.019	.032	.079	.130	.164	.176
1680.	PI+P	13.72	5.93	2.321	.787	.251	.144	.184	.255	.319	.371
5.137	PI-P	12.28	4.78	2.014	.832	.329	.159	.147	.190	.233	.254
	CHEX	.58	.38	.189	.066	.017	.026	.070	.124	.166	.182
1720.	PI+P	13.05	5.71	2.284	.802	.273	.156	.182	.245	.309	.362
4.992	PI-P	12.45	4.79	1.931	.764	.299	.149	.143	.191	.240	.263
	CHEX	.60	.37	.174	.058	.015	.025	.064	.111	.149	.168
1760.	PI+P	12.65	5.50	2.211	.795	.292	.191	.225	.283	.325	.349
4.853	PI-P	12.87	4.71	1.852	.728	.289	.154	.159	.210	.252	.263
	CHEX	.63	.36	.161	.054	.015	.026	.068	.117	.151	.163
1800.	PI+P	12.34	5.35	2.102	.755	.301	.213	.243	.293	.329	.346
4.723	PI-P	13.30	4.73	1.777	.675	.270	.157	.173	.228	.267	.270
	CHEX	.63	.36	.150	.049	.014	.023	.061	.107	.139	.152

K[MEV/C]	COS THETA									DSIG/DOM[MB/SR]	
DOM/DT	.00	-.10	-.20	-.30	-.40	-.50	-.60	-.70	-.80	-.90	-1.00
1174.	.24	.28	.36	.45	.52	.54	.50	.43	.47	.83	1.72
8.074	.30	.33	.34	.37	.42	.53	.70	.88	.97	.77	.08
	.10	.05	.05	.08	.14	.21	.27	.30	.31	.38	.73
1210.	.27	.34	.45	.55	.62	.63	.58	.49	.51	.90	2.03
7.764	.30	.33	.33	.34	.36	.44	.56	.70	.77	.61	.11
	.08	.05	.05	.09	.15	.20	.24	.23	.23	.34	.77
1235.	.29	.38	.52	.64	.72	.73	.65	.55	.55	.99	2.23
7.562	.29	.32	.34	.34	.36	.41	.51	.63	.69	.55	.13
	.07	.04	.05	.09	.15	.20	.23	.21	.20	.32	.81
1280.	.32	.45	.61	.75	.85	.87	.77	.61	.58	1.09	2.48
7.223	.29	.32	.34	.34	.36	.38	.41	.48	.53	.49	.18
	.07	.04	.05	.10	.16	.21	.22	.20	.17	.28	.86
1324.	.38	.55	.74	.91	1.00	.99	.85	.65	.61	1.17	2.78
6.918	.29	.33	.35	.36	.37	.37	.39	.43	.47	.45	.28
	.08	.06	.07	.11	.16	.20	.21	.18	.17	.29	.89
1360.	.44	.61	.80	.97	1.07	1.07	.95	.73	.61	1.09	2.72
6.686	.27	.31	.33	.35	.36	.36	.35	.36	.38	.38	.28
	.10	.08	.09	.12	.17	.21	.22	.20	.17	.26	.85
1400.	.49	.67	.87	1.04	1.13	1.12	.96	.72	.58	1.03	2.60
6.445	.26	.30	.33	.34	.35	.35	.34	.32	.32	.32	.30
	.12	.10	.11	.13	.17	.20	.21	.18	.16	.27	.81
1430.	.50	.68	.88	1.05	1.16	1.16	1.01	.74	.56	.93	2.49
6.275	.25	.30	.33	.35	.36	.35	.33	.31	.31	.32	.37
	.14	.12	.11	.13	.16	.19	.21	.19	.16	.24	.71
1473.	.53	.70	.89	1.05	1.15	1.12	.93	.65	.48	.79	2.18
6.046	.26	.29	.31	.33	.33	.31	.28	.23	.21	.21	.32
	.16	.14	.13	.14	.15	.18	.19	.17	.14	.22	.64
1505.	.523	.684	.860	1.013	1.091	1.049	.878	.624	.440	.655	1.928
5.886	.258	.284	.305	.319	.320	.300	.257	.204	.167	.181	.307
	.167	.150	.137	.133	.141	.159	.171	.159	.134	.187	.554
1550.	.498	.627	.774	.913	.997	.970	.806	.554	.364	.521	1.588
5.674	.257	.275	.290	.298	.295	.276	.239	.187	.138	.144	.293
	.175	.154	.134	.123	.126	.140	.150	.134	.107	.158	.434
1590.	.476	.580	.707	.836	.919	.900	.752	.511	.306	.385	1.249
5.498	.257	.271	.284	.291	.283	.258	.216	.159	.106	.110	.259
	.179	.160	.138	.120	.118	.132	.146	.131	.097	.122	.327
1640.	.43	.51	.62	.73	.81	.78	.62	.39	.22	.28	.86
5.292	.25	.26	.26	.26	.26	.23	.19	.13	.77-01	.84-01	.21
	.17	.15	.13	.10	.99-01	.11	.11	.93-01	.66-01	.96-01	.22
1680.	.42	.47	.54	.62	.68	.66	.53	.34	.19	.22	.64
5.137	.25	.25	.24	.23	.22	.20	.15	.10	.58-01	.59-01	.18
	.17	.15	.12	.92-01	.85-01	.94-01	.99-01	.84-01	.61-01	.80-01	.16
1720.	.40	.43	.47	.53	.59	.59	.48	.30	.16	.16	.44
4.992	.26	.24	.23	.22	.21	.19	.14	.86-01	.39-01	.42-01	.14
	.17	.15	.11	.84-01	.73-01	.82-01	.91-01	.76-01	.51-01	.67-01	.12
1760.	.36	.38	.41	.45	.49	.48	.39	.25	.12	.11	.33
4.853	.25	.23	.21	.19	.17	.15	.12	.72-01	.29-01	.32-01	.13
	.15	.13	.10	.75-01	.65-01	.72-01	.78-01	.65-01	.42-01	.54-01	.10
1800.	.35	.35	.37	.41	.44	.42	.34	.21	.10	.88-01	.25
4.723	.25	.22	.19	.17	.15	.13	.10	.59-01	.18-01	.32-01	.10
	.15	.13	.10	.75-01	.63-01	.67-01	.70-01	.57-01	.38-01	.59-01	.90-01

KCMEV/C] DOM/DT		COS THETA 1.00	.90	.80	.70	.60	.50	.40	.30	DSIG/DOM[MB/SR] .20	.10
1840.	PI+P	12.07	5.15	2.090	.791	.328	.223	.240	.281	.313	.327
4.598	PI-P	13.96	4.77	1.746	.644	.253	.154	.177	.236	.278	.280
	CHEX	.69	.35	.141	.045	.014	.023	.058	.099	.132	.146
1880.	PI+P	11.94	5.09	2.019	.750	.320	.229	.251	.294	.323	.328
4.480	PI-P	14.64	4.80	1.662	.583	.230	.149	.181	.251	.302	.301
	CHEX	.66	.35	.132	.044	.017	.027	.060	.099	.126	.133
1920.	PI+P	11.86	4.93	1.938	.736	.328	.240	.256	.291	.312	.311
4.368	PI-P	15.21	4.83	1.600	.541	.215	.154	.193	.258	.298	.291
	CHEX	.68	.32	.116	.036	.017	.028	.056	.088	.113	.124
1980.	PI+P	11.90	4.88	1.917	.728	.325	.232	.240	.271	.289	.283
4.210	PI-P	16.26	4.85	1.494	.468	.185	.146	.192	.260	.303	.296
	CHEX	.59	.29	.103	.037	.022	.033	.056	.081	.102	.110
2030.	PI+P	11.97	4.71	1.824	.710	.334	.248	.251	.267	.271	.258
4.086	PI-P	16.76	4.78	1.385	.414	.168	.146	.196	.257	.291	.279
	CHEX	.58	.26	.092	.035	.029	.042	.059	.074	.086	.093
2070.	PI+P	12.09	4.70	1.748	.645	.302	.240	.261	.289	.294	.269
3.992	PI-P	17.20	4.78	1.294	.350	.138	.131	.190	.263	.305	.291
	CHEX	.51	.24	.084	.034	.034	.048	.062	.075	.085	.087
2150.	PI+P	12.52	4.65	1.656	.593	.275	.219	.236	.255	.254	.231
3.816	PI-P	17.65	4.70	1.160	.271	.109	.122	.181	.243	.273	.254
	CHEX	.42	.21	.074	.036	.039	.053	.062	.063	.062	.061
2200.	PI+P	12.95	4.61	1.556	.537	.253	.211	.233	.254	.252	.225
3.713	PI-P	18.05	4.56	1.077	.242	.101	.119	.175	.233	.261	.240
	CHEX	.34	.19	.068	.031	.037	.053	.062	.062	.059	.057
2280.	PI+P	13.79	4.51	1.435	.478	.234	.207	.230	.244	.231	.193
3.560	PI-P	18.40	4.43	.990	.207	.091	.114	.165	.217	.239	.217
	CHEX	.26	.19	.059	.029	.043	.061	.062	.052	.044	.041
2340.	PI+P	14.49	4.49	1.340	.427	.212	.192	.210	.216	.201	.169
3.453	PI-P	18.62	4.29	.928	.196	.095	.118	.162	.203	.218	.194
	CHEX	.20	.19	.053	.024	.039	.057	.059	.050	.041	.036
2460.	PI+P	15.81	4.45	1.179	.335	.173	.170	.188	.192	.171	.130
3.257	PI-P	18.71	4.12	.842	.172	.094	.117	.150	.176	.179	.151
	CHEX	.14	.20	.038	.019	.047	.066	.059	.038	.023	.016
2560.	PI+P	16.76	4.37	1.057	.288	.162	.158	.158	.151	.133	.103
3.110	PI-P	19.10	3.98	.787	.164	.093	.107	.126	.140	.139	.116
	CHEX	.15	.20	.028	.018	.053	.067	.054	.032	.017	.009
2750.	PI+P	18.077	4.092	.84	.22	.15	.16	.15	.13	.11	.73-01
2.863	PI-P	19.806	3.729	.66	.15	.97-01	.11	.11	.99-01	.80-01	.56-01
	CHEX	.224	.191	.17-01	.17-01	.51-01	.61-01	.47-01	.29-01	.17-01	.92-02
3000.	PI+P	18.958	3.722	.71	.19	.13	.13	.11	.90-01	.61-01	.32-01
2.591	PI-P	21.466	3.392	.54	.14	.10	.10	.89-01	.62-01	.40-01	.25-01
	CHEX	.296	.146	.90-02	.16-01	.33-01	.34-01	.27-01	.18-01	.98-02	.39-02
3400.	PI+P	20.361	3.133	.54	.16	.12	.93-01	.61-01	.32-01	.14-01	.57-02
2.249	PI-P	23.659	2.769	.36	.10	.86-01	.80-01	.56-01	.27-01	.11-01	.52-02
	CHEX	.285	.086	.98-02	.22-01	.24-01	.14-01	.69-02	.42-02	.30-02	.17-02
3650.	PI+P	21.634	2.822	.41	.14	.94-01	.60-01	.34-01	.17-01	.91-02	.51-02
2.077	PI-P	24.618	2.541	.29	.77-01	.64-01	.47-01	.25-01	.14-01	.79-02	.28-02
	CHEX	.304	.064	.85-02	.21-01	.19-01	.10-01	.39-02	.18-02	.92-03	.63-03
4000.	PI+P	23.110	2.464	.31	.10	.71-01	.46-01	.19-01	.44-02	.11-02	.97-03
1.876	PI-P	25.826	2.112	.21	.75-01	.64-01	.40-01	.15-01	.43-02	.12-02	.48-03
	CHEX	.287	.048	.11-01	.19-01	.11-01	.41-02	.99-03	.32-03	.47-03	.59-03

K[MEV/C] DOM/DT	COS THETA .00	-.10	-.20	-.30	-.40	-.50	-.60	-.70	-.80	DSIG/DOM[MB/SR] -.90	-1.00
1840.	.32	.32	.32	.35	.37	.35	.26	.15	.76-01	.69-01	.18
4.598	.25	.21	.17	.15	.14	.11	.80-01	.40-01	.11-01	.23-01	.86-01
	.14	.12	.86-01	.60-01	.50-01	.53-01	.55-01	.45-01	.34-01	.58-01	.80-01
1880.	.31	.30	.30	.32	.33	.31	.23	.13	.65-01	.57-01	.13
4.480	.25	.19	.15	.14	.13	.10	.69-01	.32-01	.78-02	.21-01	.66-01
	.12	.10+00	.72-01	.51-01	.45-01	.49-01	.50-01	.38-01	.30-01	.57-01	.73-01
1920.	.29	.27	.25	.26	.26	.24	.18	.11	.51-01	.43-01	.10
4.368	.25	.19	.15	.12	.10+00	.79-01	.52-01	.21-01	.28-02	.23-01	.46-01
	.12	.99-01	.71-01	.48-01	.39-01	.43-01	.45-01	.35-01	.28-01	.56-01	.64-01
1980.	.26	.23	.21	.20	.21	.19	.14	.82-01	.38-01	.31-01	.59-01
4.210	.24	.18	.12	.95-01	.79-01	.62-01	.38-01	.12-01	.16-02	.27-01	.27-01
	.10	.83-01	.57-01	.39-01	.37-01	.42-01	.41-01	.30-01	.25-01	.52-01	.48-01
2030.	.23	.20	.18	.17	.17	.15	.11	.62-01	.28-01	.25-01	.41-01
4.086	.23	.16	.11	.78-01	.63-01	.49-01	.29-01	.71-02	.42-02	.35-01	.14-01
	.88-01	.72-01	.50-01	.34-01	.31-01	.36-01	.35-01	.26-01	.23-01	.54-01	.38-01
2070.	.23	.18	.16	.15	.14	.13	.96-01	.55-01	.27-01	.25-01	.35-01
3.992	.23	.15	.96-01	.65-01	.52-01	.41-01	.25-01	.65-02	.44-02	.37-01	.74-02
	.78-01	.60-01	.40-01	.29-01	.31-01	.37-01	.33-01	.22-01	.20-01	.52-01	.33-01
2150.	.20	.16	.13	.12	.11	.93-01	.70-01	.43-01	.25-01	.28-01	.55-01
3.816	.20	.13	.75-01	.47-01	.34-01	.25-01	.13-01	.27-02	.11-01	.51-01	.52-03
	.56-01	.46-01	.33-01	.25-01	.28-01	.33-01	.31-01	.20-01	.19-01	.53-01	.32-01
2200.	.19	.15	.12	.11	.95-01	.81-01	.60-01	.38-01	.23-01	.26-01	.83-01
3.713	.18	.11	.64-01	.38-01	.26-01	.18-01	.88-02	.23-02	.16-01	.57-01	.70-03
	.51-01	.41-01	.30-01	.24-01	.27-01	.32-01	.29-01	.18-01	.18-01	.50-01	.37-01
2280.	.15	.11	.94-01	.82-01	.72-01	.63-01	.51-01	.37-01	.29-01	.36-01	.14
3.560	.16	.94-01	.47-01	.24-01	.17-01	.13-01	.63-02	.31-02	.19-01	.60-01	.54-02
	.38-01	.30-01	.20-01	.18-01	.25-01	.31-01	.27-01	.15-01	.15-01	.50-01	.53-01
2340.	.13	.10	.80-01	.66-01	.56-01	.51-01	.48-01	.43-01	.36-01	.41-01	.19
3.453	.14	.84-01	.41-01	.20-01	.15-01	.11-01	.55-02	.38-02	.20-01	.48-01	.11-01
	.32-01	.24-01	.16-01	.14-01	.20-01	.28-01	.27-01	.17-01	.16-01	.46-01	.66-01
2460.	.86-01	.56-01	.44-01	.39-01	.38-01	.40-01	.46-01	.48-01	.42-01	.52-01	.27
3.257	.10	.54-01	.23-01	.11-01	.88-02	.69-02	.36-02	.58-02	.21-01	.49-01	.22-01
	.13-01	.10-01	.76-02	.86-02	.14-01	.21-01	.21-01	.13-01	.11-01	.45-01	.84-01
2560.	.71-01	.46-01	.34-01	.30-01	.29-01	.32-01	.43-01	.49-01	.41-01	.47-01	.27
3.110	.78-01	.42-01	.19-01	.89-02	.59-02	.39-02	.22-02	.51-02	.19-01	.37-01	.27-01
	.64-02	.52-02	.48-02	.57-02	.97-02	.17-01	.21-01	.16-01	.12-01	.21-01	.80-01
2750.	.41-01	.20-01	.14-01	.17-01	.24-01	.33-01	.42-01	.48-01	.44-01	.38-01	.24
2.863	.33-01	.16-01	.76-02	.72-02	.81-02	.60-02	.34-02	.60-02	.14-01	.16-01	.32-01
	.42-02	.15-02	.20-02	.53-02	.99-02	.14-01	.15-01	.13-01	.12-01	.18-01	.66-01
3000.	.13-01	.59-02	.76-02	.13-01	.17-01	.18-01	.25-01	.34-01	.25-01	.17-01	.15
2.591	.14-01	.57-02	.45-02	.78-02	.86-02	.56-02	.43-02	.59-02	.75-02	.55-02	.23-01
	.12-02	.64-03	.20-02	.49-02	.65-02	.60-02	.63-02	.76-02	.59-02	.54-02	.41-01
3400.	.24-02	.39-02	.74-02	.95-02	.90-02	.87-02	.12-01	.15-01	.11-01	.67-02	.85-01
2.249	.20-02	.25-02	.63-02	.73-02	.44-02	.17-02	.45-03	.66-03	.11-02	.45-02	.14-01
	.86-03	.69-03	.11-02	.15-02	.13-02	.16-02	.42-02	.59-02	.43-02	.86-02	.25-01
3650.	.27-02	.16-02	.27-02	.49-02	.58-02	.64-02	.71-02	.98-02	.12-01	.21-02	.78-01
2.077	.10-02	.71-03	.12-02	.26-02	.19-02	.87-03	.10-02	.71-03	.81-03	.13-02	.12-01
	.74-03	.76-03	.80-03	.87-03	.84-03	.19-02	.28-02	.31-02	.46-02	.30-02	.22-01
4000.	.16-02	.18-02	.14-02	.16-02	.19-02	.23-02	.50-02	.86-02	.96-02	.28-02	.68-01
1.876	.45-03	.57-03	.95-03	.16-02	.15-02	.90-03	.94-03	.88-03	.65-03	.21-02	.13-01
	.13-02	.16-02	.82-03	.46-03	.65-03	.79-03	.14-02	.22-02	.29-02	.21-02	.18-01

Table 2.3.1.3. Coefficients of a Legendre expansion, Eq. (A.4.21), of the differential cross sections as calculated from the solution KH 78. Only the first 10 (dimensionless) coefficients are given.

LEGENDRE EXPANSION COEFFICIENTS FOR DIFFERENTIAL CROSS SECTIONS PI+P->PI+P

K GEV/C	A 0	A 1	A 2	A 3	A 4	A 5	A 6	A 7	A 8	A 9	A 10
.247	1.297	-.282	1.140	-.004	.003	-.000	.000	.000	.000	.000	.000
.267	1.785	-.109	1.649	-.002	.002	.000	.000	.000	.000	.000	.000
.280	1.972	.052	1.873	.001	.000	.000	.000	.000	.000	.000	.000
.290	2.043	.165	1.980	-.000	-.000	.000	.000	.000	.000	.000	.000
.295	2.069	.264	2.022	-.001	-.001	.000	.000	.000	.000	.000	.000
.301	2.073	.363	2.051	-.001	-.003	.000	.000	.000	.000	.000	.000
.305	2.063	.426	2.059	-.001	-.004	.000	.000	.000	.000	.000	.000
.310	2.035	.502	2.062	-.001	-.006	.000	.000	.000	.000	.000	.000
.320	1.966	.652	2.043	.001	-.008	.000	.000	.000	.000	.000	.000
.331	1.885	.712	1.947	.003	-.008	.000	.000	.000	.000	.000	.000
.351	1.694	.948	1.816	.026	-.023	-.001	.000	.000	.000	.000	.000
.378	1.440	1.037	1.609	.017	-.012	.000	-.000	.000	.000	.000	.000
.408	1.248	1.044	1.445	-.004	.005	.001	-.000	.000	.000	.000	.000
.427	1.146	1.080	1.316	.046	-.026	.000	-.000	.000	.000	.000	.000
.456	.980	1.127	1.153	.029	-.049	-.001	.001	.000	.000	.000	.000
.490	.869	1.138	1.002	-.023	-.038	.002	-.000	.000	.000	.000	.000
.532	.743	1.151	.914	.036	-.035	-.024	-.004	.002	-.000	.000	.000
.573	.643	1.063	.780	.045	-.030	.017	-.003	.002	-.000	.000	.000
.614	.578	1.003	.651	.013	-.085	.034	.013	.020	.002	-.002	.000
.658	.522	.915	.546	-.024	-.101	.054	.048	.034	.002	-.003	.000
.675	.492	.854	.493	-.048	-.112	.060	.059	.032	-.001	-.003	.000
.705	.442	.766	.442	-.031	-.104	.045	.061	.042	.009	.002	-.000
.725	.420	.714	.403	-.049	-.121	.035	.055	.052	.019	.003	-.002
.750	.381	.617	.373	-.030	-.121	.037	.052	.059	.018	.005	-.007
.777	.354	.548	.349	-.016	-.151	.045	.065	.045	.012	-.005	-.005
.800	.335	.517	.382	.035	-.129	.026	.017	.019	-.008	-.002	-.001
.822	.319	.477	.360	.046	-.140	.022	.007	.012	.026	.031	.008
.851	.330	.467	.395	.090	-.136	.024	.025	.010	.000	.004	-.007
.875	.367	.485	.440	.120	-.119	.031	.021	.008	-.011	.013	.006
.895	.426	.540	.509	.150	-.144	-.001	.022	.005	.003	.013	-.002
.923	.521	.638	.612	.158	-.147	-.018	.024	.009	-.002	.000	.006
.954	.627	.766	.747	.159	-.154	-.052	.028	.016	.013	.006	.006
.975	.690	.842	.819	.131	-.169	-.060	.061	.061	.035	-.015	.001
1.000	.758	.978	1.014	.238	-.055	-.064	.049	-.007	-.038	-.033	-.001
1.030	.788	1.028	1.054	.174	-.047	-.091	.106	.031	-.003	-.026	-.002
1.055	.823	1.090	1.156	.159	-.020	-.134	.113	.036	-.005	-.010	-.011
1.080	.856	1.123	1.238	.241	.044	-.145	.168	.022	-.042	-.017	-.018
1.113	.914	1.186	1.367	.331	.121	-.192	.219	.024	.006	.007	-.003
1.154	1.008	1.244	1.576	.509	.279	-.202	.318	.023	.014	.028	.010
1.174	1.065	1.258	1.682	.613	.334	-.219	.361	.023	.042	.065	.018
1.210	1.174	1.323	1.913	.859	.548	-.184	.513	.062	.066	.033	.020
1.235	1.257	1.349	2.120	1.117	.785	-.124	.615	.048	.031	.013	.002
1.280	1.407	1.468	2.548	1.640	1.261	.043	.853	.072	.053	.037	-.026
1.324	1.550	1.574	2.980	2.285	1.907	.276	1.122	.127	.075	.050	-.001
1.360	1.649	1.703	3.266	2.864	2.372	.624	1.364	.202	.195	.087	.029
1.400	1.750	1.898	3.612	3.522	2.948	.968	1.552	.256	.187	.091	.016
1.430	1.803	2.014	3.785	3.909	3.188	1.171	1.674	.272	.260	.155	.091
1.473	1.849	2.245	3.988	4.415	3.642	1.502	1.810	.343	.227	.142	.066
1.505	1.886	2.488	4.224	4.824	3.962	1.821	1.890	.357	.164	.003	-.014
1.550	1.830	2.535	4.115	4.796	3.872	1.857	1.777	.279	.136	.049	.013
1.590	1.788	2.607	4.041	4.838	3.838	1.990	1.757	.275	.131	.039	-.012
1.640	1.726	2.731	3.980	4.783	3.744	2.007	1.666	.325	.090	.095	-.010
1.680	1.682	2.816	3.951	4.730	3.774	2.210	1.662	.408	.161	.119	.006
1.720	1.639	2.863	3.887	4.634	3.648	2.191	1.538	.378	.171	.123	-.024
1.760	1.598	2.938	3.849	4.506	3.611	2.263	1.523	.402	.162	.125	.015

K GEV/C	A 0	A 1	A 2	A 3	A 4	A 5	A 6	A 7	A 8	A 9	A 10
1.800	1.579	2.977	3.831	4.467	3.639	2.350	1.545	.451	.157	.099	-.018
1.840	1.544	3.047	3.834	4.369	3.598	2.329	1.493	.513	.201	.166	.032
1.880	1.543	3.107	3.870	4.397	3.676	2.437	1.527	.532	.181	.149	.015
1.920	1.517	3.160	3.907	4.357	3.723	2.544	1.580	.619	.261	.167	.027
1.980	1.521	3.296	4.076	4.466	3.843	2.675	1.632	.679	.301	.192	.041
2.030	1.510	3.354	4.136	4.484	3.928	2.815	1.772	.836	.403	.239	.080
2.070	1.524	3.428	4.236	4.586	4.120	3.034	1.879	.859	.394	.235	.071
2.150	1.546	3.579	4.512	4.861	4.446	3.333	2.129	1.077	.527	.274	.106
2.200	1.570	3.677	4.682	5.071	4.737	3.630	2.384	1.249	.640	.319	.129
2.280	1.607	3.834	5.003	5.388	5.153	4.095	2.837	1.579	.882	.462	.247
2.340	1.643	3.972	5.302	5.725	5.553	4.492	3.203	1.866	1.099	.538	.293
2.460	1.717	4.252	5.883	6.366	6.315	5.295	3.963	2.394	1.486	.763	.438
2.560	1.776	4.476	6.312	6.903	6.905	5.899	4.556	2.909	1.875	.956	.533
2.750	1.870	4.811	6.905	7.688	7.810	6.975	5.632	3.825	2.564	1.428	.823
3.000	1.936	5.161	7.469	8.486	8.640	7.911	6.602	4.811	3.347	2.053	1.286
3.400	2.037	5.601	8.228	9.580	9.884	9.294	8.115	6.441	4.800	3.273	2.217
3.650	2.134	5.947	8.862	10.497	11.027	10.561	9.418	7.729	5.980	4.246	2.914
4.000	2.258	6.363	9.599	11.549	12.335	12.055	10.993	9.332	7.481	5.558	4.005
5.000	2.521	7.253	11.263	14.097	15.742	16.146	15.558	14.178	12.319	10.180	8.080
6.000	2.804	8.138	12.804	16.392	18.812	19.966	20.031	19.128	17.546	15.464	13.184

LEGENDRE EXPANSION COEFFICIENTS FOR DIFFERENTIAL CROSS SECTIONS PI-P->PI-P

K GEV/C	A 0	A 1	A 2	A 3	A 4	A 5	A 6	A 7	A 8	A 9	A 10
.247	.132	.060	.119	.012	.002	.000	.000	.000	.000	.000	.000
.267	.188	.070	.183	.010	.002	.000	.000	.000	.000	.000	.000
.280	.212	.064	.212	.007	.001	.000	.000	.000	.000	.000	.000
.290	.224	.060	.227	.002	.001	.000	.000	.000	.000	.000	.000
.295	.228	.055	.233	-.001	.001	.000	.000	.000	.000	.000	.000
.301	.231	.052	.238	-.005	.001	.000	.000	.000	.000	.000	.000
.305	.231	.049	.238	-.008	.000	.001	.000	.000	.000	.000	.000
.310	.228	.046	.236	-.011	-.000	.001	.000	.000	.000	.000	.000
.320	.227	.043	.235	-.018	-.002	.001	.000	.000	.000	.000	.000
.331	.222	.038	.221	-.022	-.002	.001	.000	.000	.000	.000	.000
.351	.209	.034	.195	-.034	-.004	.002	.000	.000	.000	.000	.000
.378	.187	.030	.157	-.048	-.006	.002	.000	.000	.000	.000	.000
.408	.181	.033	.122	-.065	-.007	.003	.000	.000	.000	.000	.000
.427	.189	.064	.082	-.068	-.020	.006	.001	.000	.000	.000	.000
.456	.200	.082	.108	-.079	-.012	.010	.002	.000	.000	.000	.000
.490	.249	.152	.155	-.082	-.008	.013	.002	.000	.000	.000	.000
.532	.281	.196	.235	-.095	-.001	.003	-.000	-.000	-.000	.000	.000
.573	.351	.297	.354	-.102	.012	.013	.003	.000	.000	.000	.000
.614	.432	.433	.496	-.068	.014	.051	.008	.005	.010	.003	.001
.658	.555	.649	.805	.029	.078	.130	.063	.035	.025	.005	.002
.675	.612	.742	.948	.075	.094	.149	.064	.038	.032	.006	.002
.705	.708	.969	1.179	.211	.149	.178	.097	.041	.046	.004	.000
.725	.738	1.058	1.216	.246	.125	.174	.115	.042	.055	.005	.001
.750	.735	1.079	1.201	.315	.074	.146	.096	.024	.038	.014	-.001
.777	.673	.978	1.059	.344	.011	.148	.119	.030	.047	.029	.004
.800	.624	.868	.963	.404	.013	.183	.128	.043	.033	.030	.018
.822	.600	.749	.884	.458	.004	.259	.180	.064	.050	.034	.025
.851	.635	.684	.941	.616	.059	.418	.239	.101	.064	.033	.029
.875	.729	.713	1.161	.863	.180	.605	.257	.133	.070	.046	.041
.895	.851	.793	1.466	1.173	.306	.792	.299	.141	.081	.052	.034
.923	1.064	.986	2.066	1.758	.621	1.145	.341	.197	.116	.050	.055
.954	1.310	1.276	2.824	2.528	1.046	1.508	.391	.226	.148	.066	.071
.975	1.460	1.424	3.278	2.966	1.280	1.746	.375	.316	.244	.106	.148
1.000	1.637	1.742	3.898	3.626	1.767	2.050	.425	.267	.198	.047	.052
1.030	1.635	1.750	3.873	3.562	1.685	1.837	.257	.276	.252	.126	.126

K GEV/C	A 0	A 1	A 2	A 3	A 4	A 5	A 6	A 7	A 8	A 9	A 10
1.055	1.585	1.778	3.779	3.417	1.633	1.643	.100	.180	.181	.106	.084
1.080	1.498	1.747	3.504	3.174	1.471	1.362	-.017	.084	.091	.061	.058
1.113	1.359	1.711	3.139	2.779	1.291	1.065	-.120	.046	.048	.036	.037
1.154	1.218	1.704	2.799	2.416	1.165	.833	-.145	.049	.041	.032	.028
1.174	1.176	1.720	2.696	2.309	1.134	.769	-.113	.097	.100	.069	.061
1.210	1.142	1.788	2.619	2.256	1.182	.730	-.024	.164	.161	.084	.062
1.235	1.146	1.856	2.679	2.363	1.313	.773	.047	.193	.159	.086	.071
1.280	1.153	1.959	2.740	2.458	1.485	.766	.148	.220	.146	.115	.042
1.324	1.167	2.010	2.850	2.654	1.720	.864	.259	.220	.135	.096	.045
1.360	1.149	2.049	2.878	2.770	1.889	.975	.401	.262	.158	.136	.066
1.400	1.144	2.078	2.923	2.902	2.062	1.112	.568	.336	.230	.173	.093
1.430	1.156	2.093	2.968	2.989	2.192	1.210	.678	.393	.262	.203	.140
1.473	1.156	2.209	3.008	3.119	2.350	1.318	.787	.410	.289	.210	.135
1.505	1.187	2.323	3.128	3.277	2.527	1.426	.846	.379	.241	.175	.099
1.550	1.206	2.401	3.189	3.367	2.651	1.551	.941	.414	.279	.183	.109
1.590	1.220	2.478	3.245	3.477	2.789	1.676	1.053	.488	.326	.213	.126
1.640	1.238	2.578	3.335	3.602	2.959	1.846	1.161	.545	.358	.240	.124
1.680	1.249	2.667	3.409	3.705	3.138	2.047	1.305	.660	.445	.310	.182
1.720	1.267	2.751	3.505	3.859	3.325	2.218	1.412	.708	.462	.306	.148
1.760	1.283	2.846	3.614	3.971	3.516	2.436	1.573	.839	.561	.367	.200
1.800	1.314	2.961	3.759	4.139	3.762	2.686	1.744	.947	.619	.399	.198
1.840	1.356	3.113	3.948	4.371	4.039	2.952	1.934	1.084	.713	.485	.245
1.880	1.397	3.243	4.130	4.608	4.357	3.284	2.173	1.233	.816	.569	.267
1.920	1.433	3.383	4.323	4.829	4.647	3.577	2.390	1.410	.908	.604	.286
1.980	1.492	3.585	4.634	5.213	5.135	4.076	2.773	1.700	1.104	.733	.339
2.030	1.517	3.690	4.818	5.436	5.428	4.412	3.060	1.933	1.239	.814	.373
2.070	1.543	3.782	4.975	5.652	5.728	4.726	3.282	2.075	1.340	.879	.385
2.150	1.571	3.920	5.260	5.978	6.126	5.159	3.655	2.366	1.483	.951	.418
2.200	1.583	3.980	5.388	6.138	6.342	5.424	3.914	2.606	1.667	1.091	.513
2.280	1.609	4.091	5.606	6.390	6.638	5.768	4.240	2.880	1.870	1.235	.607
2.340	1.624	4.169	5.737	6.566	6.820	6.005	4.501	3.120	2.058	1.349	.705
2.460	1.657	4.316	6.031	6.880	7.149	6.386	4.914	3.458	2.288	1.498	.813
2.560	1.691	4.473	6.319	7.240	7.497	6.750	5.321	3.844	2.598	1.694	.969
2.750	1.757	4.757	6.831	7.889	8.138	7.490	6.171	4.596	3.174	2.073	1.273
3.000	1.884	5.189	7.558	8.849	9.197	8.675	7.465	5.852	4.254	2.909	1.916
3.400	2.029	5.694	8.471	10.145	10.792	10.492	9.474	7.906	6.135	4.509	3.190
3.650	2.112	5.987	9.013	10.915	11.719	11.502	10.501	8.926	7.129	5.388	3.899
4.000	2.243	6.394	9.725	11.922	12.997	13.008	12.185	10.679	8.819	6.869	5.109
5.000	2.546	7.370	11.497	14.573	16.502	17.257	17.005	15.907	14.230	12.186	10.032
6.000	2.849	8.301	13.107	16.924	19.610	21.107	21.511	20.928	19.578	17.638	15.356

LEGENDRE EXPANSION COEFFICIENTS FOR DIFFERENTIAL CROSS SECTIONS PI-P->PI0 N

K GEV/C	A 0	A 1	A 2	A 3	A 4	A 5	A 6	A 7	A 8	A 9	A10
.247	.323	-.169	.275	-.015	-.000	-.000	-.000	.000	.000	.000	.000
.267	.432	-.117	.382	-.012	-.000	.000	-.000	.000	.000	.000	.000
.280	.470	-.057	.430	-.008	.000	.000	-.000	.000	.000	.000	.000
.290	.485	-.018	.452	-.004	.000	-.000	-.000	.000	.000	.000	.000
.295	.488	.019	.463	-.002	.000	-.000	-.000	.000	.000	.000	.000
.301	.488	.056	.471	.002	.001	-.000	-.000	.000	.000	.000	.000
.305	.484	.082	.473	.005	.001	-.000	-.000	.000	.000	.000	.000
.310	.478	.118	.477	.011	.001	-.000	-.000	.000	.000	.000	.000
.320	.462	.167	.472	.017	.001	-.000	-.000	.000	.000	.000	.000
.331	.443	.195	.456	.023	.002	.000	-.000	.000	.000	.000	.000
.351	.402	.288	.448	.045	.003	-.001	-.000	.000	.000	.000	.000
.378	.355	.349	.421	.060	.008	.000	-.000	.000	.000	.000	.000
.408	.327	.392	.415	.078	.017	.001	-.000	.000	.000	.000	.000
.427	.315	.425	.420	.111	.022	.001	-.001	.000	.000	.000	.000
.456	.299	.481	.402	.134	.022	.000	-.001	.000	.000	.000	.000

K GEV/C	A 0	A 1	A 2	A 3	A 4	A 5	A 6	A 7	A 8	A 9	A10
.490	.312	.519	.366	.142	.029	.000	-.002	.000	.000	.000	.000
.532	.312	.539	.356	.159	.031	-.004	-.003	.001	-.000	.000	.000
.573	.318	.544	.376	.183	.064	.005	-.002	.001	.000	.000	.000
.614	.314	.525	.408	.198	.111	.038	.035	.019	.007	.000	.001
.658	.336	.561	.523	.233	.161	.062	.050	.028	.021	.003	.001
.675	.339	.542	.526	.189	.137	.065	.047	.029	.021	.003	.001
.705	.349	.513	.551	.179	.144	.090	.041	.024	.014	.005	.003
.725	.331	.433	.489	.131	.109	.103	.024	.040	.019	.007	.006
.750	.304	.311	.413	.115	.082	.127	.013	.036	.010	.013	.004
.777	.265	.166	.308	.109	.004	.174	.026	.021	.020	.011	.004
.800	.248	.084	.298	.123	-.003	.190	.006	.035	.005	.018	.003
.822	.233	.007	.274	.144	-.044	.244	.001	.023	.047	.007	.004
.851	.237	-.034	.330	.192	-.021	.291	-.014	.032	.025	.010	.008
.875	.260	-.059	.414	.246	.002	.374	-.038	.030	.033	.006	.005
.895	.293	-.072	.492	.302	.028	.450	-.046	.030	.031	.012	.009
.923	.353	-.075	.614	.379	.081	.599	-.050	.034	.048	.018	.020
.954	.420	-.068	.735	.411	.149	.744	-.065	.054	.043	.024	.034
.975	.457	-.079	.775	.394	.149	.842	-.064	.076	.048	.017	.078
1.000	.476	-.046	.814	.390	.232	.863	-.020	.053	.021	.011	.021
1.030	.448	-.052	.703	.256	.208	.773	.004	.016	.007	.016	.048
1.055	.405	-.049	.605	.140	.211	.645	.041	.002	-.017	.009	.023
1.080	.349	-.049	.461	.050	.171	.522	.066	-.038	-.025	-.016	.027
1.113	.291	-.054	.334	-.037	.137	.348	.103	-.069	-.014	-.018	.009
1.154	.238	-.049	.229	-.087	.096	.184	.134	-.105	-.015	-.013	-.002
1.174	.222	-.036	.198	-.086	.075	.144	.141	-.124	-.017	-.009	.010
1.210	.210	-.026	.195	-.096	.071	.054	.152	-.146	-.033	-.011	-.005
1.235	.210	-.017	.199	-.100	.066	.012	.159	-.171	-.044	-.018	-.005
1.280	.216	-.007	.209	-.087	.053	-.057	.130	-.222	-.047	-.028	-.003
1.324	.223	-.012	.228	-.080	.066	-.123	.087	-.250	-.057	-.028	.005
1.360	.232	-.016	.212	-.053	.059	-.125	.056	-.269	-.043	-.039	.013
1.400	.233	-.021	.206	-.035	.101	-.136	.030	-.258	-.038	-.027	.009
1.430	.236	-.014	.187	-.007	.113	-.099	.003	-.261	-.031	-.031	.009
1.473	.241	.003	.178	.038	.173	-.084	-.007	-.249	-.027	-.015	.005
1.505	.240	.031	.175	.086	.208	-.035	-.017	-.223	-.017	-.032	.004
1.550	.233	.055	.154	.116	.242	-.001	-.037	-.221	-.026	-.017	-.003
1.590	.228	.069	.131	.152	.238	.028	-.048	-.197	-.014	-.028	-.015
1.640	.209	.100	.127	.186	.278	.070	-.031	-.151	-.014	-.006	-.018
1.680	.205	.117	.122	.205	.298	.110	-.036	-.127	.012	.012	-.023
1.720	.196	.124	.121	.218	.303	.122	-.030	-.103	.023	.004	-.030
1.760	.191	.143	.124	.222	.313	.151	-.010	-.097	.019	.012	-.022
1.800	.188	.145	.135	.233	.328	.152	-.001	-.076	.013	.010	-.033
1.840	.181	.158	.152	.233	.349	.175	.010	-.055	.025	.022	-.033
1.880	.176	.166	.164	.226	.341	.185	.029	-.056	.009	.022	-.038
1.920	.169	.159	.161	.225	.344	.189	.039	-.027	.024	.019	-.042
1.980	.158	.152	.152	.200	.308	.181	.047	-.026	.016	.022	-.052
2.030	.149	.149	.154	.182	.285	.179	.064	.002	.011	.018	-.050
2.070	.143	.144	.145	.157	.259	.172	.074	-.009	-.004	.021	-.057
2.150	.130	.128	.143	.126	.206	.141	.083	.006	-.019	.004	-.057
2.200	.124	.118	.130	.107	.187	.127	.078	-.008	-.034	-.005	-.059
2.280	.117	.114	.142	.087	.164	.108	.098	-.010	-.051	-.034	-.087
2.340	.112	.110	.146	.083	.156	.106	.092	-.033	-.064	-.061	-.097
2.460	.105	.120	.185	.079	.154	.106	.138	-.021	-.088	-.114	-.138
2.560	.102	.147	.197	.121	.150	.145	.158	-.003	-.072	-.153	-.145
2.750	.104	.167	.224	.161	.191	.199	.202	.045	-.048	-.133	-.142
3.000	.087	.172	.228	.212	.230	.231	.223	.103	.019	-.071	-.084
3.400	.071	.150	.227	.211	.229	.213	.210	.150	.098	.010	-.047
3.650	.066	.148	.224	.222	.238	.224	.229	.179	.137	.044	-.015
4.000	.062	.150	.225	.240	.253	.237	.229	.196	.160	.079	.019
5.000	.056	.147	.228	.260	.285	.284	.285	.264	.241	.185	.131
6.000	.053	.143	.228	.271	.309	.314	.317	.300	.281	.238	.197

2.3.2 Polarization and spin-rotation parameters

Tables 2.3.2.1 and 2.3.2.2 give the polarization parameter P and the spin-rotation parameter R as calculated from the Karlsruhe – Helsinki partial wave solution. Although $P=0$ at $\cos\theta = \pm 1$, it can happen that P still has a large value at $\cos\theta = \pm 0.99$ at momenta k, where a zero trajectory passes the forward or backward direction (see the figures in Sect. 2.4.3 and Fig. 3.2.1). Therefore, the tables in our "Handbook" include these $\cos\theta$-values (79 R Höhler-Kaiser-Koch-Pietarinen). These tables also give the spin-rotation parameter A, which was omitted in the present volume. (The tables for A_0 and R_0 are incorrect.) $|A|$ can easily be calculated from the tables for P and R and Eq. (A.4.1), and its sign can be taken from Fig. 3.2.2.

In the table for the R-parameter, we have added the nucleon recoil angle θ_N, because it is needed for the calculation of the spin-rotation angle β and of the other spin-rotation parameters [see Eqs. (A.4.2) and (A.4.15)].

Contour lines for the polarization and spin-rotation parameters P, R, and A are shown in Figs. 3.2.1 and 3.2.2.

Electromagnetic interactions can lead to a considerable modification of the angular dependence of P (Sect. A.4.6). In particular, a cusp-like variation of P_+ is predicted at 200 MeV/c (Fig. 3.2.3a). Experimental data in the range of these modifications would give an important additional input for partial wave analysis.

Legendre expansions of polarization parameters can be made in different ways. Table 2.3.2.3 was calculated from the KH 80 partial wave solution, using the definition in Eq. (A.4.21), i.e. the dimensionless coefficients B_n belong to the expansion of $P(d\sigma/d\Omega)/\sin\theta$.

The tables for the coefficients A_n and B_n have been cut off at n=10 and n=11 respectively, although a larger number of coefficients is required in the upper part of the momentum range.

For Table 2.3.2.1, see p. 172.

For Table 2.3.2.2, see p. 182.

For Table 2.3.2.3, see p. 194.

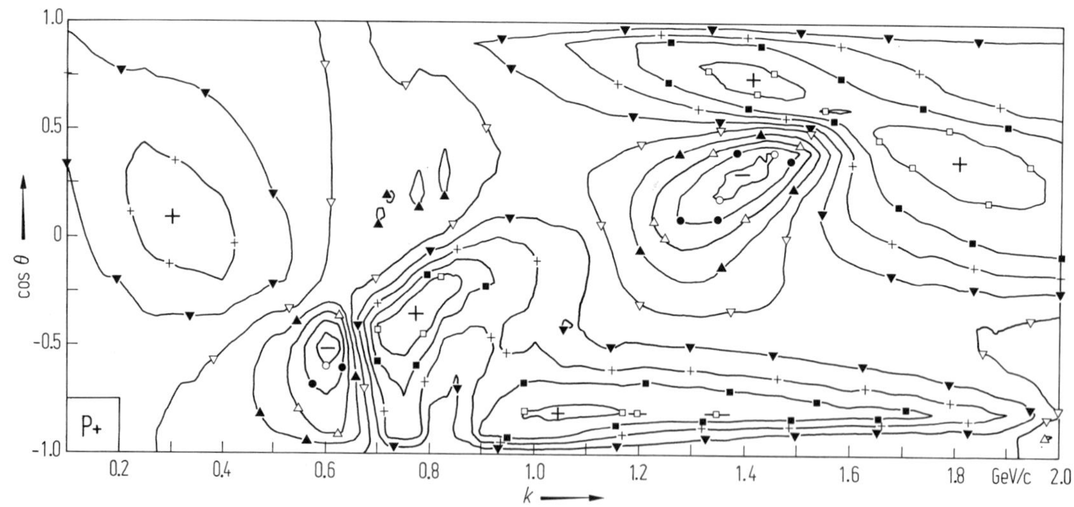

Fig. 3.2.1a.

For caption of Figs. 3.2.1a–c, see next page.

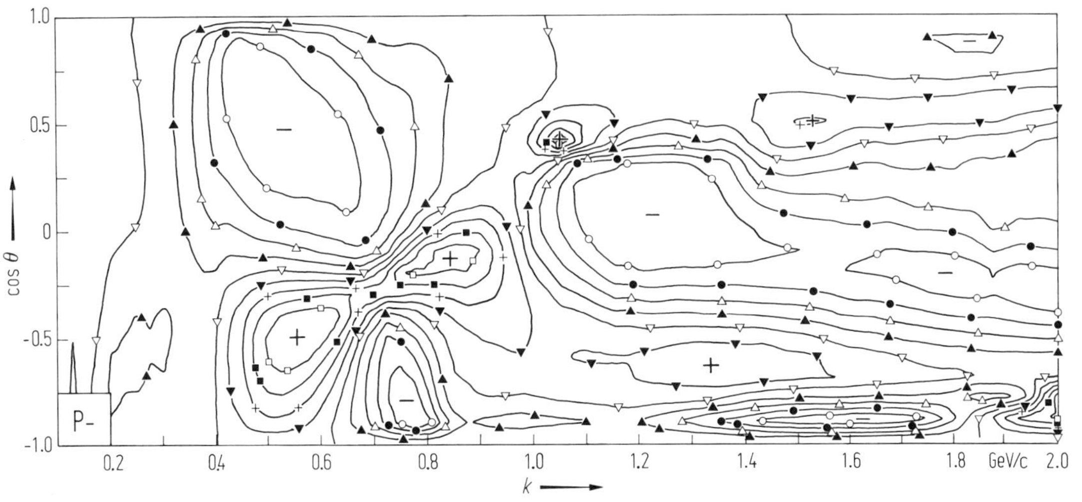

Fig. 3.2.1b.

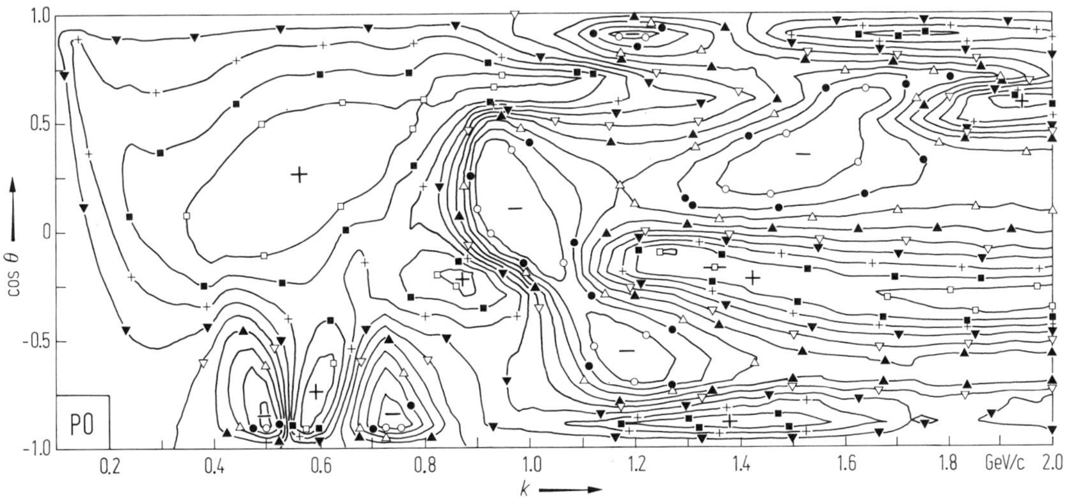

Fig. 3.2.1c.

Fig. 3.2.1a—c. Contour lines for the polarization parameters P_+, P_- and P_0 in $\pi^\pm p$ elastic and $\pi^- p$ charge-exchange scattering. The location of the maxima and minima is indicated by the symbols "+" and "−".
□ 0.8; ■ 0.6; + 0.4; ▼ 0.2; ▽ 0; ▲ −0.2; △ −0.4, ● −0.6; ○ −0.8.

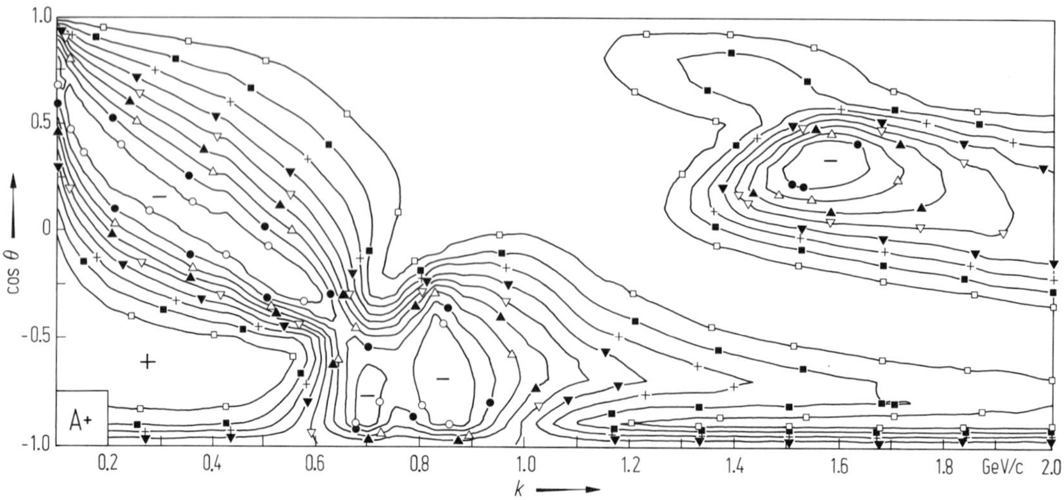

Fig. 3.2.2a.

For caption of Figs. 3.2.2 a–d, see next page.

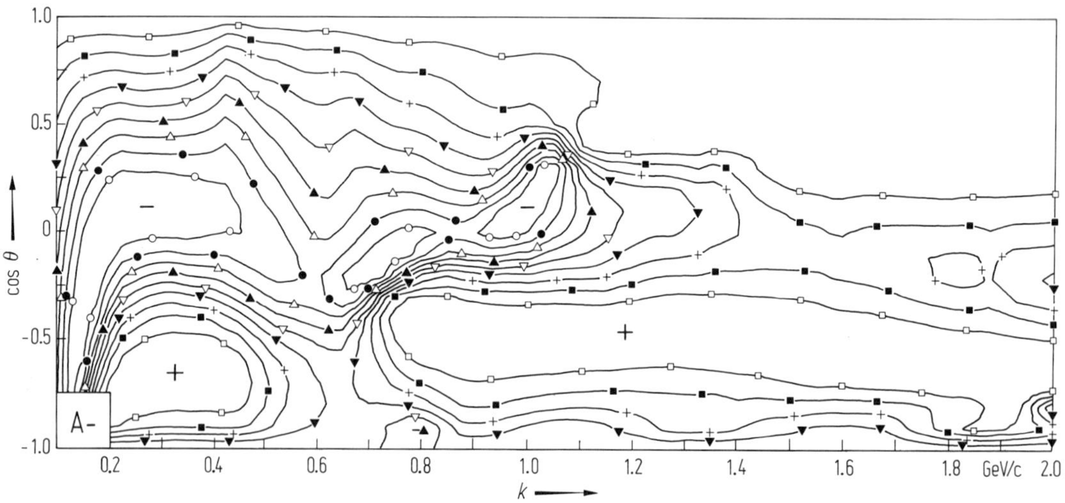

Fig. 3.2.2b.

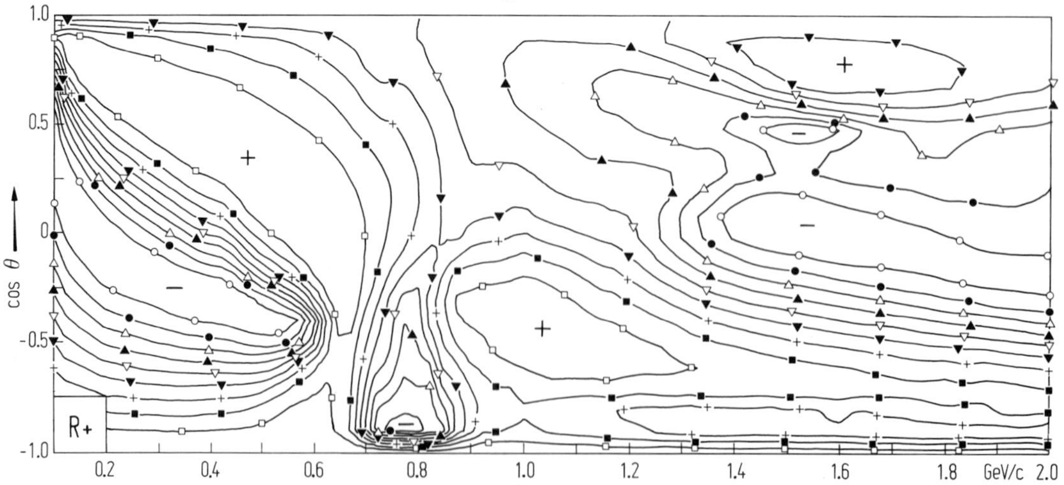

Fig. 3.2.2c.

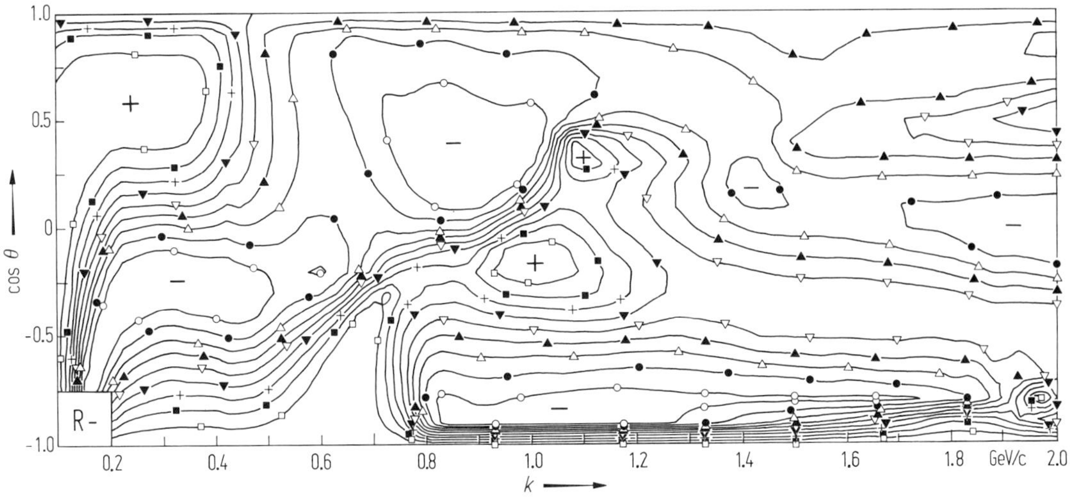

Fig. 3.2.2d.

Fig. 3.2.2a···d. Contour lines for the spin-rotation parameters R_+, A_+, R_-, A_- in elastic $\pi^{\pm}p$ scattering. The location of maxima and minima is indicated by the symbols "+" and "−".
□ 0.8; ■ 0.6; + 0.4; ▼ 0.2; ▽ 0; ▲ −0.2; △ −0.4; ● −0.6; ○ −0.8

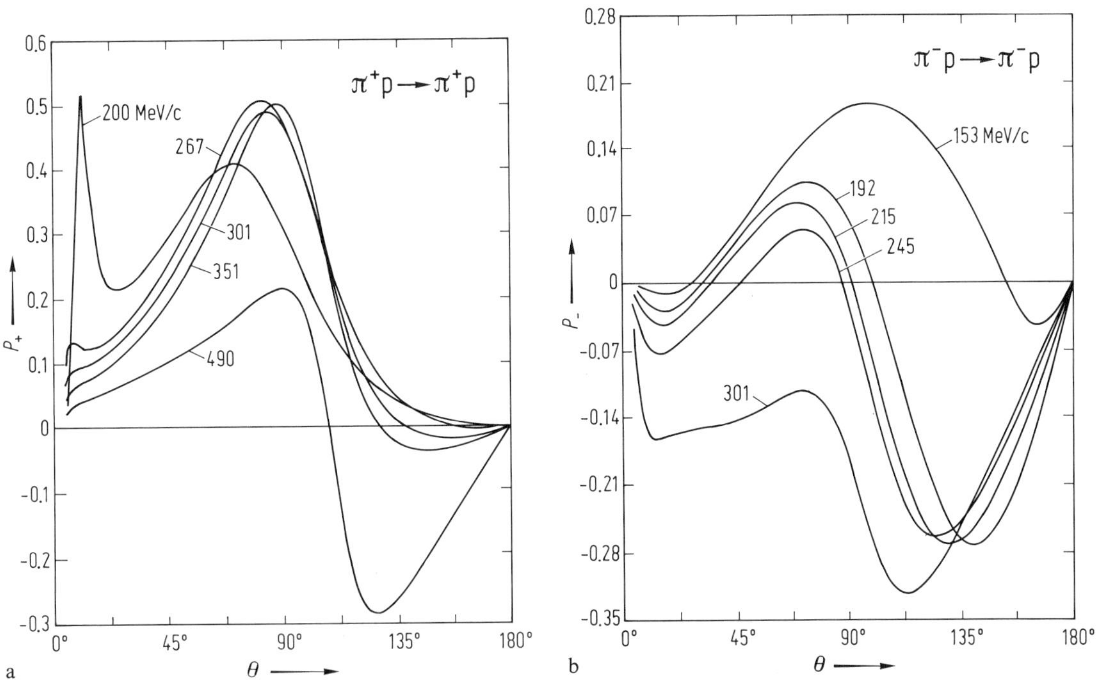

Fig. 3.2.3 a, b. Angular dependence of the polarization parameter in elastic $\pi^{\pm}p$ scattering at 153···490 MeV/c. The structures at small angles are due to the Coulomb interference term.

Table 2.3.2.1. The polarization parameters P as reconstructed from the KH partial wave solution. "THETA" is the c.m. scattering angle θ. KH 80 was used below 0.5 GeV/c and KH 78 at higher momenta.

		COS THETA								POLARIZATION	
K[MEV/C]		.95	.90	.80	.70	.60	.50	.40	.30	.20	.10
20.0	PI+P	.00	.00	.00	.00	.00	.00	.00	.00	.00	.00
	PI-P	.00	.00	.00	.00	.00	.00	.00	.00	.00	.00
	CHEX	-.00	-.00	-.00	-.00	-.00	-.00	-.00	-.00	-.00	-.00
40.0	PI+P	.00	.01	.01	.01	.01	.01	.01	.01	.01	.01
	PI-P	.00	.00	.01	.01	.01	.01	.01	.01	.01	.01
	CHEX	-.00	-.00	-.00	-.00	-.00	-.00	-.00	-.00	-.00	-.00
60.0	PI+P	.03	.04	.05	.05	.05	.04	.04	.04	.03	.03
	PI-P	.01	.02	.02	.03	.03	.03	.03	.04	.04	.04
	CHEX	-.00	-.00	-.00	-.00	-.00	-.00	-.00	-.00	-.00	-.00
79.0	PI+P	.33	.30	.23	.19	.15	.13	.11	.10	.08	.07
	PI-P	.02	.03	.04	.05	.05	.06	.07	.07	.07	.08
	CHEX	-.00	-.00	-.00	-.00	-.00	-.00	-.00	-.00	-.00	-.00
97.0	PI+P	.65	.61	.48	.38	.30	.25	.20	.17	.14	.12
	PI-P	.03	.04	.06	.07	.08	.09	.10	.11	.12	.12
	CHEX	.01	.01	.02	.01	.01	.01	.01	.01	.01	.01
112.0	PI+P	.29	.38	.43	.42	.37	.32	.27	.23	.19	.16
	PI-P	.04	.05	.07	.09	.10	.12	.13	.14	.15	.15
	CHEX	.11	.10	.08	.07	.06	.05	.04	.04	.03	.03
130.0	PI+P	.16	.23	.32	.36	.38	.36	.33	.29	.25	.21
	PI-P	.04	.06	.08	.10	.12	.13	.14	.16	.17	.17
	CHEX	.75	.56	.36	.26	.20	.15	.12	.10	.08	.07
153.0	PI+P	.12	.17	.26	.33	.38	.40	.40	.38	.34	.29
	PI-P	.04	.06	.09	.12	.14	.15	.17	.18	.19	.19
	CHEX	.32	.44	.55	.54	.48	.40	.33	.27	.22	.18
172.0	PI+P	.10	.15	.23	.30	.35	.39	.41	.40	.37	.33
	PI-P	.04	.06	.09	.11	.13	.15	.16	.17	.17	.17
	CHEX	.20	.29	.42	.51	.54	.52	.46	.40	.33	.28
185.0	PI+P	.09	.13	.20	.27	.32	.36	.39	.39	.37	.33
	PI-P	.04	.05	.08	.10	.12	.13	.14	.14	.14	.13
	CHEX	.16	.23	.35	.44	.50	.52	.50	.46	.40	.33
200.0	PI+P	.08	.12	.19	.24	.30	.34	.37	.38	.37	.34
	PI-P	.03	.05	.07	.09	.10	.11	.12	.12	.10	.08
	CHEX	.13	.20	.30	.40	.47	.52	.53	.50	.46	.39
218.0	PI+P	.08	.12	.18	.24	.30	.35	.39	.41	.41	.39
	PI-P	.03	.04	.06	.08	.10	.11	.11	.11	.09	.07
	CHEX	.13	.19	.29	.38	.46	.52	.56	.57	.54	.49
247.0	PI+P	.08	.12	.19	.25	.31	.37	.42	.46	.48	.47
	PI-P	.01	.01	.03	.04	.06	.07	.08	.08	.07	.04
	CHEX	.14	.20	.30	.38	.47	.54	.60	.64	.65	.62
267.0	PI+P	.08	.12	.18	.23	.29	.35	.41	.46	.49	.50
	PI-P	-.02	-.02	-.01	-.01	.01	.02	.03	.03	.02	-.00
	CHEX	.13	.19	.29	.37	.45	.53	.60	.66	.69	.68
280.0	PI+P	.07	.11	.16	.22	.27	.33	.39	.44	.47	.47
	PI-P	-.03	-.04	-.05	-.05	-.05	-.05	-.05	-.06	-.08	-.12
	CHEX	.13	.19	.28	.36	.44	.52	.60	.66	.69	.70
290.0	PI+P	.07	.11	.16	.22	.27	.32	.38	.43	.46	.47
	PI-P	-.04	-.06	-.07	-.07	-.07	-.07	-.07	-.08	-.10	-.13
	CHEX	.13	.19	.28	.36	.44	.51	.58	.65	.69	.70
295.0	PI+P	.07	.11	.16	.22	.27	.32	.38	.43	.47	.48
	PI-P	-.05	-.07	-.08	-.08	-.08	-.08	-.08	-.08	-.10	-.13
	CHEX	.13	.19	.28	.36	.44	.51	.59	.65	.70	.71

	COS THETA									POLARIZATION
K[MEV/C]	.00	-.10	-.20	-.30	-.40	-.50	-.60	-.70	-.80	-.90	-.95
20.0	.00	.00	.00	.00	.00	.00	.00	.00	.00	.00	.00
	.00	.00	.00	.00	.00	.00	.00	.00	.00	.00	.00
	-.00	-.00	-.00	-.00	-.00	-.00	-.00	-.00	-.00	-.00	-.00
40.0	.01	.01	.01	.01	.01	.01	.01	.00	.00	.00	.00
	.01	.01	.01	.01	.01	.01	.01	.01	.01	.01	.00
	-.00	-.00	-.00	-.00	-.00	-.00	-.00	-.00	-.00	-.00	-.00
60.0	.03	.03	.02	.02	.02	.02	.01	.01	.01	.01	.00
	.04	.04	.04	.04	.04	.04	.03	.03	.03	.02	.01
	-.00	-.00	-.00	-.00	-.00	-.00	-.00	-.00	-.00	-.00	-.00
79.0	.06	.05	.05	.04	.04	.03	.03	.02	.02	.01	.01
	.08	.08	.08	.09	.09	.09	.08	.08	.07	.05	.04
	-.00	-.00	-.00	-.00	-.00	-.00	-.00	-.00	-.00	-.00	-.00
97.0	.10	.09	.07	.06	.05	.04	.04	.03	.02	.01	.01
	.13	.13	.14	.14	.14	.15	.15	.14	.13	.11	.09
	.01	.01	.01	.01	.00	.00	.00	.00	.00	.00	.00
112.0	.14	.11	.10	.08	.07	.05	.04	.03	.02	.02	.01
	.16	.17	.18	.18	.19	.20	.20	.21	.21	.20	.17
	.02	.02	.02	.02	.01	.01	.01	.01	.01	.00	.00
130.0	.18	.15	.12	.10	.08	.06	.05	.04	.03	.02	.01
	.18	.19	.20	.20	.21	.22	.23	.24	.26	.31	.36
	.06	.05	.04	.04	.03	.02	.02	.02	.01	.01	.01
153.0	.25	.20	.17	.13	.11	.09	.07	.05	.03	.02	.01
	.20	.20	.19	.18	.17	.15	.13	.10	.06	.01	-.02
	.15	.12	.10	.08	.07	.06	.05	.04	.03	.02	.01
172.0	.28	.23	.19	.15	.12	.09	.07	.05	.03	.02	.01
	.16	.14	.12	.09	.05	.01	-.04	-.09	-.13	-.15	-.13
	.23	.19	.15	.12	.10	.08	.06	.05	.04	.02	.02
185.0	.29	.24	.19	.15	.12	.09	.06	.04	.03	.02	.01
	.11	.08	.04	-.00	-.06	-.11	-.16	-.20	-.22	-.20	-.16
	.28	.23	.18	.15	.12	.09	.07	.06	.04	.03	.02
200.0	.30	.25	.20	.15	.12	.08	.06	.04	.02	.01	.01
	.05	.00	-.05	-.11	-.17	-.22	-.26	-.29	-.28	-.23	-.17
	.33	.27	.22	.18	.14	.11	.09	.06	.05	.03	.02
218.0	.34	.29	.23	.18	.13	.09	.06	.04	.02	.01	.01
	.02	-.03	-.09	-.15	-.20	-.23	-.25	-.25	-.23	-.17	-.13
	.43	.36	.29	.23	.19	.14	.11	.08	.06	.04	.02
247.0	.43	.37	.30	.23	.17	.12	.08	.05	.03	.01	.00
	-.00	-.06	-.12	-.18	-.22	-.25	-.25	-.24	-.21	-.15	-.11
	.56	.49	.41	.33	.26	.20	.15	.11	.08	.04	.03
267.0	.47	.41	.34	.26	.19	.13	.08	.05	.02	.01	.00
	-.05	-.11	-.17	-.22	-.26	-.27	-.27	-.25	-.21	-.15	-.11
	.64	.57	.48	.39	.31	.24	.18	.13	.09	.05	.03
280.0	.45	.39	.31	.23	.16	.10	.05	.02	.00	-.01	-.01
	-.17	-.23	-.28	-.32	-.34	-.34	-.32	-.28	-.23	-.16	-.11
	.66	.59	.50	.40	.31	.23	.17	.12	.08	.04	.03
290.0	.45	.40	.32	.24	.16	.10	.05	.02	-.00	-.01	-.01
	-.18	-.24	-.29	-.32	-.34	-.33	-.30	-.27	-.22	-.15	-.11
	.68	.61	.52	.42	.33	.25	.18	.12	.08	.04	.03
295.0	.46	.40	.32	.24	.16	.09	.05	.01	-.01	-.01	-.01
	-.18	-.23	-.28	-.31	-.32	-.32	-.29	-.25	-.21	-.14	-.10
	.68	.62	.53	.43	.33	.24	.17	.12	.07	.04	.02

K[MEV/C]		COS THETA .95	.90	.80	.70	.60	.50	.40	.30	POLARIZATION .20	.10
301.0	PI+P	.07	.11	.16	.22	.27	.32	.38	.43	.47	.48
	PI-P	-.06	-.08	-.09	-.10	-.09	-.09	-.08	-.08	-.09	-.12
	CHEX	.14	.19	.28	.36	.44	.51	.58	.65	.70	.71
305.0	PI+P	.07	.11	.16	.21	.27	.32	.38	.43	.47	.49
	PI-P	-.06	-.09	-.11	-.11	-.11	-.10	-.09	-.09	-.10	-.12
	CHEX	.14	.20	.29	.36	.44	.51	.58	.65	.70	.72
310.0	PI+P	.07	.11	.16	.21	.27	.32	.38	.43	.47	.49
	PI-P	-.07	-.10	-.12	-.13	-.12	-.11	-.10	-.10	-.10	-.12
	CHEX	.14	.20	.29	.36	.44	.51	.58	.65	.70	.72
320.0	PI+P	.07	.11	.16	.21	.27	.32	.38	.43	.48	.50
	PI-P	-.09	-.12	-.15	-.15	-.15	-.14	-.12	-.11	-.11	-.12
	CHEX	.14	.20	.29	.37	.44	.51	.59	.65	.71	.74
331.0	PI+P	.07	.10	.15	.20	.25	.30	.35	.41	.45	.48
	PI-P	-.11	-.15	-.19	-.20	-.20	-.20	-.19	-.17	-.17	-.17
	CHEX	.14	.20	.28	.36	.43	.51	.58	.65	.71	.75
351.0	PI+P	.06	.09	.14	.19	.24	.29	.35	.40	.46	.49
	PI-P	-.17	-.22	-.27	-.29	-.29	-.28	-.26	-.25	-.23	-.21
	CHEX	.14	.20	.29	.37	.44	.52	.60	.68	.75	.79
378.0	PI+P	.06	.09	.13	.17	.21	.26	.30	.36	.41	.45
	PI-P	-.27	-.35	-.44	-.47	-.47	-.47	-.44	-.41	-.37	-.32
	CHEX	.14	.21	.30	.39	.46	.54	.62	.70	.77	.83
408.0	PI+P	.06	.08	.12	.15	.19	.23	.27	.32	.37	.41
	PI-P	-.41	-.54	-.65	-.69	-.68	-.65	-.61	-.55	-.47	-.38
	CHEX	.15	.21	.31	.40	.48	.57	.65	.74	.82	.88
427.0	PI+P	.05	.07	.10	.14	.17	.21	.25	.29	.34	.38
	PI-P	-.51	-.66	-.79	-.82	-.80	-.76	-.70	-.63	-.54	-.42
	CHEX	.16	.24	.35	.45	.55	.65	.74	.83	.90	.94
456.0	PI+P	.03	.05	.08	.11	.14	.18	.22	.26	.31	.35
	PI-P	-.58	-.75	-.92	-.96	-.96	-.91	-.85	-.77	-.67	-.54
	CHEX	.18	.26	.39	.50	.60	.70	.80	.88	.94	.96
490.0	PI+P	.04	.06	.08	.10	.12	.14	.16	.18	.20	.22
	PI-P	-.56	-.74	-.92	-.99	-1.00	-.97	-.93	-.86	-.76	-.63
	CHEX	.20	.29	.44	.57	.69	.80	.89	.96	.99	.97
532.0	PI+P	.03	.04	.06	.08	.10	.12	.14	.16	.18	.20
	PI-P	-.49	-.66	-.83	-.91	-.94	-.95	-.93	-.89	-.83	-.72
	CHEX	.24	.34	.50	.63	.75	.85	.93	.98	1.00	.97
573.0	PI+P	.00	.01	.02	.03	.05	.06	.07	.08	.09	.09
	PI-P	-.38	-.52	-.69	-.79	-.85	-.88	-.89	-.88	-.84	-.76
	CHEX	.22	.32	.49	.63	.76	.87	.95	.99	.99	.94
614.0	PI+P	-.01	-.02	-.02	-.02	-.01	-.01	-.01	-.01	-.01	-.02
	PI-P	-.33	-.47	-.65	-.75	-.81	-.84	-.86	-.86	-.83	-.78
	CHEX	.21	.32	.50	.65	.78	.89	.96	.99	.96	.88
658.0	PI+P	-.00	-.01	-.03	-.05	-.06	-.08	-.09	-.10	-.11	-.12
	PI-P	-.16	-.28	-.49	-.64	-.74	-.80	-.84	-.85	-.85	-.83
	CHEX	.24	.35	.52	.67	.79	.90	.96	.95	.87	.71
675.0	PI+P	.00	-.01	-.04	-.06	-.09	-.11	-.12	-.14	-.15	-.16
	PI-P	-.13	-.24	-.43	-.58	-.67	-.73	-.77	-.80	-.81	-.79
	CHEX	.25	.36	.52	.66	.79	.90	.96	.95	.86	.71
705.0	PI+P	.03	.03	.00	-.04	-.08	-.12	-.15	-.18	-.20	-.21
	PI-P	-.09	-.17	-.32	-.44	-.54	-.61	-.65	-.68	-.70	-.68
	CHEX	.22	.32	.48	.64	.78	.89	.93	.89	.80	.67

POLARIZATION

K[MEV/C]	COS THETA .00	-.10	-.20	-.30	-.40	-.50	-.60	-.70	-.80	-.90	-.95
301.0	.46	.41	.33	.24	.16	.09	.05	.01	-.01	-.02	-.02
	-.16	-.22	-.27	-.30	-.31	-.31	-.28	-.25	-.20	-.14	-.10
	.69	.63	.54	.44	.34	.25	.18	.12	.07	.04	.02
305.0	.47	.42	.34	.25	.16	.10	.04	.01	-.01	-.02	-.02
	-.16	-.21	-.26	-.29	-.30	-.30	-.27	-.24	-.19	-.13	-.09
	.70	.64	.55	.45	.34	.25	.18	.12	.07	.03	.02
310.0	.48	.42	.34	.25	.16	.09	.04	.00	-.02	-.02	-.02
	-.15	-.19	-.24	-.27	-.28	-.28	-.25	-.22	-.18	-.12	-.09
	.71	.65	.56	.45	.34	.25	.17	.11	.06	.03	.02
320.0	.48	.43	.34	.24	.15	.07	.02	-.01	-.03	-.03	-.02
	-.15	-.19	-.24	-.27	-.28	-.27	-.25	-.21	-.17	-.12	-.08
	.73	.67	.57	.45	.33	.23	.15	.09	.05	.02	.01
331.0	.48	.44	.36	.27	.17	.09	.03	-.01	-.03	-.03	-.02
	-.18	-.20	-.23	-.24	-.24	-.23	-.21	-.18	-.14	-.10	-.07
	.75	.70	.60	.48	.36	.25	.16	.09	.04	.01	.00
351.0	.50	.45	.37	.25	.15	.06	.01	-.02	-.04	-.03	-.03
	-.19	-.18	-.17	-.17	-.15	-.14	-.11	-.09	-.07	-.04	-.03
	.79	.73	.61	.45	.30	.17	.08	.02	-.01	-.03	-.02
378.0	.48	.46	.39	.28	.16	.06	-.01	-.05	-.07	-.06	-.05
	-.26	-.20	-.13	-.08	-.05	-.03	-.02	-.01	-.01	-.01	-.00
	.84	.79	.66	.48	.29	.13	.01	-.06	-.10	-.09	-.07
408.0	.45	.44	.38	.26	.13	.02	-.06	-.10	-.10	-.09	-.07
	-.27	-.15	-.03	.07	.14	.17	.16	.14	.10	.06	.04
	.88	.80	.59	.32	.05	-.14	-.25	-.28	-.27	-.21	-.15
427.0	.42	.41	.35	.24	.11	-.00	-.07	-.10	-.11	-.09	-.07
	-.29	-.13	.04	.21	.35	.43	.44	.41	.33	.23	.16
	.89	.73	.46	.12	-.19	-.39	-.49	-.51	-.46	-.35	-.25
456.0	.39	.39	.33	.19	.03	-.10	-.16	-.17	-.15	-.11	-.08
	-.38	-.19	.04	.27	.46	.56	.57	.51	.41	.28	.19
	.92	.79	.56	.24	-.12	-.42	-.62	-.70	-.65	-.50	-.36
490.0	.23	.21	.15	.02	-.14	-.25	-.28	-.27	-.22	-.16	-.11
	-.45	-.22	.07	.39	.66	.81	.82	.72	.57	.38	.26
	.90	.76	.57	.33	.07	-.21	-.48	-.73	-.89	-.88	-.73
532.0	.21	.22	.19	.09	-.13	-.33	-.40	-.38	-.32	-.24	-.17
	-.56	-.30	.06	.48	.83	.97	.91	.74	.54	.33	.22
	.90	.79	.65	.50	.36	.23	.11	.01	-.10	-.27	-.44
573.0	.10	.09	.06	-.08	-.45	-.72	-.67	-.57	-.47	-.36	-.27
	-.61	-.36	.04	.54	.92	.99	.82	.60	.39	.21	.13
	.85	.74	.64	.55	.51	.52	.60	.75	.91	.93	.78
614.0	-.02	-.03	-.07	-.22	-.71	-.97	-.76	-.61	-.52	-.44	-.37
	-.67	-.44	-.01	.59	.93	.82	.55	.33	.16	.05	.01
	.77	.65	.56	.53	.57	.69	.85	.94	.81	.48	.28
658.0	-.11	-.09	-.04	.09	.21	.01	-.17	-.24	-.27	-.30	-.30
	-.75	-.56	-.09	.55	.56	.25	.01	-.13	-.20	-.21	-.18
	.54	.40	.34	.37	.45	.49	.38	.16	-.07	-.22	-.21
675.0	-.15	-.13	-.04	.18	.44	.30	.09	-.02	-.08	-.14	-.17
	-.73	-.56	-.06	.58	.32	-.04	-.23	-.31	-.34	-.31	-.24
	.53	.38	.31	.34	.37	.27	.04	-.19	-.37	-.44	-.38
705.0	-.20	-.13	.05	.43	.85	.82	.59	.42	.32	.21	.13
	-.61	-.41	.15	.59	-.06	-.40	-.53	-.59	-.59	-.50	-.38
	.54	.47	.45	.41	.23	-.09	-.38	-.59	-.73	-.77	-.66

		COS THETA								POLARIZATION	
KEMEV/CJ		.95	.90	.80	.70	.60	.50	.40	.30	.20	.10
725.	PI+P	.06	.07	.04	-.02	-.08	-.12	-.16	-.19	-.20	-.20
	PI-P	-.08	-.15	-.29	-.40	-.49	-.55	-.60	-.63	-.64	-.62
	CHEX	.22	.31	.47	.63	.78	.88	.90	.85	.75	.64
750.	PI+P	.09	.10	.06	-.00	-.06	-.10	-.14	-.18	-.19	-.18
	PI-P	-.08	-.15	-.27	-.37	-.45	-.50	-.53	-.55	-.54	-.49
	CHEX	.21	.30	.46	.63	.77	.84	.81	.70	.57	.47
777.	PI+P	.07	.06	.01	-.06	-.11	-.15	-.18	-.21	-.22	-.19
	PI-P	-.08	-.16	-.29	-.37	-.40	-.41	-.42	-.42	-.40	-.31
	CHEX	.23	.32	.48	.66	.80	.84	.78	.65	.50	.39
800.	PI+P	.09	.08	.03	-.03	-.08	-.12	-.16	-.18	-.18	-.14
	PI-P	-.06	-.14	-.27	-.32	-.33	-.33	-.33	-.33	-.30	-.17
	CHEX	.26	.38	.56	.72	.80	.78	.66	.51	.39	.34
822.	PI+P	.10	.11	.04	-.05	-.12	-.17	-.21	-.22	-.21	-.14
	PI-P	-.03	-.10	-.22	-.25	-.24	-.23	-.24	-.24	-.22	-.08
	CHEX	.27	.38	.61	.80	.83	.72	.55	.39	.30	.29
851.	PI+P	.15	.16	.10	.01	-.06	-.12	-.16	-.17	-.14	-.04
	PI-P	.01	-.04	-.14	-.18	-.18	-.15	-.14	-.12	-.07	.12
	CHEX	.26	.40	.65	.83	.81	.58	.29	.06	-.07	-.06
875.	PI+P	.18	.20	.15	.06	-.02	-.08	-.11	-.10	-.06	.06
	PI-P	.02	-.00	-.08	-.13	-.12	-.09	-.07	-.06	-.01	.17
	CHEX	.21	.35	.62	.84	.77	.35	-.04	-.29	-.39	-.38
895.	PI+P	.19	.21	.15	.08	.02	-.02	-.03	-.03	.02	.15
	PI-P	.01	-.01	-.08	-.12	-.11	-.07	-.05	-.03	.03	.21
	CHEX	.16	.29	.59	.87	.70	-.06	-.56	-.75	-.80	-.75
923.	PI+P	.21	.23	.18	.12	.07	.02	-.00	.00	.06	.20
	PI-P	-.00	-.03	-.07	-.09	-.07	-.05	-.03	-.01	.05	.19
	CHEX	.09	.18	.49	.89	.66	-.44	-.89	-.98	-.99	-.97
954.	PI+P	.25	.27	.22	.15	.09	.04	.01	.02	.07	.19
	PI-P	-.02	-.05	-.09	-.10	-.06	-.00	.03	.03	.04	.08
	CHEX	-.01	.06	.37	.88	.66	-.47	-.84	-.93	-.96	-.98
975.	PI+P	.26	.28	.23	.17	.11	.06	.02	.01	.07	.19
	PI-P	-.02	-.05	-.10	-.11	-.06	.03	.08	.03	-.03	-.05
	CHEX	-.06	-.01	.29	.85	.74	-.28	-.71	-.85	-.91	-.95
1000.	PI+P	.28	.30	.26	.19	.12	.06	.02	.02	.08	.19
	PI-P	-.01	-.04	-.09	-.10	-.05	.14	.28	.10	-.15	-.30
	CHEX	-.11	-.07	.24	.87	.65	-.23	-.61	-.77	-.85	-.91
1030.	PI+P	.34	.37	.32	.23	.14	.07	.03	.01	.05	.14
	PI-P	.01	-.00	-.03	-.04	.03	.31	.68	-.10	-.46	-.54
	CHEX	-.19	-.16	.15	.82	.72	-.07	-.46	-.64	-.75	-.82
1055.	PI+P	.38	.41	.35	.26	.17	.10	.04	.02	.06	.15
	PI-P	.03	.04	.02	.02	.09	.42	.96	-.31	-.64	-.67
	CHEX	-.26	-.22	.14	.86	.69	.01	-.36	-.56	-.69	-.76
1080.	PI+P	.40	.44	.38	.28	.18	.09	.04	.02	.04	.13
	PI-P	.05	.07	.06	.05	.12	.39	.59	-.62	-.83	-.82
	CHEX	-.33	-.31	.03	.83	.66	.06	-.29	-.49	-.61	-.67
1113.	PI+P	.43	.49	.42	.31	.21	.11	.04	-.01	-.02	.01
	PI-P	.07	.08	.08	.10	.17	.33	.20	-.72	-.93	-.93
	CHEX	-.51	-.53	-.08	.78	.57	.07	-.24	-.42	-.53	-.55
1154.	PI+P	.45	.54	.51	.38	.24	.12	.02	-.06	-.10	-.11
	PI-P	.06	.08	.09	.12	.17	.21	-.06	-.75	-.97	-.98
	CHEX	-.76	-.81	-.28	.58	.47	.07	-.21	-.37	-.45	-.42

K[MEV/C]	COS THETA .00	-.10	-.20	-.30	-.40	-.50	-.60	-.70	-.80	-.90	POLARIZATION -.95
725.	-.16	-.06	.16	.56	.94	.86	.66	.55	.52	.55	.58
	-.53	-.27	.37	.47	-.30	-.62	-.74	-.79	-.79	-.68	-.51
	.55	.51	.50	.43	.15	-.25	-.55	-.73	-.86	-.93	-.85
750.	-.12	.03	.34	.78	1.00	.88	.71	.61	.57	.57	.62
	-.31	.12	.77	.45	-.21	-.56	-.76	-.88	-.95	-.88	-.70
	.44	.50	.58	.55	.29	-.07	-.37	-.61	-.80	-.97	-.99
777.	-.08	.20	.66	.98	.91	.71	.55	.46	.44	.52	.66
	-.07	.47	.85	.37	-.12	-.41	-.61	-.78	-.93	-.98	-.85
	.40	.54	.70	.62	.34	.07	-.15	-.33	-.52	-.79	-.93
800.	.01	.36	.84	.99	.82	.60	.41	.27	.17	.14	.19
	.18	.75	.83	.41	.03	-.22	-.39	-.53	-.67	-.86	-.89
	.42	.63	.76	.64	.39	.16	-.02	-.17	-.32	-.55	-.75
822.	.02	.43	.96	.81	.50	.34	.28	.23	.13	.00	.01
	.31	.83	.80	.43	.14	-.05	-.16	-.24	-.34	-.58	-.79
	.40	.62	.80	.66	.35	.14	.04	-.03	-.12	-.34	-.57
851.	.21	.63	.88	.71	.46	.30	.20	.17	.20	.26	.29
	.53	.92	.79	.47	.24	.11	.03	-.01	-.07	-.28	-.58
	.15	.57	.86	.71	.42	.23	.14	.11	.08	-.05	-.25
875.	.30	.61	.67	.52	.40	.36	.36	.37	.35	.33	.37
	.55	.90	.80	.52	.31	.16	.07	.01	-.04	-.19	-.45
	-.19	.30	.88	.82	.50	.30	.20	.16	.13	.04	-.08
895.	.36	.59	.64	.52	.40	.34	.35	.42	.53	.65	.68
	.56	.83	.71	.48	.32	.21	.14	.07	-.02	-.23	-.48
	-.54	.03	.81	.80	.49	.30	.22	.20	.19	.17	.13
923.	.42	.59	.60	.50	.43	.41	.46	.54	.62	.66	.59
	.43	.60	.56	.43	.31	.22	.16	.09	-.02	-.26	-.50
	-.86	-.44	.55	.76	.47	.31	.24	.22	.21	.22	.26
954.	.38	.52	.53	.43	.36	.36	.45	.59	.73	.76	.63
	.17	.28	.33	.32	.28	.24	.19	.11	-.03	-.30	-.56
	-.97	-.76	.11	.61	.40	.26	.21	.20	.21	.25	.33
975.	.37	.49	.50	.42	.36	.37	.49	.66	.81	.78	.56
	-.00	.10	.20	.26	.28	.25	.19	.10	-.04	-.31	-.57
	-.98	-.88	-.12	.50	.40	.26	.17	.14	.16	.25	.34
1000.	.33	.40	.39	.36	.36	.41	.52	.67	.81	.75	.56
	-.31	-.20	-.02	.12	.18	.19	.15	.10	-.02	-.27	-.54
	-.96	-.99	-.61	.14	.13	.07	.05	.08	.16	.32	.42
1030.	.27	.35	.33	.28	.26	.33	.48	.69	.84	.73	.52
	-.51	-.38	-.19	.00	.15	.20	.19	.12	-.01	-.26	-.53
	-.88	-.97	-.88	-.26	-.06	-.04	-.03	.01	.11	.31	.44
1055.	.27	.33	.29	.21	.17	.22	.40	.68	.86	.66	.45
	-.63	-.53	-.35	-.14	.05	.15	.17	.14	.02	-.23	-.47
	-.80	-.86	-.94	-.66	-.41	-.31	-.23	-.12	.08	.39	.52
1080.	.24	.28	.25	.21	.21	.27	.43	.70	.85	.58	.37
	-.77	-.65	-.45	-.18	.05	.19	.21	.15	.02	-.22	-.46
	-.65	-.60	-.67	-.77	-.64	-.51	-.39	-.24	.04	.44	.58
1113.	.09	.14	.14	.12	.13	.21	.40	.71	.86	.54	.31
	-.89	-.80	-.62	-.33	-.03	.16	.21	.17	.04	-.19	-.42
	-.47	-.29	-.24	-.62	-.80	-.78	-.66	-.45	-.07	.44	.55
1154.	-.06	-.01	.02	.04	.09	.19	.38	.69	.83	.47	.26
	-.97	-.92	-.77	-.49	-.14	.11	.21	.18	.07	-.15	-.38
	-.25	.14	.25	-.34	-.78	-.93	-.91	-.71	-.18	.51	.57

K[MEV/C]		COS THETA .95	.90	.80	.70	.60	.50	.40	.30	POLARIZATION .20	.10
1174.	PI+P	.47	.58	.55	.40	.25	.10	-.01	-.10	-.16	-.17
	PI-P	.05	.07	.10	.12	.15	.12	-.23	-.84	-.99	-1.00
	CHEX	-.87	-.89	-.32	.48	.44	.08	-.19	-.34	-.40	-.37
1210.	PI+P	.50	.64	.64	.48	.28	.09	-.06	-.19	-.29	-.33
	PI-P	.04	.07	.11	.12	.11	.03	-.29	-.82	-.99	-1.00
	CHEX	-.99	-.90	-.40	.26	.36	.08	-.17	-.35	-.44	-.40
1235.	PI+P	.52	.66	.69	.53	.30	.07	-.12	-.29	-.43	-.47
	PI-P	.06	.07	.09	.10	.08	-.02	-.35	-.86	-.99	-.99
	CHEX	-.99	-.86	-.44	.12	.29	.06	-.20	-.39	-.50	-.45
1280.	PI+P	.51	.68	.77	.62	.36	.07	-.19	-.42	-.61	-.63
	PI-P	.08	.08	.07	.08	.07	-.00	-.31	-.83	-.95	-.97
	CHEX	-.88	-.70	-.41	.03	.28	.09	-.21	-.45	-.57	-.53
1324.	PI+P	.48	.67	.82	.72	.43	.04	-.34	-.64	-.77	-.64
	PI-P	.10	.09	.07	.08	.10	.08	-.18	-.72	-.87	-.90
	CHEX	-.64	-.53	-.37	-.03	.26	.03	-.35	-.63	-.73	-.62
1360.	PI+P	.45	.65	.85	.81	.48	-.03	-.50	-.81	-.84	-.61
	PI-P	.10	.11	.11	.13	.13	.05	-.28	-.67	-.75	-.82
	CHEX	-.35	-.41	-.36	-.08	.21	-.04	-.45	-.71	-.80	-.67
1400.	PI+P	.42	.61	.83	.86	.57	-.07	-.68	-.88	-.69	-.39
	PI-P	.09	.11	.13	.15	.19	.19	.01	-.33	-.52	-.69
	CHEX	.15	-.09	-.28	-.17	.08	-.20	-.64	-.87	-.88	-.66
1430.	PI+P	.40	.58	.80	.88	.63	-.17	-.82	-.85	-.56	-.28
	PI-P	.06	.09	.12	.17	.21	.24	.16	-.13	-.37	-.59
	CHEX	.39	.06	-.22	-.18	.02	-.32	-.74	-.93	-.92	-.69
1473.	PI+P	.38	.53	.73	.86	.74	-.12	-.74	-.58	-.27	-.08
	PI-P	.01	.03	.07	.13	.23	.38	.30	-.06	-.33	-.54
	CHEX	.74	.33	-.18	-.33	-.19	-.52	-.90	-.99	-.91	-.64
1505.	PI+P	.35	.48	.66	.81	.77	-.08	-.57	-.27	.02	.12
	PI-P	-.03	-.03	.01	.11	.25	.43	.29	-.08	-.32	-.52
	CHEX	.87	.47	-.17	-.41	-.26	-.62	-.95	-.99	-.86	-.59
1550.	PI+P	.32	.43	.57	.71	.81	.26	-.07	.06	.19	.22
	PI-P	-.09	-.11	-.07	.06	.25	.41	.19	-.13	-.32	-.50
	CHEX	.90	.53	-.09	-.49	-.62	-.86	-.98	-.94	-.81	-.57
1590.	PI+P	.29	.39	.51	.65	.81	.55	.31	.37	.42	.39
	PI-P	-.11	-.15	-.11	.03	.21	.30	.05	-.18	-.30	-.44
	CHEX	.90	.58	-.06	-.54	-.85	-.91	-.90	-.84	-.73	-.54
1640.	PI+P	.28	.36	.45	.55	.74	.82	.66	.59	.55	.48
	PI-P	-.13	-.17	-.13	.02	.22	.29	-.01	-.22	-.31	-.44
	CHEX	.89	.61	-.01	-.59	-.99	-.70	-.77	-.78	-.68	-.49
1680.	PI+P	.26	.34	.42	.50	.67	.89	.83	.73	.64	.54
	PI-P	-.14	-.18	-.16	-.02	.19	.27	.02	-.18	-.31	-.47
	CHEX	.88	.66	-.04	-.69	-.87	-.49	-.68	-.70	-.62	-.45
1720.	PI+P	.26	.32	.39	.47	.66	.89	.86	.77	.68	.60
	PI-P	-.15	-.19	-.16	-.00	.23	.27	-.00	-.17	-.26	-.39
	CHEX	.82	.65	.03	-.59	-.50	-.21	-.60	-.69	-.62	-.44
1760.	PI+P	.25	.31	.36	.41	.59	.88	.92	.84	.75	.63
	PI-P	-.16	-.21	-.17	-.00	.22	.23	-.04	-.19	-.28	-.40
	CHEX	.76	.65	.07	-.58	-.24	.10	-.42	-.58	-.56	-.43
1800.	PI+P	.24	.30	.33	.35	.52	.82	.92	.88	.80	.70
	PI-P	-.16	-.22	-.19	-.01	.24	.25	-.03	-.18	-.25	-.37
	CHEX	.71	.63	.06	-.61	.05	.35	-.35	-.54	-.52	-.39

	COS THETA									POLARIZATION	
K[MEV/C]	.00	-.10	-.20	-.30	-.40	-.50	-.60	-.70	-.80	-.90	-.95
1174.	-.12	-.07	-.02	.04	.12	.23	.40	.66	.80	.42	.20
	-.98	-.92	-.75	-.46	-.13	.11	.20	.19	.09	-.15	-.44
	-.17	.30	.37	-.29	-.73	-.94	-.99	-.79	-.18	.60	.62
1210.	-.28	-.19	-.10	-.02	.09	.23	.43	.71	.80	.37	.16
	-.99	-.94	-.81	-.55	-.20	.10	.24	.22	.09	-.23	-.58
	-.09	.62	.64	-.11	-.60	-.86	-.99	-.88	-.20	.62	.59
1235.	-.37	-.24	-.13	-.04	.07	.21	.42	.71	.79	.33	.11
	-.98	-.92	-.77	-.52	-.19	.11	.27	.24	.08	-.30	-.67
	-.06	.79	.67	-.09	-.54	-.80	-.94	-.87	-.15	.67	.57
1280.	-.49	-.33	-.20	-.07	.06	.20	.38	.65	.79	.30	.07
	-.98	-.93	-.77	-.50	-.16	.14	.32	.29	.05	-.41	-.77
	-.11	.81	.75	.02	-.42	-.66	-.80	-.73	-.03	.75	.58
1324.	-.44	-.30	-.19	-.07	.06	.21	.39	.67	.80	.25	.01
	-.93	-.91	-.78	-.53	-.20	.14	.35	.32	.02	-.55	-.89
	-.12	.72	.80	.25	-.20	-.50	-.69	-.63	.12	.76	.53
1360.	-.36	-.22	-.12	-.03	.07	.18	.34	.60	.81	.26	.01
	-.88	-.87	-.75	-.51	-.21	.11	.35	.34	-.02	-.64	-.92
	-.19	.55	.80	.40	-.06	-.39	-.55	-.47	.18	.80	.55
1400.	-.22	-.14	-.09	-.03	.06	.18	.34	.61	.78	.22	-.02
	-.82	-.85	-.76	-.53	-.21	.11	.33	.29	-.12	-.77	-.96
	-.16	.48	.77	.52	.08	-.28	-.46	-.37	.30	.78	.52
1430.	-.14	-.09	-.06	-.01	.06	.17	.32	.58	.76	.21	-.01
	-.76	-.83	-.76	-.55	-.24	.11	.35	.29	-.19	-.84	-.96
	-.24	.35	.74	.63	.19	-.20	-.41	-.36	.23	.80	.57
1473.	-.01	-.00	.00	.02	.07	.15	.29	.57	.79	.18	-.04
	-.72	-.81	-.76	-.56	-.27	.03	.26	.25	-.19	-.89	-.95
	-.19	.34	.70	.67	.30	-.10	-.31	-.22	.42	.74	.45
1505.	.09	.04	.03	.05	.09	.15	.27	.51	.76	.20	-.04
	-.69	-.80	-.76	-.58	-.29	.01	.24	.24	-.23	-.93	-.93
	-.20	.27	.65	.72	.40	-.04	-.29	-.19	.40	.75	.48
1550.	.19	.13	.09	.07	.08	.13	.24	.49	.74	.14	-.09
	-.67	-.79	-.78	-.62	-.35	-.06	.17	.21	-.22	-.96	-.89
	-.22	.23	.63	.77	.49	.04	-.25	-.19	.40	.65	.39
1590.	.30	.19	.11	.08	.08	.12	.22	.41	.67	.14	-.11
	-.62	-.78	-.81	-.69	-.45	-.15	.11	.19	-.20	-.98	-.85
	-.25	.13	.54	.80	.62	.11	-.25	-.27	.19	.56	.37
1640.	.37	.24	.13	.07	.06	.11	.22	.45	.72	.09	-.17
	-.63	-.79	-.82	-.69	-.44	-.17	.06	.15	-.25	-.98	-.76
	-.21	.15	.53	.78	.62	.16	-.20	-.24	.26	.43	.17
1680.	.41	.28	.16	.08	.06	.09	.18	.39	.67	.13	-.18
	-.65	-.81	-.84	-.73	-.51	-.24	.01	.14	-.23	-.97	-.64
	-.21	.12	.52	.83	.68	.19	-.19	-.25	.19	.40	.15
1720.	.49	.36	.20	.08	.04	.08	.18	.39	.68	.07	-.28
	-.58	-.79	-.88	-.78	-.55	-.28	-.05	.09	-.23	-.88	-.48
	-.18	.13	.50	.84	.74	.20	-.17	-.21	.20	.19	-.09
1760.	.49	.33	.18	.08	.05	.06	.13	.30	.59	.15	-.21
	-.58	-.78	-.89	-.82	-.61	-.34	-.09	.06	-.24	-.69	-.22
	-.22	.08	.46	.84	.79	.22	-.23	-.35	.01	.24	.03
1800.	.57	.40	.23	.10	.04	.04	.10	.26	.55	.06	-.29
	-.54	-.76	-.89	-.86	-.65	-.36	-.09	.06	-.30	-.41	-.07
	-.19	.09	.46	.84	.79	.21	-.26	-.36	.07	.14	-.10

		COS THETA									POLARIZATION
K[MEV/C]		.95	.90	.80	.70	.60	.50	.40	.30	.20	.10
1840.	PI+P	.24	.29	.31	.33	.46	.74	.88	.87	.81	.73
	PI-P	-.16	-.22	-.20	.01	.29	.22	-.10	-.24	-.29	-.37
	CHEX	.63	.57	.08	-.47	.45	.50	-.30	-.53	-.52	-.40
1880.	PI+P	.24	.28	.29	.29	.44	.73	.85	.85	.81	.76
	PI-P	-.15	-.21	-.20	.03	.34	.22	-.11	-.22	-.23	-.26
	CHEX	.59	.57	.08	-.48	.61	.52	-.30	-.54	-.53	-.43
1920.	PI+P	.24	.28	.28	.27	.37	.63	.80	.83	.82	.77
	PI-P	-.13	-.19	-.18	.05	.34	.16	-.15	-.25	-.27	-.33
	CHEX	.52	.50	.08	-.34	.72	.47	-.26	-.51	-.51	-.39
1980.	PI+P	.25	.30	.29	.27	.35	.58	.74	.79	.79	.77
	PI-P	-.09	-.14	-.13	.10	.34	.06	-.23	-.30	-.31	-.35
	CHEX	.43	.48	.17	-.08	.72	.38	-.29	-.55	-.56	-.45
2030.	PI+P	.28	.32	.31	.25	.29	.51	.68	.75	.78	.77
	PI-P	-.06	-.10	-.10	.11	.29	-.05	-.29	-.32	-.32	-.34
	CHEX	.31	.37	.24	.15	.59	.29	-.22	-.52	-.56	-.44
2070.	PI+P	.29	.34	.31	.23	.25	.48	.66	.73	.76	.77
	PI-P	-.04	-.08	-.08	.12	.24	-.16	-.34	-.34	-.31	-.30
	CHEX	.23	.35	.31	.21	.53	.29	-.15	-.44	-.51	-.45
2150.	PI+P	.32	.37	.31	.19	.20	.43	.59	.66	.70	.72
	PI-P	-.02	-.05	-.06	.06	-.02	-.38	-.43	-.37	-.31	-.27
	CHEX	.04	.23	.48	.40	.36	.14	-.19	-.46	-.54	-.43
2200.	PI+P	.33	.38	.31	.17	.16	.39	.56	.65	.70	.74
	PI-P	-.02	-.04	-.06	.03	-.13	-.45	-.47	-.39	-.31	-.25
	CHEX	-.04	.22	.55	.41	.22	.06	-.18	-.41	-.51	-.43
2280.	PI+P	.35	.40	.31	.14	.13	.33	.49	.59	.68	.76
	PI-P	-.03	-.04	-.03	.04	-.28	-.58	-.54	-.42	-.29	-.19
	CHEX	-.14	.14	.63	.33	-.03	-.08	-.16	-.31	-.46	-.49
2340.	PI+P	.36	.42	.35	.16	.14	.35	.49	.56	.62	.68
	PI-P	-.04	-.04	.00	.07	-.30	-.57	-.54	-.44	-.34	-.25
	CHEX	-.13	.16	.70	.27	-.15	-.16	-.24	-.40	-.53	-.52
2460.	PI+P	.36	.43	.36	.14	.14	.34	.43	.47	.52	.60
	PI-P	-.07	-.05	.07	.10	-.47	-.70	-.62	-.47	-.31	-.18
	CHEX	-.02	.16	.70	-.11	-.37	-.30	-.28	-.32	-.45	-.58
2560.	PI+P	.36	.43	.37	.17	.22	.41	.48	.51	.53	.56
	PI-P	-.09	-.06	.10	.08	-.56	-.75	-.68	-.51	-.32	-.14
	CHEX	.10	.16	.63	-.24	-.38	-.27	-.20	-.21	-.37	-.60
2750.	PI+P	.31	.34	.28	.28	.53	.60	.57	.55	.58	.65
	PI-P	-.14	-.12	.06	-.03	-.54	-.70	-.69	-.58	-.42	-.26
	CHEX	.21	.13	.56	-.26	-.24	-.14	-.04	.09	.21	.18
3000.	PI+P	.26	.24	.17	.35	.60	.65	.65	.64	.59	.52
	PI-P	-.15	-.15	.04	-.06	-.45	-.55	-.53	-.44	-.40	-.39
	CHEX	.28	.16	.42	-.11	-.21	-.16	-.06	.12	.29	.22
3400.	PI+P	.28	.25	.10	.36	.65	.65	.61	.56	.28	-.07
	PI-P	-.10	-.07	.11	-.21	-.40	-.36	-.29	-.37	-.77	-1.00
	CHEX	.23	.25	.17	.12	.05	-.01	-.05	.16	.36	.33
3650.	PI+P	.29	.25	.09	.36	.58	.62	.65	.63	.40	.15
	PI-P	-.08	-.03	.06	-.29	-.42	-.38	-.23	-.18	-.35	-.68
	CHEX	.25	.16	.43	.23	.04	.00	.24	.24	.36	.80
4000.	PI+P	.27	.20	.11	.46	.55	.59	.71	.78	.62	.13
	PI-P	-.10	-.05	.02	-.39	-.40	-.14	.26	.54	.12	-.45
	CHEX	.20	.19	.38	.07	-.09	-.08	-.06	.58	.83	.71

	COS THETA									POLARIZATION	
K[MEV/C]	.00	-.10	-.20	-.30	-.40	-.50	-.60	-.70	-.80	-.90	-.95
1840.	.62	.45	.24	.06	-.02	-.00	.08	.25	.53	.16	-.28
	-.50	-.71	-.88	-.87	-.68	-.40	-.14	-.02	-.48	-.05	.22
	-.21	.06	.44	.85	.79	.18	-.27	-.35	.10	.17	-.06
1880.	.67	.51	.29	.11	.01	-.02	.01	.16	.42	.14	-.15
	-.36	-.55	-.77	-.82	-.67	-.43	-.18	-.00	-.44	.23	.43
	-.27	-.02	.38	.85	.81	.23	-.23	-.33	.09	.24	.09
1920.	.68	.52	.29	.08	-.03	-.06	-.01	.12	.38	.01	-.43
	-.45	-.64	-.85	-.91	-.74	-.45	-.17	-.01	-.35	.59	.63
	-.21	.04	.42	.87	.80	.14	-.31	-.35	.21	.26	.02
1980.	.70	.53	.28	.05	-.05	-.07	-.06	-.04	.03	-.24	-.51
	-.43	-.60	-.82	-.94	-.80	-.50	-.20	-.06	.57	.79	.73
	-.29	-.03	.41	.92	.77	.17	-.23	-.31	.17	.28	.09
2030.	.71	.57	.32	.05	-.13	-.20	-.22	-.18	.01	-.07	-.40
	-.40	-.55	-.79	-.95	-.82	-.47	-.14	-.04	.99	.86	.84
	-.24	.05	.50	.95	.69	.03	-.33	-.29	.44	.55	.35
2070.	.71	.56	.30	.05	-.12	-.21	-.29	-.34	-.21	-.03	-.26
	-.35	-.49	-.74	-.96	-.86	-.53	-.20	-.14	.80	.82	.79
	-.30	-.02	.49	.97	.67	.10	-.25	-.30	.40	.61	.45
2150.	.70	.56	.30	.00	-.23	-.37	-.46	-.45	-.13	.03	-.15
	-.29	-.39	-.65	-.95	-.92	-.55	-.17	-.33	.67	.77	.71
	-.21	.12	.60	.95	.55	-.02	-.29	-.16	.70	.81	.71
2200.	.72	.58	.30	-.04	-.33	-.51	-.60	-.51	-.02	.14	-.10
	-.24	-.32	-.58	-.93	-.94	-.56	-.18	-.24	.67	.74	.64
	-.21	.15	.66	.91	.39	-.14	-.35	-.13	.82	.91	.84
2280.	.75	.55	.15	-.21	-.43	-.55	-.58	-.43	.04	.14	-.05
	-.14	-.18	-.40	-.81	-.89	-.58	-.38	-.57	.52	.73	.60
	-.40	-.13	.39	.68	.23	-.15	-.27	-.05	.83	.96	.91
2340.	.69	.56	.25	-.15	-.48	-.66	-.68	-.49	.05	.32	.04
	-.18	-.15	-.23	-.57	-.76	-.46	-.10	-.38	.42	.70	.54
	-.35	-.05	.46	.66	.14	-.22	-.30	-.09	.76	.97	.88
2460.	.66	.54	.15	-.21	-.39	-.47	-.49	-.35	.13	.56	.27
	-.08	.01	.05	-.14	-.34	-.23	-.15	-.28	.39	.60	.24
	-.56	-.36	.08	.34	.08	-.15	-.23	-.14	.60	.99	.96
2560.	.60	.52	.21	-.09	-.20	-.27	-.40	-.39	.06	.38	-.19
	.01	.15	.27	.20	-.04	.07	.56	.11	.25	.21	-.32
	-.58	-.29	.03	.17	-.00	-.21	-.32	-.25	.40	.98	.79
2750.	.70	.60	.16	-.17	-.17	-.11	-.22	-.34	.01	.62	.18
	-.05	.31	.78	.73	.64	.76	.51	-.06	.18	-.26	-.97
	-.08	-.49	-.31	-.36	-.39	-.29	-.26	-.42	-.03	.93	.88
3000.	.51	.56	.24	.08	.07	-.02	-.29	-.39	-.15	.57	.27
	-.20	.51	.89	.49	.63	.92	.31	-.26	-.18	-.44	-.82
	-.36	-.72	-.98	-.92	-.72	-.46	-.37	-.47	-.28	.92	.71
3400.	.47	1.00	.91	.72	.62	.76	.53	-.04	-.86	-.18	.05
	-.87	.55	.79	.88	.98	.88	.07	.03	.14	.02	-.16
	.19	.17	-.18	-.64	-.69	.55	.63	.05	-.99	.12	.24
3650.	.19	.68	.97	.82	.77	.58	.26	.10	-.51	-.03	-.27
	-.46	-.05	.45	.82	.99	.14	-.83	-.90	-.39	.32	-.54
	.82	.82	.28	-.49	.05	.40	.24	.47	-.31	.25	.22
4000.	-.23	-.32	-.00	.24	-.21	-.43	.02	-.10	-.72	-.17	.68
	-.49	-.33	.59	.98	.51	-.78	-.49	-.18	-.97	-.05	-.13
	-.08	-.53	-.92	-.63	-.45	-.37	-.07	-.12	-.52	-.38	-.27

Table 2.3.2.2. The spin-rotation parameter R as reconstructed from the KH partial wave solution. $|A|$ follows from Eq. (A.4.1) and the sign of A from Fig. 3.2.2. THETA$-$N$=\theta_N$ (Fig. A.1.6). KH 80 was used below 0.5 GeV/c and KH 78 at lower momenta.

		COS THETA						SPIN-ROTATION PARAMETER R			
K[MEV/C]		.95	.90	.80	.70	.60	.50	.40	.30	.20	.10
20.0	PI+P	.13	.19	.27	.33	.39	.44	.48	.53	.57	.61
	PI-P	.17	.24	.34	.42	.48	.53	.58	.62	.66	.70
	CHEX	.15	.21	.30	.37	.43	.48	.53	.57	.61	.65
	THETA-N	80.90	77.08	71.56	67.21	63.43	60.00	56.78	53.72	50.76	47.86
40.0	PI+P	.04	.06	.10	.15	.19	.23	.28	.32	.37	.42
	PI-P	.21	.29	.41	.49	.56	.62	.67	.71	.75	.78
	CHEX	.13	.18	.26	.32	.38	.43	.47	.52	.56	.60
	THETA-N	80.90	77.07	71.55	67.20	63.42	59.98	56.77	53.71	50.75	47.85
60.0	PI+P	-.20	-.25	-.28	-.26	-.23	-.18	-.12	-.05	.02	.09
	PI-P	.26	.37	.51	.60	.68	.74	.79	.83	.86	.89
	CHEX	.08	.11	.17	.22	.27	.31	.36	.41	.45	.50
	THETA-N	80.89	77.06	71.54	67.18	63.40	59.96	56.75	53.69	50.73	47.83
79.0	PI+P	-.81	-.88	-.88	-.85	-.78	-.70	-.61	-.52	-.41	-.30
	PI-P	.32	.44	.60	.71	.79	.85	.90	.93	.96	.97
	CHEX	-.04	-.04	-.02	.02	.06	.11	.16	.22	.27	.33
	THETA-N	80.88	77.05	71.52	67.16	63.38	59.94	56.72	53.66	50.70	47.80
97.0	PI+P	.65	.21	-.45	-.78	-.93	-.97	-.94	-.86	-.76	-.64
	PI-P	.38	.52	.70	.81	.89	.94	.97	.99	.99	.99
	CHEX	-.24	-.29	-.30	-.27	-.22	-.15	-.09	-.01	.06	.14
	THETA-N	80.87	77.03	71.50	67.13	63.35	59.90	56.69	53.62	50.66	47.76
112.0	PI+P	.90	.91	.45	-.10	-.53	-.81	-.94	-.97	-.93	-.84
	PI-P	.41	.57	.76	.87	.94	.97	.99	.99	.97	.94
	CHEX	-.65	-.72	-.71	-.65	-.57	-.48	-.38	-.28	-.18	-.07
	THETA-N	80.86	77.02	71.48	67.11	63.32	59.87	56.66	53.59	50.63	47.73
130.0	PI+P	.77	.94	.87	.52	.06	-.37	-.70	-.89	-.97	-.95
	PI-P	.44	.61	.80	.91	.97	.99	.98	.95	.89	.82
	CHEX	-.10	-.53	-.88	-.97	-.94	-.85	-.74	-.62	-.48	-.35
	THETA-N	80.84	77.00	71.45	67.08	63.28	59.83	56.61	53.55	50.58	47.68
153.0	PI+P	.66	.86	.96	.82	.49	.07	-.34	-.67	-.88	-.96
	PI-P	.46	.63	.83	.94	.99	.98	.93	.85	.73	.58
	CHEX	.85	.89	.43	-.19	-.64	-.88	-.94	-.90	-.79	-.64
	THETA-N	80.82	76.97	71.41	67.03	63.23	59.78	56.55	53.48	50.52	47.61
172.0	PI+P	.61	.81	.96	.91	.69	.35	-.05	-.43	-.73	-.91
	PI-P	.46	.64	.85	.95	.99	.97	.89	.77	.60	.39
	CHEX	.72	.90	.86	.50	.01	-.45	-.76	-.91	-.92	-.84
	THETA-N	80.80	76.94	71.37	66.99	63.18	59.72	56.50	53.42	50.46	47.55
185.0	PI+P	.58	.77	.95	.95	.79	.50	.13	-.27	-.61	-.85
	PI-P	.46	.64	.85	.95	.99	.97	.88	.74	.55	.31
	CHEX	.66	.86	.93	.74	.35	-.10	-.51	-.79	-.92	-.91
	THETA-N	80.79	76.92	71.35	66.95	63.14	59.68	56.46	53.38	50.41	47.51
200.0	PI+P	.55	.74	.94	.97	.86	.62	.29	-.10	-.47	-.76
	PI-P	.46	.63	.84	.96	.99	.97	.87	.72	.51	.25
	CHEX	.61	.81	.95	.86	.59	.20	-.23	-.59	-.83	-.92
	THETA-N	80.77	76.90	71.31	66.91	63.10	59.64	56.41	53.33	50.36	47.46
218.0	PI+P	.52	.71	.91	.97	.90	.71	.41	.04	-.34	-.67
	PI-P	.45	.63	.85	.96	1.00	.96	.84	.64	.38	.06
	CHEX	.56	.75	.93	.92	.76	.46	.08	-.31	-.64	-.84
	THETA-N	80.75	76.87	71.27	66.87	63.05	59.58	56.34	53.27	50.29	47.39
247.0	PI+P	.49	.67	.87	.96	.94	.81	.58	.25	-.13	-.50
	PI-P	.44	.61	.83	.95	1.00	.97	.85	.64	.34	-.02
	CHEX	.50	.68	.87	.92	.85	.67	.39	.03	-.33	-.63
	THETA-N	80.71	76.82	71.20	66.78	62.95	59.48	56.24	53.16	50.18	47.27
267.0	PI+P	.46	.64	.85	.95	.95	.87	.68	.39	.03	-.35
	PI-P	.43	.60	.82	.95	1.00	.97	.86	.65	.35	-.03
	CHEX	.47	.64	.83	.91	.89	.77	.55	.24	-.11	-.45
	THETA-N	80.69	76.78	71.15	66.72	62.89	59.41	56.16	53.08	50.10	47.19

K[MEV/C]	COS THETA .00	-.10	-.20	-.30	-.40	-.50	-.60	SPIN-ROTATION -.70	PARAMETER -.80	-.90	R -.95
20.0	.65	.69	.72	.76	.80	.83	.87	.90	.93	.97	.98
	.74	.77	.80	.83	.86	.88	.91	.93	.96	.98	.99
	.69	.73	.76	.79	.82	.86	.89	.92	.94	.97	.99
	45.00	42.13	39.23	36.27	33.21	30.00	26.56	22.78	18.43	12.92	9.10
40.0	.47	.52	.57	.62	.67	.73	.78	.84	.89	.94	.97
	.82	.84	.87	.89	.91	.93	.95	.96	.98	.99	.99
	.64	.68	.72	.75	.79	.83	.86	.90	.93	.97	.98
	44.98	42.11	39.21	36.25	33.19	29.98	26.55	22.77	18.42	12.91	9.09
60.0	.16	.24	.32	.40	.48	.57	.65	.74	.83	.91	.96
	.91	.93	.95	.96	.97	.98	.99	.99	1.00	1.00	1.00
	.55	.59	.64	.68	.73	.77	.82	.86	.91	.95	.98
	44.96	42.09	39.19	36.23	33.17	29.96	26.53	22.76	18.41	12.90	9.08
79.0	-.19	-.07	.04	.16	.28	.40	.52	.64	.76	.88	.94
	.99	.99	1.00	1.00	.99	.99	.99	.99	.99	.99	.99
	.39	.45	.51	.57	.63	.70	.76	.82	.88	.94	.97
	44.93	42.06	39.16	36.20	33.14	29.94	26.51	22.73	18.39	12.89	9.07
97.0	-.50	-.36	-.21	-.06	.09	.24	.40	.55	.70	.85	.93
	.98	.97	.95	.93	.91	.90	.88	.88	.89	.93	.96
	.22	.30	.38	.46	.54	.62	.69	.77	.85	.92	.96
	44.89	42.02	39.12	36.17	33.11	29.90	26.48	22.71	18.37	12.87	9.06
112.0	-.71	-.56	-.40	-.23	-.05	.12	.30	.48	.65	.83	.91
	.91	.86	.81	.76	.71	.66	.62	.60	.61	.70	.80
	.03	.13	.23	.33	.43	.53	.63	.72	.82	.91	.95
	44.86	41.99	39.09	36.13	33.08	29.87	26.45	22.68	18.35	12.86	9.05
130.0	-.86	-.73	-.57	-.38	-.19	.01	.21	.41	.61	.81	.90
	.73	.64	.53	.41	.29	.16	.04	-.08	-.19	-.22	-.11
	-.21	-.08	.05	.18	.31	.43	.55	.67	.78	.89	.95
	44.81	41.94	39.04	36.09	33.04	29.83	26.41	22.65	18.32	12.84	9.04
153.0	-.93	-.83	-.67	-.48	-.27	-.05	.16	.38	.59	.80	.90
	.40	.20	-.01	-.23	-.45	-.66	-.85	-.97	-.98	-.67	-.19
	-.48	-.32	-.15	.01	.17	.32	.47	.61	.74	.87	.94
	44.74	41.87	38.98	36.03	32.97	29.78	26.36	22.60	18.28	12.81	9.02
172.0	-.96	-.90	-.77	-.58	-.37	-.14	.10	.33	.56	.78	.89
	.15	-.10	-.36	-.61	-.81	-.96	-1.00	-.90	-.60	-.01	.43
	-.70	-.54	-.35	-.17	.02	.20	.37	.54	.70	.85	.93
	44.68	41.81	38.92	35.97	32.92	29.72	26.31	22.56	18.24	12.78	9.00
185.0	-.95	-.94	-.83	-.66	-.44	-.20	.05	.29	.54	.77	.89
	.05	-.23	-.50	-.74	-.91	-.99	-.95	-.75	-.37	.20	.57
	-.82	-.66	-.48	-.28	-.08	.12	.31	.50	.67	.84	.92
	44.64	41.77	38.88	35.93	32.88	29.69	26.28	22.53	18.22	12.76	8.98
200.0	-.93	-.96	-.88	-.72	-.50	-.25	.00	.26	.52	.76	.88
	-.04	-.34	-.61	-.83	-.96	-.97	-.84	-.57	-.16	.37	.68
	-.89	-.76	-.59	-.39	-.17	.04	.25	.45	.64	.83	.91
	44.58	41.72	38.82	35.88	32.83	29.64	26.23	22.49	18.19	12.74	8.97
218.0	-.88	-.96	-.90	-.75	-.53	-.28	-.02	.25	.51	.76	.88
	-.27	-.59	-.83	-.97	-.97	-.84	-.59	-.26	.13	.56	.78
	-.90	-.84	-.70	-.50	-.28	-.05	.18	.40	.61	.81	.91
	44.52	41.65	38.76	35.81	32.77	29.58	26.18	22.44	18.15	12.71	8.95
247.0	-.78	-.92	-.92	-.79	-.58	-.33	-.05	.23	.50	.76	.88
	-.40	-.72	-.93	-.98	-.88	-.67	-.37	-.03	.32	.67	.83
	-.81	-.85	-.78	-.61	-.39	-.15	.10	.34	.57	.79	.90
	44.40	41.54	38.65	35.70	32.67	29.49	26.09	22.36	18.08	12.66	8.91
267.0	-.68	-.88	-.93	-.83	-.63	-.36	-.08	.21	.49	.75	.88
	-.43	-.76	-.95	-.96	-.82	-.58	-.27	.06	.39	.71	.86
	-.70	-.82	-.80	-.67	-.46	-.21	.04	.30	.55	.78	.89
	44.32	41.45	38.57	35.62	32.59	29.41	26.02	22.30	18.03	12.63	8.89

		COS THETA									SPIN-ROTATION PARAMETER R
K[MEV/C]		.95	.90	.80	.70	.60	.50	.40	.30	.20	.10
280.0	PI+P	.45	.63	.83	.94	.96	.89	.73	.46	.11	-.28
	PI-P	.43	.60	.82	.94	1.00	.98	.87	.67	.37	-.01
	CHEX	.45	.62	.81	.90	.90	.80	.61	.33	-.01	-.36
	THETA-N	80.67	76.75	71.12	66.68	62.84	59.36	56.11	53.02	50.05	47.14
290.0	PI+P	.44	.61	.82	.93	.96	.91	.76	.51	.17	-.22
	PI-P	.43	.60	.82	.94	1.00	.97	.87	.66	.36	-.03
	CHEX	.43	.59	.79	.89	.90	.83	.66	.41	.08	-.28
	THETA-N	80.66	76.74	71.09	66.65	62.81	59.32	56.07	52.98	50.00	47.09
295.0	PI+P	.44	.61	.82	.93	.96	.92	.78	.54	.20	-.19
	PI-P	.43	.60	.82	.94	.99	.97	.87	.66	.35	-.04
	CHEX	.42	.58	.78	.88	.90	.84	.68	.44	.12	-.24
	THETA-N	80.65	76.73	71.08	66.64	62.79	59.30	56.05	52.96	49.98	47.07
301.0	PI+P	.43	.60	.81	.92	.96	.92	.79	.57	.24	-.15
	PI-P	.44	.61	.82	.94	.99	.97	.87	.66	.35	-.04
	CHEX	.41	.57	.76	.87	.89	.85	.71	.48	.17	-.18
	THETA-N	80.64	76.71	71.06	66.62	62.77	59.28	56.03	52.94	49.96	47.05
305.0	PI+P	.43	.59	.80	.92	.96	.93	.81	.59	.27	-.12
	PI-P	.44	.61	.82	.94	.99	.97	.86	.66	.35	-.04
	CHEX	.40	.56	.75	.86	.89	.85	.73	.51	.21	-.15
	THETA-N	80.63	76.71	71.05	66.60	62.75	59.26	56.01	52.92	49.94	47.03
310.0	PI+P	.42	.59	.79	.91	.96	.93	.82	.61	.30	-.08
	PI-P	.44	.61	.82	.94	.99	.97	.86	.66	.35	-.04
	CHEX	.39	.55	.74	.85	.89	.85	.74	.54	.25	-.10
	THETA-N	80.63	76.70	71.04	66.59	62.73	59.24	55.99	52.90	49.92	47.00
320.0	PI+P	.41	.58	.78	.90	.95	.94	.84	.66	.37	-.01
	PI-P	.44	.61	.82	.94	.99	.97	.86	.66	.36	-.03
	CHEX	.37	.52	.71	.82	.87	.86	.77	.59	.32	-.02
	THETA-N	80.61	76.68	71.01	66.55	62.70	59.20	55.95	52.85	49.87	46.96
331.0	PI+P	.41	.57	.77	.90	.96	.95	.87	.69	.42	.05
	PI-P	.45	.62	.82	.94	.98	.95	.85	.65	.35	-.03
	CHEX	.36	.50	.69	.81	.87	.86	.79	.63	.37	.04
	THETA-N	80.60	76.65	70.98	66.52	62.66	59.16	55.90	52.81	49.82	46.91
351.0	PI+P	.39	.55	.75	.87	.94	.96	.90	.77	.54	.20
	PI-P	.45	.62	.82	.92	.96	.92	.82	.63	.35	-.01
	CHEX	.32	.46	.63	.75	.82	.84	.80	.68	.48	.18
	THETA-N	80.57	76.61	70.92	66.45	62.58	59.08	55.81	52.72	49.73	46.82
378.0	PI+P	.37	.52	.71	.85	.93	.96	.94	.84	.65	.35
	PI-P	.44	.60	.78	.86	.88	.84	.73	.56	.30	-.02
	CHEX	.27	.38	.54	.66	.74	.78	.78	.70	.55	.29
	THETA-N	80.52	76.55	70.84	66.35	62.47	58.96	55.69	52.59	49.60	46.69
408.0	PI+P	.35	.49	.69	.82	.91	.96	.96	.89	.74	.48
	PI-P	.42	.55	.68	.73	.71	.65	.54	.39	.18	-.07
	CHEX	.22	.31	.44	.55	.63	.68	.70	.67	.56	.34
	THETA-N	80.48	76.49	70.75	66.25	62.36	58.83	55.56	52.45	49.46	46.54
427.0	PI+P	.34	.48	.67	.81	.90	.96	.97	.92	.79	.55
	PI-P	.31	.42	.53	.57	.57	.52	.44	.31	.15	-.06
	CHEX	.17	.25	.36	.45	.52	.56	.57	.53	.42	.22
	THETA-N	80.45	76.44	70.69	66.18	62.28	58.75	55.47	52.36	49.37	46.45
456.0	PI+P	.32	.45	.64	.77	.87	.94	.97	.96	.87	.70
	PI-P	.10	.15	.21	.25	.26	.25	.20	.12	-.00	-.17
	CHEX	.14	.20	.29	.36	.41	.44	.45	.42	.33	.18
	THETA-N	80.40	76.38	70.60	66.07	62.16	58.62	55.33	52.22	49.22	46.30
490.0	PI+P	.28	.40	.58	.71	.82	.91	.96	.98	.95	.84
	PI-P	-.10	-.11	-.09	-.07	-.05	-.05	-.07	-.12	-.21	-.33
	CHEX	.03	.05	.09	.12	.16	.18	.19	.17	.12	.03
	THETA-N	80.34	76.30	70.49	65.94	62.02	58.47	55.17	52.05	49.05	46.13

K[MEV/C]	COS THETA							SPIN-ROTATION PARAMETER R			
	.00	-.10	-.20	-.30	-.40	-.50	-.60	-.70	-.80	-.90	-.95
280.0	-.64	-.87	-.95	-.86	-.66	-.39	-.10	.20	.48	.75	.88
	-.41	-.74	-.93	-.93	-.77	-.51	-.20	.12	.44	.73	.87
	-.65	-.81	-.82	-.71	-.50	-.26	.01	.28	.53	.77	.89
	44.26	41.40	38.51	35.57	32.54	29.37	25.98	22.27	18.00	12.60	8.87
290.0	-.59	-.85	-.95	-.87	-.68	-.41	-.11	.20	.48	.75	.88
	-.43	-.77	-.94	-.92	-.74	-.47	-.16	.16	.46	.74	.88
	-.59	-.78	-.82	-.73	-.53	-.28	-.01	.26	.52	.77	.89
	44.22	41.36	38.47	35.53	32.50	29.33	25.95	22.24	17.97	12.59	8.86
295.0	-.57	-.84	-.95	-.88	-.69	-.41	-.11	.19	.48	.75	.88
	-.45	-.78	-.94	-.92	-.73	-.46	-.14	.18	.48	.75	.88
	-.56	-.77	-.83	-.74	-.55	-.30	-.03	.25	.52	.77	.89
	44.20	41.33	38.45	35.51	32.48	29.31	25.93	22.22	17.96	12.58	8.85
301.0	-.53	-.82	-.94	-.89	-.70	-.42	-.12	.19	.48	.75	.88
	-.45	-.79	-.95	-.92	-.72	-.44	-.12	.19	.49	.76	.88
	-.52	-.75	-.83	-.75	-.57	-.32	-.04	.24	.51	.77	.89
	44.17	41.31	38.42	35.49	32.46	29.29	25.91	22.20	17.94	12.57	8.84
305.0	-.51	-.81	-.94	-.89	-.70	-.43	-.12	.19	.48	.75	.88
	-.45	-.79	-.95	-.92	-.72	-.43	-.12	.20	.49	.76	.88
	-.49	-.73	-.82	-.76	-.58	-.33	-.05	.23	.51	.76	.88
	44.15	41.29	38.41	35.47	32.44	29.27	25.89	22.19	17.93	12.56	8.84
310.0	-.48	-.79	-.94	-.90	-.71	-.44	-.13	.18	.48	.75	.88
	-.46	-.79	-.96	-.92	-.72	-.43	-.11	.21	.50	.76	.88
	-.45	-.71	-.83	-.78	-.60	-.35	-.06	.22	.50	.76	.88
	44.13	41.27	38.38	35.45	32.42	29.25	25.88	22.17	17.92	12.55	8.83
320.0	-.42	-.76	-.94	-.91	-.73	-.45	-.13	.18	.48	.75	.88
	-.45	-.79	-.96	-.92	-.71	-.41	-.09	.23	.51	.77	.89
	-.38	-.68	-.82	-.80	-.63	-.38	-.08	.21	.50	.76	.88
	44.09	41.22	38.34	35.40	32.38	29.22	25.84	22.14	17.89	12.53	8.82
331.0	-.36	-.72	-.92	-.92	-.75	-.47	-.15	.18	.48	.75	.88
	-.45	-.79	-.96	-.92	-.71	-.42	-.09	.22	.51	.77	.89
	-.32	-.63	-.80	-.80	-.64	-.40	-.10	.20	.49	.76	.88
	44.04	41.17	38.29	35.36	32.33	29.17	25.80	22.11	17.86	12.51	8.80
351.0	-.22	-.63	-.90	-.95	-.79	-.50	-.17	.17	.47	.75	.88
	-.42	-.77	-.97	-.95	-.75	-.46	-.13	.20	.49	.76	.88
	-.19	-.55	-.79	-.84	-.71	-.45	-.14	.17	.48	.75	.88
	43.94	41.08	38.20	35.27	32.25	29.09	25.73	22.04	17.81	12.47	8.77
378.0	-.05	-.49	-.83	-.96	-.83	-.54	-.19	.17	.49	.76	.89
	-.39	-.74	-.96	-.97	-.79	-.49	-.14	.20	.50	.77	.89
	-.05	-.42	-.73	-.85	-.77	-.53	-.21	.13	.46	.75	.88
	43.81	40.95	38.07	35.15	32.13	28.98	25.63	21.95	17.73	12.41	8.73
408.0	.10	-.36	-.77	-.97	-.87	-.57	-.19	.18	.50	.77	.89
	-.36	-.66	-.89	-1.00	-.91	-.66	-.31	.07	.42	.73	.87
	.02	-.39	-.76	-.94	-.85	-.58	-.22	.14	.46	.75	.88
	43.67	40.81	37.93	35.01	32.00	28.86	25.52	21.85	17.65	12.35	8.69
427.0	.18	-.28	-.73	-.97	-.90	-.61	-.22	.16	.49	.77	.89
	-.31	-.57	-.80	-.94	-.93	-.74	-.42	-.04	.34	.69	.85
	-.10	-.48	-.82	-.99	-.93	-.67	-.32	.04	.39	.71	.86
	43.58	40.72	37.84	34.92	31.92	28.78	25.44	21.79	17.60	12.31	8.66
456.0	.40	-.04	-.55	-.93	-.96	-.65	-.22	.20	.54	.80	.91
	-.37	-.60	-.81	-.93	-.88	-.65	-.30	.09	.44	.75	.88
	-.05	-.34	-.65	-.90	-.99	-.87	-.58	-.19	.23	.64	.82
	43.43	40.57	37.70	34.79	31.79	28.66	25.33	21.69	17.51	12.25	8.62
490.0	.63	.25	-.28	-.81	-.99	-.71	-.22	.24	.58	.83	.92
	-.48	-.65	-.79	-.85	-.75	-.50	-.15	.20	.52	.78	.90
	-.10	-.26	-.45	-.62	-.77	-.85	-.84	-.69	-.36	.19	.56
	43.26	40.40	37.54	34.62	31.63	28.51	25.20	21.57	17.41	12.18	8.57

K[MEV/C]		COS THETA .95	.90	.80	.70	.60	.50	SPIN-ROTATION PARAMETER R .40	.30	.20	.10
532.	PI+P	.29	.40	.56	.67	.77	.85	.92	.96	.98	.96
	PI-P	-.27	-.33	-.36	-.35	-.33	-.30	-.30	-.31	-.35	-.42
	CHEX	-.02	-.02	-.02	-.01	.01	.02	.03	.03	.01	-.04
	THETA-N	80.27	76.20	70.36	65.78	61.84	58.28	54.97	51.84	48.84	45.92
573.	PI+P	.24	.34	.49	.61	.71	.81	.88	.94	.98	.99
	PI-P	-.34	-.43	-.50	-.51	-.49	-.47	-.45	-.46	-.49	-.55
	CHEX	-.10	-.13	-.16	-.17	-.16	-.14	-.12	-.10	-.09	-.09
	THETA-N	80.20	76.10	70.22	65.62	61.66	58.09	54.77	51.64	48.63	45.70
614.	PI+P	.17	.26	.41	.53	.64	.73	.81	.88	.94	.98
	PI-P	-.40	-.51	-.57	-.56	-.54	-.53	-.52	-.52	-.54	-.58
	CHEX	-.22	-.29	-.34	-.34	-.30	-.26	-.21	-.17	-.13	-.09
	THETA-N	80.13	76.00	70.09	65.46	61.48	57.89	54.57	51.43	48.42	45.49
658.	PI+P	.11	.18	.30	.42	.54	.64	.73	.81	.87	.92
	PI-P	-.45	-.60	-.71	-.70	-.65	-.60	-.55	-.50	-.47	-.47
	CHEX	-.28	-.38	-.50	-.52	-.48	-.39	-.28	-.14	.01	.15
	THETA-N	80.05	75.89	69.94	65.29	61.29	57.69	54.35	51.20	48.19	45.26
675.	PI+P	.09	.14	.26	.37	.49	.59	.68	.75	.82	.87
	PI-P	-.44	-.59	-.73	-.75	-.72	-.68	-.63	-.57	-.51	-.47
	CHEX	-.31	-.43	-.55	-.57	-.52	-.42	-.28	-.11	.06	.23
	THETA-N	80.02	75.85	69.88	65.22	61.21	57.61	54.27	51.12	48.10	45.17
705.	PI+P	.03	.05	.13	.24	.37	.50	.60	.69	.74	.77
	PI-P	-.41	-.56	-.72	-.79	-.80	-.79	-.75	-.70	-.62	-.52
	CHEX	-.37	-.51	-.65	-.66	-.59	-.45	-.28	-.09	.12	.33
	THETA-N	79.97	75.77	69.78	65.10	61.08	57.46	54.12	50.96	47.94	45.01
725.	PI+P	.01	.02	.08	.20	.33	.45	.55	.62	.67	.69
	PI-P	-.38	-.53	-.70	-.78	-.81	-.82	-.80	-.75	-.67	-.56
	CHEX	-.41	-.56	-.70	-.70	-.62	-.47	-.29	-.08	.15	.38
	THETA-N	79.93	75.72	69.71	65.02	60.99	57.37	54.02	50.86	47.84	44.91
750.	PI+P	.00	.02	.10	.20	.31	.41	.50	.56	.59	.60
	PI-P	-.38	-.52	-.67	-.76	-.81	-.84	-.85	-.82	-.76	-.65
	CHEX	-.46	-.63	-.77	-.75	-.63	-.43	-.20	.04	.28	.52
	THETA-N	79.88	75.66	69.63	64.92	60.88	57.25	53.89	50.73	47.71	44.78
777.	PI+P	-.01	.01	.09	.18	.27	.34	.41	.46	.49	.49
	PI-P	-.39	-.52	-.67	-.75	-.82	-.88	-.91	-.90	-.84	-.72
	CHEX	-.50	-.69	-.82	-.75	-.55	-.30	-.04	.19	.40	.59
	THETA-N	79.84	75.59	69.53	64.81	60.76	57.12	53.76	50.59	47.57	44.64
800.	PI+P	-.02	-.01	.06	.12	.19	.26	.32	.37	.40	.40
	PI-P	-.40	-.54	-.68	-.77	-.85	-.91	-.94	-.94	-.89	-.77
	CHEX	-.52	-.70	-.80	-.69	-.47	-.21	.03	.24	.43	.63
	THETA-N	79.79	75.54	69.46	64.72	60.66	57.01	53.65	50.48	47.45	44.52
822.	PI+P	-.06	-.07	-.03	.06	.14	.20	.25	.29	.31	.30
	PI-P	-.40	-.55	-.70	-.78	-.86	-.92	-.96	-.97	-.94	-.82
	CHEX	-.53	-.72	-.77	-.58	-.32	-.10	.09	.28	.49	.68
	THETA-N	79.75	75.48	69.38	64.63	60.56	56.91	53.54	50.36	47.33	44.40
851.	PI+P	-.08	-.11	-.09	-.04	.02	.08	.12	.15	.16	.16
	PI-P	-.38	-.54	-.70	-.78	-.85	-.91	-.96	-.99	-.97	-.85
	CHEX	-.51	-.68	-.73	-.53	-.22	.03	.18	.30	.43	.60
	THETA-N	79.70	75.41	69.28	64.51	60.43	56.77	53.39	50.22	47.18	44.25
875.	PI+P	-.09	-.12	-.12	-.08	-.04	-.00	.03	.05	.07	.10
	PI-P	-.35	-.50	-.67	-.76	-.82	-.90	-.96	-1.00	-.98	-.86
	CHEX	-.49	-.66	-.73	-.51	-.10	.21	.32	.36	.43	.55
	THETA-N	79.66	75.35	69.20	64.42	60.32	56.66	53.27	50.10	47.06	44.13
895.	PI+P	-.10	-.13	-.14	-.12	-.09	-.06	-.03	.01	.06	.14
	PI-P	-.33	-.48	-.65	-.74	-.81	-.88	-.95	-.99	-.99	-.84
	CHEX	-.50	-.67	-.74	-.46	.16	.48	.39	.28	.27	.38
	THETA-N	79.62	75.30	69.13	64.34	60.24	56.56	53.18	49.99	46.96	44.03

K[MEV/C]	COS THETA .00	-.10	-.20	-.30	-.40	-.50	-.60	SPIN-ROTATION PARAMETER R -.70	-.80	-.90	-.95
532.	.85	.60	.14	-.52	-.98	-.72	-.10	.41	.73	.91	.96
	-.52	-.64	-.74	-.72	-.53	-.19	.18	.49	.72	.88	.95
	-.11	-.20	-.31	-.42	-.53	-.64	-.73	-.82	-.89	-.93	-.89
	43.04	40.19	37.33	34.42	31.44	28.33	25.03	21.42	17.29	12.09	8.51
573.	.95	.83	.52	-.12	-.80	-.44	.27	.65	.84	.93	.96
	-.64	-.75	-.80	-.67	-.31	.15	.52	.74	.87	.95	.98
	-.11	-.15	-.21	-.26	-.33	-.39	-.45	-.47	-.39	-.02	.37
	42.83	39.98	37.12	34.23	31.25	28.16	24.87	21.28	17.17	12.01	8.44
614.	1.00	.98	.89	.61	.09	.26	.64	.79	.84	.87	.90
	-.67	-.78	-.82	-.56	.04	.57	.83	.93	.97	.98	.99
	-.07	-.05	-.03	-.01	.04	.11	.21	.34	.48	.70	.84
	42.62	39.77	36.92	34.03	31.06	27.98	24.71	21.13	17.05	11.92	8.38
658.	.96	.98	1.00	.99	.91	.77	.75	.74	.73	.72	.76
	-.50	-.57	-.54	.02	.72	.97	1.00	.98	.97	.97	.98
	.27	.36	.47	.59	.74	.86	.89	.83	.78	.83	.90
	42.39	39.54	36.70	33.81	30.86	27.79	24.53	20.98	16.92	11.83	8.32
675.	.91	.94	.96	.97	.88	.74	.67	.63	.58	.57	.63
	-.45	-.46	-.36	.37	.94	.99	.95	.93	.92	.94	.97
	.38	.52	.67	.82	.93	.89	.75	.63	.60	.72	.84
	42.30	39.46	36.61	33.73	30.78	27.72	24.46	20.92	16.87	11.79	8.29
705.	.78	.77	.72	.63	.49	.39	.31	.19	.05	-.01	.11
	-.43	-.33	-.05	.76	.94	.84	.78	.77	.79	.86	.92
	.53	.72	.87	.88	.71	.46	.29	.22	.27	.49	.70
	42.14	39.30	36.46	33.59	30.64	27.59	24.35	20.81	16.79	11.73	8.25
725.	.68	.65	.58	.45	.30	.22	.10	-.13	-.38	-.48	-.32
	-.42	-.21	.26	.85	.70	.56	.52	.52	.58	.74	.86
	.61	.79	.86	.71	.30	-.08	-.23	-.22	-.10	.20	.50
	42.04	39.20	36.36	33.49	30.55	27.50	24.27	20.75	16.73	11.69	8.22
750.	.57	.48	.34	.15	.06	.04	-.05	-.26	-.52	-.68	-.64
	-.47	-.13	.46	.59	.35	.21	.15	.15	.24	.48	.70
	.73	.85	.75	.34	-.20	-.56	-.68	-.66	-.54	-.24	.13
	41.91	39.07	36.24	33.37	30.44	27.39	24.17	20.66	16.66	11.64	8.18
777.	.46	.34	.13	-.09	-.21	-.28	-.36	-.47	-.59	-.68	-.65
	-.50	-.02	.53	.45	.18	-.03	-.20	-.32	-.31	.04	.43
	.77	.84	.57	-.00	-.49	-.76	-.88	-.91	-.85	-.58	-.19
	41.77	38.93	36.10	33.24	30.32	27.28	24.06	20.57	16.58	11.58	8.14
800.	.34	.21	.04	-.07	-.13	-.21	-.30	-.40	-.54	-.74	-.87
	-.46	.13	.49	.34	.08	-.16	-.39	-.59	-.69	-.47	-.02
	.79	.78	.42	-.14	-.57	-.81	-.94	-.98	-.94	-.77	-.49
	41.65	38.82	35.99	33.13	30.21	27.18	23.98	20.49	16.51	11.53	8.11
822.	.25	.14	.12	.26	.22	-.00	-.27	-.46	-.53	-.50	-.57
	-.48	.18	.51	.34	.05	-.23	-.49	-.71	-.86	-.81	-.41
	.82	.79	.37	-.30	-.73	-.89	-.96	-.99	-.99	-.86	-.60
	41.54	38.71	35.88	33.03	30.11	27.09	23.89	20.41	16.45	11.49	8.07
851.	.17	.26	.46	.57	.53	.41	.24	.03	-.23	-.48	-.52
	-.44	.21	.46	.30	.05	-.21	-.45	-.67	-.85	-.95	-.77
	.80	.81	.31	-.37	-.76	-.92	-.98	-.99	-.99	-.96	-.83
	41.39	38.56	35.74	32.89	29.98	26.97	23.78	20.31	16.37	11.43	8.03
875.	.17	.40	.71	.83	.77	.60	.38	.18	.06	.04	.12
	-.45	.23	.51	.35	.09	-.18	-.43	-.64	-.80	-.93	-.89
	.73	.86	.46	-.32	-.75	-.90	-.96	-.98	-.99	-.95	-.81
	41.27	38.44	35.62	32.78	29.87	26.87	23.69	20.23	16.30	11.38	8.00
895.	.27	.51	.74	.84	.81	.71	.56	.39	.24	.24	.40
	-.35	.38	.63	.43	.13	-.16	-.40	-.61	-.78	-.91	-.88
	.63	.92	.56	-.32	-.76	-.91	-.96	-.98	-.98	-.95	-.83
	41.16	38.34	35.52	32.68	29.79	26.78	23.61	20.17	16.25	11.34	7.97

K[MEV/C]		COS THETA .95	.90	.80	.70	.60	.50	SPIN-ROTATION PARAMETER R .40	.30	.20	.10
923.	PI+P	-.11	-.14	-.17	-.16	-.13	-.09	-.04	.02	.08	.17
	PI-P	-.32	-.45	-.62	-.71	-.79	-.86	-.94	-.99	-.97	-.73
	CHEX	-.52	-.71	-.80	-.45	.56	.82	.44	.21	.14	.22
	THETA-N	79.57	75.23	69.04	64.23	60.11	56.43	53.04	49.85	46.82	43.88
954.	PI+P	-.12	-.17	-.20	-.19	-.15	-.11	-.05	.02	.09	.19
	PI-P	-.32	-.45	-.61	-.69	-.76	-.86	-.95	-1.00	-.90	-.49
	CHEX	-.55	-.76	-.88	-.47	.75	.83	.36	.11	.02	.05
	THETA-N	79.51	75.15	68.93	64.10	59.98	56.28	52.89	49.70	46.66	43.73
975.	PI+P	-.14	-.18	-.23	-.23	-.20	-.14	-.05	.04	.13	.24
	PI-P	-.33	-.46	-.62	-.70	-.77	-.87	-.98	-.97	-.76	-.25
	CHEX	-.57	-.79	-.93	-.53	.63	.72	.32	.06	-.07	-.10
	THETA-N	79.48	75.10	68.86	64.02	59.88	56.19	52.78	49.59	46.55	43.62
1000.	PI+P	-.15	-.21	-.26	-.26	-.23	-.17	-.08	.01	.11	.24
	PI-P	-.34	-.47	-.63	-.72	-.79	-.88	-.96	-.81	-.45	.03
	CHEX	-.58	-.80	-.96	-.46	.61	.52	.20	-.00	-.14	-.21
	THETA-N	79.43	75.03	68.77	63.92	59.77	56.07	52.66	49.47	46.43	43.49
1030.	PI+P	-.15	-.22	-.29	-.30	-.26	-.20	-.12	-.04	.04	.16
	PI-P	-.35	-.48	-.64	-.71	-.75	-.79	-.65	-.32	-.07	.26
	CHEX	-.60	-.83	-.99	-.55	.38	.32	.06	-.14	-.30	-.42
	THETA-N	79.38	74.96	68.67	63.80	59.64	55.93	52.52	49.32	46.28	43.34
1055.	PI+P	-.13	-.21	-.29	-.31	-.28	-.23	-.16	-.08	.00	.12
	PI-P	-.35	-.48	-.62	-.69	-.72	-.63	.17	.34	.23	.39
	CHEX	-.61	-.84	-.99	-.49	.24	.15	-.06	-.24	-.40	-.53
	THETA-N	79.33	74.90	68.59	63.70	59.53	55.81	52.40	49.19	46.15	43.22
1080.	PI+P	-.12	-.21	-.32	-.35	-.32	-.26	-.18	-.10	-.02	.08
	PI-P	-.33	-.46	-.60	-.67	-.67	-.49	.67	.68	.32	.35
	CHEX	-.58	-.81	-1.00	-.55	.08	.02	-.15	-.31	-.49	-.66
	THETA-N	79.29	74.83	68.50	63.60	59.42	55.70	52.28	49.07	46.03	43.09
1113.	PI+P	-.09	-.19	-.33	-.38	-.36	-.31	-.23	-.14	-.05	.06
	PI-P	-.31	-.42	-.56	-.62	-.61	-.39	.47	.68	.32	.28
	CHEX	-.50	-.74	-.99	-.62	-.14	-.14	-.26	-.41	-.59	-.77
	THETA-N	79.23	74.75	68.39	63.47	59.28	55.55	52.12	48.91	45.86	42.93
1154.	PI+P	-.07	-.18	-.36	-.43	-.42	-.35	-.25	-.14	-.04	.04
	PI-P	-.27	-.39	-.52	-.57	-.54	-.32	.25	.49	.24	.15
	CHEX	-.34	-.56	-.92	-.81	-.40	-.29	-.35	-.49	-.66	-.85
	THETA-N	79.15	74.65	68.25	63.31	59.10	55.36	51.92	48.71	45.66	42.73
1174.	PI+P	-.05	-.18	-.38	-.45	-.42	-.35	-.25	-.16	-.07	.02
	PI-P	-.27	-.38	-.51	-.56	-.52	-.32	.15	.32	.10	.06
	CHEX	-.26	-.46	-.87	-.87	-.48	-.35	-.41	-.54	-.70	-.87
	THETA-N	79.12	74.60	68.19	63.23	59.02	55.27	51.83	48.61	45.57	42.63
1210.	PI+P	-.01	-.12	-.35	-.48	-.48	-.39	-.27	-.14	-.05	-.01
	PI-P	-.27	-.36	-.47	-.52	-.50	-.35	.00	.20	.05	.00
	CHEX	-.10	-.27	-.73	-.96	-.68	-.46	-.43	-.52	-.67	-.86
	THETA-N	79.05	74.51	68.06	63.09	58.87	55.10	51.66	48.44	45.39	42.46
1235.	PI+P	.02	-.08	-.32	-.49	-.52	-.43	-.29	-.15	-.08	-.09
	PI-P	-.27	-.35	-.45	-.50	-.50	-.39	-.08	.08	-.04	-.06
	CHEX	-.04	-.15	-.58	-.97	-.79	-.52	-.44	-.49	-.63	-.83
	THETA-N	79.01	74.45	67.98	62.99	58.76	54.99	51.54	48.32	45.27	42.34
1280.	PI+P	.10	.03	-.23	-.46	-.54	-.48	-.33	-.20	-.17	-.28
	PI-P	-.26	-.34	-.42	-.48	-.52	-.48	-.26	-.13	-.25	-.24
	CHEX	.19	.08	-.39	-.92	-.86	-.54	-.40	-.45	-.60	-.81
	THETA-N	78.93	74.34	67.83	62.82	58.57	54.79	51.33	48.11	45.06	42.13
1324.	PI+P	.15	.11	-.12	-.40	-.57	-.54	-.39	-.28	-.38	-.57
	PI-P	-.24	-.32	-.41	-.48	-.54	-.54	-.39	-.32	-.41	-.38
	CHEX	.40	.22	-.25	-.82	-.92	-.51	-.30	-.35	-.52	-.77
	THETA-N	78.85	74.23	67.69	62.65	58.38	54.59	51.13	47.90	44.85	41.92

K[MEV/C]	COS THETA .00	-.10	-.20	-.30	-.40	-.50	-.60	SPIN-ROTATION PARAMETER R -.70	-.80	-.90	-.95
923.	.32	.56	.77	.86	.84	.76	.64	.52	.47	.57	.73
	-.08	.63	.80	.54	.18	-.15	-.41	-.62	-.78	-.90	-.86
	.45	.88	.78	-.24	-.74	-.90	-.95	-.97	-.97	-.91	-.76
	41.02	38.20	35.39	32.55	29.66	26.67	23.51	20.08	16.17	11.29	7.93
954.	.34	.57	.78	.90	.92	.87	.75	.61	.51	.60	.77
	.25	.85	.91	.62	.23	-.12	-.41	-.63	-.82	-.92	-.83
	.23	.64	.87	-.13	-.71	-.89	-.95	-.98	-.97	-.89	-.71
	40.87	38.05	35.24	32.41	29.53	26.54	23.39	19.97	16.09	11.23	7.89
975.	.40	.61	.79	.90	.93	.89	.77	.60	.48	.63	.83
	.47	.93	.94	.65	.26	-.11	-.41	-.65	-.84	-.92	-.82
	.03	.45	.80	-.05	-.65	-.87	-.96	-.99	-.98	-.86	-.63
	40.76	37.94	35.14	32.31	29.44	26.46	23.32	19.91	16.03	11.19	7.86
1000.	.43	.65	.82	.91	.93	.90	.81	.68	.57	.65	.80
	.54	.92	.97	.69	.27	-.12	-.43	-.67	-.86	-.95	-.84
	-.19	.07	.65	-.03	-.66	-.88	-.96	-1.00	-.97	-.80	-.53
	40.64	37.82	35.02	32.20	29.33	26.36	23.22	19.82	15.96	11.14	7.82
1030.	.35	.59	.79	.92	.96	.94	.85	.69	.54	.61	.78
	.64	.91	.93	.69	.30	-.09	-.42	-.68	-.88	-.96	-.83
	-.45	-.23	.45	.07	-.58	-.84	-.95	-1.00	-.97	-.72	-.33
	40.49	37.67	34.88	32.06	29.20	26.24	23.11	19.73	15.88	11.08	7.78
1055.	.30	.54	.75	.89	.96	.97	.91	.72	.50	.59	.77
	.63	.84	.88	.67	.30	-.08	-.40	-.67	-.89	-.97	-.85
	-.59	-.42	.27	.22	-.43	-.75	-.91	-.99	-.95	-.59	-.14
	40.36	37.55	34.76	31.95	29.09	26.14	23.02	19.65	15.82	11.03	7.75
1080.	.27	.50	.70	.83	.92	.95	.90	.71	.47	.55	.74
	.55	.76	.83	.66	.32	-.07	-.43	-.70	-.90	-.97	-.86
	-.76	-.53	.32	.50	-.22	-.65	-.87	-.97	-.93	-.48	.04
	40.24	37.43	34.64	31.84	28.99	26.04	22.94	19.57	15.75	10.99	7.72
1113.	.22	.43	.63	.77	.88	.94	.91	.70	.42	.54	.76
	.41	.60	.71	.62	.32	-.08	-.44	-.72	-.91	-.98	-.90
	-.87	-.62	.37	.78	.17	-.36	-.70	-.89	-.86	-.30	.28
	40.08	37.28	34.49	31.69	28.85	25.91	22.82	19.47	15.67	10.93	7.67
1154.	.16	.34	.53	.69	.83	.92	.91	.72	.39	.49	.71
	.22	.37	.51	.51	.28	-.09	-.46	-.73	-.91	-.98	-.92
	-.96	-.62	.49	.94	.52	.04	-.36	-.68	-.70	.02	.54
	39.88	37.08	34.30	31.51	28.68	25.75	22.68	19.34	15.56	10.85	7.62
1174.	.14	.32	.51	.68	.82	.90	.89	.74	.42	.43	.64
	.17	.33	.47	.45	.22	-.14	-.48	-.73	-.89	-.96	-.89
	-.98	-.59	.66	.96	.64	.25	-.16	-.53	-.60	.15	.66
	39.79	36.99	34.21	31.43	28.60	25.68	22.61	19.28	15.51	10.82	7.60
1210.	.05	.19	.38	.58	.75	.86	.87	.68	.35	.45	.69
	.07	.19	.29	.30	.14	-.16	-.48	-.73	-.89	-.93	-.79
	-.99	-.53	.69	.99	.78	.49	.15	-.23	-.33	.38	.75
	39.62	36.82	34.05	31.27	28.45	25.54	22.48	19.17	15.42	10.75	7.55
1235.	-.06	.11	.33	.54	.71	.83	.85	.67	.31	.43	.69
	.03	.14	.23	.22	.09	-.17	-.48	-.72	-.87	-.89	-.68
	-1.00	-.47	.74	.98	.82	.60	.32	-.02	-.15	.50	.81
	39.50	36.71	33.94	31.17	28.35	25.45	22.40	19.10	15.36	10.71	7.52
1280.	-.29	-.11	.15	.41	.63	.78	.84	.72	.34	.37	.62
	-.12	.04	.16	.16	.03	-.20	-.48	-.71	-.85	-.78	-.49
	-.99	-.57	.54	.88	.85	.74	.57	.30	.13	.57	.81
	39.29	36.50	33.74	30.98	28.17	25.28	22.25	18.97	15.25	10.63	7.46
1324.	-.52	-.29	-.00	.28	.52	.71	.80	.69	.29	.36	.63
	-.24	-.07	.08	.13	.06	-.14	-.41	-.69	-.85	-.69	-.27
	-.99	-.69	.20	.71	.86	.84	.71	.47	.28	.63	.83
	39.08	36.30	33.55	30.80	28.00	25.12	22.10	18.84	15.15	10.56	7.41

K[MEV/C]		COS THETA .95	.90	.80	.70	.60	.50	SPIN-ROTATION PARAMETER R .40	.30	.20	.10
1360.	PI+P	.17	.18	-.00	-.32	-.58	-.56	-.37	-.27	-.47	-.72
	PI-P	-.20	-.28	-.38	-.46	-.53	-.55	-.49	-.55	-.60	-.51
	CHEX	.59	.35	-.15	-.76	-.95	-.51	-.26	-.30	-.48	-.74
	THETA-N	78.78	74.14	67.57	62.51	58.23	54.43	50.97	47.74	44.68	41.75
1400.	PI+P	.20	.23	.11	-.20	-.56	-.66	-.48	-.47	-.72	-.91
	PI-P	-.15	-.22	-.33	-.42	-.49	-.53	-.52	-.63	-.69	-.60
	CHEX	.63	.41	-.07	-.70	-.97	-.43	-.13	-.21	-.45	-.74
	THETA-N	78.71	74.04	67.44	62.36	58.07	54.26	50.79	47.55	44.50	41.57
1430.	PI+P	.21	.26	.20	-.06	-.52	-.73	-.50	-.49	-.77	-.96
	PI-P	-.12	-.18	-.27	-.36	-.45	-.51	-.52	-.61	-.68	-.63
	CHEX	.55	.40	.02	-.60	-.98	-.35	-.04	-.14	-.40	-.70
	THETA-N	78.66	73.97	67.34	62.25	57.94	54.13	50.65	47.42	44.36	41.43
1473.	PI+P	.21	.28	.30	.11	-.41	-.90	-.65	-.60	-.83	-.99
	PI-P	-.09	-.13	-.21	-.30	-.38	-.39	-.26	-.41	-.59	-.61
	CHEX	.32	.31	.04	-.52	-.97	-.22	.07	-.09	-.37	-.67
	THETA-N	78.58	73.86	67.20	62.09	57.77	53.94	50.46	47.22	44.16	41.24
1505.	PI+P	.22	.29	.34	.19	-.36	-.96	-.68	-.61	-.81	-.98
	PI-P	-.08	-.12	-.19	-.29	-.37	-.31	-.13	-.31	-.51	-.58
	CHEX	.19	.29	.11	-.48	-.95	-.12	.11	-.07	-.34	-.63
	THETA-N	78.53	73.79	67.10	61.97	57.64	53.81	50.32	47.08	44.02	41.10
1550.	PI+P	.21	.29	.37	.29	-.20	-.96	-.64	-.59	-.79	-.95
	PI-P	-.11	-.14	-.19	-.26	-.30	-.18	-.03	-.25	-.47	-.56
	CHEX	-.02	.17	.15	-.30	-.77	.21	.19	-.08	-.35	-.60
	THETA-N	78.45	73.68	66.95	61.80	57.46	53.62	50.12	46.88	43.82	40.90
1590.	PI+P	.19	.27	.36	.31	-.13	-.81	-.56	-.54	-.71	-.89
	PI-P	-.13	-.17	-.22	-.26	-.25	-.10	-.03	-.25	-.44	-.53
	CHEX	-.12	.12	.17	-.24	-.51	.39	.24	-.06	-.33	-.56
	THETA-N	78.38	73.58	66.83	61.66	57.30	53.45	49.95	46.70	43.64	40.73
1640.	PI+P	.18	.25	.32	.27	-.09	-.58	-.48	-.51	-.68	-.85
	PI-P	-.16	-.20	-.24	-.27	-.23	-.03	.00	-.27	-.49	-.59
	CHEX	-.30	.03	.30	.13	.13	.65	.24	-.11	-.37	-.57
	THETA-N	78.29	73.47	66.67	61.47	57.10	53.24	49.73	46.48	43.43	40.51
1680.	PI+P	.15	.22	.30	.27	-.03	-.44	-.43	-.50	-.66	-.82
	PI-P	-.18	-.22	-.24	-.26	-.23	-.05	-.03	-.26	-.47	-.57
	CHEX	-.41	-.05	.34	.31	.47	.71	.17	-.17	-.39	-.55
	THETA-N	78.23	73.37	66.54	61.33	56.94	53.07	49.56	46.31	43.25	40.34
1720.	PI+P	.14	.20	.27	.24	-.04	-.40	-.44	-.51	-.65	-.78
	PI-P	-.20	-.24	-.27	-.27	-.20	.02	.00	-.27	-.49	-.62
	CHEX	-.52	-.13	.42	.57	.85	.70	.17	-.18	-.43	-.58
	THETA-N	78.16	73.28	66.42	61.18	56.78	52.91	49.39	46.14	43.08	40.17
1760.	PI+P	.13	.18	.25	.23	-.03	-.34	-.38	-.46	-.61	-.76
	PI-P	-.22	-.26	-.28	-.28	-.20	.03	-.02	-.27	-.48	-.61
	CHEX	-.60	-.20	.45	.67	.91	.61	.10	-.22	-.43	-.57
	THETA-N	78.09	73.18	66.29	61.04	56.63	52.75	49.23	45.97	42.92	40.01
1800.	PI+P	.11	.16	.24	.20	-.06	-.34	-.39	-.46	-.57	-.70
	PI-P	-.24	-.29	-.30	-.30	-.20	.06	.01	-.24	-.47	-.62
	CHEX	-.66	-.26	.48	.74	.78	.45	.00	-.27	-.44	-.55
	THETA-N	78.02	73.09	66.17	60.90	56.48	52.58	49.06	45.81	42.75	39.84
1840.	PI+P	.10	.14	.19	.15	-.10	-.37	-.44	-.49	-.58	-.68
	PI-P	-.27	-.32	-.33	-.30	-.15	.15	.03	-.26	-.49	-.64
	CHEX	-.74	-.31	.56	.88	.46	.15	-.04	-.27	-.45	-.55
	THETA-N	77.95	73.00	66.05	60.76	56.32	52.42	48.90	45.64	42.58	39.68
1880.	PI+P	.07	.10	.16	.12	-.14	-.41	-.49	-.53	-.58	-.64
	PI-P	-.29	-.36	-.36	-.31	-.10	.20	.02	-.29	-.52	-.66
	CHEX	-.78	-.33	.63	.85	.12	.03	-.12	-.34	-.49	-.56
	THETA-N	77.89	72.91	65.93	60.62	56.17	52.27	48.73	45.48	42.42	39.52

K[MEV/C]	COS THETA .00	-.10	-.20	-.30	-.40	-.50	-.60	SPIN-ROTATION -.70	PARAMETER -.80	R -.90	-.95
1360.	-.68	-.45	-.15	.16	.44	.67	.80	.74	.33	.32	.60
	-.33	-.11	.06	.13	.08	-.10	-.38	-.68	-.86	-.60	-.11
	-.97	-.83	-.13	.50	.80	.88	.82	.66	.45	.59	.79
	38.92	36.14	33.40	30.65	27.86	24.99	21.99	18.74	15.06	10.49	7.37
1400.	-.82	-.55	-.23	.09	.38	.61	.75	.71	.31	.35	.64
	-.42	-.19	.02	.15	.13	-.04	-.32	-.64	-.83	-.50	.02
	-.96	-.88	-.36	.28	.70	.88	.89	.76	.54	.63	.81
	38.74	35.97	33.23	30.49	27.71	24.85	21.86	18.62	14.97	10.43	7.32
1430.	-.88	-.64	-.33	.01	.33	.59	.74	.71	.32	.35	.64
	-.46	-.23	-.01	.13	.15	.02	-.27	-.65	-.84	-.41	.18
	-.93	-.93	-.54	.10	.62	.87	.91	.81	.59	.60	.77
	38.60	35.84	33.10	30.37	27.60	24.75	21.76	18.54	14.90	10.38	7.29
1473.	-.93	-.70	-.38	-.04	.27	.53	.71	.71	.28	.36	.70
	-.49	-.27	-.04	.12	.14	.02	-.25	-.62	-.85	-.37	.28
	-.90	-.93	-.63	-.05	.52	.85	.95	.87	.63	.66	.84
	38.41	35.65	32.92	30.20	27.44	24.60	21.63	18.42	14.81	10.31	7.24
1505.	-.95	-.73	-.43	-.10	.21	.49	.69	.73	.36	.34	.68
	-.49	-.28	-.05	.11	.14	.02	-.23	-.59	-.84	-.29	.33
	-.86	-.94	-.72	-.18	.43	.83	.95	.90	.71	.65	.82
	38.27	35.52	32.79	30.07	27.32	24.49	21.53	18.34	14.74	10.26	7.20
1550.	-.95	-.78	-.49	-.16	.17	.45	.66	.71	.34	.33	.69
	-.49	-.30	-.07	.11	.15	.04	-.20	-.55	-.84	-.21	.44
	-.81	-.90	-.77	-.32	.32	.80	.96	.92	.76	.73	.86
	38.08	35.33	32.61	29.90	27.16	24.34	21.39	18.22	14.64	10.19	7.15
1590.	-.94	-.82	-.55	-.22	.11	.41	.63	.72	.47	.36	.68
	-.48	-.31	-.06	.13	.17	.03	-.24	-.58	-.86	-.14	.52
	-.76	-.88	-.83	-.46	.23	.78	.95	.94	.89	.81	.86
	37.91	35.16	32.45	29.74	27.01	24.21	21.27	18.11	14.55	10.13	7.11
1640.	-.92	-.84	-.61	-.28	.05	.36	.59	.68	.36	.31	.66
	-.55	-.35	-.08	.12	.17	.06	-.18	-.53	-.85	.02	.62
	-.72	-.81	-.81	-.55	.10	.70	.94	.95	.89	.85	.89
	37.70	34.96	32.25	29.56	26.84	24.05	21.13	17.99	14.45	10.05	7.05
1680.	-.91	-.86	-.65	-.33	.01	.32	.57	.69	.43	.28	.64
	-.54	-.37	-.12	.08	.14	.06	-.14	-.48	-.84	.11	.73
	-.67	-.76	-.77	-.53	.11	.68	.91	.97	.96	.82	.84
	37.53	34.79	32.10	29.41	26.70	23.92	21.01	17.89	14.36	9.99	7.01
1720.	-.87	-.86	-.70	-.39	-.03	.29	.54	.67	.43	.27	.65
	-.62	-.45	-.16	.09	.16	.05	-.17	-.48	-.83	.28	.79
	-.66	-.70	-.70	-.53	.06	.67	.91	.98	.98	.77	.78
	37.37	34.63	31.94	29.26	26.56	23.79	20.90	17.79	14.28	9.94	6.97
1760.	-.87	-.88	-.73	-.43	-.08	.24	.51	.67	.52	.23	.57
	-.62	-.47	-.20	.04	.12	.03	-.19	-.47	-.76	.47	.84
	-.65	-.70	-.70	-.54	.03	.63	.86	.92	.99	.67	.64
	37.20	34.48	31.79	29.12	26.43	23.67	20.79	17.69	14.20	9.88	6.93
1800.	-.82	-.88	-.78	-.50	-.13	.23	.53	.70	.55	.22	.55
	-.67	-.54	-.27	.00	.13	.06	-.13	-.38	-.69	.58	.75
	-.60	-.63	-.65	-.54	-.02	.58	.83	.92	.97	.53	.52
	37.04	34.32	31.64	28.98	26.29	23.54	20.67	17.59	14.12	9.82	6.89
1840.	-.79	-.86	-.80	-.54	-.17	.18	.46	.66	.57	.31	.55
	-.70	-.61	-.35	-.04	.12	.07	-.13	-.43	-.57	.53	.63
	-.60	-.62	-.63	-.51	-.01	.53	.77	.88	.92	.43	.41
	36.88	34.17	31.49	28.84	26.16	23.42	20.57	17.50	14.04	9.77	6.85
1880.	-.74	-.85	-.85	-.60	-.20	.19	.51	.68	.54	.26	.51
	-.74	-.70	-.48	-.15	.09	.17	.11	-.14	-.33	.49	.45
	-.59	-.60	-.62	-.50	.00	.51	.72	.82	.81	.31	.31
	36.73	34.01	31.35	28.70	26.03	23.30	20.46	17.40	13.97	9.71	6.81

		COS THETA						SPIN-ROTATION PARAMETER R			
K[MEV/C]		.95	.90	.80	.70	.60	.50	.40	.30	.20	.10
1920.	PI+P	.05	.08	.11	.05	-.20	-.45	-.51	-.54	-.57	-.63
	PI-P	-.31	-.38	-.39	-.31	-.02	.28	.07	-.25	-.49	-.66
	CHEX	-.81	-.35	.67	.86	-.29	-.20	-.14	-.30	-.44	-.52
	THETA-N	77.82	72.82	65.80	60.48	56.02	52.11	48.57	45.31	42.26	39.36
1980.	PI+P	.03	.03	.05	.00	-.23	-.48	-.57	-.59	-.60	-.63
	PI-P	-.32	-.41	-.42	-.28	.12	.38	.09	-.26	-.51	-.67
	CHEX	-.86	-.35	.80	.70	-.56	-.31	-.12	-.24	-.38	-.45
	THETA-N	77.72	72.68	65.62	60.27	55.80	51.88	48.33	45.07	42.02	39.12
2030.	PI+P	.01	.02	.03	-.05	-.33	-.57	-.62	-.61	-.61	-.63
	PI-P	-.32	-.41	-.42	-.24	.22	.41	.09	-.26	-.52	-.70
	CHEX	-.89	-.36	.82	.49	-.79	-.42	-.11	-.16	-.32	-.43
	THETA-N	77.64	72.57	65.48	60.10	55.61	51.68	48.14	44.88	41.82	38.93
2070.	PI+P	.00	.01	.02	-.07	-.37	-.61	-.65	-.63	-.62	-.63
	PI-P	-.31	-.40	-.41	-.19	.33	.43	.05	-.30	-.54	-.71
	CHEX	-.91	-.36	.83	.31	-.85	-.47	-.17	-.18	-.30	-.39
	THETA-N	77.57	72.48	65.36	59.97	55.47	51.53	47.98	44.72	41.67	38.78
2150.	PI+P	.01	.00	-.01	-.13	-.48	-.72	-.73	-.70	-.67	-.66
	PI-P	-.28	-.37	-.40	-.15	.41	.36	-.01	-.33	-.56	-.73
	CHEX	-.89	-.31	.83	-.00	-.93	-.46	-.10	-.06	-.22	-.37
	THETA-N	77.44	72.30	65.12	59.70	55.18	51.23	47.68	44.41	41.36	38.47
2200.	PI+P	.01	.00	-.02	-.16	-.53	-.76	-.76	-.72	-.68	-.65
	PI-P	-.26	-.35	-.40	-.16	.37	.28	-.05	-.35	-.57	-.74
	CHEX	-.83	-.23	.79	-.09	-.95	-.47	-.11	-.03	-.15	-.31
	THETA-N	77.36	72.19	64.98	59.53	55.00	51.04	47.49	44.22	41.17	38.29
2280.	PI+P	.03	.01	-.05	-.26	-.68	-.86	-.82	-.75	-.66	-.61
	PI-P	-.23	-.33	-.41	-.18	.30	.16	-.14	-.40	-.60	-.75
	CHEX	-.71	-.15	.74	-.40	-.94	-.49	-.18	-.04	-.08	-.22
	THETA-N	77.23	72.02	64.75	59.27	54.72	50.75	47.19	43.92	40.88	38.00
2340.	PI+P	.04	.01	-.07	-.32	-.74	-.88	-.85	-.80	-.74	-.69
	PI-P	-.22	-.32	-.41	-.19	.23	.08	-.19	-.42	-.60	-.74
	CHEX	-.59	-.07	.66	-.56	-.85	-.36	-.05	.06	-.05	-.26
	THETA-N	77.14	71.89	64.58	59.08	54.51	50.54	46.97	43.70	40.66	37.79
2460.	PI+P	.06	.05	-.07	-.45	-.91	-.94	-.90	-.85	-.78	-.72
	PI-P	-.20	-.29	-.39	-.13	.13	-.07	-.26	-.45	-.62	-.75
	CHEX	-.41	.06	.67	-.92	-.56	-.18	.03	.14	.15	-.02
	THETA-N	76.95	71.63	64.24	58.70	54.10	50.11	46.54	43.27	40.23	37.37
2560.	PI+P	.09	.09	-.03	-.54	-.95	-.91	-.88	-.84	-.79	-.75
	PI-P	-.20	-.27	-.32	-.06	-.01	-.23	-.33	-.46	-.61	-.75
	CHEX	-.37	.11	.70	-.96	-.43	-.11	.17	.41	.51	.30
	THETA-N	76.79	71.42	63.97	58.38	53.77	49.77	46.19	42.92	39.89	37.03
2750.	PI+P	.11	.15	.01	-.65	-.85	-.78	-.82	-.83	-.79	-.74
	PI-P	-.21	-.25	-.25	-.08	-.29	-.47	-.50	-.56	-.66	-.75
	CHEX	-.27	.22	.75	-.73	-.13	.12	.27	.38	.54	.71
	THETA-N	76.50	71.02	63.45	57.80	53.15	49.13	45.54	42.28	39.25	36.41
3000.	PI+P	.07	.09	-.12	-.71	-.80	-.76	-.75	-.76	-.80	-.83
	PI-P	-.27	-.31	-.28	-.16	-.48	-.66	-.72	-.78	-.81	-.78
	CHEX	-.30	.36	.81	-.42	.11	.27	.25	.17	.23	.55
	THETA-N	76.13	70.52	62.80	57.07	52.37	48.33	44.73	41.47	38.46	35.64
3400.	PI+P	.02	-.02	-.34	-.84	-.76	-.75	-.75	-.81	-.96	-1.00
	PI-P	-.30	-.36	-.24	-.18	-.58	-.76	-.84	-.88	-.63	-.09
	CHEX	-.30	.53	-.67	-.15	.53	.77	.73	.43	.20	.31
	THETA-N	75.56	69.75	61.80	55.96	51.20	47.12	43.52	40.28	37.29	34.50
3650.	PI+P	.02	-.02	-.43	-.91	-.81	-.78	-.75	-.77	-.89	-.99
	PI-P	-.28	-.33	-.20	-.32	-.61	-.72	-.91	-.97	-.88	-.73
	CHEX	-.22	.55	-.90	-.05	.41	.49	.58	.77	.61	-.04
	THETA-N	75.21	69.28	61.21	55.29	50.50	46.41	42.82	39.58	36.61	33.84

SPIN-ROTATION PARAMETER R

K[MEV/C]	COS THETA .00	-.10	-.20	-.30	-.40	-.50	-.60	-.70	-.80	-.90	-.95
1920.	-.73	-.84	-.85	-.64	-.27	.13	.46	.69	.64	.26	.44
	-.75	-.69	-.45	-.12	.10	.10	-.04	-.22	.15	.20	.28
	-.53	-.52	-.51	-.42	.00	.45	.65	.80	.75	.14	.14
	36.57	33.86	31.20	28.56	25.90	23.19	20.35	17.31	13.89	9.66	6.77
1980.	-.72	-.84	-.89	-.70	-.32	.08	.40	.62	.63	.29	.37
	-.76	-.74	-.54	-.17	.11	.18	.11	.02	.80	-.10	-.05
	-.46	-.46	-.46	-.33	.08	.38	.52	.65	.62	.03	-.01
	36.34	33.64	30.99	28.36	25.72	23.01	20.20	17.17	13.78	9.58	6.72
2030.	-.70	-.82	-.89	-.75	-.40	.01	.35	.58	.59	.08	.15
	-.80	-.80	-.61	-.22	.11	.20	.19	.36	-.13	-.31	-.26
	-.45	-.43	-.39	-.26	.01	.20	.32	.53	.49	-.11	-.12
	36.15	33.46	30.82	28.20	25.56	22.87	20.07	17.06	13.69	9.52	6.67
2070.	-.70	-.83	-.91	-.77	-.42	-.03	.29	.49	.49	-.02	-.06
	-.81	-.83	-.67	-.25	.15	.30	.35	.62	-.48	-.47	-.45
	-.42	-.43	-.39	-.21	.06	.18	.24	.43	.47	-.16	-.22
	36.00	33.32	30.68	28.07	25.44	22.76	19.97	16.97	13.62	9.46	6.64
2150.	-.71	-.82	-.94	-.87	-.56	-.14	.18	.29	.14	-.17	.11
	-.86	-.90	-.75	-.30	.20	.42	.55	.94	-.70	-.64	-.70
	-.43	-.43	-.40	-.30	-.13	-.01	.08	.35	.33	-.21	-.19
	35.71	33.03	30.41	27.81	25.20	22.54	19.77	16.80	13.48	9.36	6.57
2200.	-.69	-.81	-.94	-.88	-.57	-.18	.07	.12	.01	-.14	.29
	-.87	-.93	-.80	-.31	.27	.54	.71	.54	-.72	-.67	-.76
	-.40	-.45	-.45	-.35	-.18	-.10	-.01	.31	.33	-.18	-.11
	35.53	32.86	30.25	27.66	25.06	22.41	19.65	16.70	13.39	9.30	6.52
2280.	-.66	-.83	-.99	-.94	-.66	-.32	-.11	-.11	-.21	-.15	.34
	-.87	-.96	-.89	-.31	.45	.69	.83	-.26	-.84	-.66	-.74
	-.37	-.54	-.73	-.65	-.31	-.12	-.01	.28	.35	-.01	.11
	35.25	32.59	29.99	27.41	24.83	22.20	19.46	16.53	13.26	9.21	6.46
2340.	-.71	-.82	-.97	-.99	-.81	-.53	-.30	-.11	.05	-.06	.30
	-.86	-.96	-.95	-.38	.53	.86	.99	-.46	-.90	-.68	-.71
	-.43	-.57	-.71	-.65	-.42	-.29	-.15	.29	.59	.08	.20
	35.04	32.39	29.80	27.23	24.66	22.04	19.32	16.41	13.16	9.14	6.41
2460.	-.73	-.83	-.92	-.94	-.92	-.83	-.69	-.51	-.14	-.05	.20
	-.86	-.97	-.95	-.24	.68	.92	.56	-.90	-.92	-.73	-.75
	-.28	-.56	-.87	-.89	-.62	-.46	-.38	-.04	.74	.17	.22
	34.64	32.00	29.42	26.88	24.33	21.74	19.05	16.18	12.97	9.01	6.31
2560.	-.78	-.85	-.87	-.85	-.93	-.95	-.71	-.31	.08	-.35	-.06
	-.87	-.97	-.90	-.26	.62	.91	.28	-.97	-.97	-.93	-.82
	-.19	-.64	-.92	-.98	-.84	-.64	-.38	.21	.91	.16	.36
	34.31	31.68	29.12	26.60	24.07	21.50	18.84	16.00	12.81	8.90	6.24
2750.	-.72	-.76	-.84	-.79	-.83	-.97	-.90	-.35	.32	-.10	.10
	-.84	-.91	-.51	.46	.73	.55	-.39	-.99	-.96	-.89	-.10
	.72	-.10	-.89	-.65	-.72	-.78	-.60	.15	1.00	.30	.26
	33.71	31.11	28.58	26.09	23.59	21.07	18.45	15.66	12.54	8.70	6.10
3000.	-.83	-.77	-.33	-.27	-.59	-.97	-.85	-.54	-.22	-.21	.05
	-.74	-.63	.46	.86	.77	.24	-.79	-.96	-.98	-.89	-.19
	.81	.67	.07	-.38	-.60	-.60	-.33	.01	.72	.38	.39
	32.97	30.40	27.90	25.45	23.01	20.53	17.97	15.24	12.20	8.46	5.93
3400.	-.86	.02	.27	.08	-.29	-.63	-.66	-.72	-.17	.87	.73
	-.33	-.02	.48	.45	.15	-.44	-.99	-.32	-.92	.90	.98
	.71	.89	.84	.57	-.15	-.79	-.54	-.76	-.11	.99	.93
	31.87	29.35	26.92	24.52	22.15	19.75	17.27	14.64	11.71	8.12	5.69
3650.	-.89	-.67	-.04	-.15	-.63	-.29	-.08	-.75	-.74	.60	.21
	.32	.76	.37	.35	.12	-.95	-.55	-.42	.76	.91	.79
	-.35	.02	.54	.05	-.99	.31	.58	-.60	-.95	.88	.81
	31.24	28.75	26.35	23.99	21.66	19.30	16.87	14.29	11.43	7.92	5.55

(Table 2.3.2.2)

K[MEV/C]		COS THETA .95	.90	.80	.70	.60	.50	SPIN-ROTATION PARAMETER R .40	.30	.20	.10
4000.	PI+P	.04	-.02	-.63	-.88	-.83	-.77	-.69	-.63	-.78	-.97
	PI-P	-.25	-.31	-.25	-.45	-.66	-.77	-.85	-.83	-.51	.17
	CHEX	-.13	.80	-.75	.42	.65	.51	.10	-.25	.20	-.38
	THETA-N	74.74	68.65	60.41	54.41	49.58	45.47	41.88	38.66	35.71	32.98
5000.	PI+P	-.01	-.16	-.96	-.59	-.70	-.92	-.94	.72	.00	.86
	PI-P	-.28	-.32	-.54	-.80	-.85	-.87	.44	-.93	-.60	.56
	CHEX	.30	.92	.31	1.00	.08	-.84	.45	-.12	-.98	.19
	THETA-N	73.48	66.98	58.31	52.12	47.20	43.08	39.51	36.34	33.48	30.83
6000.	PI+P	-.03	-.30	-.87	-.31	-.37	.54	.27	-.07	.36	-.18
	PI-P	-.28	-.31	-.87	-.69	-.23	.28	-.64	-.21	.83	-.82
	CHEX	.53	-.24	.86	.30	-.14	-.43	.63	.62	-.74	.75
	THETA-N	72.33	65.46	56.44	50.11	45.14	41.04	37.51	34.40	31.61	29.05

Table 2.3.2.3. Coefficients of a Legendre expansion, Eq. (A.4.21), for $(P d\sigma/d\Omega)/\sin\theta$ as calculated from the solution KH 80. Only the first 11 (dimensionless) coefficients are shown.

LEGENDRE EXPANSION COEFFICIENTS FOR POLARIZATION PARAMETERS PI+P→PI+P

K GEV/C	B 1	B 2	B 3	B 4	B 5	B 6	B 7	B 8	B 9	B 10	B 11
.247	.303	.088	-.004	.000	.000	.000	.000	.000	.000	.000	.000
.267	.429	.130	-.011	.000	.000	.000	.000	.000	.000	.000	.000
.280	.452	.147	-.015	.001	.000	.000	.000	.000	.000	.000	.000
.290	.447	.149	-.013	.001	.000	.000	.000	.000	.000	.000	.000
.295	.469	.169	-.012	.002	.000	.000	.000	.000	.000	.000	.000
.301	.476	.191	-.011	.002	.000	.000	.000	.000	.000	.000	.000
.305	.473	.197	-.012	.002	.000	.000	.000	.000	.000	.000	.000
.310	.460	.212	-.011	.002	.000	.000	.000	.000	.000	.000	.000
.320	.447	.226	-.013	.003	.000	.000	.000	.000	.000	.000	.000
.331	.418	.178	-.013	.003	.000	.000	.000	.000	.000	.000	.000
.351	.375	.189	-.000	.002	.000	.000	.000	.000	.000	.000	.000
.378	.289	.167	-.009	.005	.000	.000	.000	.000	.000	.000	.000
.408	.225	.148	-.017	.008	.000	.000	.000	.000	.000	.000	.000
.427	.177	.135	-.007	.008	.000	.000	.000	.000	.000	.000	.000
.456	.131	.111	-.007	.002	.000	.000	.000	.000	.000	.000	.000
.490	.084	.094	-.004	.006	.000	.000	.000	.000	.000	.000	.000
.532	.061	.069	-.003	.003	-.004	-.000	.000	-.000	.000	.000	.000
.573	.014	.037	-.013	.001	-.005	-.001	.000	-.000	.000	.000	.000
.614	-.020	.006	-.014	.002	-.002	-.000	.001	-.000	.000	-.000	.000
.658	-.036	-.019	-.006	.010	.002	-.000	-.000	.000	.000	-.000	.000
.675	-.042	-.029	-.003	.012	.004	.001	-.000	.000	.000	.000	-.000
.705	-.031	-.028	.011	.018	.007	.000	-.001	.000	-.000	.000	-.000
.725	-.021	-.025	.020	.021	.012	.000	-.002	-.001	-.000	.000	.000
.750	-.001	-.023	.024	.021	.011	.001	-.000	.000	-.000	.000	.000
.777	-.008	-.039	.013	.020	.011	.002	.001	.000	.000	.000	.000
.800	.001	-.023	.011	.021	.009	.002	.001	.000	.001	.000	.000
.822	-.007	-.023	.010	.024	.013	.004	-.000	-.000	-.000	-.001	.000
.851	.016	-.010	.020	.027	.018	.004	.001	.000	-.001	.000	-.000
.875	.042	-.007	.035	.029	.020	.007	-.000	-.000	.000	-.001	.000
.895	.070	-.008	.048	.023	.026	.006	.003	.002	.000	-.000	.000
.923	.108	-.004	.069	.029	.031	.007	.001	.003	.001	.000	.000
.954	.138	.006	.100	.033	.045	.009	.002	.004	.001	.000	.000
.975	.161	.012	.119	.035	.048	.010	-.001	.008	.001	.002	.001
1.000	.176	.032	.146	.050	.063	.016	.006	.005	.003	.002	.000
1.030	.187	.065	.176	.065	.077	.019	.007	.006	.003	.002	.000

(continued)

(Table 2.3.2.2)

SPIN-ROTATION PARAMETER R

K[MEV/C]	COS THETA .00	-.10	-.20	-.30	-.40	-.50	-.60	-.70	-.80	-.90	-.95
4000.	-.75	-.66	-.93	-.95	-.87	-.23	.24	-.34	-.52	.66	-.73
	.64	.94	.70	.14	-.28	-.50	-.30	-.60	.17	.94	-.34
	-.31	.02	.18	.60	.79	.72	.63	-.20	-.64	.78	-.96
	30.41	27.97	25.61	23.30	21.02	18.72	16.36	13.85	11.07	7.67	5.37
5000.	.21	.18	.75	-.28	-.90	.09	.82	.30	-.03	.89	-.49
	-.68	-.12	.70	-.07	-.16	-.46	-.47	-.25	.27	.29	-.17
	.43	-.99	.59	-.11	-.92	-.05	-.43	-.66	-.01	.97	.75
	28.37	26.03	23.79	21.61	19.47	17.31	15.11	12.78	10.20	7.06	4.94
6000.	-.21	.83	.20	-.88	-.93	.52	-.12	.31	.10	.47	.12
	-.97	.84	.24	-.91	-.97	.68	-.83	.69	.52	1.00	.84
	.29	.32	.16	-.78	-.77	-.33	.72	-.84	-.12	-.13	.50
	26.68	24.44	22.31	20.24	18.21	16.18	14.10	11.92	9.51	6.58	4.60

Table 2.3.2.3 (continued)

K GEV/C	B 1	B 2	B 3	B 4	B 5	B 6	B 7	B 8	B 9	B 10	B 11
1.055	.204	.099	.200	.076	.088	.021	.008	.009	.003	.003	.001
1.080	.216	.113	.217	.096	.094	.029	.011	.008	.002	.002	.000
1.113	.232	.149	.258	.121	.105	.037	.013	.011	.003	.002	.000
1.154	.261	.194	.321	.160	.123	.046	.012	.007	.001	.001	.000
1.174	.278	.211	.355	.188	.130	.054	.009	.004	-.002	.001	-.002
1.210	.313	.262	.441	.244	.157	.076	.010	.005	-.001	.001	-.001
1.235	.326	.290	.498	.287	.179	.097	.014	.010	-.000	.002	-.001
1.280	.358	.352	.593	.353	.214	.126	.022	.013	-.004	.005	-.004
1.324	.383	.388	.660	.415	.243	.158	.024	.020	-.004	.008	-.007
1.360	.401	.401	.692	.452	.270	.169	.029	.026	-.004	.007	-.004
1.400	.424	.415	.698	.480	.296	.197	.039	.033	.002	.007	-.006
1.430	.438	.414	.688	.485	.307	.203	.043	.031	.006	.005	-.002
1.473	.450	.406	.636	.476	.318	.218	.060	.047	.012	.008	-.001
1.505	.463	.396	.590	.461	.330	.232	.079	.055	.017	.013	.000
1.550	.447	.360	.495	.403	.293	.213	.070	.054	.015	.009	.002
1.590	.444	.342	.420	.356	.274	.204	.073	.052	.020	.014	.002
1.640	.435	.336	.369	.317	.253	.197	.073	.053	.022	.012	.003
1.680	.434	.335	.332	.287	.247	.192	.078	.052	.019	.010	.003
1.720	.431	.327	.300	.267	.235	.183	.072	.054	.025	.012	.004
1.760	.424	.336	.277	.242	.229	.182	.085	.050	.021	.012	.004
1.800	.422	.325	.242	.223	.226	.182	.088	.054	.024	.012	.004
1.840	.405	.326	.233	.213	.225	.177	.085	.054	.028	.014	.005
1.880	.406	.320	.212	.206	.229	.176	.092	.056	.030	.014	.005
1.920	.390	.320	.209	.207	.231	.183	.095	.060	.032	.017	.006
1.980	.376	.336	.229	.230	.247	.194	.106	.064	.038	.020	.007
2.030	.375	.357	.254	.255	.275	.211	.123	.077	.042	.021	.009
2.070	.381	.370	.267	.273	.300	.237	.143	.085	.048	.027	.011
2.150	.365	.387	.309	.319	.342	.271	.174	.111	.063	.033	.014
2.200	.365	.396	.322	.335	.366	.293	.191	.125	.073	.039	.018
2.280	.353	.408	.355	.365	.396	.327	.221	.146	.091	.053	.024
2.340	.356	.428	.394	.403	.423	.345	.242	.163	.097	.055	.025
2.460	.346	.437	.443	.444	.458	.378	.283	.190	.117	.068	.032
2.560	.349	.452	.461	.472	.465	.400	.301	.214	.128	.082	.036
2.750	.325	.408	.413	.407	.418	.374	.315	.235	.166	.112	.063
3.000	.261	.336	.341	.339	.347	.331	.303	.245	.188	.141	.097
3.400	.248	.333	.372	.390	.382	.374	.352	.302	.243	.180	.127
3.650	.242	.343	.399	.430	.431	.424	.398	.349	.287	.222	.162
4.000	.216	.323	.381	.414	.425	.422	.407	.367	.319	.259	.203
5.000	.164	.260	.326	.378	.402	.413	.408	.388	.361	.321	.276
6.000	.147	.233	.303	.359	.391	.411	.413	.405	.386	.357	.323

LEGENDRE EXPANSION COEFFICIENTS FOR POLARIZATION PARAMETERS PI-P→PI-P

K GEV/C	B 1	B 2	B 3	B 4	B 5	B 6	B 7	B 8	B 9	B 10	B 11
.247	-.006	.010	-.004	-.000	.000	.000	.000	.000	.000	.000	.000
.267	-.015	.009	-.007	-.001	.000	.000	.000	.000	.000	.000	.000
.280	-.020	.007	-.009	-.001	.000	.000	.000	.000	.000	.000	.000
.290	-.031	.004	-.010	-.001	.000	.000	.000	.000	.000	.000	.000
.295	-.027	.006	-.012	-.001	.000	.000	.000	.000	.000	.000	.000
.301	-.028	.006	-.013	-.001	.000	.000	.000	.000	.000	.000	.000
.305	-.029	.003	-.013	-.002	.000	.000	.000	.000	.000	.000	.000
.310	-.029	.001	-.014	-.002	.000	.000	.000	.000	.000	.000	.000
.320	-.037	-.002	-.014	-.002	.000	.000	.000	.000	.000	.000	.000
.331	-.042	-.013	-.014	-.002	.000	.000	.000	.000	.000	.000	.000
.351	-.045	-.024	-.015	-.002	.000	.000	.000	.000	.000	.000	.000
.378	-.062	-.040	-.018	-.002	.000	.000	.000	.000	.000	.000	.000
.408	-.078	-.063	-.021	-.002	.000	.000	.000	.000	.000	.000	.000
.427	-.086	-.091	-.025	-.004	.000	.000	.000	.000	.000	.000	.000
.456	-.102	-.116	-.029	-.005	.000	.000	.000	.000	.000	.000	.000
.490	-.131	-.169	-.031	-.004	-.000	-.000	.000	.000	.000	.000	.000
.532	-.141	-.192	-.036	-.007	-.001	-.000	-.000	-.000	.000	.000	.000
.573	-.177	-.213	-.047	-.002	-.001	-.000	-.000	-.000	.000	.000	.000
.614	-.229	-.237	-.074	.000	-.004	.000	.001	.000	-.000	.000	-.000
.658	-.290	-.224	-.100	.025	.009	.007	.002	.001	.000	.000	.000
.675	-.302	-.205	-.115	.030	.010	.007	.004	.002	.000	.000	.000
.705	-.301	-.159	-.129	.043	.010	.007	.003	.001	-.001	-.000	.000
.725	-.306	-.135	-.141	.049	.010	.006	.002	.001	-.001	.000	.000
.750	-.279	-.110	-.154	.047	.008	.007	.004	.003	-.001	-.000	.000
.777	-.223	-.090	-.148	.029	.003	.008	.007	.006	.001	.000	.000
.800	-.153	-.072	-.120	.021	.001	.006	.006	.009	.002	.001	.000
.822	-.086	-.058	-.083	.019	.001	.005	.005	.011	.002	.001	-.000
.851	-.014	-.052	-.051	.022	.004	.009	.003	.011	.001	.001	-.000
.875	.024	-.043	-.041	.030	.002	.008	.005	.011	-.001	.001	-.001
.895	.041	-.052	-.043	.034	-.010	.015	.004	.010	.002	.000	-.001
.923	.050	-.057	-.050	.036	-.030	.018	-.001	.010	.004	.001	.002
.954	.038	-.073	-.063	.029	-.061	.027	-.003	.010	.007	.001	.002
.975	.024	-.076	-.069	.031	-.076	.033	.001	.010	.007	.001	.001
1.000	-.008	-.062	-.051	.019	-.080	.041	.003	.015	.005	.003	.002
1.030	-.004	-.030	.004	.039	-.075	.052	.003	.012	.003	-.001	-.002
1.055	-.008	.016	.059	.050	-.054	.059	.002	.013	.003	-.003	-.001
1.080	-.006	.035	.099	.068	-.043	.063	.010	.009	.003	-.003	-.003
1.113	-.021	.058	.127	.062	-.033	.056	.010	.008	.005	.000	-.001
1.154	-.044	.062	.132	.048	-.029	.042	.004	.004	.002	.001	-.001
1.174	-.048	.057	.129	.048	-.029	.034	-.002	.002	-.002	.002	.000
1.210	-.064	.058	.127	.048	-.032	.034	-.008	-.004	-.005	.000	.001
1.235	-.070	.053	.120	.054	-.027	.044	-.001	.001	-.001	.002	.002
1.280	-.074	.058	.118	.063	-.024	.058	.010	.010	.005	.008	.005
1.324	-.069	.077	.122	.072	-.021	.078	.017	.017	.011	.011	.006
1.360	-.050	.099	.134	.087	-.008	.077	.014	.012	.010	.009	.006
1.400	-.038	.116	.128	.082	-.013	.074	.010	.004	.007	.007	.002
1.430	-.043	.111	.109	.066	-.028	.065	-.001	-.006	.005	.003	.001
1.473	-.068	.077	.070	.014	-.051	.034	-.010	-.011	-.000	.002	-.001
1.505	-.095	.043	.032	-.024	-.078	.013	-.019	-.014	-.000	.003	-.001
1.550	-.139	-.008	-.023	-.082	-.116	-.020	-.035	-.018	-.000	.004	.000
1.590	-.171	-.037	-.052	-.117	-.129	-.036	-.045	-.022	.001	.005	.001
1.640	-.186	-.058	-.070	-.136	-.143	-.054	-.050	-.027	-.002	.004	.001
1.680	-.200	-.074	-.082	-.153	-.149	-.064	-.050	-.028	-.005	.001	.001
1.720	-.201	-.080	-.090	-.167	-.153	-.075	-.057	-.034	-.005	.002	.001
1.760	-.208	-.094	-.100	-.177	-.160	-.087	-.061	-.037	-.007	.000	.001
1.800	-.209	-.108	-.114	-.189	-.169	-.096	-.064	-.038	-.007	.001	.001
1.840	-.210	-.114	-.116	-.190	-.171	-.107	-.067	-.037	-.004	.002	.002
1.880	-.188	-.116	-.123	-.187	-.164	-.111	-.070	-.034	-.000	.003	.003
1.920	-.188	-.106	-.098	-.169	-.154	-.103	-.062	-.033	-.001	.004	.004
1.980	-.168	-.083	-.060	-.127	-.122	-.087	-.049	-.023	.002	.005	.004

K GEV/C	B 1	B 2	B 3	B 4	B 5	B 6	B 7	B 8	B 9	B 10	B 11
2.030	-.144	-.065	-.030	-.090	-.089	-.068	-.036	-.016	.006	.007	.004
2.070	-.131	-.054	-.013	-.062	-.064	-.052	-.025	-.007	.010	.010	.006
2.150	-.111	-.050	.003	-.029	-.029	-.032	-.014	-.001	.012	.011	.005
2.200	-.103	-.054	.003	-.022	-.022	-.030	-.014	-.001	.010	.010	.003
2.280	-.092	-.062	-.009	-.022	-.022	-.041	-.028	-.011	.001	.005	-.001
2.340	-.091	-.063	-.011	-.020	-.032	-.052	-.041	-.023	-.012	-.006	-.007
2.460	-.087	-.080	-.036	-.038	-.054	-.084	-.081	-.059	-.042	-.026	-.021
2.560	-.088	-.091	-.063	-.056	-.081	-.107	-.112	-.086	-.066	-.044	-.033
2.750	-.123	-.148	-.136	-.126	-.150	-.162	-.156	-.122	-.091	-.059	-.038
3.000	-.125	-.166	-.171	-.169	-.184	-.188	-.173	-.142	-.105	-.068	-.040
3.400	-.072	-.101	-.107	-.108	-.120	-.129	-.127	-.115	-.098	-.073	-.050
3.650	-.057	-.080	-.091	-.094	-.103	-.109	-.112	-.108	-.098	-.079	-.061
4.000	-.067	-.101	-.124	-.133	-.140	-.142	-.143	-.138	-.128	-.107	-.083
5.000	-.081	-.124	-.159	-.181	-.196	-.204	-.209	-.209	-.201	-.186	-.163
6.000	-.089	-.140	-.178	-.206	-.225	-.238	-.246	-.250	-.247	-.237	-.219

LEGENDRE EXPANSION COEFFICIENTS FOR POLARIZATION PARAMETERS PI-P->PI0 N

K GEV/C	B 1	B 2	B 3	B 4	B 5	B 6	B 7	B 8	B 9	B 10	B 11
.247	.106	.019	.003	.001	.000	.000	.000	.000	.000	.000	.000
.267	.156	.034	.003	.001	.000	.000	.000	.000	.000	.000	.000
.280	.169	.042	.004	.001	.000	.000	.000	.000	.000	.000	.000
.290	.177	.045	.006	.001	.000	.000	.000	.000	.000	.000	.000
.295	.181	.051	.008	.002	-.000	.000	.000	.000	.000	.000	.000
.301	.183	.058	.009	.002	-.000	.000	.000	.000	.000	.000	.000
.305	.183	.062	.009	.002	-.000	.000	.000	.000	.000	.000	.000
.310	.179	.069	.010	.003	-.000	.000	.000	.000	.000	.000	.000
.320	.180	.076	.009	.003	-.000	.000	.000	.000	.000	.000	.000
.331	.175	.071	.010	.003	-.000	.000	.000	.000	.000	.000	.000
.351	.162	.085	.014	.003	-.000	.000	.000	.000	.000	.000	.000
.378	.149	.091	.014	.004	-.000	.000	.000	.000	.000	.000	.000
.408	.141	.103	.016	.005	-.000	.000	.000	.000	.000	.000	.000
.427	.138	.121	.023	.007	-.000	.000	.000	.000	.000	.000	.000
.456	.148	.129	.031	.006	-.000	-.000	.000	.000	.000	.000	.000
.490	.185	.148	.036	.006	-.000	-.000	.000	.000	.000	.000	.000
.532	.210	.151	.052	.007	-.001	.000	.000	.000	.000	.000	.000
.573	.217	.141	.055	.005	-.001	.000	.000	-.000	.000	.000	.000
.614	.208	.134	.060	.005	-.004	.001	.001	.000	.000	-.000	.000
.658	.192	.158	.072	.019	-.004	.004	.004	.002	-.000	.000	.000
.675	.186	.168	.065	.020	-.009	.005	.004	.002	-.000	.000	.000
.705	.165	.176	.042	.027	-.015	.006	.004	.004	.000	.000	.000
.725	.143	.173	.021	.030	-.020	.004	.004	.004	.001	.000	.000
.750	.108	.149	.004	.037	-.023	.004	.003	.005	.001	.000	.000
.777	.091	.111	-.000	.030	-.020	.003	.005	.005	.000	-.000	.000
.800	.081	.085	.005	.036	-.014	.004	.002	.005	-.000	.000	-.000
.822	.071	.057	.012	.029	-.009	.003	-.000	.005	.000	.001	.001
.851	.064	.025	.032	.029	-.002	.002	-.001	.006	-.002	.001	-.000
.875	.056	.004	.045	.031	-.000	.002	-.002	.004	-.002	.000	.000
.895	.045	-.018	.063	.030	.003	-.004	-.003	.004	-.002	-.000	.000
.923	.027	-.041	.080	.028	-.000	-.012	-.005	.003	-.000	-.001	.001
.954	.001	-.057	.089	.016	-.006	-.026	-.009	.001	-.000	-.001	.001
.975	-.016	-.058	.084	.012	-.013	-.037	-.009	.000	.001	.000	-.002
1.000	-.043	-.052	.081	-.007	-.007	-.050	-.009	-.001	-.002	-.001	-.002
1.030	-.061	-.039	.062	-.018	-.011	-.059	-.008	-.004	.001	-.002	-.004
1.055	-.074	-.018	.048	-.033	-.000	-.068	-.005	-.004	-.002	-.001	-.005
1.080	-.077	-.006	.030	-.038	.005	-.068	-.003	-.003	-.001	-.001	-.005
1.113	-.081	.011	.005	-.041	.007	-.068	-.003	-.002	-.001	-.000	-.003
1.154	-.074	.016	-.013	-.043	.011	-.066	-.002	-.001	-.001	.000	-.001
1.174	-.069	.015	-.016	-.044	.012	-.064	-.003	-.001	-.001	.001	-.001

K GEV/C	B 1	B 2	B 3	B 4	B 5	B 6	B 7	B 8	B 9	B 10	B 11
1.210	-.063	.005	-.021	-.045	.010	-.062	-.002	-.000	-.001	-.001	-.001
1.235	-.063	-.003	-.020	-.045	.010	-.061	-.002	.001	.000	-.001	-.002
1.280	-.057	-.012	-.016	-.044	.010	-.057	.001	.005	.002	-.002	-.001
1.324	-.049	-.026	-.008	-.034	.013	-.054	.004	.010	.004	.000	-.001
1.360	-.046	-.033	-.001	-.023	.016	-.050	.009	.015	.005	.001	-.000
1.400	-.035	-.037	.014	-.004	.027	-.040	.015	.019	.005	.001	-.001
1.430	-.037	-.037	.023	.010	.031	-.037	.020	.020	.006	.001	-.000
1.473	-.026	-.039	.035	.030	.039	-.022	.025	.029	.007	.001	-.001
1.505	-.021	-.034	.042	.042	.045	-.013	.031	.032	.009	.003	-.000
1.550	-.020	-.029	.044	.053	.043	-.006	.032	.033	.008	.003	-.000
1.590	-.017	-.020	.041	.058	.041	-.000	.035	.030	.006	.003	.001
1.640	-.007	-.010	.041	.061	.038	.008	.033	.032	.007	.004	.000
1.680	-.005	-.006	.041	.061	.037	.012	.036	.033	.007	.004	-.000
1.720	-.000	-.000	.038	.060	.033	.013	.032	.032	.006	.003	-.000
1.760	-.000	.006	.039	.058	.032	.013	.034	.029	.006	.003	-.000
1.800	.005	.008	.036	.056	.030	.014	.032	.029	.005	.003	-.001
1.840	.003	.009	.036	.050	.027	.013	.029	.028	.005	.002	-.000
1.880	.000	.009	.039	.047	.025	.012	.029	.025	.005	.001	-.002
1.920	.003	.008	.033	.041	.022	.009	.025	.024	.004	.002	-.001
1.980	-.001	.007	.033	.036	.015	.003	.020	.018	.002	.000	-.002
2.030	.003	.004	.029	.026	.014	-.004	.014	.014	.000	-.001	-.002
2.070	.003	.003	.028	.021	.009	-.007	.012	.009	-.002	-.002	-.003
2.150	.006	-.002	.023	.010	.008	-.015	.003	.002	-.004	-.004	-.003
2.200	.005	-.004	.022	.006	.008	-.017	.001	-.001	-.006	-.006	-.003
2.280	-.001	-.007	.020	-.001	.007	-.017	-.000	-.008	-.007	-.007	-.002
2.340	-.005	-.008	.020	.001	.009	-.017	-.002	-.008	-.008	-.007	-.003
2.460	-.007	-.010	.018	.000	.016	-.011	.003	-.010	-.003	-.005	-.001
2.560	-.007	-.003	.012	.007	.015	-.000	.003	-.003	-.002	-.001	.000
2.750	.001	.005	.012	.007	.019	.008	.011	.004	.005	.004	.005
3.000	.001	.010	.011	.011	.018	.014	.013	.009	.009	.007	.006
3.400	.006	.009	.010	.011	.010	.010	.009	.006	.007	.005	.003
3.650	.008	.009	.010	.011	.011	.011	.010	.009	.009	.007	.006
4.000	.004	.008	.008	.009	.010	.009	.009	.006	.007	.005	.006
5.000	.003	.005	.005	.007	.007	.007	.007	.007	.006	.005	.004
6.000	.001	.003	.003	.005	.005	.005	.005	.005	.004	.004	.004

2.4 Further results derived from πN partial waves

2.4.1 Parameters of pion-nucleon resonances

2.4.1.1 The notion of a resonance

In partial wave analysis, one frequently finds Argand diagrams of partial wave amplitudes which show a "resonance-like behaviour": as the energy increases, the complex vector $T_{\ell\pm}$ moves in a counter-clockwise direction along a more or less deformed circle (see Fig. A.3.1 and Figs. 2.2.1···2.2.21).

The interpretation is that of a scattering resonance. In the scattering process, an unstable excited state of the nucleon is formed and decays within a time of the order of $\tau = \hbar/\Gamma \approx 10^{-23}$ s. Of course, one has to consider the possibility that the "resonance-like" shape of the Argand diagrams is produced by other effects. Therefore, one has to check whether the criteria for the existence of a resonance are fulfilled.

These criteria are derived from resonance models. For instance, the properties of resonances can be studied in potential scattering (72 R Nussenzveig).

Another example for resonance scattering, which is nearer to our application, is a Lee model (54 L Lee). This model describes 3 particles (V, N, θ), which are coupled in such a way that the only possible transition is $V \leftrightarrow N + \theta$. Resonance scattering $N + \theta \to V \to N + \theta$ occurs, if $m_V > m_N + m_\theta$ and if the VNθ-coupling is not too strong. V corresponds to a nucleon resonance, for instance Δ(1232), and N, θ to N, π.

For a vanishing decay coupling, the excited state V is a stable particle. Its discrete eigenvalue lies at $E=m_V$ in the continuous spectrum of the N+θ states and the S-function has a pole on the real E-axis (E=energy).

If the V→N+θ decay coupling is turned on, the pole moves to a complex position in the second sheet. Its distance from the real E-axis determines the width of the scattering resonance. If $|V\rangle$ is chosen as the initial state, the time dependence of its decay follows an exponential law in a very good approximation, and this law is a consequence of the existence of the complex pole.

For increasing values of the coupling constant, the pole moves to the left and finally arrives on the real E-axis below threshold, describing again a stable particle (see 58 L Höhler for details).

The association of poles of the S-matrix with particles and resonances is one of the fundamental features of S-matrix theory (66 R Chew). The particle-like properties of a resonance are well-pronounced, if the pole is located near the physical region. If the distance becomes larger, they fade away and the notion of a resonance looses gradually its usefulness.

Furthermore, the typical properties of a resonance are modified, if the complex pole is located near a threshold branchpoint. An example is the πN resonance S 11 N(1535), which lies just above the η-production threshold. Another well-known example is the complex pole, which belongs to the "virtual" ^{1}S-state in NN-scattering near threshold.

From the above mentioned models and others, which include inelastic processes, one can derive the following criteria for the existence of a resonance:

i) It should be possible to represent the energy dependence of the elastic partial wave amplitude by a Breit-Wigner-type formula and a background or, equivalently, an analytic continuation of the partial wave amplitude into the lower half plane should lead to clear evidence for a pole.

ii) Effects of the resonance should also exist in other channels, including electromagnetic interactions. For instance, the Δ(1232) isobar is dominant also in the reactions $\gamma p \to p\pi^0$ and $\gamma p \to \gamma p$. Of course, in special cases the coupling to other channels can be suppressed by selection rules.

The first criterion includes the following properties, which are related to each other:

(a) In its Argand diagram, $T_{\ell\pm}$ moves along a (frequently deformed) "circle" in a *counter-clockwise direction* (we shall omit the index $\ell\pm$ in the following).

(b) The resonant phase shift is *rapidly increasing* by an amount of the order of π. In simple cases, $\delta=\pi/2$ lies at the center of the rapid variation, but this is not true if background effects are present [see the Argand diagram of the S 31 Δ(1650) resonance, Fig. 2.2.2].

(c) In order to study the rapid variations of T and of δ in detail, one can calculate the "speed" $|\mathrm{d}T/\mathrm{d}W|$ as a function of W. For a resonant amplitude, the speed must have a well-pronounced maximum. Its W-value is an estimate for the resonance mass M and its height gives $2/\Gamma$, where Γ is the resonance width (see Sect. 2.4.1.2).

The magnitude of the speed has the following physical interpretation (55 L Wigner, 64 R Goldberger-Watson). If the scattering process is described by wave packets, one can calculate a "time delay" by comparing the free motion of the incident particle with the motion in the scattering process. The interaction with the scatterer leads to a time delay

$$Q=-i\hbar\frac{\mathrm{d}S}{\mathrm{d}W}S^{-1}=2\hbar\frac{\mathrm{d}\delta}{\mathrm{d}W}. \tag{4.1.1}$$

The first version of the formula is valid also in the multichannel case, where Q becomes the lifetime matrix.

One can expect large positive values of Q and therefore of $\mathrm{d}\delta/\mathrm{d}W$ and $|\mathrm{d}T/\mathrm{d}W|$ in resonance scattering, where the incident particle and the target particle form an unstable particle, which can have an appreciable lifetime (in the time scale of 10^{-23} s). However, in a *causal theory*, there is no mechanism which can produce a large negative value of Q, i.e. a *rapidly decreasing phase shift* corresponding to a behaviour of T in the Argand diagram, in which the vector moves rapidly in a *clockwise* direction.

A quantitative investigation for non-relativistic S-wave scattering has been performed by Wigner (55 L), who derived the following "causal inequality" for a case, where the interaction vanishes outside a sphere of radius a (units $\hbar=1$)

$$\frac{\mathrm{d}\delta}{\mathrm{d}q}\geqq-\left(a+\frac{1}{2q}\right), \tag{4.1.2}$$

where q is the c.m. momentum. See 72 R Nussenzveig for a detailed treatment.

In the case of πN partial waves, a general bound for $\mathrm{d}\delta/\mathrm{d}W$ cannot be given. A partial wave $T(s)$ is compatible with causality, if it fulfills the partial wave dispersion relation, assuming a "reasonable" behaviour of the unknown contributions of the distant cuts (Sect. A.7.1).

Another aspect of the time delay was investigated by SMITH (60 L), who used the time-independent scattering theory in order to calculate the excess of particles in the interaction region over the number which would be there if the interaction were turned off. This problem is related to that of the calculation of the "Q-factor" of an electromagnetic resonator. The result is equivalent to Eq. (4.1.1) (see also the summary in 70 R DALITZ-MOORHOUSE). – Reviews on the decay of unstable states: 77 R and 78 R FONDA.

Resonance-like phenomena

For the discussion of partial wave amplitudes, it is of interest to know that there exist several phenomena which have some but not all properties of a resonance. They are not classified as resonances, because the amplitude does not have a pole in the second sheet.

i) The model of CALUCCI, FONDA, and GHIRARDI

One could think that a rapid increase of the phase shift by an amount of the order π indicates always the presence of a nearby pole or at least of a nearby singularity. However, this is not true. CALUCCI et al. have been able to construct a model with the following properties (CALUCCI 68 L-1, 68 L-2, 72 R FONDA, 73 L FONDA).

In the neighborhood of a given energy $E=E_R$, the phase shift is rapidly rising from 0 to π in a way which approaches a step function, if one of the parameters goes to infinity. The S-function, which belongs to this phase shift, is analytic in the energy plane except for a left hand cut from $-\infty$ to $-\mu$ and a right hand cut from 0 to $+\infty$. The left hand cut is of a logarithmic type and the right hand cut of a square root type. The S-function goes to 1 for $E\to\infty$ in any direction in the E-plane and in any of the Riemannian sheets. There are no other singularities, in particular no poles, besides the above mentioned branch points.

It is possible to approximate a Breit-Wigner phase shift in a finite region around the resonance energy to any desired accuracy. Therefore, this model can lead approximately to the same exponential decay law which follows for a Breit-Wigner resonance.

The work of CALUCCI et al. shows that a typical "narrow resonance" behaviour of $\delta(W)$ and the corresponding time delay cannot only be described by a resonance pole but also in other less transparent ways. If one had very accurate data, one could expect to be able to distinguish both cases by an analytic continuation. MIYAKOSHI and KANAZAWA (77 L) made a calculation of this type and found indeed a characteristic difference. But their method has still to be tested in more difficult cases, where the resonance is inelastic and has a large background.

Up to now, a physical mechanism which belongs to the S-matrix of CALUCCI et al. is not known. There is no reason to apply this parametrization in practice, but its existence is a warning. It shows that uncertainties of pole positions determined by analytic continuation can have unexpected errors, if the discontinuity along the cut behaves similarly as in the ansatz of CALUCCI et al.

ii) SCHMID loops

SCHMID (68 L, 69 L) considered partial wave projections, Eq. (A.3.8), of invariant πN amplitudes for reggeized ϱ-exchange and found that the shape of the Argand diagrams is similar to that of resonant amplitudes. However, the W-dependence of the "speed" $|dA_\ell/dW|$, etc. is different. It does not show the pronounced maximum which is characteristic for a resonance.

It is clear that the invariant Regge amplitude has no pole in the s-plane and no structure in its s-dependence at fixed t. The loops of the different partial wave amplitudes are highly correlated and cancel in the sum of the partial wave expansion.

SCHMIDs result is a crude approximation, because the Regge pole model for ϱ-exchange is used at relatively low energies (1.0⋯2.4 GeV/c), and not only for near-forward scattering, but also at large angles where other mechanisms are dominant.

The relation between the SCHMID loops and the Argand diagrams of partial wave analysis can be discussed as follows. Above about 4 GeV/c, the Argand diagrams of the isospin odd invariant amplitudes at fixed-t are approximately straight lines, as expected from a reggeized ϱ-exchange model according to Eqs. (A.9.16) and (A.9.17). At lower momenta, the amplitudes can be decomposed into the continuation of the straight lines (the Regge background) and a second component, which has a resonance structure in the s-dependence at fixed t. (The Argand diagrams are shown in Sect. 6.2 of the "Handbook", 79 R HÖHLER-KAISER-KOCH-PIETARINEN; structures are seen up to about 3⋯4 GeV/c.)

Since the partial wave projection is a linear operation, one can add the projection of the Regge background, which gives the SCHMID loops, and that of the difference between the total amplitude and the background, which shows relatively large resonance structures.

The latter contribution modifies the SCHMID loops in such a way that the criteria for a resonance are fulfilled.

For a quantitative discussion, it would be better to consider a decomposition into three components. First, Im A and Im B are split into "elastic" and "inelastic" parts as defined in Sect. A.5.1.2. Then, Re A_{el} and Re A_{in} etc. are calculated from the corresponding imaginary parts and the dispersion relation. Finally, the "inelastic part" is split into the reggeized ϱ-contribution and the rest, which has relatively large resonance structures up to 3⋯4 GeV/c (75 L JAKOB).

SCHMIDs first paper has raised a controversial discussion. References are given in his second paper (69 L).

iii) The diffraction background

In the isospin even invariant amplitude, the main background contribution comes from diffraction. For a semiquantitative discussion, one can describe the diffraction peak by an ansatz

$$\frac{d\sigma^+}{dt} \approx (\mathrm{Im}\, F^+_{++})^2 \approx \frac{(\sigma^+)^2}{16\pi} e^{Bt}, \tag{4.1.3}$$

where σ^+ and B are energy independent.

In order to have a simple partial wave decomposition, we use instead the same ansatz for the no-flip amplitude

$$\mathrm{Im}\, G^+(s,t) = \frac{q\sigma^+}{4\pi} e^{Bt/2}. \tag{4.1.4}$$

The partial wave projection of this expression reads

$$\mathrm{Im}\, \hat{T}_\ell(s) = \frac{q^2\sigma^+}{4\pi}\left(\frac{\pi}{2Bq^2}\right)^{1/2} e^{-Bq^2} I_{\ell+1/2}(Bq^2), \tag{4.1.5}$$

where the isospin index of the partial waves is omitted and

$$(2\ell+1)\hat{T} \equiv (\ell+1)T_{\ell+} + \ell T_{\ell-}; \tag{4.1.6}$$

$I_{\ell+1/2}$ is a Bessel function with imaginary argument.

As mentioned in Sect. A.3.2, Im $\hat{T}_\ell$ agrees approximately with the impact parameter transform

$$\mathrm{Im}\, \hat{T}_\ell(s) \approx \frac{\sigma^+}{8\pi}\Gamma(s,b^2) = \frac{\sigma^+}{8\pi B}\exp(-b^2/2B) = \frac{\sigma^+}{8\pi B}\exp[-(\ell+1/2)^2/2Bq^2] \tag{4.1.7}$$

for q^2-values which are not too small.

The partial waves Im $\hat{T}_\ell$ start at zero and are monotonously increasing at higher energies towards $\sigma^+/8\pi B$, which amounts to 0.27 for $B = 9\ \mathrm{GeV}^{-2}$, $\sigma^+ = 24$ mb. Total absorption corresponds to the value 0.5.

For increasing values of ℓ, the maximum of the slope (d/ds)Im $\hat{T}_\ell$ is located at increasing energies. This is a geometrical effect, since the main contribution to the projection integral comes from small t. The slope $(d/dt)P_\ell(1+t/2q^2)$ at $t=0$ is $\ell(\ell+1)/4q^2$. If we consider a fixed value of ℓ, and q^2 small and increasing, we obtain an appreciable contribution of the corresponding partial wave, if q^2 is so large that the slope $(d/dt)P_\ell$ becomes comparable with the fixed slope in t of the diffraction amplitude.

The steepest increase of Im $\hat{T}_\ell$ occurs approximately in the energy range of the peripheral resonances, at least for $\ell = 3$ to 5.

It is interesting to consider the effect of this background contribution to the integral over the physical cut in the partial wave dispersion relation, Eq. (A.7.1). The energy dependence of Im f_ℓ = Im T_ℓ/q is similar to that of a strongly inelastic resonance. The rapid increase of the imaginary part leads to a positive contribution from the principal value integral to the real part, and this gives a "resonance-like" structure in the Argand diagram.

Although the model is rather crude, there is no doubt that a more elaborate diffraction model would lead to a similar result. The situation is analogous to that in the case of the isospin odd background: again we have a background effect which leads to a (strongly deformed) "resonance-like circle" in the Argand diagram. The criteria for a resonance are fulfilled only, if one adds the contribution from the difference between the amplitude reconstructed from partial waves and the assumed diffraction amplitude.

Our discussion is incomplete insofar as we have considered only the no-flip amplitude, which leads to a result for a combination of partial waves, Eq. (4.1.6). Partial wave analysis shows that there is a repulsive background in some partial waves (see the figures in Sect. 2.2.2), and an attractive one in others. The question is how the latter is related to the mechanism of the higher resonances.

In the "dual absorption model" of HARARI and others (71 L, 72 L), it is assumed that the isospin even amplitude can be decomposed into a "pomeron" contribution and a secondary term, and that the latter is "dual" to the s-channel resonances. The above remark shows that it is difficult to see, how one can separate the "pomeron" contribution to a partial wave from the resonant contribution.

The discussion of the partial wave dispersion relation is related to an investigation by BALL and FRAZER (61 L), who studied resonance-like phenomena, which are produced by a rapidly increasing absorption. However, these authors used a dispersion relation for the complex phase shift. Further details on the partial wave projection of the diffraction amplitude are given in Sect. 2.6.3 and in 73 L HÖHLER-1.

A thorough treatment of the notion of a resonance in elementary particle physics can be found in the book of BRANSDEN and MOORHOUSE (73 R). See also 59 R BRENIG-HAAG, 64 R GOLDBERGER-WATSON, 70 R DALITZ-MOORHOUSE, the textbook 79 R BOEHM and, for resonance phenomena in dispersion theory, 62 R AMATI-FUBINI and 63 R FULTON. The practical criteria for resonances are discussed in the Review of Particle Properties (80 R BRICMAN).

2.4.1.2 Parametrizations of resonances

All determinations of resonance parameters are based on the assumption that the resonance can be described by a pole of the partial wave amplitude in the lower half of the second sheet of the W- or s-plane (Breit-Wigner ansatz). In the following, we omit the lower index of the partial wave amplitudes $T_{\ell\pm}$. The simplest Breit-Wigner formula reads

$$T_R(W)=\frac{R_1}{M-W-i\Gamma/2}. \tag{4.1.8}$$

The resonance is characterized by 4 real parameters: the location of the pole in the W-plane: $M-i\Gamma/2$ and the complex residue R_1. $\mathrm{Im}\,T_R$ and $|T_R|^2$ have a peak at $W=M$, whose full width at half height is given by Γ.

In the case of elastic scattering, $R_1=\Gamma/2$ and T_R describes a circle of radius 1/2 in a counter-clockwise direction as the energy increases. (Fig. A.3.1). The "speed" has a well-pronounced maximum at $W=M$, its height being

$$\left|\frac{dT_R}{dW}\right|_{W=M}=\frac{2}{\Gamma}. \tag{4.1.9}$$

If T_R is considered as a function of s, the Breit-Wigner parametrization reads

$$T_R(s)=\frac{R}{M^2-s-iM\Gamma}. \tag{4.1.10}$$

$R=M\Gamma$ in the case of elastic scattering.

For small values of Γ, the location of the pole and the interpretation of Γ are approximately the same as in the case of $T_R(W)$, Eq. (4.1.8).

A comparison with the properties of partial waves listed in Sect. A.7.1 shows that the above Breit-Wigner formulas are too simple because

i) they do not have the correct cut structure,

ii) they do not fulfill the reality condition Eq. (A.7.20).

In order to take into account the left hand cut contributions, one can add a "background" term, which also includes the tails of other resonances. The "physical cut" is usually introduced by assuming a suitable dependence of R and Γ on the c.m. momentum q.

According to the reality condition, Eq. (A.7.20), resonance poles can occur only in pairs at complex conjugate locations. However, one should note that the pole at $s=M^2-iM\Gamma$ lies much nearer to the physical region than the pole at $s=M^2+iM\Gamma$. The latter lies in the upper half of the second sheet. Starting from physical s-values, it can be reached only if one goes around the branch point. Therefore, its effect can be considered as part of the non-resonant background.

The background amplitude has to be introduced in such a way that unitarity is secured. The usual procedure is to take a product of two unitary S-functions

$$S=1+2iT=S_R S_B \tag{4.1.11}$$

and to define the resonance and background parts of the T-functions by

$$S_R=1+2iT_R; \quad S_B=1+2iT_B. \tag{4.1.12}$$

This leads to

$$T = T_R S_B + T_B; \tag{4.1.13}$$

(see for instance 76 L CUTKOSKY and 73 L LICHTENBERG, where earlier references are given).

As long as a theory of the pion-nucleon resonances does not exist, the deviation $T - T_R$ of the physical amplitude from the simple resonance formula of Eqs. (4.1.8) or (4.1.10) can be taken into account in different ways. In general, a modified version of the simplest Breit-Wigner formula Eq. (4.1.10) is used

$$T_R(s) = \frac{R(q)}{M^2 - s - \mathrm{i}M\Gamma(q)} \tag{4.1.14}$$

and the rest of the deviation is described by the background term T_B.

This method offers the possibility to include the elastic phase space factor in $R(q)$ and the sum of the elastic and all inelastic phase space factors in $\Gamma(q)$.

Recent discussions of multichannel resonance parametrizations are due to CUTKOSKY (79 L-2) and NOVOSELLER (80 L). — A duality diagram prediction for the resonance background was derived by DICKEY (75 L).

Parameters of nucleon resonances

The question is now, how to define parameters of the resonances, which can be used for a comparison of the results of different partial wave analyses. There are two conventions:

i) Most of the tables list the parameters M, Γ and Γ_{el} or $x_{el} = \Gamma_{el}/\Gamma$.

The *resonance mass* M is defined to be the (real) c.m. energy W at which the real part of the denominator of the resonance term T_R, Eq. (4.1.14), vanishes. One should note that a general statement on the location of this point on the "resonance circle" cannot be made, because there is a complex factor RS_B in the numerator of the first term in Eq. (4.1.13).

The *resonance width* Γ is defined by $\Gamma(q)$, taken at $W = M$.

Assuming that the energy dependence of the other terms is negligible, one can estimate the resonance parameters by calculating the "speed" from Eq. (4.1.9). Its maximum occurs at $W = M$ and has the value $4R_1/\Gamma^2$; $R_1 = \Gamma_{el}/2$.

Although $R(q)$ can have complex values, it is usually chosen real, because a phase of $R(q)$ can be included in the background factor S_B in Eq. (4.1.13). Introducing $R = M\Gamma_{el}$, the resonant term reads

$$T_R(s) = \frac{M\Gamma_{el}(q)}{M^2 - s - \mathrm{i}M\Gamma(q)}; \qquad x_{el}(q) = \frac{\Gamma_{el}(q)}{\Gamma(q)}. \tag{4.1.15}$$

The third resonance parameter is the *elasticity* x_{el}, taken at $W = M$.

If one considers the errors of the parameters M, Γ, and x_{el}, one should keep in mind that an additional uncertainty comes from the fact that the choice for $\Gamma(q)$ and $\Gamma_{el}(q)$ or $x_{el}(q)$ as well as the parametrization of $S_B(s)$ are to some extent arbitrary.

ii) Another set of resonance parameters is given by the location of the complex pole of $T(s)$ and by the value of the complex residue. The advantage is that these parameters are defined in a unique way. However, there is also the disadvantage that one has to perform an analytic continuation into the second sheet over an appreciable distance.

The continuation can be carried out in several different ways:

(a) One fits a parametrization of the type of Eqs. (4.1.13) and (4.1.15) to the experimental partial wave amplitude and calculates the location of the pole and the value of its residue.

For this purpose, one can use a parametrization in which the width parameter in the denominator of T_R is taken to be a constant. The necessary flexibility is obtained by the choice of $x_{el}(q)$ and $S_B(q)$. The advantage is that the location of the pole is directly given by two of the parameters of the fit (73 L LICHTENBERG).

Another method based on a Breit-Wigner type parametrization is due to VASAN (76 L). BALL et al. (72 L) used an N/D-method.

(b) A second class of papers is devoted to "model-independent" determinations of the location of the pole. Polynomial fits were used to the zero trajectories (Sect. 2.4.3) by SPEARMAN (74 L) and to the K-matrix by BALL and GOBLE (75 L).

Improved versions of expansion methods, which take into account a stabilizing condition were developed by NOGOVA and PISUT (73 L), LICHARD (75 L) and by CAPRINI, CIULLI, POMPONIOU and SABBA-STEFANESCU (72 L), SABBA-STEFANESCU (73 L), KANAZAWA and MIYAKOSHI (77 L), and MIYAKOSHI and KANAZAWA (77 L).

The problem of determining the pole position occurs also in other reactions, for instance in pion-pion scattering. The last mentioned method (b) was applied in this case by KIMURA et al. (78 L) and by LANG (79 L), who used PIETARINENs smoothness condition (72 L-3) as a stabilizing lever.

Parametrizations used in the CMU-LBL and KH analyses

CMU-LBL analyses

In the version of 1976, the authors used essentially a parametrization as described in Eqs. (4.1.13) and (4.1.15). A much greater effort was spent for the 1979 version, where the partial waves are fitted using a multichannel coupled-resonance scheme. This approach is similar to the use of a K-matrix formalism, but it is easier to build in the correct analyticity properties. Data on production channels were included, but their errors were stretched by a factor of 3. Full details can be found in the paper 79 L CUTKOSKY-2.

Further work of CUTKOSKY et al. (80 L-1) is in progress. In particular it was found that the recently proposed R-matrix approach (79 L JAFFE and LOW, 79 L ROIESNEL) does not provide an economical parametrization as the background repulsion is too large (earlier work on this method: 64 L FESHBACH and LOMON).

The KH 78 analysis

The ansatz Eq. (4.1.15) was used for $T_R(s)$, taking into account higher two-particle thresholds (ηn, πΔ, ϱN, etc.) and a diffractive background (78 L PIETARINEN-1, 79 R HÖHLER-KAISER-KOCH-PIETARINEN).

A new approach to multichannel resonance parametrization was recently proposed by NOVOSELLER (80 L). The method is simpler than the K-matrix formalism, but it still has several weaknesses. For instance there is no procedure for overlapping resonances and for the incorporation of the general structure demanded by analyticity.

Literature on resonance formulas

The first discussion on the shape of Argand diagrams of resonant amplitudes can be found in ADAIRs paper (59 L), which is based on the scattering formalism of WIGNER and EISENBUD (47 L) and includes the connection with the forward dispersion relation.

Starting from ADAIRs work (59 L), DONNACHIE (67 R) describes the resonance parametrization used in the CERN partial wave analysis.

In their review, DALITZ and MOORHOUSE (70 R) discuss the earlier work on the parametrization of multichannel resonances. They conclude that, due to the complexity of branch cuts without physical significance, the eigenphase representation of the S-matrix is not useful. Following GOEBEL-MCVOY (67 R), and MICHAEL (66 L), they derive another representation which is used for the discussion of Argand diagrams of resonant amplitudes.

A thorough treatment of resonance formulas can also be found in the books BRANDSEN-MOORHOUSE (73 R), PERL (74 R) and PILKUHN (67 R and 79 R).

2.4.1.3 Results for the parameters of nucleon resonances

In the following we shall present tables of resonance parameters which were determined in recent partial wave analyses. Some reviews on hadron spectroscopy: 68 R BARBARO-GALTIERI, 73 R DONNACHIE, 79 R PROTOPOPESCU-SAMIOS, 80 R LIPKIN. The new review of the Particle Data Group (82 R ROOS) appeared after the completion of the manuscript. We have used 80 R BRICMAN.

Table 2.4.1.1. Resonance parameters derived from the KH 78 partial wave solution. In the first column the name of the excited nucleon state contains a nominal mass value, which agrees with the value in the table of the Particle Data Group (80 R BRICMAN). – The SU(6)×O(3) classification is explained in Sect. 2.4.1.5. It gives only a first orientation. Mixing is ignored and only the dominant state is listed. In some cases the assignment is not yet well established. – The errors of the resonance parameters are of interest mainly relatively to each other. There are additional uncertainties and, in some cases, important error correlations. – The values of the lab.momentum k and of s were calculated for $W=M$. – The impact parameter b is defined by $b=(J+1/2)/q$, Eq. (6.3.8). – The last column gives the usual classification by assigning stars. Only 4-star resonances are well established. 3-star resonances are good candidates which are still in need of clarification. 2-star objects are open to considerable question. We have listed the 1-star objects separately, because they belong to weak structures which can disappear, if the data are fitted in a different way. Further information on the status of a resonance follows from the Argand plots in Sect. 2.2.2 and from the "Data Card Listings" in the Review of Particle Properties (p. 175 ff.) (80 R BRICMAN). Column 3: $\ell\pm$ indicates $J=\ell\pm 1/2$.

Name	ℓ 2I 2J	$\ell\pm$	J^P	SU(6)×O(3)	Mass $M\pm\Delta M$ MeV	Full width $\Gamma\pm\Delta\Gamma$ MeV	Elasticity $x_{el}\pm\Delta x_{el}$	b fm	k GeV/c	s GeV²	Stars
N(1470)	P11	1−	$1/2^+$	($\underline{56}$, 0^+)*	1410 ± 12	135 ± 10	0.51 ± 0.05	0.53	0.563	1.988	4
N(1520)	D13	2−	$3/2^-$	($\underline{70}$, 1^-)	1519 ± 4	114 ± 7	0.54 ± 0.03	0.87	0.737	2.307	4
N(1535)	S11	0+	$1/2^-$	($\underline{70}$, 1^-)	1526 ± 7	120 ± 20	0.38 ± 0.04	0.43	0.749	2.329	4
N(1650)	S11	0+	$1/2^-$	($\underline{70}$, 1^-)	1670 ± 8	180 ± 20	0.61 ± 0.04	0.35	0.997	2.789	4
N(1670)	D15	2+	$5/2^-$	($\underline{70}$, 1^-)	1679 ± 8	120 ± 15	0.38 ± 0.03	1.05	1.013	2.819	4
N(1688)	F15	3−	$5/2^+$	($\underline{56}$, 2^+)	1684 ± 3	128 ± 8	0.65 ± 0.02	1.04	1.022	2.836	4
N(1700)	D13	2−	$3/2^-$	($\underline{70}$, 1^-)	1731 ± 15	110 ± 30	0.08 ± 0.03	0.66	1.108	2.996	2
N(1710)	P11	1−	$1/2^+$	($\underline{70}$, 0^+)	1723 ± 9	120 ± 15	0.12 ± 0.04	0.33	1.094	2.969	3
N(1810)	P13	1+	$3/2^+$	($\underline{56}$, 2^-)	1710 ± 20	190 ± 30	0.14 ± 0.03	0.67	1.070	2.924	4
N(1990)	F17	3+	$7/2^+$	($\underline{70}$, 2^+)	2005 ± 150	350 ± 100	0.04 ± 0.02	1.02	1.657	4.020	2
N(2040)	D13	2−	$3/2^-$	($\underline{70}$, 3^-)	2081 ± 20	265 ± 40	0.06 ± 0.02	0.48	1.823	4.331	2
N(2100)	D15	2+	$5/2^-$	($\underline{70}$, 3^-)	2228 ± 30	310 ± 50	0.07 ± 0.02	0.62	2.161	4.964	2
N(2190)	G17	4−	$7/2^-$	($\underline{70}$, 3^-)	2140 ± 12	390 ± 30	0.14 ± 0.02	0.92	1.956	4.580	4
N(2200)	G19	4+	$9/2^-$	($\underline{70}$, 3^-)	2268 ± 15	300 ± 40	0.10 ± 0.02	1.06	2.257	5.144	3
N(2220)	H19	5−	$9/2^+$	($\underline{56}$, 4^+)	2205 ± 10	365 ± 30	0.18 ± 0.015	1.10	2.107	4.862	4
N(2600)	I1, 11	6−	$11/2^-$	($\underline{70}$, 5^-)	2577 ± 50	400 ± 100	0.05 ± 0.01	1.06	3.056	6.641	3
N(2700)	K1, 13	7−	$13/2^+$	($\underline{56}$, 6^+)	2612 ± 45	350 ± 50	0.04 ± 0.01	1.22	3.153	6.823	2
N(2000)	F15	3−	$5/2^+$	($\underline{70}$, 2^+)	1882 ± 10	95 ± 20	0.04 ± 0.02	0.85	1.401	3.542	1
	P11	1−	$1/2^+$	($\underline{70}$, 2^+)	2050 ± 20	200 ± 30	0.10 ± 0.04	0.25	1.754	4.203	1
N(2100)	S11	0+	$1/2^-$	($\underline{56}$, 1^-)	1880 ± 20	95 ± 30	0.09 ± 0.05	0.28	1.397	3.534	1
N(2800)	G19	4+	$9/2^-$	($\underline{70}$, 5^-)	2792 ± 100	240 ± 100	0.02 ± 0.015	0.80	3.672	7.795	1

(continued)

Table 2.4.1.1 (continued)

Name	ℓ 2I 2J	$\ell \pm$	J^P	SU(6)×O(3)	Mass $M \pm \Delta M$ MeV	Full width $\Gamma \pm \Delta\Gamma$ MeV	Elasticity $x_{el} \pm \Delta x_{el}$	b fm	k GeV/c	s GeV^2	Stars
Δ(1232)	P33	1+	$3/2^+$	$(\underline{56}, 0^+)$	1233 ± 2	116 ± 5	1.0	1.73	0.300	1.520	4
Δ(1650)	S31	0+	$1/2^-$	$(\underline{70}, 1^-)$	1610 ± 7	139 ± 18	0.35 ± 0.06	0.38	0.891	2.592	4
Δ(1670)	D33	2−	$3/2^-$	$(\underline{70}, 1^-)$	1680 ± 70	230 ± 80	0.20 ± 0.03	0.70	1.015	2.822	4
Δ(1890)	F35	3−	$5/2^+$	$(\underline{56}, 2^+)$	1905 ± 20	260 ± 20	0.15 ± 0.02	0.83	1.448	3.629	4
Δ(1910)	P31	1−	$1/2^+$	$(\underline{56}, 2^+)$	1888 ± 20	280 ± 50	0.24 ± 0.06	0.28	1.413	3.565	4
Δ(1950)	F37	3+	$7/2^+$	$(\underline{56}, 2^+)$	1913 ± 8	224 ± 10	0.38 ± 0.02	1.10	1.464	3.660	4
Δ(1960)	P33	1+	$3/2^+$	$(\underline{56}, 2^+)$	1868 ± 10	220 ± 80	0.14 ± 0.04	0.57	1.373	3.489	3
Δ(1960)	D35	2+	$5/2^-$	$(\underline{56}, 1^-)$	1901 ± 15	195 ± 60	0.04 ± 0.03	0.83	1.439	3.614	2
Δ(2420)	H3, 11	5+	$11/2^+$	$(\underline{56}, 4^+)$	2416 ± 17	340 ± 28	0.08 ± 0.015	1.16	2.627	5.837	3
Δ(2750)	I3, 13	6+	$13/2^-$	$(\underline{70}, 7^-)$	2794 ± 80	350 ± 100	0.04 ± 0.015	1.12	3.678	7.806	2
Δ(2950)	K3, 15	7+	$15/2^+$	$(\underline{56}, 6^+)$	2990 ± 100	330 ± 100	0.04 ± 0.02	1.18	4.282	8.940	2
Δ(1690)	P33	1+	$3/2^+$	$(\underline{56}, 0^+)$*	1522 ± 13	220 ± 40	0.21 ± 0.06	0.86	0.742	2.316	1
Δ(1900)	S31	0+	$1/2^-$	$(\underline{56}, 1^-)$	1908 ± 30	140 ± 40	0.08 ± 0.04	0.28	1.454	3.640	1
Δ(2160)	G37	4−	$7/2^-$	$(\underline{70}, 3^-)$	2215 ± 60	400 ± 100	0.05 ± 0.02	0.87	2.130	4.906	1
Δ(2300)	H39	5−	$9/2^+$	$(\underline{56}, 4^+)$	2217 ± 80	300 ± 100	0.03 ± 0.02	1.09	2.135	4.915	1
	D35	2+	$5/2^-$	$(\underline{70}, 3^-)$	2305 ± 26	300 ± 70	0.04 ± 0.02	0.62	2.348	5.313	1
	F37	3+	$7/2^+$	$(\underline{56}, 4^+)$	2425 ± 60	300 ± 80	0.07 ± 0.04	0.70	2.651	5.881	1
Δ(2500)	G39	4+	$9/2^-$	$(\underline{70}, 5^-)$	2468 ± 50	480 ± 100	0.06 ± 0.03	0.94	2.763	6.091	1

Table 2.4.1.2. Resonance parameters derived from the CMU – LBL 79 partial wave solution. This table is the most recent result (80 L Cutkosky-1, July 1980). It differs from the table in 79 L Cutkosky-2 (see Table 2.4.1.4).

The first column gives the name of the resonance, which is used for identification. The mass value is a nominal one and not a best average (80 R Bricman, quoted as "PDG" = "Particle Data Group"). The "quality" is denoted in the usual star notation (see the caption of Table 2.4.1.1). The additional letters in the classification of the CMU – LBL group have the following meaning. H or Q: half or a quarter of the resonance structure lies above the range of the analysis. The accuracy of the determination of the parameters is reduced, because it was necessary to use SACLAY partial waves. E.: the partial wave amplitudes have large errors or are erratic in the neighborhood of the resonance. This affects the accuracy of the determination of parameters. There may also be additional real structure. The errors (90% confidence level) are listed below the values of the parameters in the fourth and subsequent columns. R is the residue of the pole and θ the phase angle.

Name PDG	Quality PDG	Quality CMU-LBL	Mass MeV M / ΔM	Width MeV Γ / $\Delta\Gamma$	Elasticity x_{el} / Δx_{el}	Pole position MeV Re W / ΔRe W	Pole position MeV $-$Im W / Δ Im W	Residue MeV $\lvert R\rvert$ / $\Delta\lvert R\rvert$	Residue deg θ / $\Delta\theta$
N(939) P11	4								
N(1470) P11	4	4E	1440	340	0.68	1375	90	52	−100
			30	70	0.04	30	20	5	35
N(1710) P11	4	4E	1700	90	0.20	1690	40	8	175
			50	30	0.04	20	10	2	35
P11		1E	2125	260	0.12	2120	120	14	35
			75	100	0.03	40	40	7	25
N(1810) P13	4	4E	1700	125	0.10	1680	60	8	−160
			50	70	0.04	30	20	2	30
N(1688) F15	4	4	1680	120	0.62	1667	55	34	25
			10	10	0.05	5	5	2	5
N(1990) F17	3	4E	1970	350	0.06	1900	130	9	− 60
			50	120	0.02	30	30	3	30
N(2220) H19	4	4Q	2230	500	0.15	2160	240	45	− 45
			80	150	0.03	80	50	20	25
N(1535) S11	4	4	1550	240	0.50	1510	130	120	15
			40	80	0.10	50	40	40	45
N(1650) S11	4	4	1650	150	0.65	1640	75	60	− 75
			30	40	0.10	20	15	10	25
N(2100) S11	1	1E	2180	350	0.18	2150	175	40	0
			80	100	0.08	70	50	20	90
N(1520) D13	4	4	1525	120	0.58	1510	57	35	− 12
			10	15	0.03	5	5	2	5
N(1700) D13	4	4	1675	90	0.11	1660	45	6	0
			25	40	0.05	30	20	3	50
D13		1	1880	180	0.10	1880	80	10	100
			100	60	0.04	100	40	5	80
N(2040) D13	2	3	2060	300	0.14	2050	100	30	0
			80	100	0.07	70	40	20	100
N(1670) D15	4	4	1675	160	0.38	1660	70	31	− 30
			10	20	0.05	10	5	5	10
N(2100) D15	2	3Q	2180	400	0.10	2100	180	20	− 90
			80	100	0.03	60	40	10	50
N(2190) G17	4	4EQ	2200	500	0.12	2100	200	25	− 30
			70	150	0.06	50	80	10	50
N(2200) G19	4	4Q	2250	480	0.10	2150	180	20	− 50
			80	120	0.02	50	50	6	20
Δ(1910) P31	4	4	1910	225	0.19	1880	100	20	− 90
			40	50	0.03	30	20	4	30
Δ(1232) P33	4	4	1232	120	1.00	1210	50	53	− 47
			3	5	0.00	1	1	2	1
Δ(1690) P33	3	4	1600	300	0.18	1550	100	17	−150
			50	100	0.04	40	30	4	30
Δ(1960) P33	2	3	1920	300	0.20	1900	150	24	−150
			80	100	0.05	80	50	4	30
Δ(1890) F35	4	4	1910	400	0.14	1830	140	25	− 50
			30	100	0.03	40	30	8	20
F35		1Q	2200	400	0.07	2150	175	16	150
			125	125	0.04	100	50	5	90
Δ(1950) F37	4	4	1950	340	0.39	1890	130	50	− 33
			15	50	0.04	15	20	7	8
F37		1H	2350	300	0.08	2350	130	12	− 90
			100	100	0.04	100	50	6	60

continued

Table 2.4.1.2 (continued)

Name PDG	Quality PDG	Quality CMU-LBL	Mass MeV M / ΔM	Width MeV Γ / $\Delta\Gamma$	Elasticity x_{el} / Δx_{el}	Pole position MeV Re W / ΔRe W	Pole position MeV $-$Im W / Δ Im W	Residue MeV $\|R\|$ / $\Delta\|R\|$	Residue deg θ / $\Delta\theta$
Δ(2300) H39	1	2H	2400	425	0.06	2370	210	10	− 20
			125	150	0.02	80	80	4	30
Δ(2420) H3,11	3	4H	2400	450	0.08	2360	210	18	− 30
			125	150	0.03	100	50	6	40
Δ(1650) S31	4	4	1620	140	0.25	1600	60	15	−110
			20	20	0.03	15	10	2	20
Δ(1900) S31	2	3	1890	170	0.10	1870	90	10	20
			50	50	0.03	40	25	3	40
S31		2	2150	200	0.08	2140	100	7	− 60
			100	100	0.02	80	40	2	90
Δ(1670) D33	4	4	1710	280	0.12	1675	110	13	− 20
			30	80	0.03	25	20	3	25
D33		2	1940	200	0.05	1900	100	8	135
			100	100	0.02	100	30	3	45
Δ(1960) D35	3	4	1940	320	0.14	1890	130	18	− 20
			30	60	0.04	50	30	6	40
D35		2H	2400	400	0.20	2400	200	15	− 70
			125	150	0.10	125	75	8	70
G37		2Q	2200	450	0.06	2100	170	8	− 70
			80	100	0.02	50	40	3	40
Δ(2500) G39	1	2Q	2300	330	0.05	2260	160	8	− 25
			100	100	0.02	60	80	4	15

Table 2.4.1.3. Resonance parameters derived from HENDRY's partial wave analysis (80 L HENDRY). The loops in his Argand diagrams are qualitatively similar to those at lower energies. Therefore, it is reasonable to attempt an interpretation in terms of resonances. The notation "+" and "○" indicates, whether a resonance-like structure is seen in the preliminary KH solution (80 L KOCH-2). The notation in the first and second columns is the same as in Table 2.4.1.5.

P η τ		Mass $M \pm \Delta M$ MeV	Full width $\Gamma \pm \Delta\Gamma$ MeV	Elasticity $x_{el} \pm \Delta x_{el}$	KH
N_α + + +	H19	2300 ± 100	450 ± 150	0.12 ± 0.04	+
	K1, 13	3000 ± 100	900 ± 150	0.07 ± 0.02	+
	M1, 17	2800 ± 200	1600 ± 200	0.040 ± 0.015	+
N_β − − +	G19	2200 ± 100	350 ± 100	0.09 ± 0.02	+
N_γ − + −	G17	2140 ± 40	270 ± 50	0.16 ± 0.04	+
	I1, 11	2700 ± 100	900 ± 100	0.08 ± 0.02	+
	L1, 15	3500 ± 200	1300 ± 200	0.055 ± 0.02	+
	N1, 19	4100 ± 200	1900 ± 300	0.030 ± 0.015	+
N_δ + − −					○
Δ_α + + +	H39	2450 ± 100	500 ± 200	0.08 ± 0.02	+
	K3, 13	3200 ± 200	1000 ± 300	0.045 ± 0.02	○
Δ_β − − +	G39	2200 ± 100	450 ± 200	0.10 ± 0.03	+
	I3, 13	2650 ± 100	500 ± 100	0.05 ± 0.01	+
	L3, 17	3300 ± 200	1100 ± 300	0.03 ± 0.01	+
	N3, 21	4100 ± 300	1600 ± 500	0.018 ± 0.01	○
Δ_γ − + −	G37	2280 ± 80	400 ± 150	0.09 ± 0.02	+
	I3, 11	2850 ± 150	700 ± 200	0.06 ± 0.02	○
Δ_δ + − −	H3, 11	2400 ± 60	460 ± 100	0.11 ± 0.02	+
	K3, 15	2850 ± 100	700 ± 200	0.03 ± 0.01	+
	M3, 19	3700 ± 200	1300 ± 400	0.025 ± 0.01	+

Table 2.4.1.4. Comparison of resonance masses obtained in different partial wave analyses [76 L Ayed,(SACLAY), 79 R Höhler (KH 78), 79 L Cutkosky-2 (CMU – LBL 79), 80 L Hendry]. In column "KH", we have added preliminary results for resonances at higher energies (80 L Koch-2). The errors in this column include only part of the total uncertainty.
The notation in the first column is explained in Table 2.4.1.5. The third column gives the name of the resonance as defined in 80 R Bricman. The "brightness parameter" B (79 L Cutkosky-2) is a measure of the average contribution of the resonance to $\pi^{\pm}p$ elastic and charge-exchange scattering amplitudes $B = B_0 - \ln|(x_{el}/q)\sqrt{2J+1}\sqrt{2I+5}|$, where $B_0 = 3.2$ is chosen such that $B = 0$ for Δ(1232). A larger brightness corresponds to a fainter resonance. – In the last but one column, numbers in brackets refer to cases, where the Particle Data Group lists a well-established bump, but does not give quantum numbers. (See our comments below.) The classification of the CMU – LBL group is listed in Table 2.4.1.2.
The resonance parameters in column "CMU – LBL" are those published in 1979 (79 L Cutkosky-2) together with the results for the partial waves. They are somewhat different from those presented by Cutkosky at the Toronto Conference (80 L Cutkosky-1) which are listed in Table 2.4.1.2. Mass values in brackets have been obtained by using in addition SACLAY partial waves (76 Ayed).

		PDG MeV	SACLAY MeV	KH MeV	CMU – LBL MeV	Hendry MeV	B	PDG	KH
				Even parity states					
N_α	P11 (1)	939	939	939	939				
N_α	(2)	1470	1413	1410 ± 12	1450 ± 30		1.6	4	4
N_α	(2′)		1532						
N_α	(3)	1710	1730	1723 ± 9	1710 ± 60		3.2	4	3
N_α	(4)			2050 ± 20					1
N_δ	P13	1810	1696	1710 ± 20	1740 ± 80		2.9	4	4
N_α	F15 (1)	1688	1679	1684 ± 3	1680 ± 15		1.4	4	4
N_α	(2)	2000	2025	1882 ± 10				2	1
N_δ	F17	1990	2049	2005 ± 150	1970 ± 80		3.9	3	2
N_α	H19	2220	2249	2205 ± 10	(2250)	2300	2.8	4	4
N_α	K1, 13	2700		2612 ± 45		3000		1	2
N_α	M1, 17			≈3500		3800			
N_α	O1, 21			≈4000					
Δ_α	P31	1910	1789	1888 ± 20	1920 ± 50		3.2	4	4
Δ_δ	P33 (1)	1232	1231	1233 ± 2	–		0	4	4
Δ_δ	(2)	1690		1522 ± 13	1640 ± 50		2.5	3	1
Δ_δ	(3)	1960	1900	1868 ± 10	1960 ± 80		3.0	2	3
Δ_α	F35	1890	1869	1905 ± 20	1920 ± 30		2.8	4	4
Δ_δ	F37 (1)	1950	1928	1913 ± 8	1950 ± 20		1.7	4	4
Δ_δ	(2)			2425 ± 60					1
Δ_α	H39 (1)	2300		2217 ± 80				1	1
Δ_α	(2)					2450			
Δ_δ	H3, 11	2420	2392	2416 ± 17		2400		3	3
Δ_α	K3, 13					3200			
Δ_δ	K3, 15	2950		2990 ± 100		2850		1	2
Δ_δ	M3, 19	3230		≈3500		3700		(3)	

(continued)

Table 2.4.1.4 (continued)

		PDG MeV	SACLAY MeV	KH MeV	CMU−LBL MeV	HENDRY MeV	B	PDG	KH
				Odd parity states					
N_β	S11 (1)	1535	1519	1526 ± 7	1540 ± 20		2.1	4	4
N_β	(2)	1650	1673	1670 ± 8	1640 ± 30		1.9	4	4
N_β	(3)	2100		1880 ± 20				1	1
N_β	(4)		2280						
N_γ	D13 (1)	1520	1525	1519 ± 4	1525 ± 15		1.5	4	4
N_γ	(2)	1700	1710	1731 ± 15	1670 ± 25		3.4	4	2
N_γ	(3)				1830 ± 50		4.1		
N_γ	(4)	2040	2029	2081 ± 20	2100 ± 80		3.6	2	2
N_β	D15 (1)	1670	1660	1679 ± 8	1680 ± 15		2.0	4	4
N_β	(2)	2100	2076	2228 ± 30				2	2
N_γ	G17	2190	2141	2140 ± 12	(2150)	2140	3.0	4	4
N_β	G19 (1)	2200	2133	2268 ± 15	(2200)	2200	3.4	4	3
N_β	(2)	2800		2792 ± 100				1	1
N_γ	I1, 11	2600		2577 ± 50		2700		3	3
N_γ	L1, 15			≈ 3100		3500			
N_γ	N1, 19			3500–4000		4100			
Δ_β	S31 (1)	1650	1623	1610 ± 7	1620 ± 20		2.6	4	4
Δ_β	(2)	1900	2001	1908 ± 30	1850 ± 35		4.0	2	1
Δ_β	(3)				2150 ± 100		4.0		
Δ_γ	D33 (1)	1670	1723	1680 ± 70	1730 ± 30		3.1	4	4
Δ_γ	(2)				2010 ± 100		4.3		
Δ_β	D35 (1)	1960	1894	1901 ± 15	1930 ± 20		3.1	3	2
Δ_β	(2)			2305 ± 26					1
Δ_γ	G37	2160		2215 ± 60	(2200)	2280	4	(3)	1
Δ_β	G39 (1)		2170			2200			
Δ_β	(2)	2500		2468 ± 50				1	1
Δ_γ	I3, 11					2850			
Δ_β	I3, 13	2750		2794 ± 80		2650		1	2
Δ_β	L3, 17	3230		≈ 3300 ?		3300		(3)	
Δ_β	N3, 21					4100			

Discussion of the resonances in the mass region $M \lesssim 2400$ MeV

***I*=1/2**

The Particle Data Group ("PDG") lists 12 4-star resonances in addition to the nucleon. There are one or two 4-star resonances in each partial wave up to H 19, except for F 17, whose resonance has only 3 stars because its Argand diagram is not yet well determined (Fig. 2.2.13).

***I*=3/2**

The PDG lists 6 4-star resonances, one occurring in each partial wave up to F 37 except for D 35, which has a 3-star resonance. Its shape has even a 4-star quality in the CMU-LBL solution, whereas the KH solution suggests a less pronounced structure (Fig. 2.2.10). Furthermore, the recent results of the CMU-LBL and KH groups justify the promotion of H 3,11(2420) to a 4-star status.

If one compares the parameters of the 4-star and 3-star resonances, one should note that it does not make sense to compare the errors. The CMU-LBL group has spent a much greater effort in determining the errors of the partial waves and of the resonance parameters, whereas the errors of the KH group are those of the fit, other sources of errors being ignored.

However, this does not mean that the values of the CMU-LBL group for the masses, widths and elasticities are more accurate than those of the KH group. In my opinion, the discrepancies between the parameters give a realistic estimate of the present uncertainty.

One reason for discrepancies is for instance the much stronger dispersion constraint of the KH solution which should, in general, improve the result. In particular, some discrepancies between the partial waves (Figs. 2.2.1···2.2.21) are probably due to the fact that the CMU-LBL analysis used partial waves of the SACLAY analysis (1973) near the upper end of their energy range, whereas the fixed-t method of the KH group made it possible to use the most recent data at higher energies.

Another reason for discrepancies is a problem with the charge-exchange data of the Rutherford group (76 BROWN and 78 BROWN), which was noticed by both the CMU-LBL and KH groups. These data cannot be well fitted together with the elastic data and the isospin and dispersion constraints, although the isospin bounds are not violated. Both groups had to make a compromise, which led to somewhat different resonance parameters. The final decision can only come from a new experiment.

It is important to distinguish two types of discrepancies between the resonance parameters of different partial wave analyses:

i) Some of the discrepancies arise from differences between the results for the partial waves (Figs. 2.2.1···2.2.21). It will be of interest to reconstruct from both partial wave solutions the measurable quantities, to determine the kinematic regions of the largest discrepancies and to perform in these regions a careful comparison with the experimental data. It is planned to carry out this investigation at Karlsruhe.

ii) Other discrepancies are a consequence of the fact that the notion of a broad, inelastic resonance is not well-defined, unless one assumes a model which includes the background amplitude. Since a fundamental theory of the pion-nucleon system does not exist, one can have several equally justified descriptions of the same partial wave, which give different values of the parameters M, Γ, and x_{el}, and different decompositions into resonance and background contributions. For this reason several authors argued that it is preferable to use the pole position and the residue as resonance parameters. This would be true, if one had partial waves with infinite accuracy. However, the analytic continuation of error affected data depends on assumptions (75 R CIULLI-POMPONIU-SABBA-STEFANESCU). One has not only to exclude the artificial case of CALUCCI et al. (Sect. 2.4.1.1) but also others, which can falsify the result of the usual determinations of the pole position. It seems that a systematic study of this problem for broad inelastic resonances has not yet been performed.

It would be useful to compare the information on resonances obtained in different pàrtial wave analyses also in a way which is less model dependent than the determination of the parameters M, Γ and of the pole parameters. First, the partial wave amplitudes are fitted by a suitable flexible expansion in the whole energy range. Then, the maxima of the speed $|dT_{\ell\pm}/dW|$ are calculated and parameters M_{sp}, $(2\,x_{el}/\Gamma)_{sp}$ are determined from their location and height. These parameters are more directly related to the experimental data than all others and therefore, it should be easier to find the origin of discrepancies.

The overall agreement between the resonance parameters of the CMU-LBL and KH groups is good although the methods differ considerably. In Table 2.4.1.4 we also have listed the SACLAY results in order to show that the situation of the 4- and 3-star resonances is stable. Most of the masses are near the present values. The mass of P 31(1910) is different and the SACLAY group has not seen the P 33(1690) resonance, which also is a weak effect in the KH 78 solution.

The largest discrepancies between CMU-LBL and KH resonance parameters occur in the following cases:

mass: D 13(1700); P 33(1690); F 37(1950); quality: D 35(1960);

width: P 11(1470); S 11(1535); F 37(1950).

In the 1450···1800 MeV mass region this is probably due to the above mentioned problem with the charge-exchange data. A discrepancy in the 1900 MeV region is not surprising, because one has to disentangle the contributions of many resonances (9 in the case of the CMU-LBL solution, see Table 2.4.1.7).

As discussed in Sect. 2.1.8, there is at present no reason to consider a "splitting" of the P 11(1470) resonance. But the somewhat erratic behaviour of the partial wave shows that new accurate data would be desirable.

Comments on structures at high energies in the list of the Particle Data Group

The Baryon Table of the Particle Data Group (80 R BRICMAN, p. 27) contains four 3-star objects, whose quantum numbers are not specified.

Δ(2160)

The KH analysis led to structures G 37(2215) and H 39(2217) which are so weak that they could as well have been omitted. The July 80 version of the CMU-LBL analysis quotes a G 37(2200) 2-star object, but part of the structure lies outside the range of the analysis. In my opinion, a 1-star or at most 2-star classification would be adequate.

Δ(2850)

The total cross section difference shows a well-established structure (Fig. 2.1.7), which is interpreted as being due to I 3,13(2750) and K 3,15(2950) in the KH analysis. The structures in the s-dependence at fixed large t-values of elastic differential cross sections (78 JENKIN) support the discussion in terms of resonances (80 R KOCH). HENDRY found resonances not only in these two partial waves but also in I 3,11 at $M = 2850$ MeV. The compatibility of his fit with the data 78 JENKIN has not yet been checked.

Δ(3230)

The evidence in the total cross section data 66 CITRON has a 1-star quality, since the structure is broad and very small. Furthermore, a systematic error was noticed by GIACOMELLI (70 R). Figure 2.2.7 shows that even if this correction is taken into account, there remains a fairly large discrepancy with the data of FOLEY et al. (67). — The interpretation of a structure in the π^+p backward cross section in this energy region is strongly model dependent and not conclusive (75 LENNOX, see our Fig. 6.7.1). The KH analysis finds a candidate for a peripheral resonance in this range in the partial wave L 3,17, which belongs to the Δ_β trajectory. HENDRY lists in addition a K 3,13(3200) resonance (Δ_α).

N(3030)

A small bump is clearly visible in the total cross section difference of the data 66 CITRON. The KH analysis has a peripheral candidate L 1,15(3100), whereas HENDRY has instead a candidate K 1,13(3000). In the KH-analysis, the K 1,13 resonance lies at a lower mass (near 2600 MeV).

Higher peripheral resonances

The list of impact parameters of the nucleon resonances in Table 2.4.1.1 shows that almost all values of $b = (J + 1/2)/q$ are smaller than 1.2 fm, the only exception being the lowest resonance Δ(1232). It is remarkable that a considerable part of the well-established resonances has b-values in the range 1.0···1.2 fm ("peripheral resonances"). The nucleon has obviously a characteristic length parameter of about 1.1 fm for strong interactions.

Figure 4.1.1 shows CHEW-FRAUTSCHI plots of the well-established resonances and the best candidates. The classification in terms of quantum numbers naturality (or "natural parity") and signature (Table A.9.1) was chosen, because it is needed in Regge theory (Sect. A.9.2.2) (61 L CHEW).

In these plots, the peripheral resonances are those with the smallest mass for a given spin J or the highest spin for a given mass. It is seen that the resonances of the N_α and Δ_δ families are peripheral and the N_γ-resonances are almost peripheral.

Candidates for the continuation of these trajectories to higher energies are listed in Table 2.4.1.5. It is remarkable that there is a reasonable agreement between the results of HENDRY (80 L) and of the preliminary KH-analysis (80 L KOCH-2). For I-waves and higher angular momenta, there are 9 cases, in which candidates for resonances were found in both analyses. HENDRY lists 3 additional candidates: I 3,11; K 3,13; N 3,21, but he has not seen a structure in O 1,21 which is mentioned in KOCHs list.

One could argue that HENDRYs method is biased in favour of peripheral resonances because of the second term in his ansatz Eq. (1.12.1). Therefore, it is reassuring that most of the candidates were also found in a quite different way, taking into account the new wide-angle data of the Argonne group (78 JENKIN).

At large fixed values of t, for instance $t = -2.5$ GeV^2, these data show large structures of the s-dependence, which are obviously a continuation of similar structures at lower energies caused by well-established resonances. The structure in the range of the resonances H 3,11(2420) and K 3,15(2950) is shown in KOCHs review (80 R).

The question is now whether the structures in the Argand diagrams of the high partial waves are caused by resonances, which can be compared with quark model predictions.

One cannot expect an answer from partial wave analysis alone for the following reasons.

The results at intermediate energies show that the elasticity x_{el} of the resonances is in general decreasing and the width is increasing as the energy increases, as expected from the growing number of inelastic channels. Of course, it is an experimental question whether this trend continues at higher energies. A recent search for narrow structures in π^-p total cross sections at 2···14 GeV/c (80 BARREL) and in π^-p differential cross sections at large angles ($-0.4 < \cos\theta < 0.4$) at 6···13 GeV/c (80 and 80 L BAILLON) led to a negative result. However, the large-angle data suggest that broad structure in s at fixed angles, similar to those in the data of JENKINS et al. (78), exist also at higher energies.

Presumably, one can expect a decreasing size of the resonance structures in the Argand diagram as the energy increases, the highest resonance which has a well-pronounced structure being H 3,11(2420).

The problem is now that the relative magnitude of the background in comparison with the resonance structure is increasing and that the main background terms (diffraction and reggeized ϱ-exchange) also lead to loops in Argand diagrams as discussed above (Sect. 2.4.1.1).

One could think that the isospin odd background (ϱ-exchange) is negligible, for instance above 6 GeV/c, because the elastic cross section is so much larger than the charge-exchange cross section, but this is not true. The table of partial waves (Table 2.2.2.1) shows that the approximate equality of the isospin $I=1/2$ and 3/2 partial waves is valid only for the lower partial waves and mainly for the imaginary parts.

We conclude that the separation of resonance and background becomes increasingly difficult at higher energies. The main reason is not the small size of the resonance structure, but the fact that the background contribution to the partial waves can also have a "resonance-like" shape. Only relatively strong and narrow resonances can be separated in an almost model-independent way. In our case, resonance parameters can only be determined, if one assumes a model which includes the resonance and at least the dominant part of the background.

Two observations are of interest:

i) If one compares the shapes of the resonance structures in the Argand diagrams of well-established resonances and of the candidates, for instance in the partial waves K 1,13; K 3,15; L 1,15; M 1,17 etc., one finds a remarkable similarity, except for a scale factor and the change expected from the higher inelasticity (80 R Koch).

ii) In a Chew-Frautschi-plot, Fig. 4.1.1, the higher peripheral resonances lie on a smooth continuation of the straight lines connecting the well-established N_α and Δ_δ resonances. The existing crude mass estimates suggest a continuation along the curve $b=(J+1/2)/q=1.1$ fm rather than along the usual straight line $J/s=$ const (80 L Hendry).

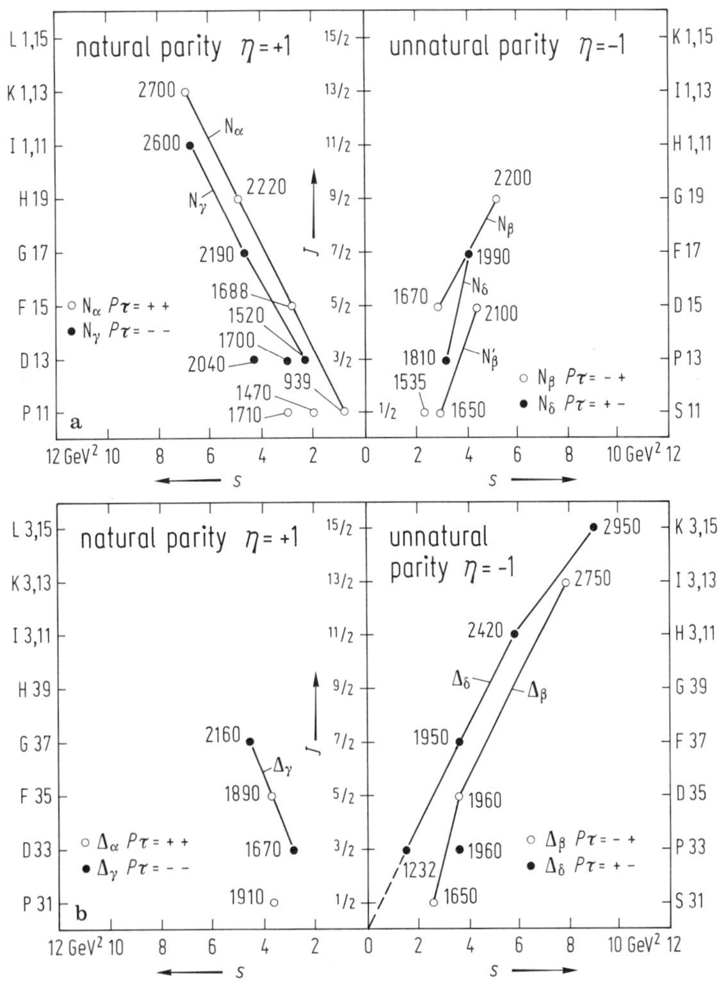

Fig. 4.1.1. Chew-Frautschi-plots of nucleon trajectories. The notation is explained in Table A.9.1. See also Table 2.4.1.5, which includes several candidates for resonances. s = resonance mass squared.

Higher non-peripheral resonances

The preliminary version of the new KH-analysis, which includes several important new data sets, led to resonance-like structures in the partial waves S 31, P 33, D 13 in the mass regions 2700, 2800, and 2600 MeV, respectively (80 L Koch-2, 80 L Höhler-2).

We add a table in which the resonances are ordered with respect to the quantum numbers parity, naturality and signature, because this leads to resonance families which occur in Regge models (Sect. A.9.2.2) and in quark models.

Table 2.4.1.5. Classification of nucleon resonances according to parity (P), naturality (η) and signature (τ). See also Table A.9.1. The 4-star resonances of the classification of the Particle Data Group are marked with full lines and the 3-star resonances with dashed lines. Dotted lines indicate either 2-star resonances or the candidates mentioned by Koch (80 L-2). 1-star objects of the Particle Data Group are omitted. Three additional candidates found by Hendry are marked with small circles. There is no evidence for resonances in those cases where the quantum numbers are not marked in the above mentioned way (I1,13, etc.).

	$P\ \eta\ \tau$						
N_α	+ + +	P11 (939) P11 (1470) P11 (1710)	F15 (1688) F15 (2000)	H19 (2220)	K1, 13 (2700)	M1, 17	O1, 21
N_β	− − +	S11 (1535) S11 (1650)	D15 (1670) D15 (2100)	G19 (2200)	I1, 13	L1, 17	N1, 21
N_γ	− + −		D13 (1520) D13 (1700) D13 (2040)	G17 (2190)	I1, 11 (2600)	L1, 15	N1, 19
N_δ	+ − −	P13 (1810)	F17 (1990)	H1, 11	K1, 15	M1, 19	O1, 23
Δ_α	+ + +	P31 (1910)	F35 (1890)	H39	K3, 13	M3, 17	O3, 21
Δ_β	− − +	S31 (1650) S31 (1900)	D35 (1960)	G39	I3, 13 (2750)	L3, 17	N3, 21
Δ_γ	− + −		D33 (1670)	G37 (2160)	I3, 11	L3, 15	N3, 19
Δ_δ	+ − −	P33 (1232) P33 (1690) P33 (1960)	F37 (1950)	H3, 11 (2420)	K3, 15 (2950)	M3, 19	O3, 23

2.4.1.4 Nucleon resonances in other reactions than πN→πN

In general, we consider in this volume only elastic and charge-exchange pion-nucleon scattering and total pion-nucleon cross sections. However, in the following we shall mention briefly some other reactions, which also are of interest for the study of nucleon resonances, because one can derive additional information.

i) Production of nucleon resonances

If a resonance is weakly coupled to the πN channel, it is useful to study its production in final states of inelastic πN scattering or of other reactions, analyzing the invariant mass of one of the outgoing nucleons combined with outgoing pions. In this way, one can determine its main decay channels. A clear resonance signal in these channels can help to confirm the existence of those resonances, which cause a weak structure in elastic scattering.

Of special interest is, of course, the possibility to detect in this way nucleon resonances which are decoupled from the πN channel. The Data Card Listings of the PDG (80 R Bricman) mention the following cases

P 13(1540), N(3245), N(3690), N(3755);   P 31(1550).

The P 13 and P 31 states were found in an analysis of the reaction $\pi N \to N\pi\pi$ (77 L and 78 L LONGACRE). Since this analysis involves simplifying assumptions and approximations, one has to wait for an independent confirmation.

There exists a large number of investigations in which nucleon resonances are produced in other strong interaction reactions, the initial channel being NN, $\bar{N}N$, KN etc. An interesting case is the peripheral production of a nucleon resonance by one-pion exchange, which is related to the formation of this resonance in πN scattering.

In general, the analyses of these experiments are more important for an understanding of the production mechanism than for an improved knowledge of the resonance properties. A complete list of references can be found in the Data Card Listings of the Particle Data Group.

A summary of the present status of production experiments and their analysis is given on p. 188 in 80 R BRICMAN.

ii) Inelastic resonance decay couplings

Another additional information follows from a study of the different inelastic decay couplings of the nucleon resonances, which can be compared with theoretical predictions.

The simplest cases are the inelastic two-body reactions

$$\pi^- p \to K^0 \Lambda; \quad \pi^\pm p \to K\Sigma; \quad \pi^- p \to \eta n,$$

which can be analyzed in analogy to πN partial wave analysis. The residues of the resonance poles are proportional to the decay couplings.

A detailed analysis of the first reaction has been performed in 77 L BAKER and 80 L SAXON. The authors used the BARRELET method (Sect. 2.1.11) and, in addition, a conventional energy dependent fit. It turned out that only the resonances S 11(1650), P 11(1710), P 31(1810), and D 13(1700) are required, but not the 4-star resonances D 15(1670) and F 15(1688), which also lie in the range of the analysis (threshold: 1613 MeV). The decoupling of the latter resonances is a prediction from $SU(6)_W$. See also the discussion in 80 R KELLY.

The most recent analysis of $\pi^\pm p \to K\Sigma$ in all charge states is due to LIVANOS et al. (80 L). An analysis of the reaction $\pi^- p \to n\eta$ in the range $1500 < W < 2250$ MeV can be found in 79 L BAKER. See also Sect. 2.1.13.

A great effort has been spent in analyses of the reactions

$$\pi^- p \to \pi^- \pi^0 p; \quad \pi^+ p \to \pi^+ \pi^0 p;$$
$$\pi^- p \to \pi^- \pi^+ n; \quad \pi^+ p \to \pi^+ \pi^+ n,$$

at energies below $W = 2000$ MeV, which are much more difficult. All fits use an isobar model, assuming that the reaction proceeds by the formation of a quasi-two-body state, which decays into the three final particles. The authors usually consider the quasi-two-body states $\pi\Delta(1232)$, ϱN, εN and sometimes πN(1470). The approximation of the $I = J = 0$ ππ-interaction by an ε-meson is very crude (Sect. 2.4.6.2), and it is difficult to estimate the neglected effects of other interactions. See 79 L AITCHISON for an improved version of the isobar model.

The results for the decay couplings are of great interest for a comparison with predictions from quark models. See 76 R BARNHAM, 80 R KELLY, and 80 R BRICMAN, p. 182 for reviews of the experimental situation. The status of the theory is discussed in 75 L HEY, 76 R HEY, 76 R LITCHFIELD-CASHMORE-HEY, 79 R HEY, 80 R HEY, 82 R LE YAOUANC, 82 L GAVELA.

iii) Photon couplings of nucleon resonances

It is an important property of resonances that they should be observable in several channels. The most detailed studies of nucleon resonances in another channel have been performed in single-pion photoproduction

$$\gamma p \to \pi^0 p \quad \text{and} \quad \gamma p \to \pi^+ n.$$

The partial wave analysis of these reactions can be carried out in a similar way as for πN-scattering, but it is much more difficult, because one has 12 independent invariant amplitudes. Instead of three accurately measurable reactions one has only two. The only advantage is that there is an additional type of experiment in which the photon is polarized.

In all analyses of photoproduction data the results of πN partial wave analyses are used as an additional input. In the elastic region, it follows from unitarity that the phase of a multipole amplitude agrees with the corresponding scattering phase shift. At high energies, the masses and widths of the nucleon resonances are used as starting values in the fits.

The least biased method is an energy-independent analysis, but it can be carried out only up to the inelastic threshold, because one has to use the πN scattering phase shifts. Several authors extended the analysis to higher energies, imposing fixed-t analyticity as an additional constraint.

A third method is an energy-dependent fit, using an ansatz which contains the nucleon Born terms, Breit-Wigner formulas for the resonances and a smooth background. This ansatz can be made flexible enough to give an excellent fit to the data.

The problem of this method is that some of the amplitudes could nevertheless be quite wrong because, due to the large number of amplitudes, there are many possibilities to fit the available data. In my opinion, an isobar model analysis is of interest only, if it is combined with a fixed-t analysis, as it was done by some authors.

Significant results for the photon couplings of nucleon resonances are known for about 10 resonances. They are important for a comparison with predictions from quark models.

The present status of pion photoproduction experiments is summarized in 80 R KAJIKAWA. The results of analyses are mentioned in the Data Card Listings of the Particle Data Group (80 R BRICMAN), which also include a summary (see also 78 R MOORHOUSE). Two recent analyses were presented at the Toronto Conference (80 L ARAI and 80 L CRAWFORD).

The excitation of nucleon resonances by virtual photons has been studied in detail in pion electroproduction experiments

$$e^- p \to e^- p\pi^0 \quad \text{and} \quad e^- p \to e^- n\pi^+ .$$

The analysis is even more difficult than that of photoproduction due to the larger number of invariant amplitudes, but it has the advantage that one can investigate the dependence on the "photon mass" and the contribution of the longitudinal multipoles.

The electroproduction of η near threshold is of interest because it leads to information on the resonance S 11(1535).

A summary of the present status is given in 80 R FOSTER.

2.4.1.5 Models for nucleon resonances

A detailed discussion of the models for nucleon resonances does not belong to the topics of this volume. We only want to mention briefly some recent developments and to give references, where a full treatment can be found. Furthermore, Table 2.4.1.6 will be explained, in which the nucleon resonances are classified according to SU(6) × O(3) supermultiplets, because the notation is used in many discussions of results of partial wave analyses. Finally, Table 2.4.1.7 is presented in order to call attention to the fact that the locations of the complex poles of relatively large groups of resonances agree within the errors.

The SU(6) × O(3) classification

In the non-relativistic quark shell models it is assumed that three spin 1/2 quarks move in a deep potential well and that the confining forces are independent of flavour (u, d or s) and spin degrees of freedom. The formalism includes u, d, and s-quarks, but only u and d-quarks are needed for the nucleon resonances.

In the SU(6) limit, all quark masses are equal. As a zero order approximation for subsequent applications of perturbation methods, one assumes harmonic forces between the quarks. This leads to a Hamiltonian, whose eigenvalue problem can be solved exactly

$$H = \frac{1}{2m}(p_1^2 + p_2^2 + p_3^2) + \tfrac{1}{2}K(r_{12}^2 + r_{23}^2 + r_{31}^2). \tag{4.1.16}$$

One introduces new coordinates

$$\boldsymbol{R} = \tfrac{1}{3}(\boldsymbol{r}_1 + \boldsymbol{r}_2 + \boldsymbol{r}_3); \quad \boldsymbol{\varrho} = \frac{1}{\sqrt{2}}(\boldsymbol{r}_1 - \boldsymbol{r}_2); \quad \boldsymbol{\lambda} = \frac{1}{\sqrt{6}}(\boldsymbol{r}_1 + \boldsymbol{r}_2 - 2\boldsymbol{r}_3). \tag{4.1.17}$$

Aside from the center-of-mass motion, one obtains two independent oscillator degrees of freedom. The $\boldsymbol{\varrho}$-coordinate is antisymmetric under the exchange of 1 and 2 and the $\boldsymbol{\lambda}$-coordinate is symmetric.

The total orbital angular momentum $\boldsymbol{L}$, the total angular momentum $\boldsymbol{J}$ and the parity P are defined by

$$\boldsymbol{L} = \boldsymbol{l}_\varrho + \boldsymbol{l}_\lambda; \quad \boldsymbol{J} = \boldsymbol{L} + \boldsymbol{S}; \quad P = (-1)^{\ell_\varrho + \ell_\lambda}. \tag{4.1.18}$$

where $\boldsymbol{S}$ is the total quark spin ($S = 1/2$ or $3/2$). S, L, ℓ_ϱ and ℓ_λ are the quantum numbers belonging to $\boldsymbol{S}^2$, $\boldsymbol{L}^2$, $\boldsymbol{l}_\varrho^2$ and $\boldsymbol{l}_\lambda^2$, respectively.

The energy of a state is determined by the total number $\mathcal{N}$ of excitation quanta of the two oscillators

$$\mathcal{N}=(\ell_\varrho+2n_\varrho+\ell_\lambda+2n_\lambda-4). \quad (4.1.19)$$

There are rotational excitations ($L>0$) and radial excitations (n_ϱ or $n_\lambda>1$). The ground state has $L=0$, $n_\varrho=n_\lambda=1$, $\mathcal{N}=0$.

The state vector of the three-quark system can be written

$$|3q\rangle=\underbrace{|\text{flavour}\rangle\,|\text{spin}\rangle\,|\text{space}\rangle}_{SU(6)\times O(3)}\,\underbrace{|\text{colour}\rangle}_{\text{singlet}}, \quad (4.1.20)$$

where |flavour⟩ denotes the general case, where u, d, and s-quarks are admitted. The antisymmetry condition for the left hand side leads to the conclusion that the SU(6) × O(3) part must be symmetric.

Taking together the SU(6) wave function and the internal space wave function, the excited SU(6) × O(3) supermultiplet may be specified by the notation (α, $L^P_{\mathcal{N}}$), where α denotes the SU(6) representation and $\mathcal{N}$ the number of excitation quanta, Eq. (4.1.19). For each value of $\mathcal{N}$ there are in general several supermultiplets which are separated in energy, if corrections to the harmonic forces are taken into account.

Another notation used in Table 2.4.1.6 is $^{2S+1}$N and $^{2S+1}\Delta$, where N and Δ refer to the SU(3) octet and decuplet respectively, of which only the nucleon N, the isobar Δ(1232) and their excited states occur in our case. A star ( )* denotes the first radial excitation. In the first column, we give L_S and L_M, where L is expressed in the usual way by S, P, D, F, ...; the subscript S and M indicate the symmetry of the SU(6) multiplet with respect to permutations of the labels of the three quarks. The $\underline{56}$-representation has complete symmetry (S), the $\underline{70}$-representation has mixed symmetry (M) and the antisymmetric $\underline{20}$-representation does not occur in our table.

The lowest state in the harmonic oscillator quark model has $\mathcal{N}=0$, $L=0$, $n_\varrho=n_\lambda=1$. Its space configuration is $(1s)_\varrho(1s)_\lambda$ and by symmetry, the supermultiplet is $(\underline{56}, 0^+_0)$. The decomposition in terms of SU(3) representations and quark spin reads

$$\underline{56}={}^4\underline{10}+{}^2\underline{8}, \quad (4.1.21)$$

corresponding to the Δ(1232) and N(939) states.

The first excited supermultiplet $\mathcal{N}=1$ has the space configurations $(1s)_\varrho(1p)_\lambda$ and $(1p)_\varrho(1s)_\lambda$. These have M symmetry, so they can only occur in the $\underline{70}$-representation. Hence, the only supermultiplet belonging to $\mathcal{N}=1$ is $(\underline{70}, 1^-_1)$.

For larger $\mathcal{N}$, the number of supermultiplets is rapidly growing. Most of the well-established states belong to the set of *minimal supermultiplets*

$$(\underline{56}, L^+_{L+2n}) \text{ for } L \text{ even}; \quad (\underline{70}, L^-_{L+2n}) \text{ for } L \text{ odd}, \quad n=0,1,2,\dots \quad (4.1.22)$$

The best known example for a state which does not fit into this scheme is the resonance Δ(1960) D 35, which is usually classified in the supermultiplet $(\underline{56}, 1^-_3)$.

Table 2.4.1.6. Classification of nucleon resonances according to SU(6) × O(3). The notation is explained in the text. As mentioned in the caption of Table 2.4.1.1, we have listed only the dominant state and ignore the mixing. In some cases, the assignment could as well be chosen in a different way. – 4-star resonances of the classification of the Particle Data Group are marked by full lines, 3-star resonances by dashed lines and weaker indications for a resonant behaviour by dotted lines. The mass values and classifications of the CMU–LBL results are taken from 79 L Cutkosky-2, who lists in addition a S31 (2150) in (70, 1^-)*. Mass values in brackets were determined using in addition SACLAY 73 partial waves. More recent mass values derived from the CMU-LBL 79 solution are listed in Table 2.4.1.2.

		^{2}N		^{4}N or $^4\Delta$				$^2\Delta$	
		P11		P33					
	SACLAY	939		1231					
$(\underline{56}, 0_0^+)$	KH	939		1233					
S_S	CMU–LBL	939		1234					
	PDG	939		1232					
		S11	D13	S11	D13	D15		S31	D33
	SACLAY	1519	1525	1673	1710	1660		1623	1723
$(\underline{70}, 1_1^-)$	KH	1526	1519	1670	1731	1679		1610	1680
P_M	CMU–LBL	1540	1525	1640	1670	1680		1620	1730
	PDG	1535	1520	1650	1700	1670		1650	1670
		P13	F15	P31	P33	F35	F37		
	SACLAY	1696	1679	1789	1900	1869	1928		
$(\underline{56}, 2_2^+)$	KH	1710	1684	1888	1868	1905	1913		
D_S	CMU–LBL	1740	1680	1920	1960	1920	1950		
	PDG	1810	1688	1910	1960	1890	1950		
		P13	F15	P11	P13	F15	F17	P33	F35
	SACLAY					2025	2049		
$(\underline{70}, 2_2^+)$	KH			2050		1882	2005		
D_M	CMU–LBL						1970		
	PDG					2000	1990		
		P11		P33					
	SACLAY	1413							
$(\underline{56}, 0_2^+)$*	KH	1410		1522					
S_S	CMU–LBL	1450		1640					
	PDG	1470		1690					
		P11		P11	P13			P31	
	SACLAY	1730							
$(\underline{70}, 0_2^+)$	KH	1723							
S_M	CMU–LBL	1710							
	PDG	1710							
		S11	D13	S31	D33	D35			
	SACLAY			2001		1894			
$(\underline{56}, 1_3^-)$	KH	1880		1908		1901			
P_S	CMU–LBL		1830	1850	2010	1930			
	PDG			1900		1960			
		D15	G17	D13	D15	G17	G19	D35	G37
	SACLAY			2029	2076	2141	2133		
$(\underline{70}, 3_3^-)$	KH			2081	2228	2140	2268	2305	2215
F_M	CMU–LBL			2100		(2150)	(2200)		(2200)
	Hendry					2140	2200		2280
	PDG			2040	2100	2190	2200		2160

(continued)

Table 2.4.1.6 (continued)

		^2N		^4N or $^4\Delta$				$^2\Delta$	
		F17	$\underline{\text{H19}}$	F35	F37	H39	$\underline{\text{H3, 11}}$		
$(\underline{56}, 4_4^+)$ G_S	SACLAY		2249				2392		
	KH		2205		2425	2217	2416		
	CMU–LBL		(2250)						
	HENDRY		2300			2450	2400		
	PDG		2220			2300	2420		
		G19	I1, 11	G17	G19	$\underline{\text{I1, 11}}$	I1, 13	G39	I3, 11
$(\underline{70}, 5_5^-)$ H_M	KH				2792	2577		2468	
	HENDRY					2700			2850
	PDG					2600		2500	

Summary on the quark-shell model and references

The non-relativistic quark-shell model leads to a simple description of the main features of the baryon spectrum. In a first step, one considers harmonic forces between the quarks and obtains a harmonic oscillator spectrum. Additional non-harmonic forces lead to a separation of the SU(6) × O(3) supermultiplets in mass. One can distinguish rotational excitations (related to Regge recurrences) and radial excitations (pulsations). Details and further references can be found in GREENBERG (67 L), DIVGI (68 L), HORGAN (73 L, 74 L, 82 L), DALITZ (73 R, 75 R, 76 R, 79 L), GROMES (76 L, 77 L, 79 L, 80 L), ISGUR (79 L), HEY (78 R, 79 R, 80 R), CLOSE (79 R, Part 5), BOWLER (81 L), DALITZ (82 R).

Mass splittings within the SU(6) × O(3) multiplets are treated by perturbation methods, using "QCD-inspired" interactions (75 L DE RUJULA). A very successful ansatz for the interaction was proposed and investigated by ISGUR and KARL (77 L, 78 L-1 and 2, 79 L-1 and 2, 80 L, 81 R ISGUR), who chose the hyperfine interaction and neglected the spin-orbit forces. See GROMES (80 L) and LE YAOUANC et al. (77 L, 78 L) for discussions of the hyperfine interaction and 80 L REINDERS for the spin-orbit force. In the supermultiplets $(\underline{70}, 1^-)$, $(\underline{56}, 2^+)$, and $(\underline{56}, 1^-)$ the spin-orbit force is remarkably small.

According to the CMU-LBL analysis (Table 2.4.1.2), the mass splitting between the two Δ-states of the $(\underline{70}, 1^-)$ supermultiplet is well-established (the evidence for a splitting is weak in the KH 78 solution). In the perturbed harmonic oscillator model the degeneracy is not lifted by the usual perturbations. One needs a "three-body spin-orbit interaction" (79 L GROMES, 80 L REINDERS, 80 R KELLY).

It was a long-standing problem of the quark-shell model that the number of states found in analyses of the data was much smaller than the number of predicted states. For some time, it seemed that only the "minimal supermultiplets" of Eq. (4.1.22) are coupled to the πN system. These states follow from models in which the nucleon has a quark-diquark structure (78 R HENDRY-LICHTENBERG, 79 R VAN DE WALLE). However, the evidence for non-minimal supermultiplets is now improving [$(\underline{56}, 1^-)$, $(\underline{70}, 0^+)$, $(\underline{70}, 2^+)$], see Table 2.4.1.6 and 80 L CUTKOSKY-2. Furthermore, FORSYTH and CUTKOSKY (81 L-3) found evidence against diquarks from baryon resonance decays.

The perturbation of the harmonic oscillator Hamiltonian leads to a mixing between the different states which was investigated by several authors. In the ISGUR-KARL model there is a substantial mixing between all five supermultiplets in the $\mathcal{N} = 2$ band: $(\underline{56}, 0^+)$, $(\underline{70}, 0^+)$, $(\underline{56}, 2^+)$, $(\underline{70}, 2^+)$, $(\underline{20}, 1^+)$. KONIUK and ISGUR (80 L-1 and 2) were able to explain qualitatively, why more than half of the $\mathcal{N} = 2$ states are not seen in πN partial wave analysis. They calculated the partial decay widths into the πN-channel from a model for the coupling of the excited state to the πN-state, assuming that the pion couples to one of the incoming quarks, the other two acting as "spectators". Earlier work in this field is due to FAIMAN and HENDRY (68 L, 69 L), HORGAN (76 R) and others. See the review 73 R LIPKIN, 74 R ROSNER, 77 R MITRA-SOOD, and 82 L GAVELA.

FORSYTH and CUTKOSKY (81 L-1, see also FORSYTH 81 L-2) investigated quantitative fits of decay models to the widths of $(\underline{70}, 1^-)$ and $(\underline{56}, 1^-)$ resonances. Since the above spectator model turned out to be unsatisfactory, they assumed that the parameters of the interaction depend on the orbital and spin states of the spectator quarks.

Electromagnetic interactions of baryons have been studied by ISHIDA et al. (79-1 and 2).

Instead of using an "explicit" quark model, one can also perform an algebraic $SU(6)_W$ analysis of the resonance decay. In this approach, meson emission is treated via a PCAC or vector dominance assumption. A review of earlier work and updated tables can be found in 76 R LITCHFIELD-CASHMORE-HEY. Other reviews: 76 R HEY, 74 R ROSNER, 77 R MITRA-SOOD.

There exists a large number of further papers on quark-shell models of the baryons. We mention some additional references such that most of the literature can be traced from the references mentioned in this subsection: BÖHM (79 L, 80 L, 82 L), 79 L NAKAMURA, 78 L and 80 L REINDERS, 78 L LE YAOUANC, 82 L KALMAN. An introduction into this field is given in CLOSEs book (79 R). Recent developments are summarized in his lecture at the KfK Summer School at Bad Liebenzell (82 R CLOSE).

Bag models

In comparison with the success of the quark shell model, the progress in the description of baryon resonances by bag models (74 L and 75 L CHODOS, 76 R JOHNSON "M.I.T. bag") is rather slow, one of the reasons being that calculations are much more unwieldy than in the quark-shell model. The present status has been reviewed by DE TAR (79 R), DE GRAND (80 R), JAFFE (79 R, 80 R), and HELLER (82 R). HELLERs lectures include a discussion of the application to nucleon-nucleon scattering.

We mention some of the recent papers: DE TAR (81 L) investigates the question how chiral symmetry in a bag model can be restored by introducing an external pion field ("hybrid pion-quark-bag model"; see also 79 L BARNHILL-2); CARLSON et al. (82 L) and LIAN-SOU et al. (81 L) study effects of the center-of-mass motion. The "little" and "cloudy" bags will be discussed in Sect. 2.5.1.7.

A phenomenological version of the bag model was developed by DE SWART et al. (79 L and 80 L MULDERS, 80 R DE SWART). The authors describe the ground state by a M.I.T. bag and assume that the fine structure of the multiplet follows from colour-magnetic forces. Orbitally excited states are described by a string-like bag, which has a single quark at one end and a quark pair (diquark) at the other end. They are held together by the colour-electric force. This picture leads to linear Regge trajectories (76 L JOHNSON).

Another bag model for baryons has been proposed by PREPARATA and SZEGOE (77 R, 78 L). Their quark fields have properties different from those assumed in QCD. It turns out that "geometrodynamical" baryons exist only in three spatial configurations of the quark-diquark type. Aside from $(\underline{56}, 0^+)$ and $(\underline{70}, 1^-)$ only two further multiplets are allowed below a mass of 2 GeV: $(\underline{56}, 2^+)$ and $(\underline{70}, 2^+)$. The above discussion of the quark-shell model shows that there is substantial evidence for additional multiplets and that the suppression of other states has found a reasonable explanation in the ISGUR-KARL model.

Multiquark baryons

It is usually assumed that baryons are states of three quarks: (qqq). However, some of the excited baryon states could possibly be states of 4 or 5 quarks and one or two antiquarks ($q^4\bar{q}$ or $q^5\bar{q}^2$). In the first investigation of this problem JAFFE (77 L) used the M.I.T. bag model. His work was continued by STROTTMANN (79 L), who developed a group-theoretical formalism for these cases. He claims that his results for the $\frac{1}{2}^-$-baryons are in good agreement with experiments, whereas the comparison of known resonances with the predictions for the $\frac{3}{2}^-$ and $\frac{5}{2}^-$-states is less auspicious. The group-theoretical part of STROTTMANNs work has recently been criticized by BICKERSTAFF (82 L).

The work of ROIESNEL (79 L) is based on the "P-matrix"-formalism of JAFFE and LOW (79 L), which was inspired by the R-matrix theory of nuclear reactions (47 L WIGNER, 58 R LANE-THOMAS). In practice, the calculation can be carried out only at low energies, where the inelasticity is negligible. In this case P is a function of the c.m. momentum q, which is simply related to the phase shift δ and to the S-function, Eq. (A.7.28)

$$P = q\cot(bq+\delta); \qquad S = e^{2i\delta}. \tag{4.1.23}$$

A pole of P corresponds to a state whose wave function vanishes at $r=b$, in agreement with the boundary condition for the internal wave function in a bag model. Outside a relative separation b, the πN system is supposed to be free. The location of the pole of P and its residue are calculated from the phase shift and compared with the prediction from the bag model for $q^4\bar{q}$-states. For the πN S-waves, the agreement is reasonable, but the evidence is not strong due to possible corrections from a second channel.

Multiquark baryons were also studied by HOEGAASEN and SORBA (80 L) and by DE SWART et al. (80 R). A detailed treatment of the baryon masses and decay properties is given in 80 L MULDERS.

JAFFE (77 L) calculated the lightest $q^2\bar{q}^2$ mesons in the M.I.T. bag model. He suggested that the ππ S-wave resonance "ε(700)" belongs to this category. The doubtful status of this resonance is discussed in 80 R BRICMAN on p. S 134. See also 81 R JAFFE and 79 L ESTABROOKS.

The present status of the various models for hadrons has been treated in QUIGGS review (81 R). Very recently, it has become possible to predict meson and baryon masses from Monte Carlo computations in lattice QCD (see for instance 82 L FUCITO and 82 L HASENFRATZ) and from QCD sum rules (81 L IOFFE).

Dynamical models of nucleon resonances, which are based on partial wave dispersion relations, are mentioned in Sect. A.12 (A.7.1).

2.4.1.6 Degeneracy of the complex pole positions

A fairly large number of papers deals with a comparison between the "experimental masses" of the nucleon resonances (listed in Tables 2.4.1.1···2.4.1.6) and predictions derived from various quark models for stable excited states. Even if the decay couplings are calculated, the effect of the decay on the "mass" of the resonance is not taken into account [cf. for instance the "self-energy" contribution to the mass of an unstable V-particle in the Lee-model (58 L Höhler)].

It would certainly be preferable to perform the comparison for the complex pole positions which, however, can only be predicted from a quark model, if one is able to describe the scattering reaction. This is much more difficult than the calculation of the excited states. The work of Jaffe and Low (79 L) is a first step in this direction.

It is of interest to note that the values for the real parts of the pole position are shifted in comparison with the values of the mass parameters M by an amount in the range 0···100 MeV, i.e. the "mass splitting" between different states for these two quantities can be considerably different.

In Table 2.4.1.7 we have ordered the resonances of Table 2.4.1.2 according to the real part of the pole position. It is surprising to see that the pole positions agree within the errors for large groups of resonances. For instance, there are 18 N-resonances. 6 of them have their pole near (1670−50i) MeV and 7 near (2120−170i) MeV.

The first question is, of course, if all these resonances really exist. If several strong resonances occur at almost the same energy, it could happen that structures in non-resonant partial waves are produced either by systematic errors of the data or even by the method of partial wave analysis. This point deserves further study as far as the one- and two-star objects are concerned. But the degeneracy remains impressive even if one considers only the 3- and 4-star resonances.

The table is important for an understanding of the smoothness of the zero trajectories (Sect. 2.4.3). Originally, it was not clear how the trajectories can pass the resonance surfaces of non-peripheral resonances, which do not have a sufficient number of Legendre zeros. Now, it is seen that these resonances are in general degenerated (within the errors) with resonances of higher angular momentum. If a group of resonances belongs to the same complex pole, the number of Legendre zeros is determined by that of the highest partial wave, whose zeros are only shifted by the contributions of the lower partial waves.

Table 2.4.1.7. Nucleon resonances ordered according to the real part of the pole position. The parameters are those of Table 2.1.4.2. See the remark in the caption of Table 2.4.1.1 concerning the classification in the last column.

Name	Re W MeV	−Im W MeV	Quality	SU(6)×O(3)
P11(1470)	1375± 30	90±20	4	(56, 0⁺)*
S11(1535)	1510± 50	130±40	4	(70, 1⁻)
D13(1520)	1510± 5	57± 5	4	(70, 1⁻)
S11(1650)	1640± 20	75±15	4	(70, 1⁻)
D13(1700)	1660± 30	45±20	4	(70, 1⁻)
D15(1670)	1660± 10	70± 5	4	(70, 1⁻)
F15(1688)	1667± 5	55± 5	4	(56, 2⁺)
P13(1810)	1680± 30	60±20	4	(56, 2⁺)
P11(1710)	1690± 20	40±10	4	(70, 0⁺)
D13	1880±100	80±40	1	(56, 1⁻)
F17(1990)	1900± 30	130±30	4	(70, 2⁺)
D13(2040)	2050± 70	100±40	3	(70, 3⁻)
G17(2190)	2100± 50	200±80	4	(70, 3⁻)
D15(2100)	2100± 60	180±40	3	(70, 3⁻)
P11	2120± 40	120±40	1	
S11(2100)	2150± 70	175±50	1	
G19(2200)	2150± 50	180±50	4	(70, 3⁻)
H19(2220)	2160± 80	240±50	4	(56, 4⁺)
P33(1232)	1210± 1	50± 1	4	(56, 0⁺)
P33(1690)	1550± 40	100±30	4	(56, 0⁺)*
S31(1650)	1600± 15	60±10	4	(70, 1⁻)
D33(1670)	1675± 25	110±20	4	(70, 1⁻)
F35(1890)	1830± 40	140±30	4	(56, 2⁺)
S31(1900)	1870± 40	90±25	3	(56, 1⁻)
P31(1910)	1880± 30	100±20	4	(56, 2⁺)
D35(1960)	1890± 50	130±30	4	(56, 1⁻)
F37(1950)	1890± 15	130±20	4	(56, 2⁺)
P33(1960)	1900± 80	150±50	3	(56, 2⁺)
D33	1900±100	100±30	2	(56, 1⁻)
G37(2160)	2100± 50	170±40	2	(70, 3⁻)
S31	2140± 80	100±40	2	(70, 1⁻)*
F35	2150±100	175±50	1	
G39	2260± 60	160±80	2	
F37	2350±100	130±50	1	(56, 4⁺)
H3,11(2420)	2360±100	210±50	4	(56, 4⁺)
H39(2300)	2370± 80	210±80	2	(56, 4⁺)
D35	2400±125	200±75	2	

The degeneracy of the pole positions raises the question, whether the notion of resonances with well-defined quantum numbers is really the basic phenomenon at intermediate energies. It seems that πN scattering can be well described by assuming *a few complex poles of the invariant amplitudes* instead of poles in many partial waves at slightly different positions. This would be similar to a VENEZIANO-type model for πN scattering. Unfortunately, earlier attempts to develop such a model were not successful (68 L IGI, 69 L CHU. 69 L BERGER, 69 L LOVELACE, 70 L FENSTER, 70 L HENDRY, 73 L ODORICO, 73 L PAL), one of the unsolved problems being the unitarization.

Another difficulty of the quark models is that the predicted density of states increases rapidly with energy. The description of pion-nucleon scattering becomes extremely complicated in the same energy region where the s- and t-dependence of the scattering amplitudes becomes very simple. Furthermore, the highest nucleon resonances are tiny structures superimposed on a large background. They are apparently related to geometrical effects occurring in the scattering process. It is difficult to believe that they can be explained as excited states of the nucleon, ignoring the dominant diffractive part of the scattering mechanism.

Conclusion

The improved versions of the non-relativistic quark-shell model have led to an important progress in baryon spectroscopy. However, as pointed out by CUTKOSKY (80 L-2), it would be wrong to assume that the physics of baryon resonances is a closed field and presents no longer interesting problems to theorists. There are still unsolved fundamental questions. Some of them can be investigated only in collaboration with experimentalists, who improve the accuracy of the existing data.

2.4.2 Meson-nucleon coupling constants

2.4.2.1 Determination of the pion-nucleon coupling constant f^2

If the pion-nucleon coupling constant shall be determined, one has to separate the nucleon-exchange contribution from the pion-nucleon scattering amplitude. First results were obtained within the framework of the static model of CHEW and LOW (56 L), where the coupling constant was derived from an effective range plot for the P 33 amplitude. However, there was a considerable uncertainty due to the approximations of this model (see below).

The fixed-t dispersion relations (Sect. A.6.3) led to an exact definition of the problem. The remaining uncertainties arise only from experimental data and from the unknown high energy region, but the latter are unimportant.

Nucleon pole terms are exactly defined also in other single-variable dispersion relations (Sect. A.6), but these follow from the Mandelstam representation, which is a conjecture, whereas fixed-t dispersion relations can be derived from axiomatic field theory. Furthermore, the other dispersion relations have contributions from a t-channel cut, whose branch point lies near the position of the nucleon poles. The discontinuity along this cut is not as well known as that of the left hand cut in the fixed-t dispersion relation.

Methods based on forward dispersion relations for the amplitudes $C^{\pm}$

We start with the forward dispersion relation for the invariant amplitude C (Sect. A.6.4) because, due to the optical theorem, Eq. (A.5.5), the dispersion integrals can be evaluated more accurately than in all other cases.

$\mathrm{Re}\,C(\omega)$ is calculated from a partial wave solution. Of course, f^2 can be determined only if the nucleon pole terms C_N, Eqs. (A.6.46) and (A.6.48), are not small in comparison with $|C|$.

It turns out that $C_N^+(\omega)$ is large, but slowly varying. If one splits off a constant part g^2/m, which cannot be separated from the subtraction constant, the rest is too small for an accurate determination of f^2.

The situation is more favorable for $C^-(\omega)$ at low momenta. For instance at 200 MeV/c, $|C^-|$ is only twice as large as C_N^-. However, data at intermediate and high energies are not sensitive to the value of f^2, the ratio $|C^-|/C_N^-$ being about 10 at 400 MeV/c and 6 in the dip of $|C^-|$ near 1.3 GeV/c (see Fig. 2.1.14).

A difficulty for the determination of f^2 lies in the fact that dispersion methods are frequently used in phase shift analyses, i.e. one starts with an assumption for the value of f^2. The CERN and SACLAY analyses and also the analyses KH 78 and 80 and CMU-LBL 79 used evaluations of the forward dispersion relations as part of their input. Furthermore, the CERN method includes partial wave dispersion relations, the CMU-LBL method has a constraint from dispersion relations along certain hyperbolas and the Karlsruhe-Helsinki method has even stronger dispersion constraints. Both, the CMU-LBL and the KH analyses start with $f^2 = 0.079$.

The important point is now that the constraints are enforced only within certain errors. As a consequence, one finds small discrepancies, if $\mathrm{Re}\,C$ as reconstructed from partial waves is compared with the input. The question is now, whether these discrepancies are systematic in the sense that the agreement is improved, if a correction is applied to f^2.

We denote the input value of $\mathrm{Re}\,C^-$ by a subscript "Table" and the value calculated from phase shifts by the subscript "Ph". Then, we can calculate at each energy where partial waves are available

$$\Delta f^2 \approx \frac{\omega}{8\pi}\{(\mathrm{Re}\,C^-)_{\mathrm{Ph}} - (\mathrm{Re}\,C^-)_{\mathrm{Table}}\}, \tag{4.2.1}$$

assuming that the difference comes entirely from the nucleon pole term.

For the KH 80 solution and Table 80/1 (80 R HÖHLER-KAISER) one obtains as an average over the range $112\cdots408$ MeV/c the correction $\Delta f^2 = -0.001$. – We have not yet included a discussion of the small corrections $(\mathrm{Im}\,C^-)_{\mathrm{Ph}} - (\mathrm{Im}\,C^-)_{\mathrm{Table}}$ and of possible effects of the $\Delta^{++} - \Delta^0$ mass difference. — The same calculation can be done for the CMU-LBL solution, but it starts above the momentum range where the amplitude C^- is sensitive to the value of f^2.

At high energies, the sensitivity of $\mathrm{Re}\,C^-$ to the value of f^2 is strongly enhanced, if one uses a *subtracted* dispersion relation, Eq. (A.6.50). However, this is not helpful for a determination of f^2, because there is a strong correlation with the value of the subtraction constant, which is proportional to the S-wave scattering length a^-, whose value is not known with a sufficient accuracy. The connection between a^- and f^2 is the same as in the GOLDBERGER-MIYAZAWA-OEHME sum rule, Eq. (A.6.49).

It is possible to determine $\mathrm{Re}\,C^-(\omega)$ up to a sign without using a phase shift analysis, if suitable charge-exchange cross section data are available for an extrapolation to $t=0$. $|\mathrm{Re}\,C^-|$ follows by inserting the forward cross section and the difference between the $\pi^\pm$p total cross sections into the formula

$$\frac{\mathrm{d}\sigma_0}{\mathrm{d}t} = \frac{1}{8\pi}\{(\mathrm{Re}\,C^-/k)^2 + (\sigma_- - \sigma_+)^2/4\}. \tag{4.2.2}$$

This method has the advantage that an error for $|\mathrm{Re}\,C^-|$ is easily obtained and that an input for f^2 is not necessary.

In practice, the determination of the forward cross section can be difficult, even if data are available in a large angular range, because the cross sections sometimes show a rapid variation even between $\cos\theta = 0.95$ and 1 (79 L HÖHLER-1). Therefore, it would be desirable to perform a special experiment for the charge-exchange cross section very near $\theta = 0°$, similarly to the experiment of BORGEAUD et al. (64). The result would also be helpful for the determination of the higher partial waves and of the shape of resonant amplitudes.

Another experiment which would not only lead to an improved f^2-determination but also to more accurate phase shifts is a measurement of the elastic cross sections at low energies and near-forward angles, where data do not exist yet (see Table 1.2.2). The knowledge of the Coulomb interference structure would help considerably to determine accurate real parts at and near $t=0$.

f^2-determination from the dispersion relation for the amplitude B

First we compare the nucleon pole term of the forward dispersion relation for $B^+(\omega)$ with $|B^+|$ as determined from partial wave analysis. Table 2.2.3.1 shows that the magnitude of

$$B_N^+(\omega) = \frac{g^2}{m}\frac{\omega}{\omega_B^2 - \omega^2} \approx -\frac{g^2}{m\omega} = -16\pi f^2 m/\mu^2\omega \tag{4.2.3}$$

is comparable with $|B^+|$ or even larger from threshold up to about 3 GeV/c. Then $\mathrm{Im}\,B^+$ becomes roughly constant, but $|\mathrm{Re}\,B^+|$ remains smaller than $|B_N^+|$, which continues its decrease. For small negative values of t, the situation is similar. Obviously $B^+(\nu,t)$ is very well suited for a determination of f^2, whereas B^- cannot be used, because its nucleon pole term is smaller and decreases more rapidly.

Instead of evaluating the dispersion relation for B^+, we prefer to work with a combination of the relations for the $\pi^\pm$p amplitudes and to present a plot which shows whether discrepancies are due to π^+p- or π^-p-data. We combine the unsubtracted dispersion relation for B^+ with the subtracted one for B^- and use the notation $\tilde{B}^-$ for the amplitude minus the nucleon pole term B_N^-

$$y_\pm \equiv (\nu_B \pm \nu)\left[\mp \mathrm{Re}\,B_\pm(\nu,t) \pm \frac{\nu}{\pi}\,⨍_{\nu_{\mathrm{th}}}^{\infty}\left\{\frac{\mathrm{Im}\,B_+}{\nu' \mp \nu} + \frac{\mathrm{Im}\,B_-}{\nu' \pm \nu}\right\}\frac{\mathrm{d}\nu'}{\nu'}\right] = \frac{g^2}{m} + \tilde{B}^-(0,t)(\nu_B \pm \nu), \tag{4.2.4}$$

where $\nu_{\mathrm{th}} = \mu + t/4m$. We calculate $B_\pm$ from phase shifts and plot the left hand side $(y_\pm)$ as a function of $x \equiv \nu_B \pm \nu$, choosing y_+ if $x>0$ and y_- if $x<0$. Then the points on the right hand side belong to $\mathrm{Re}\,B_+$ and to the sensitive principal value integral over $\mathrm{Im}\,B_{+}$, and the points on the left hand side belong to the corresponding quantities for the elastic π^-p amplitude (64 L HUEPER).

The consistency with the dispersion relation demands that all points must lie on a straight line, whose intercept with the ordinate gives g^2/m. The intercept with the abscissa determines $\tilde{B}^-(0,t)$. The plot shows directly, how well g^2 and $\tilde{B}^-(0,0)=b_{00}^-$ (Table 2.4.7.1) are determined.

The most recent evaluation of Eq. (4.2.4) has been performed by KOCH and PIETARINEN (80 L-1). Figure 4.2.1 shows their result for $t=0$. For $t<0$ the useful range becomes smaller on the low energy side, because one has to remain within the small Lehmann ellipse of the expansion for $\mathrm{Re}\,B$.

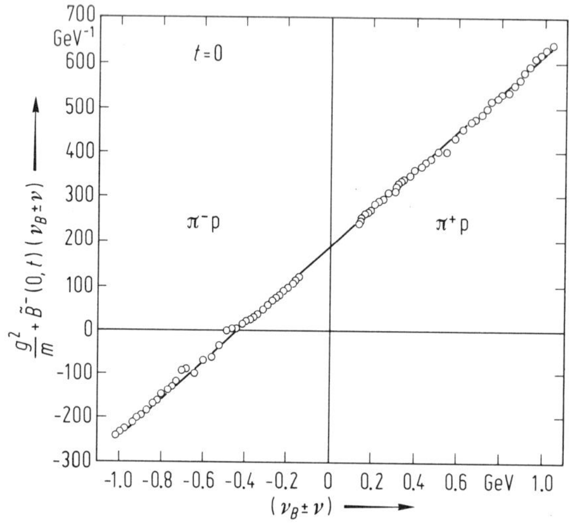

Fig. 4.2.1. Determination of the pion-nucleon coupling constant from Eq. (4.2.4). The straight line belongs to $f^2=0.079$.

The best value for f^2 varies between 0.0797 for $t=0$ and 0.0781 for $t=-0.3\,\mathrm{GeV}^2$. This result and the earlier one for C^- show that there is no indication for the necessity of corrections to the input value $f^2=0.079$.

It is remarkable that the present value $f^2=0.079$ is almost the same as results obtained one or two decades ago, when the accuracy of the data was much poorer. In their review, HAMILTON and WOOLCOCK (63 R) quoted $f^2=0.081\pm0.002$. A larger uncertainty was found, when the data available in 1969 were used (69 L HÖHLER). More recent analyses are quoted in the compilation 79 R NAGELS. Some of them are mentioned in the following.

Other methods

The number of determinations of f^2 is fairly large. Instead of evaluating the dispersion integrals one can apply expansion methods, treating f^2 as an adjustable parameter. One can take PIETARINENs expansion, (Sect. A.6.3.4) or those of LANGBEIN (75 L, 78 L) and of LICHARD (77 L). The difference between the results gives an estimate of the error of the methods, if the same partial wave analysis is used.

LANGBEINs f^2-determination (78 L-2) is based on an expansion of the amplitudes in terms of rational functions (75 L LANGBEIN). He claims that it is an advantage to perform a simultaneous fit to all 4 invariant amplitudes, but this is not true because, as discussed above, the dependence of some amplitudes on the nucleon pole term is weak. The error of his result is derived from a very crude estimate of the error of the invariant amplitudes which, in my opinion, cannot be taken seriously.

LICHARD et al. (77 L) applied the method described in 71 L PRESNAJDER and 72 L PAZMAN, which starts with a mapping of the cut ν^2-plane onto the unit disk of the Z-plane and uses the "most probable analytic representation" of the data. The amplitudes are expanded in terms of $Z^n w(Z)$, where $w(Z)$ depends on the errors and the density of data points. Unfortunately, the power of this method cannot yet be exploited because the information on the input errors is not available.

WOOLCOCK (74 L) gives a careful discussion of the different sources for errors of f^2-determinations, including electromagnetic effects and the unphysical region contribution due to the reaction $\pi^-p\to\gamma n$ [see also SAMARANAYAKE and WOOLCOCK (72 L)]. In his more recent paper SAMARANAYAKE (79 L) uses outdated phase shifts and the numerical results are therefore not of interest.

HITE et al. (75 L) determined f^2 from interior dispersion relations, claiming that their method is more accurate than others. I think that this method has two weak points which can be improved:

i) the nearby part of the left hand cut should be evaluated instead of treating it as part of the discrepancy,

ii) the invariant amplitudes should be fitted not only to the amplitudes reconstructed from phase shifts but simultaneously to the data, as it is done in the KH method for the fixed-t amplitudes.

It is of interest to ask whether an improved version of the CHEW-LOW method can lead to an accurate determination of f^2. Figure A.7.1 shows the branch cuts of the P 33 partial wave. The nucleon-exchange pole of the CHEW-LOW model is an approximation to the "short nucleon cut". The important point is now that the amplitude has a fairly large discontinuity along the front of the "circle cut" which intersects the short nucleon cut. The physical origin of this contribution is the exchange of two pions in a $I=J=0$ state. It can be calculated from the known $\pi\pi N\bar{N}$ partial wave f_+^0 (Sect. 2.4.6).

If the P 33 phase shift is fitted to a CHEW-LOW effective range formula

$$\frac{q^3}{\omega\mu^2}\cot\delta(\mathrm{P}\,33)=\frac{3}{4f_{\mathrm{CL}}^2}(1-\omega/\omega_{\mathrm{R}}), \tag{4.2.5}$$

one finds an effective coupling constant $f_{\mathrm{CL}}^2\approx 0.10$ which is about 25% higher than the expected value 0.079, because it includes the attractive $\pi\pi$ interaction in the $I=J=0$ state (67 R HAMILTON-1, 72 R HAMILTON-TROMBORG).

A new evaluation of the left hand cut contributions to the P 33 partial wave has been performed by HUTT and KOCH (80 L HUTT, 81 L HUTT).

A slightly different "relativistic CHEW-LOW formula" was used by PILKUHN (79 R, Fig. 7.2.2) in order to determine the Δ(1232) resonance parameters from the KH 80 partial wave solution.

Determination of f^2 from Current Algebra constraints

BROWN, PARDEE, and PECCEI (71 L) performed several "determinations of f^2", using Current Algebra constraints and amplitudes calculated by earlier authors from data available at that time. They did not notice that the earlier authors had to assume an input value for f^2 and that some of their data were outdated in the meantime. Since Current Algebra constraints for on-shell amplitudes contain terms which are not well known, there is no chance that f^2-determinations can compete with those based on dispersion relations. It is better to solve the Current Algebra relations for other quantities which are less well known.

f^2 from pion-photoproduction and from NN-scattering

The πNN vertex occurs also in pion-photoproduction and in NN-scattering. Therefore, it is of interest to ask whether f^2 can be determined with a good accuracy from these reactions.

In pion-photoproduction nucleon exchange gives a large contribution to the reaction $\gamma p\to n\pi^+$ near threshold, where the multipole E_{0+} is strongly dominating. In Table 2.4.2.1 we have listed the most recent results for the extrapolation of E_{0+} to threshold.

Table 2.4.2.1. Multipole E_{0+} for π^+-photoproduction at threshold.

$E_{0+}\cdot 10^2\,\mu^{-1}$	
2.83 ± 0.05	70 L ADAMOVICH } Quoted from the compilation 79 R NAGELS
2.98	NOELLE (1971) } Quoted from the compilation 79 R NAGELS
2.75 ± 0.12	78 L MCCLASKEY, from interior dispersion relations
2.76	Nucleon pole term of fixed-t dispersion relations ($f^2=0.079$)

It is seen that the nucleon pole term (which includes the magnetic terms) agrees within the errors with the experimental value. The question is whether reliable theoretical estimates can be given for the additional contributions.

In dispersion theory, Δ(1232)-exchange usually plays an important role, but in the present case its effect at threshold is negligible. There is also a t-channel effect in the isoscalar amplitude (ϱ-exchange) which is difficult to calculate, but a comparison between the experimental values of E_{0+} at threshold for $\gamma p\to n\pi^+$ and $\gamma n\to p\pi^-$ shows that this term must be small (see for instance HÖHLER 64 L and 65 R).

The nucleon-exchange dominance follows from the KROLL-RUDERMANN theorem and more recent Low Energy Theorems for pion photoproduction (see 73 R ALFARO et al., p. 240ff. and 79 R AMALDI-FUBINI-FURLAN, p. 52ff.). MACMULLAN and SCADRON (79 L) started from a Ward identity and used Current Algebra and PCAC in a way analogous to the treatment of pion-nucleon scattering which leads to Eq. (A.8.31) (see also 81 R SCADRON, Sect. 5.5 and 80 L NATH). The conclusion is that if one uses $f^2=0.079$, there is no serious deviation from the prediction, but it is difficult to give an error.

In *nucleon-nucleon* scattering, one-pion-exchange is known to be responsible for the long range interaction. The problem is to separate its contribution from two-pion exchange, which was treated by some authors in a crude approximation. Some recent results are $g^2/4\pi=13.1\pm0.4$ (74 L GERSTEN) and 15.3 ± 0.3 (74 L VIOLLIER). The most reliable result follows from an analysis of new pp-scattering data at 150···515 MeV (78 L BUGG).

$$g^2/4\pi=14.06\pm0.65 \ (\pi^0 p\bar{p} \text{ coupling}). \tag{4.2.6a}$$

It has to be compared with $g^2/4\pi=14.3\pm0.2$ from πN scattering.

One should notice that the pion mass squared enters in the relation between f^2 and g^2. This gives a 7% difference between charged and neutral pions. The value for πN scattering is essentially the $n\bar{p}\pi^+$ coupling constant, since the charge-exchange data are less accurate. Earlier results are listed in the compilation 79 R NAGELS.

Conclusion

The determination of the πN coupling from πN scattering data leads to

$$f^2=0.079\pm0.001; \quad g^2/4\pi=14.3\pm0.2. \tag{4.2.6b}$$

Other determinations from pp-scattering and from pion photoproduction are compatible with this value. Reliable results for the charge-dependence are not yet available.

2.4.2.2 Other meson-nucleon coupling constants

From pion-nucleon scattering amplitudes one can determine the meson-nucleon coupling constants for those mesons which can be exchanged in the t-channel: ϱ(770), f(1270), g(1700), h(2040). One has to choose a suitable one-dimensional dispersion relation in such a way that the contribution of the dispersion integral over the s-channel cut can be evaluated and that the partial wave expansion along the t-channel cut converges.

The first investigations of this type (62 L ATKINSON, 69 R SHIRKOV-SEREBRYAKOV-MESCHCHERYAKOV, 69 R LOVELACE) were carried out for backward dispersion relations in order to determine the contributions from ϱ and "ε" exchange. More accurate results were derived by ENGELS et al. (70-1 and 70-2) from CERN 68 phase shifts. STRAUSS (70 L) made a first attempt to separate the different contributions by considering fixed-θ dispersion relations for near-backward angles.

Some problems of the earlier investigations could be understood, when accurate determinations of the lowest t-channel partial waves f_+^0 and $f_\pm^1$ became available (Sect. 2.4.6): a description of ϱ- and "ε"-exchange by narrow resonances is bad, since both partial waves $\mathrm{Im}\, f_+^0$ and $\mathrm{Im}\, f_\pm^1$ have a peculiar shape.

The situation is similar as in the cases of the isovector nucleon form factors (Sect. 2.5.2) and of the two-pion contribution to the imaginary part of the $N\bar{N}$ amplitude along its unphysical cut (Sect. 2.5.3): since f_+^0 and $f_\pm^1$ are known on the nearby part of the cut, one can use these amplitudes as part of the input and hope that the contributions of the higher partial waves can be approximated by pole terms.

Investigations of this type were carried out independently by KAISER et al. (76 L KAISER, 77 L BORIE, 78 L KAISER) and by the HEDEGAARD-JENSEN (77 L). Dispersion relations along hyperbolas in the MANDELSTAM plane were used in both cases and both sets of curves included the backward hyperbola.

KAISER et al. chose the interior dispersion relation (Sect. A.6.7.2). The t-channel partial wave expansion converges, if the fixed lab. angle θ_π lies in the range $180°>\theta_\pi>95°$. The important point is that the contributions of the different t-channel partial waves can easily be separated by studying the dependence on the parameter a in Eq. (A.1.30). The KH 78 partial wave solution was used in the last paper and preliminary versions in the earlier ones.

In most of his calculations, HEDEGAARD-JENSEN used the hyperbolas $su=b$. For large s, these curves approach the line $u=0$, which lies within the physical region near the backward direction (Fig. A.6.5). However, near threshold these hyperbolas lie in general outside the physical region (Fig. A.6.11), except for the backward hyperbola, Eq. (A.1.32). Therefore, one has to use the partial wave expansion for the invariant amplitudes in the unphysical region in order to evaluate the s-channel dispersion integral. This is a disadvantage in comparison with the "interior" dispersion relation, where the path for the s-channel dispersion integral lies entirely in the physical region.

HEDEGAARD-JENSEN ("HJ") used phase shifts determined by NIELSEN and OADES (74 L-2) from CERN 72 and LONDON 73 phase shifts, taking into account fixed-t analyticity. (The phase shifts used by KAISER et al. include more recent data.) The separation of the different t-channel partial waves can be made in a similar way as in the case of the interior dispersion relation.

Results for the contributions of the different t-channel partial waves

$J=0$ exchange

In this case the results are relatively weak, because the convergence of the dispersion integral is slow and there is no well-pronounced resonance. HJ studied this exchange in detail, giving estimates for the $\pi\pi$-scattering length a_0^0 and for the contribution from the range $t>40\mu^2$. It would be of interest to reconsider the evaluation, using the new results for the amplitudes.

$J=1$ exchange

The compatibility with the results for $\operatorname{Im} f_\pm^1$ (Sect. 2.4.6) is reasonable. Small discrepancies, which possibly indicate non-ϱ contributions, are comparable with the errors.

Results for higher partial waves are given, assuming that a description by a single pole term is adequate

$$\operatorname{Im} f_\pm^J(t)=\pi R_\pm^{\mathrm{X}}\,\delta(t-t_{\mathrm{X}}). \tag{4.2.7}$$

X denotes the meson resonance. t_{X} is calculated from the masses of f(1270), g(1700), h(2040). $R_\pm^{\mathrm{X}}$ is essentially a product of the $\mathrm{X}\pi\pi$ and XNN coupling constants.

$J=2$ exchange

Both analyses see a clear signal for f-meson exchange. The fNN coupling constants are defined as in the "Compilation" (76 R NAGELS)

$$R_+^{\mathrm{f}}=\frac{8}{15\pi\mu}G_{\mathrm{f}\pi\pi}G_{\mathrm{fNN}}^{(1)}\left[1+\frac{4m^2-m_{\mathrm{f}}^2}{4m^2}G_{\mathrm{fNN}}^{(2)}/G_{\mathrm{fNN}}^{(1)}\right];\quad R_-^{\mathrm{f}}=\frac{4\sqrt{6}}{15\pi m\mu}G_{\mathrm{f}\pi\pi}G_{\mathrm{fNN}}^{(1)}, \tag{4.2.8}$$

where $G_{\mathrm{f}\pi\pi}$ can be obtained from the f-width (79 R PILKUHN, p. 189)

$$\frac{G_{\mathrm{f}\pi\pi}^2}{4\pi}=\frac{5}{8}\frac{m_{\mathrm{f}}^2\mu^2}{q_{\mathrm{f}}^5}\Gamma_{\mathrm{f}\pi\pi}=0.031\pm0.004;\qquad q_{\mathrm{f}}^2=(m_{\mathrm{f}}^2/4)-\mu^2. \tag{4.2.9}$$

One should note that different conventions for all three coupling constants have been used in the literature. [There is a printing error in 77 L BORIE and in Eq. (12.2) of PILKUHNs book.]

The results for the fNN coupling constants are

	77L HJ	77L BORIE	Tensor dominance	
$[G_{\mathrm{fNN}}^{(1)}]^2/4\pi$	4.0 ±1.0	2.2±0.9	0.35	(4.2.10)
$G_{\mathrm{fNN}}^{(2)}/G_{\mathrm{fNN}}^{(1)}$	0.00±0.07	0.6±0.9	0.	

Earlier determinations were performed by STRAUSS (70 L) and ENGELS (70 L-2). As usual in this field, the errors cannot be calculated in a systematic way and should therefore not be taken too seriously.

The results are of interest for a comparison with the prediction from tensor dominance models: 70 L JONES, 71 L RENNER, 72 R RENNER, 73 L KAFIEF, 79 R GENZ (p. 506). The prediction for $G_{\mathrm{fNN}}^{(2)}$ is confirmed, but the results for $G_{\mathrm{fNN}}^{(1)}$ are much larger than predicted: $G_{\mathrm{fNN}}^{(1)}=G_{\mathrm{f}\pi\pi}m/2\mu$.

One could think that the fNN-coupling constants could be estimated from reggeized f-exchange at high energies. However, it is not possible to separate the f-exchange contribution from the "pomeron" contribution in a generally accepted way (Sect. 2.6.3).

In a Feynman graph calculation for f-exchange one obtains for the contribution to the amplitudes A^+ and B^+ (70 L ACHUTHAN)

$$\begin{aligned}A_{\mathrm{f}}^+&=\left[\frac{16}{3\mu}G_{\mathrm{f}\pi\pi}G_{\mathrm{fNN}}^{(1)}q_{\mathrm{f}}^2\left(1-\frac{p_{\mathrm{f}}^2}{m^2}\frac{G_{\mathrm{fNN}}^{(2)}}{G_{\mathrm{fNN}}^{(1)}}\right)-\frac{16}{\mu}G_{\mathrm{f}\pi\pi}G_{\mathrm{fNN}}^{(2)}\left[\frac{s-u}{4m}\right]^2\right]\frac{1}{m_{\mathrm{f}}^2-t};\\ B_{\mathrm{f}}^+&=\frac{16}{\mu}G_{\mathrm{f}\pi\pi}G_{\mathrm{fNN}}^{(1)}\frac{s-u}{4m}\frac{1}{m_{\mathrm{f}}^2-t};\qquad p_{\mathrm{f}}^2=(m_{\mathrm{f}}^2/4)-m^2.\end{aligned} \tag{4.2.11}$$

At high energies, the second term in the first line is dominant. It vanishes, if $G^{(2)}_{fNN}=0$. The first term is proportional to $B^+/(s-u)$ and therefore negligible. The smallness of the A^+-amplitude in high energy near-forward scattering will be discussed in Sect. 2.6.3.8.

$J=3$ exchange

Again, there is a clear signal in both analyses. We list only the residues as defined in 76 R NAGELS

$$R^{(1)}_g=-28\pi(m_g^2/4-\mu^2)R^g_+\,;\qquad R^{(2)}_g=28\pi(m_g^2/4-m^2)(m_g^2/4-\mu^2)R^g_-/\sqrt{12} \tag{4.2.12}$$

	77 L HJ	77 L BORIE
$R^{(1)}_g$	$(-10\pm3)\mu^{-1}$	$(-12.1\pm2.1)\mu^{-1}$
$R^{(2)}_g$	$(-12\pm3)\mu^{-1}$	$(-8.0\pm2.4)\mu^{-1}$.

(4.2.13)

BORIE and KAISER also give values for the gNN coupling constants.

$J=4$ and 5 exchanges

In both analyses, there is evidence for $J=4$ and $J=5$ exchange contributions. The residues are comparable with the errors or only slightly larger.

The work of JACOB and HITE (75 L)

These authors employed interior dispersion relations in order to determine the t-channel $J=1$ and $J=3$ exchanges, using a pole term ansatz in both cases. As discussed in connection with the determination of $f^1_\pm(t)$ in Sect. 2.4.6, a pole term ansatz for ϱ-exchange is misleading, because of the enhancement on the left wing of the ϱ-peak, which is due to a nucleon-exchange contribution.

ϱNN- and ωNN-coupling constants

The numerical determination of the $\pi\pi N\bar{N}$ partial waves $f^1_\pm$ is treated in Sect. 2.4.6. These partial waves have the quantum numbers of the ϱ-meson, but they are not dominated by ϱ-exchange. Not even the right hand cut contribution in their partial wave dispersion relation can be quantitatively approximated by a ϱ-pole term or a Breit-Wigner type formula (see the discussion in Sect. 2.5.2.2). Nevertheless one can ask for the value of the ϱNN coupling constants.

This question was investigated in collaboration with E. PIETARINEN (75 L HÖHLER-4), starting from the work of KROLL, LEE, and ZUMINO (67 L). The result for the coupling constants essentially agrees with that of GUSTAFSON, NIELSEN, and OADES (unpublished, quoted in 79 R NAGELS, p. 241; see Sect. 2.4.6.3). Our values are (error ≈15%)

$$(G^{\mathrm{v}}_{\varrho NN})^2/4\pi\approx2.2\,;\qquad G^{\mathrm{T}}_{\varrho NN}/G^{\mathrm{v}}_{\varrho NN}\approx6.6\,, \tag{4.2.14}$$

whereas GUSTAFSON et al. obtained 6.0 for the tensor/vector coupling ratio (see also Sect. A.8.1.3).

As far as ϱ-exchange in near-forward high energy scattering is concerned, the situation is much better than for f-exchange, because ϱ-exchange is the leading process. It is possible to check at least in a crude approximation, whether the coupling constants Eq. (4.2.14) are compatible with the Regge residue at the pole position (see Sect. 2.6.5.4).

The ωNN-coupling constants can be estimated from analyses of isoscalar nucleon form factors (Sect. 2.5.2) and of the unphysical cut in nucleon-nucleon forward scattering (Sect. 2.5.3).

Some information on coupling constants between mesons and nucleons or nucleon resonances can be derived from the behaviour of zero trajectories at "double poles": Eq. (4.3.6).

BECKER et al. (80 L) made an attempt to determine the ratio of the πNN and ηNN coupling constants. In order to avoid $I_u=3/2$ exchanges in near-backward scattering, they measured the reactions $K^-p\to\Lambda\pi^0$ and $K^-p\to\Lambda\eta$. It is expected that in both cases the exchange of the nucleon trajectories N_α and N_γ is dominant (Sect. A.9.2.3). A comparison of the data leads to $G^2_{\eta NN}/G^2_{\pi NN}=0.26\pm0.10$, which is compatible with the (somewhat smaller) SU(3) prediction.

2.4.3 Zero trajectories

2.4.3.1 Introduction

Zero trajectories are two-dimensional surfaces in the four-dimensional space of the complex MANDELSTAM variables s, t along which a scattering amplitude vanishes: $A(s,t)=0$. One reason why zeros of scattering amplitudes attracted attention is the VENEZIANO model (68 L), which describes the amplitudes by a simple pattern of zeros and poles.

ODORICO (72 L, 73 L) performed the first detailed search for dip structures in differential cross section data, which indicate the presence of nearby zero trajectories. His "linear zero model" is an adaptation of the VENEZIANO model to pion-nucleon scattering. BARRELET (72 L) investigated zeros of transversity amplitudes and developed a method for phase shift analysis, in which zero trajectories play an important role (Sect. 2.1.11).

Since that time, zero trajectories were used not only in πN-scattering (D. CHEW, Sect. 2.1.11) and $\bar{N}N\pi\pi$-scattering (80 R PENNINGTON), but also in ππ-scattering (76 R MARTIN-MORGAN-SHAW, 79 L BECKER), kaon-nucleon-scattering (for instance 75 L VAN HORN, 80 L ATKIN), αα- and $p^{12}C$-scattering (75 L HEEMSKERK, 76 L KOK), pion photoproduction (76 L LOOS, 80 L VON GEHLEN), Compton scattering and weak axial pion production (80 L VON GEHLEN).

In collaboration with I. SABBA-STEFANESCU (79 L HÖHLER-2) we have studied the singularities of zero trajectories. Our group has also investigated the zero trajectories of invariant πN amplitudes, using the KH 78 and CMU-LBL 79 partial wave analyses (80 L BORIE). Further problems related to zero trajectories and applications are treated in SABBA-STEFANESCUs thesis (78 L) and in his recent papers (80 L-2, 81 L, 82 L-2).

2.4.3.2 General properties of zero trajectories

We assume that the invariant amplitude $A(s,t)$ has the analytic properties of the MANDELSTAM representation (Sect. A.6.2) and consider a point s_0, t_0 on a zero trajectory, i.e. $A(s_0,t_0)=0$. The question is, under which conditions the equation $A(s,t)=0$ can be solved for t, giving the description $t=\tau(s)$ for the zero trajectory in the neighborhood of the point s_0, t_0.

The answer follows from the implicit function theorem for two complex variables. s_0, t_0 must be a regular point of $A(s,t)$ and $A(s_0,t)$ must have a single zero at $t=t_0$. The derivative of $t=\tau(s)$ is given by

$$\frac{dt}{ds}=-\frac{\partial A/\partial s}{\partial A/\partial t}. \tag{4.3.1}$$

The zero trajectory $t=\tau(s)$ is locally described by a *conformal mapping* of the s-plane onto the t-plane, if $\partial A/\partial s \neq 0$ in addition to $\partial A/\partial t \neq 0$ at s_0, t_0.

One can distinguish two types of singularities of zero trajectories:

i) In general, $t=\tau(s)$ is singular at points where the invariant amplitude has a singularity. $A(s,t)$ has the cut structure and the nucleon poles described in the MANDELSTAM representation (Sect. A.6.2). In addition, it has poles in the second sheet of the s-plane, which belong to nucleon resonances, and poles in the second sheet of the t-plane, which belong to meson resonances. All poles are of first order.

ii) Furthermore, the zero trajectories may be singular at points where $\partial A/\partial t$ vanishes together with $A(s,t)$. In general this happens only at isolated points in the complex s, t space.

In the following, we shall discuss the behaviour of zero trajectories in the neighborhood of these singularities.

Poles

In the complex s, t space, the poles lie on *pole surfaces* $s=M_s^2$ and $t=M_t^2$, where $M=m_{res}-i\Gamma/2$. Considering an s-channel nucleon resonance as an example, $A(s,t)$ can be written

$$A(s,t)=\frac{r(t)}{s-M_s^2}+\text{background}. \tag{4.3.2}$$

$r(t)$ is a polynomial in $z=\cos\theta=1+t/2q^2$, which is usually expressed in terms of Legendre polynomials and their derivatives. The order of the polynomial depends on the spin of the resonance and the choice of the amplitude.

We define *Legendre zeros* as points $s=M_s^2$, $t=t_L$, which lie on the pole surface and fulfill the relation

$$\lim_{s\to M_s^2}(s-M_s^2)A(s,t_L)=r(t_L)=0, \tag{4.3.3}$$

i.e. at the Legendre zero, the amplitude has a singularity of the type 0/0, but it is important to notice that the zeros of the numerator and denominator belong to *different variables.*

In the theory of functions of two complex variables (see for instance 70 R BEHNKE-THULLEN, p. 94) it is shown that in every neighborhood of a Legendre zero the amplitude $A(s,t)$ assumes every finite value including zero. Since $r(t)$ has only simple zeros, we conclude that exactly one zero trajectory must pass a pole surface at each of its Legendre zeros.

The Legendre zeros are *singular* points of $A(s,t)$, since one cannot assign a value to A. However, the trajectory is regular at $s=M_s^2$, $t=t_L$. There is no resonance-like structure in $\tau(s)$ when s goes through the value M_s^2 or through $\mathrm{Re}\,M_s^2$ along the real s-axis.

A singularity of another type occurs at *double poles*, which are points in the complex s, t space at which two pole surfaces intersect each other, for instance $s=M_s^2$ and $u=M_u^2$. We assume that the two pole terms occur additively in the amplitude

$$A(s,t)=\frac{r_s(t)}{s-M_s^2}+\frac{r_u(t)}{u-M_u^2}+\beta(s,t), \tag{4.3.4}$$

where $\beta(s,t)$ denotes the background. Then, in the expression

$$(s-M_s^2)(u-M_u^2)A(s,t)=(u-M_u^2)r_s(t)+(s-M_s^2)r_u(t)+(s-M_s^2)(u-M_u^2)\beta(s,t), \tag{4.3.5}$$

the right hand side is zero at the double pole and in its neighborhood at points, where the relation

$$\frac{u-M_u^2}{s-M_s^2}=-\frac{r_u(t_d)}{r_s(t_d)}, \quad \text{evaluated at the double pole} \tag{4.3.6}$$

is fullfilled. According to Eq. (4.3.5), Eq. (4.3.6) describes locally a zero trajectory of $A(s,t)$. The double pole is a singular point of $A(s,t)$ and a regular one for $t=\tau(s)$.

The assumption Eq. (4.3.4) is true for the usual resonance poles. If one assumes another term, $r_3(t)/[(s-M_s^2)(u-M_u^2)]$, Eq. (4.3.2) would have *a pole in its numerator function.* However, $r(t)$ must be a polynomial, if the pole term belongs to a resonance of finite integer or half integer spin. Therefore, $r_3(t)$ must be identically zero.

It follows from Eq. (4.3.6) that the direction of the zero trajectory at the double pole is determined by the ratio of the residues, i.e. by the coupling constants of the pion-nucleon-resonance vertex.

Intersection of zero trajectories

We now consider singularities of $t=\tau(s)$ at points where $A(s,t)$ is regular, but $\partial A/\partial t$ vanishes together with $A(s,t)$ (see 79 L HÖHLER-2).

A general representation of $A(s,t)$ in the neighborhood of a point s_0, t_0 where $A(s,t)$ is regular and vanishes is given by WEIERSTRASS' "preparation theorem" (70 R BEHNKE-THULLEN, p. 90, 65 R OSGOOD, p. 71). We shall discuss only the special case where the polynomial in $(t-t_0)$ is of second order

$$\begin{aligned} A(s,t)&=a(s,t)\,\Omega(s,t);\\ a(s,t)&=(t-t_0)^2+\alpha_1(s)(t-t_0)+\alpha_2(s). \end{aligned} \tag{4.3.7}$$

The $\alpha_i(s)$ are regular at s_0, $\alpha_i(s_0)=0$ and $\Omega(s,t)$ is regular at (s_0,t_0) and has no zeros.

It is seen that $\partial A(s_0,t_0)/\partial t=0$, i.e. we have a singularity of the above mentioned type. For a given value of $s\neq s_0$, one finds two solutions for t, i.e. in the neighborhood of s_0 there exist two function elements $t=\tau_i(s)$ which describe two trajectories. At $s=s_0$ these functions $\tau_i(s)$ have a square root type branch point.

In the simplest extension of the VENEZIANO model to πN-scattering (ODORICO 72 L and 73 L), the trajectories are parallel to $t=$const and $u=$const. It corresponds to

$$a(s,t)=-(t-t_1)(u-u_1); \qquad t_1 \text{ and } u_1 \text{ real}, \tag{4.3.8}$$

if we consider only two trajectories. This ansatz belongs to a trivial special case (it is reducible), in which branch points do not occur.

Realistic zero trajectories are more complicated, because they must pass Legendre zeros and double poles, which lie at complex s-values. The simplest generalization of Eq. (4.3.8) is the nonreducible ansatz

$$a(s,t)=c^2-(t-t_1)(u-u_1); \quad \Omega=1, \tag{4.3.9}$$

where c^2 is a complex constant. Now there exist two branch points. At a large distance from these points the trajectories are similar to those described by Eq. (4.3.8), but near the branch points, which define the location of the "intersection region", there is a characteristic perturbation.

As mentioned above, a zero trajectory can be associated with a conformal mapping of the s-plane onto the t-plane. For the description of two trajectories in the intersection region, one has to consider the mapping onto two Riemannian sheets. It is interesting to notice that the mapping which belongs to Eq. (4.3.9) is a generalized JOUKOVSKY airfoil transformation

$$-(s-s_1)=t-t_1+\frac{c^2}{t-t_1}, \tag{4.3.10}$$

and therefore, a plot of the trajectories in the t-plane for real s shows a similarity with streamlines.

In collaboration with I. SABBA-STEFANESCU, we have studied in detail the behaviour of the experimental zero trajectories (calculated from partial wave solutions) in the intersection region. It can be described in a semiquantitative way already with the simple ansatz Eq. (4.3.9). In some cases, the branch points lie very near to the physical s, t plane. A general quadratic ansatz in s and t for $a(s,t)$ is sufficient for a quantitative description (79 L HÖHLER-2).

OEHME (71 R-2) studied a similar mathematical problem for Regge trajectories.

2.4.3.3 Zero trajectories of invariant amplitudes

There are two possibilities to calculate projections of zero trajectories from amplitudes derived from the results of partial wave analyses. One can calculate

i) zeros at real s in the complex t-plane, or

ii) zeros at real t in the complex s-plane. In this case one can use either the KH 78 fixed-t solution or amplitudes at fixed-t which have been determined by other methods from partial wave solutions.

Zeros at real s

Results for zero trajectories derived from the CMU-LBL 79 and KH 78 partial wave solutions are shown in Figs. 4.3.1···4.3.3. The following points are of interest:

i) Although the zeros are rather sensitive to details of a partial wave solution, the agreement between the trajectories calculated from the CMU-LBL and KH solutions is reasonable.

ii) The trajectories have a tendency to run along t=const and u=const except at intersections, where they behave as discussed in connection with Eq. (4.3.9). They remain at a close distance to the physical s,t-plane unless they are forced to go to a larger distance, because they must pass Legendre zeros or double poles. The deviation from the real s,t-plane is caused by the finite width of the resonances, i.e. by unitarity effects.

iii) It is surprising that the zero trajectories have a simple shape, since they have to pass many resonance surfaces, and this is possible only at the Legendre zeros (in the s,t-range of the physical region). In several cases, the mass differences between adjacent resonances is small, but the t_L-values of their Legendre zeros differ widely and even the number of Legendre zeros is different. One would expect rapid variations of the zero trajectories in energy, which are not observed.

One could argue that the spacing of the data in energy is not close enough, or that the variation occurs only for complex s near the poles and is washed out at real s. In any case, it would be of interest to have data closely spaced in energy in the range of a resonance family, for instance in π^+p-scattering around at total c.m. energy of 1900 MeV.

Furthermore, one should check whether the mass splitting is really well established in cases where the mass differences within a group of adjacent resonances are not much larger than the errors. Maybe an equally good fit can be found, if one admits only *a single pole in the invariant amplitude* as suggested by dual models (cf. Table 2.4.1.7). Then, the numerator in Eq. (4.3.2) is a superposition of Legendre functions belonging to different spins, and the zero trajectories have a smooth s-dependence. — It is difficult to see, how the simplicity of the zero trajectories can be explained in quark models.

iv) ODORICO (73 L) expected that the zero trajectories of the four invariant amplitudes $A^\pm$ and $B^\pm$ are different from each other, in particular because strong diffraction effects occur only in B^+ and not in A^+. Our figures show, however, *a remarkable similarity in all four cases:* A^+, A^-, B^+, B^-, although for instance the nucleon poles do not occur in $A^\pm$ and the ϱ-pole does not occur in A^+, B^+. The reason for the similarity is not yet known. Figure 4.3.3 shows the zero trajectories for the elastic A and B amplitudes.

v) The zero trajectories of the invariant amplitudes C^+ and C^- are different from those of the amplitudes $A^\pm$ and $B^\pm$ and also from each other. At intermediate and high energies, C^- has a zero trajectory along $t \approx -0.2\,\text{GeV}^2$, which is related to the zero of the difference $d\sigma_-/dt - d\sigma_+/dt$ ("cross-over effect", Sect. 2.6.5.3). Plots of the zero trajectories for the amplitudes $C^\pm$ are shown in 78 L BORIE-3.

vi) One can ask for the continuation of those zero trajectories which leave the physical region in the backward direction in Figs. 4.3.1 and 4.3.2. They approach either the Δ(1232) Legendre zero of the $A^\pm$-amplitudes or the nucleon-Δ(1232) double pole of the $B^\pm$-amplitudes. Their continuation beyond these points can be determined from the KH78 fixed-t amplitudes, an advantage being that the amplitudes are real below threshold. (It is useful to consider amplitudes which are even in ν: C^+, C^-/ν etc.)

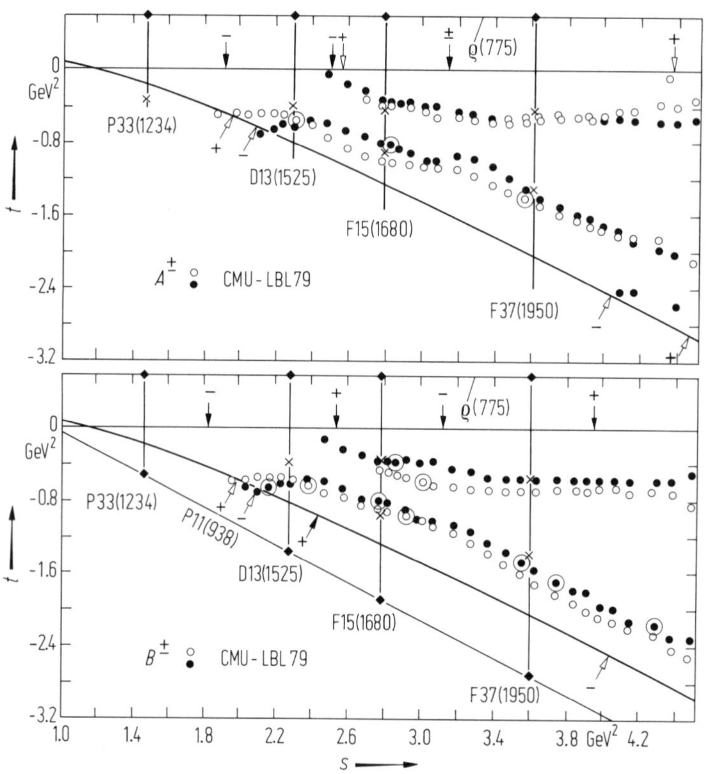

Fig. 4.3.1. Projections of zero trajectories of invariant amplitudes $A^\pm$ and $B^\pm$ onto the physical s, t-plane as calculated from the CMU-LBL 79 partial wave solution. Intersections of the trajectories with the s, t-plane are denoted by ◉ or ◎. Legendre zeros: × ; double poles: ◆. Arrows pointing to the lines $\theta=0°$ and $\theta=180°$ show the location of dips of the modulus. "+" and "−" indicate if the dip occurs in the isospin even or odd amplitude. Well-pronounced dips are denoted by "open" arrows ↗.

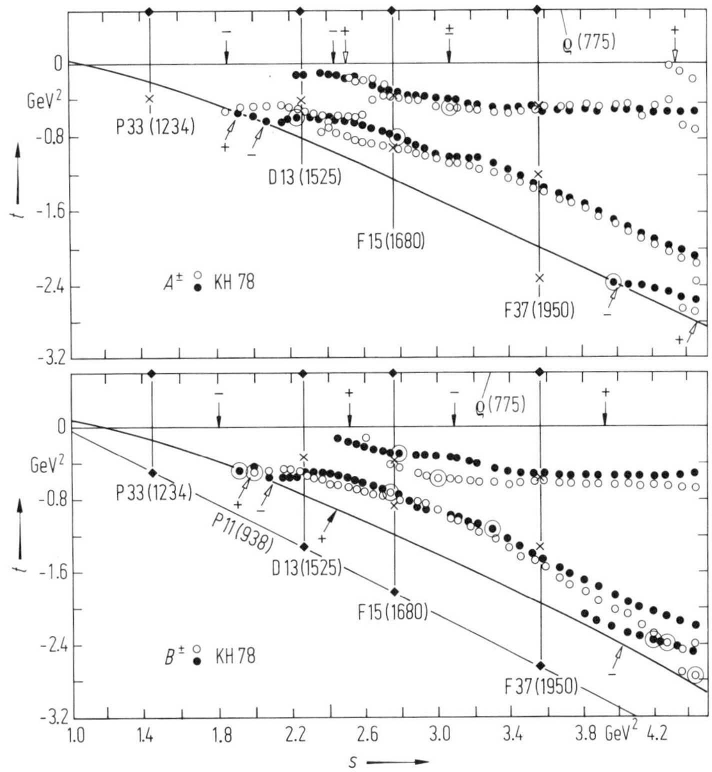

Fig. 4.3.2. This figure differs from Fig. 4.3.1 only by the use of the KH 78 partial wave solution for the calculation of the zeros and dips. The agreement is good except for some details.

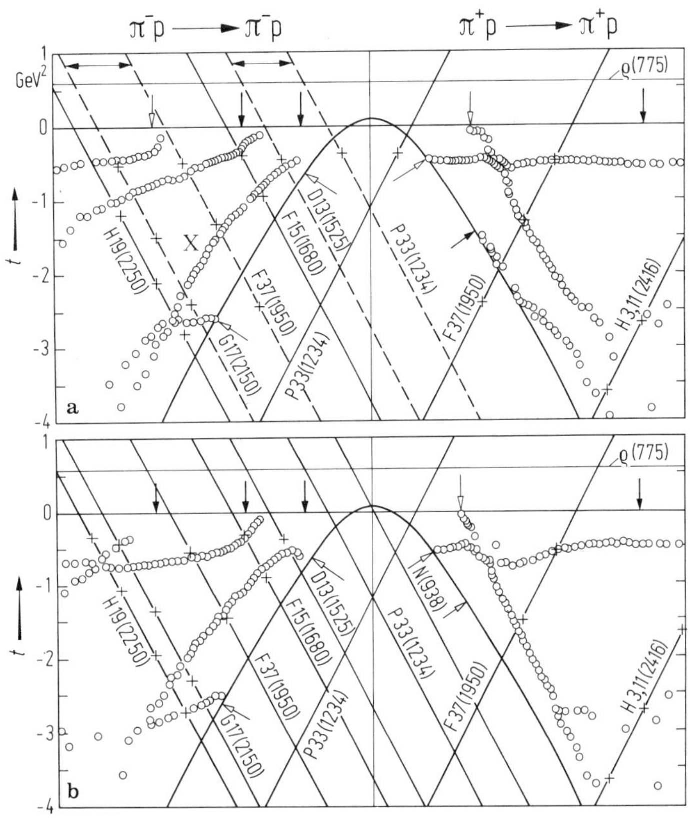

Fig. 4.3.3. Projections of zero trajectories of the invariant amplitudes A (upper part) and B (lower part) for elastic π^+p scattering (right hand side) and π^-p scattering (left hand side). The zeros were calculated from the KH 78 solution. The right hand side shows the projections of trajectories $t=\tau_i(s)$ and the left hand side projections of $t=\tau_i(u)$, so there is no simple analytic continuation from one side to the other, except for almost real trajectories. The arrows have the same meaning as in Fig. 4.3.1 and "+" denotes the Legendre zeros. The horizontal arrows ↔ in the upper left figure show the pairs of resonances which approximately cancel each other in ODORICOs model (73 L). See 80 L BORIE for a detailed discussion.

In the neighborhood of the point $\nu = t = 0$, the location of zeros can also be obtained from the subthreshold expansion (Sect. 2.4.7.1), adding the nucleon pole term. The most interesting case is the zero trajectory of D^+

$$D^+(\nu, t) \approx \frac{g^2}{m} \frac{\nu_B^2}{\nu_B^2 - \nu^2} + d_{00}^+ + d_{01}^+ t + d_{10}^+ \nu^2 . \tag{4.3.11}$$

The CHENG-DASHEN point $\nu = 0$, $t = 2\mu^2$ (i.e. $\nu_B = 0$) is exceptional insofar as it is a double pole and at the same time a Legendre zero. Therefore *two* zero trajectories go through this point.

The dominant contribution in Eq. (4.3.11) is the pseudovector nucleon Born term. It has a twofold zero along the line $t = 2\mu^2$, $\nu_B = 0$ (Fig. 4.3.4). If one uses the approximation of Eq. (4.3.11), it turns out that the additional small terms have a surprisingly large effect. The zero trajectory has now two separated branches which are connected by a "bridge" at complex ν_B-values (Fig. 4.3.4). The splitting of the two branches is proportional to the small sigma term (Sect. 2.5.1.3).

Since the validity of the approximation Eq. (4.3.11) is limited, we have calculated the curves in Fig. 4.3.4 from the expansion in terms of $\pi\pi N\bar{N}$ partial waves and, at $t < 0$, from the KH 78 fixed-t amplitudes.

One of the trajectories is directed towards the s-channel threshold, as noticed already by CHENG and DASHEN (71 L). It is responsible for the small S-wave scattering length. The "reason" for this behaviour is that this trajectory can pass the Δ-resonance surface only at its Legendre zero ($t_L = (-0.083 + 0.04i)\,\mathrm{GeV}^2$).

Another branch of the trajectory goes to negative t-values, remaining at small negative values of ν^2 up to $t \approx -0.5\,\mathrm{GeV}^2$, where it turns around, passes the NΔ-double pole and enters the physical region from the backward direction. It is remarkable that two distinct trajectories of the physical region join each other in the unphysical region at the CHENG-DASHEN point.

Further results on zero trajectories in the unphysical region can be found in 78 L BORIE-3. The most important Legendre zeros and double poles are listed in Tables 2.4.3.1 and 2.4.3.2, respectively (see next pages).

In 80 L BORIE it was shown that the "linear zero model" of ODORICO (73 L) is not satisfactory. There are also problems with more recent models (77 R FUKUGITA-IGI).

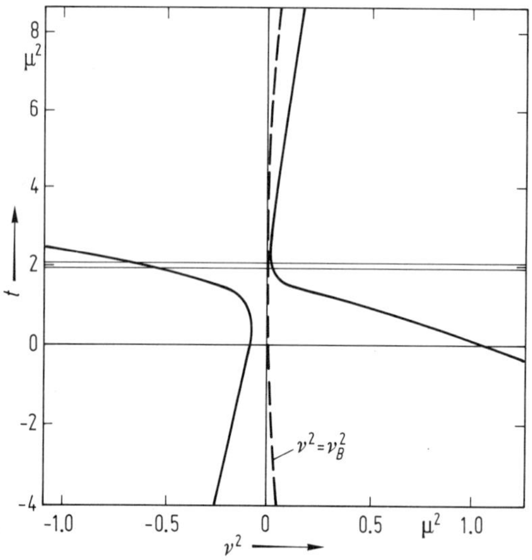

Fig. 4.3.4. Zero trajectories of the invariant amplitude D^+ in the ν^2, t-plane. Near the CHENG-DASHEN point, the trajectories can be calculated from Eq. (4.3.11). The second, third and fourth term are known to be "small" from the discussion of the sigma term. However, if these terms are omitted, the zero trajectories are strongly modified and go over into a double zero along the line $t = 2\mu^2$, which belongs to $\nu_B^2 = 0$ in the numerator of the first term (the pseudovector nucleon Born term).

Zeros at real t

Zeros at real values of t in the complex s-plane have been determined by SABBA-STEFANESCU (78 L). Since a direct calculation of zeros from PIETARINENs expansion is not practical due to the large number of coefficients, he tried several other methods and found that the most convenient one is a simple search for minima of the modulus. Earlier authors had used dispersion relations, but this is rather cumbersome. (71 L SHAFEE).

Above threshold, zeros can be determined in the t-plane at real s, and also in the s-plane at real t. It is interesting to connect the results by an analytic continuation. This was done by SABBA-STEFANESCU for the amplitude C^- at small t near 1 GeV/c, where a zero of the fixed-t amplitude shows a complicated behaviour, and also for the amplitude C_+ (79 L HÖHLER-2, Figs. 5 and 6).

Table 2.4.3.1 t-values in GeV^2 of the Legendre zeros of the most important resonances [from Z. Physik C, Particles and Fields **4** (1980) 333]. Tables 2.4.3.1 and 2.4.3.2 are based on the list of pole positions in 79 L Cutkosky-2, which is an earlier version of Table 2.4.1.2.

	Amplitude $A(s,t)$				Amplitude $B(s,t)$			
Odd parity								
N(1525) D13	−0.43+0.08i				−0.39+0.07i			
N(1670) D13	−0.66+0.06i				−0.60+0.05i			
N(1680) D15	−0.54+0.09i	−1.61+0.22i			−0.64+0.12i			2.70+0.02i
N(2150) G17	−0.53+0.12i	−1.50+0.34i	−2.39+0.54i		−0.46+0.10i	−1.36+0.30i	−2.29+0.51i	
N(2200) G19	−0.46+0.09i	−1.38+0.26i	−2.39+0.47i	−3.56+0.60i	−0.56+0.12i	−1.57+0.32i	−2.54+0.51i	2.98+0.03i
Δ(1730) D33	−0.70+0.25i				−0.63+0.21i			
Δ(1930) D35	−0.86+0.16i		−2.38+0.39i			−1.07+0.22i		2.62+0.04i
Even parity								
N(1740) P13		−1.48+0.20i						2.10+0.05i
N(1680) F15	−0.37+0.05i	−0.93+0.13i			−0.33+0.04i	−0.89+0.12i		
N(2250) H19	−0.39+0.11i	−1.17+0.33i	−2.06+0.57i	−2.75+0.76i	−0.34+0.09i	−1.06+0.28i	−1.93+0.52i	−2.69+0.73i
Δ(1234) P33	−0.38+0.09i							2.32+0.01i
Δ(1920) F35	−0.57+0.15i	−1.41+0.36i			−0.50+0.12i	−1.33+0.33i		
Δ(1960) P33		−1.09+0.27i						1.95+0.10i
Δ(1950) F37	−0.48+0.10i	−1.32+0.29i	−2.43+0.44i		−0.57+0.13i	−1.45+0.33i		2.88+0.02i
Δ(2416) H3, 11	−0.42+0.15i	−1.32+0.45i	−2.45+0.83i	−3.47+1.19i	−0.51+0.19i	−1.52+0.55i	−2.68+0.95i	−3.62+1.27i
				−4.64+1.41i				3.04+0.05i

Table 2.4.3.2. Location of the most important double poles [from Z. Physik C, Particles and Fields **4** (1980) 333].

	N(939)	Δ(1234)	ϱ(776)	f(1271)
	t-values in GeV^2		u-values in GeV^2	
Odd parity				
N(1525) D13	−1.35+0.17i	−1.93+0.29i	−1.07+0.29i	−2.08+0.40i
N(1540) S11	−1.21+0.37i	−1.79+0.49i	−0.92+0.49i	−1.93+0.60i
N(1640) S11	−1.76+0.23i	−2.34+0.35i	−1.47+0.34i	−2.49+0.46i
N(1680) D15	−1.84+0.25i	−2.42+0.37i	−1.55+0.36i	−2.57+0.48i
N(2150) G17	−3.51+0.65i	−4.09+0.77i	−3.22+0.77i	−4.23+0.88i
Δ(1620) S31	−1.63+0.19i	−2.21+0.31i	−1.33+0.31i	−2.35+0.42i
Δ(1730) D33	−1.92+0.49i	−2.50+0.61i	−1.62+0.61i	−2.64+0.72i
Δ(1930) D35	−2.70+0.43i	−3.28+0.55i	−2.42+0.55i	−3.43+0.66i
Even parity				
N (939) P11	$2\mu^2=0.04$	−0.54+0.12i	0.33+0.12i	−0.69+0.23i
N(1450) P11	−0.95+0.24i	−1.52+0.36i	−0.65+0.36i	−1.67+0.47i
N(1680) F15	−1.85+0.19i	−2.43+0.31i	−1.56+0.30i	−2.57+0.42i
N(1740) P13	−1.97+0.27i	−2.55+0.39i	−1.68+0.38i	−2.69+0.50i
N(2250) H19	−3.79+0.87i	−4.37+0.99i	−3.50+0.99i	−4.52+1.1i
Δ(1234) P33	−0.54+0.12i	−1.12+0.24i	−0.25+0.24i	−1.26+0.35i
Δ(1920) P31	−2.57+0.37i	−3.15+0.49i	−2.27+0.49i	−3.30+0.60i
Δ(1920) F35	−2.54+0.50i	−3.12+0.62i	−2.25+0.61i	−3.26+0.72i
Δ(1950) F37	−2.65+0.47i	−3.22+0.59i	−2.35+0.58i	−3.37+0.70i
Δ(2416) H3, 11	−4.82+1.45i	−5.40+1.57i	−4.53+1.56i	−4.63+1.56i

Below threshold, the only information on zeros comes from the analysis of zeros of the fixed-t amplitudes.

In special cases, the number of zeros of invariant amplitudes at fixed-t is limited by rigorous results based on the phase representation (Sect. A.6.4.3), and, in the isospin even case, on positivity in the range $0<t<4\mu^2$ (63 L Sugawara, 64 L Jin, 75 L Burkhardt).

The amplitude $D^+(\nu,t)$ at fixed t in the range $0<t<4\mu^2$ has at most 4 zeros in the physical sheet of the ν-plane, i.e. two zeros in the ν^2-plane (Fig. 4.3.4). At $t=0$, there are only two zeros in the physical sheet of the ν-plane, because the other two have entered the second sheet at a small positive value of t through the branch point $\nu=\mu+t/4m$.

Except near threshold, it is difficult to determine the continuation of zero trajectories into the second sheet of the s-plane.

2.4.3.4 Zero trajectories of transversity amplitudes

The zeros of transversity amplitudes have been discussed in Sect. 2.1.11, where it is explained why they are of interest for phase shift analysis. These trajectories can be determined from experimental data or from partial wave solutions.

A detailed investigation of zero trajectories for π^+p scattering and for charge-exchange scattering was performed by D. Chew (76 L, 78 L, 79 L-1, 80 L). Barrelet et al. (77) determined zero trajectories for π^-p elastic scattering from experimental data. See also the Figures in 79 L Höhler-1.

Zero trajectories of transversity amplitudes obtained from partial wave solutions were studied by I. Sabba-Stefanescu (78 L, 81 L). He is mainly interested in tests of the compatibility with two-variable analyticity, in particular near the above mentioned singularities, and in analytic continuations into the unphysical region.

2.4.3.5 Zero trajectories of other amplitudes

VON GEHLEN and PFEIL (80 L) studied the direction of zero trajectories at double poles, which belong to the intersection of the u-channel nucleon pole surface with the pole surfaces of different s-channel resonances. According to Eq. (4.3.6), the result is related to the ratio of coupling constants.

The model was chosen as simple as possible, since the authors were mainly interested in the general features and not in quantitative results. They work with s-channel helicity amplitudes from which the half-angle factors are removed

$$f_+ \equiv \frac{qF_{++}}{\sqrt{\pi}\cos(\theta/2)} = f_1 + f_2; \quad f_- \equiv \frac{qF_{+-}}{\sqrt{\pi}\sin(\theta/2)} = f_1 - f_2. \tag{4.3.12}$$

Only π^+p-resonances with angular momentum $J=\ell+1/2$ and parity $P=(-1)^{\ell+1}$ are considered. The ansatz for the π^+p-amplitudes takes into account only the u-channel nucleon pole term and a simple Breit-Wigner formula for the resonance contribution

$$f_\pm = \frac{r^u_\pm}{u-m^2} - \frac{\Gamma M \operatorname{Im} f^{\text{res}}_{\ell+}}{s-M^2+\mathrm{i}\Gamma M}\{P'_{\ell+1}(z) \mp P'_\ell(z)\}. \tag{4.3.13}$$

$\operatorname{Im} f^{\text{res}}_{\ell+}$ is the value of the partial wave at the resonance energy $s=M^2$, where the real part vanishes.

The ansatz is simplified by omitting $\mathrm{i}\Gamma M$ in the denominator. Furthermore, $z=\cos\theta$ is replaced by its unphysical value z_0 at the double pole. Finally, Γ is replaced by the elastic width [Eq. (4.1.15)]

$$\Gamma \cdot \operatorname{Im} f^{\text{res}}_{\ell+} = \Gamma_{\text{el}}/q. \tag{4.3.14}$$

This leads to the following relation between the elastic width and the direction $\xi_\pm = \mathrm{d}t/\mathrm{d}s$ of the trajectories at the double pole

$$\Gamma_{\text{el}} M(1+\xi_\pm) = -\frac{q r^u_\pm}{P'_{\ell+1}(z_0) \mp P'_\ell(z_0)}. \tag{4.3.15}$$

An application to the NΔ(1232)-double pole gives $\xi_+ = -0.23$, $\xi_- = -0.28$. Within the crude approximation, this agrees with ODORICOs model.

The assumption of a dominant F 37-resonance is not well justified, since there are two other 4-star resonances with almost the same mass.

It is encouraging that the simple model leads to results for pion-nucleon scattering, which are compatible with those of more accurate calculations. Therefore, the authors applied similar methods to pion-photoproduction, Compton scattering and to weak axial pion production, where more accurate calculations are much more difficult or not yet possible. — The problems discussed in connection with the $I=1/2$ amplitudes are probably related to difficulties of ODORICOs model for the π^-p-trajectories (Fig. 4.3.3, see the discussion in 80 L BORIE).

Zero trajectories for amplitudes at real t in the s-plane were also calculated in 79 L LANGBEIN. However, he considered invariant amplitudes from which the nucleon pole terms are substracted and therefore, directly compare his results with those of other authors.

Finally, we mention a study of zero trajectories for πN helicity amplitudes with definite u-channel isospin (75 L BUTTIMORE and SPEARMAN). The results are of interest for the theoretical interpretation of the interaction mechanism in πN near-backward scattering.

Conclusion on zero trajectories

Further work on zero trajectories will be useful for the improvement of partial wave analyses.

Although all attempts to develop a unitarized dual model have failed, it remains an interesting problem to represent the scattering amplitude by its poles and zeros. It will be a critical test for new models of the pion-nucleon system to check, whether they can reproduce the simple pattern of zero trajectories.

2.4.4 The saturation of isospin bounds

2.4.4.1 Introduction

Originally, it was believed that isospin invariance (Sects. A.2.2 and A.6.1) is an exact law for the strong interaction part of the pion-nucleon scattering amplitude. An accurate test is difficult, because the experimental data include electromagnetic effects, and a complete theory for the separation of the strong and electromagnetic parts of the amplitude does not exist. Even if one applies the dispersion method of the NORDITA group (Sect. 2.1.17) to very accurate data, the strong interaction part of the amplitude (the "hadronic" part) can be derived only up to a certain contribution, which cannot be calculated in a reliable way.

Within the framework of Quantum Chromodynamics("QCD"), SU(2) invariance is not an exact symmetry of the strong pion-nucleon interaction. It is broken by the u−d quark mass difference. A recent determination of the ratio of the d- and u-quark masses from a fit to the baryon and meson mass spectrum (81 L GASSER) gives: $m_d/m_u = 1.8 \pm 0.2$.

The relatively large mass difference of the u- and d-quarks does not lead to a strong violation of the SU(2) symmetry, because both quark masses are small in comparison with a characteristic mass parameter $\Lambda \approx 500$ MeV. Λ^{-1} is a critical length, corresponding roughly to the size of hadrons in treatments of the quark confinement (77 R FRITZSCH).

The u−d quark mass difference contributes to the mass difference within the isospin multiplets in addition to the electromagnetic self energy. This picture is very successful and solves for instance the long-standing problem of the neutron-proton mass difference [GASSER and LEUTWYLER (75 L)].

WEINBERG (77 L) pointed out that an isospin breaking term due to the u−d quark mass difference cannot occur in the effective Lagrangian for low energy pion-pion scattering. Therefore, it is of great interest to search in pion-nucleon scattering for isospin conservation violating effects. He predicted a 30% difference between the $\pi^0 n \to \pi^0 n$ and $\pi^0 p \to \pi^0 p$ S-wave scattering lengths, but it is difficult to see, how this violation of charge-symmetry can be measured. Numerical estimates for the other scattering lengths have not yet been calculated.

Since isospin invariance is not an exact law, it is of interest to check whether the experimental data at low or high energies indicate a violation. The main problem is that one needs a high precision of the absolute normalization of the data. Furthermore, near the forward direction and, at high energies, also near the backward direction one has to separate the electromagnetic effects. – The evidence of charge-symmetry breaking is discussed in 82 R VAN OERS.

2.4.4.2 Isospin bounds for backward scattering

We start with backward scattering, because the isospin bound Eq. (A.10.12) turns out to be a strong constraint

$$(\sqrt{\sigma_+} - \sqrt{\sigma_-})^2 \leqq 2\sigma_0 \leqq (\sqrt{\sigma_+} + \sqrt{\sigma_-})^2; \quad \sigma \equiv d\sigma(180°)/d\Omega. \tag{4.4.1a}$$

If this bound is saturated, the isospin triangle, which can be constructed from the backward cross sections, is degenerated and the backward amplitudes are parallel or antiparallel to each other (Fig. A.10.2).

TØRNQVIST (68 L) studied this bound and found appreciable violations of isospin invariance. Furthermore, the bound seemed to be approximately saturated in a large energy range. This led to the speculation of an approximate "phase degeneracy" $\phi_1 \approx \phi_3$ between the phases of the isospin 1/2 and 3/2 backward amplitudes (KORKEA-AHO and TØRNQVIST 71 L). The phase difference can be calculated from the backward cross sections, Eq. (A.10.16).

Since a "phase degeneracy" looks very strange from a dynamical point of view, we investigated the bound and the phase difference, using CERN and SACLAY phase shifts (HÖHLER and W. SCHMIDT 72 L-2). It turned out that not even in a crude approximation the phase difference is near to zero in the momentum range 0···2 GeV/c, although the isospin bound is indeed almost saturated in a large part of this range.

For this discussion, it is useful to consider the isospin triangle belonging to the backward amplitudes

$$F^{3/2} = F^{1/2} + 3F_0/\sqrt{2}. \tag{4.4.1b}$$

It is clear that this isospin triangle is almost degenerated, if the ratio $|F^{1/2}/F^{3/2}|$ is either small or large in comparison with 1, *independently of the angle* between the amplitudes $F^{1/2}$ and $F^{3/2}$.

It follows that the near-degeneracy occurs either at $I=1/2$ or 3/2 resonances or at energies where the amplitude has a nearby complex zero. The dynamical property, which causes the approximate saturation in a large momentum range, is the location of the poles and zeros of the backward amplitudes.

The present situation of the bound for the backward cross sections is shown in Fig. 4.4.1. One should note that the ordinate has a *logarithmic* scale. In a linear scale, the large deviations from saturation near the zeros of the lower bound would barely be visible. Above 2.5 GeV/c, the approximate saturation is not fulfilled any more.

The gap between the upper and lower bounds is narrow at momenta, where the π^+p- or π^-p-backward cross section has a deep dip (at 0.8 and 2.2 GeV/c, respectively). In principle, these regions could be used for a test of isospin invariance but, with the present data, a discrepancy would probably be interpreted as a normalization or extrapolation error.

The extrapolation to 180° (and to 0° in charge-exchange scattering) is more difficult than assumed by many authors. If one has data in a large range $-0.95<\cos\theta<0.95$, one could think that a Legendre fit, Eq. (A.4.21), can be used for an accurate extrapolation. However, in some momentum ranges, the variation of the cross sections in the range $1>|\cos\theta|>0.95$ is large and shows a curvature, so that the extrapolation has an appreciable uncertainty. This is related to the contribution of high partial waves, which cannot be determined from data in the range $|\cos\theta|<0.95$. (79 L HÖHLER-1).

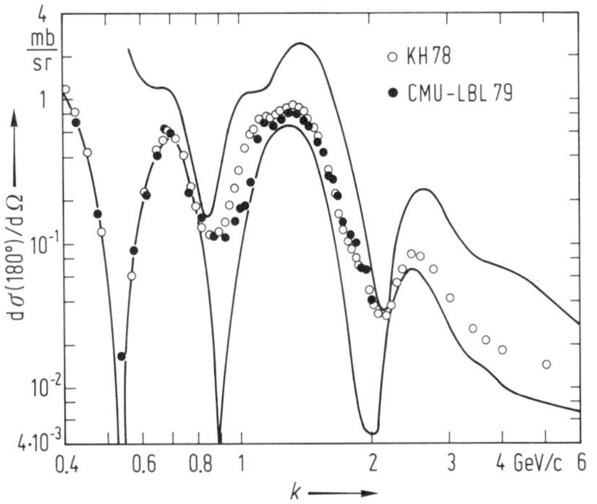

Fig. 4.4.1. The isospin bound, Eq. (4.4.1a), for the charge-exchange backward cross sections. The solid lines show the bounds as calculated from the KH 78 partial wave solution. It is seen that the charge-exchange cross section calculated from partial waves almost saturates the lower bound in parts of the momentum range. There are two momenta, which are of special interest for tests of isospin invariance (or of the normalizations), because the upper and lower bounds come close together.

2.4.4.3 Isospin bounds for cross sections and polarizations

Isospin bounds can be derived for different sets of experimental data (DONCEL, MICHEL, and MINNAERT 72 L). At low and intermediate energies, the most interesting case is that differential cross section and polarization data are given for all three reactions. One can use the bound Eq. (A.10.9) or an equivalent version (Sect. A.10.2).

In order to determine the most favorable kinematic regions for tests of isospin invariance (or of the consistency of the normalizations), one can calculate "phase contours" in a (s,t) or $(k,\cos\theta)$-plane, along which one of the isospin bounds Eq. (A.10.9) is saturated. The result is shown in Fig. 4.4.2 as following from the KH 78 partial wave solution.

Along these "phase contours", the difference of the phases of the $I=1/2$ and $3/2$ transversity amplitudes is 0 or π, and one of the isospin triangles belonging to Eq. (A.10.1) is degenerated. A more detailed version of the figures shows lines of constant area of the isospin triangles of the transversity amplitudes (Figs. 8.1 and 8.2 in 79 R HÖHLER-KAISER-KOCH-PIETARINEN).

The plots depend in a rather sensitive way on the experimental input of the phase shift analysis. Therefore, it will be of interest to calculate the phase contours also from the CMU-LBL 79 solution and from the amalgamated data. The result will help to find the origin of discrepancies between the KH 78 and CMU-LBL 79 partial waves.

The phase contours either go to $k=0$ or to $k=\infty$ or they form closed loops. If they reach $\cos\theta=\pm1$ in the plot belonging to one of the transversity amplitudes $F(\pm)$, they have a continuation in the other one. They pass through all points in the $(k,\cos\theta)$-plane, where one of the transversity amplitudes has a zero.

The condition for the equality (up to π) of the phases of the $I=1/2$ and $3/2$ transversity amplitudes defines a 3-dimensional surface in the 4-dimensional complex s,t-space. The above phase contours are the intersection of this surface with the physical s,t-plane.

The 3-dimensional surface contains the resonance pole surfaces, because the dominance of a resonant amplitude leads to a degeneracy of the isospin triangle. Furthermore, the zero trajectories of the transversity amplitudes (Sect. 2.4.3) also lie within this surface. We conclude that zeros and poles near the physical s,t-plane have an important influence on the pattern of the phase contours in Fig. 4.4.2.

ERVANNE et al. (75 L) attempted to use the bound of DONCEL et al. (72 L) in the case where spin-rotation data are not available. The bound for the first expression in Eq. (A.10.5) is equivalent to one of the two inequalities

$$[\sqrt{-\Lambda(-)} \pm \sqrt{-\Lambda(+)}]^2 = \text{real and} \geqq 0. \tag{4.4.2}$$

The sign ambiguity cannot be resolved without further information. Lines along which this "most stringent bound" is saturated in the sense that it is valid with the equality sign were calculated and discussed by ERVANNE, KORKEA-AHO, and TØRNQVIST (75 L). (The reality condition is an important difference in comparison with our contour lines in Fig. 4.4.2.)

In discussions with I. SABBA-STEFANESCU we came to the conclusion that the bound Eq. (4.4.2) is of little interest for the following reason. In general, this bound can be "saturated", i.e. it can be valid with the equality sign, although none of the bounds Eq. (A.10.9) is saturated. Isospin invariance is *not* violated, if one starts with data which "saturate" the bound and considers arbitrary small changes. An exception is the case where $\Lambda(+) = \Lambda(-) = 0$. — On the other hand, a small variation can lead to a violation of isospin invariance, if one starts with a case which is far from "saturation".

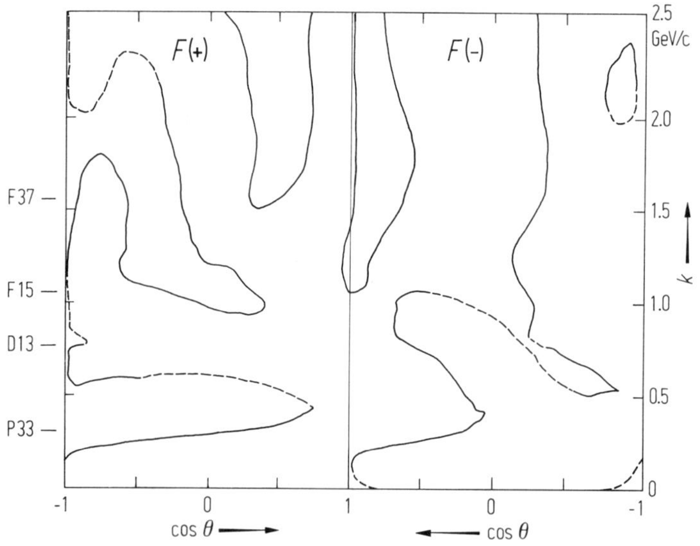

Fig. 4.4.2. Saturation of the isospin bound for differential cross sections and polarizations, Eq. (A.10.9). The plots show lines along which one of the isospin triangles Fig. A.10.1 for the transversity amplitudes $F(+)$ or $F(-)$ is degenerated. The upper bound is saturated along the solid lines and the lower bound along the dashed lines. — The figure can be thought to be drawn on the surface of a cylinder, the k-direction being parallel to the axis and the unit circle in the $w = \exp(i\theta)$-plane being the basis. If one of the curves reaches the boundary $\theta = 0°$ or 180°, it has a smooth continuation in the other part of the figure.

2.4.4.4 Isospin bounds at high energies

Figure 4.4.2 suggests that some of the phase contour lines run towards higher energies. The question is whether they still exist in the 10 or 100 GeV region.

i) One of the lines starts with a loop at 1 GeV/c and remains near $\cos\theta = 1$ in the plot for $F(-)$ (Fig. 4.4.2). The following argument suggests that this line belongs to a saturation of one of the isospin bounds, which is valid in the asymptotic region.

Figure 4.4.3 shows an Argand diagram for the t-dependence of the isospin even and odd transversity amplitudes at 10 GeV/c. These amplitudes are well known at $t = 0$ (Sect. 2.2.1). Their t-dependence is shown as following from the KH 78 partial wave solution, which agrees with amplitude analyses of the type discussed in Sect. 2.1.16.

The phase of the isospin even transversity amplitude varies only slowly, whereas the phase of the isospin odd amplitude varies rapidly. The complex vector turns to the left and, at $t = -0.04\,\text{GeV}^2$, it becomes parallel to the isospin even amplitude. This corresponds to a degeneracy of the isospin triangle and the saturation of one of the isospin bounds is a consequence.

The important point is now that the t-dependence remains similar up to the highest energies where data are available (200 GeV/c). The variation of the phase of the isospin even amplitude becomes even smaller, because the Re/Im ratio decreases as the energy increases. — The rapid rotation of the isospin odd transversity amplitude is a consequence of the large contribution of the flip amplitude in charge-exchange scattering.

One does not need details of the Regge model (Sect. 2.6.5.4), but only the assumption that the amplitude B^- gives the dominant contribution to $d\sigma_0/dt$ at and somewhat beyond the peak at $t = -0.035\,\text{GeV}^2$. Then, the decrease of $|B^-|$ at increasing energy is known and $\arg B^-$ can be obtained by an extension of the methods used in Sect. A.6.4.4 to $t<0$. The amplitude H^- follows from Eq. (A.2.32), using the dominance of B^- over C^- in this t-range.

The saturation of the isospin bound at high energies and $t \approx -0.04\,\mathrm{GeV}^2$ (Fig. 4.4.4) was studied in 80 L Höhler-1, using the bound Eq. (A.10.14). This bound is of interest for the extrapolation of $(P_+ - P_-)/\sqrt{-t}$ to $t=0$, which is needed for the determination of the forward amplitude B^-. One should note that it is necessary to take into account the electromagnetic corrections to the published polarization data (Sect. 2.1.17.4).

ii) The two phase contour lines adjacent to the line described in i) can also be followed up to higher energies. They remain near $t=-0.5\,\mathrm{GeV}^2$, where the charge-exchange cross section has its dip, which usually is ascribed to a zero of the flip amplitude. The saturation of the isospin bounds at $t \approx -0.5\,\mathrm{GeV}^2$ is useful for an adjustment of the normalizations in amplitude analysis.

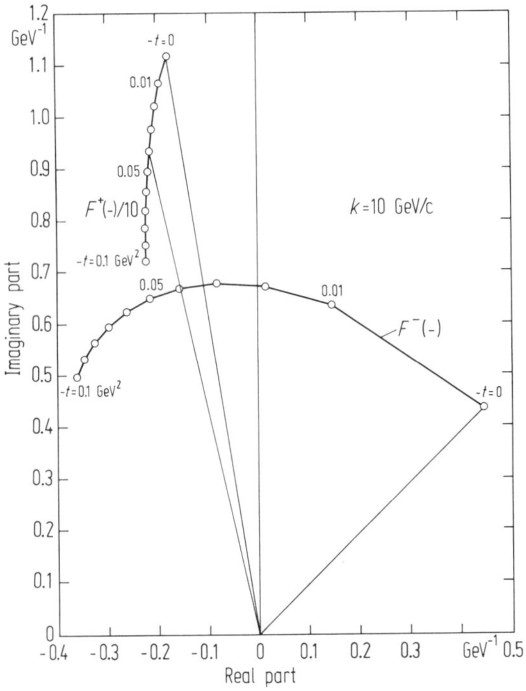

Fig. 4.4.3. Saturation of an isospin bound at high energies and $t=-0.04\,\mathrm{GeV}^2$. The Argand diagram for the transversity amplitude $F^+(-)$ shows a slow variation of the phase at small $|t|$. However, there is a very rapid variation of the phase of the charge-exchange amplitude $F^-(-)=G^- - \mathrm{i}H^-$. At $t=0$ the amplitude H^- is zero for kinematical reasons [Eq. (A.2.32)] but, as $-t$ increases, H^- becomes dominant already at $t<-0.03\,\mathrm{GeV}^2$. The direction of $\mathrm{i}H^-$ is almost orthogonal to that of G^- due to the approximate validity of the Regge law. The bound is saturated if the direction of $F^+(-)$ coincides with that of $F^-(-)$. The plot shows the amplitudes of the KH 78 solution.

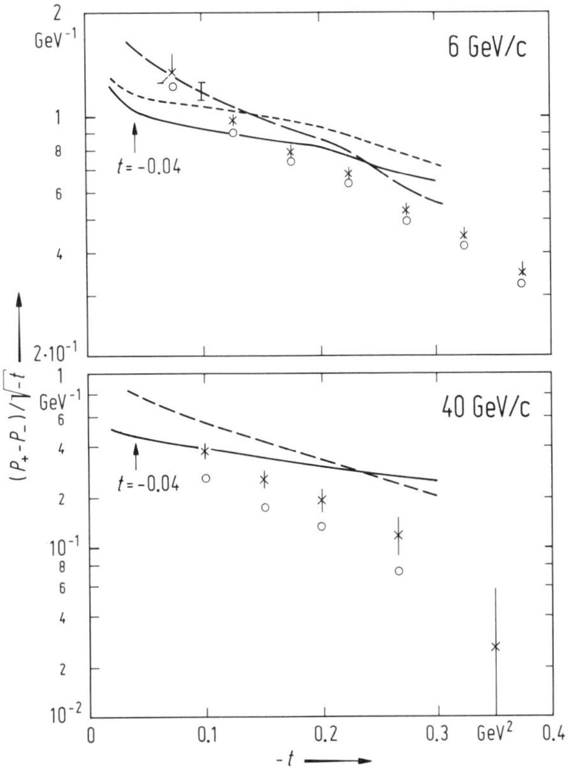

Fig. 4.4.4. An experimental test for the saturation of an isospin bound at $t=-0.04\,\mathrm{GeV}^2$. The isospin bound for $(P_+ - P_-)/\sqrt{-t}$ follows from Eq. (6.5.25) [or from Eq. (A.10.14) and $P^+ \approx 0$]. The solid lines show our evaluation at 6 and 40 GeV/c, using a fit of the type Eq. (6.5.16) for $\Delta(t)$. The dashed line is the contribution of the term which does not depend on Δ. Its uncertainty is marked at 6 GeV/c and $t=-0.1\,\mathrm{GeV}^2$. The symbols ⚹ show the quantity $(P_+ - P_-)/\sqrt{-t}$ as calculated from the published data (70 Borghi, 75 Gaidot, 76 Gaidot). If the electromagnetic corrections are taken into account (78 L Borie), one obtains the points denoted by ○ which are lower. — If the isospin bound is valid, the experimental points must lie *below* the solid lines. As explained in Sect. 2.6.5.3 and Fig. 4.4.3 one expects a saturation of the bound at $t \approx -0.04\,\mathrm{GeV}^2$. — The corrected data at 40 GeV/c are compatible with the saturation. The 6 GeV/c data have a t-dependence which leads to a violation of the bound. Presumably, this is due either to the fit assumed for $\Delta(t)$ or to a systematic error of the $P_\pm$-data. The dotted line shows the bound as calculated from the KH 78 partial wave analysis which is based on charge-independence. Similar figures for data at 10 and 14 GeV/c are shown in 80 L Höhler-1.

2.4.4.5 Further isospin bounds

A variety of other isospin bounds was discussed in ROYs review (72 R). We mention only the bound Eq. (A.10.15) for the integrated elastic and charge-exchange cross sections. ROY noticed a saturation which is much better than expected from the dominance of $I=3/2$ scattering alone.

The "saturation" was explained in the above mentioned paper (72 L HÖHLER-2). One can cancel large terms which occur on both sides of the inequality. This leads to an equivalent version

$$\tfrac{1}{9}(\sigma_-^{el} - 2\sigma_0^{el} + \tfrac{1}{3}\sigma_+^{el})^2 \leqq \tfrac{16}{81}\sigma_{el}^{1/2}\sigma_{el}^{3/2}; \qquad \sigma_{el}^{1/2} = \tfrac{3}{2}(\sigma_-^{el} + \sigma_0^{el}) - \tfrac{1}{2}\sigma_+^{el}, \tag{4.4.3}$$

which shows the deviation from saturation more clearly. Now, there is no evidence for a saturation except near threshold, where all amplitudes become almost real.

Another isospin bound will be mentioned in connection with Eq. (A.5.45).

2.4.5 Invariant amplitudes in the unphysical region and near threshold

2.4.5.1 Analytic continuation at fixed *t*

Figure A.6.3 shows that the invariant amplitudes in the unphysical region between the s- and u-channel backward directions can be determined by analytic continuation at fixed t. One can employ either dispersion relations or an expansion in terms of suitable analytic functions.

i) Dispersion relations are practical only for the determination of the (real) amplitudes $C^{\pm}$ at $t=0$ between the s- and u-channel thresholds. In this case, the experimental information on the imaginary part is much better than that on the real part due to the optical theorem.

ii) In order to apply an *expansion method*, one starts with a partial wave solution and reconstructs the invariant amplitudes in the physical range, and their imaginary parts along the unphysical section of the fixed-t line above threshold (Fig. A.6.5). Then, the coefficients of the expansion are fitted to the invariant amplitudes and the result gives directly the extrapolation. One has to ensure that the expansion has a reasonable behaviour at high energies, above the range where partial waves are available. Furthermore, t must lie in the range where the partial wave series converges (Sect. 6.2.4). At larger negative values of t, an extrapolation into part of the unphysical region is possible, if discrepancy methods are employed. Expansion methods were developed and used for this purpose by PIETARINEN (Sects. 2.1.7 and A.6.3.4), by LANGBEIN (79 L and earlier papers) and LICHARD (76 L-1, 77 L). See also the papers on analytic continuation quoted in Sect. A.12 (A.6.3).

iii) PIETARINENs results for the expansions are more reliable than those of other authors, because he determined the coefficients in a *simultaneous fit* to the partial wave solution and to the data, using an iteration procedure (Sect. 2.1.7).

iv) Since the double spectral functions A_{us} and B_{us} are non-zero only for $t < -85\mu^2 = -1.65\,\mathrm{GeV}^2$, it is possible in principle to calculate $\mathrm{Im}\,A$ and $\mathrm{Im}\,B$ from other expansions (at fixed s), using the values in the physical region for the determination of the coefficients (68 L CUTKOSKY, 69 L CIULLI). However, it turns out that the uncertainty grows rapidly with the distance from the physical region. A truncated Legendre expansion still gives a first approximation up to about $t=-1\,\mathrm{GeV}^2$. The extrapolations will be discussed in the forthcoming second part of our paper 80 L BORIE in connection with the continuation of zero trajectories into the unphysical region.

v) Invariant amplitudes near threshold. It is not possible to perform an accurate determination of πN amplitudes near threshold from data, because the electromagnetic effects are increasing as the energy decreases and it is difficult to work with slow pions. Therefore, the invariant amplitudes and the partial wave amplitudes near threshold have to be determined by analytic continuations from the results obtained at higher energies. Aside from the fixed-t dispersion relation one can use the partial wave dispersion relation (Sect. A.7.1). At threshold, the fixed-s dispersion relation (Sect. A.6.6) leads to an interesting information (Sect. 2.4.7).

vi) Invariant amplitudes at complex s-values. As long as one remains in the first sheet of the s-plane, the invariant amplitudes can be calculated from fixed-t dispersion relations or from PIETARINENs expansion (Sect. A.6.3.4). Another possibility is to start from the continuation of the partial waves into the complex s-plane and to calculate the continuation of the invariant amplitude from the partial wave expansion. This method has the advantage that one can study the continuation into the second sheet, passing the real s-axis between threshold and the first inelastic threshold (Sects. A.6.4.6 and A.7.3). A detailed discussion of the analytic continuation can be found in OEHMEs review (61 R). This method was used by BARUT (65 L) and by SAWADA (67 L) for an investigation of the "quasipole" in the second sheet. — The continuation to resonance poles in the second sheet has been treated in Sect. 2.4.1.2.

2.4.5.2 Analytic continuation along other paths

For tests of predictions from Current Algebra, PCAC and quark models (Sect. 2.5.1) it is of great interest to continue the invariant amplitudes to the region $0<t<4\mu^2$ and $\nu\approx 0$ (Fig. A.3.4), in particular to the CHENG-DASHEN point $\nu=0$, $t=2\mu^2$. This can be done by means of a suitable one-dimensional dispersion relation. In all cases the path intersects the line $t=4\mu^2$, i.e. there is a contribution from a t-channel cut.

Some authors ignored the existing information on the discontinuity along the nearby part of the t-channel cut or took into account only the value of the S-wave scattering length (for instance 74 L HITE, 75 L CHAO, 76 L MOIR, 79 L LANGBEIN).

It is clear that the extrapolation becomes more reliable if this information is included, because one exploits an additional experimental input: the $\pi\pi$-phase shifts (Sect. 2.4.6). The evaluation of the nearby part of the t-channel cut was included in the work of BORIE (77 L), HEDEGAARD-JENSEN (77 L) and KOCH (78 L and 82 L-1). At $t>4\mu^2$, the invariant amplitudes can be calculated from t-channel partial waves (Sect. 2.4.6) in the region where this expansion converges (Sect. A.6.2.4).

The paths chosen by the above mentioned authors were hyperbolas or other curves in the Mandelstam plane, along which the invariant amplitudes can be calculated from πN partial waves in the range of the s-channel cut. Another possibility is to perform the extrapolation in two steps, starting with the determination of the amplitude in the region between the s- and u-channel backward directions (Sect. 2.4.5.1) and evaluating a fixed-ν dispersion relation (Sect. A.6.10) for the continuation to $t>0$ (Fig. 4.6.2, 77 L HÖHLER-2).

2.4.5.3 Errors of invariant amplitudes in the unphysical region

The situation for an analytic continuation at fixed-t is more favorable than in all other cases, because the invariant amplitudes are known in a *large energy interval* in the s- and u-channels and one performs an *interpolation* over a much shorter distance. In the physical sheet, the only singularities are the nucleon poles, whose residue is known. The "quasipole" of BARUT (65 L) and SAWADA (67 L) in the second sheet does not produce a large effect and furthermore, its properties follow from those of the nucleon poles (Sect. A.6.4.6).

One could argue that for the crossing-even amplitudes C^+, C^-/ν, etc. one has to consider an *extrapolation* to ν^2-values below threshold. This is true, but again the distance to $\nu^2=0$ is short and there are no unknown singularities in the neighborhood.

The difference between the amplitude at a point in the low energy region and a point at the same t-value in the unphysical region is given by a rapidly converging dispersion integral. Therefore, the uncertainty of the extrapolated amplitude is not much larger than that of the amplitude in the low energy region, as long as t is not too large.

The error of the amplitudes in the unphysical region follows from several sources:

i) Errors of the amplitudes in the physical region

There exist mathematical methods for the treatment of statistical errors of the data, but in general the systematic errors are more serious, in particular the discrepancies between different data sets, which are sometimes much larger than the error estimates of the authors.

ii) Deviations from two-variable analyticity

As discussed in Sect. 2.1, none of the existing partial wave analyses leads to invariant amplitudes, which have exactly the correct analyticity properties of the Mandelstam representation. Therefore, the application of one-dimensional dispersion relations along different paths will lead to somewhat different results for the extrapolation into the unphysical region. In a recent paper, KOCH (82 L-1) has investigated this question for the extrapolation of the amplitude D^+ to the CHENG-DASHEN point. The fluctuation of the results obtained using a variety of paths is a good estimate of the uncertainty, as far as extrapolations based on the KH 80 partial wave solution are concerned (see Sect. 2.5.1.3).

iii) Errors of the t-channel input

The sensitivity of the extrapolation to the CHENG-DASHEN point due to changes of the input for the S-wave $\pi\pi$ scattering length has been studied by KOCH (82 L-1).

iv) Errors due to charge-dependent effects

As discussed in Sects. 2.1.17 and 2.4.4, the hadronic amplitudes in the physical region still have small uncertainties, because a complete theory of the charge-dependent effects does not yet exist. This contributes to the error of the extrapolation.

Invariant amplitudes at $t>0$

The extrapolation of the amplitudes $D^{\pm}$ to the CHENG-DASHEN point $\nu=0$, $t=2\mu^2$ is treated in Sects. 2.4.7 and 2.5.1. In general, the extrapolation to positive t-values is more difficult than that to comparable negative t-values, because one approaches the t-channel cut.

One can reconstruct the invariant amplitudes from s-channel and t-channel partial waves in the regions where these expansions converge. (Sect. A.6.2.4). Other extrapolations employ one-dimensional dispersion relations along hyperbolas or other curves.

It would be desirable to study the discrepancies between the different methods and to summarize the results by preparing a table, since the experimental information in the physical region of the t-channel ($\bar{N}N\to\pi\pi$ scattering) is rapidly increasing. As soon as reliable amplitudes in this region are available, (see Sect. 2.5.4), it will be of interest to investigate interpolations between the s- and t-channel amplitudes. Another application is a study of zero trajectories at positive t which will be helpful for amplitude analysis, since these trajectories must fulfill certain constraints due to the presence of double poles (Sect. 2.4.3).

2.4.6 $\pi\pi N\bar{N}$ partial waves in the unphysical region

2.4.6.1 $\pi\pi N\bar{N}$ partial waves at $t\leqq 0$

As discussed in Sect. 2.4.5.1, it is possible to determine the invariant πN amplitudes in the unphysical region between the s- and u-channel backward directions in the range $0\geqq t\gtrsim -0.5\,\mathrm{GeV}^2$. From this result, one can calculate the projection integrals for $\pi\pi N\bar{N}$ partial waves, using the formulas given in Sect. A.3.4.1. The integration limits are shown in Fig. A.3.4. It is equivalent to apply the Froissart-Gribov formulas of Sect. A.7.4.1.

The projection of the nucleon pole terms $[f^J_{\pm N}(t)]$, whose singularities lie within the range $-1<Z<+1$, is calculated separately (Sect. A.7.4.2). We shall frequently use the difference

$$\tilde{f}^J_{\pm}(t)\equiv f^J_{\pm}(t)-f^J_{\pm N}(t), \qquad (4.6.1)$$

because these amplitudes are slowly variable even in the range $0<t<4\mu^2$, where $f^J_{\pm N}(t)$ has a very rapid variation (Fig. 4.6.1).

Starting from the CERN 67/68 phase shift solution, we calculated $\pi\pi N\bar{N}$-partial waves up to $J=3$ (68 L HÖHLER-2, 70 L STRAUSS). The method was improved and the input updated in subsequent papers, (75 L HÖHLER-4, 77 L PIETARINEN), but the changes are not large. The most recent result is based on the KH 78 partial wave solution (Table 2.4.6.1). It agrees with the table in 79 R HÖHLER-KAISER-KOCH-PIETARINEN, except for the P-waves, where a preliminary result was printed instead of the final one, but we have used the final result in our calculations since 1978. Plots of the partial waves are shown in Fig. 4.6.1.

Independent determinations of $\pi\pi N\bar{N}$-partial waves have been performed by NIELSEN and OADES (72 L-2, 74 L-1). Subsequent work of these authors in collaboration with GUSTAFSON was not published, but results for the extrapolation to $t>0$ will be mentioned in Sect. 2.4.6.3.

In Fig. 4.6.2 we show plots of invariant amplitudes from which the nucleon pole term is subtracted, in comparison with the lowest $\pi\pi N\bar{N}$-partial waves. Because of the threshold behaviour (Sect. A.3.5.3), the lowest partial wave is dominant at $t=4\mu^2$. The second term of the expansion remains small up to about $t=-8\mu^2$, where the Δ-exchange contribution becomes important. The magnitudes of the third terms of the expansions of $\tilde{C}^+$ and $\tilde{C}^-$ are accurately known at $t=0$, where they can be calculated from total cross sections [cf. Eqs. (4.7.28) and (4.7.36)].

For Table 2.4.6.1, see p. 247.

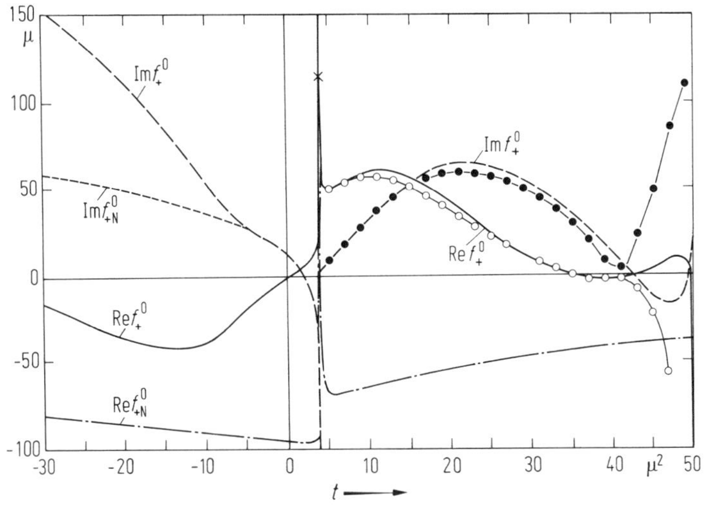

Fig. 4.6.1a.

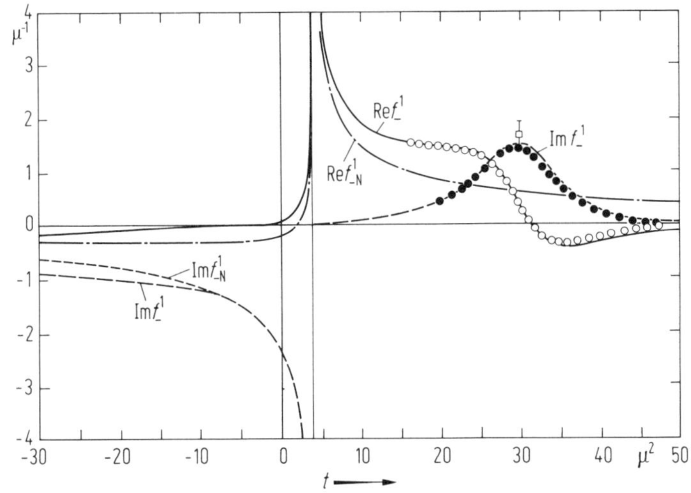

Fig. 4.6.1b.

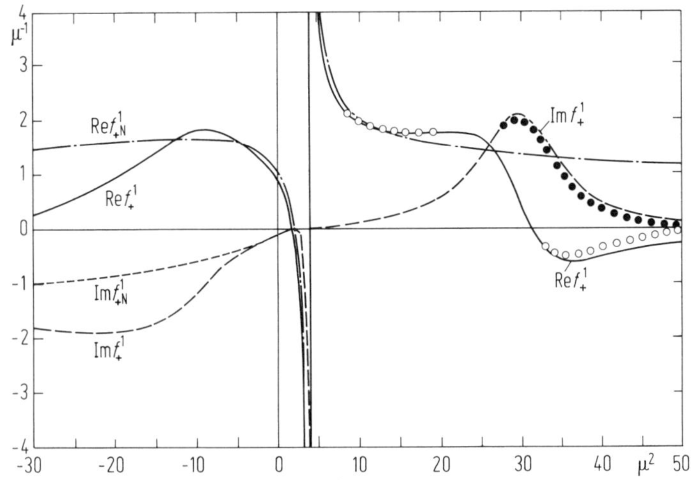

Fig. 4.6.1c.

Fig. 4.6.1a⋯c. Real and imaginary parts of the lowest $\pi\pi N\bar{N}$ partial waves as calculated from the KH 78 partial wave solution. ● and ○: results of Nielsen and Oades (unpublished, 1978); ⏀: an earlier result based on a different choice of the D_1-function (75 L Höhler-4). ×: value of $\mathrm{Re}\,f_+^0$ at $t=4\mu^2$.

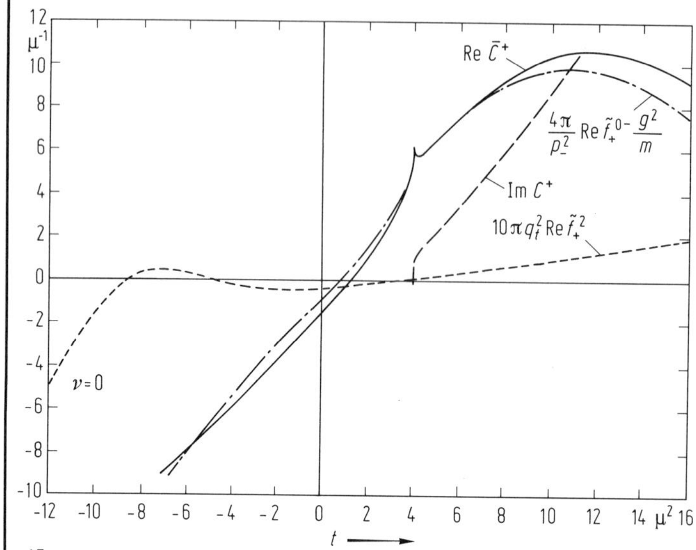

Fig. 4.6.2a.

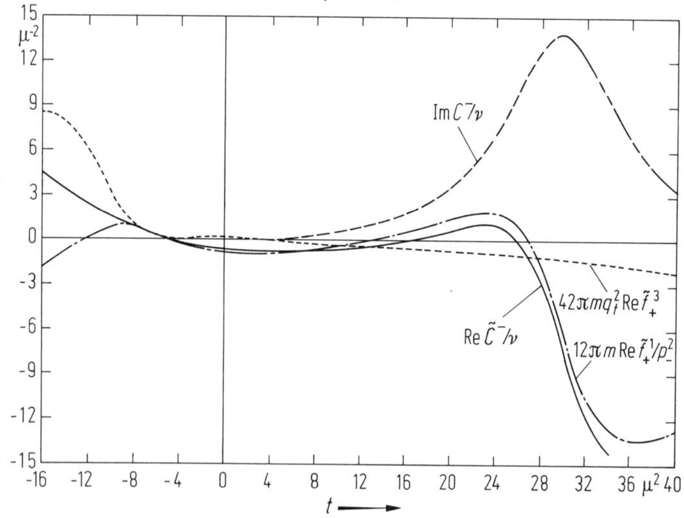

Fig. 4.6.2b.

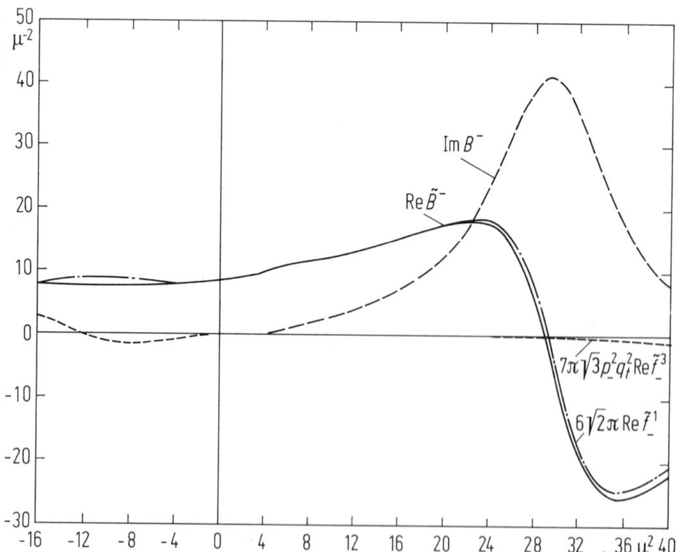

Fig. 4.6.2c.

Fig. 4.6.2a…c. Real and imaginary parts of the invariant amplitudes $\bar{C}^+$, $\tilde{C}^-/\nu$ and $\tilde{B}^-$ at $\nu=0$ as calculated from the KH 78 partial wave solution. In addition the plots show the contributions of the leading t-channel partial waves. Details of the neighborhood of $t=4\mu^2$ are seen more clearly in Fig. A.3.5.

Table 2.4.6.1. $\pi\pi N\bar{N}$ partial wave amplitudes. A twittle denotes the amplitude minus the nucleon pole term contribution. J and D are the numerator and denominator functions, respectively, and δ is the $\pi\pi$-phase shift in degrees.

t n.u.	t GeV^2	$\mathrm{Re}\, f^0_+$	$\mathrm{Im}\, f^0_+$	$\lvert f^0_+\rvert$	$\mathrm{Re}\, \tilde{f}^0_+$	$\tilde{J}^0_+$	f^0_{+N}	δ^0_0	$\lvert D_0\rvert$
		amplitudes in natural units, δ in degrees							
40.0	.779	−1.99	12.48	12.64	38.65	12.27	−40.64	99.0	.971
38.0	.740	−1.86	21.75	21.83	39.88	20.33	−41.74	94.9	.931
36.0	.701	−.54	30.34	30.34	42.35	27.11	−42.89	91.0	.893
35.0	.682	.52	34.36	34.37	44.01	30.07	−43.48	89.1	.875
34.0	.662	1.84	38.20	38.25	45.92	32.78	−44.09	87.2	.857
32.0	.623	5.19	45.29	45.59	50.51	37.47	−45.33	83.5	.822
30.0	.584	9.47	51.52	52.39	56.09	41.32	−46.62	79.6	.789
28.0	.545	14.65	56.77	58.63	62.62	44.40	−47.97	75.5	.757
26.0	.506	20.65	60.83	64.24	70.04	46.78	−49.39	71.2	.728
25.0	.487	23.92	62.34	66.77	74.04	47.73	−50.12	69.0	.715
24.0	.468	27.34	63.46	69.10	78.22	48.52	−50.88	66.7	.702
22.0	.429	34.47	64.41	73.05	86.92	49.64	−52.45	61.8	.679
20.0	.390	41.68	63.42	75.89	95.79	50.15	−54.11	56.7	.661
18.0	.351	48.47	60.34	77.40	104.34	50.06	−55.87	51.2	.647
16.0	.312	54.23	55.18	77.36	111.97	49.35	−57.75	45.5	.638
15.0	.292	56.51	51.87	76.71	115.25	48.75	−58.73	42.5	.636
14.0	.273	58.30	48.14	75.61	118.06	47.98	−59.76	39.5	.635
12.0	.234	60.15	39.69	72.06	122.05	45.92	−61.91	33.4	.637
10.0	.195	59.38	30.43	66.73	123.58	43.11	−64.19	27.1	.646
8.0	.156	55.99	21.06	59.83	122.50	39.54	−66.51	20.6	.661
6.0	.117	50.87	12.19	52.31	118.95	35.66	−68.07	13.5	.682
5.0	.097	49.72	8.00	50.36	116.43	34.96	−66.72	9.1	.694
4.1	.080	72.10	3.50	72.18	114.99	51.00	−42.89	2.8	.707
4.0	.078	116.37	.00	116.37	116.37	82.40	.00	.0	.708

t n.u.	t GeV^2	$\mathrm{Re}\, f^0_+$	$\mathrm{Im}\, f^0_+$	$\lvert f^0_+\rvert$	$\mathrm{Re}\, \tilde{f}^0_+$	$\mathrm{Im}\, \tilde{f}^0_+$	$\mathrm{Re}\, f^0_{+N}$	$\mathrm{Im}\, f^0_{+N}$	D_0
		amplitudes in natural units							
3.9	.076	37.56	−68.14	77.81	112.45		−74.89	−68.13	.744
3.5	.068	14.89	−24.02	28.26	108.44		−93.55	−24.03	.795
3.0	.058	9.84	−11.30	14.99	105.31		−95.47	−11.31	.837
2.5	.049	6.88	−4.62	8.29	102.79		−95.92	−4.61	.871
2.0	.039	4.59	.01	4.59	100.59		−96.01	.00	.901
1.5	.029	2.58	3.56	4.40	98.54		−95.95	3.56	.928
1.0	.019	.82	6.50	6.55	96.65		−95.83	6.49	.954
.5	.010	−.86	8.99	9.03	94.81		−95.67	9.01	.977
.0	.000	−2.45	11.22	11.49	93.02		−95.48	11.22	1.000

t n.u.	t GeV^2	$Re\, f^0_+$	$Im\, f^0_+$	$\|f^0_+\|$	$Re\, \bar{f}^0_+$	$Im\, \bar{f}^0_+$	$Re\, f^0_{+N}$	$Im\, f^0_{+N}$	D_0
		amplitudes in natural units							
.0	.000	-2.45	11.22	11.49	93.02	.00	-95.48	11.22	1.000
-2.0	-.039	-8.69	18.19	20.16	85.91	-.03	-94.60	18.22	1.082
-4.0	-.078	-15.71	24.10	28.77	77.94	.57	-93.65	23.53	1.154
-6.0	-.117	-23.95	30.87	39.07	68.73	2.95	-92.68	27.92	1.219
-8.0	-.156	-32.56	40.39	51.88	59.14	8.71	-91.71	31.69	1.280
-10.0	-.195	-38.83	52.98	65.69	51.90	17.96	-90.74	35.02	1.336
-12.0	-.234	-41.61	66.35	78.32	48.16	28.33	-89.77	38.02	1.390
-14.0	-.273	-41.91	78.74	89.20	46.91	37.99	-88.82	40.75	1.440
-16.0	-.312	-40.79	90.18	98.97	47.09	46.91	-87.88	43.27	1.489
-18.0	-.351	-38.86	100.55	107.80	48.08	54.94	-86.94	45.61	1.535
-20.0	-.390	-36.49	110.58	116.45	49.53	62.79	-86.02	47.80	1.579
-22.0	-.429	-33.15	120.13	124.62	51.96	70.28	-85.11	49.85	1.622
-24.0	-.468	-29.33	128.76	132.06	54.88	76.98	-84.21	51.78	1.663
-26.0	-.506	-25.38	136.90	139.23	57.94	83.28	-83.32	53.61	1.703
-28.0	-.545	-21.13	144.66	146.19	61.32	89.30	-82.44	55.35	1.742
-30.0	-.584	-16.66	151.93	152.84	64.92	94.93	-81.57	57.01	1.779
-32.0	-.623	-12.07	158.87	159.33	68.64	100.29	-80.72	58.59	1.815
-34.0	-.662	-7.28	165.50	165.66	72.58	105.40	-79.87	60.10	1.850
-36.0	-.701	-2.34	171.67	171.68	76.69	110.12	-79.03	61.54	1.885
-38.0	-.740	2.51	177.40	177.41	80.72	114.47	-78.20	62.93	1.918
-40.0	-.779	7.20	182.89	183.03	84.59	118.62	-77.39	64.27	1.951
-42.0	-.818	11.81	188.24	188.61	88.39	122.69	-76.58	65.55	1.982
-44.0	-.857	16.36	193.41	194.10	92.14	126.62	-75.78	66.79	2.013
-46.0	-.896	20.74	198.40	199.48	95.73	130.42	-74.99	67.98	2.044
-48.0	-.935	24.89	203.34	204.86	99.10	134.21	-74.21	69.13	2.073
-50.0	-.974	28.96	208.36	210.36	102.39	138.11	-73.44	70.25	2.102
-52.0	-1.013	33.11	213.32	215.87	105.78	141.99	-72.67	71.33	2.130
-54.0	-1.052	37.34	217.89	221.07	109.25	145.52	-71.92	72.37	2.158
-56.0	-1.091	41.29	221.77	225.58	112.46	148.38	-71.17	73.39	2.185
-58.0	-1.130	44.39	224.89	229.23	114.83	150.52	-70.43	74.37	2.212
-60.0	-1.169	46.10	227.55	232.17	115.80	152.22	-69.70	75.32	2.238
-62.0	-1.208	46.07	230.20	234.77	115.05	153.95	-68.98	76.25	2.264
-64.0	-1.247	44.23	233.28	237.43	112.49	156.13	-68.27	77.15	2.289
-66.0	-1.286	40.52	236.94	240.38	108.08	158.91	-67.56	78.03	2.314
-68.0	-1.325	34.71	241.07	243.56	101.57	162.19	-66.86	78.88	2.338
-70.0	-1.364	26.06	245.41	246.79	92.23	165.70	-66.17	79.71	2.362

t n.u.	t GeV^2	Re f^1_-	Im f^1_-	$\lvert f^1_-\rvert$	Re $\tilde{f}^1_-$	$\tilde{f}^1_-$	f^1_{-N}	δ^1_1	$\lvert D_1\rvert$
		amplitudes in natural units, δ in degrees							
40.0	.779	-.324	.308	.447	-.809	.256	.485	136.5	.574
38.0	.740	-.385	.451	.593	-.889	.282	.504	130.5	.475
36.0	.701	-.426	.668	.792	-.951	.310	.525	122.5	.391
35.0	.682	-.423	.812	.916	-.959	.325	.536	117.5	.355
34.0	.662	-.390	.981	1.055	-.937	.341	.548	111.7	.323
32.0	.623	-.169	1.358	1.368	-.743	.374	.573	97.1	.274
30.0	.584	.310	1.626	1.655	-.292	.411	.602	79.2	.249
28.0	.545	.888	1.576	1.809	.254	.453	.634	60.6	.250
26.0	.506	1.293	1.274	1.815	.622	.499	.671	44.6	.275
25.0	.487	1.407	1.100	1.786	.716	.524	.691	38.0	.293
24.0	.468	1.477	.938	1.750	.765	.551	.713	32.4	.315
22.0	.429	1.538	.671	1.678	.776	.611	.762	23.6	.364
20.0	.390	1.554	.480	1.627	.734	.680	.820	17.2	.418
18.0	.351	1.568	.346	1.606	.677	.762	.890	12.5	.475
16.0	.312	1.598	.250	1.617	.620	.862	.978	8.9	.533
15.0	.292	1.623	.212	1.637	.593	.921	1.030	7.5	.563
14.0	.273	1.657	.180	1.667	.567	.987	1.090	6.2	.592
12.0	.234	1.761	.126	1.766	.520	1.151	1.242	4.1	.652
10.0	.195	1.940	.085	1.942	.478	1.382	1.462	2.5	.712
8.0	.156	2.267	.051	2.267	.441	1.750	1.826	1.3	.772
6.0	.117	3.017	.024	3.017	.406	2.512	2.611	.5	.832
5.0	.097	4.018	.012	4.018	.390	3.467	3.629	.2	.863
4.1	.080	8.872	.003	8.872	.374	7.897	8.498	.0	.890
4.0	.078	13.836	.000	13.836	.372	12.359	13.464	.0	.893

t n.u.	t GeV^2	Re f^1_-	Im f^1_-	$\lvert f^1_-\rvert$	Re $\tilde{f}^1_-$	Im $\tilde{f}^1_-$	Re f^1_{-N}	Im f^1_{-N}	D_1
		amplitudes in natural units							
3.9	.076	8.429	-12.007	14.671	.368		8.061	-12.008	.896
3.5	.068	1.715	-6.569	6.789	.359		1.356	-6.568	.908
3.0	.058	.805	-4.732	4.800	.352		.454	-4.732	.922
2.5	.049	.496	-3.876	3.908	.345		.151	-3.876	.936
2.0	.039	.339	-3.356	3.373	.339		.000	-3.356	.949
1.5	.029	.244	-2.995	3.005	.334		-.090	-2.995	.962
1.0	.019	.180	-2.727	2.733	.330		-.150	-2.727	.975
.5	.010	.134	-2.516	2.520	.326		-.192	-2.517	.988
.0	.000	.101	-2.347	2.349	.324		-.223	-2.347	1.000

t n.u.	t GeV^2	Re f^1_- (amplitudes in natural units)	Im f^1_-	$\lvert f^1_-\rvert$	Re $\tilde{f}^1_-$	Im $\tilde{f}^1_-$	Re f^1_{-N}	Im f^1_{-N}	D_1
.0	.000	.101	-2.347	2.349	.324	-.000	-.223	-2.347	1.000
-2.0	-.039	.024	-1.889	1.889	.317	-.001	-.293	-1.888	1.048
-4.0	-.078	-.005	-1.616	1.616	.320	-.006	-.325	-1.610	1.093
-6.0	-.117	-.011	-1.436	1.436	.331	-.018	-.342	-1.418	1.136
-8.0	-.156	-.009	-1.314	1.314	.343	-.040	-.351	-1.274	1.177
-10.0	-.195	-.008	-1.236	1.236	.348	-.074	-.356	-1.162	1.217
-12.0	-.234	-.013	-1.183	1.183	.346	-.113	-.359	-1.070	1.255
-14.0	-.273	-.030	-1.142	1.142	.330	-.149	-.360	-.993	1.292
-16.0	-.312	-.046	-1.109	1.110	.313	-.180	-.359	-.928	1.328
-18.0	-.351	-.068	-1.082	1.084	.290	-.210	-.358	-.872	1.363
-20.0	-.390	-.092	-1.056	1.060	.264	-.234	-.356	-.822	1.396
-22.0	-.429	-.120	-1.032	1.039	.235	-.254	-.354	-.778	1.429
-24.0	-.468	-.148	-1.003	1.014	.204	-.264	-.352	-.739	1.461
-26.0	-.506	-.168	-.972	.986	.181	-.268	-.349	-.704	1.492
-28.0	-.545	-.183	-.946	.964	.163	-.275	-.346	-.672	1.522
-30.0	-.584	-.200	-.924	.946	.143	-.282	-.343	-.642	1.552
-32.0	-.623	-.216	-.903	.928	.124	-.287	-.340	-.616	1.581
-34.0	-.662	-.230	-.883	.912	.107	-.292	-.337	-.591	1.609
-36.0	-.701	-.245	-.864	.898	.089	-.297	-.334	-.568	1.636
-38.0	-.740	-.259	-.846	.885	.071	-.299	-.331	-.547	1.663
-40.0	-.779	-.273	-.827	.871	.055	-.300	-.327	-.527	1.689
-42.0	-.818	-.284	-.808	.857	.041	-.300	-.324	-.508	1.715
-44.0	-.857	-.293	-.792	.844	.028	-.301	-.321	-.491	1.741
-46.0	-.896	-.303	-.776	.833	.015	-.301	-.318	-.475	1.765
-48.0	-.935	-.312	-.760	.822	.003	-.301	-.315	-.459	1.790
-50.0	-.974	-.319	-.744	.810	-.007	-.299	-.312	-.445	1.814
-52.0	-1.013	-.323	-.730	.798	-.015	-.298	-.308	-.431	1.837
-54.0	-1.052	-.327	-.718	.789	-.021	-.299	-.305	-.419	1.860
-56.0	-1.091	-.330	-.708	.781	-.028	-.302	-.302	-.406	1.883
-58.0	-1.130	-.334	-.700	.775	-.034	-.305	-.299	-.395	1.905
-60.0	-1.169	-.337	-.692	.769	-.041	-.308	-.296	-.384	1.927
-62.0	-1.208	-.339	-.684	.763	-.045	-.311	-.294	-.373	1.949
-64.0	-1.247	-.339	-.678	.758	-.048	-.315	-.291	-.363	1.970
-66.0	-1.286	-.337	-.676	.755	-.049	-.322	-.288	-.354	1.991
-68.0	-1.325	-.336	-.678	.756	-.051	-.333	-.285	-.344	2.012
-70.0	-1.364	-.336	-.683	.761	-.054	-.347	-.282	-.336	2.032

t n.u.	t GeV^2	Re f_+^1	Im f_+^1	$\lvert f_+^1\rvert$	Re $\tilde{f}_+^1$	$\tilde{J}_+^1$	f_{+N}^1	δ_1^1	$\lvert D_1\rvert$
		amplitudes in natural units, δ in degrees							
40.0	.779	-.537	.510	.740	-1.756	.424	1.219	136.5	.574
38.0	.740	-.604	.708	.931	-1.843	.442	1.238	130.5	.475
36.0	.701	-.638	1.001	1.188	-1.898	.465	1.259	122.5	.391
35.0	.682	-.621	1.193	1.345	-1.891	.477	1.270	117.5	.355
34.0	.662	-.561	1.414	1.521	-1.843	.491	1.282	111.7	.323
32.0	.623	-.236	1.892	1.907	-1.541	.522	1.306	97.1	.274
30.0	.584	.419	2.199	2.239	-.913	.556	1.332	79.2	.249
28.0	.545	1.169	2.074	2.381	-.192	.596	1.361	60.6	.250
26.0	.506	1.659	1.635	2.329	.267	.640	1.392	44.6	.275
25.0	.487	1.784	1.394	2.265	.375	.664	1.410	38.0	.293
24.0	.468	1.850	1.174	2.192	.423	.690	1.428	32.4	.315
22.0	.429	1.880	.821	2.051	.412	.747	1.468	23.6	.364
20.0	.390	1.854	.573	1.941	.340	.811	1.514	17.2	.418
18.0	.351	1.823	.403	1.867	.254	.887	1.569	12.5	.475
16.0	.312	1.808	.283	1.830	.172	.975	1.635	8.9	.533
15.0	.292	1.810	.237	1.825	.135	1.027	1.675	7.5	.563
14.0	.273	1.820	.197	1.831	.100	1.084	1.720	6.2	.592
12.0	.234	1.873	.134	1.878	.039	1.224	1.834	4.1	.652
10.0	.195	1.993	.087	1.995	-.012	1.420	2.005	2.5	.712
8.0	.156	2.257	.051	2.258	-.053	1.743	2.310	1.3	.772
6.0	.117	3.032	.024	3.032	-.084	2.524	3.116	.5	.832
5.0	.097	4.458	.014	4.458	-.097	3.846	4.555	.2	.863
4.1	.080	20.286	.006	20.286	-.104	18.058	20.391	.0	.890
4.0	.078	63.896	.000	63.896	-.102	57.073	63.998	.0	.893

t n.u.	t GeV^2	Re f_+^1	Im f_+^1	$\lvert f_+^1\rvert$	Re $\tilde{f}_+^1$	Im $\tilde{f}_+^1$	Re f_{+N}^1	Im f_{+N}^1	D_1
		amplitudes in natural units							
3.9	.076	-32.285	-29.275	43.582	-.107		-32.178	-29.275	.896
3.5	.068	-6.447	-1.627	6.649	-.114		-6.333	-1.626	.908
3.0	.058	-2.266	-.255	2.280	-.118		-2.148	-.254	.922
2.5	.049	-.838	-.034	.838	-.120		-.717	-.034	.936
2.0	.039	-.120	-.000	.120	-.120		.000	-.000	.949
1.5	.029	.310	-.016	.310	-.119		.428	-.016	.962
1.0	.019	.595	-.048	.597	-.116		.711	-.048	.975
.5	.010	.799	-.085	.803	-.111		.910	-.086	.988
.0	.000	.952	-.124	.960	-.104		1.056	-.124	1.000

t n.u.	t GeV^2	Ref^1_+ (amplitudes in natural units)	Imf^1_+	$\lvert f^1_+ \rvert$	Re$\tilde{f}^1_+$	Im$\tilde{f}^1_+$	Ref^1_{+N}	Imf^1_{+N}	D_1
.0	.000	.952	-.124	.960	-.104	-.000	1.056	-.124	1.000
-2.0	-.039	1.323	-.273	1.351	-.057	-.007	1.380	-.266	1.048
-4.0	-.078	1.550	-.427	1.607	.029	-.045	1.521	-.382	1.093
-6.0	-.117	1.722	-.619	1.830	.134	-.141	1.588	-.478	1.136
-8.0	-.156	1.848	-.890	2.051	.229	-.331	1.619	-.559	1.177
-10.0	-.195	1.836	-1.223	2.206	.206	-.594	1.631	-.629	1.217
-12.0	-.234	1.694	-1.497	2.261	.064	-.807	1.630	-.690	1.255
-14.0	-.273	1.509	-1.684	2.261	-.112	-.940	1.621	-.744	1.292
-16.0	-.312	1.311	-1.812	2.237	-.296	-1.021	1.608	-.792	1.328
-18.0	-.351	1.118	-1.883	2.190	-.473	-1.049	1.591	-.834	1.363
-20.0	-.390	.946	-1.924	2.144	-.625	-1.051	1.571	-.873	1.396
-22.0	-.429	.781	-1.942	2.093	-.769	-1.035	1.550	-.908	1.429
-24.0	-.468	.631	-1.934	2.035	-.896	-.995	1.527	-.939	1.461
-26.0	-.506	.504	-1.913	1.979	-1.001	-.945	1.504	-.968	1.492
-28.0	-.545	.390	-1.886	1.926	-1.091	-.892	1.481	-.994	1.522
-30.0	-.584	.291	-1.850	1.873	-1.166	-.832	1.457	-1.018	1.552
-32.0	-.623	.209	-1.812	1.824	-1.224	-.772	1.433	-1.040	1.581
-34.0	-.662	.140	-1.779	1.784	-1.269	-.718	1.409	-1.060	1.609
-36.0	-.701	.077	-1.749	1.751	-1.308	-.670	1.385	-1.079	1.636
-38.0	-.740	.020	-1.721	1.721	-1.341	-.625	1.362	-1.096	1.663
-40.0	-.779	-.030	-1.698	1.698	-1.369	-.587	1.338	-1.111	1.689
-42.0	-.818	-.080	-1.682	1.684	-1.395	-.556	1.315	-1.126	1.715
-44.0	-.857	-.135	-1.669	1.675	-1.428	-.530	1.292	-1.139	1.741
-46.0	-.896	-.196	-1.652	1.664	-1.466	-.501	1.270	-1.151	1.765
-48.0	-.935	-.256	-1.627	1.647	-1.504	-.465	1.248	-1.162	1.790
-50.0	-.974	-.311	-1.595	1.625	-1.537	-.423	1.226	-1.173	1.814
-52.0	-1.013	-.359	-1.561	1.602	-1.563	-.379	1.204	-1.182	1.837
-54.0	-1.052	-.402	-1.527	1.579	-1.585	-.337	1.183	-1.191	1.860
-56.0	-1.091	-.443	-1.493	1.558	-1.605	-.295	1.162	-1.198	1.883
-58.0	-1.130	-.482	-1.457	1.534	-1.624	-.251	1.142	-1.206	1.905
-60.0	-1.169	-.518	-1.414	1.506	-1.640	-.202	1.122	-1.212	1.927
-62.0	-1.208	-.547	-1.364	1.470	-1.649	-.146	1.102	-1.218	1.949
-64.0	-1.247	-.563	-1.309	1.425	-1.646	-.086	1.083	-1.224	1.970
-66.0	-1.286	-.564	-1.251	1.373	-1.628	-.023	1.064	-1.229	1.991
-68.0	-1.325	-.547	-1.193	1.313	-1.593	.040	1.045	-1.233	2.012
-70.0	-1.364	-.509	-1.137	1.245	-1.536	.100	1.027	-1.237	2.032

t n.u.	t GeV^2	Re f^2_-	Im f^2_-	$\|f^2_-\|$	Re $\tilde{f}^2_-$	J^2_-	f^2_{-N}	δ^0_2	$\|D_2\|$
		amplitudes in natural units, δ in degrees							
40.0	.779	.0129	.00142	.0130	.00131	.0096	.0116	6.3	.736
38.0	.740	.0136	.00131	.0137	.00105	.0102	.0125	5.5	.748
36.0	.701	.0144	.00120	.0144	.00080	.0110	.0136	4.8	.761
35.0	.682	.0148	.00115	.0149	.00068	.0114	.0142	4.4	.767
34.0	.662	.0153	.00110	.0154	.00056	.0119	.0148	4.1	.774
32.0	.623	.0165	.00101	.0165	.00032	.0130	.0162	3.5	.787
30.0	.584	.0179	.00093	.0179	.00008	.0143	.0178	3.0	.801
28.0	.545	.0196	.00084	.0196	−.00016	.0160	.0197	2.5	.815
26.0	.506	.0216	.00076	.0216	−.00041	.0179	.0220	2.0	.829
25.0	.487	.0228	.00072	.0228	−.00053	.0191	.0234	1.8	.836
24.0	.468	.0242	.00069	.0242	−.00066	.0204	.0248	1.6	.843
22.0	.429	.0274	.00061	.0274	−.00094	.0235	.0283	1.3	.857
20.0	.390	.0315	.00053	.0315	−.00123	.0275	.0328	1.0	.871
18.0	.351	.0370	.00046	.0370	−.00155	.0328	.0386	.7	.885
16.0	.312	.0445	.00038	.0445	−.00190	.0400	.0464	.5	.899
15.0	.292	.0493	.00035	.0493	−.00209	.0447	.0514	.4	.906
14.0	.273	.0552	.00031	.0552	−.00229	.0504	.0575	.3	.913
12.0	.234	.0717	.00023	.0717	−.00274	.0665	.0745	.2	.927
10.0	.195	.0997	.00016	.0997	−.00325	.0938	.1030	.1	.941
8.0	.156	.1563	.00010	.1563	−.00384	.1490	.1601	.0	.954
6.0	.117	.3222	.00004	.3222	−.00455	.3113	.3268	.0	.966
5.0	.097	.6274	.00002	.6274	−.00496	.6099	.6323	.0	.972
4.1	.080	3.5566	.00001	3.5566	−.00535	3.4761	3.5619	.0	.977

t n.u.	t GeV^2	Re f^2_-	Im f^2_-	$\|f^2_-\|$	Re $\tilde{f}^2_-$	Im $\tilde{f}^2_-$	Re f^2_{-N}	Im f^2_{-N}	D_2
		amplitudes in natural units							
3.9	.076	−4.5541	−8.9363	10.0298	−.0054		−4.5487	−8.9364	.979
3.5	.068	−1.9516	−.7701	2.0981	−.0057		−1.9459	−.7701	.981
3.0	.058	−1.0377	−.1844	1.0540	−.0059		−1.0318	−.1844	.984
2.5	.049	−.7019	−.0502	.7037	−.0062		−.6957	−.0502	.986
2.0	.039	−.5283	−.0000	.5283	−.0065		−.5218	−.0000	.989
1.5	.029	−.4224	.0232	.4230	−.0068		−.4156	.0232	.992
1.0	.019	−.3511	.0350	.3529	−.0071		−.3440	.0350	.995
.5	.010	−.3000	.0415	.3029	−.0075		−.2925	.0415	.997
.0	.000	−.2616	.0450	.2654	−.0075		−.2537	.0450	1.000

t n.u.	t GeV^2	$Re f^2_-$	$Im f^2_-$	$\|f^2_-\|$	$Re \tilde{f}^2_-$	$Im \tilde{f}^2_-$	$Re f^2_{-N}$	$Im f^2_{-N}$	D_2
		amplitudes in natural units							
.0	.000	-.2616	.0450	.2654	-.0079	-.0000	-.2537	.0450	1.000
-2.0	-.039	-.1728	.0480	.1793	-.0100	.0003	-.1627	.0477	1.010
-4.0	-.078	-.1299	.0469	.1381	-.0128	.0016	-.1171	.0453	1.020
-6.0	-.117	-.1053	.0464	.1151	-.0156	.0044	-.0897	.0421	1.030
-8.0	-.156	-.0891	.0479	.1012	-.0175	.0089	-.0716	.0390	1.040
-10.0	-.195	-.0760	.0508	.0915	-.0173	.0147	-.0588	.0362	1.049
-12.0	-.234	-.0636	.0535	.0831	-.0144	.0198	-.0492	.0336	1.058
-14.0	-.273	-.0519	.0546	.0753	-.0100	.0232	-.0418	.0314	1.067
-16.0	-.312	-.0414	.0541	.0681	-.0053	.0247	-.0360	.0294	1.076
-18.0	-.351	-.0324	.0525	.0617	-.0011	.0249	-.0313	.0276	1.084
-20.0	-.390	-.0248	.0502	.0560	.0026	.0242	-.0275	.0260	1.092
-22.0	-.429	-.0185	.0474	.0509	.0057	.0229	-.0242	.0245	1.100
-24.0	-.468	-.0134	.0445	.0465	.0081	.0213	-.0215	.0232	1.108
-26.0	-.506	-.0091	.0415	.0425	.0100	.0195	-.0192	.0220	1.116
-28.0	-.545	-.0057	.0386	.0390	.0114	.0177	-.0172	.0209	1.124
-30.0	-.584	-.0030	.0359	.0360	.0125	.0160	-.0154	.0199	1.131
-32.0	-.623	-.0007	.0333	.0333	.0132	.0144	-.0139	.0189	1.139
-34.0	-.662	.0012	.0309	.0309	.0138	.0129	-.0126	.0181	1.146
-36.0	-.701	.0027	.0287	.0288	.0141	.0114	-.0114	.0172	1.153
-38.0	-.740	.0038	.0266	.0269	.0142	.0102	-.0103	.0165	1.160
-40.0	-.779	.0048	.0249	.0253	.0142	.0091	-.0094	.0158	1.167
-42.0	-.818	.0057	.0232	.0239	.0143	.0081	-.0086	.0151	1.174
-44.0	-.857	.0065	.0217	.0226	.0143	.0072	-.0078	.0145	1.180
-46.0	-.896	.0071	.0202	.0214	.0142	.0063	-.0071	.0139	1.187
-48.0	-.935	.0076	.0190	.0204	.0141	.0056	-.0065	.0134	1.193
-50.0	-.974	.0080	.0178	.0196	.0140	.0050	-.0060	.0129	1.200
-52.0	-1.013	.0085	.0168	.0188	.0140	.0044	-.0055	.0124	1.206
-54.0	-1.052	.0090	.0158	.0182	.0140	.0039	-.0050	.0119	1.212
-56.0	-1.091	.0094	.0147	.0175	.0140	.0033	-.0046	.0115	1.218
-58.0	-1.130	.0098	.0137	.0169	.0140	.0026	-.0042	.0111	1.224
-60.0	-1.169	.0101	.0128	.0163	.0140	.0021	-.0038	.0107	1.230
-62.0	-1.208	.0104	.0119	.0158	.0139	.0015	-.0035	.0103	1.236
-64.0	-1.247	.0107	.0110	.0153	.0139	.0011	-.0032	.0100	1.242
-66.0	-1.286	.0110	.0102	.0150	.0139	.0006	-.0029	.0096	1.248
-68.0	-1.325	.0113	.0094	.0146	.0140	.0000	-.0027	.0093	1.253
-70.0	-1.364	.0115	.0085	.0143	.0140	-.0005	-.0025	.0090	1.259

t n.u.	t GeV^2	Re f^2_+	Im f^2_+	$\|f^2_+\|$	Re $\tilde{f}^2_+$	J^2_+	f^2_{+N}	δ^0_2	$\|D_2\|$
		amplitudes in natural units, δ in degrees							
40.0	.779	.0567	.0062	.0570	.0241	.0420	.0326	6.3	.736
38.0	.740	.0579	.0056	.0582	.0236	.0435	.0343	5.5	.748
36.0	.701	.0594	.0050	.0596	.0231	.0453	.0363	4.8	.761
35.0	.682	.0602	.0047	.0604	.0229	.0464	.0373	4.4	.767
34.0	.662	.0611	.0044	.0613	.0227	.0474	.0385	4.1	.774
32.0	.623	.0632	.0039	.0634	.0223	.0499	.0409	3.5	.787
30.0	.584	.0657	.0034	.0658	.0220	.0527	.0437	3.0	.801
28.0	.545	.0686	.0030	.0687	.0217	.0560	.0469	2.5	.815
26.0	.506	.0721	.0025	.0721	.0215	.0598	.0506	2.0	.829
25.0	.487	.0741	.0024	.0741	.0213	.0619	.0527	1.8	.836
24.0	.468	.0762	.0022	.0763	.0212	.0643	.0550	1.6	.843
22.0	.429	.0813	.0018	.0813	.0210	.0697	.0603	1.3	.857
20.0	.390	.0875	.0015	.0875	.0208	.0762	.0667	1.0	.871
18.0	.351	.0953	.0012	.0954	.0205	.0844	.0748	.7	.885
16.0	.312	.1056	.0009	.1056	.0203	.0950	.0853	.5	.899
15.0	.292	.1121	.0008	.1121	.0202	.1016	.0919	.4	.906
14.0	.273	.1197	.0007	.1197	.0200	.1093	.0996	.3	.913
12.0	.234	.1403	.0005	.1403	.0197	.1301	.1205	.2	.927
10.0	.195	.1739	.0003	.1739	.0194	.1635	.1545	.1	.941
8.0	.156	.2401	.0001	.2401	.0189	.2290	.2212	.0	.954
6.0	.117	.4438	.0001	.4438	.0183	.4287	.4254	.0	.966
5.0	.097	.8846	.0000	.8846	.0180	.8599	.8667	.0	.972
4.1	.080	9.5289	.0000	9.5289	.0173	9.3134	9.5117	.0	.977

t n.u.	t GeV^2	Re f^2_+	Im f^2_+	$\|f^2_+\|$	Re $\tilde{f}^2_+$	Im $\tilde{f}^2_+$	Re f^2_{+N}	Im f^2_{+N}	D_2
		amplitudes in natural units							
3.9	.076	-30.2740	11.9464	32.5459	.0177		-30.2917	11.9473	.979
3.5	.068	-.8474	2.0036	2.1755	.0174		-.8648	2.0035	.981
3.0	.058	-.0798	.5005	.5068	.0171		-.0968	.5004	.984
2.5	.049	.0059	.1375	.1377	.0167		-.0107	.1376	.986
2.0	.039	.0165	-.0000	.0165	.0165		-.0000	-.0000	.989
1.5	.029	.0122	-.0634	.0646	.0160		-.0038	-.0635	.992
1.0	.019	.0052	-.0958	.0959	.0157		-.0105	-.0958	.995
.5	.010	-.0022	-.1130	.1130	.0151		-.0173	-.1130	.997
.0	.000	-.0087	-.1220	.1223	.0147		-.0234	-.1220	1.000

t n.u.	t GeV^2	Re f_+^2 amplitudes in natural units	Im f_+^2	$\lvert f_+^2 \rvert$	Re $\tilde{f}_+^2$	Im $\tilde{f}_+^2$	Re f_{+N}^2	Im f_{+N}^2	D_2
.0	.000	-.0087	-.1220	.1223	.0147	.0000	-.0234	-.1220	1.000
-2.0	-.039	-.0302	-.1274	.1309	.0103	-.0003	-.0405	-.1271	1.010
-4.0	-.078	-.0467	-.1137	.1230	.0030	.0043	-.0498	-.1181	1.020
-6.0	-.117	-.0585	-.0914	.1085	-.0034	.0159	-.0551	-.1073	1.030
-8.0	-.156	-.0605	-.0611	.0860	-.0025	.0360	-.0581	-.0971	1.040
-10.0	-.195	-.0453	-.0325	.0557	.0145	.0554	-.0598	-.0879	1.049
-12.0	-.234	-.0195	-.0182	.0267	.0411	.0616	-.0606	-.0798	1.058
-14.0	-.273	.0040	-.0185	.0190	.0648	.0541	-.0608	-.0726	1.067
-16.0	-.312	.0211	-.0261	.0335	.0817	.0402	-.0606	-.0662	1.076
-18.0	-.351	.0316	-.0370	.0486	.0918	.0236	-.0602	-.0605	1.084
-20.0	-.390	.0368	-.0478	.0603	.0964	.0076	-.0596	-.0554	1.092
-22.0	-.429	.0388	-.0579	.0697	.0977	-.0071	-.0589	-.0508	1.100
-24.0	-.468	.0381	-.0670	.0771	.0961	-.0203	-.0580	-.0467	1.108
-26.0	-.506	.0357	-.0744	.0826	.0929	-.0315	-.0571	-.0429	1.116
-28.0	-.545	.0328	-.0808	.0872	.0889	-.0414	-.0561	-.0395	1.124
-30.0	-.584	.0289	-.0867	.0914	.0841	-.0503	-.0551	-.0364	1.131
-32.0	-.623	.0240	-.0915	.0946	.0781	-.0580	-.0541	-.0335	1.139
-34.0	-.662	.0188	-.0949	.0968	.0719	-.0641	-.0531	-.0308	1.146
-36.0	-.701	.0138	-.0975	.0985	.0659	-.0691	-.0521	-.0284	1.153
-38.0	-.740	.0087	-.0998	.1002	.0598	-.0736	-.0511	-.0262	1.160
-40.0	-.779	.0034	-.1015	.1015	.0535	-.0774	-.0501	-.0241	1.167
-42.0	-.818	-.0021	-.1024	.1024	.0470	-.0802	-.0491	-.0222	1.174
-44.0	-.857	-.0073	-.1025	.1028	.0407	-.0822	-.0481	-.0204	1.180
-46.0	-.896	-.0122	-.1022	.1030	.0349	-.0835	-.0471	-.0187	1.187
-48.0	-.935	-.0169	-.1016	.1030	.0292	-.0845	-.0461	-.0172	1.193
-50.0	-.974	-.0214	-.1008	.1030	.0238	-.0851	-.0452	-.0157	1.200
-52.0	-1.013	-.0259	-.0996	.1029	.0184	-.0852	-.0443	-.0144	1.206
-54.0	-1.052	-.0302	-.0978	.1024	.0132	-.0847	-.0433	-.0131	1.212
-56.0	-1.091	-.0341	-.0957	.1016	.0084	-.0838	-.0425	-.0120	1.218
-58.0	-1.130	-.0375	-.0934	.1006	.0040	-.0825	-.0416	-.0109	1.224
-60.0	-1.169	-.0406	-.0908	.0995	.0001	-.0810	-.0407	-.0098	1.230
-62.0	-1.208	-.0434	-.0882	.0983	-.0035	-.0793	-.0399	-.0089	1.236
-64.0	-1.247	-.0458	-.0854	.0970	-.0067	-.0775	-.0391	-.0080	1.242
-66.0	-1.286	-.0479	-.0825	.0954	-.0096	-.0753	-.0383	-.0071	1.248
-68.0	-1.325	-.0496	-.0793	.0935	-.0120	-.0730	-.0375	-.0063	1.253
-70.0	-1.364	-.0506	-.0759	.0912	-.0138	-.0704	-.0368	-.0056	1.259

t n.u.	t GeV^2	$Re f^3_-$ amplitudes in natural units, δ in degrees	$Im f^3_-$	$\lvert f^3_-\rvert$	$Re \tilde{f}^3_-$	$\tilde{f}^3_-$	f^3_{-N}	δ^1_3	$\lvert D_3\rvert$
40.0	.779	.208-03	.109-05	.208-03	-.113-03	.238-03	.321-03	.30	1.145
38.0	.740	.257-03	.119-05	.257-03	-.103-03	.290-03	.359-03	.27	1.128
36.0	.701	.312-03	.127-05	.312-03	-.933-04	.347-03	.405-03	.23	1.114
35.0	.682	.342-03	.131-05	.342-03	-.887-04	.379-03	.431-03	.22	1.107
34.0	.662	.375-03	.134-05	.375-03	-.842-04	.413-03	.459-03	.20	1.101
32.0	.623	.449-03	.138-05	.449-03	-.756-04	.489-03	.525-03	.18	1.089
30.0	.584	.538-03	.141-05	.538-03	-.673-04	.580-03	.605-03	.15	1.079
28.0	.545	.646-03	.142-05	.646-03	-.593-04	.691-03	.705-03	.13	1.069
26.0	.506	.780-03	.142-05	.780-03	-.514-04	.828-03	.832-03	.10	1.061
25.0	.487	.860-03	.141-05	.860-03	-.475-04	.910-03	.908-03	.09	1.057
24.0	.468	.951-03	.139-05	.951-03	-.436-04	.100-02	.995-03	.08	1.053
22.0	.429	.117-02	.136-05	.117-02	-.357-04	.123-02	.121-02	.07	1.046
20.0	.390	.147-02	.130-05	.147-02	-.278-04	.153-02	.150-02	.05	1.040
18.0	.351	.189-02	.122-05	.189-02	-.195-04	.196-02	.191-02	.04	1.035
16.0	.312	.251-02	.112-05	.251-02	-.109-04	.259-02	.252-02	.03	1.029
15.0	.292	.294-02	.106-05	.294-02	-.637-05	.302-02	.294-02	.02	1.027
14.0	.273	.348-02	.994-06	.348-02	-.168-05	.356-02	.348-02	.02	1.025
12.0	.234	.512-02	.839-06	.512-02	.836-05	.522-02	.511-02	.01	1.020
10.0	.195	.832-02	.649-06	.832-02	.195-04	.845-02	.830-02	.00	1.016
8.0	.156	.161-01	.417-06	.161-01	.323-04	.163-01	.161-01	.00	1.012
6.0	.117	.468-01	.143-06	.468-01	.474-04	.472-01	.467-01	.00	1.009
5.0	.097	.126+00	.136-07	.126+00	.560-04	.127+00	.126+00	.00	1.007
4.1	.080	.172+01	.000	.172+01	.626-04	.173+01	.172+01	.00	1.006

t n.u.	t GeV^2	$Re f^3_-$ amplitudes in natural units	$Im f^3_-$	$\lvert f^3_-\rvert$	$Re \tilde{f}^3_-$	$Im \tilde{f}^3_-$	$Re f^3_{-N}$	$Im f^3_{-N}$	D_3
3.9	.076	-.792+01	-.136+00	.792+01	.739-04		-.792+01	-.136+00	1.0056
3.5	.068	-.383+00	.634+00	.740+00	.775-04		-.383+00	.634+00	1.0050
3.0	.058	-.660-01	.253+00	.262+00	.809-04		-.660-01	.253+00	1.0042
2.5	.049	-.146-01	.141+00	.142+00	.853-04		-.147-01	.141+00	1.0034
2.0	.039	.899-04	.920-01	.920-01	.899-04		-.697-09	.920-01	1.0027
1.5	.029	.535-02	.653-01	.655-01	.100-03		.525-02	.653-01	1.0020
1.0	.019	.734-02	.491-01	.496-01	.107-03		.723-02	.491-01	1.0013
.5	.010	.801-02	.384-01	.392-01	.118-03		.790-02	.384-01	1.0006
.0	.000	.811-02	.309-01	.320-01	.129-03		.798-02	.309-01	.9999

t n.u.	t GeV^2	Ref^3_- amplitudes in natural units	Imf^3_-	$\lvert f^3_-\rvert$	Re$\tilde{f}^3_-$	Im$\tilde{f}^3_-$	Ref^3_{-N}	Imf^3_{-N}	D_3
.0	.000	.811-02	.309-01	.000	.129-03	-.391-06	.798-02	.309-01	.9999
-2.0	-.039	.701-02	.156-01	.171-01	.196-03	-.196-04	.682-02	.156-01	.9973
-4.0	-.078	.578-02	.928-02	.109-01	.263-03	-.914-04	.552-02	.937-02	.9949
-6.0	-.117	.481-02	.596-02	.766-02	.302-03	-.204-03	.451-02	.616-02	.9926
-8.0	-.156	.405-02	.394-02	.565-02	.291-03	-.353-03	.376-02	.429-02	.9905
-10.0	-.195	.336-02	.261-02	.425-02	.183-03	-.501-03	.317-02	.311-02	.9885
-12.0	-.234	.273-02	.175-02	.324-02	.827-05	-.573-03	.272-02	.232-02	.9867
-14.0	-.273	.219-02	.121-02	.250-02	-.168-03	-.548-03	.236-02	.176-02	.9850
-16.0	-.312	.176-02	.892-03	.197-02	-.298-03	-.467-03	.206-02	.136-02	.9834
-18.0	-.351	.144-02	.701-03	.160-02	-.379-03	-.359-03	.181-02	.106-02	.9819
-20.0	-.390	.120-02	.581-03	.133-02	-.413-03	-.251-03	.161-02	.832-03	.9805
-22.0	-.429	.102-02	.501-03	.114-02	-.418-03	-.156-03	.144-02	.656-03	.9792
-24.0	-.468	.889-03	.444-03	.994-03	-.401-03	-.747-04	.129-02	.519-03	.9780
-26.0	-.506	.791-03	.395-03	.884-03	-.372-03	-.142-04	.116-02	.409-03	.9769
-28.0	-.545	.709-03	.354-03	.792-03	-.343-03	.322-04	.105-02	.322-03	.9758
-30.0	-.584	.642-03	.325-03	.719-03	-.314-03	.740-04	.956-03	.251-03	.9749
-32.0	-.623	.592-03	.302-03	.665-03	-.279-03	.109-03	.871-03	.193-03	.9740
-34.0	-.662	.554-03	.278-03	.620-03	-.243-03	.132-03	.796-03	.146-03	.9731
-36.0	-.701	.519-03	.256-03	.579-03	-.211-03	.148-03	.730-03	.107-03	.9724
-38.0	-.740	.489-03	.238-03	.544-03	-.182-03	.163-03	.671-03	.754-04	.9717
-40.0	-.779	.465-03	.225-03	.516-03	-.153-03	.176-03	.618-03	.489-04	.9710
-42.0	-.818	.447-03	.213-03	.496-03	-.122-03	.186-03	.570-03	.269-04	.9704
-44.0	-.857	.435-03	.201-03	.479-03	-.923-04	.192-03	.527-03	.865-05	.9698
-46.0	-.896	.424-03	.188-03	.464-03	-.640-04	.195-03	.488-03	-.656-05	.9693
-48.0	-.935	.416-03	.176-03	.452-03	-.371-04	.195-03	.453-03	-.192-04	.9689
-50.0	-.974	.410-03	.165-03	.442-03	-.105-04	.194-03	.421-03	-.297-04	.9684
-52.0	-1.013	.408-03	.152-03	.435-03	.160-04	.191-03	.392-03	-.384-04	.9681
-54.0	-1.052	.406-03	.139-03	.429-03	.413-04	.184-03	.365-03	-.456-04	.9677
-56.0	-1.091	.405-03	.123-03	.424-03	.644-04	.175-03	.341-03	-.515-04	.9674
-58.0	-1.130	.403-03	.108-03	.417-03	.847-04	.164-03	.318-03	-.564-04	.9671
-60.0	-1.169	.400-03	.922-04	.411-03	.103-03	.153-03	.298-03	-.603-04	.9669
-62.0	-1.208	.397-03	.773-04	.405-03	.118-03	.141-03	.279-03	-.635-04	.9666
-64.0	-1.247	.394-03	.625-04	.399-03	.132-03	.128-03	.261-03	-.659-04	.9665
-66.0	-1.286	.389-03	.476-04	.392-03	.144-03	.115-03	.245-03	-.679-04	.9663
-68.0	-1.325	.383-03	.326-04	.384-03	.152-03	.102-03	.230-03	-.693-04	.9662
-70.0	-1.364	.372-03	.182-04	.372-03	.155-03	.885-04	.217-03	-.703-04	.9660

t n.u.	t GeV^2	$Re f^3_+$	$Im f^3_+$	$\lvert f^+_3 \rvert$	$Re \tilde{f}^3_+$	J^3_+	$\tilde{f}^3_{+N}$	δ^1_3	$\lvert D_3 \rvert$
		amplitudes in natural units, δ in degrees							
40.0	.779	.675-03	.353-05	.675-03	-.262-03	.773-03	.937-03	.30	1.145
38.0	.740	.779-03	.362-05	.779-03	-.245-03	.879-03	.102-02	.27	1.128
36.0	.701	.894-03	.365-05	.894-03	-.231-03	.995-03	.112-02	.23	1.114
35.0	.682	.956-03	.366-05	.956-03	-.224-03	.106-02	.118-02	.22	1.107
34.0	.662	.102-02	.365-05	.102-02	-.219-03	.113-02	.124-02	.20	1.101
32.0	.623	.117-02	.360-05	.117-02	-.208-03	.128-02	.138-02	.18	1.089
30.0	.584	.134-02	.352-05	.134-02	-.200-03	.145-02	.154-02	.15	1.079
28.0	.545	.155-02	.341-05	.155-02	-.193-03	.166-02	.174-02	.13	1.069
26.0	.506	.179-02	.326-05	.179-02	-.188-03	.190-02	.198-02	.10	1.061
25.0	.487	.194-02	.317-05	.194-02	-.185-03	.205-02	.212-02	.09	1.057
24.0	.468	.210-02	.308-05	.210-02	-.184-03	.221-02	.228-02	.08	1.053
22.0	.429	.249-02	.287-05	.249-02	-.181-03	.260-02	.267-02	.07	1.046
20.0	.390	.299-02	.263-05	.299-02	-.179-03	.311-02	.317-02	.05	1.040
18.0	.351	.366-02	.236-05	.366-02	-.179-03	.379-02	.384-02	.04	1.035
16.0	.312	.462-02	.206-05	.462-02	-.180-03	.475-02	.480-02	.03	1.029
15.0	.292	.525-02	.190-05	.525-02	-.181-03	.539-02	.543-02	.02	1.027
14.0	.273	.604-02	.173-05	.604-02	-.182-03	.619-02	.622-02	.02	1.025
12.0	.234	.835-02	.137-05	.835-02	-.185-03	.852-02	.854-02	.01	1.020
10.0	.195	.126-01	.986-06	.126-01	-.188-03	.128-01	.128-01	.00	1.016
8.0	.156	.227-01	.587-06	.227-01	-.191-03	.229-01	.228-01	.00	1.012
6.0	.117	.624-01	.191-06	.624-01	-.191-03	.630-01	.626-01	.00	1.009
5.0	.097	.178+00	.190-07	.178+00	-.187-03	.179+00	.178+00	.00	1.007
4.1	.080	.477+01	.243-06	.477+01	-.180-03	.480+01	.477+01	.00	1.006

t n.u.	t GeV^2	$Re f^3_+$	$Im f^3_+$	$\lvert f^+_3 \rvert$	$Re \tilde{f}^3_+$	$Im \tilde{f}^3_+$	$Re f^3_{+N}$	$Im f^3_{+N}$	D_3
		amplitudes in natural units							
3.9	.076	-.229+01	.262+02	.263+02	-.161-03		-.229+01	.262+02	1.0056
3.5	.068	.664+00	.422+00	.787+00	-.242-03		.664+00	.422+00	1.0050
3.0	.058	.125+00	.340-01	.130+00	-.213-03		.125+00	.340-01	1.0042
2.5	.049	.283-01	.309-02	.284-01	-.202-03		.285-01	.309-02	1.0034
2.0	.039	-.204-03	.339-06	.204-03	-.204-03		-.324-17	.320-09	1.0027
1.5	.029	-.104-01	.850-03	.104-01	-.191-03		-.102-01	.850-03	1.0020
1.0	.019	-.141-01	.214-02	.143-01	-.191-03		-.139-01	.214-02	1.0013
.5	.010	-.153-01	.324-02	.156-01	-.179-03		-.151-01	.324-02	1.0006
.0	.000	-.153-01	.408-02	.158-01	-.165-03		-.151-01	.408-02	.9999

t n.u.	t GeV^2	Ref^3_+ amplitudes in natural units	Imf^3_+	$\lvert f^+_3\rvert$	Re$\tilde{f}^3_+$	Im$\tilde{f}^3_+$	Ref^3_{+N}	Imf^3_{+N}	D_3
.0	.000	-.153-01	.408-02	.158-01	-.165-03	.000	-.151-01	.408-02	.9999
-2.0	-.039	-.125-01	.561-02	.137-01	-.104-03	-.609-04	-.124-01	.567-02	.9973
-4.0	-.078	-.967-02	.573-02	.112-01	-.455-04	-.220-03	-.962-02	.595-02	.9949
-6.0	-.117	-.761-02	.531-02	.928-02	-.116-03	-.481-03	-.750-02	.579-02	.9926
-8.0	-.156	-.631-02	.466-02	.785-02	-.399-03	-.833-03	-.592-02	.549-02	.9905
-10.0	-.195	-.580-02	.422-02	.718-02	-.108-02	-.927-03	-.472-02	.515-02	.9885
-12.0	-.234	-.552-02	.430-02	.699-02	-.172-02	-.505-03	-.380-02	.480-02	.9867
-14.0	-.273	-.507-02	.464-02	.687-02	-.198-02	.168-03	-.308-02	.447-02	.9850
-16.0	-.312	-.443-02	.497-02	.666-02	-.192-02	.805-03	-.251-02	.417-02	.9834
-18.0	-.351	-.371-02	.519-02	.638-02	-.166-02	.131-02	-.204-02	.388-02	.9819
-20.0	-.390	-.297-02	.526-02	.604-02	-.131-02	.164-02	-.166-02	.362-02	.9805
-22.0	-.429	-.228-02	.522-02	.570-02	-.934-03	.184-02	-.135-02	.338-02	.9792
-24.0	-.468	-.165-02	.510-02	.536-02	-.562-03	.194-02	-.109-02	.316-02	.9780
-26.0	-.506	-.109-02	.490-02	.502-02	-.223-03	.194-02	-.868-03	.296-02	.9769
-28.0	-.545	-.619-03	.466-02	.470-02	.646-04	.189-02	-.684-03	.277-02	.9758
-30.0	-.584	-.212-03	.440-02	.440-02	.315-03	.180-02	-.527-03	.260-02	.9749
-32.0	-.623	.141-03	.412-02	.412-02	.534-03	.168-02	-.394-03	.244-02	.9740
-34.0	-.662	.432-03	.383-02	.385-02	.712-03	.153-02	-.280-03	.229-02	.9731
-36.0	-.701	.661-03	.353-02	.359-02	.844-03	.137-02	-.182-03	.216-02	.9724
-38.0	-.740	.836-03	.324-02	.334-02	.934-03	.120-02	-.987-04	.203-02	.9717
-40.0	-.779	.964-03	.296-02	.311-02	.991-03	.105-02	-.268-04	.191-02	.9710
-42.0	-.818	.105-02	.270-02	.290-02	.102-02	.898-03	.351-04	.181-02	.9704
-44.0	-.857	.111-02	.247-02	.271-02	.102-02	.768-03	.883-04	.171-02	.9698
-46.0	-.896	.115-02	.227-02	.254-02	.101-02	.655-03	.134-03	.161-02	.9693
-48.0	-.935	.117-02	.208-02	.239-02	.997-03	.557-03	.174-03	.152-02	.9689
-50.0	-.974	.119-02	.191-02	.225-02	.978-03	.468-03	.208-03	.144-02	.9684
-52.0	-1.013	.119-02	.175-02	.211-02	.952-03	.384-03	.237-03	.136-02	.9681
-54.0	-1.052	.118-02	.160-02	.199-02	.919-03	.309-03	.262-03	.129-02	.9677
-56.0	-1.091	.116-02	.147-02	.187-02	.880-03	.244-03	.283-03	.123-02	.9674
-58.0	-1.130	.114-02	.135-02	.177-02	.840-03	.191-03	.301-03	.116-02	.9671
-60.0	-1.169	.112-02	.125-02	.168-02	.803-03	.147-03	.316-03	.110-02	.9669
-62.0	-1.208	.110-02	.115-02	.159-02	.771-03	.106-03	.329-03	.105-02	.9666
-64.0	-1.247	.108-02	.106-02	.152-02	.744-03	.658-04	.340-03	.995-03	.9665
-66.0	-1.286	.107-02	.968-03	.144-02	.718-03	.231-04	.348-03	.945-03	.9663
-68.0	-1.325	.105-02	.878-03	.137-02	.692-03	-.213-04	.355-03	.899-03	.9662
-70.0	-1.364	.103-02	.789-03	.130-02	.667-03	-.655-04	.361-03	.855-03	.9660

t n.u.	t GeV^2	Re Γ_1	Im Γ_1	$\lvert\Gamma_1\rvert$	Re $\tilde{\Gamma}_1$	J_1	Γ_{1N}	δ_1^1	$\lvert D_1\rvert$
		amplitudes in natural units, δ in degrees							
40.0	.779	-.037	.036	.052	-.173	.030	.135	136.5	.574
38.0	.740	-.041	.048	.064	-.180	.030	.138	130.5	.475
36.0	.701	-.044	.068	.081	-.185	.032	.142	122.5	.391
35.0	.682	-.043	.082	.093	-.186	.033	.143	117.5	.355
34.0	.662	-.039	.098	.106	-.184	.034	.145	111.7	.323
32.0	.623	-.017	.136	.137	-.166	.037	.149	97.1	.274
30.0	.584	.031	.163	.166	-.122	.041	.153	79.2	.249
28.0	.545	.091	.161	.185	-.067	.046	.157	60.6	.250
26.0	.506	.135	.133	.189	-.028	.052	.162	44.6	.275
25.0	.487	.148	.116	.188	-.017	.055	.165	38.0	.293
24.0	.468	.157	.100	.186	-.010	.059	.168	32.4	.315
22.0	.429	.168	.073	.183	-.006	.067	.174	23.6	.364
20.0	.390	.173	.054	.182	-.008	.076	.181	17.2	.418
18.0	.351	.179	.039	.183	-.011	.087	.190	12.5	.475
16.0	.312	.185	.029	.188	-.014	.100	.200	8.9	.533
15.0	.292	.190	.025	.191	-.016	.108	.206	7.5	.563
14.0	.273	.195	.021	.196	-.017	.116	.213	6.2	.592
12.0	.234	.210	.015	.210	-.020	.137	.230	4.1	.652
10.0	.195	.234	.010	.234	-.022	.166	.255	2.5	.712
8.0	.156	.277	.006	.277	-.023	.214	.300	1.3	.772
6.0	.117	.393	.003	.393	-.023	.327	.416	.5	.832
5.0	.097	.601	.002	.601	-.023	.519	.624	.2	.863
4.1	.080	2.942	.001	2.942	-.022	2.619	2.964	.0	.890
4.0	.078	9.498	.000	9.498	-.021	8.484	9.520	.0	.893

t n.u.	t GeV^2	Re Γ_1	Im Γ_1	$\lvert\Gamma_1\rvert$	Re $\tilde{\Gamma}_1$	Im $\tilde{\Gamma}_1$	Re Γ_{1N}	Im Γ_{1N}	D_1
		amplitudes in natural units							
3.9	.076	-5.040	-4.264	6.601	-.022		-5.018	-4.264	.896
3.5	.068	-1.002	-.155	1.014	-.022		-.980	-.155	.908
3.0	.058	-.352	.018	.353	-.022		-.330	.018	.922
2.5	.049	-.131	.033	.135	-.022		-.110	.033	.936
2.0	.039	-.021	.027	.034	-.021		-.000	.027	.949
1.5	.029	.045	.015	.048	-.020		.065	.015	.962
1.0	.019	.088	.003	.088	-.019		.107	.004	.975
.5	.010	.119	-.008	.119	-.017		.136	-.008	.988
.0	.000	.142	-.018	.143	-.015		.157	-.018	1.000

t n.u.	t GeV^2	$Re\,\Gamma_1$ (amplitudes in natural units)	$Im\,\Gamma_1$	$\lvert\Gamma_1\rvert$	$Re\,\tilde{\Gamma}_1$	$Im\,\tilde{\Gamma}_1$	$Re\,\Gamma_{1N}$	$Im\,\Gamma_{1N}$	D_1
.0	.000	.142	-.018	.143	-.015	-.000	.157	-.018	1.000
-2.0	-.039	.195	-.055	.202	-.006	-.001	.201	-.054	1.048
-4.0	-.078	.225	-.087	.242	.009	-.007	.216	-.080	1.093
-6.0	-.117	.248	-.122	.276	.027	-.021	.221	-.101	1.136
-8.0	-.156	.263	-.166	.311	.043	-.048	.220	-.118	1.177
-10.0	-.195	.259	-.218	.338	.042	-.086	.217	-.132	1.217
-12.0	-.234	.236	-.261	.352	.024	-.118	.212	-.143	1.255
-14.0	-.273	.207	-.290	.357	.001	-.137	.206	-.153	1.292
-16.0	-.312	.177	-.311	.358	-.022	-.150	.199	-.162	1.328
-18.0	-.351	.147	-.324	.356	-.045	-.155	.192	-.169	1.363
-20.0	-.390	.120	-.332	.353	-.065	-.157	.185	-.175	1.396
-22.0	-.429	.094	-.337	.350	-.084	-.157	.178	-.180	1.429
-24.0	-.468	.071	-.337	.344	-.101	-.153	.171	-.185	1.461
-26.0	-.506	.051	-.335	.339	-.114	-.147	.165	-.188	1.492
-28.0	-.545	.033	-.333	.334	-.125	-.141	.158	-.192	1.522
-30.0	-.584	.017	-.329	.329	-.134	-.134	.151	-.195	1.552
-32.0	-.623	.004	-.325	.325	-.141	-.128	.145	-.197	1.581
-34.0	-.662	-.008	-.322	.322	-.147	-.123	.139	-.199	1.609
-36.0	-.701	-.019	-.318	.319	-.152	-.118	.133	-.201	1.636
-38.0	-.740	-.029	-.315	.317	-.156	-.114	.127	-.202	1.663
-40.0	-.779	-.039	-.313	.315	-.160	-.110	.121	-.203	1.689
-42.0	-.818	-.048	-.311	.314	-.163	-.107	.116	-.204	1.715
-44.0	-.857	-.057	-.309	.314	-.167	-.105	.110	-.204	1.741
-46.0	-.896	-.067	-.307	.314	-.172	-.103	.105	-.205	1.765
-48.0	-.935	-.076	-.304	.313	-.176	-.099	.100	-.205	1.790
-50.0	-.974	-.085	-.300	.312	-.180	-.095	.095	-.205	1.814
-52.0	-1.013	-.093	-.296	.310	-.183	-.091	.090	-.205	1.837
-54.0	-1.052	-.099	-.292	.308	-.185	-.087	.086	-.204	1.860
-56.0	-1.091	-.105	-.288	.307	-.187	-.084	.081	-.204	1.883
-58.0	-1.130	-.112	-.284	.305	-.189	-.081	.077	-.204	1.905
-60.0	-1.169	-.117	-.280	.303	-.190	-.077	.073	-.203	1.927
-62.0	-1.208	-.122	-.275	.300	-.191	-.072	.069	-.202	1.949
-64.0	-1.247	-.125	-.269	.297	-.190	-.068	.065	-.202	1.970
-66.0	-1.286	-.125	-.264	.292	-.187	-.063	.061	-.201	1.991
-68.0	-1.325	-.124	-.260	.288	-.182	-.060	.058	-.200	2.012
-70.0	-1.364	-.121	-.257	.284	-.175	-.058	.054	-.199	2.032

t n.u.	t GeV^2	Re Γ_2	Im Γ_2	$\lvert\Gamma_2\rvert$	Re $\tilde{\Gamma}_2$	J_2	Γ_{2N}	δ_1^1	$\lvert D_1\rvert$
		amplitudes in natural units, δ in degrees							
40.0	.779	-.192	.182	.265	-.399	.152	.207	136.5	.574
38.0	.740	-.231	.271	.356	-.449	.169	.218	130.5	.475
36.0	.701	-.258	.404	.479	-.487	.187	.230	122.5	.391
35.0	.682	-.256	.492	.555	-.492	.197	.236	117.5	.355
34.0	.662	-.236	.595	.640	-.479	.207	.242	111.7	.323
32.0	.623	-.103	.825	.831	-.359	.227	.257	97.1	.274
30.0	.584	.188	.986	1.004	-.085	.250	.273	79.2	.249
28.0	.545	.537	.953	1.094	.246	.274	.291	60.6	.250
26.0	.506	.780	.768	1.094	.468	.301	.312	44.6	.275
25.0	.487	.847	.662	1.075	.523	.315	.324	38.0	.293
24.0	.468	.887	.563	1.051	.551	.331	.336	32.4	.315
22.0	.429	.920	.401	1.003	.555	.365	.365	23.6	.364
20.0	.390	.926	.286	.969	.527	.405	.399	17.2	.418
18.0	.351	.930	.205	.952	.490	.452	.440	12.5	.475
16.0	.312	.944	.148	.956	.453	.510	.492	8.9	.533
15.0	.292	.958	.125	.966	.435	.543	.523	7.5	.563
14.0	.273	.977	.106	.982	.418	.582	.558	6.2	.592
12.0	.234	1.036	.074	1.038	.387	.677	.648	4.1	.652
10.0	.195	1.138	.050	1.139	.360	.811	.779	2.5	.712
8.0	.156	1.326	.030	1.326	.334	1.024	.991	1.3	.772
6.0	.117	1.740	.014	1.740	.310	1.449	1.430	.5	.832
5.0	.097	2.240	.007	2.240	.298	1.933	1.942	.2	.863
4.1	.080	3.331	.001	3.331	.286	2.965	3.045	.0	.890
4.0	.078	.285	.000	.285	.285	.255	.000	.0	.893

t n.u.	t GeV^2	Re Γ_2	Im Γ_2	$\lvert\Gamma_2\rvert$	Re $\tilde{\Gamma}_2$	Im $\tilde{\Gamma}_2$	Re Γ_{2N}	Im Γ_{2N}	D_1
		amplitudes in natural units							
3.9	.076	11.000	-4.227	11.785	.282		10.718	-4.227	.896
3.5	.068	2.214	-4.490	5.006	.276		1.938	-4.490	.908
3.0	.058	.922	-3.364	3.488	.271		.651	-3.364	.922
2.5	.049	.482	-2.774	2.816	.265		.216	-2.774	.936
2.0	.039	.260	-2.399	2.413	.260		.000	-2.399	.949
1.5	.029	.128	-2.133	2.137	.256		-.128	-2.133	.962
1.0	.019	.039	-1.932	1.932	.252		-.213	-1.932	.975
.5	.010	-.024	-1.772	1.772	.248		-.272	-1.772	.988
.0	.000	-.070	-1.641	1.642	.244		-.315	-1.641	1.000

t n.u.	t GeV^2	Re Γ_2 amplitudes in natural units	Im Γ_2	$\lvert\Gamma_2\rvert$	Re $\tilde{\Gamma}_2$	Im $\tilde{\Gamma}_2$	Re Γ_{2N}	Im Γ_{2N}	D_1
.0	.000	-.070	-1.641	1.642	.244	-.000	-.315	-1.641	1.000
-2.0	-.039	-.178	-1.281	1.293	.230	.000	-.408	-1.281	1.048
-4.0	-.078	-.229	-1.056	1.081	.217	.002	-.446	-1.058	1.093
-6.0	-.117	-.256	-.894	.929	.207	.008	-.463	-.902	1.136
-8.0	-.156	-.269	-.763	.809	.199	.020	-.469	-.783	1.177
-10.0	-.195	-.264	-.656	.707	.204	.034	-.469	-.690	1.217
-12.0	-.234	-.245	-.576	.626	.220	.038	-.465	-.613	1.255
-14.0	-.273	-.228	-.517	.565	.232	.032	-.460	-.549	1.292
-16.0	-.312	-.209	-.472	.517	.244	.022	-.453	-.495	1.328
-18.0	-.351	-.195	-.441	.482	.250	.007	-.445	-.448	1.363
-20.0	-.390	-.186	-.415	.454	.252	-.008	-.437	-.407	1.396
-22.0	-.429	-.179	-.393	.432	.250	-.023	-.429	-.370	1.429
-24.0	-.468	-.175	-.372	.411	.245	-.034	-.420	-.338	1.461
-26.0	-.506	-.169	-.352	.390	.242	-.043	-.411	-.309	1.492
-28.0	-.545	-.162	-.336	.374	.240	-.053	-.403	-.283	1.522
-30.0	-.584	-.158	-.325	.361	.236	-.065	-.394	-.260	1.552
-32.0	-.623	-.156	-.313	.350	.229	-.075	-.385	-.238	1.581
-34.0	-.662	-.155	-.303	.340	.222	-.084	-.377	-.219	1.609
-36.0	-.701	-.154	-.293	.331	.215	-.092	-.369	-.201	1.636
-38.0	-.740	-.154	-.283	.322	.206	-.098	-.360	-.185	1.663
-40.0	-.779	-.154	-.272	.313	.198	-.102	-.353	-.170	1.689
-42.0	-.818	-.153	-.261	.302	.192	-.105	-.345	-.156	1.715
-44.0	-.857	-.151	-.251	.292	.186	-.108	-.337	-.143	1.741
-46.0	-.896	-.148	-.241	.283	.182	-.110	-.330	-.131	1.765
-48.0	-.935	-.144	-.233	.274	.178	-.113	-.322	-.120	1.790
-50.0	-.974	-.140	-.226	.266	.175	-.116	-.315	-.110	1.814
-52.0	-1.013	-.136	-.220	.259	.172	-.120	-.309	-.100	1.837
-54.0	-1.052	-.132	-.216	.253	.170	-.124	-.302	-.092	1.860
-56.0	-1.091	-.128	-.213	.248	.167	-.129	-.295	-.083	1.883
-58.0	-1.130	-.124	-.211	.245	.165	-.135	-.289	-.076	1.905
-60.0	-1.169	-.121	-.209	.242	.162	-.141	-.283	-.068	1.927
-62.0	-1.208	-.118	-.209	.240	.159	-.147	-.277	-.062	1.949
-64.0	-1.247	-.115	-.210	.240	.156	-.155	-.271	-.055	1.970
-66.0	-1.286	-.113	-.214	.242	.152	-.164	-.265	-.049	1.991
-68.0	-1.325	-.113	-.219	.247	.146	-.176	-.259	-.044	2.012
-70.0	-1.364	-.117	-.226	.255	.137	-.188	-.254	-.038	2.032

2.4.6.2 ππ-scattering phase shifts

For the extrapolation of $\pi\pi N\bar{N}$ partial waves to positive t-values, it is necessary to use the experimental information on ππ-scattering phase shifts at $t \lesssim 50\mu^2 \approx 1\,\mathrm{GeV}^2$. Therefore, we give a short summary of the present status. Further details can be found in J. L. PETERSENs contribution in Sect. 9 of the "Compilation" (79 R NAGELS), in his lecture notes (71 R, 77 R PETERSEN) and in the book 76 R MARTIN-MORGAN-SHAW. Our notation is defined in Sect. A.7.6.

The main source of information on ππ phase shifts is the reaction πp→Nππ at high momenta of the incoming pion (>2 GeV/c) such that, at small momentum transfer, the pion pair is produced in a peripheral process, which can approximately be described by one-pion exchange (Fig. 4.6.3). The analysis is difficult because of background effects, in particular at low energies of the ππ-system.

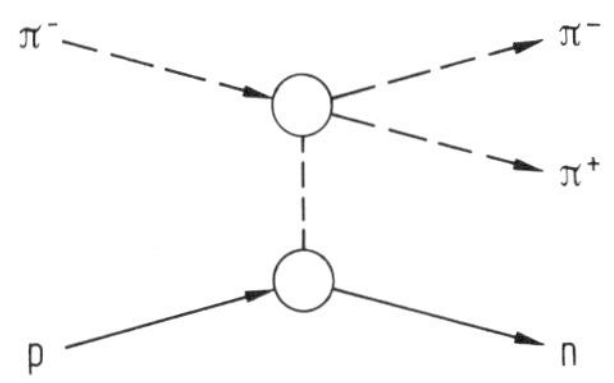

Fig. 4.6.3. Determination of ππ-scattering amplitudes from data for the reaction πN→ππN.

The $I=J=1$ (P-wave) phase shift can be determined in a weakly model-dependent way from data for the reaction $e^+e^- \to \pi^+\pi^-$ at energies, where elastic unitarity is a good approximation. A complication arises from ϱ−ω interference. (See the discussion in Sect. 2.5.2.)

A model-independent result for a combination of low energy phase shifts can be obtained from data for the K_{e4}-decay (77 L ROSSELET).

Independent determinations of the S-wave scattering lengths were made from $\pi^-p \to \pi^-\pi^+n$-data near threshold, but the analysis is model-dependent (80 L BJORK). Other recent papers on the reaction πp→Nππ at low energies: 79 L ALEKSEEVA, BELKOV (79 L-2 and 80 L), 82 L BOLOKHOV. MUKHIN (80 L) presented a new result for the S-wave ππ-phase shift from threshold up to 1 GeV and BELKOV (79 L-1) discussed the S-wave scattering length, his result being $a_0^0 = 0.23 \pm 0.05\,\mu^{-1}$.

In the upper part of the range which is of interest to us ($500 < \sqrt{t} < 950$ MeV), the results of different phase shift analyses are in good agreement with each other. At higher values up to about 1800 MeV, the main characteristics are also well established (79 L BECKER, 78 L and 79 L CORDEN, 78 L MARTIN and PENNINGTON, 75 L and 77 L FROGGATT and PETERSEN).

Unfortunately, there are serious problems in the threshold region, which is most important for our applications. The analysis of the $\pi^-p \to \pi^+\pi^-n$-data of the CERN-MUNICH group (74 L MÄNNER) suggests values for the S-, P-, and D-wave scattering lengths, which are strongly incompatible with the dispersion analysis of all data (75 L BASDEVANT) and also with other recent determinations.

Therefore, we have used an essentially unique phase shift solution derived by FROGGATT and PETERSEN (77 L), who analyzed extrapolated moments obtained by ESTABROOK and MARTIN (75 L) from the 17 GeV/c $\pi^-p \to \pi^-\pi^+n$ data of the CERN-MUNICH group and also the K_{e4}-data of ROSSELET et al. (77 L). This work is based on ROY-equation fits, using the method of BASDEVANT, FROGGATT, and PETERSEN (74 L) (Tables 2.4.6.2 and 2.4.6.3 and Fig. 4.6.4).

Many authors have discussed the status of the scalar meson resonances (ε, S* etc.), because it is difficult to determine their parameters and to find a description in terms of a quark model (see for instance JAFFE (81 R, 77 L), 81 L GIDAL, 81 L IRVING, 79 L ESTABROOKS, 82 L DELBOURGO, 82 L WEINSTEIN).

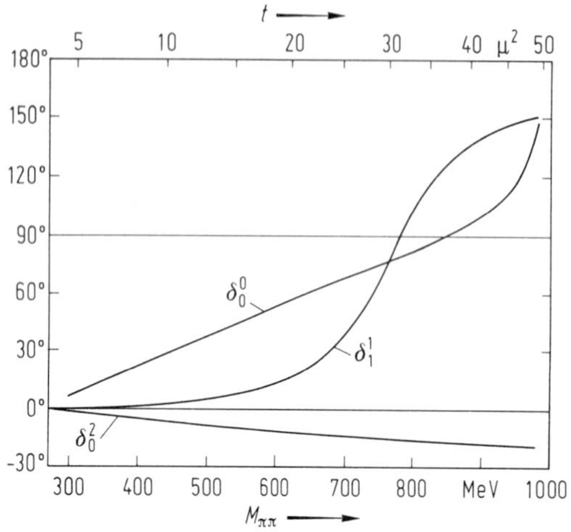

Fig. 4.6.4. Pion-pion scattering phase shifts δ_ℓ^I as functions of $M_{\pi\pi}=\sqrt{t}$. The phaseshifts were calculated by FROGGATT and J. L. PETERSEN (77 L) from experimental results derived from $\pi^- p \to \pi^- \pi^+ n$ data, using analyticity constraints and assuming the value $a_0^0=0.30\mu^{-1}$ for the S-wave scattering length.

The rapid variation of δ_0^0 near $M_{\pi\pi}=1000$ MeV is caused by the opening of the strongly coupled $K\bar{K}$-channel (threshold: 987 MeV) and also by the S*(980) resonance. Many authors ascribed the increase of δ_0^0 below this energy region to a "σ" or "ε"-resonance. However, the existence of this resonance is not established. — There is evidence for a broad ε(1300) resonance (width about 300 MeV). See the discussion in 80 R BRICMAN, p. S 134 and 79 L ESTABROOKS.

Table 2.4.6.2. ππ-phase shifts in degrees as calculated by FROGGATT and PETERSEN (77 L) ($t=M_{\pi\pi}^2$). This table is based on $a_0^0=0.30\,\mu^{-1}$ and $a_1^1=0.031\,\mu^{-3}$ and on the D- and F-wave phase shifts listed in Table 2.4.6.3. Notation: δ_J^I (I=isospin, J=angular momentum). See Fig. 4.6.4.

t [μ^2]	$M_{\pi\pi}$ [MeV]	δ_0^0	δ_1^1	δ_0^2
4.62	300	7.1	0.1	−0.6
	320	10.7	0.4	−1.2
5.93	340	14.0	0.7	−1.8
	360	17.1	1.0	−2.4
7.41	380	20.2	1.5	−3.1
	400	23.3	2.0	−3.7
9.05	420	26.4	2.5	−4.3
	440	29.6	3.2	−5.0
10.9	460	32.7	4.0	−5.6
	480	35.8	4.9	−6.2
12.8	500	39.0	5.9	−6.8
	520	42.1	7.1	−7.4
15.0	540	45.1	8.5	−8.0
	560	48.2	10.1	−8.6
17.3	580	51.2	12.1	−9.1
	600	54.1	14.4	−9.6
19.7	620	57.0	17.3	−10.1
	640	59.9	20.9	−10.6
22.4	660	62.7	25.4	−11.1
	680	65.5	31.2	−11.5
25.1	700	68.3	38.7	−12.0
	720	71.1	48.4	−12.4
28.1	740	73.9	60.6	−12.8
	760	76.7	74.9	−13.2
31.2	780	79.6	90.0	−13.6
	800	82.5	103.8	−13.9
34.5	820	85.6	115.3	−14.3
	840	88.8	124.3	−14.6
38.0	860	92.4	131.3	−14.9
	880	96.3	136.7	−15.2
41.6	900	100.8	141.0	−15.5
	920	106.3	144.5	−15.8
45.3	940	113.6	147.3	−16.1
	960	124.5	149.7	−16.4
49.3	980	146.9	151.8	−16.6

Table 2.4.6.3. D- and F-wave phase shifts in degrees. These values were used in the calculation of Table 2.4.6.2.

$M_{\pi\pi}$[MeV]	δ_2^0	δ_2^2	δ_3^1
400	0.07±0.01	0.00±0.01	0.002±0.001
600	0.9 ±0.1	−0.1 ±0.1	0.04 ±0.02
800	3.5 ±0.5	−0.5 ±0.2	0.2 ±0.1
1000	11 ±2	−1.0 ±0.6	0.8 ±0.2

Near threshold, the phase shifts can be described by effective range expansions

$$\operatorname{Re} A_J^I/q_{\pi\pi}^{2J} = a_J^I + b_J^I q_{\pi\pi}^2 + O(q_{\pi\pi}^4). \qquad (4.6.2)$$

In 79 R Nagels, J. L. Petersen gives the following values of these parameters (which are not quite consistent with the phase shifts in Table 2.4.6.2, see also 77 R Petersen).

Table 2.4.6.4. $\pi\pi$ scattering lengths and effective ranges from Roy-equation fits constrained by K_{e4}-data (units: powers of μ^{-1}).

S-waves	D-waves
$a_0^0 = 0.26 \pm 0.05$; $b_0^0 = 0.25 \pm 0.03$	$a_2^0 = (17 \pm 3) \cdot 10^{-4}$
$a_0^2 = -0.028 \pm 0.012$; $b_0^2 = -0.082 \pm 0.008$	$a_2^2 = (1.3 \pm 3) \cdot 10^{-4}$
$2a_0^0 - 5a_0^2 = 0.657 \pm 0.052$	
P-wave	**F-wave**
$a_1^1 = 0.038 \pm 0.002$	$a_3^1 = (0.6 \pm 0.2) \cdot 10^{-4}$

For the $I = J = 1$ partial wave (the P-wave), one can use a Chew-Mandelstam parametrization, whose parameters are given in Table A.7.3.

Soft pion theory

Since the Current Algebra model for πN scattering will be discussed in some detail in Sect. 2.5.1, we mention here briefly the simplest version for the case of ππ-scattering (66 L Weinberg).

One can distinguish consequences of three assumptions:

Linearity: $2a_0^0 - 5a_0^2 = 18a_1^1 = 3b_0^0 = -6b_0^2 = 6L$; the other a_ℓ^I and b_ℓ^I vanish.

Current Algebra: $L = \left(\frac{4\pi f_\pi^2}{\mu}\right)^{-1} = 0.089\,\mu^{-1}$.

In Table 2.4.6.5 we compare the predictions with the experimental values.

Table 2.4.6.5. Comparison of predictions for scattering lengths with the experimental values. Units: powers of μ^{-1}.

	$2a_0^0 - 5a_0^2$	a_1^1	b_0^0	b_0^2
Prediction	0.53	0.030	0.178	−0.089
Exp. value	0.657 ± 0.052	0.038 ± 0.002	0.25 ± 0.03	−0.082 ± 0.008

The additional assumption that the $I = 2$ part of the sigma term vanishes leads to the following values for the S-wave scattering lengths

Sigma condition: $2a_0^0 + 7a_0^2 = 0$; $\quad a_0^0 = 0.158\mu^{-1}$; $\quad a_0^2 = -0.045\mu^{-1}$. (4.6.3)

A comparison with the experimental values shows a fairly large discrepancy, but one should note that a $\pi^- p \to \pi^- \pi^+ n$ experiment near threshold led recently to the scattering length $a_0^0 = 0.177 \pm 0.017\mu^{-1}$. (80 L Bjork). Furthermore, one can expect appreciable corrections to Weinbergs prediction.

In ππ-scattering, the sigma terms σ^I are defined by

$$A(2\mu^2, \mu^2, \mu^2) = -A(0, \mu^2, \mu^2) = \tfrac{1}{3}(\sigma^0 - \sigma^2); \qquad (4.6.4)$$

$$A(\mu^2, 2\mu^2, \mu^2) = -A(\mu^2, 0, \mu^2) = \tfrac{1}{2}\sigma^2. \qquad (4.6.5)$$

Current algebra predicts $2\sigma^0 - 5\sigma^2 = 3\mu^2/8\pi f_\pi^2$, whereas WEINBERGs sigma condition demands $\sigma^2 = 0$ (77 L PETERSEN, 76 R MARTIN-MORGAN-SHAW).

The amplitude $A(s,t,u)$ is related to the amplitude F^I of Eq. (A.7.52) by

$$A(s,t,u) = \tfrac{1}{3}(F^0 - F^2). \tag{4.6.6}$$

The CHEW-MANDELSTAM coupling constant (60 L CHEW) is defined by

$$\lambda = -A(\tfrac{4}{3}, \tfrac{4}{3}, \tfrac{4}{3}). \tag{4.6.7}$$

At the ADLER point $s = t = u = \mu^2$ the amplitude vanishes as a consequence of PCAC (65 L ADLER-1)

$$A(\mu^2, \mu^2, \mu^2) = 0. \tag{4.6.7a}$$

One of the pions is "soft" (i.e. it has zero mass), since the relation $s + t + u = 4\mu^2$ is valid "on-shell".

If linearity is assumed, Eq. (4.6.7a) can be used in order to derive Eqs. (4.6.4) and (4.6.5). Furthermore, the additional assumption $\sigma^2 = 0$ leads to

$$A(s,t,u) = \frac{s - \mu^2}{16\pi f_\pi^2}. \tag{4.6.7b}$$

This relation shows the on-shell location of the zero trajectory, which belongs to the ADLER zero.

LOVELACE (68 L) pointed out that the ADLER zero leads to the condition for the ϱ-trajectory

$$\alpha(\mu^2) = 0.5 \tag{4.6.7c}$$

in a VENEZIANO model for ππ-scattering. This agrees well with the experimental result for the ϱ-trajectory as derived from πN charge-exchange scattering (Sect. 2.6.5.2).

LEHMANN (72 L) extended the low-energy theorems for massless pions by terms which are of fourth order in the pion momenta.

A detailed treatment of the zero trajectories and their role in ππ-phase shift analysis can be found in 76 R MARTIN-MORGAN-SHAW, 77 R PETERSEN and 78 L MARTIN-PENNINGTON, 79 L BECKER. Part of this summary follows J.L. PETERSENs contribution in 79 R NAGELS, 77 L PETERSEN, and 76 R MARTIN-MORGAN-SHAW. See also 75 R PAGELS.

The most recent review on pion-pion scattering is due to MUKHIN and PATARAKIN (81 R). Theoretical questions of low energy pion physics are treated in the review of VOLKOV and PERVUSHIN (76 R).

2.4.6.3 Extrapolation of ππNN̄ partial waves to $t > 0$

The problem is the following: $\mathrm{Re} f_\pm^J(t)$ and $\mathrm{Im} f_\pm^J(t)$ are known in a range $0 \geqq t \gtrsim -0.5\,\mathrm{GeV}^2$. Furthermore, from elastic unitarity, Eq. (A.5.25), it is known that

$$f_\pm^J(t) = \pm |f_\pm^J(t)| e^{i\delta_J^I}; \quad 4\mu^2 < t < 16\mu^2, \tag{4.6.8}$$

where $I = 0$ if J is even and $I = 1$ if J is odd. The range can be extended up to about $40\mu^2$, because ππ-phase shift analysis shows that the 4π-intermediate states are unimportant up to the KK̄-threshold for the S-wave and up to the ωπ-threshold for the P-wave.

We explain the extrapolation method for the $J = 1$ partial waves in order to simplify the notation.

The singularities of $f_\pm^1(t)$ are known from the partial wave dispersion relation (Sect. A.7.5). We use an ansatz of the N/D type, writing

$$f_\pm^1(t) = \frac{J_\pm(t)}{D_1(t)}, \tag{4.6.9}$$

where $J_\pm(t)$ has a left hand cut $t \leqq 3.98\mu^2$ and is real for larger t-values up to $t_r \approx 40\mu^2$. $D_1(t)$ has only the right hand cut $t \geqq 4\mu^2$ and is real for smaller t-values.

In general, $D_1(t)$ can be obtained from a phase representation

$$D_1(t) = \exp\left\{-\frac{t}{\pi}\int_{4\mu^2}^{\infty} \frac{\delta_1^1(t')\,dt'}{t'(t'-t)}\right\}, \tag{4.6.10}$$

which has, however, the difficulty that one has to make an assumption on the behaviour of $\delta_1^1(t)$ at large t.

In our example $J=1$ there is in addition the possibility to choose

$$D_1(t)=1/F_\pi(t), \tag{4.6.11}$$

where $F_\pi(t)$ is the electromagnetic pion form factor, which also has the phase $\delta_1^1(t)$ in the range where the 4π-intermediate states can be neglected [cf. Eq. (A.11.27)].

Taking for instance the second choice, Eq. (4.6.11), one can calculate $\mathrm{Re}J_\pm(t)$ and $\mathrm{Im}J_\pm(t)$ in the range $0>t>-0.5\,\mathrm{GeV}^2$. The aim is to extrapolate $J_\pm(t)$ to $t>4\mu^2$. This is simpler than an extrapolation of $f_\pm^1(t)$, because $J_\pm(t)$ is real at $t>3.98\mu^2$.

Following FRAZER and FULCO (60 L-1), one can use a dispersion relation for $J_\pm(t)$ and a discrepancy method, assuming that the distant parts of the cuts give slowly variable contributions in the t-range of interest.

For the calculation, it is preferable to work with the amplitudes Γ_1 and Γ_2, Eq. (A.3.52), because they are more rapidly decreasing than $f_\pm^1$ at large values of t according to Eq. (A.7.49). The dispersion relation for

$$J_\mathrm{i}(t)\equiv D_1(t)\Gamma_\mathrm{i}(t); \quad \mathrm{i}=1,2, \tag{4.6.12}$$

reads

$$J_\mathrm{i}(t)=\frac{1}{\pi}\int_{-\infty}^{t_\mathrm{nc}}\frac{\mathrm{Im}J_\mathrm{i}(t')}{t'-t}\mathrm{d}t'+\frac{1}{\pi}\int_{t_\mathrm{r}}^{\infty}\frac{\mathrm{Im}J_\mathrm{i}(t')}{t'-t}\mathrm{d}t', \tag{4.6.13}$$

where $t_\mathrm{nc}=3.98\mu^2$ is the branch point of the left hand cut of the partial wave dispersion relation (Sect. A.7.5), and $t_\mathrm{r}\approx 40\mu^2$ is the t-value above which the 4-pion intermediate states in the unitarity relation become important.

The discrepancy function $\Delta_\mathrm{i}^{(1)}(t)$ is defined as the contribution of the unknown parts of the cuts. It is calculated from $\mathrm{Re}J_\mathrm{i}(t)$ and the integral over $\mathrm{Im}J_\mathrm{i}(t)$ in the range, where $J_\mathrm{i}(t)$ is known.

$$\Delta_\mathrm{i}^{(1)}(t)\equiv\mathrm{Re}J_\mathrm{i}(t)-\frac{1}{\pi}\,⨍_{t_1}^{t_\mathrm{nc}}\frac{\mathrm{Im}J_\mathrm{i}(t')}{t'-t}\mathrm{d}t'=\frac{1}{\pi}\left\{\int_{-\infty}^{t_1}+\int_{t_\mathrm{r}}^{\infty}\right\}\frac{\mathrm{Im}J_\mathrm{i}(t')}{t'-t}\mathrm{d}t'. \tag{4.6.14}$$

$\mathrm{Im}J_\mathrm{i}(t)$ is known not only at negative t but also in the range $0<t<t_\mathrm{nc}$, where the only contribution to $\mathrm{Im}\Gamma_\mathrm{i}(t)$ comes from the nucleon pole term ("nc" = nucleon cut).

It is useful to choose for t_1 a larger negative value than $-0.5\,\mathrm{GeV}^2$, because this leads to a slower variation of the discrepancy function without falsifying the result in the t-range of interest.

As expected, the discrepancy function turns out to be slowly variable in the range $-0.5\,\mathrm{GeV}^2\lesssim t\leqq 0$. It can be fitted by a linear ansatz in t which is then used for the extrapolation to $t>0$. The result of this extrapolation is inserted into the dispersion relation Eq. (4.6.13) instead of the contribution of the distant parts of the cuts. Evaluating the nearby part of the cut, one obtains $\mathrm{Re}J_\mathrm{i}(t)$ in $0<t<t_\mathrm{nc}$ and the real function $J_\mathrm{i}(t)$ for $t>t_\mathrm{nc}$. Finally, $\mathrm{Re}\Gamma_\mathrm{i}$ and $\mathrm{Im}\Gamma_\mathrm{i}$ follow from $\Gamma_\mathrm{i}(t)=J_\mathrm{i}(t)/D_1(t)$.

There are two problems:

i) Near $t=t_\mathrm{nc}$, the numerical evaluation of the first integral in Eq. (4.6.14) is difficult because of the singularity of the nucleon pole term projection $\Gamma_\mathrm{iN}(t)$;

ii) the reliability of the extrapolation decreases as t increases, because one approaches the unknown cut at $t>t_\mathrm{r}$.

The first difficulty can be removed by choosing another discrepancy function, which belongs to the dispersion relation for

$$J_\mathrm{i}(t)-\Gamma_\mathrm{iN}(t)D_1(t_\mathrm{nc})=\Gamma_\mathrm{iN}(t)\,[D_1(t)-D_1(t_\mathrm{nc})]+\tilde{\Gamma}_\mathrm{i}(t)D_1(t). \tag{4.6.15}$$

The new discrepancy function reads

$$\Delta_\mathrm{i}^{(2)}(t)\equiv J_\mathrm{i}(t)-\Gamma_\mathrm{iN}(t)D_1(t_\mathrm{nc})-\frac{1}{\pi}\int_{t_1}^{0}\frac{D_1(t')\mathrm{Im}\tilde{\Gamma}_\mathrm{i}(t')}{t'-t}\mathrm{d}t'-\frac{1}{\pi}\int_{t_1}^{t_\mathrm{nc}}\frac{[D_1(t')-D_1(t_\mathrm{nc})]\mathrm{Im}\Gamma_\mathrm{iN}(t')}{t'-t}\mathrm{d}t'. \tag{4.6.16}$$

It turns out that $\Delta_\mathrm{i}^{(2)}(t)$ can be evaluated without problems. It is practically linear in t in the range $-0.5\,\mathrm{GeV}^2<t\leqq 0$. An extrapolation of the linear function is inserted into the dispersion relation and Re and $\mathrm{Im}\Gamma_\mathrm{i}(t)$ are calculated as described above.

Details of the calculation can be found in 75 L HÖHLER-4 and 77 L PIETARINEN. The D-functions are shown in Figs. 4.6.5 and 4.6.6.

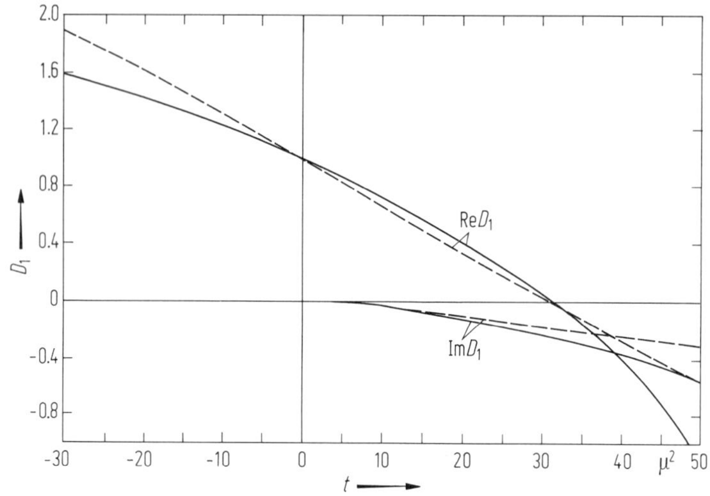

Fig. 4.6.5. Real and imaginary part of the D-function for the P-wave. Solid line: 77 L PIETARINEN, dashed line: $D_1 = 1/F_\pi$ (75 L HÖHLER-4).

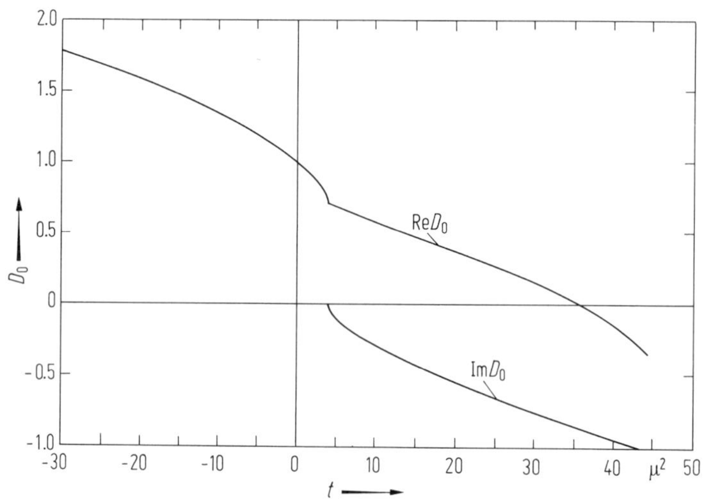

Fig. 4.6.6. Real and imaginary part of the D-function for the S-wave. (77 L PIETARINEN).

A remarkable feature of our result is the unusual threshold behaviour near $t=4\mu^2$, which is clearly seen in Figs. A.3.5 and 4.6.7 for the S- and P-waves. The reason is that the amplitude has a second branch point at a close distance ($t=t_{nc}=3.98\mu^2$), which belongs to the cut of the nucleon pole term projection (cf. Sect. A.3.5.3).

Figure 4.6.7 shows the typical P-wave threshold behaviour of $\mathrm{Im}\,F_\pi(t)$, which is much flatter than that of $\mathrm{Im}\,f_\pm^1(t)$. In the relation $\mathrm{Im}\,f_\pm^1(t)=J_\pm(t)\mathrm{Im}F_\pi(t)$, the threshold behaviour is modified by the rapid increase of $J_\pm(t)$ as t approaches the branchpoint t_{nc}.

In general, a "numerator function" is assumed to be slowly variable, because its singularities are distant (see for instance the N-function in ππ-scattering, Sect. A.7.6). The present case is different because of the small mass ratio μ/m.

The enhancement of the left wing of the ϱ-peak in $\mathrm{Im}\,f_\pm^1(t)$ was known to FRAZER and FULCO (60 L-1), but it was ignored by many subsequent authors. One of its consequences is that "effective ϱ-pole fits" led to mass values considerably below the ϱ-mass. The influence of the enhancement on their isovector nucleon form factors will be discussed in Sect. 2.5.2.

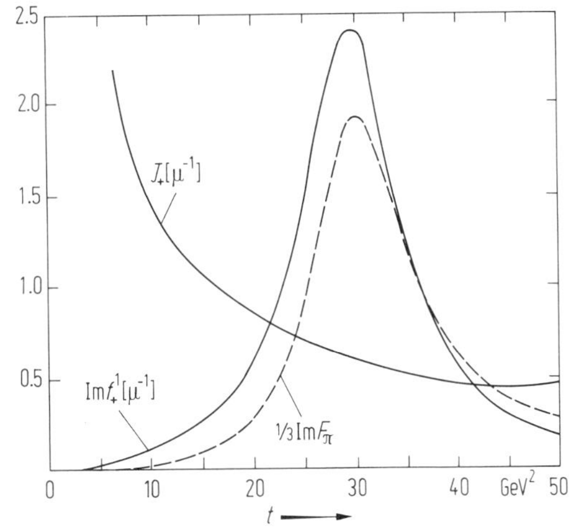

Fig. 4.6.7. t-dependence of $\mathrm{Im}\, f_+^1$. J_+ is the numerator function in Eq. (4.6.9), whose t-dependence has been neglected by some authors. The increase at small $|t|$ is due to a singularity at $t=3.98\mu^2$. It leads to an enhancement of $\mathrm{Im}\, f_+^1$ at small $|t|$ in comparison with the behaviour expected for the imaginary part of a P-wave amplitude. This enhancement is much more pronounced in Fig. 5.2.1, because the imaginary part is divided by t^2 [cf. Eq. (5.2.4)].

The second of the above mentioned problems is more serious and it is hard to see how a satisfactory solution can be found in the near future. The attempts to include inelastic channels are limited to other two-particle channels. A large effect is expected from the opening of the K$\bar{\mathrm{K}}$-channel in $f_+^0(t)$ at $t=0.97\,\mathrm{GeV}^2$ and from the S*(980) resonance, whose pole lies very near to this threshold.

In the P-waves $f_\pm^1(t)$, the most important inelastic contributions come probably from ωπ-production above its threshold at $t=0.84\,\mathrm{GeV}^2$ and from a possible ϱ'(1250) resonance in this channel. The evidence for this resonance is not yet compelling (80 R Bricman, see also 80 L Aston and 81 R Paul). A large effect in this mass region is suggested by the analysis of the isovector nucleon form factors (Sect. 2.5.2). The inelastic contribution to the P-wave was discussed in 76 L Willrodt and 76 L Budnev.

The work of Nielsen and Oades

In Fig. 4.6.1 we have also shown the results of Nielsen and Oades (private communication, 1978). These authors used backward dispersion relations for the invariant amplitudes C and B. They assumed that in a range $4\mu^2<t<40\mu^2$ the invariant amplitudes can be approximated by the leading term of their t-channel partial wave expansions (Table A.3.3). The higher terms are suppressed by threshold factors (Sect. A.3.5.3) and moreover, the lowest $J=2$ and $J=3$ resonances have a rather high mass ($t_{\mathrm{res}}=1.62\,\mathrm{GeV}^2$ for the f-meson and $2.89\,\mathrm{GeV}^2$ for the g-meson).

Once subtracted backward dispersion relations are written in the variable t for amplitudes, which are multiplied by a suitable D-function after the nucleon pole term has been subtracted. As an example, we consider $\tilde{B}^-(t)D_1(t)$. For $t>4\mu^2$, the approximation

$$\tilde{B}^-(t)D_1(t)\approx\frac{12\pi}{\sqrt{2}}\tilde{f}_-^1(t)D_1(t)=\frac{12\pi}{\sqrt{2}}\tilde{J}_-(t) \tag{4.6.17}$$

is used. The reason for the choice of this function is that one can calculate its imaginary part also on the t-channel cut

$$\mathrm{Im}\{\tilde{B}^-(t)D_1(t)\}\approx-\frac{12\pi}{\sqrt{2}}f_{-\mathrm{N}}^1(t)\,\mathrm{Im}\,D_1(t);\quad 4\mu^2<t<40\mu^2, \tag{4.6.18}$$

because the imaginary part of $f_-^1 D_1$ vanishes in this range.

The contributions of the distant parts of the cuts are again described by a discrepancy function $\Delta(t)$, which can be determined from the evaluation of the dispersion relation at $t<0$. It turns out to be slowly variable. The extrapolation to $t>0$ is used together with the dispersion relation in order to determine $J_-(t)$, from which Re and $\mathrm{Im}\, f_-^1(t)$ at $t>4\mu^2$ can be obtained.

This method has the advantage that the discontinuity along the cut $t<0$ is well known in a much larger range than in Eqs. (4.6.14) and (4.6.16). But one has to pay for it, since it is necessary to estimate the errors coming from the neglection of the higher t-channel partial waves.

The magnitude of these errors is known to some extent from the study of dispersion relations for a set of paths in the MANDELSTAM plane, which includes the backward hyperbola as a special case (70 L STRAUSS, 76 L KAISER, 77 L BORIE, 77 L HEDEGAARD-JENSEN). The dispersion integrals over negative t-values are evaluated from πN partial waves, and the results for $f_{\pm}^{0}$ and $f_{\pm}^{1}$ are used for the evaluation of the nearby part of the t-channel cut. There remains a discrepancy, which belongs to the neglected higher t-channel partial waves and to errors of the input for f_{+}^{0} and $f_{\pm}^{1}$. The different contributions can be disentangled by studying how the discrepancy depends on the parameter, which characterizes the different paths (see Sect. 2.4.2.2).

The corrections from higher partial waves are included in the results of NIELSEN and OADES shown in Fig. 4.6.1. The determination of $\pi\pi N\bar{N}$-partial waves is treated in 78 R HAMILTON-OADES, which deals mainly with kaon-nucleon scattering.

The difficulties in determinations of $\pi\pi N\bar{N}$-partial waves were also studied by EPSTEIN and MC KELLAR (74 L) but, to my knowledge, these authors have not presented results based on more recent partial wave solutions.

The work of BONNIER and GAURON (80 L)

The authors studied the question how sensitive the extrapolation of $f_{+}^{0}(t)$ to $t>4\mu^2$ is to various assumptions on the S-wave scattering length a_0^0 and to different extrapolation methods. For $a_0^0=0.30\mu^{-1}$ they found values for $|f_{+}^{0}|$ which are lower by 15% at $t=20\mu^2$ and higher by 10% at $t=30\mu^2$ than our results. At least part of the discrepancy is due to a different assumption on the ππ S-wave phase shift. For instance at $t=30\mu^2$ their interpolation gives $\delta_0^0=64°$, whereas our value 80° is considerably higher and near to that of FROGGATT-PETERSEN ($\delta_0^0=77.4°$). $|f_{+}^{0}|$ is plotted in Fig. 5.3.2.

The model of DURSO, JACKSON, and VERWEST (80 L)

These authors wanted to construct a simple model for the $\pi\pi N\bar{N}$ partial waves, which can be used for the description of two-pion intermediate states in nuclear physics.

They included Born terms for N, Δ(1232) and P 11 N(1470) exchanges with suitable form factors and described rescattering by simple separable ππ-interactions. In the S-wave, the inelasticity is taken into account by coupling a resonant $K\bar{K}$ channel to the ππ-system. The agreement with our results for the partial waves is reasonable. It would be of some interest, to compare the different contributions with their dispersion theoretic counterparts (see Sects. 2.5.1.6 and A.8.3).

Determination of ππ-scattering lengths from πN partial waves

In the sixties, it was an important problem to estimate ππ-phase shifts from the experimental information on πN scattering (67 R HAMILTON-1 and 2). More recent work in this field was done by NIELSEN et al. (70 L, 72 L-2, 74 L-1) and others [see Chap. 4.1 (d) in 76 R MARTIN-MORGAN-SHAW].

In the meantime, ππ-scattering has been investigated more directly (Sect. 2.4.6.2) and tables of ππ-phase shifts are available. Therefore, several groups used these "experimental" ππ-phase shifts in their determination of $\pi\pi N\bar{N}$-amplitudes and evaluated the nearby part of the t-channel cut in suitable dispersion relations. An analysis of the discrepancy function then shows whether the input for the s- and t-channels is compatible with the dispersion relation (77 L BORIE, 77 L HEDEGAARD-JENSEN and others). No indication for a systematic error due to wrong ππ-phase shifts was found.

HITE and JACOB (78 L) and MÜLLER, HITE, and JACOB (80 L) reconsidered the problem to determine ππ-scattering lengths from πN partial waves. They evaluated the s-channel cut of the interior dispersion relation (Sect. A.6.7.2) and performed an extrapolation towards the branch point of the t-channel cut, taking into account (in the second paper) the well-known properties of the nucleon pole term projections. As expected, they found essentially the same values which the other authors had used as part of their input.

The small values for the errors are unrealistic. In my opinion, there is no doubt that the extrapolation towards the branch point of an unknown cut cannot be as accurate as the ROY-equation fit to the ππ-scattering and K_{e4}-data (79 R NAGELS, Sect. 9). It would be an improvement to work with a variety of paths and to include the information on the discontinuity at $t>4\mu^2$, varying only a_0^0 (cf. 82 L KOCH-1).

***J*-plane singularities of $\pi\pi N\bar{N}$ amplitudes at $t=0$**

PIETARINEN (72 L-1) made an attempt to use the FROISSART-GRIBOV-representation. (Sect. A.7.4), for a continuation of $\pi\pi N\bar{N}$ partial waves $f_{\pm}^{J}(t)$ into the J-plane. He calculated the Mellin transforms of the imaginary parts of the forward amplitudes and continued these transforms in order to determine nearby singularities in J. Finally, he discussed the advantages of his method in comparison with others, which consider only fits to the high energy behaviour.

2.4.7 Low energy expansions

2.4.7.1 The subthreshold expansion at $\nu=t=0$

Since the point $\nu=t=0$ lies near the nucleon poles at $\nu=\pm\nu_B$, the invariant amplitudes show very rapid variations in its neighborhood. Therefore, the expansion has to be made for invariant amplitudes from which the nucleon pole terms have been subtracted.

For other applications, it is suitable to subtract the nucleon pole term of the fixed-t dispersion relations, which agrees with the *pseudoscalar* nucleon Born term. The result is denoted by a tilde: $\tilde{C}$ etc. In the present case we prefer to subtract the *pseudovector* nucleon Born term, because it occurs in the predictions from low energy models (Sect. 2.5.1). The result is denoted by $\bar{C}$ etc.

The relation between the different amplitudes reads (cf. Sect. A.8.1)

$$\bar{A}^+=A^+-g^2/m;\quad \bar{B}^+=\tilde{B}^+;\quad \bar{C}^+=\tilde{C}^+-g^2/m;\quad \bar{D}^+=\tilde{D}^+-g^2/m;$$

$$\bar{A}^-=A^-;\quad \bar{B}^-=\tilde{B}^-+g^2/2m^2;\quad \bar{C}^-=\tilde{C}^-+\frac{g^2}{2m^2}\frac{\nu}{1-t/4m^2};\quad \bar{D}^-=\tilde{D}^-+\frac{g^2}{2m^2}\nu; \tag{4.7.1}$$

$$g^2/m=26.70\mu^{-1};\quad g^2/2m^2=1.99\mu^{-2}.$$

We divide the crossing odd amplitudes by ν in order to have an expansion in powers of ν^2 and t

$$X(\nu^2,t)=\sum_{\mathrm{m,n}} x_{\mathrm{mn}}\nu^{2\mathrm{m}}t^{\mathrm{n}}. \tag{4.7.2}$$

For X one can insert the amplitudes $\bar{A}^+$, $\bar{A}^-/\nu$, $\bar{B}^+/\nu$, $\bar{B}^-$, $\bar{C}^+$, $\bar{C}^-/\nu$, $\bar{D}^+$, $\bar{D}^-/\nu\equiv\bar{G}^-$. The corresponding small letters are used for the coefficients: c^+_{mn} etc.

The expansion converges up to the nearest singularities, which are the s, t and u-channel thresholds. The nearest poles belong to Δ- and ϱ-exchange. Furthermore, there is a singularity near $\nu^2=0$ in the second sheet of the ν^2-plane (see Sect. A.6.4.6).

We shall list the coefficients for the amplitudes $\bar{D}^+$, $\bar{D}^-/\nu\equiv\bar{G}^-$, $\bar{B}^+/\nu$, $\bar{B}^-$, $\bar{A}^+$ and $\bar{A}^-/\nu$. Since other tables give the coefficients for $\bar{C}^+$ and $\bar{C}^-/\nu$, we mention the relation between these amplitudes

$$\bar{D}^+=\bar{C}^+-\nu\bar{B}^+\frac{t/4m^2}{1-t/4m^2};\quad \bar{G}^-\equiv\bar{D}^-/\nu=\bar{C}^-/\nu-\bar{B}^-\frac{t/4m^2}{1-t/4m^2}. \tag{4.7.3}$$

The relations between the coefficients read

$$d^+_{\mathrm{mn}}=c^+_{\mathrm{mn}}\ \text{ if }\ \mathrm{m}=0\ \text{ or }\ \mathrm{n}=0\ \text{ or both }=0;\quad d^+_{11}=c^+_{11}-b^+_{00}/4m^2;$$
$$d^-_{\mathrm{mn}}=c^-_{\mathrm{mn}}\ \text{ if }\ \mathrm{n}=0;\quad d^-_{01}=c^-_{01}-b^-_{00}/4m^2;\quad d^-_{02}=c^-_{02}-b^-_{01}/4m^2-b^-_{00}/16m^4; \tag{4.7.4}$$
$$d^-_{11}=c^-_{11}-b^-_{10}/4m^2.$$

Furthermore, some of the earlier tables are valid for expansions of amplitudes, from which the *pseudoscalar* nucleon Born term is subtracted. Our $c^+_{00}=d^+_{00}$ is obtained if one subtracts $g^2/m=26.7\mu^{-1}$ and our $d^-_{00}=c^-_{00}$ and b^-_{00} are obtained if one adds $g^2/2m^2=1.99\mu^{-2}$.

Determination of the coefficients

In principle, all coefficients except $a^+_{0\mathrm{n}}=c^+_{0\mathrm{n}}=d^+_{0\mathrm{n}}$ can be calculated from the imaginary parts of the invariant amplitudes, using forward dispersion relations or derivatives of fixed-t dispersion relations at $\nu=t=0$, from which the nucleon pole term is omitted. Therefore, the result does not depend on the value of the coupling constant f^2 and on effects of the neutron-proton mass difference.

For the determination of $c^+_{0\mathrm{n}}$ one needs in addition results from partial wave analysis for $\mathrm{Re}\,C^+$, because a subtraction is required in the dispersion relation. c^+_{00} can be determined from the following version of the forward dispersion relation

$$c^+_{00}=C^+(0,0)-g^2/m=\mathrm{Re}\,C^+(\omega,0)-\frac{g^2}{m}\frac{\mu^4/4m^2}{\omega^2-\mu^4/4m^2}-\frac{2\omega^2}{\pi}\fint_0^\infty dk'\frac{k'^2}{\omega'^2}\frac{\sigma^+(k')}{k'^2-k^2}. \tag{4.7.5}$$

The right hand side is evaluated for all values of ω where partial waves are available. The average is the best value for c^+_{00}. The deviation from constancy shows to what extent the partial wave solution is compatible with the dispersion relation.

The KH 78 partial wave solution leads to $c^+_{00}=-1.50\mu^{-1}$, whereas an updated version gives $(-1.46\pm0.07)\mu^{-1}$. The value $-1.33\mu^{-1}$ is obtained from the CMU-LBL 79 solution, if our interpolation of the total cross section data is used.

The coefficient $c^+_{01}=d^+_{01}$ follows from the dispersion relation for $\partial C^+/\partial t$ at $t=0$ (Sect. A.6.5). In this case, the consistency of the KH 78 and CMU-LBL phase shift solutions with the dispersion relation is not yet satisfactory (Fig. 4.7.1). This is due to errors in the high partial waves, whose contributions are enhanced by a factor $\approx\ell^3$. The situation is even worse for c^+_{02}.

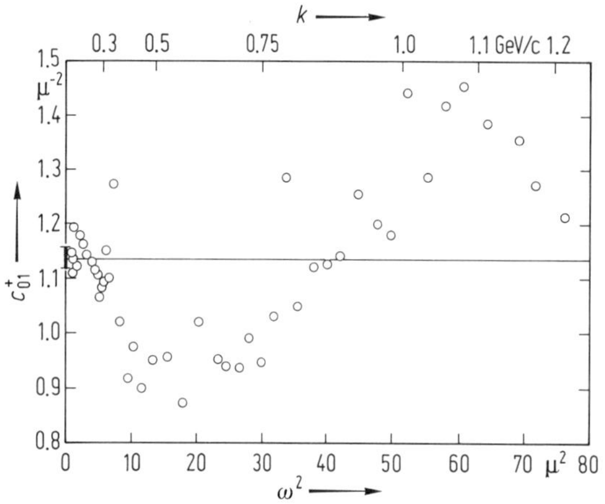

Fig. 4.7.1. Determination of the coefficient c^+_{01} of the subthreshold expansion. If the subtraction constant of the dispersion relation for $\partial C^+/\partial t$ at $t=0$ is determined at different values of ω^2, the result must be independent of ω^2. The figure shows that the compatibility of the KH 78 partial wave solution with this dispersion relation is not yet satisfactory (the same is true for the CMU-LBL 79 solution). The agreement is much better, if one uses the partial wave dispersion relation as a constraint for the high partial waves R. Koch (82 L-2 and work in progress). One should note that the scale of the ordinate starts at 0.8.

The values for x_{m0} in Table 2.4.7.1 were determined with the expansion method for the fixed-t solution of the KH 78 analysis, except for c^+_{00}, where Eq. (4.7.5) was used.

In order to improve the accuracy of the coefficients x_{mn} for $n\neq0$, we have evaluated a fixed-ν dispersion relation (72 L Engels, 74 L Nielsen-1, 77 L Höhler-2, 78 L Koch), which reads at $\nu=0$ (Sect. A.6.10)

$$\mathrm{Re}X(0,t)=\frac{1}{\pi}\left\{\int_{-\infty}^{-4m\mu}+\int_{4\mu^2}^{\infty}\right\}\frac{\mathrm{Im}X(0,t')}{t'-t}\,dt'. \tag{4.7.6}$$

$X(0,t)$ is one of the crossing-even amplitudes $\tilde{C}^+$, $\tilde{C}^-/\nu$ etc.

The nearby part of the right hand cut can be evaluated from our $\pi\pi N\bar{N}$ partial waves (Sect. 2.4.6), which include the additional experimental information from pion-pion scattering phase shifts. It is also possible to evaluate the nearby part of the left hand cut, which is dominated by Δ-exchange. Re$X(0,t)$ is taken from the fixed-t solution of the KH 78 analysis.

A smooth result for Re$X(0,t)$ near $t=0$ is obtained by using the discrepancy method. The difference between Re$X(0,t)$ and the contributions from the nearby parts of the cuts turns out to be almost a linear function in t. It is fitted by a linear t-dependence. The derivatives of Eq. (4.7.6), which are needed for the higher coefficients x_{0n}, show even a weaker dependence on the more distant parts of the cuts.

Our results for the coefficients of the subthreshold expansion are listed in Table 2.4.7.1 together with predictions for some contributions which will be explained in Sect. 2.5.1.

The numerical values of the lower coefficients are not much changed in comparison with the first version of this table (72 L Höhler-1). The recalculation by Nielsen and Oades (74 L-2) was an improvement, because the authors used phase shifts based on the new accurate low energy data of Bussey et al. (73). Applying other methods, the coefficients were also calculated by Lichard (76 L-1) and some of them were calculated by Lannbein (75 L, 79 L). Our table is based on 78 L Koch and 79 R Höhler-Kaiser-Koch-Pietarinen.

Table 2.4.7.1. Coefficients of the subthreshold expansion Eq. (4.7.2). Δ_p and Δ_{np} are the Δ(1232)-exchange pole and non-pole terms as discussed in Sects 2.5.1.5 and A.8.1.2. The non-pole terms were calculated with the parameter $Z = -1/2$. "ETC" refers to the contribution of the equal-time commutator term, which is expressed by the isovector nucleon form factors in Eq. (A.8.31). Units for the coefficients: powers of μ^{-1} (natural units). The error estimate is based on deviations from the internal consistency. The total uncertainty is somewhat larger. See Table 5.6 in 79 R Nagels for other results.

		t	ν^2	$\nu^2 t$	ν^4	t^2	value at $\nu=0$, $t=2\mu^2$
	x_{00}	x_{01}	x_{10}	x_{11}	x_{20}	x_{02}	
$\bar{D}^+$	-1.46 ± 0.10	1.14 ± 0.02	1.12 ± 0.02	0.17 ± 0.01	0.200 ± 0.005	0.036 ± 0.003	1.02
$\Delta_p + \Delta_{np}$	$23.7 - 25.2$	$0.74 + 0.03$	$0.81 - 0.13$	0.15	0.17		
$\bar{B}^+/\nu$	-3.54 ± 0.06	0.18 ± 0.01	-1.00 ± 0.02	0.08 ± 0.01	-0.31 ± 0.02	-0.01	-3.14
$\Delta_p + \Delta_{np}$	$-3.87 - 0.13$	0.16	-0.76	0.06	-0.15	-0.01	
$\bar{G} \equiv \bar{D}^-/\nu$	1.53 ± 0.02	-0.134 ± 0.005	-0.167 ± 0.005	-0.042 ± 0.003	-0.039 ± 0.002	0.010 ± 0.001	1.28
ETC	1.12	0.054	–	–	–	0.007	1.25
$\Delta_p + \Delta_{np}$	$-0.91 + 1.17$	$-0.185 - 0.002$	-0.18	-0.016	-0.04	–	
$\bar{B}^-$	10.36 ± 0.10	0.24 ± 0.01	1.08 ± 0.05	-0.055 ± 0.005	0.29 ± 0.02	0.025 ± 0.002	10.98
ETC	5.27	0.32	–	–	–	–	
$\Delta_p + \Delta_{np}$	$4.35 + 0.73$	$-0.108 - 0.002$	0.86	-0.035	0.17	–	
$\bar{A}^+$	-1.46 ± 0.10	1.14 ± 0.02	4.66	-0.01	1.20 ± 0.02	0.036 ± 0.003	1.02
$\Delta_p + \Delta_{np}$	$23.7 - 25.2$	$0.74 + 0.03$	4.68	-0.009	0.93	-0.012	
A^-/ν	-8.83 ± 0.10	-0.374 ± 0.02	-1.247 ± 0.05	0.013 ± 0.006	-0.33 ± 0.02	-0.015 ± 0.002	-9.70
ETC	-4.15	-0.266	–	–	–		
Δ	$-5.26 + 0.44$	-0.077	-1.04	0.019	-0.21		

2.4.7.2 The threshold expansion

According to Sect. A.3.5.2, the partial waves and phase shifts have the following threshold behaviour

$$\lim_{q\to 0} \operatorname{Re} f_{\ell\pm}/q^{2\ell} = \lim_{q\to 0} \delta_{\ell\pm}/q^{2\ell+1} = a_{\ell\pm}\,. \tag{4.7.7}$$

$a_{\ell\pm}$ is called the "scattering length". The "effective range parameter" $r_{\ell\pm}$ was originally defined by

$$q^{2\ell+1}\cot\delta_{\ell\pm} = \frac{1}{a_{\ell\pm}} + \tfrac{1}{2} r_{\ell\pm} q^2 + \dots, \tag{4.7.8}$$

but, for our purpose, it is preferable to choose the expansion

$$\operatorname{Re} f_{\ell\pm}(q)/q^{2\ell} = a_{\ell\pm} + b_{\ell\pm} q^2 + \dots, \tag{4.7.9}$$

because it leads to an expansion of the real parts of the invariant amplitudes in powers of q^2 and t. The formulas are given in Eqs. (A.3.66) and (4.7.32).

The question is now, how the scattering lengths and effective ranges can be determined from experimental results. There are 3 possibilities:

i) by an extrapolation of partial waves or phase shifts,
ii) by an extrapolation of invariant amplitudes,
iii) from other experiments than πN scattering (mesic atoms and PANOFSKY ratio).

Extrapolation of partial waves to threshold

In the early fifties, the authors of the first experiments attempted to determine the scattering lengths by fitting their results to effective range formulas. A somewhat different version, Eq. (4.2.6), was derived by CHEW and LOW (56 L) from their static model for the resonant partial wave P 33. It was used by many authors in subsequent investigations. The most recent determination of scattering lengths from effective-range type formulas without further theoretical constraints was made by ZIDELL, ARNDT, and ROPER (80 L-2; Sect. 2.1.10).

A weak point of this method is, of course, that one does not know the energy range, in which two or three terms of the expansion give a reasonable approximation. This question can be answered within the framework of the dispersion approach introduced by CHEW, GOLDBERGER, LOW, and NAMBU (57 L). However, in their evaluation the authors used approximations which led to larger errors than expected. An exact projection of the S-wave from fixed-t dispersion relations shows that the higher terms in the expansion of $\operatorname{Re} f_{0+}^-$ are important already at kinetic energies of the order of 10 MeV [HÖHLER and DIETZ (60 L), see also Fig. 1 in 71 L ACHUTHAN and Fig. 2.2.22].

It is remarkable that, accidentally, $\operatorname{Re} f_{0+}^-$ remains approximately constant between threshold and about 700 MeV/c, i.e. empirically a scattering length approximation seems to be valid in a large energy range, but one obtains a wrong value for the effective range b_{0+}^- (Fig. 2.2.22).

The approximate constancy of $\operatorname{Re} f_{0+}^-$ is due to a compensation between the nucleon pole term part $f_{0+,N}^-$ and the rest $\operatorname{Re} \tilde{f}_{0+}^- \equiv \operatorname{Re} f_{0+}^- - f_{0+,N}^-$, which comes mainly from the P 33 contribution (Δ-exchange). The behaviour of $\operatorname{Re} \tilde{f}_{0+}^-$ is related to the approximate crossing antisymmetry of this partial wave, which was noticed by CINI, GATTO, GOLDWASSER, and RUDERMAN (58 L). (See Fig. 1 in 69 L BAACKE-1.)

The extension of this method to the other partial waves is discussed in the reviews 70 R BAACKE and 71 R STEINER (see Sect. A.7.2). A new evaluation will be presented soon by KOCH (Karlsruhe).

The example of $f_{0+}^-(q)$ shows that accurate data at very low energies would be necessary for a "direct" determination of the scattering lengths. It is doubtful whether these can be obtained, because the importance of the Coulomb effects becomes larger as the energy decreases and, due to decay and absorption, it is difficult to perform accurate measurements with very slow pions.

We conclude that reliable results for scattering lengths can be obtained only if the extrapolation to threshold is restricted by theoretical constraints. As an alternative to the above "partial wave relation approach" (Sect. A.7.2) one can use partial wave dispersion relations (Sect. A.7.1).

This method is described in the reviews 67 R HAMILTON, 67 R DONNACHIE, 71 R HAMILTON, and in 71 L HAMILTON-1 and 2. A new evaluation of the partial wave dispersion relations, using KH 78 and 80 partial waves, has been performed by HUTT and KOCH (80 L HUTT, 81 L HUTT). In this approach, it is necessary to assume that the contributions of the distant left hand cuts are slowly variable in the physical region, and can therefore be described by a simple parametrization.

An assumption of this type is not needed in the partial wave relation method which is, however, limited to momenta below 520 MeV/c.

The investigations of partial wave dispersion relations are based on the singularities derived from the MANDELSTAM representation. In recent papers, it was suggested that quark models can lead to an additional logarithmic singularity at $t=0$, which belongs to a long range VAN DER WAALS-type force between hadrons.

SAWADA (81 L) claimed to have found evidence for this effect in the P 33 partial wave at low energies. However, none of the above mentioned authors noticed a significant deviation from consistency with the partial wave dispersion relation. In particular, HUTT and KOCH found no discrepancy although they used the same partial wave solution as SAWADA (81 L HUTT).

If further progress in QCD calculations should lead to a well-founded prediction for a new long-range force, one would have to study the possibilities to separate its effect from that of the electromagnetic interaction. It would be necessary to perform a new experiment, because the present information on elastic scattering at small t is poor.

Extrapolation of invariant amplitudes to threshold

First we consider the application of *fixed-t dispersion relations*. Equation (A.3.66) shows that the scattering lengths and effective ranges can be determined, if one knows the coefficients of expansions of $\mathrm{Re}\,C$ and $\mathrm{Re}\,B$ in powers of q^2 and t at threshold. These coefficients can be determined from forward dispersion relations as described in 63 R HAMILTON-WOOLCOCK and, using more recent data, in 72 L HÖHLER-1, 74 L WOOLCOCK, 78 L KOCH, 79 R HÖHLER-KAISER-KOCH-PIETARINEN. One possibility is to use subtracted dispersion relations for all forward amplitudes C and B and for the derivatives $\partial/\partial t$ at $t=0$. Furthermore, one needs the derivatives $\partial/\partial q^2$ of the forward dispersion relations at $q^2=0$.

These relations are evaluated by inserting the real and imaginary parts of the invariant amplitudes as calculated from a partial wave solution and from a suitable high energy ansatz. In the case of $\mathrm{Im}\,C$ at $t=0$, the high energy behaviour is given by the total cross section data. It is also known for $(\partial/\partial t)\mathrm{Im}\,C^+$ from the analysis of Coulomb interference experiments (Sects. A.6.5 and 2.6.2).

In general, errors of the high energy ansatz are not serious for the present purpose, since the subtraction constants can be determined at all energies where a partial wave solution is available. A wrong high energy input leads to a characteristic energy dependence of the results for the subtraction constants and can therefore be corrected.

Of course, one can use expansions in terms of analytic functions instead of dispersion relations. One should note that more reliable results can be expected if the expansion is fitted not only to the amplitudes calculated from phase shifts [as it was done by LANGBEIN (79 L) and by LICHARD (76 L-2)] but in addition also to the data (76 L PIETARINEN, 79 R HÖHLER-KAISER-KOCH-PIETARINEN).

Fixed-t dispersion relations are suitable for the determination of S- and P-wave scattering lengths, although the compatibility of the dispersion relation for $\partial C/\partial t$ at $t=0$ with the partial wave solutions KH 78 and CMU-LBL 79 is not yet satisfactory (Fig. 4.7.1). The reason is that the derivative

$$\frac{\partial C}{\partial t}=\frac{\pi W}{mq^3}\sum_{\ell}\{(\ell+1)^2(\ell+1-E/m)T_{\ell+}+\ell^2(\ell+E/m)T_{\ell-}\}+\frac{1}{4m^2}C \tag{4.7.10}$$

is very sensitive to the contribution of higher partial waves, which are enhanced by a factor $\approx\ell^3$, and these are not yet well determined. (Recently R. KOCH obtained encouraging results. See below.)

The situation becomes even worse, if one needs the second derivative for the coefficient of t^2 in Eq. (A.3.66), i.e. fixed-t dispersion relations are not well suited for the determination of the D-wave and higher scattering lengths.

The reason for this problem is well-known from the study of partial wave dispersion relations (67 R HAMILTON). The front of the circle in Fig. A.7.1, which belongs to t-channel exchanges, gives relatively increasing contributions as ℓ increases. Using a more recent input, the relation between the πN D-waves near threshold and the input for $\mathrm{Im}\,f_+^0$ was studied by NIELSEN and OADES (74 L-1).

In his new determination of scattering lengths, KOCH (78 L) used a *fixed-s dispersion relation* at $s=(m+\mu)^2$ (see Sect. A.6.6), which was already studied by LYTH (65 R) with the input available at that time. For πN D-waves and higher waves, one needs subtracted versions, in which only the nearby parts of the t-channel cut give appreciable contributions. The discontinuity in this region is fairly well known (Sect. 2.4.6). Another important contribution comes from the u-channel nucleon pole term.

An estimate for the behaviour of the higher partial waves between threshold and the resonance region can be derived by combining the scattering length obtained from the fixed-s dispersion relation with an evaluation of the partial wave dispersion relation [KOCH and OTTO (82 L-2)].

Of course, one can use other one-dimensional dispersion relations, whose path goes through threshold. If the evaluation can be made with a comparable reliability, the small discrepancies between the results for the scattering lengths give a realistic estimate of the errors. For instance "interior dispersion relations" (Sect. A.6.7) are a good choice.

Up to now, only the S-wave scattering lengths have been determined in this way by Moir, Jacob, and Hite (76 L). It would be of interest to improve this work, taking the new phase shifts and including the known information on the left hand cut. An evaluation of this type was made by Borie and Kaiser (77 L), but these authors did not study the consequences for the scattering lengths. Results for combinations of S- and P-wave scattering lengths are quoted in 76 L Kaiser. — J. Stahov (Karlsruhe) is studying the method of Sect. A.7.2.2, which was proposed by Hite and Steiner (73 L-1).

Determination of πN scattering lengths from pionic atoms

Investigations of X-ray transitions in pionic atoms lead to result for level shifts, from which pion-nucleus scattering lengths can be derived. These scattering lengths can be analyzed in terms of pion-nucleon scattering lengths.

In particular, the π^-d scattering length can be related to the average of the π^+p and π^-p scattering lengths (assuming charge-symmetry), i.e. to the isospin even S-wave scattering length a^+. The result is compatible with that derived from pion-nucleon scattering, but the experimental error is still fairly large. See Fig. 7 in the review 77 R Tauscher, which summarizes the experimental and theoretical situation. The more recent review 80 R Gensini treats mainly the predictions for mesic atoms, which can be derived from current algebra and PCAC. The investigation by Rasche and Woolcock (82 L) deals with the connection between the complex scattering lengths and the level shifts and widths of hadronic atoms.

Determination of the scattering length $a_1 - a_3$ from the Panofsky ratio

The assumption of time-reversal invariance for the electromagnetic interaction leads to the relation

$$\sigma(\pi^- p \to \gamma n) = 2(q_\gamma^2/q^2)\,\sigma(\gamma n \to \pi^- p), \tag{4.7.11}$$

where the total c.m. energy is the same for both reactions and q and q_γ are the c.m. momenta of π^- and γ.

We introduce two ratios of measurable quantities

$$P = \frac{\sigma(\pi^- p \to \pi^0 n)}{\sigma(\pi^- p \to \gamma n)}; \qquad R = \frac{\sigma(\gamma n \to \pi^- p)}{\sigma(\gamma p \to \pi^+ n)}. \tag{4.7.12}$$

The Panofsky ratio P is defined for stopped π^- and the ratio R for the limit of pion production at threshold. Equation (4.7.11) can now be rewritten as follows (see for instance 64 R Kallen),

$$\sigma(\pi^- p \to \pi^0 n) = 2P(q_\gamma^2/q^2)\,\sigma(\gamma n \to \pi^- p) = 2PR(q_\gamma^2/q^2)\,\sigma(\gamma p \to \pi^+ n). \tag{4.7.13}$$

The charge-exchange cross section is expressed in terms of the S-wave scattering length, taking into account the $\pi^\pm - \pi^0$ and n−p mass differences by introducing the c.m. momentum q_0 of the π^0 (for a stopped π^-, the outgoing π^0 has the velocity $0.21c$). Furthermore, we express the pion photoproduction cross section by the S-wave multipole E_{0+}

$$\sigma(\pi^- p \to \pi^0 n) = \frac{8\pi}{9}\frac{q_0}{q}(a_1 - a_3)^2; \qquad \sigma(\gamma p \to \pi^+ n) = 4\pi \frac{q}{q_\gamma}|E_{0+}^{\pi^+}|^2. \tag{4.7.14}$$

We substitute these expressions in Eq. (4.7.13) and find

$$(a_1 - a_3)^2 = (9q_\gamma/q_0)P|E_{0+}^{\pi^-}|^2 = (9q_\gamma/q_0)PR|E_{0+}^{\pi^+}|^2. \tag{4.7.15}$$

Inserting the following values (77 L Spuller, 75 L Adamovich, 79 R Nagels, Table 4.8)

$$P = 1.543 \pm 0.008; \quad R = 1.34 \pm 0.02; \quad E_{0+}^{\pi^+} = (2.83 \pm 0.05)10^{-2}\mu^{-1}; \quad 9q_\gamma/q_0 = 41.3, \tag{4.7.16}$$

one finds

$$a_1 - a_3 = (0.263 \pm 0.005)\mu^{-1}. \tag{4.7.17}$$

A careful investigation of low energy pion reactions, which includes the Coulomb interaction is due to Rasche and Woolcock (76 L-1, 76 L-2, 77 L-1, 77 L-2).

Scattering lengths and S-wave effective ranges

Table 2.4.7.2 gives the scattering lengths and the S-wave effective ranges as determined from the most recent πN low energy partial wave analysis, which takes into account the new data and analyticity constraints [KOCH and PIETARINEN (80 L)]. The results are compatible with the partial wave dispersion relations as evaluated by HUTT and KOCH (80 L) and also with the fixed-s dispersion relation at $s=(m+\mu)^2$ (78 L KOCH). Furthermore, KOCH has checked the compatibility with the partial wave relation in the case of the isospin odd S-wave (Sect. A.7.2). Figure 4.7.2 shows a test with forward dispersion relations.

Another recent determination of the scattering lengths has been performed by ZIDELL, ARNDT, and ROPER (80 L-2), who used an empirical parametrization, ignoring the theoretical information on the nearby left hand cut singularities. It is difficult to see how a convincing determination of scattering lengths can be performed in this way, because the data can be fitted by a variety of parametrizations which lead to different sets of scattering lengths (cf. Figs. 2.2.22···2.2.33 and Sect. 2.1.10).

Earlier determinations of scattering lengths are listed in the compilation 79 R NAGELS and in 80 L ZIDELL-2 (71 R HAMILTON-2, 72 L NIELSEN-1, 72 L SAMARANAYAKE-1 and 2, 73 L BUGG, 73 L CARTER-1, 74 L WOOLCOCK, 76 L LICHARD-2, 75 L PRESNAJDER, 76 L MOIR, 78 L LANGBEIN-1, 77 L BRUNET, 78 L ROWE (cf. Sect. 2.1.10), 79 R HÖHLER-KAISER-KOCH-PIETARINEN). The work of BRUNET is based on a fit to PADÉ approximants of a Lagrangian model. It is a test of this model rather than a new determination of the scattering lengths.

Table 2.4.7.2. Scattering lengths and S-wave effective ranges. Units for the scattering lengths: $\mu^{-2\ell-1}$ and for the S-wave effective ranges: μ^{-3}. a_N is the contribution of the projection of the nucleon pole term of the fixed-t dispersion relations (80 L KOCH-1). The errors refer only to that part which can be estimated from deviations from the internal consistency of the method.

$\ell\pm$	$I=1/2$ 10^3a	10^3a_N	$I=3/2$ 10^3a	10^3a_N	$3\cdot10^3a^+$	$3\cdot10^3a_N^+$	$3\cdot10^3a^-$	$3\cdot10^3a_N^-$
S1 0+	173 ±3	−1583	−101 ±4	−1998	−29 ±5	−5579	274 ±5	415
P1 1−	−81 ±2	−180	−45 ±2	−53	−170 ±4	−286	−35 ±3	−127
P3 1+	−30 ±2	−54	214 ±2	107	398 ±4	160	−244 ±3	−160
D3 2−	2.4 ±0.4	−0.84	2.1 ±0.2	1.67	6.6 ±1.0	2.51	0.30±0.45	−2.51
D5 2+	4.6 ±0.3	3.44	−5.0 ±0.2	−6.88	−5.4 ±1.0	−10.3	9.6 ±0.4	10.3
F5 3−	0.35±0.03	0.038	0.05±0.03	−0.076	0.45±0.35	−0.11	0.3 ±0.05	0.11
F7 3+	−0.07±0.02	−0.237	0.67±0.04	0.474	1.27±0.40	0.71	−0.74±0.05	−0.71
	10^3b	10^3b_N	10^3b	10^3b_N	$3\cdot10^3b^+$	$3\cdot10^3b_N^+$	$3\cdot10^3b^-$	$3\cdot10^3b_N^-$
S1 0+	−18±12	235	−58±9	94	−133±20	422	40±18	141

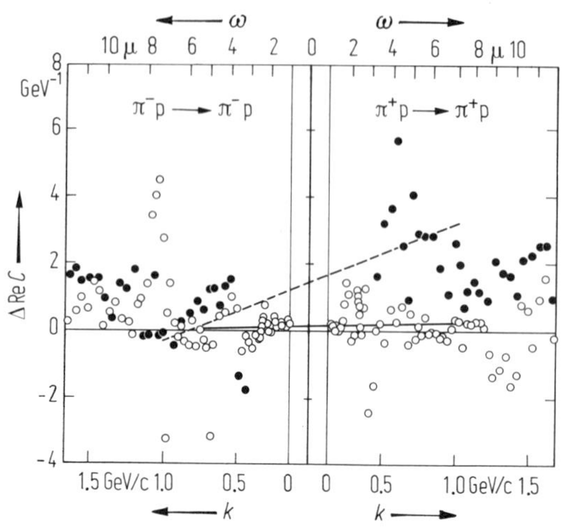

Fig. 4.7.2. Compatibility of partial wave solutions with the evaluation of the forward dispersion relations. The points show $\Delta\,\mathrm{Re}\,C = \mathrm{Re}\,C$ as calculated from partial waves minus $\mathrm{Re}\,C$ as listed in Table 2.2.1.1. On the right hand side $\Delta\,\mathrm{Re}\,C$ for π^+p scattering is plotted vs. ω and on the left hand side $\Delta\,\mathrm{Re}\,C$ for π^-p scattering vs. $-\omega$. CMU-LBL 79: ●; KH 78: ○. — Disregarding possible errors of the interpolation of the total cross sections and of $f^2 = 0.079$, one obtains from the dispersion relations: $\Delta\,\mathrm{Re}\,C_+(\omega) = \Delta C^+(\mu) - (\omega/\mu)\Delta C^-(\mu) = \Delta\mathrm{Re}\,C_-(-\omega)$. Therefore, the points should lie on a straight line, whose parameters give corrections to the values of the scattering lengths belonging to $C^\pm(\omega)$ in Table 2.2.1.1. A more detailed discussion is given in the review 80 R Höhler-Kaiser, Sect. 5, where one also can find figures and a table of the discrepancies between the interpolation of the total cross sections in Table 2.2.1.1 and the values reconstructed from partial wave solutions. Conclusion: the corrections to the scattering lengths are comparable with the errors.

Low energy behaviour of the partial waves

If one wants to judge the accuracy and the reliability of the scattering lengths, one has to study the low energy behaviour of $\mathrm{Re} f_{\ell\pm}/q^{2\ell}$.

S-waves

The *isospin even scattering length* a_{0+}^+ is determined by the subtraction constant of the forward dispersion relation for C^+. Its uncertainty is practically the same as that of c_{00}^+ in Sect. 2.4.7.1. The error given by some earlier authors is much too small. One should note that it is not claimed that the error in Table 2.4.7.2 is the total error. Instead, it shows only that part which follows from deviations from the internal consistency of the method.

A more realistic error can be obtained if one considers the spread of the values obtained for a_{0+}^+, if the extrapolation is carried out along different hyperbolas. Koch (82 L-1) gives the result

$$3a_{0+}^+ = (-0.023 \pm 0.007)\mu^{-1}. \tag{4.7.18}$$

Presumably, other determinations from the KH 80 phase shift solution will lie in this range. Unfortunately, other phase shift analyses which include the new data and take into account the analyticity constraints do not exist at present.

For the connection of the value of a_{0+}^+ with phase shifts, it is important to know the effective range b_{0+}^+, Eq. (4.7.9). Since the effective range occurs in Eq. (A.3.66), it can be calculated in combination with P-wave scattering lengths from the derivative $\partial/\partial q^2$ of the forward dispersion relation for C^+ (subtracted at threshold) and it can be expected that the result has a reasonable accuracy

$$b_{0+}^+ = -2a_{1+}^+ - a_{1-}^+ - a_{0+}^+/2m\mu + \left(1+\frac{\mu}{m}\right)\left[\frac{\mu^2 f^2}{m(\mu^2-\omega_B^2)^2} + \frac{1}{2\pi^2}\int_0^\infty \frac{\mathrm{d}k}{k^2}\{\sigma^+(\omega)-\sigma^+(\mu)\}\right]. \tag{4.7.19}$$

$\sigma^+(\mu)$ is subtracted under the integral, using Eq. (A.6.47), in order to obtain the limit $q^2 \to 0$ in a simple way. The total cross section near threshold should be compatible with Eq. (A.3.67).

Our value for b_{0+}^+ is given in Table 2.4.7.2 and the effective range approximation is plotted in Fig. 2.2.23. Zidell et al. found a steeper slope b_{0+}^+ and a *positive* value for a_{0+}^+ [$q^2 \approx T_\pi 2\mu/(1+2\mu/m)$ for $T_\pi \ll \mu$].

The *isospin odd scattering length* a_{0+}^- can be calculated from the unsubtracted forward dispersion relation for the amplitude C^-, Eq. (A.6.49), i.e. from the "Goldberger-Miyazawa-Oehme sum rule". Nowadays, the total cross section difference is known up to 300 GeV/c. The decrease can be well described by a power law and therefore, the uncertainties from the high energy behaviour are unimportant (Figs. 2.1.5 and 2.1.7). The result $3a_{0+}^- = 0.276\mu^{-1}$ is compatible with the value listed in Table 2.4.7.2.

The expression for the sum rule

$$C^-(\mu)/\mu = 4\pi\frac{m+\mu}{3m}(a_1-a_3) = \frac{8\pi f^2}{\mu^2-\omega_B^2}+4\pi J\,; \qquad J=\frac{1}{2\pi^2}\int_0^\infty \sigma^-(k)\frac{dk}{\omega}, \tag{4.7.20}$$

shows that the accurate total cross section data lead to a strong correlation between the values of f^2 and of a_1-a_3. Our value for J is (80 R Höhler-Kaiser)

$$J=\tilde{G}(\mu)/4\pi = -1.06\,\text{mb} = -0.0530\mu^{-2}. \tag{4.7.21}$$

The most accurate determination of J is to adjust the unknown high energy tail of the total cross section difference in such a way that the charge-exchange forward scattering data at high energies up to 200 GeV/c are well described (Sect. 2.6.5).

The scattering length can also be determined as the subtraction constant of the subtracted dispersion relation for C^-, but this must give of course the same result, if our values for $\text{Re}\,C^-$ and f^2 are inserted. Otherwise, one has to check the consistency with the charge-exchange scattering data at high energy. The amplitude in Carters table (73 L-2) is not compatible with these data.

There was a longstanding discrepancy between the a_{0+}^--values derived from πN scattering and from the Panofsky ratio. A comparison of the present values in Eq. (4.7.17) and in Table 2.4.7.2 shows a reasonable agreement.

Zidell et al. found a much higher value $3a_{0+}^- = (0.3016 \pm 0.0062)\mu^{-1}$. This is related to the fact that their effective range b_{0+}^- has the wrong sign (see Fig. 2.2.22).

The isospin odd S-wave effective range can be calculated in a similar way as b_{0+}^+ (71 L Achuthan)

$$b_{0+}^- = -2a_{1+}^- - a_{1-}^- + \frac{a_{0+}^-}{2\mu^2}\left(1+\frac{\mu}{m}+\frac{\mu^2}{m^2}\right)+\left(1+\frac{\mu}{m}\right)\left[-\frac{2f^2/\mu}{\mu^2-\omega_B{}^2}+\frac{\mu}{2\pi^2}\int_0^\infty\frac{dk}{k^2}\left\{\frac{\sigma^-(\omega)}{\omega}-\frac{\sigma^-(\mu)}{\mu}\right\}\right]. \tag{4.7.22}$$

Figure 2.2.22 shows that the effective range approximation is valid only in a small range above threshold, where data do not yet exist.

The shape of $\text{Re}\,f_{0+}^-(s)$ has already been discussed above. The partial wave method (Sect. A.7.2) offers probably the best possibility for obtaining a smooth result in the range where it is applicable, because the dominant contributions (N and Δ-exchange) are well known and there are no contributions from unknown cuts.

For some discussions, it is useful to have figures for the S-waves in the isospin states $I=1/2$ and 3/2 (Figs. 2.2.24 and 2.2.25). The discrepancy with the result of Zidell et al. occurs mainly in the S 11 wave. One should note that these authors presented a variety of solutions. The curves shown in Figs. 2.2.22···2.2.33 were calculated from the only solution for which the authors listed the parameters (Table VI in 80 L Zidell-1). The scattering lengths belonging to this solution differ in some cases from those listed in Table II, III, and IV of the second paper 80 L Zidell-2. Furthermore, Zidell et al. give results for the “nuclear” phases, whereas our results belong to the “hadronic” phases. The difference is discussed in Sect. 2.1.17.

The connection between low-energy scattering parameters and energy shifts for pionic hydrogen has been discussed in 82 L Rasche. See 77 R Tauscher for results derived from levels of higher pionic atoms.

P-waves

The shape of $\text{Re}\,f_{1\pm}/q^2$ is shown in Figs. 2.2.26···2.2.29. The main discrepancy with the result of Zidell et al. occurs in the P 11 wave, where the authors give the value

$$a_{11} = -(133.4 \pm 3.6)10^{-3}\mu^{-3}. \tag{4.7.23}$$

It lies far below our value $a_{11} = -(81 \pm 2)10^{-3}\mu^{-3}$, which is near to the results of other authors who also used dispersion methods. The reason for the discrepancy is probably that the parametrization of Zidell et al. is not well suited for the description of the shape of this partial wave, which has a large contribution from the nucleon pole term projection. This discrepancy is related to another one noticed by Koch (80 R): if the invariant forward amplitude B is reconstructed from the partial wave solution of Zidell et al., it turns out that it is not compatible with the forward dispersion relation. It is true that one has to employ another phase shift solution above $T_\pi = 350$ MeV, but this cannot explain the observed deviation.

The P-wave phase shifts and scattering lengths of Koch and Pietarinen (80 L) are in reasonable agreement with the results of earlier authors who used dispersion methods (72 L Engels, 72 L Nielsen-1, 73 L Carter-1, 75 L Langbein). The changes caused by the new data are comparable with the errors.

D- and F-waves

Earlier determinations of D- and F-waves led to widely differing results. The scattering lengths of NIELSEN and OADES (72 L-1) agree fairly well with the values in our Table 2.4.7.2, but some of LANGBEINs scattering lengths (75 L) have even a different sign. In their Table IV, ZIDELL et al. (80 L-2) give D-wave scattering lengths which are smaller than ours by one order of magnitude. – See also the discussion of higher partial waves in Sect. 2.1.14 and Figs. 2.2.30···2.2.33.

Conclusions on scattering lengths

The scattering lengths cannot be obtained from experimental data alone. Instead, one has to determine scattering amplitudes in a large energy range, which describe correctly the data and are compatible with the theoretical constraints. These amplitudes are then extrapolated to threshold. For D- and higher partial waves, it is important to include the information on pion-pion scattering phase shifts. This can be done by choosing a method where the t-channel cut gives a substantial contribution.

2.4.7.3 Connection between the two expansions

The isospin odd amplitude

We start with the invariant amplitude $G^-(\omega) \equiv C^-(\omega)/\omega \equiv D^-(\omega)/\omega$ at $t=0$. Both, the S-wave scattering length $\tilde{a}_{0+}^-$ and $d_{00}^- - g^2/2m^2$, are calculated from the same dispersion integral, evaluated at $\omega=\mu$ and $\omega=0$, respectively

$$\mathrm{Re}\,\tilde{G}^-(\omega) = \frac{2}{\pi} \fint_\mu^\infty \sigma^-(\omega') \frac{k' \mathrm{d}\omega'}{\omega'^2 - \omega^2}. \tag{4.7.24}$$

One can obtain accurate values of the coefficients d_{m0}^-, if the term $1/(\omega'^2 - \omega^2)$ is expanded in powers of ω^2.
The difference $\tilde{G}^-(\mu) - \tilde{G}^-(0)$ is well determined by the rapidly converging integral

$$\tilde{G}^-(\mu) - \tilde{G}^-(0) = \mu^2 \frac{2}{\pi} \int_0^\infty \mathrm{d}k\, \sigma^-(\omega)/\omega^3. \tag{4.7.25}$$

We compare the result with the first terms of the subthreshold expansion

$$\tilde{G}^-(\mu) - \tilde{G}^-(0) = -0.202\mu^{-2} = d_{10}^- \mu^2 + d_{20}^- \mu^4 + \ldots = (-0.167 - 0.039 + \ldots)\mu^{-2} \tag{4.7.26}$$

and notice that the truncation gives a reasonable approximation, although $\omega=\mu$ lies on the circle of convergence (see Fig. 4.7.3). This figure shows that the amplitude has a cusp at threshold, which can be exhibited by writing the dispersion relation as follows [cf. Eqs. (A.6.47) and (A.6.50)].

$$\mathrm{Re}\,\tilde{G}^-(\omega) = \tilde{G}^-(\mu) + k^2 \frac{2}{\pi} \fint_0^\infty \frac{\mathrm{d}k'}{k'^2 - k^2} \left\{ \frac{\sigma^-(\omega')}{\omega'} - \frac{\sigma^-(\mu)}{\mu} \right\} - \sigma^-(\mu) \sqrt{\mu^2 - \omega^2}\, \Theta(-k^2)/\mu. \tag{4.7.27}$$

The magnitude of the cusp depends on the value of $\sigma^-(\mu)$, which is 1.65 mb if calculated from Eq. (A.3.67) and our S-wave scattering lengths. See Sects. A.6.4.6 and A.7.3 for a further discussion.
Between the thresholds, $\tilde{G}^-(\omega)$ can be expanded in terms of t-channel partial waves

$$\tilde{G}^-(\omega) = 4\pi \left\{ \frac{3}{m} \mathrm{Re}\,\tilde{f}_+^1(0) + \frac{7}{2} m (5\omega^2 - 3\mu^2) \mathrm{Re}\,\tilde{f}_+^3(0) + \frac{11}{8} m^3 (63\omega^4 - 70\omega^2\mu^2 + 15\mu^4) \mathrm{Re}\,\tilde{f}_+^5(0) + \ldots \right\}$$

$$= \left\{ (-0.600 + 0.132 - 0.012 + \ldots) + (-0.220 + 0.055 + \ldots) \left(\frac{\omega}{\mu}\right)^2 - 0.050 \left(\frac{\omega}{\mu}\right)^4 \right\} \mu^{-2}. \tag{4.7.28}$$

In this formula, the Legendre expansion is used in a case where branch points lie at $\cos\theta_t = \pm 1$. It converges only on the real axis.

Finally, we give the relation between d_{00}^- and the S-wave scattering length $3a_{0+}^- = a_1 - a_3$, using Eq. (4.7.26)

$$G^-(\mu) = \tilde{G}^-(\mu) + \frac{8\pi f^2}{\mu^2 - \omega_B^2} = \frac{4\pi}{3}\frac{m+\mu}{m\mu}(a_1 - a_3); \tag{4.7.29}$$

$$\tilde{G}^-(0) = d_{00}^- - g^2/2m^2 = -0.463\mu^{-2} = 4.81(a_1 - a_3)\mu^{-1} + 0.202\mu^{-2} - 25.27 f^2 \mu^{-2}, \tag{4.7.30}$$

where $\omega_B = -\mu^2/2m$.

The t-dependence of $G^-(\nu, t)$ at $\omega = \mu$ can be expressed by the scattering lengths. In addition to the formulas Eq. (A.3.66) one needs the expansion of

$$G^- \equiv D^-/\nu \equiv C^-/\nu - B^- \frac{t/4m^2}{1 - t/4m^2} \tag{4.7.31}$$

at threshold ($\nu_{th} = \mu + t/4m$) which follows from

$$\begin{aligned} \operatorname{Re} D(\nu_{th}, t)/4\pi = {} & \left(1 + \frac{\mu}{m}\right) a_{0+} + \left\{\left(1 + \frac{3}{2}\frac{\mu}{m}\right) a_{1+} + \frac{1}{2} a_{1-} + \frac{a_{0+}}{8m^2}\right\} t \\ & + \left\{\frac{9}{8}\left(1 + \frac{5}{3}\frac{\mu}{m}\right) a_{2+} + \frac{3}{4} a_{2-} + \frac{3}{16m^2} a_{1+}\right\} t^2 + \dots \,. \end{aligned} \tag{4.7.32}$$

If one wants to express the t-dependence of $\tilde{G}^-(\nu, t)$ in terms of t-channel partial waves, one should note that the expansion Eq. (4.7.28) belongs to C^- (see Table A.3.3). According to Eq. (A.3.53) the expansion of Re $\tilde{G}^-(\nu, t)$ starts with

$$\operatorname{Re} \tilde{G}^-(\nu, t) = 12\pi \operatorname{Re} \tilde{\Gamma}_1(t) + \dots. \tag{4.7.33}$$

The relation between the derivatives $\partial \tilde{G}^-/\partial t$ and $\partial^2 \tilde{G}^-/\partial t^2$ at $t = 0$, taken at $\nu = 0$ and at threshold, is again given by rapidly converging dispersion integrals and the ω-dependence has a cusp at threshold. The variation of $\tilde{G}^-(\nu, t)$, $\tilde{C}^-(\nu, t)$, and $\tilde{B}^-(\nu, t)$ in the low energy region as following from our amplitudes is given in Table 2.4.7.3.

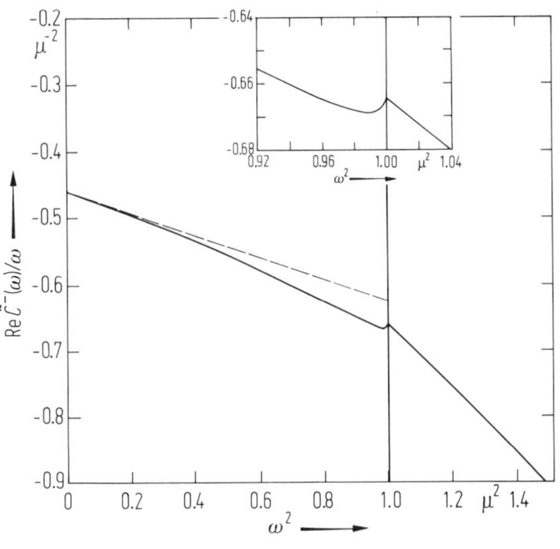

Fig. 4.7.3. The cusp of Re $\tilde{G}^-(\omega) \equiv$ Re $\tilde{C}^-(\omega)/\omega$ at $\omega = \mu$. The tilde indicates that the nucleon pole term, Eq. (A.6.48), has been subtracted. The dashed line shows the sum of the first two terms of the subthreshold expansion, Eq. (4.7.1).

Table 2.4.7.3. The amplitudes $\tilde{G}^-(\nu, t)$, $\tilde{C}^-(\nu, t)$, and $\tilde{B}^-(\nu, t)$ in the subthreshold and threshold regions. At $t = 4\mu^2$ we have added in brackets the values calculated from the $I = J = 1$ ππNN̄ partial waves. The small discrepancies show to what extent our extrapolations based on fixed-ν dispersion relations and t-channel partial wave dispersion relations are consistent with each other. The numerical determination of the amplitudes at the cusp at $t = 4\mu^2$ has some technical difficulties due to the nearby branch point at $t = 3.98\,\mu^2$. The shape of the t-dependence at $\nu = 0$ is shown in Fig. 4.6.2.

t/μ^2	$\tilde{G}^-(\nu,t)[\mu^{-2}]$		$\tilde{C}^-(\nu,t)/\nu[\mu^{-2}]$	$\tilde{B}^-(\nu,t)[\mu^{-2}]$
	$\nu=0$	$\nu=\nu_{th}$	$\nu=0$	$\nu=0$
4	−0.84		−0.62	10.05
	(−0.79)		(−0.58)	(9.92)
3	−0.80	−1.20	−0.64	9.42
2	−0.71	−1.05	−0.61	8.99
1	−0.59	−0.87	−0.55	8.65
0	−0.46	−0.66	−0.46	8.37
−1	−0.31	−0.44	−0.35	8.14

The isospin even amplitude

Because of the subtraction, the treatment of D^+ is slightly different. The quantity $\tilde{D}^+(\mu)-\tilde{D}^+(0)$ is given by the dispersion integral

$$\tilde{D}^+(\mu)-\tilde{D}^+(0)=\mu^2\frac{2}{\pi}\int_0^\infty dk\sigma^+(\omega)/\omega^2. \tag{4.7.34}$$

We compare the result with the first terms of the subthreshold expansion

$$\tilde{D}^+(\mu)-\tilde{D}^+(0)=1.47\mu^{-1}=d^+_{10}\mu^2+d^+_{20}\mu^4+\ldots=(1.12+0.20+\ldots)\mu^{-1}. \tag{4.7.35}$$

The cusp effect is larger than in the isospin odd case, because our S-wave scattering lengths lead to a larger value $\sigma^+(\mu)=4.25$ mb for the total cross section at threshold according to Eq. (A.3.67). A more detailed treatment of the threshold cusp can be found in Sects. A.6.4.6 and A.7.3.

Between the thresholds, one can use an expansion in terms of t-channel partial waves

$$\tilde{D}^+(\nu,t)=\frac{4\pi}{p_-^2}\left\{\mathrm{Re}\,\tilde{f}^0_+(t)+\frac{5}{2}(3m^2\nu^2-p_-^2q_-^2)\,\mathrm{Re}\,\tilde{f}^2_+(t)-\frac{15m}{4\sqrt{6}}\nu^2t\,\mathrm{Re}\,\tilde{f}^2_-(t)+\ldots\right\} \tag{4.7.36}$$

ich gives at $t=0$ from Tables 2.4.7.1 and 2.4.6.1, respectively

$$\tilde{D}^+(\omega)-\frac{g^2}{m}=(-1.46+1.12\omega^2+\ldots)\mu^{-1}=(-0.82-0.46+\ldots)\mu^{-1}+(1.39+\ldots)\omega^2\mu^{-1}+\ldots. \tag{4.7.37}$$

It is seen that the $J=4$ term in Eq. (4.7.36) is not yet negligible.

The amplitude at threshold is given by ($\bar{D}^+=\tilde{D}^+-g^2/m$)

$$D^+(\mu)=4\pi\left(1+\frac{\mu}{m}\right)a^+=-\frac{4\pi f^2}{m(1-\mu^2/4m^2)}+\bar{D}^+(\mu)=(-0.149+0.010)\mu^{-1}=-0.139\mu^{-1}, \tag{4.7.38}$$

and the relation between d^+_{00} and $3a^+=(a_1+2a_3)$ reads

$$d^+_{00}=\bar{D}^+(0)=-1.47\mu^{-1}+4.81(a_1+2a_3)+1.88\mu^{-1}f^2. \tag{4.7.39}$$

It is seen that the uncertainty of f^2 is unimportant in this expression.

The relation between the slopes of the t-dependence at $\nu=0$ and at threshold is analogous to that in the isospin odd case. — One can use Eq. (4.7.32) for $D^+(\nu_{th},t)$.

The variation of $\bar{D}^+(\nu,t)$ in the low energy region is shown in the following table.

Table 2.4.7.4. The amplitude $\bar{D}^+(\nu,t)$ in the subthreshold and threshold regions. The value in brackets was calculated from $\tilde{f}^0_+(4\mu^2)$. See the remarks in the caption of Table 2.4.7.3 and Fig. 4.6.2a.

t/μ^2	$\bar{D}^+(\nu,t)$ $[\mu^{-1}]$ $\nu=0$	$\nu=\nu_{th}$
4	5.92	
	(6.39)	
3	2.57	5.27
2	1.02	3.34
1	−0.28	1.61
0	−1.46	0.01
−1	−2.57	−1.49

We have discussed the energy dependence in the range $\omega=0$ to μ in some detail, because several authors believed that the accuracy of the coefficients of the subthreshold expansion, Eq. (4.7.2) and of $\bar{D}^+(0, 2\mu^2)$ can be improved by taking into account the "experimental" information on the scattering lengths [71 L ALTARELLI-1, 71 L ALTARELLI-2, OLSSON (80 L-1, 80 L-2)].

Up to now, accurate direct determinations of the scattering lengths do not exist. The so-called "experimental" values were determined from evaluations of dispersion relations as described already in the review by HAMILTON and WOOLCOCK (63 R). *The same dispersion relations* can be used for the extrapolation to the threshold region and for the determination of the coefficients of the subthreshold expansion, Eq. (4.7.2). Therefore, the on-shell part of the method of ALTARELLI et al. for the determination of $\bar{D}^+(0.2\mu^2)$ (i.e. of the sigma term) is an unnecessary detour, which introduces an additional uncertainty.

Recently, an improved version of this method was developed by OLSSON (79 L) and OLSSON and OSYPOWSKY (80 L). It leads to results equivalent to those of our earlier method (72 L HÖHLER-1), but not to the work based on the fixed-ν dispersion relation, since the effect of the curvature of $\bar{D}^+(0, t)$ is neglected.

The authors were motivated by the hope that it will be possible in the near future to determine scattering lengths and S-wave effective ranges directly from experiments at meson factories. Unfortunately, it seems that one will always need an extrapolation from the region beyond the validity of the effective range approximation (Fig. 2.2.22). The new data will be helpful for an improvement of the phase shift solution, and fixed-t or other dispersion relations will be used for the extrapolation to threshold. Then, of course, it is simpler and more transparent to apply the same dispersion relations for the extrapolation to the region, where one wants to test predictions of Current Algebra and PCAC.

2.5 Applications of πN and ππNN̄ amplitudes

2.5.1 Low-energy models for the pion-nucleon system

2.5.1.1 Introduction

In this section we shall briefly discuss various models for the pion-nucleon system at low energies. The basic idea is in most cases to represent the scattering amplitudes as sums of contributions from particle and resonance exchanges in the s, t, and u-channels. However, one cannot simply add the expressions obtained from second order perturbation theory, because this would lead to "double-counting" (see 70 L OEHME-2).

We begin with a discussion of theoretical predictions for the amplitudes D^+ and D^-, which were derived from current algebra and PCAC (Sects. 2.5.1.2 and 2.5.1.3). The next subsections are devoted to the GOLDBERGER-TREIMAN-relation and to the current algebra model, which is expected to be valid for the on-shell amplitude in the whole low energy region. Dispersion models (Sect. 2.5.1.6) are based on decompositions of fixed-t or partial wave dispersion relations, which are similar to the above representation in terms of particle and resonance exchanges. The final Sect. 2.1.5.7 contains a discussion of various models, which have been developed for applications in nuclear physics, where off-shell amplitudes are needed.

2.5.1.2 The ADLER-WEISBERGER relation

Using the PCAC relation Eq. (A.8.30) and current algebra, ADLER (65 L-2) derived the following result

$$\tilde{G}^-(0,0;0,0)=f_\pi^{-2}(1-g_A^2), \tag{5.1.1}$$

where the notation is defined by *)

$$G^-(\nu, \nu_B; \mathbf{q}^2, \mathbf{q}'^2)\equiv D^-/\nu; \qquad \nu_B=(t-\mathbf{q}^2-\mathbf{q}'^2)/4m. \tag{5.1.2}$$

The tilde indicates that the nucleon pole term has been subtracted. $\mathbf{q}^2$ and $\mathbf{q}'^2$ are the squared masses of the incoming and outgoing pions.

*) We are using ADLERs notation. The amplitude G should not be confused with the spin no-flip amplitude of Eq. (A.2.2). Other versions of Eq. (5.1.1) follow, if the GOLDBERGER-TREIMAN-relation Eq. (A.8.32) is substituted (Sect. 2.5.1.4).

If one assumes that G^- satisfies an unsubtracted dispersion relation in the variable ν, the ADLER-WEISBERGER relation Eq. (5.1.1) can be expressed in terms of a dispersion integral (65 L ADLER-2, 66 L WEISBERGER).

With the present values $f_\pi = (0.9448 \pm 0.0009)\mu$ (72 L SIRLIN) and $g_A = -1.254 \pm 0.007$ (80 R BRICMAN, p. 96), one obtains

$$\tilde{G}^-(0,0;0,0) = -(0.64 \pm 0.02)\mu^{-2}. \qquad (5.1.3)$$

(ADLER used $g_A = 1.18 \pm 0.02$). The problem is now to compare this prediction with an extrapolation of "experimental" amplitudes as determined from data. [WILKINSON (81 R) gives $g_A = 1.249 \pm 0.006$, BRYNE (82 R) $g_A = 1.257 \pm 0.009$ and ROOS (82 R) $g_A = 1.2546 \pm 0.0063$.]

FUBINI and FURLAN (68 L) chose a continuation along a parabola in the (ν, $\mathbf{q}^2 = \mathbf{q}'^2$)-plane, which connects the amplitude $\tilde{G}^-(0,0;0,0)$ with the physical amplitude at threshold. Evaluating part of the corrections, they obtained for the difference of the $I = 1/2$ and $3/2$ S-wave scattering lengths

$$a_1 - a_3 = \frac{3}{1+\mu/m}\left[L + \frac{2\mu}{3\pi}(a_1^2 - a_3^2)\right] = (0.233 + 0.010)\mu^{-1} = 0.243\mu^{-1}, \qquad (5.1.4)$$

where

$$L = \frac{\mu}{4\pi f_\pi^2} = 0.089\mu^{-1}. \qquad (5.1.5)$$

The second term is an estimate of a small unitarity correction, which is related to the threshold cusp in Eq. (4.7.27). It can be expressed by the total cross section at threshold, Eq. (A.3.67)

$$\sigma^-(\mu)/4\pi = \tfrac{1}{3}(a_1^2 - a_3^2).$$

The first term was derived in a different way by TOMOZAWA (66 L) and by WEINBERG (66 L). The discrepancy with the "experimental value" $a_1 - a_3 = (0.274 \pm 0.005)\mu^{-1}$ in Table 2.4.7.2 is not serious, since the validity of WEINBERGs approximation $\mu/m \ll 1$ has not yet been investigated in detail. In dispersion theory, the recoil effects in the S-wave amplitude f_{0+}^- at threshold turned out to be larger than expected (60 L HÖHLER).

The following remark might be of interest. We subtract from the first term in Eq. (5.1.4) the nucleon pole term contribution and compare the result with the ADLER-WEISBERGER relation Eq. (5.1.1), substituting the GOLDBERGER-TREIMAN relation Eq. (A.8.32)

$$\begin{aligned} \tilde{G}^-(\mu, -\mu^2/2m; \mu^2, \mu^2)_{TW} &= f_\pi^{-2} - \frac{g^2}{2m^2}\frac{1}{1-\mu^2/4m^2} = (1.12 - 2.00)\mu^{-2} = -0.88\mu^{-2}; \\ \tilde{G}^-(0,0;0,0)_{AW} &= f_\pi^{-2} - \frac{g^2}{2m^2}K^2(0) = (1.12 - 1.76)\mu^{-2} = -0.64\mu^{-2}. \end{aligned} \qquad (5.1.6)$$

It is an empirical fact that, according to Table 2.4.7.3, the experimental value $\tilde{G}^- = -0.66\mu^{-2}$ for the amplitude in the first line agrees within the errors with the prediction for the unphysical kinematical point in the second line. The discrepancy of the experimental value with the TOMOZAWA-WEINBERG prediction is of the same magnitude as the effect of the factor $K^2(0)$, which is related to the GOLDBERGER-TREIMAN discrepancy (Sect. 2.5.1.4).

Since TOMOZAWA and WEINBERG obtained $a_1 + 2a_3 = 0$, the discrepancies between their predictions for a_1 and a_3

$$a_1 = 2L(1+\mu/m)^{-1} = 0.155\mu^{-1}; \quad a_3 = -L(1+\mu/m)^{-1} = -0.078\mu^{-1} \qquad (5.1.7)$$

and the experimental values (Table 2.4.7.2) has essentially the same origin.

An on-shell version of the ADLER-WEISBERGER theorem was derived by BROWN, PARDEE, and PECCEI (71 L)

$$\tilde{G}^-(\nu, \nu_B; \mu^2, \mu^2) = 2f_\pi^{-2}F_{1\nu}(t) - \frac{g^2}{2m^2} + f_\pi^{-2}\left\{\left(\frac{\mu}{m}\right)^2 d + \frac{2\nu_B}{m}e + \left(\frac{\nu}{m}\right)^2 f\right\}, \qquad (5.1.8)$$

where $\nu_B = (t - 2\mu^2)/4m$. $F_{1\nu}$ is the isovector nucleon form factor (Table A.11.1).

The first term comes from the equal-time commutator and belongs to the first term in Eq. (5.1.1). The second term follows from the second term in Eq. (5.1.1) and the GOLDBERGER-TREIMAN relation. The last term is the "remainder", d, e, and f are dimensionless nonsingular functions of ν and ν_B, up to logarithms.

$F_{1v}(t)$ is well-known from the analysis of the electron-nucleon scattering data (Sect. 2.5.2). Its expansion at small t reads

$$F_{1v}(t) = 0.5 + 0.024t + 0.003t^2 + \ldots; \quad t \text{ in } \mu^2. \tag{5.1.9}$$

In order to determine the coefficient d, we calculate

$$\bar{G}^-(0,0;\mu^2,\mu^2) \equiv \tilde{G}^-(0,0;\mu^2,\mu^2) + \frac{g^2}{2m^2} = d_{00}^- + 2d_{01}^- + 4d_{02}^- = 1.30\mu^{-2};$$
$$\tilde{G}^-(0,0;\mu^2,\mu^2) = -0.69\mu^{-2}, \tag{5.1.10}$$

from Table 2.4.7.1 and find $d \approx 2.0$. This result differs considerably from the value given in 71 L Brown and 75 R Pagels, because these authors used numerical values calculated in 65 L Adler-2, which were based on the data available at that time.

The difference in comparison with Table 10.2 in 79 R Höhler-Kaiser-Koch-Pietarinen is due to a printing error: this table gives $\tilde{C}^-/\nu$ instead of $\tilde{D}^-/\nu$. A comparison with Table 2.4.7.3 shows the truncation error in Eq. (5.1.10).

In a similar way, our tables can be used in order to determine the other parameters of the remainder: $e \approx -15$, $f \approx -10$. Theoretical estimates will be discussed below.

Since the contribution of the higher $\pi\pi N\bar{N}$-partial waves ($J \geqq 3$) is small in the range of interest ($0 \leqq t < 4\mu^2$, see Fig. 4.6.2), the leading terms of Eq. (5.1.8) can be written [cf. Eq. (A.3.53)]

$$\bar{G}^- \equiv \tilde{G}^- + \frac{g^2}{2m^2} = \frac{2F_{1v}}{f_\pi^2}; \quad \bar{\Gamma}_1(t) = \tilde{\Gamma}_1 + \frac{g^2}{24\pi m^2} = \frac{1}{6\pi f_\pi^2} F_{1v}(t). \tag{5.1.11}$$

We compare the imaginary parts of both sides on the right hand cut (ignoring the correction terms)

$$\operatorname{Im}\bar{\Gamma}_1(t) = \operatorname{Im}\Gamma_1 = \frac{q_t^3}{\sqrt{t}} \Gamma_1 h_1^{1*} = \frac{1}{6\pi f_\pi^2} \frac{q_t^3}{\sqrt{t}} F_\pi \Gamma_1^*, \tag{5.1.12}$$

substituting the two-particle unitarity relations for the nucleon form factor, Eq. (A.11.13) and for the $\pi\pi N\bar{N}$ partial wave, Eq. (A.5.23). Γ_1 is defined in Eq. (A.3.52) and h_1^1 in Eq. (A.7.54). Next $h_1^1 = N/D = NF_\pi$ and $\Gamma_1 = J_1/D = J_1 F_\pi$ are substituted and we obtain

$$N = (6\pi f_\pi^2)^{-1}. \tag{5.1.13}$$

This result leads to the Weinberg value for the P-wave $\pi\pi$-scattering length (Table A.7.3), if the difference between $F_\pi(4\mu^2) = 1.16$ and $F_\pi(0) = 1$ is neglected [cf. Eq. (A.7.58)]

$$a_1^1 = h_1^1(4\mu^2)/2\mu = NF_\pi(4\mu^2)/2\mu \approx N/2\mu = \frac{\mu^{-1}}{12\pi f_\pi^2} = 0.030\mu^{-3}. \tag{5.1.14}$$

If the factor 1.16 is taken into account, the result is near the present experimental value of a_1^1, which follows from Roy equation fits constrained by the K_{e4}-data (79 R Nagels, Sect. 9.1): $a_1^1 = (0.038 \pm 0.002)\mu^{-3}$.

We conclude that the main term in Eq. (5.1.8) is compatible with the statement that the *left hand cut* contribution to $\operatorname{Re}\bar{\Gamma}_1(t)$ vanishes near $t = 2\mu^2$, i.e. the right hand cut contribution (mainly ϱ-exchange) is dominant.

This statement is related to Sakurais ϱ-dominance (69 R Sakurai). It is interesting to note that the form factor $F_{1v}(t)$ does *not* have the pronounced dipole-like structure of the spectral function, which is found in the other form factors (Sect. 2.5.2).

Sakurais ϱ-dominance has the character of a *sum rule*, but not of a model for the $J = I = 1$ $\pi\pi N\bar{N}$-partial waves, since the left hand cut contribution (mainly Δ-exchange) is in general equally important and furthermore, there is a strongly variable nucleon-exchange contribution (Fig. 4.6.2). A more detailed discussion can be found in the papers Höhler (71 L-2) and (76 R).

2.5.1.3 Adlers consistency condition and the pion-nucleon sigma term

Using PCAC, Adler derived the following "consistency condition" (65 L Adler-1)

$$A^+(0,0;0,\mu^2) = A^+(0,0;\mu^2,0) = \frac{g^2}{m} K(0). \tag{5.1.15}$$

We use the notation $A^+(\nu, \nu_B; \mathbf{q}^2, \mathbf{q}'^2)$ for the amplitude and $K(\mathbf{q}^2)$ for the πNN form factor, which is normalized to $K(\mu^2)=1$.

If current algebra is taken into account, the value of $A^+(0,0;0,0)$ can be expressed by an integral over the nucleon expectation value of the equal-time commutator of the axial current with its divergence, which is called the *sigma term* (66 L WEINBERG)

$$A^+(0,0;0,0) - \frac{g^2}{m} K^2(0) = -\frac{2}{f_\pi^2}\sigma_{NN}(0). \qquad (5.1.16)$$

There are now three problems

i) to derive a theoretical prediction for σ_{NN} $(t=0)$ from QCD,

ii) to find a relation between $A^+(0,0;0,0)$ and the *on-shell* pion-nucleon scattering amplitude. It turns out that an "experimental value" for σ_{NN} can be determined, if one knows the on-shell amplitude $A^+(\nu,t)$ or $D^+(\nu,t)$ in the unphysical range between the s- and u-channel thresholds.

iii) to extrapolate the invariant amplitudes from the physical region, where they can be calculated from partial waves, to the unphysical region mentioned in ii). This problem has already been treated in Sects. 2.4.5 and 2.4.7.

Theoretical predictions for the sigma term

The sigma commutator is defined as the equal-time commutator of the axial-vector current with its divergence, and the sigma term σ_{NN} as the expectation value of this commutator between nucleon states. In lowest order in the symmetry breaking part of the Hamiltonian, H', the expression for the sigma term reads (66 L WEINBERG, 74 R REYA, 75 R PAGELS)

$$\sigma_{NN}(t) = \tfrac{1}{3}\sum_\alpha \langle \mathbf{p}'|[{}^5Q^\alpha, [{}^5Q^\alpha, H'(0)]]|\mathbf{p}\rangle, \qquad (5.1.17)$$

where $t=(\mathbf{p}'-\mathbf{p})^2$ and ${}^5Q^\alpha$ are the weak axial charges of isospin index α.

In QCD, the chiral-symmetry breaking part H' is entirely due to the current quark mass terms

$$H' = m_u\bar{u}u + m_d\bar{d}d + m_s\bar{s}s. \qquad (5.1.18)$$

One obtains in a good approximation (see for instance 81 R DOMINGUEZ-LANGACKER)

$$\sigma_{NN}(0) = \tfrac{1}{2}(m_u + m_d)\langle \mathbf{p}|\bar{u}u + \bar{d}d|\mathbf{p}\rangle. \qquad (5.1.19)$$

In order to evaluate this matrix element, one can introduce the SU(3) octet operator $\bar{u}u+\bar{d}d+2\bar{s}s$, whose matrix elements can be expressed in terms of the baryon masses. If one accepts the validity of the "quark line" or "OZI" rule (63 L OKUBO, 77 L OKUBO, 64 L ZWEIG, 70 R MANDULA-WEYERS-ZWEIG, 66 L IIZUKA, 78 R OKUBO) in the present case, the matrix element of $\bar{s}s$ between nucleon states vanishes and one can predict the sigma term, if one knows the quark mass ratio $r_+ = (m_u+m_d)/2m_s$, which can be determined for instance from the pseudoscalar meson mass spectrum ($r_+ = 0.034 \pm 0.008$). The prediction from the simplest model is 23 ± 5 MeV. Including higher order terms in the quark mass expansion, GASSER (82 R) finds

$$\sigma_{NN}(0) = 35 \pm 5\ \text{MeV}. \qquad (5.1.20)$$

The problems connected with the application of this method are treated in 68 L GELL-MANN, 68 L GLASHOW, 69 L DASHEN, 72 R RENNER, 73 R KLEINERT, 74 R REYA, 75 L GASSER, 76 L CHENG, 80 L JAFFE, 81 R DOMINGUEZ-LANGACKER.

A careful investigation of the possibilities for corrections to the predicted value of the sigma term has recently been performed by GASSER (81 L). He concludes that an "experimental" value of about 60 MeV leads to a serious difficulty. This value is not compatible with the structure of the meson and baryon spectrum, unless the nucleon mass is smaller than 600 MeV in the chiral limit $m_u = m_d = m_s = 0$. See also the review 82 R GASSER-LEUTWYLER.

Other authors developed a formalism in which $\sigma_{NN} = 60$ MeV is acceptable. SCADRON et al. (77 L GUNION, 79 L FUCHS, 81 R SCADRON, 82 L FUCHS-2) proposed an alteration of the GELL-MANN-OAKES-RENNER scheme ("neutral PCAC") and found encouraging results for a variety of reactions. However, their value for the ratio $2m_s/(m_u+m_d)$ is 5 instead of the usual value 26 (81 L GASSER). Arguments in favor and against neutral PCAC were recently exchanged between SCADRON (81 L) and GENSINI (81 L-2). The possibility that the ratio is somewhat smaller than 26 was also discussed by FRITZSCH (77 L).

JAFFE (80 L) calculated the sigma term in a hybrid chiral bag model. He came to the conclusion that there is no discrepancy with the experimental value.

A different version of the bag model, the "cloudy bag" (Sect. 2.5.1.7) was used in a recent investigation of the sigma term by BANERJEE and MILLER (81 L). The authors derived a modified relation between the soft pion limit of the isospin even πN scattering amplitude A^+ and the expectation value of the sigma commutator. They concluded that the bag radius must be smaller than 0.7 fm, if the "experimental" sigma term is larger than 50 MeV.

Extrapolation in pion mass

Most of the authors used the prescription of CHENG and DASHEN (71 L, see also 70 L OSYPOWSKI) which is a linear approximation in $\mathbf{q}^2$ and $\mathbf{q}'^2$ for the amplitude

$$\bar{D}^+(\nu, \nu_B; \mathbf{q}^2, \mathbf{q}'^2) \equiv A^+ + \nu B^+ - \frac{\nu_B^2}{\nu_B^2 - \nu^2} \frac{g^2}{m} K(\mathbf{q}^2)\, K(\mathbf{q}'^2). \tag{5.1.21}$$

Taken on-shell, the last term differs by g^2/m from the nucleon pole term $D_N^+ = \nu B_N^+$ of the fixed-t dispersion relation, Eq. (A.6.30). g^2/m is the nucleon exchange contribution in second order perturbation theory to the amplitude A^+, if *pseudovector* coupling is used.

In terms of $\bar{D}^+$, ADLERs consistency condition and the definition of the sigma term reads

$$\bar{D}^+(0,0;0,\mu^2) = \bar{D}^+(0,0;\mu^2,0) = 0; \qquad \bar{D}^+(0,0;0,0) = -\frac{2}{f_\pi^2}\sigma_{NN}(0), \tag{5.1.22}$$

if the singular point $\nu = \nu_B = 0$ of the nucleon pole term is approached by first setting $\nu = 0$ and then $\nu_B = 0$.

The linearity assumption leads to

$$\bar{D}^+(0,0;\mu^2,\mu^2) \equiv \bar{D}(\nu=0, t=2\mu^2) = \frac{2}{f_\pi^2}\sigma_{NN}(0) + O(\mu^4/m^4). \tag{5.1.23}$$

Another method for the continuation was proposed earlier by FUBINI and FURLAN (68 L). However, the existing applications were criticized by BROWN, PARDEE, and PECCEI (71 L), because the uncertainties are too large.

See also the discussion of the extrapolation in 73 R KLEINERT.

The Ward identity approach

The continuation in pion mass can be avoided, if a relation between the sigma term and the amplitude A^+ is derived from the Ward identity for the on-shell amplitude (see Sect. 2.5.1.5). Of course, this does not solve the problem, but the corrections have now a different physical interpretation, which can be used for an approximate evaluation. We mention here only the result of BROWN, PARDEE, and PECCEI (71 L), who derived the following relation between the on-shell amplitude A^+ and the sigma term

$$A^+(\nu, \nu_B) - g^2/m = \frac{2}{f_\pi^2}\sigma_{NN}(t) + \frac{g^2}{m}\left\{\left(\frac{\mu}{m}\right)^4 a + \frac{2\nu_B}{m} b + \left(\frac{\nu}{m}\right)^2 c\right\}, \tag{5.1.24}$$

where a, b, and c are non-singular functions of ν and ν_B (up to logarithms). At $\nu = \nu_B = 0$ this agrees with the result of CHENG and DASHEN, except that the sigma term is now taken at $t = 2\mu^2$ instead of $t = 0$.

The difference was calculated by PAGELS and PARDEE (71 L)

$$\sigma_{NN}(2\mu^2) - \sigma_{NN}(0) = \frac{3 g_A^2 \mu^2}{16\pi f_\pi^2}\mu + O(\mu^4 \log\mu). \tag{5.1.25}$$

This correction amounts to 15 MeV. GASSER-LEUTWYLER [82 R, Eq. (D. 2)] found in an identical analysis 32 instead of 16 in the denominator.

In his recent paper, GASSER (81 L) has calculated non-analytic chiral corrections in a different way. Together with kinematic corrections he found an increase by 10 MeV, i.e. this amount has to be subtracted from the experimental value, following from $\bar{D}^+(0, 2\mu^2)$, before the comparison with the theoretical prediction is made. This result has no direct relation to that of PAGELS amd PARDEE [In a more recent calculation GASSER found (5 ± 2) MeV].

Another estimation of the left hand side of Eq. (5.1.25) was made by GENSINI (81 L-1). He evaluated the t-channel S-wave unitarity correction and found 7 MeV. See 72 L AUVIL and 72 L SCHNITZER for earlier discussions of the t-channel contribution.

Since the sigma term is of order $(\mu/m)^2$, it is important that the remainder in Eq. (5.1.24) is of order $(\mu/m)^4$ at the CHENG-DASHEN point. At threshold, where the scattering length $(a_1+2a_3)\propto D^+(\mu)\approx 0$ was predicted by TOMOZAWA (66 L) and by WEINBERG (66 L), the remainder is of the same order as the sigma term.

Extrapolation into the unphysical region

The problem is now to determine the value of the *on-shell amplitude* $\bar{D}^+$ at the unphysical point $\nu=0$, $t=2\mu^2$ (Fig. A.3.4). There are several possibilities for the extrapolation of the amplitude from the physical region of the s-channel to this point:

i) CHENG and DASHEN evaluated a fixed-t dispersion relation at $t=2\mu^2$, using the partial wave expansion for the extrapolation of $\mathrm{Im}\,D^+$ and, in the low energy region, also of $\mathrm{Re}\,D^+$ to the unphysical t-value. Applying a "broad area subtraction method", the authors obtained the large value $\sigma_{NN}=110$ MeV, in violent disagreement with the theoretical prediction (25 MeV; 72 R and 72 L RENNER, 68 L GELL-MANN).

ii) At that time, we had just calculated the coefficients of the expansion Eq. (4.7.2). Using CERN phase shifts, we obtained from the linear approximation $\bar{D}^+(0,2\mu^2)=d_{00}^+ + 2\mu^2 d_{01}^+$ the value $\sigma_{NN}\approx 40$ MeV [HÖHLER (71 L-3, 72 L-1)]. Essentially the same result was found by JAKOB (71 L-2), who evaluated the fixed-t dispersion relation at $t=2\mu^2$.

The discrepancy between the result of our group and that of CHENG and DASHEN was due to the fact that we used $\mathrm{Re}\,D^+$ as reconstructed from phase shifts at *all* energies, whereas CHENG and DASHEN used only the $\mathrm{Re}\,D^+$ values in the low energy region. Their method is at variance with the real parts of D^+ at higher energies.

iii) ALTARELLI, CABIBBO, and MAIANI (71 L ALTARELLI-1 and 2) developed a method for the calculation of the sigma term from scattering lengths, because they believed that these are better known than the coefficients of the subthreshold expansion.

The coefficients of the linear expansion

$$\bar{D}^+ = A\mu^2 + Bt + C(\mathbf{q}^2+\mathbf{q}'^2) + D\nu^2 + R(\nu^2, t, \mathbf{q}^2, \mathbf{q}'^2) \tag{5.1.26}$$

are determined from ADLERs consistency condition and from S- and P-wave scattering lengths, using the forward dispersion relation and estimating the Δ(1232)-exchange contribution.

It is clear that the on-shell value of R can be exactly expressed by a dispersion integral, because this is true for $\bar{D}^+$ and for the coefficients of the expansion. A calculation of this type was carried out in detail by OLSSON et al. (80 L-1 and 2). The result is an exact version of the method of ALTARELLI et al., but it is also seen that it becomes *equivalent* to our use of the subthreshold expansion, as long as the scattering lengths are determined from dispersion relations.

The on-shell part of the calculation of ALTARELLI et al. is also equivalent to the use of the expansion Eq. (A.3.66) for an extrapolation to the CHENG-DASHEN point

$$\nu=0;\quad \nu_B=0;\quad t=2\mu^2;\quad q^2=-\mu^2(1-\mu^2/4m^2);\quad s=m^2, \tag{5.1.27}$$

adding $\mathrm{Im}\,C^+$ which includes the threshold cusp. However, since only terms *linear in t* are taken into account, the *curvature* of the t-dependence of $\bar{D}^+(0,t)$ is neglected. It will be seen below that this is not justified.

We conclude that it is necessary to improve the approximations of ALTARELLI et al. But even the exact version is an unnecessary detour in comparison with the more transparent direct calculation, as long as the scattering lengths are determined from dispersion relations, and this will remain the most accurate determination also in the future (see the concluding remarks in Sect. 2.4.7).

iv) The experimental information on pion-nucleon scattering at low energies was considerably improved by the experiment 73 BUSSEY, which was the main input in the phase shift analysis 73 L CARTER-1. These new phase shifts and the CERN 72 solution were used by NIELSEN and OADES (74 L-1) for a redetermination of phase shifts which are compatible with fixed-t analyticity and for a recalculation of the coefficients of the subthreshold expansion. Due to the improved input, they were able to calculate the coefficient $d_{02}^+=c_{02}^+$, which turned out to be larger than expected.

The result for the sigma term reads

$$(f_\pi^2/2)\bar{D}^+(0,2\mu^2) = (f_\pi^2/2)(d_{00}^+ + 2\mu^2 d_{01}^+ + 4\mu^4 d_{02}^+)$$
$$= (f_\pi^2/2)(-1.45+2.36+0.14)\mu = (66\pm 9)\,\mathrm{MeV}, \tag{5.1.28}$$

It is seen that the quadratic term in t contributes about 9 MeV. The authors also gave an argument why the sigma term calculated in 73 L CARTER-2 is too large.

In a subsequent paper (74 L NIELSEN-2) the authors used their result for c_{02}^{+} and a fixed-ν dispersion relation at $\nu=0$ in order to select one of several possible parametrizations for the $\pi\pi N\bar{N}$ partial wave $\operatorname{Im} f_{+}^{0}$. The same dispersion relation had been used earlier by ENGELS (72 L) for a determination of $\bar{D}^{+}(0,t)$.

v) A fairly large number of authors developed various methods for the extrapolation procedure, but they used essentially the same phase shifts: CERN 72, SACLAY 73, 73 L CARTER and preliminary versions of KH 78 and CMU-LBL 79. The numerical results of these papers are outdated insofar as the input amplitudes do not include the information from new πN scattering data. This statement refers to the papers: 74 L HITE, 75 L CHAO, 75 L LANGBEIN-1, 75 L OLSSON, 76 L HITE, 76 L LICHARD-1, 76 L MOIR, 79 L LANGBEIN, 79 L RODRIGUEZ-VARGAS, 80 L OLSSON-2. The results agree more or less with that of NIELSEN and OADES (74 L-2).

As far as the methods are concerned, none of the papers includes the existing information on the t-channel discontinuity, although the extrapolation is made towards the t-channel branchpoint (see below). The method of RODRIGUEZ-VARGAS is less reliable than others, because a Finite Energy Sum Rule is used. It is necessary to make ad hoc assumptions in order to be able to evaluate the off-shell sum rule from on-shell partial waves and Regge fits.

The present status of attempts to derive the πN sigma term from π^{-}-nucleus scattering is reviewed in 80 R GENSINI. An application of the FUBINI-FURLAN method leads to $\sigma_{NN}=(56.9\pm6.0)$ MeV (see also 80 L GENSINI).

In all cases, the errors are highly subjective and reflect the more or less optimistic mentality of the authors, since a systematic error analysis is not feasible (cf. Sect. 2.4.5.3). Of greater interest is the spread of the results if different methods are used. Of course, one should consider only methods of comparable quality, as far as the procedure and the input amplitude is concerned.

vi) In order to improve the reliability of the extrapolation, it is useful to exploit the existing information on the discontinuity of the t-channel cut, thereby *introducing the additional experimental information on the pion-pion phase shifts.* As discussed in Sect. 2.4.6 the $\pi\pi N\bar{N}$-amplitudes can be determined up to about $t=40\mu^2$.

In 77 L HÖHLER-2, and 78 L KOCH the extrapolation of $\bar{D}^{+}(0,t)=\bar{C}^{+}(0,t)$ to $t>0$ was performed by means of a fixed-ν dispersion relation at $\nu=0$

$$\operatorname{Re} C^{+}(0,t)=\frac{1}{\pi}\left\{\int_{-\infty}^{-4m\mu}+\int_{4\mu^2}^{\infty}\right\}\frac{\operatorname{Im} C^{+}(0,t')}{t'-t}\,dt'. \tag{5.1.29}$$

This relation gives an interesting information on the shape of the (real) subtraction function $C^{+}(0,t)$ in the range of interest $(-25\mu^2<t<4\mu^2)$. On the nearby right hand cut, $t>4\mu^2$, $\operatorname{Im} C^{+}$ is *positive* (Fig. 4.6.2) and therefore, the contributions to the coefficients $c_{02}^{+}, c_{03}^{+}\ldots$ are also positive, i.e. this contribution to $C^{+}(0,t)$ has an *upward curvature.* It is calculated from the imaginary parts of the $\pi\pi N\bar{N}$ partial waves (Sect. 2.4.6). The dominant contribution comes from $\operatorname{Im} f_{+}^{0}$.

The nearby part of the left hand cut is known from the πN partial wave expansion. Its contribution to $C^{+}(0,t)$ is dominated by the Δ(1232)-exchange term. Therefore, its contribution to $C^{+}(0,t)$ has the usual resonance-like structure for real parts: its tail near $t=0$ has a *downward curvature* which is weaker, because the peak of the imaginary part lies far away [at $t=-1.24\ \text{GeV}^2$]. It is true that this is beyond the range of the partial wave expansion, but this does not invalidate our treatment, because errors on distant parts of the cuts can be tolerated, if one uses the "discrepancy" method.

We conclude that the subtraction function $C^{+}(0,t)$ has a *turning point*, which lies at $t\approx-4\mu^2$. This makes it difficult to perform an accurate extrapolation from the range $t<0$ to $t=2\mu^2$ without knowing the contribution of the nearby part of the cut $t>4\mu^2$. Furthermore, the turning point near $t=0$ indicates a slow convergence of the subthreshold expansion for $\bar{C}^{+}(0,t)$.

The extrapolation to $t>0$ is now performed by calculating the discrepancy function, which consists of $\operatorname{Re} C^{+}(0,t)$ as determined from the KH 78 fixed-t solution, minus the contributions of the nearby parts of both cuts. It is expected that this function is slowly variable in t in the range of interest. Figure 5.1.1 p. 300 shows that there is no significant deviation from a linear variation in the interval $-16\mu^2<t<0$. Therefore, we fit these points by a straight line and extrapolate $\Delta(t)$ to $t=2\mu^2$. The result is used together with the known contributions of the nearby parts of the cuts in order to calculate $\bar{C}^{+}(0,2\mu^2)$ and σ_{NN}

$$\bar{C}^{+}(0,2\mu^2)=\bar{D}^{+}(0,2\mu^2)=(1.01\pm0.1)\mu^{-1};\qquad \sigma_{NN}(2\mu^2)=(63\pm6)\ \text{MeV}. \tag{5.1.30}$$

The first terms of the subthreshold expansion read

$$\bar{D}^{+}(0,2\mu^2)=(-1.46+2\cdot1.14+4\cdot0.036)\mu^{-1}=0.964\mu^{-1}. \tag{5.1.31}$$

As expected, the convergence is slow due to the nearby turning point.

vii) In a recent paper, LANGBEIN (79 L) made an attempt to improve his earlier work by taking into account a constraint for $\bar{C}^+(0, 4\mu^2)$. He used the expansion $(q_-^2 = \mu^2 - t/4 = -q_t^2)$

$$\bar{C}^+(0,t) = \sum_{n=0}^{3} c_n q_-^n , \tag{5.1.32}$$

determining its coefficients from a fit to his result for the subtraction function at $t<0$ and from the constraint

$$\lim_{q_t\to 0} (q_t^{-1}\mathrm{Im}\,C^+(0,t)/\mathrm{Re}\,C^+(0,t)) = a_0^0 = 0.28\mu^{-1} = -c_1/c_0 . \tag{5.1.33}$$

He disregards the difficulty that this expansion is used far outside its range of convergence, which is given by the singularity at $t=3.98\mu^2$ of $C^+(0,t)$ in the second sheet (Sect. A.3.5.3; see also the analogous case in Sect. A.11.3).

The structure of the cusp at $t=4\mu^2$ is more complicated than assumed in the ansatz Eq. (5.1.32). It has been calculated from the fixed-v dispersion relation using the discrepancy method as described in vi) (79 R HÖHLER-KAISER-KOCH-PIETARINEN). Since the higher partial waves are negligible near $t=4\mu^2$, the cusp is almost the same as that in $\mathrm{Re}\,\tilde{f}_+^0$ (Fig. A.3.5). This figure also shows the unusual shape of $\mathrm{Im}\,f_+^0$ above $t=4\mu^2$.

LANGBEINs method is an unnecessary simplification of our evaluation of the fixed-v dispersion relation, The latter takes into account not only the value of a_0^0 but the complete existing information on the ππN$\bar{\text{N}}$ partial waves.

viii) Finally, we add some remarks on the uncertainties of the result Eq. (5.1.30) for σ_{NN}.

There are many possibilities for extrapolations from the physical region to the CHENG-DASHEN point. The results would be the same, if the partial wave solution would lead to invariant amplitudes which are analytic in s and t as assumed in the MANDELSTAM representation and if the extrapolation methods were perfect.

Of course, this can be true only in an approximation and therefore it is of great interest to perform the extrapolation along a variety of different paths.

The paths should be chosen such that the integral over the s-channel cut can be evaluated in a reliable way and that the ππN$\bar{\text{N}}$-partial wave expansion converges along the t-channel cut.

A calculation of this type has recently been carried out by R. KOCH (82 L-1). He chose the curves

$$(t-t_0)(v^2-v_0^2) = a^2/2 \tag{5.1.34}$$

and values of the parameters for which the curves go either through the CHENG-DASHEN point or through threshold or through both.

It turns out that the spread of the results for σ_{NN} is

$$\sigma_{NN}(2\mu^2) = (64 \pm 8)\,\mathrm{MeV} \tag{5.1.35}$$

and the spread of the scattering lengths

$$a_1 + 2a_3 = -(0.023 \pm 0.007)\mu^{-1} . \tag{5.1.36}$$

I think that this result together with the earlier one discussed in vi) rules out the possibility that values for σ_{NN} near 30 MeV can be obtained from the KH 78 and 80 partial wave solutions.

My conclusion is at variance with that of other authors (78 L BANERJEE-1, 81 R DOMINGUEZ), who performed simple fits to $\bar{C}^+(0,t)$ at $t<0$ in order to show that $\sigma_{NN}=25\cdots30$ MeV is well compatible with the experimental scattering amplitude.

The above discussion leaves no doubt that *a much greater effort* is necessary, if one wants to make useful statements on the experimental value of the sigma term.

For instance DOMINGUEZ and LANGACKER (81 R) made a linear fit in t to $\bar{C}^+(0,t)$ in the range $-10\mu^2<t<0$ and observed that the fit is still good if it is constrained by $\sigma_{NN}=30$ MeV. This agrees well with our result derived from the fixed-v dispersion relation: the tangent at $t=-4\mu^2$ gives 29 MeV. if it is extrapolated to $t=2\mu^2$! However, the point is that the tangent at $t=0$ gives 51 MeV and, due to the curvature, the final result is 63 MeV. In another fit they used LANGBEINs expansion, Eq. (5.1.32), ignoring the difficulties mentioned above in connection with this ansatz.

Furthermore, DOMINGUEZ and LANGACKER argued that there is an uncertainty of the t-channel contribution, because the model of DURSO, JACKSON, and VERWEST (80 L) leads to an amplitude $\bar{D}^+(0,t)$ in approximate agreement with our result in the range $4\mu^2<t<35\mu^2$, although it is assumed that $\sigma_{NN}=0$. However, this model was constructed for applications in nuclear physics. It is not accurate enough for the description of small details in the range $t<4\mu^2$ as can be seen by a comparison with the fixed-v dispersion relation.

Another possible difficulty was investigated by GENSINI (76 L and 77 L), who claimed that the small neutron-proton mass difference, which is usually neglected, has an appreciable influence on the low energy amplitudes and, in particular, on the result for the sigma term.

It is true, that the small mass difference leads to a remarkably large change of some quantities. For instance, in elastic $\pi^{\pm}$p scattering, the exact position of the neutron pole is $\nu_B(t=0) = -0.065\mu$ which differs by 14% from the value used in our calculations ($\nu_B = -0.074\mu$), where the mass difference is neglected.

Furthermore, the t-value where the pole lines $\nu = \pm\nu_B$ meet on the $\nu=0$ axis in the MANDELSTAM diagram is now $t=1.75\mu^2$ instead of the usual value $t=2\mu^2$.

The first question is whether our result for $D^+(0,t)$ is changed if one allows for the mass difference. In the KH-partial wave analyses, the charge-dependent effects are treated as proposed in 77 L TROMBORG, i.e. the neutron-proton mass difference is taken into account. In the expansion of the fixed-t amplitudes, the nucleon pole terms are treated separately. The exact expression for the nucleon pole term in Eq. (A.8.6) shows that there is no appreciable effect of the mass difference in the physical region. Therefore, the result for $\bar{D}^+(0,t)$ is not affected.

Furthermore, one has to repeat the derivation of ADLERs consistency condition and of the CHENG-DASHEN extrapolation for a finite n–p mass difference. It turns out that the sigma term follows from the amplitude $\bar{D}^+$ at $\nu=0$, $t=2\mu^2$ as before i.e., contrary to the result of GENSINI, the neglection of the mass difference leads to negligible corrections for σ_{NN} (77 L HÖHLER-2, 79 L LANGBEIN). Nevertheless, GENSINI has called attention to an interesting point, since the effect of the mass difference can be appreciable in other applications. Therefore we have listed the exact pole term formulas at the end of Sect. A.8.1.1.

Conclusion on the sigma term

It is necessary to continue the search for the origin of the large discrepancy between the experimental value, Eq. (5.1.35), and the theoretical prediction Eq. (5.1.20). In my opinion, it is not worthwhile to develop further methods for the analytic continuation, unless someone finds a method in which the error of $\bar{D}^+(0,t)$ can be calculated from the errors of the *experimental scattering data*. Instead, it would be important to investigate the following points:

i) An improved determination of the slope of $\bar{D}^+(0,t)$ at $t=0$

Since d^+_{02}, d^+_{03}... are positive and the error of d^+_{00} is relatively small, one can obtain a low sigma term $\sigma_{NN} \approx 30$ MeV only, if the true value of d^+_{01} is considerably smaller than our result. The accuracy of our value for d^+_{01} is limited by the fact that, at low and intermediate energies, the experimental information on $\pi^{\pm}$p scattering at small t (in the Coulomb interference region) is still very poor, except for the experiment of BAILLON et al. (74) (see Fig. 4.7.1). Accurate data in this region would also help to improve the determination of the higher partial waves and of the coupling constant f^2, the latter being of interest because of the GOLDBERGER-TREIMAN discrepancy.

ii) A check of the *t*-channel contribution

Our calculation of the curvature of $\bar{D}^+(0,t)$ depends mainly on the ππ S-wave phase shift near threshold. It is necessary to study the discrepancy between the scattering length $a^0_0 = (0.28 \pm 0.05)\mu^{-1}$ derived from the K_{e4}-decay data and ROY equation constraints (see Sect. 9.1 in 79 R NAGELS) and the new result $a^0_0 = (0.177 \pm 0.017)\mu^{-1}$ obtained from data for the reaction $\pi^- p \to \pi^- \pi^+ n$ near threshold (80 L BJORK) in a model dependent analysis. KOCH (82 L-1) has recently investigated the sensitivity of the experimental sigma term to variations of the t-channel input. Further discussions of t-channel unitarity effects: 72 L SCHNITZER, 72 L AUVIL, 81 L GENSINI-1.

iii) A check of the relation between $\bar{D}^+(0,2\mu^2)$ and $\sigma_{NN}(0)$

This question was ignored in many earlier publications. It is necessary to continue the discussions, because a widely accepted conclusion does not yet exist. A new result is discussed in 82 R GASSER-LEUTWYLER and 81 L GASSER.

2.5.1.4 The Goldberger-Treiman discrepancy

Goldberger and Treiman (58 L) derived the following relation between the pion decay constant f_π, the nucleon axial-vector coupling constant g_A and the pion-nucleon coupling constant f^2 or g^2

$$f_{\rm GT}^2 = g_{\rm GT}^2 \frac{\mu^2}{16\pi m^2} = \frac{g_A^2 \mu^2}{8\pi f_\pi^2}. \qquad (5.1.37)$$

We have added the subscript "GT", because the evaluation with the present values $g_{\rm A} = (1.254 \pm 0.007)$ (80 R Bricman) and $f_\pi = (0.9448 \pm 0.0009)\mu$ (79 R Nagels, Sect. 3.3) gives

$$f_{\rm GT}^2 = 0.070 \pm 0.001\,; \qquad \Delta_{\rm GT} \equiv 1 - \frac{\sqrt{2}\, m g_{\rm A}}{f_\pi g} = 0.06 \pm 0.01\,. \qquad (5.1.38)$$

$\Delta_{\rm GT}$ is called the Goldberger-Treiman discrepancy. It describes the discrepancy between $f_{\rm GT}^2$ and the experimental value $f^2 = 0.079$ (Sect. 2.4.2) which differ by 11%.

More recent derivations of the Goldberger-Treiman relation can be found for instance in the reviews 68 R Adler-Dashen, 75 R Pagels, and 81 R Scadron. The "field theoretic derivation" shows that g has to be replaced by $gK(0)$, where $K(\mathbf{q}^2)$ is the form factor of the pion-nucleon vertex. This version was used in Adlers consistency condition, Eq. (5.1.15).

If the discrepancy is entirely due to the πNN form factor, its relation to $K(0)$ is

$$K(0) = 1 - \Delta_{\rm GT} = 0.94\,. \qquad (5.1.38a)$$

However, different attempts to determine the πNN form factor at $\mathbf{q}^2 = 0$ led to larger values.

One possibility is to evaluate a dispersion relation for the form factor, inserting suitable intermediate states (76 L Nutt, 77 L Durso, 79 L Cass). Another one is to extract the form factor from high energy data for np and $\bar{\text{p}}$p charge-exchange scattering (74 L Bongardt, 80 L Dominguez-1) or from other reactions (80 L Dominguez-2, 81 L Dominguez). In both cases, one obtains $K(0) \approx 0.97$, i.e. one can explain only about half of the discrepancy.

In order to explain the remaining part, Scadron (81 R) suggests that it could be due to a $\mathbf{q}^2$-dependence of the pion decay constant f_π. Dominguez (82 L) ascribes it to the fact that the radial excitation of the pion has not been explicitly incorporated in this calculation. There is now experimental evidence for a heavy pion (81 L Aaron, 81 L Bonesini). Furthermore, he gives a new relation between the discrepancy $\Delta_{\rm GT}$ and the sigma term. It is fulfilled for $\sigma_{\rm NN} = 35$ MeV and the value $M = 885$ MeV for the chiral mass of the nucleon as defined in 80 L Gasser. — Dominguez (82 L) adds the statement that the uncertainty of the experimental sigma term has been considerably underestimated. In the above point viii) (Sect. 2.5.1.3) I have shown that his arguments are based on a simplified treatment which is not tenable (81 R Höhler).

2.5.1.5 Current algebra models

The success of Adler and Weisberger encouraged many authors to investigate further consequences of current algebra and PCAC for pion-nucleon scattering amplitudes. It turned out that in all cases, except those treated in Sects. 2.5.1.2 and 2.5.1.3, the result depends on properties of the axial vector-nucleon scattering amplitudes, which can be estimated only from exchange models. Furthermore, during the first years the usual procedure was to start from the scattering of zero mass pions and to treat afterwards the continuation to the physical pion mass (67 L Raman, 67 L Schnitzer, 68 R Renner).

As an example we mention the work of Goldberg and Gross (67 L Goldberg, see also 67 L Gerstein). An interesting part of their result is the following exact relation for the amplitudes A^- at $t=0$

$$A^-(\omega)/\omega = -\frac{2}{f_\pi^2} F_{2\rm v}(0) + \frac{\omega m^2}{f_\pi^2} \tilde{A}_1^-(\omega) + \frac{4m}{f_\pi^2} \tilde{A}_{11}^-(\omega)\,. \qquad (5.1.39)$$

A_1^- and A_{11}^- are invariant axial vector-nucleon scattering amplitudes in the tensor basis of these authors. The tilde indicates that the nucleon pole term contributions are subtracted.

Assuming that A^-/ω and $A_{11}(\omega)$ fulfill unsubtracted dispersion relations, Goldberg and Gross obtained the sum rule

$$F_{2\rm v}(0) = -\frac{m^2}{\pi} \int_\mu^\infty \mathrm{Im}\, A_1^-(\omega)\, \mathrm{d}\omega\,. \qquad (5.1.40)$$

The lower limit of the integral is the physical pion mass for the same reason as in the original form of the ADLER-WEISBERGER relation.

As discussed in 67 L GOLDBERG and 72 L HÖHLER-1, the isobar contribution to the integral is small in comparison with the left hand side. It would be interesting to know the mechanism for the dominant contribution (or an argument against the assumptions).

Further work in this field can be found in 68 R ADLER-DASHEN, 69 R SCHNITZER-1 and 2, 70 R WEINBERG, 71 R WEINSTEIN, 72 R LEE, 73 R ALFARO-FUBINI-FURLAN-ROSETTI, 78 R PAVER.

In order to avoid problems with the continuation in pion mass, subsequent authors developed an on-shell method, starting from a Ward identity (69 L CARRERAS, 69 L WRAY, 70 L OSYPOWSKI, 70 L HUANG, 71 L BROWN, 71 L HÖHLER-2, 73 L OLSSON, 74 L SCADRON, 75 L OLSSON).

A different approach to the same problem was used by authors who started with a "Phenomenological Lagrangian" which is applied only in the tree-graph approximation. WESS and ZUMINO (67 L WESS) considered only nucleon and ϱ-exchange. The inclusion of isobar exchange is necessary, if one wants to have a more realistic model (68 L PECCEI). The connection with the Ward identity approach is discussed in 69 R SCHNITZER-2.

If one compares the results of the above mentioned papers, it is seen that in general the πN amplitudes are represented by a sum of three contributions [see Eq. (A.8.31)].

i) equal-time commutator terms, which are expressed by isovector nucleon form factors and by the sigma term,

ii) nucleon exchange terms,

iii) isobar exchange terms, which are assumed to dominate the remainder.

Some authors considered in addition "ε-exchange" terms, a "diffractive" term (in B^+) or a subtraction constant (in D^+).

The evaluation of these contributions led to widely differing results. For instance some authors approximated the isovector nucleon form factors by ϱ-exchange terms, although the experimental data for F_{2v} demand a pronounced *dipole*-like structure of $\mathrm{Im}\,F_{2v}$ (see Sect. 2.5.2). Furthermore, strongly differing results were obtained for the isobar exchange contributions, because different Δ-propagators and different values for the πNΔ coupling constants were used (see Sect. A.8.2).

One could have expected that the comparison of the predictions with the "experimental amplitudes" would show that some results are favoured and others have to be rejected, but this is not the case. Most of the authors concluded that their method leads to a satisfactory or even good agreement with "the experiments."

In this situation it is desirable to standardize the comparison with the experimental amplitudes using a method described in 72 L HÖHLER-1.

Comparison of models with experimental amplitudes

The best kinematic region for the comparison between different low-energy models is the vicinity of the point $\nu = t = 0$, where details of the shapes of the nearest resonances are less important than near threshold. The best choice of the amplitudes is to take the invariant amplitudes $D^{\pm}$ and $B^{\pm}$. One reason is that the Low Energy Theorems of Sects. 2.5.1.2 and 2.5.1.3 are valid for $D^{\pm}$ and furthermore, each of the equal-time commutator terms σ_{NN}, F_{1v}, and G_{Mv} contributes only once.

Since nucleon pole terms occur in all models, we consider amplitudes $\bar{D}^{\pm}$, $\bar{B}^{\pm}$, from which the nucleon pole terms are subtracted. This has also the advantage that the functions $\bar{D}^{\pm}$, $\bar{B}^{\pm}$ are slowly variable and can be expanded in powers of ν^2 and t, after having divided the crossing odd amplitudes by ν. We have subtracted the *pseudovector* nucleon pole terms, because these occur in most of the models.

The Table 2.4.7.1 of the coefficients x_{mn} of the expansion Eq. (4.7.2) is a condensed version of the experimental information. Since a number of accurate experiments has been performed during the last years, one should prefer tables based on recent partial wave analyses. The use of a "world average" is misleading, since, for most of the x_{mn}, reliable error estimates do not exist and there is no reason to give a larger weight to the results of optimistic authors. It is desirable that different methods are used for the extrapolation to the subthreshold region in order to get an estimate of the uncertainties. As discussed in Sect. 2.4.7, this table cannot be improved by adding information on scattering lengths.

In Table 2.4.7.1 we also have listed the contributions of the nucleon form factor terms of Eq. (A.8.31) to x_{00}, x_{01}, and x_{02} and the contributions of the Δ-*exchange pole terms*, Eq. (A.8.19), using our value Eq. (A.8.15) for the πNΔ coupling constant. Furthermore, we have added the Δ-*exchange non-pole term* contributions, choosing the frequently used value $Z = -1/2$ in Eq. (A.8.20).

The question is now whether the prediction from Eq. (A.8.31) approximately agrees with the "experimental" coefficients, if the amplitudes with the subscript "ax" are approximated by Δ-exchange.

One should note that discrepancies are significant only if they are appreciably larger than at least 10%, because the Δ-exchange pole terms, for which all authors give the same formula, are different in their tables. For instance b_{00}^{+}, b_{10}^{+}, and b_{20}^{+} become smaller by 10%, 20%, and 40% respectively, if our effective Δ-mass value 1219 MeV is replaced by the usual mass value 1232 MeV. Furthermore, some authors find the same pole terms larger by 40%, because they calculate the πNΔ coupling constant from the Δ-width according to Eq. (A.8.17).

The errors given for the experimental values of the coefficients belong only to internal inconsistencies of our method. There are additional errors which follow from the errors of the data and from discrepancies between different experiments. For the latter it is not possible to give a reliable estimate. However, it is reassuring to see that the changes in comparison with the coefficients derived from earlier phase shift analyses (72 L Höhler-1, 74 L Nielsen-2) are relatively small.

The overall agreement between the experimental coefficients and the prediction in Table 2.4.7.1 is impressive, although there are relatively large discrepancies in some of the coefficients. However, one should note that there exists no generally accepted theoretical argument for the choice of Z. Some authors claimed to have shown that Z must be $+1/2$ (71 L Nath), whereas others prefer both $Z=\pm 1/2$ (75 L Jenkins), but there are still problems (71 L Hagen).

The approximate agreement obtained with our choice ($Z=-1/2$) is completely destroyed for $Z=+1/2$. In the latter case the Δ-exchange amplitudes agree with those derived from the Rarita-Schwinger formalism (56 R Umezawa).

In 75 L Olsson the authors treated Z and g_{Δ}^2 together with other quantities as adjustable parameters, fitted in such a way that the prediction agrees with the experimental values. Their result $Z=-0.45\pm 0.2$ and $g_{\Delta}^2/4\pi =(0.27\pm 0.01)\mu^{-2}$ agrees with our choice, Eq. (A.8.15).

In order to avoid the uncertainty following from the spin 1/2-admixture in the Δ-propagator, Scadron et al. used a dispersion method for the calculation of Δ-exchange in axial vector-nucleon scattering (74 L Scadron). In this approach the ambiguity of the Δ-exchange term is isolated in the coefficient $d_{01}^{+}=a_{01}^{+}$ as a subtraction constant. The resulting Δ-exchange non-pole terms cannot be described by Eq. (A.8.20). The agreement with the experimental coefficients is comparable with that mentioned above for $Z=-1/2$ (78 L Wilde, 81 R Scadron).

Conclusion on current algebra models

If the axial-vector terms in Eq. (A.8.31) are approximated by Δ-exchange, one obtains encouraging results for all 4 invariant amplitudes. However, the Goldberg-Gross sum rule, Eq. (5.1.40), is a warning, since it shows that isobar exchange is not always the dominant contribution.

For the treatment of isobar exchange, a generally accepted formalism does not yet exist. It is also unclear, how one should explain the remaining discrepancies with the experimental values of the coefficients in Table 2.4.7.1, which are quite large in some cases. Some authors added the exchange of higher nucleon resonances, of t-channel resonances or of a diffractive contribution, without discussing the rules which are necessary in order to avoid double counting and to separate resonances and background in strongly inelastic amplitudes.

Finally, the current algebra models cannot lead to accurate quantitative results as long as broad resonances are treated in a narrow resonance approximation. The main questions are, however, to reach an agreement on the treatment of Δ(1232)-exchange and to calculate further contributions to the axial vector amplitudes.

2.5.1.6 Dispersion models

The first dispersion model was proposed by Chew, Goldberger, Low, and Nambu (57 L Chew, frequently quoted as "CGLN"), who approximated the amplitudes $A(\nu,t)$ and $B(\nu,t)$ by writing fixed-t dispersion relations (Sect. A.6.3), taking into account only the nucleon pole term and the P33 contribution to the dispersion integral. Since the latter contribution agrees with the Δ-exchange pole term in Table 2.4.7.1, the situation is the same as in the current algebra model, except for the equal-time commutator terms, the non-pole Δ-exchange terms and the difference between the pseudovector and pseudoscalar nucleon Born terms ($g^2/m=26.7\mu^{-1}$, $g^2/2m^2=1.99\mu^{-2}$)

$$\tilde{D}^{+}(\nu,t)-\bar{D}^{+}(\nu,t)=g^2/m;\qquad \tilde{B}^{-}(\nu,t)-\bar{B}^{-}(\nu,t)=\tilde{G}^{-}(\nu,t)-\bar{G}(\nu,t)=-g^2/2m^2. \qquad (5.1.41)$$

It is remarkable that a zero order approximation for $d_{00}^{+}+g^2/m=a_{00}^{+}+g^2/m$ and for $d_{01}^{+}=a_{01}^{+}$ is obtained from the P 33 contribution to the *unsubtracted dispersion relation for the amplitude* A^{+}. If this dispersion integral exists, asymptotic s-channel helicity conservation is a consequence (70 L Höhler-1, cf. Sect. 2.6.3.8).

For some of the coefficients x_{mn} the simplest dispersion model is satisfactory, but for others it is poor. Of course one can easily find the higher resonance and background contributions to the fixed-t dispersion relation which cause the discrepancies, but the problem is another one: one would like to have a dispersion model which is similar to the current algebra model insofar as it includes in addition to N and Δ-exchange the contribution of the nearest *t-channel* singularity, which is described by the isovector form factor terms in Eq. (A.8.31).

A suitable method was developed by DIETZ (62 L, 63 L) who followed suggestions by MANDELSTAM (59 L-1) and FRAZER (60 L-3). The connection with the current algebra models was investigated by HÖHLER and STICHEL (71 L-2).

In this method, the FROISSART-GRIBOV representation for the first term of the t-channel partial wave expansion is subtracted from the fixed-t dispersion relation as described in Sect. A.8.3. The main effect is to cut off the high energy tail of the isospin odd amplitudes (reggeized ϱ-exchange) and to replace it by the $J=I=1$ $\pi\pi$N$\bar{\text{N}}$ partial wave, which includes the effect of ϱ-exchange (as a spin 1 particle). Since reggeized ϱ-exchange is expected to belong to the "inelastic part" of the amplitude as defined in Eq. (A.8.43) and Fig. A.5.1 (61 R CINI-VITALE), we have subtracted in Eq. (A.8.45) the inelastic part of $\text{Re}\, f_+^1(t)$. The (elastic) Δ-exchange pole term remains the same as before.

The subtraction of the lowest t-channel partial wave can also be made in the C^+-amplitude but, since a subtraction in this amplitude is required anyway, one obtains only a version in which the usual subtraction function is replaced by $\text{Re}\,\tilde{f}_+^0$ [Eq. (A.8.48)].

For those who are interested to improve the current algebra model, it is useful to study the relation with the dispersion model as described in our paper (71 L HÖHLER-2). Since the "experimental values" of the subthreshold coefficients are calculated from dispersion relations, one can easily see the origin of discrepancies with an approximation in which only the nucleon and isobar pole terms are retained.

As an example, we consider the coefficient d_{10}^+ which is calculated from the dispersion integral

$$d_{10}^+ = \frac{2}{\pi}\int_0^\infty \sigma^+(\omega) k^2 \mathrm{d}k/\omega^4 \, . \tag{5.1.42}$$

If the total cross section σ^+ is expressed in terms of partial waves, one finds that 75% of the result comes from the P 33 contribution. This is slightly larger than the value expected from Table 2.4.7.1, which is 0.81/1.12, because P 33 at higher energies does not belong to Δ(1232)-exchange. The S-waves give a fairly large contribution of 12%, P 11 gives 3%.

The contributions of the higher resonances are strongly suppressed by the kinematical factor k^2/ω^4, which has its maximum already at 140 MeV/c, far below the maximum of the Δ(1232) resonance (300 MeV/c).

One could be surprised that the S-wave contribution is so large, because $\text{Re}\, C^+$ at threshold is known to be very small. The reason is the large value of σ^+ at threshold (4.25 mb), which follows mainly from the large value of the *isospin odd* scattering length in the unitarity relation, Eq. (A.3.67).

The result suggests that one should consider unitarity effects at the s-channel threshold (which are taken into account in the FUBINI-FURLAN method).

Other interesting cases are those coefficients, where neither form factor terms nor Δ-nonpole terms play a role.

Furthermore, it is not clear how the subtraction function $D^+(0,t)$ or $A^+(0,t)$ should be described in a current algebra model. According to Eq. (A.8.31) there are contributions from D_{ax}^+ and A_{ax}^+, which are dominated by Δ(1232) exchange, and from $\sigma_{\text{NN}}(t)$. In old papers, the latter were described by a pole term ansatz for "σ"- or "ε"- exchange, but this is doubtful, because a well-pronounced resonance in the $\pi\pi$ S-wave does not exist below the S*(980) (80 R BRICMAN, p. 124). An estimation of t-channel exchange contributions to the axial vector amplitudes would be of interest.

The extent to which low energy πN scattering is determined by N- and Δ-exchange and the low-energy theorems was studied by RESELE and SCHMID (80 L). The treatment differs from earlier investigations mainly by the fact that the authors work with an expansion in powers of μ/m, even in cases where an exact calculation is available.

2.5.1.7 Models for applications in nuclear physics

The advent of meson factories led to an increased interest in pion-nucleus scattering. For the analysis of these reactions, one needs off-shell continuations of pion-nucleon scattering amplitudes. In principle, these can be obtained in 2 steps:

i) The analysis of pion-nucleon scattering data has led to accurate invariant and partial wave amplitudes in the physical region. Assuming MANDELSTAM analyticity, these amplitudes can be continued into unphysical regions and one can study a variety of dispersion relations. The results for the different amplitudes are available in the form of numerical tables (Sect. 2.2) or of very flexible expansions (Sect. A.6.3.4).

As described in the previous subsection, one can calculate from the dispersion relations quantitatively the contributions to an amplitude from exchanges of nucleons, isobars, ϱ-mesons etc. It is also possible to determine the parameters of "effective poles", which represent for instance isobar exchange, and also the error of this approximation.

However, one should not forget that the representation as a sum over exchange contributions and some background is not unique: it depends on the choice of the one-dimensional or partial wave dispersion relation (Sects. A.7.1, A.7.2, A.8.3, 2.5.1.6).

ii) The second step is to construct models which lead to reasonable approximations of the amplitudes obtained in i). From these models one can calculate off-shell pion-nucleon scattering amplitudes which are needed in the theory of pion-nucleus scattering. The problem is that the off-shell continuation is not unique.

A general discussion cannot be given in the present volume, but we shall mention some of the models studied in recent papers.

"Little" and "cloudy" bag models

Most of the investigations on quark bag models for nucleons deal with the excited states, i.e. with nucleon spectroscopy. First applications of these models to pion-nucleon scattering at intermediate energies are based on the P-matrix formalism of JAFFE and LOW (79 L); see for instance 79 L ROIESNEL.

Attempts to apply the M.I.T. bag model in nuclear physics led to several problems. One would like, of course, to derive the experimentally confirmed one-pion-exchange potential. In the original version of this model, the emission of a pion would have to be described by a fissioning of the bag and the creation of a $q\bar{q}$ pair. The chiral invariance of the interaction is not taken into account. Another difficulty resulted from the relatively large radius of the M.I.T. bag (about 1 fm).

The problems led to the proposal of a "little bag", which (in the first papers) has a radius of about 0.3 fm (BROWN et al. 79 L-1 and 2). In this model, pions are coupled to the bag surface in such a way as to make the axialvector current continuous across the boundary. They exist only outside the bag. This theory is very similar to the old fixed source theory of the nucleon (54 L CHEW, 56 L CHEW-LOW, 61 R THIRRING, 62 R HENLEY-THIRRING), except that the source is now described in terms of 3 quarks. More recent versions: 80 L VENTO and 82 L BROWN, 82 R RHO.

The "cloudy bag model" is another modern version of the CHEW-LOW model (THOMAS, 81 L-1 to 3, 82 L; MILLER, 80 L and 81 R; THEBERGE 80 L, 81 L, 82 L). The nucleon is described by a spherical static cavity of radius R, filled with three massless free quarks. A vacuum pressure is assumed to balance the pressure of the quark field. The new feature in comparison with the M.I.T. bag is a cloud of pions *inside and around the nucleon.* The pions are free except for emission and absorption processes at the bag surface, which are related to chiral symmetry. They can be quantized in the standard way.

The cloudy bag model has been used for a description of $\Delta(1232)$ as a resonance in πN scattering (80 L THEBERGE, 80 L MILLER). S-wave scattering has been treated by deriving a generalization of the WEINBERG effective Lagrangian (WEINBERG 66 L and 67 L). Furthermore, a calculation of the nucleon form factors led to a reasonable agreement with experiment. In all cases, it is necessary to assume a bag radius of about 0.8 fm, which practically agrees with the radius belonging to the form factor G_{Ep} (Table A.11.1). It is much larger than the radius of the "little bag" of BROWN et al.

The average number of pions in the cloud has the bound 0.9 ± 1.0, which is appreciably smaller than the bound for the CHEW-LOW model, because the bare $\Delta(1232)$ state is included in the Hamiltonian. The small number of pions suggests a rapid convergence of the perturbation expansion (81 L DODD).

A quark model for the $\pi N\Delta$ vertex has been proposed in 81 L NISKANEN and 81 L VENTO.

The coupling of pions to quark bags and the incorporation of chiral symmetry have been treated in several other recent papers, starting with the early work of CHODOS and THORN (75 L CHODOS, 78 L CALLEN, 79 R JAFFE, BARNHILL 79 L-1 and 2, 80 L, 81 L DETAR, 81 L ISRAILOV). See 81 R MILLER for a review.

Chiral Lagrangians

HAYSACK and LENDYEL (76 L and earlier papers) described the πN system at low energies by a nonlinear $SU(2) \times SU(2)$ chiral invariant Lagrangian. They used the superpropagator method for the regularization and the PADÉ method for the summation of the perturbation expansion. The results for πN phase shifts agree roughly with the phase shifts derived from experiments, although freely adjustable parameters do not occur.

Current Algebra models

These models were already treated in Sect. 2.5.1.5. Applications to off-shell amplitudes were made by SCADRON et al. In 75 L COON and 79 L COON the authors derived the three-nucleon potential of the two-pion exchange type, which is needed in nuclear structure calculations. WILDE, COON, and SCADRON (78 L) used their off-shell πN amplitudes in an investigation of the critical density for the onset of pion condensation in neutron matter.

The model of BANERJEE and CAMMARATA

In a series of papers (BANERJEE 77 L, 78 L-1 and 2, 79 L-1 and 2, 80 L, WEI, 80 L-1 and 2) the authors developed a dynamical model for low-energy pion-nucleon scattering, which is an extension of the static model of CHEW and LOW (56 L). They used the LOW equation (55 L) obtained by LSZ reduction (LEHMANN 55 L and 59 R) but, in contrast to CHEW and LOW, they retained the seagull terms and the antinucleon intermediate state contribution. (Reviews on the CHEW-LOW model: 55 R WICK, 61 R THIRRING, 62 R HENLEY-THIRRING.)

It is assumed that the isovector seagull term vanishes or is negligibly small. This term plays an important role in the Ward identity approach (Sects. A.8.2 and 2.5.1.5), where it includes the effect of ϱ-exchange in the t-channel. In the method of BANERJEE et al., the existence of a ϱ-exchange pole is possible in principle, but its effects are described in a very complicated way.

Using PCAC and a soft-pion limit, the isoscalar seagull term is expressed by the nucleon expectation value of the sigma commutator. This leads to a subtracted version of the LOW equation.

The LOW expansion of the integrals is truncated, retaining only the following intermediate states: N, Nπ and NN$\bar{N}$ ("z-graph terms"). Next, the authors use an expansion in terms of pion-nucleon partial waves. The result is an inhomogeneous non-linear integral equation for the πN amplitudes with half-off-mass-shell kinematics.

The numerical evaluation is extremely difficult and the accuracy is limited by the capacity of the computer.

A large contribution to the isospin odd amplitude comes from the z-graph terms. These terms partially replace the effect of the isovector equal-time commutator term, which is responsible for the success of the current algebra model. Unfortunately, the result depends on a parametrization of the πNN form factor at $t>4m^2$ which is very uncertain, since there are still problems with the same vertex near $t=0$ (Sect. 2.5.1.4).

The authors calculated the on-shell nucleon pole terms with the πN coupling constant $f^2=0.0725$, which follows from the GOLDBERGER-TREIMAN relation and a special assumption on $K(0)$ (Sect. 2.5.1.4), although this value is considerably lower than all reliable determinations of f^2 (Sect. 2.4.2).

The rescattering integrals in the LOW expansion are taken from threshold to infinity. One would like to see a discussion of the error which follows from the neglection of the inelasticity of the S-waves. The reliability of the treatment of the P-waves depends on a model for the off-shell extrapolation (78 L BANERJEE-2). Again, it is not clear whether one can trust in the accuracy of the evaluation of the integral.

The model contains 6 adjustable parameters which were determined by a fit to the experimental S-wave phase shifts. The agreement with the S 11 phase shift of the CBC analysis (Sect. 2.1.6) is good, but there remains an appreciable systematic deviation from the S 31 phase shift of this analysis.

The main result is a prediction for the on-shell amplitude $\bar{D}^+(0,t)$ which includes a value for the sigma term

$$\bar{D}^+(0,t)\equiv\bar{C}^+(0,t)=2f_\pi^{-2}\frac{\sigma(\pi\mathrm{N})\,(t-\mu^2)/\mu^2}{(1-t/\mu_1^2)^2(1-t/\mu_2^2)}+\dots$$
$$\mu_1=8.24\mu,\quad \mu_2=7.5\mu;\quad \sigma(\pi\mathrm{N})=25.5\,\mathrm{MeV}. \qquad (5.1.43)$$

The denominator is a rather arbitrary parametrization of the t-dependence. The factor $(t-\mu^2)$ is derived for the *soft pion amplitude*. It belongs to the cancellation of the sigma term contribution in ADLERs consistency condition Eq. (5.1.15). For the *on-shell amplitude* it is a special property of the model, since it does not occur in Eq. (5.1.24) and in the discussion leading to Eq. (5.1.25).

One should notice that $\sigma(\pi\mathrm{N})$ is somewhat different from $\sigma_{\mathrm{NN}}(2\mu^2)$ and $\sigma_{\mathrm{NN}}(0)$ as defined in Eqs. (5.1.24) and (5.1.25).

The result for the on-shell amplitude of Eq. (5.1.43) including the corrections has only a *very small quadratic term* in t, because the authors took into account only S and P-waves. At $t\approx-4\mu^2$ the agreement with our result is reasonable. If we use a *linear extrapolation* to $t=2\mu^2$ along the tangent at $t\approx-4\mu^2$, we find a sigma term of about 30 MeV. Our final result is considerably higher because of the *curvature* (Fig. 5.1.1).

It is easy to see why the authors did not find the correct curvature. They realized that their method becomes progressively less justified as t approaches $4\mu^2$. But they thought that the discontinuity of the t-channel cut starts to be significant only at much larger t-values. Our evaluation of the fixed-ν dispersion relation shows that this is not true (Sect. 2.4.7.1).

On the one hand, the dynamical model leads to some encouraging results, although this might be due to some extent to the 6 adjustable parameters and to the freedom in the form factor parametrizations.

On the other hand, the authors *grossly* overestimate the accuracy of the present version of their method. The statement that it is impossible to reproduce the experimental S-wave phase shifts with $\sigma(\pi\mathrm{N})$ differing by more than *one MeV or so* from 25 MeV reflects this misjudgment (79 L BANERJEE-2).

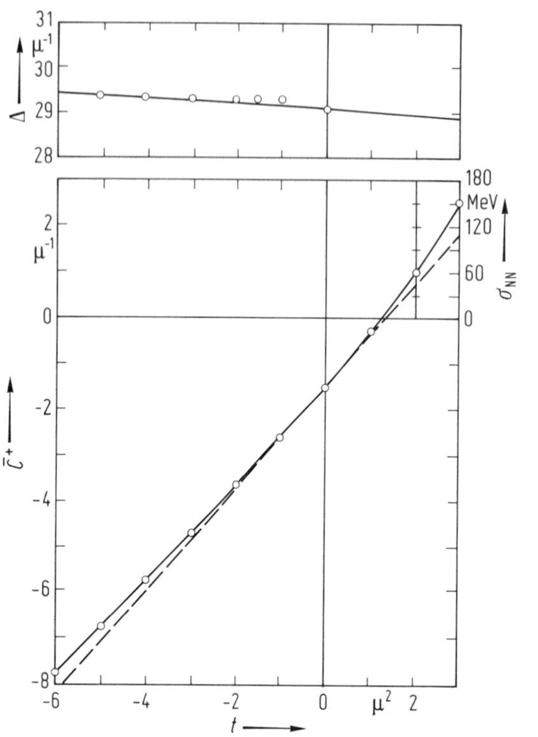

Fig. 5.1.1. Extrapolation of the amplitude $\bar{C}^{+} \equiv \bar{D}^{+}$ at $\nu=0$ from the physical region to $t=2\mu^2$. The result determines the value of the πN sigma term, Eq. (5.1.24). Solid line: the result of our calculation. (Sect. 2.4.6). It is seen that a linear extrapolation along the tangent at $t=0$ gives a lower value and that the result becomes still lower, if one uses a linear extrapolation from a range $t<-2\mu^2$. — The scale on the right hand side gives the sigma term in MeV, if it is used at $t=2\mu^2$. The cusp effect at $t=4\mu^2$ is shown in Figs. 4.6.1a and A.3.5. The curvature depends on the value assumed for the S-wave ππ-scattering length ($a_0^0=0.26\mu^{-1}$, see Table 2.4.6.2). The upper part of this figure shows the discrepancy function and its extrapolation to $t=4\mu^2$.

The model can certainly be improved by changing some assumptions and by extending it. For instance, one can include D-waves, which are expected to give a substantial contribution to the curvature of $\bar{D}^{+}(0,t)$. An important extension would be to build in a realistic strength of the t-channel cut near $t=4\mu^2$, taking into account the existing information on the ππN$\bar{N}$ partial wave f_{+}^{0} and the fairly large ππ S-wave scattering length a_0^0.

In my opinion, it is necessary to continue the test of the dynamical model by comparing it with results derived from dispersion relations, for instance with further coefficients of the subthreshold expansion, with the S-wave effective ranges and with the continuation of the πN S-wave Re$\tilde{f}_{0+}$ into the unphysical region below threshold (71 R Steiner). Because of the approximations, it is better to fit the adjustable parameters in the subthreshold region or at threshold instead of the fit to S-wave phase shifts (of which some are outdated due to new SIN results: 81 Alder).

As far as the belief of Banerjee and Cammarata in the accuracy of their prediction and the uncertainties of all dispersion calculations is concerned, one should keep in mind that the problems in the latter case are much less serious than in the dynamical model, where the rescattering integrals contain *off-shell amplitudes.* Parametrizations of form factors are chosen without giving a detailed theoretical justification, higher partial waves and some inelasticities are neglected without estimating the errors, and the reliability of the numerical solution of the complicated integral equation is not well known. Furthermore, the dispersion calculations have the advantage that they were performed independently by several authors, who used different phase shift solutions and a variety of extrapolation methods.

The more recent papers 80 L Wei-1 and 2 contain the extension to the treatment of the P-waves. A correction to the sigma term is given in 81 L Banerjee.

The model of Ernst, M. B. Johnson, and McLeod

The authors started from the Chew-Low model (56 L), extending it beyond the "one-meson approximation" to introduce the effects of inelastic channels (78 L Ernst). Next, the Chew-Low model is combined with a separable potential model (72 L Landau, 74 L Ernst) for the S-, D-, and F-waves. The no-crossing approximation is used for the P 13, P 31, and P 33 waves (80 L Ernst).

Finally, the model is generalized to include relativistic kinematics and relativistic pion-nucleon couplings. It is based on applications of N/D methods to relativistic partial wave dispersion theory. Several cases are considered: pseudoscalar and pseudovector πN coupling and a model in which the nucleon-antinucleon intermediate states are removed ("pair suppression"). The results are compared with experimental phase shifts and parameters are adjusted in order to obtain in some cases a quantitative agreement (81 L McLeod).

Comments

i) The authors investigated πN partial wave dispersion relations, but their list of references shows that they have not yet attempted to exploit the large amount of information available in the papers of Hamilton and coworkers (67 R, 71 R, 72 R, 76 L, 72 L Lea). The most recent evaluation is due to Hutt and Koch (80 L).

ii) They discuss the method of CGLN (57 L Chew), but they do not consider the exact version of this method (70 R Baacke, 71 R Steiner). In particular the *generalized potentials* (Sect. A.7.2) should be of interest, because they were derived from fixed-t dispersion relations and have therefore a good reliability (70 L Engels-3, 71 R Steiner). Engels and Fleischer (71 L) gave results for N/D calculations for the P 33 wave with effective Lagrangian input.

Separable potential models

Separable potential models offer a possibility to derive off-shell pion-nucleon amplitudes, which are needed in the theory of pion-*nucleus* scattering, from on-shell pion-nucleon scattering amplitudes. Various methods have been developed for the construction of the potential from scattering phase shifts, i.e. for the solution of the inverse scattering problem.

The literature can be traced from the early paper 72 L Landau and from the recent ones: 81 L Coronis, 81 L Fuda, 81 L Nakano, 82 L Garcilazo. Fröhlich et al. (82 L-1 and 2) presented a separable approach to the Bethe-Salpeter equation for the pion-nucleon system, which can be applied to pion-deuteron scattering.

***t*-matrix**

For applications in nuclear physics, van Orden et al. (81 L) have constructed a t-matrix from the pion-nucleon scattering amplitudes.

Padé approximants

A Padé model for pion-nucleon scattering (72 L Filkov), had serious problems (73 L Bergère). Engels and Fleischer (74 L) studied Padé approximants for Legendre expansions in order to probe the singularity structure of the scattering amplitudes. A Padé method was also used in the above mentioned papers 76 L Haysak, 76 L Lin, and 77 L Brunet. General discussions of Padé methods can be found for instance in 70 R Baker-Gammel, 71 R Zinn-Justin-1 and 2 and 72 R Basdevant.

2.5.2 Electromagnetic nucleon and pion form factors

2.5.2.1 Introduction

Phenomenological analyses of electromagnetic nucleon form factors usually start with a dispersion relation, which reads for instance for the electric (Sachs) form factor of the proton

$$G_{\mathrm{Ep}}(t)=\frac{1}{\pi}\int_{4\mu^2}^{\infty}\frac{\operatorname{Im}G_{\mathrm{Ep}}(t')}{t'-t}\,\mathrm{d}t', \tag{5.2.1}$$

Similar relations are written for the form factors G_{Mp}, $F_{1\mathrm{p}}$, $F_{2\mathrm{p}}$ and for the form factors of the neutron (see Sect. A.11.1). The dispersion relations for the Sachs form factors (G_{E}, G_{M}) and for the Dirac and Pauli form factors F_1 and F_2 are equivalent, neither of them is more fundamental than the others. However, the results can be different, if approximations are made (for instance ϱ-dominance).

The data suggest that $t^2G_{\mathrm{Mp}}\to$const as $t\to-\infty$ (Fig. 5.2.3). This leads to the "superconvergence sum rule"

$$\int_{4\mu^2}^{\infty}\operatorname{Im}G_{\mathrm{Mp}}(t)\mathrm{d}t=0. \tag{5.2.1a}$$

Since the values of the form factors at $t=0$ are known (Table A.11.1), one can perform a subtraction without introducing an unknown parameter.

The aim is now to exploit the rich experimental information on the form factors at spacelike values of t ($t<0$) in order to derive information on the behavior of the imaginary parts ("spectral functions") at $t>4\mu^2$.

A summary of the experimental data and of the methods for the determination of form factors is given in my lecture notes (76 R and 77 R HÖHLER) and in 76 L HÖHLER. New data at low energies can be found in SIMON (80 L-1, 80 L-2, and 81 L). Other new data are mentioned in the reviews 79 R ATWOOD and 79 R BERKELMAN.

Mathematically, the determination of the spectral functions from the real values of the form factors at $t<0$ is an analytic continuation from the interior of the analyticity domain to its boundary. As discussed for instance in the reviews 70 R PISUT, 75 R CIULLI-POMPONIU-SABBA-STEFANESCU and in 70 L NIELSEN, this continuation is, in a local sense, basically instable. One cannot expect to obtain reliable results for the shape of the spectral functions but only for certain "moments".

In principle, the situation is more favorable, since the nature of the branch point is known and it is reasonable to assume that there are no unexpected nearby singularities in the second sheet. However, in practice, the continuation still has a large uncertainty.

In this situation, it is useful to restrict the solution by physical assumptions. Many authors simplified the problem by assuming that the γNN-interaction is mediated by the exchange of vector mesons, which can be treated in the narrow resonance approximation. Then, the dispersion integral is replaced by a sum of pole terms. ZOVKOs review (75 R) summarizes the results obtained with this method. Its shortcomings will be discussed below.

It is remarkable that the data for G_{Ep}, G_{MP}, and G_{MN} can be described in a zero order approximation by the "dipole fit" of Eq. (A.11.9). The magnitude of the systematic deviations has been studied in 76 L HÖHLER.

Further books and reviews: 61 R DRELL-ZACHARIASEN, 61 R FRAZER, 61 R GOURDIN, 67 R WILSON, 68 R BERNSTEIN, 72 R BARTOLI-FELICETTI-SILVESTRINI, 72 R FEYNMAN, 74 R GOURDIN, 74 R RASZILLIER, 75 R WEST, 78 R DONNACHIE-SHAW-LYTH.

2.5.2.2 Dispersion theory of the nucleon isovector form factors

A strong constraint for the isovector spectral functions follows from extended unitarity in the range where it is sufficient to consider two-pion intermediate states

$$\operatorname{Im} G_{Ev}(t) = \frac{1}{m} \frac{q_t^3}{\sqrt{t}} F_\pi(t) f_+^{1*}(t); \quad 4\mu^2 < t < 16\mu^2. \tag{5.2.2}$$

See Eq. (A.11.13) for the other spectral functions. $F_\pi(t)$ is the electromagnetic pion form factor and $q_t^2 = (t/4) - \mu^2$. The 4-pion intermediate states are usually neglected up to the $\omega\pi^0$ threshold ($t = 43\mu^2$), because the 4π production in e^+e^- scattering is small at lower t-values (79 L CORDIER, 79 R SIDOROV, 81 R PAUL) and the inelasticity in ππ-scattering in the $I = J = 1$ state is also small (77 L FROGGATT, 78 L MARTIN-1).

The fundamental papers on the dispersion theory of the nucleon form factors and the application of the unitarity condition are 58 L CHEW, 58 L FEDERBUSH and 60 L FRAZER-FULCO-1. Practically all subsequent authors in this field used the formalism of FRAZER and FULCO, who developed the theory of the partial wave dispersion relations for the ππN$\bar{\text{N}}$ amplitudes and applied the N/D method to the determination of the isovector spectral functions. An important point in their analysis was that they postulated the existence of a $I = J = 1$ ππ-resonance, the ϱ-meson, which was subsequently discovered in other reactions.

For a quantitative application of the unitarity relation, Eq. (5.2.2), it is necessary to know the ππN$\bar{\text{N}}$ partial waves $f_\pm^1(t)$ for $t > 4\mu^2$. As discussed in Sect. 2.4.6, reliable results for these partial waves could only be obtained after the CERN 68 phase shift solution became available (68 L HÖHLER-2). An updated and improved version was published in collaboration with E. PIETARINEN (75 L HÖHLER-3 and 4).

As mentioned in Sects. A.11.1 and 2.4.6, it is useful to introduce the numerator functions $J_\pm(t)$ by

$$f_\pm^1(t) \equiv \frac{J_\pm(t)}{D_1(t)} = J_\pm(t) F_\pi(t), \tag{5.2.3}$$

Then, the unitarity relation reads

$$\operatorname{Im} G_{Ev}(t) = \frac{1}{m} \frac{q_t^3}{\sqrt{t}} |F_\pi(t)|^2 J_+(t), \quad \text{etc.} \tag{5.2.4}$$

The problem is now reduced to the determination of $J_\pm(t)$, since $|F_\pi(t)|^2$ follows from the data for the reaction $e^+e^- \to \pi^+\pi^-$, Eq. (A.11.30). However, these data have to be analyzed in order to remove the ω – ϱ-interference effect and to obtain a smooth interpolation, which fulfills the dispersion relation for $F_\pi(t)$ (see below and 79 L COSTA DE BEAUREGARD).

Instead of the choice $D_1(t)=F_\pi^{-1}(t)$ in Eq. (5.2.3) one could as well calculate $D_1(t)$ from the $\pi\pi$-phase shift $\delta_1^1(t)$, using a phase representation Eq. (A.7.51b). The difference between the results shows the uncertainty of the method (75 L HÖHLER-4, 77 L PIETARINEN).

A remarkable complication arises from the fact that, in contradistinction to other cases, the *numerator functions* $J_\pm(t)$ *have a rapid t-dependence*, because the left hand branch cut of the partial wave dispersion relation for $f_\pm^1(t)$ starts at $t=t_{nc}=(4-\mu^2/m^2)\mu^2=3.98\mu^2$, i.e. *very near* the branch point of the right hand cut. This is a consequence of the small mass ratio μ/m (Sect. A.7.5).

The rapid increase of $J_+(t)$ as t approaches t_{nc} is shown in Fig. 4.6.7. As a consequence, the isovector spectral functions and also $\mathrm{Im} f_\pm^1$ (Fig. 4.6.7) have an unusual threshold behaviour at $t \geqq 4\mu^2$.

$\mathrm{Im} F_\pi(t)$ behaves as expected for a P-wave (Fig. 4.6.7). But the spectral functions of the form factors and $\mathrm{Im} f_\pm^1(t)$ have a strong enhancement of the left wing of the ϱ-resonance peak.

A large effect of this enhancement can be expected if one considers the nucleon radii, which are defined by Eq. (A.11.6) and follow from

$$\langle r^2 \rangle = \frac{6}{\pi} \int_{4\mu^2}^{\infty} \frac{\mathrm{Im}\, G(t)}{t^2} \mathrm{d}t. \tag{5.2.5}$$

Figure 5.2.1 shows $2\,\mathrm{Im}\, G_{Ev}(t)/t^2$ in comparison with $\mathrm{Im} F_\pi(t)/t^2$, which has the typical P-wave behaviour. The rapid increase of $2\,\mathrm{Im}\, G_{Ev}(t)/t^2$ at $t=4\mu^2$ is due to the above mentioned enhancement.

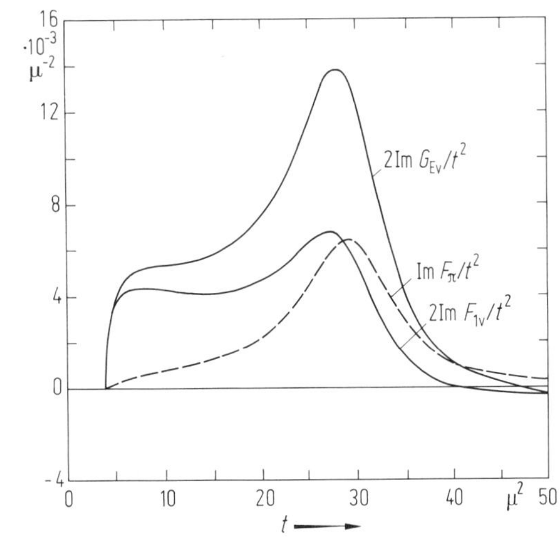

Fig. 5.2.1. Contributions to the dispersion integral for the pion and isovector nucleon radii. Equation (5.2.5) shows that $\langle r^2 \rangle$ depends on $\mathrm{Im}\, G/t^2$. It is seen that the vector dominance model is qualitatively wrong, because it ignores the large enhancement near threshold. The dipole model is also misleading, since the effect of the enhancement near threshold is described by the distant second pole. In the case of the pion, the vector dominance model gives a reasonable approximation. Some authors assumed $F_\pi(t) \approx 2F_{1v}(t)$, but this is not justified, because the spectral functions are much different.

If $\langle r^2 \rangle$ is calculated using the unitarity condition and Eq. (5.2.5), one obtains values near to the result following from the slope of the experimental form factors and Eq. (A.11.6), but almost *twice as large* as predicted from the *ϱ-dominance model*. Only the radius belonging to the isovector Dirac form factor is somewhat smaller (75 L HÖHLER-3). The radii are listed in Table 1.11.1 (76 L HÖHLER).

In generalized vector dominance models, the correct radii are obtained in the fit by considering the exchanges of one or two additional vector mesons of higher mass in addition to ϱ-exchange. The effective ϱ-mass usually comes out smaller than the true ϱ-mass. Figure 5.2.1 shows that all these fits are misleading, because an important contribution to the radii comes from the enhancement near $t=4\mu^2$, which is not related to the exchange of an additional vector meson. The empirical "dipole fit" is a special case of these exchange models.

The last remark is an example for the problems occurring in an analytic continuation from the interior to the boundary of the analyticity domain. The experimental form factor data can be well fitted by three vector meson exchange terms, but the result for the spectral function is *completely wrong* already along the nearest part of the cut.

A number of authors carried out more sophisticated analytic continuations, applying conformal mapping techniques (references are given in my lecture notes 76 R HÖHLER, see also 81 L SIMON-1). However, even these techniques do not lead to a satisfactory result, if one attempts a continuation to a boundary (the cut $t>4\mu^2$) and does not realize that a singularity lies *just behind* this boundary, at a very close distance (in the second sheet). Of course, this singularity belongs to the above mentioned enhancement and lies at $t_{nc}=3.98\mu^2$ (see Sect. A.11.3).

The continuation of form factors into the second sheet was treated by OEHME (61 L, 61 R) and by RASZILLIER (74 R).

Although the effect of the nucleon cut was described in some detail by FRAZER and FULCO (60 L-1), it was ignored in most of the analyses of the nucleon form factors (for instance in 64 L LEVINGER, 74 L HAMMER (see our comment 75 L HÖHLER-2) and 71 R PFISTER, 73 L DEO, 82 L IWAO). In a recent PADÉ-type analysis, DUBNICKA (81 L-2) found a pole which simulates the nucleon cut in the second sheet. Earlier applications of the PADÉ method led to contradictory results (76 L DUMBRAJS, 78 L BOWCOCK).

If one calculates from the form factors a charge distribution, using the inverse of Eq. (A.11.7), it is seen that the enhancement of ImG is related to the *tail* of the charge distribution, far outside the nucleon. Near $t=4\mu^2$, ImG can be approximated by the expression following from lowest order perturbation theory, which is usually called "uncorrelated two-pion exchange".

2.5.2.3 Dispersion analysis of nucleon form factor data

The discussion in the previous subsection has shown, how one can proceed in analyses of nucleon form factor data. The shape of the isovector spectral functions in the range $4\mu^2 < t \lesssim 40\mu^2$ is taken from the unitarity relation Eq. (A.11.13), using our determination of $J_{\pm}(t)$ (Sect. 2.4.6). The result is denoted by $\mathrm{Im}\,G^{2\pi}$.

It is impossible to determine details of the shape of ImG at larger t. Instead of applying sophisticated conformal mapping techniques, one can as well assume an ansatz of the type

$$\mathrm{Im}\,G_{\mathrm{Ev}} = \mathrm{Im}\,G_{\mathrm{Ev}}^{2\pi} + \sum_{\mathrm{i}} \frac{a_{\mathrm{i}}}{t_{\mathrm{i}} - t}; \quad \text{etc.,} \tag{5.2.6}$$

which contains a small number of vector meson exchange terms.

This ansatz is based on the expectation that above $t = m_{\varrho}^2$ the spectral functions are dominated by a few pronounced bumps and dips. If this is true, then one can hope to determine their location. Otherwise, the poles represent averages over fairly large t-intervals.

In the isoscalar case the situation is quite different, because there exists no evaluation of the unitarity relation near the threshold for 3-pion states, which lies at a larger distance ($t = 9\mu^2$). The usual assumption is that the spectral function is negligible below the ω-peak.

The ansatz for the isoscalar spectral function contains of course a pole term for ω-exchange. We have added a small number of further pole terms in analogy to Eq. (5.2.6).

Our analysis (76 L HÖHLER) was performed in collaboration with E. PIETARINEN, I. SABBA-STEFANESCU and with the experimental group at Mainz (F. BORKOWSKI, G. G. SIMON, V. H. WALTHER, and R. D. WENDLING).

Various fits showed that none of the adjustable poles chose a mass below the ϱ- and ω-mass, i.e. the data are compatible with our assumptions on the spectral functions at small t.

One could expect from the success of the dipole fit that the spectral functions have a pronounced bump-dip structure. This is true for $\mathrm{Im}\,G_{\mathrm{Ep}}$, G_{Mp} and G_{Mn}, but it is remarkable that the Dirac and Pauli spectral functions show the pronounced structure only in $F_{1\mathrm{s}}$ and $F_{2\mathrm{v}}$ and not in $F_{1\mathrm{v}}$ and $F_{2\mathrm{s}}$.

Since the height and the shape of the ϱ-peak and the ω-mass are given by the input, the form factor data determine fairly well the location of the effective poles for the dips in $\mathrm{Im}\,F_{1\mathrm{s}}$ and $\mathrm{Im}\,F_{2\mathrm{v}}$. The dip in $\mathrm{Im}\,F_{1\mathrm{s}}$ is found near the ϕ-mass (1020 MeV), but somewhat larger values up to about 1250 MeV cannot be excluded. It is difficult to interpret the result, since the contribution from ϕ-exchange is expected to be weak due to the OZI-rule (63 L OKUBO) and an ω′(1250) is not known. The situation is discussed in 76 L GENZ.

The dip in $\mathrm{Im}\,F_{2\mathrm{v}}$ is well determined at $t \approx 80\mu^2$, corresponding to a mass of 1250 MeV. The evidence for the existence of a resonance ϱ′(1250) is not compelling (80 R BRICMAN, Review of Particle Properties). See also 80 L ASTON and 81 R PAUL.

Since the ϱ-peak is given by the input, the residue of the ω-pole is better determined than in earlier analyses. According to the vector dominance model, it is a product of the γω- and ωNN-coupling constants

$$a_1(\omega) = (t_{\omega}/f_{\omega})\, g_1(\omega\mathrm{NN}). \tag{5.2.7}$$

Using the $\omega \to e^+e^-$ width (0.76 ± 0.17 keV from 80 R BRICMAN) which has not been remeasured since 1974, one finds

$$f_{\omega}^2/4\pi = 18.3 \pm 4; \quad g_1^2(\omega\mathrm{NN})/4\pi = 24 \pm (5 \pm 7). \tag{5.2.8}$$

The first error comes from that of f_{ω} and the second one belongs to $a_1(\omega)$, whose error depends of course on our assumptions. The ωNN-coupling constant is important in the theory of nucleon-nucleon scattering (Sect. 2.5.3).

Further details can be found in the above mentioned papers. The following points are of interest for improvements of the results:

i) In the isovector case one needs a reliable model for the discontinuity along the nearby part of the four-pion cut. In particular, it is important to know, whether ϱ′(1250) exists and can be made responsible for the dip in $\mathrm{Im}F_{2v}$. Investigations of this problem were carried out by BUDNEV et al. (76 L-1 and 2, 77 R SEREBRYAKOV) and by WILLRODT (76 L).

ii) In the isoscalar case the main questions are the origin of the dip in $\mathrm{Im}\,F_{1s}$ and the possible existence of a tail of $\mathrm{Im}\,F_{1s}$ below $t=m_\omega^2$. Improved data for the reaction $e^+e^-\to\pi^+\pi^-\pi^0$ would be helpful. The available data are shown in Fig. 8 of the review 80 R AUGUSTIN.

iii) Important constraints for the dispersion analysis would follow from accurate data for the reactions $e^+e^-\leftrightarrow\bar{N}N$ (see 79 L DELCOURT and 77 L BASSOMPIERRE) and from the masses of higher vector mesons (82 L CORDIER, 82 L DELCOURT, 82 L MANE).

iv) At present, the information on G_{Ep} is rather poor beyond $t=-1.5\,\mathrm{GeV}^2$, because the G_{Mp}-contribution becomes dominant in the ROSENBLUTH formula Eq. (A.11.15). The only possibility to determine G_{Ep} at large negative t-values is to use longitudinally polarized electrons and a polarized proton target and to measure the asymmetry on flipping the target polarization. The result is proportional to $G_{Ep}G_{Mp}$ (69 R DOMBEY). — Polarized electron-nucleon scattering experiments are being carried out mainly because one can study the interference between electromagnetic and weak interactions (see for instance 78 L CAHN, 79 L CHRISTOVA).

v) The prediction of perturbative quantum chromodynamics for the large-t behaviour of the nucleon form factors can be used as constraint. According to LEPAGE and BRODSKY (79 L and 80 L), the form factor G_{Mp} behaves as t^{-2} up to logarithmic terms and the helicity flip form factor F_{2p} falls roughly as t^{-3} (see also 80 L DUNCAN and 78 L COQUEREAUX, 82 R SIVERS).

For new analyses it is of interest to have an empirical parametrization of the results obtained for the ϱ-contribution at $t\leqq 0$, if the unitarity relation is evaluated as described in Sect. 2.5.2.2 and substituted into the dispersion relation (t in GeV^2)

$$2F_{1v}^{\varrho}(t)=\frac{0.955+0.090(1-t/0.355)^{-2}}{1-t/0.536};\qquad 2F_{2v}^{\varrho}(t)=\frac{5.335+0.962(1-t/0.268)^{-1}}{1-t/0.603};\qquad t\leqq 0. \tag{5.2.9}$$

Some results for the nucleon form factors are shown in Figs. 5.2.2···5.2.6. The discrepancies between the dispersion fits and the results derived from Rosenbluth plots are discussed in 76 L HÖHLER.

We include a table of proton and pion form factors, which is of interest for an evaluation of the electromagnetic contributions to pion-nucleon scattering amplitudes.

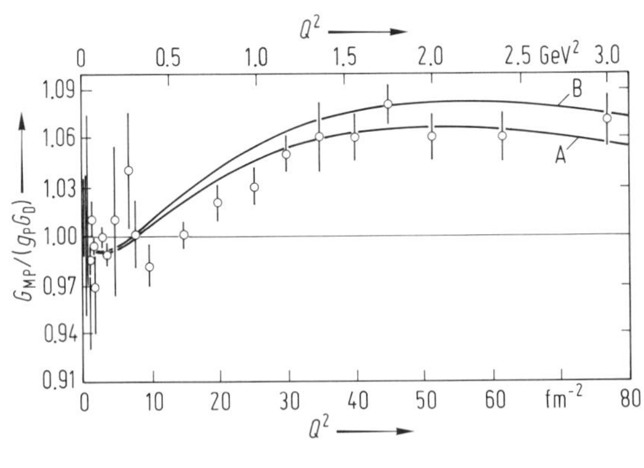

Fig. 5.2.2. The deviation of the form factor $G_{Mp}(t)$ from the dipole fit. The solid lines are examples for dispersion fits, which used Eq. (5.2.9), a pole at the ω-mass and a few additional poles. The ω-residue and the residues and masses of the other poles are treated as adjustable parameters. Fit A includes 237 and fit B 298 data points. — The points with error bars have been calculated from Rosenbluth plots. Only about 60 % of the data points are suitable for this purpose. See 76 L HÖHLER for a discussion of the discrepancy and for references of the data points. $Q^2=-t$; $1\,\mathrm{fm}^{-2}=0.03894\,\mathrm{GeV}^2=2.00\,\mu^2$.

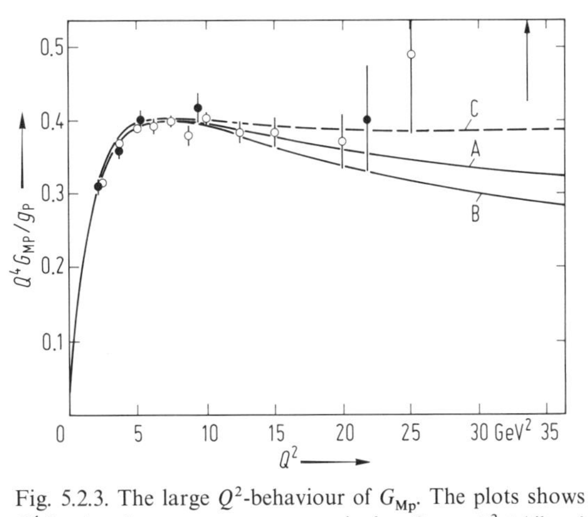

Fig. 5.2.3. The large Q^2-behaviour of G_{Mp}. The plots shows Q^4G_{Mp}/g_p, since one wants to test whether $G_{Mp}\propto t^{-2}$. "A" and "B" are the dispersion fits of Fig. 5.2.2. The experimental points have been derived from ep→ep scattering data, assuming the validity of the scaling law $G_{Ep}(t)=G_{Mp}(t)/g_p$, Eq. (A.11.19). The dashed line "C" shows the result calculated from the cross section obtained from fit A and analyzed in the same way as the SLAC data, although the scaling law is not valid in this fit. References for the data are given in 76 L HÖHLER.

Table 2.5.2.1. Proton and pion form factors. G_{Ep} is the result of our dispersion fit (76 L HÖHLER). The third column gives $F_\pi(t)$ according to solution A of HEYN and LANG (81 L, $r_\pi = 0.70$ fm). The fourth column is based on a pole fit which is adjusted to the electroproduction data at small $(-t)$. It belongs to $r_\pi = 0.68$ fm. The last column is the product squared of the second and fourth column, which occurs in Eq. (A.4.30) for the pure pion-nucleon Coulomb cross section. The error of the pion form factor is comparable with the difference between the third and fourth column. The error of G_{Ep} is smaller, because of the accurate Mainz data at small $(-t)$ (below 1%, see 76 L HÖHLER and 80 L SIMON-1). 1 GeV^2 = 25.68 fm^{-2}.

$-t$ GeV^2	G_{Ep}	F_π	F_π	$F_\pi^2 G_{Ep}^2 = f_E^2$
0.00	1.000	1.000	1.000	1.00
0.05	0.865	0.904	0.910	0.620
0.10	0.757	0.825	0.835	0.399
0.15	0.668	0.757	0.771	0.266
0.20	0.595	0.699	0.716	0.182
0.25	0.532	0.649	0.669	0.127
0.30	0.479	0.606	0.627	$0.905 \cdot 10^{-1}$
0.35	0.434	0.567	0.591	$0.656 \cdot 10^{-1}$
0.40	0.394	0.533	0.558	$0.483 \cdot 10^{-1}$
0.45	0.359	0.503	0.529	$0.361 \cdot 10^{-1}$
0.50	0.329	0.476	0.503	$0.273 \cdot 10^{-1}$
0.55	0.302	0.451	0.479	$0.209 \cdot 10^{-1}$
0.60	0.278	0.429	0.457	$0.161 \cdot 10^{-1}$
0.65	0.256	0.408	0.437	$0.126 \cdot 10^{-1}$
0.70	0.237	0.390	0.419	$0.986 \cdot 10^{-2}$
0.75	0.220	0.373	0.403	$0.782 \cdot 10^{-2}$
0.80	0.204	0.357	0.387	$0.624 \cdot 10^{-2}$
0.85	0.190	0.343	0.373	$0.501 \cdot 10^{-2}$
0.90	0.177	0.329	0.360	$0.405 \cdot 10^{-2}$
0.95	0.165	0.317	0.347	$0.329 \cdot 10^{-2}$
	76 L HÖHLER	81 L HEYN	pole term	

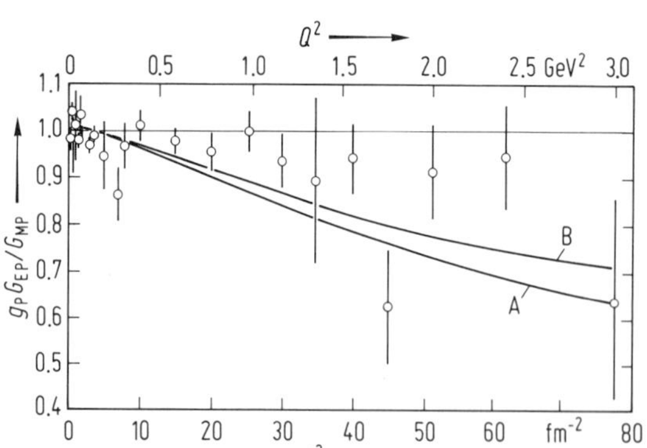

Fig. 5.2.4. Test of the scaling law. See Fig. 5.2.2 for the notation. Experiments with polarized electrons are required for a more accurate experimental determination of G_{Ep}/G_{Mp}.

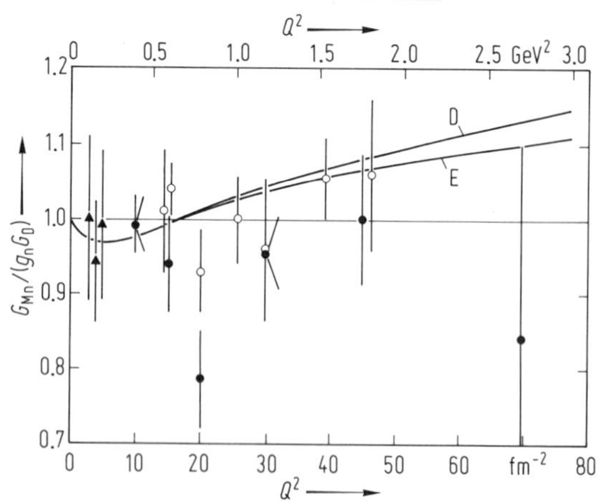

Fig. 5.2.5. Deviation of the form factor G_{Mn} from the dipole fit. "D" and "E" are dispersion fits, assuming adjustable poles in the isoscalar and isovector spectral functions in addition to Eq. (5.2.9) for ϱ and to a ω-pole. The second isovector pole has always a mass near 1.2 GeV. The second isoscalar pole chooses a mass slightly below the ϕ-mass (fit "D"), but the fit is still good if it is assumed to be located at the ϕ-mass ("E"). Details of the fits are given in 76 HÖHLER.

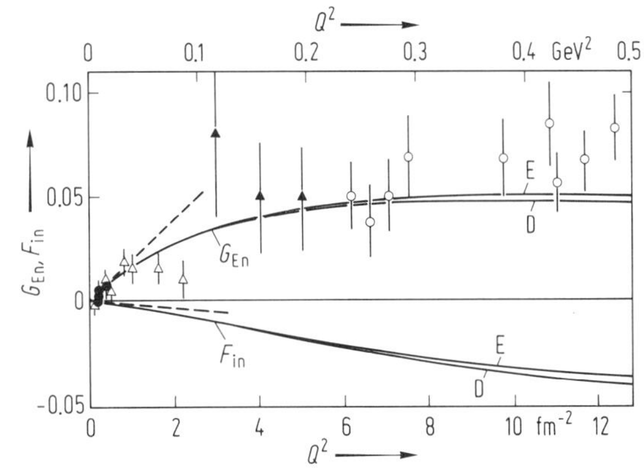

Fig. 5.2.6. The electric form factors G_{En} and F_{1n} of the neutron. The dispersion fits "D" and "E" are the same as in Fig. 5.2.5. The dashed lines are the slopes of G_{En} and of F_{1n} at $Q^2=0$. See Eq. (A.11.9) for the relation between these slopes. The experimental data are discussed in 76 L Höhler. (In the figure F_{in} has to be replaced by F_{1n}.)

Nucleon form factors derived from quark models

A fairly large number of authors attempted to derive electromagnetic properties of the nucleon (and other baryons) from quark models. One of the aims was to improve the SU (6) relation for the ratio of the proton and neutron magnetic moments: $\mu_p/\mu_n=-3/2$, and to derive the values for proton and neutron separately (82 L Böhm, 80 L Isgur; here and below, we mention only a few recent papers from which earlier references can be traced). Martinelli et al. (82 L) calculated the magnetic moments from lattice QCD.

Another frequently discussed problem is the calculation of nucleon radii, in particular of the "neutron charge radius"*) (80 L Parashar, 81 L Isgur, 81 L Dziembrowsky, 82 L Theberge-1, 81 L Thomas-3, 82 L Thomas, Barnhill (79 L-1 and 80 L), 79 R van de Walle). One should note that a dispersion analysis leads to more reliable values for the radii and also for $G_{En}(t)$ (Fig. 5.2.6) than simple vector dominance models (82 L Mickens), which have a wrong spectral function.

It might be of interest to demand that a model leads to the decomposition Eq. (A.11.9) which shows that a large part of the slope of G_{En} at $t=0$ belongs to the Foldy term and is therefore determined by the *static value* of the anomalous magnetic moment.

Some authors used quark models for a prediction not only of the slopes of the form factors at $t=0$ but also of the t-dependence (for instance 79 L Ishida, 82 L Theberge-1, 81 L Margaritisz, 82 L Stanley). In this case, it is necessary to discuss the problem resulting from the fact that the spatial charge and magnetic moment distributions agree with the Fourier transforms of the form factors only in the non-relativistic limit (62 L Sachs, Appendix; 75 L Skachkov).

Field theoretic models

We mention only the most recent model of this type which is due to Gutbrod and Hartmann (79 L), who considered skeleton diagrams for a pseudoscalar pion-nucleon interaction. Unknown parameters describing higher order corrections are fixed such that a ϱ-meson with the correct properties is generated.

The polarizability of the nucleon

The electric and magnetic polarizability of the nucleon plays a role in the nucleon Compton amplitude ($\gamma N \to \gamma N$) at low energies, in hyperfine splitting of atoms and in level shifts of hadronic atoms. See 80 L Schröder for a recent discussion.

Ericson et al. (81 L) calculated the axial nucleonic polarizability of the nucleon, i.e. the response to a pionic wave, from the coefficient c_{01}^{+} of the expansion Eq. (4.7.2).

Conclusion on nucleon form factors

i) The isovector spectral functions in the range $4\mu^2<t<40\mu^2$ can be predicted from the unitarity relation, using results derived from πN-scattering amplitudes and the pion form factor. The experimental nucleon form factors are compatible with this prediction.

*) This notion is misleading, since a neutral particle does not have a "radius" in the sense of Eq. (A.11.8). The authors take six times the slope at $t=0$ for $\langle r^2\rangle$ [cf. Eq. (A.11.6)]. — In quark models, some authors take for $\langle r^2\rangle$ the average of $\langle r^2\rangle$ of the quarks, weighted by their charges.

ii) It is not possible to determine accurate values of the spectral functions directly from form factor data, even for small values of t near $t=4\mu^2$. The difficulty is caused by a singularity at $t=3.98\,\mu^2$, which lies near the path of continuation.

iii) The usual fits of the nucleon form factor data by a sum of vector-meson exchange terms are misleading for the isovector contribution, because ϱ-exchange cannot be approximated by a pole term or by a Breit-Wigner term due to the singularity mentioned in ii).

The weak nucleon form factor

We mention some references for the axial-vector nucleon form factor $G_A(\mathbf{q}^2)$: 79 R Amaldi-Fubini-Furlan, 79 L Gault, 79 R Sehgal, 78 R Paver, 74 R Gourdin, 73 R Alfaro-Fubini-Furlan-Rosetti. It is defined in analogy to Eq. (A.11.1) by a matrix element of the weak axial-vector current between nucleon states. G_A occurs not only in the amplitude for the "elastic" neutrino reactions (for instance $\bar{\nu}_\mu + p \rightarrow \mu^+ + n$) but, together with the electromagnetic form factors, also in the amplitude for pion electroproduction near threshold, if current algebra techniques and the PCAC assumption are used. $G_A(0)=g_A$ is the axial-vector nucleon coupling constant in Eqs. (5.1.1) and (5.1.37).

2.5.2.4 The pion form factor

In this volume, the pion form factor is needed for the evaluation of the unitarity relation for $\mathrm{Im}\, f_\pm^1$ (Sect. 2.4.6) and for the isovector spectral functions of the nucleon form factors. Furthermore, it occurs in the electromagnetic corrections to pion-nucleon scattering (Sect. A.2.3).

The pion form factor is defined in Sect. A.11.2. It can be determined in several reactions:

i) For negative values of t one has to separate the relevant contribution in pion electroproduction ($e^- p \rightarrow e^- n \pi^+$).

ii) For small negative values of t, the pion form factor can be measured almost directly in pion-electron scattering, using a high energy pion beam.

iii) For small positive values of t, the pion form factor can be obtained by separating the relevant contribution to inverse electroproduction ($\pi^- p \rightarrow e^+ e^- n$).

iv) For larger positive values of t, the modulus of the pion form factor can be measured directly in $e^+e^- \rightarrow \pi^+\pi^-$, but one has to correct for the electromagnetic $\omega \rightarrow \pi^+\pi^-$ decay ($\omega - \varrho$ interference).

v) The phase of the pion form factor in the range $4\mu^2 < t < 40\mu^2$ agrees with the ππ-scattering phase in the $I=J=1$ state. In principle, this is valid only up to $16\mu^2$, but the 4π-states can be neglected up to the $\omega\pi^0$-threshold as discussed above. Equation (A.11.27) shows that under certain assumptions and approximations $F_\pi(t)$ can be determined from the phase. — The ππ-scattering phase follows from an analysis of $\pi N \rightarrow N\pi\pi$ data at high energies, where the one-pion exchange contribution is large at small angles (Sect. 2.4.6.2).

It is assumed that the pion form factor fulfills the dispersion relation

$$F_\pi(t) = \frac{1}{\pi} \int_{4\mu^2}^{\infty} \frac{\mathrm{Im}\, F_\pi(t')}{t'-t}\, dt', \tag{5.2.10}$$

The dominant singularity is the ϱ-pole. An analysis of the unitarity relation for $\mathrm{Im}\, F_\pi(t)$, Eq. (A.11.26), shows that the situation is much simpler than in the case of the isovector nucleon spectral functions. The numerator function of the $I=J=1$ ππ-scattering amplitude is slowly varying and its branch cut lies at some distance [Eq. (A.7.55)].

The simplest model for the analytic structure of $F_\pi(t)$ at low values of t is an adaptation of the Chew-Mandelstam model for ππ-scattering, which leads to a generalized effective-range formula [Frazer and Fulco (60 L-1)]. The model is frequently called "Gounaris-Sakurai fit" (68 L, 69 L), although these authors pointed out that they have made only a small modification.

Similarly to the Chew-Low effective range model, this model can accommodate a resonance. It has only two adjustable parameters. The somewhat complicated formulas are given in Sects. A.7.6 and A.11.2. Table A.11.2 lists the parameters of a fit to the pion form factor data. This model is a reasonable approximation in the range where $F_\pi(t)$ is needed for our applications.

The main problem is, of course, to determine a function $F_\pi(t)$, which fulfills the dispersion relation and is compatible with *all* data mentioned above under i) till v). We give the most recent references:

Pion electroproduction: 78 L BEBEK, 77 L BARDIN, 79 R AMALDI-FUBINI-FURLAN.
Pion-electron scattering: 77 L DALLY, 77 L ADYLOV, 82 L DALLY.
Inverse electroproduction: 77 L BEREZHNEV, 79 L ALIZADE.
$e^+e^- \to \pi^+\pi^-$: 80 L SIDOROV, 80 R AUGUSTIN, 79 R WIIK-WOLF, 78 L QUENZER, 78 L BUKIN, 77 L ESPOSITO, 76 L COSME, 75 L BOLLINI.
$\pi\pi$-scattering: see Sect. 2.4.6.2.

In the literature one can find a large number of analyses of the pion form factor (see the reviews and books: 74 R GOURDIN, 77 R PEREZ-Y-ORBA, 78 R DONNACHIE-SHAW-LYTH, 79 R AMALDI-FUBINI-FURLAN, 79 R WIIK-WOLF). The most recent detailed investigation of dispersion fits to all pion form factor data is due to HEYN and LANG (81 L). Their method follows previous work by LANG and SABBA-STEFANESCU (75 L). The authors introduce stabilizing conditions for the fit. A parametrization in the elastic region is combined with an expansion in order to have enough flexibility for the large-t region. The parametrization is a generalization of the CHEW-MANDELSTAM model. It has one additional parameter.

The result of this analysis is not quite satisfactory. The $\pi\pi$-scattering phase shifts are not included in the input. If one calculates the scattering length a_1^1, a value is found which is almost three times as large as the value obtained from $\pi\pi$-scattering data combined with crossing and analyticity constraints (Sect. 2.4.6.2). If the latter value, $a_1^1 = 0.038\,\mu^{-3}$, is enforced by the choice of the parameters, one can have an equally good fit, but the solution shows more structure.

For the calculation of the Coulomb interference term in Sect. A.2.3, it is important to know the pion radius, which is defined in Eq. (A.11.24).

From their pion electroproduction data in the range $-0.15\,\text{GeV}^2 > t > -10\,\text{GeV}^2$, BEBEK et al. (78 L) derived the radius $r_\pi = (0.711 \pm 0.018)$ fm, assuming that the data can be fitted by a simple pole term formula. The mass of the pole is 680 MeV. Of course, the error does not include the uncertainty of the parametrization. Using a more flexible parametrization, one can fit the data equally well, but the radius can be considerably different.

A more reliable determination follows from the fits to all spacelike and timelike data. HEYN and LANG found the value $r_\pi = 0.70$ fm in their best fit and 0.67 fm, if a_1^1 is fixed at $0.038\,\mu^{-3}$. In the narrow resonance approximation, a ϱ-pole gives the radius 0.62 fm, which is considerably smaller.

We mention some other recent analyses (see also: 77 L HAMMER, 77 L RASZILLIER, 78 L GENSENI, 82 L DOMINGUEZ-2).

79 L FELICETTI. Data in the timelike region are fitted to a sum of four resonance parametrizations, taking into account radiative corrections. In addition to ϱ(776), ϱ′(1266), ϱ″(1540) the authors consider a narrow resonance ϱ_A (1100), $\Gamma_{\varrho A} = 0.03$ GeV, for which some evidence has been seen in a FRASCATI-DESY-experiment (see 79 R WIIK-WOLF, p. 43). The agreement with the spacelike data is reasonable, but the authors do not show the most recent result of the HARVARD group (78 L BEBEK).

78 L QUENZER. The authors analyze data in the timelike region, using a "GOUNARIS-SAKURAI" parametrization and adding terms for the contribution of ω and of higher resonances.

77 L PANTEA, 81 L DUBNICKA-1, 81 L DUBNICKA-3. The authors employ various dispersion methods. In the second paper a PADE method is used and it is claimed that the F_π-data would be consistent with the P-wave $\pi\pi$ phase shifts. It would be useful to check, whether there is a contradiction to the result of HEYN and LANG (81 L).

For new analyses of the pion form factor, it will be of interest to take into account the prediction derived from QCD for the asymptotic behaviour (79 L FARRAR)

$$F_\pi(t) \to \frac{8\pi\alpha_s(t) f_\pi^2}{-t} \approx \frac{2 f_\pi^2}{b(-t)\ln|t|} \quad \text{as} \quad t \to \pm\infty\,; \qquad b = \frac{11 - 2N_f/3}{16\pi^2}. \tag{5.2.11}$$

N_f is the number of quark flavors and $\alpha_s(t)$ the QCD expansion parameter, which has a logarithmic variation in t.

MACHET (80 L) has used a special modulus representation to derive constraints for $F_\pi(t)$ in the timelike region from information following from perturbative QCD in the spacelike region. The problem in calculations of this type is that one does not know the t-range, in which the asymptotic formula Eq. (5.2.11) and more elaborate versions are a reasonable approximation.

More recent developments are treated in 80 L EFREMOV, 80 L LEPAGE, 80 L DITTES, 81 L BRODSKY, 81 L GUPTA, 82 R BRODSKY, 82 L IOFFE, 82 R SIVERS.

The role of the pion form factor in pion electroproduction at low energies is treated in the reviews 79 R AMALDI-FUBINI-FURLAN and 81 R SCADRON and in the papers 79 L FURLAN and 79 L MACMULLEN. This reaction is of interest because it can be described using soft-pion and PCAC low energy theorems (see also 78 R PAVER and 70 L VON GEHLEN).

We add a list of recent determinations of the pion radius.

Table 2.5.2.2. Recent results for the pion radius $r_\pi = \langle r_\pi^2 \rangle^{1/2}$. The ϱ-dominance value is $r_\varrho = 0.62$ fm.

	r_π [fm]		r_π [fm]
77 L Adylov	0.78 ± 0.10	79 L Felicetti	0.69
77 L Bardin	0.74 ± 0.12	80 L Dubnicka-1	0.66 ± 0.04
77 L Dally	0.56 ± 0.04	81 L Heyn	0.70; 0.69; 0.67
78 L Bebek	0.711 ± 0.018	82 L Dally	0.663 ± 0.023
78 L Quenzer	0.678 ± 0.008		

2.5.3 Two-pion exchange in nucleon-nucleon scattering

2.5.3.1 Introduction

Many investigations of nucleon-nucleon scattering start with the unitarity relation for the imaginary part in the *crossed channel* (antinucleon-nucleon scattering). The intermediate states of lowest mass are the one-, two-, and three-pion states. In a dispersion relation approach the exchange of these states leads to the long and intermediate range forces in nucleon-nucleon scattering, as expected from an extension of Yukawas ideas. Details of the calculation are given in 73 L Cottingham. A clear introductory treatment can be found in 76 R Brown-Jackson and 78 R Cottingham. More recent results and a description of the methods developed for the short range forces are included in the reviews 79 R Holinde and 82 R Vinh Mau-1 and 2.

The one-pion exchange contribution to NN-scattering is well-known, because the πNN coupling constant has also been determined in πN scattering (Sect. 2.4.2). At earlier times, two- and three-pion exchange was described by "one-boson exchange models", assuming that it is sufficient to consider the exchange of resonances in the narrow resonance approximation. A fit of these models to experimental NN-phase shifts led to results for "coupling constants" of the nucleon to the different meson resonances (79 R Nagels, Sect. 7.2).

One should note that these "coupling constants" of the nuclear physicists are a notion different from the coupling constants in dispersion theory, where all particles are taken on their respective mass shells [see 75 L Höhler-4, where the ϱNN-coupling is discussed in connection with the work of Kroll, Lee, and Zumino (67 L)]. An important property of a "coupling constant" is that it can be used for the same vertex in different reactions. It is difficult to see how this can be done with the "couplings at $t=0$" in Table 7.3 of the "Compilation" (79 R Nagels).

The above mentioned application of dispersion methods to NN-scattering made it possible to calculate the effect of two-pion exchange from ππNN̄-amplitudes instead of using the crude boson exchange models.

For a certain period the contact between physicists working in nuclear physics and in πN scattering was rather weak. It happened that in some cases models for ππNN̄-amplitudes were used in NN-scattering, which were incompatible with results derived in πN scattering from reliable dispersion methods. For instance the model of Brown and Durso (71 L) was used in several subsequent papers, but nowadays there is a reasonable agreement (80 L Durso).

Since the treatment of two-pion intermediate states for a general NN scattering amplitude is rather complicated, we refer to the above mentioned reviews and restrict ourselves to the simpler case of forward scattering, which also has the advantage that the test of the compatibility with the input for the ππNN̄ amplitudes is more transparent.

2.5.3.2 Two-pion-exchange in NN forward scattering

In NN forward scattering there are three independent amplitudes for each value of the isospin. We consider the "spin-averaged" forward amplitude, whose imaginary part is determined by the total cross sections.

The forward dispersion relation for nucleon-nucleon scattering is more complicated than that for pion-nucleon scattering, because below threshold one has not only the deuteron and pion poles, but also an unphysical cut. This cut belongs to the amplitude for the crossed reaction (N̄N scattering). The imaginary part of the N̄N-amplitude does not vanish below the N̄N threshold, because multipion and K̄K states occur as intermediate states in the unitarity relation. The present status of NN scattering above threshold is summarized in 82 R Vinh Mau.

Following earlier work by Bugg (68 L), Grein (77 L) used the forward dispersion relation in order to obtain information on the unknown parts of the N̄N-cut and to test the compatibility with the input for the two-pion cut.

The procedure is the usual discrepancy method. One calculates the real part of the NN-forward amplitude from phase shifts and uses in addition results derived from Coulomb interference experiments. The *discrepancy function* is obtained, if all other calculable contributions are subtracted:

i) the integral over the total cross sections,

ii) the subtraction constant as calculated from the scattering lengths,

iii) the pion pole term and (in np scattering) the deuteron pole term,

iv) the integral over the two-pion cut as far as it can be calculated from the available ππNN̄ amplitudes, which were determined as described in Sect. 2.4.6.

The discrepancy function, which is known in a large energy range, is equal to all contributions of the unphysical part of the N̄N-cut, except those already taken into account in iv).

The problem is now a similar one as in the case of the nucleon form factors: one wants to derive information on the imaginary parts along the N̄N-cut from the discrepancy function. Again, the ansatz consists of a sum of pole terms. In general, the masses are given by the masses of meson resonances, but a pole with an adjustable mass parameter was also considered. If this pole had chosen a mass value in the energy range, where the input for the ππNN̄ amplitudes is assumed to describe completely the imaginary part of the N̄N-amplitude, one would have to conclude that the ππNN̄ amplitudes derived from πN scattering are not compatible with the information on NN scattering. Fortunately, it turns out that the overall consistency is as good as can be expected.

The residues of the other pole terms are adjustable parameters, which are proportional to the squares of the meson-nucleon coupling constants. Only the fNN and gNN coupling constants were taken from the result of Borie and Kaiser (77 L) (see Sect. 2.4.2).

Since one can calculate the discrepancy functions for two reactions: pp and np scattering, the different contributions can be separated to a certain extent.

An interesting part of the result is the dip in the imaginary part of the isospin $I=0$ N̄N-amplitude (Fig. 5.3.1). In a very crude approximation, this broad dip can be described by an effective pole term whose mass parameter is 670 MeV. It occurs in the range where the f_+^0-contribution is expected to dominate ($f_\pm^1$ does not contribute to this amplitude). Obviously, this dip is the reason why earlier authors found the exchange of a "σ" or "ε" meson in this mass range. One should note that in this amplitude the effect of the ππ-interaction is relatively small in comparison with the nucleon pole term contribution, whereas the latter contribution amounts only to 1/3 of the imaginary part of the $I=1$ NN-amplitude at the ϱ-resonance (Fig. 5.3.1).

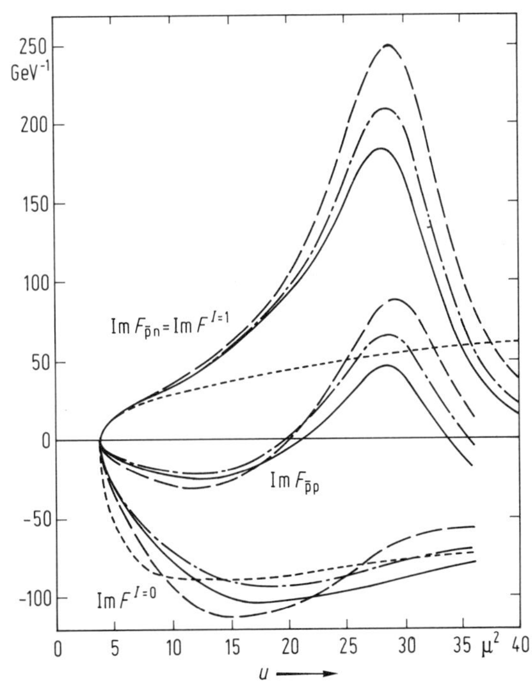

Fig. 5.3.1. Contribution of ππ-intermediate states to the imaginary parts of NN-scattering amplitudes (77 L Grein). Long-dashed lines: from a preliminary version of our ππNN̄-amplitudes (Fig. 4.6.1); dot-dashed lines: from earlier versions of the result of Nielsen and Oades (Fig. 4.6.1). The solid lines show the behaviour obtained from CERN 72 partial waves. Short-dashed lines: contribution of the ππNN̄ Born term amplitudes. – The broad dip in Im$F^{I=0}$ is approximated by σ- or ε-exchange in one-boson-exchange models.

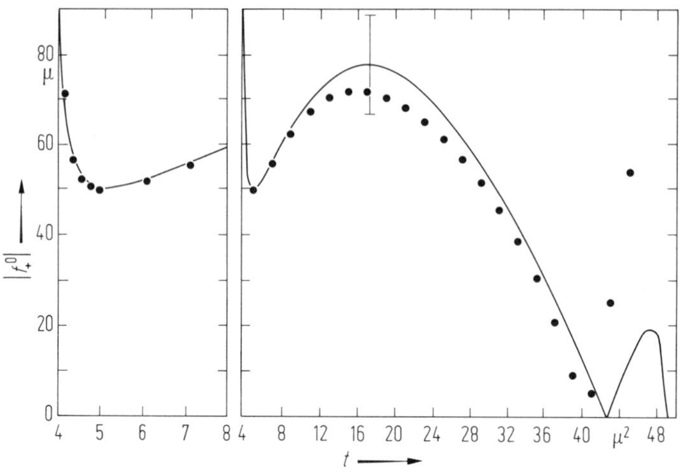

Fig. 5.3.2. The modulus of the ππN$\bar{N}$-partial wave f_+^0. See Fig. 4.6.1a for the real and imaginary parts and for the notation. $|f_+^0|$ has been discussed in several papers on NN-scattering, for instance in 80 L DURSO.

An Argonne group (78 L AUER) has recently measured total NN cross sections in pure spin states. The results made it possible to extend GREINs work to all three NN-forward amplitudes [GREIN and KROLL (78 L, 80 L and 82 L, 81 R KROLL)]. It is now easier to disentangle the different contributions. The imaginary parts of two NN- forward amplitudes depend on $|f_+^J|^2$ and $|f_-^J|^2$, respectively, and the imaginary part of the third amplitude has no two-pion contribution in its unitarity relation.

The authors conclude that the two-pion contribution as calculated from πN scattering is very well compatible with the information on the NN-amplitudes. However, they obtain a rather small value for the ωNN coupling constant

$$g_1^2(\omega NN)/4\pi = 8.1 \pm 1.5\,, \tag{5.3.1}$$

which disagrees with the value derived from the analysis of nucleon form factors, Eq. (5.2.8). I do not think that the accuracy of the value derived from form factor data is as much lower as suggested by the errors, since in this field the errors are always based on assumptions. It will be interesting to check, under which conditions a small value of the ωNN coupling constant can be reconciled with the form factor data.

Values of the ωNN-coupling constant derived from KN-scattering are given in Table XII of 78 R HAMILTON-OADES [$g_1^2(\omega\, NN) \approx 5$].

In any case, there is no doubt that the spectral functions $\mathrm{Im}\,F_{1s}$ and $\mathrm{Im}\,F_{2v}$ have a *dipole-like structure*, i.e. there are effects localized at slightly higher masses, which have a *strength comparable with that of* ϱ- *and* ω-*exchange*. The question is, whether these effects can also be found on the N$\bar{N}$-cut. If this is not the case, one would have to conclude that they are strongly enhanced by the coupling to the photon in the unitarity graph for γNN.

Another result of GREIN and KROLL is the necessity to include a three-pion exchange contribution, which is small in comparison with π- and ϱ-exchange, but comparable with the other exchanges. This contribution is centered at a relatively small mass (500···800 MeV).

A detailed investigation of uncorrelated and enhanced three-pion exchange has recently been performed by HAMILTON (81 L). He found a broad $I=1$ pseudoscalar exchange term centred at about 700 MeV mass and, at higher values of t, an $I=0$ pure tensor term and an $I=1$, $C=1$, axial term. A discussion of πϱ-exchange (77 L DURSO-1) is included.

Further work on three-pion intermediate states is necessary. The problem is not solved unless one is able to describe simultaneously the form factors in which three-pion exchange plays a role (γNN and πNN), taking into account the 3π-production data in e^+e^-- and $\bar{N}$N-scattering. As pointed out in Sect. 2.5.2.3, a large unexplained effect exists in the isoscalar spectral function $\mathrm{Im}\,F_{1s}$, unless a large 3π-exchange contribution comes from ϕ-exchange, violating the OZI-rule (78 R OKUBO).

2.5.3.3 Quark models of nucleon-nucleon scattering – Van der Waals forces

In recent years many authors tried to describe nucleon-nucleon scattering in terms of a quark model. Unfortunately, the application of QCD to distances larger than 1 fm requires non-perturbative dynamics for which reliable calculational techniques do not yet exist. The present status is described in lectures given at the KfK summer school at Bad Liebenzell (82 R Brodsky, 82 R Faessler and 82 R Heller). See also 80 R Craigie, 80 L Lomon, 82 R Brown and 82 R De Tar.

A special problem is the possible existence of a strong long-range "van der Waals"-force between hadrons. In atomic and molecular physics, van der Waals forces between two neutral systems are well-known. They are obtained in 2nd order perturbation theory from the dipole interaction part of the perturbation operator, which couples the electrons to the electromagnetic field. If R is the distance between the two atoms, the van der Waals potential decreases as R^{-6} if $a_0 \ll R \ll \lambda$ and as R^{-7} if $R \gg \lambda$. a_0 is the radius of the atom and λ is the largest wavelength associated with electric dipole excitations of the atomic ground state (48 L Casimir, 67 R Hirshfeld).

The electromagnetic interaction between two neutral atoms is to some extent analogous to the gluon-mediated interaction between two hadrons which are color-neutral. However, the hadrons cannot exchange a single colored gluon. The exchange of gluon pairs is not forbidden.

The exponentially decreasing pion-exchange mediated Yukawa-interaction is itself a residual effect of the colored gluon interaction. The question is, whether there exist in addition strong long-range forces.

A new long-range force would lead to a singularity of the invariant scattering amplitudes at $t=0$, which does not occur in dispersion theory. For instance a decrease of the potential as R^{-7} is related to a singularity of the type $t^2 \ln(-t)$.

If a new long-range force exists, it should show up somewhere in accurate experiments. Feinberg and Sucher (79 L) determined bounds for the strength of this force from various sources. Sawada (81 L-1 and 2) claimed that he has found evidence for the new force in his analysis of the partial wave dispersion relation for the πN P33 partial wave. However, no effect was seen in the careful analysis of Hutt and Koch (81 L).

Gavela et al. (79 L) generalized the treatment of the electromagnetic van der Waals force to long-range forces induced by instantaneous confining potentials. Their result is in contradiction with experimental nucleon-nucleon scattering phase shifts for a large class of confining potential.

A general discussion of the problem has recently been given by Greenberg (80 L).

2.5.4 $\bar{\text{N}}$N↔ππ amplitudes at $t > 4m^2$

As discussed in Sect. A.6.1, the reactions πN→πN and $\bar{\text{N}}$N↔ππ are closely related to each other. Therefore, I have included the formulas for the latter reactions in the Appendix (Sects. 1.4.1; 2.1.6; 2.2.4; 3.4.2; 3.5.4; 4.5; 7.5).

Up to now, the number of accurate experiments is still small, (Table 2.5.4.1), but one can expect that it will increase as soon as the new $\bar{\text{p}}$p-storage ring (Lear) is completed.

Table 2.5.4.1. Recent experiments on $\bar{p}p \to \pi\pi$. DCS = differential cross section, P= polarization parameter.

Reference	Reaction	Data	Lab. momentum range GeV/c
75 L Eisenhandler	$\bar{p}p \to \pi^-\pi^+$	DCS	0.79···2.43
77 L Carter-3	$\bar{p}p \to \pi^-\pi^+$	P and DCS	1.0 ···2.2
78 L Dulude-1	$\bar{p}p \to \pi^0\pi^0$	DCS	1.1 ···2.0

Data for the inverse process $\pi^-\pi^+ \to \bar{p}p$ have been derived from the reaction $\pi^-p \to \bar{p}pn$ at 10, 16, and 18 GeV/c. The references are given in Table 2.5.4.2. The final state $\bar{p}n$ was also detected, but in this case the production did not follow the prediction from the one-pion exchange model and therefore, one cannot derive the $\pi\pi \to \bar{p}n$ amplitude.

Since the data indicate a rich resonance structure, the partial wave analysis is rather difficult and a general agreement has not yet been reached. It seems that there are resonances in all spin states for $J=1$ to $J=5$ in the momentum range of the data. Table 2.5.4.2 lists recent analyses.

Data for the reaction $\bar{p}p \to \pi^-\pi^+$ at high energies are of interest for a line-reversal comparison with π^+p near-backward elastic scattering (Sects. 2.6.7 and A.9.2.4): 78 L BERGLUND (10 GeV/c), 77 STEIN (6 GeV/c), 76 L BURAN (6.2 GeV/c).

Table 2.5.4.2. Recent analyses of the reactions $\bar{p}p \leftrightarrow \pi\pi$. The lab. momentum range 1.1···2.0 GeV/c corresponds to the mass range 2.12···2.43 GeV.

Reference	Reaction	Mass range GeV
79 L EVANGELISTA	$\pi^-\pi^+ \to \bar{p}p$	1.9 –2.6
80 L ROZANSKA	$\pi^-\pi^+ \to \bar{p}p$	1.9 –2.8
77 L CARTER-1 and 2	$\bar{p}p \to \pi^-\pi^+$	2.02–2.58
78 L DULUDE-2	$\bar{p}p \to \pi^0\pi^0$	2.15–2.43
78 L B.R. MARTIN-MORGAN	$\bar{p}p \to \pi^-\pi^+$	2.0 –2.6
79 L A.D. MARTIN-PENNINGTON	$\bar{p}p \to \pi^-\pi^+$ and $\pi^0\pi^0$	2.1 –2.5
80 L B.R. MARTIN-MORGAN	$\bar{p}p \to \pi^-\pi^+$ and $\pi^0\pi^0$	2.1 –2.5
80 R PENNINGTON (review)	$\bar{p}p \to \pi^-\pi^+$ and $\pi^0\pi^0$	2.0 –2.6

2.6 Analysis of high energy data

2.6.1 Introduction

In order to show the energy and angular dependence of the elastic differential cross sections at high energies, we have plotted in Fig. 6.6.1 (p. 376) a selection of π^-p elastic scattering cross sections vs. $\cos\theta$. It is seen that one can distinguish three regions:

i) The forward peak, whose energy dependence at constant t is slow. This peak is due to diffraction scattering.

ii) The backward peak, whose value at $\cos\theta = -1$ decreases roughly as k^{-2}. The energy dependence at fixed u is similar to the expectation from baryon exchange Regge models.

iii) The "wide-angle" region between these peaks, in which the cross section at fixed angle decreases approximately as s^{-8}, i.e. very rapidly. A behaviour of this type is predicted from quark models.

In the following we start with the analysis of data in the Coulomb interference region. Then, we discuss fits to the data in the diffraction peak region (Sect. 2.6.3) and models of diffraction scattering (Sect. 2.6.4). Section 2.6.5 is devoted to near-forward charge-exchange scattering, Sect. 2.6.6 to wide-angle scattering and Sect. 2.6.7 to near-backward scattering. The last subsection, Sect. 2.6.8, contains a summary of total elastic and charge-exchange cross sections.

2.6.2 The Coulomb interference region

Elastic $\pi^\pm p$ differential cross section data at very small t-values are of great interest, because the interference between the strong and electromagnetic amplitudes offers a possibility to determine the real part of the strong amplitude from experimental data. The same quantity can be predicted from total cross section data and an assumption on their high energy behaviour, if one uses the forward dispersion relation (Sect. 2.2.1).

Originally, the comparison was performed in order to test the validity of the dispersion relation, which has been derived from the axioms of quantum field theory. The crucial assumption is the principle of microcausality, according to which the fields commute at spacelike separations.

Nowadays, the real part data are often used in order to derive information on the total cross sections above the experimental region, assuming that the dispersion relation is valid.

The formulas for the strong and electromagnetic contributions to the cross section are given in Sect. A.4.6. At very small t, the pure Coulomb scattering is dominant. At 100 GeV/c, its cross section is equal to the strong cross section at $t = -0.003\,\text{GeV}^2$.

The range in which the Coulomb interference term is appreciable depends on the magnitude of the real part of the strong amplitude. The real part for π^+p-scattering has a zero near 100 GeV/c and the real part for π^-p-scattering near 60 GeV/c. In general, the magnitude of the interference term becomes comparable with the experimental errors near $t = -0.02\,\text{GeV}^2$.

The analysis of data in the Coulomb interference region usually follows the prescriptions of LOCHER (67 L) and WEST and YENNIE (68 L) (see also 82 L CAHN). The data are fitted by adjusting two parameters: the ratio of the real and imaginary parts of the strong forward amplitude and the logarithmic slope at $t=0$

$$\varrho_\pm \equiv \left.\frac{\operatorname{Re} C_\pm(s,t)}{\operatorname{Im} C_\pm(s,t)}\right|_{t=0}; \qquad b_\pm \equiv \left.\frac{d}{dt}\ln\frac{d\sigma_\pm}{dt}\right|_{t=0}. \tag{6.2.1}$$

If accurate data are analyzed, one should check whether the usual neglection of the t-dependence of ϱ between $t=0$ and the Coulomb interference region is justified. The t-dependence of $\operatorname{Im} C^+$ is well known and that of $\operatorname{Re} C^+$ can be calculated from the dispersion relation for $\partial C^+/\partial t$ at $t=0$. The result depends on the assumption for the high energy behaviour of $b^+\sigma^+$, where b^+ is the logarithmic slope at $t=0$ (Sect. A.6.5). In a recent study of the dispersion relation, KOCH and MANNING (81 L-2) found that the correction is not larger than the present error bars.

If total cross section data are used in the fit, one should admit an adjustable factor in order to correct for relative normalization errors. Furthermore, the t-range of the fit should include all data of the experiment. In my opinion, both points made it necessary to reanalyze the data 69 FOLEY. Since the ϱ-values in this range nowadays are well known, we inserted the prediction from Table 2.2.1.1. It turned out that the new fit led to smaller values of the logarithmic slopes and to relative normalization errors of 3···6%. (72 L KRUBASIK, 79 L HÖHLER-4) (Table 2.6.2.2, p. 319).

A useful test of experimental data can be carried out in the following way. Taking the result of the analysis of the Coulomb interference data, one can calculate the interference term and the pure Coulomb term and subtract them from the experimental cross sections. The result must be slowly variable in the range of the experiment. Therefore, one can introduce larger t-bins in order to obtain a smaller number of experimental points, which have smaller errors than the original data.

This calculation was done using the data set 78 BURQ (Fig. 6.3.2, p. 323). The remaining structure must have an experimental reason. It is possibly related to the fact that the authors used different methods for very small and somewhat larger t-values. If the origin of the structure is found, one can obtain more accurate values of the slope (79 L HÖHLER-3).

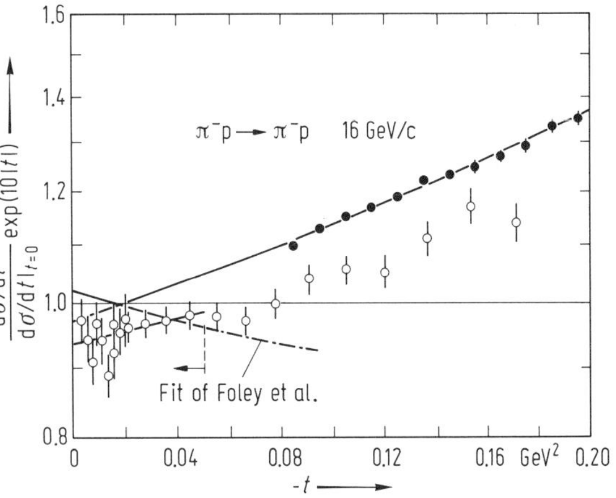

Fig. 6.2.1. π^-p elastic scattering at 16 GeV/c and small $|t|$. Data : ♦ 77 RUSS, ○ 69 FOLEY. The Coulomb and Coulomb interference contributions, which are large only at $t > -0.01\,\text{GeV}^2$, have been subtracted, using the prediction for $\varrho_- = \operatorname{Re} C_-/\operatorname{Im} C_-$ at $t=0$ from the forward dispersion relation (Table 2.2.1.1). Since the quantity plotted in the figure must be slowly variable, the data points at small $|t|$ have been combined in larger bins. The fit of FOLEY et al. includes only the data at $t > -0.05$ and the optical point. Our fit is the solid line which aims towards the points of FOLEY et al. at larger $|t|$. The slope at $t=0$ is 25% smaller than that of FOLEY et al. and there is evidence for a necessary renormalization by about 6% (Table 2.6.2.2, p. 319; 79 L HÖHLER-4).

Simultaneous analysis of $\pi^\pm$p data

If data for both reactions, π^+p and π^-p scattering, are available at the same momentum, it is an advantage to analyze the sum and the difference, as pointed out in earlier papers of our group (72 L KRUBASIK, 73 L HÖHLER-3). The new Fermilab data 80 FAJARD were recently analyzed in this way (80 L HÖHLER-1).

In the difference $d\sigma_-/dt - d\sigma_+/dt$, the Coulomb term cancels and the extrapolation of the interference term to $t=0$ determines the *isospin even combination* of the real parts: $\operatorname{Re} C^+$, since the other term $2\delta \operatorname{Im} C^-$ in Eq. (A.4.33) is negligible, i.e. the uncertainty of the BETHE phase is unimportant. It is practical to plot t times the difference and to use the fact that the slope of this quantity at $t=0$ is well known from Table 2.2.1.1. These figures give a realistic information on the uncertainty of the determination of $\operatorname{Re} C^+$.

The situation is more difficult for the sum of the cross sections

$$\frac{1}{2}\left\{\frac{d\sigma_-}{dt}+\frac{d\sigma_+}{dt}-\frac{d\sigma_0}{dt}\right\}_{\exp}=\frac{d\sigma}{dt_C}+\frac{\alpha}{|t|k}f_E(t)\{\mathrm{Re}\,C^- - 2\delta\,\mathrm{Im}\,C^+\}\left(1-\frac{t}{4m^2}\right)+\frac{d\sigma^+}{dt}. \tag{6.2.2}$$

(The small charge-exchange contribution on the left hand side is negligible at high energies.)

The problem is that, due to the factor $\mathrm{Im}\,C^+$, the term with the BETHE phase is now comparable with $\mathrm{Re}\,C^-$ or even larger. According to WEST and YENNIE (68 L), the uncertainty of the BETHE phase can easily correspond to a factor of 2.5 in the quantity $\sqrt{R^2/6}$ (their b'). For this reason, it is difficult to test the dispersion relation or to derive information on the total cross sections above the experimental range in the present case.

Instead, one can assume the validity of charge-independence and take the absolute value of $\mathrm{Re}\,C^-$ from the data for the charge-exchange forward cross section (Sect. 2.6.5) or from Table 2.2.1.1. The result is substituted into Eq. (6.2.2) and the question is, whether the data are compatible with the estimate for the BETHE phase (Table A.4.1).

This test has been performed in our paper (80 L HÖHLER-1). There is a considerable discrepancy, but it would be too early to draw conclusions, because we noticed some problems with the results of FAJARDO et al. (81) for the real/imaginary part ratios (see below).

Results for the real/imaginary part ratios

In Figs. 6.2.2 and 6.2.3 we compare the experimental results of various groups with the predictions from Table 2.2.1.1. The largest discrepancy occurs at 10···40 GeV/c in ϱ_- and it includes data of the BROOKHAVEN and SERPUCHOV groups.

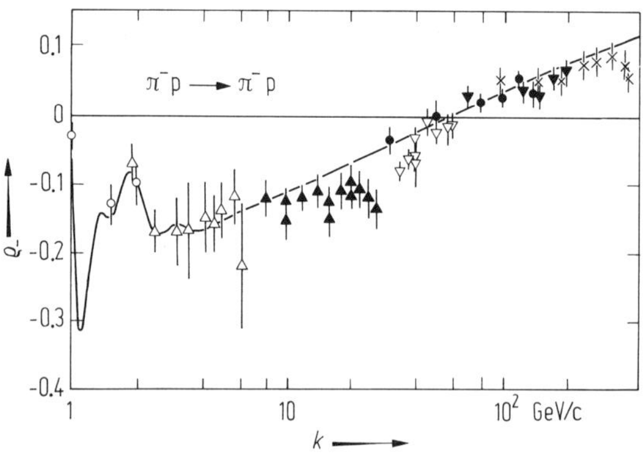

Fig. 6.2.2. Momentum dependence of the ratio $\varrho_- = \mathrm{Re}\,C_-/\mathrm{Im}\,C_-$ at $t=0$. Data: 74 BAILLO, 74 VOROBE, 69 FOLEY, 76 APOKIN (the highest point at 40 GeV/c is due to 79 ABLEEV), 78 BURQ, 81 BURQ, 81 FAJARD. Solid line: our prediction (Table 2.2.1.1). The different fits shown in Figs. 2.1.4 and 6.2.4 are useful for a discussion of the systematic deviation at the highest momenta. If the trend of the last three points on the right would be continued at higher energies, the data would not be compatible with the dispersion relation and a structureless σ^+ above 200 GeV/c.

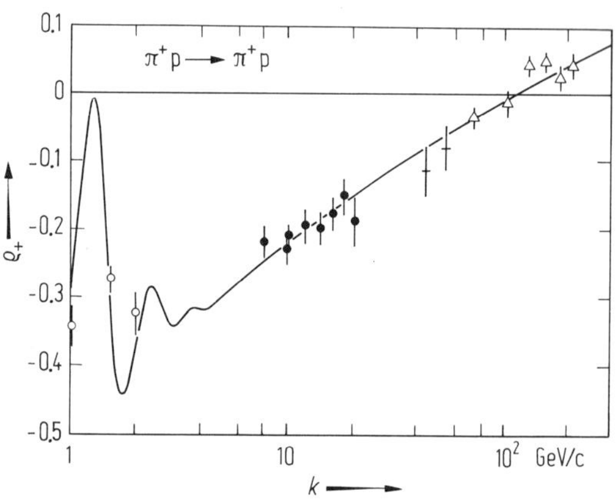

Fig. 6.2.3. Momentum dependence of the ratio $\varrho_+ = \mathrm{Re}\,C_+/\mathrm{Im}\,C_+$ at $t=0$. Data: 74 BAILLO, 69 FOLEY, 77 APOKIN, 81 FAJARD. Solid line: our prediction (Table 2.2.1.1).

However, a new measurement at SERPUCHOV (79 ABLEEV) at 40 GeV/c led to a ϱ_--value in good agreement with the prediction. As far as the data set 69 FOLEY is concerned, our reanalysis has shown that there is no discrepancy with the prediction, if one admits a relative normalization error between the differential and total cross section data and if one includes all data points of this experiment (79 L HÖHLER-4).

A slowly variable discrepancy in ϱ_- above 200 GeV/c is of a different type, because it can easily be removed by changing the assumption for the total cross sections above the experimental region. Since the high energy behaviour of σ^- is compatible with the charge-exchange forward scattering cross sections, one has to adjust the parametrization for σ^+.

In 80 R HÖHLER-KAISER, we have discussed several possibilities to extrapolate the existing σ^+-data to higher energies. Figure 6.2.4 shows the corresponding results for ϱ^+. The highest value belongs to Table 2.2.1.1 and to the curve in Fig. 6.2.2. Therefore, it is not difficult to find parametrizations which fit the total cross section data as well as the ϱ-values. In 81 Burq, the authors considered total cross sections which become constant above a certain momentum. They concluded that the increase continues at least up to 1000···2000 GeV/c.

A model independent conclusion can be drawn from the new data as follows. One calculates

$$\mathrm{Re}\,C^+ = \varrho_-\,\mathrm{Im}\,C_- - \mathrm{Re}\,C^- \tag{6.2.3}$$

from the ϱ_- data, taking $\mathrm{Im}C_-$ and $\mathrm{Re}C^-$ from Table 2.2.1.1. Then one considers the effect of a correction to the σ^+-values of the table, denoting these values by the subscript "Tab"

$$\mathrm{Re}\,C^+ - \mathrm{Re}\,C^+_{\mathrm{Tab}} = k^2 \frac{2}{\pi} \int_{k_0}^{\infty} \frac{\sigma^+ - \sigma^+_{\mathrm{Tab}}}{k'^2 - k^2}\,\mathrm{d}k'\,. \tag{6.2.4}$$

k_0 is the momentum up to which data are available. Since the correction will be appreciable only at some distance above k_0, it is sufficient to take the leading term of an expansion in powers of k^2

$$(\mathrm{Re}\,C^+ - \mathrm{Re}\,C^+_{\mathrm{Tab}})/k^2 \approx \frac{2}{\pi} \int_{k_0}^{\infty} (\sigma^+ - \sigma^+_{\mathrm{Tab}})\mathrm{d}k'/k'^2\,. \tag{6.2.5}$$

Taking the average of the left hand side over all data which give a significant value different from zero, one obtains the constraint for the necessary correction to σ^+_{Tab}.

Finally, we consider a test of results obtained from Coulomb interference experiments, in which both ϱ_+ and ϱ_- are determined. We have plotted in Fig. 6.2.5 the quantity $\sqrt{k}(\varrho_- - \varrho_+)/2$ as a function of k. The reason for this choice is the relation

$$\frac{1}{2}(\varrho_- - \varrho_+) = \frac{\sigma^-}{\sigma^+}\left\{\varrho^- - \frac{1}{2}(\varrho_- + \varrho_+)\right\} \approx \frac{\sigma^-}{\sigma^+}\varrho^-\,. \tag{6.2.6}$$

If this equation is multiplied by $\sqrt{k}$, the right hand side is approximately constant and well determined by the ϱ^--value derived from the charge-exchange reaction. Figure 6.2.5 shows that 3 of the points agree with the more accurate result and two others are too low by 2...3 standard deviations. As long as this discrepancy is not understood, one cannot hope to be able to determine the Bethe phase, since Eq. (6.2.2) determines essentially $\frac{1}{2}(\varrho_- - \varrho_+) - 2\delta$.

Table 2.6.2.1 gives the results for the ϱ-values obtained in the recent experiments at CERN and at Fermilab.

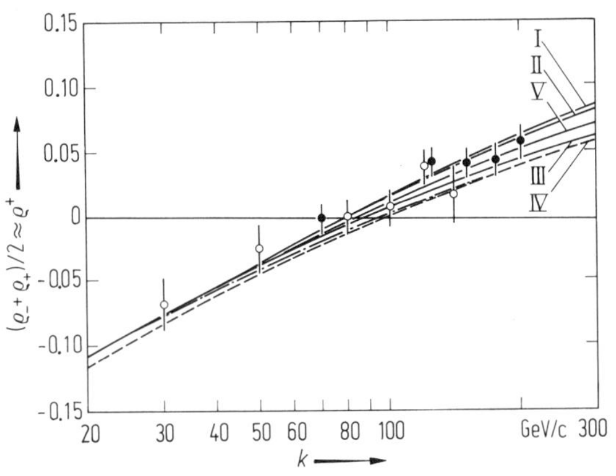

Fig. 6.2.4. The momentum dependence of ϱ^+. In order to show the sensitivity of the prediction to different high energy assumptions for σ^+, we have plotted ϱ^+ for various parametrizations. The fits I, II, III are explained in connection with Fig. 2.1.4. In fit V, σ^+ is the same as in fit I up to 1000 GeV/c and it remains constant at higher momenta. — Fit IV belongs to Lipkins analytic parametrization (dashed line "L" in Fig. 2.1.4). The dot-dashed line follows from the same σ^+, if $\mathrm{Re}C^+$ is calculated from the dispersion relation.

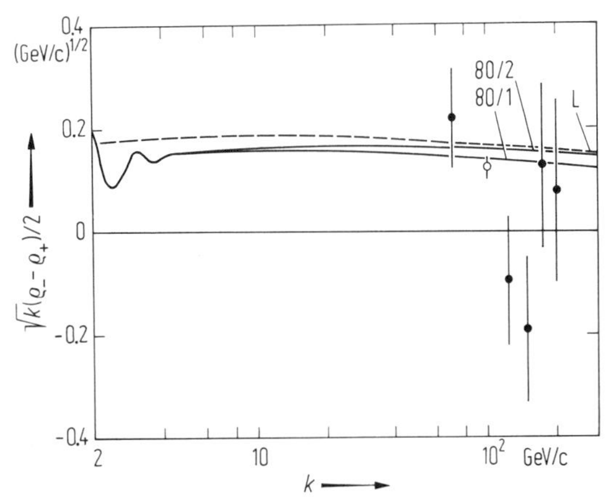

Fig. 6.2.5. The momentum dependence of $\varrho_- - \varrho_+$. The solid lines are the predictions for $\sqrt{k}(\varrho_- - \varrho_+)/2$ as following from two assumptions on the high energy behaviour of σ^- (see Fig. 2.1.7). The point ⲫ shows the relatively small uncertainty of a direct prediction from Eq. (6.2.6) and the charge-exchange data. The dashed line is Lipkins fit. — The agreement with the data 81 Fajard is good at 70, 175, and 200 GeV/c, but there is a discrepancy at 125 and 150 GeV/c.

Table 2.6.2.1. Recent results for the real/imaginary part ratio $\varrho_{\pm}$ of the forward amplitude.

k GeV/c	ϱ_-	k GeV/c	ϱ_-
30	-0.04 ± 0.02	100	0.048 ± 0.012
50	-0.003 ± 0.013	150	0.048 ± 0.012
80	0.017 ± 0.013	200	0.064 ± 0.011
100	0.023 ± 0.013	250	0.078 ± 0.013
120	0.052 ± 0.013	280	0.087 ± 0.012
140	0.03 ± 0.02	300	0.090 ± 0.012
		325	0.084 ± 0.012
		345	0.075 ± 0.012
78 BURQ		81 BURQ, fit 1	

k GeV/c	ϱ_+	ϱ_-	ϱ^+	ϱ^-	σ^+/σ^-	ϱ^+
70	-0.025 ± 0.016	0.027 ± 0.016	0.001 ± 0.001	1.58 ± 0.70	60.54	-0.008
100	-0.003 ± 0.020	–	–	–		0.018
125	0.052 ± 0.014	0.035 ± 0.017	0.043 ± 0.011	-0.59 ± 0.80	74.22	0.034
150	0.058 ± 0.014	0.027 ± 0.018	0.042 ± 0.011	-1.17 ± 0.90	78.05	0.046
175	0.035 ± 0.018	0.054 ± 0.016	0.045 ± 0.012	0.92 ± 1.1	92.19	0.056
200	0.053 ± 0.017	0.064 ± 0.020	0.059 ± 0.013	0.60 ± 1.3	98.31	0.065
	81 FAJARD		Eq. (6.2.6)			Prediction from Table 2.2.1.1

Results for the slopes at $t=0$

The logarithmic slopes b_0 at $t=0$ are determined together with the ϱ-values in the analysis of data in the Coulomb interference region. Other authors calculated values for b_0 from fits to data which start outside the Coulomb interference region. Figure 6.2.6 shows that there are large discrepancies between the published values. See also 72 R LASINSKI.

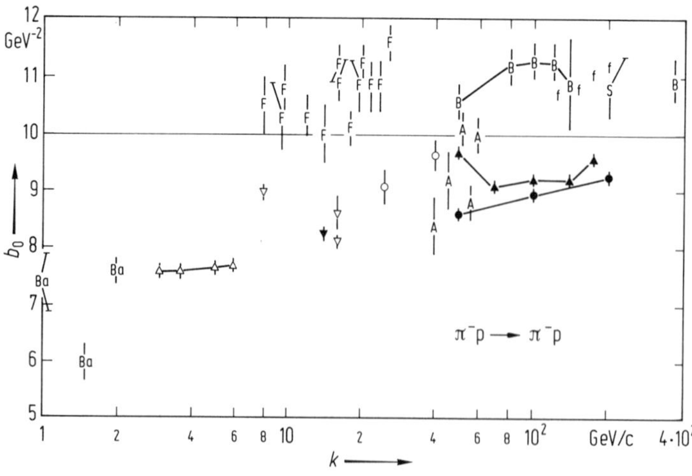

Fig. 6.2.6. Published results for the logarithmic slope of $d\sigma/dt$ at $t=0$. Values derived from data in the Coulomb interference region are denoted by letters. B: 78 BURQ and a preliminary value of 81 BURQ; S: 81 SCHIZ; f: 81 FAJARD; A: 75 APOKIN and 76 APOKIN; F: 69 FOLEY; Ba 76 BAILLO. Extrapolation from data at larger $|t|$: 74 AMBATS; 77 RUSS (two fits at 16 GeV/c): 73 DZIERB; 74 ANTIPO; 76 AKERLO; 77 AYRES.

One reason is an unexpected curvature in the shape of the diffraction peak at small t, which will be discussed in Sect. 2.6.3. Because of this effect, reliable values for the slope at $t=0$ can only be obtained from data in the Coulomb interference region.

The second reason is the above mentioned problem with the analysis of the BROOKHAVEN data (69 FOLEY). Table 2.6.2.2 gives the result of our reanalysis of all π^-p Coulomb interference data above 7 GeV/c, which extend at least up to $t=-0.04\ \mathrm{GeV}^2$. The values of ϱ_- and the total cross sections were taken from Table 2.2.1.1. The new slopes of the BROOKHAVEN data are now considerably lower.

The slopes obtained by the Fermilab group (81 FAJARD) and by the CERN group (78 BURQ, 81 BURQ) are listed in Tables 2.6.2.3 and 2.6.2.4, respectively.

Table 2.6.2.2. Results of a reanalysis of data in the Coulomb interference region (79 L HÖHLER-4). The total cross sections and the ϱ-values were taken from Table 2.2.1.1 and the formulas of Sect. A.4.6 were used. Only data sets which extend at least up to $t=-0.04\ \mathrm{GeV}^2$ were considered. N_1 is the renormalization factor of the differential cross section data with respect to the predictions for the forward cross sections in Table 2.2.1.1. The errors are of interest only relatively to each other, since errors of the total cross sections and of the ϱ-values are not included. The last column gives the results of the authors. In the case of the experiment 78 BURQ we show the published values. In a subsequent paper the authors mentioned that their b_0-values should be diminished by $0.5\ \mathrm{GeV}^{-2}$ due to radiative and other corrections (81 BURQ). One should note that our method practically reproduces the results of 78 BURQ, but it leads to much lower b_0-values for the data of FOLEY et al. (see Fig. 6.2.1). Column 5: χ^2 per degree of freedom.

Experiment $\pi^-p\to\pi^-p$	k GeV/c	b_0 GeV^{-2}	N_1	χ^2_{DF}	b_0 [GeV^{-2}] result of the authors
69 FOLEY	7.89	10.52±0.5	0.99 ±0.01	0.53	10.5±0.5
69 FOLEY	9.89	9.32±0.4	1.04 ±0.01	0.52	10.8±0.4
69 FOLEY	11.89	9.85±0.4	1.03 ±0.01	0.75	10.3±0.3
69 FOLEY	14.16	9.11±0.6	1.02 ±0.01	0.52	10.0±0.5
69 FOLEY	15.99	9.54±0.3	1.04 ±0.01	0.33	11.0±0.3
69 FOLEY	16.00	8.97±0.3	1.065±0.01	0.57	11.2±0.3
69 FOLEY	18.19	8.93±0.3	1.03 ±0.01	0.56	10.1±0.3
69 FOLEY	20.15	9.73±0.4	1.03 ±0.01	0.30	10.9±0.4
69 FOLEY	20.38	9.83±0.3	1.04 ±0.01	0.61	11.3±0.3
69 FOLEY	22.13	9.59±0.4	1.03 ±0.01	0.55	10.9±0.4
69 FOLEY	24.22	9.11±0.4	1.05 ±0.01	0.39	10.9±0.4
69 FOLEY	26.23	9.62±0.3	1.06 ±0.01	0.16	11.6±0.3
76 APOKIN	45.47	9.96±0.6	0.97 ±0.01	0.68	9.2±0.5
78 BURQ	50.0	10.8 ±0.2	1.004±0.004	1.48	10.6±0.3
76 APOKIN	50.48	9.34±0.3	1.025±0.01	0.73	10.1±0.3
76 APOKIN	56.72	7.8 ±0.4	1.01 ±0.01	1.83	8.8±0.3
76 APOKIN	59.4	9.40±0.3	1.005±0.09	0.88	10.0±0.3
78 BURQ	80.0	11.0 ±0.2	1.01 ±0.03	1.29	11.2±0.3
78 BURQ	100.0	10.96±0.2	1.02 ±0.003	1.68	11.3±0.3
78 BURQ	120.0	11.37±0.2	1.00 ±0.004	1.09	11.3±0.3
78 BURQ	140.0	10.0 ±0.5	1.03 ±0.008	0.95	10.9±0.8

Table 2.6.2.3. Logarithmic slopes at $t=0$ of the fit to the data set 81 FAJARD as obtained by the authors.

k GeV/c	$b_0(\pi^+p)$ GeV^{-2}	$b_0(\pi^-p)$ GeV^{-2}
100	10.38±0.20	–
125	10.75±0.12	10.74±0.12
150	10.74±0.08	10.91±0.10
175	10.85±0.10	11.18±0.07
200	11.03±0.06	11.21±0.06

Table 2.6.2.4. Logarithmic slopes at $t=0$ of the fit to the data sets 78 BURQ and 81 BURQ as obtained by the authors. The results of 78 BURQ are corrected by subtracting 0.5 GeV^{-2} as mentioned in 81 BURQ.

k GeV/c	b_0 GeV^{-2}	k GeV/c	b_0 GeV^{-2}
50	10.1±0.3	100	10.54±0.31
80	10.7±0.3	150	10.09±0.28
100	10.8±0.3	200	10.30±0.27
120	10.8±0.3	250	10.39±0.31
140	10.4±0.8	280	10.51±0.29
		300	10.73±0.29
78 BURQ (corrected)		325	10.90±0.30
		345	11.03±0.31
		81 BURQ	

Figure 6.2.7 shows the slopes listed in Tables 2.6.2.2, 2.6.2.3, and 2.6.2.4 and some results at lower momenta. The authors of 81 L BURQ concluded that the results for b_0 above 8 GeV/c can be fitted by an ansatz suggested by Regge pole theory

$$b_0(k)=b_0+b_1/\sqrt{k}+b_2\ln k \tag{6.2.7}$$

and they noticed that the shrinkage parameter b_2 is approximately the same at $t=-0.2$ and $t=-0.4$ GeV^2, an average value being $b_2=2\alpha'_P=0.27$ GeV^{-2} (α'_P is the slope of the pomeron trajectory).

However, the theoretical arguments in favor of the ansatz Eq. (6.2.7) are not strong. One could as well conclude that the data for b_0 in the range 80 GeV/c$\leqq k \leqq$345 GeV/c are compatible with a constant slope. Theoretical bounds for the slope depend on the energy dependence of the total cross sections (Sect. A.5.3). The k-dependence of the slope at $t=-0.2$ and -0.4 GeV^2 will be discussed in Sect. 2.6.3.

Figures 6.2.8 and 6.2.9 show the energy dependence of the slopes in the resonance region. There are some large discrepancies between the predictions of the CMU-LBL and KH-partial wave analyses, which might be due to some extent to the fact that the CMU-LBL group has included the preliminary data set 76 BARDSL. Figure 6.2.10 is an example for discrepancies between different data sets.

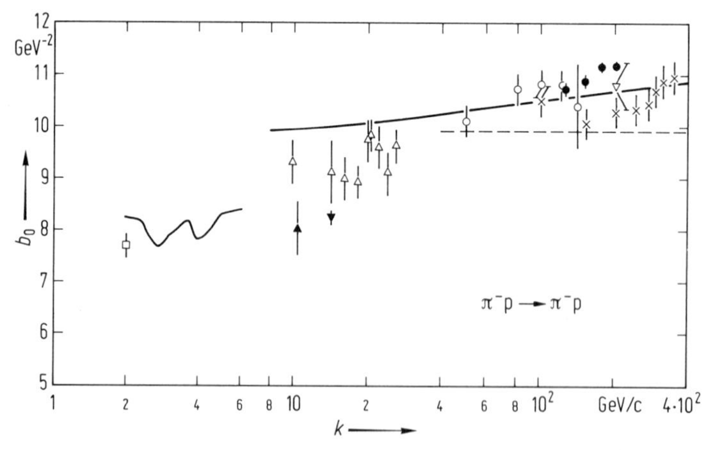

Fig. 6.2.7. New results and corrected values for the logarithmic slope of $d\sigma_-/dt$ at $t=0$. ⚹ 81 BURQ; ⧫ 81 FAJARD; ▽ 81 SCHIZ; ○ 78 BURQ (lowered by 0.5 GeV^{-2} due to corrections mentioned in 81 BURQ): △ 69 FOLEY (Table 2.6.2.2, 3rd column); □ 74 BAILLO; ▼ 77 RUSS; ▲ 75 BRANDE. Table 2.6.3.1 shows that the results for the slopes below 50 GeV/c depend on the fit and are therefore not yet well determined by the existing data. Solid line at 2···6 GeV/c: slope as calculated from the KH 78 partial wave solution. Solid line at high energies: fit proposed in 81 BURQ [Eq. (6.2.7)]. Dashed line: slope calculated from the product of pion and proton form factors squared [$f_E^2(t)$ in Eq. (6.3.2)].

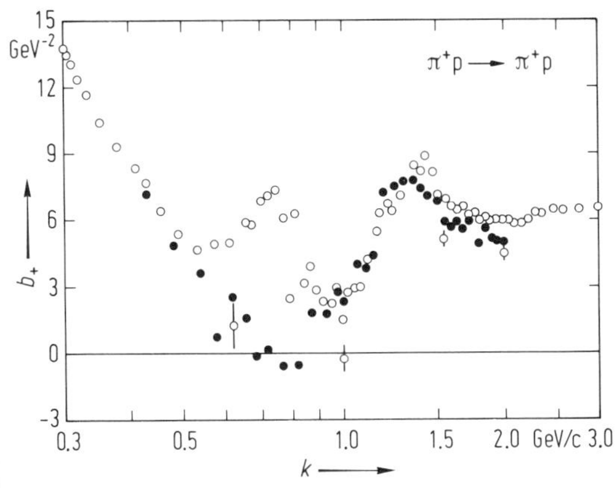

Fig. 6.2.8. Logarithmic slopes of $d\sigma_+/dt$ at $t=0$ as calculated from partial wave solutions. ● CMU-LBL 79, ○ KH 78. Slope from data in the Coulomb interference region: ϕ 74 BAILLO. There are fairly large discrepancies, because there exist only a few reliable data points in the Coulomb interference region. An example for a discrepancy between data sets is shown in Fig. 6.2.10. Furthermore, the preliminary large data set 76 BARDSL has been included in the CMU-LBL analysis but not in ours. The discrepancies are being studied by investigating dispersion relations for $\partial C/\partial t$ and for B at $t=0$ (81 L KOCH-2).

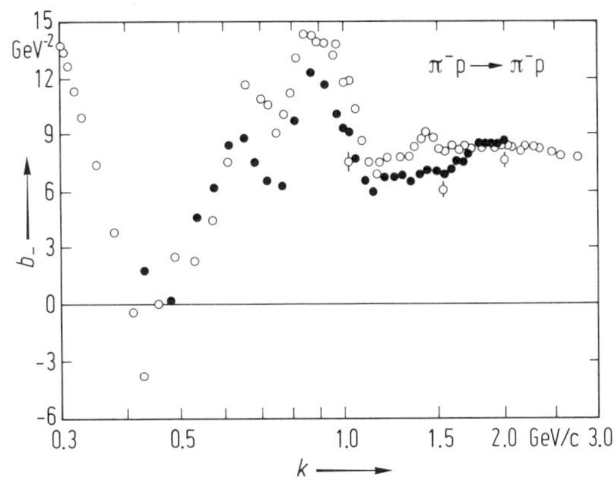

Fig. 6.2.9. Logarithmic slopes of $d\sigma_-/dt$ at $t=0$ as calculated from partial wave solutions. The notation is the same as in Fig. 6.2.8.

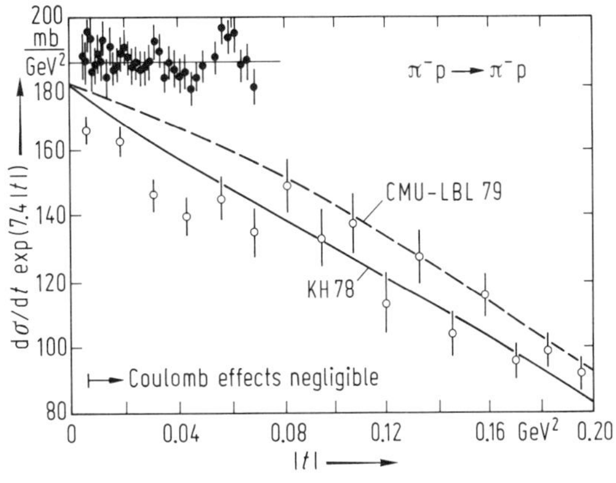

Fig. 6.2.10. Discrepancies for the slope in $d\sigma_-/dt$ at 1.0 GeV/c. Data: ♦ 74 BAILLO (1.015 GeV/c); ϕ 76 BARDSL (1.002 GeV/c); reconstruction from partial waves: CMU-LBL 79 (1.005 GeV/c); KH 78 (1.000 GeV/c). The partial wave solutions are also constrained by other data not shown in the figure. The data 76 BARDSL have not been used in the KH 78 analysis. — The momentum dependence of the slope near 1.0 GeV/c can be estimated from Fig. 6.2.9. ►

pp- and p̄p-scattering

The references for the high energy behaviour of the total cross sections, real parts and slopes in elastic pp- and p̄p-scattering can be traced from 82 L BURQ and 82 L BLOCK.

2.6.3 Diffraction scattering: fits to the data and amplitude analysis

In elastic $\pi^\pm p$ scattering at and above the energy of the higher resonances, the diffraction peak is the dominant phenomenon (Fig. 6.6.1, p. 376). We start with a discussion of the energy dependence of its shape at small $|t|$ and show results for the partial wave decomposition up to 10 GeV/c (Sect. 2.6.3.3), including the inelastic overlap function (Sect. 2.6.3.4). The high energy approximation of the impact parameter formalism (Sect. 2.6.3.5) is needed for the short summary on pp-scattering at very high energies (Sect. 2.6.3.6) and also for πN amplitude analysis and for the description of models (Sect. 2.6.4). Some results for the transition to the wide-angle region (Sect. 2.6.6) are given in Sect. 2.6.3.7. The existing information on the elastic flip amplitude is discussed in Sect. 2.6.3.8.

2.6.3.1 The shape of the diffraction peak

The diffraction peak is usually described in terms of the cross section $d\sigma/dt$ because, at high energies, the shape (i.e. the t-dependence at fixed s) and its height become almost independent of s.

The highest momentum up to which accurate elastic differential cross section data are available is 200 GeV/c (81 SCHIZ, 81 FAJARD, 81 COOL, 76 AKERLO). Since the charge-exchange cross sections at this momentum (76 BARNES) are small in comparison with the elastic cross sections, the isospin even cross section, Eq. (A.10.11), is practically given by the average of the $\pi^{\pm}p$ cross sections. It belongs to pure diffraction scattering, whereas the elastic cross sections still contain a contribution from the isospin odd amplitude, which leads for instance to the splitting of the strong $\pi^{\pm}p$ forward cross sections (4% at 200 GeV/c).

Since we are interested in small details comparable in magnitude with the statistical errors, the usual plots of $\ln d\sigma/dt$ vs. t are not suitable. We prefer to plot $d\sigma/dt$ divided by a function of t which is chosen in such a way that the ratio is almost constant in the range of interest ($0 \geqq t > -0.5\,\mathrm{GeV}^2$).

According to the CHOU-YANG model and to the Additive Quark Model (see below, Sects. 2.6.4.1 and 2.6.4.6), the shape of the πN diffraction amplitude can approximately be described by the product of the electromagnetic form factors of the nucleon and pion, aside from rescattering corrections. Following FAJARDO et al. (81), we parametrize the main part of the t-dependence of $d\sigma/dt$ by $f_E^2(t)$, where

$$f_E(t) = G_{Ep}(t)\,F_\pi(t)\,. \tag{6.3.1}$$

We have chosen G_{Ep} and not for instance the Dirac form factor F_{1p}, because we describe diffraction in terms of C^+ and this amplitude belongs to G_E, if one considers decompositions in terms of invariants like Eq. (A.2.13) and (A.11.1) [see also Eq. (A.2.80)]. F_{1p}, F_{2p}, and G_{M_p} belong to the invariant amplitudes D, A, and B, respectively [see Eq. (A.8.31)].

The experimental information on $f_E(t)$ has been summarized in Sect. 2.5.2. We take the values listed in Table 2.5.2.1. It turns out that $(d\sigma^+/dt)/f_E^2(t)$ still has an appreciable t-dependence.

Therefore, we describe the isospin even πp cross section at 200 GeV/c by the empirical parametrization

$$\left.\frac{d\sigma}{dt}\right|_{200} = \frac{d\sigma}{dt}(t=0)\,f_E^2(t)(1-0.78\,|t|)\,; \qquad t \geqq -0.5\,\mathrm{GeV}^2\,, \tag{6.3.2}$$

which follows from an overall fit to the $\pi^{\pm}p$ data 81 SCHIZ and 81 FAJARD. The forward cross section is taken from Table 2.2.1.1.

Figure 6.3.1 is a plot of the $\pi^{\pm}p$ averages of the data sets 81 SCHIZ and 81 FAJARD, divided by our parametrization. It is seen that the present experimental uncertainty is about $\pm 5\%$ and that the discrepancy cannot be corrected by a normalization factor. Both experiments have been performed by the same group at Fermilab.

In addition, we have plotted the 50 GeV/c data of AYRES et al. (77), which were corrected by a renormalization factor $N = 0.89$. This correction is necessary, if one wants to have a smooth extrapolation of the π^-p data to the data in the Coulomb interference region (78 BURQ, see Fig. 6.3.2) and also to the splitting of the $\pi^{\pm}p$ forward cross sections (Table 2.2.1.1).

We conclude from Fig. 6.3.1 that the energy dependence of $d\sigma^+/dt$ at fixed t between 50 and 200 GeV/c is smaller than the present experimental errors of the most accurate differential cross section data. At $t=0$ it amounts to 4.5% (Table 2.2.1.1), but it is not known how the shape of $d\sigma^+/dt$ is affected. It could be a slowly t-dependent increase or, for instance, a local change restricted to an interval $0 > t > -0.1\,\mathrm{GeV}^2$.

Since the experimental information on π^-p scattering at high energies is much better than that on π^+p scattering, we have summarized in Fig. 6.3.3 the variation of the shape of the π^-p forward peak between 3 and 40 GeV/c. In the scale of this figure, the variation between 40 GeV/c and 200 GeV/c is negligible.

For a fixed value of t, the ratio $(d\sigma_-/dt)/(d\sigma^+/dt)_{200}$ decreases monotonously towards the value 1. The variation is larger at $-0.1 > t > -0.5\,\mathrm{GeV}^2$ than at $t=0$ (shrinkage) and correspondingly, the slope of the t-dependence at $t=0$ is increasing with energy (Fig. 6.2.7).

Figures 6.3.1 and 6.3.3 would look appreciably different without the renormalization of some data sets. In order to demonstrate the necessity of the renormalization, we have plotted in Fig. 6.3.2 the published values of the elastic π^-p cross sections at 50 GeV/c (76 AKERLO, 77 AYRES). The data 78 BURQ were treated as follows. First, we have subtracted the Coulomb effects, using essentially the same formulas as the authors (Sect. 2.6.2). The result for the strong interaction part of the cross section must be slowly variable in t. Therefore, it is justified to introduce larger bins, which has the advantage that the errors become smaller.

The solid curve is the fit of AYRES et al. to their data in a larger t-interval than shown in the figure. The straight line is the fit of BURQ et al. At $t = -0.05\ \mathrm{GeV}^2$ the discrepancy amounts to 11 %, which is much larger than the estimate of AYRES et al. for their systematic error (3 %).

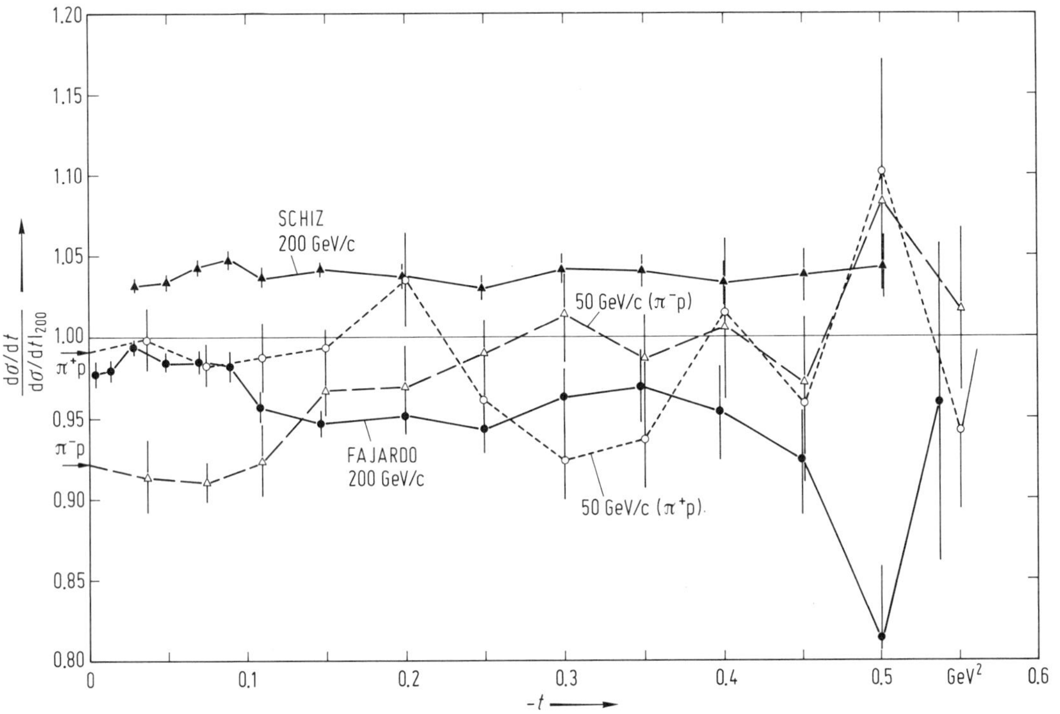

Fig. 6.3.1. $\pi^{\pm}$p cross sections at 200 and 50 GeV/c. The data are divided by the parametrization Eq. (6.3.2) of the 200 GeV/c data for $d\sigma^{+}/dt \approx (d\sigma_{+}/dt + d\sigma_{-}/dt)/2$. In addition to the $\pi^{\pm}$p averages of the data 81 SCHIZ and 81 FAJARD the plot shows the 50 GeV/c data 77 AYRES, renormalized by a factor $N = 0.89$. The arrows point to the predictions for the $\pi^{\pm}$p forward cross sections at 50 GeV/c.

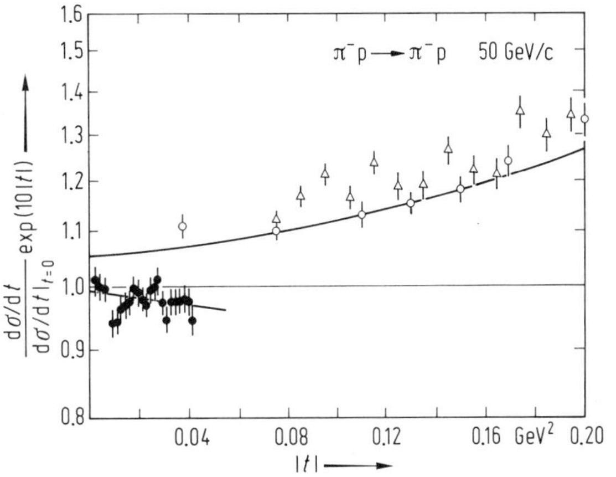

Fig. 6.3.2. π^{-}p near-forward cross sections at 50 GeV/c. The data are divided by the prediction for the forward cross section and multiplied by a factor $\exp(10|t|)$ in order to show details. ⧫ 78 BURQ, ⍋ 76 AKERLO, ⌽ 77 AYRES. Solid lines: fits of AYRES et al. and of BURQ et al. AYRES et al. give a systematic error of 3 %. The points at small $|t|$ (78 BURQ) have been obtained by subtracting the Coulomb effects and introducing larger bins. ►

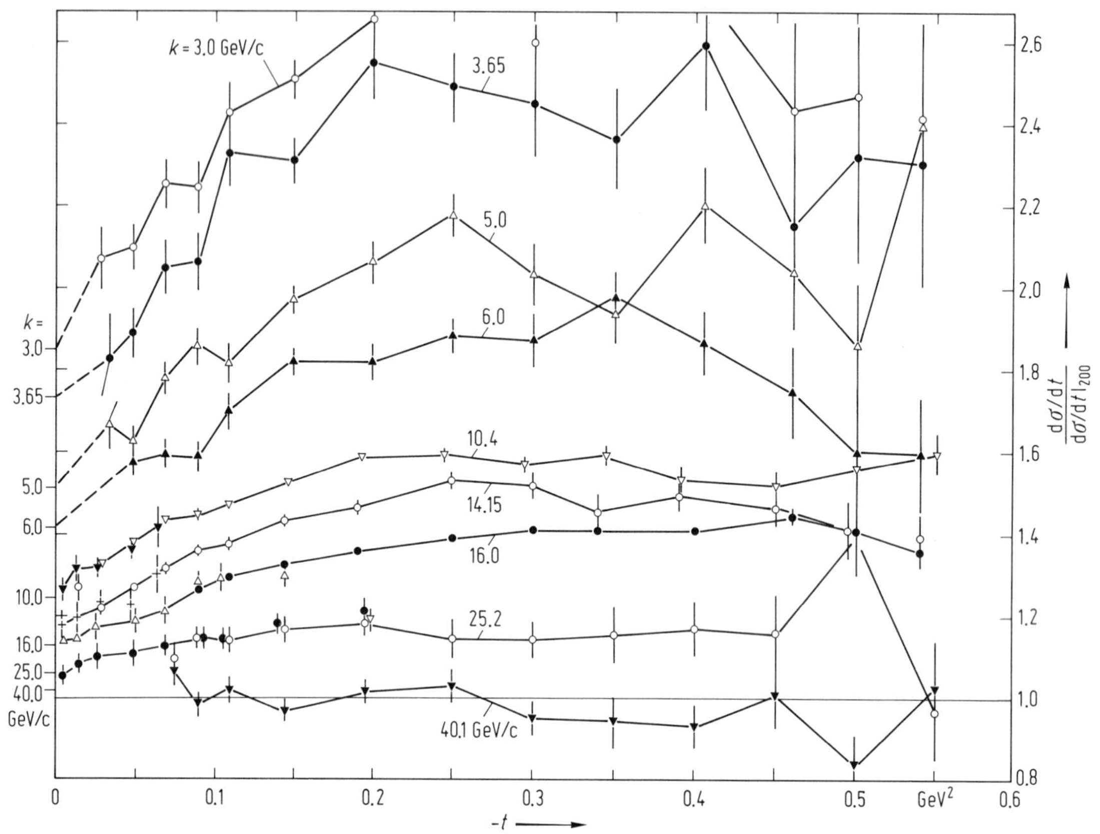

Fig. 6.3.3. Energy dependence of the π^-p cross sections between 3 and 40 GeV/c. The data are divided by the parametrization Eq. (6.3.2) of $d\sigma^+/dt$ at 200 GeV/c. 3⋯6 GeV/c: 74 AMBATS; 10.4 GeV/c: 75 BRANDE; 14.15 GeV/c: 73 DZIERB; 16.0 GeV/c: 77 RUSS; 25.2 and 40.1 GeV/c: 74 ANTIPO. At small $|t|$ we have plotted the data of FOLEY et al. (69). The Coulomb and Coulomb interference terms have been subtracted. The points belong to an average over larger bins. The scale on the left hand side shows the prediction for the forward cross section. Renormalizations: see Table 2.6.3.1.

In order to check whether a renormalization of the data of AYRES et al. can lead to a smooth transition to the data in the Coulomb interference region, we have performed two types of fits.

The first ansatz is a generalization of the usual parametrization

$$\ln\frac{d\sigma}{dt} = \sum_{m=0}^{M} a_m t^m; \qquad \text{“}t\text{-fit”}. \tag{6.3.3}$$

The second ansatz takes into account the existence of the branch cut at $t > 4\mu^2$. If it happens that, for t-values near the branch point, the discontinuity along this cut is rather large, the “t-fit” is not well suited because it is too stiff. We have used the ansatz

$$\ln\frac{d\sigma}{dt} = \sum_{m=0}^{M} c_m w^m; \qquad w = \frac{\sqrt{\alpha} - \sqrt{1 - t/4\mu^2}}{\sqrt{\alpha} + \sqrt{1 - t/4\mu^2}}; \qquad \text{“}w\text{-fit”}. \tag{6.3.4}$$

The points $t = 4\mu^2$ and $t = -\infty$ are mapped onto the points $w = +1$ and -1 respectively, and $t = -4\mu^2(\alpha - 1)$ is mapped onto $w = 0$. For our fits in the range $0 > t > -0.8\ \text{GeV}^2$, we have chosen $\alpha = 6$, mapping $w = 0$ onto the center of the interval. Our w-fit is a special case of a more general ansatz studied by BOWCOCK, COTTERILL, and QUEEN (79 L).

Table 2.6.3.1 gives the results of combined fits to data in the Coulomb interference region and to accurate data at somewhat larger $|t|$. The renormalization of the data in the Coulomb interference region is slightly different from that in Table 2.6.2.2, because the new fits are more flexible.

Table 2.6.3.1. Combined fits to $\pi^- p$ differential cross section data in the Coulomb interference region and at larger $|t|$. $\chi^2_{DF}=\chi^2$ per degree of freedom [HÖHLER, KAISER and STAUDENMAIER (79 L−4)]. The values in brackets in the column for $b(0)$ are the result of BURQ et al., N_i is the renormalization factor as determined in the fit and $0>t>t_{max}$ is the range in which the fit was made. The forward cross section of Table 2.2.1.1 was used as part of the input. — In 81 BURQ the authors mentioned that further corrections to the data reduce the slopes listed in 78 BURQ by −0.5 GeV^{-2}.

t-fit, Eq. (6.3.3)	k GeV/c	N_i	$b(0)$ GeV^{-2}	χ^2_{DF}	Order M	$\lvert t\rvert_{max}$ GeV^2	w-fit, Eq. (6.3.4) N_i	$b(0)$ GeV^{-2}	χ^2_{DF}	Order M
75 BRANDENBURG	10.4	0.98	8.05	1.00	5	1.35	0.95	10.0	1.04	5
69 FOLEY	9.89	1.06					1.06			
73 DZIERBA	14.15	1.06	8.3	0.91	3	0.76	1.06	8.3	0.96	4
69 FOLEY	14.16	1.04					1.04			
73 DZIERBA	14.15						1.01	11.3	0.89	5
69 FOLEY	14.16						1.0			
77 RUSS	16.0	1.00	9.0	0.74	4	0.78	0.96	11.3	0.68	5
69 FOLEY	16.0	1.08					1.06			
77 RUSS	16.0						1.00	8.1	0.96	4
69 FOLEY	16.0						1.09			
77 AYRES	50	0.89	10.5	1.47	4	0.75	0.88	12.9	1.43	5
78 BURQ	50	1.01	(10.6)				0.99			
76 AKERLOF	50	0.86	10.9	1.45	6	1.60	0.84	11.4	1.46	4
78 BURQ	50	1.02	(10.6)				1.00			
77 AYRES	100	0.92	11.6	1.59	5	0.80	0.92	11.7	1.62	4
78 BURQ	100	1.01	(11.3)				1.01			
76 AKERLOF	100	0.85	11.1	1.31	6	1.82	0.84	11.4	1.36	4
78 BURQ	100	1.02	(11.3)				1.02			
77 AYRES	140	0.95	10.5	1.02	5	0.75	0.97	9.5	1.13	4
78 BURQ	140	1.02	(10.9)				1.03			
77 AYRES	140						0.94	13.6	1.05	5
78 BURQ	140						1.00			

From the results of the fits one can calculate the logarithmic slope $b(t)$ which does not depend on the normalization

$$b(t) \equiv \frac{d}{dt} \ln \frac{d\sigma}{dt}. \tag{6.3.5}$$

As an example for the fits listed in Table 2.6.3.1 we show in Fig. 6.3.4 the renormalized data and the fits at 100 GeV/c.

Many earlier authors, who fitted data in the range −0.1 to −0.8 GeV^2, came to the conclusion that it is sufficient to take into account a *quadratic ansatz* in t in Eq. (6.3.3), corresponding to a linear t-dependence of $b(t)$. They used this fit in order to determine the slope $b(0)$. In some cases the data were normalized by adjusting the fit to the forward cross section as predicted from the optical theorem and the dispersion relation. The discrepancies in Fig. 6.2.1 are due to this procedure.

Our result shows that, at least at and above 50 GeV/c, $b(t)$ has a fairly strong curvature and one has to choose M=4 or 5. In Eq. (6.3.2) this curvature of $b(t)$ is described by the product of the nucleon and pion form factors $f_E^2(t)$. The structure at 10 GeV/c was discussed in 75 L CARNEGIE-2.

Figure 6.3.3 shows that, in general, $d\sigma/dt$ decreases without structures as a function of s at fixed t in the 6···16 GeV/c region. But there is a strange effect which is omitted in the figure. The 10 GeV/c data of BRANDENBURG et al. (75) lie very well on the interpolation between the data at lower and higher momenta, but the 8 GeV/c data of RUSS et al. (77) show an unexpected structure near $t=-0.2$ GeV^2, which is much larger than the errors (Fig. 6.3.5). According to BARRELET et al. (80), the total $\pi^- p$ cross section shows no indication for a resonance at this momentum. It would be of interest to check whether the strange phenomenon does really exist.

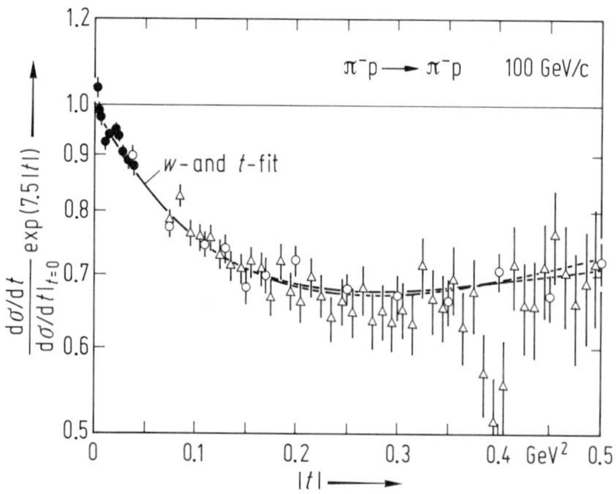

Fig. 6.3.4. Renormalized data at 100 GeV/c and the fits of Eqs. (6.3.3) and (6.3.4). ♦ 78 BURQ, △ 76 AKERLO, ○ 77 AYRES.

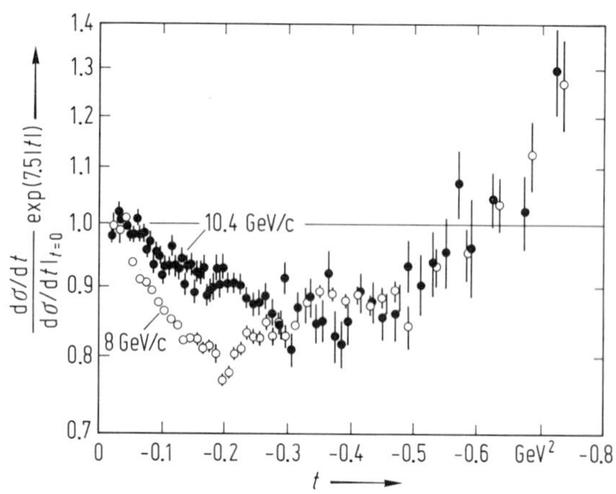

Fig. 6.3.5. An unexpected structure in the 8 GeV/c $\pi^- p$ scattering data 77 RUSS. At fixed $t=-0.2$ GeV^2, the data at 3···40 GeV/c (Fig. 6.3.3) show a smooth decrease except for the 8 GeV/c point.

2.6.3.2 The momentum dependence of C^+ and of the slope of Im C^+

Dispersion analysis

As discussed in Sect. 2.1.16, there are reasons to assume that the contribution of the amplitude A^+ is negligible at high energies in the range $0 > t \gtrsim -0.5$ GeV^2. It then follows from Eq. (1.15.7) that $|C^+|$ can be determined from differential cross section data at fixed t in a large momentum range between 3 GeV/c and 200 GeV/c. In addition, ReC^+ and ImC^+ are known from partial wave analysis below 3 GeV/c. As described in Sect. 2.1.16, one can use this information in order to determine ReC^+ and ImC^+ above 3 GeV/c, employing either dispersion relations or a suitable expansion.

The result of this calculation is shown in Fig. 6.3.6. The line belonging to $t=0$ follows from the forward dispersion relation. If one considers the t-dependence at a fixed value of the momentum k, for instance 10 GeV/c, it is seen that the imaginary part decreases much faster than the real part. As a consequence, the ratio $|\text{Re}C^+/\text{Im}C^+|$ assumes larger values than expected by many authors (Table 2.6.3.2).

Models for diffraction scattering in this momentum range are obviously too simple, if a purely imaginary amplitude is assumed. We mention only two examples: the application of HARARIs dual absorption model by DAVIER (72 L) and by KAMRAN (79 L), but there are many others.

The decrease of the ratio $|\text{Re}C^+/\text{Im}C^+|$ is the main reason for the energy dependence in Fig. 6.3.3. A calculation of the ratio Im$C^+(k,t)$/Im$C^+(k,0)$ between 3 and 200 GeV/c shows that, in the interval $-0.2 > t > -0.4$ GeV^2, the momentum dependence is not larger than the experimental normalization errors, whereas a slight decrease is noticed at $t=-0.1$ GeV^2 (see Fig. 5 in 79 R HÖHLER).

Table 2.6.3.2. The ratio $|\mathrm{Re}\, C^+/\mathrm{Im}\, C^+|$ at different momenta and t-values (79 R HÖHLER-KAISER-KOCH-PIETARINEN).

$-t$ GeV^2	$\lvert\mathrm{Re}\, C^+/\mathrm{Im}\, C^+\rvert$ $k=6$ GeV/c	$k=10$ GeV/c
0	0.20	0.16
0.1	0.32	0.22
0.2	0.42	0.26
0.3	0.44	0.32
0.4	0.45	0.36

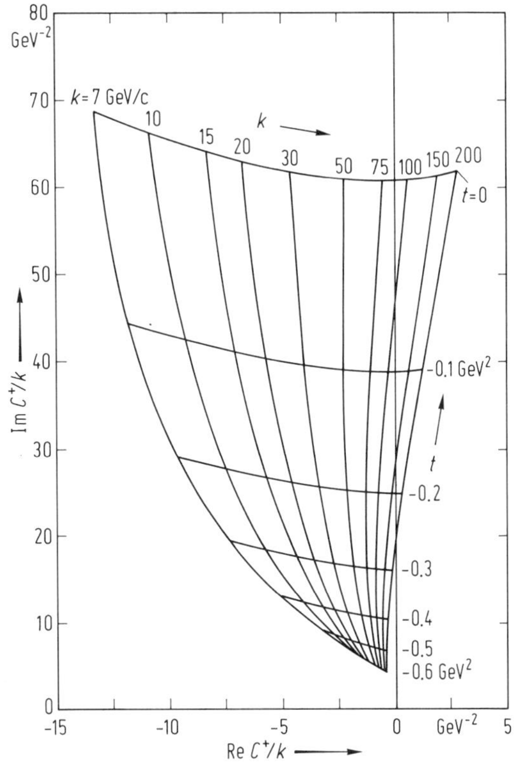

Fig. 6.3.6. An Argand diagram for the diffraction peak. The plot shows the complex amplitude C^+/k as a function of k and t as determined from data and fixed-t dispersion relations. The result depends to some extent on the high energy assumption for C^+ at fixed t.

The s- and t-dependence of the logarithmic slope

S.M. ROY (79 L) presented recently an interesting new unitarity bound for $b(t)$ [Eq. (A.5.38)] and compared it with *fits* of the authors of the accurate π^-p scattering experiments at 100 GeV/c (78 BURQ, 77 AYRES). He noticed a "substantial violation of unitarity".

However, a plot of the data shows that there is obviously a large error of the relative normalization, which can only be due to the data 77 AYRES, because the data 78 BURQ are normalized to the optical point.

The bound is fulfilled, if the data 77 AYRES are renormalized such that they join smoothly the data 78 BURQ (HÖHLER 79 L-3). This conclusion was confirmed by the results of several new experiments (81 FAJARD, 81 SCHIZ, 81 COOL) (see Figs. 6.3.2 and 6.3.4).

Several rigorous bounds for the slope and for the shape of the diffraction peak are listed in Sect. A.5.3. Since these bounds were derived from fundamental theoretical principles, the chance to find a violation is very small. Apparent violations are of interest, because they help to localize experimental errors.

It is sometimes asked: what can one learn from the fact that a bound is *almost* saturated? In the case of the almost saturated isospin bound for the backward cross sections (Sect. 2.4.4.2), the conclusion is that one of the isospin amplitudes must be small, but there is no restriction for its phase.

The MACDOWELL-MARTIN bound Eq. (A.5.33) is also nearly saturated at high energies. It is difficult to see, how the restriction for the partial waves could be applied in amplitude analysis.

The curvature of $b(t)$ in the fits of Table 6.3.1 is probably an energy-dependent phenomenon. It is well-established in the 50...200 GeV/c region. As far as it can be judged from the data of FOLEY et al. (69) and other data, it is weaker at 10...20 GeV/c. There are no accurate data in the Coulomb interference region between 2 and about 50 GeV/c from which the energy-dependence of the curvature can be determined.

A possible reason for an increase of the slope as t goes to zero was proposed by ANSELM and GRIBOV (72 L) on the basis of a Regge pole model. It would be of interest to adapt this model to the present experimental information in order to find out, whether the magnitude of the effect can be explained.

LEADER and MAOR (73 L-1) proposed an ansatz for the "break" which exploits the fact that the slope at $t=0$ is bounded *above and below* by $\ln^2 s$, if the total cross section increases $\propto \ln^2 s$.

The slope of the diffraction peak has been investigated in a series of papers by PARIDA et al. (see 80 L PARIDA for earlier references). A critical discussion was given in a comment written in collaboration with I. SABBA-STEFANESCU (82 L HÖHLER). In my opinion, our objections have not been disproved by PARIDAs reply (82 L PARIDA).

2.6.3.3 The partial wave decomposition of the diffraction peak

In Sect. 2.1.14, we have already briefly mentioned the partial wave projection of an exponentially decreasing diffraction peak [Eq. (1.14.1)]. This is, of course, a crude approximation. Now, we shall discuss realistic partial waves obtained in an updated version of the KH 78 partial wave analysis [HÖHLER and KOCH (80 L-2), see also 76 L PIETARINEN-4].

The partial wave decomposition of the s-channel helicity amplitudes is given in Eq. (A.3.9), the relation between the partial waves F^J_{++} and the usual partial waves $T_{\ell\pm}$ being [Eq. (A.3.12)]

$$2F^J_{++} = T_{(J-1/2)+} + T_{(J+1/2)-}\,. \tag{6.3.6}$$

The impact parameter transform of the s-channel helicity amplitudes is defined in analogy to Eq. (A.3.19)

$$\begin{aligned} f_{+\pm}(s,b) &= \frac{1}{4\sqrt{\pi}} \int_{-4q^2}^{0} F_{+\pm}(s,t) J_{\Delta\lambda}(b\sqrt{-t})\,\mathrm{d}t\,, \\ F_{+\pm}(s,t) &= 2\sqrt{\pi} \int_0^\infty f_{+\pm}(s,b) J_{\Delta\lambda}(b\sqrt{-t})\, b\,\mathrm{d}b\,, \end{aligned} \tag{6.3.7}$$

where $\Delta\lambda = 0$ for the index $++$ and $\Delta\lambda = 1$ for $+-$. The lower limit of the integral over $\mathrm{d}t$ is $-\infty$ in the case of the BLANKENBECLER-GOLDBERGER transform (62 L).

At high energies and not too large J, we have in analogy to Eqs. (A.3.27) and (A.3.29)

$$f_{+\pm}(s,b) \approx F^J_{+\pm}(s)\,; \qquad b = (J+1/2)/q\,; \tag{6.3.8}$$

$$\Gamma(s,b) = \frac{8\pi}{\sigma^+} \operatorname{Im} f^+_{++}(s,b)\,; \qquad \int_0^\infty \Gamma(s,b)\, b\,\mathrm{d}b = 1\,, \tag{6.3.9}$$

where σ^+ is the isospin even total cross section and f^+_{++} the isospin even combination of the impact parameter transforms: $f^+_{++}(s,b) = (f^{1/2}_{++} + 2f^{3/2}_{++})/3$.

It can be compared with the simplest approximation of an exponential decrease of $\operatorname{Im} F^+_{++}$

$$\operatorname{Im} F^+_{++}(s,t) = \frac{\sigma^+}{4\sqrt{\pi}} e^{Bt/2}\,; \qquad \operatorname{Im} f^+_{++}(s,b) = \frac{\sigma^+}{8\pi B} e^{-b^2/2B}\,. \tag{6.3.10}$$

Figure 6.3.7 shows two kinds of deviations:

i) At 10 GeV/c, the shape is roughly an exponential decrease in b^2, but there is a well established deviation. In order to check whether this deviation is outside the error bars, one can determine $\operatorname{Im} F^+_{++}(s,t)$ in a different way, using the result of amplitude analysis (Sects. 2.1.15 and 2.1.16) for $\operatorname{Im} C^+$ and neglecting $\operatorname{Im} A^+$. It turns out that there is a significant deviation of $\operatorname{Im} F^+_{++}$ from an exponential decrease, Eq. (6.3.10).

ii) Below 4 GeV/c, there is a peak at about $b^2 = 1.2\ \mathrm{fm}^2$. Its height is rapidly decreasing at increasing energy. This "peripheral peak" is closely related to a large number of 4-star resonances (see the column for the impact parameter b in Table 2.4.1.1). A similar peak occurs in the partial wave decomposition of the isospin odd helicity flip amplitude (Fig. 6.5.7, p. 372).

An ARGAND diagram for $f^+_{++}(s,b)$ can easily be plotted, if one uses the approximation of Eq. (6.3.8). It shows a remarkable simplicity. In the range 3···10 GeV/c, the points belonging to different pairs of variables (s,b) [where $b = (J+1/2)/q$] lie approximately on a single curve. For negative values of $\operatorname{Re} f^+_{++}(s,b)$ it is a straight line

$$\operatorname{Im} f^+_{++}(s,b) \approx -c_1 \operatorname{Re} f^+_{++}(s,b) + c_2\,; \qquad c_1 \text{ and } c_2 > 0\,. \tag{6.3.11}$$

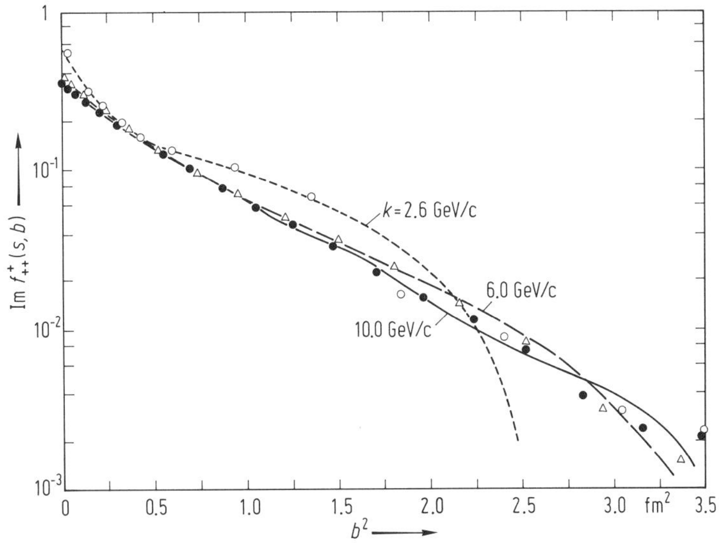

Fig. 6.3.7. Impact parameter transform of the diffraction amplitude. The points have been calculated from partial waves and Eq. (6.3.8), whereas the curves follow from the evaluation of the integral in Eq. (6.3.7) (updated KH 78 analysis).

The straight line intersects the ordinate of the ARGAND diagram at $b \approx 1.1$ fm. The points on this line belong to partial waves, which have a J-value smaller than the "peripheral" one at the same energy (see Fig. 11 in 80 R KOCH).

Partial wave decomposition of the slope of ImF^+_{++}

The slope of Im$F^+_{++}(s,t)$ at $t=0$ is sensitive to the contribution of high partial waves. Using the formula for the second derivative of the Legendre function at $z=1$

$$P''_n(1) = \tfrac{1}{8}(n^2+n)(n^2+n-2), \qquad (6.3.12)$$

one obtains for the logarithmic slope

$$\left.\frac{(\partial/\partial t)\,\mathrm{Im}\,F^+_{++}}{\mathrm{Im}\,F^+_{++}}\right|_{t=0} = \frac{\pi\sqrt{8}}{q^4\sigma^+}\sum_{J=1/2}^{\infty}\frac{\mathrm{d}}{\mathrm{d}z}[\sqrt{1+z}\,\{P'_{J+1/2}(z)-P'_{J-1/2}(z)\}]_{z=1}\,\mathrm{Im}\,F^J_{++}(s)$$

$$= \frac{\pi}{2q^4\sigma^+}\sum_{J=1/2}^{\infty}(4J^3+6J^2+J-1/2)\,\mathrm{Im}\,F^J_{++}(s)\,. \qquad (6.3.13)$$

Figure 6.3.8 shows the contributions of the different partial waves to the logarithmic slope at $t=0$. At each momentum, the points differ from those in Fig. 6.3.7 only by the scale, the J-dependent factor in Eq. (6.3.13), which enhances the contributions of high partial waves, and by a factor q. The factor q was chosen in order to get a plot, in which the area under the curve gives the logarithmic slope at $t=0$. The ordinate is proportional to $b^3\Gamma(b,s)$ in terms of the profile function, Eq. (6.3.9).

In Im$f^+_{++}(s,b)$ one can distinguish a "diffractive component", which is monotonously decreasing as b increases and a "peripheral component", which has its peak near $b=1.1$ fm (see Sect. 2.6.4.8). It is interesting to notice that, for the quantity plotted in Fig. 6.3.8, *both* have their maximum near $b=1.1$ fm. $b_{\max}=\sqrt{3B}$ in the exponential case of Eq. (6.3.10).

One can argue that both are related to the same radius of the matter distribution of the nucleon, the diffractive contribution belonging to a volume effect and the peripheral contribution to a surface effect.

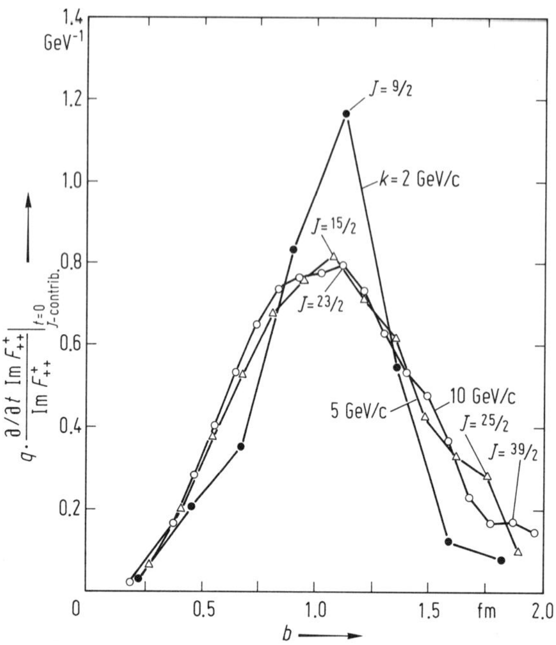

Fig. 6.3.8. Contributions of the different partial waves to the logarithmic slope at $t=0$. On the abscissa we have plotted $b=(J+1/2)/q$ instead of J. At each value of J, the ordinate gives the contribution to the second sum in Eq. (6.3.13). The logarithmic slope at $t=0$ is given by the area under the curves.

2.6.3.4 The inelastic overlap function

In some models the inelastic overlap function is the quantity for which a simple behaviour is expected (Sect. 2.6.4.4). $\mathrm{Im}\, F_{++\mathrm{in}}(s,t)$ follows from Eq. (A.3.9) if Eq. (A.3.12) and the second term of Eq. (A.5.20) are substituted. The approximate equality in Eq. (6.3.8) reads for the inelastic parts

$$\mathrm{Im}\, F^J_{++\mathrm{in}}(s) \equiv \tfrac{1}{4}\varrho^J_{++}(s) = \tfrac{1}{8}\{(1-\eta^2_{(J-1/2)+}) + (1-\eta^2_{(J+1/2)-})\} \approx \tfrac{1}{4}\varrho_{++}(s,b) \equiv \mathrm{Im}\, f_{++\mathrm{in}}(s,b); \quad b=(J+1/2)/q. \tag{6.3.14}$$

Figure 6.3.9 shows the transform of the isospin even inelastic overlap function as calculated from updated KH 78 partial waves. It is seen that, in the 4···10 GeV/c region, the approximate agreement with a Gaussian

$$\varrho^+_{++}(b) = \varrho^+_{++}(0)\, \mathrm{e}^{-b^2/2B_{\mathrm{in}}} \tag{6.3.15}$$

is now much better than in the case of $\mathrm{Im}\, f^+_{++}(s,b)$, Fig. 6.3.7. The parameters of the Gaussian are

$$\varrho^+_{++}(0) = 0.91; \quad B_{\mathrm{in}} = 0.370\,\mathrm{fm}^2 = 9.51\,\mathrm{GeV}^{-2} = 3.70\,\mathrm{mb}; \quad 10\,\mathrm{GeV/c}. \tag{6.3.16}$$

The value of $\varrho^+_{++}(0)$ is restricted by the total inelastic cross section $\sigma_{\mathrm{in}} = 4\sqrt{\pi}\, \mathrm{Im}\, F^+_{++\mathrm{in}}(s,0)$ and Eq. (6.3.18)

$$\sigma^+_{\mathrm{in}} = 2\pi \varrho^+_{++}(0) B_{\mathrm{in}} = 21.2\,\mathrm{mb}; \quad 10\,\mathrm{GeV/c}. \tag{6.3.17}$$

According to Eq. (6.3.14), the maximum value of ϱ^J_{++} is 1. If we denote the average of the two η^2-values by $[\eta^+_{++}(b)]^2 \approx (\eta^J_{++})^2 \equiv 1 - \varrho^J_{++}$, the value $\varrho^+_{++}(0)$ in Eq. (6.3.16) belongs to the absorption parameter $\eta^+_{++}(b=0) = 0.30$. — At $b=1.1$ fm, our fit Eq. (6.3.15) gives $\varrho^+_{++} = 0.18$, corresponding to the absorption parameter $\eta^+_{++} = 0.90$.

The impact parameter transform of the Gaussian, Eq. (6.3.15), leads to an exponential t-dependence for $\mathrm{Im}\, F^+_{++\mathrm{in}}(s,t)$

$$\mathrm{Im}\, F^+_{++\mathrm{in}}(s,t) = \frac{\sqrt{\pi}}{2}\, \varrho^+_{++}(s,0)\, B_{\mathrm{in}} \exp(B_{\mathrm{in}} t/2). \tag{6.3.18}$$

Figure 6.3.10 shows that, at 10 GeV/c, $\mathrm{Im}F_{++\mathrm{in}}(s,t)$ is in good agreement with the exponential decrease for about 2 orders of magnitude. At $t \approx -1\,\mathrm{GeV}^2$, the overlap function has its first zero, which is followed by oscillations at larger $-t$.

The zero of the overlap function is expected from general arguments (see 71 R Zachariasen, Sect. 3.3). An example is given in Eq. (6.3.35).

The above results are taken from our paper 80 L Höhler-2. See also 80 R Koch.

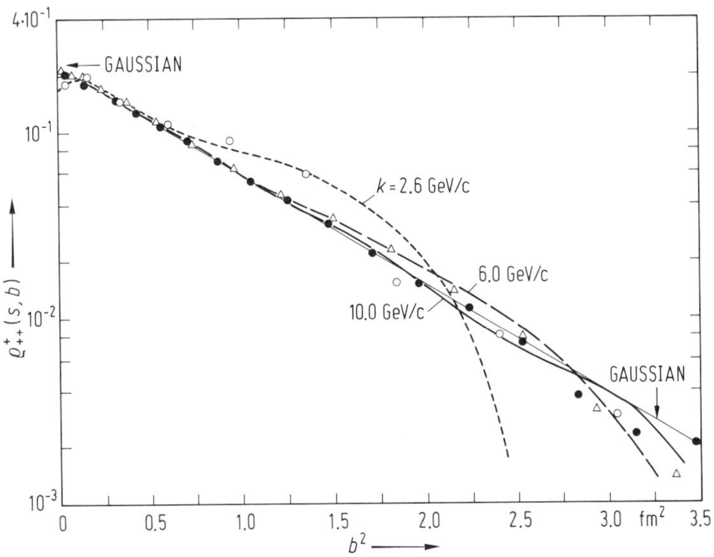

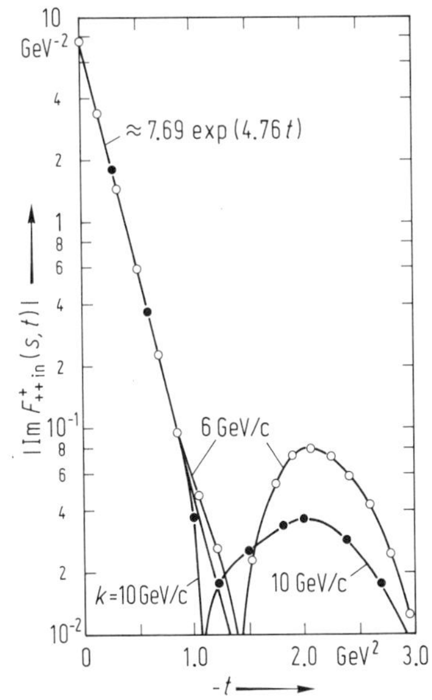

Fig. 6.3.9. Impact parameter transform of the inelastic overlap function. It is seen that the approximation of the 10 GeV/c result by a Gaussian is significantly better than in the case of Im f^+_{++} in Fig. 6.3.7, where the symbols and curves are explained. The scale of the ordinate has to be multiplied by a factor of 4 in order to obtain the normalization of Eq. (6.3.14).

Fig. 6.3.10. The inelastic overlap function Im $F^+_{++\mathrm{in}}(s,t)$ as calculated from the partial wave solution. The first two decades of the decrease are well described by an exponential, but there is a zero near $t=-1.2$ GeV/c (we have plotted the absolute value of Im $F^+_{++\mathrm{in}}$).

2.6.3.5 A high energy approximation of the impact parameter formalism

In amplitude analyses at very high energies and for the discussion of several models one needs the following approximation of the impact parameter formalism of Sect. A.3.3. The lower limit in the Fourier-Bessel integral Eq. (A.3.19), $t=-4q^2$, is replaced by $-\infty$, corresponding to the Blankenbecler-Goldberger (62 L) version of the impact parameter transform. Furthermore, it is assumed that the real parts of the amplitudes are negligible and that the two independent amplitudes are chosen in such a way that one of them is strongly dominant.

The scattering amplitude at a given energy is written $\mathrm{i}A(t)$, omitting the variable s for simplicity. The normalization is given by

$$\frac{\mathrm{d}\sigma}{\mathrm{d}t}=\pi|A(t)|^2, \qquad A(0)=\sigma/4\pi. \tag{6.3.19}$$

Applying the formalism for scattering of spinless particles, one obtains for the partial wave decomposition [cf. Eqs. (A.3.17) and (A.3.20)]

$$A(t)=\frac{1}{2q^2}\sum_{\ell=0}^{\infty}(2\ell+1)(1-S_\ell)P_\ell(\cos\theta); \qquad S_\ell=1+2\mathrm{i}T_\ell. \tag{6.3.20}$$

S_ℓ is an element of the diagonal S-matrix, which is real in our approximation.

The Fourier-Bessel transforms read, using the notation of Eqs. (A.3.34) and (A.3.38)

$$\begin{aligned} A(t)&=\int_0^\infty J_0(b\sqrt{-t})\,a(b)\,b\,\mathrm{d}b=\langle a(b)\rangle; \\ a(b)&=\frac{1}{2}\int_{-\infty}^{0} J_0(b\sqrt{-t})\,A(t)\,\mathrm{d}t=\langle A(t)\rangle. \end{aligned} \tag{6.3.21}$$

We mention the comparison of the present notation with that in Sect. A.3.3: $G(s,t)=\mathrm{i}qA(t)$; $g(s,b)=\mathrm{i}a(b)/2$.

Because of the assumptions, the different variables of Sect. A.3.3 are now related as follows: (the s-dependence is omitted in some cases)

$$T_\ell(s) \approx \frac{i}{2} a(b); \quad S_\ell(s) = \eta_\ell(s) \approx S(b) = \eta(b); \quad \chi(s,b) = \frac{i}{2} \Omega(b); \tag{6.3.22}$$

$$S(b) = e^{-\Omega(b)}; \quad a(b) = 1 - S(b) = (\sigma/4\pi)\Gamma(b), \tag{6.3.23}$$

where

$$b = (\ell + 1/2)/q. \tag{6.3.24}$$

$\Gamma(b)$ is the "profile function" and $\Omega(b)$ the "opacity" or "opaqueness" or "blackness".

The unitarity relation reads [see Eq. (A.5.22)]

$$2a(b) = a^2(b) + \varrho(b) + \dots, \tag{6.3.25}$$

where

$$\begin{aligned} \varrho(b) &= 1 - \eta^2(b) = 1 - S^2(b) = 1 - e^{-2\Omega(b)} \approx 1 - \eta_\ell^2; \\ A(t) &= A_{el}(t) + A_{in}(t); \quad A_{in}(t) = \langle \tfrac{1}{2}\varrho(b) \rangle. \end{aligned} \tag{6.3.26}$$

The question is now, how the different quantities can be calculated from each other.

If one starts with $d\sigma/dt$, one knows $|A(t)|$. It must be positive at small t because of the optical theorem

$$A(0) = \sigma/4\pi = \int_0^\infty a(b) b \, db. \tag{6.3.27}$$

If a dip occurs, it is usually assumed that $A(t)$ changes its sign and $d\sigma/dt$ at the dip is due to the contribution of the real part. Fixed-t dispersion relations can be used in order to estimate the real part and to check if it becomes negligible again beyond the dip.

$\Omega(b)$ follows from Eq. (A.3.45)

$$\Omega(b) = -\ln S(b) = -\ln(1 - \langle A \rangle) = \langle A \rangle + \tfrac{1}{2}\langle A \rangle^2 + \tfrac{1}{3}\langle A \rangle^3 + \dots. \tag{6.3.28}$$

The transform of this equation gives the "optical Born amplitude", which is defined in Eq. (A.3.32)

$$A_B(t) \equiv \langle \Omega(b) \rangle = A(t) + \tfrac{1}{2} A(t) * A(t) + \tfrac{1}{3} A(t) * A(t) * A(t) + \dots. \tag{6.3.29}$$

The theorem Eq. (A.3.41) has been used in the second step.

Some models start with an ansatz for the optical Born amplitude. Then, $A(t)$ can be calculated from

$$\begin{aligned} A(t) &= \langle 1 - S(b) \rangle = \langle 1 - e^{-\Omega} \rangle = \left\langle \Omega - \frac{1}{2!}\Omega^2 + \frac{1}{3!}\Omega^3 - \dots \right\rangle \\ &= \langle \Omega \rangle - \frac{1}{2!}\langle \Omega^2 \rangle + \frac{1}{3!}\langle \Omega^3 \rangle - \dots = A_B(t) - \frac{1}{2!} A_B(t) * A_B(t) + \frac{1}{3!} A_B(t) * A_B(t) * A_B(t) - \dots. \end{aligned} \tag{6.3.30}$$

It is seen that the optical Born amplitude $A_B(t)$ is the first approximation for $A(t)$, which is valid for small t, if the higher terms of the "multiple scattering expansion" are small.

The expansions Eqs. (6.3.29) and (6.3.30) are discussed in the first paper of CHOU and YANG (68 L), who also describe the relation to GLAUBERs theory of multiple scattering (58 R, 67 R), which can be used for instance for pion-nucleus scattering.

If one starts with the inelastic overlap function $\varrho(b)$, $a(b)$ is obtained from Eq. (6.3.25)

$$a(b) = 1 - \sqrt{1 - \varrho}. \tag{6.3.31}$$

The sign of the square root follows from the condition that one must have $a = 0$ for $\varrho = 0$. The total elastic and inelastic cross sections are given by

$$\sigma_{el} = \pi \int_{-\infty}^{0} A^2(t) dt = 2\pi \int_0^\infty a^2(b) b \, db; \quad \sigma_{in} = 2\pi \int_0^\infty \varrho(b) b \, db. \tag{6.3.32}$$

We also mention the expressions for the logarithmic slopes at $t=0$ (79 L Chou-1)

$$\frac{A'(0)}{A(0)}=\frac{1}{4}\frac{\int_0^\infty b^3 a(b)\,db}{\int_0^\infty b a(b)\,db};\qquad \frac{A'_B(0)}{A_B(0)}=\frac{1}{4}\frac{\int_0^\infty b^3 \Omega(b)\,db}{\int_0^\infty b\Omega(b)\,db}. \tag{6.3.33}$$

In many calculations the fact is exploited that the Fourier-Bessel transform of a Gaussian in $\sqrt{-t}$ is a Gaussian in b

$$\langle e^{-b^2/4\beta}\rangle \equiv \int_0^\infty e^{-b^2/4\beta} J_0(b\sqrt{-t})\,b\,db = 2\beta e^{\beta t};$$
$$\langle e^{\beta t}\rangle \equiv \tfrac{1}{2}\int_{-\infty}^0 e^{\beta t} J_0(b\sqrt{-t})\,dt = \frac{1}{2\beta}e^{-b^2/4\beta}. \tag{6.3.34}$$

A simple example is an exponentially decreasing diffraction peak

$$A(t)=(\sigma/4\pi)e^{Bt/2};\quad a(b)=(\sigma/4\pi B)e^{-b^2/2B};\quad \Gamma(b)=B^{-1}e^{-b^2/2B};\quad \sigma_{el}=\frac{\sigma^2}{16\pi B};$$
$$\Omega(b)=-\ln\{1-(\sigma/4\pi B)e^{-b^2/2B}\}=\sum_1^\infty \frac{1}{n}\left(\frac{\sigma}{4\pi B}\right)^n e^{-nb^2/2B}; \tag{6.3.35}$$
$$A_B(t)=B\sum_1^\infty n^{-2}\left(\frac{\sigma}{4\pi B}\right)^n e^{Bt/2n};\quad A_{in}(t)=(\sigma/4\pi)\,e^{Bt/2}-(\sigma_{el}/4\pi)\,e^{Bt/4};$$

$$\frac{A'_B(0)}{A_B(0)}=\frac{B}{2}\frac{\sum_1^\infty n^{-3}\left(\frac{\sigma}{4\pi B}\right)^n}{\sum_1^\infty n^{-2}\left(\frac{\sigma}{4\pi B}\right)^n}. \tag{6.3.36}$$

From $\sigma=25$ mb and $B=10\,\mathrm{GeV}^{-2}$ one obtains for the expansion parameter $\sigma/4\pi B=0.50$.

Another example is an exponential ansatz for $A_B(t)$, which was proposed by Frautschi and Margolis (68 L)

$$A_B(t)=Ke^{\beta t};\qquad A(t)=-\sum_1^\infty \frac{2\beta}{n!\,n}(-K/2\beta)^n e^{\beta t/n}. \tag{6.3.37}$$

The result for $d\sigma/dt$ shows several dips, but the model is too simple for a quantitative description. It was mentioned, because one can clearly see the effect of multiple scattering, which is less transparent in other cases, where a numerical treatment is necessary.

If one wants to calculate $A(t)$ from a more general ansatz for $A_B(t)$, one can fit $A_B(t)$ by a sum of exponentials in t. The transform $\Omega(b)$ is then a sum of Gaussians in b and this is also true for the powers of Ω. Equation (6.3.30) then gives the result for $A(t)$. See for instance 78 L Chan.

Some problems which occur in numerical calculations of the various transforms have recently been discussed by Hutt and Sabba-Stefanescu (82 L).

2.6.3.6 pp-scattering amplitudes in the ISR range

If one wants to perform an amplitude analysis of πN scattering data at high energies, it is of interest to study the analyses of pp-scattering for the following reasons:

i) The same methods can be used, since it is assumed in both cases that one of the scattering amplitudes is strongly dominant and that the real part is small or negligible, except near a dip of the cross section.

ii) The existing data support the assumption that the asymptotic behaviour of the πN amplitude is similar to that of the pp-amplitude (slow increase of the total cross section, shape of the diffraction peak, existence of a dip). pp-scattering has the advantage that data are available up to much higher energies from experiments at the Intersecting Storage Rings ("ISR") at CERN. The amplitudes derived from these data can be used in order to estimate the behaviour of the πN amplitudes above the experimentally accessible region. Furthermore, it is of interest to adapt those models to πN scattering, which have been successfully tested in pp-scattering in the ISR range.

At very high energies the amplitude analysis can be carried out as follows, using the formalism of Sect. 2.6.3.5. From $d\sigma/dt$ and Eq. (6.3.19) one obtains $|A(t)|$. The value of $A(0)=\sigma/4\pi$ is positive and it is usually assumed that $A(t)$ changes its sign at the dip of $d\sigma/dt$.

Since the real part is important near the dip, we introduce a complex-valued amplitude $f(s,t)$, where $\mathrm{Im}\, f = A(t)$. The magnitude of $\mathrm{Re}\, f$ can be determined from a fixed-t dispersion relation (75 L GREIN). As long as $(\mathrm{Re}\, f)^2 \ll (\mathrm{Im}\, f)^2$, one can as well employ a derivative analyticity relation (Sect. A.6.4.5.), which can be written in the case of a crossing even-amplitude (77 L SUKHATME-2)

$$\frac{\mathrm{Re}\, f}{\mathrm{Im}\, f} = \tan\left(\frac{\pi}{4}\frac{\mathrm{d}}{\mathrm{d}(\ln s)}\ln\frac{d\sigma}{dt}\right). \tag{6.3.38}$$

A careful investigation of this type has been carried out by NAGY et al. (79 L). Their result is well compatible with the assumption that $\mathrm{Im}\, f$ is zero at the dip and that the differential cross section near the dip is dominated by the contribution of the real part.

The most detailed analyses of the ISR data on pp-scattering are due to members of the experimental groups at the ISR: AMALDI and SCHUBERT (80 R) and NAGY et al. (79 L). AMALDI and SCHUBERT determined the scattering amplitude and the inelastic overlap function, which is called $\varrho(s,b)$ in our notation [Eq. (6.3.25)]. Their result *rules out a factorizable eikonal* (or opacity) of the type

$$\Omega(s,b) = K(s)\tilde{\Omega}(b), \tag{6.3.39}$$

which is used in several models (see below, Sects. 2.6.4.1 and 2.6.4.2).

In another recent analysis of the same data, FRANCA and HAMA (79 L) assumed an imaginary scattering amplitude with two zeros. One zero is necessary in order to describe the dip at $t=-1.4\,\mathrm{GeV}^2$ and the other one is needed, if one wants to have an always positive optical Born term. This condition is fulfilled for instance in the CHOU-YANG model where $A_B(t) \propto G^2_{Ep}(t)$ (see 68 L DURAND). The exact location of the second zero is not important. The authors concluded that there is a well-determined energy dependence of the opacity, which can be described by

$$\Omega(s,b) = \Omega_f(b) + \ln(s/s_0)\Omega_0(b). \tag{6.3.40}$$

The result confirms the conclusion of AMALDI and SCHUBERT that a factorizable opacity is not tenable.

Both AMALDI and SCHUBERT and FRANCA and HAMA criticize an earlier analysis by CHOU (78 L), giving arguments why his results are less reliable.

A quite different method was used by HENZI and VALIN (79-1 and 2). The authors started from a flexible parametrization of the inelastic overlap function, taking into account the fact that a Gaussian is a zero order approximation

$$\varrho(s,b) = P\sum_{n=0}^{\infty}\delta_{2n}\left[\frac{\gamma^2}{2B}\right]^n b^{2n} e^{n-b^2/4B_n}; \qquad B_n = B/(2n\gamma^2+1). \tag{6.3.41}$$

The adjustable parameters depend on energy. It turns out that 3 terms of the series are sufficient for a good fit in the lab. momentum range 200⋯2000 GeV/c and for t-values up to $-14\,\mathrm{GeV}^2$.

JENKOVSZKY and PACCANONI (81 L) determined the inelastic overlap function by fitting the data to a Regge dipole formula. Discrepancies of their result with that of other authors are explained by numerical problems with the small-t break and by the sensitivity to the assumption on the behaviour at large t. They point out that there are appreciable discrepancies between the results of the above mentioned analyses. BOURRELY et al. (77 L) made an eikonal-type ansatz for the pp-scattering amplitudes and fitted the differential cross sections and the polarization data in the range 5⋯2000 GeV/c.

KROLLs review on NN-scattering (81 R) includes a chapter on impact parameter analyses and tables of the results of AMALDI and SCHUBERT (81 R) for the impact parameter transforms of the pp-scattering amplitude and the inelastic overlap function (see also our Table 2.6.4.2). Tables of pp-scattering data at high energies are given in SCHUBERTs contribution to Vol. I/9a of this series. More recent results can be found in 79 L NAGY, 80 R AMALDI-SCHUBERT, 81 L FAISSLER. A dip in $\bar{p}p$ scattering was recently discovered at 50⋯200 GeV/c (82 L ASAD, 82 L KAPLAN)

Note added in proof, see p. 404.

Conclusion on pp-amplitude analyses

For a critical test of models, a comparison with the results of amplitude analyses is preferable to a direct comparison with the data, because it is a difficult task to produce a coherent and complete set of data as it was done by AMALDI and SCHUBERT (80 R).

It is important to know to what extent the scattering amplitude and its transforms are at present well determined. Therefore, it is necessary to search for the origin of the discrepancies mentioned in 81 L JENKOVSZKY, which could be due to the treatment of the input data as well as to the methods used in the analyses.

2.6.3.7 Elastic scattering at larger $|t|$

In elastic scattering at high energies, one can easily distinguish three different regions (Fig. 6.6.1): near-forward scattering, wide-angle scattering and near-backward scattering. Near-forward scattering is characterized by the "diffraction peak", which is slowly variable in energy, if t is chosen as the second variable. The wide-angle region is centered around $\theta = 90°$. The angular dependence is rather flat but there is a very rapid energy dependence at fixed θ. The data and the amplitudes are usually discussed in terms of the variables s and $\cos\theta$ (Sect. 2.6.6).

The present subsection is devoted to the transition between the near-forward and wide-angle regions. At intermediate energies, we shall include data of the wide-angle region, because $\theta = 90°$ lies at relatively small t-values. The s- and t-dependence is more complicated than in the diffraction peak and the results of amplitude analysis are still rather uncertain. Therefore we shall treat only two aspects:

i) The energy dependence of $d\sigma/dt$ at fixed t,

ii) characteristic structures of $d\sigma/dt$ vs. t.

The energy dependence of $d\sigma/dt$ at fixed t

Several plots of $\ln(d\sigma/dt)$ vs. $\ln k$ at fixed t are shown in Fig. 6.3.11. In earlier discussions of Regge pole models, similar plots have been used in order to define an "effective trajectory" by

$$\frac{d\sigma}{dt} = F(t)\,(s/s_0)^{2\alpha_{\rm eff}(t)-2}, \qquad (6.3.42)$$

although, at intermediate energies and large t, one should better use the variable $v = \omega + t/4m \approx k + t/4m$. Since in elastic scattering the slope in plots of $\ln d\sigma/dt$ vs. $\ln s$ is a function of energy, $\alpha_{\rm eff}(t)$ can only be defined locally in energy intervals.

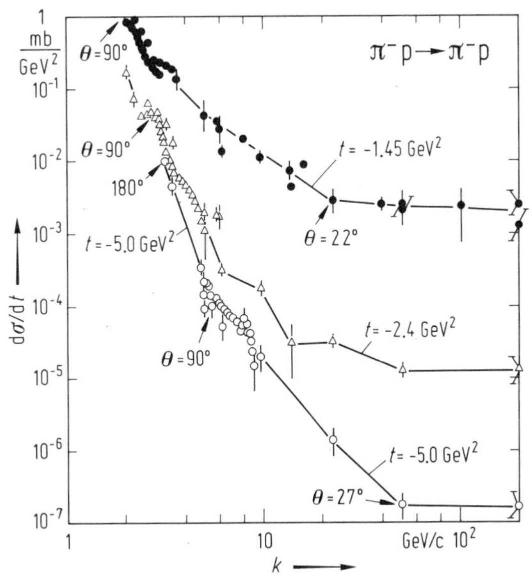

Fig. 6.3.11. $d\sigma/dt$ vs. k at fixed values of t. It is seen that the curves have a rapidly decreasing part at intermediate energies and an almost constant part at high energies. For a comparison with other figures (for instance Figs. 6.6.1 and 6.3.12) the c.m. scattering angle is given in a few cases. References for the data: see Table 1.2.2 and 81 BAKER.

For sufficiently high energies, the contributions of the low-lying trajectories become negligible and one expects to obtain an energy-independent $\alpha_{eff}(t)$, which belongs to the leading pomeron trajectory.

Figure 6.3.11 shows that the curves $\ln d\sigma/dt$ vs. $\ln k$ have two parts:

i) A rapid decrease which follows roughly a power law. The oscillations which are superimposed are due to resonances at least below 3 GeV/c. The effective α is for instance -1.75 at $t=-2.4\,\text{GeV}^2$.

ii) A practically constant behaviour at the highest momenta, i.e. the leading term has $\alpha_{eff}\approx 1.0$ even at large negative t.

At $t=-5\,\text{GeV}^2$, the curve starts with a more rapid decrease, because the first point lies on the backward peak.

It would be of interest to perform a systematic study of the t-dependence of α_{eff} in the almost linear part of the curves. Unfortunately, this is difficult because there are large discrepancies even between different new data sets and only a few of them are probably due to the overall normalization. Furthermore, there are still large gaps in the data. The inclusion of the results of BAILLON et al. (80), which are not yet available to us, will be an important progress.

The t-dependence of $d\sigma/dt$ at fixed s

In Fig. 6.3.12 we have plotted $\ln d\sigma/dt$ vs. t in order to show the characteristic structures of the cross section. The most prominent structures are the dip at $t\approx-0.6\,\text{GeV}^2$, the secondary peak at $t\approx-1.3\,\text{GeV}^2$ and the dip

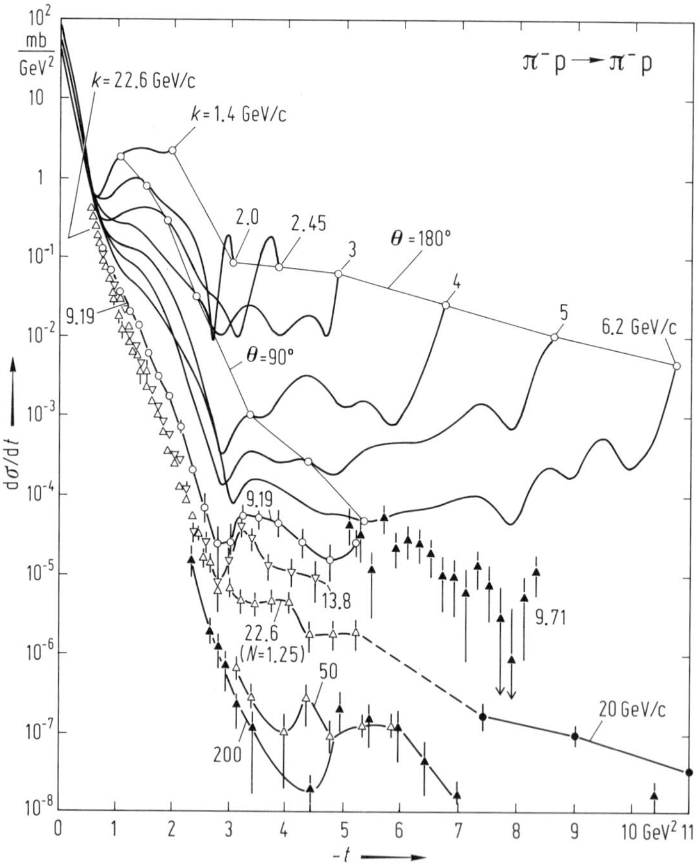

Fig. 6.3.12. $d\sigma/dt$ vs. t at different energies. The curves belonging to 1.4···6.2 GeV/c have been calculated from the preliminary updated KH 78 partial wave solution. Some of the structures are not yet well determined due to discrepancies between different data sets (see for instance Fig. 6.3.13). The points at c.m. scattering angles $\theta=90°$ and 180° are connected by straight lines. Data at higher momenta: 9.19 GeV/c: 78 JENKIN, 9.71 GeV/c: 69 OWEN, 13.8 GeV/c: 73 CORNIL, 20 GeV/c: 80 ALMAS, 22.6 GeV/c: 73 CORNIL, 50 GeV/c and 200 GeV/c: 82 ASAD and 81 BAKER (only part of the data points 81 BAKER are shown). The data set 73 CORNIL at 22.6 GeV/c has been renormalized by a factor of $N=1.25$ in order to have a smooth s-dependence at small t, where data at somewhat smaller and higher momenta can be adjusted to the forward cross section (Table 6.3.1). — The dip at $t\approx-3\,\text{GeV}^2$ is clearly visible up to 13.8 GeV/c. There is a second dip near this t-value in the lower part of the momentum range (80 TERADA, 77 VAVRA).

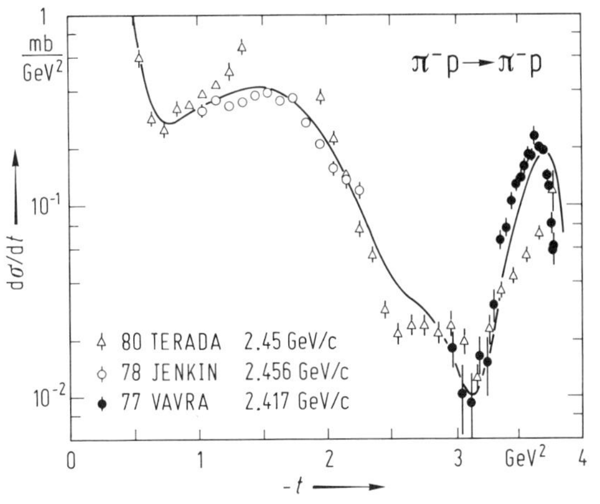

Fig. 6.3.13. Discrepancies between the results of different π^-p scattering experiments near 2.45 GeV/c. The solid line has been calculated from the updated KH 78 partial wave analysis at $k=2.45$ GeV/c, which does not include the data set 80 TERADA. The shift in comparison to the data 77 VAVRA is due to the small difference of the momenta. The data 77 VAVRA show a dip in the backward direction between $k=1.917$ and 2.600 GeV/c, but a backward peak was found by APLIN et al. (71) at 2.46 GeV/c.

at $t \approx -3.0\,\text{GeV}^2$. At some energies, the shape of the differential cross section is not yet well determined due to discrepancies between different experiments (see for instance Fig. 6.3.13).

As the energy increases, the cross sections at fixed t are monotonously decreasing until they reach the value given by the constant high energy behaviour in Fig. 6.3.11. The above mentioned dip-bump structures disappear and are therefore obviously a property of the decreasing component.

It might be that the dip at $t=-4\,\text{GeV}^2$ in the 200 GeV/c data 81 BAKER is of a different nature and belongs to the diffractive component, similarly to the dip at $t=-1.4\,\text{GeV}^2$ in pp-scattering, which is seen at all momenta, including the ISR range, above about 200 GeV/c lab. momentum. [Recently, ASAD et al. (82 L) have discovered a dip at the same t-value in $\bar{\text{p}}$p-scattering at 50 GeV/c.] One argument in favour of this assumption is that the cross section is practically constant between 50 and 200 GeV/c at $t=-5\,\text{GeV}^2$, i.e. beyond the dip.

Amplitude analyses at large $|t|$ and high energies do not yet exist. The problem is that there are neither polarization data nor spin-rotation data and that there is no reason to expect that one of the amplitudes is strongly dominant. One could try to derive some information on the phase of the invariant amplitudes from a suitable one-dimensional dispersion relation. Recent models will be mentioned in Sect. 2.6.4.

2.6.3.8 The isospin even s-channel helicity flip amplitude

Isospin even πN scattering is described by two independent amplitudes, for which one can choose for instance the invariant amplitudes (C^+, B^+) or (C^+, A^+) or the s-channel helicity amplitudes (F^+_{++}, F^+_{+-}), or the spin no-flip and flip amplitudes (G^+, H^+). In all cases, the imaginary part of the first amplitude of the pair at $t=0$ is determined by the optical theorem. The differential cross section at small $|t|$ is dominated by the contribution of this amplitude, because the contribution of the second amplitude is suppressed by a factor of t.

In order to simplify the discussion, we shall consider in the following the pair (F^+_{++}, F^+_{+-}). As mentioned in Sect. 2.1.15, experimental information on F^+_{+-}/F^+_{++} can be obtained only from polarization and spin-rotation experiments [see also Eq. (A.4.13)].

The isospin even amplitudes are related to the following combination of measurable quantities

$$P^+ \frac{d\sigma^+}{dt} \equiv \tfrac{1}{2}\left\{P_+ \frac{d\sigma_+}{dt} + P_- \frac{d\sigma_-}{dt} - P_0 \frac{d\sigma_0}{dt}\right\} = 2\,\text{Im}(F^+_{++} F^{+*}_{+-}) \equiv 2|F^+_{++}|(F^+_{+-})_\perp . \qquad (6.3.43)$$

From similar combinations of the spin-rotation data, one can determine $\varkappa = |F^+_{+-}|/|F^+_{++}|$ and the component of F^+_{+-} parallel to F^+_{++}, using Eqs. (1.15.3) and (1.15.4).

It turns out that, at high energies and $t \gtrsim -0.5\,\text{GeV}^2$, $|F^+_{+-}|^2 \ll |F^+_{++}|^2$. Furthermore, the charge-exchange cross section is negligible in comparison with the elastic cross sections and $d\sigma_-/dt - d\sigma_+/dt$ is small in comparison with $d\sigma^+/dt$. Therefore

$$P^+ \approx \tfrac{1}{2}(P_+ + P_-) \approx 2\frac{(F^+_{+-})_\perp}{|F^+_{++}|} . \qquad (6.3.44)$$

There are two possibilities for the high energy behaviour of the flip amplitude at fixed t, which are frequently discussed:

i) The ratio $|F^{+}_{+-}|/|F^{+}_{++}|$ remains finite. Then, the spin-rotation parameters S^{+} and T^{+} approach values different from zero and one.

ii) Asymptotic s-channel helicity conservation

$$|F^{+}_{+-}|/|F^{+}_{++}| \to 0 \quad \text{as} \quad s\to\infty \quad \text{at fixed } t. \tag{6.3.45}$$

The spin-rotation parameter S^{+} goes to 1 and T^{+} goes to 0. Equation (A.4.15) shows that the R- and A-parameters behave as follows

$$R^{+}\to -\cos\theta_{\rm N} \to -\left[\frac{-t}{4m^2-t}\right]^{1/2}; \qquad A^{+}\to \sin\theta_{\rm N} \quad \text{as} \quad s\to\infty. \tag{6.3.46}$$

In both cases, $P^{+}(s,t)$ goes to zero as $s\to\infty$. This follows in the first case by using the relation between the energy dependence of the modulus and the phase (Sect. A.6.4) at $t\neq 0$. It is clear that the question of helicity conservation *cannot be resolved* from P^{+}-data alone as it was claimed in the title of the paper 72 L BARGER-3.

The near-forward $\pi^{\pm}$p polarization data are approximately mirror symmetric, i.e. $P_{+}-P_{-}$ is large and positive and $P_{+}+P_{-}$ is small. The energy dependence of $P_{+}+P_{-}$ at fixed t shows, as expected, a decrease. At 3⋯14 GeV/c, $P_{+}+P_{-}$ is positive at least up to $t=-0.5\,\text{GeV}^2$. The 40 GeV/c data (75 GAIDOT, 76 GAIDOT) are *negative*, but this trend is not clearly visible in the 100 GeV/c data of KLINE et al. (80) (Fig. 6.3.14).

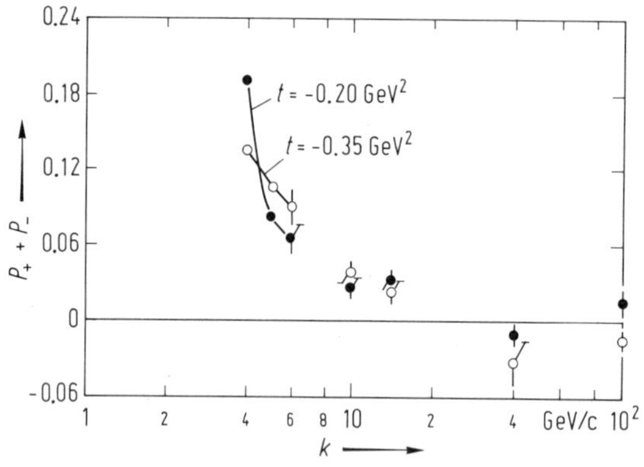

Fig. 6.3.14. Sum of the elastic polarization parameters at high energies. Data: 100 GeV/c: 80 KLINE = 77 AUER, 40 GeV/c: 75 GAIDOT and 76 GAIDOT, 6, 10, and 14 GeV/c: 72 BORGHI, 4 and 5 GeV/c: from KH 78 partial waves. The negative values at 40 GeV/c are mainly due to the behaviour of the P_{+} data (76 GAIDOT). A confirmation by another experiment would be desirable.

According to the amplitude analyses (Sect. 2.1.16), both amplitudes F^{+}_{++} and F^{+}_{+-} have a dominant imaginary part which is positive. A change of sign of P^{+} indicates that the relative phase of these two amplitudes has a zero as well as the relative phase of C^{+} and A^{+} [Eq. (A.4.19)].

If one wants to describe this effect in Regge pole theory, a third pole is required in addition to P and f. It is called P'' or σ (69 L BARGER, 76 L GIRARDI, 78 L MORROW, 78 L MARTIN-2). However, the evidence for a σ-trajectory is weak, since the change of sign in $P_{+}+P_{-}$ is not yet well established and there is no guarantee that a sum of Regge poles is sufficient for an accurate description of the amplitudes.

In dispersion theory, the behaviour of P^{+} is described in terms of the invariant amplitudes C^{+} and A^{+}. The ratio A^{+}/C^{+} is related to F^{+}_{+-}/F^{+}_{++} by Eq. (1.15.6). The data for P^{+} and the spin-rotation data at 6 and 40 GeV/c are the only constraint for the amplitude A^{+}, whose high energy behaviour is not restricted by a generally accepted theoretical condition.

As discussed in Sect. 2.1.15, the assumption $A^{+}(s,t)\to 0$ as $s\to\infty$ at fixed t has as a consequence asymptotic s-channel helicity conservation, but it is stronger. In a Regge pole model, it corresponds to a decoupling of the P- and f-Regge poles from the amplitude A^{+}. It is remarkable that the corresponding fNN-coupling constant $G^{(2)}_{\rm fNN}$ is compatible with zero [Eq. (4.2.10)]; see 70 L HÖHLER-1 and 73 L KAFIEF.

s-channel helicity conservation in other reactions has been discussed in 70 L GILMAN; see 78 R LEITH (Sect. 3.2.2) for a recent review. The best example is the photoproduction of ϱ on protons, which behaves like a diffractive process.

A better information on F^{+}_{+-} and A^{+} in the asymptotic region can only be obtained from new polarization and spin-rotation measurements in the Fermilab and SPS region. The great effort would be justified, because the results would help to understand the spin structure of diffraction scattering.

MÜLLERs review (72 R) gives a fairly complete treatment of the literature on helicity conservation available at that time, including many theoretical papers. As an example for more recent work, we mention the idea of CHOU and YANG (76 L) that a polarized proton can be described by a rotating matter distribution. A modified version was discussed by BOURRELY, SOFFER, and WU (79 L). These models lead to a prediction for the high energy behaviour of F^{+}_{+-} which, however, is uncertain due to the difficulty of the factorizable eikonal ansatz (see Sects. 2.6.4.1 and 2.6.4.2).

In a recent paper, YOUNG (79 L) considered a modified CHOU-YANG model, in which the continuous matter distribution in the nucleon is replaced by three quarks. He showed that the flip amplitude is suppressed.

Further reviews and papers on diffraction scattering are mentioned at the end of Sect. 2.6.4.

2.6.4 Models of diffraction scattering

There exists a large number of models of diffraction scattering which are described for instance in the reviews 71 R ZACHARIASEN, 76 R ABARBANEL. Most of them belong to one of the following categories:

i) *s*-channel models

An ansatz is made for one of the quantities which occur in the impact parameter formalism (Sect. 2.6.3.5), for instance for the opacity or for the overlap function. It is usually motivated by geometrical or optical ideas, if the hadron is considered as a continuous matter distribution. In other models, the eikonal for hadron-hadron scattering is expressed by contributions from the scattering between the "constituents" (quarks and gluons). Primarily, the *s*-channel models describe the shape of the diffraction peak. In some cases, it is difficult to find an extension which also describes the *s*-dependence at fixed *t* in agreement with the data.

ii) *t*-channel models

The scattering process is mediated by the exchange of objects in the *t*-channel. The leading term comes from "pomeron" exchange, a notion for which a generally accepted definition does not exist since the discovery of the increase of the total cross sections. Secondary terms belong to the exchange of meson resonances. The *t*-channel exchange terms have a well-defined *s*-dependence at fixed *t*, but the *t*-dependence at fixed *s* is arbitrary to a large extent.

We start with a discussion of the most popular *s*-channel models.

2.6.4.1 The CHOU-YANG model

CHOU and YANG (68 L, see also 80 R YANG) proposed an optical model for elastic hadron-hadron scattering, in which the hadrons are treated as spatially extended Lorentz contracted balls of "hadronic matter", which propagate through each other. During the collision, various inelastic interactions take place and lead to elastic "shadow scattering".

In addition to the spatial density distribution $\varrho_i(x, y, z)$, the authors introduce the two-dimensional density distributions of the two particles ($i=1,2$)

$$d_i(b) = \int_{-\infty}^{+\infty} \varrho_i(x, y, z)\,\mathrm{d}z\,, \tag{6.4.1}$$

where $b=\sqrt{x^2+y^2}$. ϱ_i is assumed to be spherically symmetrical.

One of the basic assumptions is now that the opacity for elastic scattering is given by the convolution integral over the product of the density functions $d_i(b)$ of the two particles

$$\Omega(b) = K \iint \mathrm{d}^2 b' d_1(\boldsymbol{b}') d_2(\boldsymbol{b}-\boldsymbol{b}'), \tag{6.4.2}$$

where K is a constant, which is adjusted to the forward amplitude as following from the optical theorem. $\boldsymbol{b}$ and $\boldsymbol{b}'$ are two-dimensional vectors in a plane perpendicular to the *z*-direction.

According to the convolution theorem, Eq. (A.3.41), the Fourier transform of $\Omega(b)$, which is by definition the optical Born amplitude $A_B(t)$, Eq. (6.3.29), is given by the product of the transforms $D_i(t)$ of the functions $d_i(b)$ [Eq. (A.3.36)]

$$\langle d_i(b)\rangle = D_i(t) = \frac{1}{2\pi}\iint \mathrm{d}^2 b\, e^{i\boldsymbol{\varkappa}\cdot\boldsymbol{b}} d_i(b); \quad \langle\Omega(b)\rangle \equiv A_B(t) = K D_1(t) D_2(t)\,. \tag{6.4.3}$$

The second basic assumption is that the hadronic matter distributions agree with the electric charge distributions. Then, their transforms are the electromagnetic form factors

$$D_1(t)=F_\pi(t)\,; \quad D_2(t)=G_{Ep}(t)\,. \tag{6.4.4}$$

The fact that the charge distribution $\varrho(x,y,z)$ calculated from the 3-dimensional Fourier transform of an electromagnetic form factor has a simple meaning only in the non-relativistic limit (62 L SACHS, Appendix; 75 L SKACHKOV) was discussed in 76 L CHOU.

The expression for the optical Born amplitude for pion-nucleon scattering reads

$$\langle\Omega(b)\rangle \equiv A_B(t)=K\,F_\pi(t)\,G_{Ep}(t)=Kf_E(t)\,. \tag{6.4.5}$$

The prediction can be tested in two different ways. One can start from the result of amplitude analysis for $\Omega(b)$ and compare the transform $\langle\Omega(b)\rangle$ with the prediction of Eq. (6.4.5). It turns out that there are some systematic deviations (77 L LOMBARD). A new analysis has been performed by HUTT and SABBA-STEFANESCU (82 L). The results are given in Tables 2.6.4.1 and 2.6.4.2 and in Figs 6.4.1 and 6.4.2.

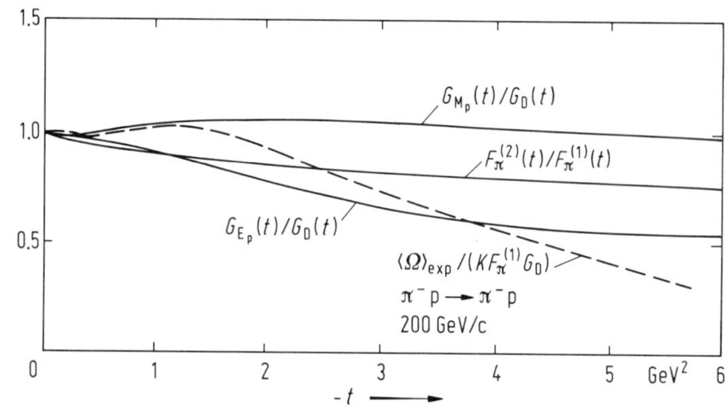

Fig. 6.4.1. Test of the CHOU-YANG model with 200 GeV/c $\pi^-p\to\pi^-p$ scattering data (82 L HUTT and SABBA-STEFANESCU). The optical Born amplitude $\langle\Omega\rangle_{exp}$ has been calculated from the data sets 81 L SCHIZ, 76 L AKERLO and 81 L BAKER, including the dip at $t=-4\,\text{GeV}^2$. $G_D(t)$ is the dipole fit, Eq. (A.11.20) and $F_\pi^{(1)}(t)=(1-t/0.505)^{-1}$, t in GeV². K has been adjusted such that the dashed curve starts at unity at $t=0$. — In order to show the dependence on the form factor input, we have also plotted G_{Mp}/G_D, G_{Ep}/G_D and $F_\pi^{(2)}/F_\pi^{(1)}$, where $F_\pi^{(2)}$ is a fit by HEYN and LANG (81 L). According to Eq. (6.4.5), the dashed line should coincide with the line for G_{Ep}/G_D. The uncertainties of G_{Mp} and G_{Ep} are shown in Figs. 5.2.2 and 5.2.4. — The difference between the $\langle\Omega\rangle_{exp}$-values at 200 and 50 GeV/c is comparable with the errors, which are discussed in the above mentioned paper.

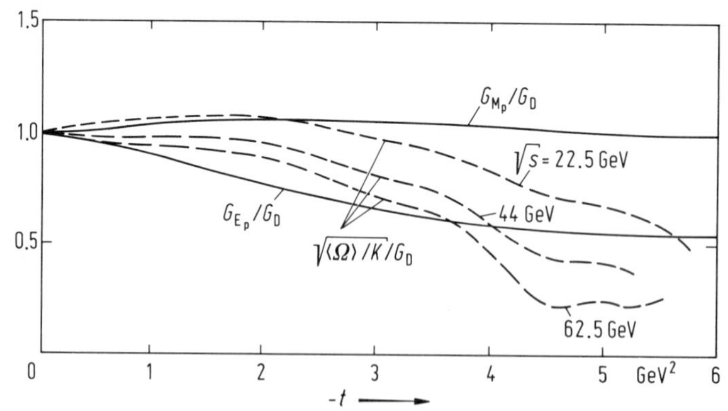

Fig. 6.4.2. Test of the CHOU-YANG model with pp→pp scattering data (82 L HUTT and SABBA-STEFANESCU). The dashed lines show $\sqrt{\langle\Omega\rangle_{exp}}/(\sqrt{K}G_D)$ as calculated from the ISR-data at $\sqrt{s}=22.5$, 44, and 62.5 GeV. The uncertainties of G_{Mp} and G_{Ep} are shown in Figs. 5.2.2 and 5.2.4. — It is seen that the CHOU-YANG model is a reasonable approximation and that the curves derived from the scattering data are similar as in the case of π^-p-scattering (Fig. 6.4.1). The failure of the factorizing eikonal ansatz, Eq. (6.3.39) in Sect. 2.6.3.6, leads to the systematic energy dependence of the dashed lines, which is a small effect, almost comparable with the uncertainties of the experimental form factors.

Table 2.6.4.1. Amplitudes for elastic πp-scattering at high energies. $A(t)$ is the imaginary part of the amplitude, $A_{in}(t)$ the inelastic contribution to it and $A_B(t)$ the (real) optical Born amplitude as defined in Eqs. (6.3.19), (6.3.26), and (6.3.29), respectively (82 L Hutt). Real parts and flip amplitudes were neglected.

Values at $t=0$ in units of $\sqrt{\text{mb}}/\text{GeV}$:

50 GeV/c	$A(0)=3.01$	$A_{in}(0)=2.61$	$A_B(0)=3.63$
200 GeV/c	$A(0)=3.08$	$A_{in}(0)=2.67$	$A_B(0)=3.70$

$-t$	50 GeV/c			200 GeV/c		
	$\frac{A(t)}{A(0)}$	$\frac{A_{in}(t)}{A_{in}(0)}$	$\frac{A_B(t)}{A_B(0)}$	$\frac{A(t)}{A(0)}$	$\frac{A_{in}(t)}{A_{in}(0)}$	$\frac{A_B(t)}{A_B(0)}$
GeV^2	10^{-3}	10^{-3}	10^{-3}	10^{-3}	10^{-3}	10^{-3}
0.05	766.	745.	790.	777.	758.	799.
0.10	600.	567.	638.	609.	579.	645.
0.15	479.	439.	525.	482.	444.	526.
0.20	389.	346.	439.	385.	343.	435.
0.25	320.	276.	373.	311.	267.	363.
0.30	267.	222.	320.	253.	209.	306.
0.35	224.	180.	277.	209.	165.	261.
0.40	189.	147.	241.	173.	131.	225.
0.45	161.	121.	211.	145.	105.	195.
0.50	137.	99.	186.	123.	85.	171.
0.55	117.	81.	164.	105.	69.	151.
0.60	100.	66.	145.	90.	56.	134.
0.65	86.	54.	129.	78.	46.	119.
0.70	74.	44.	115.	67.	38.	107.
0.75	64.	36.	103.	59.	31.	96.
0.80	55.4	28.8	92.0	51.5	26.0	86.8
0.85	47.9	23.0	82.7	45.3	21.5	78.7
0.90	41.5	18.1	74.6	40.0	17.8	71.5
0.95	36.0	14.1	67.4	35.3	14.8	65.2
1.00	31.2	10.8	61.0	31.3	12.1	59.5
1.25	15.7	1.3	38.4	17.5	4.0	38.8
1.50	8.2	-1.8	25.3	9.9	0.4	26.1
1.75	4.5	-2.5	17.6	5.6	-1.1	18.1
2.00	2.6	-2.3	12.7	3.2	-1.6	12.9
2.25	1.5	-1.9	9.3	1.8	-1.7	9.4
2.50	0.93	-1.49	6.94	0.98	-1.51	6.91
2.75	0.58	-1.14	5.25	0.52	-1.28	5.19
3.00	0.36	-0.86	4.08	0.25	-1.04	3.96
3.25	0.21	-0.66	3.24	0.10	-0.83	3.07
3.50	0.12	-0.51	2.56	0.02	-0.65	0.24
3.75	0.05	-0.40	2.00	-0.03	-0.51	1.89
4.00	0.01	-0.32	1.55	-0.06	-0.40	1.50
4.25	-0.02	-0.26	1.21	-0.07	-0.31	1.19
4.50	-0.04	-0.22	0.97	-0.07	-0.24	0.95
4.75	-0.05	-0.18	0.81	-0.08	-0.19	0.77
5.00	-0.06	-0.16	0.68	-0.08	-0.15	0.62
5.25	-0.06	-0.13	0.56	-0.07	-0.12	0.50
5.50	-0.07	-0.12	0.43	-0.07	-0.10	0.40
5.75	-0.07	-0.10	0.32	-0.07	-0.08	0.32
6.00	-0.07	-0.09	0.23	-0.06	-0.07	0.26

Table 2.6.4.2. Impact parameter transforms of elastic πp and pp-scattering amplitudes at high energies. $a(b)$, $\varrho(b)$ and $\Omega(b)$ are defined in Eqs. (6.3.21), (6.3.26), and (6.3.22), respectively. The calculation at 50 and 200 GeV/c is due to Hutt and Sabba-Stefanescu (82 L).

The pp-amplitudes were calculated from the fit of Amaldi and Schubert (80 R). Kroll (81 R) gives a more detailed table which includes an error estimate. – The transforms at 10 GeV/c belong to the amplitude Im $F^{+}_{++}(t)$ (80 L Höhler-2). They were calculated using the approximation $T_\ell(s) \approx ia(b)/2$ (82 Hutt).

b fm	b^2 fm^2	π^-p $\sqrt{s}=9.7$ GeV, $k=50$ GeV/c $a(b)$	$\varrho(b)$	$\Omega(b)$	π^-p $\sqrt{s}=19.4$ GeV, $k=200$ GeV/c $a(b)$	$\varrho(b)$	$\Omega(b)$	pp $\sqrt{s}=62.5$ GeV $a(b)$	$\varrho(b)$	$\Omega(b)$
0.000	0.000	0.6350	0.8668	1.0079	0.5991	0.8393	0.9140	0.7343	0.9294	1.3254
0.050	0.003	0.6288	0.8622	0.9911	0.5968	0.8375	0.9084	0.7324	0.9284	1.3183
0.100	0.010	0.6114	0.8490	0.9453	0.5892	0.8312	0.8896	0.7268	0.9254	1.2978
0.150	0.023	0.5857	0.8283	0.8811	0.5753	0.8197	0.8564	0.7175	0.9202	1.2641
0.200	0.040	0.5546	0.8016	0.8087	0.5560	0.8028	0.8118	0.7045	0.9127	1.2192
0.250	0.062	0.5204	0.7700	0.7347	0.5321	0.7810	0.7594	0.6879	0.9026	1.1645
0.300	0.090	0.4844	0.7341	0.6624	0.5046	0.7545	0.7023	0.6678	0.8896	1.1018
0.350	0.122	0.4472	0.6944	0.5928	0.4743	0.7236	0.6430	0.6445	0.8736	1.0342
0.400	0.160	0.4096	0.6514	0.5269	0.4421	0.6887	0.5836	0.6183	0.8542	0.9628
0.450	0.202	0.3720	0.6056	0.4652	0.4089	0.6506	0.5257	0.5895	0.8314	0.8901
0.500	0.250	0.3353	0.5582	0.4085	0.3754	0.6098	0.4706	0.5586	0.8050	0.8174
0.600	0.360	0.2670	0.4627	0.3106	0.3102	0.5242	0.3714	0.4925	0.7421	0.6776
0.700	0.490	0.2080	0.3727	0.2331	0.2508	0.4386	0.2887	0.4240	0.6678	0.5510
0.800	0.640	0.1594	0.2934	0.1736	0.1992	0.3587	0.2221	0.3569	0.5859	0.4408
0.900	0.810	0.1209	0.2271	0.1288	0.1561	0.2879	0.1698	0.2941	0.5011	0.3477
1.000	1.000	0.0910	0.1738	0.0955	0.1211	0.2275	0.1291	0.2375	0.4180	0.2706
1.100	1.210	0.0683	0.1320	0.0708	0.0930	0.1774	0.0977	0.1883	0.3406	0.2082
1.200	1.440	0.0512	0.0998	0.0526	0.0708	0.1366	0.0734	0.1468	0.2716	0.1585
1.300	1.690	0.0383	0.0752	0.0391	0.0533	0.1038	0.0548	0.1129	0.2127	0.1196
1.400	1.960	0.0286	0.0564	0.0290	0.0396	0.0776	0.0404	0.0859	0.1641	0.0896
1.500	2.250	0.0213	0.0421	0.0215	0.0290	0.0571	0.0294	0.0648	0.1253	0.0669
1.600	2.560	0.0157	0.0311	0.0158	0.0209	0.0413	0.0211	0.0489	0.0952	0.0500
1.700	2.890	0.0114	0.0227	0.0115	0.0148	0.0293	0.0149	0.0369	0.0724	0.0376
1.800	3.240	0.0082	0.0164	0.0083	0.0102	0.0204	0.0103	0.0281	0.0554	0.0285
1.900	3.610	0.0058	0.0117	0.0059	0.0070	0.0139	0.0070	0.0217	0.0428	0.0219
1.950	3.803	0.0049	0.0098	0.0049	0.0057	0.0114	0.0057	0.0191	0.0378	0.0193
2.200	4.840	0.0019	0.0038	0.0019	0.0019	0.0039	0.0019	0.0103	0.0205	0.0104
2.400	5.760	0.0008	0.0016	0.0008	0.0007	0.0015	0.0007	0.0059	0.0117	0.0059
2.600	6.760	0.0003	0.0007	0.0003	0.0003	0.0005	0.0003	0.0026	0.0052	0.0026
2.800	7.840	0.0001	0.0002	0.0001	0.0001	0.0002	0.0001	0.0002	0.0005	0.0003

b fm	b^2 fm^2	$\sqrt{s}=4.4$ GeV, $k=10$ GeV/c $a(b)$	$\varrho(b)$	$\Omega(b)$
0.093	0.009	0.6956	0.8500	1.1894
0.187	0.035	0.6398	0.8220	1.0211
0.280	0.078	0.5912	0.7948	0.8945
0.373	0.139	0.5308	0.7504	0.7567
0.466	0.217	0.4630	0.6928	0.6218
0.560	0.314	0.3790	0.6016	0.4764
0.653	0.426	0.3150	0.5240	0.3783
0.746	0.557	0.2556	0.4416	0.2952
0.839	0.704	0.2032	0.3632	0.2272
0.933	0.870	0.1528	0.2804	0.1658
1.026	1.053	0.1168	0.2188	0.1242
1.119	1.252	0.0922	0.1748	0.0967
1.212	1.469	0.0668	0.1288	0.0691
1.306	1.706	0.0458	0.0888	0.0469
1.399	1.957	0.0316	0.0620	0.0321
1.492	2.226	0.0234	0.0460	0.0237
1.585	2.512	0.0150	0.0296	0.0151
1.679	2.819	0.0078	0.0156	0.0078
1.772	3.140	0.0048	0.0096	0.0048
1.865	3.478	0.0042	0.0084	0.0042
1.958	3.834	0.0032	0.0060	0.0032
2.052	4.211	0.0016	0.0032	0.0016
2.145	4.601	0.0010	0.0020	0.0010
2.238	5.009	0.0022	0.0044	0.0022

The other possibility is to start from the experimental form factors and to calculate $A(t)$ from $A_B(t)$. This can be done in principle with Eq. (6.3.30), but in practice it is easier to represent $A_B(t)$ by a sum of exponentials and to use Eq. (6.3.37). The value of K is determined in such a way that $A(0)=\sigma/4\pi$. The interaction strength K amounts to about 5.7 GeV^{-2} for π^-p scattering at 100 GeV/c. It is considerably smaller than the value for pp-scattering ($K_{pp}=9.5$ GeV^{-2}). Therefore, the first dip in the πp cross section is expected at a larger value of $-t$. An estimate gives a dip at $t=-5$ GeV^2 and 200 GeV/c. However, the prediction is not accurate due to the uncertainty of the pion form factor (78 L CHAN).

It is very interesting to note that a prominent dip in π^-p scattering has recently been discovered by BAKER et al. (81) at 200 GeV/c and $t=-4$ GeV^2. The optical Born term derived from these data is shown in Fig. 6.4.1.

CHOU (79 L-1) has made an attempt to determine the radii of pions and kaons by fitting the CHOU-YANG formulas to the elastic scattering data at 200 GeV/c (76 AKERLO) [cf. Eq. (6.3.33)]. It is misleading to call the result a "theoretical value", since there is no theoretical argument why the matter distribution should be exactly equal to the charge distribution (see below). Furthermore, there are problems with the factorizing eikonal ansatz Eq. (6.4.7) for the energy dependence of the CHOU-YANG model. Finally, CHOUs work is based on the value $b=(9.26\pm0.06)$ GeV^{-2} for the slope of the π^-p cross section, whereas the new value of BURQ et al. (81) is (10.37 ± 0.27) GeV^{-2}. The discrepancy is related to the effect shown in Fig. 6.3.2.

For some applications it is useful to know that a closed expression for the opacity can be given in the case of pp-scattering, if one assumes a dipole fit for the nucleon form factor (68 L DURAND)

$$G_{Ep}(t)=(1-t/t_D)^{-2}; \qquad \Omega(b)=\tfrac{1}{8}\{(b\sqrt{t_D})^3 K_3(b\sqrt{t_D})\}\,\Omega(0). \tag{6.4.6}$$

$K_3(z)$ is the hyperbolic Bessel function of the second kind and $\Omega(0)$ is proportional to the constant K in Eq. (6.4.3). The authors find two zeros of $A(t)$. The first one occurs near $t=-1.5$ GeV^2 in agreement with the pp-data, but the second one near $t=-6$ GeV^2 is not observed.

At the time when the CHOU-YANG model was proposed, it was believed that the total cross sections approach a constant at very high energies. Several years later, new experiments showed that they are slowly increasing. This made it necessary to introduce an assumption on the energy dependence of the CHOU-YANG model.

HAYOT and SUKHATME (74 L) suggested to consider the interaction strength K as a function of energy which is determined by the increase of the total cross sections. The ansatz corresponds to a *factorizing eikonal*

$$\Omega(s,b)=K(s)\tilde{\Omega}(b), \tag{6.4.7}$$

where $\tilde{\Omega}(b)$ is the integral in Eq. (6.4.2). The shape of the differential cross section is now changing with energy, since the multiple scattering effects become larger as the energy increases and the dips approach $t=0$, due to the increase of the interaction strength $K(s)$.

The existence of dips is related to the alternating signs in the multiple scattering expansion, Eq. (6.3.30). They occur for instance in the model of FRAUTSCHI and MARGOLIS [Eq. (6.3.37)] which, however, is too simple.

The CHOU-YANG model has an impressive success insofar as the decrease of the pp-cross section over several orders of magnitude up to about $t=-1.2$ GeV^2 is described without an adjustable parameter. Furthermore, the existence of a dip at $t=-1.4$ GeV^2 is predicted in agreement with the data, including its energy dependence.

However, the model predicts a second dip at $t=-4\cdots-5$ GeV^2, which is not observed (see for instance 77 L SUKHATME-1, 78 L CHOU). This difficulty has recently been discussed by CHOU and YANG (79 L-2), who argue that the dip should appear at higher energies. In any case, it is an important question to find the mechanism which dominates the pp-amplitude beyond $t=-2$ GeV^2 in the ISR range.

Another problem is the fact that the factorizing eikonal ansatz Eq. (6.4.7) is ruled out according to the results of amplitude analyses (see Sect. 2.6.3.6 and Fig. 6.4.2).

An interesting extension of the CHOU-YANG model deals with the attempt to describe a polarized proton by a rotation of the hadronic matter (76 L CHOU). The scattering process now becomes spin-dependent because of the increasing opaqueness at increasing relative velocity.

The authors predict a spin-rotation parameter R at 200 GeV/c which is slightly different for π^+p and π^-p scattering and amounts to about -0.06 at $t=-0.4$ GeV^2. The experimental value at 40 GeV/c is $R=-0.28\pm0.07$ (75 PIERRA).

The generalization of the CHOU-YANG model to the helicity flip amplitude has also been studied by YOUNG (79 L), who proposed an ansatz which leads to s-channel helicity conservation, and by MAJEROTTO (75 L).

In their recent work on the charge and matter distributions in a nucleon, MIETTINEN and THOMAS (80 L) came to the conclusion that these distributions are very different. Using the experience gained in nuclear physics, they generalized the impact parameter description in such a way that diffraction dissociation is taken into account and used the result for an analysis of the pp-scattering data. Furthermore, they discussed the relation of optical models to QCD inspired pictures of hadrons.

Conclusion on the CHOU-YANG model

The model leads to a remarkably good first approximation for the shape of the diffraction peak. Relatively small effects of the energy-dependence and of the spin-dependence are not yet well understood and it is not known whether the mechanism remains dominant beyond the dip at 1.5 GeV^2 and, at small $|t|$, in the high energy limit (see the systematic variation in Fig. 6.4.2). – Further references: 81 L and 82 L CHOU.

2.6.4.2 The CHENG-WU model

The authors started from a detailed study of the high energy behaviour of quantum field theory (69 R and 70 L CHENG). A first result was the prediction that the FROISSART bound for the total cross sections (Table A.5.1) is saturated in the high energy limit. At that time, it was generally believed that the total cross sections approach a constant, but several years later, an increase of the total pp-cross section was found in an ISR experiment. However, the data are still far below the FROISSART bound. Experiments at Fermilab showed that the increase occurs also in πp and Kp scattering (76 CARROLL).

The prediction for the behaviour of the real part follows already from analyticity (Sect. 2.2.1) and should not be considered as an independent success of the model.

In order to be able to make further predictions, the authors had to add assumptions which have no support from quantum field theory. For instance, they added Regge terms and chose an optical Born amplitude in such a way that the computations are simple (73 L CHENG, WALKER, and WU-1 and 2).

We discuss only the most recent version [BOURRELY, SOFFER, and WU (79 L)]. These authors proposed the following ansatz for the eikonal (they call $-2\mathrm{i}\chi=\Omega-2\mathrm{i}\delta$ the opaqueness)

$$-2\mathrm{i}\chi=S_0(s)F(b)+R_0(s,b)\,. \tag{6.4.8}$$

The first term is the diffractive component ("pomeron exchange") and the second term is the "Regge background", which is needed to describe the data at intermediate energies (pp-data above 14 GeV/c).

The complex quantity

$$S_0(s)=\frac{s^c}{(\ln s)^{c'}}+\frac{u^c}{(\ln u)^{c'}} \tag{6.4.9}$$

is taken from the high energy behaviour of quantum field theory (70 L CHENG). The earlier ansatz for $F(b)$ is replaced by an ansatz for the optical Born amplitude which has 4 adjustable parameters

$$\langle F(b)\rangle=fG^2(t)\frac{a^2+t}{a^2-t}\,;\qquad G(t)=[(1-t/m_1^2)(1-t/m_2^2)]^{-1}\,. \tag{6.4.10}$$

$G(t)$ is similar to the electromagnetic form factor of the nucleon but, in contrast to the CHOU-YANG model, the ansatz is modified by a factor which introduces a real zero at $t=-a^2$, a being an adjustable parameter. The parametrization in Eqs. (6.4.8) and (6.4.10) was chosen because it turns out to be useful in a fit to the data. It has not been derived from quantum field theory.

The Fourier transform of the Regge term $\langle R_0(s,b)\rangle$ represents the standard even and odd signature Regge exchange contributions with an exchange degenerate trajectory, whose parameters are determined by the fit:

$$\alpha(t)=0.352+0.694\,t\,. \tag{6.4.11}$$

At high energies, the Regge term becomes negligible and the ansatz Eq. (6.4.8) is of the type of a factorizable eikonal, which is ruled out according to the results of AMALDI and SCHUBERT (80 R) and of FRANCA and HAMA (79 L) as mentioned in Sect. (2.6.3.6).

The model of BOURRELY et al. (79 L) also includes a description of polarized protons, which is somewhat different from that in the model of CHOU and YANG (76 L). Since a rigid rotation of the hadronic matter leads to a serious problem at very high energies, the authors prefer to assume a "soft rotation", in which the angular velocity ω is a function of b^2 such that $\omega(b^2)\to 0$ as $b^2\to\infty$. Predictions are made for the polarization and the R-parameter of pp-scattering in the Fermilab and ISR range. An extension to pion-nucleon scattering is possible.

Conclusion on the CHENG-WU model

The model of CHENG-WU and collaborators can be fitted to the pp-data and the $\bar{p}p$ total cross sections in the lab. momentum range of 14···2000 GeV/c, but the number of adjustable parameters is rather large (10). There are systematic deviations from the data in the ISR range due to the assumption of a factorizable eikonal. The model has an input from quantum field theory, but the ansatz for the eikonal is a parametrization which is assumed without being derived from a fundamental theory or from an intuitive picture.

2.6.4.3 Geometrical scaling

In general, a scattering amplitude depends on two variables: s and t. Various models have been proposed, in which the amplitudes approach a simple behaviour at high energies in the sense that they depend only on *a single variable* τ, which is a function of s and t ("scaling").

A scaling property is usually assumed only for the imaginary part of the amplitude, since the real part then follows from a fixed-t dispersion relation. At very high energies, the real part can as well be calculated from a "derivative analyticity relation" or, equivalently, from an analytic parametrization. However, the latter methods can be misleading at intermediate energies, because there may be other contributions of comparable magnitude, which are "non-local" in energy (see Sect. 2.2.1.3).

In the range where the real parts are not negligible, one cannot demonstrate the validity of the scaling law by using differential cross sections. Above about 50 GeV/c, the real parts are expected to be small in near-forward scattering, but the variation of the cross section at fixed $t \geqq -0.6\ \mathrm{GeV}^2$ in the range up to 200 GeV/c is comparable with the experimental errors (Fig. 6.3.1). Therefore, scaling laws in πp scattering should be studied using the results of amplitude analyses.

It is easier to test scaling laws for pp-scattering, because data are available up to very high energies (total c.m. energy 62.5 GeV, corresponding to a lab. momentum of 2100 GeV/c). In the following, we shall discuss the "geometrical scaling" model, which is compatible with the pp-scattering data (80 R AMALDI-SCHUBERT) and should therefore also be studied in the case of πN scattering.

Geometrical scaling is based on the idea that hadron-hadron interactions mainly depend on a well-defined interaction radius R, which is a function of energy. First, we consider very high ernergies, where a single scattering amplitude is strongly dominant and the real part is negligible, except near the zeros of the imaginary part.

The hypothesis of geometrical scaling can be described as follows. If one starts with the amplitude and the impact parameter transforms of Sect. 2.6.3.5 at a certain high energy $\sqrt{s_0}$, the energy dependence is given in terms of a radius $R(s)$ (73 L DIAS DE DEUS, 74 L BURAS, 75 L BARGER)

$$a(bR_0/R)\,;\quad S(bR_0/R)\,;\quad \Omega(bR_0/R),\quad \text{etc.,}\quad A(tR^2/R_0^2)R^2/R_0^2\,;$$
$$\mathrm{d}\sigma/\mathrm{d}t=\pi\frac{R^4}{R_0^4}A^2(tR^2/R_0^2),\quad \text{where}\quad R_0=R(s_0). \tag{6.4.12}$$

One can conclude from Eq. (6.3.32) and from the first Eq. (6.3.33) that the total cross sections and the logarithmic slope B at $t=0$ are proportional to $R^2(s)$

$$\sigma\propto\sigma_{\mathrm{el}}\propto\sigma_{\mathrm{in}}\propto B\propto R^2(s),\quad \sigma=\sigma_{\mathrm{el}}+\sigma_{\mathrm{in}}\,. \tag{6.4.13}$$

The first tests of geometrical scaling are the constancy of the ratios

$$\sigma_{\mathrm{el}}/\sigma\quad \text{and}\quad B/\sigma\,, \tag{6.4.14}$$

and the fact that in a plot of the following ratio at very high energies

$$\frac{\mathrm{d}\sigma(s,t)/\mathrm{d}t}{\mathrm{d}\sigma(s,0)/\mathrm{d}t}\,\text{vs.}(-t\sigma)\,, \tag{6.4.15}$$

the data at different energies should lie on the same curve.

These predictions are fulfilled in pp-scattering, where the variation of the total cross section and therefore of R^2 between a lab. momentum of 200 GeV/c (Fermilab data) and the highest energy of the ISR data amounts to 13%. In particular, it is striking to see that the plot of Eq. (6.4.15) leads to a universal curve for the dip-bump structure.

The most accurate test has been performed by AMALDI and SCHUBERT (80 R), who studied the energy dependence of the inelastic overlap function and found a good agreement with the geometrical scaling model.

A test is much more difficult in πN scattering. One has to wait for data at 1000 GeV/c, where the dip at $t=-4.0\ \mathrm{GeV}^2$ and 200 GeV/c should be shifted to $t=-3.6\ \mathrm{GeV}^2$.

If geometrical scaling is assumed, there exists a relation between the s- and t-dependence of the differential cross section, Eq. (6.4.12). We define the local slope by (75 L BARGER)

$$B(s,t)\equiv\frac{\partial}{\partial t}\ln\frac{\mathrm{d}\sigma}{\mathrm{d}t} \tag{6.4.16}$$

and an "effective Regge trajectory" by

$$2\alpha_{\mathrm{eff}}(s,t)-2\equiv\frac{\partial}{\partial\ln s}\left(\ln\frac{\mathrm{d}\sigma}{\mathrm{d}t}\right). \tag{6.4.17}$$

It then follows from Eq. (6.4.12) that these quantities are related by

$$\alpha_{\mathrm{eff}}(s,t)-1=\{1+B(s,t)t/2\}\frac{\mathrm{d}(\ln R^2)}{\mathrm{d}(\ln s)}. \tag{6.4.18}$$

It is interesting to notice that $\mathrm{d}\sigma/\mathrm{d}t$ at fixed t is *increasing with energy* if

$$t>-2/B\approx-0.2\ \mathrm{GeV}^2 \tag{6.4.19}$$

and *decreasing* if $t<-2/B$. The forward cross section is increasing and the slope is shrinking. Both effects compensate each other at $t=-2/B$.

Prediction for the real part

The energy dependence of the imaginary part A of the scattering amplitude at fixed t is determined by Eq. (6.4.12), if $R(s)$ is substituted as following for instance from Eq. (6.4.13) and the total cross section data. Therefore, one can predict the real part of the amplitude from fixed-t analyticity.

We have to generalize the ansatz Eq. (6.3.19) by introducing the real part

$$F(s,t)=\mathrm{Re}\,F(s,t)+\mathrm{i}sA(s,t)=\frac{m}{2\pi}(1-t/4m^2)\,C^+(s,t);$$
$$\mathrm{Im}\,F(s,0)=\frac{s\sigma}{4\pi}\approx\mathrm{Im}\,C^+(s,0)\frac{m}{2\pi};\quad\frac{\mathrm{d}\sigma}{\mathrm{d}t}=\pi\frac{|F(s,t)|^2}{s^2}. \tag{6.4.20}$$

For applications in πN scattering, we have added the relation to the amplitude C^+ of Eq. (1.15.7), using the high energy approximation $2mk\approx2m\omega=s-m^2-\mu^2\approx s$.

We describe the t-dependence of $\mathrm{Im}\,F(s,t)$ by a function Φ which fulfills geometrical scaling

$$\mathrm{Im}\,F(s,t)\equiv\frac{s\sigma}{4\pi}\Phi(\tau)=\mathrm{Im}\,F(s,0)\Phi(\tau);\quad\tau=-t\sigma(s). \tag{6.4.21}$$

At fixed-t, $F(s,t)$ has the same analytic properties as $C^+(s,t)$. At very high energies, its real part can be calculated for instance by fitting an analytic parametrization, Eq. (A.6.62), to $\mathrm{Im}\,F(s,t)$ or by using the prescription of Eq. (A.6.61). The derivative analyticity relation Eq. (A.6.63) (74 L BRONZAN) gives the same result

$$\frac{1}{s}\mathrm{Re}\,F(s,t)=\frac{\pi}{2}\frac{\partial}{\partial(\ln s)}\left[\frac{1}{s}\mathrm{Im}\,F(s,t)\right]=\frac{1}{s}\mathrm{Re}\,F(s,0)\frac{\mathrm{d}}{\mathrm{d}\tau}[\tau\Phi(\tau)]. \tag{6.4.22}$$

Combining the last two equations, one obtains for the differential cross section

$$\frac{\mathrm{d}\sigma}{\mathrm{d}t}(s,t)=\frac{\mathrm{d}\sigma}{\mathrm{d}t}(s,0)\left[\Phi^2(\tau)+\varrho^2(s,0)\left\{\frac{\mathrm{d}}{\mathrm{d}\tau}\{\tau\Phi(\tau)\}\right\}^2\right]\frac{1}{1+\varrho^2(s,0)}, \tag{6.4.23}$$

where $\varrho(s,0)=\mathrm{Re}\,F(s,0)/\mathrm{Im}\,F(s,0)$. The above discussion follows 78 L DIAS DE DEUS and KROLL and 81 R KROLL.

In the review 81 R KROLL one can find tables of the impact parameter transforms of pp-amplitudes and also of the scaling function Φ and of the inelastic overlap function $\varrho(s,b)$, which were calculated by AMALDI-SCHUBERT (80 R) and by DIAS DE DEUS and KROLL (78 L). The agreement is good except for minor deviations.

KROLL (81 R) and AMALDI-SCHUBERT (80 R) found a good compatibility with the geometrical scaling hypothesis. Small deviations were noticed and discussed by FRANCA and HAMA (79 L) and by NAGY et al. (79 L). It will be interesting to test this hypothesis using an amplitude analysis of the new πp-scattering data up to 200 GeV/c. An extension to intermediate energies was proposed in 77 L DIAS DE DEUS.

At intermediate energies, the different cross sections in Eq. (6.4.13) are not yet proportional to each other. It will be seen in Sect. 2.6.8 that the decrease of the total cross section σ^+ from 31 mb at 3 GeV/c to 23.5 mb at 70 GeV/c is essentially due to the decrease of the elastic cross section, which is caused to some extent by the diminution of the real parts and flip terms (see also Fig. 6.3.12). In this situation, the scaling property should be assumed only for the no-flip inelastic overlap function (75 L GREIN).

Many other scaling laws have been discussed in the literature. We mention only the interesting proposal by LEADER and PENNINGTON (73 L) to work with the variables s and n^2, where $n^2 = -\Phi/s^2$ has been introduced in order to unify the group-theoretical kinematic structure of elastic and inelastic reactions. $\Phi(s,t)$ is the Kibble function, Eq. (A.1.15). For πN scattering n^2 is given by

$$n^2 = \frac{t}{s}(4q^2 + t). \qquad (6.4.24)$$

The application to the energy dependence of the logarithmic slope at $t=0$ of the diffraction peak leads to a *constant* behaviour in the high energy limit: $b(s) = b_0 4q^2/s$, whereas the new data, in particular for pp-scattering (see Fig. 4 on p. 227 of Vol. I/9a), show an *increase*. It is remarkable that the quantity $4q^2/s$ approaches its asymptotic value 1 only slowly (0.95 at 20 GeV/c in πN scattering).

In some cases the arguments in favor of empirical scaling laws are not tenable (see the comments 75 L EICHMANN and 82 L HÖHLER).

Conditions under which a scaling law can be derived from analyticity and unitary have been given by CORNILLE and MARTIN (76 L).

2.6.4.4 The VAN HOVE model for the inelastic overlap function

At high energies, elastic scattering can be considered as the shadow of inelastic processes. Therefore, the inelastic overlap function is a good candidate for a quantity which has simple properties. VAN HOVE (64 R) investigated the overlap function $\operatorname{Im} F_{\text{in}}(s,t)$ under the assumption that the multiplicity of secondaries in the inelastic collisions is high and that they are only weakly correlated in momentum distribution. In the notation of Sect. 2.6.3.5 his result reads

$$A_{\text{in}}(t) = (\sigma_{\text{in}}/4\pi)\,e^{B_{\text{in}}t/2}; \quad \varrho(b) = (\sigma_{\text{in}}/2\pi B_{\text{in}})\,e^{-b^2/2B_{\text{in}}}. \qquad (6.4.25)$$

If the real parts are neglected and the high energy approximation of the unitarity relation in impact parameter space is used, the two parameters σ_{in} and B_{in} can be expressed by the total and the total elastic cross sections σ and σ_{el}.

The maximum value of $\varrho(0)$ is 1, corresponding to total absorption. This value leads to the "VAN HOVE limits", which are upper bounds (65 L COTTINGHAM, 74 R BARGER)

$$\sigma_{\text{el}}/\sigma \leqq 0.185; \quad B/\sigma \leqq 0.115 = 0.296\ \text{GeV}^{-2}/\text{mb}. \qquad (6.4.26)$$

B is the logarithmic slope of $d\sigma/dt$ at $t=0$. B/σ is dimensionless.

The product of the two inequalities gives the quantity for which the MAC-DOWELL-MARTIN bound (64 L) is valid [Eq. (A.5.33)]

$$\frac{B\sigma_{\text{el}}}{\sigma^2} \geqq \frac{1}{18\pi} = 0.018. \qquad (6.4.26a)$$

In πN scattering this bound is almost saturated. Therefore, the deviations from the VAN HOVE limits are not independent.

In pp-scattering, the data approach the limit from *above* and are lower in the ISR range, where AMALDI and SCHUBERT (80 R) give an average

$$\sigma_{\text{el}}/\sigma = 0.1747 \pm 0.001; \quad \text{pp-scattering}.$$

The result for πp scattering at 200 GeV/c is lower (81 SCHIZ)

$$\begin{aligned}\sigma_{el}/\sigma &= 0.136 \pm 0.006\ \pi^- p \to \pi^- p;\\ &= 0.133 \pm 0.003\ \pi^+ p \to \pi^+ p.\end{aligned} \tag{6.4.27}$$

The ISR data for pp-scattering practically agree with the bound for B/σ and the experimental values at lower momenta approach it from below.

In π^-p scattering at 200 GeV/c the ratio is $B/\sigma = 0.165 \pm 0.005$ (81 BURQ).

If one wants to calculate $A(t)$ and $d\sigma/dt$ from the VAN HOVE ansatz, one can use Eqs. (6.3.21) and (6.3.31)

$$A(t) = \int_0^\infty J_0(b\sqrt{-t})[1 - \{1 - \varrho(0)\,e^{-b^2/2B_{in}}\}^{1/2}]\,b\,db. \tag{6.4.28}$$

This relation leads to Eq. (4.105) in 71 R ZACHARIASEN, if another variable is introduced under the integral.

It was pointed out by GREIN and KROLL (75 L) that the unitarity relation for the impact parameter transforms is rather sensitive to the real part contribution, i.e. one has to include corrections to Eq. (6.3.31) even for relatively small real parts.

The unitarity relation for the impact parameter transform of the s-channel helicity amplitude, Eq. (6.3.7), reads including the real part

$$\mathrm{Im} f_{++}(s,b) \approx (\mathrm{Im} f_{++}(s,b))^2 + (\mathrm{Re} f_{++}(s,b))^2 + \tfrac{1}{4}\varrho_{++}(s,b), \tag{6.4.29a}$$

if the flip contribution to the rescattering term is neglected. If ϱ_{++} and $\mathrm{Re} f_{++}$ are known, one can solve for $\mathrm{Im} f_{++}$

$$2\,\mathrm{Im} f_{++}(s,b) \approx 1 - [1 - \varrho_{++}(s,b) - 4(\mathrm{Re} f_{++}(s,b))^2]^{1/2}, \tag{6.4.29b}$$

The total elastic cross section is given by (neglecting again f_{+-})

$$\sigma_{el}(s) = 8\pi \int_0^\infty |f_{++}(s,b)|^2 b\,db. \tag{6.4.29c}$$

The exponential ansatz Eq. (6.4.25) is a useful first approximation even at intermediate energies (Fig. 6.3.9). The new πp scattering data up to 200 GeV/c have been used for a study of the overlap function in 82 L HUTT.

In pp-scattering, HENZI and VALIN (79 L-1 and 2) found that, as expected by VAN HOVE, the inelastic overlap function in b-space is simpler than the transform of the scattering amplitude. The latter has additional structure due to "rescattering", which is given by the first term of the unitarity relation Eq. (6.3.25). The "short range expansion" of these authors, Eq. (6.3.41), leads to an inelastic overlap function $A_{in}(t)$ which has two zeros.

Beyond the second zero (at $t < -3.5$ GeV^2), $A_{in}(t)$ becomes much smaller than the total amplitude $A(t)$, i.e. $d\sigma/dt$ is *dominated by the rescattering effect*. As a consequence, the decrease becomes flatter than at small t.

The present status of the information on the overlap function in pp-scattering is discussed in 81 R KROLL.

2.6.4.5 Other s-channel models

The model of B. and F. SCHREMPP is an attempt to describe diffraction scattering as well as wide-angle scattering within the same formalism. Quantitative applications to diffractive πp and pp-scattering do not yet exist (77 L SCHREMPP, 80 L SCHREMPP). Applications to charge-exchange scattering and wide-angle scattering will be discussed in Sects. 2.6.5 and 2.6.6, respectively.

In the model of HEINES and ISLAM (81 L) it is assumed that the scattering amplitude can be represented as a sum of two complex-valued contributions which belong to diffraction scattering and to a hard scattering mechanism, respectively.

The *diffractive part* is described by an ansatz for the impact parameter transform of the scattering amplitude, which has a pole at a complex value of b^2. This agrees with the assumption in the "dual peripheral model" of B. and F. SCHREMPP for the no-flip amplitude in charge-exchange scattering: Eq. (6.5.41). It corresponds to the contribution of an s-channel Regge pole.

For the *hard scattering part* the authors use the following prescription. The t-dependence is chosen to be the same as derived from scattering at a short range complex optical potential [see Eq. (A.3.32)]. The s-dependence is assumed to agree with that of a fixed Regge cut. Finally, absorption is included by considering the sum of the eikonals of the two contributions and keeping only the term linear in the hard scattering eikonal.

The dip-bump structure follows from the destructive interference between the two contributions. In contrast to the CHOU-YANG model a second dip is not predicted. The ansatz can be interpreted in terms of a quark model.

2.6.4.6 Constituent models

One of the earliest constituent models for hadron-hadron scattering is the *Additive Quark Model* of KOKKEDEE and VAN HOVE (66 L, 69 R KOKKEDEE), which leads in the impulse approximation to the following expression for the cross section

$$\frac{d\sigma}{dt} \propto F_{\pi}^{2}(t)\, G_{Ep}^{2}(t)\, |A_{qq}(s,t)|^{2}, \qquad (6.4.30)$$

We have already substituted the electromagnetic form factors for the "hadronic form factors" as it is usually done. The use of G_{Ep} instead of the Dirac form factor F_{1p} is an additional assumption. $A_{qq}(s,t)$ is the quark-quark scattering amplitude.

Originally, the main application of the model was to predict relations between cross sections of different reactions. FAJARDO et al. (81) have recently performed a comparison with their new πp and pp-scattering data, assuming that $|A_{qq}|^2$ can be parametrized by an exponential: $\exp(-r_q^2 t/2)$, r_q being called the "radius" of the constituent quarks.

However, one should notice that Eq. (6.4.30) is only an approximation which has to be corrected by rescattering effects. There is a remarkable similarity with the CHOU-YANG model (see also 77 L BIALAS and 78 L LEVIN, where other references are given).

A detailed investigation of the multiple scattering expansion in a constituent model has been performed by WAKAIZUMI and TANIMOTO (77 L, 78 L, 82 L). The authors claim that a fit to the pp-data beyond the dip is possible only if the number of constituents is *three*, whereas an infinite number leads to a behaviour similar to that in the CHOU-YANG model.

Other multiple scattering or iteration expansions were used in 78 L MARTIN-2 and 81 L KULESHOV.

The role of gluons in elastic hadron-hadron scattering has been investigated by VAN HOVE (77 L). He starts from an intuitive ansatz for the transform $a(b)$ of the scattering amplitude in terms of a convolution integral over the transforms of the glue-glue amplitude and the glue distribution inside the proton. In t-space, this ansatz leads to a product.

$$A(t) = A_g(t)\,\gamma(t) \qquad (6.4.31)$$

for the scattering amplitude defined in Eq. (6.3.19). $A_g(t)$ is the glue-glue scattering amplitude and $\gamma(t)$ is the square of the form factor which describes the glue distribution inside the proton.

It is shown that the pp-scattering data at ISR energies in the t-range up to $-8\ \mathrm{GeV}^2$ can be fitted, if one assumes the difference between two Gaussians for the transform of $A_g(t)$. One wants to have a rather flat behaviour near $b=0$ in order to be able to produce the zero of $A_g(t)$ at the dip of $d\sigma/dt$. The square of the gluonic proton form factor turns out to be well approximated by a dipole fit.

The identification of the active constituent with the glue does not follow from the elastic scattering data but from the study of inelastic processes.

Further references are given in the review 79 R HENDRY-LICHTENBERG.

2.6.4.7 General remarks on t-channel models

There exists a very large number of papers in which πN scattering at high energies is described in terms of Regge pole and cut models (see 69 L BARGER and the reviews 71 R SQUIRES, 71 R ZACHARIASEN, 72 R CHIU, 72 R HARA, 72 R SCHREMPP, 73 R FOX-QUIGG, 75 R SCHREMPP, 76 R KANE-SEIDL, 77 R COLLINS, 77 R IRVING-WORDEN, 81 R COLLINS-MARTIN). Most of the authors found an "excellent fit to the data", but it is difficult to decide whether this can be taken as a good argument in favor of the assumptions of their model. The problem is that all of these models have a great flexibility and that the data show a relatively simple behaviour: a smooth t-dependence and a weak s-dependence at fixed t. In fact, the s-dependence of $d\sigma/dt$ at small $|t|$ is comparable with the experimental errors in the range 40···200 GeV/c (cf. Figs. 6.3.1 and 6.3.3).

In the t-channel approach the main problem in elastic πN scattering is to understand the leading terms of the high energy expansion, which are usually called P- and f-exchange, respectively. "P" refers to pomeron-exchange and "f" to the exchange of a reggeized f(1270) meson, which was called "P′-exchange" in earlier papers (62 L IGI).

Originally, it was thought that P- and f-exchange are different effects. HARARI (68 L) and FREUND (68 L) proposed a *two-component duality* model, according to which P-exchange in the t-channel is "dual" to the s-channel background, whereas the exchange of f and of the other "ordinary" trajectories is "dual" to the s-channel resonances.

(P+f) Regge pole fits were discussed by many authors, for instance by ZARMI (71 L) and by BARGER et al. (72 L-8).

CARLITZ et al. (71 L) investigated the hypothesis of an "f-dominance of the pomeron", according to which scattering at small t can be described by a pomeron which couples to the hadrons via an f-meson. This hypothesis was used by COLLINS et al. (78 L-1, see also 82 R COLLINS-MARTIN) and studied by IRVING et al. (79 L).

More recently, ROSENZWEIG and CHEW (75 L) proposed their *Pomeron-f identity model.* They assumed that there is only one high-lying trajectory which is *curved.* It goes through the point $\alpha=2$ at $t=m_f^2$ and is also responsible for the effects of the Pomeron (see also 78 R CHEW-ROSENZWEIG). Further papers on this question are mentioned in 79 L SUKHATME and 80 L DASH. Other aspects of the P- and f-couplings have been discussed in Sect. 2.4.2.2.

We start with a discussion of HARARIs "dual absorptive model" which combines features of both s- and t-channel models.

2.6.4.8 HARARIs dual absorptive model

A test of the HARARI-FREUND conjecture was attempted by HARARI and ZARMI (69 L). They compared the imaginary parts of the P- and f-exchange amplitudes as obtained in Regge fits, with the imaginary part of the s-channel background and of the sum over all s-channel resonances respectively, and concluded that their result was in agreement with two-component duality.

Recently, this calculation was repeated by DASH, JONES, and MARTIN (80 L), taking into account new results from Regge fits and phase shift analysis. These authors concluded that there is a *large discrepancy with two-component duality.*

Unfortunately, two-component duality is not a well-defined notion, because the decomposition of an amplitude into resonance and background contributions is *arbitrary to a large extent*, as mentioned already by HARARI and ZARMI (69 L). This is in particular true for the strongly inelastic resonances at higher energies, where the rapid increase of the absorption in a partial wave can lead to a loop in the Argand diagram, if analyticity is taken into account (see Sect. 2.4.1.1). Furthermore, the separation of P- and f-exchange in fits of the data to Regge pole models also has a considerable uncertainty, because there are many possibilities to parametrize the increasing total cross sections.

Next we shall briefly describe the *dual absorptive model* of HARARI and coworkers (71 L, 72 L), because it was frequently used in discussions. The model is based on the decomposition of the no-flip isospin even amplitude $F(s,t)$ into two components

$$F(s,t)=P(s,t)+R(s,t)\,, \tag{6.4.32}$$

which are characterized by the b-dependence of the impact parameter transforms of the imaginary parts.

The *diffractive component* $P(s,t)$ (also called "pomeron-exchange" contribution because of the relation to Regge theory) has an imaginary part whose shape is of the "central" type: it has a maximum at $b=0$ and decreases without structure rapidly enough such that one can define a "radius". The simplest parametrization of a central component follows from a scattering at a *black sphere.* We use the notation of Sect. 2.6.3.5, identifying $P(s,t)$ with $\mathrm{i}A(t)$

$$\begin{aligned} a(b)&=1-S(b)=1, \quad \text{if} \quad b<R\,;\\ &=0, \quad \text{if} \quad b>R\,;\\ A(t)&=R^2\frac{J_1(R\sqrt{-t})}{R\sqrt{-t}}=\frac{R^2}{2}\left(1-\frac{R^2}{8}(-t)+\cdots\right). \end{aligned} \tag{6.4.33}$$

HARARI et al. chose instead a *Gaussian shape* of $a(b)$, which has the special property of leading to a structureless $A(t)$

$$a(b)=1-S(b)=\mathrm{e}^{-b^2/2B}; \quad A(t)=B\,\mathrm{e}^{Bt/2}. \tag{6.4.34}$$

The *peripheral component* $R(s,t)$ is characterized by a peak of the imaginary part of the impact parameter transform at $b\approx 1.1$ fm. It is assumed to represent either the sum of the s-channel resonances or, according to "duality", the sum of the t-channel exchanges, except the Pomeron.

The simplest ansatz for a "surface interaction" is

$$a(b)=1-S(b)=\delta(R^2-b^2); \quad A(t)=\tfrac{1}{2}J_0(R\sqrt{-t}). \tag{6.4.35}$$

HARARI and SCHWIMMER (72 L) used a more realistic version which has a finite width of the peak

$$a(b)=\frac{1}{2B}\mathrm{e}^{-(R^2+b^2)/4B}I_0(Rb/2B)\approx\frac{1}{2}(\pi RBb)^{-1/2}\mathrm{e}^{-(R-b)^2/4B}; \quad \text{for} \quad b>2B/R;$$
$$A(t)=\mathrm{e}^{Bt}J_0(R\sqrt{-t}). \tag{6.4.36}$$

The treatment of the s-channel helicity flip amplitude is analogous, but the Bessel function J_0 has to be replaced by J_1 in the Fourier-Bessel transformation which leads from $A(t)$ to $a(b)$, and therefore also in Eqs. (6.4.35) and (6.4.36), where the hyperbolic Bessel function I_0 has to be replaced by I_1 [see Eq. (6.3.7)].

As far as the real part is concerned, the authors suggested to calculate it from the fixed-t dispersion relation. But this has not been done, because the model does not specify the energy dependence of the two components and furthermore, it is assumed to be valid only in the range 2···20 GeV/c.

Semi-quantitative applications have been made for various reactions, in some cases with a remarkable success. In two calculations for πN scattering the real part has been neglected. This is not justified (Table 2.6.3.2, p. 327) and leads to fairly large errors (72 L DAVIER, 79 L KAMRAN). See also the criticism of BARGER and HALZEN (72 L-6 and 8).

Figure 6.3.7 shows the impact parameter transform of the isospin even no-flip amplitude as derived from our analysis (p. 329). In the resonance region there is a broad peak at $b\approx 1.1$ fm, which shows the magnitude of $R(s,t)$, Eq. (6.4.32), in comparison with $P(s,t)$. The height of the peak decreases rapidly as the momentum increases and at about 5 GeV/c it is comparable with the errors. This agrees with the fact that there are no significant indications for peripheral (or other) nucleon resonances at higher momenta.

Since $P(s,t)$ does not contribute to charge-exchange scattering, the partial wave decomposition shows in this case directly the transform of the peripheral component $R(s,t)$. According to Fig. 6.5.7 (p. 372) the transform of the dominant flip amplitude has the typical "peripheral" shape and its height and momentum dependence are similar as in the isospin even case. This is expected from HARARIs model, where the dominant t-channel reggeon exchanges are f- and ϱ-exchange respectively, since the no-flip fNN and the flip ϱNN coupling constants have the same order of magnitude (Sect. 2.4.2). A quantitative estimate depends on assumptions on the residue functions, for which a justification is difficult, in particular for f-exchange.

The smallness of the peripheral component is seen from the fact that the total elastic πp cross section at 10 GeV/c is 90 times larger than the total charge-exchange cross section (Tables 2.6.5.2, p. 362 and 2.6.8.1, p. 400).

Comment

HARARIs dual absorptive model describes in a semiquantitative way interesting relations between many different reactions. However, there exists no successful *quantitative* application to πN scattering. It is not clear if the peripheral component (mainly f-exchange), which is rapidly decreasing in the range 1.5···5 GeV/c, is also responsible for the decrease of $d\sigma^+/dt$ between 5 and 50 GeV/c.

2.6.4.9 Recent work on Regge pole phenomenology

The most detailed recent investigation of the application of Regge pole theory to hadron-hadron scattering is due to P.D.B. COLLINS et al. (74 L, 76 L-1, 76 L-2, 78 L-1, 78 L-2, 78 L-3, 82 R COLLINS-MARTIN). The authors wanted to give a simple description of the main features of πp, Kp, pp, and p̄p scattering in the range $s>15$ GeV^2 ($k>7.5$ GeV/c for πp scattering).

The scattering amplitudes are described as sums over a few factorizing Regge pole terms (P, f, ϱ, ω, A_2) and some corrections. The ansatz differs from the usual one insofar as the pomeron intercept $\alpha_P(0)$ is treated as an adjustable parameter. The fit leads to the Pomeron trajectory

$$\alpha_P(t)=\alpha_P(0)+\alpha'_P t\,;\quad \alpha_P(0)=1.06\,;\quad \alpha'_P=0.24\ \mathrm{GeV}^{-2}. \tag{6.4.37}$$

It is true that the amplitude violates the FROISSART bound (Table A.5.1). But the authors argue that this occurs at extremely high energies ($s>10^8$ GeV^2) and that in the energy range of the available experimental data there are no signs of the unitarity corrections which ensure that the FROISSART bound is satisfied.

All Regge trajectories are assumed to be linear. Aside from the pomeron trajectory they are determined by the particle masses.

In order to reduce the large flexibility of the residue functions, COLLINS and GAULT (76 L-1) introduced the *reggeon-photon coupling analogy*, according to which the couplings of the natural parity trajectories P, f, ϱ, ω, A_2 to any vertex are related to those of the photon.

In πN scattering the ansatz for the contribution of a Regge pole exchange (R=P, f, ϱ) to the s-channel helicity amplitudes has the following factorized form (78 L COLLINS-1)

$$\begin{aligned}4\sqrt{\pi}\, sF^R_{++}&=-2\{mg^R_{1NN}(t)+g^R_{2NN}(t)\}\,g^R_{1\pi\pi}(t)\,R\{s,\alpha(t)\}\,;\\ 4\sqrt{\pi}\, sF^R_{+-}&=-g^R_{1NN}(t)\,g^R_{1\pi\pi}(t)\,R\{s,\alpha(t)\}\,;\end{aligned} \tag{6.4.38}$$

$$R\{s,\alpha(t)\}\equiv\frac{\alpha(t)}{\{\alpha(t)\}!}\,\frac{e^{-i\pi\alpha(t)}+\tau}{2\sin\pi\alpha(t)}\,s^{\alpha(t)}. \tag{6.4.39}$$

$\tau=\pm1$ for isospin even/odd amplitudes is the signature of the trajectory (see Sect. A.9.1.3).

According to the reggeon-photon analogy the P, f, and ω couplings are proportional to the isoscalar photon couplings and the ϱ, A_2 couplings to the isovector photon couplings.

The s-channel helicity amplitudes F_{++}, F_{+-} belong to the electromagnetic nucleon form factors F_1 and F_2, respectively, if one compares the decomposition in terms of invariants [cf. Eqs. (A.2.13) and (A.11.1)] (76 L COLLINS-1). The reggeon-photon analogy relates two well-known facts to each other: the smallness of the form factor ratio F_{2s}/F_{1s} (Table A.11.1) and the approximate s-channel helicity conservation, i.e. the smallness of F^+_{+-}/F^+_{++} (see Sects. 2.1.15 and 2.6.3.8). For simplicity the authors neglect the amplitude F^+_{+-} in their fit.

The reggeon-photon coupling analogy has the following difficulty. In hadron-hadron scattering, ω and ϱ are the only vector mesons taken into account in t-channel exchanges. But it is known from electron-nucleon scattering that the isoscalar and isovector γNN-vertex cannot be well described by ω- and ϱ-exchange respectively. A second pole is required in both cases (Sect. 2.5.2.3). Its residue must have a magnitude comparable with that of the ϱ- and ω-poles, because a dipole fit is approximately valid.

One could consider the introduction of ω'- and ϱ'-trajectories, but the relation of the dips of the spectral functions to higher vector mesons is not yet understood. Therefore, the authors preferred to produce the correct shape of the form factors at $t<0$ by assuming ω- and ϱ-dominance respectively, and a pole at $t=0.9$ GeV^2 in the t-dependence of the couplings.

Because of the reggeon-photon analogy, this pole also occurs in the PNN- and fNN-couplings, for which the "f-dominated pomeron hypothesis" of CARLITZ et al. (71 L) is used. The ansatz $\varrho(t)=\varrho\exp(at)$ is made for the ratio of the P- and f-couplings, where ϱ and a are adjustable parameters.

According to the prescription of the authors, no t-dependence is assumed in the ω-, ϱ-, and A_2-couplings. One could have expected that the dipole character of the spectral function $\mathrm{Im}\,F_{2v}$ and the different behaviour of $\mathrm{Im}\,F_{1v}$ (Sect. 2.5.2.3) are taken into account.

The quark model values are used for the magnetic moments of the nucleon. This eliminates the P-, f-, and ω-flip couplings and gives $4/m=4.26$ for the tensor/vector ratio of the ϱNN-couplings, for which our value is 6.6 [Eq. (4.2.14)]. $f_\varrho=f_\omega/3$ is assumed corresponding to ideal mixing for the vector meson-gamma couplings.

An extrapolation of the meson couplings to t=meson mass squared leads to decay widths which are too large by almost a factor of two. This is not surprising, since the assumptions on the t-dependence, which include the $\alpha(t)$ factors in Eq. (6.4.39), are rather crude and not uniquely determined by the basic idea of the model.

In order to obtain reasonable fits to the data, the authors found it necessary to introduce in the ω-, ϱ-, and A_2-exchange no-flip amplitudes rather strong "absorption" effects, which are described by an empirical parametrization. We mention only the case of ϱ-exchange, where the replacement

$$\alpha_\varrho(t)\to\alpha_\varrho(0)\,x(1+t/t_0)\,e^{c_\varrho t} \tag{6.4.40}$$

is made in the first factor of Eq. (6.4.39). x, t_0, and c_ϱ are adjustable parameters.

At very high energies, the only surviving term in the above Regge pole model belongs to pomeron exchange. In order to be able to describe the dip in pp-scattering at $t=-1.4\,\mathrm{GeV}^2$, the authors added a P⊗P cut. The cut resulting from an eikonal model would be much too strong. Therefore, a parametrization of the t-dependence of the Gribov vertex is assumed and two parameters are fitted to the data. The weakness of the cut is seen from the fact that the cross section at the second maximum, where the cut contribution dominates, is 10^6 times smaller than the forward cross section.

The branch point trajectory

$$\alpha_C(t)=2\alpha_P(0)-1+\tfrac{1}{2}\alpha'_P t=1.12+0.12\,t; \quad t \text{ in GeV}^2 \tag{6.4.41}$$

accounts for the shrinkage found in the large $|t|$ region.

Some problems of this model are discussed in 78 L Collins-3. For instance, the prediction deviates from the data, if a P⊗P⊗P cut is added.

The dip in πp scattering can be treated in a similar way. In 74 L Collins the authors predicted a dip between $t=-1.5$ and $-2.5\,\mathrm{GeV}^2$, which is considerably lower than the location of the dip in the experiment 81 Baker ($t=-4\,\mathrm{GeV}^2$). However, it is probably not difficult to adjust the model to the data by modifying the parameters in the Gribov vertex function.

Comment

It is remarkable that a large amount of data of various reactions can approximately be described by a Regge pole model with factorizable residues, which are determined to some extent by the reggeon-photon analogy. However, it is necessary to introduce several adjustable parameters and functions and therefore, it is not clear up to what accuracy the reggeon-photon analogy and the restriction to a few Regge poles have been tested.

It would be useful to perform a *quantitative* comparison of the model with the isospin even combination of the πN data, taking the parameters of the Pomeron-exchange amplitude from the analysis of the pp-scattering data in the ISR range. The effect of the P⊗P-cut is much smaller in πN scattering than in pp-scattering, because the ratio of the cross sections at the secondary peak and at $t=0$ is 100 times smaller. Correspondingly, the dip occurs at a larger t-value ($t=-4\,\mathrm{GeV}^2$) than in the case of pp-scattering ($t=-1.4\,\mathrm{GeV}^2$).

According to the model of Collins et al., P- and f-exchange should be sufficient to fit the πN data in the momentum range 10···200 GeV/c and a t-range from $t=0$ to a value somewhat below the location of the dip. For the comparison with the data, plots of $\ln d\sigma/dt$ vs. $\ln s$ at fixed t are of special interest, since the model predicts a simple shape for the transition to the asymptotic behaviour where pomeron exchange is dominant (see Fig. 6.3.11, p. 335).

2.6.4.10 Lipkins two-component pomeron

In his study of various total cross sections for hadron-hadron scattering, Lipkin noticed regularities in the discrepancies between the data and predictions based on two-component duality with exact exchange degeneracy, universality and SU (3) symmetry (75 L Lipkin, 78 L Lipkin). These regularities are described by his ansatz for the "two-component pomeron". The new component has the same energy dependence for πN, KN, and NN scattering.

For the isospin even and odd pion-nucleon total cross sections Lipkins ansatz reads

$$\begin{aligned}\sigma^+(k)&=13(k/20)^{0.13}+8.8(k/20)^{-0.2}+2.63(k/20)^{-0.5},\\ \sigma^-(k)&=0.88(k/20)^{-0.5}; \quad \sigma \text{ in mb}, \ k \text{ in GeV/c}.\end{aligned} \tag{6.4.42}$$

The real parts of the forward amplitudes are calculated from analytic parametrizations which belong to this fit (see Table A.6.4). The prediction for the real parts was used by Fajardo et al. (81) for a comparison with their data. Their figure shows deviations of Lipkins predictions from ours. These are due to two reasons:

i) Lipkins fit deviates systematically from the new total cross section data of the Fermilab group above 200 GeV/c (Fig. 2.1.4) and also from the experimental total cross section difference at lower momenta (Fig. 2.1.7);

ii) Lipkins determination of the real part from his analytic parametrization has a systematic error due to effects "non-local" in energy (see Sect. 2.2.1.3 and 80 R Höhler-Kaiser, p. 13).

2.6.4.11 Diffraction in electromagnetic processes

The proposal for a reggeon-photon coupling analogy (Sect. 2.6.4.9) was motivated by many earlier results which showed that rather strong similarities exist between diffractive hadron-hadron scattering and the reactions $\gamma N \to \gamma N$ and $\gamma N \to VN$, where V denotes a vector meson. We shall discuss briefly some of the most important properties of these reactions.

The total πN cross section and the vector dominance model

It is a surprising experimental fact that the isospin even total πN cross section $\sigma^+(\pi N)$ and the average $\sigma(\gamma N)$ of the total γp and γn cross sections are proportional to each other in a good approximation, except for deviations near threshold and near $W = 1.5$ GeV (see Fig. 1 in 73 L GENZ)

$$\sigma(\gamma N) \approx \tfrac{1}{220}\sigma^+(\pi N). \tag{6.4.43}$$

The photon energy E_γ and the pion energy ω in the lab. system are chosen in such a way that the total c.m. energy W is the same in both reactions

$$E_\gamma = \omega + \frac{\mu^2}{2m} = T_\pi + E_{\text{th}}; \quad W^2 = 2m\omega + m^2 + \mu^2 = 2mE_\gamma + m^2. \tag{6.4.44}$$

$T_\pi = \omega - \mu$ is the kinetic energy of the pion and E_{th} the threshold energy for photopion production (≈ 150 MeV for $\pi^\pm$ and ≈ 145 MeV for π^0).

The total γp cross section has been measured up to $E_\gamma = 185$ GeV at Fermilab (78 L CALDWELL). References for data at lower energies are given in 80 R BRICMAN (p. 54). The difference between the total γp and γn cross sections is rather small (see Fig. 34 in 78 R BAUER-SPITAL-YENNIE-PIPKIN).

According to the optical theorem, Eq. (6.4.43) is a relation between the imaginary parts of the γp forward Compton amplitude ($\gamma p \to \gamma p$) and the πN forward diffraction amplitude. In terms of t-channel exchanges, both reactions are dominated at high energies by pomeron exchange. The empirical relation Eq. (6.4.43) is therefore a relation between the $\gamma\gamma P$ and $\pi^0\pi^0 P$ couplings.

At intermediate energies one has to consider in addition f-exchange in both reactions. A_2-exchange contributes to γp and γn scattering but it cancels in the sum of the amplitudes.

Many discussions of Eq. (6.4.43) start from the *vector dominance model*. In the simplest version it is assumed that in hadronic reactions at high energies a photon acts like a superposition of vector mesons V = ϱ, ω, and ϕ, which undergo conventional hadronic interactions. Improved versions are discussed in detail in the review 80 R BAUER-SPITAL-YENNIE-PIPKIN. See also 72 R FEYNMAN.

If this model is applied to the imaginary parts of the forward amplitudes of the reactions $\gamma N \to \gamma N$ and $\gamma N \to VN$, one obtains ($\alpha = 1/137$; the real parts are neglected; σ is the total cross section)

$$\begin{aligned} \sigma(\gamma N) &= \alpha \sum_V (4\pi/f_V^2)\sigma(VN) \\ &= \sum_V \left[16\pi\alpha(4\pi/f_V^2)\frac{d\sigma}{dt}(\gamma N \to VN)\right]^{1/2}. \end{aligned} \tag{6.4.45}$$

The $\gamma - V$ coupling strength can be determined from the production of vector mesons in e^+e^--storage rings, which gives the partial decay width [Eq. (A.11.37) includes a correction factor]

$$\Gamma(V \to e^+e^-) \approx \frac{1}{3}\alpha^2 m_V \frac{4\pi}{f_V^2}, \qquad V = \varrho, \omega, \phi, \psi, \cdots \tag{6.4.46}$$

or from photoproduction experiments. The predictive power of the vector dominance model follows from the assumption that f_V is independent of the four-momentum squared of the photon, at least in the large range from 0 to m_V^2.

Our notation for the coupling f_V agrees with that in the "Compilation" (79 R NAGELS). Instead of f_V, $\bar{f}_V$ is used in 78 R BAUER-SPITAL-YENNIE-PIPKIN (see their Table I) and g_V in 77 R PEREZ-Y-ORBA. LEITH (78 R) and others prefer $\gamma_V = f_V/2$.

Since $\sigma(\mathrm{VN})$ cannot be measured directly, it is estimated using the Additive Quark Model which gives (69 R Kokkedee, Chaps. 16 and 17)

$$\begin{aligned} &\sigma(\varrho^0 p) = \sigma(\omega p) = \sigma^+(\pi N);\\ &\sigma(\phi p) = \tfrac{1}{2}\{\sigma(K^- p) + \sigma(K^+ p) + \sigma(K^- n) + \sigma(K^+ n)\} - \sigma^+(\pi N). \end{aligned} \tag{6.4.47}$$

If the small ϕ-contribution is neglected, Eqs. (6.4.45) and (6.4.47) show how a relation of the type of Eq. (6.4.43) can be derived.

It turns out that the "Compton Sum Rule" Eq. (6.4.45) explains 75···85% of the total γp cross section, if ϱ, ω, and ϕ are taken into account. The ϱ alone provides 65···70%. The origin of the remainder is not yet well understood, although it is clear that there must be contributions from higher vector mesons (78 R Leith, 78 R Bauer-Spital-Yennie-Pipkin, p. 402; 79 L Gotsman). Collins et al. (78 L-2) obtained a good fit to $\sigma(\gamma p)$ which, however, is based on assumptions for the contribution of the daughter sequence (ϱ', ϱ'', ω', ω'', ···) of the leading vector mesons, which are not confirmed by an independent information.

Up to now, we have discussed only part of a possible derivation of Eq. (6.4.43), since this relation is not only valid at high energies but also for instance at the peak of the $\Delta(1232)$ resonance. Genz and Schmidt (73 L) proposed a conjecture which relates the isovector photon forward Compton amplitude with the isospin even πN scattering amplitude for zero mass pions. The scaling factor 1/220 is determined in the low energy region, using the value of the πN sigma term.

Dispersion relation for the $\gamma p \to \gamma p$ forward scattering amplitude

The total γp cross section can be used in order to calculate the real part of the spin-averaged Compton forward scattering amplitude $T_{\|}(\nu)$ from the dispersion relation (70 L Damashek)

$$\operatorname{Re} T_{\|}(\nu) = -\frac{\alpha}{m} + \frac{\nu^2}{2\pi^2} \int_{\nu_0}^{\infty} \frac{\sigma(\nu')}{\nu'^2 - \nu^2} d\nu'; \quad \nu \equiv E_\gamma, \tag{6.4.48}$$

where E_γ is the photon energy in the laboratory system [see the second Eq. (6.4.44)]. The normalization of the (laboratory frame) amplitude $T_{\|}$ follows from

$$\left.\frac{d\sigma}{dt}\right|_{t=0} = \frac{\pi}{\nu^2}\{|T_{\|}|^2 + |T_{\perp}|^2\}; \quad \operatorname{Im} T_{\|} = \frac{\nu}{4\pi}\sigma(\nu). \tag{6.4.49}$$

$\sigma(\nu)$ is the spin-averaged total γp cross section which has been measured up to 185 GeV. The constant $-\alpha/m = -3.0\ \mu\text{b GeV}$ is the "Thomson limit" of the scattering amplitude as $\nu \to 0$ and ν_0 is the threshold energy for photopion production.

It is clear that the prediction for the real part is similar to that in the case of the isospin even πN forward amplitude because of the relation Eq. (6.4.43).

An experimental determination of $\operatorname{Re} T_{\|}$ at 2.2 GeV was performed by Alvensleben et al. (73 L). The authors studied the reaction $\gamma p \to p e^+ e^-$ in the case where the invariant mass of the pair and the momentum transfer to the recoil proton are almost zero. The real part of the Compton amplitude follows from the asymmetry due to the interference between the Compton and Bethe-Heitler amplitudes. The result is compatible with the prediction.

Several authors investigated the question whether the asymptotic expansion of $T_{\|}(\nu)$ contains a real constant C which would be related to a "fixed pole" in Regge theory. In practice, the value of C is determined from the sum rule (70 L Damashek, 77 R Collins)

$$C = T_{\|}(0) + \frac{2}{\pi}\int_0^{\nu_0} \frac{d\nu}{\nu} \operatorname{Im} T_{\|}^{R}(\nu) + \frac{2}{\pi}\int_{\nu_0}^{\infty} \frac{d\nu}{\nu} \operatorname{Im}\{T_{\|}^{R}(\nu) - T_{\|}(\nu)\}. \tag{6.4.50}$$

$T_{\|}^{R}(\nu)$ includes all Regge pole contributions with $\alpha(0) > 0$. The second integral has no contribution from the range above a certain energy, where the sum of Regge pole terms is fitted to the data.

Some of the authors concluded that $C = T_{\|}(0) = -\alpha/m$ is compatible with the data. However, it is difficult to see how a significant result can be obtained, if a larger variety of parametrizations of the amplitude at high energies is considered. At that time, it was not yet known that the total γp cross section is increasing above 50 GeV.

Caldwell et al. (78 L) fitted their data above 35 GeV to a *linear* increase in ν. However, this was done only in order to demonstrate that the increase exists. There is no experimental reason to doubt that Eq. (6.4.43) is valid up to the highest energy where data are available.

A sum rule similar to Eq. (6.4.50) has been discussed for πN scattering in Sect. 9.6.2 of the review 80 R Höhler-Kaiser, Sect. 9.6.2. A good fit to the data can be obtained with the assumption that the constant term is zero. It would be of some interest to consider the difference between this sum rule and Eq. (6.4.50), assuming that the high energy part is the same up to a factor.

The forward peaks in the reactions $\gamma N \rightarrow \gamma N$ and $\gamma N \rightarrow VN$

Experiments on Compton scattering at high energies (81 L Breakstone, 77 L Crigee, 78 R Leith) show a shape of $d\sigma/dt$ which is very similar to the πN diffraction peak. It is not yet known if this includes the details discussed in Sect. 2.6.3.1.

The similarity is also observed in the photoproduction of ϱ, ω, and ϕ. There are indications for a steepening of the slope at small $|t|$ (78 R Leith).

A detailed discussion of Compton scattering can be found in the review 76 R Rollnik-Stichel, Caprini et al. (82 L) have recently performed a new dispersion analysis.

2.6.4.12 The transition to the wide-angle region

In pp-scattering it is well-known that cross section data at intermediate energies ($s > 15\,\mathrm{GeV}^2$ corresponding to lab. momenta above 7 GeV/c) can be described by a law derived from the dimensional-counting rule (Sect. 2.6.6)

$$\frac{d\sigma}{dt} = f(\theta)\, s^{-10}, \qquad (6.4.51)$$

if $t > 2.5\,\mathrm{GeV}^2$, corresponding to c.m. scattering angles $\theta > 38°$ (73 L Landshoff).

The shape of the function $f(\theta)$ as following from the fit is such that $d\sigma/dt$ *at fixed t* falls sharply with increasing energy. It would be interesting to use fixed-t analyticity in order to discuss the consequence for the phase of the amplitude (assuming that a single amplitude is dominant). In models from which Eq. (6.4.51) is derived (Sect. 2.6.6), it is usually assumed that the amplitude is *real*.

Donnachie and Landshoff (79 L) started from the observation that the rapid decrease of $d\sigma/dt$ at fixed t becomes slower at very high energies, $d\sigma/dt$ finally being energy-independent (see Fig. 1 in Vol. I/9a, p. 228). The πp cross section behaves in a similar way (Fig. 6.3.11).

The rather sudden change suggests that a new dynamical mechanism becomes dominant, but it is not known if this mechanism contributes to the same amplitude which is dominant at lower energies or to another one.

The onset of the new mechanism can as well be seen in plots of $d\sigma/dt$ at different energies vs. t, if one looks at the behaviour at fixed t and increasing s (Fig. 6.3.12). From this figure it is clear that there are two cases: one can choose a t-value to the left or to the right or the dip, which occurs at $t = -4\,\mathrm{GeV}^2$ in π^-p-scattering. On the left of the dip one arrives obviously at the amplitude for diffraction scattering, whose energy dependence at high energies is comparable with the experimental errors.

Donnachie and Landshoff (79 L) were only interested in the other case, because in pp-scattering the dip occurs already at $t = -1.4\,\mathrm{GeV}^2$. They found that the pp-data at $t < -3\,\mathrm{GeV}^2$ in the ISR range can be fairly well described by a law

$$\frac{d\sigma}{dt} = \mathrm{const} \cdot t^{-8}, \qquad (6.4.52)$$

which had been proposed earlier (74 L Landshoff) from a model, in which quark-quark scattering is mediated by gluon-exchange between almost on-shell quarks. In pp-scattering, 3 gluons are exchanged in such a way that each of the quarks is coupled to one of the gluons ("triple scattering"), see also 81 R Landshoff.

This model has recently been criticized by Collins and Gault (82 L), who pointed out that the triple scattering amplitude is *real* and can therefore not interfere destructively with the almost imaginary diffraction amplitude to produce the dip at $t = -1.4\,\mathrm{GeV}^2$. The phase of the diffraction amplitude is known from fixed-t analyticity and the fact that the differential cross sections are almost independent of energy.

Another problem of the model is that a recent experiment on $\bar{p}$p-scattering at 50 GeV/c (82 L Asad) shows almost the same shape of the differential cross section as pp-scattering in the ISR-range, including the dip structure. This fact leads to the conclusion that at $t < -2\,\mathrm{GeV}^2$ and very high energies, the pp-scattering amplitude is *even* under charge conjugation whereas the triple scattering term is *odd*.

Collins and Gault (82 L) continued the discussion of their own model (Sect. 2.6.4.9) pointing out that the dip of the pp-scattering cross section is filled in at lower energies by the $f-\omega$ contribution, which is purely real, if f and ω are exchange degenerate. Due to a change of the relative sign of the f- and ω-contributions, the sum has a rotating phase $\exp(-i\pi\alpha_R)$ in the case of $\bar{p}p$ scattering and this leads to a predominantly imaginary amplitude (cf. 77 R Collins, Sect. 6.8.h). The effect is to deepen the dip in agreement with the experimental result.

In πN scattering the transition from the diffraction region to the wide-angle region is not yet understood. Figures 6.3.3, 6.3.11, and 6.3.12 suggest that there exist two components which are similar to those discussed in Sect. 2.6.4.8:

i) The diffractive component

This component becomes dominant above 20···50 GeV/c at fixed t in the range from $t=0$ up to at least $t\approx -1.0\,\mathrm{GeV}^2$. It is characterized by a slow variation of $d\sigma/dt$ at fixed t with energy, which is comparable with the variation of the forward cross section.

Presumably, one can neglect the "flip"-amplitude at least up to $t\approx -0.5\,\mathrm{GeV}^2$, but it is not known whether the neglection of F_{+-} or of A or of H is preferable. This leads to a small value of the polarization parameter.

However, the inverse is not true, since the polarization vanishes if no-flip and flip amplitudes are comparable and have the same energy dependence (and therefore also the same phase, cf. Sect. A.6.4.4).

The dip structure at $t=-4\,\mathrm{GeV}^2$ and 200 GeV/c and the slower decrease beyond the dip suggest that the diffraction amplitude can be decomposed into two contributions. One of them follows from the usual diffraction mechanism. It is strongly dominant at small $|t|$ and decreases rapidly as $-t$ increases. The second contribution becomes dominant beyond the dip, where the decrease is much slower. Its physical origin is not yet known. Various possibilities have been discussed above. The two contributions interfere destructively.

The similarities between the diffraction peaks and the dip structures in pp-, $\bar{p}$p-, and πp-scattering suggest that the mechanism is the same in all three cases.

From fixed-t analyticity and the slow energy dependence of $d\sigma/dt$ at fixed t one can conclude that the diffraction amplitude is almost purely imaginary on both sides of the dip. The simplest interpretation of the dip is that the imaginary part has a zero and that $d\sigma/dt$ near the dip is dominated by the real part (79 L Nagy).

ii) The peripheral component

This component is responsible for the rapid decrease of $d\sigma/dt$ at fixed t in Figs. 6.3.3, 6.3.11, and 6.3.12. It can be described in different ways.

The structures in the s- and t-dependence belong to the pattern of s-channel resonances and zero trajectories discussed in Sect. 2.4.3. The peripheral resonances are related to the geometrical picture of scattering at intermediate energies (Sect. 2.6.4.8). Another description is based on the exchange of reggeons in the t-channel (Sect. 2.6.4.9). In elastic π^-p scattering f- and ϱ-exchange are expected to give the dominant contributions at high energies.

Figure 6.6.2 (p. 377) shows a comparison between the elastic and charge-exchange π^-p cross sections. It is seen that they have a comparable magnitude in the t-range where the diffractive component is small in comparison with the peripheral component.

There exists no reliable information on the question whether one of the amplitudes (for instance F_{++} or F_{+-}) is dominating in the peripheral region. In the case of ϱ- and f-exchange, one of the coupling constants is suppressed in comparison with the other one and the corresponding residue functions are small near $t=0$. But there is no guarantee that they remain small at large negative t.

Furthermore, it would be too simple to assume that the absolute phase of the peripheral amplitude is given by the Regge phase, because there could be contributions "non-local in energy" (Sect. 2.2.1.3). It would be useful to calculate the absolute phase of an amplitude whose modulus is given by the behaviour shown in Fig. 6.3.11.

In particular, it would be of interest to study the energy dependence of the absolute phase for an amplitude at $\theta=90°$. At low and intermediate energies this amplitude belongs to the resonance and Regge region, but at high energies the hard scattering models of Sect. 2.6.6.2 predict a real amplitude and s-channel helicity conservation (see also 77 L Höhler-1). The data points in Fig. 6.6.3 and the contour lines in Fig. 6.6.4 do not indicate where the transition between the two mechanisms occurs.

Theoretical investigations of this question have been performed by Blankenbecler et al. (73 L, 74 L) and by Coon et al. (78 L). See also the review 76 R Sivers-Brodsky-Blankenbecler and the most recent paper of this group (80 L Lepage) on exclusive processes in perturbative QCD.

Spin effects in hadronic collisions at large momentum transfer have recently been reviewed by Bourrely, Leader, and Soffer (80 R) and by Szwed (82 R).

2.6.4.13 Theoretical papers on diffraction scattering

Up to now we have discussed phenomenological models of diffraction scattering, which have been invented in order to describe the experimental results. These models play an important role, since they can be used as a guide in attempts to derive the main features of diffraction scattering from fundamental principles.

The early work on multiperipheral models and Regge models is reviewed in 61 R CHEW, 61 R OEHME, 63 R FRAUTSCHI, 63 R FUBINI, 64 R OEHME and many other articles. Further developments can be found for instance in 68 R COLLINS-SQUIRES, 68 L GRIBOV, 71 R ABARBANEL, 71 R ZACHARIASEN, 75 R ABARBANEL-BRONZAN-SUGAR-WHITE, 76 R ABARBANEL, 76 R BAKER-TER MARTIROSYAN, 77 R COLLINS, 78 R MOSHE. OEHME et al. (71 R, 72 L, 74 L, 75 L, 76 L, 76 L PARANJAPE) studied general singular Regge surfaces ($J=\alpha(t)$) of the continued partial wave amplitude in connection with requirements of elastic t-channel unitary.

LOW (75 L) proposed a model for the "bare" (or lowest order) pomeron formulated in terms of the M.I.T. bag (74 L CHODOS), which leads to constant total cross sections and to predictions for the elastic scattering amplitudes including flip terms (see also 75 L NUSSINOV). There are interesting relations to the CHOU-YANG model.

GRIBOV (76 L) studied the pomeron interaction in the preasymptotic region on the basis of the complex angular momentum theory. In his model the total cross sections rise as $(\ln \ln E)^2$ and the interaction radii as $\ln \ln E$.

More recent developments can be traced from the Proceedings of a Conference at Marseille (78 R BOURRELY) and from papers by BARTELS (80 L), BAUMEL (82 L) and by WHITE (81 L, 81 R, 82 R), who points out that diffraction scattering provides potentially the most detailed experimental test of QCD.

WHITE (82 R) compares the experimental situation with the prediction for a "Critical Pomeron" in Reggeon Field Theory (75 R ABARBANEL-BRONZAN-SUGAR-WHITE) (the name refers to an analogy with the critical phenomenon in statistical mechanics). Aside from logarithmic terms, the contribution to the scattering amplitude is that of a single, even-signatured Regge pole with intercept $\alpha_P(0)=1$. This agrees in a first approximation with the high energy behaviour of the elastic hadron-hadron scattering reactions.

The slope $\alpha'_P(0)$ of the trajectory can be determined from the lab. momentum dependence of the logarithmic slope b of $d\sigma/dt$ at $t=0$ [Eq. (6.2.1)]

$$b(k)=c+2\alpha'_P(0)\ln k\,. \tag{6.4.53}$$

GIACOMELLI (76 R) obtained $\alpha'_P(0)=0.30\,\mathrm{GeV}^{-2}$ from pp-scattering data. BURQ et al. (82 L) added to the parametrization a third term, which describes the interference with a secondary contribution and is needed because of the curvature of $b(k)$ in a plot against $\ln k$. Their result is lower [Eq. (6.2.7)]

$$\alpha'_P(0)=(0.13\pm 0.025)\mathrm{GeV}^{-2}\,. \tag{6.4.54}$$

Furthermore, it is consistent with the value derived by the same authors from π^-p scattering data which are, however, not accurate enough for a strong test (Fig. 6.2.7).

The description by a single Regge pole is also favored by the fact that factorization seems to be valid up to at least 10% accuracy.

The important point is, of course, to perform a *quantitative test* of the Critical Pomeron model. In elastic scattering, the problem is to check whether the data show the predicted logarithmic terms in the energy dependence of the total and differential cross sections and the predicted scaling properties.

In this model the total cross sections at very high energies are increasing as $(\ln s)^{0.26}$. This is much slower than the increase $\propto(\ln s)^2$, which has been found in fits to the pp- and πp-total cross sections (Fig. 2.1.4).

A weakening of the increase of the total cross sections not far above the present experimental region can be determined indirectly from experimental data for the real parts at $t=0$. There should be a systematic discrepancy in comparison with a prediction from the dispersion relation based on an increase $\propto(\ln s)^2$.

Unfortunately, the experimental results for the real/imaginary part ratio ϱ of the pp-forward amplitude show a considerable disagreement around 100 GeV/c (82 L BURQ) and a new value in the ISR region is lower than the earlier ones (quoted in 82 L WHITE, Fig. 24).

The situation in π^-p scattering has been discussed in Sect. 2.6.2 in connection with Fig. 6.2.2. WHITE suggests that ϱ_- seems to have stopped increasing and may even be decreasing towards zero, but this would not be compatible with analyticity and a reasonable energy dependence of the total cross section. The sensitivity of ϱ to changes of the high energy behaviour of σ is shown in Fig. 6.2.4.

If all Critical Pomeron phenomena are confirmed, then strong interactions should be described by a SU(3) gauge theory containing the maximum number of quarks consistent with asymptotic freedom (82 R WHITE).

2.6.5 Near-forward charge-exchange scattering

2.6.5.1 The shape of the $\pi^- p \to \pi^0 n$ near-forward cross section

Accurate determinations of the shape have been performed in three recent experiments: 79 APEL, 81 BINON, 81 DAUM. The statistics is lower in the experiment 76 BARNES, but this experiment is the only one which covers momenta up to 200 GeV/c.

The data in the range $0 \geqq t > -0.4\,\text{GeV}^2$ can be parametrized by the empirical ansatz

$$\frac{d\sigma}{dt} = \left.\frac{d\sigma}{dt}\right|_{t=0} \{1 - g(s)c(s)t\}\, e^{c(s)t}. \tag{6.5.1}$$

The simplest interpretation of this expression is that the two terms in the bracket belong to the spin no-flip and flip contributions, respectively (74 BOLOTO). Then, g is the ratio of the integrated flip and no-flip differential cross sections.

However, there is no theoretical reason why this should be true, since the flip and no-flip contributions can be separated only if one has spin-rotation data [see Eq. (A.4.13)].

In 79 APEL the authors fitted their data and the data sets 76 BARNES and 65 STIRLI in the range $10 < k < 200\,\text{GeV/c}$ and $0 > t > -0.2\,\text{GeV}^2$ to the parametrization of Eq. (6.5.1). They found a logarithmic increase of $c(s)$

$$c(s) = c_0 + c_1 \ln(s/s'_0); \quad s'_0 = 10\,\text{GeV}^2; \quad c_0 = (12.7 \pm 0.3)\,\text{GeV}^{-2}; \quad c_1 = (1.57 \pm 0.12)\,\text{GeV}^{-2}. \tag{6.5.2}$$

It describes a "shrinkage" of the near-forward peak.

The energy dependence of $g(s)$ also can be fitted by a logarithmic term, but in this case one obtains a decrease at higher energies. It turns out that the product $g(s)c(s)$ in Eq. (6.5.1) has only a weak energy dependence. Therefore, a reasonable fit to the data above 10 GeV/c is possible already with the simpler ansatz

$$\frac{d\sigma}{dt} = \left.\frac{d\sigma}{dt}\right|_{t=0} (1 - At)\, e^{ct}, \tag{6.5.3}$$

where A is a constant and c has the s-dependence of Eq. (6.5.2).

Table 2.6.5.1. Parameters of the fit Eq. (6.5.1). $A = gc$.

	k GeV/c	A GeV^{-2}	c GeV^{-2}	$\lvert t\rvert_{max}$ GeV^2
79 APEL	10	32.9	13.8	0.2
	200	32.4	17.4	0.2
81 BINON	40	31.8±1.0	15.8±0.2	0.3
81 DAUM	63	30.2±1.8	16.0±0.2	0.4

Equation (6.5.3) describes part of the characteristic structure of the near-forward cross section. At $t = 0$ it starts with an increase towards a peak at $t \approx -0.035\,\text{GeV}^2$. Then, it decreases almost exponentially towards a dip at $t \approx -0.6\,\text{GeV}^2$, which is followed by a broad secondary peak. At 40 GeV/c, the first peak is higher than the forward value by a factor of 1.22 and the dip is lower than the peak by a factor of about 700. Due to the shrinkage, which is also observed at larger values of $-t$ (see below), the shape depends on energy. The dip-bump structure becomes a shoulder above 50 GeV/c. The energy dependence of the shape is shown in Fig. 6.5.1 (see p. 361).

Another parametrization, which was used by many authors, follows from Regge theory, assuming that it is sufficient to consider the exchange of a single Regge pole (ϱ-exchange). According to Eq. (A.9.23), a high energy approximation reads

$$\frac{d\sigma}{dt} = F(t)\,(s/s_0)^{2\alpha(t)-2}. \tag{6.5.4}$$

If the corrections to the approximations $\nu=\omega+t/4m\approx\omega=(s-m^2-\mu^2)/2m\approx s/2m$ are appreciable, one can use a more accurate version by substituting Eqs. (A.9.16) and (A.9.17) into Eq. (A.4.17) (63 L SINGH). However, the non-vanishing polarization shows that there are secondary terms in addition to ϱ-exchange, and their effects can be comparable with the first mentioned kinematical correction.

Equation (6.5.4) can be tested by checking whether plots of $\ln(d\sigma/dt)$ vs. $\ln s$ give straight lines. It turns out that this is true to a remarkable extent in the range 6⋯200 GeV/c, $0\geqq t\geqq -1.3\,\mathrm{GeV}^2$. Some systematic deviations will be discussed below.

At each value of t, the fits give a result for $\alpha(t)$ and for $F(t)$. It turns out that $\alpha(t)$ is approximately linear at least in the range $0>t\gtrsim -0.4\,\mathrm{GeV}^2$, where the parametrization Eq. (6.5.3) is valid (79 APEL)

$$\alpha(t)=\alpha(0)+\alpha' t;\qquad \alpha(0)=0.48\pm 0.01;\qquad \alpha'\approx 0.90\,\mathrm{GeV}^{-2}. \tag{6.5.5}$$

We substitute Eq. (6.5.5) into Eq. (6.5.4) and find

$$\frac{d\sigma}{dt}=\left.\frac{d\sigma}{dt}\right|_{t=0} T(t)\,e^{\bar{c}(s)t};\qquad \bar{c}(s)=2\alpha'\ln(s/s_0);\qquad T(t)=F(t)/F(0). \tag{6.5.6}$$

It is seen that the earlier ansatz Eq. (6.5.3) is a special case of the Regge formula, if a linear Regge trajectory is assumed and if $d\sigma/dt$ at $t=0$ in Eq. (6.5.3) fulfills the Regge law. One should notice that the result for the shape of $T(t)$ depends on the parameter s_0, which is not determined by Regge theory. We learn from the success of Eq. (6.5.3) that, in the range $0>t>-0.4\,\mathrm{GeV}^2$, $T(t)$ can be approximated by a linear function of t, if s_0 is chosen in a suitable way. $c_1/2$ corresponds to α' in Eq. (6.5.6).

The parametrization Eq. (6.5.3) can only be used up to about $t=-0.4\,\mathrm{GeV}^2$, because it cannot describe the dip-bump structure at larger $|t|$. BARNES et al. (76) fitted a more flexible ansatz to their data, choosing in Eq. (6.5.4)

$$F(t)=\beta_1\,e^{b_1 t}-t(t-t_0)^2(\beta_2+\beta_3 t+\beta_4 t^2)\,e^{b_2 t} \tag{6.5.7}$$

and

$$\alpha(t)=\alpha_1+\alpha_2 t+\alpha_3 t^2, \tag{6.5.8}$$

because a non-linear trajectory was required in the larger t-range.

The parametrization of BARNES et al. is a good approximation to the s- and t-dependence of the charge-exchange cross sections in the range 20⋯200 GeV/c, $0>t>-1.4\,\mathrm{GeV}^2$ ($\chi^2=107$ for 128 degrees of freedom).

Since many error bars are too small to be visible in plots of $\ln d\sigma/dt$ vs. t, we have plotted in Fig. 6.5.2 the ratio of the data and the cross sections calculated from the parametrization of BARNES et al. Furthermore, the data of some experiments are too densely spaced for our scale. Since the ratio is slowly variable in t, it is justified to plot the average of the data over larger bins.

Figure 6.5.2 (see p. 361) gives information on two questions:

i) on discrepancies between different data sets,

ii) on systematic deviations from the simplest version of the Regge law, Eq. (6.5.4).

63 GeV/c

A comparison between the new ACCMOR data (81 DAUM) and the data 76 BARNES shows a serious discrepancy. The data 76 BARNES are higher by a factor of about 1.4 near $t=0$ and they are lower by a factor of about 2 beyond $t=-0.6\,\mathrm{GeV}^2$.

In Fig. 6.5.2, all ACCMOR data have been multiplied by a factor of 1.4. There are two arguments why the correction is applied to the ACCMOR data and not to the data 76 BARNES:

i) The latter data agree with the data 79 APEL at 20 and 40 GeV/c within the quoted errors, which are much smaller than the discrepancy.

ii) Even if the data 76 BARNES would not exist, the small value of DAUM et al. for the forward cross section (Table 2.6.5.2) could not be accepted, because it leads to a real part of the forward amplitude, which violates strongly the forward dispersion relation, if one inserts the known total cross section difference.

At $t<-0.6\,\mathrm{GeV}^2$, the plots of $\ln d\sigma/dt$ vs. $\ln s$ at fixed t show a break above 40 GeV/c, if the data of DAUM et al. are correct, whereas the data of BARNES et al. show a smaller systematic deviation from the Regge behaviour in the sense that the points at $k>60$ GeV/c lie above the straight lines. This discrepancy can only be resolved by a new experiment.

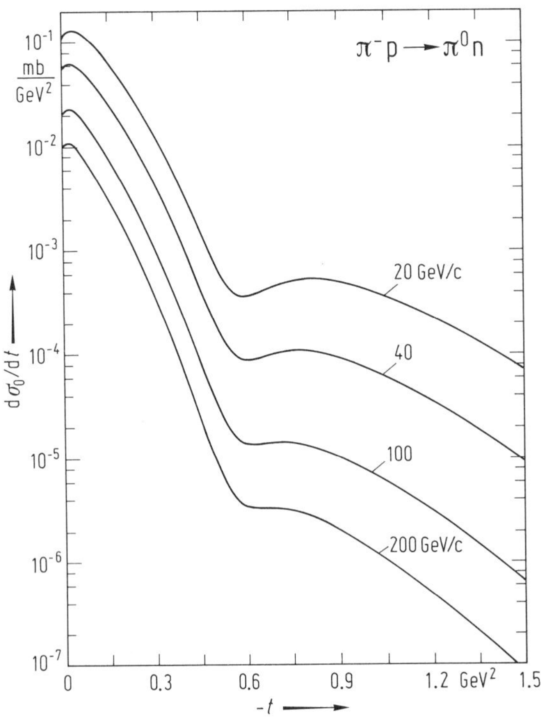

Fig. 6.5.1. $d\sigma_0/dt$ vs. t as following from the fit of Eqs. (6.5.7) and (6.5.8) to the data of BARNES et al. (76).

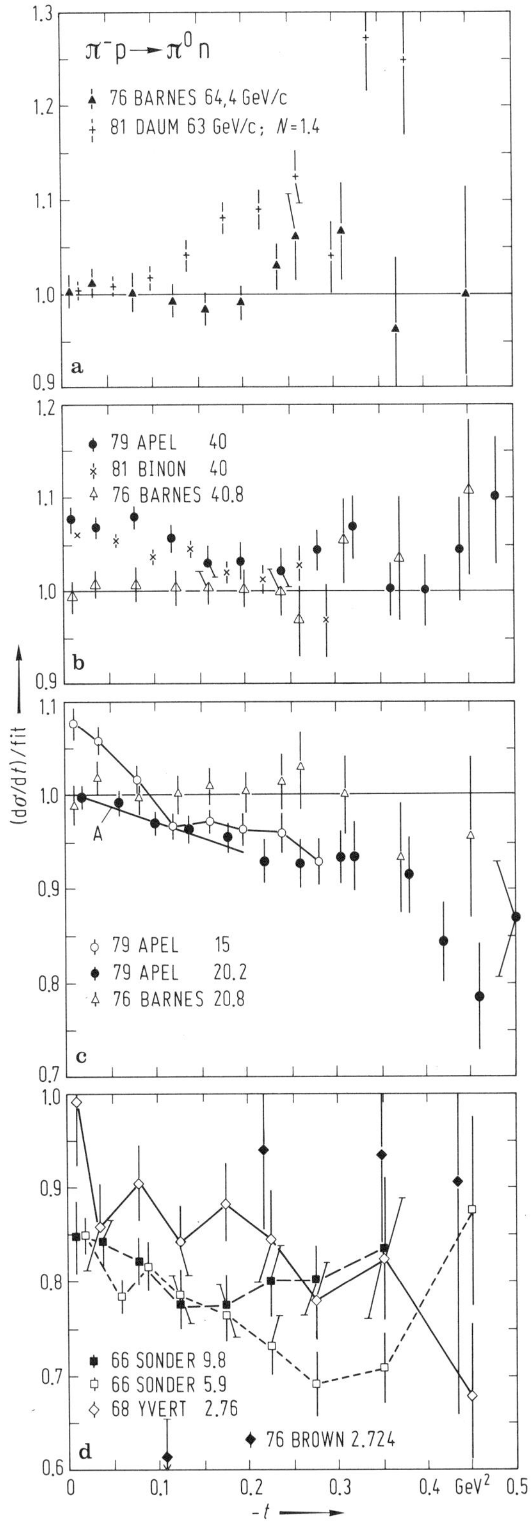

Fig. 6.5.2a…d. Various data sets divided by the fit of BARNES et al. (76), Eqs. (6.5.7) and (6.5.8).
a) Data near 60 GeV/c. The data 81 DAUM have been renormalized by a factor of $N=1.4$. They disagree with the data 76 BARNES.
b) Data near 40 GeV/c. There is a systematic difference between the Serpuchov data (79 APEL, 81 BINON) and the Fermilab data (76 BARNES) at $t > -0.15\,\mathrm{GeV}^2$.
c) Data at 15…20 GeV/c. Again the slope of the data 79 APEL at $t=0$ is steeper than that of the data 76 BARNES. At 15 GeV/c, the data near $t=0$ of APEL et al. are higher than the fit, but the data at 9.8 and 5.9 GeV/c are lower (Fig. d).
d) Data at 2.7…10 GeV/c. The first point of 76 BROWN lies at 0.56.

Table 2.6.5.2. Forward charge-exchange cross sections and integrated differential cross sections (see also Fig. 6.8.5 on p. 399).

k GeV/c	$d\sigma_0/dt$ µb GeV^{-2}	$k\, d\sigma_0/dt$ mb GeV^{-1}	σ_0 µb	Reference
199.3	9.61	1.92	1.44±0.06	76 Barnes
150.2	12.88	1.93	2.02±0.09	76 Barnes
100.7	19.51	1.96	3.20±0.14	76 Barnes
64.4	30.9	1.99	5.61±0.25	76 Barnes
63	23.4 ± 0.7	1.47±0.04	4.3 ±0.5	81 Daum
40.8	49.8	2.03	9.8 ±0.5	76 Barnes
40.0	54.7 ± 2.7	2.19±0.11	–	81 Binon
40.0	54.9 ± 1.6	2.20±0.06	10.6 ±0.5	79 Apel
30.0	70 ± 2	2.10±0.06	14.3 ±0.7	79 Apel
25.0	84 ± 2	2.10±0.05	17.8 ±0.8	79 Apel
20.8	100	2.09	22.6 ±1.1	76 Barnes
20.2	104 ± 3	2.10±0.06	22.8 ±1.1	79 Apel
18.2	132 ± 3	2.40	24.4 ±2.6	66 Sonder
15.0	152 ± 5	2.28±0.08	34.2 ±1.8	79 Apel
13.3	184 ± 8	2.45	37.2 ±2.2	66 Sonder
9.8	201 ± 8	2.0 ±0.1		66 Sonder
5.9	340 ± 10	2.0		66 Sonder
5.74	350 ± 20	2.0		68 Yvert
5.55	380 ± 40	2.1		68 Yvert
5.38	490 ± 50	2.6 ±0.3		68 Yvert
5.21	470 ± 50	2.5		68 Yvert
5.01	440 ± 40	2.2		68 Yvert
4.82	460 ± 30	2.2		68 Yvert
4.64	450 ± 30	2.1		68 Yvert
4.45	470 ± 50	2.1 ±0.2		68 Yvert
4.25	570 ± 50	2.4		68 Yvert
4.06	570 ± 30	2.3		68 Yvert
3.86	510 ± 40	2.0		68 Yvert
3.66	670 ± 30	2.5 ±0.1	191±21	68 Yvert
3.52	580 ± 80	2.0		68 Yvert
3.38	680 ± 50	2.3		68 Yvert
3.22	850 ± 60	2.7		68 Yvert
3.07	840 ± 80	2.6 ±0.3	260±30	68 Yvert
2.92	830 ±120	2.4		68 Yvert
2.76	780 ± 60	2.2		68 Yvert
2.55	800 ± 70	2.0 ±0.2		68 Yvert

40 GeV/c

Near $t=0$ the Serpuchov data (79 Apel, 81 Binon) are about 7% higher than the Fermilab data (76 Barnes), but the agreement is good beyond $t=-0.15\,\text{GeV}^2$. The systematic errors are about 5% in all three cases.

15 and 20 GeV/c

Near $t=0$ the data 79 Apel lie above the fit at 15 GeV/c, on the fit at 20.2 GeV/c and again above the fit at 40 GeV/c. In all three cases the slope at $t=0$ is steeper than that of the fit. The line denoted by "A" is the fit of Apel et al.

2.7···10 GeV/c

The figure shows that the experimental situation is not satisfactory, because the variation between 1.7 and 10 GeV/c is comparable with the errors. New accurate data at 2···20 GeV/c would be of great interest, because one could deduce the second term of the asymptotic expansion, which has been discussed in many models. Furthermore, these data would show structures due to nucleon resonances, which are the continuations of well-known resonance families to higher masses (see Sect. 2.4.1).

For the application of the optical theorem to the charge-exchange forward amplitude, the difference of the $\pi^{\pm}$p total cross sections is required. Again, the experimental information at 6···20 GeV/c is not satisfactory (Fig. 2.1.7) and this will remain so for some time, because only the π^-p total cross section has been remeasured in a recent experiment (80 Barrel).

2.6.5.2 Validity of the reggeized ϱ-exchange model

The simplest Regge model for π^-p charge-exchange scattering is the reggeized ϱ-exchange model (Sect. A.9.1). It leads to two predictions which can easily be tested:

i) The differential cross section is given by Eq. (6.5.4). Therefore, a plot of $\ln d\sigma/dt$ vs. $\ln s$ at fixed t should give straight lines. The slopes of these lines determine $\alpha(t)$.

ii) The Regge trajectory $\alpha(t)$ should aim towards the points $\mathrm{Re}\,\alpha=1$ at $t=m_\varrho^2=(0.776)^2\,\mathrm{GeV}^2$ and $\mathrm{Re}\,\alpha=3$ at $t=m_g^2=(1.700)^2\,\mathrm{GeV}^2$.

The ϱ-trajectory has the signature $\tau=-1$ and, according to Eq. (A.9.20), the next resonance on this trajectory has spin 3. Since $\alpha(t)$ has a branch cut at $t=4\mu^2$, $\mathrm{Re}\,\alpha(t)$, at $t>4\mu^2$ is not a smooth continuation of the real trajectory $\alpha(t)$ at $t<4\mu^2$. However, a detailed investigation has shown that a strong effect is possible only if $\alpha(4\mu^2)<0.5$, whereas in our case $\alpha(4\mu^2)\approx 0.55$. Furthermore, the cusp effects at threshold are very small in potential scattering (see 68 R Collins-Squires, p. 74).

The comparison of both predictions with data has recently been discussed by Apel et al. (79). Between $t=0$ and the dip region ($t\approx -0.6\,\mathrm{GeV}^2$) there are no significant deviations from the power law, Eq. (6.5.4), in the range 15···200 GeV/c (see also Fig. 6.5.2). $\alpha(t)$ almost agrees with the straight line through $\mathrm{Re}\,\alpha=1$, $t=m_\varrho^2$ and $\mathrm{Re}\,\alpha=3$, $t=m_g^2$, $\alpha(t)=0.48+0.87t$, t in GeV^2 (see Fig. 6.5.3).

These results are a strong argument in favor of the approximate validity of the reggeized ϱ-exchange model in π^-p charge-exchange scattering. Up to now, there are no generally accepted demonstrations that other predictions derived from Regge poles or cuts are valid in πN scattering.

In particular, it is important to notice that the experimental Regge trajectory (Fig. 6.5.3) has no significant structure in the dip region ($t\approx -0.6\,\mathrm{GeV}^2$). This fact disagrees with predictions from several models, for instance from the Regge cut model of Fox (69 R) and the dual peripheral model of Schrempp (74 L, see 74 R Barger).

The following difficulty of the reggeized ϱ-exchange model is not yet well understood. In this model, the expressions for the differential charge-exchange forward cross section and for the difference $2\sigma^-$ of the total $\pi^{\pm}$p cross sections read with the parameters of the fit of Barnes et al. (cf. Figs. 2.1.7 and 2.1.8)

$$k\left.\frac{d\sigma_0}{dt}\right|_{t=0}=\beta k^{-0.038}; \qquad \sqrt{k}\,\sigma^-=ck^{-0.019};$$
$$\alpha=0.481; \quad \beta=2.34\,\mathrm{mb}\cdot\mathrm{GeV}^{2-2\alpha}; \quad c=\sqrt{8\pi\beta}\cos(\alpha\pi/2)=3.48\,\mathrm{mb}\cdot\mathrm{GeV}^{1-\alpha} \tag{6.5.9}$$

One obtains a prediction for σ^- which is in reasonable agreement with the data at 15···100 GeV/c, but it is too low at 100···340 GeV/c. The discrepancy is also seen in the result for the exponent of the power law fit of Carroll et al. (79): $\alpha=0.54\pm 0.03$, which is higher than the result of Barnes et al.: $\alpha=0.481\pm 0.004$.

As mentioned in connection with Fig. 2.1.8, it is necessary to consider radiative corrections, which have to be applied to the published charge-exchange data. The discrepancy is appreciably diminished, if one uses the estimate for this correction by Ginzburg et al. (78 L). In my opinion, the value $\alpha(0)=0.500$ for both cases is not yet excluded (see Figs. 2.1.7, 2.1.8, and 6.5.4). Of course, one also has to consider the possibility that a power law is not a sufficiently good approximation for the decrease of σ^- above 20 GeV/c.

It is remarkable that a power law fit to the difference of the pp and $\bar{\mathrm{p}}$p total cross sections (82 L Carboni-1 and 2) at the highest energies leads to $\alpha=0.49\pm 0.01$ (82 L Block) for the dominant ω-exchange contribution.

The apparent incompatibility of the charge-exchange data with conventional Regge asymptotic behaviour has been discussed in detail by Joynson, Leader and Nicolescu (75 L). The authors considered a possible odd-signatured analogue of the pomeron ("odderon"), which corresponds to an additive real term in the unsubtracted fixed-t dispersion relation (see Sect. A.6.3.2). – In my opinion the discrepancy is still within the uncertainties.

Lovelace (68 L) predicted the ϱ-trajectory by imposing the Adler zero of the off-shell amplitude at $s=t=u=\mu^2$ in a Veneziano model for ππ-scattering ($\alpha(\mu^2)=0.5$, see 76 R Martin-Morgan-Shaw, p. 329)

$$\alpha(t)=0.5+0.5(t-\mu^2)/(t_\varrho-\mu^2)=0.48+0.86t. \tag{6.5.9a}$$

At that time the "experimental value" of α(0) derived from data at 6⋯18 GeV/c was 0.57, but now the agreement is very good.

The discussion of the reggeized ϱ-exchange model will be continued in Sect. 2.6.5.4.

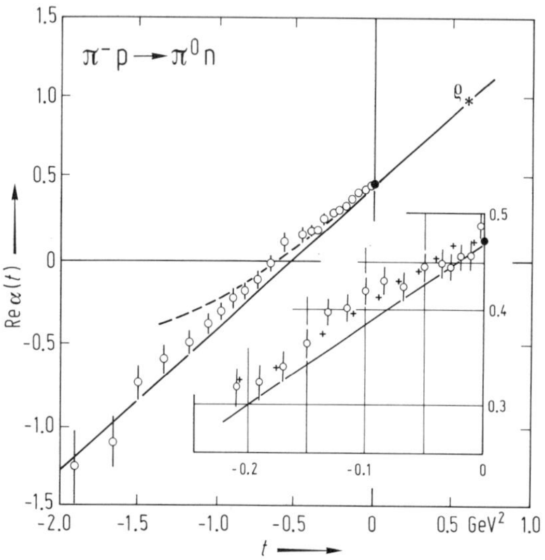

Fig. 6.5.3. The effective Regge trajectory as derived from charge-exchange scattering data at 6⋯200 GeV/c (79 Apel). The straight line goes through the points belonging to the ϱ and g-mesons. ♁: result derived from the data of Apel et al. (79). Dashed line: result of Barnes et al. for the trajectory in the range 20⋯200 GeV/c. The insert shows the small t region in greater detail. +: values for α derived from a fit to all data at 6⋯200 GeV/c.

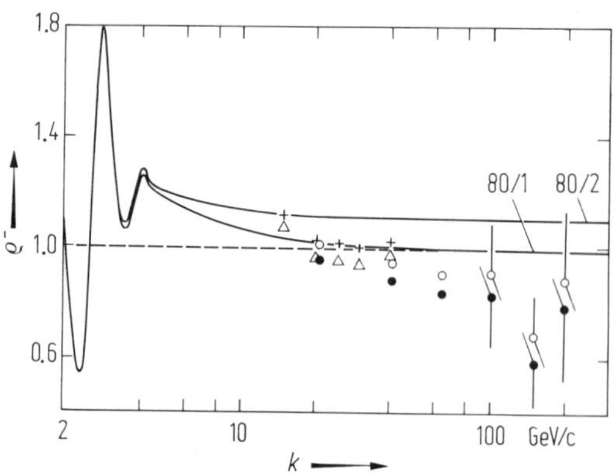

Fig. 6.5.4. The ratio $\varrho^- \equiv \mathrm{Re}\,C^-/\mathrm{Im}\,C^-$ at $t=0$ vs k. The solid line 80/1 follows from Table 2.2.1.1, where the total cross section difference is assumed to decrease as $1/\sqrt{k}$ at high energies. The solid line 80/2 belongs to a power law fit to the total cross section difference ($\alpha=0.534$; Table 80/2 in 80 R Höhler-Kaiser). The symbols show ϱ^--values calculated from $d\sigma_0/dt$ and σ^--data. ♁ and ●: Barnes (76) with and without radiative corrections, + and △: Apel (79) with and without radiative corrections. The corrections are crude estimates (78 L Ginzburg).

2.6.5.3 Determination of π⁻p charge-exchange scattering amplitudes at high energies

As mentioned above, the peak of $d\sigma_0/dt$ at the small value $t=-0.035\,\mathrm{GeV}^2=-1.8\mu^2$ strongly suggests that the helicity flip contribution is large. The ratio of the non-flip and flip contributions follows easily from $d\sigma_0/dt$-data and the spin-rotation parameter S, Eq. (A.4.13), but this quantity has not yet been measured in charge-exchange scattering. In the following, we shall discuss the different possibilities to derive information on the amplitudes from other experimental quantities for which data are available.

$d\sigma_0/dt$ and P_0

From the charge-exchange differential cross section and polarization one can calculate

$$8\pi\sqrt{s}\,\frac{P_0 d\sigma_0/dt}{\sin\theta}=\mathrm{Im}(C^{-*}B^-)=|B^-||C^-|\sin(\phi_B-\phi_C)\equiv|B^-|C_\perp^-\equiv|C^-|B_\perp^-, \tag{6.5.10}$$

where ϕ_B and ϕ_C are the phases of the complex vectors B^- and C^-.

In the reggeized ϱ-exchange model, we have $\phi_B=\phi_C$ and therefore $P_0=0$, $B_\perp^-=C_\perp^-=0$.

The extrapolation of the experimental data for the left hand side to $t=0$ gives a finite value. It can be used for the determination of $B_\perp^-$, since the forward amplitude C^- is well-known (Table 2.2.1.1).

At small $|t|$, the data for P_0 are positive in the range 2⋯40 GeV/c. It follows that $\phi_B>\phi_C$, i.e. a *clockwise rotation* leads from the direction of B^- to that of C^-.

The magnitude and the energy dependence of $C_\perp^-$ can be estimated at $t \approx -0.2\ \mathrm{GeV}^2$, where $d\sigma_0/dt$ is strongly dominated by the contribution of $|B^-|^2$, because $|C^-|^2$ is small (see Figs. 6.5.5 and 6.5.6). Equation (6.5.10) leads to

$$P_0 \approx \frac{4m}{\sqrt{-t}\,k|B^-|} C_\perp^- ; \quad t \approx -0.2\ \mathrm{GeV}^2 . \tag{6.5.10a}$$

The momentum dependence of $|B^-|$ follows from that of $d\sigma_0/dt$, i.e. $k|B^-| \propto k^\alpha$, where $\alpha = 0.33$ at $t = -0.2\ \mathrm{GeV}^2$. An estimate of the decrease of the P_0-data at fixed $t = -0.2\ \mathrm{GeV}^2$ in the range 5⋯40 GeV/c gives $P_0 \propto 1/\sqrt{k}$, within the rather larger errors of the data. It follows that $C_\perp^- \propto k^{-0.17}$, in agreement with the fact that a possible candidate for the secondary term is the contribution from a ϱ′-Regge pole, whose trajectory lies *half a unit* below the ϱ-trajectory. The mass of the ϱ′-resonance would be about 1200 MeV, near the location of the dip in the isovector spectral function of the electromagnetic nucleon form factor [Sect. 2.5.2.3; see for instance 74 L Barger for a ϱ, ϱ′-model and 81 R Paul for the status of ϱ′ (1250)]. Of course, the polarization can as well be due to a pole-cut interference.

$d\sigma_-/dt - d\sigma_+/dt$

Above 2 GeV/c the elastic $\pi^\pm p$ differential cross sections show a "cross-over" at about $t = -0.2\ \mathrm{GeV}^2$, because the $\pi^- p$ forward cross section is higher and has a larger logarithmic slope than the $\pi^+ p$ cross section. In an optical model this is a consequence of the fact that the $\pi^- p$ interaction has a larger interaction radius than the $\pi^+ p$ interaction, since it has additional channels in the final state.

Barger and Phillips (76 L) discussed this point quantitatively as follows. According to the hypothesis of "geometrical scaling", the ratio of the total cross section and the logarithmic slope at $t = 0$ should be independent of energy, since both are proportional to the square of the energy-dependent "interaction radius" [Eq. (6.4.13)]. If one adds the assumption that this constant is the same for $\pi^+ p$ and $\pi^- p$ scattering, it follows that

$$\frac{d\sigma_\pm}{dt} = \mathrm{const} \cdot (B_\pm)^2\, e^{B_\pm t} , \tag{6.5.11}$$

and one obtains for a small difference $B_- - B_+$ the result

$$t_c = -\frac{4}{B_+ + B_-} \approx -0.2\ \mathrm{GeV}^2 \tag{6.5.12}$$

for the location of the cross-over, in approximate agreement with the experimental data. It is surprising that the prediction is reasonable even at intermediate energies, where geometrical scaling is not yet valid.

The determination of a charge-exchange amplitude from the data for the cross-over effect was discussed in 64 L Höhler-2. In the expression

$$\Delta \equiv \frac{d\sigma_-}{dt} - \frac{d\sigma_+}{dt} \approx \frac{1}{4\pi}\left[\left(1 - \frac{t}{4m^2}\right)\frac{\mathrm{Re}(C^+ C^{-*})}{k^2} - \frac{t}{4m^2}\,\mathrm{Re}(B^+ B^{-*})\right], \tag{6.5.13}$$

the first term is dominant for small $|t|$. Furthermore, we can use an approximation in which the contribution of A^+ to the cross section is neglected at high energies [Eq. (1.15.7)]

$$\Delta \approx \frac{1}{4\pi k^2}\,\mathrm{Re}(C^+ C^{-*}) = \frac{1}{4\pi k^2}|C^+||C^-|\cos(\phi_+ - \phi_-) \equiv \frac{1}{4\pi k^2}|C^+| C_\parallel^- \approx \sqrt{d\sigma^+/dt}\,\frac{C_\parallel^-}{k\sqrt{\pi}} . \tag{6.5.14}$$

It is important to study the decrease of Δ already in the Coulomb interference region (72 L Krubasik). In Eq. (A.4.33), the term $2\delta\,\mathrm{Im}\,C^-$ is negligible in comparison with $\mathrm{Re}\,C^+$. We denote the experimental value of the difference of the cross sections by Δ_{exp} and obtain

$$C_\parallel^- = \frac{4\pi k^2}{|C^+|}\left\{\Delta_{\mathrm{exp}} - \frac{2\alpha}{k|t|} f_E(t)\,\mathrm{Re}\,C^+\right\} . \tag{6.5.15}$$

The interference term is appreciable in a larger t-range than expected. For instance at 10 GeV/c and $t = -0.10\ \mathrm{GeV}^2$, it contributes 0.4 mb/GeV² which is larger than the statistical error of Δ in 75 Brande.

The most accurate data for the cross-over effect in πN scattering are the 3···6 GeV/c data 74 AMBATS and the 10.4 GeV/c data 75 BRANDE. Both groups fitted their data to an ansatz (75 L CARNEGIE-1)

$$\bar{\Delta} \equiv \frac{\Delta(s,t)}{4\sqrt{d\sigma^+/dt}} = \frac{C_{\|}^-}{4k\sqrt{\pi}} = a\,e^{bt} J_0(r\sqrt{-t}), \tag{6.5.16}$$

which follows from the dual absorptive model of HARARI et al. (71 L, 72 L; see Sect. 2.6.4.8).

However, there is no evidence for a continuation of the oscillations beyond the second zero and, at 10.4 GeV/c, one can as well use the simpler ansatz (75 L CARNEGIE-1).

$$\bar{\Delta} = a(1 - t/t_c)\,e^{bt}. \tag{6.5.17}$$

The parameter a is determined by the total cross sections and the real parts at $t=0$ as calculated from the dispersion relation.

Since the location of the cross-over is sensitive to the absolute normalization of the elastic differential cross sections, an accurate determination can only be performed, if the data include the Coulomb interference region and can be adjusted to the forward cross sections, which follow from the total cross section data. Unfortunately, the accuracy of the data for the total cross section difference near 10 GeV/c is not yet satisfactory (see Fig. 2.1.7).

In 80 HÖHLER-1 we have studied the energy dependence of the logarithmic slope

$$\beta \equiv \frac{\partial}{\partial t}\ln\bar{\Delta}(s,t) = \frac{\partial}{\partial t}\ln C_{\|}^- \approx \frac{\partial}{\partial t}\ln \operatorname{Im} C^-, \quad \text{at } t=0, \tag{6.5.18}$$

including an average of the 150···200 GeV/c data 81 FAJARD. It turned out that, within the relatively large uncertainties, β remains constant in the large range 3···200 GeV/c. It is about twice as large as the slope of $\operatorname{Im} C^+$ ($\beta = 7\cdots10\ \text{GeV}^{-2}$). β is related to the logarithmic slopes $b_\pm$ and b^+ of the elastic cross sections by

$$\frac{b_- - b_+}{\Delta(t=0)} \approx \frac{\beta - b^+/2}{d\sigma^+/dt|_{t=0}}. \tag{6.5.19}$$

The result for β is of interest for an improved investigation of the dispersion relation for $\partial C^-/\partial t$ at $t=0$, which leads to a constraint in amplitude analysis (73 L JAKOB and KROLL; R. KOCH, work in progress).

The energy dependence of $\Delta(s,t)$ is considerably reduced, if we consider

$$\tilde{\Delta}(s,t) \equiv \Delta(s,t)\left[8\frac{d\sigma^+}{dt}\frac{d\sigma_0}{dt}\right]^{-1/2} \equiv \bar{\Delta}(s,t)\frac{\sqrt{2}}{\sqrt{d\sigma_0/dt}}. \tag{6.5.20}$$

If the contribution of the amplitude B^- is negligible (very near $t=0$ and, as usually assumed, at the dip near $t=-0.6\ \text{GeV}^2$), we find using Eq. (6.5.14)

$$\tilde{\Delta}(s,t) \approx \frac{C_{\|}^-}{|C^-|} = \cos(\phi_+ - \phi_-); \quad \phi_\pm = \arg C^\pm; \quad \text{if } B^- \approx 0. \tag{6.5.21}$$

Furthermore, the isospin bound Eq. (A.10.8), taken for the differential cross sections, leads to an upper bound for $|\tilde{\Delta}|$ (Fig. 3 in 80 L HÖHLER-1)

$$|\tilde{\Delta}(s,t)| \leqq 1; \quad \frac{|\Delta|}{d\sigma^+/dt} \leqq \left[8\frac{d\sigma_0/dt}{d\sigma^+/dt}\right]^{1/2}. \tag{6.5.22}$$

At 200 GeV/c and $t \approx -0.5\ \text{GeV}^2$ the bound is much smaller than the statistical errors of the most accurate data. Therefore, it can be used for a test of the relative normalization of the $\pi^\pm p$ scattering cross sections.

Since the polarization parameters are small near $t=-0.5\ \text{GeV}^2$, the bound Eq. (6.5.22) is not much different from the stronger bounds Eq. (A.10.8) for $\Sigma(\pm)$. It is possible that these bounds are saturated at high energies near $t=-0.5\ \text{GeV}^2$, because the reggeized ϱ-exchange model suggests that the Argand vectors of the isospin odd transversity amplitudes become antiparallel to those of the isospin even amplitudes. There is evidence for the saturation as far as accurate data are available (up to 10 GeV/c, see also Fig. 4.4.2).

$\boldsymbol{P_+ - P_-}$

At high energies, the data for the polarization parameters P_+ and P_- can be used in order to derive information on the charge-exchange amplitude B^-. In the expression

$$\frac{8\pi\sqrt{s}}{\sin\theta}\left[P_+\frac{d\sigma_+}{dt} - P_-\frac{d\sigma_-}{dt}\right] = \mathrm{Im}(C^+B^{-*} + C^-B^{+*}), \tag{6.5.23}$$

the second term on the right hand side is negligible for small $|t|$. Since $|C^+|$ can be determined from $d\sigma^+/dt$ data using Eq. (1.15.7), one obtains information on the component of B^- perpendicular to C^+.

In a zero order approximation, one could take C^+ purely imaginary and check, whether the s- and t-dependence of $B_\perp^- = \mathrm{Re}\,B^-$ agrees with the prediction derived from the reggeized ϱ-exchange model, Eq. (A.9.17).

The result confirms that "pomeron-ϱ interference" gives the dominant contribution to the left hand side, i.e. essentially to $P_+ - P_-$ at high energies, but there are deviations which are clearly outside the errors.

One reason for the discrepancy follows from Table 2.6.3.2: below 50 GeV/c it is not justified to neglect $\mathrm{Re}\,C^+$ (p. 327).

A second reason is the neglection of electromagnetic corrections to the invariant amplitude B (Sect. 2.1.17.4). These corrections are increasing with energy and they remain appreciable beyond the "Coulomb interference region" of the differential cross sections. At 40 GeV/c and $t = -0.2\ \mathrm{GeV}^2$ they are larger than the experimental errors (78 L Borie-2) (see Figs. 1.17.1 and 4.4.4).

In their amplitude analysis, these authors concluded that the data are not compatible with a Regge behavior of the B^--amplitude and the assumption of C^+-dominance for the diffraction amplitude. At present, we are investigating the possibility to change this unpleasant result by modifying the high energy ansatz for $C^+(\nu, t)$.

Amplitude analysis at 6 GeV/c shows that $d\sigma_0/dt$ at $t \approx -0.15\ \mathrm{GeV}^2$ is strongly dominated by the contribution of $|B^-|^2$ (Fig. 6.5.5). If this remains approximately true at higher energies, as usually assumed, the Regge behavior of $|B^-|$ follows directly from that of the cross section data and the Regge phase follows from a general argument (Sect. A.6.4.4).

The electromagnetic corrections to P_+ and P_- are also of interest in connection with an isospin bound, which is saturated at high energies near $t = -0.04\ \mathrm{GeV}^2$ (Sect. 2.4.4.4). The bound can be derived from Eq. (A.10.9) or from the following exact relation for the transversity amplitudes

$$(1-P_-)\frac{d\sigma_-}{d\Omega} - (1-P_+)\frac{d\sigma_+}{d\Omega} = |F_-(-)|^2 - |F_+(-)|^2 = 4|F^+(-)||F^-(-)|\cos\phi$$
$$= \left[8(1-P^+)\frac{d\sigma^+}{d\Omega}(1-P_0)\frac{d\sigma_0}{d\Omega}\right]^{1/2}\cos\phi\,. \tag{6.5.24}$$

ϕ is the angle between $F^+(-)$ and $F^-(-)$.

At high energies and small $|t|$, one can neglect P^+ and a good approximation for the left hand side of Eq. (6.5.24) is

$$\left[\Delta + (P_+ - P_-)\frac{d\sigma^+}{dt}\right]\frac{q^2}{\pi}.$$

Since $P_+ - P_-$ is positive, one obtains from $\cos^2\phi \leqq 1$

$$\Delta \equiv \frac{d\sigma_-}{dt} - \frac{d\sigma_+}{dt} \leqq \left[8\frac{d\sigma^+}{dt}\frac{d\sigma_0}{dt}(1-P_0)\right]^{1/2} - (P_+ - P_-)\frac{d\sigma^+}{dt}. \tag{6.5.25}$$

The bound is valid for the strong interaction part. The electromagnetic corrections to the polarization parameters are given by (78 L Borie-2)

$$\left.\frac{P_+ - P_-}{\sqrt{-t}}\right|_{\mathrm{exp}} \approx \left.\frac{P_+ - P_-}{\sqrt{-t}}\right|_{\mathrm{strong}} + \frac{16\pi\alpha}{m|t|(1-t/4m^2)^2}\,\frac{1.79\,f_E(t)}{(\mathrm{Im}\,C^+/k)}. \tag{6.5.26}$$

A term proportional to $f_E\,\mathrm{Im}\,A^+$ has been neglected. For the magnetic form factor we have substituted $2.79\,G_{Ep}(t)$ and Eq. (1.15.7) is used for $d\sigma^+/dt \approx d\sigma_-/dt \approx d\sigma_+/dt$, neglecting a factor $(\mathrm{Re}\,C^+/\mathrm{Im}\,C^+)^2 + 1$.

Equation (6.5.25) is an *upper* bound for $P_+ - P_-$. Since the correction in Eq. (6.5.26) is positive, the isospin bound is apparently violated near $t = -0.04\ \mathrm{GeV}^2$ at high energies, if the uncorrected polarization parameters are inserted (Fig. 4.4.4).

The bound is helpful for the extrapolation of $(P_+ - P_-)/\sqrt{-t}$ to $t=0$, which is needed in amplitude analysis for the determination of B^- at $t=0$. A small violation is observed at 6 GeV/c even if the corrected polarization data are used, but this is not serious because of the uncertainty of the shape of $\Delta(t)$ (80 L HÖHLER-1).

2.6.5.4 The amplitudes of the reggeized ϱ-exchange model

As discussed above, the experimental charge-exchange data at 4···200 GeV/c are compatible with the assumption that the dominant contribution comes from reggeized ϱ-exchange. A small secondary term is sufficient to produce the polarization and some deviations (except for the deviation at $t > -1\ \mathrm{GeV}^2$ above 50 GeV/c). Even in the dip region the decrease at fixed t follows the Regge law. The largest effect of the correction occurs near the zero of the amplitude C^-.

Therefore, it is of interest to consider the amplitudes of the simple ϱ-exchange model (66 L HÖHLER-3, see also 66 L DESAI), although many authors proposed much more complicated models, which include absorptive effects described by Regge cuts (see Sect. 6.5.5).

The amplitudes for reggeized ϱ-exchange are given by Eqs. (A.9.16)···(A.9.19). It is convenient to work with dimensionless residue functions

$$b_+(t) = \frac{2\pi}{m}\gamma_+^-(t)\,; \quad b_-(t) = 2\pi[0.5+\alpha(t)]^{-1} m^2 \gamma_-^-(t)\,. \tag{6.5.27}$$

The amplitudes and the differential cross section are given by

$$C_\varrho^- = b_+(t)\frac{2\sqrt{\pi}}{m(1-t/4m^2)}\varphi(\alpha)\left(\frac{2\nu}{m}\right)^{\alpha}; \tag{6.5.28}$$

$$B_\varrho^- = b_-(t)\frac{4\sqrt{\pi}}{m^2}\alpha\varphi(\alpha)\left(\frac{2\nu}{m}\right)^{\alpha-1}; \tag{6.5.29}$$

$$\frac{\mathrm{d}\sigma_0}{\mathrm{d}t} = \frac{2m^{-4}}{1-t/4m^2}|\varphi(\alpha)|^2\left(\frac{2\nu}{m}\right)^{2\alpha-2}\{b_+^2(t)\nu^2/k^2 - (t/4m^2)\alpha^2(t)b_-^2(t)\}\,, \tag{6.5.30}$$

where

$$\varphi(\alpha) \equiv \frac{\Gamma(\alpha+\frac{3}{2})}{\Gamma(\alpha+1)}\,\frac{1-\mathrm{e}^{-i\pi\alpha}}{\sin\pi\alpha}\,; \quad \frac{1-\mathrm{e}^{-i\pi\alpha}}{\sin\pi\alpha} = \tan(\pi\alpha/2) + i = i\frac{\mathrm{e}^{-i\pi\alpha/2}}{\cos(\pi\alpha/2)}\,. \tag{6.5.31}$$

At high energies, where the approximation $\nu/k \approx 1 + t/4mk \approx 1$ is valid, the expression

$$\frac{\mathrm{d}\sigma_0}{\mathrm{d}t}\left(\frac{2\nu}{m}\right)^{2-2\alpha}\frac{1-t/4m^2}{2m^{-4}}|\varphi(\alpha)|^{-2} \approx b_+^2 - \frac{t\alpha^2}{4m^2}b_-^2 \equiv V(t) \tag{6.5.30a}$$

is independent of energy. It can be calculated from $F(t)$, Eq. (6.5.7).

Assuming a linear trajectory, the denominator in the signature factor [Eq. (6.5.31), 2nd term] can be approximated by $\sin\pi\alpha(t) \approx \pi\alpha'(t_\varrho - t)$ near $t = t_\varrho$. A comparison with pole term expressions for ϱ-exchange in $C^-(\nu,t)$ and $B^-(\nu,t)$ (Table A.8.1) leads to the following results for the dimensionless residue functions [cf. Eq. (A.9.22)]

$$\begin{aligned} b_+(t_\varrho) &= \frac{2\pi m^2\alpha'}{3}\,\frac{1}{4\pi}G_{\varrho\pi\pi}G_{NN\varrho}^{V}\left\{1+\frac{t_\varrho}{4m^2}\,\frac{G_{NN\varrho}^{T}}{G_{NN\varrho}^{V}}\right\} \approx 8.5 \pm 2.5\,; \\ b_-(t_\varrho) &= \frac{2\pi m^2\alpha'}{3}\,\frac{1}{4\pi}G_{\varrho\pi\pi}G_{NN\varrho}^{V}\left\{1+\frac{G_{NN\varrho}^{T}}{G_{NN\varrho}^{V}}\right\} \approx 30 \pm 10\,. \end{aligned} \tag{6.5.32}$$

The numerical values are obtained with $G_{\varrho\pi\pi}G_{NN\varrho}^{V}/4\pi = 2.4$ and $G_{NN\varrho}^{T}/G_{NN\varrho}^{V} = 6.6$, the uncertainty being of the order of 15% (see Sect. 2.4.2.2). The uncertainty of the ratio $b_-(t_\varrho)/b_+(t_\varrho) = 3.6 \pm 0.25$ is smaller.

$b_+(0)$ follows from the charge-exchange forward cross section or from the optical theorem and Eq. (2.1.3). Assuming $\alpha(0) = 0.5$, one obtains

$$k\left.\frac{\mathrm{d}\sigma_0}{\mathrm{d}t}\right|_{t=0} = b_+^2(0)\frac{8}{\pi m^3}\,; \quad \sqrt{k}\sigma^- = b_+(0)\frac{\sqrt{32}}{m^{3/2}}\,; \quad b_+(0) = 1.37 \pm 0.15\,. \tag{6.5.33}$$

See Figs. 2.1.7 and 2.1.8 for the errors. The same method has been used in an earlier determination of $b_{\pm}(t)$ (66 L Höhler-2 and 3).

The ratio $b_{+}(t_{\varrho})/b_{+}(0) \approx 6$ is fairly large and this is one of the reasons why many authors assumed that the no-flip amplitude is strongly modified by "absorption effects". As a consequence, the energy dependence at fixed-t should be different from that in Eq. (6.5.4) in the region where the no-flip amplitude is dominant, in particular at $t=0$.

It could happen that the deviation from a power law is not large in the range of the experimental data, although there is a substantial contribution from absorption effects to the amplitude. Then, an "effective $\alpha(t)$" is determined by a fit to Eq. (6.5.4). One would expect that the results for $\alpha(t)$ near $t=0$ and $t=-0.55\,\mathrm{GeV}^2$, where the no-flip amplitude is dominant, do not lie on a smooth curve which connects the $\alpha(t)$ values in the region of flip dominance with the values $\mathrm{Re}\,\alpha=1$ and $\mathrm{Re}\,\alpha=3$ at $t=m_{\varrho}^2$ and $t=m_{g}^2$, respectively. Figure 6.5.3 shows that there is no evidence for a behaviour of this type.

If nevertheless one insists on introducing a large absorption, one has to assume that it practically has the same energy dependence as the ϱ-exchange term, i.e. absorption has the same effect as a modification of the ϱ-exchange residue function (78 L Collins-1).

Another argument in favour of absorption is the existence of a non-vanishing polarization P_0. From the P_0-data information is obtained on the component of C^- orthogonal to B^- and this is certainly a correction to the ϱ-exchange model. However, Fig. 6.5.5 shows that the extrapolation of this component to $t=0$ is small in comparison with the forward amplitude.

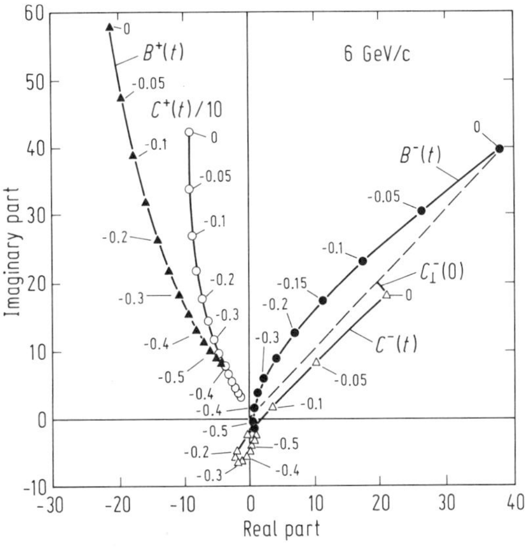

Fig. 6.5.5. Argand diagram of the isospin even and odd invariant amplitudes at 6 GeV/c as calculated from KH 78 partial waves. The t-values are given in GeV². $C_\perp^-$ (t=0) is small in comparison with $C^-(0)$, i.e. the extrapolated data for $P_0 d\sigma_0/dt/\sqrt{-t}$ do not enforce a large correction to C_{ϱ}^- at $t=0$.

Since the (real) residue functions are expected to be smooth, the rapid decrease of $b_{+}(t)$ between $t=t_{\varrho}$ and $t=0$ suggests that this function has a zero near $t=-0.15\,\mathrm{GeV}^2$, where $C_{\parallel}^-$ in Eq. (6.5.14) has its "cross-over zero". – At intermediate energies, it is well-known from partial wave and amplitude analysis that the amplitude $C^-(s,t)$ has a zero trajectory, which runs near the physical s,t-plane at an almost constant value of $t \approx -0.15\,\mathrm{GeV}^2$ (Sect. 2.4.3.3).

Since the polarization parameter P_0, for which data exist up to 40 GeV/c (80 Avvaku), does not show a dip at this t-value, it is clear that the zero of the total C^--amplitude, which includes a secondary term, lies in the complex domain, near to the real s,t-plane. In the Argand diagram (Fig. 6.5.5), $C^-(t)$ is close to its complex zero when it passes by the origin, and it is seen that this happens near the t-value where $C_{\parallel}^-$ is zero.

In the literature, it is usually pointed out that $b_{+}(t)$ cannot have a zero near $t=-0.2\,\mathrm{GeV}^2$, because this leads to difficulties with factorization. In particular, Collins (77 R, p. 204) mentions the reaction $\pi^- p \to \omega n$, which can have ϱ-exchange in the t-channel.

However, the effective Regge trajectory for this reaction in the range 20···200 GeV/c shows a *strong curvature* and does not point towards $\alpha=1$, at $t=t_{\varrho}$. Therefore, it is obvious that ϱ-exchange is not dominant in the interval $0>t>-0.5\,\mathrm{GeV}^2$. The trajectory bends towards zero as $t \to 0$. This suggests that lower lying unnatural parity exchanges become dominant (see 77 L Dahl).

In order to estimate the t-dependence of $b_-(t)$, we use the fact that $|B^-|^2$ can be determined from $d\sigma_0/dt$ in the t-range where $|C^-|^2$ is negligible, because t is near the complex zero. One obtains

$$b^-(t=-0.15\,\mathrm{GeV}^2)\approx 29\,. \tag{6.5.34}$$

A comparison with Eq. (6.5.32) shows that $b_-(t)$ remains *constant* within the relatively large errors.

It is reasonable to assume that b_- has approximately the same value at $t=0$. Then, one can decompose the total logarithmic slope of $d\sigma_0/dt$ at $t=0$ into a contribution of the derivative of $|C^-|$ and a second contribution proportional to $|B^-(t=0)|^2$ [see Eq. (A.4.17)]. The result is

$$\frac{\partial}{\partial t}\ln\frac{d\sigma_0}{dt}\bigg|_{t=0}=(16-32)\,\mathrm{GeV}^2=-16\,\mathrm{GeV}^{-2};\quad 40\,\mathrm{GeV/c}\,. \tag{6.5.35}$$

It is remarkable that the decomposition is the same as in Eq. (6.5.3), but in this parametrization the no-flip term does not have a dip at the t-value of the "cross-over zero".

Further evidence for the validity of the reggeized ϱ-exchange model follows from the fact that, according to Eq. (6.5.29), a zero of B_ϱ^- is predicted at the value $t\approx-0.55\,\mathrm{GeV}^2$, where $\alpha(t)$ vanishes. Since $|B_\varrho^-|^2$ gives the dominant contribution at and beyond the peak, the zero should lead to a dip in $d\sigma_0/dt$ and this dip is indeed seen in the data (Fig. 6.5.1).

At higher momenta the dip becomes a shoulder, but this is due to the shrinkage factor. The dip is seen more clearly in $V(t)$, Eq. (6.5.30a), which does not depend on energy.

Unfortunately, the predictive power of Regge theory for the t-dependence of the residue function $b_-(t)$ in Eq. (6.5.29) is poor. It can happen that the zero of B_ϱ^- at the "wrong-signature" point $\alpha=0$ is cancelled by a fixed pole, which stems from the presence of the third double spectral function (A_{us} in Fig. A.6.3; see Sect. 6.3c in 77 R Collins). However, the simplest interpretation of the data is to assume that $\alpha=0$ leads to $B_\varrho^-=0$, but C_ϱ^- remains different from zero.

One should note that many authors prefer to use s-channel helicity amplitudes for discussions of Regge theory, whereas the invariant amplitudes C^- and B^- correspond to t-channel helicity amplitudes [cf. Eqs. (A.2.24) and (A.2.35)].

The vanishing of B_ϱ^- at $t=-0.55\,\mathrm{GeV}^2$ makes it possible to calculate C^- at this t-value from $d\sigma_0/dt$. A smooth interpolation from $t=m_\varrho^2$ to $t=0$ and $t=-0.55\,\mathrm{GeV}^2$ leads to a decomposition of $d\sigma_0/dt$ into contributions from $|C^-|^2$ and $|B^-|^2$ (Fig. 6.5.6).

One of the corrections to the simple ϱ-exchange model follows from the fixed-t analyticity constraint, which modifies the real part (69 L Baacke, 73 L and 74 L Argyres, 73 L Devenish, 74 L Joynson, 77 L Svec, see also the Karlsruhe-Helsinki partial wave analysis, Sect. 2.1.7).

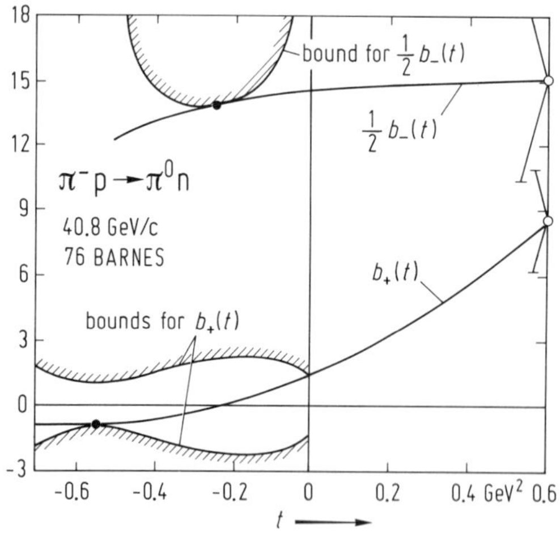

Fig. 6.5.6. A crude estimate for the contribution of the amplitudes C^- and B^- to charge-exchange scattering (66 L Höhler-3). The bounds follow from the data and Eq. (6.5.30a), putting $b_+\equiv0$ and $b_-\equiv0$, respectively. — The values at $t=t_\varrho=0.6\,\mathrm{GeV}^2$ are those of Eq. (6.5.32) (the overlap of the error bars is due to the fact that b_+ has been plotted together with $b_-/2$). — The curve for $b_+(t)$ is a smooth connection of the point near $t=0.6\,\mathrm{GeV}^2$ with the points at $t=0$ and $t=-0.55\,\mathrm{GeV}^2$, assuming that ϱ-exchange is dominant and that B^- vanishes at the zero of $\alpha(t)$. The curve for $b_-(t)$ at $t<0$ follows from the $d\sigma_0/dt$-data and $b_+(t)$.

2.6.5.5 Other models

The reaction $\pi^- p \to \pi^0 n$ has been discussed in a very large number of papers. We refer to the reviews and books 68 R COLLINS-SQUIRES, 69 R FOX, 71 R COLLINS-2, 71 R SQUIRES, 72 R SCHREMPP, 73 R FOX-QUIGG, 75 R SCHREMPP, 77 R COLLINS and 77 R IRVING-WORDEN and add only a few remarks.

As mentioned above, the main indication for the presence of a non-ϱ contribution comes from the non-vanishing polarization P_0. Unfortunately, it is difficult to detect the corresponding correction in the data for $d\sigma_0/dt$ and for the total cross section difference, because the experimental uncertainties in the 5···20 GeV/c region are too large (Figs. 2.1.7 and 2.1.8). At momenta above 50 GeV/c and $t < -0.6\,\text{GeV}^2$, a correction to the ϱ-exchange amplitudes is necessary, because the shrinkage is smaller than predicted.

One possibility to describe the non-ϱ contribution is to assume a second Regge pole ϱ′. BARGER and PHILLIPS (74 L) showed that their earlier parametrization (68 L RARITA, 69 L BARGER) can be adjusted to the data in the 5···18 and 20···100 GeV/c ranges available at that time (67 SONDER and a preliminary version of 76 BARNES) and also to the polarization data. Their ϱ′-trajectory is linear: $\alpha_{\varrho'}(0)=0$, $\alpha'_{\varrho'}=0.29\,\text{GeV}^{-2}$. Both ϱ′-residue functions are zero at $t=0$.

The introduction of an "effective pole" ϱ′ is of interest, because the exchange of a second object with the quantum numbers of the ϱ-meson is suggested by the dipole-like behaviour of the isovector nucleon form factors (Sect. 2.5.2.3).

Regge cuts

In Regge theory it is shown that one has to assume the existence of Regge cuts in order to satisfy t-channel unitarity. GRIBOV (68 L) developed a "Reggeon calculus" from which one can derive the discontinuity across J-plane cuts. However, it is not yet possible to deduce quantitative predictions for the Regge cut contributions in specific reactions (77 R COLLINS, Sect. 8.3; 75 R ABARBANEL-BRONZAN-SUGAR-WHITE).

Therefore, a variety of approximate methods has been proposed which are based on intuitive prescriptions for eikonal and absorption models. The idea of taking into account the effect of "absorption" is certainly reasonable, but it is difficult to say which of the existing phenomenological models will survive, if any.

The flexibility of the residue functions of Regge poles and of the prescriptions for the treatment of absorption is so large that a good fit to the t-dependence of the data is not impressive. Unfortunately, a large energy range and a high precision of the data are required to test the predictions for the s-dependence, because the Regge cut contributions usually depend on $\log s$.

The problems of Regge cut models are discussed in the reviews 71 R COLLINS-1, 72 R SCHREMPP, 73 R FOX-QUIGG, 77 R COLLINS, 77 R IRVING-WORDEN. A more optimistic view is presented in 73 L HARTLEY and in the article of KANE and SEIDL (76 R). It would be of interest to check whether their model is compatible with fixed-t analyticity.

The coherent droplet model

BYERS and YANG (66 L) proposed an eikonal-type description of charge-exchange scattering. The ansatz includes an absorption factor. This model has not been used in recent years, although similar geometrical pictures of elastic scattering are popular (the CHOU-YANG model, see Sect. 2.6.4.1).

The dual absorptive model

Other geometrical models have been combined with duality by HARARI et al. (71 L, 72 L). Their work is based on the ansatz (see also Sect. 2.6.4.8)

$$\text{Im}\,F_{++} \propto \text{"}J_0(R\sqrt{-t})\text{"}; \quad \text{Im}\,F_{+-} \propto \text{"}J_1(R\sqrt{-t})\text{"}; \quad R \approx 1\,\text{fm}. \tag{6.5.36}$$

The quotation marks indicate that the Bessel functions should be taken in a semiquantitative sense. An important point is that the position of the first zero is different for the no-flip and flip amplitudes. Using $R=1$ fm, one finds $t_0=-0.23\,\text{GeV}^2$ and $-0.60\,\text{GeV}^2$ respectively, in approximate agreement with the result of amplitude analysis.

There is no prediction for the real parts. In this model, they have to be calculated from fixed-t dispersion relations.

In the impact parameter representation (Sect. A. 3.3), the transforms have a peripheral peak at $b=R$. Figure 6.5.7 shows the peripheral "peak" of $\text{Im}\,F^{-}_{+-}(s,b)$ as following from the KH 78 partial wave analysis. Up to about 2.5 GeV/c, the peak belongs to the well-pronounced peripheral resonances. At higher momenta it becomes broader and flatter and approaches the shape following from the reggeized ϱ-exchange model.

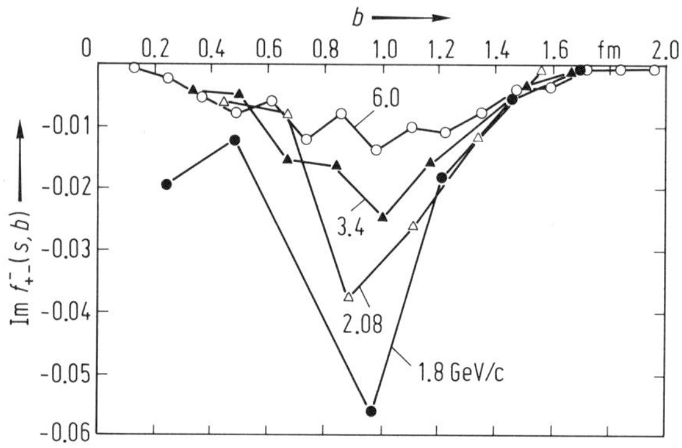

Fig. 6.5.7. Impact parameter transform of the imaginary part of the charge-exchange flip amplitude. The points have been calculated from the updated KH 78 analysis using Eq. (6.3.8).

The dual peripheral model

Several groups proposed a direct channel peripheral model as an alternative to the t-channel exchange approach (71 L Chu, 72 L Imachi and earlier papers of these authors; Schrempp 73 L-1, 73 L-2, 74 L).

The dominant s-channel resonances are peripheral (Table 2.4.1.3) and can be described by an effective trajectory

$$\mathrm{Re}\,\alpha(s)+1/2=qR\,; \qquad R\approx 1\,\mathrm{fm}. \tag{6.5.37}$$

We shall discuss only the method of B. and F. Schrempp, who assumed an analytic form for $\alpha(s)$

$$\alpha(s)\approx -\frac{R}{2\pi}\sqrt{-s}\,\log(-s/s_0)\,; \qquad |s| \text{ large} \tag{6.5.38}$$

which approximately agrees with Eq. (6.5.37) at high energies, where $s\approx 4q^2$.

The expression for an s-channel Regge pole amplitude (Sect. A.9.2) is approximated for large s and fixed negative t using

$$\frac{P_{\alpha(s)}(-\cos\theta)}{\sin\pi\alpha(s)}\approx -\mathrm{i}H_0^{(1)}\{b_0(s)\sqrt{-t}\}, \tag{6.5.39}$$

where $H_n^{(1)}$ is the Hankel function of the first kind and $b_0(s)=\{\alpha(s)+1/2\}/q$.

For large values of its argument, the Hankel function can be approximated by

$$\mathrm{i}H_n^{(1)}(b_0(s)\sqrt{-t})\approx\{b_0(s)\sqrt{-t}\}^{-1/2}\,\mathrm{e}^{-\mathrm{Im}\,b_0(s)\sqrt{-t}}\,\mathrm{e}^{\mathrm{i}\,\mathrm{Re}\,b_0(s)\sqrt{-t}}\,\mathrm{e}^{\mathrm{i}\pi/4}\,\mathrm{e}^{-\mathrm{i}n\pi/4}. \tag{6.5.40}$$

It is seen that $\mathrm{Im}\,b_0$ determines the exponential decrease of the amplitude and $\mathrm{Re}\,b_0$ is the rotating phase. The approximation is valid for $t<-0.25\,\mathrm{GeV}^2$ and $n\leqq 2$.

For $\mathrm{Im}\,b_0 \ll \mathrm{Re}\,b_0=R\approx 1\,\mathrm{fm}$, the imaginary part of the amplitude is similar to Harari's ansatz Eq. (6.5.36).

The approximation Eq. (6.5.39) leads to an interesting simplification of the impact parameter transform [Eq. (6.3.7)] of the amplitude, which belongs to an s-channel Regge pole exchange. One obtains

$$f_{++}(s,b)\propto\frac{1}{b^2-b_0^2(s)}\,; \qquad f_{+-}(s,b)\propto\frac{b/b_0(s)}{b^2-b_0^2(s)}, \tag{6.5.41}$$

i.e. the s-channel Regge pole in the J-plane leads to a pole of the impact parameter transform in the b^2-plane.

The energy dependence of the peripheral peak, for instance in Fig. 6.5.7, is described by a constant $\mathrm{Re}\,b_0\approx 1\,\mathrm{fm}$ and an increasing $\mathrm{Im}\,b_0(s)$, which gives the decrease of the peak and its broadening.

Using assumptions on the s-dependence of the residue functions, and certain approximations, B. and F. Schrempp derived a formula which is similar to the reggeized ϱ-exchange amplitude F^-_{+-} [without the factor $\alpha(t)$]. In the no-flip case, the result simulates a t-channel Regge cut with the same trajectory, which turns out to be linear in $\sqrt{-t}$. Furthermore, it is possible to deduce the "derivative relation" (71 L Högaasen, 75 L Schrempp, 75 R Schrempp)

$$F_{+-}(s,t)=C(s)\frac{\partial}{\partial\sqrt{-t}}F_{++}(s,t), \tag{6.5.42}$$

which is expressed in Eq. (6.5.41) by the factor b in the numerator of $f_{+-}(s,b)$.

The dual peripheral model was applied to many non-diffractive two-body reactions and had an encouraging success. The predictive power for the t-dependence is much larger than in the case of a t-channel Regge model.

In this section, we are only interested in the question whether the model is suitable for a *quantitative* description of the reaction $\pi^- p \to \pi^0 n$ at high energies. B. and F. SCHREMPP (74 L) had demonstrated that they could obtain a reasonable fit to the data available at that time. However, they predicted that the dip at $t=-0.6\,\mathrm{GeV}^2$ will become more pronounced at higher energies. This is in contradiction with the data 76 BARNES, according to which the dip-bump structure becomes a shoulder above 60 GeV/c (Fig. 6.5.1).

It is not clear, whether the model leads to flip dominance or to no-flip dominance at the dip ($t \approx -0.6\,\mathrm{GeV}^2$). The interesting plot of $\sqrt{-t}\,d\sigma/dt$ vs. $\sqrt{-t}$ is discussed under the assumption that the flip term is also dominating at the dip, whereas a comparison with amplitude analyses at 6 GeV/c shows a dominating no-flip term, in agreement with the usual opinion (74 L SCHREMPP). In the latter case, one would expect a difficulty of the model with the experimental t-channel Regge trajectory, because the predicted effective α for the flip term is considerably higher than that for the no-flip term.

APEL et al. (79) tested the model in the 15···40 GeV/c range as far as the ansatz for $\mathrm{Im}\,b_0(s)$ is concerned and found a good agreement. It would be of interest to check, whether the above mentioned qualitative discrepancy with the data of BARNES et al. (76) can be removed.

2.6.5.6 Superconvergence, finite energy sum rules and duality

In the sixties, the $\pi^- p$ charge-exchange amplitude played an important role for the development of the notions of superconvergence and of Finite Energy Sum Rules ("FESR"). An analogous treatment of the isospin even πN amplitude is now outdated, because it was based on the assumption of asymptotically constant total cross sections.

We start with the IGI-MATSUDA sum rule (67 L) for the charge-exchange forward amplitude $C^-(\omega)$. It is assumed that the reggeized ϱ-exchange amplitude $C_\varrho(\omega)$ [Eq. (A.9.9)]

$$C_\varrho(\omega) = 2\pi\beta\{P_\alpha(\omega/\mu) - P_\alpha(-\omega/\mu)\}/\sin\pi\alpha \tag{6.5.43}$$

is the leading term of the asymptotic expansion. It fulfills the dispersion relation

$$\{P_\alpha(\omega/\mu) - P_\alpha(-\omega/\mu)\}/\sin\pi\alpha = \frac{2\omega}{\pi}\int_\mu^\infty d\omega' \frac{P_\alpha(\omega'/\mu)}{\omega'^2-\omega^2}, \tag{6.5.44}$$

which is similar to the unsubtracted dispersion relation for $C^-(\omega)$. We consider $\hat{C} \equiv C^- - C_\varrho$. Its dispersion relation reads

$$\mathrm{Re}\,\hat{C}(\omega) = C_N^-(\omega) + \frac{2\omega}{\pi}\,⨍_\mu^\infty d\omega' \frac{\mathrm{Im}\,\hat{C}(\omega')}{\omega'^2-\omega^2}; \qquad \mathrm{Im}\,\hat{C}(\omega) = k\sigma^- - 2\pi\beta P_\alpha(\omega/\mu). \tag{6.5.45}$$

The numerical value of β follows from Eqs. (A.9.13) and (2.1.3) or from another fit of C_ϱ to the high energy behaviour of the data for $k\sigma^-$.

In order to test whether the second term of the asymptotic expansion vanishes faster than ω^{-1}, IGI and MATSUDA derived a consequence of the condition

$$\omega|\hat{C}(\omega)| \to 0 \quad \text{for} \quad \omega \to \infty. \tag{6.5.46}$$

An integral over the contour shown in Fig. A.6.7 leads in the limit of an infinite radius of the circle to the IGI-MATSUDA sum rule

$$\int_\mu^\infty \mathrm{Im}\,\hat{C}(\omega)\,d\omega = \int_\mu^\infty \{k\sigma^- - 2\pi\beta P_\alpha(\omega/\mu)\,d\omega = 4\pi^2 f^2. \tag{6.5.47}$$

The upper limit can be replaced by the energy above which the data are fitted by $\sigma^- = \mathrm{Im}\,C_\varrho/k$.

An evaluation shows that the σ^--data are not accurate enough for a good test (Fig. 2.1.7), but the sum rule is at least approximately fulfilled and this is a remarkable result.

One could be surprised that the Regge pole amplitude Eq. (6.5.43) occurs in Eq. (6.5.47), since it was derived for the asymptotic behavior and the continuation to lower energies is not unique. However, the sum rule could have as well been written for another amplitude $\hat{C} = C^- - \bar{C}_\varrho$, where $\bar{C}_\varrho$ agrees asymptotically with C_ϱ

$$\bar{C}_\varrho(\omega) = c_\varrho \omega k^{\alpha-1}\left(\tan\frac{\alpha\pi}{2} + i\right) = c_\varrho\omega(-ik)^{\alpha-1}/\cos\frac{\alpha\pi}{2}. \tag{6.5.48}$$

$\bar{C}_\varrho(\omega)$ fulfills the same dispersion relation. The parameter c_ϱ follows from β and Eq. (A.9.13).

Figure 2.1.7 shows that the transition from the resonance region to the asymptotic region (ϱ-dominance) is rather uncertain, due to experimental errors and discrepancies between two data sets. One can try to parametrize the second term of the asymptotic expansion by an effective Regge pole, which is usually called ϱ'.

$$\operatorname{Im}\tilde{C}(\omega)=c_{\varrho}\omega k^{\alpha_{\varrho}-1}+c_{\varrho'}\omega k^{\alpha_{\varrho'}-1}. \tag{6.5.49}$$

We assume that $\omega|C^- - \tilde{C}|\to 0$ in the high energy limit and find the sum rule

$$4\pi^2 f^2=\int_{\mu}^{\infty}(\operatorname{Im}C^- - c_{\varrho}\omega k^{\alpha_{\varrho}-1}-c_{\varrho'}\omega k^{\alpha_{\varrho'}-1})\mathrm{d}\omega$$

$$\approx\int_{\mu}^{\Omega}k\sigma^-\mathrm{d}\omega - c_{\varrho}\frac{\Omega^{\alpha_{\varrho}+1}}{\alpha_{\varrho}+1}-c_{\varrho'}\frac{\Omega^{\alpha_{\varrho'}+1}}{\alpha_{\varrho'}+1}. \tag{6.5.50}$$

Ω is the energy above which the data are fitted by the parametrization Eq. (6.5.49). (We have neglected μ^2 in comparison with Ω^2.) $\alpha_{\varrho'}$ can be <-1 or >-1.

One can test the sum rule Eq. (6.5.50) by fitting the data to Eq. (6.5.49) under the constraint that the sum rule is valid. The result is shown in Fig. 2.1.7.

Equation (6.5.50) is an example of a "Finite Energy Sum Rule" (67 L Logunov). A detailed investigation of these sum rules was performed by Dolen, Horn, and Schmid (68 L), who considered both amplitudes C^- and B^- and t-values different from zero.

The FESR can be generalized by considering $\omega^{2n}C^-(\omega)$ instead of $C^-(\omega)$ or by using modified dispersion relations (Sect. A.6.4.2) which lead to "Continuous Moment Sum Rules" (CMSR) [Liu and Okubo (67 L)].

The different sum rules played an important role in earlier amplitude analyses, since phase shifts can be used for the determination of residue functions of the Regge poles in a first approximation (68 L Dolen). Furthermore, the sum rules can be used as a constraint in fits to the high energy data [Barger and Phillips (69 L)].

Nowadays, the invariant amplitudes in a limited t-range near the forward direction can be determined by a simultaneous fit to data at all energies, if Pietarinens expansion is used (Sect. 2.1.16). The direct application of the analyticity constraint is not only more transparent. One avoids a bias due to the choice of the parametrization of the residue functions and difficulties resulting from the cut-off at a relatively low momentum. Constraints from fixed-t dispersion relations were also used in 73 L and 74 L Argyres and 75 R Contogouris.

It can still be of interest to study superconvergence sum rules similar to that of Igi and Matsuda. An interesting question is for instance the following. According to the dispersion relation, Eq. (A.6.35), each resonance gives a contribution to the dispersion integral, which decreases as $1/\nu$ at high energies. Therefore, one could expect a contribution of the order $1/\nu$ in the high energy expansion of $\operatorname{Re}C^-(\nu,t)$.

The most accurate data are available at $t=0$. A discussion similar to that for C^+ in Eq. (2.1.11) leads to the result that the coefficient of $1/\omega$ is zero within the errors (80 R Höhler-Kaiser, Sect. 9.6.1).

This remark is related to the notion of "duality", which was derived by Dolen et al. (68 L) from a study of Finite Energy Sum Rules. In Eq. (6.5.50), one can add further Regge poles and approximate the integral by a sum over resonance contributions. Then the sum rule leads to the statement that the sum over s-channel resonance terms (including the nucleon exchange term $4\pi f^2$) is equal to the sum over t-channel Regge pole terms.

Jakob (75 L) investigated consequences of the assumption that the t-channel exchanges in πN charge-exchange scattering are "dual" to the *inelastic parts* of the s- and u-channel resonances as suggested by the decomposition of Cini and Fubini (61 R Cini-Vitale) (Fig. A.5.1).

Ferrari et al. (73 L and 74 L) studied duality for the pion-nucleon forward amplitude in the complex energy plane and Arci et al. (80 L) reconsidered the Finite Energy Sum Rule for the amplitude B^-.

During a certain period the notion of "duality" was very popular, but it turned out that it is difficult to give a precise definition and to perform a quantitative test. For instance there is no unique way in which a resonant part of the amplitude can be separated from the background. The mathematical content of "duality" has been discussed by Oehme (70 L-2).

A special model which fulfills the requirements of duality was constructed by Veneziano (68 L) for the reaction $\pi\pi\to\pi\omega$. An adaptation to $\pi\pi$-scattering was proposed by Lovelace (68L) and a smoothed version by Burkhardt et al. (70 L). Extensions to πN scattering were attempted by many authors, but a quantitative success was not achieved (69 L Berger, 69 L Chu, 69 L Lovelace, 70 L Fenster, 70 L Hendry, 68 L Igi, 73 L Pal). However, one should keep in mind that the zero trajectory structure of "experimental" invariant πN amplitudes has a remarkable similarity with predictions from Veneziano-type models (73 L Odorico, see Sect. 2.4.3).

As far as "duality" is concerned, it seems that it is not a basic law of strong interaction processes and it is difficult to find successful quantitative applications. Duality is useful as a "mnemonic for many of the basic facts of two-body processes and as a simplifying model for more complex ones" (77 R COLLINS). A detailed discussion can also be found in 69 R HORN, 70 R KUGLER, 71 R OEHME-3, 77 R IRVING-WORDEN. FUKUGITA and IGI (77 R) presented a more optimistic view of duality. They covered all aspects of this notion and gave a long list of references (more than 500 entries). A detailed review of dispersion sum rules is due to FERRO-FONTAN, QUEEN and VIOLINI (72 L). Their list of references has more than 600 entries. A considerable part of the results is out-dated, because the increase of the total cross sections at high energies was not yet known at that time.

2.6.5.7 Charge-exchange scattering at intermediate energies

In recent years, several accurate π^-p charge-exchange scattering experiments have been carried out in the range 0.6···4 GeV/c (76 BROWN, 78 BROWN, 79 MINOWA, 79 SUZUKI, 80 ADACHI). The data of the Rutherford group have been used in the CMU-LBL 79 and KH 78 partial wave analyses, whereas the data of the KYOTO group at the KEK accelerator have been taken into account only in preliminary updated versions.

In both analyses the inclusion of the data of BROWN et al. has led to difficulties in the attempts to fit simultaneously the elastic scattering data and the dispersion constraints. Therefore, we have performed a detailed analysis of the charge-exchange data, using Legendre expansions for $d\sigma/d\Omega$ and for $P(d\sigma/d\Omega)/\sin\theta$ and calculating the zero trajectories from the Legendre expansions as well as from the partial wave solution (79 L HÖHLER-1).

It turned out that some of the Legendre coefficients show a strange behaviour. For instance the coefficient A_{10} of $d\sigma/d\Omega$ has a peak at $k = 1.0$ GeV/c (Fig. 6.5.8). This could be due to an unexpected large contribution of the $J = 15/2$ partial wave (or to a systematic experimental error). Furthermore, the zero trajectories are not as smooth as they should be, in particular at 0.6···1 GeV/c.

A test of the isospin bounds, Eq. (A.10.9), led to the result that they are compatible with the data.

Unfortunately, the constraints from forward and backward dispersion relations are not as useful as they could be, because the data in the angular range $-0.95 < \cos\theta < 0.95$ cannot be extrapolated in a reliable way to $\cos\theta = \pm 1$ due to rapid variations of the cross sections at $|\cos\theta| > 0.95$ in part of the momentum range.

New accurate data in the range of the near-forward and near-backward peaks and dips would also be important for an improved determination of the higher partial waves.

Charge-exchange forward scattering at all energies has been discussed in Sect. 2.2.1.

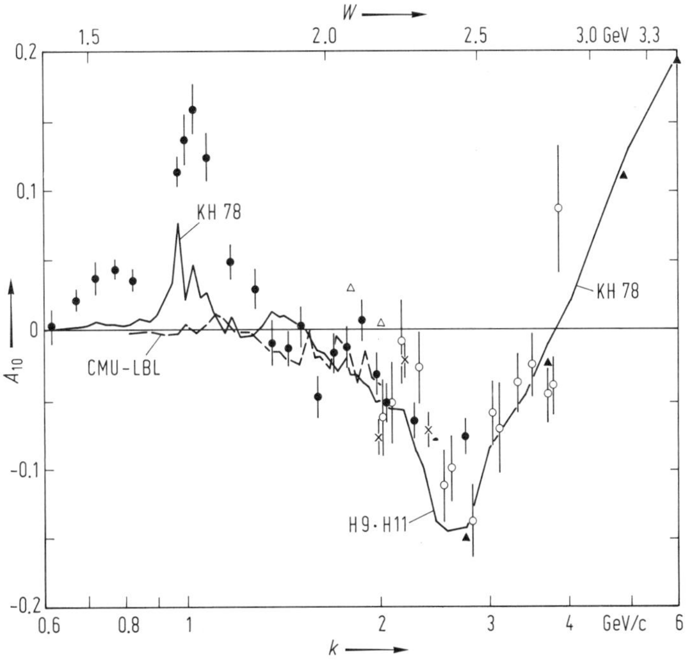

Fig. 6.5.8. Momentum dependence of the Legendre coefficient A_{10} in charge-exchange scattering. Solid line: calculation from the KH 78 analysis; dashed line: calculation from the CMU-LBL 79 analysis. Values from differential cross section data: ● 76 BROWN; × 73 NELSON; △ 75 FELTES; ▲ 68 YVERT; 70 BORIGH; 74 BROCKE; 75 MARZO; ○ from the experiment 80 CROUCH. The large dip is mainly due to the interference of H 9 and H 11 waves. The data 76 BROWN lead to an unexpected peak at 1 GeV/c. (79 L HÖHLER-1).

2.6.6 Wide-angle scattering

2.6.6.1 Experimental results on wide-angle scattering at high energies

In elastic scattering at high energies, the angular range between the forward and backward peaks is called the "wide-angle" (or "large-angle") region. Experimental data for π^-p scattering are available up to 30 GeV/c (80 ALMAS). In two recent experiments data at many momenta have been measured at 2···9 GeV/c (78 JENKIN) and at 5.8···13 GeV/c (80 BAILLO). Furthermore, there are earlier data at 5 GeV/c (70 BRABSO, 73 EIDE), 6.2 GeV/c (76 BURAN), and 5.9 and 9.8 GeV/c (69 OWEN) (Fig. 6.6.1).

In π^+p scattering, there are data at 20 GeV/c (80 ALMAS), 10 GeV/c (75 BAGLIN), 1.9···6 GeV/c (78 JENKIN) and 5 GeV/c (70 BRABSO, 73 EIDE). New $\pi^{\pm}p$ data at 50 GeV/c up to $|t|=8\,\mathrm{GeV}^2$: ASAD.

Wide-angle data for charge-exchange scattering exist only up to 5.9 GeV/c (74 BROCKE, Fig. 6.6.2).

The shape of the elastic π^-p cross sections at wide angles is shown in Fig. 6.6.1. Because of the scaling law, Eq. (6.6.1), we have plotted in Figs. 6.6.3a and b the energy dependence of $d\sigma/dt$ at fixed t/s. At intermediate energies, both figures show broad structures.

At and below 2.5 GeV/c, these structures are caused by the well-known nucleon resonances. In partial wave analyses at higher momenta, the structures of $d\sigma/dt$ also lead to a resonance-like behaviour in some of the partial waves, but the magnitude of the loops in the Argand diagrams becomes gradually smaller as the energy increases. The candidates for the highest resonances are included in Table 2.4.1.4.

As discussed in Sect. 2.4.3, the resonances are closely related to zero trajectories. One obtains some information on the projection of these trajectories onto the physical s,t-plane by studying the valleys in a three-dimensional plot of $d\sigma/dt$ above the s,t-plane (73 L ODORICO). The projections of the minima at fixed s for many s-values onto the s,t-plane have a tendency to run along t=const or u=const, but there are perturbations around points where two of these lines intersect each other (see Sect. 2.4.3.2).

The most impressive dip starts at $k\approx 2$ GeV/c in the backward direction (Figs. 6.7.1 and 6.7.2). It remains roughly at $t=-3\,\mathrm{GeV}^2$ in a large momentum range at least up to 14 GeV/c (see Figs. 6.3.12, 6.6.1, and Fig. 13 in 80 L BAILLON). A systematic study of the dip structure is difficult because of gaps in the data and large discrepancies between different experimental results (see for instance Fig. 6.3.13).

BAILLON et al. (80 L) spent a great effort in searching for narrow resonances between 5.8 and 13 GeV/c, but the result was negative.

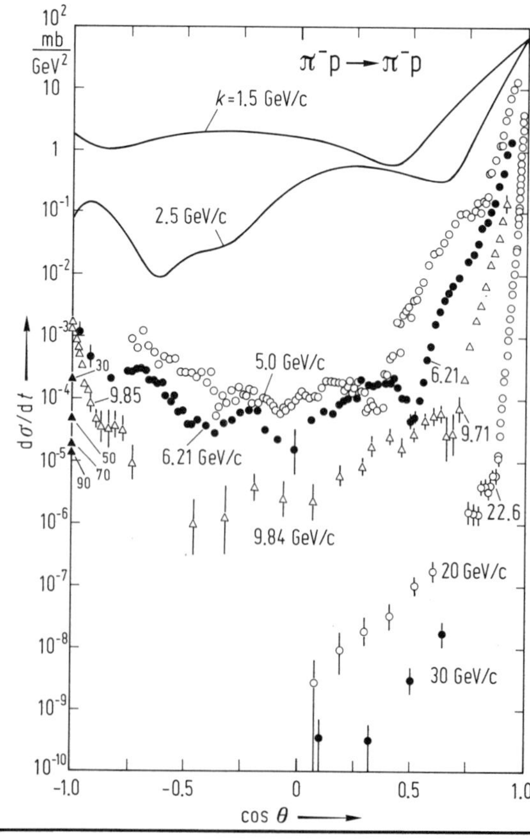

Fig. 6.6.1. π^-p elastic scattering cross sections at high energies and all angles. Solid lines: reconstruction from partial wave analysis (KH 78). 5 GeV/c: 73 EIDE; 6.21 GeV/c: 76 BURAN; 9.71, 9.84, and 9.85 GeV/c: 69 OWEN; 20 and 30 GeV/c: 80 ALMAS; 22.6 GeV/c: 73 CORNIL; backward points: 79 BAKER.

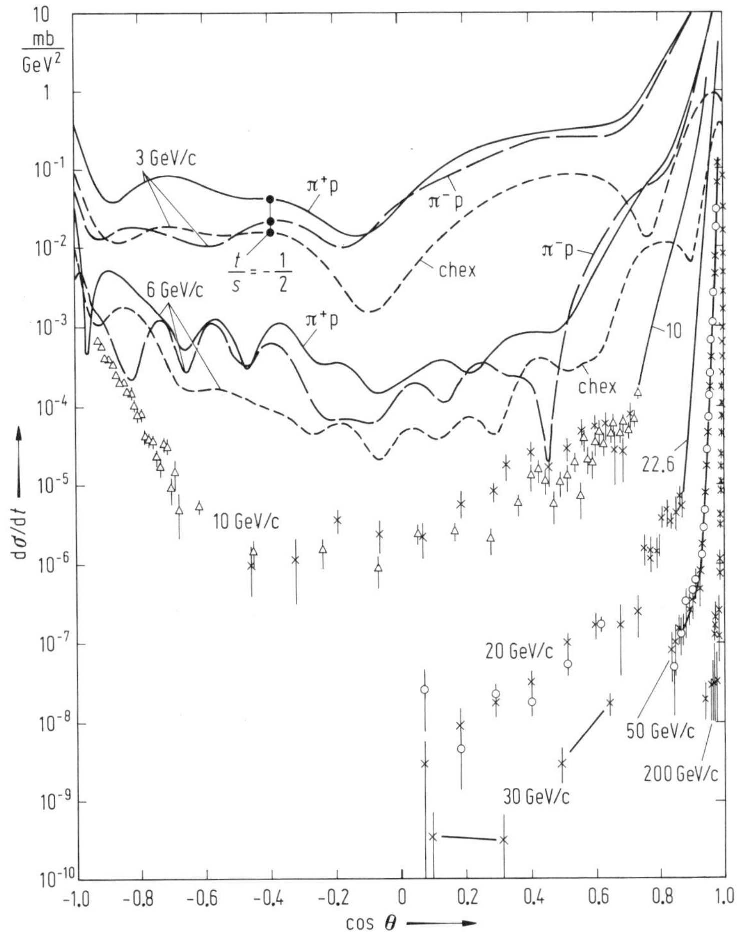

Fig. 6.6.2. Comparison between $\pi^{\pm}$p elastic and π^-p charge-exchange cross sections at high energies and all angles. The curves at 3 and 6 GeV/c belong to the updated KH 78 partial wave solution. π^-p elastic scattering data: ⚹ see Fig. 6.6.1. In addition we have plotted at 50 GeV/c the data 82 Asad and at 200 GeV/c the data 81 Baker (taken from their graph). π^+p data: △ 75 Baglin (10 GeV/c), ○ 82 Asad (50 GeV/c). In regions where the data points lie very close to each other, they have been replaced by a line. — The wide-angle charge-exchange cross sections at 3 GeV/c and 6 GeV/c follow the data sets 79 Suzuki and 74 Brocke, respectively. — It is seen that in the wide-angle region the three cross sections have the same order of magnitude. — The structures in the cosθ-dependence at fixed s are related to those in Figs. 6.3.11 and 6.6.3. $t/s = -1/2$, which belongs asymptotically to $\theta = 90°$, is marked at 3 GeV/c.

2.6.6.2 Hard scattering models

It is assumed that the basic processes in wide-angle scattering are pointlike interactions between the constituents of the hadrons. The *Dimensional Scaling Laws* (or *Constituent Counting Rules*) have been derived from the assumption that in high energy scattering the hadrons behave as collinear systems of constituents and that the interaction between these constituents, which leads to wide-angle scattering, is scale-invariant (Brodsky 73 L and 75 L, 73 L Matveev).

In the special case of pion-nucleon scattering the prediction reads

$$\frac{d\sigma}{dt} = s^{-8} f(t/s); \qquad t/s = -\frac{4q^2}{s}\frac{1}{2}(1-\cos\theta). \tag{6.6.1}$$

At very high energies, fixed t/s corresponds to fixed c.m. scattering angle θ, but one should note that the limit $4q^2/s \to 1$ is approached only slowly. $t/s = -1/2$ belongs to $\theta = 90°$ in the high energy limit, to $\theta = 102°$ at 5 GeV/c and to 123° at 2 GeV/c. Nevertheless the comparison of Eq. (6.6.1) with experimental data is usually made at fixed angle.

Figure 6.6.3 shows that the prediction is fulfilled in a zero order approximation, since $d\sigma/dt$ varies between 5 and 30 GeV/c by almost a factor of 10^6. However, the data show a significant structure, which is the continuation of the structure due to resonances and zero trajectories at lower momenta.

One should note that our figures start at 3 GeV/c in a t-region, where one expects the Regge model to be valid. It is difficult to estimate where the hard scattering mechanism becomes dominant (Fig. 6.6.4). This problem has been investigated by BLANKENBECLER et al. (73 L, 74 L). See also the review 76 R SIVERS-BRODSKY-BLANKENBECLER and 78 L COON.

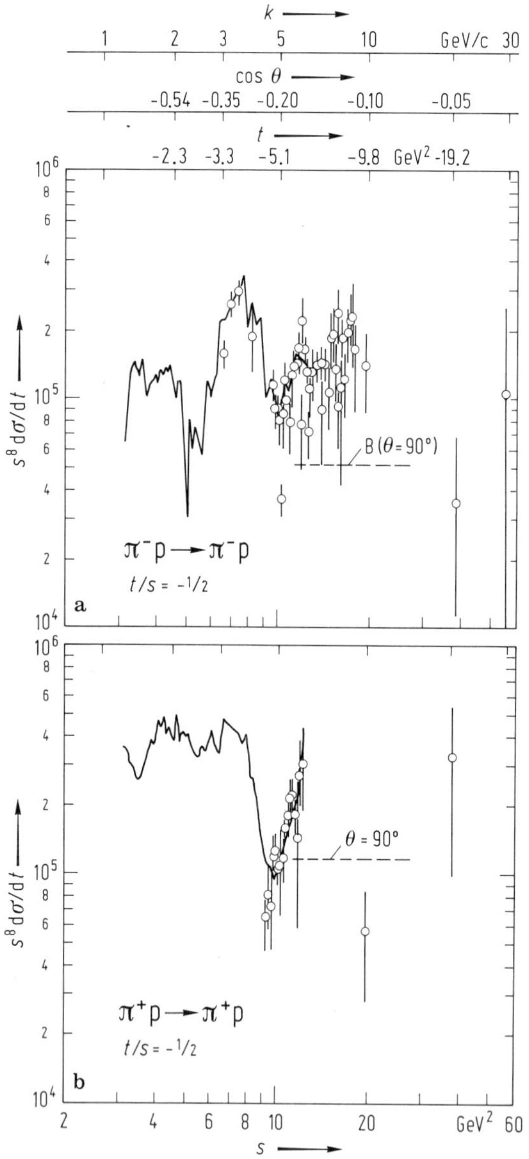

Fig. 6.6.3. $\pi^\pm p$ cross sections at fixed $t/s = -0.5$. In the range of most of the experimental data the plots differ considerably from plots at fixed angle $\theta = 90°$, because the second equation in Eq. (6.6.1) approaches its asymptotic behaviour only slowly (see the $\cos\theta$-scale). — The t-scale shows that most of the wide-angle data also belong to the "large-t" region discussed in Sects. 2.6.3.7 and 2.6.4.12. — In the plots of $d\sigma/dt$ at fixed-t (Fig. 6.3.11), the data of the present figures lie in the range of the rapid decrease. — The solid lines below 6 GeV/c have been calculated from the updated KH 78 partial wave analysis. Data at high energies: 20 and 30 GeV/c: 80 ALMAS; 10 GeV/c, $\pi^+ p$: 75 BAGLIN; 9.84 GeV/c, $\pi^- p$: 69 OWEN. Most of the other data are from 78 JENKIN; see Table 1.2.2 for a complete list. The large $\pi^- p$-data set 80 BAILLO is not yet available to us. See the figures in 80 L BAILLON. — The horizontal dashed line denoted by "B" is the fit to the data of BAILLON et al. at $\theta = 90°$, Eq. (6.6.4) (80 L BAILLON). The dashed line in the lower figure is the prediction for the $\pi^+ p$-data at 90° from this fit and Eq. (6.6.2).

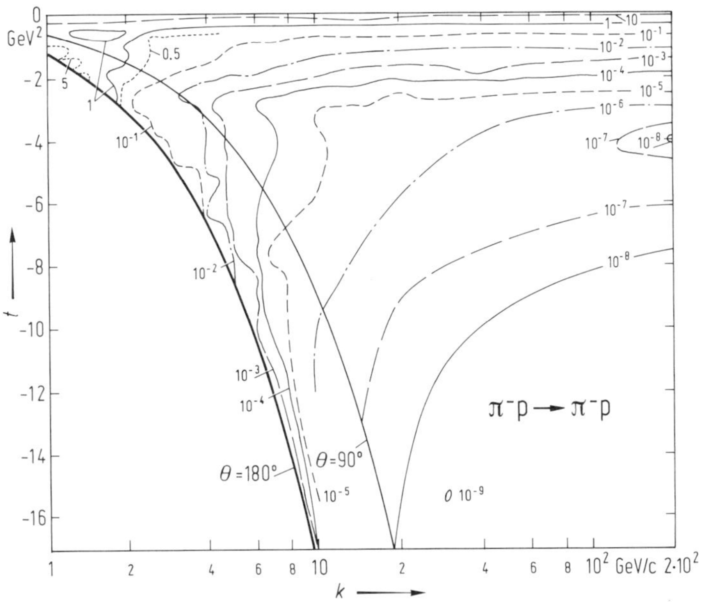

Fig. 6.6.4. Contour lines of π^-p elastic cross sections (in mb/GeV²) at high energies. The figure shows the relations between structures in the energy dependence at fixed t, fixed u and fixed θ.

A prediction for the function $f(t/s)$ can be made if one specifies the mechanism of the interaction. In the *Constituent Interchange Model* ("CIM"), Gunion, Brodsky, and Blankenbecler (73 L) assumed that the scattering proceeds by an interchange of quarks between the colliding particles. At high energies, where $f(t/s)$ is a function of $z=\cos\theta$, the prediction reads

$$s^{-8}\frac{d\sigma}{dt}=\sigma_0\frac{1+z}{(1-z)^4}[4\alpha(1+z)^{-2}+\beta]^2 \quad \text{for} \quad \pi^+p\to\pi^+p;$$
$$=\sigma_0\frac{1+z}{(1-z)^4}[4\beta(1+z)^{-2}+\alpha]^2 \quad \text{for} \quad \pi^-p\to\pi^-p. \qquad (6.6.2)$$

For the standard quark configuration the parameters have the values $\alpha=2$, $\beta=1$.

In this model, s-channel helicity is conserved at high energies and fixed θ. The invariant amplitudes are real. They are related by $B\approx -A/m\approx C/2m$ and B gives the dominant contribution to F_{++}, Eq. (A.2.29), and to $d\sigma/dt$

$$\frac{d\sigma}{dt}\approx\frac{1}{16\pi}\frac{-u}{s}|B|^2; \qquad B_+=\frac{1}{t^2}\left(\frac{\alpha}{u^2}+\frac{\beta}{s^2}\right). \qquad (6.6.3)$$

An earlier derivation of s-channel helicity conservation in wide-angle scattering at high energies is due to Gross and Wess (70 L), who studied consequences of conformal invariance.

A comparison of the CIM-prediction with $\pi^\pm$p scattering data has been made by the authors of recent experiments in the wide-angle scattering region (78 Jenkins, 80 L Baillon, 80 Almas). The result of Baillon et al. for the only adjustable parameter σ_0 in Eq. (6.6.2)

$$\sigma_0=(560\pm 20)\text{mb GeV}^{14}=\frac{s^8}{36}\frac{d\sigma_-(90^\circ)}{dt} \qquad (6.6.4)$$

agrees approximately with the values derived from the data sets 80 Almas and 75 Baglin, but Baillon mentions a "complete disagreement" with the fit of Jenkins et al. (78). Details of this discrepancy cannot be shown, because Baillon et al. have not yet distributed a table of their data.

The ratio of the π^-p and π^+p cross sections does not agree with the prediction. As far as the data of Jenkins et al. are concerned, one can argue that the CIM-prediction is not valid below 6 GeV/c.

Another model for wide-angle scattering has recently been proposed by PREPARATA and SOFFER (80 L). The authors considered the exchange of mesons and of baryons in a quark model, generalizing a method proposed in an earlier paper (79 L) for nucleon-nucleon scattering. This model also predicts that the wide-angle amplitudes are real at very high energies. The fit to the data at 10···30 GeV/c is not much better than that to the CIM-prediction, although the model of PREPARATA et al. has three adjustable parameters. It is remarkable that the cross section varies as s^{-9} instead of s^{-8}. In contrast to other models the helicity flip amplitude is not negligible.

Since 1978 there has been an increasing activity in attempts to describe wide-angle scattering (and form factors at large negative t) by perturbative Quantum Chromodynamics ("QCD"). Since the incoming and outgoing particles are hadrons, it is necessary to include some assumptions on the treatment of non-perturbative bound state effects. A summary of the present status can be found in BRODSKYs lectures at the KfK summer school (82 R) and in the reviews 82 R LEPAGE, 82 R SIVERS. Details are given in 80 L LEPAGE; see also 78 L BLANKENBECLER, 79 L PARISI, 80 L LANDSHOFF.

A systematic treatment of wide-angle scattering processes in QCD is very difficult, one of the reasons being that one has to consider a very large number of Feynman diagrams for quarks and gluons. The approach of BRODSKY and LEPAGE leads to a relatively small modification of the Counting Rule Eq. (6.6.1) by the occurrence of the QCD coupling constant α_S which has a logarithmic t-dependence. The total hadron helicity is conserved up to corrections of order $m/\sqrt{-t}$.

It is obvious that, due to the large number of diagrams, quantitative predictions for the angular dependence, like Eq. (6.6.2), are not yet possible. The small oscillatory deviations from the Counting Rule prediction, Eq. (6.6.1) (see below), are not explained. It is not yet clear whether the rapidly varying spin correlations observed in polarized pp-scattering can be understood in this model.

See also the discussion of the large-t behaviour in Sect. 2.6.4.12.

2.6.6.3 Geometrical models

OREAR (64 L) noticed that the wide-angle data for pp-scattering as well as the few data available at that time for πp scattering depend on energy and angle in such a way that the cross section as a function of the scaling variable $p_T = q\sin\theta$ (transverse momentum) is given by

$$\frac{d\sigma}{dt} = A e^{-ap_T}, \tag{6.6.5}$$

where $a \approx 1.2$ fm, $1/a \approx 160$ MeV/c.

More accurate data showed that in both cases the exponential law gives only the average behaviour and that "break" structures are superimposed. HENDRY (74 L) found evidence for the statement that the breaks occur at fixed values of t. For πp scattering, the location agrees with the zeros of the Bessel function $J_1(R\sqrt{-t})$, where $R = 0.9$ fm.

B. and F. SCHREMPP (75 L-2 and 75 R) generalized their dual peripheral model (Sect. 2.6.5) for applications in wide-angle scattering by substituting p_T for $\sqrt{-t}$. The argument is that p_T is approximately equal to $\sqrt{-t}$ and $\sqrt{-u}$ for near-forward and near-backward scattering, respectively

$$p_T^2 = q^2\sin^2\theta = -t(1+t/4q^2) = (-t/4q^2)(u_b - u) \to \frac{tu}{s} \quad \text{as} \quad s \to \infty. \tag{6.6.6}$$

u_b is the backward value of u. It is small at high energies [Eq. (A.1.32)]. They obtain Eq. (6.6.5), modified by a modulating factor $(1 + C\cos(Rp_T + \varphi))$, similar to that following from the rotating phase in Eq. (6.5.40).

It turns out that in pp-scattering the earlier "breaks" are well described by the oscillations deduced from the rotating phase (data are shown up to $p_T = 3.0$ GeV/c).

In a subsequent paper, B. and F. SCHREMPP (80 L) concluded that a modified OREAR-type law

$$\frac{d\sigma}{dt} = A\frac{1}{p_T} e^{-ap_T} \tag{6.6.7}$$

is valid in pp-scattering even in the FNAL-ISR range for $p_T \gtrsim 1.5$ GeV/c as far as data are available (up to $p_T = 3.8$ GeV/c). However, in the region $2.0 < p_T < 3.8$ GeV/c the data can *as well* be described by a power-law: $d\sigma/dt \propto (p_T^2)^{-N}$, where N = 8 to 10, similar to Eq. (6.6.1).

The new model describes scattering as a tunneling phenomenon, which takes place at the interface between the highly excited hadronic matter and the surrounding vacuum. The power law follows from the geometry of the stringlike "confining walls" and not from the interactions of pointlike constituents.

Another geometrical approach was proposed by ELVEKJAER and J. L. PETERSEN (75 L), who started from their modified impact parameter representation of amplitudes, which are considered as functions of the longitudinal and transverse momentum p_L and p_T (Sect. A.3.3.3). They noticed that the ansatz Eq. (6.6.5) has a transform with singularities on the imaginary b-axis at $b = \pm iR$, where $R = a/2$.

A more general ansatz, which has several poles in the b^2-plane, was studied by MARTIN and PESCHANSKI (78 L-2). See also Eq. (6.5.41).

A comparison with new data shows that the exponential law Eq. (6.6.5) agrees with the cross sections at $\theta = 90°$ ($p_L = 0$) only up to about $p_T = 2$ GeV/c, expect for the above mentioned oscillatory structure. The data 80 ALMAS extend the range up to $p_T = 3.7$ GeV/c (at 30 GeV/c). At this p_T-value, the prediction from the exponential law Eq. (6.6.5) is too low by 3···4 orders of magnitude, i.e. it is not tenable.

The *quasipotential approach* was developed by LOGUNOV and TAVKHELIDZE (63 L), who constructed a velocity-dependent generalized complex potential for hadron-hadron scattering, using methods of quantum field theory. A review of applications to high energy scattering can be found in TAVKHELIDZEs report (72 R) and a detailed application to πN scattering in 73 L DZHGARKAVA.

Subsequent work is mentioned by GOLOSKOKOV (78 L, 79 L, 81 L, 82 L), who applied this quasi-optical approach to wide-angle πN scattering. Their quasipotential consists of two terms: a long range component described by a Gaussian whose parameters are fixed by the small-angle scattering data, and a short range potential for the hard interactions, chosen to guarantee the asymptotic validity of the Constituent Counting Rule.

For pp-scattering, a two-component model for diffraction and the hard scattering mechanism has recently been proposed by HEINES and ISLAM (81 L) (see Sect. 2.6.4.5).

2.6.6.4 Statistical models

The basic ideas of the statistical model are due to FERMI (50 L). It is assumed that the incoming particles form a highly excited intermediate state, which comes to an equilibrium and decays into many channels in proportion to phase space.

The connection with a similar situation in nuclear physics has been reviewed by FRAUTSCHI (72 L). In πN scattering, the level density of the nucleon resonances is increasing with energy. Above 2 GeV/c, one has a situation in which many resonances are overlapping within their widths. Dips and peaks in differential cross sections are assumed to be due to fluctuations in the level density and in the coupling strengths and not to individual resonances.

FRAUTSCHI suggested that the peaks and dips in πN elastic scattering between 1.5 and 5 GeV/c can be interpreted as "ERICSON fluctuations" (66 R ERICSON-MAYER-KUCKUK) from which one can derive information on the level densities. The usual assumption is an exponential increase with $\sqrt{s}$: $\varrho(s) \propto \exp(\sqrt{s}/T_0)$. The "hadronic temperature" T_0 is of the order of 150 MeV.

A detailed statistical model for πN scattering at $\theta = 90°$ was proposed by EILAM, GELL, MARGOLIS, and MEGGS (73 L). The authors assumed an incoherent superposition of a large number of resonance contributions. They used a statistical bootstrap model for the couplings and for the density of states, assuming that ϱ increases exponentially in $(\sqrt{s}/T_0)$. The prediction is similar to Eq. (6.6.5)

$$\left.\frac{d\sigma}{dt}\right|_{90°} \propto \exp(-\sqrt{s}/T_0) \propto 1/\varrho(s), \tag{6.6.8}$$

since $s \approx 4q^2$ at high energies. Therefore, the prediction for the 30 GeV/c data 80 ALMAS is too low by several orders of magnitude.

Several groups of authors made an attempt to search directly for ERICSON fluctuations in πN scattering data.

SCHMIDT et al. (73 L) analyzed the 5 GeV/c $\pi^{\pm}$p data 73 EIDE as follows. The elastic events were divided into two momentum groups, which are separated by 15···20 MeV in the c.m. system. At each value of t, an asymmetry parameter was defined by

$$A(t) = \frac{(d\sigma/dt)_u - (d\sigma/dt)_l}{(d\sigma/dt)_u + (d\sigma/dt)_l}, \tag{6.6.9}$$

where the index "u" or "l" denotes the cross sections in the upper and lower momentum bins, respectively.

Plots of $A(t)$ for π^+p- and π^-p-scattering show that the π^-p-data are consistent with $A=0$, whereas a significant asymmetry is found in the π^+p data in the range $-3>t>-7\,\mathrm{GeV}^2$. The cross section varies up to a factor of 3 in the small range $\Delta\sqrt{s}=30$ MeV corresponding to $\Delta k=100$ MeV/c.

A much more detailed investigation of the asymmetry has recently been performed by JENKINS et al. (78). At 5 GeV/c, $A(t)$ has been determined for an interval $\Delta\sqrt{s}=30$ MeV. There is a clear disagreement with the result of SCHMIDT et al. (73 L).

JENKINS et al. examined their large data set and found that significant asymmetries occurred in general only for the larger c.m. energy interval 100 MeV, corresponding to $\Delta k\approx 330$ MeV/c at 5 GeV/c. In contrast, there is no asymmetry in the pp-data of this experiment.

JENKINS et al. also calculated the correlation function which is defined by ($\sigma\equiv d\sigma/dt$)

$$C=\left\langle\frac{(\sigma-\sigma_{sm})^2}{\sigma_{sm}^2}\right\rangle. \tag{6.6.10}$$

σ_{sm} is an estimate of the smooth non-fluctuating part of the cross section. First the correlation is calculated using fits to constant t data. Then, for each energy interval, the correlation functions of each-t-value are averaged. The average is taken over an energy range $\Delta\sqrt{s}$ which is large in comparison with the widths of the resonances. There are significant correlations, but the result differs strongly from FRAUTSCHIs prediction (72 L).

Conclusion

There exist fluctuations and correlations in the wide-angle $\pi^\pm$p scattering data above the energy range of the well-established resonances, but they differ strongly from the predictions of statistical models. Presumably one of the reasons is that the resonances have a tendency to occur in groups, which have the same position of the complex pole (Table 2.4.1.7). Furthermore, the fact that important resonances lie on Regge trajectories has an influence on the level density. — The prediction of an exponential decrease at $\theta=90°$ is not tenable.

2.6.7 Near-backward scattering

2.6.7.1 Experimental data for near-backward scattering at high energies

Above 20 GeV/c, near-backward elastic $\pi^\pm$p scattering has been investigated at Serpuchov (72 BABAEV, 77 BABAEV) and at Fermilab (79 BAKER). In the neighborhood of $u=0$, the data are usually fitted to an exponential parametrization

$$\frac{d\sigma}{du}=\frac{d\sigma(u=0)}{du}\,e^{Bu}=\frac{d\sigma(180°)}{du}\,e^{B(u-u_b)}, \tag{6.7.1}$$

where

$$u_b=\frac{(m^2-\mu^2)^2}{s};\qquad (m^2-\mu^2)^2=0.741\,\mathrm{GeV}^4. \tag{6.7.2}$$

In the high energy limit, u_b approaches 0. At 6 GeV/c, $u=0$ belongs to the lab. scattering angle $\theta_\pi=148°$, the c.m. scattering angle $\theta=171°$, $\cos\theta=-0.9885$ and $u_b=+0.061\,\mathrm{GeV}^2$.

Experimentally, the exponential decrease is observed in a range $u_b>u\gtrsim-0.1\,\mathrm{GeV}^2$ in π^+p scattering. The range is limited by the fact that the π^+p differential cross section has a dip at $u\approx-0.15\,\mathrm{GeV}^2$, which occurs in a very large momentum range at an approximately constant value of u (72 L ODORICO). It is clearly visible at all momenta above 1.2 GeV/c up to 70 GeV/c.

In π^-p-scattering, the exponential decrease is valid in a larger range, which extends at least up to $u=-0.4\,\mathrm{GeV}^2$ at 30 GeV/c and up to $-0.70\,\mathrm{GeV}^2$ at 10 GeV/c. A structure near $u=-1.0\,\mathrm{GeV}^2$ was found in the experiments 69 OWEN, 77 JACHOL, and 81 GHIDIN.

The parameters of exponential fits to the data at high energies are listed in Tables 2.6.7.1 and 2.6.7.2.

Figures 6.7.1 and 6.7.2 show $k^2 d\sigma/du$ as a function of the pion lab. momentum k for $u=0$ and for $\theta=180°$. The data points have been calculated from the fits of the authors. Below 3 GeV/c, the cross sections have been reconstructed from the KH 78 partial wave solution. The following points are of interest:

i) The figures show that the Serpuchov points disagree strongly with the Fermilab results. The following discussion is based on the Fermilab data (79 BAKER).

ii) It is seen that there are no well-established structures above 5 GeV/c. The data above 5 GeV/c are compatible with a power law

$$\frac{d\sigma(u=0)}{du} = ck^{-n}. \tag{6.7.3}$$

A best fit gives (79 Baker)

$$\pi^+ p \to \pi^+ p: \quad n_+ = 2.33 \pm 0.07;$$
$$c_+ = 877 \mu b (GeV)^{n_+ - 2};$$
$$\pi^- p \to \pi^- p: \quad n_- = 2.08 \pm 0.06;$$
$$c_- = 197 \mu b (GeV)^{n_- - 2}. \tag{6.7.4}$$

The slopes in π^+p scattering are much larger than those in π^-p-scattering, because of the above mentioned dip at $t = -0.15\,\mathrm{GeV}^2$. Due to the large slope in π^+p-scattering, the ratio of $d\sigma/du$ at $u=0$ and at $\theta = 180°$ is appreciably different from 1 even at the highest energies of our figure, where u_b is very small.

The energy dependence of the slopes is compatible with a law $B = 2\ln k + \mathrm{const}$, but the accuracy is not good enough to exclude for instance a constancy above 10 GeV/c (Fig. 6.7.3).

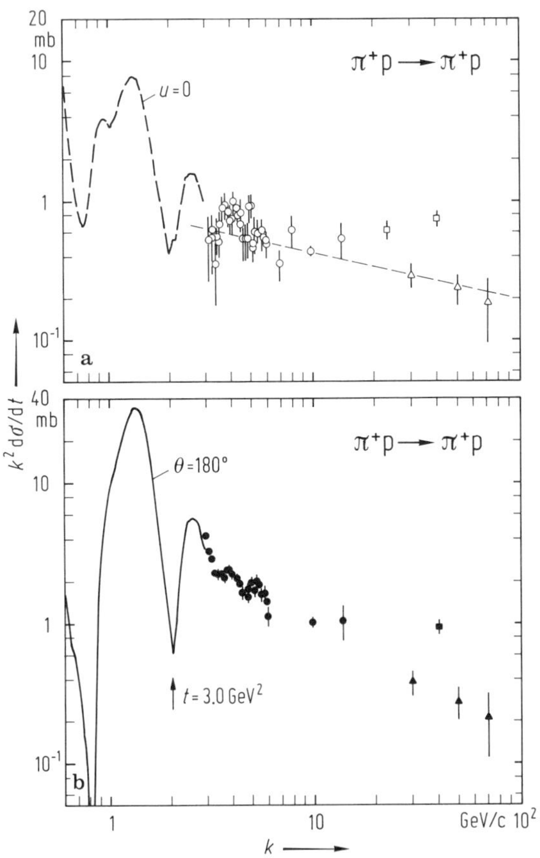

Fig. 6.7.1. $k^2 d\sigma/dt$ for elastic π^+p scattering at $u=0$ and $\theta=180°$. Dashed line and solid line below 3 GeV/c: reconstruction of the cross section from KH 78 partial waves. Straight dashed line: fit of Baker et al. (79), Eqs. (6.7.3) and (6.7.4). Data: △▲ 79 Baker, □■ Babaev. Other data: see Tables 2.6.7.1 and 2.6.7.2. The arrow shows the dip at $t \approx 3\,\mathrm{GeV}^2$, which is seen at higher momenta at approximately the same t-value.

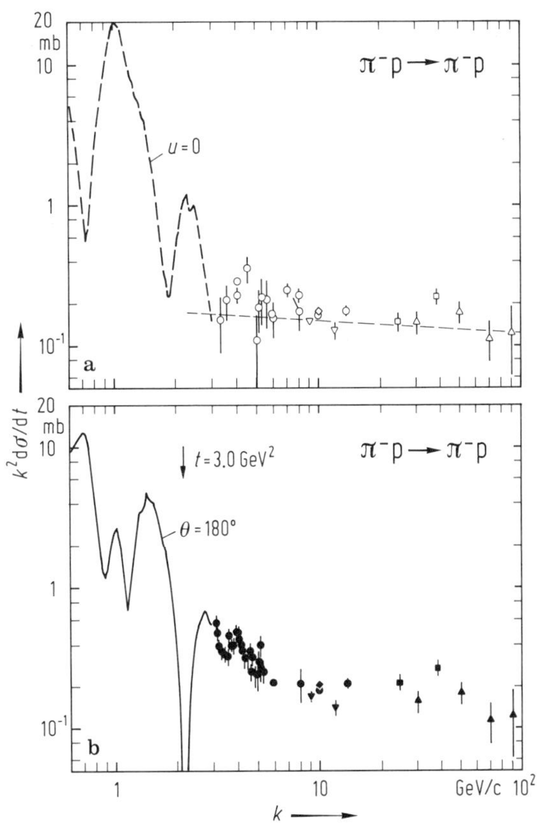

Fig. 6.7.2. $k^2 d\sigma/dt$ for elastic π^-p scattering at $u=0$ and $\theta=180°$. Dashed line and solid line below 3 GeV/c: reconstruction of the cross section from KH 78 partial waves. Straight dashed line: fit of Baker et al. (79), Eqs. (6.7.3) and (6.7.4). Data: △▲ 79 Baker, □■ 77 Babaev, ▽▼ 77 Jachol, ◇◆ 81 Ghidin. Other data: see Table 2.6.7.1.

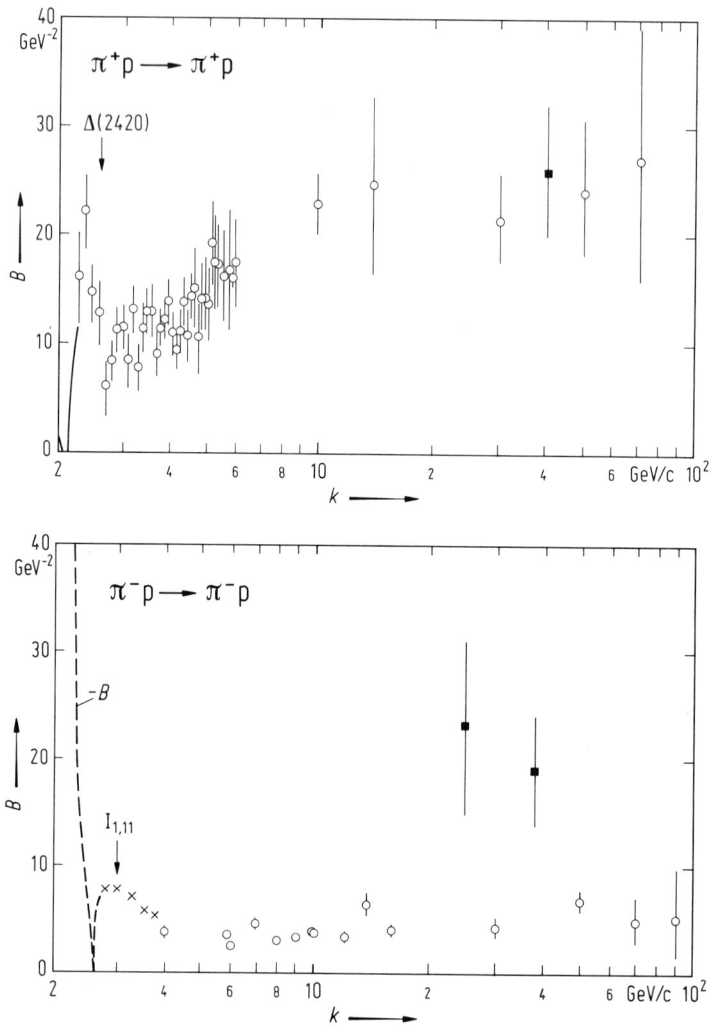

Fig. 6.7.3. Energy dependence of the slope of the backward peak. The experimental points are those of Tables 2.6.7.1 and 2.6.7.2. The solid line at the lowest momenta has been calculated from the updated KH 78 partial wave analysis. In the case of π^-p scattering the slopes of the Serpuchov data (■) are much higher than those of the Fermilab data. — The logarithmic increase predicted by a simple Regge law [Eq. (6.7.18)] is compatible with the data, but the data are not accurate enough to exclude for instance a constancy above 10 GeV/c. — In the case of π^-p scattering the experimental information at intermediate energies is rather poor. Above 2.5 GeV/c the values reconstructed from the KH 78 partial wave analysis are not satisfactory. The points denoted by × have been determined by exponential fits to the data 77 Vavra and 76 Hoffma. The dashed line is an estimate of the rapid variation, which is related to the intersection of a zero trajectory with the backward hyperbola in the Mandelstam plane (Fig. 4.3.3). Below 2.65 GeV/c the dashed line shows $-B$.

Table 2.6.7.1. Parameters of fits to near-backward scattering data at high energies. The data 72 BABAEV belong to $\pi^- n \to \pi^- n$.

$\pi^- p \to \pi^- p$

k GeV/c	$d\sigma(u=0)/du$ μb/GeV²	B GeV⁻²	Reference
90	0.0155 ± 0.0081	$5.2\,^{+4.7}_{-3.2}$	79 BAKER
70	0.0233 ± 0.0075	$5.0\,^{+2.2}_{-1.8}$	79 BAKER
50	0.070 ± 0.013	$6.89^{+0.99}_{-0.86}$	79 BAKER
37.8	0.160 ± 0.022	19 ± 5	77 BABAEV
30	0.167 ± 0.032	$4.47^{+0.83}_{-0.73}$	79 BAKER
24.7	0.262 ± 0.037	23 ± 8	77 BABAEV
16.25	0.45 ± 0.11	–	69 OWEN
16	0.91 ± 0.13	4.23 ± 0.40	68 ANDERS
13.7	0.95 ± 0.07	6.62 ± 0.94	69 OWEN
12	0.91 ± 0.12	3.6 ± 0.35	78 JACHOL
10	1.76 ± 0.18	3.9 ± 0.3	81 GHIDIN
9.9	1.70 ± 0.06	4.05 ± 0.16	69 OWEN
9.0	1.88 ± 0.13	3.5 ± 0.2	78 JACHOL
8.0	3.75 ± 0.35	3.16 ± 0.24	68 ANDERS
7.0	5.3 ± 0.3	4.7 ± 0.4	71 BAKER
6.0	4.2 ± 0.3	2.7 ± 0.3	78 SHARFM
5.9	4.94 ± 0.12	3.72 ± 0.08	69 OWEN
4.0	14.2 ± 3.0	3.86 ± 0.40	78 SCHARR

$\pi^+ p \to \pi^+ p$

k GeV/c	$d\sigma(u=0)/du$ μb/GeV²	B GeV⁻²	Reference
70	0.037 ± 0.018	$27\,^{+16}_{-11}$	79 BAKER
50	0.091 ± 0.023	$24.0\,^{+6.7}_{-5.6}$	79 BAKER
40	0.444 ± 0.053	26 ± 6	72 BABAEV
30	0.317 ± 0.063	$21.4\,^{+4.1}_{-3.6}$	79 BAKER
23	1.10 ± 0.02	–	72 BABAEV
13.9	2.69 ± 0.74	24.7 ± 8.0	69 OWEN
9.9	4.27 ± 0.30	22.9 ± 2.7	69 OWEN
5.9	15.1 ± 0.4	16.1 ± 0.5	69 OWEN

Table 2.6.7.3. Parameters of an exponential fit, Eq. (6.7.1), to charge-exchange cross sections in the range $u_b > u > -0.2\,\mathrm{GeV}^2$ (75 MARZO).

k GeV/c	$d\sigma(u=0)/du$ μb/GeV²	B GeV⁻²
2.6	93.8 ± 2.2	7.1 ± 0.3
3.5	37.6 ± 1.0	6.2 ± 0.4
4.3	17.4 ± 0.5	5.8 ± 0.4
6.0	7.2 ± 0.3	6.9 ± 0.6
8.0	3.1 ± 0.4	7.6 ± 2.3

Table 2.6.7.2. Differential cross sections and slopes B for $\pi^+ p$ backward scattering at 2···6 GeV/c according to LENNOX et al. (76 LENNOX). – In Table III of this paper the columns for $d\sigma/d\Omega$ and for $-\cos\theta_{cm}$ are not correct. I am grateful to P. KOEHLER and K.P. PRETZL for a discussion

k GeV/c	$u(180°)$ GeV²	W MeV	$d\sigma(180°)/du$ μb/GeV²	B GeV⁻²
2.0	0.1590	2159	$183\,^{+42}_{-34}$	12.1 ± 10.2
2.1	0.1528	2202	$74\,^{+24}_{-18}$	-6.2 ± 11.5
2.2	0.1478	2244	$266\,^{+54}_{-45}$	8.1 ± 8.0
2.3	0.1418	2286	$638\,^{+55}_{-51}$	16.3 ± 3.9
2.4	0.1370	2326	$979\,^{+55}_{-52}$	22.2 ± 3.2
2.5	0.1324	2366	$982\,^{+47}_{-46}$	14.8 ± 2.3
2.6	0.1281	2405	$955\,^{+57}_{-53}$	13.0 ± 2.7
2.7	0.1241	2444	$847\,^{+38}_{-37}$	6.2 ± 2.1
2.8	0.1203	2482	$716\,^{+19}_{-18}$	8.5 ± 1.5
2.9	0.1167	2520	$615\,^{+27}_{-25}$	11.3 ± 2.0
3.0	0.1134	2556	$460\,^{+23}_{-22}$	11.6 ± 2.0
3.1	0.1103	2592	$337\,^{+21}_{-20}$	8.7 ± 2.3
3.2	0.1073	2628	$280\,^{+11}_{-15}$	13.2 ± 2.1
3.3	0.1044	2664	$208\,^{+13}_{-12}$	7.9 ± 2.1
3.4	0.1017	2699	$191\,^{+12}_{-11}$	11.5 ± 2.2
3.5	0.0992	2733	$182\,^{+12}_{-11}$	13.0 ± 2.1
3.6	0.0968	2767	$171\,^{+13}_{-12}$	13.1 ± 2.4
3.7	0.0945	2801	$153\,^{+10}_{-9}$	9.2 ± 1.9
3.8	0.0923	2834	$165\,^{+7}_{-7}$	11.5 ± 1.4
3.9	0.0902	2867	$161\,^{+7}_{-7}$	12.3 ± 1.6
4.0	0.0881	2900	$140\,^{+10}_{-9}$	14.0 ± 2.0
4.1	0.0862	2932	$138\,^{+7}_{-7}$	11.2 ± 1.7
4.2	0.0844	2964	$118\,^{+8}_{-7}$	9.6 ± 1.8
4.3	0.0826	2995	$108\,^{+9}_{-8}$	11.2 ± 2.0
4.4	0.0809	3026	$97.6^{+5.4}_{-5.0}$	13.9 ± 2.2
4.5	0.0793	3057	$80.4^{+8.2}_{-7.6}$	10.9 ± 2.3
4.6	0.0777	3088	$83.3^{+4.8}_{-4.5}$	14.5 ± 2.1
4.7	0.0762	3118	$79.6^{+7.2}_{-6.8}$	15.2 ± 3.7
4.8	0.0748	3148	$66.3^{+6.3}_{-6.0}$	10.8 ± 3.5
4.9	0.0734	3177	$75.7^{+7.5}_{-6.9}$	14.1 ± 3.4
5.0	0.0721	3207	$76.8^{+6.9}_{-6.3}$	14.2 ± 3.8
5.1	0.0708	3236	$68.8^{+5.9}_{-5.3}$	13.7 ± 3.4
5.2	0.0695	3265	$62.9^{+6.0}_{-5.4}$	19.3 ± 3.8
5.3	0.0683	3293	$69.9^{+7.0}_{-6.8}$	17.5 ± 4.3
5.4	0.0672	3322	$64.3^{+6.2}_{-5.8}$	17.3 ± 3.6
5.6	0.0649	3378	$50.3^{+6.1}_{-5.3}$	16.3 ± 4.2
5.8	0.0629	3433	$47.8^{+5.8}_{-5.1}$	16.8 ± 5.5
6.0	0.0610	3487	$37.3^{+5.2}_{-4.5}$	17.5 ± 3.9

π^-p charge-exchange near-backward scattering

The differential cross sections for π^-p charge-exchange near-backward scattering have been measured up to 13.8 GeV/c (70 BORIGH, 71 SCHNEI, 75 MARZO), but the experiment by MARZO et al. is the only one in which the appropriate radiative corrections have been applied (Sect. 2.1.17.3). Therefore, the results cannot be compared in a simple way with those of the earlier experiments.

In the experiment 72 KISTIA, the cross section at $\theta \approx 180°$ was measured at many momenta in the range 1.8···6 GeV/c. The data are shown in Fig. 4.4.3. The agreement with the results of other authors is reasonable except for a large discrepancy with 70 CHASE.

According to 75 MARZO, the data show a backward peak, which can be fitted by an exponential law, Eq. (6.7.1), in the range $u_b > u > -0.2\,\text{GeV}^2$. The parameters are listed in Table 2.6.7.3 (see p. 385).

The momentum dependence of the cross sections at $u=0$ between 4.3 and 8.0 GeV/c can be fitted by a power law, $d\sigma(u=0)/du \propto s^{-n_0}$, and $n_0 = 2.96 \pm 0.12$ (75 MARZO), the decrease being steeper than the above mentioned fit to the π^+p data ($n_+ = 2.33$) which, however, was made in a much larger momentum range. KISTIAKOWSKI et al. (72) gave a power law with $n_0 = 1.86 \pm 0.15$ for a fit to the *backward* cross sections between 4 and 6 GeV/c.

MARZO et al. (75) found a dip in the differential charge-exchange cross section at $u \approx -0.3\,\text{GeV}^2$ and a shoulder or a secondary bump at $u \approx -0.7\,\text{GeV}^2$.

The dips at constant u in the elastic and charge-exchange cross sections are related to zero trajectories, which have been discussed in Sect. 2.4.3.

Near-backward polarization data

The experimental information on polarization parameters at high momenta and near-backward angles is very poor. The highest momentum where elastic polarization measurements have been made is 6 GeV/c.

The 6 GeV/c P_+-data (72 DICK) have small values between the backward direction and the dip region ($u = -0.15\,\text{GeV}^2$). Then, the polarization parameter goes to fairly large negative values (-0.6) and remains almost constant. Data at somewhat lower momenta (3.25···4.0 GeV/c, 73 SCHEID, 73 BRADAM) show a similar shape and magnitude.

The 6 GeV/c P_--data (73 DICK) have a maximum of about 0.4 at $u = -0.3\,\text{GeV}^2$ and are then decreasing to small negative values in the range $-0.5 < u < -0.9\,\text{GeV}^2$. The next data at lower momenta lie already in the resonance region (80 FUKUSH, 2.22···3.5 GeV/c; 81 AUER, 2.9 and 3.3 GeV/c). A comparison of the shape at 3.5 GeV/c with that at 6 GeV/c is difficult due to large errors.

The information is even poorer in the charge-exchange reaction, where near-backward polarization data have not yet been measured above the resonance region. The experiment 79 MINOWA includes P_0 data up to 3.0 GeV/c and even these data go only up to $u = -0.24\,\text{GeV}^2$ ($\cos\theta > -0.85$).

2.6.7.2 Dispersion analysis of backward scattering data

Backward scattering cross sections are of special interest, because one can measure directly the modulus of the invariant amplitude C

$$\frac{d\sigma(180°)}{d\Omega} = \frac{d\sigma(180°)}{du}\frac{q^2}{\pi} = \left(\frac{E}{4\pi W}\right)^2 |C|^2 . \tag{6.7.5}$$

The analytic properties of the backward amplitude C as a function of energy can be used in order to determine the *absolute phase*.

In practice, one can proceed as follows. The backward dispersion relations (Sect. A.6.9) have a right hand (s-channel) and a left hand (t-channel) cut. On the right hand cut, ReC and ImC are known from partial wave analysis in a certain range. At higher momenta, ImC is taken from a Regge model, whose parameters are derived from a fit to the backward cross sections.

One can use the fits Eqs. (6.7.3), (6.7.4) above 5 GeV/c and KH 78 partial waves at lower momenta. If an expansion method is used (Sect. 2.6.7.7), one can easily take into account in addition accurate backward cross section data, which are not well enough represented by the preliminary partial waves at 2.5···5 GeV/c.

The nearby part of the left hand cut can be evaluated by substituting the $\pi\pi N\bar{N}$ partial waves of Table 2.4.6.1. The more distant part of the left hand cut has to be treated with a discrepancy method. Above the $\bar{p}p$-threshold, some information on the $\pi\pi N\bar{N}$ backward amplitude is available from a partial wave analysis of the $\bar{p}p \to \pi\pi$ data (Table 2.5.4.2). Finally, one can use line-reversal (Sects. 2.6.7.5 and A.9.2.4) for an estimate of the very distant part of the t-channel cut (see below).

Evaluations of backward dispersion relations were performed by several authors since the pioneering work of ATKINSON (62 L), for instance by LOVELACE et al. (69 R), ENGELS et al. (70 L-1, 70 L-2), NIELSEN, PETERSEN, and PIETARINEN (70 L-1), PIETARINEN (78-2) and CUTKOSKY et al. (79 L-1). In some cases, the input for the nearby part of the left hand cut was not yet available or it was not used.

In other investigations, backward dispersion relations occur as a special case, for instance in the study of the interior dispersion relations by KAISER et al. (76 L) and BORIE and KAISER (77 L) or in HEDEGAARD-JENSENs work (77 L).

The result for the phase of the backward amplitude can be shown by plotting Argand diagrams (70 L ENGELS-1 and 2, 79 R HÖHLER-KAISER-KOCH-PIETARINEN, Sect. 6.3).

One should note that the shape of the Argand diagrams is not only determined by the resonance poles, but also by the nearby complex zeros of the scattering amplitude, which lead to pronounced dips of the backward cross sections. The location of these dips is marked in Figs. 4.3.1 and 4.3.2.

The high energy behaviour of $|C|$ follows from the power law fit Eq. (6.7.3) and from (6.7.5)

$$|C_+| \propto s^{-0.67}; \qquad |C_-| \propto s^{-0.54}. \tag{6.7.6}$$

It follows that the dispersion integrals converge without subtraction. In the more general case of the interior dispersion relation a subtraction is required, because one uses the amplitude $p_\pm^2 C^+$ (77 L BORIE).

The nucleon pole term goes to zero as $1/s$ and therefore, it is unimportant at high energies. It is a distant singularity at fixed angle $\theta = 180°$, but the nearest singularity at high energies and fixed s.

If one wants to use a dispersion relation for near-backward amplitudes, the situation is much more complicated, since one has to deal with two independent amplitudes C and B. The expression for the differential cross section, Eq. (A.4.17) can be written

$$\frac{d\sigma}{du} = \frac{1}{16\pi}\left\{\left(1-\frac{t}{4m^2}\right)|C/k|^2 - \frac{t}{4q^2}\frac{u_b-u}{4m^2-t}|B|^2\right\}. \tag{6.7.7}$$

The high energy approximation for near-backward scattering reads

$$\frac{d\sigma}{du} \approx \frac{1}{16\pi s}\{|C|^2+(u_b-u)|B|^2\}; \qquad t \approx -4q^2 \gg 4m^2. \tag{6.7.8}$$

It is not possible to determine the amplitudes C and B separately from amplitude analysis at high energies, since one would have to use not only polarization data but also spin-rotation data. A model-independent result for the high energy behaviour of ImB can be derived by evaluating the nearby part of the left hand cut of the dispersion relation Eq. (A.6.100), using the $\pi\pi N\bar{N}$ partial waves of Table 2.4.6.1 and the right hand cut up to 5 GeV/c, using πN partial waves of the KH solution. The question is then whether the discrepancy gives a clear signal for corrections which are *increasing* as the energy increases, indicating a contribution from the high energy part of the right hand cut.

A strong contribution of the B-amplitude to near-backward scattering is excluded by the fact that one observes a backward *peak*, whereas the $|B|^2$-term in Eq. (6.7.8), taken alone, produces a backward *dip*.

Useful constraints on near-backward amplitudes follow from three different types of dispersion relations: interior dispersion relations (Sect. A.6.7.2), fixed c.m. angle dispersion relations (Sect. A.6.8) and fixed-u dispersion relations (Sect. A.6.6.).

Connection between high energy backward scattering and πN partial waves at $s=0$

As discussed in Sect. A.7.1.4, the behaviour of the s-channel πN partial waves near $s=0$ is related to that of the invariant amplitudes at high energies and near-backward angles. This problem was investigated in detail by JAKOB and STEINER (69 L). Subsequent authors (72 L ELVEKJAER, 75 L DIU) claimed to have found large inconsistencies in the P11 partial wave and discussed this problem as a consequence of deviations from the Regge behaviour at high energies, which were suggested by the Serpuchov data (72 BABAEV).

As mentioned above, the new data 81 BAKER are compatible with a Regge behaviour. No inconsistencies were noticed in the most recent analyses of the πN partial waves (80 L HUTT, 82 L KOCH-2).

2.6.7.3 Isospin analysis of near-backward scattering data

Since near-backward scattering is assumed to be dominated by u-channel baryon exchanges, it is useful to work with amplitudes which have a well-defined u-channel isospin I_u. We denote the $I_u=1/2$ and $3/2$ amplitudes by an upper index N and Δ respectively. The relation between these amplitudes and the amplitudes $F_\pm$, F_0 of the measurable reactions reads (Table A.2.1)

$$F_+=\tfrac{1}{3}(2F^{\mathrm{N}}+F^{\Delta}); \quad F_-=F^{\Delta}; \quad F_0=\frac{\sqrt{2}}{3}(F^{\mathrm{N}}-F^{\Delta}). \tag{6.7.9}$$

In this subsection, we denote the differential cross sections $d\sigma/du$ by σ. Then, one can derive the following relations (72 L Barger-1)

$$\begin{aligned}&\sigma^{\Delta}=\sigma_-; \quad \sigma^{\mathrm{N}}=\tfrac{1}{2}[3(\sigma_++\sigma_0)-\sigma_-]; \quad \sigma_+=\tfrac{4}{9}(\sigma^{\mathrm{N}}+D)+\tfrac{1}{9}\sigma^{\Delta};\\ &\sigma_0=\tfrac{2}{9}(\sigma^{\mathrm{N}}-2D)+\tfrac{2}{9}\sigma^{\Delta}; \quad D\equiv\sum\mathrm{Re}(F^{\mathrm{N}}F^{\Delta *})=\tfrac{3}{4}(\sigma_+-2\sigma_0+\tfrac{1}{3}\sigma_-).\end{aligned} \tag{6.7.10}$$

The sum in D is taken over the s-channel helicity non-flip and flip amplitudes. σ^{N} and σ^{Δ} are the cross sections belonging to pure $I_u=1/2$ and $3/2$ amplitudes. Results for the separation of N- and Δ-exchange contributions are given in 72 L Barger-1 and 2.

At $\theta=180°$, the helicity non-flip amplitude vanishes and the interference term D gives information on the phase difference of the N- and Δ-amplitudes in analogy to Eq. (A.10.16). The following convention is used for the phases in order to obtain quantities which can be interpreted directly in Regge models

$$F^{\mathrm{N}}_{+-}=-|F^{\mathrm{N}}_{+-}|\exp[-\mathrm{i}(\alpha^{\mathrm{N}}-1/2)\pi/2]; \quad F^{\Delta}_{+-}=-\mathrm{i}|F^{\Delta}_{+-}|\exp[-\mathrm{i}(\alpha^{\Delta}-1/2)\pi/2]. \tag{6.7.11}$$

One obtains

$$D(u_b)=(\sigma^{\mathrm{N}}\sigma^{\Delta})^{1/2}\sin[(\alpha^{\Delta}-\alpha^{\mathrm{N}})\pi/2]; \quad \sin[(\alpha^{\Delta}-\alpha^{\mathrm{N}})\pi/2]=\frac{3(\sigma_+-2\sigma_0+\sigma_-/3)}{4(\sigma^{\mathrm{N}}\sigma^{\Delta})^{1/2}} \quad \text{at} \quad u=u_b. \tag{6.7.12}$$

The polarization data can be treated in an analogous way, but P^{N} cannot be extracted, because near-backward P_0 data do not yet exist. Therefore, one can discuss only the N−Δ interference term

$$X=\tfrac{1}{4}(9P_+\sigma_+-P_-\sigma_-). \tag{6.7.13}$$

Furthermore, one of the isospin bounds of Sect. A.10.2 can be modified in such a way that it is applicable to P^{N} (72 L Barger-1).

Isospin bounds for backward scattering have been discussed in Sect. 2.4.4.2.

2.6.7.4 Baryon-exchange Regge pole analyses of near-backward data

The formalism for the reggeized u-channel baryon-exchange contributions to πN near-backward scattering is described in detail in 69 R Barger-Cline. We have listed the main formulas in Sect. A.9.2.3. Equation (A.9.49) gives the expression for the differential cross section and Eqs. (A.9.46) and (A.9.48) define the amplitudes. The trajectories α and the residues are functions of $W_u=\sqrt{u}$, Eq. (A.9.43).

At high energies, $d\sigma/du$ has the following s-dependence, if we consider only a single trajectory

$$\frac{d\sigma}{du}=A(u)s^{2\alpha(\sqrt{u})-2}. \tag{6.7.14}$$

According to Eq. (6.7.10), the π^-p elastic cross section depends only on Δ-trajectories, because N-exchange is not possible. The dominant contribution comes from the Δ_δ-trajectory (Fig. 4.1.1), which starts with the resonance Δ(1232). The fit of Baker et al. (79), Eq. (6.7.4), to π^-p data in the range from 5 to 90 GeV/c gives $2\alpha^{\Delta}-2=-2.08\pm0.06$ or

$$\alpha^{\Delta}(0)=-0.04\pm0.03, \tag{6.7.15}$$

in agreement with the expectation for the Δ_δ-trajectory ($2mk\approx s$ at high energies). See the extrapolation of the Δ_δ-trajectory in Fig. 4.1.1.

Since charge-exchange backward cross sections have been measured only up to 13.8 GeV/c, one cannot disentangle the Δ- and N-contributions to the π^+p cross section. Furthermore, it could be that the N_γ-contribution is not negligible in comparison with the N_α-contribution. The fit of BAKER et al. (79) gives an "effective α" derived from the π^+p data

$$\alpha_{\text{eff}}(0) = -0.16 \pm 0.04 \quad \text{from} \quad d\sigma_+/du. \tag{6.7.16}$$

It can be compared with the value expected from a linear extrapolation of the N_α-trajectory: −0.37, which would lead to a decrease $\propto k^{-2.74}$. One can conclude that there is no evidence for a strong N_α-dominance in π^+p scattering.

MARZO et al. (75) determined an effective α from a fit to their charge-exchange data in the range 4.3···8.0 GeV/c.

$$\alpha_{\text{eff}}(0) = -0.48 \pm 0.06, \tag{6.7.17}$$

Regge theory predicts a shrinkage of the backward slopes, corresponding to an energy dependence of B in Eq. (6.7.1). If we assume a linear u-dependence as suggested by the CHEW-FRAUTSCHI-plots of N_α and Δ_δ (Fig. 4.1.1), one finds $\alpha'(0) \approx 1.0\,\text{GeV}^{-2}$ and

$$B(s) = 2\alpha'(0)\ln s + \text{const}. \tag{6.7.18}$$

BAKER et al. (79) have shown that there is some evidence for shrinkage, which is compatible with the expected behaviour, but due to the errors the test is still weak (Fig. 6.7.3).

It is difficult to perform a Regge analysis of the invariant amplitude B, which is needed if $\theta < 180°$, because the result is sensitive to the parametrization of the trajectory and residue functions. If both do not contain a $\sqrt{u}$ term, the Regge pole does not contribute to this amplitude [see Eqs. (A.9.46) and (A.9.48)].

Parity doublets

The generalized MAC-DOWELL symmetry relation, Eq. (A.9.38) requires that the trajectories for u-channel baryon exchanges of given signature τ satisfy the relation

$$\alpha^{(+)}(\sqrt{u}) = \alpha^{(-)}(-\sqrt{u}); \qquad u > 0, \tag{6.7.19}$$

where the superscripts (±) refer to the naturality $\eta = P\tau$.

In Fig. 4.1.1, the natural ($\eta = 1$) and unnatural ($\eta = -1$) parity trajectories are plotted back to back. The MAC-DOWELL pairs are states of equal signature, opposite parity and equal J, which lie in the left hand and right hand plot, respectively.

Since the trajectories are approximately linear in u, it follows from Eq. (6.7.19) that both trajectories are described by the ansatz

$$\alpha^{(\pm)}(\sqrt{u}) = \alpha(0) + \alpha' u. \tag{6.7.20}$$

The resonances are expected to appear in pairs of equal mass, whose location is symmetric to the line $s = 0$ in Fig. 4.1.1.

The plots in Fig. 4.1.1 show that there are some examples of mass degenerate parity doublets; for instance D 15(1670) and F 15(1688) and G 19(2200) and H 19(2220), but there is no partner for the nucleon and for many of the 4-star resonances.

A more general ansatz

$$\alpha^{(\pm)}(\sqrt{u}) = \alpha(0) \pm \alpha_1\sqrt{u} + \alpha_2 u \tag{6.7.21}$$

makes the trajectories in Fig. 4.1.1 curved and splits the degeneracy, but this does not solve the problem, unless the trajectory remains on one side below the physical J-values. If this ansatz is used, Eq. (6.7.14) has to be replaced by

$$\frac{d\sigma}{du} = A(u)\, s^{\alpha(\sqrt{u}) + \alpha(-\sqrt{u}) - 2}. \tag{6.7.22}$$

The term $\alpha_1\sqrt{u}$ cancels in the sum.

A satisfactory and generally accepted answer to the problem of parity doublets is not yet known. One can assume that zeros in residue functions kill the unwanted states (69 R BARGER-CLINE) or one can introduce a branch point at $J = \alpha(0)$ and place the unwanted state on the unphysical side of the cut [70 L CARLITZ, see also

77 R COLLINS, p. 290 and the application of effective action techniques to the CARLITZ-KISLINGER fermion Reggeization scheme by KARSTEN (79 L)].

If the $\sqrt{u}$-term is present, polarization can arise from a single trajectory.

Exchange degeneracy

The $\Delta J=2$ spacing of the Regge trajectories is due to the presence of exchange forces. If it happens that these forces are small, then trajectories with the same isospin and naturality and opposite signature are almost degenerate, the new spacing being $\Delta J=1$. See 70 R MANDULA-WEYERS-ZWEIG, for a review on exchange degeneracy.

From Fig. 4.1.1 one can conclude that an approximate exchange degeneracy is valid for the pairs N_α, N_γ and Δ_β, Δ_δ.

As mentioned in connection with Eq. (A.9.24), the necessity to treat the even and odd angular momenta differently follows from the existence of the u-channel integral. If an approximate exchange degeneracy holds, one can conclude that the discontinuity along the u-channel cut is small.

Determination of $\mathrm{Im}\,\alpha(\sqrt{u})$ from resonance widths

Unless trajectories cross each other, $\alpha(\sqrt{u})$ will have a cut at threshold $\sqrt{u}=m+\mu$. The magnitude of $\mathrm{Im}\,\alpha(\sqrt{u})$ at the position of the nucleon resonances can be estimated from the total widths, assuming $(\mathrm{Re}\,\alpha/\mathrm{Im}\,\alpha)^2 \gg 1$ [Eq. (A.9.37)] (69 R BARGER-CLINE, p. 114)

$$\mathrm{Im}\,\alpha(\sqrt{u}=M)=\frac{\Gamma}{2}\left.\frac{\mathrm{d}\,\mathrm{Re}\,\alpha}{\mathrm{d}\sqrt{u}}\right|_{\sqrt{u}=M}. \tag{6.7.23}$$

We substitute the linear ansatz Eq. (6.7.20) and find

$$\mathrm{Im}\,\alpha(\sqrt{u}=M)=\alpha'\Gamma M, \tag{6.7.24}$$

where $\alpha' \approx 1.0\,\mathrm{GeV}^{-2}$ for the trajectories N_α and Δ_δ (Fig. 4.4.1). Figure 6.7.4 shows a plot of $\mathrm{Im}\,\alpha$ against M^2 for the Regge trajectories Δ_δ, N_α and N_γ.

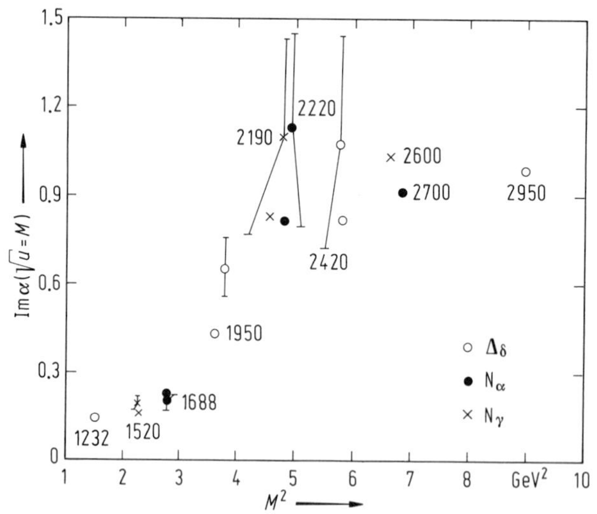

Fig. 6.7.4. Plot of $\mathrm{Im}\,\alpha(\sqrt{u}=M)$ against M^2 for the trajectories N_α, N_γ, and Δ_δ. The symbols with error bars belong to the CMU-LBL partial wave solution and the others to the KH 78 solution. $\mathrm{Im}\,\alpha$ has been calculated from Eq. (6.7.24), assuming $\alpha'=1.0\,\mathrm{GeV}^{-2}$ and using Tables 2.4.1.1 and 2.4.1.2. — In earlier versions of this plot, the experimental points suggested a linear increase. One should note that the determination of the width of the higher resonances is sensitive to the separation of the background.

Extrapolation to the particle poles

The extrapolation of the residue functions $\gamma^{(\pm)}(\sqrt{u})$ in Eq. (A.9.46) to $u>0$ is constrained by the following condition. The Regge amplitude reduces to the lowest order Feynman graph amplitude (or Breit-Wigner amplitude), if $\sqrt{u}$ is near to the mass of the nucleon (or a nucleon resonance). The residue at $\sqrt{u}=M$ is related to the square of the πNN or πNN* coupling constants, respectively.

Details are given in the book 69 R BARGER-CLINE, p. 132. A more recent application can be found in 75 L STORROW.

The dip of $d\sigma_+/du$ at $u=-0.15\,\mathrm{GeV}^2$

As mentioned above, the π^+p cross section has a near-backward dip which occurs in the data at constant $u\approx-0.15\,\mathrm{GeV}^2$ from about 2 GeV/c (77 OTT, 77 STEIN, 71 SIDWELL, see also Fig. 4.3.3) up to 50 GeV/c (79 BAKER). In the smaller range where differential cross section data for all three reactions are available, BARGER et al. (72 L-1 and 2) used Eq. (6.7.10) for a determination of the purely $I_u=1/2$ cross section σ^N and of the interference term D (see also 73 L HAYOT).

At all momenta above 4 GeV/c the result for σ^N at $u=-0.15\,\mathrm{GeV}^2$ is zero within the errors and the interference term D is also zero. The conclusion that both $I_u=1/2$ amplitudes vanish at this u-value leads to the prediction that the polarizations P_+ and P_- are equal, since both come from the interference between the Δ-exchange no flip and flip terms. This was confirmed by the data of DICK et al. (72 and 73).

One should note that, in contrast to near-forward scattering, a non-vanishing polarization can arise from a *single* baryon Regge exchange. The reason is that parity doublets may have non-degenerate trajectories (i.e. terms odd in $\sqrt{u}$) and therefore different Regge phases.

The vanishing of both $I_u=1/2$ amplitudes is a consequence of the properties of the expression for $R[\alpha(W_u),s]$, Eq. (A.9.46), if one assumes a N_α trajectory which is linear in u

$$\alpha(\sqrt{u})=-0.37+\alpha' u; \quad \alpha'\approx 1\,\mathrm{GeV}^{-2}; \quad N_\alpha\text{-trajectory}. \tag{6.7.25}$$

For $u=-0.13\,\mathrm{GeV}^2$, the trajectory passes $\alpha=-1/2$, where $1/\Gamma(\alpha+\frac{1}{2})$ is zero. The signature factor is finite because $\tau=+1$.

In general, R vanishes at negative half-integer values of α at which $(-1)^{\alpha-1/2}=-\tau$ ("wrong signature, nonsense values of α"). However, since $\sqrt{u}$ becomes imaginary at this point, an appreciable dip of the cross section will occur only if the $\sqrt{u}$-term in Eq. (6.7.21) is small.

According to BARGER et al. (73 L) the dip location differs systematically from $u=$const and is more accurately described by

$$u_{\mathrm{dip}}-u_{\mathrm{b}}\approx-0.20\,\mathrm{GeV}^2; \quad u_{\mathrm{b}}=(m^2-\mu^2)^2/s. \tag{6.7.26}$$

Fixed-$(u-u_{\mathrm{b}})$ dips occur in geometrical models (see also 75 L BUTTIMORE).

Further Regge pole analyses of near-backward πN scattering data

Investigations which do not yet take into account the polarization data at 6 GeV/c (72 DICK, 73 DICK) are outdated. As an example for a more recent analysis in terms of Regge poles N_α, N_β, and Δ_δ we mention the work of STORROW and WINBOW (75 L) and BIRSA et al. (77 L). The new Fermilab data 79 BAKER are taken into account in the analyses 81 L GLUSHKO and 81 L MIR. Other papers, in particular those which introduce Regge cuts and absorption corrections, are discussed in the books and reviews 72 R CHIU, 73 R FOX-QUIGG, 77 R COLLINS, 77 R IRVING-WORDEN and 77 R LYUBIMOV. (See also 71 L BERGER, 73 L-1 and 2 STORROW, and 74 L DONNACHIE.)

2.6.7.5 Line-reversed reactions

Starting from a reaction $a+b\to c+d$, one can define "line-reversed" reactions $\bar{c}+b\to\bar{a}+d$ and $\bar{d}+b\to c+\bar{a}$. In baryon-exchange Regge models, the contribution of a baryon trajectory pair of signature τ to the amplitudes of the original and the line-reversed reaction differs only by a factor τ [see Eq. (A.9.52)].

If one assumes that the reactions can be described by the exchange of a single trajectory pair, the differential cross sections agree up to a kinematic factor, which practically only depends on s. The comparison has to be performed at the same values of s and u, i.e. at the same total c.m. energy (see Sect. A.9.2.4).

In pion-nucleon scattering, two cases have been investigated

$$\pi^- p \to p\pi^- \longleftrightarrow \bar{p}p \to \pi^+\pi^- ;$$

$$\pi^+ p \to p\pi^+ \longleftrightarrow \bar{p}p \to \pi^-\pi^+ .$$

We have applied the usual convention in this field: in ab→cd, the incoming particle is a. Particle c is scattered into the forward direction and particle d into the backward direction.

The reactions $\pi^- p \to p\pi^-$ and $\bar{p}p \to \pi^+\pi^-$

SHARFMAN et al. (78) have measured both reactions at 6 GeV/c in the range $u_b > u > -1.5\,\mathrm{GeV}^2$ under nearly identical conditions to minimize systematic errors in the comparison.

The π^-p-cross section drops out smoothly without apparent structure to $u = -1.0\,\mathrm{GeV}^2$. It can be well fitted by an exponential, Eq. (6.7.1).

The $\bar{p}p \to \pi^+\pi^-$ cross section is also compatible with an exponential decrease, in contrast to previous experiments (76 L BURAN). The slope agrees approximately with that in π^-p backward scattering.

In order to compare the magnitudes of the two cross sections, one has to scale the data to the same value of s and to apply the factor which relates the two cross sections (Sect. A.9.2.4; see 67 L BARGER for a detailed treatment). The ratio of the π^-p and $\bar{p}$p cross sections is 2.5. It is in reasonable agreement with the prediction.

The reactions $\pi^+ p \to p\pi^+$ and $\bar{p}p \to \pi^-\pi^+$

Both reactions have been measured in the same experiment by STEIN et al. (77) at 6 GeV/c. The π^+p near-backward cross section shows the above mentioned narrow dip in agreement with earlier experiments. In the overlap region, the agreement of the new $\bar{p}p \to \pi^-\pi^+$ data with the data of BURAN et al. (76 L) is good.

The shape of the $\bar{p}p \to \pi^-\pi^+$ cross section differs considerably from the line-reversal prediction, which is based on the assumption that the contribution of the trajectory N_α is dominant. There is a shoulder or a shallow dip at $u = -0.4\,\mathrm{GeV}^2$ and an essentially flat behaviour up to $u = -1.3\,\mathrm{GeV}^2$. The ratio of the two cross sections agrees roughly with the prediction. The discrepancy between the shapes of the two cross sections has been discussed in 73 L BARGER using earlier data. The proposed law Eq. (6.7.26) leads to a shift of the dip in the reaction $p\bar{p} \to \pi^-\pi^+$.

SHARFMAN et al. (78) noticed that the shape of the $\bar{p}p \to \pi^-\pi^+$ cross section has a remarkable similarity to that of the $\pi^- p \to n\pi^0$ cross section, if plotted against $u - u_b$ although the two reactions are not related by line-reversal.

BERGLUND et al. (78 L) used their data for $\bar{p}p \to \pi^-\pi^+$ at 10 GeV/c (which have a lower accuracy) and the $\pi^+ p \to \pi^+ p$ data 69 OWEN for a comparison and found a similar disagreement.

2.6.7.6 Dominance of peripheral partial waves in backward scattering

Many authors based their analyses of πN backward scattering data on the idea that, at intermediate energies, the backward peak is dominated by the contribution of the "peripheral" resonances. We shall mention only a few typical investigations.

The ℓ-dominance sum rule

HALZEN, OLSSON, and YOKOSAWA (76 L) noticed that the near-backward $\pi^\pm$p differential cross sections and polarization parameters in the momentum range 1.6⋯3.5 GeV/c fulfill semiquantitatively the rule

$$P\frac{d\sigma}{d\Omega} = \lambda(s)\frac{\partial}{\partial\theta}\frac{d\sigma}{d\Omega}. \tag{6.7.27}$$

They derived this relation from the assumption that the partial waves $f_{\ell\pm}$ belonging to a single value of ℓ dominate the spin no-flip and flip amplitudes G and H in the near-backward angular range and that $f_{\ell+}$ and $f_{\ell-}$ are not 90 degrees out of phase.

Using results of partial wave analysis, the authors demonstrated that the G and H-amplitudes as reconstructed from partial waves at 1.6 and 2.4 GeV/c approximately agree with the "dominant" contribution of the $\ell = 3$ and $\ell = 4$ partial waves $f_{3\pm}$ and $f_{4\pm}$, respectively, in the range $u_b > u > -0.8\,\mathrm{GeV}^2$. They pointed out that the phase condition is fulfilled due to the fact that the spectrum of nucleon resonances contains pairs of nearly degenerate resonances (N, N), (N, Δ), and (Δ, Δ).

The reason for the approximate validity of Eq. (6.7.27) can be described as follows. In elastic scattering, the imaginary parts of all partial waves are added with the same sign at $\theta = 0°$, whereas a strong cancellation occurs at $\theta = 180°$ due to a factor $(-1)^{\ell}$ which follows from the Legendre functions. If a certain partial wave is enhanced due to a peripheral resonance, the modification of the forward peak is relatively small, but in the backward direction the effect of the enhancement is larger than the sum of all other partial waves, due to the cancellations.

Dominance of peripheral partial waves

Another (more realistic) model contains peripheral partial waves within a range of ℓ-values peaking at $\ell = \ell_0$. ℓ_0 is expected to increase as $\sqrt{s}$.

In forward scattering, peripheral models are usually described in terms of impact parameter transforms (Sects. A.3.3 and 2.6.3.3) and Bessel functions are obtained for the t-dependence (Sect. 2.6.4.8). Several authors made an analogous ansatz for the backward peak, for instance

$$\mathrm{Im}\, F^{\mathrm{N}}_{+-}(u) = A\, \mathrm{e}^{-B^{\mathrm{N}}(u_{\mathrm{b}}-u)} J_0(R\sqrt{u_{\mathrm{b}}-u})\,. \tag{6.7.28}$$

If one assumes the usual value for the interaction radius $R = 1\,\mathrm{fm} \mathrel{\hat{=}} 5\,\mathrm{GeV}^{-1}$, the first zero of the Bessel function gives a zero of $\mathrm{Im}\,F^{\mathrm{N}}_{+-}(u)$ at $u_{\mathrm{b}} - u = 0.25\,\mathrm{GeV}^2$, which corresponds in the 3 GeV region to the zero of σ^{N} at $u = -0.15\,\mathrm{GeV}^2$. The slope of the Bessel function at $u = u_{\mathrm{b}}$ gives a contribution $R^2/2 \approx 13\,\mathrm{GeV}^{-2}$ to B, Eq. (6.7.1), which is of the same order of magnitude as the backward slope of the $\pi^+\mathrm{p}$ cross section at $3 \cdots 4\,\mathrm{GeV/c}$.

If one wants to use these ideas for data analysis, it is necessary to add a number of further assumptions, which are to some extent arbitrary (see for instance 73 L TAKAHASHI and KOHSAKA and 80 L MATSUI). The author of the second paper presented amplitudes which lead to reasonable fits to all data at 3.5 and 6 GeV/c. However, he had to use *17 adjustable parameters* at each momentum and therefore, the success of his fit is not a strong argument in favour of the model.

The glory effect

One can substitute the kinematical formula

$$(u_{\mathrm{b}} - u)^{1/2} = 2q\cos(\theta/2) \approx q(\pi - \theta) \quad \text{for} \quad \theta \approx \pi \tag{6.7.29}$$

into the argument of the Bessel function in Eq. (6.7.28). Omitting the exponential factor and other terms, one obtains in a crude approximation

$$\mathrm{d}\sigma/\mathrm{d}u \propto |J_0(qR(\pi - \theta)|^2 \quad \text{for} \quad \theta \approx \pi\,. \tag{6.7.30}$$

A relation of this type is well-known from classical optics. If visible light is scattered off water droplets with radius R (for instance in a cloud), one can observe the "glory effect", which is a strong enhancement of backward scattering, the angular distribution being given by Eq. (6.7.30) (74 R BRYANT-JARMIE).

A theoretical explanation for this effect was given by NUSSENZVEIG (72 R) in terms of surface waves creeping around the surface and enhancing backward scattering.

The analogy between the glory effect and pion-nucleon backward scattering has been discussed by BRYANT and JARMIE (68 L) and, more recently, by VIANO (74 L), LYUBIMOV (77 R) and SCHREMPP (80 R), where further references can be found.

The glory effect is closely related to Regge poles (72 R NUSSENZVEIG, pp. 357⋯359). VIANO (74 L) approximates the near-backward amplitude by a single s-channel Regge pole.

The backward peak was also studied within the quasipotential approach of LOGUNOV and TAVKHELIDZE (74 L DZHGARKAVA).

Derivative relations for helicity amplitudes

A study of models for impact parameter amplitudes and of the dual peripheral model (Sect. 6.5.5) led to the result that, for a variety of reactions, the s-channel partial waves are related by a "derivative relation" which is given for πN scattering by Eq. (6.5.42) or, for the impact parameter transform by

$$f_{+-}(s, b) = c(s)\, b f_{++}(b, s)\,. \tag{6.7.31}$$

In a systematic investigation B. and F. SCHREMPP (75 L-1) came to the conclusion that the relation Eq. (6.5.42) can be valid only in two cases: for large s and $-t \ll s$ and for large s and $-u \ll s$. The second case belongs to near-backward scattering.

ADER (75 L) combined the derivative relation with the assumption that the no-flip amplitudes for $I_u = 1/2$ and 3/2 are well described by N_α- and Δ_δ-exchanges, respectively. He obtained a reasonable fit to the 3···13 GeV/c data, but there remain some discrepancies.

Interference models

Figures 6.7.1 and 6.7.2 suggest that the near-backward amplitudes can be decomposed into a Regge-type background amplitude and resonance contributions, which are large at 2···3 GeV/c and become negligible above 5 GeV/c. The problem is that a simple addition of these amplitude is expected to lead to "double-counting", since part of the Regge amplitude contributes to the resonance structures.

The model of MA and SHAW (76 L, 75 LENNOX) used the prescription to add Breit-Wigner terms for the resonances to *the non-signatured part* of the Regge amplitude, i.e. the term proportional to the signature τ in the expression for R, Eq. (A.9.46), is omitted. The motivation was the somewhat vague concept of "duality", according to which the sum of the direct channel resonances is represented by the signatured part of the Regge amplitude.

The aim of these investigations was to determine the parameters of the higher resonances. Nowadays, the experimental information is much better, in particular in the large angle region. The resonance parameters can be determined in a more reliable way by a partial wave analysis which includes data at all angles. The interesting question is now to use the result of partial wave analysis in order to study the properties of the background.

2.6.7.7 Expansion methods for the evaluation of dispersion integrals

Since the discussion in Sect. A.6.3.4 is rather short, we shall treat here some aspects which are of interest not only for the evaluation of the backward dispersion relation but also for instance for that of the partial wave dispersion relation of Sect. A.7.1.

We consider the dispersion relation

$$F(s) = \frac{1}{\pi} \int_{s_{th}}^{\infty} \frac{ds'}{s'-s} \operatorname{Im} F(s'). \tag{6.7.32}$$

The method is also applicable, if there is in addition a term due to left hand cuts, which is real at $s \geqq s_{th}$.

The transformation

$$w = \frac{b^2 - (s/s_{th} - 1)}{b^2 + (s/s_{th} - 1)}; \qquad s - s_{th} = s_{th} b^2 \frac{1-w}{1+w} \tag{6.7.33}$$

maps the s-range of the cut ($s_{th} < s < \infty$) onto the interval $1 > w > -1$. b^2 is a real, positive parameter. It allows to choose the point $s = (b^2+1)s_{th}$, which is mapped onto the center $w=0$ of the w-interval.

In terms of the new variable the dispersion relation for $f(w) \equiv F(s)$ reads

$$f(w) = \frac{1}{\pi} \int_{-1}^{+1} dw' \left[\frac{1}{w'-w} - \frac{1}{w'+1}\right] \operatorname{Im} f(w'). \tag{6.7.34}$$

In the range of the cut $F(s)$ is defined as $F(s+i0)$ and this goes over into $f(w-i0)$.

We exploit the fact that the integral can be done analytically if $\operatorname{Im} f(w)$ is given by $\sqrt{1-w^2}\,U_{n-1}(w)$, where $U_{n-1}(w)$ is a Chebyshev Polynomial of the second kind [72 R ABRAMOWITZ-STEGUN, Eq. (22.13.4)]

$$-\frac{1}{\pi} \int_{-1}^{+1} \frac{\sqrt{1-w'^2}\,U_{n-1}(w')}{w'-(w-i0)} dw' = T_n(w) \begin{matrix} -\sqrt{w^2-1}\,U_{n-1}(w); & |w|>1; \\ +i\sqrt{1-w^2}\,U_{n-1}(w); & |w|<1. \end{matrix} \quad w \text{ real}; \tag{6.7.35}$$

$T_n(w)$ is a Chebyshev Polynomial of the first kind.

In order to evaluate the dispersion integral in Eq. (6.7.32), one can expand $\operatorname{Im} F(s)$ as follows

$$\operatorname{Im} F(s) = \sum_{n=1}^{\infty} a_n \sqrt{1-w^2}\,U_{n-1}(w). \tag{6.7.36}$$

The real part above threshold is then given by

$$\operatorname{Re} F(s) = \sum_{n=1}^{\infty} a_n \{T_n(w) - T_n(-1)\}; \qquad s > s_{th}; \qquad -1 < w < 1. \tag{6.7.37}$$

The expansion can as well be used in order to determine the analytic function $F(s)$, if the input includes $|F(s)|$ and $\mathrm{Re}\,F(s)$ at a sequence of s-values. One has to determine best values of the coefficients of a truncated expansion, taking into account the errors of the input.

The relation to other methods is seen, if one substitutes $w=\cos\phi=(z+z^{-1})/2$ and uses

$$\sin\phi\, U_{n-1}(\cos\phi)=\sin(n\phi); \qquad T_n(\cos\phi)=\cos(n\phi). \tag{6.7.38}$$

With $z=w-\sqrt{w^2-1}$, one obtains for $z=e^{i\phi}$ and real w below the cut ($|w|<1$)

$$T_n(w)+i\sqrt{1-w^2}\,U_{n-1}(w)=e^{in\phi}\equiv z^n \tag{6.7.39}$$

and Eqs. (6.7.36) and (6.7.37) become a Fourier expansion

$$\mathrm{Re}\,F(s)=\sum_{n=1}^{\infty} a_n\{\cos(n\phi)-\cos(n\pi)\}; \qquad \mathrm{Im}\,F(s)=\sum_{n=1}^{\infty} a_n\sin(n\phi); \tag{6.7.40}$$

$$F(s)=\sum_{n=1}^{\infty} a_n(z^n-e^{in\pi}). \tag{6.7.41}$$

The second term in $\mathrm{Re}\,F(s)$ enforces that $\mathrm{Re}\,F(s)\to 0$ as $s\to\infty$.

The power series can be derived directly, if one considers the conformal mapping

$$z=\frac{b-\sqrt{1-s/s_{th}}}{b+\sqrt{1-s/s_{th}}}, \tag{6.7.42}$$

which is essentially the same as in Eq. (A.6.41). The first sheet of the s-plane is mapped onto the interior of the unit circle in the z-plane. $z=-1$ belongs to $s=\infty$, $z=+1$ to $s=s_{th}$ and the upper half of the unit circle to $s+i0$ along the cut. The remaining part of the real s-axis ($-\infty<s<s_{th}$) is mapped onto the segment $-1<z<+1$ along the real axis of the z-plane (Fig. A.6.8). For real $s>s_{th}$ the real part of z in Eq. (6.7.42) agrees with w in Eq. (6.7.33).

The analytic properties are correctly described by the power series in Eq. (6.7.41), because all singularities lie outside the unit circle, i.e. in the second sheet of the s-plane.

The expansion can only be used, if it has the same behaviour as the physical amplitude in the limits $s\to s_{th}$ and $s\to\infty$. If s is the usual Mandelstam variable and q the c.m. momentum, one finds for the factor $\sqrt{1-w^2}$ in Eq. (6.7.36)

$$\sqrt{1-w^2}\to\frac{2}{b\sqrt{m\mu}}q \quad \text{as} \quad s\to s_{th}; \qquad \sqrt{1-w^2}\to(m+\mu)bq^{-1} \quad \text{as} \quad s\to\infty. \tag{6.7.43}$$

Since

$$U_n(1)=(-1)^n U_n(-1)=n+1 \tag{6.7.44}$$

is finite, the truncated expansion for $\mathrm{Im}\,F(s)$ has the behaviour of Eq. (6.7.43) in the two limits. This is correct for instance for the isospin even combination of the s-channel partial waves f_{0+}.

A method for more general cases was developed by LOVELACE [69 R; details are given in Appendix 2 of DONNACHIEs review (67 R)]. It is based on the fact that the dispersion integral can be evaluated in closed form, if the imaginary part of $F(s)$ is expanded in terms of Gegenbauer Polynomials, which are called Ultraspherical Polynomials in 68 R ABRAMOWITZ-STEGUN, Sect. 22.

LOVELACE uses the kinetic energy in the lab. system T_π instead of s in order to have a dispersion integral which starts at $T_\pi=0$

$$\frac{1}{\pi}\int_0^{\infty}\frac{h_n^{\ell}(T'_\pi)}{T'_\pi-T_\pi}\,dT'_\pi. \tag{6.7.45}$$

h_n^{ℓ} is proportional to the Gegenbauer Polynomial $C_{n-1}^{\ell+1}$

$$h_n^{\ell}(T_\pi)=\left[\frac{2\sqrt{\tau T_\pi}}{\tau+T_\pi}\right]^{2\ell+1} C_{n-1}^{\ell+1}\left(\frac{\tau-T_\pi}{\tau+T_\pi}\right), \tag{6.7.46}$$

where τ is an arbitrary positive constant. The evaluation of the integral leads to hypergeometric functions which reduce to polynomials. Its behaviour in the limits $T_\pi \to 0$ and $T_\pi \to \infty$ agrees with that of the s-channel partial waves $f_{\ell\pm}$

$$\mathrm{Im}\, h_n^\ell(T_\pi) = O(T_\pi^{\ell+1/2}) \quad \text{as} \quad T_\pi \to 0; \qquad \mathrm{Im}\, h_n^\ell(T_\pi) = O(T_\pi^{-\ell-1/2}) \quad \text{as} \quad T_\pi \to \infty. \tag{6.7.47}$$

The special case $\ell = 0$ leads to the above Fourier series.

Earlier papers on expansion methods: 61 L CIULLI, 61 L FRAZER, 62 L ATKINSON, 62 L CIULLI-1 and 2, 62 L LOVELACE, 64 L LEVINGER. Applications to the t-channel cut in $\pi\pi N\bar{N}$-amplitudes and isovector nucleon form factors are doubtful, if the singularity at $t = 4\mu^2 - \mu^4/m^2$ is ignored (see Sects. 2.4.6.3 and 2.5.2.2).

We also mention some of the more recent papers on expansion methods and related statistical problems: 69 L CIULLI-1 and 2, 69 L and 71 R CUTKOSKY, 70 L NIELSEN, 71 L PRESNAJDER, 73 L CUTKOSKY, 73 L and 73 R CIULLI, PIETARINEN (72 L-1 and 2, 73 L, 76 L-1, 78 L-1 and 2) 69 L PISUT, and three reviews: 70 R PISUT, 74 R HAMILTON-PETERSEN and 75 R CIULLI-POMPONIU-SABBA-STEFANESCU. Although some of the other methods are preferable from a mathematical point of view, our group has preferred PIETARINENs method, because it requires less computer time due to its simplicity. See also the references to Sect. A.6.3 in Sect. A.12.

Final remark on backward scattering

There exists a fairly large number of models for πN near-backward scattering. Most of them led to a reasonable fit to the data available to the authors. In some cases, the test was extended to a variety of other reactions.

It is difficult to judge the success of the models, since they depend on more or less arbitrary assumptions and some of them use a large number of adjustable parameters.

The main problem in the analysis of near-backward πN scattering at high energies is the poor experimental information. There exist no charge-exchange cross sections above 14 GeV/c, no $P_\pm$-data above 6 GeV/c and no P_0-data above 2 GeV/c. A better knowledge of the near-backward structures in the cross sections and polarizations above 2 GeV/c would also be very helpful for the analysis of the higher resonances. The $P_\pm$-data at 6 GeV/c still show a considerable variation in the u-dependence, whose s-dependence is not known.

As far as theoretical constraints are concerned, it will be interesting to impose analyticity along lines u = const and along suitable hyperbolas in the Mandelstam plane (Sects. A.6.6. and A.6.7) and to check the compatibility with models at intermediate energies.

2.6.8 Total elastic and charge-exchange cross sections

Results for integrated $\pi^\pm p$ elastic and $\pi^- p$ charge-exchange cross sections are of interest for several reasons:

i) The ratio of the elastic and total cross sections σ_{el}/σ can be compared with predictions from various models, for instance from the geometrical scaling model (Sect. 2.6.4.3) and from models which start from the inelastic overlap function, Sect. 2.6.4.4. In both cases it is of interest to calculate the total inelastic cross sections

$$\sigma_{in}(\pi^+ p) = \sigma(\pi^+ p) - \sigma_{el}(\pi^+ p); \qquad \sigma_{in}(\pi^- p) = \sigma(\pi^- p) - \sigma_{el}(\pi^- p) - \sigma_0(\pi^- p), \tag{6.8.1}$$

where σ_0 denotes the integrated charge-exchange cross section (Table 2.6.5.2).

ii) The quantity $(\sigma_{el}/\sigma^2)t$ occurs in some models as a scaling variable. Furthermore, the ratio σ_{el}/σ^2 determines the s-dependence in the rigorous bounds of MACDOWELL-MARTIN and of ROY and SINGH (Sect. A.5.3).

Figures 6.8.1 and 6.8.2 show a comparison between the variations of the total and the total inelastic $\pi^\pm p$ cross sections above 2 GeV/c. It is remarkable that a considerable part of the decrease of the total cross sections is due to that of the elastic cross sections. For instance in $\pi^+ p$ scattering, the decrease between the peak at 2.5 GeV/c and 200 GeV/c amounts to 6.94 mb for the total cross section and to 5.49 mb for σ_{el}.

The relatively small variation of the inelastic cross sections between 5 GeV/c and 200 GeV/c (4 % for $\pi^+ p$ and 11 % for $\pi^- p$) shows that the variation of the "interaction radii" $R^2(s)$ as defined in Eq. (6.4.13) is small.

The approximate constancy of $\sigma_{in}(\pi^+ p)$ is in agreement with the assumption in the work of HENZI (67 L). In a Regge pole model it is related to the fact that the f- and ϱ-exchange terms have opposite signs.

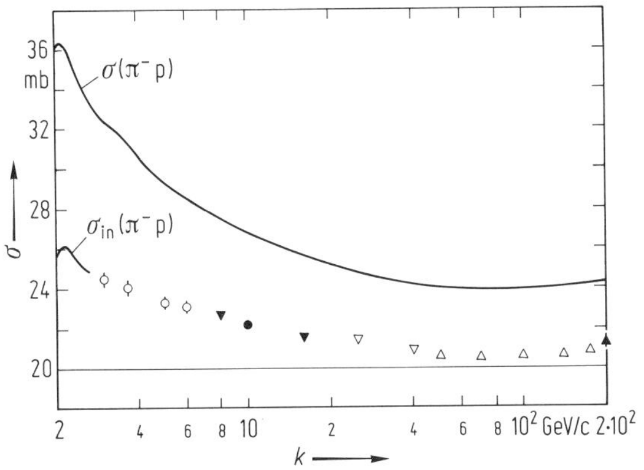

Fig. 6.8.1. π^-p total and total inelastic cross sections. σ_{el} from: ⟟ 74 AMBATS, ▼ 77 RUSS, ● BRANDE, ▽ 74 ANTIPO, △ 77 AYRES, ▲ 81 SCHIZ. Lower solid line: σ_{in} from KH 78 partial waves.

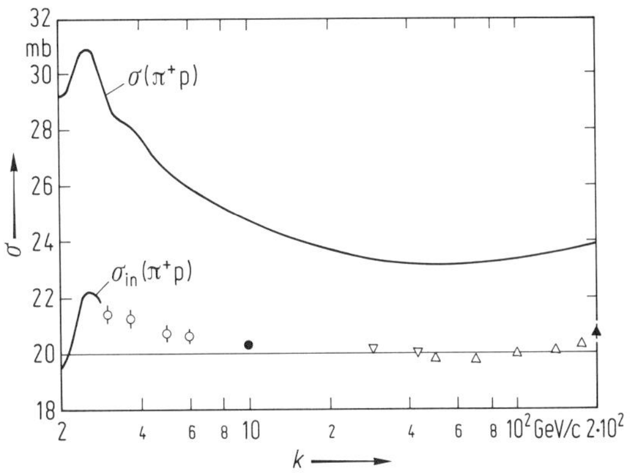

Fig. 6.8.2. π^+p total and total inelastic cross sections. The notation is the same as in Fig. 6.8.1.

The numerical values of the total elastic cross sections at high energies are listed in Table 2.6.8.1. In addition to the published values we have given results obtained in our fits. As discussed in Sect. 2.6.3.1, we have used data at small $|t|$ including the Coulomb interference region in order to adjust the normalizations and to describe the shape at $t > -0.1\,\text{GeV}^2$ in a correct way. Due to the rapid decrease of the differential cross section, an appreciable part of the total elastic cross section comes from this range. In most of the experimental papers the total elastic cross section is calculated from a fit of the type of Eq. (6.3.3) with M = 2, which is too simple.

It is seen that $\sigma_{in}(\pi^\pm p)$ has a minimum at 50⋯70 GeV/c and increases slowly at higher momenta. The accuracy of the total elastic cross sections is not good enough to decide, if the increase also occurs for this quantity.

pp-scattering data are available up to much higher energies due to the experiments at the ISR. Figure 6.8.3 shows that the total elastic-cross section is varying in such a way that the ratio σ_{el}/σ becomes constant above 100 GeV/c.

$\sigma_{in}(\pi^- p)$ is larger than $\sigma_{in}(\pi^+ p)$, because it has a larger number of channels in the final state, in particular strange particle channels.

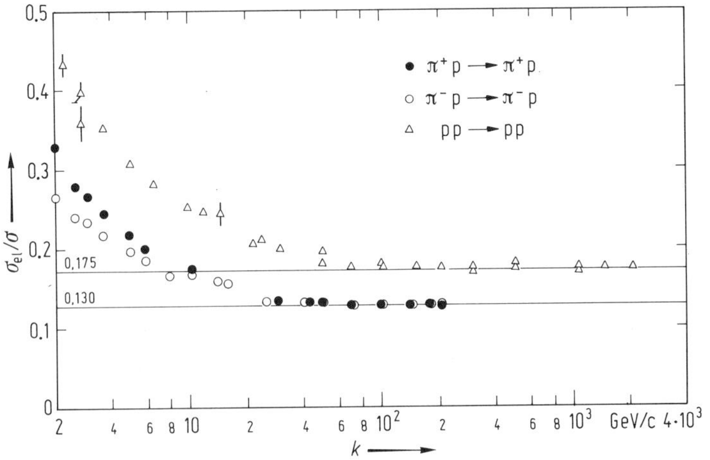

Fig. 6.8.3. Momentum dependence of the ratios σ_{el}/σ for πp and pp-scattering. k is the lab. momentum of the incoming particle. In πp scattering the data are those of our fit in Table 2.6.8.1. The statistical errors are not shown, since they would hardly be visible in our scale. It is seen that the normalization is not yet satisfactory between 5 and 50 GeV/c. — The pp-scattering data were taken from Tables 1.1.1.1.1 and 1.1.1.2.1 of Landolt-Börnstein, Vol. I/9a. Only a few of the error bars are shown. The shapes of the curves for πp and pp-scattering are similar, so one can expect that the ratio remains constant at higher energies also for πp scattering.

It is of interest to consider the decomposition of the difference of the total $\pi^\pm p$ cross section $\sigma_- - \sigma_+ \equiv 2\sigma^-$ into the elastic and inelastic contributions

$$\sigma^- = \sigma_{el}^- + \sigma_{in}^-$$
$$\sigma_{el}^- \equiv \tfrac{1}{2}\{\sigma_{el}(\pi^- p) + \sigma_0(\pi^- p) - \sigma_{el}(\pi^+ p)\}; \quad \sigma_{in}^- \equiv \tfrac{1}{2}\{\sigma_{in}(\pi^- p) - \sigma_{in}(\pi^+ p)\}, \tag{6.8.2}$$

because it gives the "elastic" and "inelastic" contributions to the imaginary part of the forward amplitude, Eq. (A.5.18),

$$\operatorname{Im} C^-(s,0) = \operatorname{Im} C_{el}^-(s,0) + \operatorname{Im} C_{in}^-(s,0). \tag{6.8.3}$$

ϱ-exchange belongs to the inelastic part, because the coupling of ϱ to four pions is negligible and the elastic part has at least four pions in the t-channel intermediate state (Fig. A.5.1). One expects therefore that the inelastic part is dominating at high energies and this turns out to be true (Fig. 6.8.4).

Above the resonance region the elastic part is small and comparable with its errors. From the known difference of the $\pi^\pm p$ forward cross sections and the assumption that the differential cross sections decrease exponentially and have a cross-over at $t \approx -0.2\,\text{GeV}^2$, one obtains the crude estimate $\sigma_{el}^- = 0.03$ mb at 200 GeV/c. Partial wave analysis gives 0.09 mb at 6 GeV/c, i.e. the decrease seems to be rather slow.

The momentum dependence of the integrated charge-exchange cross section is shown in Fig. 6.8.5. It is interesting to note that the extrapolation of the Regge law to lower momenta approximately agrees with the integral over the near-forward peak.

Figures 6.8.6a and b show the decomposition of the total $\pi^\pm p$ cross sections into elastic and inelastic contributions at low and intermediate energies.

If isospin invariance is assumed, the three integrated cross sections are constrained by the following bound (72 R Roy)

$$\{\sigma_{el}(\pi^- p) - \sigma_{el}(\pi^+ p)\}^2 < 4\sigma_0\{\sigma_{el}(\pi^- p) + \sigma_{el}(\pi^+ p) - \sigma_0\}. \tag{6.8.4}$$

At high energies, the bound is not much higher than the data (Fig. 6.8.4).

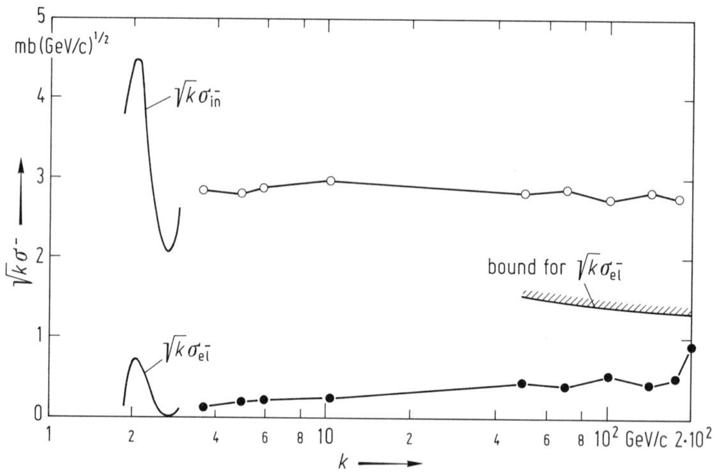

Fig. 6.8.4. Momentum dependence of $\sqrt{k}\sigma_{in}^-$ and $\sqrt{k}\sigma_{el}^-$. The factor $\sqrt{k}$ has been applied, because the sum of the two quantities: $\sqrt{k}\sigma^-$ is approximately independent of k (Fig. 2.1.7). Above 3 GeV/c, $\sqrt{k}\sigma_{el}^-$ is not well determined. The errors are comparable with the magnitude of σ_{el}^-. — The isospin bound, Eq. (6.8.4) is fulfilled (σ_0 is negligible in this range). The decrease of σ_{in}^- is compatible with a power law and $\alpha \approx 0.5$.

Fig. 6.8.5. The integrated charge-exchange cross section. $\omega\sigma_0$ has been plotted in order to have a slow energy dependence at high energies, where the reggeized ϱ-exchange contribution is dominant. The extrapolation of the Regge law to lower energies almost agrees with the integral over the cross section of the near-forward peak. ($t > -0.5\,\mathrm{GeV}^2$.) The experimental points belong to the data listed in Table 2.6.5.2.

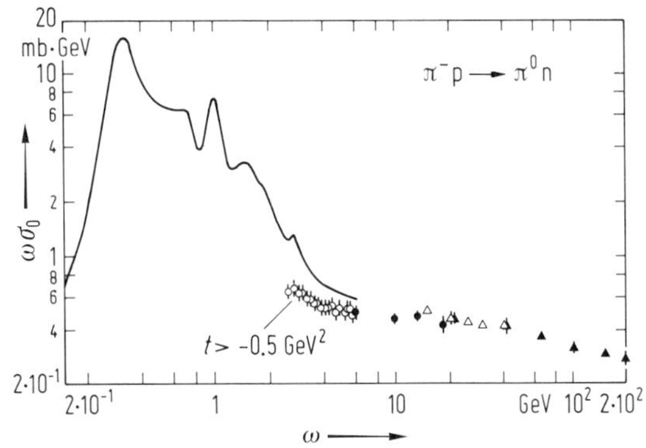

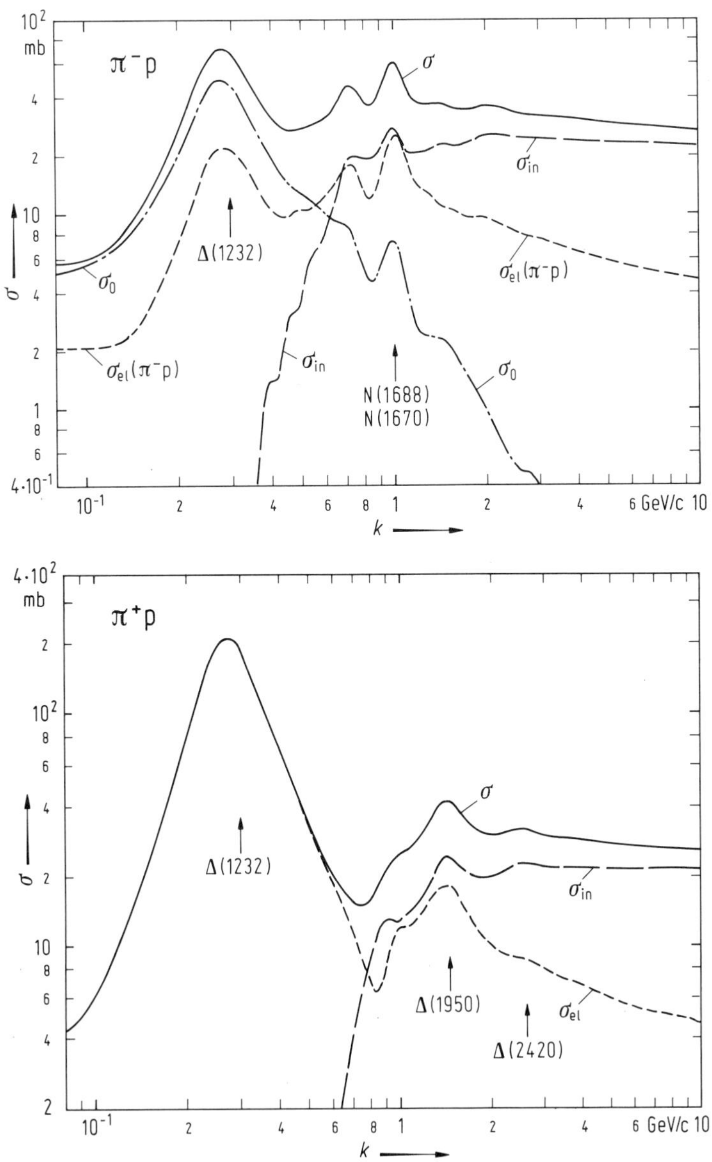

Fig. 6.8.6. Decomposition of the total $\pi^{\pm}$p cross sections into elastic and inelastic contributions at intermediate and low energies (from KH 78 partial waves).

Table 2.6.8.1. Total elastic and total inelastic $\pi^{\pm}p$ cross sections at high energies. In addition to the published values we have listed the results of our fit which is described in Sect. 2.6.3. The formal error of the fit is of little interest, because the main uncertainty comes from the possibility to chose different shapes for $d\sigma/dt$ at small $|t|$. The numbers in brackets follow from the values given for σ_{el}/σ.

$\pi^{+}p$ scattering

k GeV/c	Reference	fit of the authors σ_{el} mb	fit of the authors σ_{el}/σ	our fit σ_{el} mb	our fit σ_{el}/σ	our fit σ_{in} mb	our fit σ mb
200	81 SCHIZ	3.17±0.06	0.133±0.003	3.11	0.130	20.77	23.88
200	76 AKERLO	(3.11)	0.130±0.012				
175	77 AYRES	3.37±0.10	0.143±0.004	3.11	0.131	20.62	23.73
140	77 AYRES	3.39±0.12	0.145±0.005	3.06	0.130	20.46	23.52
100	76 AKERLO	(3.14)	0.135±0.012	3.03	0.130	20.25	23.28
100	77 AYRES	3.30±0.15	0.142±0.007				
70	77 AYRES	3.35±0.12	0.145±0.005	3.06	0.132	20.10	23.16
50	76 AKERLO	(3.33)	0.144±0.012	3.11	0.134	20.04	23.15
50	77 AYRES	3.29±0.11	0.142±0.005				
43	76 ANTIPO	3.13±0.04	0.135±0.002	3.15	0.136	20.04	23.19
29	76 ANTIPO	3.20±0.04	0.137±0.002	3.17	0.135	20.23	23.40
10.4	75 BRANDE			4.39	0.178	20.33	24.72
6.0	74 AMBATS	5.33±0.22	0.205±0.008	5.27	0.203	20.72	25.99
5.0	74 AMBATS	5.79±0.24	0.218±0.009	5.77	0.217	20.81	26.58
3.65	74 AMBATS	6.88±0.29	0.244±0.010	6.91	0.245	21.25	28.16
3.0	74 AMBATS	7.84±0.33	0.268±0.011	7.82	0.267	21.46	29.28
2.56	KH 78			8.60	0.279	22.22	30.82
2.03	KH 78			9.61	0.329	19.58	29.19
1.505	KH 78			17.30	0.429	23.03	40.33
1.00	KH 78			11.77	0.486	12.43	24.20

$\pi^{-}p$ scattering

k GeV	Reference	fit of the authors σ_{el} mb	fit of the authors σ_{el}/σ	our fit σ_{el} mb	our fit σ_{el}/σ	our fit σ_{in} mb	our fit σ mb	our fit σ_0 mb
200	81 SCHIZ	3.32±0.15	0.136±0.006	3.25	0.133	21.10	24.35	
200	76 AKERLO	(3.14)	0.129±0.012					
175	77 AYRES	3.38±0.11	0.140±0.005	3.19	0.132	21.04	24.23	
140	77 AYRES	3.36±0.12	0.140±0.005	3.13	0.130	20.94	24.07	
100	76 AKERLO	(3.23)	0.135±0.012	3.14	0.131	20.80	23.94	
100	77 AYRES	3.28±0.11	0.137±0.005					
70	77 AYRES	3.39±0.14	0.141±0.006	3.16	0.132	20.79	23.95	
50	76 AKERLO	(3.42)	0.142±0.012	3.23	0.134	20.84	24.08	0.01
50	77 AYRES	3.48±0.11	0.145±0.005					
40.1	74 ANTIPO	3.32±0.06	0.137±0.002			20.93	24.25	0.01
25.2	74 ANTIPO	3.35±0.06	0.135±0.002			21.44	24.87	0.02
16.0	77 RUSS	4.08±0.05	0.159±0.002	4.07	0.159	21.53	25.64	0.03
14.15	73 DZIERB			4.19	0.162	21.70	25.93	0.04
10.4	75 BRANDE			4.50	0.168	22.17	26.72	0.05
8.0	77 RUSS	4.75±0.05	0.173±0.002	4.60	0.167	22.84	27.51	0.07
6.0	74 AMBATS	5.30±0.22	0.186±0.008	5.35	0.188	23.07	28.52	0.10
5.0	74 AMBATS	5.85±0.24	0.200±0.007	5.82	0.199	23.33	29.28	0.13
3.65	74 AMBATS	6.97±0.29	0.223±0.009	6.83	0.218	24.23	31.27	0.21
3.0	74 AMBATS	7.57±0.31	0.233±0.010	7.61	0.235	24.48	32.44	0.35
2.56	KH 78			8.19	0.243	25.02	33.70	0.49
2.03	KH 78			9.65	0.265	25.75	36.35	0.95
1.505	KH 78			10.88	0.303	22.78	35.86	2.20
1.00	KH 78			25.43	0.420	27.66	60.48	7.39

2.6.9 Survey index for the experimental data in the near-backward *u*-range

+, −, 0 belong to the differential cross sections and P, M, N to the polarization parameters. Cf. Table 1.2.2 in Vol. I/9b1

```
K [GEV/C]                                             REF
   +0.16              U                   -0.80           PTS
2.056 *   0  0 0  0  0  0 0  0  0 0  0  0   0  * 76BROWN   13
2.057 *  P P  P  P PP  P P P   P          P  PP* 75MARTIN  13
2.060 *        -   -   -   -   -   -   -   -  -  * 80TERADA   8
2.070 *-- --   - -   -   -   -         - -   -   * 71APLIN   12
2.070 *       ++ + ++ + + ++ + ++ + ++ + ++ + + + * 74ABE     20
2.070 *              +  +   +   +  + +  +  +     * 71ALBROW   8
2.070 *                                       M  * 70HILL     1
2.070 *                     M   M   M  MM  M  M* 71HILL     7
2.070 *               P  P   P   P  P P  P  P    * 71ALBROW   8
2.100 * - --  - -   - - - -     - - -             * 71APLIN   13
2.100 *++                                        * 75LENNOX   5
2.100 *0                                         * 72KISTIA   1
2.110 *       + ++ + + ++ + + ++ + ++ + + ++ + +* 74ABE     20
2.110 * P  P P  P   P  P   P    P    P   P     P* 71BURLES  11
2.111 *------- ----- -- - - - - -                * 77VAVRA   20
2.113 *       P  P P P P  P P   P  P    P P  P   * 75MARTIN  12
2.150 * ---- - -  -   -  - - -  -   - -   -      * 71APLIN   13
2.150 *----                                      * 72MEANLY  12
2.150 *0                                         * 72KISTIA   1
2.160 *     P P   P P P  P  P  P   P  P          * 75MARTIN  10
2.168 *          N      N         N         N  N * 79MINOWA   4
2.172 *            0  0   0   0  0   0  0   0    * 79SUZUKI   8
2.180 *  + + + ++ + + ++ + + + ++ + + + ++        * 71SIDWEL  19
2.190 *         + + + + ++ + + + ++ + + + ++ + +* 74ABE     18
2.190 *          +    +     +    +    +     +    * 71ALBROW   6
2.190 *       0 00 0  0  0  0  0  0  0  0 0  0   * 73NELSON  13
2.190 *          P    P     P    P    P     P    * 71ALBROW   6
2.200 * --- - -  -   -  - - -  -    - -       -  * 71APLIN   13
2.200 *+++                                       * 75LENNOX   5
2.200 *0                                         * 72KISTIA   1
2.205 *----- ----- -- - - - - -  -  - -          * 77VAVRA   21
2.219 *     P  P  P P P P  P  P     P            * 75MARTIN   9
2.220 *    M  M   M    M   M    M      M       M* 80FUKUSH   8
2.240 * --- - -  -   -  - - -  -    - -          * 71APLIN   12
2.250 *0                                         * 72KISTIA   1
2.260 *        -   -   -   -   -   -   -   -   -* 80TERADA   9
2.267 *    0  0  0  0  0  0  0  0  0  0  0  0    * 76BROWN   12
2.267 *  N  N   N   N   N  N      N   N   N   N * 78BROWN   10
2.269 *     P  P  P P P P  P  P    P  P          * 75MARTIN  10
2.270 *        M            M                    * 70HILL     2
2.280 * + ++ + + + + + + ++ + + + + + + + ++     * 71SIDWEL  20
2.280 *   M  M  M   M    M   M    MM   M   M   M* 71HILL    11
2.290 *  -  -   -   -    -  -  -   -   -  -   -* 72HAGOPI  11
2.290 * --- -  - -   -   -- -  -       -  -      * 71APLIN   12
```

```
K [GEV/C]                                             REF
   +0.16              U                   -0.80           PTS
2.300 *+++                                       * 75LENNOX   5
2.300 *          + + + + + + + ++ + + + + + + + * 74ABE     16
2.300 *          +       +       +       +     +* 71ALBROW   5
2.300 *0                                         * 72KISTIA   1
2.300 *          P       P       P       P     P* 71ALBROW   5
2.309 *---- ---- --- - - - - -  -- -             * 77VAVRA   20
2.310 *   P P  P  P   P     P       P            * 71BURLES   7
2.316 *      P P  P  P P P  P                    * 75MARTIN   7
2.340 * --- --  - -  - - - - -         - -       * 71APLIN   12
2.350 *0                                         * 72KISTIA   1
2.360 *          N       N       N       N       * 79MINOWA   4
2.368 *      P P  P  P P P  P  P                 * 75MARTIN   8
2.370 *             0   0   0   0  0   0   0     * 79SUZUKI   7
2.380 * --- --  - -  - - - - -          - -      * 71APLIN   12
2.380 *  - - - - - - - - - - - - -  -   -  -  -  * 70CRITTE  15
2.380 * ++ + + + + + + + + + + + + + + + + + +  * 71SIDWEL  20
2.390 *          +     +       +       +     + * 71ALBROW   5
2.390 *     0 00  0 0 0   0   0   0  0  0 0  0* 73NELSON  13
2.390 *          P     P       P       P     P * 71ALBROW   5
2.395 *       P    P P  P P   P P                * 75MARTIN   7
2.400 *+++                                       * 75LENNOX   5
2.400 *0                                         * 72KISTIA   1
2.412 *---- - ------ --- - - - - - - - - -      * 77VAVRA   22
2.450 *    -   -   -   -   -   -   -   -   -   - * 80TERADA   9
2.450 *   P P  P  P P  P P  P  P   P P P P P     * 75MARTIN  14
2.460 * --- - - -  -   -  - -  -   -          -  * 71APLIN   12
2.460 *   M   M   M    M   M    M      M      M* 80FUKUSH   8
2.480 *                                -         * 72ALBROW   1
2.480 *                                M         * 72ALBROW   1
2.490 * --- - - -  -   -  - -  -   -          -  * 71APLIN   12
2.500 *------                                    * 72MEANLY  11
2.500 *  - -  - - - - - - - -  -   -   -   -   -* 70CRITTE  14
2.500 *++++                                      * 75LENNOX   5
2.500 *0                                         * 72KISTIA   1
2.500 *PPPP PPPPP PP PP P P  P                   * 73SCHEID  16
2.511 *   P P    P  P P P  P   P  P   P P   P   * 75MARTIN  12
2.515 *------- ----- --- - - - - - - - - -      * 77VAVRA   22
2.566 *          N       N       N       N     N* 79MINOWA   4
2.569 *         0   0    0   0   0   0   0      * 79SUZUKI   7
2.570 *     P  P  P  P P  P  P  P   P            * 75MARTIN   9
2.594 *   P  P P   P P  P P  P           P       * 75MARTIN   9
2.600 *------ ---- ---- - - - - - - - - - -      * 77VAVRA   22
2.600 * +++                                      * 75LENNOX   5
2.600 * 0000 0 0 0 00 0    0     0           0 * 75MARZO   13
```

K [GEV/C]	REF	PTS
2.600	72KISTIA	1
2.650	70CRITTE	12
2.650	80TERADA	7
2.651	75MARTIN	10
2.700	75LENNOX	5
2.700	72KISTIA	1
2.710	80FUKUSH	7
2.724	76BROWN	10
2.730	72ALBROW	2
2.730	71ALBROW	5
2.730	72ALBROW	2
2.730	71ALBROW	5
2.750	71SIDWEL	17
2.750	72KISTIA	1
2.750	73SCHEID	16
2.767	79SUZUKI	6
2.770	70BATON	9
2.800	72MEANLY	9
2.800	75LENNOX	6
2.800	72KISTIA	1
2.801	77VAVRA	22
2.850	70CRITTE	14
2.850	80TERADA	7
2.850	75MARZO	9
2.900	75LENNOX	6
2.900	72KISTIA	1
2.930	76AUER	10
2.930	81AUER	10
2.930	73SCHEID	16
2.950	72KISTIA	1
2.960	79MINOWA	3
2.965	79SUZUKI	7
3.000	75HOFFMA	14
3.000	72BRABSO	12
3.000	70CRITTE	12
3.000	75LENNOX	6
3.000	71SIDWEL	15
3.000	72KISTIA	1
3.000	70CHASE	11
3.006	77VAVRA	22
3.050	80TERADA	7
3.050	72KISTIA	1
3.100	75LENNOX	5
3.100	72KISTIA	1
3.200	75LENNOX	6
3.200	72KISTIA	1
3.250	76HOFFMA	15
3.250	74BASHIA	5

K [GEV/C]	REF	PTS
3.250	71SIDWEL	14
3.250	76AUER	10
3.250	81AUER	10
3.250	73SCHEID	10
3.260	80TERADA	8
3.300	75LENNOX	6
3.300	72KISTIA	1
3.400	75LENNOX	6
3.400	72KISTIA	1
3.480	80TERADA	6
3.500	76HOFFMA	15
3.500	72MEANLY	3
3.500	75LENNOX	6
3.500	74BASHIA	6
3.500	71SIDWEL	12
3.500	75MARZO	26
3.500	72KISTIA	1
3.500	76BIRSA	6
3.500	80FUKUSH	4
3.500	73BRADAM	10
3.600	75LENNOX	6
3.600	72KISTIA	1
3.630	71MACNAU	4
3.700	75LENNOX	6
3.700	72KISTIA	1
3.750	76HOFFMA	12
3.750	74BASHIA	6
3.750	71SIDWEL	11
3.750	72KISTIA	1
3.750	73SCHEID	8
3.800	75LENNOX	6
3.800	72KISTIA	1
3.850	72KISTIA	1
3.900	75LENNOX	6
3.900	71BASTIE	3
3.900	72KISTIA	1
3.920	76ALITTI	2
3.950	72MEANLY	3
3.950	72KISTIA	1
4.000	76HOFFMA	17
4.000	72BRABSO	12
4.000	78SCHARR	8
4.000	75LENNOX	6
4.000	74BASHIA	6
4.000	71SIDWEL	11
4.000	72KISTIA	1
4.000	71SCHNEI	8
4.000	70CHASE	12

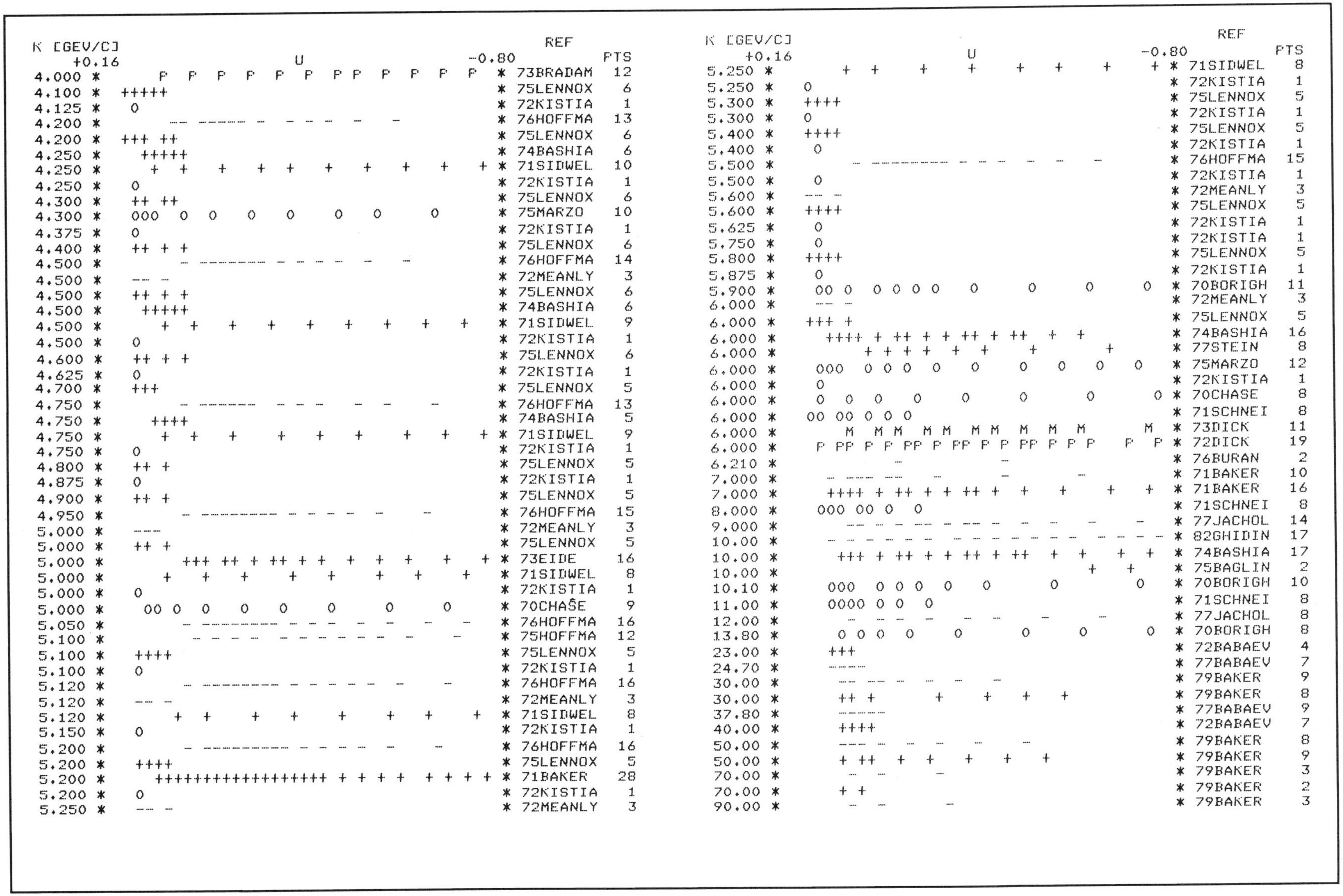

K [GEV/C]	REF	PTS
4.000	73BRADAM	12
4.100	75LENNOX	6
4.125	72KISTIA	1
4.200	76HOFFMA	13
4.200	75LENNOX	6
4.250	74BASHIA	6
4.250	71SIDWEL	10
4.250	72KISTIA	1
4.300	75LENNOX	6
4.300	75MARZO	10
4.375	72KISTIA	1
4.400	75LENNOX	6
4.500	76HOFFMA	14
4.500	72MEANLY	3
4.500	75LENNOX	6
4.500	74BASHIA	6
4.500	71SIDWEL	9
4.500	72KISTIA	1
4.600	75LENNOX	6
4.625	72KISTIA	1
4.700	75LENNOX	5
4.750	76HOFFMA	13
4.750	74BASHIA	5
4.750	71SIDWEL	9
4.750	72KISTIA	1
4.800	75LENNOX	5
4.875	72KISTIA	1
4.900	75LENNOX	5
4.950	76HOFFMA	15
5.000	72MEANLY	3
5.000	75LENNOX	5
5.000	73EIDE	16
5.000	71SIDWEL	8
5.000	72KISTIA	1
5.000	70CHAŠE	9
5.050	76HOFFMA	16
5.100	75HOFFMA	12
5.100	75LENNOX	5
5.100	72KISTIA	1
5.120	76HOFFMA	16
5.120	72MEANLY	3
5.120	71SIDWEL	8
5.150	72KISTIA	1
5.200	76HOFFMA	16
5.200	75LENNOX	5
5.200	71BAKER	28
5.200	72KISTIA	1
5.250	72MEANLY	3

K [GEV/C]	REF	PTS
5.250	71SIDWEL	8
5.250	72KISTIA	1
5.300	75LENNOX	5
5.300	72KISTIA	1
5.400	75LENNOX	5
5.400	72KISTIA	1
5.500	76HOFFMA	15
5.500	72KISTIA	1
5.600	72MEANLY	3
5.600	75LENNOX	5
5.625	72KISTIA	1
5.750	72KISTIA	1
5.800	75LENNOX	5
5.875	72KISTIA	1
5.900	70BORIGH	11
6.000	72MEANLY	3
6.000	75LENNOX	5
6.000	74BASHIA	16
6.000	77STEIN	8
6.000	75MARZO	12
6.000	72KISTIA	1
6.000	70CHASE	8
6.000	71SCHNEI	8
6.000	73DICK	11
6.000	72DICK	19
6.210	76BURAN	2
7.000	71BAKER	10
7.000	71BAKER	16
8.000	71SCHNEI	8
9.000	77JACHOL	14
10.00	82GHIDIN	17
10.00	74BASHIA	17
10.00	75BAGLIN	2
10.10	70BORIGH	10
11.00	71SCHNEI	8
12.00	77JACHOL	8
13.80	70BORIGH	8
23.00	72BABAEV	4
24.70	77BABAEV	7
30.00	79BAKER	9
30.00	79BAKER	8
37.80	77BABAEV	9
40.00	72BABAEV	7
50.00	79BAKER	8
50.00	79BAKER	9
70.00	79BAKER	3
70.00	79BAKER	2
90.00	79BAKER	3

Notes added in proof

Addendum to Sect. 2.6.3.6

The high energy behaviour of the pp-forward amplitude was recently investigated by Caprini et al. (80 L), who used the total cross section and real part data and a rather sophisticated method for the application of the analyticity constraint. The authors concluded that the data are not inconsistent with a total cross section which is asymptotically constant. – This result follows immediately from the conventional dispersion integral, because the real parts up to a given energy are not sensitive to the behaviour of the total cross section above a much higher energy as long as the Froissart bound is respected [see Eq. (6.2.5)]. – It is a disadvantage of the method of Caprini et al. that the best analytic approximant does not have the expected smoothness but shows wiggles (Fig. 2 in 80 L Caprini).

At intermediate energies the conventional method is somewhat complicated due to the left hand cut (77 L Grein, 81 R Kroll), but at very high energies it is sufficient to employ an analytic parametrization. Block and Cahn (82 L) studied various fits of this type, including the latest $\bar{p}p$-data from the ISR (82 L Carboni-1 and 2 and real parts). They found that all data above $s=25\ \mathrm{GeV}^2$ can be fitted by an ansatz for the total cross sections, which has a constant term, a term $\propto \ln^2(s/s_0)$, and a power law $\propto s^{\alpha-1}$ for the total cross section difference. The exponent $\alpha=0.48\pm0.01$ does not differ significantly from 1/2. – There is no evidence for an unconventional asymptotic behaviour, for instance for an "odderon"-exchange (cf. the remarks in Sect. 2.6.5.2 and 75 L Joynson).

Addendum to Sect. 3.1

82 R Dominguez
Present status of the Goldberger-Treiman relation
C. A. Dominguez: Valparaiso preprint and talk at the Black Forest Meeting **1982**
2.5.1.4

82 R Gasser-Leutwyler
Quark masses
J. Gasser, H. Leutwyler: Physics Rep. **87** (1982) 77
2.5.1

Addendum to Sect. 3.2

71 L Alcock
The two-pion exchange contribution to meson-nucleon scattering in the low and intermediate energy region
J. W. Alcock, W. N. Cottingham: Nucl. Phys. B **31** (1971) 443
2.1.9; 2.1.14

72 Alcock
The peripheral partial waves in pion-nucleon scattering from two-pion exchange
J. W. Alcock, W. N. Cottingham: Nucl. Phys. B **41** (1972) 141
2.1.9; 2.1.14

69 L Alvarez-Estrada
Causality and electromagnetic form factors
R. F. Alvarez-Estrada: Ann. of Phys. **54** (1969) 276
2.5.2; A.11.3

73 L Alvarez-Estrada
On the rigorous determination of the spin 0 – spin 1/2 scattering amplitudes from data and unitarity
R. F. Alvarez-Estrada: Nucl. Phys. B **54** (1973) 237
2.1.2

81 L Alvarez-Estrada
Rigorous approach to elastic meson-nucleon scattering in non-relativistic quantum field theory I and II
R. F. Alvarez-Estrada: Ann. Inst. Henri Poincaré **34** (1981) 253, 277
2.5.1.7

82 L Alvarez-Estrada
Further studies of convergence in the cloudy bag model
R. F. Alvarez-Estrada, A. W. Thomas: CERN preprint TH 3316 (1982)
2.5.1.7

82 L Breakstone
Measurement of $\bar{p}p$ and pp elastic scattering at ISR energies
A. Breakstone et al.: submitted to the Intern. Conf. on High Energies (Paris) **1982**
2.6.3.6

58 L Fronsdal
On the theory of higher spin fields
C. Fronsdal: Suppl. Nuovo Cim. **9** (1958) 416
A.8.1.2; 2.5.1.5

82 L Kaplan
$\bar{p}p$ and pp elastic scattering at 100 and 200 GeV/c
D. H. Kaplan et al.: Phys. Rev. D **26** (1982) 723
2.6.3.6

80 L Kriesche
Coulomb corrections in non-relativistic π^+p scattering
H. Kriesche, H. Zankel: J. Phys. G: **6** (1980) 853
2.1.17

3 Bibliography

In this section the reader can find a list of books, reviews and lecture notes (Sect. 3.1), of original papers and preprints (Sect. 3.2) and of papers on pion-nucleon scattering experiments (Sect. 3.3). This last list is an updated version of two pages of Sect. 1.2.1 (Vol. I/9b1).

It has been my aim to give a fairly complete compilation of books, reviews and lecture notes, including closely related fields. But the search turned out to be difficult, even if one uses a literature reference retrieval system. One of the reasons is that summer school lecture notes and reviews based on talks given at conferences are sometimes published directly by the organizers.

For those topics which have been treated in detail in this volume, I have included all relevant papers published during the last decade, which I have been able to find, and some of the earlier ones. For other topics I have only given a "guide to the literature", mentioning recent papers and preprints from which earlier work can be traced.

In the last line of each entry the subsections are given to which the reference belongs.

Section A.12 contains a list of references ordered according to the subsections of the Appendix.

Further references added in proof: see p. 404 and p. 423 (Sect. 3.1), p. 477 (Sect. 3.2), and p. 479 (Sect. 3.3).

3.1 Books, reviews and lecture notes

71R ABARBANEL
EIKONAL APPROXIMATIONS IN COLLISION THEORY
H.D.I.ABARBANEL IN"TOPICS IN STRONG INTERACTIONS"(BOULDER 1971),P.3;EDS.A.O.BARUT,W.E. BRITTIN;COLORADO UNIV.PRESS(1972)
2.6.4

75R ABARBANEL-BRONZAN-SUGAR-WHITE
REGGEON FIELD THEORY:FORMULATION AND USE
H.D.I.ABARBANEL,J.D.BRONZAN,R.L.SUGAR,A.R.WHITE: PHYSICS REP.21(1975)119
2.6.4

76R ABARBANEL
DIFFRACTION SCATTERING OF HADRONS
H.D.I.ABARBANEL:REV.MOD.PHYS.48(1976)435
2.6.4

68R ADLER-DASHEN
CURRENT ALGEBRAS AND APPLICATIONS TO PARTICLE PHYSICS
S.L.ADLER,R.F.DASHEN;BENJAMIN(1968)
A.8.2; 2.5.1

73R ALFARO-FUBINI-FURLAN-ROSETTI
CURRENTS IN HADRON PHYSICS
V.DE ALFARO,S.FUBINI,G.FURLAN,C.ROSETTI;NORTH-HOLLAND(1973)
A.8.2; 2.5.1; 2.4.2

79R AMALDI-FUBINI-FURLAN
PION-ELECTROPRODUCTION
E.AMALDI,S.FUBINI,G.FURLAN:SPRINGER TRACTS MOD.PHYS. 83(1979)1
2.5.2; A.11.2

80R AMALDI-SCHUBERT
IMPACT PARAMETER INTERPRETATION OF PP-SCATTERING
U.AMALDI,K.R.SCHUBERT:NUCL.PHYS.B166(1980)301
2.6.3; 2.6.4

62R AMATI-FUBINI
DISPERSION RELATION METHOD IN STRONG INTERACTIONS
D.AMATI,S.FUBINI:ANN.REV.NUCL.SCI.12(1962)359
A.6

73R ATKINSON
THEORETICAL INVESTIGATIONS OF PHASE-SHIFT ANALYSIS
D.ATKINSON:LECTURE NOTES IN PHYSICS 17(1973)61
2.1.1; 2.1.3

76R ATKINSON
THE CONTINUUM AMBIGUITY IN PION-NUCLEON PHASE SHIFT ANALYSIS
D.ATKINSON IN"PROC.TOPICAL CONF.ON BARYON RESONANCES"(OXFORD),P.189;RUTHERFORD LAB.(1976)
2.1.3

79R ATWOOD
LEPTON-NUCLEON SCATTERING
W.ATWOOD:SLAC SUMMER INSTITUTE ON PARTICLE PHYSICS(1979)
2.5.2

80R AUGUSTIN
SELECTED TOPICS IN E+E- PHYSICS RESULTS
J.E.AUGUSTIN IN"QUARKS AND LEPTONS"(CARGESE 1979);EDS. M.LEVY ET AL.;PLENUM PRESS(1980)
2.5.2

70R BAACKE-STEINER
PION-NUCLEON PARTIAL WAVE RELATIONS FROM FIXED-T
DISPERSION RELATIONS
J.BAACKE,F.STEINER:FORTSCHR.PHYSIK 18(1970)67
A.7.2; 2.4.7

76R BAKER – TER-MARTIROSYAN
GRIBOV'S REGGEON CALCULUS:ITS PHYSICAL BASIS AND
IMPLICATIONS
M.BAKER,K.A.TER-MARTIROSYAN:PHYSICS REP.28(1976)1
2.6.4

68R BARBARO-GALTIERI
BARYON RESONANCES
A.BARBARO-GALTIERI IN"ADVANCES IN PARTICLE PHYSICS",
VOL.2,P.175;EDS.R.L.COOL,R.E.MARSHAK;
INTERSCIENCE(1968)
2.4.1

69R BARGER-CLINE
PHENOMENOLOGICAL THEORIES OF HIGH ENERGY SCATTERING
V.D.BARGER,D.B.CLINE;BENJAMIN(1969)
A.9; 2.6.7

74R BARGER
REACTION MECHANISMS AT HIGH ENERGY
V.BARGER IN"XVII INTERNATIONAL CONFERENCE ON HIGH
ENERGY PHYSICS"(LONDON),P.I-193(1974)
2.6.3; 2.6.4; 2.6.5; 2.6.6; 2.6.7

76R BARNHAM
PI N -> N PI PI, A REVIEW OF THE EXPERIMENTAL
SITUATION
K.W.J.BARNHAM IN"PROC.TOPICAL CONF.ON BARYON
RESONANCES"(OXFORD),P.109;RUTHERFORD LAB.(1976)
2.4.1

72R BARTOLI-FELICETTI-SILVESTRINI
ELECTROMAGNETIC STRUCTURE OF HADRONS
B.BARTOLI,F.FELICETTI,V.SILVESTRINI:REVISTA NUOVO CIM.
2(1972)241
2.5.2; A.11

65R BARTON
INTRODUCTION TO DISPERSION TECHNIQUES IN FIELD THEORY
G.BARTON;BENJAMIN(1965)
2.5.2; A.7.3; A.11.3

64R BARUT
THE FRAMEWORK OF S-MATRIX THEORY
A.O.BARUT IN"STRONG INTERACTIONS AND HIGH ENERGY
PHYSICS"(SCOTT.UNIV.SUMMER SCHOOL),P.81;OLIVER AND
BOYD(1964)
A.2; A.3; A.6

67R BARUT
THEORY OF THE SCATTERING MATRIX
A.O.BARUT;MACMILLAN(1967)
A

72R BASDEVANT
THE PADE APPROXIMATION AND ITS PHYSICAL APPLICATIONS
J.L.BASDEVANT:FORTSCHR.PHYSIK 20(1972)283
2.5.1

78R BAUER-SPITAL-YENNIE-PIPKIN
THE HADRONIC PROPERTIES OF THE PHOTON IN HIGH-ENERGY
INTERACTIONS
T.H.BAUER,R.D.SPITAL,D.R.YENNIE,F.M.PIPKIN:REV.MOD.
PHYS.50(1978)261
2.6.4

70R BEHNKE-THULLEN
THEORIE DER FUNKTIONEN MEHRERER KOMPLEXER
VERAENDERLICHER
H.BEHNKE,F.THULLEN;SPRINGER(1970)
2.4.3

62R BERESTETSKI
DYNAMICAL PROPERTIES OF ELEMENTARY PARTICLES AND THE
THEORY OF THE SCATTERING MATRIX
V.B.BERESTETSKI:SOV.PHYS.USPECHI 5(1962)7
A.6

79R BERKELMAN
DIRECT EXPERIMENTAL EVIDENCE FOR CONSTITUENTS IN THE
NUCLEON FROM ELECTROMAGNETIC SCATTERING EXPERIMENTS
K.BERKELMAN:ORBIS SCIENTIAE CONFERENCE,CORAL
GABLES(1979)
2.5.2

68R BERNSTEIN
ELEMENTARY PARTICLES AND THEIR CURRENTS
J.BERNSTEIN;FREEMAN(1968)
A.8.2; 2.5.2; 2.5.1; A.11

67R BERTOCCHI
THEORETICAL ASPECTS OF HIGH-ENERGY PHENOMENOLOGY
L.BERTOCCHI IN"PROC.1967 HEIDELBERG CONF.",P.197;ED.H.
FILTHUTH
A.9

56R BETHE-DE HOFFMANN-SCHWEBER
MESONS AND FIELDS I AND II
H.A.BETHE,F.DE HOFFMANN,S.S.SCHWEBER;ROW,PETERSEN AND
CO(1956)
A

64R BJORKEN-DRELL
RELATIVISTIC QUANTUM MECHANICS
J.D.BJORKEN,S.D.DRELL;MC GRAW HILL(1964)
A

65R BJORKEN-DRELL
RELATIVISTIC QUANTUM FIELDS
J.D.BJORKEN,S.D.DRELL;MC GRAW HILL(1965)
A

79R BOEHM
QUANTUM MECHANICS
A.BOEHM;SPRINGER(1979)
2.4.1

58R BOGOLJUBOV-MEDVEDEV-POLIVANOV
PROBLEME DER THEORIE DER DISPERSIONSBEZIEHUNGEN
N.N.BOGOLJUBOV,B.V.MEDVEDEV,M.K.POLIVANOV:FORTSCHR.
PHYSIK 6(1958)169
A.6.3

76R BOGOLJUBOV-SHIRKOV
INTRODUCTION TO THE THEORY OF QUANTIZED FIELDS
N.N.BOGOLJUBOV,D.V.SHIRKOV;WILEY-INTERSCIENCE(1976)
A

60R BOHR
CAUSALITY AND DISPERSION RELATIONS
A.BOHR IN"LECTURES IN THEORETICAL PHYSICS"(BOULDER);
EDS.W.E.BRITTIN,B.W.DOWNS,J.DOWNS;INTERSCIENCE(1961)
A.6.4

80R BOURRELY-LEADER-SOFFER
POLARIZATION PHENOMENA IN HADRONIC REACTIONS
C.BOURRELY,E.LEADER,J.SOFFER:PHYSICS REP.59(1980)96
A.4; A.9; 2.6.4; 2.6.7

75R BOWCOCK-BURKHARDT
PRINCIPLES AND PROBLEMS OF PHASE-SHIFT ANALYSIS
J.E.BOWCOCK,H.BURKHARDT:REP.PROGR.PHYS.38(1975)1099
2.1.1; 2.1.2; 2.1.3

73R BRANSDEN-MOORHOUSE
THE PION-NUCLEON SYSTEM
B.H.BRANSDEN,R.G.MOORHOUSE;PRINCETON UNIV.PRESS(1973)
2; A

59R BRENIG-HAAG
ALLGEMEINE QUANTENTHEORIE DER STOSSPROZESSE
W.BRENIG,R.HAAG:FORTSCHR.PHYS.7(1959)183
2.4.1

80R BRICMAN
REVIEW OF PARTICLE PROPERTIES
C.BRICMAN ET AL.CERN REPRINT=R.L.KELLY ET AL.:REV.MOD.
PHYS.52(1980)1;NEW VERSION:82R ROOS
2.4.1

82R BRODSKY
QUANTUM CHROMODYNAMICS AT NUCLEAR DIMENSIONS
S.J.BRODSKY:SPRINGER TRACTS MOD.PHYS.100(1982)81
2.5.2; 2.5.3; 2.6.6

82R BRODSKY-HUANG-LEPAGE
QUANTUM CHROMODYNAMICS AND HADRONIC INTERACTIONS AT
SHORT DISTANCES
S.J.BRODSKY,T.HUANG,G.P.LEPAGE IN"PROC.SUMMER
INSTITUTE ON PARTICLE PHYSICS:THE STRONG
INTERACTIONS";SLAC REPORT NO.245(1982)
2.5.2

76R BROWN-JACKSON
THE NUCLEON-NUCLEON INTERACTION
G.E.BROWN,A.D.JACKSON;NORTH-HOLLAND(1976)
2.5.3

82R BROWN
NUCLEON-NUCLEON FORCES FROM BAGS,QUARKS AND BOSON
EXCHANGES
G.E.BROWN:PROGR.NUCL.PART.PHYS.8(1982)147
2.5.3

74R BRYANT-JARMIE
THE GLORY
H.C.BRYANT,N.JARMIE:SCIENT.AMERICAN 231(1974)60
2.6.7

69R BURKHARDT
DISPERSION RELATION DYNAMICS
H.BURKHARDT;NORTH-HOLLAND(1969)
A.6

73R BURKHARDT
SERIOUS AMBIGUITIES IN INELASTIC PHASE SHIFT ANALYSIS
H.BURKHARDT IN"HADRON INTERACTIONS AT LOW
ENERGIES"(TRIANGLE MEETING 1973);EDS.D.KRUPA,J.PISUT;
VEDA,BRATISLAVA(1975)
2.1.3

82R BYRNE
WEAK INTERACTIONS OF THE NEUTRON
J.BYRNE:REP.PROGR.PHYS.45(1982)115
2.5.1

71R CARRUTHERS
SPIN AND ISOSPIN IN PARTICLE PHYSICS
P.A.CARRUTHERS;GORDON AND BREACH(1971)
A.2.2

69R CENCE
PION-NUCLEON SCATTERING
R.J.CENCE:PRINCETON UNIV.PRESS(1969)
2

69R CHENG
THEORY OF HIGH ENERGY DIFFRACTION SCATTERING
H.CHENG IN"HIGH ENERGY COLLISIONS"P.329;ED.C.N.YANG
ET AL.;GORDON AND BREACH(1969)
2.6.4

59R CHEW
THE PION-NUCLEON INTERACTION AND DISPERSION RELATIONS
G.F.CHEW:ANN.REV.NUCL.SCI.9(1959)29
A.8.3

60R CHEW
DOUBLE DISPERSION RELATIONS AND UNITARITY AS A BASIS
FOR A DYNAMIC THEORY OF STRONG INTERACTIONS
G.F.CHEW IN"RELATIONS DE DISPERSION ET PARTICULES
ELEMENTAIRES"(LES HOUCHES 1960),P.455;EDS.C.DEWITT,R.
OMNES;HERMANN,PARIS(1960)
A.6

61R CHEW
S-MATRIX THEORY OF STRONG INTERACTIONS
G.F.CHEW;BENJAMIN(1961). INCLUDES REPRINTS
A.6; 2.6.4

65R CHEW
THE ANALYTIC S-MATRIX
G.F.CHEW IN"HIGH ENERGY PHYSICS"(LES HOUCHES 1965),
P.189;EDS.C.DEWITT,M.JACOB;GORDON AND BREACH(1965)
A.6; A.9

66R CHEW
THE ANALYTIC S-MATRIX
G.F.CHEW;BENJAMIN(1966)
A.6; A.9; 2.4.1

78R CHEW-ROSENZWEIG
DUAL TOPOLOGICAL UNITARIZATION:AN ORDERED APPROACH TO
HADRON THEORY
G.F.CHEW,C.ROSENZWEIG:PHYSICS REP.41(1978)263
2.6.4

72R CHIU
EVIDENCE FOR REGGE POLES AND HADRON COLLISION
PHENOMENA AT HIGH ENERGIES
C.B.CHIU:ANN.REV.NUCL.SCI.22(1972)255
A.9; 2.6.4; 2.6.7

61R CINI-VITALE
ANALYTIC PROPERTIES OF SCATTERING AMPLITUDES AND
APPLICATIONS TO PION PHYSICS
M.CINI,B.VITALE IN"LECTURES ON HIGH ENERGY
PHYSICS"(HERCEGNOVI 1961);ED.B.JAKSIC;GORDON AND
BREACH(1965)
A.6.2; A.8.3

73R CIULLI
STABILITY PROBLEMS IN ANALYTIC CONTINUATION
S.CIULLI:LECTURE NOTES IN PHYSICS 17(1973)70
A.6.3; 2.4.5; 2.5.2

75R CIULLI-POMPONIU-SABBA-STEFANESCU
ANALYTIC EXTRAPOLATION TECHNIQUES AND STABILITY
PROBLEMS IN DISPERSION RELATION THEORY
S.CIULLI,C.POMPONIU,I.SABBA-STEFANESCU:PHYSICS REP.
17(1975)133
A.6.3; 2.4.5; 2.5.2

79R CLOSE
INTRODUCTION TO QUARKS AND PARTONS
F.E.CLOSE;ACADEMIC PRESS(1979)
2.4.1

82R CLOSE
THE PAULI PRINCIPLE AND QCD FOR QUARKS AND NUCLEONS
IN HADRONS AND NUCLEI
F.E.CLOSE:SPRINGER TRACTS MOD.PHYS.100(1982)57
2.4.1

68R COLLINS-SQUIRES
REGGE POLES IN PARTICLE PHYSICS
P.D.B.COLLINS,E.J.SQUIRES: SPRINGER TRACTS MOD.PHYS.
45(1968)
A.6.2; A.9; 2.6.4; 2.6.5; 2.6.7

71R COLLINS-1
HOW IMPORTANT ARE REGGE CUTS?
P.D.B.COLLINS:SPRINGER TRACTS MOD.PHYS.
60(1971)204
A.9

71R COLLINS-2
REGGE THEORY AND PARTICLE PHYSICS
P.D.B.COLLINS:PHYSICS REP.1(1971)103
A.9; A.6; 2.6.5; 2.6.7

75R COLLINS-GAULT
A BIBLIOGRAPHY OF REGGE PHENOMENOLOGY
P.D.B.COLLINS,F.D.GAULT;DEP.OF PHYSICS,DURHAM UNIV.,
U.K.
A.9; 2.6.4; 2.6.5

77R COLLINS
REGGE THEORY AND HIGH ENERGY PHYSICS
P.D.B.COLLINS;CAMBRIDGE UNIV.PRESS(1977)
A.9; A.6; 2.6.4; 2.6.5; 2.6.7

82R COLLINS-MARTIN
HADRON REACTION MECHANISMS
P.D.B.COLLINS,A.D.MARTIN:REP.PROGR.PHYS.45(1982)335
2.6.4

72R CONTE-DE BOOR
ELEMENTARY NUMERICAL ANALYSIS
S.D.CONTE,C.DE BOOR;MC GRAW HILL(1972)
2.4.3

75R CONTOGOURIS
AMPLITUDE ANALYSES FROM INCOMPLETE DATA USING FIXED-T DISPERSION RELATIONS
A.P.CONTOGOURIS:ORSAY REPORT LPTPE 75/18(1975)
2.1.16

61R CORINALDESI-CINI
INTRODUCTION TO PION PHYSICS
E.CORINALDESI,M.CINI IN"LECTURES ON HIGH ENERGY PHYSICS"(HERCEGNOVI 1961);ED.B.JAKSIC;GORDON AND BREACH(1965)
A.1; A.2; A.3

78R COTTINGHAM
MESONS AND TWO-NUCLEON POTENTIAL
W.N.COTTINGHAM:REP.PROGR.PHYS.41(1978)1509
2.5.3

80R CRAIGIE
QCD:A THEORY OF THE NUCLEAR FORCE
N.S.CRAIGIE:TRIESTE REPORT IC/80/62(1980)
2.5.3

71R CUTKOSKY
ANALYTIC APPROXIMATION THEORY APPLIED TO ANALYSIS OF EXPERIMENTAL DATA
R.E.CUTKOSKY IN"TOPICS IN STRONG INTERACTIONS"(BOULDER 1971),P.423;EDS.A.O.BARUT,W.E.BRITTIN; COLORADO UNIV.PRESS(1972)
2.1.9; 2.6.3; 2.6.7

80R CUTKOSKY
PION-NUCLEON DATA,DISPERSION RELATIONS AND RESONANCES
R.E.CUTKOSKY IN"LOW AND INTERMEDIATE ENERGY KAON-NUCLEON PHYSICS"(ROME 1980),P.31;EDS.E.FERRARI,G. VIOLINI;REIDEL(1980)
2.1.9; 2.4.1

62R DALITZ
STRANGE PARTICLES AND STRONG INTERACTIONS
R.H.DALITZ;OXFORD UNIV.PRESS(1962)
2.1.13

70R DALITZ
WHAT IS A RESONANCE?
R.H.DALITZ,R.G.MOORHOUSE:PROC.ROY.SOC.LOND.A318(1970) 279
2.4.1

73R DALITZ
THE THREE QUARK SHELL MODEL AND THE SPECTRUM OF BARYON STATES
R.H.DALITZ IN"HADRON INTERACTIONS AT LOW ENERGIES"(TRIANGLE MEETING 1973),P.11;ED.M.BLAZEK; VEDA,BRATISLAVA(1975)
2.4.1

76R DALITZ
QUARK MODELS,OLD AND NEW
R.H.DALITZ IN"FUNDAMENTALS IN QUARK MODELS"(SCOTT. UNIV.SUMMER SCHOOL);EDS.I.M.BARBOUR,A.T.DAVIES(1976)
2.4.1

82R DALITZ
QUARKS AND LIGHT HADRONS
R.DALITZ:PROGR.PART.NUCL.PHYS.8(1982)7(ERICE 1981)
2.4.1

72R DASHEN
LECTURES ON CHIRAL SYMMETRY BREAKING
R.DASHEN IN"PROC.OF THE SCHOOL ENRICO FERMI"(VARENNA 1972);ACADEMIC PRESS(1972)
2.5.1; A.8.2

80R DE GRAND
THINKING ABOUT BARYON EXCITED STATES WITH BAGS
T.A.DE GRAND IN"BARYON 1980,PROC.TORONTO CONF.",P.209
2.4.1

80R DE SWART
MULTIQUARK BARYONS
J.J.DE SWART IN"BARYON 1980,PROC.TORONTO CONF.",P. 405(1980)
2.4.1

79R DE TAR
THE M.I.T. BAG MODEL
C.DE TAR IN"QUANTUM FLAVORDYNAMICS,QUANTUM CHROMODYNAMICS AND UNIFIED THEORIES"P.393;EDS.K.T. MAHANTHAPPA,J.RANDA;PLENUM PRESS(1979)
2.4.1

69R DOMBEY
SCATTERING OF POLARIZED LEPTONS AT HIGH ENERGY
N.DOMBEY:REV.MOD.PHYS.41(1969)236
2.5.2; A.11.1

81R DOMINGUEZ-LANGACKER
PRESENT STATUS OF THE PION-NUCLEON SIGMA TERM
C.A.DOMINGUEZ,P.LANGACKER:PHYS.REV.D24(1981)1905
2.5.1

67R DONNACHIE
PION-NUCLEON PHASE SHIFT ANALYSIS
A.DONNACHIE IN"PARTICLE INTERACTIONS AT HIGH ENERGIES";EDS.T.W.PREIST,L.L.J.VICK;OLIVER AND BOYD(1967)
A.7.1; 2.1.5; 2.4.1; 2.6.7

73R DONNACHIE
PARTIAL WAVE ANALYSIS AND BARYON RESONANCES
A.DONNACHIE:REP.PROGR.PHYS.36(1973)695
2.1; 2.4.1

78R DONNACHIE-SHAW-LYTH
FORM FACTORS AND ELECTROPRODUCTION
A.DONNACHIE,G.SHAW,D.H.LYTH IN"ELECTROMAGNETIC INTERACTIONS OF HADRONS";VOL.1,P.221;EDS.A.DONNACHIE, G.SHAW;PLENUM PRESS(1978)
2.5.2

70R DRECHSLER
COMPLEX ANGULAR MOMENTUM THEORY IN PARTICLE PHYSICS
W.DRECHSLER:FORTSCHR.PHYSIK 18(1970)305
A.9

61R DRELL-ZACHARIASEN
ELECTROMAGNETIC STRUCTURE OF NUCLEONS
S.D.DRELL,F.ZACHARIASEN;OXFORD UNIV.PRESS(1961)
2.5.2; A.11

82R DUMBRAJS
COMPILATION OF COUPLING CONSTANTS AND LOW-ENERGY PARAMETERS
O.V.DUMBRAJS ET AL.:TO APPEAR IN PHYSICS DATA(1982)
2.4.2; 2.4.7

81R EBERT-VOLKOV
MESON INTERACTIONS WITHIN NONLINEAR CHIRAL THEORIES
D.EBERT,M.K.VOLKOV:FORTSCHR.PHYSIK 29(1981)35
2.4.6; 2.5.1

66R EDEN-LANDSHOFF-OLIVE-POLKINGHORNE
THE ANALYTIC S-MATRIX
R.J.EDEN,P.V.LANDSHOFF,D.I.OLIVE,J.C.POLKINGHORNE; CAMBRIDGE UNIV.PRESS(1966)
A.6

67R EDEN-1
HIGH ENERGY COLLISIONS OF ELEMENTARY PARTICLES
R.J.EDEN;CAMBRIDGE UNIV.PRESS(1967)
A; A.6; A.9; A.5.3

67R EDEN-2
PROPERTIES OF SCATTERING AMPLITUDES AT HIGH ENERGIES
R.J.EDEN IN"PARTICLE INTERACTIONS AT HIGH ENERGIES"; EDS.T.W.PREIST,L.L.J.VICK;OLIVER AND BOYD(1967)
A.5.3; A.6.4

71R EDEN
THEOREMS ON HIGH ENERGY COLLISIONS OF ELEMENTARY PARTICLES
R.J.EDEN:REV.MOD.PHYS.43(1971)15
A.5.3; A.5.4

73R EDEN
OPTIMIZATION OF COLLISION AMPLITUDES UNDER CONSTRAINTS
R.J.EDEN:LECTURE NOTES IN PHYSICS 17(1973)43
A.5.3

66R ERICSON-MAYER-KUCKUK
FLUCTUATIONS IN NUCLEAR REACTIONS
T.ERICSON,T.MAYER-KUCKUK:ANN.REV.NUCL.SCI.16(1966)183
2.6.6

69R FELD
MODELS OF ELEMENTARY PARTICLES
B.T.FELD;BLAISDELL(1969)
A

72R FERRO-FONTAN-QUEEN-VIOLINI
DISPERSION SUM RULES FOR STRONG AND ELECTROMAGNETIC INTERACTIONS
C.FERRO-FONTAN,N.M.QUEEN,G.VIOLINI:RIVISTA NUOVO CIM. 2(1972)357
2.2.1

72R FEYNMAN
PHOTON-HADRON INTERACTIONS
R.P.FEYNMAN;BENJAMIN(1972)
2.6.4; 2.5.2

72R FISCHER
USE OF EXTRAPOLATION TECHNIQUE IN ANALYSES OF BINARY REACTIONS
J.FISCHER IN"PROC.INT.SEMINAR ON BINARY REACTIONS", DUBNA D-6004(1972)
A.6.3

80R FISCHER-SALY
THE POMERANCHUK THEOREM AND ITS MODIFICATIONS
J.FISCHER,R.SALY:FORTSCHR.PHYSIK 28(1980)237
A.5.4

81R FISCHER
GENERAL LAWS OF HADRON SCATTERING AT VERY HIGH ENERGIES
J.FISCHER:PHYSICS REP.76(1981)159
2.6.4; A.5.3; A.5.4; 2.2.1

58R FOLDY
NEUTRON-ELECTRON INTERACTION
L.L.FOLDY:REV.MOD.PHYS.30(1958)471
A.11.1

72R FONDA
ON THE CONNECTION BETWEEN PARTICLES AND POLES OF THE S-MATRIX
L.FONDA:FORTSCHR.PHYSIK 20(1972)135
2.4.1

77R FONDA
CRITICAL DISCUSSION ON THE DECAY OF QUANTUM UNSTABLE SYSTEMS
L.FONDA:FORTSCHR.PHYSIK 25(1977)101
2.4.1

78R FONDA
DECAY THEORY OF UNSTABLE SYSTEMS
L.FONDA,G.C.GHIRARDI,A.RIMINI:REP.PROGR.PHYS.41(1978)
587
2.4.1

80R FOSTER
ELECTROPRODUCTION OF NUCLEON RESONANCES
F.FOSTER IN"BARYON 1980,PROC.TORONTO CONF.",P.79
2.4.1

69R FOX
SKELETONS IN THE REGGE CUPBOARD
G.C.FOX IN"HIGH ENERGY COLLISIONS",P.367;ED.C.N.YANG
ET AL.;GORDON AND BREACH(1969)
A.9; 2.6.5

73R FOX-QUIGG
PRODUCTION MECHANISMS OF 2-2 SCATTERING PROCESSES AT
INTERMEDIATE ENERGIES
G.C.FOX,C.QUIGG:ANN.REV.NUCL.SCI.23(1973)219
A.9; 2.6.3; 2.6.5; 2.6.4

63R FRAUTSCHI
REGGE POLES AND S-MATRIX THEORY
S.C.FRAUTSCHI;BENJAMIN(1963)
A.9; 2.6.4

61R FRAZER
THE ELECTROMAGNETIC STRUCTURE OF PIONS AND NUCLEONS
W.R.FRAZER IN"DISPERSION RELATIONS"(SCOTT.UNIV.SUMMER
SCHOOL),P.227;ED.G.R.SCREATON;OLIVER AND BOYD(1961)
2.5.2; A.11

77R FRITZSCH
THE WORLD OF FLAVOUR AND COLOUR
H.FRITZSCH IN"PROC.ENRICO FERMI SCHOOL,COURSE
LXXI"(VARENNA 1977)P.351;ED.M.BALDO CEOLIN;NORTH
HOLLAND(1979)
2.4.4

63R FROISSART
HIGH-ENERGY PROPERTIES OF THE MANDELSTAM
REPRESENTATION
M.FROISSART IN"THEORETICAL PHYSICS",TRIESTE SEMINARS
1962,P.379;INTERNATIONAL ATOMIC ENERGY AGENCY,
VIENNA(1963)
A.6.2

65R FROISSART-OMNES
INTRODUCTION TO STRONG INTERACTION PHYSICS
M.FROISSART,R.OMNES IN"HIGH ENERGY PHYSICS"(LES
HOUCHES 1965),P.91;EDS.C.DEWITT,M.JACOB;GORDON AND
BREACH(1965)
A.6

61R FUBINI
THE USE OF ONE-DIMENSIONAL REPRESENTATIONS IN PION
PHYSICS
S.FUBINI IN"DISPERSION RELATIONS"(SCOTT.UNIV.SUMMER
SCHOOL),P.259;ED.G.R.SCREATON;OLIVER AND BOYD(1961)
A.8.3

63R FUBINI
MULTIPERIPHERAL MODEL FOR HIGH ENERGY INTERACTIONS
S.FUBINI IN"STRONG INTERACTIONS AND HIGH ENERGY
PHYSICS"(SCOTT.UNIV.SUMMER SCHOOL 1963),P.259;ED.R.G.
MOORHOUSE;OLIVER AND BOYD(1963)
2.6.4

77R FUKUGITA-IGI
PHENOMENOLOGICAL DUALITY
M.FUKUGITA,K.IGI:PHYSICS REP.31(1977)238
2.6.5; 2.4.3

63R FULTON
RESONANCES IN STRONG INTERACTION PHYSICS
TH.FULTON IN"ELEMENTARY PARTICLE PHYSICS AND FIELD
THEORY",P.1;ED.K.W.FORD;BENJAMIN(1963)
2.4.1

66R GASIOROWICZ
ELEMENTARY PARTICLE PHYSICS
S.GASIOROWICZ;WILEY(1966)
A

82R GASSER-LEUTWYLER
QUARK MASSES
PHYSICS REP.,IN PRINT
2.5.1

80R GENSINI
MESIC ATOMS AND THE SIGMA TERM
P.M.GENSINI IN"LOW AND INTERMEDIATE ENERGY KAON-
NUCLEON PHYSICS"(ROME 1980),P.339,EDS.E.FERRARI,G.
VIOLINI;REIDEL(1980)
2.5.1

79R GENZ
STRONG AND ELECTROMAGNETIC DECAYS OF THE NEW PARTICLES
H.GENZ:ACTA PHYS.AUSTR.,SUPPL.XXI(1979)559
2.4.2

82R GERASIMOV
EFFECTIVE INTERACTIONS OF QUARKS AND STATIC
PROPERTIES OF HADRONS
S.B.GERASIMOV:PROGR.PART.NUCL.PHYS.8(1982)207
2.4.1

70R GIACOMELLI
TOTAL CROSS SECTION MEASUREMENTS
G.GIACOMELLI:PROGR.NUCL.PHYS.12(1970)PART 2
2.2.1

76R GIACOMELLI
TOTAL CROSS SECTIONS AND ELASTIC SCATTERING AT HIGH ENERGIES
G.GIACOMELLI:PHYSICS REP.23(1976)124
2.2.1; 2.6.4; 2.6.6; 2.6.7; 2.6.8

80R GIACOMELLI
HIGH ENERGY HADRONIC INTERACTIONS
G.GIACOMELLI IN"GAUGE THEORIES AND EXPERIMENTS AT HIGH ENERGIES"(SCOTT.UNIV.SUMMER SCHOOL 1980);EDS. K.C.BOWLER,D.G.SUTHERLAND;SUSSP
2.6.4

77R GILMAN
HADRON SPECTROSCOPY
F.J.GILMAN IN"PROC.SUMMER INSTITUTE ON QUARK SPECTROSCOPY AND HADRON DYNAMICS";ED.M.C.ZIPF; STANFORD(1977)
2.4.1

58R GLAUBER
HIGH ENERGY COLLISIONS
R.GLAUBER IN"LECTURES IN THEORETICAL PHYSICS",VOL.I, P.315;BOULDER(1958)
2.6.3; A.3.3

67R GLAUBER
MULTIPLE DIFFRACTION THEORY OF HIGH ENERGY COLLISIONS
R.J.GLAUBER IN"HIGH ENERGY PHYSICS AND NUCLEAR STRUCTURE";NORTH HOLLAND(1967)
2.6.3; A.3.3

60R GOLDBERGER
INTRODUCTION TO THE THEORY AND APPLICATIONS OF DISPERSION RELATIONS
M.L.GOLDBERGER IN"RELATIONS DE DISPERSION ET PARTICULES ELEMENTAIRES"(LES HOUCHES 1960);EDS.C. DEWITT,R.OMNES;HERMANN,PARIS(1960)
A.6.3

64R GOLDBERGER-WATSON
COLLISION THEORY
M.L.GOLDBERGER,K.M.WATSON;WILEY(1964)
A.2; A.3; A.6.3; 2.4.1

61R GOURDIN
DISPERSION RELATIONS FOR PROBLEMS INVOLVING PHOTONS
M.GOURDIN IN"LECTURES ON HIGH ENERGY PHYSICS"(HERCEGNOVI 1961);ED.B.JAKSIC;GORDON AND BREACH(1965)
2.5.2

66R GOURDIN
DIFFUSION DES ELECTRONS DE HAUTE ENERGIES
M.GOURDIN;MANON ET CIE (1966)
A.11.1

74R GOURDIN
WEAK AND ELECTROMAGNETIC FORM FACTORS OF HADRONS
M.GOURDIN:PHYSICS REP.11(1974)30
A.11; 2.5.2; A.7.6

72R GUSTAFSON-HAMILTON
THE DYNAMICS OF SOME PI N RESONANCES
G.GUSTAFSON,J.HAMILTON:SPRINGER TRACTS MOD.PHYS. 61(1972)49
2.4.1; A.7.1

79R HAAN
CONSEQUENCES OF CROSSED CHANNEL STRUCTURE FOR HIGH ENERGY ELASTIC SCATTERING
O.HAAN:FORTSCHR.PHYSIK 27(1979)35
A.7.1

64R HAGEDORN-1
RELATIVISTIC KINEMATICS
R.HAGEDORN;BENJAMIN(1964)
A.1

64R HAGEDORN-2
INTRODUCTION TO FIELD THEORY AND DISPERSION RELATIONS
R.HAGEDORN;PERGAMON PRESS(1964)
A.6

59R HAMILTON
THE THEORY OF ELEMENTARY PARTICLES
J.HAMILTON;CLARENDON PRESS OXFORD(1959)
A

60R HAMILTON
DISPERSION RELATIONS FOR ELEMENTARY PARTICLES
J.HAMILTON:PROGR.NUCL.PHYS.8(1960)143
A.6.4

63R HAMILTON-WOOLCOCK
DETERMINATION OF PI N PARAMETERS AND PHASE SHIFTS FROM DISPERSION RELATIONS
J.HAMILTON,W.S.WOOLCOCK:REV.MOD.PHYS.35(1963)737
2.4.2; 2.4.7

66R HAMILTON
DYNAMICS OF THE PI N SYSTEM
J.HAMILTON:ACTA PHYS.AUSTR.SUPPL.3(1966)229
A.7.1

67R HAMILTON-1
PION-NUCLEON INTERACTIONS
J.HAMILTON IN:"HIGH ENERGY PHYSICS",VOL.1,P.193;ED. E.H.S.BURHOP;ACADEMIC PRESS(1967)
A.7

67R HAMILTON-2
THE DYNAMICS OF ELEMENTARY PARTICLES AND THE PI N INTERACTION
J.HAMILTON,NORDITA LECTURE NOTES, COURSE A AND B (1967-68)
A.6; A.7; 2.4.2

68R HAMILTON
ON ZERO MASS PIONS
J.HAMILTON:NORDITA LECTURE NOTES(1968)
2.5.1
71R HAMILTON-1
NEW METHODS IN THE ANALYSIS OF PI N SCATTERING
J.HAMILTON:SPRINGER TRACTS MOD.PHYS.57(1971)41
A.7.1; 2.4.6
71R HAMILTON-2
PION-NUCLEON SCATTERING THEORY
J.HAMILTON;PROC.CERN SIN SCHOOL(ENGADIN 1971)P.101
2.4.7
72R HAMILTON-TROMBORG
PARTIAL WAVE AMPLITUDES AND RESONANCE POLES
J.HAMILTON,B.TROMBORG;OXFORD UNIV.PRESS(1972)
A.7; 2.4.1; 2.4.2
74R HAMILTON-PETERSEN
NEW DEVELOPMENTS IN DISPERSION THEORY
J.HAMILTON,J.LYNG PETERSEN:NORDITA LECTURES,VOL.
1(1973-74)
2.1.7; 2.1.16
75R HAMILTON
ELECTROMAGNETIC CORRECTIONS TO HADRON SCATTERING
J.HAMILTON:FORTSCHR.PHYSIK 23(1975)211
A.2.3; 2.1.17
78R HAMILTON-OADES
NEW DEVELOPMENTS IN DISPERSION THEORY:STRONG
INTERACTIONS OF KAONS AND SU(3)PROPERTIES
J.HAMILTON,G.C.OADES:NORDITA LECTURE NOTES,
VOL.2A(1974-77)
2.4.6; 2.5.3
72R HARA
REGGE POLE MODEL AND HADRON PHYSICS
Y.HARA:PROGR.THEOR.PHYS.SUPPL.51(1972)96
A.9; 2.6.4
82R HELLER
BAG MODELS AND NUCLEAR FORCES
L.HELLER:SPRINGER TRACTS MOD.PHYS.100(1982)145
2.4.1
78R HENDRY-LICHTENBERG
THE QUARK MODEL
A.W.HENDRY,D.B.LICHTENBERG:REP.PROGR.PHYS.41(1978)1707
2.4.1; 2.6.4
62R HENLEY-THIRRING
ELEMENTARY QUANTUM FIELD THEORY
E.M.HENLEY,W.THIRRING;MC GRAW HILL(1962)
2.5.1
72R HERNDON
PION-NUCLEON PARTIAL WAVE AMPLITUDES
D.J.HERNDON,A.BARBARO-GALTIERI,A.H.ROSENFELD:BERKELEY
REPORT UCRL 20030(1972)
2.1
76R HEY
THE STATUS OF SU(6)W
A.J.G.HEY IN"PROC.INT.CONF.ON BARYON RESONANCES"
(OXFORD 1976),P.463
2.4.1
78R HEY
QUARKS AND SYMMETRIES
A.J.G.HEY IN"ELECTROMAGNETIC INTERACTIONS OF HADRONS",
VOL.1,P.1;EDS.A.DONNACHIE,G.SHAW;PLENUM PRESS(1978)
2.4.1
79R HEY
PARTICLE SYSTEMATICS
A.J.G.HEY IN"PROC.INT.CONF.HIGH ENERGY
PHYSICS"(GENEVA 1979),P.523
2.4.1
80R HEY
QUARK MODELS OF BARYONS
A.J.G.HEY IN"BARYON 1980,PROC.TORONTO CONF.",P.223
2.4.1
60R HILGEVOORD
DISPERSION RELATIONS AND CAUSAL DESCRIPTION
J.HILGEVOORD;NORTH HOLLAND (1960)
A.6.4
67R HIRSHFELD
INTERMOLECULAR FORCES
J.O.HIRSHFELD:ADVANCES CHEM.PHYS.12;INTERSCIENCE(1967)
2.5.3
69R HITE
RECENT DEVELOPMENTS OF THE REGGE POLE MODEL
G.E.HITE:REV.MOD.PHYS.41(1969)669
A.9
65R HOEHLER
SPECIAL MODELS AND PREDICTIONS FOR PION
PHOTOPRODUCTION
G.HOEHLER:SPRINGER TRACTS MOD.PHYS.39(1965)55
2.4.2
72R HOEHLER
REVIEW OF NEAR-FORWARD PI N SCATTERING AMPLITUDES AT
HIGH ENERGIES
G.HOEHLER IN"PROC.INT.SEMINAR ON BINARY REACTIONS";
DUBNA D-6004(1972)
2.6.2; 2.6.3

73R HOEHLER-1
DETERMINATION OF PION-NUCLEON SCATTERING AMPLITUDES FROM DATA AND FIXED-T ANALYTICITY
G.HOEHLER IN'HADRON INTERACTIONS AT LOW ENERGIES'(TRIANGLE MEETING 1973),P.11;ED.M.BLAZEK; VEDA,BRATISLAVA(1975)
2.1.7

73R HOEHLER-2
PION-NUCLEON SCATTERING AMPLITUDES AT INTERMEDIATE AND HIGH ENERGIES
G.HOEHLER:LECTURE NOTES IN PHYSICS 17(1973)21
2.6.3; 2.6.5; 2.2.1

76R HOEHLER
DISPERSION THEORY OF NUCLEON FORM FACTORS
G.HOEHLER:LECTURE NOTES IN PHYSICS 56(1976)159
A.11.1; 2.5.2; 2.5.1

77R HOEHLER
DETERMINATION OF PI N SCATTERING AMPLITUDES AND NUCLEON FORM FACTORS
G.HOEHLER IN'HADRON STRUCTURE'(2ND ADRIATIC MEETING); EDS.M.MARTINIS,N.ZOVKO(1977)
2.5.2; 2.1.7

79R HOEHLER
HIGH ENERGY PI N PHASE SHIFT ANALYSIS
G.HOEHLER IN 'PROC.INT.CONF.ON HIGH ENERGY PHYSICS'VOL.2,P.594;CERN(1979)
2.1.7

79R HOEHLER-KAISER-KOCH-PIETARINEN
HANDBOOK OF PION-NUCLEON SCATTERING
G.HOEHLER,F.KAISER,R.KOCH,E.PIETARINEN:PHYSICS DATA 12-1(1979)
A

80R HOEHLER-KAISER
REVIEW AND TABLES OF PION-NUCLEON FORWARD AMPLITUDES
G.HOEHLER,F.KAISER:KFK 3027,KERNFORSCHUNGSZENTRUM KARLSRUHE(1980)
A.6.4; 2.2.1; 2.6.5

81R HOEHLER
PION-NUCLEON SCATTERING AND QUARK MODELS
G.HOEHLER IN'PARTICLE PHYSICS'(DUBROVNIK 1980),P.343, EDS.I.ANDRIC,I.DADIC,N.ZOVKO;NORTH-HOLLAND(1981)
A.8.2; 2.5.1

63R HOFSTADTER
ELECTRON SCATTERING AND NUCLEAR AND NUCLEON STRUCTURE
R.HOFSTADTER (ED.);COLLECTION OF REPRINTS;BENJAMIN (1960)
A.11

81R HOLINDE
TWO-NUCLEON FORCES AND NUCLEAR MATTER
K.HOLINDE:PHYSICS REP.68(1981)123
2.5.3

76R HORGAN
THE THREE QUARK OSCILLATOR MODEL FOR THE BARYON SPECTRUM
R.R.HORGAN IN'PROC.CONF.ON BARYON RESONANCES,OXFORD (1976)',P.435
2.4.1

69R HORN
FINITE ENERGY SUM RULES
D.HORN IN'PARTICLE PHYSICS';ED.P.URBAN;SPRINGER(1969)
2.6.5

78R IGARASHI
CONSTITUENT REARRANGEMENT MODEL AND HADRON REACTIONS
Y.IGARASHI ET AL.:PROGR.THEOR.PHYS.SUPPL.63(1978)49
2.6.4; 2.6.6; 2.6.7

77R IRVING-WORDEN
REGGE PHENOMENOLOGY
A.C.IRVING,R.P.WORDEN:PHYSICS REP.34(1977)117
A.9; 2.6.4; 2.6.5; 2.6.7

79R ISGUR
SOFT QCD:LOW ENERGY HADRON PHYSICS WITH CHROMODYNAMICS
N.ISGUR IN'THE NEW ASPECTS OF SUBNUCLEAR PHYSICS'(ERICE 1978),P.107;ED.A.ZICHICHI;PLENUM PRESS(1979)
2.4.1

81R ISGUR
HADRONIC STRUCTURE WITH QCD
N.ISGUR IN'PARTICLES AND FIELDS - 1981'(SANTA CRUZ 1981);EDS.C.A.HEUSCH,W.T.KIRK;AIP CONF.PROC.
2.4.1

68R ISLAM
IMPACT PARAMETER DESCRIPTION OF HIGH-ENERGY SCATTERING
M.M.ISLAM IN'LECTURES IN THEORETICAL PHYSICS'(BOULDER) ;EDS.A.O.BARUT,W.E.BRITTIN;GORDON AND BREACH(1968)
A.3.3

71R ISLAM
EIKONAL DESCRIPTION OF HIGH-ENERGY PARTICLE SCATTERING
M.M.ISLAM IN'TOPICS IN STRONG INTERACTIONS'(BOULDER 1971);EDS.A.O.BARUT,W.E.BRITTIN;BOULDER UNIV. PRESS(1972)
2.6.3; A.3.3

80R ITZYKSON-ZUBER
QUANTUM FIELD THEORY
C.ITZYKSON,J.B.ZUBER;MAC GRAW HILL(1980)
A

61R JACKSON
INTRODUCTION TO DISPERSION RELATION TECHNIQUES
J.D.JACKSON IN'DISPERSION RELATIONS';ED.G.R.SCREATON;
OLIVER AND BOYD(1961)
A.6

74R JACKSON
INTRODUCTION TO HADRONIC INTERACTIONS AT HIGH ENERGIES
J.D.JACKSON IN'PHENOMENOLOGY OF PARTICLES AT HIGH
ENERGIES'(SCOTTISH UNIV.SUMMER SCHOOL);EDS.R.L.
CRAWFORD,R.JANNINGS(1974)
2.6.3; 2.6.4; 2.2.1

61R JACOB
THE HELICITY FORMALISM
M.JACOB IN'LECTURES ON HIGH ENERGY PHYSICS'
(HERCEGNOVI 1961);ED.B.JAKSIC;GORDON AND
BREACH(1965)
A.2

64R JACOB-CHEW
STRONG INTERACTION PHYSICS
M.JACOB,G.F.CHEW;BENJAMIN(1964)
A.6

79R JAFFE
THE BAG
R.L.JAFFE IN'THE SUBNUCLEAR SERIES',VOL.17;ED.
A.ZICHICHI;PLENUM PRESS(1980)
2.4.1; 2.5.1

80R JAFFE
APPLICATIONS OF THE BAG MODEL
R.L.JAFFE:ACTA PHYS.AUSTR.SUPPL.22(1980)269
2.4.1

81R JAFFE
MULTIQUARK HADRONS
R.L.JAFFE IN'PROC.1981 SYMPOSIUM ON LEPTON AND PHOTON
INTERACTIONS AT HIGH ENERGIES'(BONN),P.395;ED.W.
PFEIL(1981)
2.4.1

69R JOHNSON-STEINER
LOW ENERGY PION-NUCLEON EXPERIMENTS AND PHASE SHIFT
ANALYSES
C.JOHNSON,H.STEINER IN'PION-NUCLEON SCATTERING'
(IRVINE 1967),P.1;WILEY-INTERSCIENCE(1969)
2.1

76R JOHNSON
INTRODUCTION TO M.I.T.BAG MECHANICS
K.JOHNSON IN'FUNDAMENTALS OF QUARK MODELS'(SCOTTISH
UNIV.SUMMER SCHOOL)P.245;EDS.I.M.BARBOUR,A.T.
DAVIES(1976)
2.4.1

79R JOOS
INTRODUCTION TO QUARK CONFINEMENT IN QCD
H.JOOS:ACTA PHYS.AUSTR.SUPPL.XXI(1979)407;SCHLADMING
LECTURE
2.4.1

80R KAJIKAWA
RECENT PION PHOTOPRODUCTION AND DEUTERON PHOTO-
DISSOCIATION EXPERIMENTS
R.KAJIKAWA IN'BARYON 1980,PROC.TORONTO CONF.',P.43;ED.
N.ISGUR
2.4.1

64R KALLEN
ELEMENTARY PARTICLE PHYSICS
G.KALLEN;ADDISON-WESLEY(1964)
A; 2

76R KANE-SEIDL
AN INTERPRETATION OF TWO-BODY HADRON REACTIONS
G.L.KANE,A.SEIDL:REV.MOD.PHYS.48(1976)309
2.6.4; 2.6.5

76R KASCHLUHN-MUELLER-PREUSKER
REGGE MODELS AND DISPERSION RELATIONS FOR PARTIAL
WAVES
F.KASCHLUHN,M.MUELLER-PREUSKER:FORTSCHR.PHYSIK
24(1976)477
A.7.1

80R KELLY
NUCLEON AND DELTA RESONANCES
R.KELLY IN'BARYON 1980,PROC.TORONTO CONF.',P.149;ED.N.
ISGUR
2.4.1

73R KLEINERT
BROKEN SCALE INVARIANCE
H.KLEINERT:FORTSCHR.PHYSIK 21(1973)1,SEE ALSO ACTA
PHYS.AUSTR.SUPPL.IX(1972)533
2.5.1

80R KOCH
THE PRESENT STATUS OF PION-NUCLEON PHASE SHIFT
ANALYSIS
R.KOCH IN'LOW AND INTERMEDIATE ENERGY KAON-NUCLEON
PHYSICS'(ROME 1980),P.1;EDS.E.FERRARI,G.VIOLINI;
REIDEL(1980)
2.1.7; 2.4.1; 2.6.3

69R KOKKEDEE
THE QUARK MODEL
J.J.KOKKEDEE;BENJAMIN(1969)
2.4.1; 2.6.4

76R KROLL
NUCLEON-NUCLEON SCATTERING AT HIGH ENERGIES
P.KROLL:FORTSCHR.PHYSIK 24(1976)565
2.6.4

81R KROLL
PHENOMENOLOGY OF NN-SCATTERING
P.KROLL:PHYSICS DATA 22-1(1981)
2.6.3; 2.6.4; 2.5.3

70R KUGLER
DUALITY
M.KUGLER IN"DEVELOPMENTS IN HIGH ENERGY
PHYSICS"(SCHLADMING SCHOOL)P.443;ED.P.URBAN;
SPRINGER(1970)
2.6.5

81R LANDSHOFF
PHENOMENOLOGY OF THE STANDARD MODEL SU(3)C X SU(2) X
U(1)
P.V.LANDSHOFF IN"PROC.INT.CONF.ON HIGH ENERGY
PHYSICS",LISBON(1981)
2.6.4; 2.6.6

58R LANE-THOMAS
R-MATRIX THEORY OF NUCLEAR REACTIONS
A.M.LANE,R.G.THOMAS:REV.MOD.PHYS.30(1958)257
2.4.1

72R LASINSKI
ENERGY DEPENDENCE OF THE ELASTIC DIFFRACTION SLOPES
T.LASINSKI,R.LEVISETTI,B.SCHWARZSCHILD,P.UKLEJA:NUCL.
PHYS.B37(1972)1
2.6.2

82R LE YAOUANC-OLIVER-PENE-RAYNAL
HADRONIC TRANSITION PROCESSES AT LOW AND MEDIUM
ENERGIES(DECAY OF HADRONS)
A.LE YAOUANC,L.OLIVER,O.PENE,J.C.RAYNAL
IN"INTRODUCTION TO THE QUARK MODEL OF ELEMENTARY
PARTICLES",VOL.2;ED.D.FLAMM
2.4.1

66R LEADER
PRESENT PHENOMENOLOGICAL STATUS OF REGGE POLE THEORY
E.LEADER:REV.MOD.PHYS.38(1966)476
A.9

72R LEE
CHIRAL DYNAMICS
B.W.LEE IN"CARGESE LECTURES IN PHYSICS"(VOL.5,1970);
ED.D.BESSIS;GORDON AND BREACH(1972)
A.8.2; 2.5.1

59R LEHMANN
SCATTERING MATRIX AND FIELD OPERATORS
H.LEHMANN:SUPPL.NUOVO CIM.14(1959)153
A.6

78R LEITH
HIGH-ENERGY PHOTOPRODUCTION:DIFFRACTIVE PROCESSES
D.W.G.S.LEITH IN"ELECTROMAGNETIC INTERACTIONS OF
HADRONS I",P.345;EDS.A.DONNACHIE,G.SHAW;PLENUM
PRESS(1978)
2.6.4

82R LEPAGE
HADRONIC WAVE FUNCTIONS IN QCD
G.P.LEPAGE:PREPRINT CLNS-82/522;BANFF SUMMER
INSTITUTE,ALBERTA(1981)
2.6.6

80R LICHTENBERG-ROSEN
DEVELOPMENTS IN THE QUARK THEORY OF HADRONS
D.B.LICHTENBERG,S.P.ROSEN;HADRONIC PRESS,NONAMTUM,
MASS.(1980)
2.5.1; 2.4.1

69R LINDENBAUM
FORWARD DISPERSION RELATIONS - THEIR VALIDITY AND
PREDICTIONS
S.J.LINDENBAUM IN"PION-NUCLEON SCATTERING"(IRVINE
1967);EDS.G.L.SHAW,D.Y.WONG;WILEY(1969)
2.2.1

73R LINDENBAUM
PARTICLE INTERACTIONS AT HIGH ENERGIES
S.J.LINDENBAUM;CLARENDON PRESS(1973)
A.6.3

73R LIPKIN
QUARKS FOR PEDESTRIANS
H.J.LIPKIN:PHYSICS REP.8(1973)175
2.4.1

80R LIPKIN
WHO NEEDS BARYON SPECTROSCOPY?
H.J.LIPKIN IN"BARYON 80";PROC.4TH INT.CONF.ON BARYON
RESONANCES(TORONTO),P.35;ED.N.ISGUR
2.4.1

76R LITCHFIELD-CASHMORE-HEY
SU(6)W AND DECAYS OF BARYON RESONANCES
P.J.LITCHFIELD,R.J.CASHMORE,A.J.G.HEY IN"PROC.CONF.ON
BARYON RESONANCES",OXFORD(1976),P.477
2.4.1

67R LOVELACE
NUCLEON RESONANCES AND LOW ENERGY SCATTERING
C.LOVELACE IN"PROC.INT.CONF.ON ELEMENTARY PARTICLES",
HEIDELBERG;ED.H.FILTHUT;NORTH-HOLLAND(1968)
2.1.5

69R LOVELACE
PION-NUCLEON PHASE SHIFT ANALYSIS AND
PHENOMENOLOGICAL DISPERSION RELATIONS
C.LOVELACE IN"PION-NUCLEON SCATTERING";EDS.G.SHAW,D.Y.
WONG;WILEY(1969)
2.1.5; A.6.9; 2.6.7

72R LOVELACE
BARYON RESONANCES AND RELATED PHENOMENOLOGY
C.LOVELACE IN"PROC.INT.CONF.ON HIGH ENERGY PHYSICS",
BATAVIA;EDS.J.D.JACKSON,A.ROBERTS;P.73(1972)
2.1.5; 2.4.1

67R LOW
HIGH-ENERGY BEHAVIOUR OF SCATTERING AMPLITUDES
F.E.LOW IN"ELEMENTARY PARTICLE PHYSICS AND SCATTERING
THEORY",(BRANDEIS SCHOOL 1967);GORDON AND BREACH(1970)
2.6.4

65R LYTH
FIXED VARIABLE DISPERSION RELATIONS FOR THE PION-
NUCLEON SYSTEM
D.H.LYTH:REV.MOD.PHYS.37(1965)709
A.6.3; A.6.6

77R LYUBIMOV
BACKWARD SCATTERING OF PIONS BY NUCLEONS
V.A.LYUBIMOV:SOV.PHYS.USPECHI 20(1977)691
2.6.7

62R MANDELSTAM
DISPERSION RELATIONS IN STRONG COUPLING PHYSICS
S.MANDELSTAM:REP.PROGR.PHYS.25(1962)99
A.6.2; A.6

63R MANDELSTAM
REGGE POLES AND STRIP APPROXIMATION
S.MANDELSTAM IN"THEORETICAL PHYSICS",TRIESTE SEMINARS
1962,P.401;INTERNATIONAL ATOMIC ENERGY AGENCY,
VIENNA(1963)
A.6.2

70R MANDULA
PATTERNS OF EXCHANGE DEGENERACY
J.MANDULA,J.WEYERS,G.ZWEIG:ANN.REV.NUCL.SCI.20(1970)
289
A.9; 2.6.3; 2.6.7; 2.5.2

69R MARTIN
SCATTERING THEORY:UNITARITY,ANALYTICITY AND CROSSING
A.MARTIN:LECTURE NOTES IN PHYSICS 3(1969)
A.5.1; A.6

70R MARTIN-CHEUNG
ANALYTIC PROPERTIES AND BOUNDS OF SCATTERING
AMPLITUDES
A.MARTIN,F.CHEUNG;GORDON AND BREACH(1970)
A.6; A.5.3

70R MARTIN-SPEARMAN
ELEMENTARY PARTICLE THEORY
A.D.MARTIN,T.D.SPEARMAN;NORTH-HOLLAND(1970)
A

73R MARTIN
ANALYTICITY, UNITARITY AND SCATTERING AMPLITUDES
A.MARTIN IN"PARTICLE PHYSICS"(LES HOUCHES 1971),P.173,
EDS.C.DEWITT,C.ITZYKSON,GORDON AND BREACH(1973)
A.6; A.5

74R MARTIN
RECONSTRUCTION OF SCATTERING AMPLITUDES FROM
DIFFERENTIAL CROSS SECTIONS
A.MARTIN IN"PARTICLE PHYSICS";ED.S.M.MARTINIS,S.
PALLUA,N.ZOVKO;NORTH-HOLLAND(1974)
2.1.2

76R MARTIN
SCALING AND SPIN EFFECTS IN THE DIFFRACTION PEAK
A.MARTIN IN"HADRON STRUCTURE"(2ND ADRIATIC MEETING);
EDS.M.MARTINIS,N.ZOVKO(1977)
2.6.4

76R MARTIN-MORGAN-SHAW
PION-PION INTERACTIONS IN PARTICLE PHYSICS
B.R.MARTIN,D.MORGAN,G.SHAW;ACADEMIC PRESS(1976)
2.4.6; A.7.6; 2.6.5

74R MESHCHERYAKOV
RIGOROUS RESULTS OF LOW-ENERGY MODELS OF THE ANALYTIC
S-MATRIX THEORY
V.A.MESHCHERYAKOV IN"PARTICLE PHYSICS",ADRIATIC
MEETING 1973,P.33;EDS.M.MARTINIS,S.PALLUA,N.
ZOVKO(1974)
A.8.3; 2.5.1

75R MICHAEL
REGGE PHENOMENA
C.MICHAEL:DARESBURY LECTURE NOTE NO.14(1975)
A.9

81R MILLER-THOMAS-THEBERGE
PIONIC CORRECTIONS IN THE M.I.T.BAG MODEL
G.A.MILLER,A.W.THOMAS,S.THEBERGE:COMM.NUCL.PART.PHYS.
10(1981)101
2.5.1

77R MITRA-SOOD
RELATIVISTIC HADRON COUPLINGS:A UNIFIED FRAMEWORK
A.N.MITRA,S.SOOD:FORTSCHR.PHYSIK 25(1977)649
2.4.1

69R MOORHOUSE
PION-NUCLEON INTERACTIONS
R.G.MOORHOUSE:ANN.REV.NUCL.SCI.19(1969)301
A

78R MOORHOUSE
ELECTROMAGNETIC EXCITATION AND DECAY OF HADRON
RESONANCES
R.G.MOORHOUSE IN"ELECTROMAGNETIC INTERACTIONS OF
HADRONS,VOL.1",P.83;EDS.A.DONNACHIE,G.SHAW;PLENUM
PRESS(1978)
2.4.1

78R MOSHE
RECENT DEVELOPMENTS IN REGGEON FIELD THEORY
M.MOSHE:PHYSICS REP.37(1978)256
2.6.4

72R MUELLER
HELICITY CONSERVATION AND RELATED PROBLEMS
H.J.W.MUELLER IN"TOPICS IN PARTICLES AND FIELDS";ED.H.
H.ALY;DEKKER(1972)
2.6.3

81R MUELLER
PERTURBATIVE QCD AT HIGH ENERGIES
A.H.MUELLER:PHYSICS REP.73(1981)238
2.5.2

68R MUIRHEAD
THE PHYSICS OF ELEMENTARY PARTICLES
H.MUIRHEAD;PERGAMON(1968)
A

81R MUKHIN-PATARAKIN
THE PION-PION INTERACTION
K.N.MUKHIN,O.O.PATARAKIN:SOV.PHYS.USPECHI 24(1981)161
2.4.6

76R NAGELS
COMPILATION OF COUPLING CONSTANTS AND LOW-ENERGY
PARAMETERS
M.M.NAGELS ET AL:NUCL.PHYS.B109(1976)1 AND PHYSICS
DATA 4-1(1976)
2.4.2; 2.4.7

79R NAGELS
COMPILATION OF COUPLING CONSTANTS AND LOW-ENERGY
PARAMETERS
M.M.NAGELS ET AL.:NUCL.PHYS.B147(1979)189 AND PHYSICS
DATA 4-2(1979).NEXT EDITION:82R DUMBRAJS
2.4.2; 2.4.7

63R NISHIJIMA
FUNDAMENTAL PARTICLES
K.NISHIJIMA;BENJAMIN(1963)
A

69R NISHIJIMA
FIELDS AND PARTICLES
K.NISHIJIMA;BENJAMIN(1969)
A.6; 2.5.1

72R NUSSENZVEIG
CAUSALITY AND DISPERSION RELATIONS
H.M.NUSSENZVEIG;ACADEMIC PRESS(1972)
A.6; 2.4.1; 2.6.7

61R OEHME
THEORY OF ELEMENTARY PARTICLE REACTIONS
R.OEHME IN"LECTURES ON HIGH ENERGY PHYSICS"
(HERCEGNOVI 1961);ED.B.JAKSIC;GORDON AND
BREACH(1965)
2.5.2; 2.6.4; A.7.3

64R OEHME-1
COMPLEX ANGULAR MOMENTUM IN ELEMENTARY PARTICLE
SCATTERING
R.OEHME IN"STRONG INTERACTIONS AND HIGH ENERGY
PHYSICS";ED.R.G.MOORHOUSE;OLIVER AND BOYD(1964)
A.9

64R OEHME-2
HIGH-ENERGY SCATTERING AND DISPERSION THEORY
R.OEHME IN"DISPERSION RELATIONS AND THEIR CONNECTION
WITH CAUSALITY"(VARENNA SCHOOL),P.167;ACADEMIC PRESS
(1964)
A.9

71R OEHME-1
HIGH ENERGY SCATTERING OF HADRONS
R.OEHME IN"PHYSIKALISCHE UND PHILOSOPHISCHE
BETRACHTUNGEN ZUM 70.GEBURTSTAG VON W.HEISENBERG;ED.H.
P.DUERR;VIEWEG(1971)
A.9

71R OEHME-2
COMPLEX ANGULAR MOMENTUM
R.OEHME:SPRINGER TRACTS MOD.PHYS.57(1971)132
A.9; 2.6.5; 2.6.4

71R OEHME-3
DUALITY AND REGGE THEORY
R.OEHME:SPRINGER TRACTS MOD.PHYS.57(1971)119
2.6.5; A.9

78R OKUBO
A SURVEY OF QUARK-LINE RULE
S.OKUBO:PROGR.THEOR.PHYS.SUPPL.63(1978)1
2.5.2

60R OMNES
DEMONSTRATION DES RELATIONS DE DISPERSION
R.OMNES IN"RELATIONS DE DISPERSION ET PARTICULES
ELEMENTAIRES"(LES HOUCHES 1960),P.319;EDS.C.DEWITT,R.
OMNES:HERMANN,PARIS(1960)
A.6.3

63R OMNES-FROISSART
MANDELSTAM THEORY AND REGGE POLES
R.OMNES,M.FROISSART;BENJAMIN(1963)
A.6.2; A.9

66R OMNES
REGGE POLES IN HIGH ENERGY SCATTERING
R.L.OMNES:ANN.REV.NUCL.SCI.16(1966)263
A.9

70R OMNES
INTRODUCTION TO PARTICLE PHYSICS
R.OMNES;WILEY-INTERSCIENCE(1970)
A.6

65R OSGOOD
LEHRBUCH DER FUNKTIONENTHEORIE II
W.F.OSGOOD;CHELSEA PUBL.CO.(1965)
2.4.3

75R PAGELS
DEPARTURES FROM CHIRAL SYMMETRY
H.PAGELS:PHYSICS REP.16(1975)221
A.8.2; 2.4.6; 2.5.1

80R PARTICLE DATA GROUP
SEE 80R BRICMAN

81R PAUL
SPECTROSCOPY OF RHO,OMEGA AND PHI FAMILIES
E.PAUL IN'PROC.1981 SYMPOSIUM ON LEPTON AND PHOTON
INTERACTIONS AT HIGH ENERGIES'(BONN),P.301;ED.W.
PFEIL(1981)
2.5.2; 2.4.6; 2.6.5

78R PAVER
CURRENT ALGEBRA AND ELECTROMAGNETIC HADRON
INTERACTIONS
N.PAVER IN'ELECTROMAGNETIC INTERACTIONS OF HADRONS',
VOL.2,P.353;EDS.A.DONNACHIE,G.SHAW;PLENUM PRESS(1978)
2.4.2; 2.5.1; 2.5.2

73R PENNINGTON
ZEROS IN PI PI SCATTERING
M.R.PENNINGTON IN'AIP CONF.PROC.NO 13;PI PI SCATTERING'
(TALLAHASSEE);P.89;EDS. P.K.WILLIAMS,V.HAGOPIAN(1973)
2.6.4; 2.4.3

80R PENNINGTON
TWO BODY ANNIHILATION PROCESSES IN ANTINUCLEON-
NUCLEON INTERACTIONS
M.R.PENNINGTON IN'PROC.5TH EUROPEAN SYMPOSIUM ON
NUCLEON-ANTINUCLEON INTERACTIONS'(BRESSANONE 1980),
P.485
2.5.4; A.3.5; 2.4.3

77R PEREZ-Y-ORBA-RENARD
THE PHYSICS OF ELECTRON-POSITRON COLLIDING BEAMS
J.P.PEREZ-Y-ORBA,F.M.RENARD:PHYSICS REP.31(1977)1
2.5.2; 2.6.4

74R PERL
HIGH ENERGY HADRON PHYSICS
M.L.PERL;WILEY-INTERSCIENCE(1974)
A; 2

71R PETERSEN
MESON-MESON SCATTERING
J.L.PETERSEN:PHYSICS REP.2(1971)155
2.4.6

77R PETERSEN
THE PI PI INTERACTION
J.L.PETERSEN,CERN-YELLOW 77-04(1977)
2.4.6

71R PFISTER
DER ELEKTROMAGNETISCHE FORMFAKTOR DES PROTONS BEI
ZEITARTIGEM IMPULSUEBERTRAG
H.PFISTER:FORTSCHR.PHYSIK 19(1971)1
2.5.2

70R PHILLIPS
LECTURES ON REGGE PHENOMENOLOGY
R.J.N.PHILLIPS:ACTA PHYS.AUSTR.SUPPL.VII(1970)214
(SCHLADMING)
A.9

67R PILKUHN
THE INTERACTION OF HADRONS
H.PILKUHN;NORTH-HOLLAND(1967)
A; 2

79R PILKUHN
RELATIVISTIC PARTICLE PHYSICS
H.PILKUHN;SPRINGER(1979)
A; 2

70R PISUT
ANALYTIC EXTRAPOLATIONS AND THE DETERMINATION OF PION-
PION PHASE SHIFTS
J.PISUT:SPRINGER TRACTS MOD.PHYS.55(1970)43
2.5.2; 2.4.6; A.6.3

79R PREDAZZI
STRONG INTERACTIONS AT VERY HIGH ENERGIES OR WHERE IS
ASYMPTOPIA?
E.PREDAZZI:REVISTA NUOVO CIM.2(1979)NO.11
2.6.4

77R PREPARATA
QUARK-GEOMETRODYNAMICS:A NEW APPROACH TO HADRONS
AND THEIR INTERACTIONS
G.PREPARATA IN 'THE WHYS OF SUBNUCLEAR PHYSICS'
(ERICE 1977),P.727; ED.A.ZICHICHI;PLENUM PRESS
2.4.1

79R PROTOPOPESCU-SAMIOS
LIGHT HADRON SPECTROSCOPY:EXPERIMENTAL AND QUARK MODEL INTERPRETATIONS
S.D.PROTOPOPESCU,N.P.SAMIOS:ANN.REV.NUCL.SCI.29(1979) 339
2.4.1

74R QUEEN-VIOLINI
DISPERSION THEORY IN HIGH ENERGY PHYSICS
N.M.QUEEN,G.VIOLINI;MAC MILLAN(1974)
A.6

81R QUIGG
MODELS FOR HADRONS
CH.QUIGG:FERMILAB-CONF.81-78 THY(LES HOUCHES LECTURES 1981)
2.4.1

76R RASCHE
ELECTROMAGNETIC EFFECTS IN LOW ENERGY PION-NUCLEON SCATTERING
G.RASCHE,W.S.WOOLCOCK:HELV.PHYS.ACTA 49(1976)435
2.1.17

74R RASZILLIER
FACTORI DE FORMA
I.RASZILLIER;EDITURA ACADEMIEI REP.SOC.ROMANIA; BUCURESTI(1974)
A.11; 2.5.2

68R RENNER
CURRENT ALGEBRAS AND THEIR APPLICATIONS
B.RENNER;PERGAMON(1968)
A.8.2; 2.5.1

70R RENNER
ASPECTS OF CHIRAL SYMMETRY
B.RENNER IN"DEVELOPMENTS IN HIGH ENERGY PHYSICS"(SCHLADMING SCHOOL)P.91;ED.P.URBAN; SPRINGER(1970)
A.8.2; 2.5.1

72R RENNER
ON THE PROBLEM OF THE SIGMA TERM IN MESON-BARYON SCATTERING
B.RENNER:SPRINGER TRACTS MOD.PHYS.61(1972)120
2.5.1

74R REYA
CHIRAL SYMMETRY BREAKING AND MESON-NUCLEON SIGMA COMMUTATORS
E.REYA;REV.MOD.PHYS.46(1974)545
A.8.2; 2.5.1

82R RHO
PIONS AND THE CHIRAL BAG
M.RHO:PROGR.NUCL.PART.PHYS.8(1982)103
2.4.1

76R ROLLNIK-STICHEL
COMPTON SCATTERING
H.ROLLNIK,P.STICHEL:SPRINGER TRACT IN MOD.PHYS. 79(1976)1
2.6.4

60R ROMAN
THEORY OF ELEMENTARY PARTICLES
P.ROMAN;NORTH-HOLLAND(1960)
A

69R ROMAN
INTRODUCTION TO QUANTIZED FIELD THEORY
P.ROMAN:WILEY(1969)
A

82R ROOS
REVIEW OF PARTICLE PROPERTIES
M.ROOS ET AL.(PARTICLE DATA GROUP):PHYS.LETT. 111B(1982)1
2.4.1

74R ROSNER
THE CLASSIFICATION AND DECAYS OF RESONANT PARTICLES
J.L.ROSNER:PHYSICS REP.11(1974)189
2.4.1

72R ROY
HIGH ENERGY THEOREMS FOR STRONG INTERACTIONS
S.M.ROY:PHYSICS REP.5(1972)126
A.5.3; A.5.4; A.10; 2.4.4; 2.6.8

69R SAKURAI
CURRENTS AND MESONS
J.J.SAKURAI;UNIVERSITY PRESS,CHICAGO(1969)
A.8.2; 2.5.1

74R SAMIOS
HADRONS AND SU(3):A CRITICAL REVIEW
N.P.SAMIOS,M.GOLDBERG,B.T.MEADOWS:REV.MOD.PHYS. 46(1974)49
2.4.1

79R SCADRON
ADVANCED QUANTUM THEORY AND ITS APPLICATIONS THROUGH FEYNMAN DIAGRAMS
M.D.SCADRON;SPRINGER(1979)
A; A.8.2

81R SCADRON
CURRENT ALGEBRA,PCAC,AND THE QUARK MODEL
M.D.SCADRON:REP.PROGR.PHYS.44(1981)213
2.5.1

69R SCHNITZER-1
HARD PIONS,A PHENOMENOLOGICAL APPROACH TO CURRENT ALGEBRA
H.J.SCHNITZER IN'THE SUBNUCLEAR SERIES',VOL.7(ERICE); ED.A.ZICHICHI;BENJAMIN(1970)
2.5.1

69R SCHNITZER-2
CONNECTIONS BETWEEN WARD IDENTITIES,PHENOMENOLOGICAL LAGRANGIANS AND ALGEBRA OF FIELDS
H.J.SCHNITZER IN'ELEMENTARY PARTICLE PHYSICS I'(BOULDER,1968);EDS.K.T.MAHANTHAPPA,W.E.BRITTIN,A.O. BARUT;GORDON AND BREACH(1969)
2.5.1; A.8.2

72R SCHREMPP
ARE REGGE CUTS STILL WORTHWHILE?
B.AND F.SCHREMPP;SPRINGER TRACTS MOD.PHYS.61(1972)68
A.9; 2.6.5; 2.6.4

75R SCHREMPP
QUASI TWO-BODY REACTIONS
F.AND B.SCHREMPP IN'PROC.INTERN.CONF.ON HIGH ENERGY PHYSICS'(PALERMO 1975)
2.6.4; 2.6.5; 2.6.6

80R SCHUBERT
DIFFERENTIAL CROSS SECTIONS FOR ELASTIC PP-SCATTERING AT HIGH MOMENTA
K.R.SCHUBERT IN'LANDOLT-BOERNSTEIN',VOL.I/9A,SECT.1.2. 2;ED.H.SCHOPPER,SPRINGER(1980)
2.6.3

61R SCHWEBER
AN INTRODUCTION TO RELATIVISTIC QUANTUM FIELD THEORY
S.S.SCHWEBER;ROW,PETERSEN AND CO.(1961)
A

79R SEHGAL
NEUTRINOS AND NUCLEON STRUCTURE
L.M.SEHGAL IN'PROC.INT.CONF.ON HIGH ENERGY PHYSICS'(GENEVA)VOL.1,P.98;CERN(1979)
2.5.2

65R SEREBRYAKOV-SHIRKOV
DISPERSION THEORY OF LOW-ENERGY SCATTERING
V.V.SEREBRYAKOV,D.V.SHIRKOV:FORTSCHR.PHYSIK 13(1965) 227
A.8.3; 2.5.1

77R SEREBRYAKOV
ISOVECTOR FORM FACTORS OF MESONS AND NUCLEONS
V.V.SEREBRYAKOV IN 'PROC.1977 EUROPEAN CONF.ON PARTICLE PHYSICS'P.771;EDS.L.JENIK,I.MONTVAY
2.5.2

69R SHIRKOV-SEREBRYAKOV-MESHCHERYAKOV
DISPERSION THEORIES
D.V.SHIRKOV,V.V.SEREBRYAKOV,V.A.MESHCHERYAKOV;NORTH-HOLLAND(1969)
A.6

76R SIVERS-BRODSKY-BLANKENBECLER
LARGE TRANSVERSE MOMENTUM PROCESSES
D.SIVERS,S.J.BRODSKY,R.BLANKENBECLER:PHYSICS REP. 23(1976)1
2.6.6; 2.6.4

82R SIVERS
WHAT CAN WE COUNT ON?
D.SIVERS:ARGONNE REPORT ANL-HEP-PR-82-02
2.6.6; 2.5.2

70R SOMMER
RIGOROUS ANALYTIC PROPERTIES OF SCATTERING AMPLITUDES
G.SOMMER:FORTSCHR.PHYSIK 18(1970)577
A.6.3; A.5.3

63R SQUIRES
COMPLEX ANGULAR MOMENTA AND PARTICLE PHYSICS
E.J.SQUIRES;BENJAMIN(1963)
A.9

71R SQUIRES
REGGE-POLE PHENOMENOLOGY
E.J.SQUIRES:SPRINGER TRACTS MOD.PHYS.57(1970)71
A.9; 2.6.4; 2.6.5

70R STEINER
ON THE GENERALIZED PI N POTENTIAL.A NEW REPRESENTATION FROM FIXED-T DISPERSION RELATIONS
F.STEINER:FORTSCHR.PHYSIK 18(1970)43
A.7.2; 2.5.1

71R STEINER
PARTIAL WAVE CROSSING RELATIONS FOR MESON-BARYON SCATTERING
F.STEINER:FORTSCHR.PHYS.19(1971)115
A.7.2; 2.4.7

71R SUGAR
THE RELATIVISTIC EIKONAL MODEL
R.L.SUGAR IN'TOPICS IN STRONG INTERACTIONS'(BOULDER, 1971),P.47;EDS.A.O.BARUT,W.E.BRITTIN;BOULDER UNIV. PRESS(1972)
2.6.4

65R SUGAWARA
LECTURES ON DISPERSION THEORY
M.SUGAWARA:HELSINKI REPORT,RESEARCH INST.FOR THEOR. PHYSICS(1965)
A.6

82R SZWED
SPIN EFFECTS IN HADRONIC COLLISIONS AT LARGE MOMENTUM TRANSFERS
J.SZWED:RAPORT NO.1172/PH,UNIV.OF KRAKOW(1982)
2.6.6

77R TAUSCHER
HADRONIC ATOMS
L.TAUSCHER IN"EXOTIC ATOMS".PROC.ERICE SCHOOL,P.145; EDS.G.FIORENTINI,G.TORELLI(1977)
2.4.7

72R TAVKHELIDZE
DYNAMICS OF TWO-BODY PROCESSES AT HIGH ENERGIES
A.TAVKHELIDZE IN"PROC.XV.INT.CONF.ON HIGH ENERGY PHYSICS"(KIEV,SEPT.1970),P.367;ED.V.SHELEST;NAUKOVA DUMKA PUBL.,KIEV(1972)
A.9; 2.6.4; 2.6.6

61R THIRRING
INTRODUCTION TO THE STATIC MODEL
W.THIRRING IN"LECTURES ON HIGH ENERGY PHYSICS" (HERCEGNOVI 1961);ED.B.JAKSIC;GORDON AND BREACH(1965)
2.5.1

80R THOMAS-LANDAU
PION-DEUTERON AND PION-NUCLEON SCATTERING
A.W.THOMAS,R.H.LANDAU:PHYSICS REP.58(1980)121
2.5.1

56R UMEZAWA
QUANTUM FIELD THEORY
H.UMEZAWA;NORTH-HOLLAND(1956)
A.8.1; 2.5.1

79R VAN DE WALLE
DIQUARKS
R.T.VAN DE WALLE IN"THE SUBNUCLEAR SERIES"VOL.17 (ERICE LECTURES);ED.A.ZICHICHI;PLENUM PRESS (1980)
2.4.1; 2.5.2

72R VAN HIEU
ANALYTICAL PROPERTIES OF THE SCATTERING AMPLITUDE AND ASYMPTOTIC THEOREMS
N.VAN HIEU IN"PROC.XV.INT.CONF.ON HIGH ENERGY PHYSICS"(KIEV,SEPT.1970),P.564;ED.V.SHELEST;NAUKOVA DUMKA PUBL.,KIEV(1972)
A.5.3; A.6

64R VAN HOVE
HIGH-ENERGY COLLISIONS OF STRONGLY INTERACTING PARTICLES
L.VAN HOVE:PHYS.MOD.PHYS.36(1964)655
2.6.4

67R VAN HOVE
VERY HIGH-ENERGY SCATTERING OF STRONGLY INTERACTING PARTICLES
L.VAN HOVE IN"PARTICLE INTERACTIONS AT HIGH ENERGIES"; EDS.T.W.PREIST,L.L.J.VICK;OLIVER AND BOYD(1967)
A.9

68R VAN HOVE
TOPICS ON REGGE-POLE THEORY OF HIGH-ENERGY SCATTERING
L.VAN HOVE;CERN 68-31;ACADEMIC TRAINING PROGRAM,1968
A.9

82R VAN OERS
ON THE EVIDENCE OF CHARGE SYMMETRY BREAKING
W.T.H.VAN OERS:COMM.NUCL.PART.PHYS.10(1982)251
2.4.4

74R VENEZIANO
DUAL MODELS OF STRONG INTERACTIONS AND THEIR PHYSICAL MOTIVATIONS
G.VENEZIANO:PHYSICS REP.9(1974)199
2.4.3; 2.6.5

82R VINH MAU-1
THE NN AND N ANTI N INTERACTIONS
R.VINH MAU:NUCL.PHYS.A374(1982)3C
2.5.3

82R VINH MAU-2
NUCLEON-ANTINUCLEON FORCES AND ANNIHILATION IN THE POTENTIAL APPROACH
R.VINH MAU IN"WORKSHOP ON PHYSICS AT LEAR"(ERICE 1982)
2.5.3

76R VOLKOV-PERVUSHIN
LOW ENERGY PION PHYSICS
M.K.VOLKOV,V.N.PERVUSHIN:SOV.PHYS.USPECHI 19(1976)894
2.4.6

66R WATSON
A TREATISE ON THE THEORY OF BESSEL FUNCTIONS
G.N.WATSON;CAMBRIDGE UNIV.PRESS(1966)

70R WEINBERG
DYNAMIC AND ALGEBRAIC SYMMETRIES
S.WEINBERG IN"LECTURES ON ELEMENTARY PARTICLES AND QUANTUM FIELD THEORY";EDS.S.DESER,M.GRISARU,H. PENDLETON;M.I.T.PRESS,CAMBR.(1970)
2.5.1

71R WEINSTEIN
CHIRAL SYMMETRY
M.WEINSTEIN:SPRINGER TRACTS MOD.PHYS.60(1971)32
2.5.1

66R WERLE
RELATIVISTIC THEORY OF REACTIONS
J.WERLE;NORTH-HOLLAND(1966)
A.2; A.3; A.4

75R WEST
ELECTRON SCATTERING FROM ATOMS, NUCLEI AND NUCLEONS
G.B.WEST:PHYSICS REP.18(1975)264
A.11.1; 2.5.2

82R WHITE-1
DIFFRACTION IN ACCELERATORS,COLLIDERS AND QCD
A.R.WHITE;FERMILAB-CONF.81-16 THY
2.6.4

82R WHITE-2
THE POMERON IN QCD
A.R.WHITE:ANN.OF PHYS.,IN PRINT;CERN TH 2976(1980)
2.6.4

55R WICK
INTRODUCTON TO SOME RECENT WORK IN MESON THEORY
G.C.WICK:REV.MOD.PHYS.27(1955)339
A.8.3; 2.5.1

64R WIGNER
CAUSALITY,R-MATRIX AND COLLISION MATRIX
E.P.WIGNER IN"DISPERSION RELATIONS AND THEIR CONNECTION WITH CAUSALITY"(VARENNA SCHOOL);P.40; ACADEMIC PRESS (1964)
A.6.4

79R WIIK-WOLF
ELECTRON-POSITRON INTERACTIONS
B.H.WIIK,G.WOLF:SPRINGER TRACTS MOD.PHYS.86(1979)
2.5.2

81R WILKINSON
THE NEUTRON LIFETIME
D.H.WILKINSON:PROGR.PART.NUCL.PHYS.6(1981)325
2.5.1; A.8.2

67R WILSON
SOME FEATURES OF ELECTROMAGNETIC INTERACTIONS
R.WILSON IN"PARTICLE INTERACTIONS AT HIGH ENERGIES"; EDS.T.W.PREIST,L.L.J.VICK;OLIVER AND BOYD(1967)
A.11; 2.5.2

64R WONG
DISPERSION RELATIONS AND APPLICATIONS
D.E.WONG IN"DISPERSION RELATIONS AND THEIR CONNECTION WITH CAUSALITY"(VARENNA SCHOOL),P.68;ACADEMIC PRESS (1964)
A.6; A.11

80R YANG
THE VERY HIGH ENERGY SCATTERING AND THE SHAPE OF HADRONS
C.N.YANG IN "LECTURES ON FRONTIERS OF PHYSICS",KOREAN PHYSICAL SOCIETY(1980)
2.6.3; 2.6.4

72R YNDURAIN
RIGOROUS CONSTRAINTS,BOUNDS,AND RELATIONS FOR SCATTERING AMPLITUDES
F.J.YNDURAIN:REV.MOD.PHYS.44(1972)645
A.5.3

71R ZACHARIASEN
THEORETICAL MODELS OF DIFFRACTION SCATTERING
F.ZACHARIASEN:PHYSICS REP.2(1971)5
2.6.4; A.3.3; 2.6.3

71R ZINN-JUSTIN-1
COURSE ON PADE-APPROXIMANTS
J.ZINN-JUSTIN:SPRINGER TRACTS MOD.PHYS.57(1971)248
2.5.1

71R ZINN-JUSTIN-2
STRONG INTERACTION DYNAMICS WITH PADE APPROXIMANTS
J.ZINN-JUSTIN:PHYSICS REP.1(1971)55
2.5.1

75R ZOLIN
ELASTIC SCATTERING OF HADRONS AT HIGH ENERGIES
L.S.ZOLIN ET AL.:SOV.PHYS.USPECHI:18(1975)712
2.6.4

75R ZOVKO
HADRON ELECTROMAGNETIC STRUCTURE AND HEAVY VECTOR MESONS
N.ZOVKO:FORTSCHR.PHYSIK 23(1975)185
2.5.2

Addendum to Sect. 3.1

71R ATKINSON
SOME CONSEQUENCES OF UNITARITY AND CROSSING. EXISTENCE AND ASYMPTOTIC THEOREMS
D.ATKINSON:SPRINGER TRACTS MOD.PHYS.57(1971)1
A.5.3

71R EINHORN-BLANKENBECLER
BOUNDS ON SCATTERING AMPLITUDES
M.B.EINHORN,R.BLANKENBECLER:ANN.OF PHYS.67(1971)480
A.5.3

76R IAGOLNITZER
S-MATRIX THEORY; AN INTRODUCTION
D.IAGOLNITZER;NORTH HOLLAND (1976)
A.6

69R NISHIJIMA
FIELDS AND PARTICLES
K.NISHIJIMA;BENJAMIN (1969)
A.6; 2.5.1

71R SINGH
HIGH ENERGY BOUNDS FOR SCATTERING AMPLIDUDES
V.SINGH:FIELDS AND QUANTA 1(1971)151
A.5.3

3.2 Original papers and preprints

81L AARON
ANALYSIS OF 3-PION SYSTEMS
R.AARON,R.S.LONGACRE:PHYS.REV.D24(1981)1207
2.5.1

78L ABAEV
PHASE SHIFT ANALYSIS OF PION-NUCLEON SCATTERING
V.V.ABAEV,S.P.KRUGLOV,Y.A.MALOV:LENINGRAD REPORT NO. 438(1978)
2.1.8

70L ACHUTHAN
A MODEL RELATING THE POMERON TO THE F-MESON
T.ACHUTHAN,H.G.SCHLAILE,F.STEINER:NUCL.PHYS.B24(1970) 398
2.4.2

71L ACHUTHAN
EFFECTIVE RANGE PARAMETER IN S-WAVE PION-NUCLEON SCATTERING
P.ACHUTHAN,G.E.HITE,G.HOEHLER:Z.PHYSIK 242(1971)167
2.4.7

66L ADACHI
AN IMPACT PARAMETER FORMALISM
T.ADACHI ET AL.:PROGR.THEOR.PHYS.35(1966)463,485,576; 36(1966)745 AND SUPPL.(1965),316
A.3.3

59L ADAIR
HIGH ENERGY MAXIMA IN THE PI N CROSS SECTIONS
R.K.ADAIR: PHYS.REV.113(1959)338
2.4.1

70L ADAMOVICH
DIFFERENTIAL CROSS SECTIONS FOR PHOTOPRODUCTION OF MESONS ON PROTONS AT 165-230 MEV
M.I.ADAMOVICH ET AL.:SOV.J.NUCL.PHYS.7(1970)360
2.4.2

75L ADAMOVICH
MEASUREMENT OF THE RATIO OF THE PHOTOPRODUCTION YIELDS OF CHARGED PIONS IN DEUTERIUM NEAR THRESHOLD
M.I.ADAMOVICH,V.G.LARIONOVA,S.P.KHARLAMOV:SOV.J.NUCL. PHYS.20(1975)28
2.4.7

73L ADER
DETERMINATION OF PARITY-DOUBLET CONTRIBUTIONS IN BARYON-EXCHANGE REACTIONS
J.P.ADER,C.MEYERS,P.SALIN:NUCL.PHYS.B58(1973)621
2.6.7; A.9.2

75L ADER
BACKWARD MESON-BARYON SCATTERING AND GEOMETRICAL PROPERTIES OF BARYON EXCHANGE
J.P.ADER:NUCL.PHYS.B98(1975)154
2.6.7

65L ADLER-1
CONSISTENCY CONDITION ON THE STRONG INTERACTION IMPLIED BY A PARTIALLY CONSERVED AXIAL-VECTOR CURRENT
S.L.ADLER:PHYS.REV.137(1965)B1023
2.4.6; 2.5.1; A.8.2

65L ADLER-2
SUM RULES FOR THE AXIAL-VECTOR COUPLING CONSTANT RENORMALIZATION IN BETA-DECAY
S.L.ADLER:PHYS.REV.140(1965)B736
2.5.1; A.8.2

77L ADYLOV
A MEASUREMENT OF THE ELECTROMAGNETIC SIZE OF THE PION FROM ELASTIC PION SCATTERING AT 50 GEV/C
G.T.ADYLOV ET AL.:NUCL.PHYS. B128(1977)461
2.5.2

79L AITCHISON
MEDIUM-ENERGY N PI PI DYNAMICS
I.J.R.AITCHISON,J.J.BREHM:PHYS.REV.D20(1979)1119,1131
2.4.1

73L ALCOCK-1
HIGH PARTIAL WAVES IN PHASE SHIFT ANALYSIS OF PI PLUS PROTON SCATTERING
J.W.ALCOCK,W.N.COTTINGHAM:NUCL.PHYS.B62(1973)533
2.1.9; 2.1.14

73L ALCOCK-2
THE PERIPHERY OF THE PROTON
J.W.ALCOCK,W.N.COTTINGHAM,C.MICHAEL:NUCL.PHYS. B67(1973)445
2.1.9; 2.1.14

75L ALCOCK
REGGE EXTRAPOLATION TO THE RESONANCE REGION
J.W.ALCOCK,Y.A.CHAO,R.E.CUTKOSKY:NUCL.PHYS.B84(1975) 503
2.1.9

82L ALDER
MEASUREMENT OF THE POLARIZATION PARAMETER IN PI MINUS P ELASTIC AND CHARGE-EXCHANGE SCATTERING BETWEEN 98 AND 310 MEV
J.C.ALDER ET AL.:SIN PREPRINT;PHYS.REV.IN PRINT
2.1.8

79L ALEKSEEVA
PHASE ANALYSIS OF THE CHARGED PION-NEUTRAL PION INTERACTIONS IN THE REACTION PI P -> 2 PI P
E.A.ALEKSEEVA ET AL.:JETP LETT.29(1979)100
2.4.6

79L ALIZADE
INVERSE PION ELECTROPRODUCTION AT THE PION ENERGY 164 MEV
V.V.ALIZADE ET AL.:YAD.FIZ.30(1979)142
2.5.2

72L ALMEHED
NEW PION-NUCLEON PHASE SHIFT ANALYSIS
S.ALMEHED,C.LOVELACE:NUCL.PHYS.B40(1972)157
2.1.5

71L ALTARELLI-1
THE SIGMA TERM AND LOW ENERGY PI N SCATTERING
G.ALTARELLI,N.CABIBBO,L.MAIANI:NUCL.PHYS.B34(1971)621
2.5.1; 2.4.7

71L ALTARELLI-2
MAGNITUDE OF THE SIGMA TERM,CHIRAL SYMMETRY AND SCALE INVARIANCE
G.ALTARELLI,N.CABIBBO,L.MAIANI:PHYS.LETT.B35(1971)415
2.4.7; 2.5.1

73L ALVENSLEBEN
EXPERIMENTAL VERIFICATION OF THE KRAMERS-KRONIG RELATION AT HIGH ENERGY
H.ALVENSLEBEN ET AL.:PHYS.REV.LETT.30(1973)328
2.6.4

77L ANDREWS
ESTIMATES OF PROTON RADIUS FROM LAMB SHIFT IN HYDROGEN
D.A.ANDREWS,R.GOLUB,G.NEWTON:J.PHYS.G3(1977)L91
2.5.2

72L ANSELM
ZERO PION MASS LIMIT IN INTERACTIONS AT VERY HIGH ENERGIES
A.A.ANSELM,V.N.GRIBOV:PHYS.LETT.40B(1972)487
2.6.3

76L ANTILLE
NEW METHODS FOR MEASURING SPIN-ROTATION PARAMETERS
J.ANTILLE,L.DIEK,A.GONIDEC,K.KURODA,L.MADANSKY:CERN 76-05(MARCH 1976)
2.3.2

73L ARAI
ON THE SATURATION OF THE DIFFRACTION PEAK TO THE UNITARITY BOUND
M.ARAI:PHYS.LETT.:43B(1973)41
A.5.3

82L ARAI
PHENOMENOLOGICAL ANALYSIS OF PION PHOTOPRODUCTION USING FIXED-T DISPERSION RELATIONS
I.ARAI,H.FUJII:NUCL.PHYS.B194(1982)251
2.4.1

66L ARBAB
DIP IN PION-NUCLEON CHARGE EXCHANGE SCATTERING
F.ARBAB,C.B.CHIU:PHYS.REV.147(1966)1045
2.6.5

80L ARCI
AN INDICATION FOR A'3-COMPONENT DUALITY'IN FESR FOR PION-NUCLEON SCATTERING
E.ARCI,E.FERRARI,G.VIOLINI:NUOVO CIM.57A(1980)77
2.6.5

73L ARGYRES
REAL PARTS OF AMPLITUDES AND POLARIZATION OF PI N CHARGE-EXCHANGE SCATTERING
E.N.ARGYRES,A.P.CONTOGOURIS,M.SVEC:PHYS.REV.D8(1973) 184
2.6.5

74L ARGYRES
THE VECTOR AND TENSOR EXCHANGE CONTRIBUTIONS TO PION-NUCLEON SCATTERING
E.N.ARGYRES,A.P.CONTOGOURIS,S.RAY,M.SVEC:ANN.OF PHYS. 85(1974)283
2.1.16; 2.6.5

67L ARNOLD
AN OPTICAL POTENTIAL FOR HIGH ENERGY PHYSICS
R.C.ARNOLD:PHYS.REV.153(1967)1523
A.3.3

82L ASAD
DIP IN ANTI P P ELASTIC SCATTERING
Z.ASAD ET AL.:PHYS.LETT.108B(1982)51
2.6.4; 2.6.3

80L ASTON
OBSERVATION OF AN OMEGA-PI STATE OF MASS 1.25 GEV PROCUCED BY PHOTONS
D.ASTON ET AL.:PHYS.LETT.92B(1980)211
2.5.2; 2.4.6

80L ATKIN
THE USE OF ZEROS IN STUDYING K N PHYSICS
G.ATKIN IN'LOW AND INTERMEDIATE ENERGY KAON-NUCLEON PHYSICS':;EDS.E.FERRARI,G.VIOLINI;REIDEL PUBL.CO.(1980)
2.4.3

62L ATKINSON
PREDICTION OF PION PHASE SHIFTS
D.ATKINSON:PHYS.REV.128(1962)1908
A.6.9; 2.4.6; 2.6.7

73L ATKINSON-1
CRICHTON'S PHASE-SHIFT AMBIGUITY
D.ATKINSON,P.W.JOHNSON,M.MEHTA,M.DE ROO:NUCL.PHYS.
B55(1973)125
2.1.2

73L ATKINSON-2
CONSTRUCTION OF A UNITARY ANALYTIC SCATTERING
AMPLITUDE II
D.A.ATKINSON,G.MAHOUX,F.J.YNDURAIN:NUCL.PHYS.B66(1973)
429
2.1.3

76L ATKINSON
CONSTRUCTION OF A UNITARY ANALYTIC SCATTERING
AMPLITUDE (PI+P SCATTERING)
D.ATKINSON,A.C.HEEMSKERK,S.D.SWIERSTRA:NUCL.PHYS.
B109(1976)322
2.1.3

78L ATKINSON
CRICHTON AMBIGUITIES WITH INFINITELY MANY PARTIAL
WAVES
D.ATKINSON,L.P.KOK,M.DE ROO:PHYS.REV.17(1978)2492
2.1.2

77L AUBERSON
AN INEQUALITY ON THE SLOPE OF THE DIFFRACTION PEAK IN
ELASTIC COLLISION OF PARTICLES OF ARBITRARY SPIN
G.AUBERSON,A.MARTIN,G.MENNESSIER:PHYS.LETT.67B(1977)75
A.5.3

78L AUER
OBSERVATION OF STRUCTURES IN PP TOTAL CROSS SECTION
DIFFERENCES OF PURE HELICITY STATES IN THE MASS RANGE
2.1-2.4 GEV
I.P.AUER ET AL.:PHYS.REV.LETT.41(1978)354
2.5.3

77L AUGENSTEIN
PION-NUCLEON SCATTERING DATA
K.H.AUGENSTEIN,G.HOEHLER,E.PIETARINEN,H.M.
STAUDENMAIER:PHYSICS DATA 1-2(1977)AND MAGNETIC TAPE
2.1.7

72L AUVIL
PI PI SCATTERING AND THE PI N SIGMA TERM
P.R.AUVIL,J.J.BREHM,S.C.PRASAD:PHYS.REV.D6(1972)3526
2.5.1

70L AYED
PION-NUCLEON PHASE SHIFT ANALYSIS UP TO 2.8 GEV
R.AYED,P.BAREYRE,G.VILLET:PHYS.LETT.31B(1970)598
2.1.5

76L AYED
ANALYSE EN DEPHASAGES DE LA DIFFUSION ELASTIQUE PION-
NUCLEON
R.AYED;CEA-N-1291,SACLAY,AND THESIS,UNIV.OF PARIS(1976)
2.1.5

69L BAACKE-1
THE PI N PARTIAL WAVES AT LOW ENERGIES AND IN THE
UNPHYSICAL REGION
J.BAACKE,G.HOEHLER,F.STEINER:Z.PHYS.221(1969)134
A.7.2; 2.4.7

69L BAACKE-2
DETERMINATION OF REGGE-POLE RESIDUE FUNCTIONS FROM
FIXED-T DISPERSION RELATIONS
J.BAACKE:NUCL.PHYS.B9(1969)324
2.6.5

73L BAACKE
TEST OF FINITE-CONTOUR DISPERSION RELATIONS AT FIXED-
T IN PI N SCATTERING
J.BAACKE,J.ENGELS:NUCL.PHYS.B51(1973)434
A.6.4

80L BAILLON
SEARCH FOR NARROW RESONANCES IN PI MINUS P LARGE
ANGLE ELASTIC SCATTERING AND IN TOTAL CROSS SECTIONS
P.BAILLON IN "BARYON 80",PROC.4TH INT.CONF.ON BARYON
RESONANCES(TORONTO);ED.N.ISGUR,P.375
2.6.6; 2.2.1

76L BAKER
BARRELET ZEROS IN PARTIAL WAVE ANALYSIS
R.D.BAKER:RUTHERFORD REPORT RL-76-013(1976)
2.1.11

77L BAKER
A PARTIAL WAVE ANALYSIS OF THE WORLD DATA FOR THE
REACTION PI MINUS P -> KO LAMBDA
R.D.BAKER ET AL.:NUCL.PHYS.B126(1977)365
2.4.1

79L BAKER
POLARIZED CROSS SECTIONS FOR THE REACTION PI MINUS P
TO ETA NEUTRON AND PARTIAL WAVE ANALYSIS
R.D.BAKER ET AL.:NUCL.PHYS.B156(1979)93
2.4.1

61L BALL
POSSIBLE EXPLANATION OF HIGHER PI N RESONANCES IN
TERMS OF INELASTIC THRESHOLDS
J.S.BALL,W.R.FRAZER:PHYS.REV.LETT.7(1961)204
2.4.1

72L BALL
DETERMINATION OF THE BASIC PARAMETERS OF THE 3-3
RESONANCE
J.BALL ET AL.:PHYS.REV.LETT.28(1972)1143
2.4.1
75L BALL
DETERMINATION OF THE DELTA ++ DELTA 0 MASS DIFFERENCE
J.BALL,R.L.GOBLE:PHYS.REV.D11(1975)1171
2.1.17; 2.4.1
77L BANERJEE
LOW-ENERGY PION-NUCLEON SCATTERING
M.K.BANERJEE,J.B.CAMMARATA:PHYS.REV.D16(1977)1334
2.5.1
78L BANERJEE-1
PION-NUCLEON SIGMA COMMUTATOR
M.K.BANERJEE,J.B.CAMMARATA:PHYS.REV.D18(1978)4078
2.5.1
78L BANERJEE-2
THEORY OF THE LOW-ENERGY PION-NUCLEON INTERACTION
M.K.BANERJEE,J.B.CAMMARATA:PHYS.REV.C17(1978)1125
2.5.1
79L BANERJEE-1
PION-NUCLEON INTERACTION AT LOW ENERGY
M.K.BANERJEE:MARYLAND PREPRINT ORO 5126-60 AND TALK
AT HOUSTON CONF. ON MESON-NUCLEAR PHYSICS(1979)
2.5.1
79L BANERJEE-2
PARTIAL CONSERVATION OF AXIAL-VECTOR CURRENT,
DYNAMICAL THEORY,AND DISPERSION RELATIONS
M.K.BANERJEE,J.B.CAMMARATA:PHYS.REV.D19(1979)145
2.5.1
80L BANERJEE
TESTS FOR MODELS OF THE PION-NUCLEON AMPLITUDE
M.K.BANERJEE,J.B.CAMMARATA:MARYLAND PREPRINT ORO 5126-
93
2.5.1
81L BANERJEE
PARTIAL CONSERVED AXIAL CURRENT,NONCANONICALITY AND
THE SIGMA COMMUTATOR
M.K.BANERJEE,G.A.MILLER:SEATTLE PREPRINT,1981
2.5.1
77L BARDIN
A PI PLUS ELECTROPRODUCTION COINCIDENCE EXPERIMENT
AND THE PION RADIUS
G.BARDIN ET AL.:NUCL.PHYS.B120(1977)45
2.5.2
67L BARGER
ANTI N N ANNIHILATION INTO TWO MESONS AT HIGH ENERGIES
V.BARGER,R.J.N.PHILLIPS:PHYS.LETT.25B(1967)415
2.6.7; A.9.2
69L BARGER
MESON REGGE EXCHANGES FROM SIMULTANEOUS ANALYSIS OF
PI N SCATTERING DATA AND DISPERSION SUM RULES
V.BARGER,R.J.N.PHILLIPS:PHYS.REV.187(1969)2210
2.1.16; 2.6.3; 2.6.5
70L BARGER-1
EMPIRICAL MERGER OF GEOMETRICAL AND REGGE PICTURES
FOR ELASTIC HADRON SCATTERING
V.BARGER,D.CLINE:NUCL.PHYS.B23(1970)227
2.6.4
70L BARGER-2
UNCONVENTIONAL ASYMPTOTICS
V.BARGER,R.J.N.PHILLIPS:PHYS.LETT.31B(1970)643
2.2.1
71L BARGER
DEDUCTION FROM NEW PI N CHARGE-EXCHANGE DATA AT LARGE
MOMENTUM TRANSFER
V.BARGER,R.J.N.PHILLIPS:NUCL.PHYS.B33(1971)22
2.6.5
72L BARGER-1
SEPARATION OF N AND DELTA EXCHANGES IN PI N SCATTERING
V.BARGER,M.G.OLSSON:PHYS.REV.D5(1972)2736
2.6.7
72L BARGER-2
EXCHANGE STRUCTURE OF N AND DELTA BACKWARD PEAKS
V.BARGER,F.HALZEN,M.G.OLSSON:NUCL.PHYS.B49(1972)206
2.6.7
72L BARGER-3
RESOLUTION OF THE PI N HELICITY CONSERVATION QUESTION
FROM POLARIZATION DATA ALONE
V.BARGER,F.HALZEN:PHYS.REV.LETT.28(1972)194
2.6.3
72L BARGER-4
EMPIRICAL SYSTEMATICS OF PI N AMPLITUDES
V.BARGER,F.HALZEN:PHYS.REV.D6(1972)1918
2.1.15
72L BARGER-5
RIVAL MODELS FOR TOTAL CROSS SECTIONS
V.BARGER,R.J.N.PHILLIPS:NUCL.PHYS.B40(1972)205
2.2.1
72L BARGER-6
SUCCESSES AND FAILURES OF DUAL ABSORPTION MODELS
V.BARGER,F.HALZEN:NUCL.PHYS.B43(1972)62
2.6.4

72L BARGER-7
ON THE NATURE OF ABSORPTION
V.BARGER,K.GEER,F.HALZEN:NUCL.PHYS.B44(1972)475
2.6.4

72L BARGER-8
THE STRUCTURE OF ELASTIC SCATTERING:ARE TENSOR EXCHANGES PERIPHERAL?
V.BARGER,K.GEER,F.HALZEN:NUCL.PHYS.B49(1972)302
2.6.4

73L BARGER
LINE-REVERSAL OF BARYON-EXCHANGE REACTIONS
V.BARGER,F.HALZEN,R.J.N.PHILLIPS:NUCL.PHYS.B57(1973) 401
2.6.7

74L BARGER
COMPATIBILITY OF OLD AND NEW PI N CHARGE-EXCHANGE RESULTS
V.BARGER,R.J.N.PHILLIPS:PHYS.LETT.53B(1974)195
2.6.5

75L BARGER
TEST OF GEOMETRICAL SCALING AND GENERALIZATIONS
V.BARGER,J.LUTHE,R.J.N.PHILLIPS:NUCL.PHYS.B88(1975)237
2.6.4

76L BARGER
GEOMETRICAL ASPECTS OF HIGH ENERGY ELASTIC SCATTERING
V.BARGER,R.J.N.PHILLIPS:RUTHERFORD REPORT RL-75-176
2.6.4; 2.6.5

79L BARNHILL
PCAC EFFECTS IN M.I.T. BAG MODEL
M.V.BARNHILL,W.K.CHENG,A.HALPRIN:PHYS.REV.D20(1979)727
2.4.1; 2.5.1

80L BARNHILL
STATIC ELECTROWEAK PROPERTIES AND PCAC IN THE M.I.T. BAG MODEL
M.V.BARNHILL III,A.HALPRIN:PHYS.REV.21(1980)1916
2.5.2; 2.5.1

72L BARRELET
A NEW POINT OF VIEW IN THE ANALYSIS OF TWO-BODY REACTIONS
E.BARRELET:NUOVO CIM.8A(1972)331
A.2.1; 2.1.11; 2.4.3

76L BARRELET
COMMENTS ON THE METHOD OF ZEROS
E.BARRELET IN 'PROC.TOPICAL CONF.ON BARYON RESONANCES'(OXFORD),P.175;RUTHERFORD LAB.,1976
2.1.11

80L BARTELS
HIGH ENERGY BEHAVIOUR IN A NON-ABELIAN GAUGE THEORY
J.BARTELS:NUCL.PHYS.B151(1979)293 AND 175(1980)365
2.6.4

65L BARUT
RELATION BETWEEN THE COUPLING CONSTANT AND THE SINGULARITY IN THE SECOND SHEET
A.O.BARUT,T.SAWADA:NUOVO CIM.37(1965)1531 (SEE ALSO P.1126)
2.4.5; A.7.3; A.6.4

74L BASDEVANT
CONSTRUCTION OF PHENOMENOLOGICAL PI PI AMPLITUDES
J.L.BASDEVANT,C.D.FROGGATT,J.L.PETERSEN:NUCL.PHYS. B72(1974)413
2.4.6

75L BASDEVANT
COMMENTS ON LOW-ENERGY PI PI P-WAVE
J.J.BASDEVANT ET AL.:NUCL.PHYS.B98(1975)285
2.4.6

77L BASSOMPIERRE
FIRST DETERMINATION OF PROTON FORM FACTORS AT THE THRESHOLD OF THE TIMELIKE REGION
G.BASSOMPIERRE ET AL.:PHYS.LETT.68B(1977)477
2.5.2

82L BAUMEL
CROSS SECTIONS AND ASYMPTOTIC ENERGIES IN THE P ANTI P COLLIDER
J.BAUMEL,M.FEINGOLD,M.MOSHE:NUCL.PHYS.B198(1982)13
2.6.4

59L BAZ
ENERGY DEPENDENCE OF SCATTERING CROSS SECTIONS NEAR THRESHOLD
A.I.BAZ:SOV.PHYS.JETP 6(1959)709
2.1.13

78L BEBEK
ELECTROPRODUCTION OF SINGLE PIONS AND A MEASUREMENT OF THE PION FORM FACTOR UP TO Q SQUARED=10 GEV
C.J.BEBEK ET AL.: PHYS.REV.D17(1978)1693
2.5.2

79L BECKER
MODEL-INDEPENDENT PARTIAL WAVE ANALYSIS OF THE PI+ PI- SYSTEM PRODUCED IN THE REACTION PI P -> N PI PI AT 17.2 GEV/C
H.BECKER ET AL.:NUCL.PHYS.B151(1979)46
2.4.6; 2.4.3

80L BECKER
MEASUREMENTS OF THE BARYON-EXCHANGE REACTIONS K- P -> LAMBDA ETA AND K-P ->LAMBDA PI
H.BECKER ET AL.:NUCL.PHYS.B167(1980)292
2.4.2

68L BEG
FIELD-CURRENT IDENTITIES AND NUCLEON RADII
M.M.B.BEG,J.BERNSTEIN,M.TAUSNER:PHYS.REV.173(1968)1523
2.5.2

79L BELKOV-1
PION-PION S-WAVE SCATTERING LENGTHS
A.A.BELKOV,S.A.BUNYATOV:SOV.J.NUCL.PHYS.29(1979)666
2.4.6

79L BELKOV-2
PI PI PHASES FROM THE REACTION PI- P -> PI+ PI- N NEAR THRESHOLD
A.A.BELKOV ET AL.:JETP LETT.29(1979)597
2.4.6

80L BELKOV
A STUDY OF PI P -> N+2PI ZERO NEAR THRESHOLD
A.A.BELKOV ET AL.:YAD.FIZ.31(1980)181
2.4.6

77L BEREZHNEV
DETERMINATION OF NUCLEON AND PION FORM FACTORS BY ANALYZING INVERSE ELECTROPION-PRODUCTION AT 275 MEV
S.F.BEREZHNEV ET AL.:SOV.J.NUCL.PHYS.26(1977)181
2.5.2

69L BERGER
PION-NUCLEON AND KAON-NUCLEON SCATTERING IN THE VENEZIANO MODEL
E.L.BERGER,G.C.FOX:PHYS.REV.188(1969)2120
2.6.5

71L BERGER
COMPARATIVE EVALUATION OF THEORIES OF PI N BACKWARD SCATTERING
E.L.BERGER,G.C.FOX:NUCL.PHYS.B26(1971)1
2.6.7

73L BERGERE
PION-NUCLEON PHASE SHIFTS AND PADE-APPROXIMANTS
M.C.BERGERE,J.M.DROUFFE:NUCL.PHYS.B53(1973)191
2.5.1

78L BERGLUND
A STUDY OF THE REACTION ANTIPROTON PROTON -> PI- PI+ AT 10 GEV
A.BERGLUND ET AL.:NUCL.PHYS.B137(1978)276
2.5.4; 2.6.7

78L BHAKAR
IMPACT PARAMETER REPRESENTATION FOR FORWARD AND BACKWARD NN SCATTERING WITH SPIN
B.S.BHAKAR:PHYS.REV.D17(1978)1323
2.6.7

77L BHANDARI
PION-NUCLEON S11 PARTIAL WAVE NEAR THE ETA-NEUTRON PRODUCTION THRESHOLD
R.BHANDARI,Y.CHAO:PHYS.REV.D15(1977)192
2.1.13

77L BIALAS
ELASTIC PP CROSS SECTIONS AND MULTIPLE SCATTERING OF QUARKS
A.BIALAS ET AL.:ACTA PHYS.POLON.89(1977)855
2.6.4

82L BICKERSTAFF
PHASE CONSISTENCY AND DISSOCIATED WAVE FUNCTIONS OF QQQQ Q-BAR BARYONS
R.P.BICKERSTAFF:PHYS.REV.D25(1982)279
2.4.1

77L BILENKAYA
ELECTROMAGNETIC PROTON FORM FACTOR IN THE VECTOR DOMINANCE MODEL WITH MODIFICATIONS
S.I.BILENKAYA,N.B.SKACHKOV,L.L.SOLOVTSOV:SOV.J.NUCL.PHYS.26(1977)556
2.5.2

67L BILENKII
RECONSTRUCTION OF THE MESON-NUCLEON SCATTERING MATRIX IN EXPERIMENTS WITH A POLARIZED TARGET
S.M.BILENKII,R.M.RYNDIN:SOV.J.NUCL.PHYS.4(1967)457
2.1.1

60L BINCER
ELECTROMAGNETIC STRUCTURE OF THE NUCLEON
A.M.BINCER:PHYS.REV.118(1960)855
A.11.1

73L BINNIE
STUDY OF NARROW MESONS NEAR THEIR THRESHOLDS
D.M.BINNIE ET AL.:PHYS.REV.D8(1973)2789
2.1.13

77L BIRSA
ANALYSIS OF BACKWARD PI N SCATTERING DATA AT 3.5 GEV/C
R.BIRSA ET AL.:NUCL.PHYS.B122(1977)397
2.6.7

80L BJORK
MEASUREMENT OF PI N TO PI- PI+ N NEAR THRESHOLD AND CHIRAL SYMMETRY BREAKING
C.W.BJORK ET AL.:PHYS.REV.LETT.44(1980)62
2.4.6

62L BLANKENBECLER
BEHAVIOR OF SCATTERING AMPLITUDES AT HIGH ENERGIES, BOUND STATES AND RESONANCES
R.BLANKENBECLER,M.L.GOLDBERGER:PHYS.REV.126(1962)766
2.6.3; A.3.3

73L BLANKENBECLER
THE CONNECTION BETWEEN REGGE BEHAVIOUR AND FIXED-ANGLE SCATTERING
R.BLANKENBECLER,S.J.BRODSKY,J.F.GUNION,R.SAVIT:PHYS.REV.D8(1973)4117
2.6.4; 2.6.6

74L BLANKENBECLER
PHYSICAL EFFECTS OF HADRONIC BREMSSTRAHLUNG
R.BLANKENBECLER,S.J.BRODSKY,J.F.GUNION,R.SAVIT:PHYS.REV.D10(1974)2153
2.6.4; 2.6.6

78L BLANKENBECLER
MAGNITUDE OF LARGE TRANSVERSE-MOMENTUM CROSS SECTIONS
R.BLANKENBECLER,S.J.BRODSKY,J.F.GUNION:PHYS.REV.D18(1978)900
2.6.6

82L BLOCK
FORWARD HADRONIC PROTON-PROTON AND ANTIPROTON-PROTON ELASTIC SCATTERING AMPLITUDES
M.M.BLOCK,R.N.CAHN:CERN PREPRINT TH 3307(1982)
2.6.3; 2.6.5

73L BLUVSTEIN
CONNECTION BETWEEN THE ELECTROMAGNETIC FORM FACTORS OF HADRONS AND SIDEWISE DISPERSION RELATIONS
R.E.BLUVSTEIN,A.A.CHESHKOV,V.M.DUBOVIK:NUCL.PHYS.B64(1973)407
A.11.1

77L BOEHM
WAVE FUNCTIONS FOR RELATIVISTIC THREE-PARTICLE SYSTEMS
M.BOEHM,R.F.MEYER:ACTA PHYS.AUSTR.47(1977)249
2.4.1

79L BOEHM
SATURATION OF CONFINING FORCES IN A RELATIVISTIC QUARK MODEL FOR THE BARYONS
M.BOEHM,R.F.MEYER:ANN.OF PHYS.120(1979)360
2.4.1

80L BOEHM
THE MASSES OF THE NON-STRANGE BARYON RESONANCES IN A QUARK MODEL WITH SPIN-SPIN AND TENSOR INTERACTIONS
M.BOEHM:Z.PHYSIK C3(1980)321
2.4.1

82L BOEHM
SPIN-DEPENDENT QUARK-QUARK INTERACTION AND BARYON MAGNETIC MOMENTS
M.BOEHM,R.HUERTA,A.ZEPEDA:PHYS.REV.D25(1982)223
2.4.1; 2.5.2

75L BOLLINI
THE PION FORM FACTOR IN THE TIMELIKE RANGE 1.4-9 GEV SQUARED
D.BOLLINI:NUOVO CIM.LETT.14(1975)418
2.5.2

82L BOLOKHOV
ANALYSIS OF LOW ENERGY DATA ON THE PI N -> PI PI N REACTION
A.A.BOLOKHOV,V.V.VERESHCHAGIN,S.G.SHERMAN:DUBNA PREPRINT NO.750(1982)
2.4.6

81L BONESINI
EVIDENCE FOR A NEW PSEUDOSCALAR MESON(PI')
M.BONESINI ET AL.:PHYS.LETT.103B(1981)75
2.5.1

74L BONGARDT
DETERMINATION OF THE ONE-PION EXCHANGE POTENTIAL FROM PN AND ANTI P P CHARGE-EXCHANGE SCATTERING
K.BONGARDT,H.PILKUHN,H.G.SCHLAILE:PHYS.LETT.52B(1974)271
2.5.1

74L BONNEAU
REMARKS ON BOUNDS OF THE PION FORM FACTOR
G.BONNEAU ET AL.:PHYS.REV.D10(1974)3136
2.5.2

74L BONNIER
CONSTRAINTS IMPOSED BY DATA ABOVE 1 GEV ON THE LOW-ENERGY PI PI AMPLITUDES
B.BONNIER,N.JOHANNESSON:NUCL.PHYS.B83(1974)440
2.4.6

75L BONNIER
ANALYSIS OF PCAC CONSTRAINTS IN A DISPERSIVE APPROACH TO OFF-SHELL PI PI SCATTERING
B.BONNIER,N.JOHANNESSON:NUOVO CIM.29A(1975)565
2.4.6

80L BONNIER
EXTRAPOLATION OF THE PION-PION-NUCLEON-ANTINUCLEON S-WAVE BELOW 1 GEV
B.BONNIER,P.GAURON:NUOVO CIM.57A(1980)261
2.4.6

77L BORIE
A STUDY OF PI PI -> N ANTI N PARTIAL-WAVE AMPLITUDES
USING INTERIOR DISPERSION RELATIONS
E.BORIE,F.KAISER:NUCL.PHYS.B126(1977)173
A.6.7; 2.4.2; 2.4.6; 2.6.7

78L BORIE-1
TABLES OF RADIATIVE CORRECTIONS TO PION NUCLEON AND
NUCLEON NUCLEON SCATTERING
E.BORIE:Z.NATURFORSCH.33A(1978)1436
2.1.17

78L BORIE-2
ELECTROMAGNETIC EFFECTS IN PI N SCATTERING AT HIGH
ENERGIES
E.BORIE,H.P.JAKOB:PHYS.LETT.78B(1978)323
A.2.3; 2.1.7; 2.6.5

78L BORIE-3
ZERO TRAJECTORIES OF INVARIANT PION-NUCLEON AMPLITUDES
E.BORIE,W.GAMPP,G.HOEHLER,R.KOCH,I.SABBA-STEFANESCU:
KARLSRUHE PREPRINT TKP 78-18.THE FIRST PART IS
PUBLISHED:80L BORIE
2.4.3

80L BORIE
ZERO TRAJECTORIES OF INVARIANT PI N AMPLITUDES.I
E.BORIE,W.GAMPP,G.HOEHLER,R.KOCH,I.SABBA-STEFANESCU:
Z.PHYSIK C4(1980)333
2.4.3

73L BOURRELY
ANALYTIC PARAMETRIZATION OF HIGH ENERGY FORWARD
SCATTERING AMPLITUDES
C.BOURRELY,W.FISCHER,Z.SEKARA:NUCL.PHYS.B67(1973)452
2.2.1

77L BOURRELY-1
GLOBAL DESCRIPTION OF PP SCATTERING AMPLITUDES FROM 5
TO 2000 GEV
C.BOURRELY:J.PHYS.G3(1977)295
2.6.3

77L BOURRELY-2
COULOMB-NUCLEAR INTERFERENCE POLARIZATION IN HADRONIC
REACTIONS AT HIGH ENERGIES
C.BOURRELY,J.SOFFER:LETT.NUOVO CIM.19(1977)569
2.1.17

79L BOURRELY
NEW IMPACT PICTURE FOR LOW- AND HIGH-ENERGY PP
ELASTIC SCATTERING
C.BOURRELY,J.SOFFER,T.T.WU:PHYS.REV.D19(1979)3249
2.6.4; 2.6.3

60L BOWCOCK
EFFECT OF A PION-PION SCATTERING RESONANCE ON LOW
ENERGY PION-NUCLEON SCATTERING
J.BOWCOCK,W.N.COTTINGHAM,D.LURIE:NUOVO CIM.16(1960)
918 AND 19(1961)142
A.8.3; 2.5.1

78L BOWCOCK
EXTRAPOLATION OF THE PROTON FORM FACTOR USING PADE-
TYPE APPROXIMANTS
J.E.BOWCOCK,N.M.C.DACUNHA,N.M.QUEEN:REV.ROUM.PHYS.
23(1978)549
2.5.2

79L BOWCOCK
SLOPE OF THE FORWARD DIFFRACTION PEAK
J.E.BOWCOCK,P.A.COTTERILL,N.M.QUEEN:NUOVO CIM.
49A(1979)561
2.6.3

79L BRAMON
UNITARITY AND GLUONIC CORRECTIONS TO THE PION FORM
FACTOR AROUND THE PHI PEAK
A.BRAMON,A.VARIAS:PHYS.REV.D20(1979)2262
2.5.2

71L BRANDT
CONSTRAINTS FOR PARTIAL WAVES FROM REGGE ASYMPTOTICS
IN PI N SCATTERING IN P11 STATE
W.BRANDT,F.KASCHLUHN:FYS.CAS.SAV.21(1971)35
A.7.1

72L BRANDT
FINITE ENERGY SUM RULE FOR PARTIAL WAVES
W.BRANDT,F.KASCHLUHN:NUCL.PHYS.B40(1972)457
A.7.1

73L BRANDT
DETERMINATION OF CDD POLES IN THE PI N J=I=1/2
PARTIAL WAVE
W.BRANDT,F.KASCHLUHN,M.MUELLER-PREUSSKER:ACTA PHYS.
SLOVEN.23(1973)141
A.7.1

81L BREAKSTONE
ELASTIC PHOTON-PROTON SCATTERING IN THE 50-130 GEV
RANGE
A.M.BREAKSTONE ET AL.:PHYS.REV.LETT.47(1981)1778
2.6.4

73L BRODSKY
SCALING LAWS AT LARGE TRANSVERSE MOMENTUM
S.J.BRODSKY,G.FARRAR:PHYS.REV.LETT.31(1973)1153
2.6.6

75L BRODSKY
SCALING LAWS FOR LARGE MOMENTUM-TRANSFER PROCESSES
S.J.BRODSKY,G.R.FARRAR:PHYS.REV.D11(1975)1309
2.6.6

81L BRODSKY
WEAK AND ELECTROMAGNETIC FORM FACTORS OF BARYONS AT LARGE MOMENTUM TRANSFER
S.J.BRODSKY,G.P.LEPAGE,S.A.A.ZAIDI:PHYS.REV.D23(1981) 1152
A.11.1; 2.5.2

74L BRONZAN
OBTAINING REAL PARTS OF SCATTERING AMPLITUDES DIRECTLY FROM CROSS SECTION DATA USING DERIVATIVE DISPERSION RELATIONS
J.B.BRONZAN,G.L.KANE,U.P.SUKHATME:PHYS.LETT.49B(1974) 272
2.2.1; 2.6.4

71L BROWN-1
THE ADLER-WEISBERGER THEOREM REEXAMINED
L.S.BROWN,W.J.PARDEE,R.D.PECCEI:PHYS.REV.D4(1971)2801
A.8.2; 2.4.2; 2.5.1

71L BROWN-2
SOFT PIONEERING DETERMINATION OF THE INTERMEDIATE RANGE NUCLEON-NUCLEON INTERACTION
G.E.BROWN,J.W.DURSO:PHYS.LETT.35B(1971)120
2.5.3

79L BROWN-1
THE LITTLE BAG
G.E.BROWN,M.RHO:PHYS.LETT.82B(1979)177
2.5.3; 2.5.1

79L BROWN-2
LITTLE BAG DYNAMICS
G.E.BROWN,M.RHO,V.VENTO:PHYS.LETT.84B(1979)383
2.5.3; 2.5.1

82L BROWN
QUARKS AND NUCLEI
G.E.BROWN:NUCL.PHYS.A374(1982)63C
2.5.1

76L BRUNET
POSSIBLE AMBIGUITIES IN PI N PHASE-SHIFT ANALYSIS
R.BRUNET,L.GAUTIER,P.WINTERNITZ:PHYS.REV.D13(1976)1390
2.1.2

77L BRUNET
DYNAMICAL FIT TO LOW-ENERGY PI N PHASE SHIFTS AND DETERMINATION OF THRESHOLD PARAMETERS
R.C.BRUNET:PHYS.REV.D16(1977)85
2.4.7

68L BRYANT
NUCLEAR GLORY SCATTERING
H.C.BRYANT,N.JARMIE:ANN.PHYS.47(1968)127
2.6.7

76L BUDNEV-1
A TWO-CHANNEL PROBLEM WITH LINEAR UNITARITY
V.M.BUDNEV,V.V.SEREBRYAKOV:SOV.J.NUCL.PHYS.23(1976) 1306
2.5.2; 2.4.6

76L BUDNEV-2
ELECTROMAGNETIC FORM FACTORS FOR THE COUPLED PI PI AND PI OMEGA STATES
V.M.BUDNEV.N.M.BUDNEV,V.V.SEREBRYAKOV:SOV.J.NUCL.PHYS. 24(1976)74
2.5.2; 2.4.6

68L BUGG
MESON-NUCLEON COUPLING CONSTANTS FROM NUCLEON-NUCLEON FORWARD DISPERSION RELATIONS
D.V.BUGG:NUCL.PHYS.B5(1968)29
2.5.3; 2.4.2

73L BUGG-1
COULOMB CORRECTIONS TO ELASTIC PI N SCATTERING
D.V.BUGG:NUCL.PHYS.B58(1973)397
2.1.17

73L BUGG-2
NEW VALUES FOR PION-NUCLEON SCATTERING LENGTHS AND COUPLING CONSTANT
D.V.BUGG,A.A.CARTER,J.R.CARTER:PHYS.LETT.44B(1973)278
2.4.7; 2.4.2

78L BUGG
PROTON-PROTON ELASTIC SCATTERING FROM 150 TO 515 MEV
D.V.BUGG:J.PHYS.G4(1978)1025
2.4.2

76L BUJAK
IS THERE ANY USE FOR DERIVATIVE ANALYTICITY RELATIONS?
A.BUJAK,O.DUMBRAJS:JOURN.PHYS.G2(1976)L129
2.2.1

78L BUKIN
PION FORM FACTOR MEASUREMENT BY E+ E- -> PI+ PI- IN THE ENERGY RANGE 0.78 TO 1.34 GEV
A.D.BUKIN ET AL.:PHYS.LETT.73B(1978)226
2.5.2

76L BURAN
ANTIPROTON-PROTON ANNIHILATION INTO PI+ PI- AT 62 GEV
T.BURAN ET AL.:NUCL.PHYS.B116(1976)51
2.5.4; 2.6.7

74L BURAS
SCALING LAW FOR ELASTIC CROSS SECTIONS FROM GEOMETRICAL SCALING
A.J.BURAS,J.DIAS DE DEUS:NUCL.PHYS.71B(1974)481
2.6.4

69L BURKHARDT
ASYMPTOTIC UNITARITY AND FIXED-ANGLE SCATTERING
H.BURKHARDT:NUOVO CIM.60A(1969)160
2.6.6

70L BURKHARDT
SMOOTHED VENEZIANO MODELS AND APPLICATIONS TO PION-PION SCATTERING
H.BURKHARDT,W.N.COTTINGHAM,K.K.PHUA:NUOVO CIM.3A(1970)584
2.6.5

75L BURKHARDT
UNIQUE INELASTIC PHASE SHIFTS FROM GENERAL STRUCTURE IN TWO VARIABLES
H.BURKHARDT,A.MARTIN:NUOVO CIM.29A(1975)141
2.1.3; 2.4.3

82L BURQ
EXPERIMENTAL RESULTS ON PP FORWARD ELASTIC SCATTERING AND POSSIBILITY OF UNIVERSAL SHRINKAGE
J.P.BURQ ET AL.:PHYS.LETT.109B(1982)124
2.6.4; 2.6.2

75L BUTTIMORE
PION-NUCLEON BACKWARD SCATTERING:THE LOCATION OF THE AMPLITUDE ZEROS
N.H.BUTTIMORE,D.SPEARMAN:NUCL.PHYS.B84(1975)531
2.4.3; 2.6.7

78L BUTTIMORE
SPIN-DEPENDENT PHENOMENA INDUCED BY ELECTROMAGNETIC-HADRONIC INTERFERENCE AT HIGH ENERGIES
N.H.BUTTIMORE,E.GOTSMAN,E.LEADER:PHYS.REV.D18(1978)694
2.1.17; A.2.3; A.4.6

66L BYERS
PI- P CHARGE-EXCHANGE SCATTERING AND A COHERENT DROPLET MODEL OF HIGH ENERGY EXCHANGE PROCESSES
N.BYERS,C.N.YANG:PHYS.REV.142(1966)976
2.6.5

76L BYSTRICKY
DISCRETE TWO-VARIABLE EXPANSIONS OF PHYSICAL SCATTERING AMPLITUDES
J.BYSTRICKY,F.LEHAR,J.PATERA,P.WINTERNITZ:PHYS.REV.D13(1976)1276
2.1.4

61L CABIBBO
ELECTRON-POSITRON COLLIDING BEAM EXPERIMENTS
N.CABIBBO,R.GATTO:PHYS.REV.124(1961)1577
A.11.1

78L CAHN
POLARIZED ELECTRON-NUCLEON SCATTERING IN GAUGE THEORIES OF WEAK AND ELECTROMAGNETIC INTERACTIONS
R.N.CAHN,F.J.GILMAN:PHYS.REV.D17(1978)1313
2.5.2

82L CAHN
COULOMBIC-HADRONIC INTERFERENCE IN AN EIKONAL MODEL
R.CAHN:CERN PREPRINT TH 3292(1982)
2.6.2; 2.1.17

78L CALDWELL
MEASUREMENT OF THE PHOTON TOTAL CROSS SECTION ON PROTONS FROM 18 TO 185 GEV
D.O.CALDWELL ET AL.:PHYS.REV.LETT.40(1978)1222
2.6.4

78L CALLEN
THE STRUCTURE OF HADRONS IN QCD
C.G.CALLEN,R.F.DASHEN,D.Y.GROSS:PHYS.LETT.78B(1978)307
2.5.1

68L CALUCCI-1
CORRESPONDENCE BETWEEN UNSTABLE PARTICLES AND POLES IN S-MATRIX THEORY
G.CALUCCI,L.FONDA,G.C.GHIRARDI:PHYS.REV.166(1968)1719
2.4.1

68L CALUCCI-2
CORRESPONDENCE BETWEEN UNSTABLE PARTICLES AND POLES IN S-MATRIX THEORY:THE EXPONENTIAL DECAY LAW
G.CALUCCI,G.C.GHIRARDI:PHYS.REV.169(1968)1339
2.4.1

72L CAPRINI
BIAS-FREE METHOD FOR THE DETECTION OF BOUND,ANTI-BOUND AND RESONANT STATES FROM DATA
I.CAPRINI,S.CIULLI,C.POMPONIU,I.SABBA-STEFANESCU:PHYS.REV.D5(1972)1658
2.4.1

79L CAPRINI
A GUIDE TO ANALYTIC EXTRAPOLATIONS
I.CAPRINI ET AL.:COMP.PHYSICS:18(1979)305;SEE ALSO CERN TH-2267 AND 2268(1977)
A.6.3; 2.6.7

80L CAPRINI
HIGH ENERGY BEHAVIOUR OF PP SCATTERING
I.CAPRINI,A.SARARU,M.SARARU,C.VERZEGNASSI:NUOVO CIM.58A(1980)285
2.2.1; 2.6.3

82L CAPRINI
MODEL-INDEPENDENT DISPERSION APPROACH TO PROTON
COMPTON SCATTERING
I.CAPRINI,I.GUIASU,E.E.RADESCU:PHYS.REV.D25(1982)1808
2.6.4

82L CARBONI-1
MEASUREMENT OF ANTIPROTON-PROTON TOTAL CROSS SECTION
AND ELASTIC SCATTERING AT CERN
G.CARBONI ET AL.:PHYS.LETT.108B(1982)145
2.6.3

82L CARBONI-2
EVIDENCE FOR A RISE IN THE ANTIPROTON-PROTON TOTAL
CROSS SECTION
G.CARBONI ET AL.:PHYS.LETT.113(1982)87
2.6.3

70L CARLITZ
FERMION REGGEIZATION WITHOUT PARITY DOUBLING
R.CARLITZ,M.KISLINGER:PHYS.REV.LETT.24(1970)186
2.6.7

71L CARLITZ
DYNAMICALLY BASED COUPLING SCHEME FOR THE
POMERANCHUKON
R.CARLITZ,M.B.GREEN,A.ZEE:PHYS.REV.D4(1971)3439
2.6.4; 2.4.2

82L CARLSON
A QUARK MODEL FOR BARYONS BASED ON QCD
J.CARLSON,J.KOGUT,PANDHARIPANDE:URBANA PREPRINT ILL-
(TH)-82-23(1982)
2.4.1

75L CARNEGIE-1
THE IMAGINARY PART OF THE HELICITY NON-FLIP C=-1
EXCHANGE AMPLITUDE IN ELASTIC SCATTERING BETWEEN
6 AND 14 GEV/C
R.K.CARNEGIE ET AL.:PHYS.LETT.58B(1975)371
2.6.3

75L CARNEGIE-2
THE STRUCTURE OF THE FORWARD ELASTIC CROSS SECTION IN
10-14 GEV RANGE
R.K.CARNEGIE ET AL.:PHYS.LETT.59B(1975)313
2.6.3

62L CARRASSI
ON RELATIVISTIC EFFECTS IN LOW-ENERGY PION-NUCLEON
SCATTERING
M.CARRASSI,G.PASSATORE:NUOVO CIM.26(1962)1254 AND
27(1963)1156
A.8.3; 2.5.1

69L CARRERAS
ON-MASS-SHELL CURRENT ALGEBRA AND PION-NUCLEON
SCATTERING
C.CARRERAS,A.DE RUJULA:NUOVO CIM.64(1969)1014
2.5.1

73L CARTER-1
PI P PHASE SHIFTS FROM 88 TO 310 MEV
J.R.CARTER,D.V.BUGG,A.A.CARTER:NUCL.PHYS.B58(1973)378
2.1.6; 2.4.7

73L CARTER-2
A NEW VALUE FOR THE SIGMA COMMUTATOR IN PI N
SCATTERING
A.A.CARTER,D.V.BUGG,J.R.CARTER:LETT.NUOVO CIM.8(1973)
669
2.5.1

73L CARTER-3
TABULATION OF PION-NUCLEON FORWARD AMPLITUDES UP TO
30 GEV
A.A.CARTER,J.R.CARTER:RUTHERFORD REPORT RL-73-
024(1973)
2.2.1

77L CARTER-1
EVIDENCE FOR MESONS WITH SPINS 3,4 AND 5 IN THE MASS
REGION 2020 TO 2580 MEV
A.A.CARTER ET AL.:PHYS.LETT.67B(1977)117
2.5.4

77L CARTER-2
DIFFERENTIAL CROSS SECTION ZEROS FOR THE PROCESS
ANTIPROTON-PROTON -> TWO PIONS
A.A.CARTER:PHYS.LETT.67B(1977)122
2.5.4; 2.4.3

77L CARTER-3
MEASUREMENT OF THE POLARIZATION PARAMETER FOR
ANTIPROTON-PROTON ANNIHILATION INTO CHARGED PION PAIRS
A.A.CARTER ET AL.:NUCL.PHYS.B127(1977)202
2.5.4

48L CASIMIR
THE INFLUENCE OF RETARDATION ON THE LONDON-VAN DER
WAALS FORCES
H.B.G.CASIMIR,D.POLDER:PHYS.REV.73(1948)360
2.5.3

79L CASS
THE PI NN VERTEX FUNCTION AND OFF-SHELL PI N
SCATTERING
A.CASS,B.H.J.MC KELLAR:NUCL.PHYS.B166(1980)399
2.5.1

77L CELMASTER
BARYON RESONANCES IN A GLUON-PERTURBED LINEAR POTENTIAL
PHYS.REV.D15(1977)1391
2.4.1; 2.1.17

78L CHAN
IMPLICATION OF THE CHOU-YANG MODEL IN PI P ELASTIC SCATTERING
S.C.CHAN,S.Y.LO,H.B.LOW,K.K.PHUA:PHYS.REV.D17(1978)802
2.6.3

80L CHAN
CONSTRUCTION OF THE PHASE OF THE SCATTERING AMPLITUDE BY ITERATION OF BOUNDS FROM UNITARITY INTEGRAL
S.K.CHAN,I.A.SAKMAR:J.MATH.PHYS.21(1980)735
2.1.2; A.5.1

73L CHAO
EXTRAPOLATION OF THE PI N AMPLITUDE TO THE PI PI CUT
Y.A.CHAO,R.K.P.ZIA:PHYS.REV.D8(1973)4085
2.4.6

75L CHAO
EVALUATION OF THE PION-NUCLEON SIGMA TERM
Y.A.CHAO,R.E.CUTKOSKY,R.L.KELLY,J.W.ALCOCK:PHYS.LETT. 57B(1975)150
2.5.1

70L CHENG
LOGARITHMIC FACTORS IN THE HIGH-ENERGY BEHAVIOUR OF QUANTUM ELECTRODYNAMICS
H.CHENG,T.T.WU:PHYS.REV.D1(1970)2775
2.6.4

71L CHENG
IS SU(2) X SU(2) A BETTER SYMMETRY THAN SU(3)?
T.P.CHENG,R.DASHEN:PHYS.REV.LETT.26(1971)594
A.8.2; 2.5.1; 2.4.7

73L CHENG-1
IMPACT PICTURE OF ELASTIC SCATTERING FROM 20 TO 5000 GEV
H.CHENG,J.K.WALKER,T.T.WU:PHYS.LETT.44B(1973)97
2.6.3; 2.6.4

73L CHENG-2
RATIO OF REAL TO IMAGINARY PARTS FOR PP,PI P AND K P FORWARD ELASTIC SCATTERING AMPLITUDES FROM 50 TO 5000 GEV
H.CHENG,J.K.WALKER,T.T.WU:PHYS.LETT.44B(1973)283
2.6.4; 2.6.2

76L CHENG
ZWEIG RULE AND THE PI N SIGMA TERM
T.P.CHENG:PHYS.REV.D13(1976)2161
2.5.1

54L CHEW
RENORMALIZATION OF MESON THEORY WITH A FIXED EXTENDED SOURCE
G.F.CHEW:PHYS.REV.94(1954)1748
2.5.1

56L CHEW
EFFECTIVE-RANGE APPROACH TO THE LOW-ENERGY P-WAVE PION-NUCLEON INTERACTION
G.F.CHEW,F.E.LOW:PHYS.REV.101(1956)1570
2.5.1; 2.4.2; 2.4.7

57L CHEW
APPLICATION OF DISPERSION RELATIONS TO LOW ENERGY MESON-NUCLEON SCATTERING
G.F.CHEW,M.L.GOLDBERGER,F.LOW,Y.NAMBU("CGLN"):PHYS. REV.106(1957)1337
A.6.3; 2.5.1

58L CHEW
ELECTROMAGNETIC STRUCTURE OF THE NUCLEON IN LOCAL-FIELD THEORY
G.F.CHEW,R.KARPLUS,S,GASIOROWICZ,F.ZACHARIASEN:PHYS. REV.110(1958)265
2.5.2

60L CHEW
THEORY OF THE LOW-ENERGY PION-PION INTERACTION
G.F.CHEW,S.MANDELSTAM:PHYS.REV.119(1960)467
A.7.6; 2.4.6

61L CHEW
PRINCIPLE OF EQUIVALENCE FOR ALL STRONGLY INTERACTING PARTICLES WITHIN THE S-MATRIX FRAMEWORK
G.F.CHEW,S.C.FRAUTSCHI:PHYS.REV.LETT.7(1961)394 AND 8(1962)41
2.4.1; A.9.2

74L CHEW
BARYON-ANTIBARYON THRESHOLD RESONANCES
G.F.CHEW,J.KOPLIK:NUCL.PHYS.B79(1974)365
A.3.5; 2.5.4

76L CHEW
BARRELET ZEROS AND ELASTIC PI+ P PARTIAL WAVES
D.M.CHEW,M.URBAN IN"PROC.TOPICAL CONF.ON BARYON RESONANCES"(OXFORD),P.156;RUTHERFORD LAB.,1976
2.1.11; 2.4.3

78L CHEW
PI MINUS CHARGE-EXCHANGE AMPLITUDE ZEROS FROM A BARRELET ANALYSIS BETWEEN 1.4 AND 2.3 GEV
D.M.CHEW:PHYS.REV.D18(1978)2368
2.4.3; 2.1.11

79L CHEW-1
NEW DELTA BARYON PARAMETERS FROM A BARRELET ANALYSIS
OF ELASTIC PI+ P DATA BETWEEN 0.6 AND 2.3 GEV
D.M.CHEW:BERKELEY PREPRINT LBL-5306(1979)
2.4.1; 2.1.11; 2.4.3

79L CHEW-2
VERIFICATION OF THE BARRELET METHOD FOR PI N
AMPLITUDE ANALYSIS
D.M.CHEW:BERKELEY PREPRINT LBL-10075(1979)
2.1.11; 2.4.3

80L CHEW
NEW DELTA PARAMETERS FROM A SIMPLER,CHEAPER AND MORE
RELIABLE PION-NUCLEON PARTIAL WAVE ANALYSIS
D.CHEW IN 'BARYON 1980' PROC. 4TH INT.CONF.ON BARYON
RESONANCES IN TORONTO,P.123;ED.N.ISGUR;
2.4.1; 2.1.11; 2.4.3

81L CHEW
IS THERE A CONTINUUM AMBIGUITY FOR ELASTIC PION-
NUCLEON AMPLITUDES?
D.M.CHEW,R.P.ELY:PHYS.LETT.103B(1981)141
2.1.3; 2.1.11; 2.4.3

74L CHODOS
NEW EXTENDED MODEL OF HADRONS
A.CHODOS,R.J.JAFFE,K.JOHNSON,C.B.THORN,V.F.WEISSKOPF:
PHYS.REV.D9(1974)3471
2.4.1

75L CHODOS
CHIRAL INVARIANCE IN A BAG THEORY
A.CHODOS,C.B.THORN:PHYS.REV.D12(1975)2733
2.5.1; 2.4.1

68L CHOU
MODEL OF ELASTIC HIGH-ENERGY SCATTERING
T.T.CHOU,C.N.YANG:PHYS.REV.170(1968)1591
A.3.3; 2.6.4; 2.6.3

76L CHOU
HADRONIC MATTER CURRENT DISTRIBUTION INSIDE A
POLARIZED NUCLEUS AND A POLARIZED HADRON
T.T.CHOU,C.N.YANG:NUCL.PHYS.B107(1976)1
2.6.4; 2.6.3

78L CHOU
IMPACT PARAMETER ANALYSIS OF PP ELASTIC SCATTERING
T.T.CHOU:NUCL.PHYS.B139(1978)260
2.6.4; 2.6.3

79L CHOU-1
RADII OF PIONS AND KAONS
T.T.CHOU:PHYS.REV.D19(1979)3327
2.6.3

79L CHOU-2
ELASTIC HADRON-HADRON SCATTERING AT ULTRAHIGH
ENERGIES AND EXISTENCE OF MANY DIPS
T.T.CHOU,C.N.YANG:PHYS.REV.D19(1979)3268
2.6.4; 2.6.3

81L CHOU
DIP MOVEMENT IN PP AND ANTI P P ELASTIC COLLISIONS
T.T.CHOU,C.N.YANG:PHYS.REV.LETT.46(1981)764 AND
1350(E)
2.6.4

82L CHOU
ASYMPTOTIC BEHAVIOUR AT VERY HIGH ENERGIES FOR HADRON-
HADRON SCATTERING
T.T.CHOU,T.T.WU,C.N.YANG:PHYS.REV.D26(1982)328
2.6.4

72L CHOUDHARY
ANALYTICITY OF PARTIAL WAVE SCATTERING AMPLITUDES AND
CONFORMAL MAPPING METHODS
A.R.CHOUDHARY,R.B.JONES:J.PHYS.A5(1972)981
A.7.1

79L CHRISTOVA
MODEL INDEPENDENT ANALYSIS OF THE WEAK INTERACTION IN
ELECTRON-NUCLEON SCATTERING
E.C.CHRISTOVA,S.T.PETCOV:PHYS.LETT.84B(1979)250
2.5.2

69L CHU
VENEZIANO-TYPE REPRESENTATION FOR PI N AMPLITUDES
S.CHU,B.R.DESAI:PHYS.REV.188(1969)2215
2.6.5

71L CHU
ANALYSIS OF PI P ELASTIC SCATTERING AT ALL ANGLES
S.Y.CHU,A.W.HENDRY:PHYS.REV.D4(1971)2743,3282
2.6.4; 2.6.5; 2.6.7

76L CHUNG
RIGOROUS LOWER AND UPPER BOUNDS FOR THE SLOPE
PARAMETERS
B.K.CHUNG:NUCL.PHYS.B105(1976)178
A.5.3

58L CINI
APPARENT INCONSISTENCY AMONG MEASUREMENTS OF
THRESHOLD PION INTERACTIONS
M.CINI,R.GATTO,E.L.GOLDWASSER,M.RUDERMAN:NUOVO CIM.
10(1958)243
2.4.7

60L CINI
THEORY OF LOW-ENERGY SCATTERING IN FIELD THEORY
M.CINI,S.FUBINI:ANN.OF PHYS.10(1960)352
A.8.3

61L CIULLI
A CONVERGENT SET OF INTEGRAL EQUATIONS FOR SINGLET PP SCATTERING
S.CIULLI,J.FISCHER:NUCL.PHYS.24(1961)465
A.6.3; 2.6.7

62L CIULLI
ON THE EXTRAPOLATION OF THE EXPERIMENTAL SCATTERING AMPLITUDE TO THE SPECTRAL FUNCTION REGION
I.CIULLI,S.CIULLI,J.FISCHER:NUOVO CIM.23(1962)1129
A.6.3; 2.5.2; 2.6.7

69L CIULLI-1
A STABLE AND CONVERGENT EXTRAPOLATION PROCEDURE FOR SCATTERING AMPLITUDES(I)
S.CIULLI:NUOVO CIM.61A(1969)787
A.6.3; 2.4.5; 2.5.2; 2.6.7

69L CIULLI-2
A STABLE AND CONVERGENT EXTRAPOLATION PROCEDURE FOR SCATTERING AMPLITUDES(II).LEAST SQUARES AND INTER-POLATION IN ROOTS OF UNITY
S.CIULLI:NUOVO CIM.62(1969)301
A.6.3; 2.4.5; 2.5.2; 2.6.7

70L CIULLI
WEIGHT FUNCTIONS AND SUBTRACTIONS IN DISPERSION THEORY
S.CIULLI,J.FISCHER:NUCL.PHYS.B24(1970)537
A.6.4

73L CIULLI
OPTIMAL ANALYTIC INTERPOLATION FOR THE SCATTERING AMPLITUDE FROM CUTS TO INTERIOR POINTS
S.CIULLI,G.NENCIU:J.MATH.PHYS.14(1973)1675
2.6.7; A.6.3

68L COHEN-TANNOUDJI
KINEMATICAL SINGULARITIES,CROSSING AND KINEMATICAL CONSTRAINTS FOR 2-BODY HELICITY AMPLITUDES
G.COHEN-TANNOUDJI,A.MOREL,H.NAVELET:ANN.OF PHYS. 46(1968)239
A.3.5

81L COLBY
ON THE USE OF SPECIAL FUNCTIONS IN THE EXTRAPOLATION OF SCATTERING DATA
P.C.COLBY:NUCL.PHYS.A370(1981)77
2.4.5; A.6.3

74L COLLINS
THE POMERON IN ELASTIC MESON-BARYON SCATTERING
P.D.B.COLLINS,F.D.GAULT,A.MARTIN:NUCL.PHYS.B83(1974) 241
2.6.4

76L COLLINS-1
THE RELATION BETWEEN REGGEON AND PHOTON COUPLINGS
P.D.B.COLLINS,F.D.GAULT:NUCL.PHYS.B112(1976)483
2.6.4

76L COLLINS-2
DIFFRACTIVE BARYON PRODUCTION AND THE RELATION BETWEEN REGGEON AND PHOTON COUPLINGS
P.D.B.COLLINS,F.D.GAULT:NUCL.PHYS.B113(1976)34
2.6.4

78L COLLINS-1
ELASTIC SCATTERING AND THE REGGEON-PHOTON COUPLING ANALOGY
P.D.B.COLLINS,A.D.M.WRIGHT:J.PHYS.G4(1978)1223
2.6.4; 2.6.5

78L COLLINS-2
PREDICTION OF ELASTIC AND TOTAL CROSS SECTIONS USING THE REGGEON-PHOTON COUPLING ANALOGY
P.D.B.COLLINS,A.D.M.WRIGHT:J.PHYS.G4(1978)L201
2.6.4

78L COLLINS-3
REGGE CUTS AND LARGE T PROTON-PROTON SCATTERING
P.D.B.COLLINS,F.D.GAULT:PHYS.LETT.73B(1978)330
2.6.4

82L COLLINS
LARGE T ELASTIC PP AND ANTIPROTON-PROTON SCATTERING AND THE TRIPLE-SCATTERING MECHANISM
P.D.B.COLLINS,F.D.GAULT:PHYS.LETT.112B(1982)255
2.6.4

71L COMMON
RIGOROUS BOUNDS ON AVERAGED CROSS SECTIONS FOR THE SCATTERING OF PIONS
A.K.COMMON,F.J.YNDURAIN:NUCL.PHYS.B26(1971)167
A.5.3

76L COMMON
IMPROVED LOWER BOUND TO THE ABSORPTIVE PART IN THE UNPHYSICAL REGION AT HIGH ENERGIES
A.K.COMMON:NUCL.PHYS.B106(1976)176
A.5.3

75L COON
THE THREE-BODY FORCE,OFF-SHELL PI N SCATTERING AND BINDING ENERGIES IN NUCLEAR MATTER
S.COON,M.D.SCADRON,B.BARRETT:NUCL.PHYS.A242(1975)467
2.5.1

78L COON
PHENOMENOLOGY OF HIGH MOMENTUM TRANSFER ELASTIC PROCESSES
D.D.COON,J.F.GUNION,TRAN THANH VAN,R.BLANKENBECLER: PHYS.REV.D18(1978)1451
2.6.4; 2.6.6

79L COON
THE TWO-PION EXCHANGE THREE-NUCLEON POTENTIAL AND NUCLEAR MATTER
S.A.COON ET AL.:NUCL.PHYS.A317(1979)242
2.5.1

78L COQUEREAUX
RATIO OF MAGNETIC TO ELECTRIC PROTON FORM FACTORS IN QCD
R.COQUEREAUX,E.DE RAFAEL:PHYS.LETT.76B(1978)475
2.5.2

78L CORDEN
NEW DATA ON THE PI PLUS PI MINUS ANGULAR DISTRIBUTION IN THE MASS RANGE 0.8-2.2 GEV
M.J.CORDEN ET AL.:NUCL.PHYS.B137(1978)221
2.4.6

79L CORDEN
AN AMPLITUDE ANALYSIS OF PI PI SCATTERING FROM NEW DATA
M.J.CORDEN ET AL.:NUCL.PHYS.B157(1979)250
2.4.6

79L CORDIER
CROSS SECTION OF THE REACTION E+E- -> 2PI+ 2PI- FOR C.M.ENERGIES 890 TO 1100 MEV
A.CORDIER ET AL.:PHYS.LETT.81B(1979)389
2.5.2

82L CORDIER
STUDY OF THE REACTION E+E- -> 4 PIONS IN THE 1.4-2.18 GEV ENERGY RANGE
A.CORDIER ET AL.:PHYS.LETT.109B(1982)129
2.5.2

74L CORNILLE
CONSTRAINTS ON THE PHASES OF HELICITY AMPLITUDES DUE TO POSITIVITY
H.CORNILLE,A.MARTIN:NUCL.PHYS.B77(1974)141
A.5.3

75L CORNILLE
WHY IS THE DIFFRACTION PEAK A PEAK?
H.CORNILLE,A.MARTIN:NUCL.PHYS.B101(1975)411
A.5.3

76L CORNILLE-1
SCALING OF THE DIFFRACTION PEAK IN COLLISIONS OF PARTICLES OF ARBITRARY SPINS
H.CORNILLE,A.MARTIN:NUCL.PHYS.B115(1976)163
2.6.4; A.5.3

76L CORNILLE-2
SCALING LAWS TAKING INTO ACCOUNT POSITIVITY PROPERTIES
H.CORNILLE:PHYS.REV.D14(1976)1693
2.6.4; A.5.3

81L CORONIS
SEPARABLE POTENTIAL MODEL FOR THE OFF-SHELL PI N AMPLITUDE
C.CORONIS,R.H.LANDAU:PHYS.REV.C 24(1981)605
2.5.1

76L COSME
MULTIPION PRODUCTION BELOW 1.1 GEV BY E+E- ANNIHILATION
G.COSME ET AL.:PHYS.LETT.63B(1976)349
2.5.2

79L COSTA DE BEAUREGARD
STUDY OF OMEGA AND PHI EFFECTS ON THE PION FORM FACTOR
B.COSTA DE BEAUREGARD,B.PIRE,T.N.TRUONG:PHYS.REV.D19(1979)274
2.5.2

65L COTTINGHAM
IMPACT PARAMETER EXPANSION OF HIGH-ENERGY ELASTIC-SCATTERING AMPLITUDES
W.N.COTTINGHAM,R.F.PEIERLS:PHYS.REV.137(1965)B147
2.6.4; A.3.3

73L COTTINGHAM
NUCLEON-NUCLEON INTERACTION FROM PION-NUCLEON PHASE SHIFT ANALYSIS
W.N.COTTINGHAM ET AL.:PHYS.REV.D8(1973)800
2.5.3

72L COZZIKA
THE PION-NUCLEON SCATTERING AMPLITUDES AT 6 AND 16 GEV
G.COZZIKA ET AL.:PHYS.LETT.40B(1972)281
2.1.15

80L CRAWFORD
THE GLASGOW GAMMA,N,N* AND GAMMA,N,DELTA PARTIAL WAVE ANALYSIS
R.L.CRAWFORD IN"BARYON 1980,PROC.TORONTO CONF.",P.107
2.4.1

70L CREUTZ
NONCAUSAL DISPERSION RELATIONS AND A FUNDAMENTAL LENGTH
M.CREUTZ,R.JAFFE:PHYS.REV.D2(1970)2359
2.2.1

66L CRICHTON
PHASE SHIFT AMBIGUITIES FOR SPIN-INDEPENDENT SCATTERING
J.H.CRICHTON:NUOVO CIM.45A(1966)256
2.1.2

77L CRIEGEE
SMALL-ANGLE COMPTON SCATTERING ON HYDROGEN AND DEUTERIUM
L.CRIEGEE ET AL.:NUCL.PHYS.B121(1977)31
2.6.4

64L CURTIS
EXPANSIONS OF INTEGRALS OF BESSEL FUNCTIONS OF LARGE ORDER
J.CURTIS:J.MATH.PHYS.5(1964)561
A.3.3

60L CUTKOSKY
SINGULARITIES AND DISCONTINUITIES OF FEYNMAN AMPLITUDES
R.E.CUTKOSKY:J.MATH.PHYS.1(1960)429
A.6.2

68L CUTKOSKY
OPTIMIZED POLYNOMIAL EXPANSION FOR SCATTERING AMPLITUDES
R.E.CUTKOSKY,B.B.DEO:PHYS.REV.174(1968)1859
2.1.9; A.6.3

69L CUTKOSKY
THEORY OF REPRESENTATION OF SCATTERING DATA BY ANALYTIC FUNCTIONS
R.E.CUTKOSKY:ANN.PHYS.54(1969)350
2.1.9; A.6.3; 2.6.7

70L CUTKOSKY
ANALYSIS OF LOW-ENERGY K P SCATTERING DATA BY A NEW METHOD
R.E.CUTKOSKY,B.B.DEO:PHYS.REV.D1(1970)2547
2.1.9; A.6.3

73L CUTKOSKY
CONSTRUCTION OF REPRODUCING KERNELS FOR ANALYTIC HILBERT SPACES
R.E.CUTKOSKY:J.MATH.PHYS.14(1973)1231
2.1.9; A.6.3

76L CUTKOSKY
PION-NUCLEON PARTIAL WAVE ANALYSIS FROM 0.8 TO 2.0 GEV
R.E.CUTKOSKY ET AL. IN 'PROC.TOPICAL CONF.ON BARYON RESONANCES'(OXFORD),P.49;RUTHERFORD LAB.(1976)
2.1.9; 2.4.1

77L CUTKOSKY
BARYON SPECTRUM IN A LINEAR STRING MODEL
R.E.CUTKOSKY,R.E.HENDRICK:PHYS.REV.D16(1977)786,793
2.4.1

79L CUTKOSKY-1
PION-NUCLEON PARTIAL WAVE ANALYSIS
R.E.CUTKOSKY ET AL.:PHYS.REV.D20(1979)2840
2.1.9

79L CUTKOSKY-2
PION-NUCLEON PARTIAL WAVE AMPLITUDES
R.E.CUTKOSKY,C.P.FORSYTH,R.E.HENDRICK,R.L.KELLY:PHYS.REV.D20(1979)2839
2.1.9; 2.4.1

79L CUTKOSKY-3
ISOSPIN VIOLATION IN PI N SCATTERING FROM PI ZERO-ETA MIXING
R.E.CUTKOSKY:PHYS.LETT.88B(1979)339
2.1.13

80L CUTKOSKY-1
PION-NUCLEON PARTIAL WAVE ANALYSIS
R.E.CUTKOSKY IN 'BARYON 1980'PROC.4TH INT.CONF.ON BARYON RESONANCES IN TORONTO,P.19;ED.N.ISGUR
2.1.9; 2.4.1

80L CUTKOSKY-2
ODD-PARITY BARYONS:PROGRESS AND PROBLEMS
R.E.CUTKOSKY:CMU PREPRINT,SUBM.TO LOS ALAMOS CONF.(DEC.1980)
2.4.1

77L DAHL
REACTION PI MINUS PROTON TO OMEGA NEUTRON AT 20-200 GEV
O.I.DAHL ET AL.:PHYS.REV.LETT.38(1977)54
2.6.5

79L DALITZ
HIGH LYING BARYONIC MULTIPLETS IN THE HARMONIC QUARK SHELL MODEL
R.H.DALITZ,L.J.REINDERS:OXFORD PREPRINT(1979)
2.4.1

77L DALLY
DIRECT MEASUREMENT OF THE PI MINUS FORM FACTOR
E.B.DALLY ET AL.:PHYS.REV.LETT.39(1977)1176
2.5.2

82L DALLY
ELASTIC SCATTERING MEASUREMENT OF THE NEGATIVE PION RADIUS
E.B.DALLY ET AL.:PHYS.REV.LETT.48(1982)375
2.5.2

70L DAMASHEK
FORWARD COMPTON SCATTERING
M.DAMASHEK,F.J.GILMAN:PHYS.REV.1(1970)1319
2.6.4

80L DASH
TWO-COMPONENT DUALITY AND FLAVORING IN THE P+F MODEL
J.W.DASH,S.T.JONES,A.MARTIN:PHYS.REV.D22(1980)765
2.6.4

64L DASHEN
S-MATRIX METHOD FOR CALCULATION OF ELECTROMAGNETIC CORRECTIONS TO STRONG INTERACTIONS
R.F.DASHEN,S.C.FRAUTSCHI:PHYS.REV.135(1964)B1190,1196
2.1.17

65L DASHEN
GENERAL S-MATRIX METHODS FOR CALCULATION OF PERTURBATION ON THE STRONG INTERACTION
R.F.DASHEN,S.C.FRAUTSCHI:PHYS.REV.137(1965)B1318
2.1.17

69L DASHEN
EXPERIMENTAL TESTS FOR THEORIES OF CHIRAL-SYMMETRY BREAKING
R.DASHEN,M.WEINSTEIN:PHYS.REV.188(1969)2330
2.5.1

71L DASS
BOUNDS ON POLARIZATION IN PI N SCATTERING FROM ISOSPIN INVARIANCE
G.V.DASS,J.FROYLAND,F.HALZEN,A.MARTIN,C.MICHAEL,S.M.ROY:PHYS.LETT.36B(1971)339
2.4.4; A.10.2

80L DAUMENS
DISCRETE TWO-VARIABLE EXPANSIONS OF SCATTERING AMPLITUDES FOR PARTICLES WITH SPIN
M.DAUMENS,P.WINTERNITZ:PHYS.REV.D21(1980)1919
2.1.4

72L DAVIER
POMERON AND F-EXCHANGE AMPLITUDES IN ELASTIC PI N SCATTERING
M.DAVIER:PHYS.LETT.40B(1972)369
2.6.3; 2.6.4

72L DE LESQUEN
MEASUREMENT OF SPIN-ROTATION PARAMETERS IN PION-NUCLEON SCATTERING AT 6 AND 16 GEV
DE LESQUEN ET AL.:PHYS.LETT.40B(1972)277
A.4.2; A.4.3

75L DE RUJULA
HADRON MASSES IN GAUGE THEORY
A.DE RUJULA,H.GEORGI,S.L.GLASHOW:PHYS.REV.D12(1975)147
2.1.17; 2.4.1

81L DE TAR
QUARK-BAG MODEL WITH LOW-ENERGY PION INTERACTIONS
C.E.DE TAR:PHYS.REV.D24(1981)752,762
2.4.1; 2.5.1

82L DE TAR
COMPOUND BAGS AND HADRON-HADRON INTERACTIONS
C.DE TAR:HELSINKI PREPRINT HU-TFT-82-6
2.5.3

72L DEAN
AMBIGUITIES IN SCATTERING AMPLITUDES RESULTING FROM INSUFFICIENT DATA
N.W.DEAN,P.LEE:PHYS.REV.D5(1972)2741
2.1.1

82L DELBOURGO
DYNAMICAL SYMMETRY BREAKING AND THE SIGMA MASS IN QCD
R.DELBOURGO,M.D.SCADRON:PHYS.REV.LETT.48(1982)379
2.4.6

79L DELCOURT
STUDY OF THE REACTION E+E- -> P ANTI P IN THE TOTAL ENERGY RANGE 1925 - 2180 MEV
B.I.DELCOURT ET AL.:PHYS.LETT.86B(1979)395
2.5.2

82L DELCOURT
STUDY OF THE REACTION E+E- -> RHO PI AND OTHERS FOR TOTAL ENERGY RANGES BETWEEN 1.4 AND 2.18 GEV
B.DELCOURT ET AL.:PHYS.LETT.113B(1982)93
2.5.2

73L DEO
PROTON ELECTROMAGNETIC FORM FACTOR
B.B.DEO,M.K.PARIDA:PHYS.REV.D8(1973)2939
2.5.2

74L DEO
CALCULATION OF THE ANOMALOUS MAGNETIC MOMENTS OF THE NUCLEON
B.B.DEO,L.P.SINGH:PHYS.REV.D10(1974)308
A.11.1

66L DESAI
PI N CHARGE-EXCHANGE SCATTERING AND THE RHO-TRAJECTORY
B.R.DESAI:PHYS.REV.142(1966)1255
A.9.1; 2.6.5

77L DESHPANDE
HADRON ELECTROMAGNETIC MASS DIFFERENCES
N.G.DESHPANDE ET AL.:PHYS.REV.D15(1977)1885
2.1.17

73L DEVENISH
TWO-COMPONENT DUALITY AND FIXED-T DISPERSION RELATIONS FOR PI N CHARGE-EXCHANGE
R.C.E.DEVENISH,B.R.MARTIN:PHYS.REV.D8(1973)3126
2.6.5

73L DIAS DE DEUS
GEOMETRIC SCALING,MULTIPLICITY DISTRIBUTIONS AND CROSS SECTIONS
J.DIAS DE DEUS:NUCL.PHYS.59B(1973)231
2.6.4

77L DIAS DE DEUS
ON THE SYSTEMATICS OF ELASTIC SCATTERING AT HIGH AND INTERMEDIATE ENERGY
J.DIAS DE DEUS,P.KROLL:NUOVO CIM.37A(1977)67
2.6.4

78L DIAS DE DEUS
DIPS,ZEROS AND LARGE T BEHAVIOUR OF THE ELASTIC AMPLITUDE
J.DIAS DE DEUS,P.KROLL:ACTA PHYS.POLON.B9(1978)157
2.6.4

75L DICKEY
TEST OF DUALITY DIAGRAM PREDICTION FOR RESONANCE BACKGROUND
J.O.DICKEY:NUCL.PHYS.B90(1975)501 AND PROC.11TH RENCONTRE DE MORIOND(1976);ED.J.TRAN THANH VAN(1976)
2.4.1

62L DIETZ
EINFLUSS DER PION-PION-WECHSELWIRKUNG AUF PION-NUKLEON STREUPHASEN
K.DIETZ:Z.PHYSIK 170(1962)212
A.8.3; 2.5.1

63L DIETZ
THE EFFECT OF A PION-PION INTERACTION ON LOW-ENERGY PION-NUCLEON SCATTERING
K.DIETZ:NUOVO CIM.27(1963)141
2.5.1; A.8.3

80L DITTES
PION FORM FACTOR BEYOND THE LEADING ORDER OF PERTURBATIVE QCD
F.M.DITTES,A.V.RADYUSHKIN,DUBNA PREPRINT JINR-EZ-80-688
2.5.2

75L DIU
NON-REGGE BEHAVIOUR IN PI N BACKWARD SCATTERING AT HIGH ENERGY
B.DIU,P.TCHANG:NUOVO CIM.32A(1976)185
2.6.7; A.7.1

68L DIVGI
REANALYSIS OF THE LOWEST-MASS NEGATIVE PARITY BARYON RESONANCES
D.R.DIVGI,O.W.GREENBERG:PHYS.REV.175(1968)2024,2027
2.4.1

81L DODD
CLOUDY BAG MODEL:CONVERGENT PERTURBATION EXPANSION FOR THE NUCLEON
L.R.DODD,A.W.THOMAS,R.F.ALVAREZ-ESTRADA:PHYS.REV.D24(1981)1961;SEE ALSO TH.3316-CERN(1982)
2.5.1

68L DOLEN
FINITE ENERGY SUM RULES AND APPLICATIONS IN PI N CHARGE-EXCHANGE
R.DOLEN,D.HORN,C.SCHMID:PHYS.REV.166(1968)1768
2.6.5

80L DOMINGUEZ-1
DETERMINATION OF THE PI NN VERTEX FUNCTION FROM NP AND PP CHARGE-EXCHANGE SCATTERING
C.A.DOMINGUEZ,B.J.VERWEST:PHYS.LETT.89B(1980)333
2.5.1

80L DOMINGUEZ-2
EXTRACTION OF THE PI NN FORM FACTOR FROM CHARGED PION PHOTOPRODUCTION
C.A.DOMINGUEZ,R.B.CLARK:PHYS.REV.C21(1980)1944
2.5.1

81L DOMINGUEZ
PI NN,PI N DELTA AND PI PI RHO FORM FACTORS FROM QUASI-TWO-BODY HADRONIC REACTIONS
C.A.DOMINGUEZ:PHYS.REV.C24(1981)2611
2.5.1

82L DOMINGUEZ-1
CHIRAL-SYMMETRY BREAKING CORRECTIONS TO THE GOLDBERGER TREIMAN RELATION
C.A.DOMINGUEZ:PHYS.REV.D25(1982)1937
2.5.1

82L DOMINGUEZ-2
ELECTROMAGNETIC FORM FACTOR OF THE PIONS,VECTOR MESONS OR QUARKS?
C.DOMINGUEZ:PHYS.REV.D25(1982)3084
2.5.2

72L DONCEL
CONSTRAINTS ON SPIN-ROTATION PARAMETERS DUE TO ISOSPIN CONSERVATION
M.G.DONCEL,L.MICHEL,P.MINNAERT:PHYS.LETT.38B(1972)42
A.10; 2.4.4

64L DONNACHIE-1
SEMIPHENOMENOLOGICAL SOLUTIONS OF PI N PARTIAL-WAVE DISPERSION RELATIONS
A.DONNACHIE,J.HAMILTON:PHYS.REV.133(1964)B1053
A.7.1

64L DONNACHIE-2
PREDICTION OF P- D- AND F-WAVE PION-NUCLEON SCATTERING
A.DONNACHIE,J.HAMILTON,A.T.LEA:PHYS.REV.135(1964)B515
A.7.1

65L DONNACHIE-1
VERY SHORT-RANGE INTERACTION IN PION-NUCLEON SCATTERING
A.DONNACHIE,J.HAMILTON:PHYS.REV.138(1965)B678
A.7.1

65L DONNACHIE-2
THE QUANTUM NUMBERS OF THE NUCLEON ISOBARS
A.DONNACHIE,J.HAMILTON:ANN.OF PHYS.31(1965)410
A.7.1

68L DONNACHIE
EVIDENCE FROM PI P PHASE SHIFT ANALYSIS FOR NINE MORE POSSIBLE NUCLEON RESONANCES
A.DONNACHIE,R.KIRSOPP,C.LOVELACE:PHYS.LETT.26B(1968) 161
2.1.5

74L DONNACHIE
PION-NUCLEON SCATTERING IN THE BACKWARD HEMISPHERE
A.DONNACHIE,P.R.THOMAS:NUOVO CIM.19A(1974)279
2.6.7

79L DONNACHIE
ELASTIC SCATTERING AT LARGE T
A.DONNACHIE,P.V.LANDSHOFF,Z.PHYS.C2(1979)55
2.6.4

78L DONOGHUE
ADLER-WEISBERGER RELATION AND THE QUARK MODEL
J.F.DONOGHUE,D.WYLER:PHYS.REV.D17(1978)280
2.5.1

65L DRELL
ANOMALOUS MAGNETIC MOMENT OF THE ELECTRON,MUON AND NUCLEON
S.D.DRELL,H.R.PAGELS:PHYS.REV.140(1965)B397
A.11.1

68L DRELL
DIFFERENCE OF THE NUCLEON AND PION RADII
S.D.DRELL,D.J.SILVERMAN:PHYS.REV.LETT.20(1968)1325
A.11.1

81L DUBNICKA-1
THE DATA ON THE PION FORM FACTOR AND P-WAVE ISOVECTOR PI PI PHASE SHIFT ARE DEFINITLY CONSISTENT
S.DUBNICKA,V.A.MESHCHERYAKOV,J.MILKO:J.PHYS.G7(1981) 605

81L DUBNICKA-2
WHY THE RHO-POLE POSITION COULD NOT BE FOUND FROM SPACE-LIKE REGION DATA ON NUCLEON FORM FACTORS
S.DUBNICKA,D.KRUPA,V.A.MESHCHERYAKOV:SUBM.TO ACTA PHYS.SLOV.(1981)
2.5.2

81L DUBNICKA-3
THE PION CHARGE RADIUS OBTAINED IN THE RECENT SOVIET-AMERICAN EXPERIMENT SEEMS TO BE QUESTIONABLE
S.DUBNICKA,A.Z.DUBNICKOVA,ST.DUBNICKA:HELSINKI PREPRINT HU-TFT-81-27
2.5.2

78L DULUDE-1
OBSERVATION OF STRUCTURE IN THE REACTION ANTIPROTON-PROTON -> TWO PI ZERO
R.S.DULUDE ET AL.:PHYS.LETT 79B(1978)335
2.5.4

78L DULUDE-2
MEASUREMENT OF ANTIPROTON-PROTON -> TWO PI ZERO IN THE T-AND U-MESON REGION
R.S.DULUDE ET AL.:PHYS.LETT.79B(1978)329
2.5.4

75L DUMBRAJS
BOUNDS ON COUPLING CONSTANTS PROVIDED BY KNOWLEDGE OF MODULI AND ZEROS OF FORWARD SCATTERING AMPLITUDES
O.V.DUMBRAJS,M.STASZEL:NUCL.PHYS.B88(1975)52
2.4.2; 2.4.3

76L DUMBRAJS
EXTRAPOLATION OF THE PROTON FORM FACTOR BY MEANS OF PADE APPROXIMANTS
O.DUMBRAJS:REV.ROUM.PHYS.21(1976)273
2.5.2

80L DUNCAN
ASYMPTOTIC BEHAVIOUR OF COMPOSITE-PARTICLE FORM FACTORS AND THE RENORMALIZATION GROUP
A.DUNCAN,A.MUELLER:PHYS.REV.D21(1980)1636
2.5.2

68L DURAND
DIFFRACTION MODEL FOR HIGH-ENERGY PP-SCATTERING
L.DURAND,R.LIPES:PHYS.REV.LETT.20(1968)637
2.6.4; 2.6.3

77L DURSO-1
ISOBARS,TRANSITION POTENTIALS AND SHORT RANGE REPULSION IN THE NN-INTERACTION
J.W.DURSO,M.SAARELA,G.E.BROWN,A.D.JACKSON:NUCL. PHYS.A278(1977)445
2.5.3

77L DURSO-2
THE PI NN VERTEX FUNCTION
J.W.DURSO,A.D.JACKSON,B.J.VERWEST:NUCL.PHYS.A282(1977) 404
2.5.1

80L DURSO
MICROSCOPIC MODELS OF PSEUDOPHYSICAL ANTINUCLEON NUCLEON -> PION PION AMPLITUDES
J.W.DURSO,A.D.JACKSON,B.J.VERWEST:NUCL.PHYS.A345(1980) 471
2.5.3; 2.4.6; 2.5.1

73L DZHGARKAVA
PI N SCATTERING AT HIGH ENERGIES IN THE QUASIPOTENTIAL APPROACH
M.I.DZHGARKAVA ET AL:NUCL.PHYS.B67(1973)232
2.5.4; 2.6.6

74L DZHGARKAVA
QUASIPOTENTIAL APPROACH AND BACKWARD PI N SCATTERING
AT HIGH ENERGY
M.I.DZHGARKAVA ET AL.:NUCL.PHYS.B79(1974)396
2.6.7

81L DZIEMBOWSKI
DIQUARK CLUSTERING AND THE NEUTRON CHARGE RADIUS
Z.DZIEMBOWSKI,W.J.METZGER,R.T.VAN DE WALLE:Z.PHYSIK
C10(1981)231
2.5.2

61L EFREMOV
PION-NUCLEON SCATTERING AT LOW ENERGIES
A.V.EFREMOV,V.A.MESHCHERYAKOV,D.V.SHIRKOV:SOV.PHYS.
JETP 12(1961)308
A.6.8

80L EFREMOV
FACTORIZATION AND ASYMPTOTIC BEHAVIOUR OF PION FORM
FACTOR IN QCD
A.V.EFREMOV,A.V.RADYUSHKIN:PHYS.LETT.94B(1980)245
2.5.2

74L EICHMANN
CAN DISPERSION INTEGRALS BE REPLACED BY DIFFERENTIAL
OPERATORS?
G.EICHMANN,J.DRONKERS:PHYS.LETT.52B(1974)428
2.2.1; A.6.4

75L EICHMANN
COMMENT ON A SCALING LAW FOR PION-NUCLEON SCATTERING
G.EICHMANN,F.KAISER:PHYS.REV.D11(1975)3352
2.6.4

73L EILAM
STATISTICAL APPROACH TO DIFFERENTIAL CROSS SECTIONS
NEAR 90 DEGREES
G.EILAM,Y.GELL,B.MARGOLIS,W.J.MEGGS:PHYS.REV.D8(1973)
2871
2.6.6

75L EISENHANDLER
MEASUREMENT OF DIFFERENTIAL CROSS SECTIONS FOR
ANTIPROTON-PROTON ANNIHILATION INTO CHARGED PION PAIRS
E.EISENHANDLER ET AL.:NUCL.PHYS.B96(1975)109
2.5.4

72L ELVEKJAER
SHORT RANGE PI N P-WAVE INTERACTION. COMPARISON WITH
HIGH-ENERGY BACKWARD REGGE MODELS
F.ELVEKJAER:NUCL.PHYS.B47(1972)253
2.6.7; A.7.1

75L ELVEKJAER
GEOMETRICAL DESCRIPTION OF BINARY REACTIONS AND LARGE
TRANSVERSE MOMENTUM PHENOMENA
F.ELVEKJAER,J.L.PETERSEN:NUCL.PHYS.B94(1975)100
2.6.6; A.3.3

77L ENGELMANN
REAL PARTS OF PION-PROTON FORWARD SCATTERING
AMPLITUDES
T.R.ENGELMANN,R.E.HENDRICK:PHYS.REV.D16(1977)2891
2.2.1

70L ENGELS-1
N, DELTA AND RHO-EXCHANGE IN PION-NUCLEON SCATTERING
J.ENGELS,G.HOEHLER:NUCL.PHYS.B15(1970)365
A.8.1; 2.6.7; 2.4.2; A.6.9

70L ENGELS-2
MESON-EXCHANGES IN THE ISOSPIN ZERO T-CHANNEL OF PION-
NUCLEON SCATTERING
J.ENGELS:NUCL.PHYS.B25(1970)141
2.6.7; 2.4.2; A.6.9

70L ENGELS-3
GENERALIZED POTENTIAL AND UNIQUENESS IN THE
CALCULATION OF THE PION-NUCLEON P33-STATE
J.ENGELS,V.FRANK:NUCL.PHYS.B22(1970)429
2.5.1; A.7.1

71L ENGELS
N/D CALCULATIONS FOR THE P33 PI N PARTIAL WAVE WITH
EFFECTIVE LAGRANGIAN INPUT
J.ENGELS,J.FLEISCHER:NUOVO CIM.2A(1971)384
A.7.1; A.8.2; 2.5.1

72L ENGELS
APPLICATION OF SUBTRACTED FIXED-T DISPERSION
RELATIONS TO PI N SCATTERING
J.ENGELS:NUCL.PHYS.B36(1972)73
2.1.7; 2.4.7; 2.5.1

74L ENGELS
LEGENDRE AND PADE-APPROXIMANTS IN PI N SCATTERING
J.ENGELS,J.FLEISCHER:NUOVO CIM.24A(1974)73
2.5.1

74L EPSTEIN
DETERMINATION OF N ANTI N -> PI PI HELICITY AMPLITUDES
G.N.EPSTEIN,B.H.J.MC KELLAR:PHYS.REV.D10(1974)2169
2.4.6

81L ERICSON
THE AXIAL POLARIZABILITY OF NUCLEONS AND NUCLEI
M.ERICSON,A.FIGUREAU:J.PHYS.G7(1981)1197
2.5.2

74L ERNST
ENERGY-DEPENDENT SEPARABLE POTENTIALS
D.J.ERNST,J.T.LONDERGAN,E.J.MONIZ,R.M.THALER:PHYS.REV.
C10(1974)1708
2.5.1

78L ERNST
PION-NUCLEON FORM FACTOR IN CHEW-LOW THEORY
D.J.ERNST,M.B.JOHNSON:PHYS.REV.C17(1978)247
2.5.1

80L ERNST
SIMPLE MODELS OF PION-NUCLEON SCATTERING
D.J.ERNST,M.B.JOHNSON:PHYS.REV.C22(1980)651
2.5.1

75L ERVANNE
THE PUZZLE OF NEAR SATURATION OF ISOSPIN BOUNDS
J.ERVANNE,M.KORKEA-AHO,N.A.TORNQVIST:PHYSICA SCRIPTA
12(1975)305
2.4.4

77L ESPOSITO
MOMENTUM ANALYSIS OF KAON AND PION PAIRS PRODUCED
FROM TIMELIKE PHOTONS AT 1.6 GEV
B.ESPOSITO:PHYS.LETT.67B(1977)239
2.5.2

75L ESTABROOKS
PI PI PARTIAL WAVES FROM 0.6 TO 1.8 GEV
P.ESTABROOKS,A.D.MARTIN:NUCL.PHYS.B95(1975)322
2.4.6

79L ESTABROOKS
WHERE AND WHAT ARE THE SCALAR MESONS?
P.ESTABROOKS:PHYS.REV.D19(1979)2678
2.4.6

79L EVANGELISTA
ANALYSIS OF THE REACTION PI MINUS PROTON ->
ANTIPROTON-PROTON-NEUTRON AT 10 AND 16 GEV
C.EVANGELISTA ET AL.:NUCL.PHYS.B153(1979)253
2.5.4

68L FAIMAN
HARMONIC OSCILLATOR MODEL FOR BARYONS
D.FAIMAN,A.W.HENDRY:PHYS.REV.173(1968)1720 AND
180(1969)1609
2.4.1

81L FAISSLER
LARGE-ANGLE PP ELASTIC SCATTERING AT 201 AND 400 GEV/C
W.FAISSLER ET AL.:PHYS.REV.D23(1981)33
2.6.6

79L FARRAR
PION FORM FACTOR
G.FARRAR,D.JACKSON:PHYS.REV.LETT.43(1979)246
2.5.2

58L FEDERBUSH
ELECTROMAGNETIC STRUCTURE OF THE NUCLEON
P.FEDERBUSH,M.L.GOLDBERGER,S.B.TREIMAN:PHYS.REV.
112(1958)642
2.5.2

79L FEINBERG
IS THERE A STRONG VAN DER WAALS FORCE BETWEEN HADRONS?
G.FEINBERG,J.SUCHER:PHYS.REV.20(1979)1717
2.5.3

79L FELICETTI
RESONANT SPACE- AND TIMELIKE PION FORM FACTOR
F.FELICETTI,Y.SRIVASTAVA:PHYS.LETT.83B(1979)109
2.5.2

70L FENSTER
PION-NUCLEON SCATTERING WITH VENEZIANO TERMS
S.FENSTER,K.C.WALI:PHYS.REV.D1(1970)1409
2.6.5

50L FERMI
HIGH ENERGY EVENTS
E.FERMI:PROGR.THEOR.PHYS.5(1950)570
2.6.6

73L FERRARI
DUALITY OF THE SCATTERING AMPLITUDE IN THE COMPLEX
ENERGY PLANE
E.FERRARI:NUOVO CIM.17A(1973)505
2.6.5

74L FERRARI
ASYMPTOTIC CROSSING EVEN PI N FORWARD AMPLITUDE AND
DUALITY IN THE COMPLEX ENERGY PLANE
E.FERRARI,M.LUSIGNOLI:NUOVO CIM.23A(1974)51
2.6.5

79L FERRARI
HIGH ENERGY BEHAVIOUR OF PI N SCATTERING PARTIAL
WAVES FROM AMPLITUDE ANALYSIS
E.FERRARI:NUOVO CIM.49A(1979)1
2.1.7

64L FESHBACH
THE BOUNDARY CONDITION MODEL OF STRONG INTERACTIONS
H.FESHBACH,E.L.LOMON:ANN.OF PHYS.29(1964)19
2.4.1; 2.5.1

72L FILKOV
SOLUTION OF DISPERSION RELATIONS FOR PI N SCATTERING
AMPLITUDES BY PADE-APPROXIMANTS
L.V.FILKOV,B.B.PALYUSHER:NUCL.PHYS.B42(1972)541
2.5.1

73L FISCHER
THE USE OF UNITARITY BOUNDS FOR A STABLE EXTRAPOLATION OF LOW ENERGY DATA
J.FISCHER:COMM.MATH.PHYS.30(1973)249
A.6.3; 2.6.7

76L FISCHER
RIGOROUS PHASE-MODULUS CORRELATIONS FOR FORWARD AMPLITUDES AT HIGH ENERGIES
J.FISCHER,P.KOLAR,I.VRKOC:PHYS.REV.D(1976)133
A.6.4

78L FISCHER-1
HIGH-ENERGY STATUS OF DERIVATIVE ANALYTICITY RELATIONS
J.FISCHER,P.KOLAR:PHYS.REV.D17(1978)2168
A.6.4; 2.2.1

78L FISCHER-2
POMERANCHUK-TYPE THEOREMS FOR TOTAL AND ELASTIC CROSS SECTIONS
J.FISCHER,R.SALY,I.VRKOC:PHYS.REV.D18(1978)4271
A.5.4

64L FLETCHER
R.FLETCHER,C.M.REEVES:COMPUTER JOURN.7(1964)151
2.1.7

73L FONDA
NO-POLE S-MATRIX FIT TO THE DELTA(1232)RESONANCE
L.FONDA,G.C.GHIRARDI,G.L.SHAW:PHYS.REV.D8(1973)353
2.4.1

81L FORSYTH-1
MASSES AND WIDTHS OF ODD-PARITY N AND DELTA RESONANCES
C.P.FORSYTH,R.E.CUTKOSKY:PHYS.REV.LETT.46(1981)576
2.4.1

81L FORSYTH-2
A QUARK MODEL OF THE NON-STRANGE BARYONS
C.P.FORSYTH:THESIS;CARNEGIE-MELLON UNIVERSITY, PITTSBURGH(1981)
2.4.1

81L FORSYTH-3
EVIDENCE AGAINST DIQUARKS FORM BARYON RESONANCE DECAYS
C.P.FORSYTH,R.E.CUTKOSKY:NUCL.PHYS.B178(1981)35
2.4.1

79L FRANCA
ENERGY DEPENDENCE OF THE EIKONAL IN PP-SCATTERING
H.M.FRANCA,Y.HAMA:PHYS.REV.D19(1979)3261
2.6.3

60L FRAUTSCHI
PION-NUCLEON SCATTERING IN THE MANDELSTAM REPRESENTATION
S.C.FRAUTSCHI,J.D.WALECKA:PHYS.REV.120(1960)1487
A.7.1

68L FRAUTSCHI
MULTIPLE SCATTERING CORRECTION TO POMERON-EXCHANGE
S.FRAUTSCHI,B.MARGOLIS:NUOVO CIM.56(1968)1155
2.6.3

72L FRAUTSCHI
ERICSON FLUCTUATIONS AND THE BOHR MODEL IN HADRON PHYSICS
S.FRAUTSCHI:NUOVO CIM.12A(1972)133
2.6.6

60L FRAZER-1
PARTIAL-WAVE DISPERSION RELATIONS FOR THE PROCESS PI PI -> NUCLEON ANTINUCLEON
W.R.FRAZER,J.R.FULCO:PHYS.REV.117(1960)1603,1609
A.3.4; A.4.5; A.7.5; 2.4.6; 2.5.2

60L FRAZER-2
PARTIAL-WAVE DISPERSION RELATIONS FOR PION-NUCLEON SCATTERING
W.R.FRAZER,J.R.FULCO:PHYS.REV.119(1960)1420
A.7.1

60L FRAZER-3
APPLICATION OF THE MANDELSTAM REPRESENTATION TO PI N SCATTERING
W.R.FRAZER IN"PROC.1960 INTERN.CONF.ON HIGH ENERGY PHYSICS",P.282;EDS.E.C.G.SUDARSHAN,J.H.TINLOT,A.C. MELISSINOV;UNIV.OF ROCHESTER
2.5.1; A.8.3

61L FRAZER
APPLICATIONS OF CONFORMAL MAPPING TO THE PHENOMENOLOGICAL REPRESENTATION OF SCATTERING AMPLITUDES
W.R.FRAZER:PHYS.REV.123(1961)2180
2.6.7

68L FREUND
FINITE-ENERGY SUM RULES AND BOOTSTRAPS
P.G.O.FREUND:PHYS.REV.LETT.20(1968)235
2.6.4

77L FRITZSCH
CALCULATING THE CABIBBO ANGLE
H.FRITZSCH:PHYS.LETT.70B(1977)436
2.5.1

82L FROEHLICH-1
SEPARABLE BETHE-SALPETER EQUATION KERNELS FOR PION-NUCLEON SCATTERING
J.FROEHLICH,K.SCHWARZ,L.STREIT,H.F.K.ZINGL:PHYS.REV. C25(1982)2591
2.5.1

82L FROEHLICH-2
REDUCTION OF SEPARABLE BETHE-SALPETER EQUATION KERNELS TO QUASIPOTENTIAL EQUATIONS
J.FROEHLICH,K.SCHWARZ,H.F.K.ZINGL:UNI GRAZ UTP 05/82(1982)
2.5.1

75L FROGGATT
PHASE SHIFT ANALYSIS OF PI PLUS PI MINUS SCATTERING BETWEEN 1.1 AND 1.8 GEV BASED ON FIXED MOMENTUM TRANSFER ANALYTICITY
C.D.FROGGATT,J.L.PETERSEN:NUCL.PHYS.B91(1975)454(ERR. B104(1976)186)
2.4.6

77L FROGGATT
PHASE SHIFT ANALYSIS OF PI PLUS PI MINUS SCATTERING BETWEEN 1.0 AND 1.8 GEV BASED ON FIXED MOMENTUM TRANSFER ANALYTICITY
C.D.FROGGATT,J.L.PETERSEN:NUCL.PHYS.B129(1977)89
2.4.6; 2.5.2

61L FROISSART
ASYMPTOTIC BEHAVIOUR AND SUBTRACTIONS IN THE MANDELSTAM REPRESENTATION
M.FROISSART:PHYS.REV.123(1961)1053
A.5.3; A.6.2

68L FRYE
ANALYSIS OF PARTIAL-WAVE DISPERSION RELATIONS
G.FRYE,R.L.WARNOCK:PHYS.REV.130(1968)478
A.7.1

68L FUBINI
DISPERSION THEORY OF LOW-ENERGY LIMITS
S.FUBINI,G.FURLAN:ANN.OF PHYS.48(1968)478
2.5.1

79L FUCHS
CURRENT QUARK MASSES AND STRUCTURE FUNCTIONS
N.H.FUCHS,M.D.SCADRON:PHYS.REV.D20(1979)2421
2.5.1

82L FUCHS-1
SU(2) MASS DIFFERENCES OF HADRONS:A FINE TUNING OF CHIRAL SYMMETRY BREAKING
N.H.FUCHS,M.D.SCADRON:PURDUE PREPRINT(1982)
2.1.17

82L FUCHS-2
A UNIFIED QUANTUM CHROMODYNAMICS PICTURE OF BROKEN CHIRAL SYMMETRY
N.H.FUCHS,M.D.SCADRON:PURDUE PREPRINT(1982)
2.5.1

82L FUCITO
HADRON SPECTROSCOPY IN LATTICE QCD
F.FUCITO ET AL.:CERN PREPRINT TH 3288(1982)
2.4.1

81L FUDA
INVERSION PROCEDURE FOR A PION-NUCLEON T MATRIX
M.G.FUDA:PHYS.REV.C24(1981)614
2.5.1

79L FURLAN
WEAK FORM FACTORS FROM PION ELECTROPRODUCTION AT THRESHOLD
G.FURLAN,N.PAVER,C.VERZEGNASSI:NUOVO CIM.49A(1979)26
2.5.2

79L GANGAL
IMPROVED UNITARITY BOUND ON THE SLOPE PARAMETER
A.D.GANGAL,A.S.VENGURLEKAR:PHYS.REV.D19(1979)1465
A.5.3

82L GARCILAZO
ENERGY-INDEPENDENT SEPARABLE POTENTIALS FOR THE PION-NUCLEON P33-CHANNEL
M.GARCILAZO,L.MATHELITSCH,B.J.VERWEST:TEXAS A&M UNIV. PREPRINT(1982)
2.5.1

75L GASSER
IMPLICATIONS OF SCALING FOR THE PROTON-NEUTRON MASS DIFFERENCE
J.GASSER,H.LEUTWYLER:NUCL.PHYS.B94(1975)269
2.1.17; 2.4.4; 2.5.1

80L GASSER
APPROACHING THE CHIRAL LIMIT IN QCD
J.GASSER,A.ZEPEDA:NUCL.PHYS.B174(1980)445
2.5.1

81L GASSER
HADRON MASSES AND THE SIGMA COMMUTATOR IN CHIRAL PERTURBATION THEORY
J.GASSER:ANN.OF PHYS.136(1981)62
2.5.1; A.8.2

79L GAULT
EXTRACTION OF AXIAL VECTOR FORM FACTOR USING CURRENT ALGEBRA,PCAC AND THRESHOLD ELECTROPRODUCTION AMPLITUDES
F.D.GAULT,M.D.SCADRON:NUCL.PHYS.B157(1979)517
2.5.2

79L GAVELA
VAN DER WAALS-LIKE FORCES BETWEEN HADRONS INDUCED BY COLOR CONFINING POTENTIALS
M.B.GAVELA ET AL.:PHYS.LETT.82B(1979)431
2.5.3

```
80L GAVELA
    STANDARD SOLUTION TO THE ROPER-RESONANCE PUZZLE
    M.B.GAVELA ET AL.:PHYS.REV.D21(1980)182
    2.4.1
82L GAVELA
    PIONIC DECAYS AND SATURATION OF CURRENT ALGEBRA SUM
    RULES IN AN EXPANSION OF THE QUARK SHELL MODEL
    M.B.GAVELA ET AL.:PHYS.REV.D25(1982)1921
    2.4.1
61L GELL-MANN
    FORM FACTORS AND VECTOR MESONS
    M.GELL-MANN,F.ZACHARIASEN:PHYS.REV.124(1961)953
    A.8.1; A.8.3; 2.5.1
64L GELL-MANN
    ELEMENTARY PARTICLES OF CONVENTIONAL FIELD THEORY AS
    REGGE POLES.III
    M.GELL-MANN ET AL.:PHYS.REV.133(1964)B145
    A.9; A.3.1
68L GELL-MANN
    BEHAVIOR OF CURRENT DIVERGENCES UNDER SU(3)XSU(3)
    M.GELL-MANN,R.J.OAKES,B.RENNER:PHYS.REV.175(1968)2195
    2.5.1
76L GENSINI
    OPTIMAL TECHNIQUES FOR THE EXTRAPOLATION OF SU(2)
    VIOLATING PI N AMPLITUDES
    P.GENSINI,N.PAVER,C.VERCEGNASSI:LETT.NUOVO CIM.
    15(1976)409
    2.5.1
77L GENSINI
    PION-NUCLEON AMPLITUDES AROUND THE ZERO-MOMENTUM POINT
    P.M.GENSINI:SLAC-PUB-1944 AND 1967
    2.5.1; 2.4.7
78L GENSINI
    ANALYTIC REPRESENTATION FOR THE PION FORM FACTOR AT
    ALL Q SQUARED
    P.M.GENSINI:PHYS.REV.D17(1978)1368
    2.5.2
80L GENSINI
    MESON-NUCLEUS SCATTERING LENGTHS AND SIGMA-TERMS.THE
    PI-NUCLEUS SYSTEM
    P.M.GENSINI:NUOVO CIM.60A(1980)221
    2.5.1
81L GENSINI-1
    SIGMA TERMS ON AND OFF THE MESON MASS SHELL
    P.M.GENSINI:J.PHYS.G7(1981)1177
    2.5.1
81L GENSINI-2
    CURRENT QUARK-MASS RATIO:MESONIC COMPARED WITH
    BARYONIC DATA
    P.M.GENSINI:J.PHYS.G7(1981)1315
    2.5.1
73L GENZ
    ON A PROPORTIONALITY OF HADRONIC AND ELECTROMAGNETIC
    AMPLITUDES
    H.GENZ,W.SCHMIDT:NUCL.PHYS.B59(1973)629
    2.6.4
76L GENZ
    VALIDITY OF SU(3) AND VIOLATION OF ZWEIG'S RULE FOR
    VECTOR MESON-NUCLEON COUPLING CONSTANTS
    H.GENZ,G.HOEHLER:PHYS.LETT.61B(1976)389
    2.5.2
76L GERDT
    UNIFORMIZATION OF THE FORWARD SCATTERING AMPLITUDE AT
    HIGH ENERGY
    V.P.GERDT,V.I.INOZEMTSEV,V.A.MESHCHERYAKOV:SOV.J.NUCL.
    PHYS.24(1976)90
    2.2.1; A.6.4
67L GERSTEIN
    PION-NUCLEON SPIN-FLIP SUM RULE
    I.S.GERSTEIN:PHYS.REV.161(1967)1631
    2.5.1
69L GERSTEN
    AMBIGUITIES OF COMPLEX PHASE-SHIFT ANALYSIS
    A.GERSTEN:NUCL.PHYS.B12(1969)537
    2.1.11; A.1.3; A.2.1
74L GERSTEN-1
    ZEROS OF THE SCATTERING AMPLITUDE IN THE COMPLEX
    ANGLE PLANE
    A.GERSTEN:NUCL.PHYS.A219(1974)317
    2.1.2; 2.1.11
74L GERSTEN-2
    PION-NUCLEON COUPLING CONSTANT
    A.GERSTEN:PHYS.REV.D10(1974)2876
    2.4.2
81L GIDAL
    OBSERVATION OF S* -> PI+ PI- IN PSI DECAY
    G.GIDAL ET AL.:SLAC-PUB-2793
    2.4.6
71L GIFFON
    RECONSTRUCTION OF AMPLITUDES FOR PI N SCATTERING FROM
    DATA
    M.GIFFON,J.JOSEPH,J.MEYER:NUOVO CIM.5A(1971)478
    2.1.15
```

72L GIFFON-1
POLARIZATION PARAMETERS AND RECONSTRUCTION OF AMPLITUDES FOR PI N SCATTERING
M.GIFFON:NUOVO CIM.7A(1972)705
2.1.15

72L GIFFON-2
A NEW METHOD FOR THE DETERMINATION OF PI N SCATTERING AMPLITUDES
M.GIFFON,R.LAVERRIERE:NUOVO CIM.11A(1972)917
2.1.15

57L GILBERT
NEW DISPERSION RELATIONS FOR PION-NUCLEON SCATTERING
W.GILBERT:PHYS.REV.108(1957)1078
A.6.4

70L GILMAN
HELICITY CONSERVATION IN DIFFRACTION SCATTERING
F.J.GILMAN,J.PUMPLIN,A.SCHWIMMER,L.STODOLSKY:PHYS. LETT.31B(970)387
2.6.3; 2.1.15

78L GINZBURG
RADIATIVE CORRECTIONS IN CHARGE-EXCHANGE AND BACKWARD HADRON SCATTERING
I.F.GINZBURG,G.L.KOTKIN,V.G.SERBO:PHYS.LETT.80B(1978) 101
2.1.17; 2.2.1; 2.6.5

76L GIRARDI
LOW-LYING REGGE TRAJECTORY AND POLARIZATIONS IN PI P SCATTERING
G.GIRARDI,H.NAVELET:PHYS.REV.D14(1976)280
2.6.3

68L GLASHOW
BREAKING CHIRAL SYMMETRY
S.L.GLASHOW,S.WEINBERG:PHYS.REV.LETT.20(1968)224
2.5.1

81L GLUSHKO
HIGH ENERGY PI MINUS P BACKWARD SCATTERING AND THE STRUCTURE OF THE DELTA-SPECTRUM
N.I.GLUSHKO,N.A.KOBYLINSKY,V.V.TIMOKHIN:KIEV PREPRINT ITP-81-73E
2.6.7

67L GOEBEL
EIGENPHASES AND THE GENERALIZED BREIT-WIGNER APPROXIMATION
G.J.GOEBEL,K.W.MC VOY:PHYS.REV.164(1967)1932
2.4.1

67L GOLDBERG
CURRENT ALGEBRAS,REGGE POLES AND THE ISOVECTOR MAGNETIC MOMENT OF THE NUCLEON
H.GOLDBERG,F.GROSS:PHYS.REV.162(1967)1350
2.5.1; 2.5.2

55L GOLDBERGER-1
CAUSALITY CONDITIONS AND DISPERSION RELATIONS
M.L.GOLDBERGER:PHYS.REV.99(1955)979
A.6.4

55L GOLDBERGER-2
APPLICATION OF DISPERSION RELATIONS TO PION-NUCLEON SCATTERING
M.L.GOLDBERGER,H.MIYAZAWA,R.OEHME:PHYS.REV.99(1955)986
A.6.4; 2.2.1

58L GOLDBERGER
DECAY OF THE PI MESON
M.L.GOLDBERGER,S.B.TREIMAN:PHYS.REV.110(1958)1178
2.5.1

63L GOLDBERGER
USE OF INTENSITY CORRELATIONS TO DETERMINE THE PHASE OF A SCATTERING AMPLITUDE
M.L.GOLDBERGER,H.W.LEWIS,K.M.WATSON:PHYS.REV.132(1963) 2764
2.1.1

78L GOLOSKOKOV
CORRECTIONS TO THE AMPLITUDE OF LARGE-ANGLE SCATTERING
S.V.GOLOSKOKOV,A.V.KOUDINOV,S.P.KULESHOV:DUBNA PREPRINT E2-11539
2.6.6

79L GOLOSKOKOV-1
PREASYMPTOTIC EFFECTS IN LARGE-ANGLE PI P SCATTERING
S.V.GOLOSKOKOV,A.V.KOUDINOV,S.P.KULESHOV:LETT.NUOVO CIM.24(1979)110
2.6.6

79L GOLOSKOKOV-2
NONAUTOMODEL CORRECTIONS IN LARGE-ANGLE SCATTERING OF PARTICLES WITH SPIN
S.V.GOLOSKOKOV,A.V.KORDINOV,S.P.KULESHOV:YAD.FIZ. 29(1979)1070
2.6.6

81L GOLOSKOKOV
HIGH ENERGY PP SCATTERING AND ENERGY DEPENDENCE OF EIKONAL PHASE
S.V.GOLOSKOKOV,S.P.KULESHOV,O.V.SELJUGIN:YAD.FIZ. 34(1981)235
2.6.3; 2.6.6

82L GOLOSKOKOV
PREDICTION OF MESON-CLOUD MODEL FOR ELASTIC HADRON SCATTERING
S.V.GOLOSKOKOV,S.P.KULESHOV,O.V.SELJUGIN:SUBM.TO PHYS. LETT.;DUBNA PREPRINT E2-82-109
2.6.6

79L GOTSMAN
JETS AS A SOURCE FOR THE OBSERVED INCREASE OF THE GAMMA-PROTON CROSS SECTION
E.GOTSMAN,D.SILVERMAN,A.SONI:PHYS.REV.D20(1979)2220
2.6.4

68L GOUNARIS
FINITE WIDTH CORRECTIONS TO VECTOR MESON DOMINANCE PREDICTION FOR RHO -> E+ E-
G.J.GOUNARIS,J.J.SAKURAI:PHYS.REV.LETT.21(1968)244
2.5.2; A.7.6; A.11.2

69L GOUNARIS
EFFECTS OF FINITE WIDTH OF RHO-MESON ON SOME VECTOR DOMINANCE PREDICTIONS
G.J.GOUNARIS:PHYS.REV.181(1969)2066
2.5.2; A.11.2

74L GRASSBERGER
FINITE ENERGY BOUNDS FOR PI N SCATTERING
P.GRASSBERGER,H.KUEHNELT,D.SCHWELA:NUCL.PHYS.B75(1974) 493
A.5.3

67L GREENBERG
SYMMETRIC QUARK MODEL OF BARYON RESONANCES
O.W.GREENBERG,M.RESNIKOFF:PHYS.REV.163(1967)1844
2.4.1

71L GREENBERG
FINITE ENERGY SUM RULES ON HYPERBOLAS
O.W.GREENBERG,J.C.SANDUSKY:NUOVO CIM.6A(1971)617,627
A.6.7

80L GREENBERG
COLOR VAN DER WAALS FORCES?
O.W.GREENBERG IN'RECENT DEVELOPMENTS IN HIGH-ENERGY PHYSICS',P.67;EDS.A.PERLMUTTER,L.F.SCOTT;PLENUM PR. (1980)AND 17.KARPACZ SCHOOL
2.5.3

75L GREIN
GEOMETRICAL SCALING - A PROPERTY OF THE INELASTIC OVERLAP FUNCTION
W.GREIN,P.KROLL:PHYS.LETT.58B(1975)79
2.6.4; 2.6.3

77L GREIN
ANALYSIS OF THE NN AND N ANTI N FORWARD SCATTERING AMPLITUDES INCLUDING INFORMATION FROM PI PI AND PI N SCATTERING
W.GREIN:NUCL.PHYS.B131(1977)255
2.5.3

78L GREIN
DISPERSION THEORETIC ANALYSIS OF THE PROTON-PROTON HELICITY AMPLITUDES AT T=0
W.GREIN,P.KROLL:NUCL.PHYS.B137(1978)173
2.5.3

80L GREIN
TWO AND THREE PION-CUT CONTRIBUTIONS TO NN-SCATTERING
W.GREIN,P.KROLL:NUCL.PHYS.A338(1980)332
2.5.3

82L GREIN
AMPLITUDE ANALYSIS OF FORWARD NUCLEON-NUCLEON SCATTERING
W.GREIN,P.KROLL:NUCL.PHYS.A377(1982)505
2.5.3

62L GRIBOV
PARTIAL WAVES WITH COMPLEX ANGULAR MOMENTA AND ASYMPTOTIC BEHAVIOUR OF SCATTERING AMPLITUDES
V.N.GRIBOV:SOV.PHYS.JETP 14(1962)1395
A.7.4; A.9

68L GRIBOV
REGGEON DIAGRAM TECHNIQUES
GRIBOV:SOV.PHYS.JETP 26(1968)414
2.6.4

76L GRIBOV
THEORY OF THE HEAVY POMERON
V.N.GRIBOV:NUCL.PHYS.B106(1976)189
2.6.4

73L GRISS
A CRITICISM OF THE ACCELERATED-CONVERGENCE EXPANSION PARTIAL WAVE ANALYSIS IN K+P SCATTERING
M.L.GRISS,G.C.FOX:PHYS.REV.D7(1973)74
2.1.9

76L GROMES
LINEARLY RISING QUARK POTENTIAL AND NON-STRANGE BARYON SPECTRUM
D.GROMES,I.O.STAMATESCU:NUCL.PHYS.B112(1976)213
2.4.1

77L GROMES
ON THE EFFECTIVE QUARK POTENTIAL IN BARYONS
D.GROMES:NUCL.PHYS.B130(1977)18
2.4.1

79L GROMES
BARYON SPECTRUM AND THE FORCES BETWEEN QUARKS
D.GROMES,I.O.STAMATESCU:Z.PHYS.C3(1979)43
2.4.1

80L GROMES
HYPERFINE INTERACTIONS IN BARYONS
D.GROMES IN"BARYON 1980";PROC.4TH INT.CONF.ON BARYON RESONANCES(TORONTO),P.195;ED.N.ISGUR
2.4.1

70L GROSS
SCALE INVARIANCE,CONFORMAL INVARIANCE AND HIGH-ENERGY BEHAVIOUR OF SCATTERING AMPLITUDES
D.J.GROSS,J.WESS:PHYS.REV.D2(1970)753
2.6.6

73L GUNION
LARGE-ANGLE SCATTERING AND THE INTERCHANGE FORCE
J.F.GUNION,S.J.BRODSKY,R.BLANKENBECLER:PHYS.REV.D8(1973)287
2.6.6

77L GUNION
CHIRAL SYMMETRY BREAKING AND THE QUARK MODEL
J.F.GUNION,P.C.MCNAMEE,M.D.SCADRON:NUCL.PHYS.B123(1977)445
2.5.1

81L GUPTA
OPTIMIZED PERTURBATION THEORY:THE PION FORM FACTOR
R.GUPTA IN"PERTURBATIVE QCD";AIP CONFERENCE PROCEEDINGS,VOL.74;EDS.D.W.DUKE,J.F.OWENS(1981)
2.5.2

70L GUSTAFSON
A DYNAMICAL MODEL FOR THE PI N RESONANCE S11(1525)
G.GUSTAFSON:NUCL.PHYS.B22(1970)605
2.4.1; A.7.1

71L GUSTAFSON
A DYNAMICAL MODEL FOR THE PI N RESONANCES S31,S11,D33
G.GUSTAFSON:NUCL.PHYS.B31(1971)461
2.4.1; A.7.1

72L GUSTAFSON
A DYNAMICAL MODEL FOR THE PI N RESONANCES P13,P31,F35
G.GUSTAFSON:NUCL.PHYS.B42(1972)205
2.4.1; A.7.1

77L GUSTAFSON
BARYON DECAYS AND SU(3)BREAKING
G.GUSTAFSON,J.HAMILTON,H.NIELSEN,G.C.OADES,B.TROMBORG:PHYS.SCRIPTA 15(1977)289
2.4.1

79L GUTBROD
ON THE ISOVECTOR FORM FACTOR OF THE NUCLEON IN MESON THEORY
F.GUTBROD,G.HARTMANN:Z.PHYSIK C2(1979)153
2.5.2

77L HAAN
APPROXIMATE CROSSING SUM RULES FOR PARTIAL WAVE AMPLITUDES
O.HAAN,K.H.MUETTER:NUCL.PHYS.B131(1977)232
A.7.1

71L HAGEN
NEW INCONSISTENCIES IN THE QUANTIZATION OF SPIN 3/2-FIELDS
C.R.HAGEN:PHYS.REV.D4(1971)2204
2.5.1

72L HAHN
FINITE ENERGY BOUNDS ON SCATTERING AMPLITUDES
B.D.HAHN,D.P.HODGKINSON:NUCL.PHYS.B46(1972)232
A.5.3

71L HALZEN
AMPLITUDE ANALYSIS OF PI N SCATTERING AT 6 GEV
F.HALZEN,C.MICHAEL:PHYS.LETT.36B(1971)367
2.1.15

76L HALZEN
BACKWARD POLARIZATION AS A DIRECT EXPERIMENTAL MEASURE OF PERIPHERALITY
F.HALZEN,M.G.OLSSON,A.YOKOSAWA:NUCL.PHYS.B113(1976)269
2.6.7

61L HAMILTON-1
LOW-ENERGY PION SCATTERING
J.HAMILTON,T.D.SPEARMAN:ANN.OF PHYS.12(1961)172
A.7.1

61L HAMILTON-2
EVIDENCE FOR PION-PION INTERACTIONS FROM S-WAVE PION-NUCLEON SCATTERING
J.HAMILTON,P.MENOTTI,T.D.SPEARMAN,W.S.WOOLCOCK:NUOVO CIM.20(1961)519
A.7.1

62L HAMILTON-1
S-WAVE PION-NUCLEON SCATTERING AND PION-PION INTERACTIONS
J.HAMILTON,T.D.SPEARMAN,W.S.WOOLCOCK:ANN.OF PHYS.17(1962)1
A.7.1

62L HAMILTON-2
PION-NUCLEON SCATTERING AND PION-PION INTERACTIONS
J.HAMILTON,P.MENOTTI,G.C.OADES,L.L.J.VICK:PHYS.REV.128(1962)1881
A.7.1

66L HAMILTON
PION-NUCLEON SCATTERING AND CONSERVATION LAWS
J.HAMILTON:PHYS.LETT.20(1966)687
A.6.4

67L HAMILTON
ON ZERO MASS PIONS
J.HAMILTON:NUCL.PHYS.B1(1967)449
A.8.2; 2.5.1

71L HAMILTON
PREDICTION OF S-WAVE PION-NUCLEON SCATTERING
J.HAMILTON,J.L.PETERSEN:NUCL.PHYS.B29(1971)29,51
A.7.1

73L HAMILTON
COULOMB CORRECTIONS IN NON-RELATIVISTIC SCATTERING
J.HAMILTON,I.OVERBOE,B.TROMBORG:NUCL.PHYS.B60(1973)443
2.1.17

76L HAMILTON
THE THEORY OF P-WAVE STATES
J.HAMILTON,B.TROMBORG:NUOVO CIM.33A(1976)605
A.7.1; 2.5.1

81L HAMILTON
UNCORRELATED AND ENHANCED 3PI-EXCHANGE IN NUCLEON-NUCLEON SCATTERING
J.HAMILTON:NORDITA PREPRINT 81-45
2.5.3

74L HAMMER
NEUTRON-PROTON FORM FACTORS
C.L.HAMMER,T.A.WEBER,V.S.ZIDELL:PHYS.REV.D9(1974)158
2.5.2

77L HAMMER
PION CHARGE RADIUS AND THE PION FORM FACTOR
C.L.HAMMER ET AL.:PHYS.REV.D15(1977)696
2.5.2

68L HARARI
POMERANCHUK TRAJECTORY AND ITS RELATION TO LOW-ENERGY SCATTERING AMPLITUDES
H.HARARI:PHYS.REV.LETT.20(1968)1395
2.6.4

69L HARARI
DUALITY AND THE POMERANCHUK SINGULARITY
H.HARARI,Y.ZARMI:PHYS.REV.187(1969)2230
2.6.4

71L HARARI
ELASTIC HADRONIC PROCESSES, DUALITY AND ABSORPTION
H.HARARI:ANN.OF PHYS.63(1971)432
2.6.4; 2.6.5

72L HARARI
PROPERTIES OF HADRONIC AMPLITUDES IN AN ABSORPTIVE MODEL
H.HARARI,A.SCHWIMMER:PHYS.REV.D5(1972)2780
2.6.4; 2.6.5

73L HARTLEY
TOWARD A GENERAL DESCRIPTION OF TWO-BODY HADRON REACTIONS
B.J.HARTLEY,G.L.KANE:NUCL.PHYS.B57(1973)157
2.6.4; 2.6.5

82L HASENFRATZ
HADRON SPECTRUM CALCULATIONS IN LATTICE QCD
P.HASENFRATZ:CERN PREPRINT TH 3349(1982)
2.4.1

73L HAYOT
AMPLITUDES AND MODEL FOR PION-NUCLEON BACKWARD SCATTERING AT 6 GEV
F.HAYOT,A.MOREL:PHYS.REV.D8(1973)223
2.6.7

74L HAYOT
TEST OF A GENERALIZED CHOU-YANG MODEL
F.HAYOT,U.P.SUKHATME:PHYS.REV.D10(1974)2183
2.6.4

76L HAYSAK
INTERACTION OF LOW-ENERGY PIONS WITH NUCLEONS IN NONLINEAR CHIRAL SU(2)XSU(2) DYNAMICS
M.I.HAYSAK,V.I.LENGYEL IN"HADRON CONSTITUENTS AND SYMMETRIES"(SMOLENICE 1976),P.313;EDS.J.LANIK,M.NOGA; VEDA,BRATISLAVA(1978)
2.5.1

74L HECHT
PI N SCATTERING AMPLITUDES AT 1.5-14 GEV AND SMALL MOMENTUM TRANSFER
H.HECHT,P.KROLL,H.P.JAKOB:NUCL.PHYS.B71(1974)1
2.1.16

77L HEDEGAARD-JENSEN
STUDY OF PION PION -> NUCLEON ANTINUCLEON PARTIAL WAVE AMPLITUDES USING HYPERBOLIC DISPERSION RELATIONS
N.HEDEGAARD-JENSEN:NUCL.PHYS.B119(1977)27
A.6.7; 2.4.2; 2.4.6; 2.6.7

75L HEEMSKERK
DISCRETE AMBIGUITIES IN PHASE SHIFT ANALYSIS
A.C.HEEMSKERK:NUCL.PHYS.A244(1975)15
2.1.3; 2.4.3

81L HEINES
HIGH ENERGY PP ELASTIC SCATTERING AND EVIDENCE FOR A NUCLEON CORE OF VALENCE QUARKS
G.W.HEINES,M.M.ISLAM:NUOVO CIM.61A(1981)149
2.6.4; 2.6.6

76L HELLER
RELATIVISTIC DESCRIPTION OF DIRECTLY INTERACTING
PIONS AND NUCLEONS
L.HELLER:AIP CONF.PROC.NO 33(1976)93
2.5.1

75L HENDRICK
REAL PARTS OF FORWARD ELASTIC PI P, KP,ANTI P P AND
PP SCATTERING AMPLITUDES
R.E.HENDRICK,B.LAUTRUP:PHYS.REV.D11(1975)529
2.2.1

70L HENDRY
ON THE VENEZIANO FORMULATION FOR PION-NUCLEON
SCATTERING
A.W.HENDRY,S.T.JONES,H.W.WYLD:NUCL.PHYS.B15(1970)389
2.6.5

73L HENDRY
ARE REGGE AND OPTICAL MODELS EQUIVALENT,BY AN
AMPLITUDE ROTATION?
A.W.HENDRY:PHYS.REV.D7(1973)910
2.6.4

74L HENDRY
EVIDENCE FOR COHERENT EFFECTS IN LARGE ANGLE HADRON-
HADRON SCATTERING
A.W.HENDRY:PHYS.REV.D10(1974)2300
2.6.6

78L HENDRY
ANALYSIS OF PION-NUCLEON SCATTERING UP TO 10 GEV/C
A.W.HENDRY:PHYS.REV.LETT.41(1978)222
2.1.12

80L HENDRY
HIGH MASS PION-NUCLEON RESONANCES
A.HENDRY IN'BARYON 1980'PROC.4TH INT.CONF.ON BARYON
RESONANCES IN TORONTO',P.113;ED.N.ISGUR(1980)
2.1.12; 2.4.1

81L HENDRY
PION-NUCLEON SCATTERING UP TO 10 GEV
A.W.HENDRY:ANN.OF PHYS.136(1981)1
2.1.12

67L HENZI
OVERLAP FUNCTION ANALYSIS OF PI P SCATTERING AT HIGH
ENERGIES
R.HENZI:NUOVO CIM.52(1967)772
2.6.3; 2.6.8

68L HENZI
POLARIZATIONS IN THE OVERLAP FUNCTION ANALYSIS OF
PI P SCATTERING
R.HENZI:NUOVO CIM.57(1968)300
2.6.4

79L HENZI-1
ON ELASTIC PP DIFFRACTION SCATTERING AND ITS ENERGY
DEPENDENCE
R.HENZI,P.VALIN:NUCL.PHYS.B148(1979)513
2.6.4; 2.6.3

79L HENZI-2
RESCATTERING AS THE ORIGIN OF THE 'BREAKS' IN ELASTIC
PP-DIFFRACTION SCATTERING
R.HENZI,P.VALIN;MONTREAL PREPRINT,SUBM.TO THE INTERN.
CONF.ON HIGH ENERGY PHYSICS IN GENEVA(1979)
2.6.4; 2.6.3

75L HEY
SU(6)W AND DECAYS OF BARYON RESONANCES
A.J.G.HEY,P.J.LITCHFIELD,R.J.CASHMORE:NUCL.PHYS.
B95(1975)516
2.4.1

81L HEYN
ANALYSIS OF THE PION FORM FACTOR IN THE SPACE-LIKE
AND TIME-LIKE REGION
M.F.HEYN,C.B.LANG:Z.PHYSIK C7(1981)169
A.11.1; 2.5.2

81L HIROSHIGE
AN APPROXIMATE PHASE SHIFT ANALYSIS OF PI N SCATTERING
N.HIROSHIGE,H.KAMO,W.WATARI:PROGR.THEOR.PHYS.65(1981)
1763
2.1.7

82L HIROSHIGE
ON THE POLARIZATION OF PION-NUCLEON SCATTERING AT
LARGE ANGLES
N.HIROSHIGE,W.WATARI:OSAKA PREPRINT(1982)
2.1.7

73L HITE-1
NEW DISPERSION RELATIONS AND THEIR APPLICATIONS
G.E.HITE,F.STEINER:NUOVO CIM.18A(1973)237
A.6.7

73L HITE-2
EXTENSION OF BOUNDARY DISPERSION RELATIONS
G.E.HITE,R.J.JACOB,F.STEINER:PHYS.REV.D6(1973)3333
A.6.7

74L HITE
AN INTERIOR DISPERSION RELATION DETERMINATION OF THE
PI N SIGMA TERM
G.E.HITE,R.J.JACOB:PHYS.LETT.53B(1974)200
2.5.1

75L HITE
DETERMINATION OF THE PION-NUCLEON COUPLING CONSTANT
G.E.HITE,R.J.JACOB,D.C.MOIR:PHYS.REV.D12(1975)2677
2.4.2

76L HITE
TEST OF THE PCAC LIMIT IN PI N SCATTERING
G.E.HITE,R.J.JACOB,M.D.SCADRON:PHYS.REV.D14(1976)1306
2.5.1

78L HITE
THE I=0,S-WAVE PI PI SCATTERING LENGTH
G.E.HITE,R.J.JACOB:NUCL.PHYS.B134(1978)291
2.4.6

77L HO-KIM
RADIATIVE PION-PROTON SCATTERING
Q.HO-KIM,J.P.LAVINE:NUCL.PHYS.A285(1977)407
2.1.17

76L HODGKINSON
AMALGAMATION OF PI P SCATTERING DATA
D.P.HODGKINSON,R.L.KELLY,R.E.CUTKOSKY,J.C.SANDUSKY
IN'PROC.TOPICAL CONF.ON BARYON RESONANCES'(OXFORD),
P.41;RUTHERFORD LAB.(1976)

71L HOEGAASEN
PHYS.NORVEG. 5(1971)219
2.6.5

58L HOEHLER
UEBER DIE EXPONENTIALNAEHERUNG BEIM TEILCHENZERFALL
G.HOEHLER:Z.PHYSIK 152(1958)546
2.4.1

60L HOEHLER
BERECHNUNG DER PION-NUCLEON STREUPHASEN AUS
DISPERSIONSRELATIONEN
G.HOEHLER,K.DIETZ:Z.PHYS.160(1960)453
2.4.7; 7.2.1; A.8.3

64L HOEHLER-1
ANALYSIS OF THE PHOTOPRODUCTION OF POSITIVE PIONS
G.HOEHLER,W.SCHMIDT:ANN.OF PHYS.28(1964)34
2.4.2

64L HOEHLER-2
DETERMINATION OF THE PI N CHARGE-EXCHANGE AMPLITUDE
FROM DATA AT HIGH ENERGIES
G.HOEHLER,N.ZOVKO:Z.PHYSIK 181(1964)293
2.6.5

66L HOEHLER-1
HIGH ENERGY BEHAVIOUR OF THE PI N FORWARD CHARGE-
EXCHANGE SCATTERING AMPLITUDE
G.HOEHLER,J.BAACKE,R.STRAUSS:PHYS.LETT.21(1966)223
2.6.5

66L HOEHLER-2
ANALYSIS OF PI N CHARGE EXCHANGE SCATTERING FROM 4 TO
18 GEV
G.HOEHLER,J.BAACKE,H.SCHLAILE,P.SONDEREGGER:PHYS.LETT.
20(1966)79
2.6.5

66L HOEHLER-3
PARAMETERS OF A REGGE POLE MODEL FOR PI N CHARGE
EXCHANGE SCATTERING
G.HOEHLER,J.BAACKE,G.EISENBEISS:PHYS.LETT.22(1966)203
2.6.5

68L HOEHLER-1
ON THE ROLE OF RHO-EXCHANGE IN LOW ENERGY PI N
SCATTERING
G.HOEHLER,J.BAACKE,F.STEINER:Z.PHYSIK 214(1968)381
2.5.1; 2.4.6; 2.5.2; A.8.3

68L HOEHLER-2
PREDICTION OF THE ISOVECTOR NUCLEON FORM FACTORS FROM
PI N SCATTERING AND PION FORM FACTOR DATA
G.HOEHLER,R.STRAUSS,H.WUNDER:SUBM.TO VIENNA CONF.
(1968)
2.4.6; 2.5.2

69L HOEHLER
A NEW DETERMINATION OF THE PI N COUPLING CONSTANT
G.HOEHLER,H.SCHLAILE,R.STRAUSS:Z.PHYSIK 229(1969)217
2.4.2

70L HOEHLER-1
IS A SUBTRACTION REALLY NECESSARY IN THE ISOSPIN EVEN
PI N DISPERSION RELATION?
G.HOEHLER,R.STRAUSS:Z.PHYS.232(1970)205
A.6.4; 2.6.3; 2.1.15; 2.5.1

70L HOEHLER-2
DISCREPANCIES BETWEEN PI N CHARGE EXCHANGE FORWARD
SCATTERING DATA AND THE DISPERSION RELATION PREDICTION
G.HOEHLER,R.STRAUSS:Z.PHYS.240(1970)377
2.6.5; 2.2.1

71L HOEHLER-1
ENERGY DEPENDENCE OF THE SLOPE OF THE PI N
DIFFRACTION AMPLITUDE
G.HOEHLER,H.P.JAKOB:LETT.NUOVO CIM.2(1971)485
2.6.2; 2.5.1

71L HOEHLER-2
WARD IDENTITIES,DISPERSION RELATIONS AND DUALITY FOR
PI N SCATTERING AT LOW ENERGIES
G.HOEHLER,P.STICHEL:Z.PHYS.245(1971)187
A.8.3; 2.5.1

71L HOEHLER-3
THE MAGNITUDE OF THE SIGMA COMMUTATOR TERM IN CURRENT
ALGEBRA
G.HOEHLER,H.P.JAKOB,R.STRAUSS:PHYS.LETT.35B(1971)445
2.5.1

72L HOEHLER-1
A CRITICAL TEST OF MODELS FOR THE LOW ENERGY PI N AMPLITUDE
G.HOEHLER,H.P.JAKOB,R.STRAUSS:NUCL.PHYS.B39(1972)237
A.8; A.3.5; 2.4.7; 2.5.1

72L HOEHLER-2
ON THE SATURATION OF SOME PI N ISOSPIN BOUNDS
G.HOEHLER,W.SCHMIDT:PHYS.LETT.38B(1972)237
2.4.4

72L HOEHLER-3
ISOSPIN BOUNDS AND ZEROS OF PION-NUCLEON SCATTERING AMPLITUDES
G.HOEHLER,F.KAISER,H.KOHL,W.SCHMIDT:LETT.NUOVO CIM. 5(1972)413
2.4.4; A.10.3

73L HOEHLER-1
IMPACT PARAMETER EXPANSION OF PI N SCATTERING AMPLITUDES
G.HOEHLER,H.P.JAKOB:Z.PHYS.261(1973)371
A.3.3; 2.6.3

73L HOEHLER-2
DISCREPANCY BETWEEN CERN AND SACLAY PHASE SHIFTS FOR PION-NUCLEON SCATTERING
G.HOEHLER,F.KAISER,F.WAGNER:LETT.NUOVO CIM.8(1973)232
2.1.5

73L HOEHLER-3
DOES ASYMPTOPIA BEGIN AT 5 GEV IN PI N SCATTERING?
G.HOEHLER,H.P.JAKOB:LETT.NUOVO CIM.6(1973)389
2.6.2

74L HOEHLER
THE REAL PART OF THE PI N DIFFRACTION AMPLITUDE
G.HOEHLER,H.P.JAKOB:Z.PHYS.268(1974)75
2.6.3; 2.1.16

75L HOEHLER-1
REAL PARTS OF PI N FORWARD AMPLITUDES AT HIGH ENERGIES
G.HOEHLER,H.P.JAKOB,F.KAISER:PHYS.LETT.58B(1975)348
2.2.1

75L HOEHLER-2
COMMENTS ON FITS TO NUCLEON FORM FACTOR DATA
G.HOEHLER,H.D.KIEHLMANN,W.SCHMIDT:PHYS.REV.D11(1975) 2667
2.5.2

75L HOEHLER-3
ELECTROMAGNETIC RADII OF NUCLEON AND PION
G.HOEHLER,E.PIETARINEN:PHYS.LETT.53B(1975)471
2.5.2

75L HOEHLER-4
THE RHO NN VERTEX IN VECTOR-DOMINANCE MODELS
G.HOEHLER,E.PIETARINEN:NUCL.PHYS.B95(1975)210
2.4.2; 2.4.6; 2.5.2; 2.5.3

76L HOEHLER
ANALYSIS OF ELECTROMAGNETIC NUCLEON FORM FACTORS
G.HOEHLER ET AL.:NUCL.PHYS.B114(1976)505
2.5.2

77L HOEHLER-1
DISPERSION RELATIONS FOR PI N AMPLITUDES AT FIXED C.M. ANGLES
G.HOEHLER,R.KOCH,E.PIETARINEN:KARLSRUHE PREPRINT TKP 77-12
A.6.8; 2.1.7; 2.6.4

77L HOEHLER-2
PI N LOW-ENERGY PARAMETERS FROM A NEW AMPLITUDE ANALYSIS AND FIXED-NU DISPERSION RELATIONS
G.HOEHLER,R.KOCH,E.PIETARINEN:KARLSRUHE PREPRINT TKP 77-11;SEE 79R HOEHLER-KAISER-KOCH-PIETARINEN
2.4.7; 2.5.1; 2.4.5; A.8.1

79L HOEHLER-1
ANALYSIS OF PI N CHARGE-EXCHANGE SCATTERING DATA
G.HOEHLER,H.P.JAKOB,F.KAISER,G.BRANDENBURGER:KFK 2735; KERNFORSCHUNGSZENTRUM KARLSRUHE,1979
2.6.5; 2.4.3

79L HOEHLER-2
SINGULARITIES OF ZERO TRAJECTORIES OF SCATTERING AMPLITUDES
G.HOEHLER,I.SABBA-STEFANESCU:Z.PHYSIK C2(1979)253
2.4.3

79L HOEHLER-3
IS THE PI P DIFFRACTION PEAK COMPATIBLE WITH THE UNITARITY BOUNDS?
G.HOEHLER,F.KAISER,H.M.STAUDENMAIER:PHYS.LETT. 86B(1979)383
2.6.2; 2.6.3

79L HOEHLER-4
THE SHAPE OF THE PI MINUS P DIFFRACTION PEAK
G.HOEHLER,F.KAISER,H.M.STAUDENMAIER;KARLSRUHE PREPRINT TKP 79-4
2.6.2; 2.6.3

80L HOEHLER-1
ANALYSIS OF NEAR-FORWARD PI N SCATTERING DATA AT HIGH ENERGIES
G.HOEHLER,M.HUTT,R.KOCH,H.M.STAUDENMAIER;KARLSRUHE PREPRINT TKP 80-10
2.6.2; 2.1.15; 2.4.4; 2.6.5

80L HOEHLER-2
PION-NUCLEON AMPLITUDES AT 2-10 GEV
G.HOEHLER,R.KOCH;KARLSRUHE PREPRINT TKP 80-2
2.1.7; 2.4.1; 2.6.3

82L HOEHLER
SLOPE PARAMETERS AND ZERO TRAJECTORIES IN PI MINUS-PROTON SCATTERING
G.HOEHLER,I.SABBA-STEFANESCU:PHYS.REV.D25(1982)892
AND KARLSRUHE REPORT TKP 81-4
2.4.3; 2.6.3

75L HOFF
PI PLUS P POLARIZATION DATA AND THE STRONGLY-CORRELATED-RESONANCES MODEL
G.T.HOFF,J.N.PAPPADEMOS:PHYS.REV.D12(1975)309
2.4.1

73L HORGAN
BARYON SPECTROSCOPY AND THE QUARK SHELL MODEL I
R.HORGAN,R.H.DALITZ:NUCL.PHYS.B66(1973)135
2.4.1

74L HORGAN
BARYON-SPECTROSCOPY AND THE QUARK SHELL MODEL II
R.HORGAN:NUCL.PHYS.B71(1974)514
2.4.1

82L HORGAN
MASS SPECTRUM OF LOW LYING DELTA RESONANCES IN HARMONIC OSCILLATOR-LIKE QUARK MODELS
R.R.HORGAN:Z.PHYSIK C13(1982)25
2.4.1

70L HUANG
RHO MESON IN LOW-ENERGY PI N SCATTERING
J.C.HUANG,J.R.URANI:PHYS.REV.D1(1970)877
2.5.1

64L HUEPER
CONSISTENCY OF NEW PI N PHASE SHIFT ANALYSES
R.HUEPER:Z.PHYSIK 181(1964)426
2.4.2

80L HUTT
EVALUATION OF PI N PARTIAL WAVE DISPERSION RELATIONS USING NEW PHASE SHIFT RESULTS
M.HUTT,R.KOCH:KARLSRUHE PREPRINT TKP 80-25(1980)
A.7.1

81L HUTT
IS THERE EVIDENCE FOR STRONG VAN DER WAALS FORCES IN PI N SCATTERING?
M.HUTT,R.KOCH:PHYS.LETT.105B(1981)227
A.7.1

82L HUTT-1
AN ANALYSIS OF PI P ELASTIC SCATTERING AT 50 AND 200 GEV/C
M.HUTT,I.SABBA-STEFANESCU:KARLSRUHE PREPRINT TKP 82/12
2.6.3; 2.6.4

82L HUTT-2
A LITERATURE REFERENCE RETRIEVAL SYSTEM FOR PERSONAL COMPUTERS
M.HUTT:KARLSRUHE PREPRINT TKP 82-11
3.1

62L IGI
PI N SCATTERING LENGTH AND SINGULARITIES IN THE J-PLANE
K.IGI:PHYS.REV.LETT.9(1962)76
2.2.1; A.9.1; 2.6.4

67L IGI
NEW SUM RULES AND SINGULARITIES IN THE J-PLANE
K.IGI,S.MATSUDA:PHYS.REV.LETT.18(1967)625
2.2.1; 2.6.5; A.9.1

68L IGI
CROSSING SYMMETRIC REGGE AMPLITUDES FOR PI N SCATTERING
K.IGI:PHYS.LETT.28B(1968)330
2.6.5; A.9.1

66L IIZUKA
A SYSTEMATICS AND PHENOMENOLOGY OF MESON FAMILY
J.IIZUKA:PROGR.THEOR.PHYS.SUPPL.37(1966)21
2.5.2

72L IMACHI
HADRON REACTIONS AND URBARYON REARRANGEMENT
M.IMACHI ET AL.:PROGR.THEOR.PHYS.48(1972)210,47(1972)1958,45(1971)1849
2.6.5

81L IOFFE
CALCULATION OF BARYON MASSES IN QCD
B.L.IOFFE:NUCL.PHYS.B188(1981)317,B191(1981)591
2.4.1

82L IOFFE
PION FORMFACTOR AT INTERMEDIATE MOMENTUM TRANSFER IN QCD
B.L.IOFFE,A.V.SMILGA:PHYS.LETT.114B(1982)353
2.5.2

79L IRVING
POMERON AND F-EXCHANGE IN ELASTIC AND INELASTIC SCATTERING
A.C.IRVING,H.R.SEPANGI,A.D.MARTIN,T.SHIMADA:NUCL.PHYS.B149(1979)101
2.6.4

81L IRVING
ON THE NATURE OF THE S* AND RELATED SCALAR RESONANCES
A.C.IRVING,A.D.MARTIN,P.J.DONE:Z.PHYS.C10(1981)45
2.4.6

77L ISGUR
HYPERFINE INTERACTIONS IN NEGATIVE PARITY BARYONS
N.ISGUR,G.KARL:PHYS.LETT.72B(1977)109
2.4.1

78L ISGUR-1
SYMMETRY BREAKING IN BARYONS
N.ISGUR,G.KARL:PHYS.LETT.74B(1978)353
2.4.1

78L ISGUR-2
P-WAVE BARYONS IN A QUARK MODEL
N.ISGUR,G.KARL:PHYS.REV.D18(1978)4187
2.4.1

79L ISGUR-1
POSITIVE-PARITY EXCITED BARYONS IN QUARK MODEL WITH HYPERFINE INTERACTIONS
N.ISGUR,G.KARL:PHYS.REV.D19(1979)2653
2.4.1

79L ISGUR-2
GROUND STATE BARYONS IN A QUARK MODEL WITH HYPERFINE INTERACTIONS
N.ISGUR,G.KARL:PHYS.REV.D20(1979)1191
2.4.1

80L ISGUR
GROUND-STATE BARYON MAGNETIC MOMENTS
N.ISGUR,G.KARL:PHYS.REV.D21(1980)3175
2.4.1; 2.5.2

81L ISGUR
NEUTRON CHARGE FORMFACTOR IN A QUARK MODEL WITH HYPERFINE INTERACTIONS
N.ISGUR,G.KARL,D.W.SPRUNG:PHYS.REV.D23(1981)163
2.5.2

79L ISHIDA-1
ELECTROMAGNETIC INTERACTIONS OF HADRONS IN THE RELATIVISTIC HARMONIC OSCILLATOR QUARK MODEL
S.ISHIDA,K.TAKEUCHI,S.TSURUTA,M.WATANABE:PHYS.REV. D20(1979)2906
2.5.2

79L ISHIDA-2
RADIATIVE DECAY OF BARYON RESONANCES IN A RELATIVISTIC HARMONIC OSCILLATOR MODEL
S.ISHIDA,M.WATANABE:PROGR.THEOR.PHYS.61(1979)1412
2.4.1

76L ISLAM
IMPACT PARAMETER REPRESENTATION FROM THE SOMMERFELD-WATSON TRANSFORM
M.M.ISLAM:NUCL.PHYS.B104(1976)511
A.3.3

81L ISRAILOV
PION-NUCLEON SCATTERING IN THE CHIRAL BAG MODEL
Z.Z.ISRAILOV,M.M.MUSAKHANOV:PHYS.LETT.104B(1981)173
2.5.1

82L IWAO
CHARGE FORM FACTOR OF NEGATIVE PION IN RELATION TO THAT OF PROTON
S.IWAO:LETT.NUOVO CIM.34(1982)197
2.5.2

68L JACKSON
KINEMATIC SINGULARITIES AND THRESHOLD RELATIONS FOR HELICITY AMPLITUDES
J.D.JACKSON,G.E.HITE:PHYS.REV.169(1968)1248
A.3.5

59L JACOB
ON THE GENERAL THEORY OF COLLISIONS FOR PARTICLES WITH SPIN
M.JACOB,G.C.WICK:ANN.OF PHYS.7(1959)404
A.2; A.3

75L JACOB
SPIN CONTENT OF CROSSED-CHANNEL CONTRIBUTIONS AND INTERIOR DISPERSION RELATIONS
R.J.JACOB,G.E.HITE:PHYS.REV.D11(1975)2466
2.4.6; 2.4.2

82L JACOB
CURRENT ALGEBRA,PI PI SCATTERING LENGTHS AND THE S*(980) BACKGROUND CONTRIBUTION
R.J.JACOB,M.D.SCADRON:ARIZONA UNIV.PREPRINT(1982)
2.4.6

77L JAFFE
MULTIQUARK HADRONS
R.J.JAFFE:PHYS.REV.D15(1977)267,281
2.4.6

79L JAFFE
CONNECTION BETWEEN QUARK-MODEL EIGENSTATES AND LOW-ENERGY SCATTERING
R.L.JAFFE,F.E.LOW:PHYS.REV.D19(1979)2105
2.4.1; 2.5.1

80L JAFFE
THE SIGMA TERM REVISITED
R.L.JAFFE:PHYS.REV.D21(1980)3215
2.5.1

69L JAKOB
ON THE BEHAVIOR OF THE PI N PARTIAL WAVE AMPLITUDES
AT S=0
H.P.JAKOB,F.STEINER:Z.PHYSIK 228(1969)353
A.7.1; 2.6.7

71L JAKOB-1
APPLICATION OF PARTIAL WAVE CROSSING RELATIONS TO PI
N SCATTERING
H.P.JAKOB,E.KRUBASIK,F.STEINER:FORTSCHR.PHYS.19(1971)
27
A.7.2

71L JAKOB-2
DETERMINATION OF THE SIGMA COMMUTATOR TERM FROM PI N
SCATTERING
H.P.JAKOB:CERN PREPRINT TH 1446(1971)
2.5.1

73L JAKOB
THE SLOPE OF THE PI N CHARGE-EXCHANGE AMPLITUDE AT T=0
H.P.JAKOB,P.KROLL:NUCL.PHYS.B58(1973)269
2.6.5

75L JAKOB
UNITARISIERTE SUMMENREGELN UND ANWENDUNG AUF PION-
NUKLEON LADUNGSAUSTAUSCHSTREUUNG
H.P.JAKOB:THESIS,UNIVERSITY OF KARLSRUHE,1975
2.6.5

75L JENKINS
SOME THEORETICAL ASPECTS OF DELTA(1936)-EXCHANGE IN
LOW-ENERGY PI N SCATTERING
J.D.JENKINS:DURHAM PREPRINT(1975)
A.8.2; 2.5.1

81L JENKOVSZKY
PROTON'S OPACITY IN A DUAL MODEL
L.L.JENKOVSZKY,F.PACCANONI:KIEV PREPRINT IFPD 31-
81(1981)
2.6.3

64L JIN-1
CONNECTION BETWEEN THE ASYMPTOTIC BEHAVIOUR AND THE
SIGN OF THE DISCONTINUITY IN ONE-DIMENSIONAL
DISPERSION RELATIONS
Y.S.JIN,A.MARTIN:PHYS.REV.135(1964)B1369
A.6.3

64L JIN-2
NUMBER OF SUBTRACTIONS IN FIXED-T DISPERSION RELATIONS
Y.S.JIN,A.MARTIN:PHYS.REV.135(1964)B1375
A.6.3; 2.4.3

65L JIN
PHASE REPRESENTATION AND HIGH ENERGY BEHAVIOUR OF THE
FORWARD AMPLITUDE
Y.S.YIN,S.W.MAC DOWELL:PHYS.REV.138(1965)B1279
A.6.4

73L JOHNSON
PI N AMPLITUDE ANALYSIS AT 3.6 AND 6.0 GEV
P.JOHNSON ET AL.:PHYS.REV.LETT.30(1973)242
2.1.15

76L JOHNSON
STRINGLIKE SOLUTIONS TO THE BAG MODEL
K.JOHNSON,C.B.THORN:PHYS.REV.D13(1976)1934
2.4.1

71L JONES
S-CHANNEL HELICITY CONSERVATION AND THE STRESS-TENSOR
H.F.JONES,A.SALAM:PHYS.LETT.34B(1971)149
2.4.2

74L JOYNSON
A PHASE CORRECTED REGGE POLE MODEL FOR PION-NUCLEON
CHARGE-EXCHANGE SCATTERING
D.JOYNSON:NUCL.PHYS.B79(1974)381
2.6.5

75L JOYNSON
NON-REGGE AND HYPER-REGGE EFFECTS IN PI N CHARGE-
EXCHANGE SCATTERING AT HIGH ENERGIES
D.JOYNSON,E.LEADER,B.NICOLESCU,C.LOPEZ:NUOVO CIM.
30A(1975)345
2.6.5

73L KAFIEF
HELICITY CONSERVATION AND TENSOR-MESON DOMINANCE
Y.N.KAFIEF,V.V.SEREBRYAKOV:NUCL.PHYS.B52(1973)141
2.4.2; 2.6.3

76L KAISER
HIGHER SPIN EXCHANGE IN PI PI -> NUCLEON ANTINUCLEON
SCATTERING
F.KAISER,E.BORIE,G.HOEHLER:PHYS.LETT.62B(1976)441
2.4.2; A.6.7; 2.4.6; 2.6.7

78L KAISER
EVALUATION OF PI N INTERIOR DISPERSION RELATIONS WITH
KH 78 PHASE SHIFTS
F.KAISER:KARLSRUHE PREPRINT TKP 78-23,1978
2.4.2

82L KALMAN
CONSISTENT QUARK MODEL WITH HYPERFINE INTERACTIONS
FOR THE GROUND AND LOW-LYING EXCITED BARYON STATES
C.S.KALMAN,R.L.HALL:PHYS.REV.D25(1982)217
2.4.1

79L KAMRAN
DUAL ABSORPTIVE MODEL AND PI P ELASTIC SCATTERING UP TO 280 GEV
M.KAMRAN:NUOVO CIM.51A(1979)43
2.6.3

82L KAMRAN
PP ELASTIC POLARIZATION AND POMERON FLIP REVISITED
M.KAMRAN,M.A.BUTT:J.PHYS.G8(1982)33
2.6.4

77L KANAZAWA
ON A POLE SEARCH PROBLEM OF SCATTERING AMPLITUDES
A.KANAZAWA ET AL.:PROGR.THEOR.PHYS.57(1977)295
2.4.1

82L KAPLAN
ANTIPROTON-PROTON AND PROTON-PROTON ELASTIC SCATTERING AT 100 AND 200 GEV/C
D.H.KAPLAN ET AL.:FERMILAB-PUB-82/40-EXP
2.6.3

79L KARSTEN
ON LATTICE GAUGE THEORIES AND BACKWARD PION-NUCLEON SCATTERING
L.H.KARSTEN:THESIS,UNIV.OF AMSTERDAM(1979)
2.6.7

67L KASCHLUHN
N/D APPROACH TO PI N SCATTERING IN THE P11 STATE
F.KASCHLUHN:NUCL.PHYS.B3(1967)139
A.7.1

65L KATO
ANALYTICAL PROPERTIES OF TWO-CHANNEL S-MATRIX
M.KATO:ANN.OF PHYS.31(1965)130
2.1.13

66L KAWARABAYASHI
PCAC AND THE DECAYS OF VECTOR MESONS
K.KAWARABAYASHI,M.SUZUKI:PHYS.REV.LETT.16(1966)255,384
A.8.2

72L KELLY
AMPLITUDE ANALYSIS AT 6 GEV/C
R.L.KELLY:PHYS.LETT.39B(1972)635
2.1.16

74L KELLY
SPIN-ROTATION EFFECTS IN MESON-NUCLEON SCATTERING
R.L.KELLY,J.C.SANDUSKY,R.E.CUTKOSKY:PHYS.REV.D10(1974) 2309
A.4.2; A.4.3

79L KELLY
AMALGAMATION OF MESON-NUCLEON SCATTERING DATA
R.L.KELLY,R.E.CUTKOSKY:PHYS.REV.D20(1979)
2.1.9

67L KHALFIN
REAL PART OF SCATTERING AMPLITUDE AT HIGH ENERGIES AND DISPERSION SUM RULES
L.A.KHALFIN:SOV.PHYS.JETP LETT.6(1967)107
A.6.4

65L KHURI
FORWARD SCATTERING AMPLITUDE AND UNIVALENT FUNCTIONS
N.N.KHURI,T.KINOSHITA:PHYS.REV.140(1965)B706;137(1965) B720
A.6.4

79L KIMURA
BIAS-FREE POLE SEARCH ON THE S* RESONANCE
N.KIMURA,T.MIYAKOSHI,A.KANAZAWA:PROGR.THEOR.PHYS. 61(1979)145
2.4.1

78L KOCH
ANWENDUNG VON DISPERSIONSMETHODEN ZUR ANALYSE DER PION-NUKLEON STREUUNG
R.KOCH:THESIS,UNIVERSITY OF KARLSRUHE,1978
2.1.7; 2.4.2; 2.4.7

80L KOCH-1
LOW ENERGY PION-NUCLEON PARTIAL WAVE ANALYSIS
R.KOCH,E.PIETARINEN:NUCL.PHYS.A336(1980)331
2.1.7; 2.4.2; 2.4.7

80L KOCH-2
THE KARLSRUHE-HELSINKI PI N ELASTIC PARTIAL WAVE ANALYSIS
R.KOCH IN 'BARYON 1980',PROC. 4TH INT.CONF.ON BARYON RESONANCES IN TORONTO;ED.N.ISGUR
2.1.7; 2.4.1

81L KOCH
CONSISTENCY OF PION-NUCLEON PARTIAL WAVE RELATIONS WITH THE FORWARD DISPERSION RELATION FOR THE T-DERIVA-TIVE OF THE AMPLITUDE C PLUS
R.KOCH,U.MANNING:KARLSRUHE PREPRINT TKP 81-14
2.4.7; 2.6.2; A.6.5

82L KOCH-1
A NEW DETERMINATION OF THE PION-NUCLEON SIGMA TERM USING HYPERBOLIC DISPERSION RELATIONS
R.KOCH:Z.PHYSIK C,IN PRINT
2.5.1; 2.4.5; 2.4.7

82L KOCH-2
ANALYSIS OF THE LONG RANGE PION-NUCLEON INTERACTION USING PARTIAL WAVE DISPERSION RELATIONS
R.KOCH,J.OTTO:KARLSRUHE PREPRINT TKP 82-5(1982)
A.7.1; 2.4.7; 2.1.7

```
76L KOK
    AMBIGUOUS SETS OF PARTIAL WAVE AMPLITUDES CAN
    INTERSECT
    L.P.KOK,M.DE ROO:NUCL.PHYS.B111(1976)39
    2.1.3; 2.4.3
66L KOKKEDEE
    QUARK MODEL AND HIGH ENERGY SCATTERING
    J.J.J.KOKKEDEE,L.VAN HOVE:NUOVO CIM.42A(1966)711
    2.6.4
80L KONIUK-1
    AN ANALYSIS OF BARYON COUPLINGS IN A QUARK MODEL WITH
    QCD
    R.KONIUK,N.ISGUR:PHYS.REV.LETT.44(1980)845
    2.4.1
80L KONIUK-2
    BARYON DECAYS IN A QUARK MODEL WITH CHROMODYNAMICS
    R.KONIUK,N.ISGUR:PHYS.REV.D21(1980)1868
    2.4.1
71L KORKEA-AHO
    ISOSPIN INVARIANCE AND PI N BACKWARD SCATTERING
    M.KORKEA-AHO,N.A.TORNQVIST:NUCL.PHYS.B33(1971)239
    2.4.4; 2.6.7
67L KROLL
    NEUTRAL VECTOR MESONS AND THE HADRONIC
    ELECTROMAGNETIC CURRENT
    N.M.KROLL,T.D.LEE,B.ZUMINO:PHYS.REV.157(1967)1376
    2.4.6; 2.5.2; 2.5.3
72L KRUBASIK
    AN AMPLITUDE ANALYSIS OF PI N COULOMB INTERFERENCE
    DATA
    E.KRUBASIK:NUCL.PHYS.B44(1972)558
    2.6.2
81L KULESHOV
    GLAUBER REPRESENTATION AND HIGH-ENERGY ELASTIC PP-
    SCATTERING
    S.P.KULESHOV,V.K.MITRJUSHKIN,P.K.RASHIDOV:SUBM.TO
    HADRONIC JOURN.;DUBNA E2-81-50
    2.6.4
73L KUPSCH
    PROPERTIES OF IMPACT PARAMETER REPRESENTATION
    J.KUPSCH,I.O.STAMATESCU:NUOVO CIM.15A(1973)663
    A.3.3
74L KUPSCH
    IMPACT PARAMETER AMPLITUDES
    J.KUPSCH,I.O.STAMATESCU:NUOVO CIM.20A(1974)317
    A.3.3
59L LANDAU
    ON ANALYTIC PROPERTIES OF VERTEX PARTS IN QUANTUM
    FIELD THEORY
    L.D.LANDAU:NUCL.PHYS.13(1959)181
    A.6.2
72L LANDAU
    ABSORPTIVE SEPARABLE POTENTIALS CONSTRUCTED FROM PI N
    DATA
    R.LANDAU,F.TABAKIN:PHYS.REV.D5(1972)2746
    2.5.1
73L LANDSHOFF
    PP ELASTIC SCATTERING AT LARGE MOMENTUM TRANSFER
    P.V.LANDSHOFF,J.C.POLKINGHORNE:PHYS.LETT.44B(1973)293
    2.6.4; 2.6.6
74L LANDSHOFF
    MODEL FOR ELASTIC SCATTERING AT WIDE ANGLES
    P.V.LANDSHOFF:PHYS.REV.D10(1974)1024
    2.6.4; 2.6.6
80L LANDSHOFF
    HIGHER ORDER QCD CORRECTIONS TO EXCLUSIVE PROCESSES:
    MULTIPLE SCATTERING MECHANISM
    P.V.LANDSHOFF,D.J.PRITCHARD:Z.PHYSIK C6(1980)69
    2.6.6; 2.6.4
75L LANG
    ANALYSIS OF THE PION FORMFACTOR IN THE TIME-LIKE
    REGION
    C.B.LANG,I.SABBA-STEFANESCU:PHYS.LETT.58B(1975)450
    2.5.2
79L LANG
    DETERMINATION OF RESONANCE PARAMETERS
    C.B.LANG,A.MAS-PARAREDA:PHYS.REV.D19(1979)956
    2.4.1
75L LANGBEIN-1
    ANALYSIS OF PI N INVARIANT AMPLITUDES IN TERMS OF
    ALGEBRAIC FUNCTIONS
    W.LANGBEIN:NUCL.PHYS.94B(1975)519
    2.4.5; A.6.3; 2.4.7; 2.4.2; 2.5.1
75L LANGBEIN-2
    ARE CONFORMAL MAPPING POLYNOMIALS REALLY OPTIMAL
    ANALYTIC PARAMETRIZATIONS OF THE SCATTERING
    AMPLITUDES?
    W.LANGBEIN:NUOVO CIM.30A(1975)653
    2.4.5; A.6.3
77L LANGBEIN
    CONFORMAL MAPPING,ALGEBRAIC PARAMETRIZATION AND TWO-
    VARIABLE EXPANSION OF SCATTERING AMPLITUDES
    W.LANGBEIN:NUOVO CIM.39A(1977)313
    A.6.3
```

78L LANGBEIN-1
PI N SCATTERING LENGTHS FROM AN EVALUATION OF THE 4 INVARIANT AMPLITUDES IN TERMS OF RATIONAL ANALYTIC SERIES
W.LANGBEIN:KAISERSLAUTERN PREPRINT(1978)
2.4.7

78L LANGBEIN-2
DETERMINATION OF THE PI N COUPLING CONSTANT FROM A SIMULTANEOUS ANALYSIS OF ALL INVARIANT AMPLITUDES
W.LANGBEIN:LETT.NUOVO CIM.22(1978)85
2.4.2

79L LANGBEIN
SIMULTANEOUS ANALYSIS OF THE 4 PI N INVARIANT AMPLITUDES AT LOW ENERGIES AND EVALUATION OF THE SIGMA COMMUTATOR
W.LANGBEIN:NUOVO CIM.51A(1979)219
2.5.1; 2.4.7; 2.4.3; 2.4.5

75L LE YAOUANC
SU(6) STRONG BREAKING:STRUCTURE FUNCTIONS AND STATIC PROPERTIES OF THE NUCLEON
A.LE YAOUANC,L.OLIVER,O.PENE,J.C.RAYNAL:PHYS.REV. D12(1975)2137
2.4.1

77L LE YAOUANC
PROPERTIES OF THE BARYON GROUND STATE
A.LE YAOUANC,L.OLIVER,O.PENE,J.C.RAYNAL:PHYS.REV. D15(1977)844
2.4.1

78L LE YAOUANC
PHENOMENOLOGICAL SU(6)BREAKING OF BARYON WAVE FUNCTIONS
A.LE YAOUANC,L.OLIVER,O.PENE,J.C.RAYNAL:PHYS.REV. D18(1978)1591
2.4.1

72L LEA
EFFECTIVE RANGE FORMULAE FOR N*33
A.T.LEA,J.HAMILTON:NUCL.PHYS.B42(1972)518
2.5.1; A.7.1

73L LEADER-1
CORRELATION BETWEEN THE BREAK IN DS/DT AND THE GROWTH OF THE TOTAL CROSS SECTION
E.LEADER,U.MAOR:PHYS.LETT.43B(1973)505
2.6.3

73L LEADER-2
SCALING IN HADRONIC COLLISIONS AND THE NEW KINEMATIC VARIABLE N SQUARED
E. LEADER, M.R. PENNINGTON: PHYS.REV.D7(1973)2668
2.6.4

54L LEE
SOME SPECIAL EXAMPLES IN RENORMALIZABLE FIELD THEORY
T.D.LEE:PHYS.REV.95(1954)1329
2.4.1

74L LEE
UNITARY MULTICHANNEL SIDEWISE DISPERSION CALCULATION OF THE NUCLEON ANOMALOUS MAGNETIC MOMENTS
P.S.LEE,G.L.SHAW,D.SILVERMAN:PHYS.REV.D10(1974)2251
A.11.1

55L LEHMANN
ZUR FORMULIERUNG QUANTISIERTER FELDTHEORIEN
H.LEHMANN,K.SYMANZIK,W.ZIMMERMANN:NUOVO CIM.1(1955)205
A.6

58L LEHMANN
ANALYTIC PROPERTIES OF SCATTERING AMPLITUDES AS FUNCTIONS OF MOMENTUM TRANSFER
H.LEHMANN:NUOVO CIM.10(1958)579
A.6.2

62L LEHMANN
ON THE HIGH-ENERGY LIMIT OF ORDINARY DISPERSION RELATIONS
H.LEHMANN:NUCL.PHYS.29(1962)300
2.2.1; A.6.4

72L LEHMANN
CHIRAL INVARIANCE AND EFFECTIVE RANGE EXPANSION FOR PION-PION SCATTERING
H.LEHMANN:PHYS.LETT.41B(1972)529
2.4.6

79L LEPAGE-1
EXCLUSIVE PROCESSES IN QCD.FORM FACTORS OF BARYONS AT LARGE T
G.P.LEPAGE,S.J.BRODSKY:PHYS.REV.LETT.43(1979)545
2.5.2

79L LEPAGE-2
EXCLUSIVE PROCESSES IN QCD.FORM FACTORS OF MESONS
G.P.LEPAGE,S.J.BRODSKY:PHYS.LETT.87B(1979)359
2.5.2

80L LEPAGE
EXCLUSIVE PROCESSES IN PERTURBATIVE QCD
G.P.LEPAGE,S.J.BRODSKY:PHYS.REV.D22(1980)2157
2.6.6; 2.5.2

78L LEVIN
SMALL-ANGLE ELASTIC SCATTERING AND QUARK MODEL
E.M.LEVIN,V.M.SCHEKTER:LENINGRAD PREPRINT NO.442(1978)
2.6.4

64L LEVINGER
EXTRAPOLATION OF PROTON ELECTROMAGNETIC FORM FACTORS
J.S.LEVINGER,R.F.PEIERLS:PHYS.REV.134(1964)B1341
2.5.2; 2.6.7; A.11.1

75L LICHARD
DETERMINATION OF RESONANCE PARAMETERS FROM PARTIAL WAVE AMPLITUDES
P.LICHARD,J.PISUT IN"HADRON INTERACTIONS AT LOW ENERGIES"(TRIANGLE MEETING,NOV.1973),P.329;ED.M. BLAZEK;VEDA,BRATISLAVA(1975)
2.4.1

76L LICHARD-1
A STATISTICAL METHOD FOR REPRESENTATION OF DATA BY ANALYTIC FUNCTIONS, EXPANSIONS AROUND NU=T=0 AND THE SIGMA TERM
P.LICHARD:CZECH.J.PHYS.B26(1976)1113
2.5.1; 2.4.7; 2.4.5

76L LICHARD-2
EVALUATION OF THE PI N SCATTERING LENGTHS FROM DATA AND ANALYTICITY
P.LICHARD:ACTA PHYS.SLOV.26(1976)152
2.4.7

77L LICHARD
A GENERAL METHOD FOR TESTING THE ANALYTICITY OF SCATTERING AMPLITUDES
P.LICHARD,J.PISUT,P.PRESNAJDER:ACTA.PHYS.SLOV. 27(1977)5
A.6.3; 2.4.2; 2.4.5

73L LICHTENBERG
MASS AND WIDTH OF A BROAD RESONANCE
D.B.LICHTENBERG:LETT.NUOVO CIM.7(1973)727
2.4.1; 2.1.10

75L LICHTENBERG
THE MASSES AND WIDTHS OF THE DELTA ++ AND DELTA 0 RESONANCES
D.B.LICHTENBERG:LETT.NUOVO CIM.12(1975)616
2.1.17

76L LICHTENBERG
ELECTROMAGNETIC MASS SPLITTING OF HADRONS IN THE QUARK MODEL
D.B.LICHTENBERG:PHYS.REV.14(1976)1412
2.1.17

81L LICHTENBERG
TWO-CHANNEL DESCRIPTION OF THE RHO MESON AND DELTA BARYON
D.B.LICHTENBERG,J.G.WILLS:PHYS.LETT.100B(1981)87
2.4.1

82L LICHTENBERG
BARYON MASSES IN A RELATIVISTIC QUARK-DIQUARK MODEL
D.B.LICHTENBERG,W.NAMGUNG,E.PREDAZZI,J.G.WILLS: INDIANA UNIV.PREPRINT IUHET 74(1982)
2.4.1

76L LIN
NONLINEAR SIGMA MODEL CALCULATION OF PI N PHASE SHIFTS
W.L.LIN,R.S.WILLEY:PHYS.REV.D14(1976)196
2.5.1

77L LIOU
SOFT-PHOTON ANALYSIS OF PION-PROTON BREMSSTRAHLUNG
M.K.LIOU,W.T.NUTT:PHYS.REV.D16(1977)2176
2.1.17

78L LIOU
PION-PROTON BREMSSTRAHLUNG
M.K.LIOU,W.T.NUTT:NUOVO CIM.46A(1978)365
2.1.17

75L LIPKIN
NEW SYSTEMATICS IN HADRON TOTAL CROSS SECTIONS
H.J.LIPKIN:PHYS.REV.D11(1975)1827
2.6.4

78L LIPKIN
TWO-COMPONENT POMERON AND HADRON TOTAL CROSS SECTIONS AND REAL PARTS
H.J.LIPKIN:PHYS.REV.D17(1978)366
2.6.4

68L LIU
NEW PI N SUPERCONVERGENT SUM RULES
Y.LIU,S.OKUBO:PHYS.REV.168(1968)1712
A.6.4; 2.6.5; 2.2.1

80L LIVANOS
ENERGY DEPENDENT PARTIAL WAVE ANALYSIS OF THE REACTION PI PLUS P -> K PLUS SIGMA
PH.LIVANOS IN"BARYON 80";PROC.4TH INT.CONF.ON BARYON RESONANCES(TORONTO);ED.N.ISGUR,P.35
2.4.1

67L LOCHER
RELATIVISTIC TREATMENT OF STRUCTURE IN THE COULOMB INTERFERENCE PROBLEM
M.P.LOCHER:NUCL.PHYS.B2(1967)525
A.2.3; 2.1.17; 2.6.2

63L LOGUNOV-1
QUASI-OPTICAL APPROACH IN QUANTUM FIELD THEORY
A.A.LOGUNOV,A.N.TAVKHELIDZE:NUOVO CIM.29(1963)380
2.6.4; 2.6.6

63L LOGUNOV-2
ASYMPTOTIC RELATIONS BETWEEN CROSS SECTIONS IN LOCAL FIELD THEORY
A.A.LOGUNOV ET AL.:PHYS.LETT.7(1963)69
A.5.3

67L LOGUNOV
DISPERSION SUM RULES AND HIGH ENERGY SCATTERING
A.A.LOGUNOV,L.D.SOLOVIEV,A.N.TAVKHELIDZE:PHYS.LETT. 24B(1967)181
2.6.5; 2.2.1

77L LOMBARD
THE CHOU-YANG MODEL AND THE FORM FACTORS OF ELEMENTARY PARTICLES
R.J.LOMBARD,C.WILKIN:JOURN.PHYS.G3(1977)L5
2.6.4

80L LOMON
JOINING THE QUARK THEORY TO THE HADRON EXCHANGE THEORY OF INTERACTIONS
E.LOMON IN'AUSSIOS'(MARCH 1980);ORSAY REPORT(1980)
2.5.3

74L LOMSADZE
DISPERSION RELATIONS WITH DAMPING FUNCTIONS
Y.M.LOMSADZE,B.A.AGRANOVSKY,E.P.SABAD:NUCL.PHYS. B73(1974)536
A.6.4

77L LONGACRE
K-MATRIX FITS TO N PI -> N PI PI IN THE RESONANCE REGION
R.S.LONGACRE ET AL.:NUCL.PHYS.B122(1977)493
2.4.1

78L LONGACRE
K-MATRIX FITS TO PI N ELASTIC AND INELASTIC SCATTERING AT W=1.3 TO 2.0 GEV
R.S.LONGACRE ET AL.:PHYS.REV.D17(1978)1795
2.4.1

75L LOOS
TRANSVERSITY-AMPLITUDEN FUER DIE PHOTOPRODUKTION IM RAHMEN DER WORDENSCHEN REGGE-MODELLE
U.LOOS:DIPLOMARBEIT,UNIVERSITY OF BONN;IR-75-44
2.4.3

69L LOVE
SIDEWISE DISPERSION RELATIONS AND J=1/2 BARYON RESONANCES
A.LOVE,R.G.MOORHOUSE:NUCL.PHYS.B9(1969)577
A.11.1

62L LOVELACE
DIFFRACTION SCATTERING AND MANDELSTAM REPRESENTATION
C.LOVELACE:NUOVO CIM.25(1962)730
2.6.7; A.6.2

66L LOVELACE
NEW EVIDENCE FOR THE SIGMA MESON
C.LOVELACE,R.M.HEINZ,A.DONNACHIE:PHYS.LETT.22(1966)332
2.6.7

68L LOVELACE
A NOVEL APPLICATION OF REGGE TRAJECTORIES
C.LOVELACE:PHYS.LETT.28B(1968)264
2.4.6; 2.6.5

69L LOVELACE
REGGE FITS WITH VENEZIANO RESIDUES
C.LOVELACE:NUCL.PHYS.B12(1969)253
2.6.5

55L LOW
BOSON-FERMION SCATTERING IN THE HEISENBERG REPRESENTATION
F.E.LOW:PHYS.REV.97(1955)1392
2.5.1

75L LOW
MODEL OF THE BARE POMERON
F.E.LOW:PHYS.REV.D12(1975)163
2.6.4

66L LUMING
ON THE FOURIER-BESSEL REPRESENTATION FOR HELICITY AMPLITUDES
M.LUMING,E.PREDAZZI:NUOVO CIM.42A(1966)878
A.3.3

76L MA
N* RESONANCE PARAMETERS FROM A DUAL ANALYSIS OF BACKWARD PI- P SCATTERING
E.MA,G.L.SHAW:PHYS.REV.D13(1976)3027
2.6.7

59L MAC DOWELL
ANALYTIC PROPERTIES OF PARTIAL WAVES IN MESON-NUCLEON SCATTERING
S.W.MAC DOWELL:PHYS.REV.116(1959)774
A.7.1

64L MAC DOWELL
UNITARITY BOUNDS OF THE SCATTERING AMPLITUDES AND THE DIFFRACTION PEAK
S.W.MAC DOWELL,A.MARTIN:PHYS.REV.135(1964)B960
A.5.3

72L MAC DOWELL
ANALYTIC CONTINUATION OF REDUCED PION-NUCLEON PARTIAL WAVE AMPLITUDES
S.W.MAC DOWELL:PHYS.REV.D6(1972)3512
A.7.1

79L MAC MULLAN
LOW-ENERGY PHOTO- AND ELECTROPRODUCTION FOR PHYSICAL PIONS
J.T.MAC MULLAN,M.D.SCADRON:PHYS.REV.D20(1979)1069,1081
2.4.2; 2.5.2

81L MACHET
QCD CONSTRAINTS FOR THE PION FORM FACTOR
B.MACHET:Z.PHYSIK C8(1981)215
2.5.2

74L MAENNER
NEW RESULTS ON PI PI SCATTERING
W.MAENNER IN'PROC.IV.INT.CONF.ON EXPERIMENTAL MESON SPECTROSCOPY'(1974)
2.4.6

78L MAHARANA
BOUNDS ON THE SLOPE OF THE DIFFRACTION PEAK AND ZEROS OF THE ABSORPTIVE DIFFERENTIAL CROSS SECTIONS
J.MAHARANA:PHYS.LETT.77B(1978)102
A.5.3

76L MAHOUX
POSITIVITY PROPERTIES IN ELASTIC SCATTERING OF PARTICLES WITH ARBITRARY SPINS
G.MAHOUX:PHYS.LETT.65B(1976)139
A.5.3

75L MAJEROTTO
PP AND ANTI P P ELASTIC SCATTERING INCLUDING SPIN
W.MAJEROTTO:NUOVO CIM.25(1975)117
2.6.4

58L MANDELSTAM
DETERMINATION OF PION-NUCLEON SCATTERING AMPLITUDES FROM DISPERSION RELATIONS AND UNITARITY
S.MANDELSTAM:PHYS.REV.112(1958)1344
A.6.2

59L MANDELSTAM-1
ANALYTIC PROPERTIES OF TRANSITION AMPLITUDES IN PERTURBATION THEORY
S.MANDELSTAM:PHYS.REV.115(1959)1741
A.6.2

59L MANDELSTAM-2
CONSTRUCTION OF THE PERTURBATION SERIES FOR TRANSITION AMPLITUDES FROM THEIR ANALYTICITY AND UNITARITY PROPERTIES
S.MANDELSTAM:PHYS.REV.115(1959)1752
A.6.2

63L MANDELSTAM-1
THE REGGE FORMALISM FOR RELATIVISTIC PARTICLES WITH SPIN
S.MANDELSTAM:NUOVO CIM.30(1963)1113
A.9

63L MANDELSTAM-2
CUTS IN THE ANGULAR MOMENTUM PLANE
S.MANDELSTAM:NUOVO CIM.30(1963)1127,1148
A.9

82L MANE
OBSERVATION OF AN ISOSCALAR VECTOR MESON PHI(1.65 GEV)
F.MANE ET AL.:PHYS.LETT.112B(1982)178
2.5.2

81L MARGARITISZ
THE NUCLEON FORM FACTORS IN THE GEOMETRODYNAMICAL MODEL
T.MARGARITISZ,K.SZEGOE:HUNG.ACADEMY KFKI-1981-76
2.5.2

63L MARTIN
UNITARITY AND HIGH-ENERGY BEHAVIOUR OF SCATTERING AMPLITUDES
A.MARTIN:PHYS.REV.129(1963)1432
2.5.3

66L MARTIN
EXTENSION OF THE AXIOMATIC ANALYTICITY DOMAIN OF SCATTERING AMPLITUDES BY UNITARITY
A.MARTIN:NUOVO CIM.42(1966)930
2.6.2

78L MARTIN-1
HOW IMPOSING ANALYTICITY ON A PI PI PHASE SHIFT ANALYSIS CAN REVEAL NEW SOLUTIONS
A.D.MARTIN,M.R.PENNINGTON:ANN.OF PHYS.114(1978)1
2.4.6; 2.5.2

78L MARTIN-2
ITERATION MODELS AND ELASTIC REACTIONS BEYOND THE DIFFRACTION PEAK
A.MARTIN,R.PESCHANSKI:PHYS.REV.D18(1978)685
2.6.5; 2.6.4

78L MARTIN-3
SPECTRUM OF BROAD RESONANCES IN ANTIPROTON-PROTON -> PI PLUS PI MINUS
B.R.MARTIN,D.MORGAN IN'PROC.EUROP.SYMP. ON ANTINUCLEON-NUCLEON ANNIHILATION'(STRASBOURG 1978)
2.5.4

78L MARTIN-4
ON THE SIGN OF THE POMERON FLIP AMPLITUDE
A.MARTIN,H.NAVELET:J.PHYS.G4(1978)647
2.6.3

79L MARTIN
DETERMINATION OF AND RESONANCE STRUCTURE IN ANTIPROTON-PROTON -> TWO PION PARTIAL WAVES
A.D.MARTIN,M.R.PENNINGTON:PHYS.LETT.86B(1979)93
2.5.4

80L MARTIN
ANALYSIS OF ANTIPROTON PROTON -> PION PION IN THE RESONANCE REGION
B.R.MARTIN,D.MORGAN IN'PROC.5TH INT.EUROP.SYMP. ON NUCLEON-ANTINUCLEON INTERACTIONS'(BRESSANONE 1980);P. 397
2.5.4

82L MARTINELLI
THE PROTON AND NEUTRON MAGNETIC MOMENTS IN LATTICE QCD
G.MARTINELLI,G.PARISI,R.PETRONZIO,F.RAPVANO:CERN PREPRINT TH 3334(1982)
2.5.2

80L MATSUI
ANALYSIS OF PI N BACKWARD SCATTERING AT INTERMEDIATE ENERGY
T.MATSUI:NAGOYA PREPRINT(1980)
2.6.7

73L MATVEEV
AUTOMODELLISM IN THE LARGE ANGLE ELASTIC SCATTERING AND STRUCTURE OF HADRONS
V.MATVEEV,R.MURADYAN,A.TAVKHELIDZE:LETT.NUOVO CIM. 7(1973)719
2.6.6

78L MC CLASKEY
APPLICATION OF INTERIOR DISPERSION RELATIONS TO PION PHOTOPRODUCTION NEAR THRESHOLD
R.MC CLASKEY, R.J.JACOB,G.HITE:PHYS.REV.D18(1978)660
2.4.2

72L MC CLURE
DISPERSION RELATION CALCULATION OF THE PI N AMPLITUDE A'
J.A.MC CLURE,L.E.PITTS:PHYS.REV.D5(1972)109
2.1.16

81L MC LEOD
RELATIVISTIC GENERALIZATIONS OF SIMPLE PION-NUCLEON MODELS
R.J.MC LEOD,D.J.ERNST:PHYS.REV.C23(1981)1660
2.5.1

66L MICHAEL
RESONANCES IN THE S11 PION NUCLEON AMPLITUDE
C.MICHAEL:PHYS.LETT.21(1966)93
2.4.1

69L MICHAEL
PION-DEUTERON SCATTERING
C.MICHAEL,C.WILKIN:NUCL.PHYS.B11(1969)99
A.2.1

61L MICHEL
RELATIONS BETWEEN POLARIZATIONS DUE TO CHARGE-INDEPENDENCE
L.MICHEL:NUOVO CIM.22(1961)203
A.10.2; 2.4.4

82L MICKENS
NEUTRON ELECTRIC FORM FACTOR
R.E.MICKENS:PHYS.REV.D25(1982)273
2.5.2

80L MIETTINEN
EVIDENCE THAT HADRONIC INTERIORS HAVE A DENSER MATTER THAN CHARGE DISTRIBUTION
H.I.MIETTINEN,G.H.THOMAS:NUCL.PHYS.B166(1980)365
2.6.4

74L MILLER
PI N SCATTERING AMPLITUDE ANALYSIS
R.C.MILLER ET AL.:NUOVO CIM.22A(1974)99
2.1.15

80L MILLER
PION-NUCLEON SCATTERING IN THE BROWN-RHO BAG MODEL
G.A.MILLER,A.W.THOMAS,S.THEBERGE:PHYS.LETT.91B (1980)192
2.5.1

81L MIR
ON THE ABSOLUTE NORMALIZATION OF THE PI N BACKWARD SCATTERING DATA
K.L.MIR,J.K.STORROW:J.PHYS.G7(1981)1187
2.6.7

77L MIYAKOSHI
BIAS-FREE DETERMINATION OF DELTA(1232)POLE POSITION
T.MIYAKOSHI,A.KANAZAWA:PROGR.THEOR.PHYS.58(1977)1421
2.4.1

76L MOIR
DETERMINATION OF THE PI N SIGMA TERM
D.C.MOIR,R.J.JACOB,G.E.HITE:NUCL.PHYS.B103(1976)477
2.5.1; 2.4.7

78L MORROW
AMPLITUDE ANALYSIS OF PI P SCATTERING AND THE SIGMA TRAJECTORY
R.A.MORROW:NUOVO CIM.43A(1978)212
2.6.3

81L MUELLER
THE ISOSINGLET AND ISOVECTOR PI PI SCATTERING LENGTHS
R.MUELLER,G.E.HITE,R.J.JACOB:Z.PHYSIK C8(1981)199
2.4.6

71L MUETTER
UNITARITY CONSTRAINTS ON HELICITY FLIP AMPLITUDES IN
MESON-NUCLEON SCATTERING
K.H.MUETTER:NUCL.PHYS.B31(1971)589
A.5.3

80L MUKHIN
DETERMINATION OF PI PI S-WAVE PHASE SHIFTS IN THE
RANGE FROM THRESHOLD TO 1 GEV
K.N.MUKHIN ET AL.:SOV.PHYS.JETP LETT.32(1980)601
2.4.6

79L MULDERS
MULTIQUARK STATES:QQQ BARYON RESONANCES
P.J.MULDERS,A.T.AERTS,J.J.DE SWART:PHYS.REV.D19(1979)
2635
2.4.1

80L MULDERS
MULTIQUARK STATES IN THE BAG MODEL
P.J.G.MULDERS:THESIS;UNIVERSITY OF NIJMEGEN,1980
2.4.1

76L MYHRER
ELECTROMAGNETIC CORRECTIONS TO MASSES AND WIDTHS IN
THE BARYON DECUPLET
F.MYHRER,H.PILKUHN:Z.PHYSIK A276(1976)29
2.1.17

79L NAGY
ELASTIC PP SCATTERING AT LARGE MOMENTUM TRANSFER AT
THE ISR
E.NAGY ET AL.:NUCL.PHYS.B150(1979)221
2.6.3; 2.6.4

81L NAKANO
PI N SELF-CONSISTENT SEPARABLE MODEL
K.NAKANO:PHYS.REV.C 24(1981)561
2.5.1

71L NATH
UNIQUENESS OF THE INTERACTION INVOLVING SPIN 3/2
PARTICLES
L.M.NATH,B.ETEMADI,J.D.KIMEL:PHYS.REV.D3(1971)2153
2.5.1; A.8.1

80L NATH
PHOTOPRODUCTION OF PIONS AT LOW ENERGY
L.M.NATH,B.K.BHATTACHARYYA:Z.PHYSIK C5(1980)9
2.4.2

74L NENCIU
NEW METHOD FOR THE GENERAL SOLUTION OF PARTIAL WAVE
DISPERSION RELATIONS
G.NENCIU,G.RASCHE,W.S.WOOLCOCK:HELV.PHYS.ACTA 47(1974)
137
A.7.1

80L NENCIU
GENERAL SOLUTION OF MULTICHANNEL PARTIAL WAVE
DISPERSION RELATIONS
G.NENCIU,G.RASCHE,W.S.WOOLCOCK:HELV.PHYS.ACTA 53(1980)
134
A.7.1

70L NIELSEN
INFORMATION ON LOW-ENERGY PI PI INTERACTION
H.NIELSEN,J.LYNG PETERSEN,E.PIETARINEN:NUCL.PHYS.
B22(1970)525
A.7.1; A.6.9; 2.4.6; 2.5.2; 2.6.7

71L NIELSEN
PION-NUCLEON PHASE SHIFTS BELOW 270 MEV
H.NIELSEN:NUCL.PHYS.B30(1971)317
2.1.6

72L NIELSEN-1
PION-NUCLEON PARTIAL WAVE AMPLITUDES IN THE LOW
ENERGY REGION
H.NIELSEN,G.C.OADES:NUCL.PHYS.B49(1972)573
2.1.7; 2.4.7

72L NIELSEN-2
CONSISTENCY BETWEEN THE LOW ENERGY PION-PION
INTERACTION AND PION-NUCLEON SCATTERING
H.NIELSEN,G.C.OADES:NUCL.PHYS.B49(1972)586
2.4.6

74L NIELSEN-1
CONSTRAINTS ON THE LOW-ENERGY PI PI ISOSPIN 0 S-WAVE
FROM ELASTIC PI N SCATTERING
H.NIELSEN,G.C.OADES:NUCL.PHYS.B72(1974)321
2.4.7; 2.4.6; 2.5.1

74L NIELSEN-2
LOW ENERGY PI N PARTIAL WAVES,EXPANSION ABOUT NU=T=0
AND THE VALUE OF THE SIGMA TERM
H.NIELSEN,G.C.OADES:NUCL.PHYS.B72(1974)310
2.4.7; 2.5.1

81L NISKANEN
THE PI N DELTA COUPLING AND THE DELTA RESONANCE WIDTH
J.A.NISKANEN:PHYS.LETT.107B(1981)344
2.5.1; A.8.1

73L NOGOVA
DETERMINATION OF THE DELTA(1232)POLE POSITION
A.NOGOVA,J.PISUT:NUCL.PHYS.B61(1973)445,B65(1973)
544(ERRATUM)
2.4.1

80L NOVOSELLER
MULTICHANNEL RESONANCE PARAMETRIZATION IN THE
PRESENCE OF AN INELASTIC BACKGROUND
D.E.NOVOSELLER:NUCL.PHYS.B176(1980)153
2.4.1

75L NUSSINOV
COLOURED-QUARK VERSION OF SOME HADRONIC PUZZLES
S.NUSSINOV:PHYS.REV.LETT.34(1975)1286
2.6.4

76L NUTT
PION-NUCLEON VERTEX FUNCTION AND THE ONE-PION
EXCHANGE POTENTIAL
W.NUTT,B.LOISEAU:NUCL.PHYS.B104(1976)98
2.5.1

63L OADES
PION-NUCLEON SCATTERING AND THE J=2 T=0 PION-PION
INTERACTION
G.C.OADES:PHYS.REV.132(1963)1277
A.7.1

66L OADES
SYSTEMATIC PROPERTIES OF PION-NUCLEON SCATTERING
AMPLITUDES
G.C.OADES:SUPPL.NUOVO CIM.44(1966)217
A.6.6

68L ODORICO
DETERMINATION OF A SCATTERING AMPLITUDE FROM THE
KNOWLEDGE OF ITS MODULUS ABOVE THRESHOLD
R.ODORICO:NUOVO CIM.54A(1968)96
A.6.4; A.6.9

72L ODORICO
CRUCIAL TESTS OF VENEZIANO MODEL FROM DIP BEHAVIOUR
AT LOW ENERGY
R.ODORICO:NUCL.PHYS.B37(1972)509
2.4.3; 2.6.7

73L ODORICO
SUPPRESSION OF RESONANCE COUPLINGS AS GAUGE-LIKE
CONDITIONS FROM DUAL RESONANCE MODELS
R.ODORICO:PHYS.REV.D8(1973)3952
2.4.3; 2.6.5

55L OEHME
DISPERSION RELATIONS FOR PION-NUCLEON SCATTERING
(FLIP-AMPLITUDE)
R.OEHME:PHYS.REV.100(1955)1503
A.6.3

56L OEHME
DISPERSION RELATIONS FOR PION-NUCLEON SCATTERING (NO-
FLIP AMPLITUDE)
R.OEHME:PHYS.REV.102(1956)1174
A.6.3

61L OEHME
CONTINUATION OF SCATTERING AMPLITUDES AND FORM
FACTORS THROUGH TWO-PARTICLE BRANCH LINES
R.OEHME:PHYS.REV.121(1961)1840
A.7.3; A.11.3; 2.5.2

68L OEHME-1
SUPERCONVERGENT DISPERSION RELATIONS
R.OEHME:ANN.OF PHYS.46(1968)347
A.6.4; A.6.3

68L OEHME-2
LOW-ENERGY SCATTERING AND HIGH-ENERGY LIMITS
R.OEHME:NUOVO CIM.55A(1968)595
A.6.4

70L OEHME-1
FORWARD DISPERSION RELATIONS AND MICROCAUSALITY
R.OEHME IN'QUANTA'(G.WENTZEL FESTSCHRIFT),P.309;UNIV.
OF CHICAGO PRESS(1970)
A.6.4

70L OEHME-2
MATHEMATICAL AND PHENOMENOLOGICAL DUALITY
R.OEHME:NUCL.PHYS.B16(1970)161
2.6.5; 2.5.1

72L OEHME
RISING CROSS SECTIONS
R.OEHME:SPRINGER TRACTS MOD.PHYS.61(1972)109
2.6.4

74L OEHME
FACTORIZATION AND SHIELDING OF HARD REGGE SURFACES
R.OEHME,S.PARANJAPE:PHYS.REV.D10(1974)2229
2.6.4

75L OEHME
REGGE SURFACES AND ELASTIC UNITARITY
R.OEHME:PHYS.REV.D11(1975)1191
2.6.4

76L OEHME
ASYMPTOTIC FREEDOM,POTENTIAL SCATTERING AND HIGH-
ENERGY LIMITS
R.OEHME:PHYS.REV.D13(1976)1003
2.6.4

63L OKUBO
PHI-MESON AND UNITARY SYMMETRY MODEL
S.OKUBO:PHYS.LETT.5(1963)165
2.5.2; 2.5.1

77L OKUBO
CONSEQUENCES OF THE QUARK-LINE(OZI)RULE
S.OKUBO:PHYS.REV.D16(1977)2336
2.5.1

73L OLSSON
IMPLICATIONS FOR THE DELTA N PI INTERACTION AND SIGMA TERM FROM LOW ENERGY PI N SCATTERING
M.G.OLSSON:PHYS.REV.D7(1973)3444
2.5.1; A.8.1

75L OLSSON
SYSTEMATICS OF LOW-ENERGY PION-NUCLEON SCATTERING
M.G.OLSSON,E.T.OSYPOWSKI:NUCL.PHYS.B101(1975)136
A.8.1; 2.5.1

77L OLSSON
THE DELTA(33) PHASE SHIFT
M.G.OLSSON:NUOVO CIM.40A(1977)284
A.8.1; 2.5.1

80L OLSSON-1
A RIGOROUS DERIVATION OF THE ALTARELLI-CABIBBO-MAIANI RELATION
M.G.OLSSON,E.T.OSYPOWSKI:J.PHYS.G6(1980)423
2.5.1; 2.4.7

80L OLSSON-2
PION-NUCLEON SIGMA TERM FROM THRESHOLD DATA
M.G.OLSSON:J.PHYS.G6(1980)431
2.5.1; 2.4.7

64L OREAR
UNIVERSALITY OF TRANSVERSE MOMENTUM DISTRIBUTIONS IN HIGH ENERGY PHYSICS
J.OREAR:PHYS.LETT.13(1964)190
2.6.6

82L OREAR
HADRON-HADRON ELASTIC SCATTERING AT LARGE TRANSVERSE MOMENTUM
J.OREAR IN'RENCONTRES DE MORIOND',VOL.36,P.359;ED. TRAN THANH VAN;EDITIONS FRONTIERES(1982)
2.6.4

70L OSYPOWSKI
WARD IDENTITY TECHNIQUES AND LOW-ENERGY PION-NUCLEON SCATTERING
E.T.OSYPOWSKI:NUCL.PHYS.B21(1970)615
2.5.1

78L OYANAGI
CALCULATION OF LEGENDRE MOMENTS FROM PARTIAL WAVES
Y.OYANAGI:KEK PREPRINT 78-2
A.4.4

71L PAGELS
NON-ANALYTIC BEHAVIOUR OF THE SIGMA TERM IN PI N SCATTERING
H.PAGELS,W.PARDEE:PHYS.REV.D4(1971)3335
2.5.1

79L PAGELS
PION DECAY CONSTANT,ELECTROMAGNETIC FORM FACTOR,AND QUARK SELF-ENERGY IN QCD
H.PAGELS,S.STOKAR:PHYS.REV.D20(1979)2947
2.5.2

73L PAL
VENEZIANO TYPE AMPLITUDES WITH FORM FACTORS FOR PI N SCATTERING
B.K.PAL:PHYS.REV.D8(1973)241
2.6.5

77L PANTEA
CORRELATION BETWEEN PION RADIUS AND PI PI SCATTERING LENGTH
D.PANTEA,I.RASZILLIER:BUCHAREST CENTRAL INST.OF PHYSICS PREPRINT FT-132-1977
2.5.2

76L PARANJAPE
SHIELDED POMERON MODEL FOR HIGH ENERGY CROSS SECTIONS
S.PARANJAPE:PHYS.REV.D13(1976)1509
2.6.4

80L PARASHAR
CHARGE RADII OF NUCLEONS
D.PARASHAR:NUOVO CIM.57A(1980)268
2.5.2

80L PARIDA
CONVERGENT POLYNOMIAL EXPANSION AND SCALING IN DIFFRACTION SCATTERING III
M.K.PARIDA,N.GIRI:PHYS.REV.D21(1980)2528
2.6.3

82L PARIDA
REPLY TO'SLOPE PARAMETER AND ZERO TRAJECTORIES IN PI- P SCATTERING'
M.K.PARIDA:PHYS.REV.D25(1982)897
2.6.2; 2.6.3; 2.4.3

79L PARISI
ASYMPTOTIC PREDICTIONS FOR EXCLUSIVE PROCESSES
G.PARISI:PHYS.LETT.84B(1979)225
2.6.6

72L PAZMAN
TESTING OF ANALYTICITY OF PI N FORWARD AMPLITUDES
A.PAZMAN,J.PISUT,P.PRESNAJDER,P.LICHARD:NUCL.PHYS. B46(1972)637
A.6.3; 2.4.7

68L PECCEI
CHIRAL LAGRANGIAN CALCULATION OF PI N SCATTERING
LENGTHS
R.D.PECCEI:PHYS.REV.176(1968)1812
2.5.1

69L PETERSEN
ANALYTIC STRUCTURE OF PARTIAL WAVE AMPLITUDES FOR
INELASTIC TWO BODY COLLISIONS
J.L.PETERSEN:NUCL.PHYS.B13(1969)73
A.7.1

70L PETERSEN
CALCULATING PARTIAL WAVE AMPLITUDES ON THE LEFT-HAND
CUT
J.L.PETERSEN:NUCL.PHYS.B15(1970)549
A.7.1

77L PETERSEN
PI PI PHENOMENOLOGY AND CURRENT ALGEBRAS
J.L.PETERSEN:COPENHAGEN PREPRINT NBI-HE-77-11
2.4.6

70L PFEFFER
MESON-NUCLEON SCATTERING AND ASYMPTOTIC CONSERVATION
OF HELICITY
R.PFEFFER,W.K.CHENG,B.DUTTA-ROY,G.RENNINGER:PHYS.REV.
D2(1970)1965
2.6.3; 2.1.15

79L PICCIOTTO
POLARIZATION ASYMMETRIES IN PION PROTON RADIATIVE
SCATTERING
C.PICCIOTTO:PHYS.REV.D19(1979)3244
2.1.17

76L PIERRARD
THE PION-NUCLEON SCATTERING AMPLITUDES AT 40 GEV
J.PIERRARD ET AL.:NUCL.PHYS.B107(1976)118
2.1.15

77L PIERRARD
MESURE DU PARAMETRE DE ROTATION DE SPIN ET
DETERMINATION DES AMPLITUDES D'HELICITE A 45 GEV/C
J.PIERRARD:THESIS,ORSAY(1977)
2.1.16

72L PIETARINEN-1
PHENOMENOLOGICAL J-PLANE SINGULARITIES
E.PIETARINEN:NUCL.PHYS.B41(1972)110
A.9.1; 2.4.6

72L PIETARINEN-2
AN AMPLITUDE ANALYSIS OF PI N SCATTERING NEAR FORWARD
DIRECTIONS
E.PIETARINEN:NUCL.PHYS.B49(1972)315
2.1.7; A.6.3

72L PIETARINEN-3
DISPERSION RELATIONS AND EXPERIMENTAL DATA
E.PIETARINEN:NUOVO CIM.12A(1972)522
2.1.7; 2.2.1; A.6.3

73L PIETARINEN
APPROXIMATION OF SCATTERING AMPLITUDES USING
ANALYTICITY AND EXPERIMENTAL DATA
E.PIETARINEN:NUCL.PHYS.B55(1973)541
2.1.7; A.6.3

76L PIETARINEN-1
FIXED MOMENTUM TRANSFER ANALYSIS OF PION-NUCLEON
SCATTERING
E.PIETARINEN:NUCL.PHYS.B107(1976)21
A.6.3; 2.1.3; 2.1.7

76L PIETARINEN-2
PARTIAL WAVE ANALYSIS OF PION-NUCLEON SCATTERING
USING DISPERSION RELATIONS
E.PIETARINEN IN 'PROC.TOPICAL CONF.ON BARYON
RESONANCES'(OXFORD),P.20;RUTHERFORD LAB.,1976
2.1.7

76L PIETARINEN-3
UNIQUENESS OF FIXED MOMENTUM TRANSFER AMPLITUDES
DERIVED FROM DATA
E.PIETARINEN:PHYSICA SCRIPTA 14(1976)11
2.1.3; 2.1.7

76L PIETARINEN-4
PION-NUCLEON AMPLITUDES BETWEEN 5 AND 10 GEV
E.PIETARINEN:PHYS.LETT.61B(1976)461
2.1.7; 2.6.3

77L PIETARINEN
A CALCULATION OF PI PI -> NUCLEON-ANTINUCLEON
AMPLITUDES IN THE PSEUDOPHYSICAL REGION
E.PIETARINEN:HELSINKI PREPRINT HU-TFT-17-77
2.4.6; 2.5.2

78L PIETARINEN-1
ELASTIC PARTIAL WAVE ANALYSIS
E.PIETARINEN:HELSINKI PREPRINT HU-TFT-78-13
2.1.7

78L PIETARINEN-2
DETERMINATION OF BACKWARD PION-NUCLEON SCATTERING
AMPLITUDES
E.PIETARINEN:HELSINKI PREPRINT HU-TFT-78-13
2.6.7; 2.1.7

74L PILKUHN
COULOMB CORRECTIONS TO HADRON WIDTHS
H.PILKUHN:NUCL.PHYS.B82(1974)365
2.1.17

69L PISUT
ON ANALYTIC EXTRAPOLATION OF FORM FACTORS AND SCATTERING AMPLITUDES II:EXTRAPOLATION UP TO THE BOUNDARY
J.PISUT,P.PRESNAJDER,J.FISCHER:NUCL.PHYS.B12(1969)586
A.6.3; 2.6.7

81L POEPPING
COULOMB CORRECTIONS FOR PION-PROTON SCATTERING IN THE P33 PARTIAL WAVE
H.POEPPING,P.U.SAUER:J.PHYS.G7(1981)1349
2.1.17

58L POMERANCHUK
EQUALITY OF THE NUCLEON AND ANTINUCLEON TOTAL CROSS SECTIONS AT HIGH ENERGIES
I.JA.POMERANCHUK:SOV.PHYS.JETP 7(1958)499
A.5.4

66L PREDAZZI
INTEGRAL REPRESENTATION FOR SCATTERING AMPLITUDES
E.PREDAZZI:ANN.OF PHYS.36(1966)228
A.3.3

78L PREPARATA
BARYON STRUCTURE IN QUARK GEOMETRODYNAMICS
G.PREPARATA,K.SZEGOE:NUOVO CIM.47A(1978)303
2.4.1

79L PREPARATA
THE SPIN-DEPENDENCE OF LARGE ANGLE ELASTIC SCATTERING
G.PREPARATA,J.SOFFER:PHYS.LETT.86B(1979)304
2.6.6

80L PREPARATA
ON LARGE-ANGLE MESON-BARYON ELASTIC SCATTERING
G.PREPARATA,J.SOFFER:PHYS.LETT.93B(1980)187
2.6.6

71L PRESNAJDER
MERGING OF ANALYTICITY AND STATISTICS IN EXTRAPOLATIONS AND REPRESENTATIONS OF DATA
P.PRESNAJDER,J.PISUT:NUOVO CIM.3A(1971)603
2.4.7; A.6.3; 2.6.7

74L PRESNAJDER
CONVERGENT EXPANSIONS OF TWO-BODY SCATTERING AMPLITUDES AS FUNCTIONS OF TWO COMPLEX VARIABLES
P.PRESNAJDER,J.PISUT:NUCL.PHYS.B82(1974)313
2.1.4

75L PRESNAJDER
DETERMINATION OF LOW ENERGY PARAMETERS IN PI N SCATTERING
P.PRESNAJDER,P.LICHARD IN 'HADRON INTERACTIONS AT LOW ENERGIES' (TRIANGLE MEETING,NOV.1973),P.315;ED.M. BLAZEK;VEDA(BRATISLAVA)1975
2.4.7

76L PRESNAJDER
EXPANSION IN TWO COMPLEX VARIABLES FOR PION-PION AMPLITUDES
P.PRESNAJDER IN'HADRON CONSTITUENTS AND SYMMETRIES'(SMOLENICE 1976),P.361;EDS.J.LANIK,M.NOGA; VEDA,BRATISLAVA (1978)
2.1.4

78L QUENZER
PION FORM FACTOR FROM 480 TO 1100 MEV
A.QUENZER ET AL.:PHYS.LETT.76B(1978)512
2.5.2

77L RAINA
UNITARITY BOUNDS ON ELASTIC ABSORPTIVE AMPLITUDES, INVOLVING TOTAL CROSS SECTION,FORWARD SLOPE AND CURVATURE
A.K.RAINA,V.SINGH:PHYS.LETT.67B(1977)327
A.5.3

67L RAMAN
LOW-ENERGY PION-NUCLEON SCATTERING FROM CURRENT COMMUTATION RELATIONS AND PCAC
K.RAMAN:PHYS.REV.164(1967)1736
2.5.1

68L RARITA
REGGE POLE MODEL FOR PI P,PP AND ANTI P P SCATTERING
W.RARITA,R.J.RIDDELL,C.B.CHIU,R.J.N.PHILLIPS:PHYS.REV. 165(1968)1615
2.6.5

76L RASCHE-1
THE PARAMETRIZATION OF PI P SCATTERING EXPERIMENTS
G.RASCHE,W.S.WOOLCOCK:HELV.PHYS.ACTA 49(1976)455
2.4.7; 2.1.17

76L RASCHE-2
THE EFFECT OF RADIATIVE CAPTURE ON THRESHOLD OF PI MINUS P SCATTERING AND THE THEORY OF THE PANOFSKY RATIO
G.RASCHE,W.S.WOOLCOCK:HELV.PHYS.ACTA 49(1976)557
2.4.7; 2.1.17

77L RASCHE-1
EXTENSION OF FINAL STATE THEOREM AND ANALYSIS OF LOW ENERGY PI N SCATTERING AND PHOTOPRODUCTION EXPERIMENTS
G.RASCHE,W.S.WOOLCOCK:HELV.PHYS.ACTA 50(1977)407
2.4.7; 2.1.17

77L RASCHE-2
EXTENDED UNITARITY IN HADRONIC MULTICHANNEL SCATTERING WITH COULOMB INTERACTION PRESENT
G.RASCHE,W.S.WOOLCOCK:FORTSCHR.PHYSIK 25(1977)501
2.4.7; 2.1.17

82L RASCHE
CONNECTION BETWEEN LOW-ENERGY SCATTERING PARAMETERS AND ENERGY SHIFTS FOR PIONIC HYDROGEN
G.RASCHE,W.S.WOOLCOCK:NUCL.PHYS.A381(1982)405
2.4.7

77L RASZILLIER
PROBLEMS IN THE PHENOMENOLOGY OF THE PION ELECTROMAGNETIC FORM FACTOR
I.RASZILLIER:BUCHAREST CENTRAL INST.OF PHYSICS PREPRINTS FT-133 AND FT-140-1977
2.5.2

78L REINDERS
BARYON SPECTROSCOPY IN THE NON-RELATIVISTIC QUARK MODEL WITH ONE-GLUON EXCHANGE POTENTIAL
L.J.REINDERS:J.PHYS.G4(1978)1241
2.4.1

80L REINDERS
SPIN-ORBIT FORCES IN THE BARYON SPECTRUM
L.J.REINDERS IN"BARYON 1980,PROC.TORONTO CONF.",P.203
2.4.1

70L RENNER
EMPIRICAL TEST OF TENSOR MESON DOMINANCE
B.RENNER:PHYS.LETT.33B(1970)599
2.4.2

72L RENNER
THEORETICAL ESTIMATE OF THE MESON NUCLEON SIGMA TERM
B.RENNER:PHYS.LETT.40B(1972)
2.5.1

80L RESELE
LOW ENERGY PION-NUCLEON SCATTERING FROM N- AND DELTA-POLES AND THE LOW ENERGY THEOREMS
G.RESELE,C.SCHMID:PHYS.REV.D22(1980)121
A.8.1

66L RIAZUDDIN
ALGEBRA OF CURRENTS COMPONENT AND DECAY WIDTH OF RHO MESONS
RIAZUDDIN,FAYYAZUDDIN:PHYS.REV.147(1966)1071
A.8.2

79L RODRIGUEZ-VARGAS
THE SIGMA TERM AND THE AXIAL COUPLING CONSTANT IN PION-NUCLEON SCATTERING
A.M.RODRIGUEZ-VARGAS:NUOVO CIM.A51(1979)33
2.5.1

79L ROIESNEL
LOW-ENERGY MESON-NUCLEON SCATTERING ANALYSIS IN THE P-MATRIX FORMALISM
C.ROIESNEL:PHYS.REV.D20(1979)1646
2.5.1; 2.4.1

65L ROPER-1
ENERGY-DEPENDENT PION-NUCLEON PHASE-SHIFT ANALYSIS
L.D.ROPER,R.M.WRIGHT,B.T.FELD:PHYS.REV.138(1965)B190
2.1

65L ROPER-2
PION-NUCLEON PHASE-SHIFT ANALYSIS:0-350 MEV
L.D.ROPER,R.M.WRIGHT:PHYS.REV.138(1965)B921
2.1

75L ROSENZWEIG
RELATIONSHIP BETWEEN POMERON,REGGEON AND SU(3) SYMMETRY VIOLATION
C.ROSENZWEIG,G.F.CHEW:PHYS.LETT.58B(1975)93
2.6.4

61L ROSS
MULTICHANNEL EFFECTIVE RANGE THEORY
M.ROSS,G.SHAW:ANN.OF PHYS.13(1961)147
2.1.13

71L ROSS
DATA REPRESENTATION AND MODIFIED DISPERSION RELATIONS
G.G.ROSS:NUCL.PHYS.B31(1971)113
2.4.5; A.6.4

77L ROSSELET
EXPERIMENTAL STUDY OF 30 000 KE4-DECAYS
L.ROSSELET ET AL.:PHYS.REV.D15(1977)574
2.4.6

78L ROWE
AN ENERGY-DEPENDENT PHASE SHIFT ANALYSIS OF PI N SCATTERING BELOW 400 MEV
G.ROWE,M.SALOMON,R.H.LANDAU:PHYS.REV.C18(1978)584
2.1.10

79L ROY
APPARENT VIOLATION OF UNITARITY IN ELASTIC PI MINUS P SCATTERING
S.M.ROY:PHYS.REV.LETT.43(1979)19
2.6.3; A.5.3

82L ROYZEN
THE THEORETICAL APPROACH TO THE DESCRIPTION OF THE ELASTIC SCATTERING AT HIGH ENERGIES AND SMALL ANGLES
I.I.ROYZEN IN"RENCONTRE DE MORIOND",VOL.36,P.389;ED. TRAN THANH VAN;EDITIONS FRONTIERES(1982)
2.6.4

80L ROZANSKA
PARTIAL WAVE ANALYSIS OF THE ANTIPROTON-PROTON SYSTEM PRODUCED IN PI MINUS P SCATTERING AT 18 GEV
M.ROZANSKA ET AL.:NUCL.PHYS.B162(1980)505
2.5.4

67L RUBINSTEIN
ELECTROMAGNETIC PROPERTIES OF HADRONS IN THE QUARK MODEL
H.R.RUBINSTEIN,F.SCHECK,R.H.SOCOLOW:PHYS.REV.154(1967)1608
2.1.17

73L SABBA-STEFANESCU
ON THE ANALYTIC EXTRAPOLATION OF SCATTERING AMPLITUDES IN L SQUARED NORM
I.SABBA-STEFANESCU:NUCL.PHYS.B56(1973)287
2.4.1

78L SABBA-STEFANESCU
EINE UNTERSUCHUNG DER NULLSTELLEN-TRAJEKTORIEN DER PION NUCLEON STREUAMPLITUDEN
I.SABBA-STEFANESCU;THESIS,UNIVERSITY OF KARLSRUHE(1978)
2.4.3

80L SABBA-STEFANESCU-1
ON THE STABLE ANALYTIC CONTINUATION WITH RATIONAL FUNCTIONS
I.SABBA-STEFANESCU:J.MATH.PHYS.21(1980)175
A.6.3

80L SABBA-STEFANESCU-2
CONSTRAINTS ON PHASE-SHIFT ANALYSIS FROM TWO-VARIABLE ANALYTICITY
I.SABBA-STEFANESCU:PHYS.REV.D21(1980)3225
2.1.16; 2.4.3

81L SABBA-STEFANESCU
ZERO TRAJECTORIES AND ANALYTICITY IN PION-PROTON SCATTERING
I.SABBA-STEFANESCU:KARLSRUHE PREPRINT TKP 81-10
2.1.3; 2.4.3

82L SABBA-STEFANESCU-1
DETERMINATION OF PI N PHASE SHIFTS FROM ISOSPIN CONSTRAINTS AND FIXED T ANALYTICITY
I.SABBA-STEFANESCU:J.MATH.PHYS.23(1982)1190 AND ACTA PHYS.AUSTR.SUPPL. XXIII(1981)695
2.1.3; 2.1.7; 2.4.3

82L SABBA-STEFANESCU-2
CONSTRUCTION OF AMPLITUDES WITH MANDELSTAM ANALYTICITY FROM OBSERVABLE QUANTITIES
I.SABBA-STEFANESCU:UNIVERSITY OF KARLSRUHE PREPRINT TKP 82-13
2.1.3; 2.1.7

62L SACHS
HIGH ENERGY BEHAVIOUR OF NUCLEON ELECTROMAGNETIC FORM FACTORS
R.G.SACHS:PHYS.REV.126(1962)2256
2.5.2; A.11.1

60L SAKURAI
THEORY OF STRONG INTERACTIONS
J.J.SAKURAI:ANN.OF PHYS.11(1960)1
A.8.1; 2.5.1

72L SAMARANAYAKE-1
DETERMINATION OF PION-NUCLEON COUPLING CONSTANT AND S-WAVE SCATTERING LENGTHS
V.K.SAMARANAYAKE,W.S.WOOLCOCK:NUCL.PHYS.B48(1972)205
2.4.2; 2.4.7

72L SAMARANAYAKE-2
FORWARD DISPERSION RELATION CONSTRAINTS ON THE PI N P-WAVE AND D-WAVE SCATTERING LENGTHS
V.K.SAMARANAYAKE,W.S.WOOLCOCK:NUCL.PHYS.B49(1972)128
2.4.7

79L SAMARANAYAKE
DETERMINATION OF PION-NUCLEON COUPLING CONSTANT
V.K.SAMARANAYAKE:J.PHYS.G5(1979)657
2.4.2

79L SARMA
CUSP IN PI MINUS ELASTIC SCATTERING AT THE ETA THRESHOLD
H.N.K.SARMA ET AL.: NUCL.PHYS.B161(1979)1
2.1.13

67L SAWADA
EFFECT OF QUASIPOLE ON THE PI N FORWARD SCATTERING AMPLITUDE
T.SAWADA:PHYS.REV.158(1967)1421
2.4.5; A.6.4; A.7.3

81L SAWADA-1
STRONG VAN DER WAALS POTENTIAL IN PI N SCATTERING
T.SAWADA:NUOVO CIM.62A(1981)207
A.7.1; 2.5.3

81L SAWADA-2
ON THE UNIVERSALITY OF THE STRONG VAN DER WAALS FORCE BETWEEN HADRONS
T.SAWADA:PHYS.LETT.100B(1981)50
A.7.1; 2.5.3

80L SAXON
THE REACTION PI P -> K0 LAMBDA 0
SAXON ET AL.:NUCL.PHYS.B162(1980)522
2.4.1

74L SCADRON
MODEL-INDEPENDENT TESTS OF ON-SHELL PION-NUCLEON WARD IDENTITIES
M.D.SCADRON,L.R.THEBAUD:PHYS.REV.D9(1974)1544
2.5.1

81L SCADRON
ON THE CHOICE BETWEEN STRONG AND NEUTRAL PCAC
M.D.SCADRON:J.PHYS.G7(1981)1325
2.5.1

70L SCHLAILE
FIXED-U DISPERSIONSRELATIONEN IN DER PION-NUKLEON STREUUNG
H.G.SCHLAILE:THESIS,UNIV.OF KARLSRUHE(1970)
A.6.6

68L SCHMID
DIRECT CHANNEL RESONANCES FROM REGGE POLE EXCHANGE
C.SCHMID:PHYS.REV.LETT.20(1968)689
2.4.1

69L SCHMID
REGGE POLES AND RESONANCE CIRCLES
C.SCHMID:NUOVO CIM.61(1969)289
2.4.1

73L SCHMIDT
FLUCTUATIONS IN LARGE ANGLE PI P ELASTIC SCATTERING AT 5 GEV
F.H.SCHMIDT ET AL:PHYS.LETT.45B(1973)157
2.6.6

67L SCHNITZER
CURRENT ALGEBRA DETERMINATION OF LOW-ENERGY PION-NUCLEON SCATTERING AMPLITUDES
H.J.SCHNITZER:PHYS.REV.158(1967)1471
2.5.1

72L SCHNITZER
SIGMA TERM IN PION-NUCLEON SCATTERING
H.J.SCHNITZER:PHYS.REV.D5(1972)1482;D6(1972)1801
2.5.1

73L SCHREMPP-1
HIGH-ENERGY REACTIONS SEEN FROM THE S-CHANNEL:A COMPLEX POLE IN THE IMPACT PARAMETER PLANE
B. AND F.SCHREMPP:NUCL.PHYS.B54(1973)525
2.6.5

73L SCHREMPP-2
UNDERSTANDING AMPLITUDES FROM AN S-CHANNEL APPROACH
B.AND F.SCHREMPP:NUCL.PHYS.B60(1973)110
2.6.5

74L SCHREMPP
DUAL PERIPHERAL MODEL AT SERPUCHOV ENERGIES
B.AND F.SCHREMPP:NUCL.PHYS.B77(1974)453
2.6.5

75L SCHREMPP-1
DERIVATIVE RELATIONS FOR HELICITY AMPLITUDES
B.AND F.SCHREMPP:NUCL.PHYS.B96(1975)307
2.6.5; 2.6.7

75L SCHREMPP-2
IS LARGE ANGLE EXCLUSIVE SCATTERING CONTROLLED BY THE HADRONIC RADIUS?
B.AND F.SCHREMPP:PHYS.LETT.55B(1975)303
2.6.6

77L SCHREMPP
A SPACE (TIME) PICTURE OF HADRONIC DIFFRACTION
B. AND F. SCHREMPP:PHYS.LETT.70B(1977)88
2.6.4

80L SCHREMPP
STRONG INTERACTIONS - A TUNNELING PHENOMENON? THEORY AND APPLICATION TO HADRONIC DIFFRACTION.
B.AND F. SCHREMPP:NUCL.PHYS.B163(1980)397
2.6.4; 2.6.6

80L SCHROEDER
CALCULATION OF THE ELECTRIC POLARIZABILITIES OF PROTON AND NEUTRON
U.E.SCHROEDER:NUCL.PHYS.B166(1980)103
2.5.2

71L SHAFEE
ZEROS IN PION-NUCLEON SCATTERING AMPLITUDES
A.SHAFEE:NUCL.PHYS.B35(1971)556
2.4.3

72L SHEPARD
OPTIMIZED DATA EXTRAPOLATION OFF A BOUNDARY OF AN ANALYTIC FUNCTION
H.K.SHEPARD,C.C.SHIH:NUCL.PHYS.B42(1972)397
A.6.3; A.6.4

74L SHEPARD
OPTIMALLY-WEIGHTED DISPERSION RELATIONS
H.K.SHEPARD,C.C.SHIH:NUCL.PHYS.B77(1974)134
A.6.3; A.6.4

75L SIDHU
REAL PARTS OF FORWARD ELASTIC HADRONIC AMPLITUDES
D.P.SIDHU,U.P.SUKHATME:PHYS.REV.D11(1975)1351
2.2.1

80L SIDOROV
RECENT RESULTS OF EXPERIMENTS AT VEPP-2M
V.SIDOROV IN "PROC.1979 SYMP.ON LEPTON AND PHOTON INTERACTIONS",P.490;EDS.T.B.W.KIRK,H.D.I.ABARBANEL; FERMILAB
2.5.2

80L SIMON-1
THE STRUCTURE OF THE NUCLEONS
G.G.SIMON,F.BORKOWSKI,CH.SCHMITT,V.H.WALTHER: Z.NATURFORSCH.35A(1980)1
2.5.2

80L SIMON-2
ABSOLUTE ELECTRON-PROTON CROSS SECTIONS
G.G.SIMON,CH.SCHMITT,F.BORKOWSKI,V.H.WALTHER:NUCL.
PHYS.A333(1980)381
2.5.2

81L SIMON
ELASTIC ELECTRIC AND MAGNETIC E-D SCATTERING AT LOW
MOMENTUM TRANSFER
G.G.SIMON,CH.SCHMITT,V.H.WALTHER:NUCL.PHYS.A364(1981)
285
2.5.2

61L SINGH
THEORY OF PI N SCATTERING IN THE STRIP APPROXIMATION
TO THE MANDELSTAM REPRESENTATION
V.SINGH,B.M.UDGAONKAR:PHYS.REV.123(1961)1487
A.6.2

63L SINGH
REGGE POLES IN PI N SCATTERING AND IN PI PI->N ANTI N
V.SINGH:PHYS.REV.129(1963)1889
A.9; 2.6.5

70L SINGH
UNITARITY UPPER AND LOWER BOUNDS ON THE ABSORPTIVE
PARTS OF ELASTIC SCATTERING AMPLITUDES
V.SINGH,S.M.ROY:PHYS.REV.D1(1970)2638
A.5.3

72L SIRLIN
RADIATIVE CORRECTION TO THE GOLDBERGER-TREIMAN
RELATION
A.SIRLIN:PHYS.REV.D5(1972)436
2.5.1

75L SKACHKOV
ASYMPTOTIC BEHAVIOUR OF THE FORM FACTORS AND
INVARIANT DESCRIPTION OF SPATIAL STRUCTURE OF
PARTICLES
N.B.SKACHKOV:TEOR.I.MAT.FIZ 25(1975)313 AND DUBNA E2-
8007(1974)
2.5.2

60L SMITH
LIFE-TIME MATRIX IN COLLISION THEORY
F.T.SMITH:PHYS.REV.119(1960)2098
2.4.1

80L SMITH
MEASUREMENT OF RADIATIVE PI+ SCATTERING BELOW THE
DELTA-RESONANCE
D.E.A. SMITH ET AL.: PHYS.REV.D21(1980)1715
2.1.17

74L SOGARD
RADIATIVE CORRECTIONS IN ELASTIC HADRON SCATTERING
M.R.SOGARD:PHYS.REV.D9(1974)1486
2.1.17

67L SOMMER
EXTENSION OF THE AXIOMATIC ANALYTICITY DOMAIN:PION-
NUCLEON SCATTERING
G.SOMMER:NUOVO CIM.52A(1967)373
A.6.3

74L SPEARMAN
DETERMINATION OF THE DELTA(1236)POLE POSITION
D.SPEARMAN:PHYS.REV.D10(1974)1660
2.4.1

77L SPULLER
A REMEASUREMENT OF THE PANOFSKY RATIO
J.SPULLER ET AL.:PHYS.LETT.67B(1977)479
2.4.7

81L STAHOV
COMMENTS ON THE PI N PHASE SHIFT ANALYSIS OF THE
LENINGRAD GROUP
J.STAHOV:KARLSRUHE PREPRINT TKP 81-5
2.1.8

82L STANLEY
RELATIVISTIC FORM FACTORS FOR HADRONS WITH QUARK
MODEL WAVE FUNCTIONS
D.P.STANLEY,D.ROBSON:PHYS.REV.D26(1982)223
2.5.2

73L STORROW-1
ANALYSIS OF HIGH ENERGY BACKWARD PI N SCATTERING
J.K.STORROW,G.A.WINBOW:NUCL.PHYS.B53(1973)62
2.6.7

73L STORROW-2
AMBIGUITIES IN REGGE ANALYSES OF BACKWARD SCATTERING
J.K.STORROW,G.A.WINBOW:NUCL.PHYS.B54(1973)560
2.6.7

75L STORROW
ANALYSIS OF HIGH ENERGY BACKWARD PI N SCATTERING
INCLUDING POLARIZATION DATA
J.K.STORROW,G.A.WINBOW:J.PHYS.G1(1975)263
2.6.7

70L STRAUSS
DIE PI PI->N ANTI N PARTIALWELLEN IM UNPHYSIKALISCHEN
GEBIET
R.STRAUSS:THESIS,UNIVERSITY OF KARLSRUHE,1970
2.4.6

79L STROTTMANN
MULTIQUARK BARYONS AND THE MIT BAG MODEL
D.STROTTMANN:PHYS.REV.D20(1979)748
2.4.1

61L SUGAWARA
SUBTRACTIONS IN DISPERSION RELATIONS
M.SUGAWARA,A.KANAZAWA:PHYS.REV.123(1961)1895
A.6.3; A.6.4

63L SUGAWARA
PHASE REPRESENTATION OF ANALYTIC FUNCTIONS
M.SUGAWARA,A.TUBIS:PHYS.REV.130(1963)2127
A.6.4; 2.4.3

75L SUKHATME
EXTENSION OF THE DERIVATIVE DISPERSION RELATIONS FOR AMPLITUDE ANALYSES
U.SUKHATME,G.L.KANE,R.BLANKENBECLER,M.DAVIER:PHYS.REV. D12(1975)3431
2.2.1

77L SUKHATME-1
WHERE IS THE DIP STRUCTURE IN PP ELASTIC SCATTERING?
U.P.SUKHATME:PHYS.REV.LETT.38(1977)124
2.6.4

77L SUKHATME-2
U.SUKHATME IN"PROC.12TH RENCONTRE DE MORIOND";VOL.II, P.57;ED.J.TRAN THANH VAN(1977)
2.6.4; 2.6.3

79L SUKHATME
ELASTIC PEAK AND HADRON SIZE FROM T-CHANNEL VIEWPOINT
U.P.SUKHATME,CHUNG-I TOM,J.TRAN THANH VAN:Z.PHYSIK C1(1979)95
2.6.4

77L SVEC
PION-NUCLEON CHARGE-EXCHANGE AMPLITUDES AT FERMILAB ENERGIES
M.SVEC:LETT.NUOVO CIM.18(1977)45
2.6.5

73L TAKAHASHI
AMPLITUDE ANALYSIS OF PION-NUCLEON BACKWARD SCATTERING AT 6 GEV
Y.TAKAHASHI,Y.KOHSAKA:PROGR.THEOR.PHYS.50(1973)1334
2.6.7

80L THEBERGE
PION CORRECTIONS IN THE MIT BAG MODEL: THE (33) RESONANCE
S.THEBERGE,A.W.THOMAS,G.A.MILLER:PHYS.REV.D22(1980) 2838;ERRATUM:D23(1981)2106
2.5.1

82L THEBERGE-1
THE CLOUDY BAG MODEL:IV
S.THEBERGE,G.A.MILLER,A.W.THOMAS:CANAD.J.PHYS.60(1982) 59
2.5.1

82L THEBERGE-2
MAGNETIC MOMENT OF THE NUCLEON OCTET CALCULATED IN THE CLOUDY BAG MODEL
S.THEBERGE,A.W.THOMAS:PHYS.REV.D25(1982)284
2.5.2; 2.5.1

65L THIRRING
ELECTROMAGNETIC PROPERTIES OF HADRONS IN THE STATIC SU(6)MODEL
W.THIRRING:ACTA PHYS.AUSTR.SUPPL.II(1965)205
2.1.17

81L THOMAS-1
LOW-ENERGY PION SCATTERING IN THE CLOUDY BAG MODEL
A.W.THOMAS:J.PHYS.G7(1981)L283
2.5.1

81L THOMAS-2
A NEW SOURCE OF CHARGE SYMMETRY VIOLATION IN THE NUCLEON-NUCLEON SYSTEM
A.W.THOMAS,P.BICKERSTAFF,A.GERSTEN:PHYS.REV.D24(1981) 2539
2.5.3; 2.4.4; 2.4.2

81L THOMAS-3
CLOUDY BAG MODEL OF THE NUCLEON
A.W.THOMAS,S.THEBERGE,G.A.MILLER:PHYS.REV.D24(1981)216
2.5.1; 2.5.2

82L THOMAS
THE CLOUDY BAG MODEL
A.W.THOMAS:CZECH.J.PHYS.B32(1982)239
2.5.1; 2.5.2

56L TOLL
CAUSALITY AND THE DISPERSION RELATION:LOGICAL FOUNDATION
J.S.TOLL:PHYS.REV.104(1956)1760
A.6.4

66L TOMOZAWA
AXIAL-VECTOR COUPLING CONSTANT RENORMALIZATION
Y.TOMOZAWA:NUOVO CIM.46A(1966)707
2.5.1

68L TORNQVIST
IS ISOSPIN INVARIANCE VIOLATED IN PI N BACKWARD SCATTERING?
N.A.TORNQVIST:NUCL.PHYS.B6(1968)187
2.6.7; 2.4.4

74L TROMBORG
ELECTROMAGNETIC CORRECTIONS TO HADRON-HADRON SCATTERING
B.TROMBORG,J.HAMILTON:NUCL.PHYS.B76(1974)483
A.2.3; 2.1.17

76L TROMBORG
ELECTROMAGNETIC CORRECTIONS TO PI+P SCATTERING
B.TROMBORG,S.WALDENSTROM,I.OVERBO:ANN.OF PHYS. 100(1976)1
A.2.3; 2.1.17

77L TROMBORG
ELECTROMAGNETIC CORRECTIONS TO PI N SCATTERING
B.TROMBORG,S.WALDENSTROM,I.OVERBO:PHYS.REV.D15(1977) 725
A.2.3; 2.1.17

78L TROMBORG
ELECTROMAGNETIC CORRECTIONS IN HADRON SCATTERING,WITH APPLICATION TO PI N -> PI N
B.TROMBORG,S.WALDENSTROM,I.OVERBO:HELV.PHYS.ACTA 51(1978)584
A.2.3; 2.1.17

64L TRUEMAN
CROSSING RELATIONS FOR HELICITY AMPLITUDES
T.L.TRUEMAN,G.C.WICK:ANN.OF PHYS.26(1964)322
A.2.1

74L UCHIYAMA
RIGOROUS LOWER BOUNDS ON THE IMAGINARY PARTS OF THE SCATTERING AMPLITUDES AND THE POSITION OF ZEROS
T.UCHIYAMA:PHYS.REV.D10(1974)999
A.5.3

75L VAN HORN
ENERGY-DEPENDENT PARTIAL WAVE ANALYSIS OF K P TO LAMBDA PI ZERO BETWEEN 1540 AND 2215 MEV
A.J.VAN HORN:NUCL.PHYS.B87(1975)157
2.4.3

63L VAN HOVE
AN EXTENSION OF POMERANCHUK'S THEOREM TO DIFFRACTION SCATTERING
L.VAN HOVE:PHYS.LETT.5(1963)252;7(1963)69
A.5.3

77L VAN HOVE
THE DIFFRACTION PATTERN OF PP ELASTIC SCATTERING
L.VAN HOVE:NUCL.PHYS.B122(1977)525
2.6.4

81L VAN ORDEN
SELF CONSISTENT PION-NUCLEON T-MATRIX
J.W.VAN ORDEN,M.K.BANERJEE,D.M.SCHNEIDER,S.J.WALLACE: PHYS.REV.C23(1981)2157
2.5.1

73L VANZHA
ELECTROMAGNETIC EFFECTS IN POLARIZATION IN MESON-NUCLEON SCATTERING
A.P.VANZHA,L.I.LAPIDUS,A.V.TARASOV:SOV.J.NUCL.PHYS. 16(1973)565
2.1.17

76L VASAN
DETERMINATION OF THE POSITION AND RESIDUES OF THE DELTA ++ AND DELTA 0 POLES
S.S.VASAN:NUCL.PHYS.B106(1976)535,526
2.4.1

68L VENEZIANO
CONSTRUCTION OF A CROSSING-SYMMETRIC,REGGE-BEHAVED AMPLITUDE FOR LINEARLY RISING TRAJECTORIES
G.VENEZIANO:NUOVO CIM.57A(1968)190
2.4.3; 2.6.5

80L VENTO
CHIRAL PION DYNAMICS FOR SPHERICAL NUCLEON BAGS
V.VENTO,J.J.G.NYMAN,M.RHO,G.E.BROWN:NUCL.PHYS. A345(1980)413
2.5.1

81L VENTO
DEFORMED BAGS AND THE PI N DELTA COUPLING CONSTANT
V.VENTO,G.BAYM,A.D.JACKSON:PHYS.LETT.102B(1981)97
2.5.1; A.8.1

74L VIANO
THE GLORY EFFECT IN PION-NUCLEON SCATTERING
G.A.VIANO:NUOVO CIM.22A(1974)1
2.6.7

74L VIOLLIER
THE PP PI ZERO COUPLING CONSTANT IN PP-FORWARD DISPERSION RELATIONS
R.D.VIOLLIER,G.R.PLATTNER,K.ALDER:PHYS.LETT.48B(1974) 99
2.4.2

70L VON GEHLEN
THRESHOLD PION ELECTRO- AND PHOTOPRODUCTION
G.VON GEHLEN:UNIV.OF BONN REPORT PI 2-80(1970)
2.5.2

80L VON GEHLEN
ALGEBRAIC PREDICTION OF RESONANCE COUPLINGS FROM ASSUMPTIONS ABOUT ZERO TRAJECTORY SLOPES
G.VON GEHLEN,W.PFEIL:PHYS.REV.D21(1980)2009
2.4.3

79L VON SCHLIPPE
TABLES OF AUVIL-LOVELACE COEFFICIENTS FOR DIFF.CROSS SECTION,POLARIZATION AND SPIN-ROTATION PARAMETERS
W.VON SCHLIPPE:WEST FIELD COLLEGE(LONDON)REPORT(1979)
2.3.1; 2.3.2

77L WAKAIZUMI
AN INDICATION OF 3-QUARK STRUCTURE FROM PP ELASTIC SCATTERING
S.WAKAIZUMI,M.TANIMOTO:PHYS.LETT.70B(1977)55
2.6.3; 2.6.4

78L WAKAIZUMI
HIGH ENERGY PP ELASTIC SCATTERING AND MULTIPLE SCATTERING MODEL
S.WAKAIZUMI:PROGR.THEOR.PHYS.60(1978)1040,1930 AND 65(1981)1112
2.6.3; 2.6.4

82L WAKAIZUMI
ELASTIC ANTIPROTON-PROTON SCATTERING IN THE SPS ENERGY REGION
S.WAKAIZUMI:PHYS.LETT.113B(1982)275
2.6.3; 2.6.4

80L WEI-1
NUCLEON RECOIL EFFECT IN THE PION-NUCLEON INTERACTION
N.WEI,M.K.BANERJEE:PHYS.REV.C22(1980)2052
2.5.1

80L WEI-2
A DYNAMICAL THEORY FOR THE P-WAVE PION-NUCLEON INTERACTION
N.WEI,M.K.BANERJEE:PHYS.REV.C22(1980)2061
2.5.1

66L WEINBERG
PION SCATTERING LENGTHS
S.WEINBERG:PHYS.REV.LETT.17(1966)616
2.5.1; 2.4.6

67L WEINBERG
DYNAMICAL APPROACH TO CURRENT ALGEBRA
S.WEINBERG:PHYS.REV.LETT.18(1967)188
2.5.1

77L WEINBERG
THE PROBLEM OF MASS
S.WEINBERG:TRANSACT.NEW YORK ACAD.SCI 38(1977)185
2.5.1; 2.4.4

82L WEINSTEIN
DO MULTIQUARK HADRONS EXIST?
J.WEINSTEIN,N.ISGUR:TORONTO PREPRINT(1982)
2.4.6

66L WEISBERGER
UNSUBTRACTED DISPERSION RELATIONS AND THE RENORMALIZATION OF THE WEAK AXIAL-VECTOR COUPLING CONSTANTS
W.I.WEISBERGER:PHYS.REV.143(1966)1302
2.5.1

67L WESS
LAGRANGIAN METHOD FOR CHIRAL SYMMETRIES
J.WESS,B.ZUMINO:PHYS.REV.163(1967)1727
2.5.1

68L WEST
COULOMB INTERFERENCE IN HIGH-ENERGY SCATTERING
G.B.WEST,D.R.YENNIE:PHYS.REV.172(1968)1413
2.1.17; A.2.3; 2.6.2

80L WEST
ON THE N-P MASS DIFFERENCE IN QCD
G.B.WEST IN'RECENT DEVELOPMENTS IN HIGH-ENERGY PHYSICS'(ORBIS SCIENTIAE CONF.),P.239;EDS.A. PERLMUTTER,L.F.SCOTT;PLENUM PRESS(1980)
2.1.17

81L WHITE
THE POMERON AND HADRON THROUGH INFRARED ANALYSIS IN QCD
A.R.WHITE,CERN PREPRINT TH 3058
2.6.4

47L WIGNER
HIGHER ANGULAR MOMENTA AND LONG RANGE INTERACTION IN RESONANCE REACTIONS
E.P.WIGNER,L.EISENBUD:PHYS.REV.72(1947)29
2.4.1

48L WIGNER
ON THE BEHAVIOUR OF CROSS SECTIONS NEAR THRESHOLDS
E.P.WIGNER:PHYS.REV.73(1948)1002
2.1.13

55L WIGNER
LOWER LIMIT FOR THE ENERGY DERIVATIVE OF THE SCATTERING PHASE SHIFT
E.P.WIGNER:PHYS.REV.98(1955)145;AM.J.PHYS.23(1955)371
2.4.1

78L WILDE
PION CONDENSATION IN NEUTRON MATTER
B.H.WILDE,S.A.COON,M.D.SCADRON:PHYS.REV.D18(1978)4489
2.5.1

76L WILLRODT
INELASTIC CONTRIBUTIONS TO THE ISOVECTOR NUCLEON FORM FACTORS
J.WILLRODT:NUCL.PHYS.B116(1976)118
2.5.2; 2.4.6

82L WOLOSHYN
CONFINING POTENTIAL AND ANALYTICITY
R.M.WOLOSHYN:J.PHYS.G8(1982)1
A.6.3; 2.4.1

74L WOOLCOCK
CALCULATION OF PI N COUPLING CONSTANT AND S-WAVE
SCATTERING LENGTHS
W.S.WOOLCOCK:NUCL.PHYS.B75(1974)455
2.4.2; 2.4.7

69L WRAY
HARD PIONS,CURRENT AND FIELD COMMUTATORS AND LOW-
ENERGY PI N SCATTERING
D.A.WRAY:ANN.OF PHYS.51(1969)162
2.5.1

79L YOUNG
CHOU-YANG MODEL AND S-CHANNEL HELICITY CONSERVATION
K.YOUNG:PHYS.LETT.83B(1979)403
2.6.3

71L ZARMI
ON THE HELICITY STRUCTURE OF THE P AND P'
CONTRIBUTIONS TO TWO BODY SCATTERING AMPLITUDES
Y.ZARMI:PHYS.REV.D4(1971)3455
2.6.4

80L ZIDELL-1
PION-NUCLEON PHASE SHIFT ANALYSES:0-350 MEV
V.S.ZIDELL,R.A.ARNDT,L.D.ROPER:PHYS.REV.D21(1980)1255
2.1.10

80L ZIDELL-2
DETERMINATION OF PION-NUCLEON SCATTERING LENGTHS FROM
PARTIAL WAVE ANALYSES
V.S.ZIDELL,R.A.ARNDT,L.D.ROPER:PHYS.REV.D21(1980)1289
2.4.7

74L ZIMMERMANN
REMARKS ABOUT COULOMB CORRECTIONS TO PI P ELASTIC
SCATTERING
H.ZIMMERMANN:HELV.PHYS.ACTA 47(1974)130
2.1.17

75L ZIMMERMANN
CHARGE-INDEPENDENT ANALYSIS OF LOW-ENERGY PI N
SCATTERING DATA
H.ZIMMERMANN:HELV.PHYS.ACTA 48(1975)191
2.1.17

75L ZOVKO
DOES THE GLAUBER MODEL FAIL IN LARGE ANGLE PION-
DEUTERON SCATTERING AT HIGH ENERGIES?
N.ZOVKO:NUCL.PHYS.B91(1975)517
2.1.1

64L ZWEIG
AN SU(3)MODEL FOR STRONG INTERACTION SYMMETRY AND ITS
BREAKING
G.ZWEIG:CERN PREPRINT 8419/TH.412;REPRINTED IN 80R
LICHTENBERG-ROSEN
2.4.1; 2.5.1

ADDENDUM TO SECT.3.2

71L ALVAREZ-ESTRADA
ON THE CONSTRUCTION OF SCATTERING AMPLITUDES
FROM EXPERIMENTAL DATA AND ANALYTICITY
R.F.ALVAREZ-ESTRADA:ANN.OF PHYS.68(1971)196
2.1.16; 2.4.3; A.6.3; A.6.4

71L ARIF-UZ-ZAMAN
CONTRIBUTIONS OF SPIN J BARYON POLES TO
PI N -> PI N AND PI N -> GAMMA N AMPLITUDES
ARIF-UZ-ZAMAN:J.MATH.PHYS.12(1971)1286
A.8.1

50L ASHKIN
ON THE SCATTERING OF PI MESONS BY NUCLEONS
J.ASHKIN,A.SIMON,R.MARSHAK:PROGR.THEOR.
PHYS.5(1950)634
A.8.1

63L IGI
TWO VACUUM POLES AND PION-NUCLEON SCATTERING
K.IGI:PHYS.REV.130(1963)820
2.2.1; A.6.3; A.9.1

70L JACOBS
IMPROVED UPPER BOUND ON THE IMAGINARY PART
OF ELASTIC SCATTERING AMPLITUDES
M.A.JACOBS,A.JOSEPH,S.NUSSINOV,A.A.RANGWALA:
PHYS.REV.D2(1970)1970
A.5.3

71L NIELSEN-2
THE INVARIANT AMPLITUDES OF PION-NUCLEON
SCATTERING IN THE UNPHYSICAL REGION AND
HELICITY AMPLITUDES
H.NIELSEN:NUCL.PHYS.B33(1971)152
2.4.5; 2.4.6

70L RAVENHALL
UNITARITY UPPER BOUNDS AND ELASTIC SCATTERING
D.G.RAVENHALL,W.J.PARDEE:PHYS.REV.D2(1970)589
A.5.3

71L SAVIT
ANOTHER BOUND ON THE ABSORPTIVE PART OF THE
ELASTIC SCATTERING AMPLITUDE
R.SAVIT,R.BLANKENBECLER,M.B.EINHORN:
J.MATH.PHYS.12(1971)2092
A.5.3

72L SINGH
ASYMPTOTIC BOUNDS ON THE ABSORPTIVE PARTS OF
THE ELASTIC SCATTERING AMPLITUDES
V.SINGH,A.S.VENGURLAKAR:PHYS.REV.D5(1972)2310
A.5.3

3.3 List of pion-nucleon scattering experiments

The list is an updated version of the first part of Table 1.2.1 in Vol. I/9b1, where one can find the definitions of the abbreviations.

LIST OF EXPERIMENTS
==================

REFERENCE		P1 GEV/C	ANG. RANGE	LITERATURE	
82 ALDER	P-P0	.35-.43	M	ALDER J.C. PR IN PRINT AND SIN PREPRINT	
82 BRICK	D+T+	147.	F	D.BRICK PR D25,2794	LOW ACC
82 BURQ	D-	100.-345.	C	BURQ J.P. PL 109B,111	
82 GHIDIN	D-	10.	B	GHIDINI B.NP B195,12	
81 ASAD	D-	50.	F	ASAD Z. CERN PREPRINT EP/81-26	
81 AUER	D-	2.9,3.3	MB	AUER I.P. PR D24,1771	
81 BAKER	D-	200.	F	BAKER W.F. PRL 47,1683	
81 BEKRE1	P+	.63-.71	MB	BEKRENEV V.S. JETP LETT.34,152	
81 BEKRE2	P-	.57-.73	MB	BEKRENEV V.S. NP A364,515;EXT.OF 80 BEKREN	
81 BINON	D0	40.	F	BINON F. Z.PHYS.C 9,109	LBI/9B
81 BURQ	D-	100.-345.	C	SEE 82 BURQ	
81 COOL	D-D+	100.,200	F	COOL R.L. PR D24,2821	
81 DAUM	D0	63.	F	DAUM C. Z.PHYS.C 8,95	LBI/9B
81 FAJARD	D+D-	70.-200.	CF	FAJARDO L.A. PR D24,46;SEE 80 FAJARD	LBI/9B
81 FIDECA	D+	150.,200.	F	FIDECARO G. CERN,PRIV.COMM.	
81 GORDEE	D+D-	.4-.77	MB	GORDEEV V.A. NP A364,408,EXT.OF 79 GORDEE	LBI/9B
81 MEASDA	T0	.09,.11		MEASDAY D.F. VANCOUVER PREPRINT	
81 SADLER	D+D-	.25-.69	FB	SADLER M. SUBM.TO VERSAILLES CONF.	
81 SCHIZ	D+D-	200.	F	SCHIZ A. PR D24,26; SEE 80 SCHIZ	LBI/9B
80 ADACHI	P0	2.8-4.2	FM	ADACHI T. THESIS,KYOTO UNIV.	LB I/9B
80 ALMAS	D-D+	20.-30.	M	ALMAS R. PL 93B,199	LB I/9B
80 AVVAKU	P0	40.	F	AVVAKUMOV I.A. SERPUCHOV PREPRINT 80-94	LB I/9B
80 AZHINE	D+	32.	F	AZHINENKO I.V. YAD.FIZ.31,648	NO TABLE
80 BAILLO	D-	5.7-13.	MB	BAILLON P. PL 94B,533	NO TABLE
80 BARREL	T-	2.-14.		BARRELET E. PL 94B,541	NO TABLE
80 BEKREN	P-	.57-.69	MB	BEKRENEV V.S. YAD.FIZ.31,173	LB I/9B
80 CROUCH	D0L	1.3-3.8	FB	CROUCH H.R. PR D21,3023	LB I/9B
80 EATON	P0	.35-.41	MB	SEE 82 ALDER	
80 FAJARD	D-D+	70.-200.	CF	SEE 81 FAJARD	
80 FRASCA	D+	.23-.40	B	FRASCARIA R. PL 91B,345	NO TABLE
80 FUKUSH	P-	2.2-3.5	B	FUKUSHIMA M. NP B167,307	LB I/9B
80 JENKIN	D-D+	2.0-9.	M	JENKINS K.A. PR D21,2445,SEE 78JENKIN	2ND PUBL
80 KLINE	P-P+	100.	F	KLINE R.V. PR D22,553 SEE 77AUER	2ND PUBL
80 SCHIZ	D-D+	200.	F	SEE 81 SCHIZ	
80 TERADA	D-	2.-3.5	FB	TERADA S. NP B175,1	LB I/9B
79 ABLEEV	D-	40.	C	ABLEEV V.G. YAD.FIZ.28,1529	LB I/9B
79 ALDER	P-	.19,.35	M	SEE 81 ALDER	
79 APEL	D0	15.-40.	F	APEL W.D. NP B154,189,SEE ALSO 77APEL	LB I/9B
79 APEL	D0			AND AUGENSTEIN K.H.,THESIS,KARLSRUHE(1978)	
79 AULD	D+	.13	FB	AULD E.G. CAN.J.PHYS.57,73	LB I/9B
79 BAKER	D-D+	30.-90.	B	BAKER W.F. PRL 43,1635	LB I/9B
79 CARROL	T-T+	200.-370.		CARROL A.S. PL 80B,423	LB I/9B
79 FRANK	D-D+	.10-.18	MB	FRANK J.S. PROC.HOUSTON CONF.	NO TABLE
79 FUJISA	P+	6.,12.	F	FUJISAKI,M. NP B151,206 AND B155,544 (E)	LB I/9B
79 GORDEE	D-D+	.55-.77	MB	GORDEEV B.A. YAD.FIZ.29,657	LB I/9B
79 KARAMI	D-	1.0-1.2	M	KARAMI H. NP B154,503	LB I/9B
79 MINOWA	P0	1.9-3.0	FB	MINOWA M.,THESIS,KYOTO	LB I/9B
79 SARMA	D-	.66-.72	M	SARMA,H.N.K. NP B161,1	LB I/9B
79 SUZUKI	D0	1.9-3.0	FB	SUZUKI Y.,THESIS,KYOTO	LB I/9B
78 ALDER	P-	.41	M	ALDER,J.C. LETT.AL NUOVO CIM. 23,381	LB I/9B
78 BROWN	P0	.6-2.0	FB	BROWN R.M. NP B144,287	LB I/9B
78 BURQ	D-	30.-140.	C	BURQ J.P. PL 77B,438 AND CERN REPORT	LB I/9B
78 JENKIN	D-D+	2.0-9.	M	JENKINS K.A. PRL 40,425 SEE 80 JENKIN	LB I/9B
78 KRAVTS	D0	.52-.73	F	KRAVTSOV A.V. NP B140,279(EXTR.TO 0 DEG.)	LB I/9B
78 NEFKEN	D-D+	.38-.45	M	NEFKEN B.M.K. LOS ANGELES PREPRINT	LOW ACC
78 PEDRON	T-T+	.16-.49		PEDRONI E. NP A300,321	LB I/9B
78 SCHARR	D-	4.	B	SCHARRE D.L. PR D17,2853	LB I/9B
78 SHARFM	D-	6.	B	SHARFMAN N. PRL 40,681	NO TABLE
77 APEL	D0	40.	F	APEL W.D. PL 72B,132,JETP LETT. 26,205	2ND PUBL
77 APEL	D0			SEE 79APEL	
77 APOKIN	D+T+	42.,52.	C	APOKIN V.D. SJNP 25,51	LB I/9B
77 AUER	P-P+	100.	F	AUER P. PRL 39,313,SEE ALSO 80KLINE	LB I/9B
77 AYRES	D-D+	50.-175.	F	AYRES,D.S. PR D15,3105	LB I/9B

77	BABAEV	D-	25.,38.	B	BABAEV A. PL 67B,351,JETP LETT. 25,367	LB I/9B
77	BARREL	P-	1.2-1.4	M	BARRELET E. PRL 15,2435,PR D15,2435	LB I/9B
77	BRUNET	D-D+	39.,44.	F	BRUNETON C. NP B124,391	LB I/9B
77	DEREVS	P+	45.	F	DEREVSHCHIKOV A.A. SJNP 25,198,S.76GAIDOT	2ND PUBL
77	DUBAL	P+	.41,.43	M	DUBAL L. HELV.PHYS.ACTA 50,815	LB I/9B
77	JACHOL	D-	9.,12.	B	JACHOLSKOWSKI A. NP B126,1	LB I/9B
77	JENEFS	DO	.23-.36	MB	JENEFSKY R.F. NP A290,407	LB I/9B
77	LJUNG	D-	205.	F	LJUNG D. PR D15,3163	LOW ACC
77	OTT	D+	1.2-2.0	MB	OTT R.J. PR D16,2699	LB I/9B
77	REY	D+	4.4-6.	F	REY C.A. PR D15,59 (SLOPES AT 0 DEG.)	NO TABLE
77	RUSS	D-	8.0,16.	F	RUSS J.S. PR D15,3139	LB I/9B
77	STEIN	D+	6.	B	STEIN N.A. PRL 39,378	LB I/9B
77	VAVRA	D-	1.2-3.	MB	VAVRA K. PR D16,2687	LB I/9B
76	AKERLO	D-D+	50.-200.	F	AKERLOF C.W. PRL 35,1406,PR D14,2864	LB I/9B
76	ALITTI	D-	3.9	FB	ALITTI J. NC 33A,160	LB I/9B
76	AMSLER	P+	.19-.3	MB	AMSLER C. LETT.AL NUOVO CIM. 15,209	LB I/9B
76	ANTIPO	D+	29.,43.	F	ANTIPOV M.YU. IHEP 76-95,SERPUCHOV	LB I/9B
76	APOKIN	D-T-	33.-60.	C	APOKIN V.D. NP B106,413 ,SJNP 24,49	LB I/9B
76	AUER	P-	2.9-3.2	MB	AUER P. PRL 37,83,NP B113,279	LB I/9B
76	BAILLO	D-D+	.6-2.	C	BAILLON P. NP B105,365,SEE 74BAILLO	2ND PUBL
76	BARDSL	D-D+	.4-2.1	FB	BARDSLEY D.J. PROC.OXFORD CONF.	NO TABLE
76	BARNES	DO	20.-200	F	BARNES A.V. PRL 37,76	LB I/9B
76	BAYER	DO	.24-.39	F	BAYER W. NUCL.INST.AND METHODS 134,449	LB I/9B
76	BEKREN	P-P+	.57	MB	BEKRENEV V.S. SJNP 24,45	2ND PUB
76	BERTIN	D+	.08-.2	M	BERTIN P.Y. NP B106,341	LB I/9B
76	BIRSA	P-	3.5	B	BIRSA R. NP B117,77	LB I/9B
76	BROWN	DO	.6-2.7	FB	BROWN R.M. NP B117,12 AND B137,542 (E)	LB I/9B
76	BRUNET	R+	45.	F	BRUNETON C. SJNP 24,397,SEE 76PIERRA	LB I/9B
76	BRUNE1	R-	40.	F	BRUNETON C. SJNP 23,409,SEE 75PIERRA	2ND PUBL
76	BRUNE2	P-	40.	F	BRUNETON C. SJNP 23,405,SEE 75GAIDOT	2ND PUBL
76	BURAN	D-	6.2	MB	BURAN T. NP B111,1	LB I/9B
76	CARROL	T-T+	23.-280.		CARROL A.S. PL 61B,303	LB I/9B
76	FIREST	D-T-	360.	F	FIRESTONE A. PR D14,2902	NO TABLE
76	GAIDOT	P+	45.	F	GAIDOT A. PL 61B,103,SEE 77DEREVS	LB I/9B
76	GORDEE	D-D+	.4-.6	MB	GORDEEV V.A. SJNP 24,599	LB I/9B
76	HOFFMA	D-	3.0-5.1	B	HOFFMAN D.W. OXFORD CONF. EXT.OF 75HOFFMA	LB I/9B
76	KISTIA	DOL	1.3-3.8	FB	KISTIAKOWSKY V. TBILISI CONF.,SEE 80CROUCH	2ND PUBL
76	PIERRA	R+	45.	F	PIERRARD J. PL 61B,107,SEE 76BRUNET	2ND PUBL
75	AMSLER	P+	.35	MB	AMSLER C. PL 57B,289	LB I/9B
75	APOKIN	D-	40.,50.	C	APOKIN V.D. SJNP 21,640,PL 56B,391	LB I/9B
75	AYRES	D-D+	50.-175.	F	AYRES D.S. PRL 35,1195,SEE 77AYRES	2ND PUBL
75	BAGLIN	D+	10.	FB	BAGLIN C. NP B98,365 (EXT.OF 73BAGLIN)	LB I/9B
75	BARBER	D-	1.0-1.4	FB	BARBER P.C. NP B84,109	LB I/9B
75	BEKREN	P-P+	.57	M	BEKRENEV V.S. JETP LETT.21,282	LB I/9
75	BLAISI	D-	17.	F	BLAISING J.J. PL 58B,121	LOW ACC
75	BRANDE	D-D+	10.	F	BRANDENBURG G. PL 58B,367 AND PL 59B,313	LB I/9B
75	COMISO	DO	.24-.4	M	COMISO J.C. PR D12,738	LB I/9B
75	DEBENH	D-DO	.6-1.	B	DEBENHAM N.C. PR D12,2545	LB I/9B
75	GAIDOT	P-	40.	F	GAIDOT A. PL 57B,389	LB I/9B
75	HOFFMA	D-	3.0,5.1	B	HOFFMAN D.W. PRL 35,138	LB I/9B
75	LENNOX	D+	2.0-6.	B	LENNOX A.J. PR D11,1777	LB I/9B
75	MARTIN	P+	.6-2.6	MB	MARTIN J.F. NP B89,253	LB I/9B
75	MARZO	DO	2.6-8.	B	DE MARZO C. PL 56B,487	LB I/9B
75	MICHAE	D+	3.7,7.	MB	MICHAEL W. PRL 35,193	NO TABLE
75	PIERRA	R-	40.	F	PIERRARD J. PL 57B,393	LB I/9B
74	ABE	D+	1.5-2.	M	ABE K. PR D10,3556	LB I/9B
74	AMBATS	D-D+	3.0-6.	F	AMBATS I. PR D9,1179	LB I/9B
74	ANTIPO	D-	25.,40.	F	ANTIPOV Y.M. SJNP 18,182	LB I/9B
74	BAILLO	D-D+	.6-2.	C	BAILLON P. PL 50B,387, CERN 75-10	LB I/9B
74	BAKER	D+	2.0-6.0	B	BAKER W.F. PRL 32,251,SEE 75LENNOX	2ND PUBL
74	BASHIA	D+	3.2-10.	B	BASHIAN A. PR D9,3193	LB I/9B
74	BERTHO	D+	1.2	FB	BERTHON A. NP B81,431	LB I/9B
74	BOLOTO	DO	20.-50.	F	BOLOTOV V.N. NP B73,365,SJNP 18,538	LB I/9B
74	BROCKE	DO	5.9	M	BROCKETT W.S. PL 51B,390	LB I/9B
74	CHEZE	DO	.47-.5	B	CHEZE J.B. NP B72,365	LB I/9B
74	DEREVG	D-	50.	F	DEREVSHCHIKOV A. A. PL 48B,367	LB I/9B
74	DEREVM	D-	35.-55.	F	DEREVSHCHIKOV A.A. NP B80,442	LB I/9B
74	DOLBEA	D-	.9-1.2	FB	DOLBEAU J. NP B78,233	LB I/9B
74	FELTES	DO	1.8,2.	MB	FELTESSE J. CEA-N-1838,THESIS(PARIS)	LB I/9B

Addendum to Sect. 3.3

82	ASAD	D−D+	20,50	F	Asad, Z. et al.: CERN-EP/82-115, subm. to Phys. Lett. B
82	BURQ	D−	30–345	C	Burq, J.P. et al.: CERN-EP/82, subm. to Nucl. Phys. B

Appendix

Collection of Pion-Nucleon Scattering Formulas

1 Kinematics

1.1 Notation

As mentioned in the Introduction, we have used in Chap. 1 the general notation for two-body reactions as defined in Vols. I/7 and I/9a, whereas the formulas in Chap. 2 and in the Appendix are written in a special notation for pion-nucleon scattering, which avoids unnecessary complications and essentially agrees with that used in all theoretical papers in this field. For some quantities, the choice of the symbols is standardized, in other cases we have preferred the symbols used most frequently in the literature.

The following table shows the comparison between the two notations

Table A.1.1. Notation.

General notation Chap. 1 and Vol. I/9a	Special notation Chap. 2 and Appendix	Quantity
$m_1=m_3$	μ	pion mass
$m_2=m_4$	m	nucleon mass
$\|p_1^*\|=\|p_3^*\|$, $E_1^*=E_3^*$, θ_3^*	q, ω_q, θ	momentum, energy and scattering angle of the pion in the c.m. system
$E_2^*=E_4^*$	E	energy of the nucleon in the c.m. system
E^*	$W=\sqrt{s}$	total energy in the c.m. system
$\|p_1\|$, E_1	k, ω	momentum and energy of the incoming pion in the lab. system
$\|p_3\|$, E_3, θ_3	k_π, ω_π, θ_π	momentum, energy and scattering angle of the outgoing pion in the lab. system
$\|p_4\|$, E_4, θ_4	k_N, ω_N, θ_N	momentum, energy and scattering angle of the recoil nucleon in the lab. system
		differential cross section in the:
$d\sigma/d\Omega^*$	$d\sigma/d\Omega$	c.m. system
$d\sigma/d\Omega$	$d\sigma/d\Omega_{lab}$	lab. system

Omission of brackets: $\sin\alpha\beta=\sin(\alpha\beta)$; $a+b/xyz+c=a+[b/(xyz)]+c$

The four-vector momenta of the nucleon are denoted by $\mathbf{p},\mathbf{p}'$ and those of the pion by $\mathbf{q},\mathbf{q}'$ (Fig. A.1.1). The energy-momentum conservation law reads

$$\mathbf{p}+\mathbf{q}=\mathbf{p}'+\mathbf{q}'.$$

We are using units such that $\hbar=c=1$. With our choice of the metric, the mass shell restrictions are

$$\mathbf{p}^2=\mathbf{p}'^2=m^2;\quad \mathbf{q}^2=\mathbf{q}'^2=\mu^2. \tag{A.1.1}$$

Since only two of the kinematical variables are independent, there exists a relation between the *Mandelstam variables* s,t,u

$$s=(\mathbf{p}+\mathbf{q})^2;\quad t=(\mathbf{q}-\mathbf{q}')^2;\quad u=(\mathbf{p}'-\mathbf{q})^2;\quad s+t+u=2m^2+2\mu^2=\Sigma=92.4\mu^2=1.80\,\mathrm{GeV}^2. \tag{A.1.2}$$

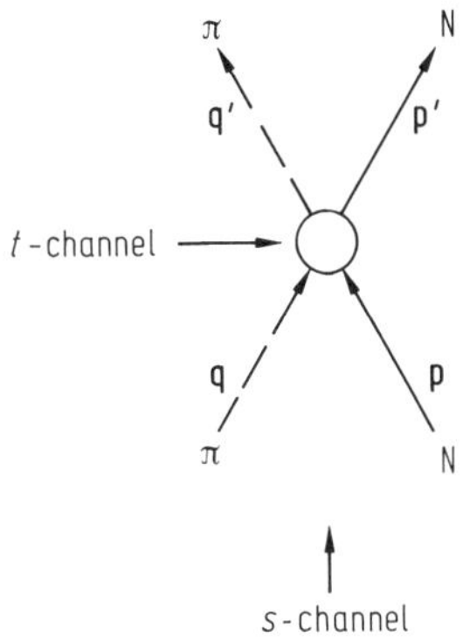

Fig. A.1.1. Notation for 4-vector momenta.

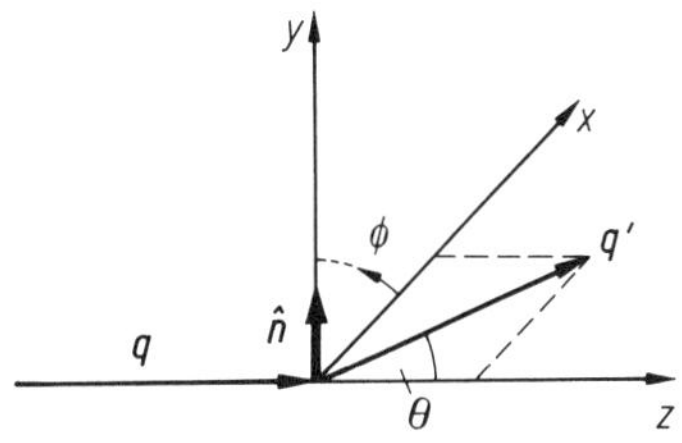

Fig. A.1.2. Definition of the polar angles. $\boldsymbol{q}$ and $\boldsymbol{q}'$ are the momentum vectors of the incoming and outgoing pions in the c.m. frame. The scattering plane is the $x-z$-plane. If the vector $\boldsymbol{q}''$ of another outgoing pion is considered, the azimuthal angle ϕ is the angle between the x-axis and the projection of $\boldsymbol{q}''$ onto the $x-y$-plane. Our scattering amplitudes do not depend on ϕ.

Crossing properties can be expressed simply in terms of the variable

$$v=(s-u)/4m=\omega+t/4m. \tag{A.1.3}$$

In some applications, it is useful to consider combinations of the four-momenta

$$\mathbf{P}=\tfrac{1}{2}(\mathbf{p}+\mathbf{p}'); \quad \mathbf{Q}=\tfrac{1}{2}(\mathbf{q}+\mathbf{q}'); \quad \boldsymbol{\varkappa}=\tfrac{1}{2}(\mathbf{q}-\mathbf{q}'), \tag{A.1.4}$$

which are related to the Mandelstam variables as follows

$$s=(\mathbf{P}+\mathbf{Q})^2; \quad t=4\boldsymbol{\varkappa}^2=2(\mu^2-\mathbf{q}\cdot\mathbf{q}'); \quad u=(\mathbf{P}-\mathbf{Q})^2, \tag{A.1.5}$$

and fulfill the relations

$$\mathbf{P}\cdot\boldsymbol{\varkappa}=\mathbf{Q}\cdot\boldsymbol{\varkappa}=0; \quad \mathbf{P}^2+\boldsymbol{\varkappa}^2=m^2; \quad \mathbf{Q}^2+\boldsymbol{\varkappa}^2=\mu^2; \quad \mathbf{P}\cdot\mathbf{Q}=mv=\tfrac{1}{2}\mathbf{q}'\cdot(\mathbf{p}+\mathbf{p}')=\tfrac{1}{2}\mathbf{q}\cdot(\mathbf{p}+\mathbf{p}'). \tag{A.1.6}$$

The usual choice for the pair of independent variables is (s,t) or (v,t), but for the discussion of low-energy theorems the pair (v, v_B) where

$$v_B=-\frac{1}{2m}\mathbf{q}\cdot\mathbf{q}'=\frac{1}{4m}(t-2\mu^2)=\frac{1}{4m}(t-\mathbf{q}^2-\mathbf{q}'^2) \tag{A.1.7}$$

is preferable.

Notation for 3-vectors: $\mathbf{q}=(q_0, \boldsymbol{q})$ etc. Normal unit vector on the scattering plane ($x-z$-plane) (Fig. A.1.2)

$$\hat{\boldsymbol{n}}=\boldsymbol{q}\times\boldsymbol{q}'/|\boldsymbol{q}\times\boldsymbol{q}'|. \tag{A.1.8}$$

$\hat{\boldsymbol{n}}$ has the direction of the y-axis.

1.2 Kinematics in the center of mass frame

Momentum conservation:

$$\boldsymbol{p}+\boldsymbol{q}=\boldsymbol{p}'+\boldsymbol{q}'=0. \tag{A.1.9}$$

The momenta and the energies are given by

$$q\equiv|\boldsymbol{q}|=|\boldsymbol{p}|=|\boldsymbol{q}'|=|\boldsymbol{p}'|; \quad \omega_q=\sqrt{\mu^2+q^2}; \quad E=\sqrt{m^2+q^2}. \tag{A.1.10}$$

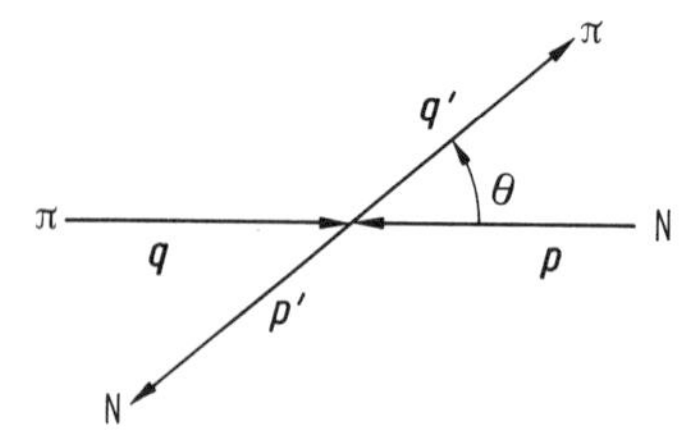

Fig. A.1.3. Variables in the center of mass frame.

1.2.1 Relations between energy and momentum variables

$$\begin{aligned}
&\sqrt{s}=W=E+\omega_q\,; && 2W(E\pm m)=(W\pm m)^2-\mu^2\,.\\
&(E+m\pm q)^2=2(E+m)(E\pm q)\,; && s-m^2+\mu^2=2W\omega_q\,;\\
&\pm q(E+m\pm q)=(E-m\pm q)(E+m)\,; && s+m^2-\mu^2=2WE\,;\\
&\sqrt{2(E\pm m)}=\sqrt{E+q}\pm\sqrt{E-q}\,; && 1\pm q/(E+m)=\sqrt{2(E\pm q)}/\sqrt{E+m}\,.
\end{aligned} \tag{A.1.11}$$

$$\begin{aligned}
\lambda(s,m^2,\mu^2)&\equiv s^2+m^4+\mu^4-2sm^2-2s\mu^2-2m^2\mu^2=s^2\left\{1-\frac{2}{s}(m^2+\mu^2)+\frac{1}{s^2}(m^2-\mu^2)^2\right\}=4sq^2\\
&=\{s-(m+\mu)^2\}\{s-(m-\mu)^2\}=\{(W+m)^2-\mu^2\}\{(W-m)^2-\mu^2\}\\
&=\{(W+\mu)^2-m^2\}\{(W-\mu)^2-m^2\}=4\{(\mathbf{p}\cdot\mathbf{q})^2-m^2\mu^2\}\,.\\
4q^2&=s-\Sigma+(m^2-\mu^2)^2/s.
\end{aligned} \tag{A.1.12}$$

For a given value of q, we have two solutions for s

$$s_\pm=2q^2+m^2+\mu^2\pm2\sqrt{q^4+q^2(m^2+\mu^2)+m^2\mu^2}\,, \tag{A.1.13}$$

which fulfill $s_+s_-=(m^2-\mu^2)^2$. The mapping of the q^2-plane onto the s-plane is described in Fig. A.1.4.

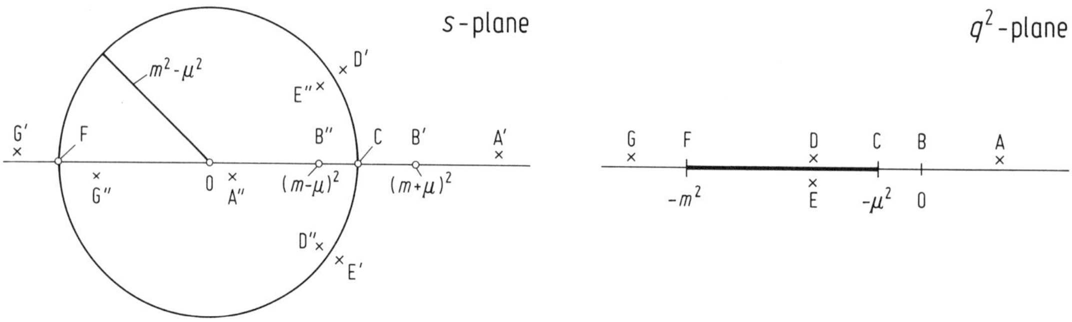

Fig. A.1.4. The mapping of the s-plane onto the q^2-plane.
The exterior of the circle of radius $m^2-\mu^2$ in the s-plane, which includes the s-channel physical region, is mapped onto the first sheet of the q^2-plane, the circle being mapped onto the segment $-m^2<q^2<-\mu^2$ along which the q^2-plane has its cut. The interior of the circle is mapped onto the second sheet. It includes part of the u-channel physical region. $q^2=0$ is mapped onto the s- and u-channel thresholds s_{th}, u_{th}. The points A′, A″ in the s-plane are mapped onto the points A in the first and second sheet of the q^2-plane respectively, etc.

1.2.2 Relations which include the scattering angle or t

$$\begin{aligned}
&\cos\theta=1+t/2q^2\,;\quad \cos\theta=-1+\frac{2[(m^2-\mu^2)^2-us]}{s^2-2s(m^2+\mu^2)+(m^2-\mu^2)^2}\,; && \cos(\theta/2)=\sqrt{1+t/4q^2}\,;\\
&\sin\theta=\sqrt{-t(1+t/4q^2)}/q\,; && \sin(\theta/2)=\sqrt{-t}/2q\,.
\end{aligned} \tag{A.1.14}$$

The quantity

$$-\Phi/t=(m^2-\mu^2)^2-su=(4q^2+t)s=4q^2s\cos^2(\theta/2) \tag{A.1.15}$$

is of interest, because the zero of the "Kibble function" Φ determines the boundary of the physical region (see Fig. A.1.5).

Transverse and longitudinal momenta

$$p_T^2=(q\sin\theta)^2=-t(1+t/4q^2)\,; \qquad p_L=q\cos\theta=q(1+t/2q^2)\,,$$
$$q^2=p_T^2+p_L^2\,; \qquad t=2\{p_L\sqrt{p_L^2+p_T^2}-(p_L^2+p_T^2)\}\,. \tag{A.1.16}$$

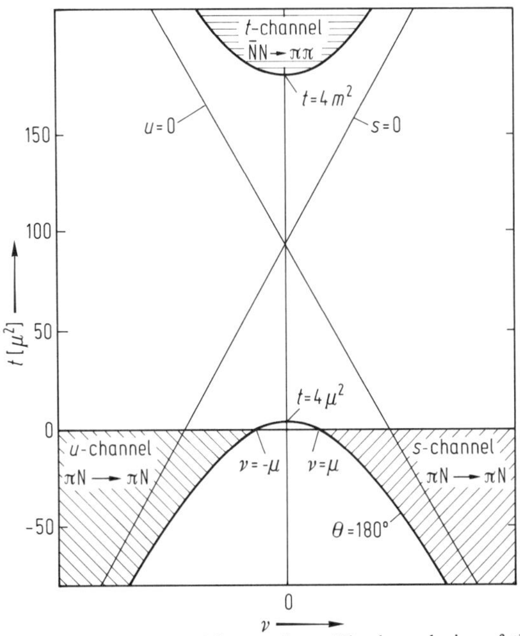

Fig. A.1.5. The Mandelstam plane. The boundaries of the physical regions are given by the curve along which the Kibble function Φ Eq. (A.1.15), is zero. The abscissa in this diagram is $\nu=(s-u)/4m$.

1.3 Relations between center of mass and laboratory frame variables

Velocity of the center of mass in the lab. frame:

$$\beta_{\rm cm}=\frac{k}{m+\omega}=\frac{q}{E}\,; \qquad \gamma_{\rm cm}=\frac{1}{\sqrt{1-\beta_{\rm cm}^2}}=\frac{E}{m}\,. \tag{A.1.17}$$

Velocity of the particles: $\beta=q/\omega_q$ or k_π/ω_π etc. Momentum conservation in the lab. system: $\boldsymbol{k}=\boldsymbol{k}_\pi+\boldsymbol{k}_{\rm N}$. Energies of the particles and energy conservation in the lab. system (T_π = kinetic energy of the pion)

$$\omega=\sqrt{\mu^2+k^2}=\mu+T_\pi\,; \quad \omega_\pi=\sqrt{\mu^2+k_\pi^2}\,; \quad \omega_{\rm N}=\sqrt{m^2+k_{\rm N}^2}\,; \quad m+\omega=\omega_{\rm N}+\omega_\pi\,. \tag{A.1.18}$$

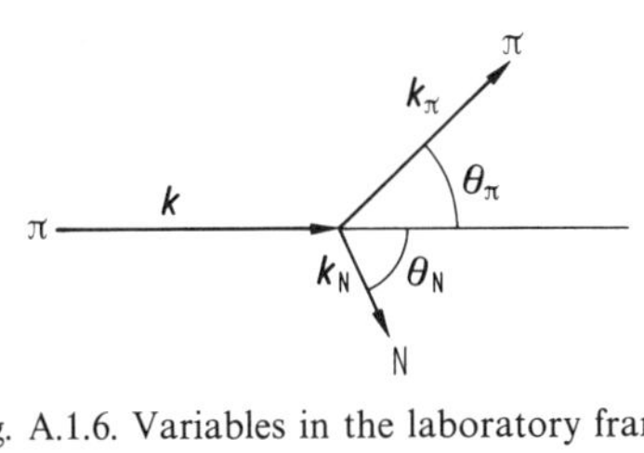

Fig. A.1.6. Variables in the laboratory frame.

1.3.1 Relations between energy and momentum variables

$$\frac{q}{k}=\frac{m}{W}=\frac{E}{m+\omega}=\frac{E\pm m}{\omega\pm W+m}=\frac{W-E}{\omega}-\frac{\mu^2}{\omega W}. \tag{A.1.19}$$

$$m\omega=EW-m^2=W^2-WE-\mu^2=W\omega_q-\mu^2=E\omega_q+q^2=(E+m)W-(W+m)m=\tfrac{1}{2}(W^2-m^2-\mu^2) \tag{A.1.20}$$

$$m(\omega\pm W\mp E)=(W\mp m)(E\pm m)\,;\qquad E\omega=m\omega_q+kq=mW-mE+kq\,. \tag{A.1.21}$$

The following table is useful for kinematical calculations on a computer:

Table A.1.2. Relations between energy and momentum variables. In our numerical calculations we have always used the proton mass and the charged pion mass ($m=938.28$ MeV, $\mu=139.57$ MeV).

	k	ω	q	s
$k=$	k	$\sqrt{\omega^2-\mu^2}$	$\frac{q}{m}\{\sqrt{m^2+q^2}+\sqrt{\mu^2+q^2}\}$	$\frac{\sqrt{(s-(m+\mu)^2)(s-(m-\mu)^2)}}{2m}$
$\omega=$	$\sqrt{\mu^2+k^2}$	ω	$\frac{q^2+\sqrt{(\mu^2+q^2)(m^2+q^2)}}{m}$	$\frac{s-m^2-\mu^2}{2m}$
$q^2=$	$\frac{k^2m^2}{m^2+\mu^2+2m\sqrt{\mu^2+k^2}}$	$\frac{m^2(\omega^2-\mu^2)}{m^2+\mu^2+2m\omega}$	q^2	$\frac{s}{4}-\frac{m^2+\mu^2}{2}+\frac{(m^2-\mu^2)^2}{4s}$
$s=$	$m^2+\mu^2+2m\sqrt{\mu^2+k^2}$	$m^2+\mu^2+2m\omega$	$\{\sqrt{m^2+q^2}+\sqrt{\mu^2+q^2}\}^2$	s
$E=$	$\frac{m^2+m\sqrt{k^2+\mu^2}}{\sqrt{m^2+\mu^2+2m\sqrt{\mu^2+k^2}}}$	$\frac{m(m+\omega)}{\sqrt{m^2+\mu^2+2m\omega}}$	$\sqrt{m^2+q^2}$	$\frac{s+m^2-\mu^2}{2\sqrt{s}}$
$\omega_q=$	$\frac{\mu^2+m\sqrt{\mu^2+k^2}}{\sqrt{m^2+\mu^2+2m\sqrt{\mu^2+k^2}}}$	$\frac{\mu^2+m\omega}{\sqrt{m^2+\mu^2+2m\omega}}$	$\sqrt{\mu^2+q^2}$	$\frac{s-m^2+\mu^2}{2\sqrt{s}}$

1.3.2 Relations which include the scattering angle or *t*

$$k_N=\sqrt{-t(4m^2-t)/4m^2}\,;\qquad \omega_N=m-t/2m\,;$$
$$k_\pi=\sqrt{k^2+\omega t/m+t^2/4m^2}\,;\qquad \omega_\pi=\omega+t/2m=\frac{m^2+\mu^2-u}{2m}. \tag{A.1.22}$$

$$\sqrt{4m^2-t}\,\sin\theta_N=2m\cos(\theta/2)\,;\qquad \tan\theta_N=(m/E)\cot(\theta/2)\,;$$
$$\sqrt{4m^2-t}\,\cos\theta_N=2E\sin(\theta/2)\,;\qquad \tan\theta_\pi=\frac{m\sin\theta}{E\cos\theta+\omega_q}\,; \tag{A.1.23}$$
$$\tan\theta_\pi=\frac{2mW\sin\theta}{(s-m^2+\mu^2)+(s+m^2-\mu^2)\cos\theta}.$$

Approximation for small-angle scattering

$$k\theta_\pi\approx q\theta\approx\sqrt{-t}\,. \tag{A.1.24}$$

Further relations

$$k_\pi^2=\lambda(u,m^2,\mu^2)/4m^2\,;\qquad k^2=\lambda(s,m^2,\mu^2)/4m^2\,; \tag{A.1.25}$$

$$k_\pi\cos\theta_\pi=k+\frac{t}{2k}\left(1+\frac{\omega}{m}\right);\qquad k_N\sin\theta_N=q\sin\theta\,. \tag{A.1.26}$$

The "triangle function" λ is defined in (A.1.12).

Relation between the differential cross sections in the center of mass and laboratory systems

$$\frac{d\Omega_{lab}}{d\Omega}=\frac{d\cos\theta_\pi}{d\cos\theta}=\frac{E}{m}\frac{\sin^3\theta_\pi}{\sin^3\theta}\left(1+\frac{\omega_q}{E}\cos\theta\right)=k\cdot k_\pi^{-3}(q^2+t\omega_q/2W). \qquad \text{(A.1.27)}$$

1.3.3 Difference and differential formulas

$$\frac{4(q^2-q'^2)}{s-s'}=1-\frac{(m^2-\mu^2)^2}{ss'}; \qquad \frac{2(E+E')}{W+W'}=1+\frac{m^2-\mu^2}{WW'};$$

$$\frac{2(E-E')}{s-s'}=(W+W')^{-1}\left(1-\frac{m^2-\mu^2}{WW'}\right); \qquad \frac{EW-E'W'}{s-s'}=\frac{m(\omega-\omega')}{s-s'}=\frac{1}{2}; \qquad \text{(A.1.28)}$$

$$\frac{1}{\nu'-\nu}=\frac{2m}{s'-s}; \qquad \frac{1}{\nu'+\nu}=\frac{2m}{s'-u}.$$

$$d\omega/k=dk/\omega=dW/q; \qquad ds=2m\,d\omega=(s/E\omega_q)dq^2. \qquad \text{(A.1.29)}$$

1.3.4 Scattering at fixed laboratory scattering angle θ_π

This case is of interest for applications of interior dispersion relations [see Sect. 2.4.2 and (A.6.7)]. θ_π is constant along the hyperbolas $(s-a)(u-a)=\lambda(a,m^2,\mu^2)$ in the Mandelstam plane (ν,t-plane). See Eq. (A.1.12) for the definition of the triangle function λ and Sect. A.1.4 for the definition of the t-channel variables p_t, q_t, θ_t.

$$\tan\theta_\pi=-\frac{2m\sqrt{-a}}{m^2-\mu^2+a}; \qquad \cos\theta=\frac{a+s}{a-s}; \qquad \cos^2\theta_t=1+\frac{at}{4(p_tq_t)^2};$$

$$4sq^2\equiv\lambda(s,m^2,\mu^2)=(a-s)t; \qquad \Phi(s,t)=-at^2 \qquad \text{(A.1.30)}$$

$$4m\nu=\{t^2-4(m^2+\mu^2)t+16m^2\mu^2+4at\}^{1/2}=4p_tq_t\cos\theta_t.$$

The Kibble function Φ is defined in Eq. (A.1.15).

Points (s',t',u') and (s,t,u) along a given hyperbola are related by

$$(t'-t)(s'-a)=-(s'-s)(s'-u); \qquad \frac{ds'}{dt'}=\frac{a-s'}{4m\nu'}. \qquad \text{(A.1.31)}$$

It is interesting to mention the special case of backward scattering, for which the formulas read

$$t=-4q^2; \qquad m\nu=E\omega_q=\omega m-q^2=\frac{s^2-(m^2-\mu^2)^2}{4s}; \qquad u_b=(m^2-\mu^2)^2/s=(E-\omega_q)^2; \qquad \text{(A.1.32)}$$

$$\left\{\frac{1}{s'-s}+\frac{1}{s'-u}\right\}ds'=\left\{\frac{1}{E'\omega_q'}+\frac{1}{q'^2-q^2}\right\}dq'^2; \qquad \left\{\frac{1}{s'-s}-\frac{1}{s'-u}\right\}ds'=\frac{dq'^2}{q'^2-q^2}\frac{E\omega_q}{E'\omega_q'}. \qquad \text{(A.1.33)}$$

1.4 Kinematics of the t-channel reaction $\pi\pi\to N\bar{N}$

The location of the physical region is shown in the Mandelstam diagram Fig. A.1.5. Amplitudes in the unphysical region $t<4m^2$ are of interest for many applications.

We consider the center-of-mass frame only. All square roots are taken of positive quantities.

Total energy: $\sqrt{t}$.

Momentum conservation: $\boldsymbol{q}_t+\boldsymbol{q}_t'=\boldsymbol{p}_t+\boldsymbol{p}_t'=0$.

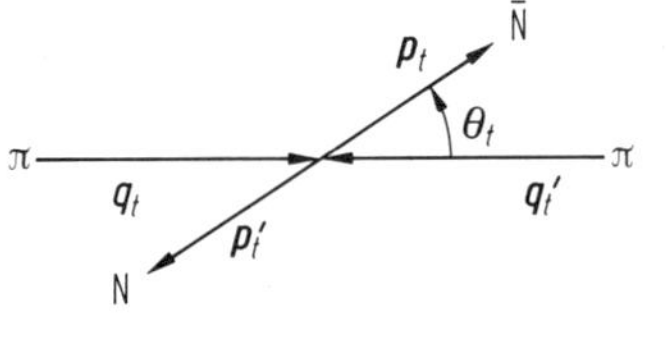

Fig. A.1.7. Variables for the t-channel reaction in the center of mass system.

1.4.1 Physical region $t \geq 4m^2$

	momentum	energy		
			$s = -p_t^2 - q_t^2 + 2p_t q_t \cos\theta_t$	
pion	$q_t = \sqrt{t/4 - \mu^2}$	$\frac{1}{2}\sqrt{t}$	$u = -p_t^2 - q_t^2 - 2p_t q_t \cos\theta_t$	(A.1.34)
nucleon	$p_t = \sqrt{t/4 - m^2}$	$\frac{1}{2}\sqrt{t}$	$t = 4(p_t^2 + m^2) = 4(q_t^2 + \mu^2)$	

Table A.1.3. Relations between s- and t-channel variables. λ is positive in the s- and t-channel physical regions. Other expressions for $\lambda(s, m^2, \mu^2)$ are given in Eqs. (A.1.12) and (A.1.25) and for Φ in Eq. (A.1.15).

	t-channel variables	s-channel variables
$\sqrt{\lambda}$	$= [(2mq_t \sin\theta_t)^2 + t(p_t - q_t \cos\theta_t)^2]^{1/2}$	$= 2qW;$
$\sqrt{\Phi}$	$= 2\sqrt{t}\, p_t q_t \sin\theta_t$	$= 2q^2 W \sin\theta;$
$\sqrt{\tau}$	$= 2\sqrt{t}\, p_t$	$= 4p_- q \sin(\theta/2)$
$s + m^2 - \mu^2$	$= -2p_t(p_t - q_t \cos\theta_t)$	$= 2WE$
$s - u$	$= 4p_t q_t \cos\theta_t$	$= 4m\nu = 4m\omega + t$

1.4.2 Unphysical region $t < 4m^2$

$$t<4m^2 \quad p_- \equiv \sqrt{m^2 - t/4} = -\mathrm{i}p_t; \qquad \Phi = -4t\{(p_- q_t)^2 + m^2\nu^2\} \tag{A.1.35}$$

$$t<4\mu^2 \quad q_- \equiv \sqrt{\mu^2 - t/4} = -\mathrm{i}q_t; \qquad p_- q_- = -p_t q_t$$

$$s = p_-^2 + q_-^2 - 2p_- q_- \cos\theta_t; \qquad \cos\theta_t = -\frac{m\nu}{p_- q_-} = -\frac{s-u}{4p_- q_-} \tag{A.1.36}$$

$$u = p_-^2 + q_-^2 + 2p_- q_- \cos\theta_t;$$

s-channel backward scattering: $\quad q_- = \omega_q, \quad p_- = E.$

If one expresses the kinematic variables in terms of Mandelstam variables s, t, u and considers a continuation from the s-channel physical region to the t-channel physical region, some signs depend on the path of continuation, i.e. they have to be fixed by a convention. We have adopted the convention of the book 70R Martin-Spearman who used the Trueman-Wick convention.

1.5 u-channel kinematics

The expressions for kinematical quantities in the u-channel are needed in discussions of the contribution of u-channel exchanges to s-channel near-backward scattering (Sect. A.9.2). They are obtained by the operation $s \leftrightarrow u$ at fixed t. $W_u = \sqrt{u}$ is the total c.m. energy for the u-channel reaction.

$$E_u = \sqrt{m^2 + q_u^2} = (u + m^2 - \mu^2)/2W_u; \qquad \omega_u = \sqrt{\mu^2 + q_u^2} = (u - m^2 + \mu^2)/2W_u; \tag{A.1.37}$$

$$q_u^2 = \frac{u}{4} - \frac{1}{2}(m^2 + \mu^2) + \frac{1}{4u}(m^2 - \mu^2)^2; \qquad W_u = E_u + \omega_u; \tag{A.1.38}$$

$$\cos\theta_u = 1 + \frac{t}{2q_u^2} = -1 + \frac{2\{(m^2 - \mu^2)^2 - us\}}{u^2 - 2u(m^2 + \mu^2) + (m^2 - \mu^2)^2}. \tag{A.1.39}$$

1.6 The brick-wall frame

In derivations of fixed-t dispersion relations one works in a special Lorentz frame, called brick-wall or Breit frame. It has the property that both the pion and the nucleon behave as if they had bounced off a brick-wall

$$\boldsymbol{p}' = -\boldsymbol{p}; \qquad \boldsymbol{q}' - \boldsymbol{q} = 2\boldsymbol{p}; \qquad |\boldsymbol{q}| = |\boldsymbol{q}'|. \tag{A.1.40}$$

For each particle, the energy of the incoming state is the same as that of the outgoing state.

At fixed t, the nucleon momentum $|\boldsymbol{p}|$ is fixed and the pion energy is a linear function of s

$$t=-4|\boldsymbol{p}|^2\,;\quad E_\pi=\sqrt{\mu^2+|\boldsymbol{q}|^2}=\frac{s+m^2-\mu^2}{2E_p}-E_p\,;\quad E_p=\sqrt{m^2+|\boldsymbol{p}|^2}\,. \tag{A.1.41}$$

For forward scattering, the brick-wall frame agrees with the laboratory frame.

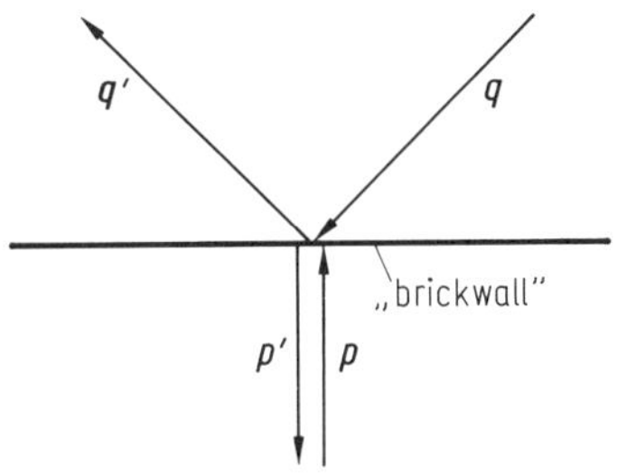

Fig. A.1.8. The brickwall frame is the center of mass frame of the initial and final nucleon. Both the nucleon and the pion behave as if they had bounced off a brickwall.

2 Scattering amplitudes

2.1 The spin structure

If one assumes Lorentz invariance and parity conservation, the most general T-matrix element for one of the elastic or charge-exchange scattering reactions contains two independent functions of the kinematic variables (s,t) which are called scattering amplitudes. It is convenient to introduce several pairs of amplitudes, each of them defined such that the formulas in a certain branch of pion-nucleon phenomenology become reasonably short and easy to survey.

2.1.1 Spin no-flip and flip amplitudes $G(s,t)$ and $H(s,t)$

The T-matrix element in the center-of-mass frame is expressed in terms of a matrix $M_{\mathrm{fi}}(s,t)$

$$\frac{\langle\mathrm{out}|\mathrm{T}|\mathrm{in}\rangle}{q/2\pi}=M_{\mathrm{fi}}(s,t)=\chi_{\mathrm{f}}^{+}(\hat{\boldsymbol{e}}')\mathsf{M}\chi_{\mathrm{i}}(\hat{\boldsymbol{e}})\,, \tag{A.2.1}$$

$$\begin{aligned}\mathsf{M}(s,t)&=f_1(s,t)+f_2(s,t)(\boldsymbol{\sigma}\cdot\hat{\boldsymbol{q}}')(\boldsymbol{\sigma}\cdot\hat{\boldsymbol{q}})\\&=G(s,t)+H(s,t)\,\mathrm{i}(\boldsymbol{\sigma}\cdot\hat{\boldsymbol{n}})=\begin{pmatrix}G & H\\ -H & G\end{pmatrix}.\end{aligned} \tag{A.2.2}$$

$\chi_{\mathrm{f}}^{+}(\hat{\boldsymbol{e}}')$ and $\chi_{\mathrm{i}}(\hat{\boldsymbol{e}})$ are rest frame Pauli spinors for the final and initial nucleons respectively, with spin quantized along the directions $\hat{\boldsymbol{e}}',\hat{\boldsymbol{e}}$. In the present case, $\hat{\boldsymbol{e}}'=\hat{\boldsymbol{e}}=e_z$ is the direction of the incoming pion. i and f take the values $\pm$ 1/2. $\boldsymbol{\sigma}=(\sigma_x,\sigma_y,\sigma_z)$ is a vector of Pauli matrices and $\hat{\boldsymbol{n}}$ is the normal to the scattering plane, Eq. (A.1.8).

The normalization is chosen such that the differential cross section in the center-of-mass system is given by

$$\frac{\mathrm{d}\sigma}{\mathrm{d}\Omega}=\frac{1}{2}\sum_{\mathrm{i,f}}|M_{\mathrm{fi}}|^2=|G|^2+|H|^2\,. \tag{A.2.3}$$

The advantages of G,H and f_1,f_2 are the simplicity of the s-channel partial wave expansion, of the expression for the cross section, Eq. (A.2.3), and of the transition to the non-relativistic limit.*)

*) In the literature one can find the name "non-relativistic amplitude" for $\mathsf{M}(s,t)$. This is misleading, since Eq. (A.2.2) is an exact relativistic equation.

2.1.2 Transversity amplitudes $F(\pm)$

The combination

$$F(\pm)\equiv G\pm \mathrm{i}H \tag{A.2.4}$$

of the spin no-flip and flip amplitudes has the property that its modulus is determined by the differential cross section and the polarization parameter P [there exist other amplitudes with the same property, see Sect. A.4.4.1 and Eq. (A.2.18)].

$$|F(\pm)|^2=\frac{\mathrm{d}\sigma}{\mathrm{d}\Omega}(1\pm P). \tag{A.2.5}$$

At fixed energy, the transversity amplitudes are usually considered either as functions of $z=\cos\theta$ or as functions of a variable w, which is connected with z by a conformal mapping

$$w=\mathrm{e}^{\mathrm{i}\theta}=z\pm\sqrt{z^2-1}\,, \tag{A.2.6}$$

$$z=\cos\theta=\frac{1}{2}\left(w+\frac{1}{w}\right); \quad \sin\theta=\frac{1}{2\mathrm{i}}\left(w-\frac{1}{w}\right). \tag{A.2.7}$$

This mapping has the property that a physical value of z (i.e. z real and $|z|\leqq 1$) is mapped onto *two* points in the w-plane, which lie on the upper and lower halves of the unit circle, respectively. Since we have two amplitudes $F(\pm\,;z)$, one can attach the values of $F(+\,;z)$ to the upper half and those of $F(-\,;z)$ to the lower half of the unit circle. It turns out that this leads to a function $F(w)$ which is analytic in the neighborhood of the unit circle, including the points ± 1.

Along the upper half, θ agrees with the scattering angle, whereas along the lower half the scattering angle is $-\theta$, if we use $|\theta|<\pi$.

The w-plane is mapped onto two sheets ($z^{\pm}$) of the z-plane: w and w^{-1} belong to the same value of z. The cuts in the z-plane are chosen along the real axis at $|z|\geqq 1$ (Fig. A.2.1, see next page).

Transversity amplitudes are useful for phase shift analysis (Sect. 2.1). One can easily see, to what extent these amplitudes are determined by the different measurable quantities (Sect. A.4.4.1). Barrelet's method starts with the determination of zeros of the transversity amplitudes in the $z^{\pm}$- or w-plane. Figures A.2.1 and A.2.2 are useful for the discussion of the energy dependence of these zeros (Sect. 2.4.3). The latter figure is well-known from hydrodynamics, where the Joukowski mapping, Eq. (A.2.7), is used for determining the streamlines of a flow around a cylinder.

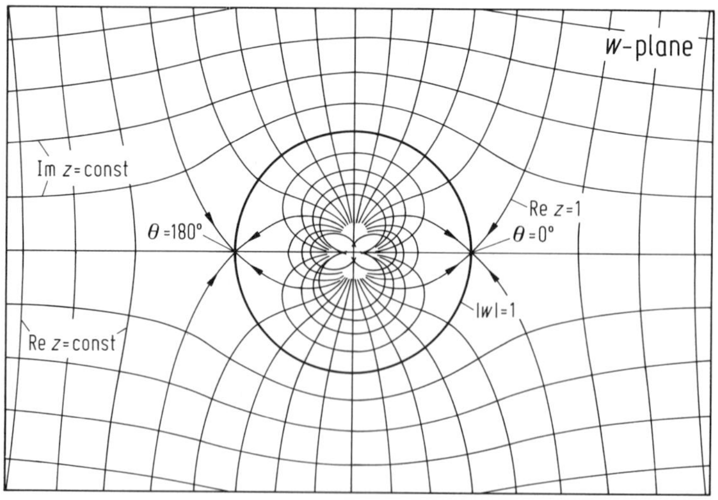

Fig. A.2.2. Lines $\mathrm{Re}\,z=$const and $\mathrm{Im}\,z=$const in the w-plane. The mapping $z=(w+w^{-1})/2$ is singular at $w=\pm 1$. $z=\cos\theta$, $w=\exp(\mathrm{i}\theta)$.

For Fig. A.2.1, see next page.

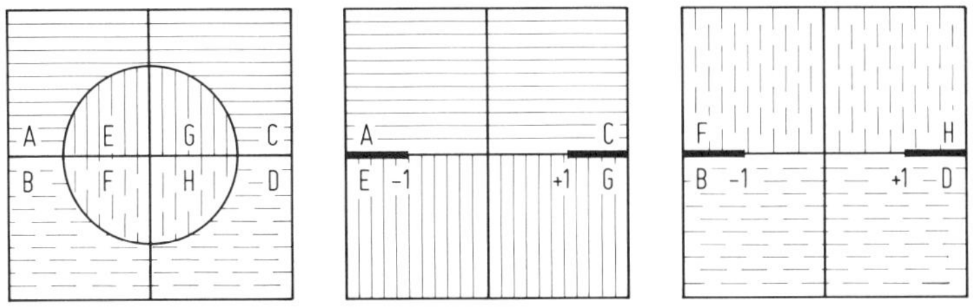

Fig. A.2.1 Conformal mapping of the w-plane onto the two sheets of the z-plane. The circle $|w|=1$ is mapped onto the real axis of the z-plane between the branch points $z=\pm 1$. The points denoted by A, B, etc. are mapped onto the points denoted by the same letters in the other plane.

2.1.3 s-channel helicity amplitudes F_{++}, F_{+-}

The helicity amplitudes are matrix elements M_{fi}, Eq. (A.2.1), with spin quantized along the directions of the outgoing and ingoing nucleon momenta: $\hat{e}'=\hat{p}'$, $\hat{e}=\hat{p}$. $f=\lambda$ and $i=\mu$ are the final and initial nucleon helicities, for which one frequently writes $\pm$ instead of $\pm 1/2$

$$M_{\lambda\mu} \equiv \frac{q}{\sqrt{\pi}} F_{\lambda\mu} = (f_1 + 4\lambda\mu f_2) d^{1/2}_{\mu\lambda}(\theta). \qquad \text{(A.2.8)}$$

The factor $q/\sqrt{\pi}$ is introduced in order to obtain the simple normalization

$$\frac{d\sigma}{dt} = |F_{++}|^2 + |F_{+-}|^2. \qquad \text{(A.2.9)}$$

The d-functions are defined in Table A.3.1. Parity invariance leads to

$$F_{++} = F_{--}, \qquad F_{+-} = -F_{-+}. \qquad \text{(A.2.10)}$$

The s-channel helicity amplitudes are frequently used in models for the high energy behaviour. $\mathrm{Im}\, F_{++}$ is dominant in the isospin even amplitude at small $|t|$. The normalization, Eq. (A.2.9), has the advantage that $\mathrm{Im}\, F^{+}_{++}(s,t)$ varies only slowly with energy at fixed t and large s.

2.1.4 s-channel laboratory frame amplitudes G_L and H_L

Up to now, we have used the T-matrix element in the center of mass frame, Eq. (A.2.1). Next, amplitudes in the laboratory frame are defined by

$$\mathrm{M}_L = G_L + H_L \mathrm{i}(\boldsymbol{\sigma}\cdot\hat{\boldsymbol{n}}); \qquad \frac{d\sigma}{d\Omega_{lab}} = |G_L|^2 + |H_L|^2. \qquad \text{(A.2.11)}$$

They are used in investigations of pion-deuteron scattering.

2.1.5 Invariant amplitudes A, B and combinations C, D

In order to obtain a manifestly Lorentz invariant description of the pion-nucleon system one has to describe the spin states by Dirac spinors. Our conventions for these spinors and for the gamma matrices are defined by the following equations

$$\{\gamma^\mu, \gamma^\nu\} = 2g^{\mu\nu}; \quad \sigma^{\mu\nu} = \frac{i}{2}(\gamma^\mu\gamma^\nu - \gamma^\nu\gamma^\mu); \quad \not{p} \equiv p_\mu \gamma^\mu;$$
$$(\not{p} - m)u(p) = 0; \quad \bar{u}(p)(\not{p} - m) = 0; \quad \bar{u}_f(p)u_i(p) = 2m\delta_{fi}. \qquad \text{(A.2.12)}$$

$g^{\mu\nu}$ is the metric tensor. The most general Lorentz invariant and parity conserving T-matrix element is given by $[\mathbf{Q} = \frac{1}{2}(\mathbf{q}+\mathbf{q}'); \boldsymbol{\varkappa} = \frac{1}{2}(\mathbf{q}-\mathbf{q}')]$

$$8\pi W M_{fi} = \bar{u}_f(p')\{A(s,t) + \not{Q} B(s,t)\} u_i(p)$$
$$= \bar{u}_f(p') \left\{ A(s,t) + \nu B(s,t) - \frac{\not{Q}\not{\varkappa}}{m} B(s,t) \right\} u_i(p). \qquad \text{(A.2.13)}$$

The second version follows from the first one and *Gordon's decomposition of the current*

$$\bar{u}_f(p')\gamma^\mu u_i(p) = \frac{1}{m}\bar{u}_f(p')\{P^\mu + i\sigma^{\mu\nu}\varkappa_\nu\}u_i(p), \qquad \text{(A.2.14)}$$

using the rule $\not{a}\not{b} = a\cdot b - i\sigma^{\mu\nu}a_\mu b_\nu$. $\mathbf{P} = \frac{1}{2}(\mathbf{p}+\mathbf{p}')$.

The invariant amplitudes $A(s,t)$ and $B(s,t)$ have the important property of being free of "kinematic singularities", which arise in the s-channel helicity amplitudes $F_{+\pm}$ at thresholds, at $s=0$ and along the physical region boundary.

For some applications, it is of interest to introduce combinations of the amplitudes $A(s,t)$ and $B(s,t)$. The amplitude

$$C(s,t) \equiv A + \frac{\omega + t/4m}{1-t/4m^2}B = A + \frac{\nu}{1-t/4m^2}B = A + \frac{m(s-u)}{4m^2-t}B, \qquad \text{(A.2.15)}$$

which is sometimes called A', includes the forward amplitude $C(s,0)$ as a special case. According to the optical theorem, $\operatorname{Im} C(s,0)$ is determined by the total cross section data [see Eq. (A.5.5)]. In elastic scattering at high energies, $\operatorname{Im} C(s,t)$ gives the dominant contribution to the diffraction peak. Finally, the pair of amplitudes C, B is proportional to the t-channel helicity amplitudes (Sect. A.2.1.6). As a consequence, $C(s,t)$ has a simpler t-channel partial wave expansion than $A(s,t)$ (Table A.3.3). Its kinematic singularity at $t=4m^2$ will be discussed in Sect. A.3.5.4.

The combination

$$D(s,t) \equiv A(s,t) + \nu B(s,t), \qquad \text{(A.2.16)}$$

which occurs in Eq. (A.2.13), is used in applications of Current Algebra and PCAC in low energy pion-nucleon scattering (Sect. A.8.2). Furthermore, the imaginary part of the isospin even amplitude, $\operatorname{Im} D^+(s,t)$, has an interesting positivity property which will be mentioned in Sect. A.5.3.

In phase shift analysis, the calculations become more transparent, if one works with amplitudes, whose modulus is determined by differential cross section and polarization data. Transversity amplitudes have this property, Eq. (A.2.5), but their analytic behaviour at fixed t is complicated, which makes it difficult to apply analyticity constraints. Therefore, Pietarinen introduced the following combinations of the invariant amplitudes

$$g(\pm) \equiv \sqrt{4m^2-t}\, C \pm \left[\frac{t\{(m^2-\mu^2)^2 - su\}}{4m^2-t}\right]^{1/2} B \qquad \text{(A.2.17)}$$

which have a simple analytic behaviour in s at fixed t and also the property Eq. (A.2.5)

$$\frac{d\sigma}{d\Omega}(1\pm P) = \frac{|g(\pm)|^2}{64\pi^2 s}. \qquad \text{(A.2.18)}$$

The square root is defined such that at fixed $t \leqq 0$ it is positive for

$$|\nu| < \bar{\nu} \equiv \sqrt{(\mu^2 - t/4)(1-t/4m^2)} = p_- q_- / m. \qquad \text{(A.2.19)}$$

It has a cut on the real ν-axis for $|\nu| \geqq \bar{\nu}$. The signs of the square root are given in Fig. A.2.3.

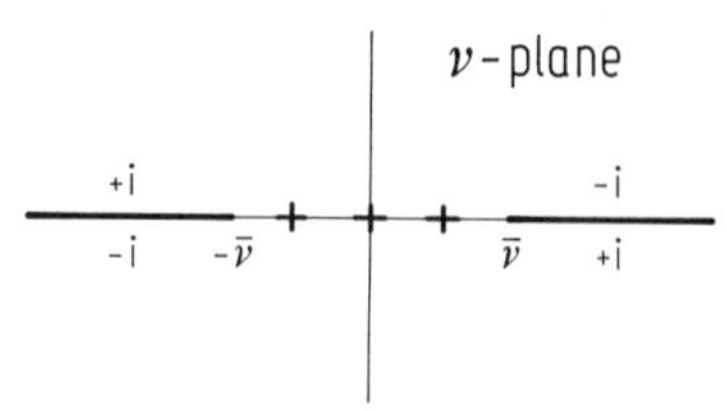

Fig. A.2.3. Cut structure of the function $g(\pm)$, Eq. (A.2.17).

2.1.6 *t*-channel amplitudes

In S-matrix theory, the amplitudes for the t-channel reactions $\bar{N}N \to \pi\pi$ and $\pi\pi \to N\bar{N}$ are analytic continuations of the invariant amplitudes $A(s,t)$, $B(s,t)$ for πN scattering. The T-matrix element in the center-of-mass frame, Eq. (A.2.1), is now written as follows

$$M^t_{\ell j}(s,t) = \chi^+_{N\ell}\{h_1(s,t)\boldsymbol{\sigma}\cdot\boldsymbol{p}_t + h_2(s,t)\boldsymbol{\sigma}\cdot\boldsymbol{q}_t\}\chi_{\bar{N}j}\,, \tag{A.2.20}$$

the normalization being given by the expression for the differential cross section in the center-of-mass system, Eq. (A.2.22), and the spins being quantized along the direction of the incoming antinucleon.

It is convenient to have a simple expression for the differential cross section for the physical reaction $\bar{N}N \to \pi\pi$. Therefore, we introduce t-channel helicity amplitudes in the c.m. frame

$$\begin{aligned} f_{++} &= \frac{m\sqrt{q_t/p_t}}{4\pi\sqrt{2t}}\{h_1 p_t + h_2 q_t \cos\theta_t\} = -\frac{\sqrt{p_t q_t}}{4\pi\sqrt{2t}}C\,, \\ f_{+-} &= \frac{m\sqrt{q_t/p_t}}{4\pi\sqrt{2t}}h_2 q_t \sin\theta_t = \frac{q_t^{3/2}}{8\pi\sqrt{2p_t}}\sin\theta_t B\,, \end{aligned} \tag{A.2.21}$$

for which the differential cross sections read

$$\frac{d\sigma(\bar{N}N \to \pi\pi)}{d\Omega} = |f_{++}|^2 + |f_{+-}|^2 = \frac{q_t^2}{4p_t^2}\frac{d\sigma(\pi\pi \to N\bar{N})}{d\Omega}\,. \tag{A.2.22}$$

t-channel helicity amplitudes are also of interest, if one describes the high energy behaviour of the s-channel amplitudes in terms of t-channel exchanges. The most important example is the reggeized ϱ-exchange model for $\pi^- p$ charge-exchange scattering.

We introduce the amplitudes

$$G_{+\pm} = \mathrm{i}\frac{\sqrt{2\pi t}}{qW}\sqrt{\frac{p_t}{q_t}}f_{+\pm} \tag{A.2.23}$$

which are normalized as follows

$$\frac{d\sigma(\pi N \to \pi N)}{dt} = |G_{++}|^2 + |G_{+-}|^2\,. \tag{A.2.22a}$$

They are simply related to the invariant amplitudes

$$\begin{aligned} 4\sqrt{\pi}\,WqG_{++} &= p_- C = -\mathrm{i}p_t C = -\mathrm{i}p_t\left(A - \frac{mq_t\cos\theta_t}{p_t}\right)B\,, \\ 4\sqrt{\pi}\,WqG_{+-} &= \frac{q^2 W}{2p_-}\sin\theta\, B = \frac{\mathrm{i}}{2}\sqrt{t}\,q_t\sin\theta_t\, B\,. \end{aligned} \tag{A.2.24}$$

Because of parity invariance, we have $G_{++} = G_{--}$ and $G_{+-} = -G_{-+}$.

2.1.7 *u*-channel amplitudes

The s-channel amplitudes f_1, f_2, Eqs. (A.2.2), (A.2.31), can be considered as functions of $W = \sqrt{s}$ and u: $f_1(s,z) \equiv f_1(W,u)$, etc. For the discussion of u-channel exchange contributions to near-backward s-channel scattering (Sect. A.9.2.3), one needs u-channel amplitudes $f_1(W_u, s)$, $f_2(W_u, s)$, where $W_u = \sqrt{u}$. These amplitudes follow from the s-channel amplitudes by applying the $s \leftrightarrow u$ crossing operation at fixed t, using Eqs. (A.2.31) and (A.6.29)

$$\begin{aligned} f_1^{\pm}(W_u, s) &= \pm\frac{(W_u+m)^2 - \mu^2}{16\pi u}\{A^{\pm}(s,t,u) - (W_u - m)B^{\pm}(s,t,u)\}\,; \\ f_2^{\pm}(W_u, s) &= \mp\frac{(W_u-m)^2 - \mu^2}{16\pi u}\{A^{\pm}(s,t,u) + (W_u + m)B^{\pm}(s,t,u)\}\,. \end{aligned} \tag{A.2.25}$$

The formula $2W_u(E_u \pm m) = (W_u \pm m)^2 - \mu^2$ [cf. Eq. (A.1.11)] was inserted in order to simplify the discussion of the *MacDowell symmetry relation* $f_1(-W_u, s) = -f_2(W_u, s)$, which follows from the fact that the invariant amplitudes are analytic functions of s, t, u.

2.1.8 Relations between different amplitudes

In numerical calculations, one starts with the amplitudes G, H or f_1, f_2, which are simply related to the partial waves [see Eqs. (A.3.1) and (A.3.4)]. The other amplitudes can be obtained from the following relations

$$\frac{A}{4\pi} = \frac{W+m}{E+m}G + \left[2m\omega + \frac{t(W+m)}{2(E+m)}\right]\frac{H}{q^2\sin\theta} = \frac{W+m}{E+m}f_1 - \frac{W-m}{E-m}f_2$$

$$= \frac{W}{k\sqrt{\pi}}\left[-\frac{\omega_q F_{++}}{\cos(\theta/2)} + \frac{\omega F_{+-}}{\sin(\theta/2)}\right]. \qquad \text{(A.2.26)}$$

$$\frac{B}{4\pi} = \frac{G}{E+m} - \left[2m - \frac{t}{2(E+m)}\right]\frac{H}{q^2\sin\theta} = \frac{f_1}{E+m} + \frac{f_2}{E-m}$$

$$= \frac{W}{k\sqrt{\pi}}\left[\frac{E}{m}\frac{F_{++}}{\cos(\theta/2)} - \frac{F_{+-}}{\sin(\theta/2)}\right]. \qquad \text{(A.2.27)}$$

$$\frac{C}{4\pi} = G\frac{W}{4m^2-t}\left(4m - \frac{t}{E+m}\right) + H\sin\theta\frac{2W(E-m)}{4m^2-t}$$

$$= \frac{k/\sqrt{\pi}}{1-t/4m^2}\{F_{++}\cos(\theta/2) + F_{+-}(E/m)\sin(\theta/2)\}. \qquad \text{(A.2.28)}$$

$$\frac{q}{\sqrt{\pi}}F_{++} = G\cos(\theta/2) - H\sin(\theta/2) = (f_1+f_2)\cos(\theta/2)$$

$$= \frac{m}{4\pi W}(A+\omega B)\cos(\theta/2) = \frac{m}{4\pi W}\left(C - t\frac{m+\omega}{4m^2-t}B\right)\cos(\theta/2) = \frac{m}{4\pi W}\left(D - \frac{t}{4m}B\right)\cos(\theta/2), \qquad \text{(A.2.29)}$$

$$\frac{q}{\sqrt{\pi}}F_{+-} = G\sin(\theta/2) + H\cos(\theta/2) = (f_1-f_2)\sin(\theta/2)$$

$$= \frac{m}{4\pi W}\left(\frac{E}{m}A + \omega_q B\right)\sin(\theta/2) = \frac{m}{4\pi W}\left(\frac{E}{m}C - \frac{4q^2+t}{4m^2-t}WB\right)\sin(\theta/2). \qquad \text{(A.2.30)}$$

$$f_1 = G + (H/\sin\theta)\cos\theta = \frac{E+m}{8\pi W}\{A + (W-m)B\} = \frac{q}{2\sqrt{\pi}}\left[\frac{F_{++}}{\cos(\theta/2)} + \frac{F_{+-}}{\sin(\theta/2)}\right],$$

$$f_2 = -H/\sin\theta = \frac{E-m}{8\pi W}\{-A + (W+m)B\} = \frac{q}{2\sqrt{\pi}}\left[\frac{F_{++}}{\cos(\theta/2)} - \frac{F_{+-}}{\sin(\theta/2)}\right]. \qquad \text{(A.2.31)}$$

$$G = f_1 + f_2\cos\theta = \frac{1}{4\pi W}\left[A\left(m - \frac{t}{4(E+m)}\right) + B\left(m\omega + \frac{t(W+m)}{4(E+m)}\right)\right]$$

$$= \frac{1}{4\pi W}\left[C\left(m - \frac{t}{4(E+m)}\right) + Bq^2\sin^2\theta\frac{W(E-m)}{4m^2-t}\right] = \frac{q}{\sqrt{\pi}}\{F_{++}\cos(\theta/2) + F_{+-}\sin(\theta/2)\}$$

$$H = -f_2\sin\theta = \frac{E-m}{8\pi W}\{A - (W+m)B\}\sin\theta = \frac{E-m}{8\pi W}\left\{C - BW\frac{4m(E+m)-t}{4m^2-t}\right\}\sin\theta \qquad \text{(A.2.32)}$$

$$= \frac{q}{\sqrt{\pi}}\{-F_{++}\sin(\theta/2) + F_{+-}\cos(\theta/2)\}.$$

The relation between the t- and s-channel helicity amplitudes $G_{+\pm}$ and $F_{+\pm}$ reads

$$\begin{pmatrix} F_{++} \\ F_{+-} \end{pmatrix} = \begin{pmatrix} \sin\chi & \cos\chi \\ \cos\chi & -\sin\chi \end{pmatrix}\begin{pmatrix} G_{++} \\ G_{+-} \end{pmatrix}, \qquad \text{(A.2.33)}$$

where the angle χ is given by

$$\sin\chi = \frac{m}{p_-}\cos(\theta/2) = \frac{2m}{\sqrt{\lambda}} q_t \sin\theta_t = \frac{2m\sqrt{\Phi}}{\sqrt{\lambda t(t-4m^2)}}\,;$$
$$\cos\chi = \frac{E}{p_-}\sin(\theta/2) = \sqrt{t/\lambda}(p_t - q_t\cos\theta_t) = -\frac{(s+m^2-\mu^2)\sqrt{t}}{\sqrt{\lambda(t-4m^2)}}\,. \qquad \text{(A.2.34)}$$

$\lambda = 4sq^2$ is positive in the physical s- and t-channel regions. In the s-channel, the angle χ agrees with the scattering angle θ_N of the recoil nucleon in the laboratory system, Eq. (A.1.23).

For the study of the "crossing relation" between s- and t-channel helicity amplitudes, one has to consider analytic continuations from one of these channels to the other one.

Amplitudes which are suitable for this purpose are obtained by multiplying $F_{+\pm}$ and $G_{+\pm}$ by a factor

$$T^s_{+\pm} = (2/\pi^{3/2})WqF_{+\pm}\,; \quad T^t_{+\pm} = (2/\pi^{3/2})G_{+\pm} = \mathrm{i}\frac{(8tp_t/q_t)^{1/2}}{\pi q W} f_{+\pm}\,. \qquad \text{(A.2.35)}$$

Since the factor is the same in both cases, the crossing matrix is the same as in Eq. (A.2.33).

We shall also need a relation between the s-channel amplitudes f_1, f_2, Eq. (A.2.2), which are now written as functions of W and u, and the u-channel amplitudes, Eq. (A.2.25). One can solve this equation for A and B and insert the result into Eq. (A.2.31)

$$f_1^\pm(W,u) = \pm\frac{E+m}{2W}\left\{(W_u - W + 2m)\frac{f_1^\pm(W_u, s)}{E_u+m} + (W_u + W - 2m)\frac{f_1^\pm(-W_u, s)}{E_u - m}\right\};$$
$$f_2^\pm(W,u) = -f_1^\pm(-W_u, u)\,. \qquad \text{(A.2.36)}$$

The amplitudes G_L, H_L in the laboratory system can be calculated from C, B

$$G_L = \frac{1}{\sqrt{J}}\sqrt{1-t/4m^2}\,\frac{C}{4\pi}\,; \quad H_L = \frac{1}{\sqrt{J}}\,\frac{k_N k\sin\theta_N}{2m\sqrt{1-t/4m^2}}\,\frac{B}{4\pi}\,. \qquad \text{(A.2.37)}$$

The expression for the Jacobian

$$J = \frac{k^2}{q^2}\frac{\mathrm{d}\cos\theta_\pi}{\mathrm{d}\cos\theta} = \frac{k^2}{q^2}\frac{\mathrm{d}\Omega_{\mathrm{lab}}}{\mathrm{d}\Omega} \qquad \text{(A.2.37a)}$$

follows from Eq. (A.1.27).

2.1.9 High energy approximations

We consider two cases:

Large s, fixed t

$$A \approx 4\sqrt{\pi}m\left\{-F_{++} + \frac{2k}{\sqrt{-t}}F_{+-}\right\}; \quad B \approx 4\sqrt{\pi}\left\{\left(1+\frac{2m^2}{s}\right)F_{++} - \frac{2m}{\sqrt{-t}}F_{+-}\right\},$$
$$C \approx \frac{4\sqrt{\pi}}{1-t/4m^2}\left\{kF_{++} + \sqrt{-t}\,\frac{s+m^2}{4m^2}F_{+-}\right\}. \qquad \text{(A.2.38)}$$

The ratio of the s-channel helicity amplitudes

$$\frac{F_{+-}}{F_{++}} = \frac{\sqrt{-t}}{2m}\,\frac{E/q}{\sqrt{1+t/4q^2}}\,\frac{\left(1+\frac{\mu^2}{m\omega}\right)\frac{m}{m+\omega} + \frac{A}{\omega B}}{1 + A/\omega B} \qquad \text{(A.2.39)}$$

tends to zero (s-channel helicity conservation) if $A/\omega B \to 0$. t-channel helicity conservation $G_{+-}/G_{++} \to 0$ belongs to $A/\omega B \to \infty$ and $F_{+-}/F_{++} \to \sqrt{-t}/2m$. Another finite value is approached: $F_{+-}/F_{++} \to (\sqrt{-t}/2m)(1+\omega B/A)^{-1}$, if $A/\omega B$ remains finite.

Large s, fixed θ

$$A\approx 4\sqrt{\pi}\left[\frac{-mF_{++}}{\cos(\theta/2)}+\frac{WF_{+-}}{\sin(\theta/2)}\right];\quad B\approx 4\sqrt{\pi}\left[\frac{F_{++}}{\cos(\theta/2)}-\frac{2m}{W}\frac{F_{+-}}{\sin(\theta/2)}\right];$$

$$C\approx A+\frac{3+\cos\theta}{1-\cos\theta}mB\approx\frac{4\sqrt{\pi}}{\sin^2(\theta/2)}(2m\cos(\theta/2)F_{++}+W\sin(\theta/2)F_{+-}). \tag{A.2.40}$$

2.2 The isospin structure

2.2.1 Isospin operators and states

The components of an isospin operator $\mathsf{I}=(\mathsf{I}_1,\mathsf{I}_2,\mathsf{I}_3)$ fulfill the same commutation relations as those of an angular momentum operator

$$[\mathsf{I}_1,\mathsf{I}_2]=i\mathsf{I}_3;\quad [\mathsf{I}_2,\mathsf{I}_3]=i\mathsf{I}_1;\quad [\mathsf{I}_3,\mathsf{I}_1]=i\mathsf{I}_2. \tag{A.2.41}$$

The charge operator Q follows from I_3 and the baryon number B

$$\mathsf{Q}=\mathsf{I}_3+\tfrac{1}{2}\mathsf{B}. \tag{A.2.42}$$

I^2 and I_3 have eigenvalues $I(I+1)$ and I_3, where I is integer or half-integer and $I_3=-I,-I+1,\dots,I$.

We consider simultaneous eigenstates of I^2 and I_3, which are orthogonal and normalized

$$\langle I,I_3|I',I_3'\rangle=\delta_{II'}\delta_{I_3I_3'}. \tag{A.2.43}$$

In order to fix the relative phases of these eigenstates, one starts with $|I,I\rangle$ and constructs the other states of the basis by repeated application of the operator I_- which is defined by

$$\mathsf{I}_\pm\equiv\mathsf{I}_1\pm i\mathsf{I}_2, \tag{A.2.44}$$

using

$$\mathsf{I}_\pm|I,I_3\rangle=\sqrt{(I\mp I_3)(I\pm I_3+1)}|I,I_3\pm 1\rangle. \tag{A.2.45}$$

The $\mathsf{I}_\pm$ fulfill the relations

$$[\mathsf{I}_3,\mathsf{I}_\pm]=\pm\mathsf{I}_\pm;\quad [\mathsf{I}_+,\mathsf{I}_-]=2\mathsf{I}_3;\quad \mathsf{I}_\mp\mathsf{I}_\pm=\mathsf{I}^2-I_3(I_3\pm 1);\quad [\mathsf{Q},\mathsf{I}_\pm]=\pm\mathsf{I}_\pm. \tag{A.2.46}$$

The isospin operators of nucleon and pion are denoted by $(1/2)\boldsymbol{\tau}$ and by $\mathbf{T}$, respectively. The components of the Pauli operator $\boldsymbol{\tau}=(\tau_1,\tau_2,\tau_3)$ have the further simple properties

$$\tau_i^2=1;\quad \tau_i\tau_j=-\tau_j\tau_i=i\tau_k,\quad \text{where}\quad i,j,k=1,2,3\quad \text{or an even permutation.} \tag{A.2.47}$$

The nucleon states are

$$|p\rangle=|\tfrac{1}{2},\tfrac{1}{2}\rangle;\quad |n\rangle=|\tfrac{1}{2},-\tfrac{1}{2}\rangle \tag{A.2.48}$$

and the phase of the pion states is chosen such that

$$|\pi^+\rangle=|1,1\rangle;\quad |\pi^0\rangle=|1,0\rangle;\quad |\pi^-\rangle=|1,-1\rangle. \tag{A.2.49}$$

As a consequence, one has to adopt the following behaviour under charge conjugation

$$\mathsf{C}|\pi^\pm\rangle=-|\pi^\mp\rangle;\quad \mathsf{C}|\pi^0\rangle=|\pi^0\rangle. \tag{A.2.50}$$

The isospin states of a πN system, for instance $|\pi^+p\rangle\equiv|\pi^+\rangle|p\rangle$, can be expressed in terms of eigenstates of the square of the total isospin

$$\mathbf{I}\equiv\tfrac{1}{2}\boldsymbol{\tau}+\mathbf{T} \tag{A.2.51}$$

and of its third component: $|I,I_3\rangle$ *)

*) Some authors write these formulas for $|N\pi\rangle$ states, taking the nucleon as the first particle. This leads to the opposite sign in front of all $|\tfrac{1}{2},I_3\rangle$ states. The signs in Table A.2.1 remain unchanged.

$$|\tfrac{3}{2},\tfrac{3}{2}\rangle=|\pi^+p\rangle\,; \quad |\tfrac{3}{2},\tfrac{1}{2}\rangle=\sqrt{\tfrac{2}{3}}|\pi^0p\rangle+\sqrt{\tfrac{1}{3}}|\pi^+n\rangle\,;$$
$$|\tfrac{3}{2},-\tfrac{1}{2}\rangle=\sqrt{\tfrac{1}{3}}|\pi^-p\rangle+\sqrt{\tfrac{2}{3}}|\pi^0n\rangle\,; \quad |\tfrac{3}{2},-\tfrac{3}{2}\rangle=|\pi^-n\rangle\,; \tag{A.2.52}$$
$$|\tfrac{1}{2},\tfrac{1}{2}\rangle=\sqrt{\tfrac{2}{3}}|\pi^+n\rangle-\sqrt{\tfrac{1}{3}}|\pi^0p\rangle\,; \quad |\tfrac{1}{2},-\tfrac{1}{2}\rangle=\sqrt{\tfrac{1}{3}}|\pi^0n\rangle-\sqrt{\tfrac{2}{3}}|\pi^-p\rangle\,.$$

In practice, one mainly needs 3 channels

$$|\pi^+p\rangle=|\tfrac{3}{2},\tfrac{3}{2}\rangle\,; \quad |\pi^-p\rangle=\sqrt{\tfrac{1}{3}}|\tfrac{3}{2},-\tfrac{1}{2}\rangle-\sqrt{\tfrac{2}{3}}|\tfrac{1}{2},-\tfrac{1}{2}\rangle\,; \quad |\pi^0n\rangle=\sqrt{\tfrac{2}{3}}|\tfrac{3}{2},-\tfrac{1}{2}\rangle+\sqrt{\tfrac{1}{3}}|\tfrac{1}{2},-\tfrac{1}{2}\rangle\,. \tag{A.2.53}$$

Projection operators for the $I=1/2$ and $I=3/2$-states read

$$\mathscr{P}(I=1/2)=\tfrac{1}{3}(1-\boldsymbol{\tau}\cdot\mathbf{T})\,; \quad \mathscr{P}(I=3/2)=\tfrac{1}{3}(2+\boldsymbol{\tau}\cdot\mathbf{T})\,. \tag{A.2.54}$$

2.2.2 Matrix representations

In some applications, it is useful to introduce matrix representations of the isospin operators and states. τ is represented by the Pauli matrices

$$\tau_1=\begin{pmatrix}0&1\\1&0\end{pmatrix};\quad \tau_2=\begin{pmatrix}0&-i\\i&0\end{pmatrix};\quad \tau_3=\begin{pmatrix}1&0\\0&-1\end{pmatrix}, \tag{A.2.55}$$

and the state of the nucleon by an isospinor

$$\chi_p=\begin{pmatrix}1\\0\end{pmatrix};\quad \chi_n=\begin{pmatrix}0\\1\end{pmatrix}. \tag{A.2.56}$$

In the case of the pion, one can again take a basis where the charge operator is diagonal, i.e. the states of the basis χ_{I_3} are eigenstates of $\mathsf{Q}=\mathsf{T}_3$

$$\chi_{+1}=\chi^{ch}_{\pi^+}=\begin{pmatrix}1\\0\\0\end{pmatrix};\quad \chi_0=\chi^{ch}_{\pi^0}=\begin{pmatrix}0\\1\\0\end{pmatrix};\quad \chi_{-1}=\chi^{ch}_{\pi^-}=\begin{pmatrix}0\\0\\1\end{pmatrix}. \tag{A.2.57}$$

The matrix representation of $\mathbf{T}$ is then given by ("ch" refers to "charge basis")

$$(\mathsf{T}^{ch}_i)_{I'_3,I_3}=\langle 1,I'_3|\mathsf{T}_i|1,I_3\rangle,$$
$$\mathsf{T}^{ch}_1=\frac{1}{\sqrt{2}}\begin{pmatrix}0&1&0\\1&0&1\\0&1&0\end{pmatrix};\quad \mathsf{T}^{ch}_2=\frac{1}{\sqrt{2}}\begin{pmatrix}0&-i&0\\i&0&-i\\0&i&0\end{pmatrix};\quad \mathsf{T}^{ch}_3=\begin{pmatrix}1&0&0\\0&0&0\\0&0&-1\end{pmatrix}. \tag{A.2.58}$$

The quantities χ_{I_3} are called "spherical tensors" (of rank 1), because they transform under a rotation in isospin space in analogy to the spherical harmonics Y^m_1 in ordinary space.

A second possibility is to choose a basis which transforms under rotations in analogy to the Cartesian vectors in ordinary space. In this basis, the physical states are

$$\chi^{Cart}_{\pi^+}=-\frac{1}{\sqrt{2}}\begin{pmatrix}1\\i\\0\end{pmatrix};\quad \chi^{Cart}_{\pi^0}=\begin{pmatrix}0\\0\\1\end{pmatrix};\quad \chi^{Cart}_{\pi^-}=\frac{1}{\sqrt{2}}\begin{pmatrix}1\\-i\\0\end{pmatrix}. \tag{A.2.59}$$

The ket-vectors of the Cartesian basis, $|\pi_\alpha\rangle, \alpha=1,2,3$, correspond to the one-particle states in field theory, where the pion is described by three real fields, which transform like a Cartesian vector under rotations in isospin space (except for phase conventions)

$$|\pi_1\rangle=\frac{1}{\sqrt{2}}[|\pi^-\rangle-|\pi^+\rangle];\quad |\pi_2\rangle=\frac{i}{\sqrt{2}}[|\pi^-\rangle+|\pi^+\rangle];\quad |\pi_3\rangle=|\pi_0\rangle. \tag{A.2.60}$$

The matrices $(\mathsf{T}_i^{\mathrm{Cart}})_{\alpha\beta} = \langle \pi_\alpha | \mathsf{T}_i | \pi_\beta \rangle$ read

$$\mathsf{T}_1^{\mathrm{Cart}} = \begin{pmatrix} 0 & 0 & 0 \\ 0 & 0 & -\mathrm{i} \\ 0 & \mathrm{i} & 0 \end{pmatrix}; \quad \mathsf{T}_2^{\mathrm{Cart}} = \begin{pmatrix} 0 & 0 & \mathrm{i} \\ 0 & 0 & 0 \\ -\mathrm{i} & 0 & 0 \end{pmatrix}; \quad \mathsf{T}_3^{\mathrm{Cart}} = \begin{pmatrix} 0 & -\mathrm{i} & 0 \\ \mathrm{i} & 0 & 0 \\ 0 & 0 & 0 \end{pmatrix}. \tag{A.2.61}$$

They can be expressed by the totally antisymmetric tensor $\varepsilon_{\mathrm{ijk}}$, which is equal to 1 for $\mathrm{i,j,k} = 1,2,3$ or even permutations, equal to -1 for odd permutations and zero otherwise. The same tensor can also be used in order to write the commutation relation Eq. (A.2.41) for $\mathsf{I}_\mathrm{i} = \frac{1}{2}\tau_\mathrm{i}$ in a compact way

$$(\mathsf{T}_\mathrm{j})_{\alpha\beta}^{\mathrm{Cart}} = \mathrm{i}\varepsilon_{\alpha \mathrm{j} \beta}, \quad \mathrm{j} = 1,2,3; \quad \tfrac{1}{2}[\tau_\alpha, \tau_\beta] = \mathrm{i}\varepsilon_{\alpha\beta\gamma}\tau_\gamma . \tag{A.2.62}$$

A combination of these two formulas and of Eq. (A.2.54) for the projection operators gives

$$\tfrac{1}{2}[\tau_\alpha, \tau_\beta] = -\boldsymbol{\tau}\cdot\mathbf{T}_{\alpha\beta} = [2\mathscr{P}(I=1/2) - \mathscr{P}(I=3/2)]_{\alpha\beta}; \quad \delta_{\alpha\beta} = [\mathscr{P}(I=1/2) + \mathscr{P}(I=3/2)]_{\alpha\beta} . \tag{A.2.63}$$

2.2.3 Isospin invariance

Isospin invariance demands that the commutator between the scattering operator and the components of the total isospin operator vanishes: $[\mathsf{S}, \mathsf{I}_\mathrm{i}] = 0$. In the following, we shall assume that isospin invariance is valid.

The consequences for the scattering amplitude are easily seen, if the initial and final states are decomposed as linear combinations of states $|I, I_3\rangle$ with well-defined quantum numbers of the total isospin. The application of the *Wigner-Eckart theorem* for tensors of rank 0 gives for the S- or T-matrix element.

$$\langle I, I_3 | \mathsf{S} | I', I'_3 \rangle = S(I)\delta_{II'}\delta_{I_3 I'_3} . \tag{A.2.64}$$

Therefore, the amplitudes of all 10 pion-nucleon scattering reactions are determined by only 2 independent S-matrix elements. Each of the amplitudes of Sect. A.2.1 carries an isospin index $I = 1/2$ or $3/2$. For each pair of the kinematical variables (s, t) there exist 4 independent complex-valued amplitudes. The relations between the different amplitudes follow from Eqs. (A.2.52). A list will be given in Table A.2.1.

Another approach is similar to the procedure in Sect. A.2.1. Now we write the most general ansatz for the T- or M-matrix which fulfills isospin invariance. The dependence on the charge state of the nucleon is expressed by the τ_i-operators, whereas the dependence on the pion charge state is treated by considering matrix elements of the M-operator taken with the *Cartesian basis*. There are only two independent matrices which can be constructed from the τ_i

$$\tfrac{1}{2}[\tau_\alpha, \tau_\beta]_+ = \delta_{\alpha\beta} \quad \text{and} \quad \tfrac{1}{2}[\tau_\alpha, \tau_\beta] . \tag{A.2.65}$$

Suppressing the indices for the nucleon spin states, we find for the generalization of Eq. (A.2.13) to the case of πN reactions in all possible charge states

$$8\pi W M_{\alpha\beta} = \bar{u}(p')\{A_{\alpha\beta} + \not{Q} B_{\alpha\beta}\} u(p), \tag{A.2.66}$$

where

$$A_{\alpha\beta}(s,t) = A^+(s,t)\delta_{\alpha\beta} + A^-(s,t)\tfrac{1}{2}[\tau_\alpha, \tau_\beta] . \tag{A.2.67}$$

A similar relation is valid for $B_{\alpha\beta}$.

The first equation (A.2.63) shows that the ansatz (A.2.67) can as well be written in such a way that the two terms belong to the invariants under rotation in isospin space: $\mathbf{1}$ and $\boldsymbol{\tau}\cdot\mathbf{T}$. The upper index in $A^\pm$, $B^\pm$, refers to the symmetric and antisymmetric behaviour of the two terms under $\alpha \leftrightarrow \beta$. The relation to the $s \leftrightarrow u$ crossing operation will be mentioned in Eq. (A.6.28).

In order to obtain the connection of $A^\pm$ with the amplitudes for a well-defined isospin I, one has to insert the expressions (A.2.63) for the projection operators. A^+ belongs to $\pi^0 p \to \pi^0 p$ and A^- to $\pi^- p \to \pi^0 n$.

Before presenting the relations between the different isospin amplitudes in Table A.2.1, we want to introduce two further amplitudes A^{N} and A^{Δ} which are needed in investigations of πN near-backward scattering at high energies. In this kinematical range, the dominant contributions come from the reggeized exchange of the nucleon ($I = 1/2$) and of the isobar $\Delta(1232)$ ($I = 3/2$) in the u-channel (see Fig. A.2.4 and Sect. A.9.2. The upper index N or Δ indicates that $I_u = 1/2$ or $3/2$, respectively.

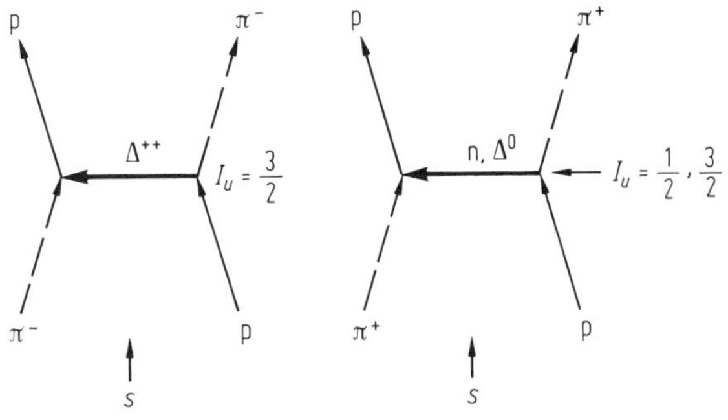

Fig. A.2.4. $I_u = 1/2$ and 3/2 exchange in $\pi^{\pm}$p elastic scattering. These contributions dominate the near-backward amplitudes.

Table A.2.1. Relations between πN isospin amplitudes. A can be any of the amplitudes in Sect. A.2.1. Lower index: $\pm, 0$ refers to $\pi^{\pm}p \to \pi^{\pm}p$ and $\pi^-p \to \pi^0 n$. Upper index: 1/2 or 3/2 gives I; N or Δ gives I_u; + or − belongs to the isospin even or odd combination in Eq. (A.2.67). The Eqs. (A.2.52) can be used for the extension of this table to other $\pi N \to \pi N$ reactions.

Amplitudes for the reactions $\pi^{\pm}p \to \pi^{\pm}p$ *and* $\pi^-p \to \pi^0 n$ and $I = 1/2$ amplitude

$$A_+ = A^{3/2} \qquad = A^+ - A^- = \tfrac{1}{3}(2A^{\mathrm{N}} + A^{\Delta}),$$

$$A_- = \tfrac{1}{3}(2A^{1/2} + A^{3/2}) \qquad = A^+ + A^- = A^{\Delta},$$

$$A_0 = \frac{-\sqrt{2}}{3}(A^{1/2} - A^{3/2}) = -\sqrt{2}A^- = \frac{\sqrt{2}}{3}(A^{\mathrm{N}} - A^{\Delta}) = \frac{1}{\sqrt{2}}(A_+ - A_-),$$

$$A^{1/2} = \tfrac{1}{2}(3A_- - A_+) \qquad = A^+ + 2A^- .$$

Isospin even and odd combinations

$$A^+ = \tfrac{1}{2}(A_- + A_+) = \tfrac{1}{3}(A^{1/2} + 2A^{3/2}); \qquad A^- = \tfrac{1}{2}(A_- - A_+) = -\frac{1}{\sqrt{2}}A_0 = \tfrac{1}{3}(A^{1/2} - A^{3/2}).$$

s − u crossing relations

$$\begin{pmatrix} A^{1/2} \\ A^{3/2} \end{pmatrix} = \tfrac{1}{3}\begin{pmatrix} -1 & 4 \\ 2 & 1 \end{pmatrix}\begin{pmatrix} A^{\mathrm{N}} \\ A^{\Delta} \end{pmatrix}; \qquad \begin{pmatrix} A^{\mathrm{N}} \\ A^{\Delta} \end{pmatrix} = \tfrac{1}{3}\begin{pmatrix} -1 & 4 \\ 2 & 1 \end{pmatrix}\begin{pmatrix} A^{1/2} \\ A^{3/2} \end{pmatrix}.$$

A graphical representation of the relation $A_+ = A_- + \sqrt{2}A_0$ between the amplitudes for the three accurately measurable reactions is called "isospin triangle" (Fig. A.2.5). It is also useful for the discussion of isospin bounds (Sect. A.10).

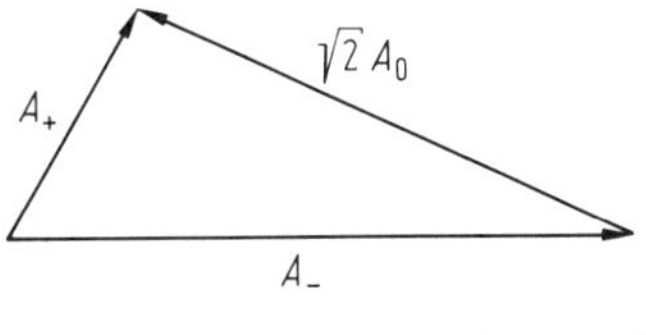

Fig. A.2.5. The isospin triangle. A can be any of the amplitudes introduced in Sect. A.2.

2.2.4 Isospin relations for $\bar{N}N \to \pi\pi$ scattering

First we discuss the isospin states for antinucleons, where a sign has to be fixed by convention. In order to have the simple assignment

$$|\bar{n}\rangle = |\tfrac{1}{2}, \tfrac{1}{2}\rangle; \quad |\bar{p}\rangle = |\tfrac{1}{2}, -\tfrac{1}{2}\rangle \tag{A.2.68}$$

one has to accept a minus sign, when the charge conjugation operator C is applied

$$\mathsf{C}|p\rangle = |\bar{p}\rangle; \quad \mathsf{C}|n\rangle = -|\bar{n}\rangle. \tag{A.2.69}$$

The operation "G-conjugation" is defined by

$$\mathsf{G} \equiv \mathsf{C}e^{-i\pi \mathsf{I}_2}. \tag{A.2.70}$$

The second factor is a rotation through π about the 2-axis of isospace, its action on a state being

$$e^{-i\pi \mathsf{I}_2}|I, I_3\rangle = (-1)^{I-I_3}|I, -I_3\rangle. \tag{A.2.71}$$

It is chosen in order to obtain an operator which commutes with all components of the isospin operator: $[\mathsf{G}, \mathsf{I}_i] = 0$.

Since $\mathsf{G}|\pi\rangle = -|\pi\rangle$, the initial state in the reaction $\bar{N}N \to \pi\pi$ must be an eigenstate of G belonging to the eigenvalue ("G-parity") $+1$. The resulting selection rules can only be discussed after we have introduced the partial wave decomposition of the $\bar{N}N \to \pi\pi$ amplitude (Sect. A.3.4).

The initial isospin states of the $\bar{N}N$ system are superpositions of the states $|1, I'_3\rangle$ and $|0, 0\rangle$

$$\begin{aligned} |\bar{p}p\rangle &= \frac{1}{\sqrt{2}}\{|1,0\rangle + |0,0\rangle\}; \quad |\bar{p}n\rangle = |1, -1\rangle; \\ |\bar{n}n\rangle &= \frac{1}{\sqrt{2}}\{|1,0\rangle - |0,0\rangle\}; \quad |\bar{n}p\rangle = |1, 1\rangle. \end{aligned} \tag{A.2.72}$$

We also need the inverse relations

$$|1,0\rangle = \frac{1}{\sqrt{2}}\{|\bar{p}p\rangle + |\bar{n}n\rangle\}; \quad |0,0\rangle = \frac{1}{\sqrt{2}}\{|\bar{p}p\rangle - |\bar{n}n\rangle\}. \tag{A.2.73}$$

The two-pion states form a triplet and a singlet

$$\begin{aligned} |1,1\rangle &= \frac{1}{\sqrt{2}}\{|\pi^+\pi^0\rangle - |\pi^0\pi^+\rangle\}; \quad |1,0\rangle = \frac{1}{\sqrt{2}}\{|\pi^+\pi^-\rangle - |\pi^-\pi^+\rangle\}; \\ |1,-1\rangle &= \frac{1}{\sqrt{2}}\{|\pi^0\pi^-\rangle - |\pi^-\pi^0\rangle\}; \end{aligned} \tag{A.2.74}$$

$$|0,0\rangle = \frac{1}{\sqrt{3}}\{|\pi^+\pi^-\rangle + |\pi^-\pi^+\rangle - |\pi^0\pi^0\rangle\}. \tag{A.2.75}$$

The isotriplet state is antisymmetric and the isosinglet state is symmetric under the exchange of the two pions.

The expansion of the two-pion states reads

$$\begin{aligned} |\pi^+\pi^0\rangle &= \frac{1}{\sqrt{2}}\{|2,1\rangle + |1,1\rangle\}; \quad |\pi^+\pi^-\rangle = \frac{1}{\sqrt{6}}|2,0\rangle + \frac{1}{\sqrt{2}}|1,0\rangle + \frac{1}{\sqrt{3}}|0,0\rangle; \\ |\pi^0\pi^0\rangle &= \sqrt{\tfrac{2}{3}}|2,0\rangle - \frac{1}{\sqrt{3}}|0,0\rangle; \quad |\pi^-\pi^0\rangle = \frac{1}{\sqrt{2}}\{|2,-1\rangle - |1,-1\rangle\}. \end{aligned} \tag{A.2.76}$$

The $I=2$ terms do not contribute in our case.

The connection between the isospin amplitudes in the s- and t-channels is given by the $s-t$ crossing relation

$$\begin{pmatrix} A^{1/2} \\ A^{3/2} \end{pmatrix} = \begin{pmatrix} \frac{1}{\sqrt{6}} & 1 \\ \frac{1}{\sqrt{6}} & -\frac{1}{2} \end{pmatrix} \begin{pmatrix} A^0 \\ A^1 \end{pmatrix}, \tag{A.2.77}$$

from which follows that A^+ and A^- have a well-defined quantum number I_t

$$A^+ = \frac{1}{\sqrt{6}} A^0; \quad A^- = \tfrac{1}{2} A^1. \tag{A.2.78}$$

We obtain for those reactions in which the target particle is a proton

$$\begin{aligned} A(\bar{n}p \to \pi^+\pi^0) &= \frac{1}{\sqrt{2}} A^1 &&= \sqrt{2} A^- \quad = -A_0 = -A(\bar{n}p \to \pi^0\pi^+), \\ A(\bar{p}p \to \pi^+\pi^-) &= \tfrac{1}{2} A^1 + \frac{1}{\sqrt{6}} A^0 = A^- + A^+ = A_-, \\ A(\bar{p}p \to \pi^0\pi^0) &= -\frac{1}{\sqrt{6}} A^0 &&= -A^+. \end{aligned} \tag{A.2.79}$$

2.3 Electromagnetic contributions

In analyses of experimental data, one has to consider the electromagnetic interaction between nucleon and pion, which gives an important contribution in certain kinematic regions. We shall list only the formulas for high energy near-forward scattering and refer to the literature for the more complicated cases of high energy near-backward scattering and low energy scattering (Sects. A.12 and 2.1.17).

In the one-photon exchange approximation the electromagnetic contributions to the invariant amplitudes C and B for elastic $\pi^\pm$p-scattering read

$$\begin{aligned} C_\pm^{\mathrm{em}}(s,t) &= \mp C_{\mathrm{em}}(s,t) = \mp \frac{8\pi\alpha}{(-t)} \frac{\nu}{1-t/4m^2} f_{\mathrm{E}}(t); \\ B_\pm^{\mathrm{em}}(s,t) &= \mp B_{\mathrm{em}}(s,t) = \mp \frac{8\pi\alpha}{(-t)} f_{\mathrm{M}}(t). \end{aligned} \tag{A.2.80}$$

$\alpha = e^2/\hbar c \approx 1/137$, and $f_{\mathrm{E}}(t), f_{\mathrm{M}}(t)$ are products of the electromagnetic Sachs form factors of the proton and the pion form factor (see Sects. A.11.1 and A.11.2).

$$f_{\mathrm{E}}(t) = G_{\mathrm{Ep}}(t) F_\pi(t); \quad f_{\mathrm{M}}(t) = G_{\mathrm{Mp}}(t) F_\pi(t). \tag{A.2.81}$$

At very small values of $(-t)$, it is sufficient to take the first two terms of an expansion

$$\begin{aligned} f_{\mathrm{E}}(t) &= 1 - (r_\pi^2 + r_{\mathrm{Ep}}^2)(-t)/6; \quad & f_{\mathrm{M}}(t) &= g_{\mathrm{p}}[1 - (r_\pi^2 + r_{\mathrm{Mp}}^2)(-t)/6]; \\ &= 1 - 4.97(-t); & &= 2.793[1 - 5.02(-t)]; \quad t \text{ in GeV}^2. \end{aligned} \tag{A.2.82}$$

r_π, r_{Ep}, r_{Mp} are the pion and nucleon "radii" and g_{p} is the magnetic moment of the proton (see Table A.11.1 for the numerical values).

Higher order corrections are usually taken into account only for C^{em} by adding a phase factor

$$\begin{aligned} C_\pm^{\mathrm{em}}(s,t) &= \mp C_{\mathrm{em}}(s,t) e^{\pm 2i\delta(t)}; \\ 2\delta(t) &\approx -\alpha\{\ln(-tR^2/6) + 0.577\}, \quad \text{where} \quad R^2 = 3b + r_\pi^2 + r_{\mathrm{EP}}^2. \end{aligned} \tag{A.2.83}$$

$b/2$ is the logarithmic slope of $\operatorname{Im} C$ at $t=0$, i.e. b agrees approximately with the slope of the diffraction peak.

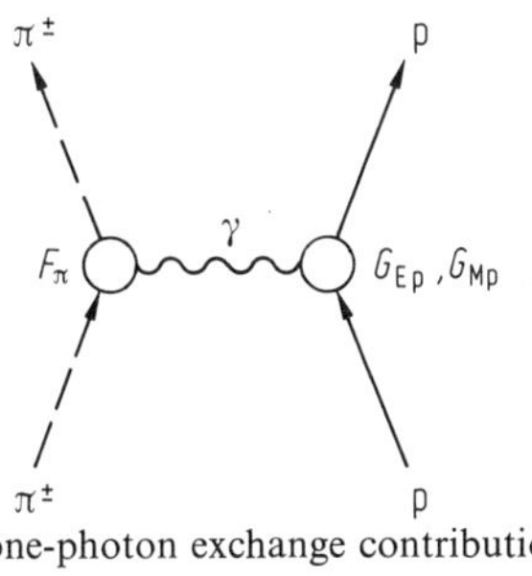

Fig. A.2.6. The one-photon exchange contribution to elastic $\pi^\pm$p scattering.

3 Expansions of scattering amplitudes

3.1 s-channel partial wave expansion

The expansions of the spin no-flip and spin flip amplitudes G and H in terms of partial wave amplitudes $T_{\ell\pm}$ read ($z=\cos\theta$)

$$G(s,z)=\frac{1}{q}\sum_{\ell=0}^{\infty}\{(\ell+1)T_{\ell+}(s)+\ell T_{\ell-}(s)\}P_\ell(z);$$
$$H(s,z)=\frac{1}{q}\sum_{\ell=1}^{\infty}\{T_{\ell+}(s)-T_{\ell-}(s)\}\sin\theta P'_\ell(z), \tag{A.3.1}$$

where $\ell\pm$ indicates that the total angular momentum is $J=\ell\pm1/2$. The $P_\ell(z)$ are Legendre polynomials and $P'_\ell(z)$ their derivatives.

The dimensionless partial wave amplitudes $T_{\ell\pm}$ can be expressed by phase shifts $\delta_{\ell\pm}$ and absorption parameters $\eta_{\ell\pm}$

$$T_{\ell\pm}\equiv qf_{\ell\pm}=\frac{1}{2\mathrm{i}}\{\eta_{\ell\pm}\exp(2\mathrm{i}\delta_{\ell\pm})-1\}. \tag{A.3.2}$$

Unitarity requires that δ be real and $0\leqq\eta\leqq1$. The partial wave amplitude $f_{\ell\pm}(s)$ is preferable in investigations of the analytic properties.

The logarithmic slope of $\mathrm{Im}\,G$ at $t=0$

$$\left.\frac{\partial\,\mathrm{Im}\,G(s,t)/\partial t}{\mathrm{Im}\,G(s,t)}\right|_{t=0}=\frac{\pi}{\sigma q^4}\sum_{\ell=1}^{\infty}\ell(\ell+1)\{(\ell+1)\,\mathrm{Im}\,T_{\ell+}+\ell\,\mathrm{Im}\,T_{\ell-}\} \tag{A.3.3}$$

is sensitive to the contributions of high partial waves, since these are enhanced by a factor $\approx\ell^3$. $\mathrm{Im}\,G(s,0)=q\sigma/4\pi$, where σ is the total cross section.

Sometimes, it is an advantage to work with the amplitudes $f_1(s,z)=G+H\cot\theta$ and $f_2(s,z)=-H/\sin\theta$, Eq. (A.2.2), whose partial wave expansion reads

$$f_1(s,z)=\sum_{\ell=1}^{\infty}\{f_{(\ell-1)+}-f_{(\ell+1)-}\}P'_\ell(z)=\sum_{J=1/2}^{\infty}\{P'_{J+1/2}(z)f_{\ell+}-P'_{J-1/2}(z)f_{\ell-}\};$$
$$f_2(s,z)=\sum_{\ell=1}^{\infty}(f_{\ell-}-f_{\ell+})P'_\ell(z)\quad=\sum_{J=1/2}^{\infty}\{P'_{J+1/2}(z)f_{\ell-}-P'_{J-1/2}(z)f_{\ell+}\}. \tag{A.3.4}$$

Starting from $G(s,z)$, $H(s,z)$ or from $f_1(s,z)$, $f_2(s,z)$, one can calculate the partial waves from the projection formula

$$T_{\ell\pm}(s)/q\equiv f_{\ell\pm}(s)=\tfrac{1}{2}\int_{-1}^{+1}\mathrm{d}z\{f_1(s,z)P_\ell(z)+f_2(s,z)P_{\ell\pm1}(z)\}. \tag{A.3.5}$$

A graphical representation of the energy dependence of $T_{\ell\pm}(s)$ is provided by its "Argand diagram", which is a plot of this amplitude in a complex plane (Fig. A.3.1). In the case of an elastic Breit-Wigner-type resonance without background, the locus of $T_{\ell\pm}$ moves in a counter-clockwise sense around a circle of radius 1/2, whose center lies at $T_{\ell\pm}=\mathrm{i}/2$. The absorption parameter $\eta_{\ell\pm}$ is equal to one and the "speed" $\mathrm{d}\delta_{\ell\pm}/\mathrm{d}W$ has its maximum at the resonance energy ($W=M_{\mathrm{res}}$), which belongs to $\delta_{\ell\pm}=90°$ and $T_{\ell\pm}=\mathrm{i}$. In general, the resonances are inelastic and have a substantial background (Fig. A.3.1b). Parametrizations for $T_{\ell\pm}$ are given in Sect. 2.4.1.2.

Next we consider the partial wave expansions of the invariant amplitudes. The sums start at $\ell=0$ for terms $\propto T_{\ell+}$ and at $\ell=1$ for terms $\propto T_{\ell-}$; $[p_-^2=m^2-t/4]$.

$$A(s,t)=\frac{4\pi m\omega}{q^3}\sum_\ell\left\{T_{\ell+}\left[\left(1+\frac{\omega_q}{\omega}\right)P'_\ell+\left(1-\frac{\omega_q}{\omega}\right)P'_{\ell+1}\right]-T_{\ell-}\left[\left(1+\frac{\omega_q}{\omega}\right)P'_\ell+\left(1-\frac{\omega_q}{\omega}\right)P'_{\ell-1}\right]\right\};$$
$$B(s,t)=\frac{4\pi E}{q^3}\sum_\ell\left\{-T_{\ell+}\left[\left(1+\frac{m}{E}\right)P'_\ell-\left(1-\frac{m}{E}\right)P'_{\ell+1}\right]+T_{\ell-}\left[\left(1+\frac{m}{E}\right)P'_\ell-\left(1-\frac{m}{E}\right)P'_{\ell-1}\right]\right\}; \tag{A.3.6}$$
$$C(s,t)=\frac{2\pi W}{p_-^2}\frac{E}{q}\sum_\ell\left\{(\ell+1)T_{\ell+}\left[\left(1+\frac{m}{E}\right)P_\ell-\left(1-\frac{m}{E}\right)P_{\ell+1}\right]+\ell T_{\ell-}\left[\left(1+\frac{m}{E}\right)P_\ell-\left(1-\frac{m}{E}\right)P_{\ell-1}\right]\right\}.$$

The projection formula reads

$$16\pi W f_{\ell\pm}(W)=(E+m)\{A_\ell+(W-m)B_\ell\}+(E-m)\{-A_{\ell\pm1}+(W+m)B_{\ell\pm1}\}, \tag{A.3.7}$$

where

$$A_\ell(s)=\int_{-1}^{+1} \mathrm{d}z P_\ell(z)A(s,t); \quad B_\ell(s)=\int_{-1}^{+1} \mathrm{d}z P_\ell(z)B(s,t). \tag{A.3.8}$$

$-2q^2(1-z)$ has to be inserted for t.

Finally, it is of interest to list the partial wave expansions of the s-channel helicity amplitudes (see Table A.3.1 for the d-functions).

$$F_{+\pm}(s,t)=\frac{2\sqrt{\pi}}{q^2}\sum_{J=1/2}^{\infty}(J+\tfrac{1}{2})d^J_{\pm+}(\theta)F^J_{+\pm}(s). \tag{A.3.9}$$

The inversion formula reads

$$F^J_{+\pm}(s)=\frac{q^2}{2\sqrt{\pi}}\int_{-1}^{+1}\mathrm{d}z\, d^J_{\pm+}(\theta)F_{+\pm}(s,t). \tag{A.3.10}$$

Under the integral $-2q^2(1-z)$ has to be substituted for t.

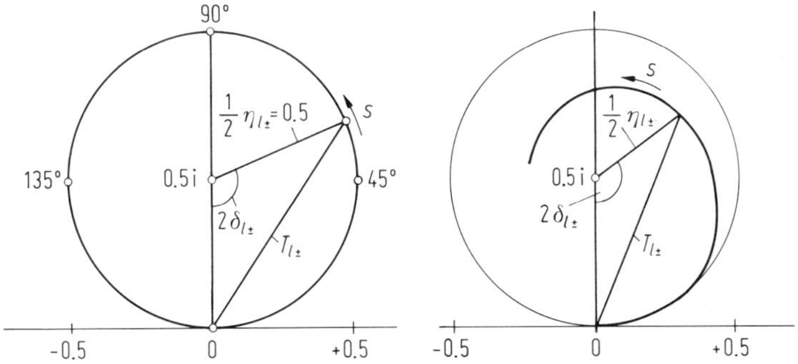

Fig. A.3.1. Argand diagrams of partial wave amplitudes. Figure A.3.1a shows the Argand diagram of an elastic resonance ($\eta=1$) without background. The vector moves in a counter-clockwise direction as the energy increases. The angles refer to $\delta_{\ell\pm}$. Figure A.3.1b shows an inelastic resonance. Argand diagrams of πN partial wave amplitudes are presented in Sect. 2.2.2.

Table A.3.1. d-functions. Our $d^J_{+\pm}$ agree with $d^J_{1/2,\pm1/2}$ in the "Review of particle properties" (80 R Brigman, p. S. 34). $z=\cos\theta$, $P'_J(z)=\mathrm{d}P_J(z)/\mathrm{d}z$.

$$(J+\tfrac{1}{2})d^J_{++}(\theta)=\cos(\theta/2)\{P'_{J+1/2}(z)-P'_{J-1/2}(z)\};$$
$$(J+\tfrac{1}{2})d^J_{-+}(\theta)=\sin(\theta/2)\{P'_{J+1/2}(z)+P'_{J-1/2}(z)\};$$
$$d^J_{+-}=-d^J_{-+}; \quad d^J_{++}=d^J_{--}; \quad d^J_{++}(\pi)=0; \quad d^J_{-+}(\pi)=(-1)^{J-1/2};$$
$$d^{1/2}_{++}=\cos(\theta/2); \quad d^{1/2}_{-+}=\sin(\theta/2);$$
$$d^{3/2}_{++}=\tfrac{1}{2}\cos(\theta/2)[3\cos\theta-1]; \quad d^{3/2}_{-+}=\tfrac{1}{2}\sin(\theta/2)[3\cos\theta+1].$$

Parity conservation leads to

$$F^J_{++}=F^J_{--}; \quad F^J_{-+}=F^J_{+-}. \tag{A.3.11}$$

The partial waves $T_{\ell\pm}$ of Eq. (A.3.2) have a definite parity $P=-(-1)^\ell$. They are linear combinations of the $F^J_{+\pm}$

$$T_{\ell+}=F^J_{++}+F^J_{+-}, \text{ where } J=\ell+1/2; \quad T_{\ell-}=F^J_{++}-F^J_{+-}, \text{ where } J=\ell-1/2;$$
$$2F^J_{+\pm}=T_{(J-1/2)+}\pm T_{(J+1/2)-}. \tag{A.3.12}$$

In Regge models for fermion exchange in the u-channel (Sect. A.9.2), one needs *parity-conserving helicity amplitudes* which are proportional to the amplitudes f_1 and f_2. In order to understand the name, we use Eq. (A.2.31) and insert the partial wave expansions (A.3.4) or (A.3.9)

$$f_{1,2}(s,z)=\frac{q}{2\sqrt{\pi}}\left\{\frac{F_{++}}{\cos(\theta/2)}\pm\frac{F_{+-}}{\sin(\theta/2)}\right\}=\sum_J\{P'_{J+1/2}f_{\ell\pm}-P'_{J-1/2}f_{\ell\mp}\}. \quad \text{(A.3.13)}$$

The upper signs belong to f_1 and the lower signs to f_2.

In addition to the usual space-parity P we introduce the quantum number *naturality* (or *normality*) η which is defined by

$$P=\eta(-1)^{J-1/2} \text{ for } \pi\text{N states}; \qquad P=\eta(-1)^J \text{ for } \pi\pi\text{- or } \text{N}\bar{\text{N}}\text{-states}. \quad \text{(A.3.14)}$$

States for which $\eta=+1$ and -1 are called *natural and unnatural parity states* respectively. For πN states, the naturality is given by the difference between the total and orbital angular momentum: $J-\ell=-\eta/2$. See Table A.3.2 for examples (in Sect. A.3.2 we use $-\varepsilon$ instead of η).

The naturality of the first term in the sum of Eq. (A.3.13) is $\eta=-1$ for f_1 and $\eta=+1$ for f_2. The name "parity conserving" is derived from the fact that the latter term is dominant in the Regge model.

3.2 Expansion of transversity amplitudes

We use the notation of Sect. A.2.1.2. The naturality is denoted by $\varepsilon=-\eta$, where η is defined in Eq. (A.3.14).

Omitting the energy variable in the argument, the expansions of G and H, Eq. (A.3.1), give for the transversity amplitude $F(\pm;z)$, Eq. (A.2.4), ($T_{\ell\pm}$ is written $T_{J,\varepsilon}$ using $J=\ell+\varepsilon/2$)

$$qF(\pm;z)=\sum_{J,\varepsilon}[(J+\tfrac{1}{2})P_\ell(z)\pm i\varepsilon\sqrt{1-z^2}\,P'_\ell(z)]T_{J,\varepsilon}. \quad \text{(A.3.15)}$$

This expansion is written in terms of w, Eq. (A.2.6),

$$qF(w)=\sum_{J,\varepsilon}R_{J,\varepsilon}T_{J,\varepsilon};$$

$$R_{J,\varepsilon}(w)=\left(J+\frac{1}{2}\right)P_{J-\varepsilon/2}\left(\frac{w+w^{-1}}{2}\right)+\varepsilon\frac{w-w^{-1}}{2}P'_{J-\varepsilon/2}\left(\frac{w+w^{-1}}{2}\right). \quad \text{(A.3.16)}$$

$R_{J,\varepsilon}(w)$ has the following properties:

1. Normalization: $\int_0^{2\pi} d\theta|\sin\theta|R^*_{J',\varepsilon'}(e^{i\theta})R_{J,\varepsilon}(e^{i\theta})=4\left(J+\frac{1}{2}\right)\delta_{JJ'}\delta_{\varepsilon\varepsilon'}$;
2. From reality of coefficients: $R^*_{J,\varepsilon}(w)=R_{J,\varepsilon}(w^*)$;
3. Minami-transformation: $R_{J,-\varepsilon}(w)=w^{-1}R_{J,\varepsilon}(w^{-1})$;
4. Recurrence relation: $(2J-1)R_{J,\varepsilon}(w)=R_{J-1,-\varepsilon}(w)+2Jw^{\varepsilon}R_{J-1,\varepsilon}(w)$;
5. $w^{J-\varepsilon/2}R_{J,\varepsilon}$ is a polynomial in w^2 of order $J-1/2$. $R_{J,\varepsilon}$ has $J-1/2$ pairs of complex conjugate zeros in the w-plane. They lie $^{\text{outside}}_{\text{inside}}$ the unit circle $|w|=1$ if $\varepsilon \lessgtr 0$, and approach the unit circle if J increases.

The following table gives explicit expressions for the lowest functions $R_{J,\varepsilon}(w)$ and the location of their zeros.

Table A.3.2. The functions $R_{J,\varepsilon}(w)$ for the lowest J-values and the location of their zeros in the w-plane and in the two sheets $z^{\pm}$ of the z-plane. All 4 combinations of the signs have to be taken in the column "w-plane". The relation to the zeros in the z-plane follows from Fig. A.2.1. In the $z^{\pm}$-planes, the real parts of the zeros lie between -1 and $+1$ and have the same property of interlacing as the zeros of Legendre polynomials.

$\ell 2J$	ε	$R_{J,\varepsilon}(w)$	Location of the zeros		
			w-plane	z^+-plane	z^--plane
S1	$+1$	1	—	—	—
P1	-1	w^{-1}	—	—	—
P3	$+1$	$\frac{1}{2}(3w+w^{-1})$	$\pm i/\sqrt{3}$	$-i/\sqrt{3}$	$i/\sqrt{3}$
D3	-1	$\frac{1}{2}(3w^{-2}+1)$	$\pm\sqrt{3}i$	$i/\sqrt{3}$	$-i/\sqrt{3}$

Table A.3.2, continued

Table A.3.2, continued

$\ell 2J$	ε	$R_{J,\varepsilon}(w)$	Location of the zeros		
			w-plane	z^+-plane	z^--plane
D5	+1	$\frac{1}{8}(15w^2+6+3w^{-2})$	$\pm 0.35 \pm 0.57i$	$\pm 0.57-0.35i$	$\pm 0.57+0.35i$
F5	−1	$\frac{1}{8}(3w+6w^{-1}+15w^{-3})$	$\pm 0.79 \pm 1.27i$	$\pm 0.57+0.35i$	$\pm 0.57-0.35i$
F7	+1	$\frac{1}{16}(35w^3+15w+9w^{-1}+5w^{-3})$	$\pm 0.70i$	$-0.36i$	$0.36i$
			$\pm 0.53 \pm 0.50i$	$\pm 0.76-0.22i$	$\pm 0.76+0.22i$
G7	−1	$\frac{1}{16}(5w^2+9+15w^{-2}+35w^{-4})$	$\pm 1.42i$	$0.36i$	$-0.36i$
			$\pm 0.99 \pm 0.93i$	$\pm 0.76+0.22i$	$\pm 0.76-0.22i$

3.3 Impact parameter representations

3.3.1 The Fourier-Bessel transform

The partial wave expansion, Eq. (A.3.1), is well suited for a description of the amplitudes at low and intermediate energies, where a relatively small number of partial waves give the main contribution. The ℓ-dependence at fixed energy is not smooth, because certain partial waves are strongly enhanced due to nucleon resonances.

Above the resonance region, one expects that the partial wave expansions of the helicity amplitudes show a smooth dependence on angular momentum. Furthermore, the number of significant partial waves becomes larger as the energy increases. This situation is favourable for an alternative description of the scattering amplitudes, in which the sum over partial waves is replaced by an integral transform called "impact parameter representation". It is based on ideas related to optical models and to the eikonal method in potential scattering.

This representation is usually treated for the spinless case, but it can be applied as well to the no-flip amplitude in πN scattering, because the partial wave expansion, Eq. (A.3.1), agrees with that in the spinless case if we introduce

$$(2\ell+1)\hat{T}_\ell(s) \equiv (\ell+1)T_{\ell+} + \ell T_{\ell-} . \qquad \text{(A.3.17)}$$

There exist several versions of impact parameter representations of which we prefer the Cottingham-Peierls formalism, because it allows us to work with exact formulas.

The impact parameter representation for the no-flip amplitude and the inversion formula are given by the Fourier-Bessel transforms

$$G(s,t) = 2q \int_0^\infty g(s,b) J_0(b\sqrt{-t})\, b\,db \quad \text{for} \quad 0 \geqq t \geqq -4q^2; \qquad G(s,t) \equiv 0 \quad \text{for} \quad t < -4q^2; \qquad \text{(A.3.18)}$$

$$g(s,b) = \frac{1}{4q} \int_{-4q^2}^{0} G(s,t) J_0(b\sqrt{-t})\,dt . \qquad \text{(A.3.19)}$$

Inserting the Legendre expansion (A.3.1)

$$G(s,t) = \frac{1}{q} \sum_\ell (2\ell+1)\hat{T}_\ell(s) P_\ell(z) \qquad \text{(A.3.20)}$$

into (A.3.19), one obtains the Neumann series expansion

$$g(s,b) = \frac{1}{qb} \sum_\ell (2\ell+1)\hat{T}_\ell(s) J_{2\ell+1}(2bq) . \qquad \text{(A.3.21)}$$

Its inverse

$$\hat{T}_\ell(s) = \int_0^\infty g(s,b) J_{2\ell+1}(2bq)\, d(2bq) \qquad \text{(A.3.22)}$$

follows from Eq. (A.3.18) and the projection formula

$$\hat{T}_\ell(s) = \frac{q}{2} \int_{-1}^{+1} G(s,t) P_\ell(z)\,dz = \frac{1}{4q} \int_{-4q^2}^{0} G(s,t) P_\ell\left(1+\frac{t}{2q^2}\right) dt . \qquad \text{(A.3.23)}$$

The well-known approximate relation between $\hat{T}_\ell(s)$ and $g(s,b)$ can be derived by inserting the expression (66R Watson, p. 158)

$$P_\ell(\cos\theta)=J_0(\eta)+\sin^2(\theta/2)\left\{\frac{J_1(\eta)}{\eta}-J_2(\eta)+\frac{\eta}{6}J_3(\eta)\right\}+O(\sin^4(\theta/2))\,; \tag{A.3.24}$$

where

$$\eta=2bq\sin(\theta/2)=b\sqrt{-t}\,;\qquad b=\frac{\ell+1/2}{q}. \tag{A.3.25}$$

For a restricted class of functions $g(s,b)$, one obtains ($g'=\partial g/\partial b^2$ etc.)

$$\hat{T}_\ell(s)=g(s,b)-\frac{1}{4q^2}\left\{g'+4b^2g''+\frac{4}{3}b^4g'''\right\}+O(q^{-4}). \tag{A.3.26}$$

$O(q^{-4})$ refers only to the contribution from the last term in (A.3.24). The validity of the approximate relation

$$\hat{T}_\ell(s)\approx g(s,b)\,;\qquad b=(\ell+1/2)/q\,, \tag{A.3.27}$$

depends on the behaviour of $g(s,b)$. Furthermore, one has to specify which of the quantities s, b is held fixed and which goes to infinity. We shall discuss the application to diffraction scattering in Sect. 2.6.3. See Fig. A.3.2 for the classical interpretation of b.

If one starts from Eq. (A.3.22), the approximation (A.3.27) follows if $J_{2\ell+1}(2bq)$ is replaced by $\delta(2\ell+1-2bq)$, which is the first term of an expansion derived in 64L Curtis. For *small* ℓ, this seems to be rather crude, but a direct evaluation of the Neumann series (A.3.21) for $\ell=0$ and 1 shows that Eq. (A.3.27) is approximately valid

$$\hat{T}_0(s)=g(s,0)\,;\qquad \hat{T}_1(s)=g(s,0)+\frac{2}{q^2}\frac{\partial g}{\partial b^2}\bigg|_{b=0}\approx g(s,\sqrt{2}/q)\,. \tag{A.3.28}$$

It is remarkable that, in diffraction scattering, $\mathrm{Im}\,\hat{T}_0(s)$ is not only determined by the transform $\mathrm{Im}\,g(s,b)$ at $b=0$ but, according to Eq. (A.3.23), also by the scattering amplitude $\mathrm{Im}\,G(s,t)$ at *small* t.

In the literature, it is usually stated that the leading term in Eq. (A.3.24)

$$P_\ell(\cos\theta)\approx J_0\{(2\ell+1)\sin(\theta/2)\}=J_0(b\sqrt{-t}) \tag{A.3.24a}$$

is an approximation for small angles and *large* ℓ. Figure A.3.3 shows to what extent the approximation is already valid at *small* ℓ.

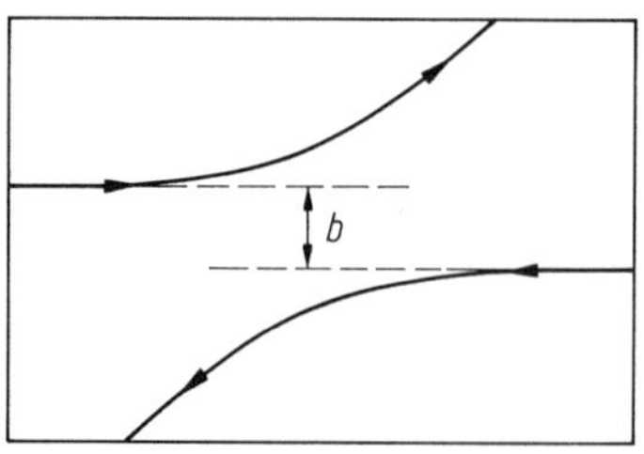

Fig. A.3.2. The impact parameter. A well-known semiclassical estimate of the impact parameter in optical models is $\hbar\ell=\hbar qb$. In our formalism, the partial wave amplitude is approximately equal to the impact parameter transform if $\ell+1/2=bq$.

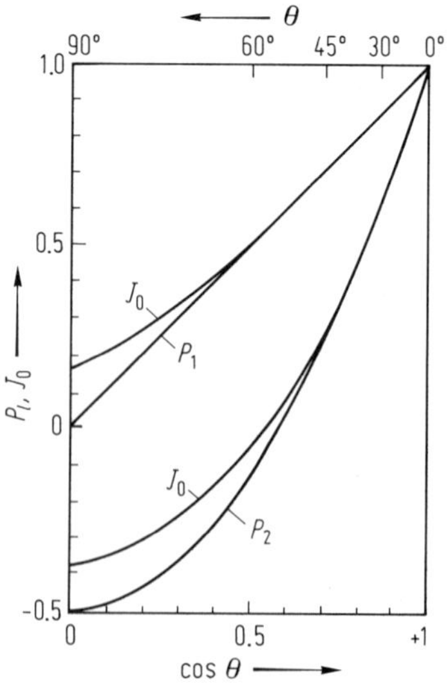

Fig. A.3.3. Validity of the approximation (A.3.24a) for $\ell=1$ and 2. It is seen that the approximation is very good up to $\theta=45°$.

In diffraction scattering, the real part of the amplitude is frequently neglected and the shape of the imaginary part is described by the *profile function*, which differs from $\operatorname{Im} g(s,b)$ only by the normalization

$$\Gamma(s,b)=\frac{8\pi}{\sigma}\operatorname{Im} g(s,b)\,; \qquad \int_0^\infty \Gamma(s,b)b\,\mathrm{d}b=1\,, \tag{A.3.29}$$

where σ is the total cross section. Γ does not depend on s in the simplest models (see Sect. 2.6.4).

If one assumes an exponential decrease of $\operatorname{Im} G(s,t)$, the profile function can be evaluated in terms of an exponential and of Lommel functions V_0, V_1

$$\begin{aligned}&\operatorname{Im} G(s,t)=(q\sigma/4\pi)\exp(Bt/2)\,;\\&\Gamma(s,b)=2q^2\left\{\int_0^\infty-\int_1^\infty\right\}J_0(2bqu)\mathrm{e}^{-2Bq^2u^2}\mathrm{d}u^2\\&\quad=B^{-1}\mathrm{e}^{-b^2/2B}-\mathrm{e}^{-2Bq^2}\{V_0(4Bq^2\mathrm{i},2bq)+\mathrm{i}V_1(4Bq^2\mathrm{i},2bq)\}B^{-1}\,.\end{aligned} \tag{A.3.30}$$

The first term belongs to the lower limit $-\infty$ instead of $-4q^2$ in Eq. (A.3.19), the latter value being chosen in the BLANKENBECLER-GOLDBERGER representation. The second term is negligible for small $2bq$.

Finally, we mention the definition of some quantities which are used in applications of the impact parameter representation. The *eikonal* χ corresponds to the complex phase shift [cf. Eq. (A.3.2)], Ω is called "*opacity*" or "opaqueness"

$$\begin{aligned}&g(s,b)=\frac{1}{2\mathrm{i}}(\mathrm{e}^{2\mathrm{i}\chi}-1)\,;\quad \chi(s,b)=\delta(s,b)-\frac{\mathrm{i}}{2}\ln\eta(s,b)=\delta(s,b)+\frac{\mathrm{i}}{2}\Omega(s,b)\,;\\&S(s,b)=1+2\mathrm{i}g(s,b)=\mathrm{e}^{2\mathrm{i}\chi(s,b)}=\mathrm{e}^{-\Omega}\mathrm{e}^{2\mathrm{i}\delta}\,.\end{aligned} \tag{A.3.31}$$

$S(s,b)$ corresponds to the diagonal S-matrix element $S_{\ell\pm}=1+2\mathrm{i}T_{\ell\pm}$.

The complex *optical potential* $V(s,r)$ and the *optical Born amplitude* $G_B(s,t)$ are defined by

$$\begin{aligned}&V(s,r)=\frac{2q}{\pi}\int_r^\infty\frac{\partial\chi(s,b)}{\partial b}\frac{\mathrm{d}b}{\sqrt{b^2-r^2}}=-\frac{1}{2\pi}\int_{-4q^2}^0 G_B(s,t)\frac{\sin(r\sqrt{-t})}{r\sqrt{-t}}\sqrt{-t}\,\mathrm{d}t\,;\\&G_B(s,t)=2q\int_0^\infty J_0(b\sqrt{-t})\chi(s,b)b\,\mathrm{d}b=-2\int_0^\infty\frac{\sin(r\sqrt{-t})}{r\sqrt{-t}}V(s,r)r^2\mathrm{d}r\,;\\&\chi(s,b)=-\frac{1}{q}\int_b^\infty\frac{V(s,r)}{\sqrt{r^2-b^2}}r\,\mathrm{d}r=-\frac{1}{q}\int_0^\infty V(s,\sqrt{b^2+z^2})\mathrm{d}z=\frac{1}{4q}\int_{-4q^2}^0 G_B(s,t)J_0(b\sqrt{-t})\mathrm{d}t\,.\end{aligned} \tag{A.3.32}$$

The expression for G_B in terms of V is the same as the one obtained in lowest order perturbation theory in non-relativistic potential scattering, except for a factor $2m_{\mathrm{red}}/\hbar^2$ (m_{red} = reduced mass). The integrals can be evaluated analytically in the case

$$V(r)=\frac{e^{-\alpha\sqrt{r^2+c^2}}}{\sqrt{r^2+c^2}}\,;\qquad \chi(b)=-\frac{1}{q}K_0(\alpha\sqrt{b^2+c^2})\,;\qquad G_B(t)=-\frac{-2c}{\sqrt{\alpha^2-t}}K_1(c\sqrt{\alpha^2-t})\,. \tag{A.3.33}$$

3.3.2 Two-dimensional Fourier transforms

We introduce the following notation for the Fourier-Bessel transform

$$X(t)\equiv\langle x(b)\rangle=\int_0^\infty x(b)J_0(b\sqrt{-t})b\,\mathrm{d}b \tag{A.3.34}$$

and insert an integral representation for the Bessel function

$$J_0(z)=\frac{1}{2\pi}\int_{-\pi}^{\pi}\mathrm{e}^{\mathrm{i}z\cos\phi}\mathrm{d}\phi\,. \tag{A.3.35}$$

The result can be written as a two-dimensional Fourier transform

$$X(t)\equiv\langle x(b)\rangle=\frac{1}{2\pi}\iint_{-\infty}^{\infty}\mathrm{d}^2b\,x(b)\mathrm{e}^{\mathrm{i}\boldsymbol{\varkappa}\cdot\boldsymbol{b}}\,, \tag{A.3.36}$$

where the vectors $\boldsymbol{\varkappa}$ and $\boldsymbol{b}$ are defined by (see Fig. A.1.2 for the $x-y$-plane)

$$\boldsymbol{\varkappa}=(\varkappa_x,\varkappa_y)\,;\quad \boldsymbol{b}=(b_x,b_y)\,;\quad \varkappa^2=-t\,;\quad \boldsymbol{\varkappa}\cdot\boldsymbol{b}=\varkappa b\cos\phi\,. \tag{A.3.37}$$

The inverse of Eq. (A.3.36) reads

$$x(b)\equiv\langle X(t)\rangle=\frac{1}{2\pi}\int\!\!\int_{-\infty}^{\infty}\mathrm{d}^2\varkappa X(t)\mathrm{e}^{-\mathrm{i}\boldsymbol{\varkappa}\cdot\boldsymbol{b}}=\frac{1}{2}\int_{-\infty}^{0}X(t)J_0(b\sqrt{-t})\mathrm{d}t\,. \tag{A.3.38}$$

Inserting Eq. (A.3.34) into Eq. (A.3.38), one obtains

$$\int_{-\infty}^{0}\mathrm{d}tJ_0(b\sqrt{-t})J_0(b'\sqrt{-t})=4\delta(b'^2-b^2)\,. \tag{A.3.39}$$

The convolution integrals are defined by

$$X*Y=\frac{1}{2\pi}\int\!\!\int_{-\infty}^{\infty}X(\boldsymbol{b}-\boldsymbol{b}')Y(\boldsymbol{b}')\mathrm{d}^2b'\,;\qquad x*y=\frac{1}{2\pi}\int\!\!\int_{-\infty}^{\infty}x(\boldsymbol{\varkappa}-\boldsymbol{\varkappa}')y(\boldsymbol{\varkappa}')\mathrm{d}^2\varkappa'\,. \tag{A.3.40}$$

The transform of a convolution integral is the product of the transforms of the two functions

$$\langle X*Y\rangle=\langle X\rangle\langle Y\rangle\,;\qquad \langle x*y\rangle=\langle x\rangle\langle y\rangle\,. \tag{A.3.41}$$

We add the application of this formalism to a purely imaginary diffraction amplitude, which belongs to an energy independent diffraction peak

$$G(s,t)=\mathrm{i}qa(t)\,. \tag{A.3.42}$$

In this case, the eikonal and the optical Born term are also purely imaginary: $\chi=(\mathrm{i}/2)\Omega$ and $G_\mathrm{B}=\mathrm{i}qa_\mathrm{B}(t)$. One finds from Eqs. (A.3.32) and (A.3.34) that $a_\mathrm{B}(t)$ is the transform of the opacity

$$\langle\Omega(b)\rangle=a_\mathrm{B}(t)\,, \tag{A.3.43}$$

and from Eqs. (A.3.31) and (A.3.18)

$$S(b)=1+2\mathrm{i}g(s,b)=1-\langle a(t)\rangle\,, \tag{A.3.44}$$

ignoring the integral in the range $-4q^2>t>-\infty$.

Now, one can derive a second expression for $\langle\Omega(b)\rangle$

$$\begin{aligned}\Omega(b)&=-\ln S(b)=-\ln(1-\langle a\rangle)=\langle a\rangle+\tfrac{1}{2}\langle a\rangle^2+\tfrac{1}{3}\langle a\rangle^3+\dots\,;\\ \langle\Omega(b)\rangle&=a(t)+\tfrac{1}{2}a(t)*a(t)+\tfrac{1}{3}a(t)*a(t)*a(t)+\dots\,.\end{aligned} \tag{A.3.45}$$

Comparing this equation with Eq. (A.3.43), it is seen that the optical Born term $a_\mathrm{B}(t)=G_\mathrm{B}(s,t)/\mathrm{i}q$ can be expressed as a sum over convolution integrals over the scattering amplitude $a(t)=G(s,t)/\mathrm{i}q$. See Sect. 2.6.3 for a further discussion of the diffraction peak.

3.3.3 Use of longitudinal and transverse momentum variables

The impact parameter representation (A.3.18), (A.3.19) has some problems which are avoided, if one introduces the impact parameter b as the variable conjugate to the transverse momentum p_T, and takes for the second variable the longitudinal momentum in the c.m. system p_L instead of s [see Eq. (A.1.16)].

The Fourier-Bessel transform is defined for the s-channel helicity amplitude F_{++} (Sect. 2.1.3)

$$f^0_{++}(p_\mathrm{L},b)=\frac{1}{2\sqrt{\pi}}\int_0^\infty\mathrm{d}p_\mathrm{T}^2J_0(p_\mathrm{T}b)F_{++}(p_\mathrm{L},p_\mathrm{T})\,. \tag{A.3.46}$$

The lower limit is $p_\mathrm{T}=q\sin\theta=0$, i.e. the integral starts in the forward or backward direction. The path remains in the physical region and approaches $\theta=90°$ in the high energy limit, whereas in Eq. (A.3.19) we have applied a cut-off at $t=-4q^2$, because the path leaves the physical region. The approximate relation

$$f^0_{++}(p_\mathrm{L},b)\approx F^J_{++}(s) \tag{A.3.47}$$

is valid for small b in the case of diffraction scattering.

Unitarity relations in the impact parameter representation will be treated in Sect. A.5.1.3.

3.4 *t*-channel partial wave expansion

t-channel partial waves are not only of interest for phase shift analyses of the reaction $\bar{\mathrm{N}}\mathrm{N}\to\pi\pi$. In fact, they were used up to now mainly in order to describe processes, in which $\pi\pi$-intermediate states in the *t*-channel couple to a nucleon, for instance in the reactions $\mathrm{NN}\to\mathrm{NN}$ and $\mathrm{e}^-\mathrm{N}\to\mathrm{e}^-\mathrm{N}$. Therefore, we begin with the range $t<4\mu^2$ and shall discuss afterwards the continuation to larger *t*.

3.4.1 $\pi\pi\to\mathrm{N}\bar{\mathrm{N}}$-partial wave expansion in the range $t<4\mu^2$

In this range, we use the following kinematical quantities of Sect. A.1.4

$$p_-=\sqrt{m^2-t/4}\,;\quad q_-=\sqrt{\mu^2-t/4}\,;\quad Z=\frac{m\nu}{p_-q_-}=-\cos\theta_t\,. \tag{A.3.48}$$

The partial wave expansion and its inversion read

$$A(\nu,t)=\frac{8\pi}{p_-^2}\sum_{J=0}^{\infty}\left(J+\frac{1}{2}\right)(p_-q_-)^J\left\{P_J(Z)\tilde{f}_+^J(t)-\frac{mZ}{\sqrt{J(J+1)}}P_J'(Z)\tilde{f}_-^J(t)\right\};$$

$$\tilde{B}(\nu,t)=8\pi\sum_{J=1}^{\infty}\frac{J+\frac{1}{2}}{\sqrt{J(J+1)}}(p_-q_-)^{J-1}P_J'(Z)\tilde{f}_-^J(t); \tag{A.3.49}$$

$$\tilde{C}(\nu,t)=\frac{8\pi}{p_-^2}\sum_{J=0}^{\infty}\left(J+\frac{1}{2}\right)(p_-q_-)^J P_J(Z)\tilde{f}_+^J(t).$$

$$\begin{aligned}\tilde{f}_+^J(t+\mathrm{i}0)&=\frac{1}{8\pi}\int_{-1}^{+1}\mathrm{d}Z P_J(Z)\left\{\frac{p_-^2}{(p_-q_-)^J}A(\nu-\mathrm{i}0,t)+\frac{mZ}{(p_-q_-)^{J-1}}\tilde{B}(\nu-\mathrm{i}0,t)\right\}\\&=\frac{1}{8\pi}\frac{p_-^2}{(p_-q_-)^J}\int_{-1}^{+1}\mathrm{d}Z P_J(Z)\tilde{C}(\nu-\mathrm{i}0,t);\end{aligned} \tag{A.3.50}$$

$$\tilde{f}_-^J(t+\mathrm{i}0)=\frac{1}{8\pi}\frac{\sqrt{J(J+1)}}{2J+1}\frac{1}{(p_-q_-)^{J-1}}\int_{-1}^{+1}\mathrm{d}Z\{P_{J-1}(Z)-P_{J+1}(Z)\}\tilde{B}(\nu-\mathrm{i}0,t).$$

Since the nucleon poles lie near to $\nu=0$ (Fig. A.3.4), we have written the expansion for the amplitudes $\tilde{C}, \tilde{B}, \tilde{f}_\pm^J$, from which the nucleon pole term contributions C_N, B_N, $f_{\pm\mathrm{N}}^J$ are subtracted.

In the argument of the invariant amplitudes in Eq. (A.3.50) one has to insert $\nu=p_-q_-Z/m$. This formula shows that $t+\mathrm{i}0$ belongs to $\nu-\mathrm{i}0$, since Z is real along the integration path.

Along the cuts of $f_\pm^J(t)$ (see Sect. A.7.5), we have made the usual choice, taking the boundary value $f_\pm^J(t+\mathrm{i}0)$ which is also listed in our tables. However, one should notice that we have also adopted the usual choice to take the boundary value $\nu+\mathrm{i}0$ in earlier formulas, for instance in Eq. (A.3.6), and to list it in our tables of the invariant amplitudes (Sect. 2.2.3). Therefore, one has to add a minus sign in front of the imaginary parts of the invariant amplitudes, if Eqs. (A.3.49) and (A.3.50) are used together with other formulas or with tables.

The integration limits belong to backward scattering in the *s*-channel ($Z=1$) and in the *u*-channel ($Z=-1$). They are denoted by B and A in Fig. A.3.4.

In the triangle between $s=(m+\mu)^2$, $u=(m-\mu)^2$, and $t=4\mu^2$, the invariant amplitudes are real (see Sect. A.6.2.3). The projection formula (A.3.50) shows that in the range $t<0$ the partial waves $\tilde{f}_\pm^J(t)$ are complex-valued, because part of the integration path lies outside the triangle. An additional contribution to the imaginary part is due to the nucleon pole term.

Obviously, within the above mentioned triangle the contributions of all imaginary parts must cancel on the right hand side of the partial wave expansion (A.3.49). Cancellations of this type are well-known for Legendre series. A simple example is

$$\sum_{n=0}^{\infty}(-1)^n\{P_{2n}(z)+P_{2n+1}(z)\}=\begin{matrix}1/\sqrt{z}\\0\end{matrix}\quad\text{for}\quad\begin{matrix}0<z<1,\\-1<z<0.\end{matrix} \tag{A.3.51}$$

The series converges for real z in the range $-1<z<+1$ and represents $\mathrm{Re}(1/\sqrt{z})$. It diverges if $\mathrm{Im}\,z\neq0$.

It is useful to list the first terms of the partial wave expansion (A.3.49) explicitly, taking into account the selection rule which follows from the generalized Pauli principle for identical bosons in the reaction $\bar{N}N \to \pi\pi$ (Table A.3.3). The exchange of the two pions in an orbital state with definite J gives a factor $(-1)^J$. Since the singlet isospin $\pi\pi$-state is symmetric and the triplet isospin state is antisymmetric (Sect. A.2.2.4), only the combinations (J even, $I_t=0$) and (J odd, $I_t=1$) are allowed. According to Eq. (A.2.78), this means that the expansion of the isospin even amplitudes A^+, B^+, C^+ contains only partial waves $f_\pm^J$ having even J, whereas only odd J partial waves occur in the expansions of A^-, B^-, C^-.

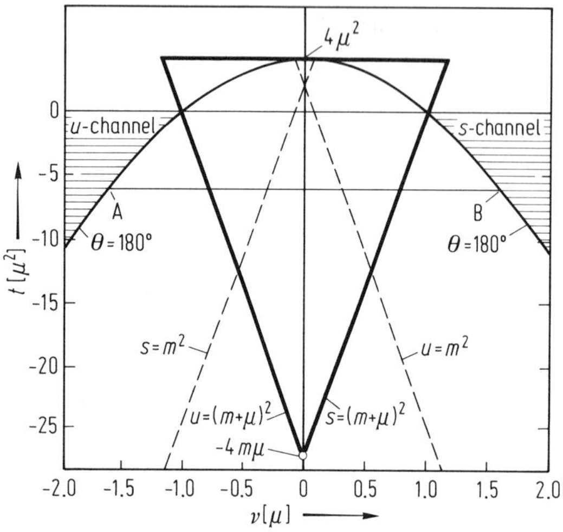

Fig. A.3.4. The Mandelstam plane in the low energy region. The invariant amplitudes are real within the triangle limited by thick lines. The dashed lines show the location of the nucleon poles. A ⋯ B is the integration range in the projection integral (A.3.50) for t-channel partial waves.

Table A.3.3. t-channel partial wave expansion of the isospin even and odd invariant amplitudes. The amplitudes are boundary values $f_+^J(t+\mathrm{i}0)$, $A^+(\nu-\mathrm{i}0, t)$ etc. Cf. the discussion of Eq. (A.3.50). The notation $\tilde{C}$ etc. indicates that the nucleon pole term contribution is subtracted.

$$(p_-^2/4\pi)A^+(\nu,t)=\tilde{f}_+^0(t)+\frac{5}{2}(3m^2\nu^2-(p_-q_-)^2)\tilde{f}_+^2(t)-\frac{15}{\sqrt{6}}m^3\nu^2\tilde{f}_-^2(t)+\ldots$$

$$(p_-^2/4\pi)A^-(\nu,t)=3m\nu\left\{\tilde{f}_+^1(t)-\frac{m}{\sqrt{2}}\tilde{f}_-^1(t)\right\}+\ldots$$

$$(1/4\pi)\tilde{B}^+(\nu,t)=(15m\nu/\sqrt{6})\tilde{f}_-^2(t)+\ldots$$

$$(1/4\pi)\tilde{B}^-(\nu,t)=\frac{3}{\sqrt{2}}\tilde{f}_-^1(t)+\frac{7\sqrt{3}}{4}\{5m^2\nu^2-(p_-q_-)^2\}\tilde{f}_-^3(t)+\ldots$$

$$(p_-^2/4\pi)\tilde{C}^+(\nu,t)=\tilde{f}_+^0(t)+\frac{5}{2}\{3m^2\nu^2-(p_-q_-)^2\}\tilde{f}_+^2(t)+\ldots$$

$$(p_-^2/4\pi)\tilde{C}^-(\nu,t)=3m\nu\tilde{f}_+^1(t)+\frac{7}{2}m\nu\{5m^2\nu^2-3(p_-q_-)^2\}\tilde{f}_+^3(t)+\ldots$$

In the dispersion theory of the nucleon form factor (Sects. 2.5.2 and A.11.1) one needs the following combinations of the $J=1$ partial waves

$$\Gamma_1(t)=\frac{m}{p_-^2}\left\{f_+^1-\frac{t}{4\sqrt{2}m}f_-^1\right\};\quad \Gamma_2(t)=-\frac{m}{p_-^2}\left\{f_+^1-\frac{m}{\sqrt{2}}f_-^1\right\};$$

$$f_+^1(t)=m\left\{\Gamma_1+\frac{t}{4m^2}\Gamma_2\right\};\quad f_-^1(t)=\sqrt{2}(\Gamma_1+\Gamma_2). \tag{A.3.52}$$

They are the leading terms in the expansions of the amplitudes A^- and $\tilde{D}^-$

$$A^-(\nu,t)=-12\pi\nu\tilde{\Gamma}_2(t)+\ldots;\quad \tilde{D}^-(\nu,t)=12\pi\nu\tilde{\Gamma}_1(t)+\ldots. \tag{A.3.53}$$

3.4.2 $\pi\pi\to N\bar{N}$ partial wave expansion in the range $t>4\mu^2$

An expansion in the range $4\mu^2<t<4m^2$ occurs, if one wants to evaluate the left hand cut of certain dispersion relations. $Z=-\cos\theta_t$ is now purely imaginary and the variables p_- and q_t are real and positive. In Table A.3.2 one has to replace q_-^2 by $-q_t^2$. It is seen that the coefficients of $f_\pm^J(t)$ are real quantities.

In the physical range for $\bar{N}N\to\pi\pi$ scattering: $t>4m^2$ one uses the variables

$$p_t=\sqrt{t/4-m^2}\,;\quad q_t=\sqrt{t/4-\mu^2}\,;\quad \cos\theta_t=\frac{m\nu}{p_t q_t},$$

replacing p_-q_- by $-p_tq_t$, p_-^2 by $-p_t^2$ and Z by $-\cos\theta_t$. The partial wave expansion of the invariant amplitudes reads

$$\begin{aligned} B(\nu,t)&=8\pi\sum_J\frac{J+\frac{1}{2}}{\sqrt{J(J+1)}}(p_tq_t)^{J-1}P_J'(\cos\theta_t)f_-^J(t)\,;\\ C(\nu,t)&=-\frac{8\pi}{p_t^2}\sum_J\left(J+\frac{1}{2}\right)(p_tq_t)^JP_J(\cos\theta_t)f_+^J(t)\,. \end{aligned}\tag{A.3.54}$$

The expansion of the helicity amplitudes, Eq. (A.2.21), is usually written in terms of partial waves $T_\pm^J(t)$

$$\begin{aligned} f_{++}(t,\cos\theta_t)&=\frac{1}{\sqrt{2p_tq_t}}\sum_J\left(J+\frac{1}{2}\right)T_+^J(t)P_J(\cos\theta_t)\,;\\ f_{+-}(t,\cos\theta_t)&=\frac{1}{\sqrt{2p_tq_t}}\sum_J\frac{J+\frac{1}{2}}{\sqrt{J(J+1)}}T_-^J(t)\sin\theta_tP_J'(\cos\theta_t)\,, \end{aligned}\tag{A.3.55}$$

which are related to the reduced partial waves $f_\pm^J(t)$ as follows

$$T_+^J(t)=\frac{q_t}{p_t}\frac{2(p_tq_t)^J}{\sqrt{t}}f_+^J(t)\,;\qquad T_-^J(t)=\frac{q_t}{p_t}(p_tq_t)^Jf_-^J(t)\,.\tag{A.3.56}$$

The amplitudes $T_\pm^J(t)$ are proportional to the diagonal elements of the S-matrix

$$T_\pm^J(t)=-\mathrm{i}\sqrt{q_t/p_t}\,S_\pm^J(t)\,.\tag{A.3.57}$$

Therefore, the unitarity condition leads to a bound, which will be used in the discussion of subtractions in the partial wave dispersion relation (Sect. A.7.5).

3.5 Low energy expansions and threshold behaviour

3.5.1 Expansion at $\nu=t=0$

We consider invariant amplitudes from which the rapidly varying nucleon pole terms are subtracted

$$\tilde{B}(\nu,t)\equiv B(\nu,t)-B_N(\nu,t)\,;\qquad \tilde{C}(\nu,t)\equiv C(\nu,t)-C_N(\nu,t)\,.\tag{A.3.58}$$

Explicit expressions for the nucleon pole terms are given in Sect. A.6.3.1. $\tilde{A}(\nu,t)\equiv A(\nu,t)$, since the nucleon pole terms do not contribute to this amplitude.

It follows from crossing symmetry that the expansion can be written

$$X(\nu^2,t)=\sum_{\mathrm{m,n}}x_{\mathrm{mn}}\nu^{2\mathrm{m}}t^{\mathrm{n}}\,,\tag{A.3.59}$$

where $X(\nu^2,t)$ is one of the following crossing even amplitudes: A^+, A^-/ν, $\tilde{B}^+/\nu$, $\tilde{B}^-$, $\tilde{C}^-/\nu$, $\tilde{D}^-/\nu$. One can as well subtract the pseudovector nucleon Born term [Eq. (4.7.1)] and expand for instance (Sect. 2.4.7.1)

$$\bar{C}^+=\tilde{C}^+-g^2/m\,;\qquad \bar{D}^+=\tilde{D}^+-g^2/m\,.\tag{A.3.60}$$

The region of convergence follows from the location of the nearest singularities, which are the s- and u-channel thresholds at $\nu_{\text{th}}^2=(\mu+t/4m)^2$ and $t=4\mu^2$, where the amplitudes have a cusp-like behaviour (see Sects. 2.4.6, 2.4.7.3, and A.3.5.3).

3.5.2 Expansions at the s-channel threshold

The threshold behaviour can be derived from Froissart-Gribov-type representations of the $T_{\ell\pm}(q)$ in terms of invariant amplitudes (Sect. A.7.1.1)

$$a_{\ell\pm}=\lim_{q\to 0}(\operatorname{Re}T_{\ell\pm}/q^{2\ell+1});\qquad (a_{\ell\pm})^2=\lim_{q\to 0}(\operatorname{Im}T_{\ell\pm}/q^{4\ell+2}); \tag{A.3.61}$$

the $a_{\ell\pm}$ are called "scattering lengths". In terms of phase shifts one obtains the "effective range approximation"

$$q^{2\ell+1}\cot\delta_{\ell\pm}=\frac{1}{a_{\ell\pm}}+\frac{1}{2}r_{\ell\pm}q^2+\ldots;\qquad q^{-(2\ell+1)}\tan\delta_{\ell\pm}=a_{\ell\pm}-\frac{1}{2}a_{\ell\pm}^2r_{\ell\pm}q^2+\ldots. \tag{A.3.62}$$

The second version is an expansion of the K-matrix element, since $K_{\ell\pm}=\tan\delta_{\ell\pm}$, see Eq. (A.7.28). The name "effective range" for $r_{\ell\pm}$ comes from an application in potential scattering.

Sometimes another expansion is used for the S-wave

$$\operatorname{Re}(T_{0+}/q)=a_{0+}+b_{0+}q^2+O(q^4), \tag{A.3.63}$$

the relation between the coefficients being

$$b_{0+}=-a_{0+}^2(a_{0+}+\tfrac{1}{2}r_{0+}). \tag{A.3.64}$$

Since the branch point at threshold is of the square root type, $T_{\ell\pm}$ can be expanded in powers of iq

$$T_{\ell\pm}(q)=q^{2\ell+1}\sum_{n=0}^{\infty}c_n(iq)^n. \tag{A.3.65}$$

The real coefficients are related to those of the effective range expansion (A.3.62), if elastic unitarity is used: $\operatorname{Im}T_{\ell\pm}=|T_{\ell\pm}|^2$. Inserting the real part of (A.3.65) into Eq. (A.3.6), one obtains an expansion of $\operatorname{Re}C$ and $\operatorname{Re}B$ in powers of q^2 and t, which is valid for positive q^2 (see Sect. A.7.3 for the continuation to negative q^2)

$$\begin{aligned}\frac{1-\frac{t}{4m^2}}{4\pi\left(1+\frac{\mu}{m}\right)}\operatorname{Re}C(q^2,t)&=a_{0+}+\left(a_{1+}+\frac{1}{2}a_{1-}-\frac{a_{0+}}{8m^2}\right)t+\frac{3}{4}\left(\frac{3}{2}a_{2+}+a_{2-}-\frac{a_{1+}}{4m^2}\right)t^2\\&\quad+\frac{5}{4}\left(a_{3+}+\frac{3}{4}a_{3-}-\frac{3}{16m^2}a_{2+}\right)t^3+\left(b_{0+}+a_{1-}+2a_{1+}+\frac{a_{0+}}{2m\mu}\right)q^2+\ldots,\\\frac{1}{8\pi m}\operatorname{Re}B(q^2,t)&=\left(a_{1-}-a_{1+}+\frac{a_{0+}}{4m^2}\right)+\frac{3}{2}\left(a_{2-}-a_{2+}+\frac{a_{1+}}{4m^2}\right)t+\frac{15}{8}\left(a_{3-}-a_{3+}+\frac{a_{2+}}{4m^2}\right)t^2+\ldots.\end{aligned} \tag{A.3.66}$$

Using the optical theorem, a similar expansion for $\operatorname{Im}C$ at $t=0$ gives an expression for the momentum dependence of the total cross sections in terms of S-wave scattering lengths and effective ranges (A.3.64)

$$\begin{aligned}\frac{\sigma^+}{4\pi}&=(a_{0+}^+)^2+2(a_{0+}^-)^2+2q^2\{a_{0+}^+b_{0+}^++2a_{0+}^-b_{0+}^-+\frac{1}{24}(a_1^4+2a_3^4)\}+\ldots\\\frac{\sigma^-}{4\pi}&=(a_{0+}^-)^2+2a_{0+}^+a_{0+}^-+2q^2\{a_{0+}^-b_{0+}^-+a_{0+}^-b_{0+}^++a_{0+}^+b_{0+}^-+\frac{1}{24}(a_1^4-a_3^4)\}+\ldots.\end{aligned} \tag{A.3.67}$$

a_1 and a_3 are the isospin 1/2 and 3/2 S-wave scattering lengths and upper indices $\pm$ denote the isospin even and odd combinations.

For a study of the cusp phenomenon at threshold, one can use the above expansions in order to continue the amplitudes at fixed t to values of ν below threshold, where q becomes purely imaginary and $T_{\ell\pm}/q$, B, and C are real. The cusp and the relation between the coefficients of the expansions (A.3.66) and (A.3.59) are discussed in Sect. 2.4.7.3. The cusp is also treated in Sect. A.7.3.

3.5.3 Behaviour near $t=4\mu^2$

The Froissart-Gribov representation shows that $f_\pm^J(4\mu^2)$ is real, finite and non-vanishing (Sect. A.7.4.3). For $t>4\mu^2$, it follows from the extended unitarity relation (A.5.25) that $f_\pm^J(t)$ is complex-valued, its phase being the same as the $\pi\pi$-scattering phase $\delta_J^I(t)$ with the same J; $I=0$ or 1 for even and odd values of J, respectively

$$f_\pm^J(t)=\pm|f_\pm^J(t)|\exp(\mathrm{i}\delta_J^I);\quad 4\mu^2<t<16\mu^2. \tag{A.3.68}$$

The $\pi\pi$-phase shifts have the usual threshold behaviour $\delta_J^I(t)/q_t^{2J+1}\to a_J^I$ for $q_t\to 0$. Therefore, one could think that the scattering length approximation $\mathrm{Im} f_\pm^J(t)\propto a_J^I q_t^{2J+1}$ is useful in a certain range above $t=4\mu^2$.

Of course, this is true in principle, but in praxis it turns out that the threshold behaviour is strongly distorted by a logarithmic branch point at $t=4\mu^2-\mu^4/m^2=3.98\mu^2$, which belongs to the partial wave projection of the nucleon pole term (Sect. A.7.4.2). The contribution of the pole term projection $f_{\pm N}^J$ can be shown explicitly

$$\frac{\mathrm{Im} f_\pm^J(t)}{q_t^{2J+1}}=\mathrm{Re} f_\pm^J(t)\frac{\tan\delta_J^I}{q_t^{2J+1}}=[f_{\pm N}^J(t)+\mathrm{Re}\tilde{f}_\pm^J(t)]\,[a_J^I+O(q_t^2)]\,. \tag{A.3.69}$$

Near $t=4\mu^2$, $f_{\pm N}^J(t)$ is varying very rapidly (Sect. A.7.4.2), whereas a numerical evaluation of the partial wave dispersion relation (Sect. A.7.5) shows that $\mathrm{Re}\tilde{f}_\pm^J(t)$ varies only slowly (Fig. A.3.5). — Using a method analogous to the treatment of the s-channel partial wave in Sect. A.7.3, one can continue $f_\pm^J(t)$ into the second sheet. It turns out that the branch point at $t=t_{\mathrm{nc}}=3.98\mu^2$ ("nc" = nucleon pole term cut) is present not only in $(f_\pm^J)^{\mathrm{II}}$ but also in $(\tilde{f}_\pm^J)^{\mathrm{II}}$, since the subtraction of $f_{\pm N}^J$ cancels only the branch point in the first sheet.

One should note that for $t>4\mu^2$ we have $\mathrm{Im}\tilde{f}_\pm^J(t)\equiv\mathrm{Im} f_\pm^J(t)$. The distortion of the threshold behaviour occurs also in $\mathrm{Im}\tilde{f}_\pm^J(t)$, since it is related to the branch cut in the second sheet at $t<t_{\mathrm{nc}}$, which is the same in both cases. The dispersion integrals over the right hand cut $t>4\mu^2$ give the same cusp-like contributions to $\mathrm{Re} f_\pm^J(t)$ and to $\mathrm{Re}\tilde{f}_\pm^J(t)$.

The behaviour of $f_\pm^J(t)$ near $t=4\mu^2$ can also be studied by using an ansatz $f_\pm^J(t)=N_\pm^J(t)/D^J(t)$, where D^J has only a right hand cut at $t>4\mu^2$ and $N_\pm^J$ has only a left hand cut at $t<t_{\mathrm{nc}}$ (Sect. A.7.5). Since D^J is calculated from the $\pi\pi$-phase shift, it has the normal threshold behaviour. In general, it is assumed that $N(t)$ is varying slowly on the right hand cut, because the left hand cut is far away. However, in the present case $N_\pm^J(t)$ has a logarithmic branch point at $t=t_{\mathrm{nc}}$ and is therefore rapidly varying near $t=4\mu^2$. As a consequence, the cusp at $t=4\mu^2$ differs considerably from the usual threshold cusps.

Finally, we discuss the behaviour of the invariant amplitudes near $\nu=0$, $t=4\mu^2$, taking $C^+(\nu,t)$ as an example. The t-dependence of $C^+(0,t)$ is governed by the dispersion relation at fixed ν (Sect. A.6.10)

$$\mathrm{Re}\,C^+(0,t)=\frac{1}{\pi}\left\{\int_{-\infty}^{-4m\mu}+\int_{4\mu^2}^{\infty}\right\}\frac{\mathrm{Im}\,C^+(0,t')}{t'-t}\,\mathrm{d}t'\,. \tag{A.3.70}$$

The nucleon pole term, Eq. (A.6.34), does not contribute since $C_N^+(0,t)\equiv 0$.

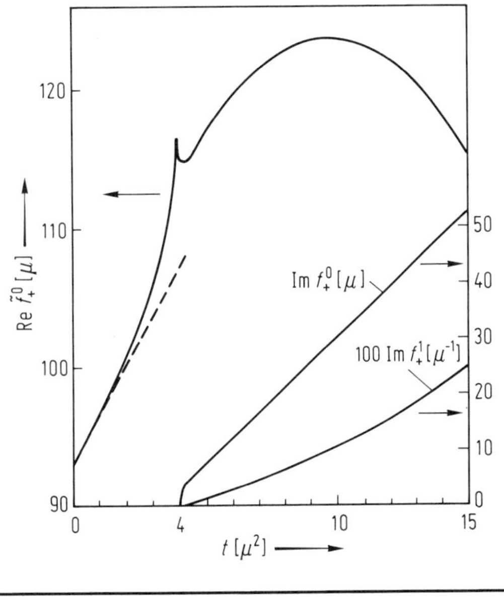

Fig. A.3.5. The behaviour of $\pi\pi N\bar{N}$ partial wave amplitudes near $t=4\mu^2$. The second branch point at $t=3.98\mu^2$ cannot be shown in the scale of the abscissa. The imaginary parts of the S-wave $\tilde{f}_+^0$ and of the P-wave $\tilde{f}_+^1$ have an unusual threshold behaviour. The dashed line is the tangent at $t=0$. Near $t=4\mu^2$ the invariant amplitudes $\tilde{C}^+$ and $\tilde{C}^-$ are determined by the partial waves f_+^0 and f_+^1, because the higher partial waves are strongly suppressed by threshold factors.

The partial wave expansion of Table A.3.3 shows that the higher partial waves are suppressed by threshold factors which are powers of q_t^2

$$C^+(0,t) \equiv \tilde{C}^+(0,t) = \frac{4\pi}{p_-^2}\left[\tilde{f}_+^0(t) + \frac{5}{2}p_-^2 q_t^2 \tilde{f}_+^2(t) + \ldots\right]. \tag{A.3.71}$$

We conclude that near $t=4\mu^2$ the shape of $\mathrm{Re}\,C^+(0,t)$ and $\mathrm{Im}\,C^+(0,t)$ is almost the same as that of $\mathrm{Re}\tilde{f}_+^0(t)$ and $\mathrm{Im}\tilde{f}_+^0(t)$. In particular, $\mathrm{Im}\,C^+(0,t)$ has the same distorted threshold behaviour as $\mathrm{Im}\tilde{f}_+^0(t)$. This is related to the fact that $C^+(0,t)$ also has a singularity at $t=t_{\mathrm{nc}}$ in the second sheet. The analytic continuation is performed in analogy to the continuation near the s-channel threshold in Sect. A.6.4.6.

Because of this singularity, expansions of partial waves $f_\pm^J(t)$ or $\tilde{f}_\pm^J(t)$ or of invariant amplitudes similar to those at the s-channel threshold (Sect. A.3.5.2) are not of interest, because the region of convergence is very small.

3.5.4 Expansion at the t-channel threshold $t=4m^2$

A simple description of $\bar{\mathrm{N}}\mathrm{N}\leftrightarrow\pi\pi$ scattering is obtained in the nonrelativistic range above $t=4m^2$, if one considers the *Russel-Saunders expansion* of the helicity states. First, the spins of $\bar{\mathrm{N}}$ and N are coupled to give a total spin S. Then, S is coupled to the orbital momentum L, whose magnetic quantum number is zero. We introduce combinations of the helicity states (given by λ_{N} and $\lambda_{\bar{\mathrm{N}}}$) which have a well-defined parity $\pm$

$$|J,\lambda_{\bar{\mathrm{N}}},\lambda_{\mathrm{N}}\rangle_\pm \equiv \frac{1}{\sqrt{2}}[|J,\lambda_{\bar{\mathrm{N}}},\lambda_{\mathrm{N}}\rangle \pm |J,-\lambda_{\bar{\mathrm{N}}},-\lambda_{\mathrm{N}}\rangle], \tag{A.3.72a}$$

and find for the expansion in terms of $|J,L,S\rangle$ states

$$\begin{aligned}\left|J,\frac{1}{2},\frac{1}{2}\right\rangle_+ &= \sqrt{\frac{J}{2J+1}}|J,J-1,1\rangle - \sqrt{\frac{J+1}{2J+1}}|J,J+1,1\rangle;\\ \left|J,\frac{1}{2},-\frac{1}{2}\right\rangle_+ &= \sqrt{\frac{J+1}{2J+1}}|J,J-1,1\rangle + \sqrt{\frac{J}{2J+1}}|J,J+1,1\rangle.\end{aligned} \tag{A.3.72b}$$

Next, the partial wave amplitudes $T_\pm^J$ of Eq. (A.3.56) are expressed in terms of amplitudes $T^J(L,S)$, which belong to transitions starting from or leading to antinucleon-nucleon states with well-defined quantum numbers for orbital momentum L and total spin S

$$\begin{aligned}T_+^J &= \frac{1}{\sqrt{2}}\left\{\sqrt{\frac{J}{2J+1}}T^J(J-1,1) - \sqrt{\frac{J+1}{2J+1}}T^J(J+1,1)\right\};\\ T_-^J &= \frac{1}{\sqrt{2}}\left\{\sqrt{\frac{J+1}{2J+1}}T^J(J-1,1) + \sqrt{\frac{J}{2J+1}}T^J(J+1,1)\right\}.\end{aligned} \tag{A.3.73a}$$

The transitions to or from singlet states, which have $S=0$, $L=J$, are forbidden, because their parity differs from that of the $\pi\pi$-state with the same J.

We give explicit formulas for the lowest values of L, using the notation $T^J(L,S) \equiv T(^{2S+1}L_J)$

$$\begin{aligned}T_+^0 &= -\frac{1}{\sqrt{2}}T(^3\mathrm{P}_0);\\ T_+^1 &= \frac{1}{\sqrt{6}}T(^3\mathrm{S}_1) - \frac{1}{\sqrt{3}}T(^3\mathrm{D}_1); \quad T_-^1 = \frac{1}{\sqrt{3}}T(^3\mathrm{S}_1) + \frac{1}{\sqrt{6}}T(^3\mathrm{D}_1);\\ T_+^2 &= \frac{1}{\sqrt{5}}T(^3\mathrm{P}_2) - \sqrt{\frac{3}{10}}T(^3\mathrm{F}_2); \quad T_-^2 = \sqrt{\frac{3}{10}}T(^3\mathrm{P}_2) + \frac{1}{\sqrt{5}}T(^3\mathrm{F}_2).\end{aligned} \tag{A.3.73b}$$

A kinematic constraint

The important point is now that the amplitudes $T^J(L,S)$ behave as p_t^L near threshold. Therefore, the second terms in Eqs. (A.3.72) and (A.3.73) are smaller by two orders in p_t. A comparison of the first terms leads to the following relation between the $T^J_\pm$, which vary both as p_t^{J-1} (except for $J=0$)

$$T^J_+ = \sqrt{\frac{J}{J+1}}\, T^J_- + O(p_t^2); \quad J \neq 0. \tag{A.3.74}$$

It is also of interest to write this equation in terms of the reduced partial waves $f^J_\pm$, Eq. (A.3.56), which have finite non-vanishing values at threshold, except for f^0_+, which varies as p_t^2 according to the first relation (A.3.73b)

$$f^J_+(4m^2) = m\sqrt{\frac{J}{J+1}}\, f^J_-(4m^2) + O(p_t^2); \quad f^0_+ \propto p_t^2. \tag{A.3.75}$$

This result can also be derived directly from the projection formula (A.3.50). It ensures that Γ_1 and Γ_2 in Eq. (A.3.52) remain finite.

Another version of the kinematic constraint at the t-channel threshold follows if we consider the limit of the combination $T^t_{++} + \mathrm{i}T^t_{+-}$ of helicity amplitudes at fixed ν as $p_t \to 0$

$$T^t_{++} + \mathrm{i}T^t_{+-} = \frac{-\mathrm{i}}{2\pi^2 Wq}\{p_t A - m q_t(\cos\theta_t + \mathrm{i}(\sqrt{t}/2m)\sin\theta_t)B\} \to \frac{-\mathrm{i}}{2\pi^2 Wq} p_t A(s,t) = O(p_t), \tag{A.3.76}$$

where the relation

$$\sin\theta_t = \{1 - (m\nu/p_t q_t)^2\}^{1/2} = \mathrm{i}\cos\theta_t + O(p_t) \tag{A.3.77}$$

was used. Since the invariant amplitudes A and B are finite at threshold, the left hand side of Eq. (A.3.76) is of order p_t. $C^-(\nu, t)$ has a pole at $t = 4m^2$ and C^+ is finite.

Next, $T^t_{++} + \mathrm{i}T^t_{+-}$ is expanded in terms of partial waves $f^J_\pm$, using Eqs. (A.2.35), (A.3.55), and (A.3.56). The leading term for small p_t (i.e. large $\cos\theta_t$) is proportional to $p_t^{-1}[f^J_+(4m^2) - f^J_-(4m^2)m\sqrt{J}/\sqrt{J+1}]$. Because of the earlier result, this combination of helicity amplitudes is of order p_t, i.e. the bracket must vanish and one obtains Eq. (A.3.75). Note that the above formula is written for the reaction $\pi\pi \to \mathrm{N\bar{N}}$: If one considers instead $\pi\pi \to \mathrm{\bar{N}N}$, interchanging the final particles, the constraint becomes $T^t_{++} - \mathrm{i}T^t_{+-} = O(p_t)$.

Selection rules

Up to now, we have mentioned in this section the selection rule following from parity invariance, which is consistent only with "natural parity" states having $P = (-1)^J$. Further selection rules follow from the conservation of charge parity for $\mathrm{\bar{p}p}$ and $\mathrm{\bar{n}n}$ initial or final states

$$P = (-1)^{L+1} = (-1)^J; \quad C = (-1)^{L+S} = (-1)^J. \tag{A.3.78}$$

Selection rules based on isospin invariance can be applied to the reaction $\mathrm{\bar{n}p} \to \pi^+\pi^0$, which has a well defined isospin $I_t = 1$. Only odd values of J lead to the correct symmetry property of the final state under the exchange of the two bosons. – Some transitions are excluded by G-parity conservation

$$G = (-1)^{L+S+I_t} = 1. \tag{A.3.79}$$

A summary of the selection rules is given in Table A.3.4.

Table A.3.4. Selection rules in $\mathrm{\bar{N}N} \to \pi\pi$. Notation for the $\mathrm{\bar{N}N}$-states: ${}^{2S+1}L_J$. The transitions marked by a minus sign are forbidden.

Reaction	${}^1S_0$	${}^3P_0$	${}^3S_1$	${}^1P_1$	${}^3P_1$	${}^3D_1$	${}^3P_2$	${}^1D_2$	${}^3D_2$	${}^3F_2$	I_t
$\mathrm{\bar{p}p} \to \pi^+\pi^-$	–	+	+	–	–	+	+	–	–	+	0, 1
$\mathrm{\bar{p}p} \to \pi^0\pi^0$	–	+	–	–	–	–	+	–	–	+	0
$\mathrm{\bar{n}p} \to \pi^+\pi^0$	–	–	+	–	–	+	–	–	–	–	1

Threshold behaviour of cross sections

Now, it is possible to discuss the behaviour of the $\mathrm{\bar{N}N} \to \pi\pi$ cross section near threshold. In the reaction $\mathrm{\bar{n}p} \to \pi^+\pi^0$, which has a well-defined isospin $I_t = 1$, the lowest partial waves are the $\pi\pi\mathrm{N\bar{N}}$ P-waves $f^1_\pm$. The expansion of the helicity amplitudes $f_{+\pm}$ starts as follows

$$f^1_{++}(t,\cos\theta_t)=\frac{3}{\sqrt{p_t}}\frac{\sqrt{2}}{\sqrt{t}}q_t^{3/2}\cos\theta_t f^1_+(t)+\ \dots;$$
$$f^1_{+-}(t,\cos\theta_t)=\frac{3}{2\sqrt{p_t}}q_t^{3/2}\sin\theta_t f^1_-(t)+\ \dots \qquad \text{(A.3.80)}$$

and the cross section reads

$$\frac{d\sigma}{d\Omega}(\bar{\text{n}}\text{p}\to\pi^+\pi^0)=\frac{9}{8}\frac{q_t^3}{p_t}\left[\frac{8}{t}\cos^2\theta_t|f^1_+|^2+\sin^2\theta_t|f^1_-|^2\right]+\dots$$
$$=\frac{9}{8}\frac{q_t^3}{p_t}\left\{|f^1_-|^2+\cos^2\theta_t\left[\left|\frac{\sqrt{2}}{m}f^1_+\right|^2-|f^1_-|^2-|f^1_+|^2\frac{2p_t^2}{m^4(1+p_t^2/m^2)}\right]\right\}+\dots. \qquad \text{(A.3.81)}$$

If the kinematic constraint Eq. (A.3.75) is inserted, it is seen that the cross section becomes isotropic for small p_t. Moreover, if threshold is approached, it increases as $1/p_t$ in agreement with the "$1/v$-law" for exoergic reactions.

In the reaction $\bar{\text{p}}\text{p}\to\pi^0\pi^0$, the isospin is $I_t=0$ and the lowest partial waves are the $\pi\pi\text{N}\bar{\text{N}}$ S- and D-waves

$$\frac{1}{\sqrt{6}}f^0_{++}(t,\cos\theta_t)=\sqrt{p_t}\sqrt{\frac{q_t}{2t}}\left\{\frac{f^0_+}{p_t^2}+5q_t^2P_2(\cos\theta_t)f^2_+ +\dots\right\};$$
$$\frac{1}{\sqrt{6}}f^0_{+-}(t,\cos\theta_t)=\frac{15}{2\sqrt{12}}\sqrt{p_t}q_t^{5/2}\sin\theta_t\cos\theta_t f^2_- +\ \dots. \qquad \text{(A.3.82)}$$

The expression for the cross section reads

$$\frac{d\sigma}{d\Omega}(\bar{\text{p}}\text{p}\to\pi^0\pi^0)=\frac{p_tq_t}{2t}\left|\frac{f^0_+}{p_t^2}+\frac{5}{2}q_t^2(3\cos^2\theta_t-1)f^2_+\right|^2+\frac{75}{16}p_tq_t^5\sin^2\theta_t\cos^2\theta_t|f^2_-|^2+\ \dots. \qquad \text{(A.3.83)}$$

The $\cos^4\theta_t$ terms cancel, if the kinematic constraint Eq. (A.3.75) is inserted. Because of the threshold behaviour of the S-wave ($f^0_+\propto p_t^2$), the contributions of the S- and D-waves to the cross section start both $\propto p_t$.

The threshold behaviour of the reaction $\bar{\text{p}}\text{p}\to\pi^+\pi^-$ is determined by the P-wave of the $I_t=1$ contribution, i.e. the cross section varies as p_t^{-1}.

4 Measurable quantities

For each pair of kinematic variables, for instance (s,t) or $(k,\cos\theta)$, there exist 4 measurable quantities: the differential cross section $d\sigma/dt$, the polarization parameter P and the spin-rotation parameters A and R, which are related by

$$P^2+R^2+A^2=1. \qquad \text{(A.4.1)}$$

They determine the two independent amplitudes up to a common phase factor, which depends on s and t. Instead of A and R one can introduce the spin-rotation angle β which is defined by

$$R=-\sqrt{1-P^2}\cos(\theta+\theta_\text{N}-\beta);\quad A=\sqrt{1-P^2}\sin(\theta+\theta_\text{N}-\beta);\quad -R+iA=\sqrt{1-P^2}e^{i(\theta+\theta_\text{N}-\beta)}. \qquad \text{(A.4.2)}$$

4.1 Description of spin states

The spin orientation of a nucleon target or a nucleon beam is described by a 2×2 density matrix $\mathsf{\rho}$. The mean value of any operator A for the statistical ensemble is

$$\langle \mathsf{A}\rangle=\text{Tr}(\mathsf{A}\mathsf{\rho}). \qquad \text{(A.4.3)}$$

The mean value of the Pauli spin operator $\boldsymbol{\sigma}$ defines a polarization vector $\boldsymbol{P}$, which describes the spin orientation of the ensemble of nucleons as well as this is done by the density matrix (I is the unit matrix)

$$\langle\boldsymbol{\sigma}\rangle\equiv\boldsymbol{P}=\text{Tr}(\mathsf{\rho}\boldsymbol{\sigma});\quad \mathsf{\rho}=\tfrac{1}{2}(\mathsf{I}+\boldsymbol{P}\cdot\boldsymbol{\sigma}). \qquad \text{(A.4.4)}$$

A definite spin orientation belongs to a pure state and has $|\boldsymbol{P}|=1$, whereas an unpolarized target has $|\boldsymbol{P}|=0$. All quantities are defined in the rest frame of the nucleon.

The relation between the density matrices of the incoming and outgoing states can be expressed in terms of s-channel helicity amplitudes, Sect. A.2.1.3

$$\rho^{\text{out}}=\mathsf{F}\rho^{\text{in}}\mathsf{F}^{+}\,; \quad \mathsf{F}=\begin{pmatrix} F_{++} & F_{+-} \\ F_{-+} & F_{--} \end{pmatrix}, \tag{A.4.5}$$

where $F_{++}=F_{--}$ and $F_{+-}=-F_{-+}$ due to parity invariance. F^{+} is the hermitian conjugate matrix.

The polarization of the outgoing beam is given by $[\mathrm{Tr}(AB)=\mathrm{Tr}(BA)]$

$$\boldsymbol{P}_{\mathrm{f}}\frac{\mathrm{d}\sigma}{\mathrm{d}t}=\mathrm{Tr}(\boldsymbol{\sigma}\rho^{\text{out}})=\mathrm{Tr}(\boldsymbol{\sigma}\mathsf{F}\rho^{\text{in}}\mathsf{F}^{+})\,; \quad \frac{\mathrm{d}\sigma}{\mathrm{d}t}=\mathrm{Tr}\rho^{\text{out}}=\langle \mathsf{F}^{+}\mathsf{F}\rangle\,. \tag{A.4.6}$$

4.2 Differential cross section measurements

If the target is *unpolarized* ($\rho^{\text{in}}=\mathsf{I}/2$), and the polarization of the final state is not measured, one obtains the "unpolarized differential cross section" $\mathrm{d}\sigma/\mathrm{d}t$

$$\frac{\mathrm{d}\sigma}{\mathrm{d}t}=\frac{1}{2}\mathrm{Tr}(\mathsf{F}\mathsf{F}^{+})=|F_{++}|^2+|F_{+-}|^2\,. \tag{A.4.7}$$

The differential cross section $\mathrm{d}\sigma/\mathrm{d}t$ is related to the cross section $\mathrm{d}\sigma/\mathrm{d}\Omega_{\text{lab}}$ measured in the laboratory system by

$$\frac{q^2}{\pi}\frac{\mathrm{d}\sigma}{\mathrm{d}t}=\frac{\mathrm{d}\sigma}{\mathrm{d}\Omega}=\frac{\mathrm{d}\sigma}{\mathrm{d}\Omega_{\text{lab}}}\left[\frac{\sin^3\theta_\pi}{\sin^3\theta}\,\frac{E+\omega_q\cos\theta}{m}\right], \tag{A.4.8}$$

the factor in the bracket being $\mathrm{d}\cos\theta_\pi/\mathrm{d}\cos\theta$, Eq. (A.1.27). If the target is *polarized*, ρ^{in} is given by

$$\rho^{\text{in}}=\tfrac{1}{2}(\mathsf{I}+\boldsymbol{P}_{\mathrm{T}}\cdot\boldsymbol{\sigma})\,, \tag{A.4.9}$$

and the differential cross section in the c.m. system reads

$$\left(\frac{\mathrm{d}\sigma}{\mathrm{d}t}\right)_{\text{pol}}=\frac{\mathrm{d}\sigma}{\mathrm{d}t}\{1+P(\hat{\boldsymbol{n}}\cdot\boldsymbol{P}_{\mathrm{T}})\}\,, \tag{A.4.10}$$

where P is the polarization parameter and $\boldsymbol{P}_{\mathrm{f}}=\hat{\boldsymbol{n}}P$.

P is usually determined by changing the sign of the target polarization $\boldsymbol{P}_{\mathrm{T}}$, which is chosen parallel to the normal unit vector $\hat{\boldsymbol{n}}$ on the scattering plane.

Another possibility is to keep $\boldsymbol{P}_{\mathrm{T}}$ fixed and to reverse the sign of $\hat{\boldsymbol{n}}=(\boldsymbol{q}\times\boldsymbol{q}')/|\boldsymbol{q}\times\boldsymbol{q}'|$. One measures $(\mathrm{d}\sigma/\mathrm{d}t)_{\text{pol}}$ in a plane orthogonal to $\boldsymbol{P}_{\mathrm{T}}$ at two values of the azimuthal angle $\phi=0$ and π, which belong to the same value θ of the scattering angle of the pion (Fig. A.6.2, p. 527).

This experiment shows how a left-right asymmetry can arise in πN scattering, although the amplitudes depend only on $\cos\theta$ and not on the azimuthal angle.

4.3 Polarization measurements

If the polarization $\boldsymbol{P}_{\mathrm{f}}$ of the recoil nucleon is measured by analyzing the azimuthal distribution in a second scattering process, it is of little interest to choose an unpolarized target, because one can determine only the polarization parameter P, which is easier obtained from Eq. (A.4.10). In this case, it follows from parity invariance that $\boldsymbol{P}_{\mathrm{f}}$ is directed along $\hat{\boldsymbol{n}}$.

The result again depends only on P, if the target is polarized along $\hat{\boldsymbol{n}}$

$$\boldsymbol{P}_{\mathrm{f}}=\frac{P+|\boldsymbol{P}_{\mathrm{T}}|}{1+P|\boldsymbol{P}_{\mathrm{T}}|}\hat{\boldsymbol{n}}\,. \tag{A.4.11}$$

The interesting case is to choose the target polarization in the scattering plane ($\boldsymbol{P}_{\mathrm{T}}\cdot\hat{\boldsymbol{n}}=0$). Now, a measurement of suitable components of $\boldsymbol{P}_{\mathrm{f}}$ determines the spin-rotation parameters A and R (Fig. A.4.1)

$$\boldsymbol{P}_{\mathrm{f}}\cdot\hat{\boldsymbol{k}}_{\mathrm{N}}=R(\hat{\boldsymbol{k}}\cdot\boldsymbol{P}_{\mathrm{T}})+A(\hat{\boldsymbol{k}}\times\hat{\boldsymbol{n}})\cdot\boldsymbol{P}_{\mathrm{T}}\,; \quad \boldsymbol{P}_{\mathrm{f}}\cdot(\hat{\boldsymbol{k}}_{\mathrm{N}}\times\hat{\boldsymbol{n}})=-A(\hat{\boldsymbol{k}}\cdot\boldsymbol{P}_{\mathrm{T}})+R(\hat{\boldsymbol{k}}\times\hat{\boldsymbol{n}})\cdot\boldsymbol{P}_{\mathrm{T}}\,; \quad \boldsymbol{P}_{\mathrm{f}}\cdot\hat{\boldsymbol{n}}=P\,. \tag{A.4.12}$$

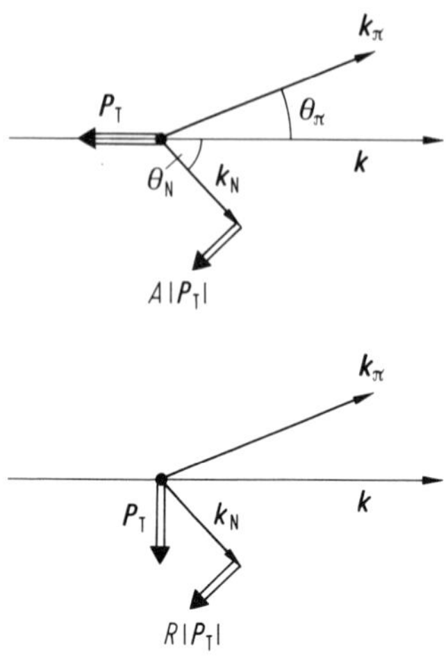

Fig. A.4.1. Spin-rotation parameters A and R.
$\boldsymbol{k}$, $\boldsymbol{k}_\pi$, and $\boldsymbol{k}_N$ are the momenta of the incoming and outgoing pion and of the recoil nucleon in the laboratory system. The target polarization $\boldsymbol{P}_T$ is assumed to lie in the scattering plane. $A|\boldsymbol{P}_T|$ and $R|\boldsymbol{P}_T|$ are the components $\boldsymbol{P}_f\cdot(\hat{\boldsymbol{k}}_N\times\hat{\boldsymbol{n}})$ of the recoil nucleon polarization $\boldsymbol{P}_f$ which are measured by analyzing the azimuthal distribution in a subsequent scattering on carbon.

4.4 Relations between measurable quantities and amplitudes

4.4.1 Amplitudes G, H and s-channel helicity amplitudes

These relations are very simple for the s-channel helicity amplitudes F_{++}, F_{+-} and for the spin no-flip and flip amplitudes G and H

$$
\begin{aligned}
\frac{d\sigma}{dt} &= |F_{++}|^2+|F_{+-}|^2\,; & \frac{d\sigma}{d\Omega} &= |G|^2+|H|^2=\frac{q^2}{\pi}\frac{d\sigma}{dt}\,;\\
S\frac{d\sigma}{dt} &= |F_{++}|^2-|F_{+-}|^2\,; & U\frac{d\sigma}{d\Omega} &= |G|^2-|H|^2\,;\\
T\frac{d\sigma}{dt} &= 2\,\mathrm{Re}(F_{++}F^*_{+-})\,; & V\frac{d\sigma}{d\Omega} &= 2\,\mathrm{Re}(GH^*)\,;\\
P\frac{d\sigma}{dt} &= 2\,\mathrm{Im}(F_{++}F^*_{+-})\,; & P\frac{d\sigma}{d\Omega} &= 2\,\mathrm{Im}(GH^*)\,.
\end{aligned}
\tag{A.4.13}
$$

The different pairs of spin-rotation parameters are related as follows [θ_N is the angle of the recoil nucleon in the lab. system, Eq. (A.1.23)]

$$S+iT=(U+iV)e^{i\theta}=(-R+iA)e^{-i\theta_N}\,. \tag{A.4.14}$$

In terms of helicity amplitudes the spin-rotation parameters R and A, which are simply related to experimental quantities according to Eq. (A.4.12), read

$$
\begin{aligned}
R\frac{d\sigma}{dt} &= -(|F_{++}|^2-|F_{+-}|^2)\cos\theta_N+2\,\mathrm{Re}(F_{++}F^*_{+-})\sin\theta_N\,;\\
A\frac{d\sigma}{dt} &= (|F_{++}|^2-|F_{+-}|^2)\sin\theta_N+2\,\mathrm{Re}(F_{++}F^*_{+-})\cos\theta_N\,.
\end{aligned}
\tag{A.4.15}
$$

Transversity amplitudes (Sect. 2.1.2) have the advantage that their modulus can be determined from $d\sigma/d\Omega$ and P-data alone

$$
\begin{aligned}
&F(\pm)\equiv G\pm iH\,; \qquad |F(\pm)|^2=\frac{d\sigma}{d\Omega}(1\pm P)\,;\\
&F(+)F(-)^*=\frac{d\sigma}{d\Omega}(U+iV)\,; \qquad \arg F(+)-\arg F(-)=-\beta\,.
\end{aligned}
\tag{A.4.16}
$$

Geometrically, the spin-rotation angle β can be interpreted as the angle between the complex vectors $\boldsymbol{F}(+)$ and $\boldsymbol{F}(-)$ (Fig. A.4.2).

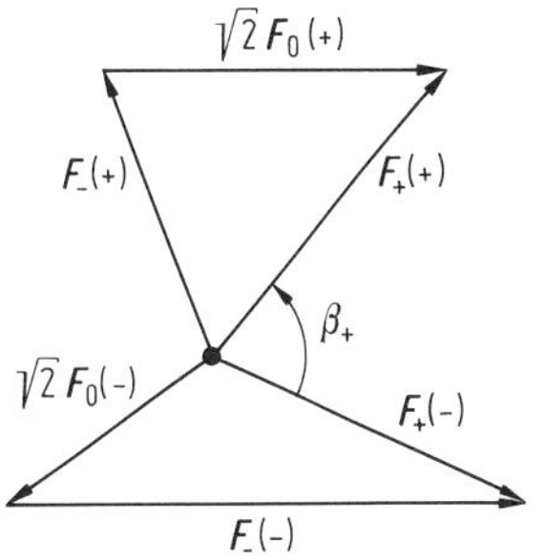

Fig. A.4.2. Isospin triangles for transversity amplitudes. The lengths of the sides of the triangles are determined by differential cross section and polarization experiments. The relative orientation of the triangles is determined by one of the spin-rotation angles, for instance β_+, except for reflections. Data for two suitable spin-rotation parameters remove all discrete ambiguities. An overall phase angle of all amplitudes is not measurable in strong interaction experiments, but it is theoretically significant (Sect. 2.1.1). The arrow for β_+ should be reversed.

4.4.2 Invariant amplitudes and *t*-channel helicity amplitudes

The relations between measurable quantities and invariant amplitudes are somewhat more complicated than Eq. (A.4.13). We have included these formulas, because they are needed in applications of dispersion methods and of Regge models.

$$\frac{d\sigma}{dt}=|G_{++}|^2+|G_{+-}|^2=\frac{1}{16\pi}\left\{\left(1-\frac{t}{4m^2}\right)\left|\frac{C}{k}\right|^2-\frac{t(1+t/4q^2)}{4m^2-t}|B|^2\right\};$$

$$16\pi\frac{d\sigma}{dt}=\left\{\frac{\sin^2(\theta/2)}{s}+\frac{1}{k^2}\right\}|C|^2+\frac{\sin^2\theta}{4(\sin^2(\theta/2)+m^2/q^2)}|B|^2; \tag{A.4.17}$$

$$16\pi\frac{d\sigma}{dt}=\left(1-\frac{t}{4m^2}\right)\left|\frac{C}{k}\right|^2\left\{1-\frac{k^2}{\omega^2}\frac{1+t/4q^2}{(1+t/4m\omega)^2}\frac{t}{4m^2}\left[\left[1-\frac{A_{\parallel}}{|C|}\right]^2+\frac{A_{\perp}^2}{|C|^2}\right]\right\},$$

where $\mathrm{Re}(CA^*)=|C|A_{\parallel}$, $\mathrm{Im}(CA^*)=|C|A_{\perp}$ can be determined from spin-rotation and polarization data. They give a small correction at high energies in elastic scattering.

In the special case of backward scattering we have

$$\frac{d\sigma}{d\Omega}=\frac{m^2}{u}\frac{d\sigma}{d\Omega_{\mathrm{lab}}}=\left(\frac{E}{4\pi W}\right)^2|C|^2;\qquad \theta=180^\circ. \tag{A.4.18}$$

The formulas for the polarization parameter and the spin-rotation parameter R read

$$P\frac{d\sigma}{dt}=-2\,\mathrm{Im}(G_{++}G_{+-}^*)=\sin\theta\,\mathrm{Im}(A^*B)/16\pi W=\sin\theta\,\mathrm{Im}(C^*B)/16\pi W$$

$$=\frac{\sin\theta}{16\pi W\omega}\frac{1-t/4m^2}{1+t/4m\omega}\mathrm{Im}(CA^*); \tag{A.4.19}$$

$$\left\{R-\frac{\omega+m}{k}\sqrt{\frac{-t}{4m^2-t}}\right\}\frac{d\sigma}{dt}=-\sqrt{\frac{-t}{4m^2-t}}\,\frac{1+t/4q^2}{8\pi k}\left\{\mathrm{Re}(B^*C)-\frac{t(\omega+m)}{4m^2-t}|B|^2\right\}. \tag{A.4.20}$$

4.4.3 Legendre expansions

At low and intermediate energies experimental data for charge-exchange cross sections, for elastic cross sections beyond the Coulomb interference range and for polarization parameters are frequently fitted to truncated Legendre expansions

$$\frac{d\sigma}{d\Omega}=\frac{1}{q^2}\sum_{n=0}^{N}A_n(k)P_n(\cos\theta);\qquad P\frac{d\sigma}{d\Omega}/\sin\theta=\frac{1}{q^2}\sum_{n=1}^{N}B_n(k)P_n'(\cos\theta). \tag{A.4.21}$$

The coefficients A_n, B_n can be expressed in terms of partial waves $T_{\ell\pm}$. The neglection of partial waves beyond $\ell=\ell_{\max}$ corresponds to the choice $N=2\ell_{\max}$. For example, one obtains for $\ell_{\max}=1$

$$A_0=|T_{0+}|^2+|T_{1-}|^2+2|T_{1+}|^2;\qquad B_1=2\,\mathrm{Im}\{T_{0+}T_{1+}^*-T_{0+}T_{1-}^*\};$$

$$A_1=2\,\mathrm{Re}\{T_{0+}(T_{1-}+2T_{1+})^*\};\qquad B_2=2\,\mathrm{Im}\{T_{1-}T_{1+}^*\}. \tag{A.4.22}$$

$$A_2=2|T_{1+}|^2+4\,\mathrm{Re}(T_{1-}T_{1+}^*);$$

4.5 $\bar{N}N \to \pi\pi$ and $\pi\pi \to N\bar{N}$ cross sections

The differential cross sections can be expressed in terms of t-channel helicity amplitudes f_{++}, f_{+-}, Eq. (A.2.21) or invariant amplitudes

$$\frac{d\sigma}{d\Omega}(\bar{N}N \to \pi\pi) = \frac{q_t^2}{4p_t^2}\frac{d\sigma}{d\Omega}(\pi\pi \to \bar{N}N) = |f_{++}|^2 + |f_{+-}|^2$$

$$= \frac{q_t p_t}{\pi}\frac{d\sigma}{ds}(\bar{N}N \to \pi\pi) = \frac{q_t p_t}{32\pi^2 t}\left\{|C|^2 + \frac{t}{4}\frac{q_t^2}{p_t^2}\sin^2\theta_t |B|^2\right\}. \tag{A.4.23}$$

At threshold ($t = 4m^2$), the invariant amplitudes A^+, B^+/v, A^-/v, B^- have non-vanishing finite values. C^- has a kinematic singularity, but $C^- p_t^2/v$ is finite and C^+ is also finite at $v = 0$.

In the t-channel region, it is useful to express the relation between C, A, and B in terms of t-channel variables

$$C = A - m\cos\theta_t \frac{q_t}{p_t} B. \tag{A.4.24}$$

For a $I_t = 1$ reaction, we find for the cross section at $v = 0$ and $p_t \to 0$

$$\frac{d\sigma}{d\Omega}(\bar{n}p \to \pi^+\pi^0) = \frac{q_t^3}{64\pi^2 p_t}|B^-|^2 + O(p_t). \qquad \cos\theta_t = 0. \tag{A.4.25}$$

The cross section goes to infinity as $p_t \to 0$ ($1/v$-law).

The situation is different for a $I_t = 0$ reaction

$$\frac{d\sigma}{d\Omega}(\bar{p}p \to \pi^0\pi^0) = \frac{q_t p_t}{32\pi^2 t}|C^+|^2 + O(p_t^3); \qquad \cos\theta_t = 0. \tag{A.4.26}$$

See Sect. A.3.5.4 for a description in terms of partial waves.

For a complete description of the relation (A.4.23), we have to add that the total center of mass energy is the same on both sides and that the following angles θ are equal: (we consider $\bar{p}p \to \pi^+\pi^-$ as an example).

i) On the left hand side, θ is the angle of the outgoing π^+ with respect to the incoming $\bar{p}$,

ii) on the right hand side, θ is the angle of the outgoing $\bar{p}$ with respect to the incoming π^+.

The relation between the unpolarized cross sections follows from time-reversal invariance.

4.6 Electromagnetic contributions to cross sections and polarizations

We only include the treatment at high energies, using the formulas of Sect. A.2.3. The expression (A.4.17), 3rd line, is simplified by assuming that $A_{\|}$ and $A_{\perp}$ are small in comparison with $|C|$. This is justified by the result of amplitude analysis at those momenta, where spin-rotation data are available (see Sect. 2.1.15). Furthermore, we use approximations for high energy, small-$|t|$ scattering and obtain

$$\frac{d\sigma}{dt} = \frac{1}{16\pi k^2}\left(1 - \frac{t}{4m^2}\right)^2 |C|^2. \tag{A.4.27}$$

The factor $(1 - t/4m^2)^2$ is not negligible, because it contributes $1/2m^2 = 0.57\,\text{GeV}^{-2}$ to the logarithmic slope at $t = 0$, which is about $10\,\text{GeV}^{-2}$.

We denote the strong amplitude by C and the sum of the strong and the electromagnetic amplitudes by C^{exp}

$$C_{\pm}^{\text{exp}}(s,t) = C_{\pm}(s,t) \mp C_{\text{em}}(s,t)e^{\pm 2i\delta}. \tag{A.4.28}$$

C_{em} and the Bethe phase δ are defined in Eqs. (A.2.80) and (A.2.83). Now, we insert Eq. (A.4.28) into Eq. (A.4.27) and obtain

$$\frac{d\sigma}{dt_{\text{exp}}} = \frac{d\sigma}{dt_C} + \frac{d\sigma}{dt_{\text{Int}}} + \frac{d\sigma}{dt}, \tag{A.4.29}$$

where (t in GeV^2)

$$\frac{d\sigma}{dt_C}=\frac{4\pi\alpha^2}{t^2}f_E^2(t)=\frac{2.606\cdot 10^{-4}}{t^2}f_E^2(t)\frac{\mathrm{mb}}{\mathrm{GeV}^2}; \tag{A.4.30}$$

$$\frac{d\sigma_\pm}{dt_{\mathrm{Int}}}=-\frac{\alpha}{|t|k}f_E(t)\,\mathrm{Im}\,C_\pm(\pm\varrho_\pm+2\delta)(1-t/4m^2) \tag{A.4.31}$$

$$=-\frac{\alpha}{|t|k}f_E(t)(\pm\mathrm{Re}\,C_\pm+2\delta\,\mathrm{Im}\,C_\pm)(1-t/4m^2). \tag{A.4.32}$$

$d\sigma/dt$ is the strong interaction part of the cross section.

In the *difference* of the elastic cross sections, the Coulomb term cancels and the interference term mainly depends on the *sum* of $\mathrm{Re}\,C_\pm$

$$\left\{\frac{d\sigma_-}{dt}-\frac{d\sigma_+}{dt}\right\}_{\mathrm{exp}}=\frac{2\alpha}{|t|k}f_E(t)(\mathrm{Re}\,C^+-2\delta\,\mathrm{Im}\,C^-)(1-t/4m^2)+\left\{\frac{d\sigma_-}{dt}-\frac{d\sigma_+}{dt}\right\}. \tag{A.4.33}$$

Numerical values of the Bethe phase are given in Table A.4.1. They follow from Eq. (A.2.83) and $r_{\mathrm{Ep}}^2=(0.836\,\mathrm{fm})^2$, $r_\pi^2=(0.68\,\mathrm{fm})^2$, $B=11.0\,\mathrm{GeV}^{-2}$, $R^2/6=10.5\,\mathrm{GeV}^{-2}$.

Table A.4.1. Numerical values of the Bethe phase.

$-t$	0.001	0.002	0.01	0.02	0.05	0.10	0.20	GeV^2
2δ	0.029	0.024	0.012	0.007	0.000	−0.004	−0.009	

Electromagnetic corrections to polarization parameters are most important in the difference of the $\pi^\pm p$ elastic scattering data

$$P_-d\sigma_-/dt-P_+d\sigma_+/dt=\frac{\sin\theta}{8\pi W}\{\mathrm{Im}(C^{+*}B^-+C^{-*}B^+)-B_{\mathrm{em}}\,\mathrm{Im}\,C^++C_{\mathrm{em}}\,\mathrm{Im}\,B^+\}. \tag{A.4.34}$$

The correction to the isospin even combination reads

$$P_-d\sigma_-/dt+P_+d\sigma_+/dt-P_0d\sigma_0/dt=\frac{\sin\theta}{8\pi W}\{\mathrm{Im}(C^{+*}B^+)-B_{\mathrm{em}}\,\mathrm{Im}\,C^-+C_{\mathrm{em}}\,\mathrm{Im}\,B^-\}. \tag{A.4.35}$$

In this case the correction is rather small.

5 Unitarity

5.1 Unitarity for πN scattering amplitudes

Starting from the unitarity of the S-operator and the definition of the T-operator, one obtains

$$S^+S=SS^+=I;\quad S=I+iT;\quad T-T^+=iT^+T. \tag{A.5.1}$$

We take matrix elements of the last relation with our incoming and outgoing πN states

$$\langle f|T|i\rangle-\langle f|T^+|i\rangle=i\sum_n\int d\tau_n\langle f|T^+|n\rangle\langle n|T|i\rangle. \tag{A.5.2}$$

The intermediate states $|n\rangle$ are real physical states which can be reached from the incoming and outgoing πN states, i.e. the conservation laws must be fulfilled. $d\tau_n$ contains the phase space factor.

Since we want to write this relation in special cases in terms of the amplitudes defined in Sect. A.2, we have to specify the description of the states $|i\rangle$, $|f\rangle$. We choose $|\theta,\phi,\lambda;\pi,N\rangle$, where θ,ϕ is given by the direction of the pion in the c.m. system. λ is the helicity and π, N refers to the charge state.

First, we suppress the notation for the charge state, considering for instance π^+p elastic scattering. The amplitude $\langle f|T|i\rangle$ is proportional to the helicity amplitude

$$\langle f|T|i\rangle=\langle\theta,\phi,\lambda'|T|0,0,\lambda\rangle=\frac{q^2}{2\pi^{3/2}}F_{\lambda'\lambda}(\theta). \tag{A.5.3}$$

$F_{\lambda'\lambda}(\theta)$ does not depend on the azimuthal angle ϕ.

5.1.1 Forward scattering

In forward scattering we have $|i\rangle = |f\rangle$ and Eq. (A.5.2) reads

$$2\,\mathrm{Im}\langle \mathrm{i}|\mathrm{T}|\mathrm{i}\rangle = (q^2/\pi^{3/2})\,\mathrm{Im}\,F_{++}(s,\theta=0°) = \sum_{\mathrm{n}} \int \mathrm{d}\tau_{\mathrm{n}} |\langle \mathrm{n}|\mathrm{T}|\mathrm{i}\rangle|^2 = (q^2/4\pi^2)\sigma\,. \tag{A.5.4}$$

This relation is called the *optical theorem*, σ being the total cross section. We write this theorem also for the other amplitudes

$$\mathrm{Im}\,C(s,0) = k\sigma\,; \quad \mathrm{Im}\,G(s,z=1) = (q/4\pi)\sigma\,; \quad \mathrm{Im}\,F_{++}(s,t=0) = \sigma/4\sqrt{\pi}\,. \tag{A.5.5}$$

Since the optical theorem is a linear relation, it is valid not only for the elastic amplitudes but also for the isospin even and odd combinations

$$\sigma_{\pm} \equiv \sigma(\pi^{\pm}\mathrm{p})\,; \quad \sigma^{\pm} \equiv \tfrac{1}{2}(\sigma_{-} \pm \sigma_{+})\,. \tag{A.5.6}$$

In particular, we obtain for the charge-exchange amplitude

$$\mathrm{Im}\,C_0(s,0) = -\sqrt{2}\,\mathrm{Im}\,C^{-}(s,0) = k(\sigma_{+} - \sigma_{-})/\sqrt{2}\,. \tag{A.5.7}$$

5.1.2 Elastic unitarity

Below the inelastic threshold the sum in the unitarity relation (A.5.2) includes only πN intermediate states, the summation being carried out over the helicity and charge states. The evaluation of this part is also of interest at higher energies. We denote this contribution by a subscript "el".

We use Eq. (A.5.3) in order to write Eq. (A.5.2) in terms of amplitudes, taking into account the elastic part only

$$[F_{\lambda'\lambda}(\theta) - F^{*}_{\lambda\lambda'}(-\theta)]_{\mathrm{el}} = \mathrm{i}\frac{q^2}{2\pi^{3/2}} \sum_{\lambda''} \int \mathrm{d}\Omega'' F^{*}_{\lambda''\lambda'}(\bar{\theta}) F_{\lambda''\lambda}(\theta'')\,. \tag{A.5.8}$$

The integral goes over the direction of the intermediate state θ'', ϕ''. The amplitude $F_{\lambda''\lambda'}(\bar{\theta})$ belongs to scattering from the direction θ'', ϕ'' to the direction θ, ϕ.

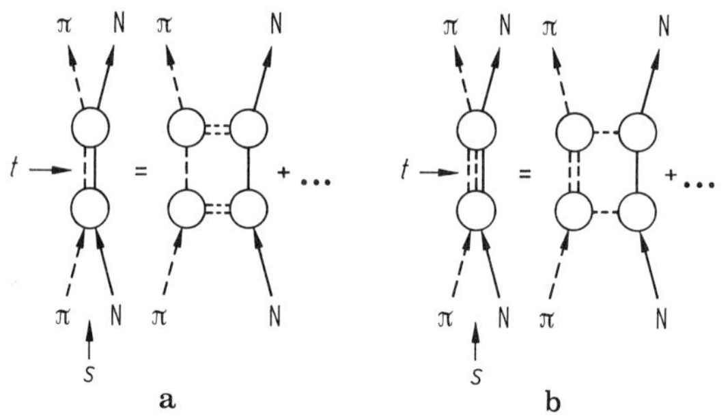

Fig. A.5.1. Intermediate states in the unitarity relation for πN scattering. a) elastic unitarity. The intermediate state is a πN state, including the possibility of charge-exchange. If we consider t-channel unitarity, the lowest intermediate state has 4 pions because of G-parity invariance.

b) The lowest inelastic intermediate state has one nucleon and two pions. If we consider t-channel unitarity, the lowest intermediate state has two pions. This case includes ϱ-exchange, because ϱ does not decay into 4 pions.

If the helicity amplitudes $F_{+\pm}$ are expressed in terms of invariant amplitudes, Eqs. (A.2.29), (A.2.30), it is seen that $F_{++}(-\theta) = F_{++}(\theta)$ and $F_{+-}(-\theta) = -F_{+-}(\theta) = F_{-+}(\theta)$. Therefore, Eq. (A.5.8) can be written

$$\mathrm{Im}\,F_{+\pm}(\theta)|_{\mathrm{el}} = (q^2/4\pi^{3/2}) \sum_{\lambda} \int \mathrm{d}\Omega'' F^{*}_{\lambda+}(\bar{\theta}) F_{\lambda\pm}(\theta'')\,. \tag{A.5.9}$$

The relation between $z = \cos\theta$, $z'' = \cos\theta''$, $\bar{z} = \cos\bar{\theta}$ and the azimuthal angle ϕ'' of the intermediate state reads (writing z' for $\bar{z}$)

$$z' = zz'' + [(1-z^2)(1-z''^2)]^{1/2}\cos\phi''\,. \tag{A.5.10}$$

It can be used in order to replace the integral over $d\Omega''$ by an integral over $dz'dz''$: $d\Omega'' = (2\Theta(K)/\sqrt{K})dz'dz''$ [see Eq. (A.5.12)].

The last remark is applied in the following version of the elastic unitarity relation for the amplitudes $G(s,z)$ and $f_2(s,z) = -H/\sin\theta$. According to Eq. (A.2.31), these amplitudes depend only on $\cos\theta$ and not on $\sin\theta$.

$$\frac{2\pi}{q}\operatorname{Im}G_{\mathrm{el}}(s,z) = \int_{-1}^{1} dz' \int_{-1}^{1} dz'' \frac{\Theta(K)}{\sqrt{K}}\{G^*(z'')G(z') + (z - z'z'')f_2^*(z'')f_2(z')\};$$
$$\frac{2\pi}{q}\operatorname{Im}H_{\mathrm{el}}(s,z)/\sin\theta = -\int_{-1}^{1} dz' \int_{-1}^{1} dz'' \frac{\Theta(K)}{\sqrt{K}}\cdot\left\{f_2^*(z'')f_2(z') + \frac{z''-zz'}{1-z^2}f_1^*(z'')f_2(z') + \frac{z'-zz''}{1-z^2}f_2^*(z'')f_1(z')\right\}; \qquad \text{(A.5.11)}$$

$$K(z,z',z'') = 1 - z^2 - z'^2 - z''^2 + 2zz'z''; \qquad \Theta(K) = \begin{matrix}1\\0\end{matrix} \text{ for } \begin{matrix}K>0\\K<0\end{matrix}. \qquad \text{(A.5.12)}$$

For a study of the limiting cases of forward and backward scattering $z \to \pm 1$, one needs the formulas

$$\pi = \int_{-1}^{1} dz''\Theta(K)/\sqrt{K}; \qquad \pi zz' = \int_{-1}^{1} dz''\Theta(K)z''/\sqrt{K}. \qquad \text{(A.5.13)}$$

The partial wave expansion of $\Theta(K)/\sqrt{K}$ is of interest, if one wants to derive the connection between Eq. (A.5.11) and the unitarity relation for partial waves, Eq. (A.5.20)

$$\Theta(K)/\sqrt{K} = -\tfrac{1}{2}\mathrm{i}\,\mathrm{disc}(1/\sqrt{-K}) = \tfrac{1}{2}\pi\sum_{n=0}^{\infty}(2n+1)P_n(z)P_n(z')P_n(z'');$$
$$1/\sqrt{-K} = \sum_{n=0}^{\infty}(2n+1)P_n(z')Q_n(z)P_n(z''). \qquad \text{(A.5.14)}$$

Due to the factor $\Theta(K)$, the integral in Eq. (A.5.11) is extended over an ellipse in the $z'z''$-plane (Fig. A.5.2). For $\theta = 90°$, the ellipse is a circle and for forward or backward scattering it degenerates into a straight line.

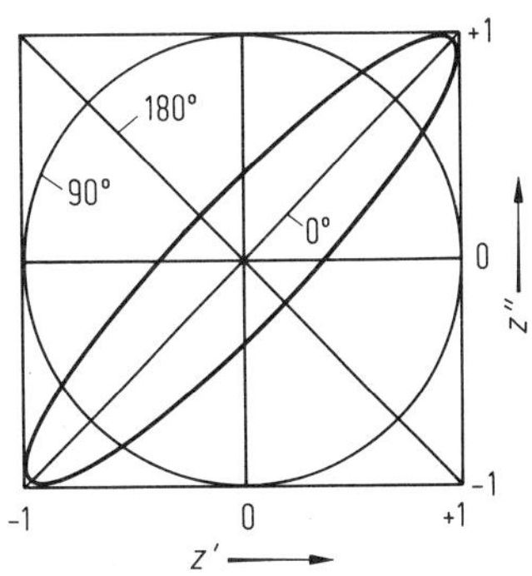

Fig. A.5.2. The integration range in the elastic unitarity relation. The factor in front of the bracket under the integral in Eq. (A.5.11) is non-zero within an ellipse, which degenerates into a straight line at $z = +1$ and $z = -1$ and becomes a circle at $z = 0$.

As an example, we mention the evaluation of the integral (A.5.11) for an exponentially decreasing diffraction peak

$$G(s,z) = \mathrm{i}(q\sigma/4\pi)\exp(bt/2); \qquad H(s,z) \equiv 0. \qquad \text{(A.5.15)}$$

$$\operatorname{Im}G_{\mathrm{el}}(s,z) = \frac{q^3\sigma^2}{64\pi^3}\int d\Omega'' e^{b(t'+t'')/2}$$
$$= \frac{q\sigma^2}{16\pi^2 b}\{2(1+z)\}^{-1/2}e^{-2bq^2}\sinh\{bq^2\sqrt{2(1+z)}\}$$
$$\approx \frac{q\sigma^2}{64\pi^2 b}e^{bt/4}; \qquad |t| \ll 8q^2. \qquad \text{(A.5.16)}$$

The unitarity relations are more complicated for the charge-exchange amplitude $G_0 = -\sqrt{2}G^-$, because the charge-exchange can occur in the first or in the second step

$$\frac{2\pi}{q}\operatorname{Im} G_{\mathrm{el}}^-(s,z) = \int_{-1}^{1} \mathrm{d}z' \int_{-1}^{1} \mathrm{d}z'' \frac{\Theta(K)}{\sqrt{K}} \cdot [\{G^{+*}(z'')G^-(z') + G^{-*}(z'')G^+(z') + G^{-*}(z'')G^-(z')\} + (z - z'z'')\{f_2^{+*}(z'')f_2^-(z') + f_2^{-*}(z'')f_2^+(z') + f_2^{-*}(z'')f_2^-(z')\}]\,. \tag{A.5.17}$$

In the special case of forward scattering ($z=1$), one obtains a contribution to the optical theorem, which is expressed in terms of the integrated differential cross sections σ^{el}

$$\operatorname{Im} C_{\mathrm{el}}^-(s,0) = 4\pi\frac{k}{q}\operatorname{Im} G_{\mathrm{el}}^-(s,1) = \frac{k}{2}(\sigma_-^{\mathrm{el}} + \sigma_0^{\mathrm{el}} - \sigma_+^{\mathrm{el}})\,. \tag{A.5.18}$$

The quantity $\operatorname{Im} G_{\mathrm{el}}^+$ is used in the theory of diffraction scattering in order to define the *inelastic overlap function* $\operatorname{Im} G_{\mathrm{in}}^+$ by

$$\operatorname{Im} G^+(s,z) \equiv \operatorname{Im} G_{\mathrm{el}}^+(s,z) + \operatorname{Im} G_{\mathrm{in}}^+(s,z)\,. \tag{A.5.19}$$

It is interesting to discuss the possible t-channel exchanges which belong to the unitarity diagram for $\operatorname{Im} G_{\mathrm{el}}$ etc. Because of G-parity conservation (Sect. A.2.2.4), the number of pions at the left hand vertices must be even. We conclude that two-pion exchange does not contribute to the "elastic" part of the amplitude (Fig. A.5.1). This fact will be used in Sect. A.8.3.

5.1.3 Unitarity for partial waves and impact parameter transforms

For the partial waves, unitarity has the consequence that $T_{\ell\pm}$ can be described by Eq. (A.3.2) with real phase shifts $\delta_{\ell\pm}$ and $0 \leqq \eta_{\ell\pm} \leqq 1$. From Eq. (A.3.2) one can derive for the isospin $I=\frac{1}{2}$ and $\frac{3}{2}$ partial waves

$$\operatorname{Im} T_{\ell\pm} = |T_{\ell\pm}|^2 + \tfrac{1}{4}(1 - \eta_{\ell\pm}^2)\,. \tag{A.5.20}$$

The first term on the right hand side belongs to πN intermediate states and the second term to all others.

If this relation is inserted into the partial wave expansion Eq. (A.3.1), one obtains the decomposition of Eq. (A.5.19) and similar decompositions of all other amplitudes.

In order to be able to write a unitarity relation for the impact parameter transforms of Sect. 3.3., we introduce

$$\varrho(s,b) = \frac{1}{q}\int_{-4q^2}^{0} J_0(b\sqrt{-t})\operatorname{Im} G_{\mathrm{in}}(s,t)\mathrm{d}t \tag{A.5.21}$$

and find

$$\operatorname{Im} g(s,b) = |g(s,b)|^2 + \frac{1}{4}\varrho(s,b) + q^{-2}\frac{\partial}{\partial b^2}\left[b^4\left|\frac{\partial g}{\partial b^2}\right|^2\right] + O(q^{-4})\,. \tag{A.5.22}$$

In addition to the terms expected from Eqs. (A.5.20) and (A.3.27), there are correction terms, which vanish at fixed b in the high energy limit.

5.2 Unitarity for $\pi\pi$N$\bar{\mathrm{N}}$-amplitudes

Extended unitarity leads to the relation

$$\operatorname{Im} f_\pm^J(t) = f_\pm^J(t)T_J^{I*}(t)\,; \quad 4\mu^2 < t < 16\mu^2\,, \tag{A.5.23}$$

where the $\pi\pi$-partial waves $T_J^I(t)$ are related to the (real) phase shifts $\delta_J^I(t)$ by

$$T_J^I(t) = \exp(\mathrm{i}\delta_J^I)\sin\delta_J^I\,. \tag{A.5.24}$$

The isospin quantum number is $I=0$ if J is even and $I=1$ if J is odd. It follows that

$$f_\pm^J(t) = \pm|f_\pm^J(t)|\exp(\mathrm{i}\delta_J^I)\,. \tag{A.5.25}$$

In these relations the amplitudes are understood as limits $t+\mathrm{i}0$. Applications of extended unitarity to electromagnetic form factors will be mentioned in Sect. A.11.

5.3 Unitarity bounds

The unitarity condition for isospin $I=1/2$ and $3/2$ πN partial waves, Eq. (A.5.20), leads to the inequality $\operatorname{Im} T_{\ell\pm} \geqq |T_{\ell\pm}|^2$ which implies that $\operatorname{Im} T_{\ell\pm} > 0$ and $|T_{\ell\pm}| \leqq 1$.

$$1 \geqq \operatorname{Im} T_{\ell\pm} \geqq |T_{\ell\pm}|^2 > 0. \qquad \text{(A.5.26)}$$

This relation has two aspects:

boundedness: $1 \geqq \operatorname{Im} T_{\ell\pm} \geqq |T_{\ell\pm}|^2$; positivity: $\operatorname{Im} T_{\ell\pm} > 0$.

Consequences of the positivity condition are usually derived for the spinless case. An extension to πN scattering is simple for the spin no-flip amplitude $G(s,z)$, because the Legendre expansion is the same, if one introduces the combination $\hat{T}_{\ell\pm}$, Eq. (A.3.17), of the partial wave amplitudes [Eq. (A.3.20)].

Since $|P_\ell(\cos\theta)| \leqq 1$ for $-1 \leqq \cos\theta \leqq 1$, it follows that

$$|\operatorname{Im} G(s,z)| \leqq \operatorname{Im} G(s,0); \quad -1 \leqq z \leqq +1. \qquad \text{(A.5.27)}$$

A much stronger positivity condition can be derived from Eq. (A.5.30)

$$\left|\left(\frac{\partial}{\partial z}\right)^n \operatorname{Im} G(s,z)\right| \leqq \left(\frac{\partial}{\partial z}\right)^n \operatorname{Im} G(s,1); \quad -1 \leqq z \leqq +1. \qquad \text{(A.5.28)}$$

It is remarkable that positivity properties can be proved also for the invariant amplitude $D(s,t)$, Eq. (A.2.16), for elastic scattering, whose Legendre expansion contains derivatives of Legendre polynomials

$$\left|\left(\frac{\partial}{\partial t}\right)^n \operatorname{Im} D(s,t)\right| \leqq \left(\frac{\partial}{\partial t}\right)^n \operatorname{Im} D(s,0); \quad 0 \geqq t \geqq -4q^2. \qquad \text{(A.5.29)}$$

A similar relation is valid in the physical region of the u-channel.

The positivity properties are consequences of Eq. (A.5.26) and of the following relations for the Legendre polynomials

$$\begin{aligned} &\left(\frac{d}{dz}\right)^n P_\ell(1) > 0; \quad \left|\left(\frac{d}{dz}\right)^n P_\ell(z)\right| \leqq \left(\frac{d}{dz}\right)^n P_\ell(1); \quad -1 \leqq z \leqq +1; \\ &\left|\left(\frac{d}{dz}\right)^n \{P'_{\ell+1}(z) - P'_\ell(z)\}\right| \leqq \left(\frac{d}{dz}\right)^n \{P'_{\ell+1}(1) - P'_\ell(1)\}; \quad 0 \leqq z \leqq +1. \end{aligned} \qquad \text{(A.5.30)}$$

Rigorous upper bounds for the elastic forward amplitudes and for cross sections in the asymptotic region are listed in Table A.5.1. These bounds were derived from axiomatic quantum field theory, taking into account unitarity.

Table A.5.1. Upper bounds for elastic forward amplitudes and forward cross sections in the limit $s \to \infty$. $\sigma_{\rm el}$ is the integrated elastic cross section. ε is an arbitrarily small positive number.

$\operatorname{Im} C(s,0)/k = \sigma < \frac{4\pi}{\mu^2 - \varepsilon} \ln^2 s$; $|C(s,0)| < \text{const} \cdot s \cdot \ln^2 s$;	Froissart-Martin
$|A(s,0)| < \text{const } s \ln^3 s$; $|B(s,0)| < \text{const} \cdot \ln^3 s$;	Sommer
$\left.\frac{d\sigma}{dt}\right|_{t=0} = \frac{\sigma^2}{16\pi}[1 + \{\operatorname{Re} C(s,0)/\operatorname{Im} C(s,0)\}^2] \leqq \frac{\sigma_{\rm el}}{16\mu^2} \ln^2(s/s_0^2 \sigma_{\rm el})$;	Roy-Singh

Bounds for the slope of the diffraction peak

The diffraction peak is parametrized as follows

$$\operatorname{Im} G(s,t) = \operatorname{Im} G(s,0) e^{b(s,t)t/2}. \qquad \text{(A.5.31)}$$

The logarithmic slope at $t=0$ is defined by

$$b(s,0) = b_0(s) = 2 \frac{d}{dt} \ln \operatorname{Im} G(s,t) \Big|_{t=0}. \qquad \text{(A.5.32)}$$

The MacDowell-Martin *bound* reads

$$b_0(s) \geqq \frac{2}{9}\left[\frac{\sigma^2}{4\pi\sigma_{el}} - \frac{1}{q^2}\right]. \tag{A.5.33}$$

It is interesting to note that the leading term of the bound is almost saturated, if we insert σ_{el} for an exponentially decreasing diffraction peak

$$\operatorname{Im} G(s,t) = (q\sigma/4\pi)\mathrm{e}^{Bt/2}; \quad \sigma_{el} = \sigma^2/16\pi B; \quad b_0(s) = B \geqq \frac{8}{9}B. \tag{A.5.34}$$

Other bounds for $b_0(s)$ are

$$\sigma(s) \leqq 16\pi b_0(s) \leqq (2\pi/\mu^2)\ln^2(s/s_0^2\sigma). \tag{A.5.35}$$

At present, the experimental data suggest that $\sigma(s)$ is increasing $\propto \ln^2 s$ (Sect. 2.2.1). Then, it follows from Eqs. (A.5.35) and (A.5.33) that $b_0(s)$ and $\sigma_{el}(s)$ are also increasing $\propto \ln^2 s$ in the asymptotic region.

A bound for the shape of the diffraction peak

A generalization of the MacDowell-Martin bound is due to Singh and Roy

$$\left|\frac{\operatorname{Im} G(s,t)}{\operatorname{Im} G(s,0)}\right| \leqq 1 - \frac{\varrho}{9} + \frac{3}{8}\left(\frac{\varrho}{9}\right)^2 - \ldots; \quad \varrho = \frac{-t\sigma^2}{4\pi\sigma_{el}}. \quad s \to \infty. \tag{A.5.36}$$

Another bound gives a restriction for the t-dependence of the logarithmic slope $b(t)$. Notation: $\sigma(t) \equiv \mathrm{d}\sigma(t)/\mathrm{d}t$,

$$b(t) = \frac{\mathrm{d}}{\mathrm{d}t}\ln\sigma(t); \quad \sigma(t) = \sigma(0)\exp\left\{\int_0^t b(t')\mathrm{d}t'\right\}. \tag{A.5.37}$$

$$b(t) \geqq b(0)\left[\frac{\sigma(0)}{\sigma(t)}\right]^{1/2}\left\{\frac{3}{2}\frac{\sigma(t_1)}{\sigma(0)}\left[\frac{b(t_1)}{b(0)}\right]^2 - \frac{1}{2}\right\}, \tag{A.5.38}$$

where $t_1 = t/3$ in a very good approximation (Roy).

Behaviour of the partial wave amplitudes for large ℓ

The following inequality can be derived from unitarity, using the domain of convergence of the partial wave series for $\operatorname{Im} G$ and polynomial boundedness of the amplitude ($< s^N$)

$$|\hat{T}_\ell|^2 \leqq \operatorname{Im} \hat{T}_\ell(s) < \mathrm{const} \cdot \exp\{-\ell(1-\varepsilon(s))4\mu/\sqrt{s} + N\ln s\}, \tag{A.5.39}$$

where $\varepsilon \to 0$ as $s \to \infty$. It is seen that $\operatorname{Re} \hat{T}_\ell$ and $\operatorname{Im} \hat{T}_\ell$ decrease exponentially with ℓ and become negligible, when s is large and

$$\ell > L = C\sqrt{s}\ln s. \tag{A.5.40}$$

Therefore, the partial wave expansion can be truncated

$$qG(s,t) = \sum_{\ell=0}^{L}(2\ell+1)\hat{T}_\ell(s)P_\ell(1+t/2q^2) + R_L(s,t). \tag{A.5.41}$$

The absolute value of the remainder is bounded by a polynomial: $|R_L(s,t)| < s^{-M}$. M can be made arbitrarily large by choice of C in Eq. (A.5.40). An estimate for the sum in the last equation, taken at $t=0$, gives the Froissart-Martin bound of Table A.5.1.

5.4 Pomeranchuk-type theorems

Pomeranchuk proved the relation $\{\sigma_{\pm} \equiv \sigma(\pi^{\pm} p) = \text{total cross section}\}$

$$\lim_{s\to\infty} (\sigma_{-} - \sigma_{+}) = 0, \tag{A.5.42}$$

using the following assumptions
i) the forward dispersion relation is valid,
ii) the high energy limits of $\sigma_{\pm}$ exist and are finite,
iii) the modulus of $C_{\pm}(s,0)/s$ is bounded by a finite constant $c_{\pm}$.

Nowadays, the data suggest that the total cross sections are not bounded. The question is, whether the relation (A.5.42) can be proven from other assumptions.

Pomeranchuks relation cannot be derived from axiomatic field theory, even if unitarity is included. In addition to assumption i), one can assume for instance

$$\lim_{s\to\infty} \frac{\mathrm{Re}\, C_{-}(s,0) - \mathrm{Re}\, C_{+}(s,0)}{s \ln s} = 0 \tag{A.5.43}$$

and the existence of the limit in Eq. (A.5.42).

If the cross sections $\sigma_{\pm}$ are increasing as $s\to\infty$, the ad hoc assumption Eq. (A.5.43) can be avoided, if analyticity of the amplitude in t and unitarity are assumed. Instead of Pomeranchuks relation one obtains

$$\lim_{s\to\infty} \sigma_{+}/\sigma_{-} = 1. \tag{A.5.44}$$

If isospin invariance is assumed, the following relation can be proved

$$|\lim_{s\to\infty} (\sigma_{-} - \sigma_{+})| \leqq \frac{\pi^{3/2}}{\sqrt{2}\mu} \lim_{s\to\infty} \sqrt{\sigma_0^{\mathrm{el}}} \tag{A.5.45}$$

provided the limits exist. σ_0^{el} is the integrated differential cross section of the reaction $\pi^- p \to \pi^0 n$.

At present the charge-exchange cross sections up to 200 GeV/c can be described fairly well by the reggeized ϱ-exchange model. If this model is inserted into Eq. (A.5.45), it turns out that the inequality is not fulfilled at high energies. This is not serious, since the inequality has been derived only for the limit and not for the cross sections at finite energies. In our case both sides go to zero, i.e. Eq. (A.5.45) is valid with the equality sign.

6 Symmetries and analytic properties of invariant amplitudes

6.1 Behaviour of invariant amplitudes under symmetry operations

The S-matrix theory of strong interactions is based on the validity of Lorentz invariance and invariance under the operations time-reversal (T), space-inversion (P) and particle-antiparticle conjugation (C). Furthermore, one assumes that the invariant scattering amplitudes [in our case $A(s,t)$ and $B(s,t)$] contain only the singularities demanded by unitarity and that they describe several reactions related by the crossing operation.

The crossing operation

Starting from a reaction $\pi + N \to \pi' + N'$, the crossing operation is carried out by transposing one of the incoming and one of the outgoing particles to the other side of the reaction, simultaneously changing them into their antiparticles and reversing the sign of their four-momenta.

This operation is sometimes called *line-reversal*, in particular in connection with Regge models (see Sect. A.9.2.4).

Up to a phase factor, two crossed reactions have the same invariant amplitudes. As an example, we consider $\pi^+ p \to \pi^+ p$. One of the crossing operations leads to a relation with the $\pi^- p \to \pi^- p$ amplitude (Fig. A.6.1)

$$A_{+}(\mathbf{q}, \mathbf{p}; \mathbf{q}', \mathbf{p}') = A_{-}(-\mathbf{q}', \mathbf{p}; -\mathbf{q}, \mathbf{p}'), \tag{A.6.1}$$

where the four-momenta as defined in Fig. A.1.1 are listed in the argument. As a consequence of the Hall-Wightman *theorem*, a Lorentz invariant function of four-momenta is an analytic function of the invariants that can be formed from these momenta. Therefore, we can write

$$A_+(s,t,u) = A_-(u,t,s)\,. \tag{A.6.2}$$

If s, t or u are chosen in the range of a cut, it is necessary to specify whether the boundary value is taken from above or below [$\pm i0$ prescription, see Eq. (A.6.27)]. If the variables in $A_+(s,t,u)$ lie in the physical region, one obtains the A_--amplitude in the unphysical region.

An application of crossing to the pions and also to the nucleons shows that the TCP-conjugated reaction $\pi^- + \bar{p} \to \pi^- + \bar{p}$ has the same invariant amplitude. The bars denote the antiparticles.

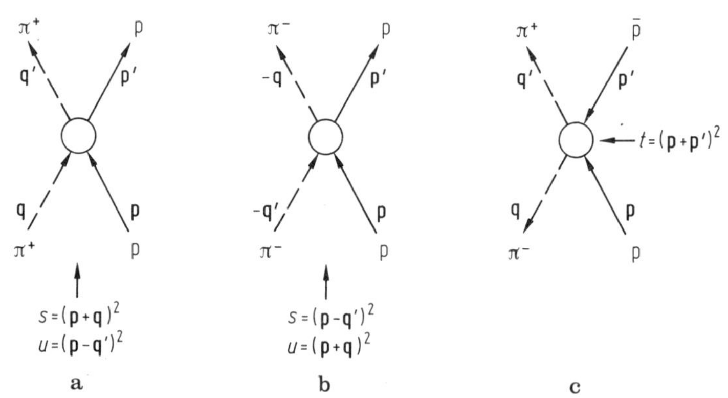

Fig. A.6.1. The crossing operation. We start from $\pi^+ p \to \pi^+ p$ (Fig. a). Crossing of the pions leads to the u-channel reaction $\pi^- p \to \pi^- p$ (Fig. b) and crossing of one pion and one nucleon leads to the t-channel reaction $\bar{p}p \to \pi^+\pi^-$ (Fig. c).

Table A.6.1. List of reactions which are described by the same pair of invariant amplitudes $A(s,t)$, $B(s,t)$. Further reactions are obtained by a repeated application of symmetry or crossing operations.

related by symmetry operation		original reaction	related by crossing operation
T	$\pi' + N' \to \pi + N$	$\pi + N \to \pi' + N'$	$\bar{\pi}' + N \to \bar{\pi} + N'$
C	$\bar{\pi} + \bar{N} \to \bar{\pi}' + \bar{N}'$		$\pi + \bar{N}' \to \pi' + \bar{N}$
TCP	$\bar{\pi}' + \bar{N}' \to \bar{\pi} + \bar{N}$		$\pi + \bar{\pi}' \to \bar{N} + N'$
			$\bar{N}' + N \to \pi' + \bar{\pi}$

Other relations between scattering amplitudes of different reactions follow from isospin invariance (Table A.2.1). A special case is *charge symmetry*, which belongs to a rotation in isospin space such that I_3 is changed into $-I_3$. In πN reactions p and n are interchanged simultaneously with π^+ and π^-, whereas π^0 remains unchanged. Example: the amplitudes for $\pi^- p \to \pi^0 n$ and for $\pi^+ n \to \pi^0 p$ are the same.

Time-reversal

As mentioned above, two reactions which are related by time-reversal have the same invariant amplitudes. However, it is not true that the same experiment carried out for both reactions leads always to the same result, i.e. the *principle of detailed balance* is not valid. This can be seen from the expression for the T-matrices of time-reversed reactions

$$\langle \boldsymbol{q}', \boldsymbol{p}', M'; \pi'N'|T|\boldsymbol{q}, \boldsymbol{p}, M; \pi N\rangle = \langle -\boldsymbol{q}, -\boldsymbol{p}, -M; \pi N|T|-\boldsymbol{q}', -\boldsymbol{p}', -M'; \pi'N'\rangle, \tag{A.6.3}$$

where πN and π'N$'$ denote the charge states. The momentum vectors are the spatial parts of the four-vectors in Fig. A.1.1 and $M, M' = \pm 1/2$ are the eigenvalues of the nucleon spin operator along a given direction. As an example we consider the matrix element in the c.m. system for fixed values of energy, scattering angle θ and quantum numbers $M = M' = 1/2$. There exist two possibilities for the scattering process $\pi N \to \pi' N'$ in a given scattering plane (Fig. A.6.2) which lead in general to different cross sections, if the target is polarized [see Eq. (A.4.10) and the subsequent discussion of the "left-right asymmetry"]. If we start with scattering to the left in $\pi N \to \pi' N'$, Eq. (A.6.3) shows the equality with the matrix element for scattering to the right in the inverse reaction $\pi' N' \to \pi N$.

The cross sections for the two reactions are equal, if the target is unpolarized and the polarization of the final state is not measured *(principle of semi-detailed balance)*. Time-reversal in $\pi\pi N\bar{N}$-scattering is discussed in Sect. A.4.5.

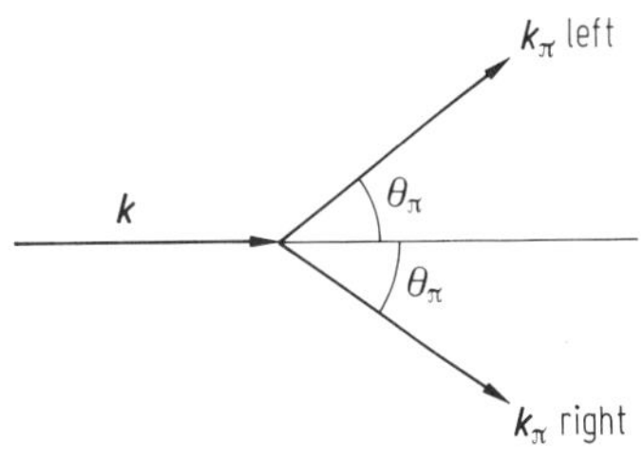

Fig. A.6.2. Left-right asymmetry. We consider two possibilities for a scattering process, in which the scattering plane and the scattering angle are the same, but the outgoing pion goes to the left in one case and to the right in the other case. The invariant amplitudes are the same but, if the target is polarized, Eq. (A.4.10) shows that the cross sections are different, because the two $\hat{\boldsymbol{n}}$-vectors have opposite directions.

Constraints derived from symmetries

It happens that the application of a symmetry operation to a reaction leads to the same reaction for different values of the kinematical variables (example: P- and T-invariance for elastic scattering). In this case one obtains a restriction for the T-matrix.

The most general Lorentz-invariant ansatz for the scattering amplitude contains in addition to the terms in Eq. (A.2.2) two others: $f_3(\boldsymbol{\sigma}\cdot\hat{\boldsymbol{q}})$ and $f_4(\boldsymbol{\sigma}\cdot\hat{\boldsymbol{q}}')$. P-invariance implies that $f_3=f_4=0$, whereas T-invariance gives the weaker restriction $f_3=f_4$.

The crossing operation leads to the same reaction, if the pions are crossed in $\pi^0 p \to \pi^0 p$. A second possibility is to start with $\pi^- p \to \pi^0 n$ and to use in addition to crossing also charge-symmetry and time-reversal invariance. In these cases one concludes that the invariant amplitudes must be symmetric or antisymmetric, respectively, under the exchange of s and u at fixed t. The same restriction follows from the generalized Pauli principle for for the two pions in the reactions $\bar{p}p \to \pi^0\pi^0$ and $\bar{p}n \to \pi^0\pi^-$.

6.2 The Mandelstam representation

6.2.1 The double-dispersion relation

The usual working hypothesis is that the invariant amplitudes are analytic functions of two complex variables, the only singularities being those derived from unitarity and from the mass spectrum (nucleon poles). All singularities occur at real values of the variables s, t, u. The Mandelstam representation, which is also called a double or two-dimensional dispersion relation, is a concise statement of these postulated analytic properties.

Although the Mandelstam representation is not used directly in phenomenological investigations, the main formulas are listed in the following, because they are needed in applications of single variable and partial wave dispersion relations.

It is assumed that, in agreement with the situation in perturbation theory, the invariant amplitudes $A(s,t)$ and $B(s,t)$ describe three reactions, related to each other by the crossing operation. We call them s, t or u-channel reactions, if s, t or u are the square of the total energy in the respective c.m. frame. Example (Fig. A.6.1):

s-channel: $\pi^+ p \to \pi^+ p$; t-channel: $\pi^+\pi^- \to \bar{p}p$; u-channel: $\pi^- p \to \pi^- p$.

The physical regions of these reactions are shown in Fig. A.6.3. The same amplitudes describe further reactions, related to the above ones by T- and C-invariance (Table A.6.1) and by isospin invariance.

In order to have simple properties under the crossing operation, the Mandelstam representation is written for the isospin even and odd amplitudes (Table A.2.1)

$$A^{\pm}(s,t,u)=\frac{1}{\pi^2}\iint\frac{A^{\pm}_{st}(s',t',\tilde{u})}{(s'-s)(t'-t)}\mathrm{d}s'\mathrm{d}t'+\frac{1}{\pi^2}\iint\frac{A^{\pm}_{tu}(\tilde{s},t',u')}{(u'-u)(t'-t)}\mathrm{d}u'\mathrm{d}t'+\frac{1}{\pi^2}\iint\frac{A^{\pm}_{us}(s',\tilde{t},u')}{(s'-s)(u'-u)}\mathrm{d}s'\mathrm{d}u'. \tag{A.6.4}$$

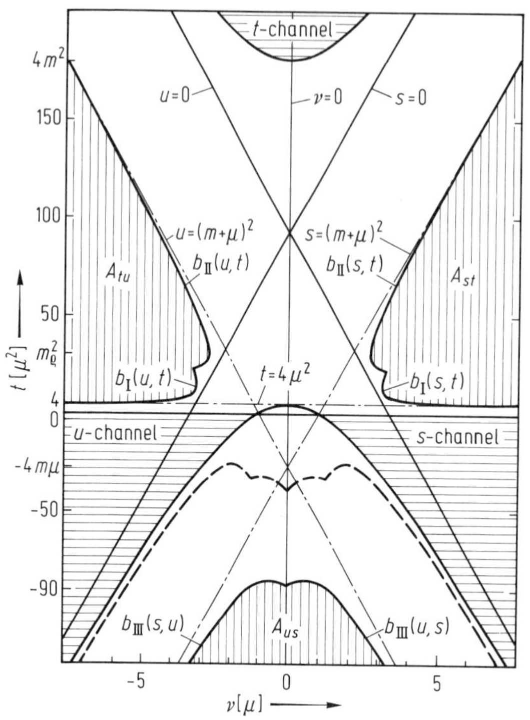

Fig. A.6.3. Mandelstam diagram: t vs. $\nu=(s-u)/4m$. Horizontally hatched: physical regions. Vertically hatched: regions of non-vanishing double spectral functions. Dot-dashed lines: limits of regions where the absorptive parts are non-zero: $s>(m+\mu)^2$, $t>4\mu^2$, $u>(m+\mu)^2$. Dashed lines: reflection of the boundary of the double spectral function region at fixed s in the $\cos\theta$-plane and its crossing symmetric counterpart. The partial wave expansions for ImA and ImB converge in the region between this boundary and the dashed line. See Table A.6.2.

Along the cuts, the value of $A(s,t,u)$ is not unique. The physical amplitude $A(s,t,u)$ is defined to be the boundary value taken from above, written as $s+i0$, $t+i0$, $u+i0$ for the s-, t-, and u-channel, respectively.

The relation for the amplitude B is analogous, but it has in addition the nucleon pole terms

$$B_N^{\pm}(s,t,u)=\frac{g^2}{m^2-s}\mp\frac{g^2}{m^2-u}. \tag{A.6.5}$$

Notation: Although only two of the variables s, t, u are independent because of the relation $s+t+u=2m^2+2\mu^2\equiv\Sigma$, all three of them will be used in order to exhibit the symmetry of the formulas. In the arguments of the amplitudes, the first, second and third position always belongs to s, t, u in this order: $B(x,y,z)$ means B taken at $s=x$, $t=y$, $u=z$. The label $\sim$ indicates that this variable is considered to be the dependent one. Its value is given by Σ minus the sum of the other two variables.

Equation (A.6.4) has been obtained on the assumption that the scattering amplitude tends to zero if either s or t tends to infinity. A more general ansatz includes one-dimensional dispersion integrals which can be considered as subtraction constants in one variable but dispersion integrals over the other one (similarly for the amplitude B)

$$\frac{1}{\pi}\int_{(m+\mu)^2}^{\infty}\frac{a_s(s')}{s'-s}ds'+\frac{1}{\pi}\int_{4\mu^2}^{\infty}\frac{a_t(t')}{t'-t}dt'+\frac{1}{\pi}\int_{(m+\mu)^2}^{\infty}\frac{a_u(u')}{u'-u}du'. \tag{A.6.6}$$

The double spectral functions A_{st}, A_{tu}, A_{us} and the single spectral functions a_s, a_t, a_u are real and this is also true for the spectral functions of the amplitude B. Crossing symmetry demands that

$$\begin{aligned}&A_{us}(s,t,u)=\pm A_{us}(u,t,s); \quad B_{us}(s,t,u)=\mp B_{us}(u,t,s);\\&A_{st}(s,t,u)=\pm A_{tu}(u,t,s); \quad B_{st}(s,t,u)=\mp B_{tu}(u,t,s);\\&a_s^{\pm}(s')=\pm a_u^{\pm}(s').\end{aligned} \tag{A.6.7}$$

The limits of the integrals in Eqs. (A.6.4) and (A.6.6), i.e. the boundaries of the regions where the spectral functions are nonvanishing, follow from unitarity. It is sufficient to consider two-particle intermediate states in all channels. Since these boundaries depend only on the masses, they are the same for $A(s,t)$ and $B(s,t)$. In the Mandelstam diagram, Fig. A.6.3, they are symmetric with respect to the line $\nu=0$ (or $s=u$) due to crossing symmetry, Eq. (A.6.7).

A_{st} and B_{st} are different from zero only within the region bounded by the curves which are shown in Fig. A.6.3

$$b_{\mathrm{I}}(s,t) \equiv (t-4\mu^2)\{s-(m+2\mu)^2\}\{s-(m-2\mu)^2\}-16\mu^4\{s+3(m^2-\mu^2)\}=0\,;$$
$$b_{\mathrm{II}}(s,t) \equiv (t-16\mu^2)\{s-(m+\mu)^2\}\{s-(m-\mu)^2\}-64\mu^4 s = 4(t-16\mu^2)sq^2-64\mu^4 s=0\,. \tag{A.6.8a}$$

The boundaries of A_{tu} and B_{tu} follow from crossing symmetry. Finally, those of A_{us}, B_{us} are given by

$$b_{\mathrm{III}}(s,u) \equiv \{u-(m+2\mu)^2\}\{u-(m-2\mu)^2\}\{s-(m+\mu)^2\}\{s-(m-\mu)^2\}-16\mu^2 m^2 us$$
$$+16\mu^2(m^2-\mu^2)^2(s+u-m^2-2\mu^2)=0 \tag{A.6.8b}$$

and its $s-u$ crossing symmetric counterpart (Fig. A.6.3).

6.2.2 Single-dispersion relations and absorptive parts

From the Mandelstam representation, one can derive the following single variable dispersion relations ($s_{\mathrm{th}}=u_{\mathrm{th}}=(m+\mu)^2$)

$$\textit{fixed-t} \qquad A(s,t,u)=\frac{1}{\pi}\int_{s_{\mathrm{th}}}^{\infty}\frac{A_s(s',t,\tilde{u})}{s'-s}\,ds'+\frac{1}{\pi}\int_{u_{\mathrm{th}}}^{\infty}\frac{A_u(\tilde{s},t,u')}{u'-u}\,du'\,;$$
$$\textit{fixed-s} \qquad A(s,t,u)=\frac{1}{\pi}\int_{4\mu^2}^{\infty}\frac{A_t(s,t',\tilde{u})}{t'-t}\,dt'+\frac{1}{\pi}\int_{u_{\mathrm{th}}}^{\infty}\frac{A_u(s,\tilde{t},u')}{u'-u}\,du'\,; \tag{A.6.9}$$
$$\textit{fixed-u} \qquad A(s,t,u)=\frac{1}{\pi}\int_{s_{\mathrm{th}}}^{\infty}\frac{A_s(s',\tilde{t},u)}{s'-s}\,ds'+\frac{1}{\pi}\int_{4\mu^2}^{\infty}\frac{A_t(\tilde{s},t',u)}{t'-t}\,dt'\,.$$

The absorptive parts A_s, A_t, A_u are analytic functions of one variable, which fulfill the following dispersion relations

$$A_s(s,t,u)=\frac{1}{\pi}\int_{T(s)}^{\infty}\frac{A_{st}(s,t',\tilde{u})}{t'-t}\,dt'+\frac{1}{\pi}\int_{U(s)}^{\infty}\frac{A_{us}(s,\tilde{t},u')}{u'-u}\,du'\,;$$
$$A_t(s,t,u)=\frac{1}{\pi}\int_{S(t)}^{\infty}\frac{A_{st}(s',t,\tilde{u})}{s'-s}\,ds'+\frac{1}{\pi}\int_{U(t)}^{\infty}\frac{A_{tu}(\tilde{s},t,u')}{u'-u}\,du'\,; \tag{A.6.10}$$
$$A_u(s,t,u)=\frac{1}{\pi}\int_{T(u)}^{\infty}\frac{A_{tu}(\tilde{s},t',u)}{t'-t}\,dt'+\frac{1}{\pi}\int_{S(u)}^{\infty}\frac{A_{us}(s',\tilde{t},u)}{s'-s}\,ds'\,.$$

They can be written in such a way that in each relation both integrals are taken over the same variable.

Similar relations are valid for the amplitude $B(s,t,u)$ and for its absorptive parts B_s, B_t, B_u, except for additional nucleon pole terms in the dispersion relations (A.6.9). The fixed-t relation has the same nucleon pole terms as the Mandelstam representation, Eq. (A.6.5), whereas the fixed-u and fixed-s relations have only pole terms at $s=m^2$ and $u=m^2$, respectively. The question of subtractions, i.e. of possible terms of the type mentioned in Eq. (A.6.6) will be discussed later.

The lower limits of the integrals in Eq. (A.6.10) are given by

$$T(s)=\mathrm{Min}\{T_{\mathrm{I}}(s),T_{\mathrm{II}}(s)\}\,;\qquad U(s)=\mathrm{Min}\{U_{\mathrm{III}}(s),S_{\mathrm{III}}(s)\}\,;$$
$$S(t)=\mathrm{Min}\{S_{\mathrm{I}}(t),S_{\mathrm{II}}(t)\}\,;\qquad U(t)=\mathrm{Min}\{S_{\mathrm{I}}(t),S_{\mathrm{II}}(t)\}\,; \tag{A.6.11}$$
$$T(u)=\mathrm{Min}\{T_{\mathrm{I}}(u),T_{\mathrm{II}}(u)\}\,;\qquad S(u)=\mathrm{Min}\{U_{\mathrm{III}}(u),S_{\mathrm{III}}(u)\}\,,$$

where the different functions $T_{\mathrm{I}}(s)$, etc. follow by solving the implicit equations (A.6.8) for the different variables

$$t=T_{\mathrm{I}}(s) \quad \text{and} \quad s=S_{\mathrm{I}}(t) \quad \text{from} \quad b_{\mathrm{I}}(s,t)=0\,;$$
$$t=T_{\mathrm{II}}(s) \quad \text{and} \quad s=S_{\mathrm{II}}(t) \quad \text{from} \quad b_{\mathrm{II}}(s,t)=0\,; \tag{A.6.12}$$
$$u=U_{\mathrm{III}}(s) \quad \text{and} \quad s=S_{\mathrm{III}}(u) \quad \text{from} \quad b_{\mathrm{III}}(s,u)=0\,.$$

The limits of the integrals are given by the points where the lines $s=$const for A_s, $t=$const for A_t and $u=$const for A_u in Fig. A.6.3 intersect the boundary curves of the regions where the double spectral functions are different from zero. If Eq. (A.6.10) is inserted into Eq. (A.6.9), one obtains the Mandelstam representation Eq. (A.6.4).

6.2.3 Further properties of absorptive parts and invariant amplitudes

i) Real analyticity

The absorptive parts A_s, A_t, A_u are real between the cuts of the single variable dispersion relations (A.6.10), since the double spectral functions are real. Furthermore, they vanish for $s<(m+\mu)^2$, $t<4\mu^2$, and $u<(m+\mu)^2$, respectively.

As a consequence, the invariant amplitudes are real between the cuts of their single variable dispersion relations (A.6.9), i.e. they are real in the triangle $t<4\mu^2$, $s<(m+\mu)^2$, $u<(m+\mu)^2$, which is located near the center of the Mandelstam diagram Fig. A.6.3, the lowest t-value being $t=-4m\mu\approx-27\mu^2$ (see also Fig. A.3.4).

If a function $f(z)$ is analytic and takes real values along some segment of the real axis, it is said to be "real analytic" and the Schwarz reflection principle can be used for the continuation from the upper to the lower half plane

$$f(z^*)=f^*(z)\,. \qquad \text{(A.6.13)}$$

We conclude that the invariant amplitudes and the absorptive parts are real analytic functions. For instance for a given real t-value the relation

$$A(s^*,t,\tilde{u})=A^*(s,t,\tilde{u}) \qquad \text{(A.6.14)}$$

is valid, and for $t>4\mu^2$

$$A_t(s^*,t,\tilde{u})=A_t^*(s,t,\tilde{u})\,. \qquad \text{(A.6.15)}$$

ii) Discontinuities

We consider as an example the fixed-t dispersion relation, Eq. (A.6.9), for a real value of t and a complex value of s. The boundary values $A(s\pm\mathrm{i}0,t,\tilde{u})$ along the cut $s_{\mathrm{th}}\leqq s<\infty$ are obtained by inserting

$$\frac{1}{s'-s\mp\mathrm{i}0}=\mathscr{P}\frac{1}{s'-s}\pm\mathrm{i}\pi\delta(s'-s) \qquad \text{(A.6.16)}$$

under the integral. "$\mathscr{P}$" indicates that one has to take the principal value of the integral, which is written: $⨍$.

The value of t is chosen in the range $4\mu^2>t>-4m\mu$, where A_s and A_u are real and the two cuts in the fixed-t dispersion relation do not overlap. Then, the first term in Eq. (A.6.16) gives the same contribution to $\mathrm{Re}\,A$, in both cases $s\pm\mathrm{i}0$. The second term shows that

$$\mathrm{Im}\,A(s\pm\mathrm{i}0,t,\tilde{u})=\pm A_s(s,t,u)\,. \qquad \text{(A.6.17)}$$

The quantity

$$\mathrm{disc}_s A(s,t,u)\equiv\frac{1}{2\mathrm{i}}[A(s+\mathrm{i}0,t,\tilde{u})-A(s-\mathrm{i}0,t,\tilde{u})]\,, \qquad \text{(A.6.18)}$$

which is called the "discontinuity of $A(s,t,u)$" along the cut, is seen to agree with the absorptive part $A_s(s,t,u)$.

Equation (A.6.10) shows that the double spectral functions are the discontinuities of the absorptive parts.

The u-channel cut of the fixed-t relation (A.6.9) is a "left hand cut" $-\infty<s<(m-\mu)^2-t=s_L$ in the s-plane. Using again Eq. (A.6.16), one finds for s-values on this cut

$$\mathrm{Im}\,A(s+\mathrm{i}0,t,\tilde{u})=\mathrm{Im}\,A(\tilde{s},t,u-\mathrm{i}0)=-A_u(s,t,u)\,, \qquad \text{(A.6.19)}$$

because $s+u=\Sigma-t$ is real. If Eq. (A.6.19) is inserted, the fixed-t dispersion relation can be written in a simplified way

$$\mathrm{Re}\,A(s,t)=\frac{1}{\pi}\left\{\int_{-\infty}^{s_L}+\int_{s_{\mathrm{th}}}^{\infty}\right\}\frac{ds'}{s'-s}\,\mathrm{Im}\,A(s,t)\,. \qquad \text{(A.6.20)}$$

Figure A.6.4 shows the discontinuities in the case that the cuts of the fixed-t dispersion relation overlap ($t<-4m\mu$).

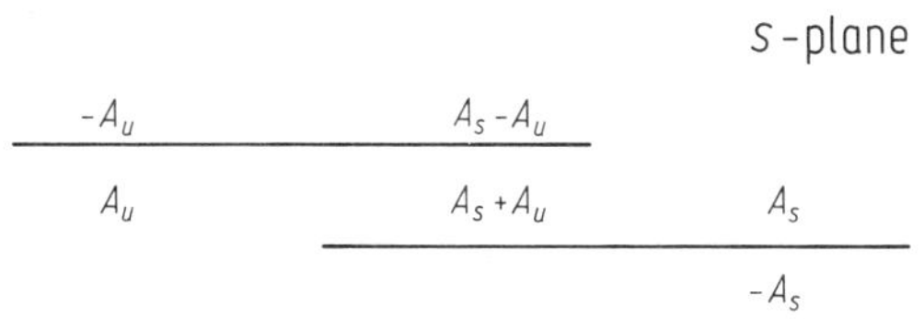

Fig. A.6.4. Boundary values $\mathrm{Im}\,A(s \pm \mathrm{i}0, t)$ in the case of overlapping s- and u-channel cuts ($t < -4m\mu$, see Fig. A.6.3). The situation is similar in the ν-plane.

iii) Crossing relation

Crossing symmetry leads to the relations

$$\begin{aligned} &A_s^{\pm}(s,t,u) = \pm A_u^{\pm}(u,t,s)\,; \quad && B_s^{\pm}(s,t,u) = \mp B_u^{\pm}(u,t,s)\,; \\ &A_t^{\pm}(s,t,u) = \pm A_t^{\pm}(u,t,s)\,; \quad && B_t^{\pm}(s,t,u) = \mp B_t^{\pm}(u,t,s)\,, \end{aligned} \tag{A.6.21}$$

which also follow from Eq. (A.6.10) if Eq. (A.6.7) is inserted. They can be used in order to simplify the fixed-t dispersion relation (A.6.9). We write $A(s,t)$ for the physical amplitude instead of $A(s+\mathrm{i}0,t)$ and $u = -s-t+\Sigma$

$$\begin{aligned} \mathrm{Re}\,A^{\pm}(s,t) &= \frac{1}{\pi} \int\!\!\!\!\!\!-_{s_{\mathrm{th}}}^{\infty} \mathrm{Im}\,A^{\pm}(s',t) \left\{ \frac{1}{s'-s} \pm \frac{1}{s'-u} \right\} \mathrm{d}s'\,; \\ \mathrm{Re}\,B^{\pm}(s,t) &= \frac{g^2}{m^2-s} \mp \frac{g^2}{m^2-u} + \frac{1}{\pi} \int\!\!\!\!\!\!-_{s_{\mathrm{th}}}^{\infty} \mathrm{Im}\,B^{\pm}(s',t) \left\{ \frac{1}{s'-s} \mp \frac{1}{s'-u} \right\} \mathrm{d}s'\,. \end{aligned} \tag{A.6.22}$$

6.2.4 Convergence of partial wave expansions

i) s-channel partial wave expansion

The singularities of $\mathrm{Re}\,A(s,t)$ and $\mathrm{Re}\,B(s,t)$ at fixed s follow from the fixed-s dispersion relation (A.6.9), and those of the absorptive parts $A_s(s,t,u)$ and $B_s(s,t,u)$ from the dispersion relation (A.6.10).

A Legendre expansion converges in the $z = \cos\theta$ plane in an ellipse which goes through the nearest singularity and has its foci at $z = \pm 1$. It can be shown that the domain of convergence of the expansion Eq. (A.3.6) is also an ellipse.

The locations of the nearest singularities of B and B_s in t at fixed s are listed in Table A.6.2, which gives also the range where the partial wave expansion converges. The parameters of the ellipses are determined by the t-channel singularities. The semimajor and semiminor axes of the ellipses a and b are related by $a^2 = b^2 + 1$, because the distance of the foci is 2.

If one performs a suitable conformal mapping, one can find other expansions, whose domain of convergence extends to a larger distance beyond the backward direction, up to the u-channel singularities.

Table A.6.2. Nearest singularities of invariant amplitudes at fixed $s > (m+\mu)^2$. The pole at $u = m^2$ does not occur in the amplitude $A(s,\ t)$ and in the $\pi^- p \to \pi^- p$ amplitude $B_-(s,\ t)$. $A_s = \mathrm{Im}\ A$ and $B_s = \mathrm{Im}\ B$ between the branch points, if the cuts do not overlap ($4\mu^2 > t > -4m\mu$). See Eqs. (A.6.11), (A.6.12), and (A.6.8) and Fig. A.6.5 for $T(s)$ and $U(s)$.

Nearest singularities of $B(s,t)$; $s > (m+\mu)^2$	
$t = 4\mu^2$	$\cos\theta = 1 + 2\mu^2/q^2$
$u = m^2$	$\cos\theta = -1 - (s - 4q^2 - m^2 - 2\mu^2)/2q^2$
$u = (m+\mu)^2$	$\cos\theta = -1 - (s - 4q^2 - \{m-\mu\}^2)/2q^2$
Legendre exp. converges if	$-1 - 2\mu^2/q^2 < \cos\theta < 1 + 2\mu^2/q^2$; $\quad -4(\mu^2+q^2) < t < 4\mu^2$
Nearest singularities of $B_s(s,t,u)$; $s > (m+\mu)^2$	
$t = T(s)$	$\cos\theta = 1 + T(s)/2q^2$
$u = U(s)$	$\cos\theta = -1 - (s - 4q^2 - U(s))/2q^2$
Legendre exp. converges if	$-1 - T(s)/2q^2 < \cos\theta < 1 + T(s)/2q^2$; $-4q^2 - T(s) < t < T(s)$

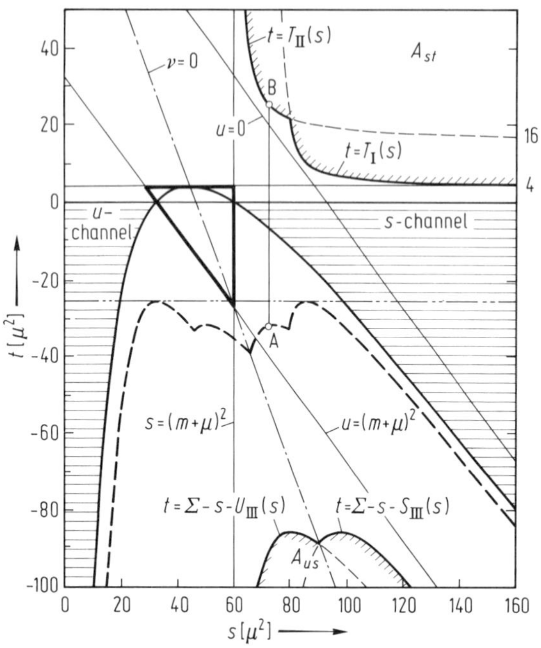

Fig. A.6.5. Convergence of the partial wave expansion (A.3.6) for ImA and ImB. The expansion converges in the s-channel range ($s>(m+\mu)^2$) between the boundary of the double spectral function A_{st} and the dashed curve (example: between the points A and B). In the fixed-t dispersion relation, a convergent expansion is needed for *all* s: the line t=const must lie between $t=4\mu^2$ and the dash-double dot line. – The figure also shows that the projection integral (A.3.8) can be evaluated in the unphysical region $0<s<(m-\mu)^2$, because the invariant amplitudes are known from crossing symmetry, Eq. (A.6.29). $A(s,t)$ and $B(s,t)$ are real in the triangle $s<(m+\mu)^2$, $t<4\mu^2$, $u<(m+\mu)^2$. See Eq. (A.6.12) for the boundaries.

ii) *t*-channel partial wave expansion

The singularities of $A(s,t)$ and $B(s,t)$ at fixed t follow from the fixed-t dispersion relation (A.6.9) and those of the absorptive parts A_t and B_t from Eq. (A.6.10). Because of crossing symmetry, it is preferable to use the variables $\nu=(s-u)/4m$. $\pm\nu_B$ and $\pm\nu_{\text{th}}$ denote the locations of the nucleon poles and of the thresholds.

The boundary of the region where the double spectral functions A_{st}, B_{st} are non-vanishing follows from Eq. (A.6.8a)

$$\nu=N(t)=\text{Min}\,\{(N_{\text{I}}(t), N_{\text{II}}(t)\}\,;$$

$$N_{\text{I}}(t)=\frac{t}{4m}+\frac{2\mu^2}{m}\left\{\frac{3}{4}+\frac{2\mu^2}{t-4\mu^2}+\left[\frac{m^2}{\mu^2}+\frac{4m^2+\mu^2}{t-4\mu^2}+\frac{4\mu^4}{(t-4\mu^2)^2}\right]^{1/2}\right\};\qquad (A.6.23)$$

$$N_{\text{II}}(t)=\frac{t}{4m}+\frac{\mu^2}{m}\left\{\frac{16\mu^2}{t-16\mu^2}+\left[\frac{m^2}{\mu^2}+\frac{16(m^2+\mu^2)}{t-16\mu^2}+\frac{(4\mu)^4}{(t-16\mu^2)^2}\right]^{1/2}\right\}.$$

Table A.6.3 summarizes the location of the nearest singularities and gives the result for the semiaxes a and b of the ellipse of convergence in the range $t>4\mu^2$.

Table A.6.3. Nearest singularities of the invariant amplitudes at fixed $t>4\mu^2$. The nucleon poles do not occur in $A(s,t)$. The location of the singularity determines the major semiaxis of the ellipse at $t>4m^2$ and the minor semiaxis at $4\mu^2<t<4m^2$. See also Figs. A.6.6 and A.6.10b.

Nearest singularities of $B(s,t)$	$t>4m^2$	$4\mu^2<t<4m^2$
$\nu=\pm\nu_B=\pm\dfrac{t-2\mu^2}{4m}$	$\cos\theta_t=\pm\dfrac{\nu_B m}{p_t q_t}=\pm a$	$\cos\theta_t=\mp\mathrm{i}\dfrac{\nu_B m}{p_- q_t}=\mp\mathrm{i}b$
$\nu=\pm\nu_{\text{th}}=\pm\left(\mu+\dfrac{t}{4m}\right)$	$\cos\theta_t=\pm\dfrac{\nu_{\text{th}} m}{p_t q_t}$	$\cos\theta_t=\mp\mathrm{i}\dfrac{\nu_{\text{th}} m}{p_- q_t}$
second semiaxis	$b=\sqrt{a^2-1}$	$a=\sqrt{b^2+1}$

Table A.6.3, continued

Table A.6.3, continued

Nearest singularity of $B_t(s, t, u)$	$t>4m^2$	$4\mu^2<t<4m^2$
$\nu=\pm N(t)$	$\cos\theta_t=\pm\frac{N(t)m}{p_t q_t}=\pm a$ $N(t)\approx\nu_{\text{th}}$	$\cos\theta_t=\mp i\frac{N(t)m}{p_- q_t}=\mp ib$

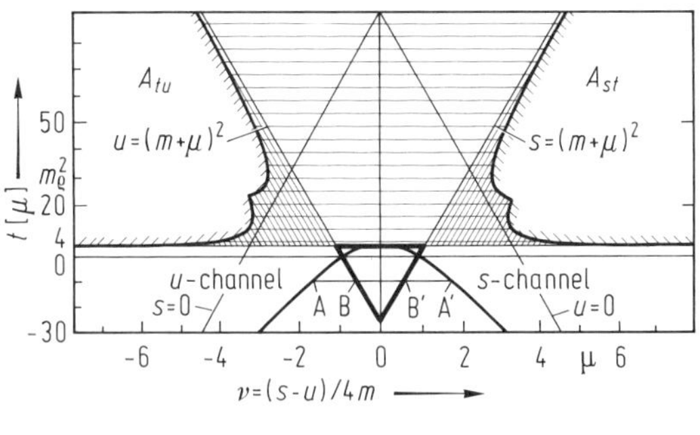

Fig. A.6.6. Convergence of the t-channel partial wave expansion. The expansion for A_t, Sect. A.3.4.2, converges in the hatched regions. $\operatorname{Im} A=A_t$ in the horizontally hatched region, where A_s and A_u vanish. In the double hatched regions, either A_t and A_s or A_t and A_u are both different from zero. As a consequence, the right and left hand cuts overlap for instance in a fixed-u dispersion relation ($u\approx 0$, Sect. A.6.6) and in certain hyperbolic dispersion relations (Sect. A.6.7). $\operatorname{Im} A$ follows in analogy to Fig. A.6.4 — A ··· A' is the integration range in the projection integral Eq. (A.3.50). The invariant amplitude is real in the central part B···B'.

In the case of B, the nearest singularity is the nucleon pole. Since its contribution is known explicitly, it can be subtracted, the expansion being carried out for $\tilde{B}\equiv B-B_N$.

At $t>50\mu^2$, the nearest singularities in $\tilde{B}$ and B_t have practically the same distance from the center of the ellipse ($\nu=0$, see Fig. A.6.6).

At $t>4m^2$ the nearest singularity determines the semimajor axis a of the ellipse and in the range $4\mu^2<t<4m^2$, it determines the semiminor axis b. If one calculates the second axis in each case ($a^2=b^2+1$), one can determine the boundary of the region of convergence for the expansion of A_t and B_t in a ν^2, t-plane (Fig. A.6.10b)

$$p_t^2 q_t^2/m^2 - N^2(t) < \nu^2 < N^2(t)\,. \tag{A.6.24}$$

The result is needed for evaluations of hyperbolic dispersion relations (Sect. A.6.7).

The s-channel partial wave dispersion relation is usually studied in the s-plane or in the W-plane. An important part of the t-channel cut lies on the circumference of the circle $|s|=m^2-\mu^2$, where $\cos\theta_t$ and $\nu=p_t q_t\cos\theta_t/m$ are complex. The expansion of A_t and B_t, in terms of $f_\pm^J$ converges on the circle in the angular range $-66°<\arg s<66°$ (see Fig. A.7.1). In the W-plane the angular range is $-33°<\arg W<33°$.

For $t>4\mu^2$, $\operatorname{Im} A=A_t$, and $\operatorname{Im} B=B_t$, if $s<(m+\mu)^2$ and $u<(m+\mu)^2$. In the remaining part of the half-plane $t>4\mu^2$, which is double-hatched in Fig. A.6.6, the cuts overlap similarly to the situation in Fig. A.6.4 [see for example Eq. (A.6.82)].

At $4\mu^2>t>0$, the semimajor axis of the ellipse of convergence for A and $\tilde{B}$ is determined by $\nu=\pm\nu_{\text{th}}$. The singularities approach the foci of the ellipse as $t\to 0$, and for $t\leqq 0$ an ellipse of convergence does not exist any more. The expansion converges only for real ν (see the remarks in Sect. A.3.4.1). The s- and u-channel cuts overlap if $t<-4m\mu$.

6.3 Fixed-*t* dispersion relations

6.3.1 General properties

In applications of the fixed-t dispersion relation (A.6.9), it is useful to work with the variable $\nu=(s-u)/4m$, because the crossing operation $s\leftrightarrow u$ at fixed t is then simply a reflection $\nu\leftrightarrow-\nu$. The analytic properties of the invariant amplitudes $A(\nu,t)$ and $B(\nu,t)$ are not only known from the Mandelstam representation. In the range $4\mu^2>t>-18\mu^2$ they can be derived from axiomatic field theory.

At fixed-t, both amplitudes are holomorphic in the complex ν-plane, except for cuts along the real axis at $|\nu|>\nu_{\text{th}}$ and poles of the amplitude $B(\nu,t)$ at $\nu=\pm\nu_B$

$$\nu_{\text{th}}=\mu+\frac{t}{4m}; \quad \nu_B=\frac{1}{4m}(t-2\mu^2)\,. \tag{A.6.25}$$

The amplitudes are real analytic

$$A(\nu^*, t) = A^*(\nu, t); \quad B(\nu^*, t) = B^*(\nu, t), \tag{A.6.26}$$

even if the cuts overlap ($\nu_{th} < 0$, $t < -4m\mu \approx -0.52\,\text{GeV}^2 = 27\mu^2$).

In the ν-plane, physical s-channel amplitudes are boundary values on the right hand cut, taken from above: $A(\nu + i0, t)$, whereas physical u-channel amplitudes are boundary values on the left hand cut, taken from *below*: $A(\nu - i0, t)$, because $u + i0$ belongs to $\nu - i0$ (t is real)

$$\begin{aligned} A_-(\nu + i0, t) &= A_+(-\nu - i0, t) = A_+^*(-\nu + i0, t); \\ B_-(\nu + i0, t) &= -B_+(-\nu - i0, t) = -B_+^*(-\nu + i0, t). \end{aligned} \tag{A.6.27}$$

For the isospin even and odd amplitudes, the crossing operation leads to the same reaction (see Sect. A.6.1), i.e. we have "crossing symmetry" which implies that the amplitudes are even or odd in ν

$$\begin{aligned} A^\pm(\nu + i0, t) &= \pm A^\pm(-\nu - i0, t) = \pm A^{\pm *}(-\nu + i0, t); \\ B^\pm(\nu + i0, t) &= \mp B^\pm(-\nu - i0, t) = \mp B^{\pm *}(-\nu + i0, t). \end{aligned} \tag{A.6.28}$$

The amplitude $C(\nu, t)$ behaves under the crossing operation in the same way as $A(\nu, t)$.

If the variables s, t, u are used, crossing symmetry is expressed by

$$A^\pm(s, t, u) = \pm A^\pm(u, t, s); \quad B^\pm(s, t, u) = \mp B^\pm(u, t, s); \tag{A.6.29}$$

the physical amplitudes being boundary values taken from above.

In terms of the variable ν, the fixed-t dispersion relation (A.6.9) reads (aside from subtractions)

$$\begin{aligned} \operatorname{Re} A^\pm(\nu, t) &= \frac{1}{\pi} \, ⨍_{\nu_{th}}^{\infty} d\nu' \, \operatorname{Im} A^\pm(\nu' + i0, t) \left[\frac{1}{\nu' - \nu} \pm \frac{1}{\nu' + \nu}\right]; \\ \operatorname{Re} B^\pm(\nu, t) &= B_N^\pm(\nu, t) + \frac{1}{\pi} \, ⨍_{\nu_{th}}^{\infty} d\nu' \, \operatorname{Im} B^\pm(\nu' + i0, t) \left[\frac{1}{\nu' - \nu} \mp \frac{1}{\nu' + \nu}\right]; \\ B_N^\pm(\nu, t) &= \frac{g^2}{2m}\left(\frac{1}{\nu_B - \nu} \mp \frac{1}{\nu_B + \nu}\right); \quad \frac{g^2}{4\pi} = \left(\frac{2m}{\mu}\right)^2 f^2. \end{aligned} \tag{A.6.30}$$

ν_{th} and ν_B are defined in Eq. (A.6.25). These dispersion relations can be derived either from Eq. (A.6.22) or from a Cauchy integral in the ν-plane, using the contour shown in Fig. A.6.7.

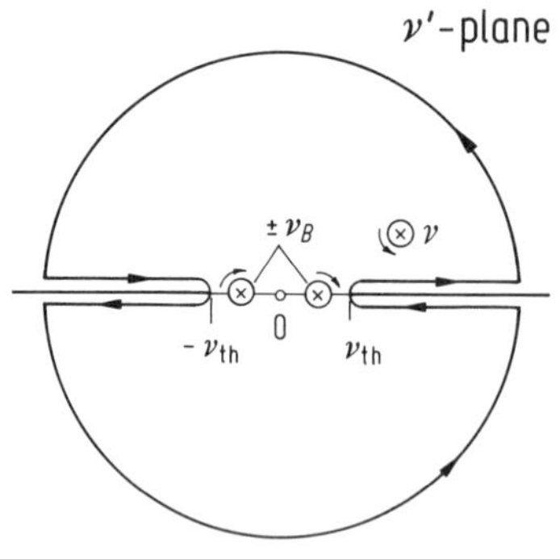

Fig. A.6.7. Contour of integration in the complex ν'-plane. If Cauchy's theorem is written for the function $B(\nu', t)/(\nu' - \nu)$ at fixed t over the small circle around ν, and the contour is blown up as shown in this figure, one obtains the dispersion relation (A.6.30), if ν is chosen real and Eq. (A.6.16) is used.

The nucleon pole term B_N agrees with the result of second order perturbation theory, if pseudoscalar coupling is used and the experimental value of the coupling constant is inserted. The coupling strength is frequently expressed by the pseudovector coupling constant f^2($f^2 = 0.079$, $g^2/4\pi = 14.3$, see Sects. A.8.1.1 and 2.4.2).

Formally, the pole term follows from the dispersion integrals Eqs. (A.6.30) and (A.6.22), if one inserts

$$\operatorname{Im} B_N^\pm(\nu + i0, t) = \frac{g^2}{2m} \pi\delta(\nu - \nu_B) = g^2 \pi \delta(s - m^2), \tag{A.6.31}$$

and replaces the lower limit by $\nu = 0$.

Sometimes it is preferable to rewrite the dispersion relations in terms of s or ω, using

$$\frac{d\nu'}{\nu'-\nu}=\frac{ds'}{s'-s}=\frac{d\omega'}{\omega'-\omega}; \quad \frac{d\nu'}{\nu'+\nu}=\frac{ds'}{s'-u}=\frac{d\omega'}{\omega'+\omega+t/2m}, \tag{A.6.32}$$

$$B_N^{\pm}(s,t)=g^2\left[\frac{1}{m^2-s}\mp\frac{1}{m^2-u}\right]=-\frac{g^2}{2m}\left[\frac{1}{\omega+\dfrac{\mu^2}{2m}}\mp\frac{1}{\omega-\dfrac{\mu^2-t}{2m}}\right]. \tag{A.6.33}$$

The lower limits $s_{th}=(m+\mu)^2$ and $\omega_{th}=\mu$, which correspond to ν_{th}, are independent of t.

Dispersion relations for the amplitudes $C^{\pm}(\nu,t)$ can be obtained by combining those for $A^{\pm}$ and $B^{\pm}$

$$\begin{aligned} \operatorname{Re} C^+(\nu,t)&=C_N^+(\nu,t)+\frac{2\nu^2}{\pi}\fint_{\nu_{th}}^{\infty}\frac{d\nu'}{\nu'}\frac{\operatorname{Im} C^+(\nu',t)}{\nu'^2-\nu^2}+C^+(0,t);\\ C_N^+(\nu,t)&=-\frac{g^2}{m}\frac{\nu^2}{(\nu^2-\nu_B^2)(1-t/4m^2)}; \end{aligned} \tag{A.6.34}$$

$$\begin{aligned} \operatorname{Re} C^-(\nu,t)&=C_N^-(\nu,t)+\frac{2\nu}{\pi}\fint_{\nu_{th}}^{\infty}d\nu'\frac{\operatorname{Im} C^-(\nu',t)}{\nu'^2-\nu^2};\\ C_N^-(\nu,t)/\nu&=-\frac{g^2}{m}\frac{\nu_B}{(\nu^2-\nu_B^2)(1-t/4m^2)}=\frac{g^2}{1-t/4m^2}\left(\frac{1}{m^2-s}+\frac{1}{m^2-u}\right). \end{aligned} \tag{A.6.35}$$

The "subtraction function" $C^+(0,t)$ is given by

$$C^+(0,t)=A^+(0,t)=\frac{2}{\pi}\fint_{\nu_{th}}^{\infty}\frac{d\nu'}{\nu'}\operatorname{Im} A^+(\nu',t), \tag{A.6.36}$$

if the integral exists. Otherwise it can be determined by calculating $\operatorname{Re} C^+$ and also the integral in Eq. (A.6.34) from phase shifts (Sect. 2.4.5).

Another possibility is to derive the dispersion relation for $C^{\pm}(\nu,t)$ directly from the analytic properties, assuming a subtraction in the case of C^+. The result is the same, except for the nucleon term, which differs from C_N^+ insofar as its numerator is taken at $\nu=\nu_B$

$$\hat{C}_N^+(\nu,t)=-\frac{g^2}{m}\frac{\nu_B^2}{(\nu^2-\nu_B^2)(1-t/4m^2)}=C_N^+(\nu,t)+\frac{g^2}{m(1-t/4m^2)}=C_N^-\nu_B/\nu\,. \tag{A.6.37}$$

In contradistinction to $C_N^+(\nu,t)$, the expression for $\hat{C}_N^+(\nu,t)$ has the property of a "pole term" to have a constant numerator. At $t=0$ it agrees with the pseudovector Born term Eq. (A.8.1). Both nucleon terms can be used equally well in applications, but $C^+-C_N^+$ is large and $C^+-\hat{C}_N^+$ is small at $\nu=0$ and $t\approx 0$.

6.3.2 Subtractions

Theoretical information on the necessity of subtractions comes from rigorous bounds for the high energy behaviour of the elastic amplitudes. A summary is given in Table A.5.1.

At $t=0$, at most two subtractions are required for the amplitudes A^+ and C^+ and no subtraction for B^+. The total cross section data show a slight increase at high energies (Sect. 1.3), i.e. the subtractions are needed for the amplitude $C^+(\nu,0)$.

In the case of the amplitude $A^+(\nu,t)$, the experimental information on the high energy behaviour comes from spin-rotation and polarization data. Because of the fairly large uncertainties of these data, the necessity of a subtraction has not yet been demonstrated beyond doubt (see Sect. 2.6.3.8). The subtracted dispersion relation for $A^+(\nu,t)$ reads

$$\operatorname{Re} A^+(\nu,t)=A^+(0,t)+\frac{2\nu^2}{\pi}\fint_{\nu_{th}}^{\infty}\frac{d\nu'}{\nu'}\frac{\operatorname{Im} A^+(\nu',t)}{\nu'^2-\nu^2}. \tag{A.6.38}$$

Because of crossing symmetry, the real parts $\operatorname{Re} A^+$, $\operatorname{Re} B^+/\nu$, $\operatorname{Re} C^+$, $\operatorname{Re} A^-/\nu$, $\operatorname{Re} B^-$, $\operatorname{Re} C^-/\nu$ and the corresponding dispersion integrals (A.6.30) are functions of ν^2. Therefore, the above mentioned two subtractions are in fact only *one* subtraction and introduce only *one* "subtraction constant" (which is a function of t).

The bounds are also valid for A^-, B^-, C^-, which are differences of elastic amplitudes (Table A.2.1). However, the data suggest that the possible subtractions are not needed, because all three amplitudes remain far below the bounds. They approximately behave as follows (Sect. 2.6.5)

$$|A^-(\nu,0)| \propto \sqrt{\nu}\,; \quad |C^-(\nu,0)| \propto \sqrt{\nu}\,; \quad |B^-(\nu,0)| \propto 1/\sqrt{\nu}\,. \tag{A.6.39}$$

In principle, the dispersion relations could have additional real contributions which are polynomials in ν. However, these are excluded by the bounds of Table A.5.1 except for a real term $c^-(t)\nu$ in the relation for $C^-(\nu,t)$. The charge-exchange data show no evidence for this term (Sect. 2.6.5). It is forbidden in Regge theory.

At $t \neq 0$ the bounds are written for the amplitude $D^+(\nu,t) = A^+ + \nu B^+$, which has the same positivity property as the elastic amplitude in the spinless case. One subtraction in ν^2 is sufficient for all t below $4\mu^2$.

6.3.3 Evaluation of the fixed-*t* dispersion relation

In order to evaluate the fixed-t dispersion relation, it is necessary to calculate the imaginary parts from phase shifts. The forward dispersion relation for C is an exception, because one can use the optical theorem (A.5.5).

The partial wave expansions of the imaginary parts of the invariant amplitudes are given by Eq. (A.3.6). In $\operatorname{Im} A(\nu',t)$ one has to calculate the energy variable from $\omega' = \nu' - t/4m$.

If $t \neq 0$, the low energy part of the integration range lies outside the physical region and therefore, it is necessary to discuss the convergence of this expansion (Fig. A.6.5).

The dispersion integral is carried out at fixed t for all s above threshold. According to Table A.6.2, the convergence of the Legendre expansion for $\operatorname{Im} A$ and $\operatorname{Im} B$ is secured, if t is larger than the highest point of the curve (Fig. A.6.5)

$$t = -4q^2 - T(s)\,. \tag{A.6.40}$$

This point belongs to $t = T_1(s)$ and lies near $t = -27\mu^2 = -0.51\,\text{GeV}^2$ and $s = 85\mu^2$, corresponding to $k = 400\,\text{MeV/c}$.

If the semiaxis of Lehmann's large ellipse is determined using rigorous methods, one obtains the smaller range: $4\mu^2 > t > -18\mu^2$. However, there is no serious argument against the application of the expansion up to $t = -27\mu^2$.

According to the Mandelstam representation, one can apply the fixed-t dispersion relation even at lower t-values, if one uses another expansion which has a larger region of convergence. A line $t = \text{const}$ in the Mandelstam diagram reaches the boundary of the double spectral function A_{us} at $t = -85\mu^2 = -1.65\,\text{GeV}^2$.

6.3.4 The expansion method

In the case of forward amplitudes $C(\nu,0)$, the imaginary parts are known much more accurately than the real parts and therefore, it is reasonable to calculate the principal value integrals directly.

However, in all other cases the situation is different: one wants to determine amplitudes with the correct analyticity and crossing properties, which agree as far as possible with amplitudes reconstructed from phase shifts. For this purpose, the best method is to use PIETARINENs expansion.

The crossing odd amplitudes are divided by ν in order to be able to work in a ν^2-plane with functions which have only a right hand cut at $\nu^2 > \nu_{\text{th}}^2$. This cut is opened by a conformal mapping

$$Z(\nu^2,t) = \frac{\nu_1 - \sqrt{\nu_{\text{th}}^2 - \nu^2}}{\nu_1 + \sqrt{\nu_{\text{th}}^2 - \nu^2}}\,, \tag{A.6.41}$$

which maps the physical sheet of the ν^2-plane onto the interior of the unit circle in the Z-plane. ν_1 is a real parameter. The expansion is a power series in Z, for instance

$$C^+(\nu,t) = C_{\text{N}}^+(\nu,t) + H^+(Z,t) \sum_{\text{m}=0}^{\text{M}} c_{\text{m}}^+ Z^{\text{m}}\,, \tag{A.6.42}$$

where H^+ is adjusted to an assumption for the asymptotic behaviour of the amplitude at high energies.

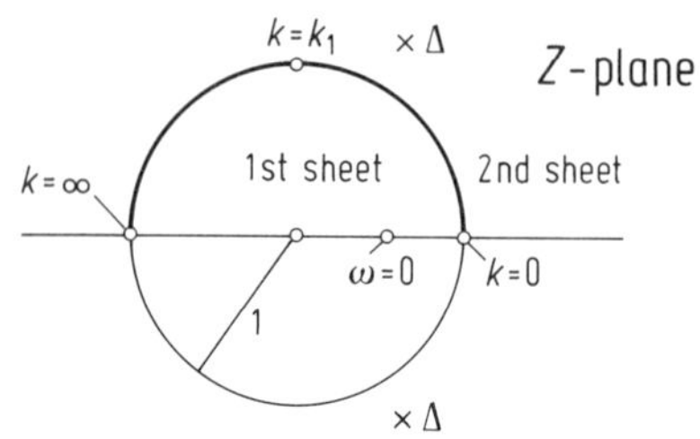

Fig. A.6.8. The conformal mapping (A.6.43). The cut at $\omega^2 > \mu^2$ of the ω^2-plane is mapped onto the circumference of the unit circle in the Z-plane. Threshold belongs to $Z = +1$ and $\omega^2 = \infty$ to $Z = -1$. The first sheet of the ω^2-plane is mapped onto the interior of the circle and the second sheet onto the exterior. The crosses show the location of the s- and u-channel $\Delta(1232)$ resonance poles. $k_1 = 0.5\,\text{GeV/c}$.

In the special case of forward scattering, we have $(\mu^2-\omega^2)^{1/2}=ik$ and

$$Z=(k_1+ik)/(k_1-ik); \quad 1+Z=2k_1/(-i\omega)+\ldots \text{for large } \omega. \tag{A.6.43}$$

The physical range $0<k<\infty$ is mapped onto the upper half of the circumference of the unit circle in the Z-plane. $Z=1$ belongs to threshold and $Z=-1$ to $k=\infty$ (Fig. A.6.8).

A reggeized ϱ-exchange term, which is usually assumed for the high energy behaviour of $C_\varrho^-(\nu,0)=C_\varrho^-(\omega)$, Eq. (A.9.16), can be written in terms of Z as follows

$$C_\varrho^-(\omega)/\omega=\gamma\frac{1-\exp(-i\pi\alpha)}{\sin\pi\alpha}\left(\frac{\omega}{\omega_0}\right)^{\alpha-1}=\frac{\gamma}{\cos(\pi\alpha/2)}\left(\frac{2k_1}{\omega_0}\right)^{\alpha-1}(1+Z)^{1-\alpha}. \tag{A.6.44}$$

The smoothing and the cut-off of the expansions of the type (A.6.42) is performed by means of a convergence test function

$$\chi_T^2=\lambda\sum_{m=0}^{M}c_m^2(m+1)^3. \tag{A.6.45}$$

The sum of the usual χ^2 and χ_T^2 is minimized in the fit. If M is chosen large enough, the result for the amplitude is practically independent of M, because the cut-off is enforced by including χ_T^2.

The expansion method is an important tool in phase shift analysis. It is hard to see, how fixed-t analyticity could be incorporated as a constraint without using a method of this type. See Sect. 2.6.7.7 for other expansions.

6.4 Forward dispersion relations

6.4.1 The basic formulas

At $t=0$ we have $\nu=\omega=\sqrt{\mu^2+k^2}$, $\nu_B=\omega_B=-\mu^2/2m$. We omit the variable t and write the forward amplitude $C(\omega)$.

The dispersion relations for the forward amplitudes can be evaluated using total cross section data, because $\operatorname{Im}C^\pm(\omega)=k\sigma^\pm$ according to the optical theorem Eq. (A.5.5).

The dispersion relation for the isospin even amplitude C^+ reads for a subtraction at $\omega=0$ and at threshold, respectively

$$\begin{aligned}\operatorname{Re}C^+(\omega)&=C^+(0)+C_N^+(\omega)+\frac{2}{\pi}\omega^2\int_\mu^\infty\frac{\sigma^+(\omega')}{\omega'^2-\omega^2}\frac{k'd\omega'}{\omega'}; \quad C_N^+(\omega)=-\frac{g^2}{m}\frac{\omega^2}{\omega^2-\omega_B^2}\\ &=C^+(\mu)+C_N^+(\omega)-C_N^+(\mu)+\frac{2k^2}{\pi}\int_0^\infty\frac{\sigma^+(k')}{k'^2-k^2}dk'; \quad C^+(\mu)=4\pi a^+(1+\mu/m).\end{aligned} \tag{A.6.46}$$

$a^+=(a_1+2a_3)/3$ is the isospin even S-wave scattering length. In the last integral one can write $\sigma^+(k')-\sigma^+(k)$ instead of $\sigma^+(k')$, because

$$\int_0^\infty\frac{dk'}{k'^2-k^2}=\begin{matrix}0\\ \pi/(2\sqrt{-k^2})\end{matrix}\quad \text{if}\quad k^2\begin{matrix}>0\\ <0.\end{matrix} \tag{A.6.47}$$

The dispersion relation for the isospin odd amplitude $C^-(\omega)$ reads

$$\operatorname{Re}C^-(\omega)=C_N^-(\omega)+\frac{2}{\pi}\omega\int_0^\infty\frac{\sigma^-(k')}{k'^2-k^2}\frac{k'^2dk'}{\omega'}; \quad C_N^-(\omega)=\frac{g^2\mu^2}{2m^2}\frac{\omega}{\omega^2-\omega_B^2}. \tag{A.6.48}$$

At $\omega=\mu$ one obtains the GOLDBERGER-MIYAZAWA-OEHME *sum rule*

$$\frac{C^-(\mu)}{\mu}=4\pi\frac{m+\mu}{3m\mu}(a_1-a_3)=\frac{8\pi f^2}{\mu^2-\omega_B^2}+4\pi J; \quad J=\frac{1}{2\pi^2}\int_0^\infty\frac{\sigma^-(k)}{\omega}dk. \tag{A.6.49}$$

Some authors preferred the subtracted dispersion relation

$$\operatorname{Re}C^-(\omega)=\frac{\omega}{\mu}C^-(\mu)-\frac{8\pi f^2k^2\omega}{(\omega^2-\omega_B^2)(\mu^2-\omega_B^2)}+k^2\omega\frac{2}{\pi}\int_0^\infty\frac{\sigma^-(k')}{k'^2-k^2}\frac{dk'}{\omega'}, \tag{A.6.50}$$

but this leads to increasing difficulties at high energies, unless the parameters fulfill the sum rule (A.6.49), and this is equivalent to using the unsubtracted relation.

6.4.2 Modified dispersion relations

Dispersion relations are written for the product of an invariant forward amplitude and a function $g(\omega)$ which is real analytic, has (at most) the same cuts as the invariant amplitude and is bounded at infinity.

GILBERTs *inverse dispersion relations* follow from the ansatz $g(\omega)=(\mu^2-\omega^2)^{-1/2}=\mathrm{i}/k$. A subtraction is required in the relation for $C^+(\omega)$.

$$\sigma^+(\omega)=\sigma^+(\mu)+\frac{8\pi f^2\omega_B}{(\mu^2-\omega_B^2)^{3/2}}\frac{k^2}{\omega^2-\omega_B^2}-\frac{2k^2}{\pi}\fint_0^\infty\frac{\mathrm{d}k'}{k'^2}\frac{\mathrm{Re}\,C^+(\omega')-\mathrm{Re}\,C^+(\mu)}{k'^2-k^2};$$

$$\sigma^-(\omega)=-\frac{8\pi f^2}{\sqrt{\mu^2-\omega_B^2}}\frac{\omega}{\omega^2-\omega_B^2}-\frac{2\omega}{\pi}\fint_0^\infty\frac{\mathrm{d}k'}{k'^2-k^2}\frac{\mathrm{Re}\,C^-(\omega')}{\omega'}. \qquad \text{(A.6.51)}$$

We mention also the inverse dispersion relations for A^+ and B^+

$$\mathrm{Im}\,A^+(\omega)=-\frac{2k}{\pi}\fint_0^\infty\frac{\mathrm{d}k'}{k'^2-k^2}\mathrm{Re}\,A^+(\omega');$$

$$\mathrm{Im}\,B^+(\omega)=-\frac{8\pi f^2}{\omega_B\sqrt{\mu^2-\omega_B^2}}\frac{\omega k}{\omega^2-\omega_B^2}-\frac{2\omega k}{\pi}\fint_0^\infty\frac{\mathrm{d}k'}{\omega'}\frac{\mathrm{Re}\,B^+(\omega')}{k'^2-k^2}. \qquad \text{(A.6.52)}$$

The more flexible ansatz

$$g(\omega)=\frac{e^{\mathrm{i}\pi\beta}}{\omega(\omega^2-\mu^2)^\beta}=\omega^{-1}(\mu^2-\omega^2)^{-\beta} \qquad \text{(A.6.53)}$$

leads to a dispersion relation, in which $\mathrm{Re}\,C^+$ as well as $\mathrm{Im}\,C^+$ are needed for the evaluation of the integral

$$\mathrm{Im}[g(\omega)C^+(\omega)]=(\cos\pi\beta\,\mathrm{Im}\,C^+ +\sin\pi\beta\,\mathrm{Re}\,C^+)/k^{2\beta}\omega\,. \qquad \text{(A.6.54)}$$

$k^{2\beta}(\omega^2-\mu^2)^\beta$ is chosen real and positive, if the cut $\mu<\omega<\infty$ is approached from above. $\mathrm{Im}(gC^+)\equiv 0$ for $|\omega|<\mu$ except for the pole contributions. β is an arbitrary real constant less than unity.

6.4.3 The phase representation

The phase of the amplitude $C^+(\omega)$ is defined by $\delta(\omega)\equiv 0$ for $\omega^2\leqq\mu^2$ and

$$C^+(\omega)/|C^+(\omega)|=e^{\mathrm{i}\delta(\omega)}\quad\text{if}\quad C^+(\mu)\geqq 0\quad\text{and}\quad =-e^{\mathrm{i}\delta(\omega)}\quad\text{if}\quad C^+(\mu)<0\,. \qquad \text{(A.6.55)}$$

Assuming the usual analytic properties for $C^+(\omega)$ and in addition a finite limit $\delta(\infty)$ and a finite number of zeros in the ω-plane, one obtains the "phase representation"

$$C^+(\omega)=\frac{P_n(\omega^2)}{\omega^2-\omega_B^2}\exp\left[\frac{2\omega^2}{\pi}\int_\mu^\infty\frac{\delta(\omega')\mathrm{d}\omega'}{\omega'(\omega'^2-\omega^2)}\right], \qquad \text{(A.6.56)}$$

where P_n is a polynomial whose order is determined by the number of zeros. The parameters are restricted by the residue at the nucleon pole

$$f^2=-\frac{P_n(\omega_B^2)}{2\omega_B}\exp\left[\frac{2\omega_B^2}{\pi}\int_\mu^\infty\frac{\delta(\omega')\mathrm{d}\omega'}{\omega'(\omega'^2-\omega_B^2)}\right]. \qquad \text{(A.6.57)}$$

Aside from logarithmic terms, the asymptotic behaviour is given by

$$C^+(\omega)\propto\omega^{2n-2-2\delta(\infty)/\pi}\exp\{\mathrm{i}\delta(\infty)\}\,. \qquad \text{(A.6.58)}$$

Inserting the Froissart bound (Table A.5.1) and the limit $|\delta(\infty)|=\pi/2$ and using the fact that the analysis leads to a negative value for the scattering length, one finds that the number of zeros must be 2. The location of the zeros is $\omega=\pm 0.92\mu i$ (see Sect. 2.4.3).

A similar representation can be written for the amplitude $C^-(\omega)$, but the number of zeros is not yet uniquely determined by the existing data (7 or 11).

6.4.4 Parametrizations for forward amplitudes at high energies

One starts with an ansatz for the momentum dependence of the total cross sections $\sigma^{\pm}(k)$ at high energies and inserts it into the dispersion relation, from which the nucleon pole term is omitted

$$C^{+}_{\text{par}}(\omega)=k^2\frac{2}{\pi}\int_0^\infty\frac{\sigma^{+}_{\text{par}}(k')}{k'^2-k^2}dk'\,;\quad C^{-}_{\text{par}}(\omega)=\omega\frac{2}{\pi}\int_0^\infty\frac{k'^2\sigma^{-}_{\text{par}}(k')/\omega'}{k'^2-k^2}dk'\,. \tag{A.6.59}$$

The result is a real analytic function of ω with the correct crossing property Eq. (A.6.28).

As an example we quote $C^{+}_{\text{par}}(\omega)$ for a power law $\sigma^{+}_{\text{par}}(k)=k^{\alpha-1}$

$$C^{+}_{\text{par}}(\omega)=-\frac{k^\alpha+(-k)^\alpha}{\sin\pi\alpha}\,;\qquad \operatorname{Im}k\neq 0\,. \tag{A.6.60}$$

In applications one needs the boundary values taken from above along the cut $\mu<\omega<\infty$. Table A.6.4 lists this boundary value for Eq. (A.6.60) and for several other parametrizations which are frequently used in fits to the total cross section data at high energies. The coefficients are omitted.

Table A.6.4. Parametrizations of forward amplitudes at high energies and different ways of writing the factor of k^α in the first two parametrizations. — The term $\pi^2/4$ in the last parametrization does not contribute to the dispersion integral [see Eq. (A.6.47)].

The ansatz for $\sigma^{-}(k)$ practically agrees with $k^{\alpha-1}$, because at high energies $\omega=k\sqrt{1-\mu^2/k^2}\approx k$ in a very good approximation.

$$\sigma^{+}_{\text{par}}(k)=k^{\alpha-1}\,;\quad C^{+}_{\text{par}}(\omega+\mathrm{i}0)=k^\alpha[-\cot(\pi\alpha/2)+\mathrm{i}]=-(-\mathrm{i}k)^\alpha/\sin(\pi\alpha/2)\,;$$

$$\sigma^{-}_{\text{par}}(k)=\omega k^{\alpha-2}\,;\quad C^{-}_{\text{par}}(\omega+\mathrm{i}0)=\frac{\omega}{k}k^\alpha[\tan(\pi\alpha/2)+\mathrm{i}]=\frac{\omega}{k}\mathrm{i}(-\mathrm{i}k)^\alpha/\cos(\pi\alpha/2)\,;$$

$$\sigma^{+}_{\text{par}}(k)=\ln k\,;\quad C^{+}_{\text{par}}(\omega+i0)=\frac{\pi}{2}k+\mathrm{i}k\ln k=-(-\mathrm{i}k)\ln(-\mathrm{i}k)\,;$$

$$\sigma^{+}_{\text{par}}(k)=\ln^2 k\,;\ C^{+}_{\text{par}}(\omega+\mathrm{i}0)=\pi k\ln k+\mathrm{i}k\ln^2 k=-(-\mathrm{i}k)\ln^2(-\mathrm{i}k)+\mathrm{i}k\pi^2/4.$$

$$C^{+}(\omega):\quad -\cot(\pi\alpha/2)+\mathrm{i}=-\frac{1+\mathrm{e}^{-\mathrm{i}\pi\alpha}}{\sin\pi\alpha}=-\frac{\mathrm{e}^{-\mathrm{i}\pi\alpha/2}}{\sin(\pi\alpha/2)}$$

$$C^{-}(\omega):\quad \tan(\pi\alpha/2)+\mathrm{i}=\frac{1-\mathrm{e}^{-\mathrm{i}\pi\alpha}}{\sin\pi\alpha}=\frac{\mathrm{i}\mathrm{e}^{-\mathrm{i}\pi\alpha/2}}{\cos(\pi\alpha/2)}$$

The formulas in Table A.6.4 are used in Sect. 2.2.1. In other applications, one can ignore the threshold effects and consider simple analytic functions of ω which are constructed from a real analytic function $f(z)=f^*(z^*)$ as follows ($\omega\approx k$ at high energies)

$$C^{+}(\omega+\mathrm{i}0)=f(-\mathrm{i}\omega)\,;\qquad C^{-}(\omega+\mathrm{i}0)=\mathrm{i}f(-\mathrm{i}\omega)\,. \tag{A.6.61}$$

These functions $C^{\pm}(\omega+\mathrm{i}0)$ have the correct behaviour under crossing, Eq. (A.6.28) but one should note that they are not real analytic in the ω-plane. For instance $f(z)=z^\alpha$ gives the correct boundary value (Table A.6.4), but Eq. (A.6.60) is the function which fulfills the dispersion relation for complex ω.

A linear combination of terms

$$f(-\mathrm{i}\omega)=(-\mathrm{i}\omega)^\alpha[\ln(-\mathrm{i}\omega)]^\beta \tag{A.6.62}$$

is general enough to describe the high energy behaviour of $C^{\pm}(\omega)$. Furthermore, it includes the amplitudes derived from Regge pole and dipole models and from special Regge cut parametrizations.

The asymptotic phase of $C^{\pm}(\omega+\mathrm{i}0)$ can also be determined from crossing symmetry and a corollary to the *Phragmén-Lindeloef theorem*, which reads:

Let $f(z)$ be an analytic function of z, regular and bounded in $\operatorname{Im}z>0$. If $f(z)$ tends to the limits L_1 and L_2 along the rays $z=x+\mathrm{i}0$ as $x\to+\infty$ and $-\infty$, then we must have $L_1=L_2$.

6.4.5 Derivative analyticity relations

The following relations were proposed as an alternative to the usual dispersion relations (A.6.46) and (A.6.48)

$$\operatorname{Im} C^{+}(\omega) = -\tan\left(\frac{\pi}{2}\frac{\mathrm{d}}{\mathrm{d}\xi}\right)\operatorname{Re} C^{+}(\omega); \quad \operatorname{Re} C^{-}(\omega) = \tan\left(\frac{\pi}{2}\frac{\mathrm{d}}{\mathrm{d}\xi}\right)\operatorname{Im} C^{-}(\omega), \tag{A.6.63}$$

where $\xi = \ln\omega$. The differential operator is approximated by a finite number of terms in the expansion

$$\tan\left(\frac{\pi}{2}\frac{\mathrm{d}}{\mathrm{d}\xi}\right) = \frac{\pi}{2}\frac{\mathrm{d}}{\mathrm{d}\xi} + \frac{1}{3}\left\{\frac{\pi}{2}\right\}^{3}\frac{\mathrm{d}^{3}}{\mathrm{d}\xi^{3}} + \ldots. \tag{A.6.64}$$

A formal relation between the differential operator and the principal value integral of the dispersion relation reads

$$\tan\left(\frac{\pi}{2}\frac{\mathrm{d}}{\mathrm{d}\eta}\right) = \frac{2}{\pi} k \, \mathrm{P}\!\!\int_{0}^{\infty} \frac{\mathrm{d}k'}{k'^{2} - k^{2}} \exp\left\{(\eta - \eta')\frac{\mathrm{d}}{\mathrm{d}\eta}\right\}, \tag{A.6.65}$$

where $\eta = \ln k$.

One obtains the correct result in the case of the simple ansatz (A.6.62), but the "derivative dispersion relation" fails if the amplitude has resonance poles and therefore it cannot be applied at intermediate or low energies (Sect. 2.2.1.3).

6.4.6 Analytic continuation into the second sheet

We consider the continuation of the invariant amplitudes through the cut $\mu < \omega < \omega_{\mathrm{in}}$ into the second sheet. $\omega_{\mathrm{in}} = 2.2\mu$ is the first inelastic threshold. Since the continuation of the partial waves will be used, we start with the no-flip amplitude $G(s, \cos\theta)$ which has the simple expansion Eq. (A.3.1)

$$G(s, \cos\theta) = \sum_{\ell=0}^{\infty} \{(\ell+1) f_{\ell+}(s) + \ell f_{\ell-}(s)\} P_{\ell}(\cos\theta); \quad f_{\ell\pm} = T_{\ell\pm}/q. \tag{A.6.66}$$

We insert the decomposition

$$f_{\ell\pm}(s) = d_{\ell\pm}(s) + \mathrm{i}q(s) a_{\ell\pm}(s) \tag{A.6.67}$$

of Eq. (A.7.26) and find

$$G(s, \cos\theta) = d(s, \cos\theta) + \mathrm{i}q(s) a(s, \cos\theta), \tag{A.6.68}$$

where $d(s, \cos\theta)$ and $a(s, \cos\theta)$ are the contributions of all $d_{\ell\pm}$ and $a_{\ell\pm}$ to the sum in Eq. (A.6.66), respectively.

For physical $\cos\theta$-values and s-values in the elastic region these functions describe $\operatorname{Re} G$ and $\operatorname{Im} G$

$$d(s, \cos\theta) = \operatorname{Re} G(s + \mathrm{i}0, \cos\theta); \quad a(s, \cos\theta) = \operatorname{Im} G(s + \mathrm{i}0, \cos\theta)/q(s + \mathrm{i}0). \tag{A.6.69}$$

In Eq. (A.6.68) the effect of the cut is completely taken into account by the factor $q(s)$. $d(s, \cos\theta)$ and $a(s, \cos\theta)$ do not have a cut at $\mu < \omega < \omega_{\mathrm{in}}$. They are regular near this part of the real s-axis except for the nucleon and resonance poles and a singularity related to the nucleon poles, which will be discussed later.

The continuation of $G(s, \cos\theta)$ into the second sheet is performed by inserting into Eq. (A.6.66) the continuation $f^{\mathrm{II}}_{\ell\pm}(s)$ of the partial waves as given in Eq. (A.7.24)

$$\begin{aligned} G^{\mathrm{II}}(s, \cos\theta) &= \sum_{\ell=0}^{\infty} \{(\ell+1) f^{\mathrm{II}}_{\ell+}(s) + \ell f^{\mathrm{II}}_{\ell-}(s)\} P_{\ell}(\cos\theta) \\ &= d(s, \cos\theta) - \mathrm{i}q(s) a(s, \cos\theta). \end{aligned} \tag{A.6.70}$$

The second equation results from $f^{\mathrm{II}}_{\ell\pm}(s) = d_{\ell\pm}(s) - \mathrm{i}q(s) a_{\ell\pm}(s)$, Eq. (A.7.26).

Now, we specialize to the case of forward scattering. $\sigma(\omega)$ denotes the total cross section

$$G(s, 1) = G(\omega) = C(\omega) m/4\pi W = d(\omega) + \mathrm{i}\frac{q(\omega)}{4\pi}\sigma(\omega). \tag{A.6.71}$$

$\sigma(\omega)$ is an analytic function, which is regular in the neighborhood of threshold. Outside the physical range, it is defined by the continuation of $a_{\ell\pm}$, Eq. (A.7.27)

$$\sigma(\omega)=4\pi\sum_{\ell=0}^{\infty}\left\{(\ell+1)\frac{f_{\ell+}^2(\omega)}{1+2iq(\omega)f_{\ell+}(\omega)}+\ell\,\frac{f_{\ell-}^2(\omega)}{1+2iq(\omega)f_{\ell-}(\omega)}\right\}. \tag{A.6.72}$$

The cusp in $\operatorname{Re}G(\omega)$ follows from the fact that $d(\omega)$ is regular at $\omega=\mu$ and $iq\sigma$ is imaginary above and real below $\omega=\mu$

$$\begin{aligned}&\operatorname{Re}G(\omega)=d(\omega);\quad \operatorname{Im}G(\omega)=\frac{m}{4\pi W}\sqrt{\omega^2-\mu^2}\,\sigma(\omega)\quad\text{for}\quad\omega>\mu;\\&\operatorname{Re}G(\omega)=d(\omega)-\frac{m}{4\pi W}\sqrt{\mu^2-\omega^2}\,\sigma(\omega);\quad \operatorname{Im}G(\omega)=0\quad\text{for}\quad\omega<\mu.\end{aligned} \tag{A.6.73}$$

A remarkable essential singularity lies near $\omega=0$ in the second sheet (Fig. A.6.9) at the same ω^2-value where the nucleon pole lies in the first sheet ($\omega^2=\omega_B^2=\mu^4/4m^2$). This singularity follows from the contribution of the u-channel nucleon pole term to the partial waves in the expansion Eq. (A.6.70) for G^{II}.

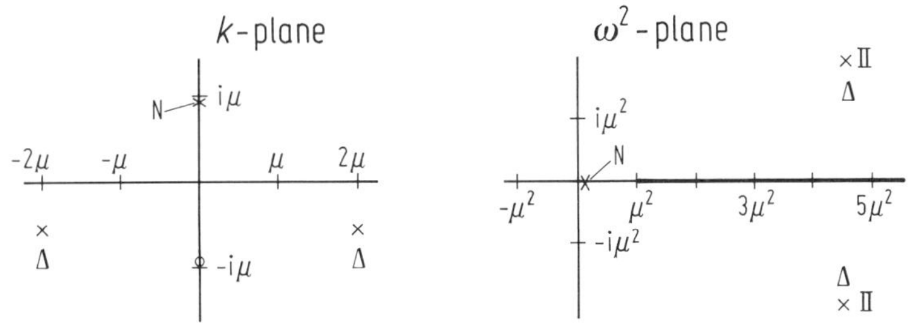

Fig. A.6.9. Location of the singularities in the low energy region at $t=0$. Nucleon pole: $\omega_B=-\mu^2/2m=-0.074\mu$; $\Delta(1232)$ pole: $\omega_\Delta=(2.15\text{–}0.46i)\mu$ corresponding to $W_\Delta=(1210\text{–}50i)$ MeV. The isobar poles lie in the second sheet of the ω^2-plane. The small circle near $k=-i\mu$ shows the location of a "quasi-pole" (65L Barut), which follows from the nucleon pole term projection in $f_{\ell\pm}$, if Eq. (A.7.24) is substituted in Eq. (A.6.70). In the ω^2-plane it lies in sheet II at the location of the nucleon pole in the physical sheet.

For some applications, it is of interest to consider expansions which converge in the physical low energy region as well as in part of the second sheet. The simplest case is an expansion of $d_{\ell\pm}(s)$ and $a_{\ell\pm}(s)$ in powers of q^2, which is equivalent to the expansion Eq. (A.3.65). The first coefficients follow from the scattering lengths and the effective ranges. One obtains G^{II} by inserting the expansions into Eq. (A.6.70).

For a direct expansion of the invariant amplitudes it is useful to choose $\tilde{C}^+(\omega)$ and $\tilde{C}^-(\omega)/\omega$, because these functions have no other singularity in the first sheet than the cut $\mu^2<\omega^2<\infty$, since the nucleon pole terms are omitted. In terms of the variable $k=\sqrt{\omega^2-\mu^2}$ the amplitudes are regular in the neighborhood of threshold, the nearest singularities being a cut starting at the inelastic threshold, the $\Delta(1232)$ resonance pole and the above mentioned singularity at $\omega^2=\omega_B^2$ in the second sheet, which lies at $k=-i\sqrt{\mu^2-\mu^4/4m^2}$. A possible ansatz is

$$C^+(k)=\sum_{n=0}^{\infty}a_n(-ik-\mu)^n. \tag{A.6.74}$$

Another possibility is to start with the conformal mapping of Eq. (A.6.41), Fig. A.6.8, but one has to choose another expansion than the power series Eq. (A.6.42), which does not converge at $|Z|>1$.

Up to now we have considered continuations in the threshold region. Another important problem is to determine the location of resonance poles in the second sheet (Sect. 2.4.1.2).

6.5 The dispersion relation for $\partial C^+/\partial t$ at $t=0$

The dispersion relation for the derivative of C^+ at $t=0$ is of interest, because experimental information on the imaginary part is available up to 200 GeV/c from the analyses of data in the Coulomb interference region.

Starting from the fixed-t dispersion relation (A.6.34) for $C^+(\nu,t)$, we insert $\nu=\omega+t/4m$, carry out the integration over $\omega'=\nu'-t/4m$, take the derivative at $t=0$ and *fixed* ω and obtain

$$\left.\frac{\partial \operatorname{Re} C^+}{\partial t}\right|_{t=0}=\left.\frac{\partial C_N^+}{\partial t}\right|_{t=0}+\left.\frac{\partial A^+(0,t)}{\partial t}\right|_{t=0}+\omega^2\frac{2}{\pi}\,⨍_{\mu}^{\infty}\frac{d\omega'}{\omega'(\omega'^2-\omega^2)}\frac{\partial}{\partial t}\operatorname{Im} C^+\left.\left(\omega'+\frac{t}{4m},t\right)\right|_{t=0}$$
$$+\frac{\omega}{2m\pi}\frac{1}{\pi}\int_{\mu}^{\infty}\frac{d\omega'}{\omega'^2}\frac{2\omega'+\omega}{(\omega+\omega')^2}\operatorname{Im} C^+(\omega',0)\,. \tag{A.6.75}$$

The slope of the subtraction function $A^+(0,t)$ is a real parameter which has to be determined from this equation. The derivative of the nucleon term is given by

$$\left.\frac{\partial C_N^+}{\partial t}\right|_{t=0}=-\frac{g^2}{4m^3}\frac{\omega}{\omega^2-\omega_B^2}\left\{\omega-\frac{\mu^2}{\omega+\omega_B}\right\}. \tag{A.6.76}$$

We define a slope parameter b_0 at $t=0$ by

$$\left.\frac{\partial \operatorname{Im} C^+}{\partial t}\right|_{t=0}=\frac{1}{2}k\sigma^+(\omega)b_0(\omega)\,, \tag{A.6.77}$$

which agrees with the slope in the usual ansatz

$$\operatorname{Im} C^+(\nu,t)=k\sigma^+(\omega)\exp\{\tfrac{1}{2}b_0(\omega)t\}\,, \tag{A.6.78}$$

and insert Eq. (A.6.77) into the principal value integral in Eq. (A.6.75)

$$\omega^2\frac{1}{\pi}\,⨍_{\mu}^{\infty}\frac{\sigma^+(\omega')b_0(\omega')}{\omega'^2-\omega^2}\frac{k'd\omega'}{\omega'}\,. \tag{A.6.79}$$

Upper bounds for $\sigma^+(\omega)$ and $b_0(\omega)$ are listed in Sect. A.5.3.

The dispersion relation (A.6.75) is of interest in Sects. 2.4.5, 2.5.1, and 2.6.2.

6.6 Fixed-u and fixed-s dispersion relations

The fixed-u dispersion relations read

$$\operatorname{Re} A(s,t,u)=\frac{1}{\pi}\,⨍_{4\mu^2}^{\infty}\frac{A_t(\tilde{s},t',u)}{t'-t}dt'+\frac{1}{\pi}\,⨍_{s_{th}}^{\infty}\frac{A_s(s',\tilde{t},u)}{s'-s}ds'\,;$$
$$\operatorname{Re} B(s,t,u)=\frac{g^2}{m^2-s}+\frac{1}{\pi}\,⨍_{4\mu^2}^{\infty}\frac{B_t(\tilde{s},t',u)}{t'-t}dt'+\frac{1}{\pi}\,⨍_{s_{th}}^{\infty}\frac{B_s(s',\tilde{t},u)}{s'-s}ds'\,. \tag{A.6.80}$$

A_s and B_s are determined from the partial wave expansion Eq. (A.3.6), whose region of convergence is discussed in Sect. A.6.2.4. Figure A.6.5 shows that the convergence is secured for all $s>s_{th}$ along a line $u=$const in a small interval around $u=0$.

For $u\lesssim-2.7\mu^2$ the lines $u=$const intersect the double spectral function region of A_{st} and B_{st}. The limitation at positive u is not directly determined by a singularity. It follows from the length of the semimajor axis of the ellipse of convergence. According to Table A.6.2, the expansion converges up to $t=-4q^2-T(s)$. This corresponds to an u-value which has its minimum at high energies: $u=4\mu^2$, i.e. the dispersion integral over the right hand cut can be evaluated in the range

$$-2.7\mu^2<u<+4\mu^2\,. \tag{A.6.81}$$

The t-channel partial wave expansion of Sect. A.3.4 can be used for the determination of A_t, B_t. According to Table A.6.3, it converges in the range of Eq. (A.6.81).

One should note that the imaginary parts are related to the absorptive parts as follows:

$$\begin{aligned} &\operatorname{Im} A(s+i0, t, u) = A_s(s, t, u), \quad \text{if} \quad t<4\mu^2 \quad \text{and} \quad s>s_{th} \\ &\operatorname{Im} A(s, t-i0, u) = -A_t(s, t, u), \quad \text{if} \quad s<s_{th}=(m+\mu)^2 \quad \text{and} \quad t>4\mu^2 \\ &\operatorname{Im} A(s+i0, t+i0, u) = A_s(s, t+i0, u) + A_t(s-i0, t, u), \quad \text{if} \quad s>s_{th} \quad \text{and} \quad t>4\mu^2 . \end{aligned} \tag{A.6.82}$$

A subtraction is not required, since the decrease of the backward cross sections at very high energies shows that the elastic backward amplitudes decrease as $|C_\pm| \propto s^{-0.67}$ and $s^{-0.54}$, respectively [Eq. (6.7.6)].

If a single u-channel Regge pole exchange dominates in the asymptotic region $s\to\infty$, the asymptotic behaviour for $t\to\infty$ is determined by line-reversal (Sect. A.9.2.4). For instance the $\pi^+ p \to \pi^+ p$ backward scattering amplitude is equal to the amplitude for $\bar{p}p \to \pi^- \pi^+$ at the same value of u, i.e. for backward angles of the π^+ with respect to the incoming $\bar{p}$. A more general relation between the asymptotic behaviour at $s\to\infty$ and $t\to\infty$ follows, if we substitute $-s-u+\Sigma$ for t and apply the Phragmén-Lindeloef theorem in the s-plane (Sect. A.6.4.4). See also Sect. A.7.1.4.

A constraint for the contribution of the unknown distant part of the left hand cut can be obtained from a "crossing sume rule" which demands that the crossing odd amplitudes vanish at $\nu=0$.

Results for fixed-s dispersion relations can be derived by applying the crossing operation.

Furthermore, the fixed-s dispersion relations can be used for the determination of the parameters of an expansion at threshold, $s=(m+\mu)^2$. In particular, it is of interest to evaluate the contribution of the nearby part of the t-channel cut (Sect. 2.4.7.2).

6.7 Hyperbolic dispersion relations

6.7.1 General properties

Starting from the Mandelstam representation, one can derive one-dimensional dispersion relations not only for straight lines (s or t or $u=$const) in the Mandelstam diagram, but also for suitable curves which are chosen in such a way that the absorptive parts can be evaluated in a large part of the integration range. An interesting choice are the $s\leftrightarrow u$ crossing symmetric hyperbolas ($\Sigma=2m^2+2\mu^2$)

$$(s-a)(u-a)=b \quad \text{or} \quad (4m\nu)^2=(t-\Sigma+2a)^2-4b, \tag{A.6.83}$$

which do not enter the double spectral function regions (Fig. A.6.10). Two special cases were studied in detail

$$a=0; \quad \nu^2=\frac{p_-^2 q_-^2}{m^2}+\frac{(m^2-\mu^2)^2-b}{4m^2}, \tag{A.6.84a}$$

$$b=\lambda(m^2, a, \mu^2); \quad \nu^2=\frac{t^2}{16m^2}-\frac{t}{4m^2}(m^2+\mu^2-a)+\mu^2 . \tag{A.6.84b}$$

A third case will be mentioned in Sect. A.7.2.2. See Eq. (A.1.12) for λ.

In the following, we shall treat only the second case, which has the advantage that the s-channel part of the integration path remains in the physical region ("interior dispersion relation"). Figure A.6.11 shows that, at low energies, the paths of Eq. (A.6.84a) lie in general at a considerable distance outside the physical region (Sect. 2.4.2.2).

6.7.2 Interior dispersion relations

A number of kinematic relations, which can be derived from Eqs. (A.6.83) and (A.6.84b), can be found in Sect. A.1.3.4. From these relations and from the Tables A.6.2 and A.6.3 one obtains the following results:

i) all curves pass the s-channel (or u-channel) thresholds (Fig. A.6.10),

ii) along each of the curves the pion scattering angle in the laboratory system θ_π has a constant value which is given by Eq. (A.1.30),

iii) The Legendre expansion of the absorptive part in the s-channel integral converges for all s and all a belonging to $0°<\theta_\pi<180°$,

iv) the Legendre expansion of the absorptive part in the t-channel integral converges for all t, if $0>a>-0.7\,\text{GeV}^2$, corresponding to $180°>\theta_\pi>95°$.

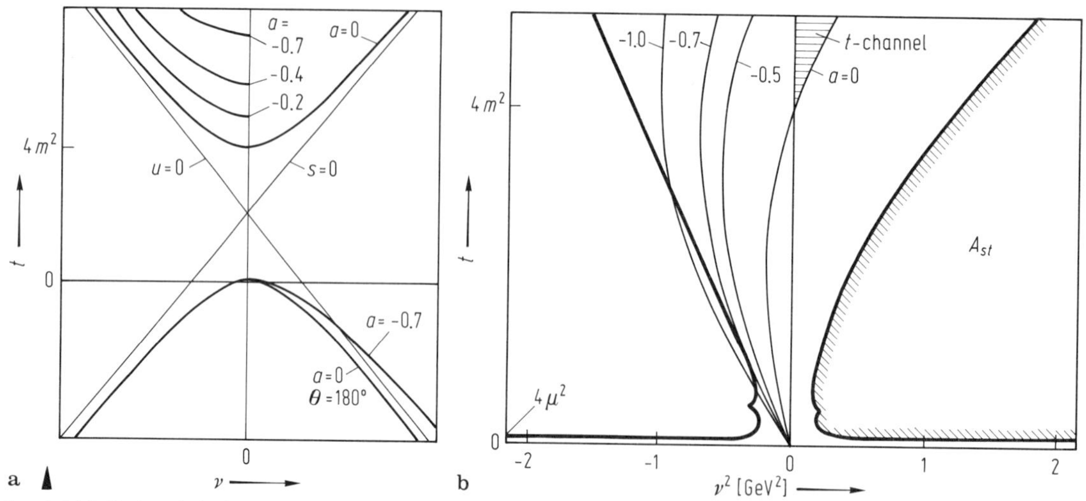

Fig. A.6.10. Paths of the interior dispersion relation in the Mandelstam plane. In the s-channel range, the curves $a=$const belong to a constant value of the scattering angle θ_π in the laboratory system, i.e. the centre-of-mass scattering angle θ goes to 180° in the high energy limit. — The t-channel partial wave expansions of $\mathrm{Im}\,A$ and $\mathrm{Im}\,B$ converge, if a lies in the range $0>a>-0.7\,\mathrm{GeV}^2$, corresponding to $180°>\theta_\pi>95°$. In Fig. A.6.10a the paths have two branches, because ν becomes purely imaginary in the intermediate t-range. Figure A.6.10b shows these paths in a ν^2-t-plane. In the range $4m^2>t>4\mu^2$, the boundary of the double spectral function A_{st} determines the semiminor axis of the ellipse of convergence. The convergence at $\nu^2<0$ is limited by the thick line, which follows from the semimajor axis. See Sect. A.6.2.4ii).

Fig. A.6.11. Paths of hyperbolic dispersion relations in the low energy region. ▶

The figure shows some hyperbolas of the family $a=0$, Eq. (A.6.84a), which belong to different b-values. It is seen that in most cases the partial wave expansion for $\mathrm{Im}\,A$, $\mathrm{Im}\,B$ is needed far outside the physical region. The lines $u=0$ and $\theta=180°$ are special cases. — All paths of the "interior dispersion relation", Eq. (A.6.48b), go through the s-channel threshold and remain within the physical region for larger values of s. In the high energy limit they approach $u=a$, $\theta=180°$.

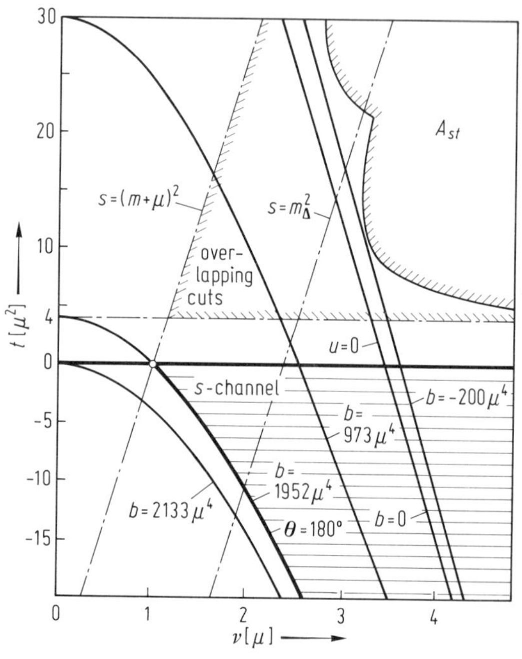

The interior dispersion relation is written for crossing symmetric amplitudes and reads (tilde: this variable is calculated from the others and Eqs. (A.6.83), (A.6.84b)

$$\mathrm{Re}\,M(s,t,a)=M_{\mathrm{N}}(t,a)+\frac{1}{\pi}\int_{4\mu^2}^{\infty}\frac{\mathrm{Im}\,M(\tilde{s},t',a)}{t'-t}\,\mathrm{d}t'+\frac{1}{\pi}\int_{(m+\mu)^2}^{\infty}\mathrm{Im}\,M(s',\tilde{t},a)\left[\frac{1}{s'-s}+\frac{1}{s'-u}-\frac{1}{s'-a}\right]\mathrm{d}s'. \qquad (\text{A.6.85})$$

For the amplitude M the following expressions can be used

$$A^+,B^-,\hat{B}^+=\frac{t-t_{\mathrm{N}}}{s-u}B^+,\quad \hat{C}^+=(4m^2-t)C^+,\quad \hat{C}^-=\frac{4m^2-t}{s-u}C^-. \qquad (\text{A.6.86})$$

The factors $(4m^2-t)$ are necessary in order to remove the kinematic singularity of the amplitudes $\hat{C}^\pm$ (see Sect. A.3.5.4). t_{N} is the location of the nucleon pole. The factor $(t-t_{\mathrm{N}})$, which removes the large contribution of the nucleon pole term, is applied only for practical reasons.

One subtraction is necessary in the dispersion relation for $\hat{C}^+$.

$\mathrm{Im}\, M(s', \tilde{t}, a)$ is calculated from Eq. (A.3.6), inserting s' and $\tilde{t} = 4s'q'^2/(a-s')$. $\mathrm{Im}\, M(\tilde{s}, t', a)$ is obtained from Table A.3.3, inserting v from Eq. (A.1.30).

The nucleon pole terms read (subtraction in $\hat{C}^+$ at $t=0$)

$$B^-_{\mathrm{N,IDR}} = g^2\left(\frac{1}{m^2-s} + \frac{1}{m^2-u} - \frac{1}{m^2-a}\right); \quad \hat{C}^+_{\mathrm{N,IDR}} = g^2 \frac{mt(t_{\mathrm{N}} - 2\mu^2)^2}{t_{\mathrm{N}}(m^2-s)(m^2-u)};$$
$$t_{\mathrm{N}} = \mu^2(4m^2 - \mu^2)/(m^2 - a). \tag{A.6.87}$$

6.8 Fixed center-of-mass angle dispersion relations

All single-variable dispersion relations considered up to now are written for paths in the Mandelstam plane, which behave asymptotically for large s as $t=$const or $u=$const. For the analysis of data at intermediate and high energies, it is of interest to have also a dispersion relation, whose path remains in the "wide-angle region", i.e. for instance near $\theta = 90°$.

If the c.m. scattering angle θ is kept constant, the path in the Mandelstam plane is given by

$$s_\pm(q^2) = m^2 + \mu^2 + 2q^2 \pm 2K(q^2); \quad K(q^2) = \{(m^2+q^2)(\mu^2+q^2)\}^{1/2} \tag{A.6.88}$$

$$t(q^2) = -2q^2(1-z); \tag{A.6.89}$$

$$t(s) = -\frac{1-z}{2s}\{s-(m+\mu)^2\}\{s-(m-\mu)^2\}. \tag{A.6.90}$$

The path has two branches (Fig. A.6.12)

$$\begin{aligned} &\text{maximum of the lower branch: } t = 2\mu^2(1-z); \quad s = m^2 - \mu^2; \\ &\text{minimum of the upper branch: } t = 2m^2(1-z); \quad s = \mu^2 - m^2. \end{aligned} \tag{A.6.91}$$

The dispersion relations are written in the q^2-plane. The mapping of the s-plane onto the two sheets of the q^2-plane, Eq. (A.6.88), is explained in Fig. A.1.4, Sect. A.1.2.1. The cut of $K(q^2)$ is chosen at $-m^2 < q^2 < -\mu^2$ and we demand that $K(q^2) > 0$ for real $q^2 > -\mu^2$.

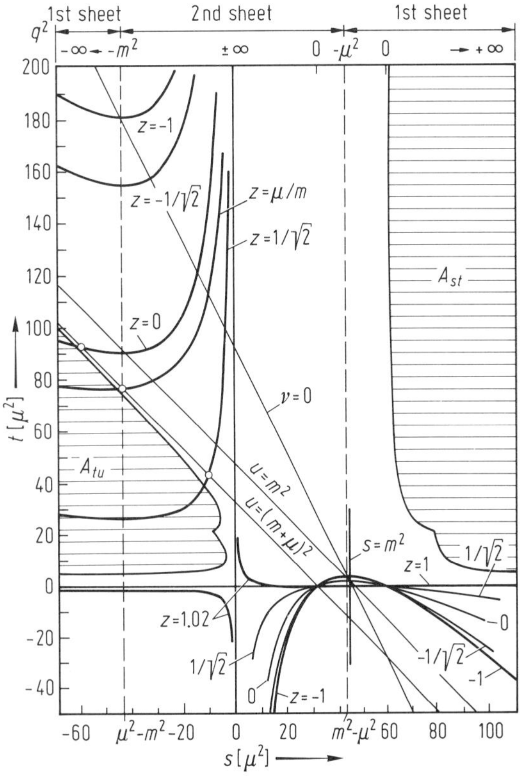

Fig. A.6.12. Paths of fixed center-of-mass angle dispersion relations in the Mandelstam plane. The lower branch has a maximum at $s = m^2 - \mu^2$ and the upper branch has a minimum at $s = \mu^2 - m^2$. In the t-range between the maximum and the minimum, s and v are complex-valued along the path. — The nucleon poles lie at the intersections of lines $z=$const and $s=m^2$ or $u=m^2$. — The u-channel cuts start at the intersections of lines $z=$const and $u=(m+\mu)^2$. The transition to forward scattering, $z=+1$, is singular at $s=0$. The curve for $z=0.98$ is the reflection at the abscissa of the curve for $z=1.02$. — On top of the figure, we have listed some q^2-values which belong to the s-values of the abscissa.

The first sheet of the q^2-plane is called the physical sheet, because it contains the s-channel physical region. The crossing even amplitudes at fixed z are denoted by $L_\pm(q^2)$, where L can be $A^+, C^+, B^-, A^-/\nu, C^-/\nu, B^+/\nu$. In $L_\pm$, the subscript + refers to the first or physical sheet and – to the second sheet.

Both $L_\pm(q^2)$ have a kinematic cut at $-m^2<q^2<-\mu^2$. It is related to the fact that for instance the values of $L_+(q^2\pm i\varepsilon)$ belong to two points in the s-plane which lie far apart: just outside the upper and lower half of the circle $|s|=m^2-\mu^2$ (see Fig. A.1.4).

The kinematic cut is cancelled in the combinations

$$L_1(q^2)\equiv\frac{1}{2}\{L_+(q^2)+L_-(q^2)\};\quad L_2(q^2)\equiv\frac{1}{2K(q^2)}\{L_+(q^2)-L_-(q^2)\}. \tag{A.6.92}$$

The dispersion relations are obtained from Cauchy integrals in the q^2-plane for $L_1(q^2)$ and $L_2(q^2)$. The singularities are: the nucleon poles, the physical cut $s>(m+\mu)^2$, the t-channel cut $t>4\mu^2$ and the u-channel cut $u>(m+\mu)^2$.

The s- and u-channel nucleon poles are located at

$$q^2_{s\mathrm{N}}=-\mu^2+\frac{\mu^4}{4m^2};\quad q^2_{u\mathrm{N}}=\frac{m^2+(1+z)\mu^2}{2(1-z^2)}\left\{-1\pm\left\{1-\frac{\mu^2(1-z^2)(4m^2-\mu^2)}{\{m^2+\mu^2(1+z)\}^2}\right\}^{1/2}\right\}. \tag{A.6.93}$$

We give only the expressions for the nucleon pole terms of the amplitude L_+ on the physical sheet. The subscripts sN and uN refer to the s- and u-channel pole terms.

$$B^-_{+s\mathrm{N}}=g^2\frac{\mu^2}{2m^2}\frac{1-\mu^2/2m^2}{q^2_{s\mathrm{N}}-q^2}M_+(q^2,q^2_{s\mathrm{N}});\quad M_\pm(q^2,q'^2)\equiv\frac{1}{2}\left\{1\pm\frac{K(q^2)}{K(q'^2)}\right\};$$

$$B^+_{+s\mathrm{N}}=-\frac{2m\nu}{q^2_{s\mathrm{N}}(1-z)+\mu^2}B^-_{+s\mathrm{N}}; \tag{A.6.94}$$

$$C^-_{+s\mathrm{N}}=\frac{\nu B^-_{+s\mathrm{N}}}{1+q^2_{s\mathrm{N}}(1-z)/2m^2};\quad C^+_{+s\mathrm{N}}=-\frac{q^2_{s\mathrm{N}}(1-z)+\mu^2}{q^2_{s\mathrm{N}}(1-z)+2m^2}mB^-_{+s\mathrm{N}}.$$

The u-channel nucleon pole term leads to two poles. We list only the nearby pole, which belongs to the + sign in Eq. (A.6.93). The other pole lies at $q^2<-m^2$.

$$B^-_{+u\mathrm{N}}=\frac{R(z)}{q^2_{u\mathrm{N}}-q^2}M_\pm(q^2,q^2_{u\mathrm{N}});\quad M_\pm \text{ for } -\tfrac{1}{2}\begin{matrix}<z<1\\ >z>-1\end{matrix};\quad R(z)=g^2\frac{q^2_{u\mathrm{N}}z-\mu^2/2}{2q^2_{u\mathrm{N}}(1-z)+\mu^2(1+z)+m^2};$$

$$B^+_{+u\mathrm{N}}=-\frac{2m\nu}{q^2_{u\mathrm{N}}(1-z)+\mu^2}B^-_{+u\mathrm{N}}; \tag{A.6.95}$$

$$C^-_{+u\mathrm{N}}=\frac{\nu B^-_{+s\mathrm{N}}}{1+q^2_{u\mathrm{N}}(1-z)/2m^2};\quad C^+_{+u\mathrm{N}}=-\frac{q^2_{u\mathrm{N}}(1-z)+\mu^2}{q^2_{u\mathrm{N}}(1-z)+2m^2}mB^-_{+u\mathrm{N}}.$$

Cuts of $L_\pm(q^2)$:

The *physical cut* $s>s_{\mathrm{th}}$ leads to a cut in $L_+(q^2)$ at $q^2>0$, but not to a cut in $L_-(q^2)$.

The *t-channel cut* at $t>4\mu^2$ leads to cuts in both $L_\pm(q^2)$ at $-\infty<q^2<a_t(z)=-2\mu^2/(1-z)$. In order to locate these cuts in the s-plane, one has to solve Eq. (A.6.90) for s and to insert $t>4\mu^2$. One finds cuts along half of the real axis: $-\infty<s<0$ and, if $z^2<1-2\mu^2/m^2$, in addition a cut along part of the circle $|s|=m^2-\mu^2$. For $z=-1$, the range $4\mu^2<t<4m^2$ gives a cut on the full circle. The cuts in $L_\pm(q^2)$ derived from the u-channel cut $u>u_{\mathrm{th}}$ are listed in Table A.6.5 (see also Fig. A.6.13).

Table A.6.5. Location of the u-channel cuts in the q^2-plane. $a_u=-(m^2+\mu^2-2m\mu z)/(1-z^2)$ is the q^2-value, where the lines $z=$const and $u=(m+\mu)^2$ in the Mandelstam plane, Fig. A.6.12, intersect each other.

q^2-plane			$L_\pm$	Mandelstam plane
U1	$1>z>-1$	$0<q^2<\infty$	L_-	lower branch; $0<s<(m-\mu)^2$
U2	$1>z>\mu/m$	$-\infty<q^2<-m^2$	L_+	upper branch; from $-\infty$ to the intersection of the lines $z=$const and $u=(m+\mu)^2$
	$-1<z<\mu/m$	$a_u(z)<q^2<-m^2$ $-\infty<q^2<a_u$	L_- L_+	

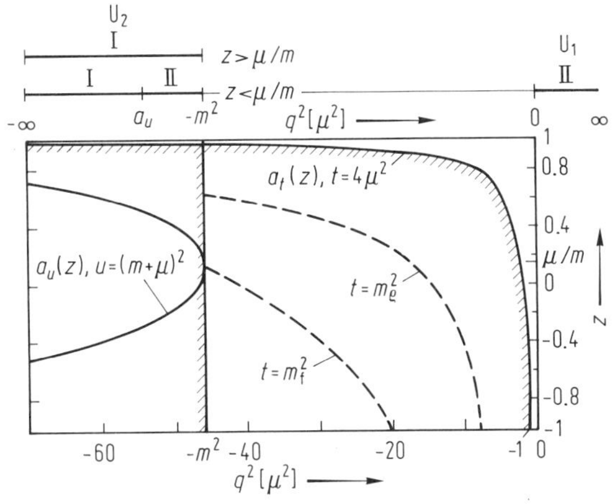

Fig. A.6.13. The z-dependence of the cuts in the q^2-plane. The nearest cut is the t-channel cut. The dashed lines show the location of the maxima of the ϱ- and f-exchange contributions to the imaginary parts of the invariant amplitudes. The u-channel cuts which belong to U_2 (Table A.6.5), start at $t=-4m^2$. — The s-channel nucleon pole and one of the u-channel nucleon poles are located very near $t=-\mu^2$ for all z-values. The second u-channel pole lies beyond $t=-4m^2$.

The dispersion relations for $L_1(q^2)$, $L_2(q^2)$ are combined to give the final result for $L_+(q^2)$

$$L_+(q^2)=L_1(q^2)+K(q^2)L_2(q^2). \qquad \text{(A.6.96)}$$

The dispersion relation for $L_+(q^2)$ reads

$$\operatorname{Re} L_+(q^2)=L_{+\mathrm{N}}+L_{\mathrm{S}}+L_{\mathrm{T}}+L_{\mathrm{U}1}+L_{\mathrm{U}2}, \qquad \text{(A.6.97)}$$

where $L_{+\mathrm{N}}$ is the sum of the s- and u-channel nucleon pole terms, Eqs. (A.6.94) and (A.6.95), and

$$\begin{aligned}
L_{\mathrm{S}}&=\frac{1}{\pi}\oint_0^{\infty}\frac{\mathrm{d}q'^2}{q'^2-q^2}M_+(q^2,q'^2)\operatorname{Im}L_+(q'^2);\\
L_{\mathrm{T}}&=\frac{1}{\pi}\oint_{-\infty}^{a_t}\frac{\mathrm{d}q'^2}{q'^2-q^2}\{M_+(q^2,q'^2)\operatorname{Im}L_+(q'^2)+M_-(q^2,q'^2)\operatorname{Im}L_-(q'^2)\};\\
L_{\mathrm{U}1}&=\frac{1}{\pi}\oint_0^{\infty}\frac{\mathrm{d}q'^2}{q'^2-q^2}M_-(q^2,q'^2)\operatorname{Im}L_-(q'^2);\\
L_{\mathrm{U}2}&=\frac{1}{\pi}\oint_{-\infty}^{a_{u+}}\frac{\mathrm{d}q'^2}{q'^2-q^2}M_+(q^2,q'^2)\operatorname{Im}L_+(q'^2)+\Theta\left(z-\frac{\mu}{m}\right)\frac{1}{\pi}\oint_{a_u}^{-m_2}\frac{\mathrm{d}q'^2}{q'^2-q^2}M_-(q^2,q'^2)\operatorname{Im}L_-(q'^2).
\end{aligned} \qquad \text{(A.6.98)}$$

Θ is the step function and $M_\pm$ is defined in Eq. (A.6.94). The limits of the integrals are

$$a_t=-2\mu^2/(1-z); \qquad a_{u+}=\begin{matrix}-m^2 & \text{for} & z>\mu/m;\\ a_u & \text{for} & z<\mu/m;\end{matrix} \qquad a_u=-\frac{m^2+\mu^2-2m\mu z}{1-z^2}. \qquad \text{(A.6.99)}$$

Figure A.6.13 shows the location of the different cuts in a $z-q^2$-plane. It is seen that only the t-channel cut starts near the physical region, whereas the u-channel cuts start far away, in particular for near-backward scattering and even at $\theta=90°$.

$\operatorname{Im}L_+(q^2)$ on the physical cut and part of the discontinuities on the other cuts can be calculated from the s- and t-channel partial wave expansions.

Figure A.6.12 shows where the line $z=\text{const}$ enters the double spectral function region. The usefulness of the dispersion relation depends on the validity of the assumption that distant cuts give slowly variable contributions.

6.9 Backward dispersion relations

Dispersion relations for invariant amplitudes at fixed angle $\theta=\theta_\pi=180°$ are an important special case of the fixed-angle dispersion relations discussed in Sects. A.6.7 and A.6.8. The amplitude is usually considered in the q^2-plane, where it has a right hand cut in the range $0<q^2<\infty$ and a left cut at $-\infty<q^2<-\mu^2$.

On the right hand cut, the modulus of the amplitude follows directly from experimental backward cross section data. Furthermore, the real and imaginary parts of the amplitude can be calculated from πN phase shifts.

On the left hand cut, the amplitude has a convergent $\pi\pi N\bar{N}$ partial wave expansion and, for $q^2<-m^2$, its modulus can be obtained from experimental $\bar{N}N\pi\pi$ backward scattering data.

As discussed in Sect. A.7.1.4, the asymptotic behaviour at $q^2\to-\infty$ is the same as that at $q^2\to+\infty$ as a consequence of the Phragmén-Lindeloef theorem. On the right hand cut, the modulus of the amplitude is known up to large q^2-values from πN backward cross sections ($k\leqq 90$ GeV/c).

The backward dispersion relations can be written as follows

$$\mathrm{Re}X(q^2)=X_N(q^2)+\frac{1}{\pi}\int_0^\infty \frac{\mathrm{Im}X(q'^2)}{q'^2-q^2}\mathrm{d}q'^2+\frac{1}{\pi}\int_{-\infty}^{-\mu^2}\frac{\mathrm{Im}X(q'^2)}{q'^2-q^2}\mathrm{d}q'^2. \qquad \text{(A.6.100)}$$

For $X(q^2)$ one can substitute one of the amplitudes listed in Table A.6.6, where one can also find the expression for the nucleon pole term.

Table A.6.6. Amplitudes and nucleon pole terms which can be substituted into Eq. (A.6.100). $q_N^2=\mu^4/4m^2-\mu^2$.

$X(q^2)=$	$\frac{C^+}{4\pi};\ \frac{m\omega_q}{E}\frac{B^+}{4\pi}$	$\frac{m}{E\omega_q}\frac{B^+}{4\pi};$	$\frac{E}{\omega_q m}\frac{C^-}{4\pi};\ \frac{B^-}{4\pi}$
$X_N(q_N^2-q^2)=$	$\frac{\mu^2}{m}f^2$	$\frac{4m}{\mu^2}f^2$	$2\left(1-\frac{\mu^2}{2m^2}\right)f^2$

Sum rules

i) The *crossing sum rule* follows if one demands that the backward dispersion relation leads to vanishing isospin odd amplitudes A^-, C^-, B^+ at $q^2=-\mu^2$, which corresponds to $\nu=0$, where these amplitudes are zero due to crossing antisymmetry.

ii) *The threshold sum rule* is based on the fact that the forward and backward amplitudes are equal at the s-channel threshold. The sum rule follows by equating the forward and backward dispersion relations. A generalization is to equate the backward dispersion relation and the fixed-t relation for an amplitude at $t=-4q^2$.

6.10 Fixed-ν dispersion relations

Similarly as in the case of the fixed-θ dispersion relations, the integration path enters the double spectral function region. Nevertheless, the dispersion relation is of interest for a determination of a smooth interpolation, assuming again that distant cuts give slowly varying contributions.

We consider only the special case of dispersion relations at $\nu=0$, which corresponds to $\cos\theta_t=0$ (Fig. A.6.3)

$$\mathrm{Re}X(0,t)=\frac{1}{\pi}\left\{\int_{-\infty}^{-4m\mu}+\int_{4\mu^2}^{\infty}\right\}\frac{\mathrm{Im}X(0,t')}{t'-t}\mathrm{d}t', \qquad \text{(A.6.101)}$$

where X is one of the amplitudes $\tilde{C}^+$, $\tilde{C}^-/\nu$, $\tilde{B}^+/\nu$, $\tilde{B}^-$, A^+, A^-/ν. The tilde indicates that the nucleon pole term of the fixed-t dispersion relations (A.6.30), (A.6.34), (A.6.35) is subtracted. At $\nu=0$ the locations of the s- and u-channel poles coincide ($t=2\mu^2$).

On the nearby part of the right hand cut, ImX can be calculated from $\pi\pi N\bar{N}$ partial waves. The s-channel partial wave expansion converges only on a small interval of the left hand cut ($-27\mu^2>t>-40\mu^2$), where both absorptive parts, A_s and A_u give a contribution (see Fig. A.6.4). Nevertheless, it is useful to work with the truncated expansion up to somewhat larger negative t-values in order to include the effect of Δ(1232)-exchange in the s- and u-channels. The isobar pole lines meet on the $\nu=0$ axis at $t=-63\mu^2\approx-1.2\,\mathrm{GeV}^2$. The double spectral function region begins at $t=-87\mu^2\approx-1.7\,\mathrm{GeV}^2$, $s=u=90\mu^2$, which corresponds to $k=430$ MeV/c (Fig. A.6.5).

7 Analytic properties of partial wave amplitudes

7.1 s-channel partial wave dispersion relations

7.1.1 The singularities of the partial wave amplitudes

The analytic properties of the partial waves $f_{\ell\pm}(s)$ are derived from a FROISSART-GRIBOV-type representation which is obtained by inserting the fixed-s dispersion relation Eq. (A.6.9) into the projection formula Eq. (A.3.8), taking into account the dispersion relation for the absorptive parts, Eq. (A.6.10), and the properties of the mapping $q^2(s)$ (Fig. A.1.4).

The singularities are listed in Table A.7.1 and their location is shown in Figs. A.7.1 and A.7.2.

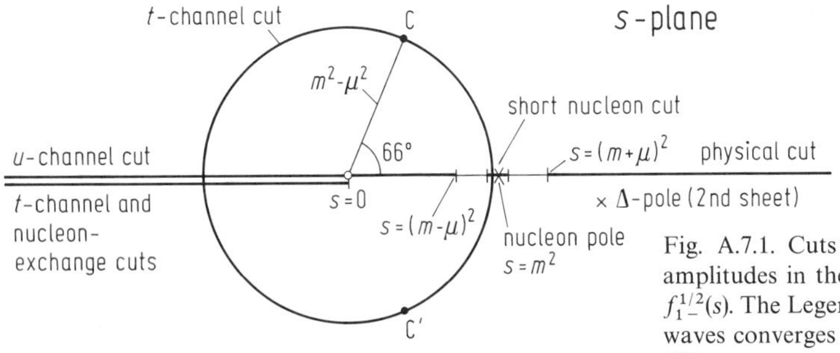

Fig. A.7.1. Cuts and poles of the s-channel partial wave amplitudes in the s-plane. The nucleon pole occurs only in $f_{1-}^{1/2}(s)$. The Legendre expansion in terms of t-channel partial waves converges only on the front of the circle in the range CC′.

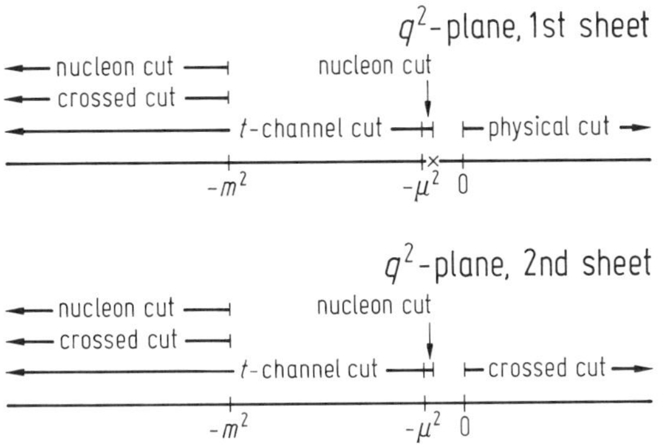

Fig. A.7.2. Cuts and poles of the s-channel partial wave amplitudes in the q^2-plane. The nucleon pole occurs only in $f_{1-}^{1/2}$. The kinematic cut of the function $s(q^2)$ is chosen at $-m^2<q^2<-\mu^2$. The figure is not drawn in scale.

The dispersion relation for the partial wave amplitude $f_{\ell\pm}(s)$ reads

$$\mathrm{Re}\, f_{\ell\pm}(s)=\frac{1}{\pi}\int_{s_{\mathrm{th}}}^{\infty}\frac{\mathrm{Im}\, f_{\ell\pm}(s')}{s'-s}\,ds'+\frac{1}{\pi}\int_{\mathrm{l.h.c.}}\frac{\mathrm{disc}_s f_{\ell\pm}(s')}{s'-s}\,ds'+\text{pole term}. \tag{A.7.1}$$

disc_s is defined in Eq. (A.6.18). The principal value integral over the physical cut is written explicitly, but we omit the complicated formulas for the discontinuities on the unphysical "left hand cuts" (abbreviated "l.h.c.").

The partial wave projection of the s-channel nucleon pole term $g^2/(m^2-s)$, Eq. (A.6.5), is different from zero only for the S11 and P11 waves

$$f_{\mathrm{N},0+}^{1/2}(s)=-\frac{g^2}{4\pi}\frac{W-m}{s-m^2}\frac{3}{4}\frac{(W+m)^2-\mu^2}{s}; \tag{A.7.2a}$$

$$f_{\mathrm{N},1-}^{1/2}(s)=-\frac{g^2}{4\pi}\frac{W+m}{s-m^2}\frac{3}{4}\frac{(W-m)^2-\mu^2}{s}. \tag{A.7.2b}$$

Both amplitudes have a kinematic pole at $s=0$, but only Eq. (A.7.2b) has a pole at $s=m^2$. The nucleon pole term in Eq. (A.7.1) follows, if $(s-m^2)f_{\mathrm{N},1-}^{1/2}$ is evaluated at $s=m^2$

$$f_{\mathrm{p},1-}^{1/2}(s)=\frac{g^2}{4\pi}\frac{3\mu^2}{2m}\frac{1}{s-m^2};\quad F_{\mathrm{p},1-}^{1/2}(s)=-\frac{g^2}{4\pi}\frac{6m}{4m^2-\mu^2}\frac{1}{s-m^2}. \tag{A.7.2c}$$

The second version belongs to the *reduced partial wave amplitude*

$$F_{\ell\pm}(s) \equiv f_{\ell\pm}(s)/q^{2\ell}, \tag{A.7.3}$$

which fulfills a dispersion relation of the same type as Eq. (A.7.1), since the division by q^2 does not introduce new singularities. $q=0$ at the s- and u-channel thresholds. At the s-channel threshold, $F_{\ell\pm}$ is finite and equal to the scattering length (Sect. A.3.5.2). There is a difficulty for $\ell>0$ at the u-channel threshold, because $\operatorname{Im} F_{\ell\pm}$ goes to infinity, but it is possible to evaluate the dispersion integral.

The advantage of the reduced partial wave amplitudes is that the contributions from the distant left and right hand cuts are suppressed.

Table A.7.1. Singularities of s-channel partial wave amplitudes. In addition, all partial waves have a kinematic cut at $s<0$, which follows from terms depending on $W=\sqrt{s}$ in the projection formula. See Sect. 7.1.4 for the singularity at $s=0$.

Branch cuts	q^2-plane	Sheet	s-plane
physical cut (s-channel)	$0 \leqq q^2 < \infty$	I	$(m+\mu)^2 \leqq s < \infty$
crossed cut (u-channel)	$0 \leqq q^2 < \infty$	II	$0 \leqq s \leqq (m-\mu)^2$
	$-\infty < q^2 \leqq -m^2$	II	$-(m^2-\mu^2) \leqq s \leqq 0$
	$-\infty < q^2 \leqq -m^2$	I	$-\infty < s \leqq -(m^2-\mu^2)$
t-channel cut ($t \geqq 4\mu^2$)	$-m^2 \leqq q^2 \leqq -\mu^2$		$\lvert s\rvert = m^2-\mu^2$
	$-\infty < q^2 \leqq -m^2$	II	$-(m^2-\mu^2) \leqq s \leqq 0$
	$-\infty < q^2 \leqq -m^2$	I	$-\infty < s \leqq -(m^2-\mu^2)$
nucleon cuts ($u=m^2$)			
long range part (short cut)	$-\mu^2 \leqq q^2 \leqq \dfrac{-\mu^2(1-\mu^2/4m^2)}{1+2\mu^2/m^2}$	I	$(m^2-\mu^2) \leqq s \leqq m^2+2\mu^2$
	$-\mu^2 \leqq q^2 \leqq -\mu^2\left(1-\dfrac{\mu^2}{4m^2}\right)$	II	$\left(m-\dfrac{\mu^2}{m}\right)^2 \leqq s < m^2-\mu^2$
short range part	$-\infty < q^2 \leqq -m^2$	II	$-(m^2-\mu^2) \leqq s < 0$
	$-\infty < q^2 \leqq -m^2$	I	$-\infty < s \leqq -(m^2-\mu^2)$
nucleon pole			
($s=m^2$) only in $f_{1-}^{1/2}(s)$	$q^2 = -\mu^2(1-\mu^2/4m^2)$	I	$s=m^2$

7.1.2 The generalized πN potential

The dispersion relation for the reduced partial waves can be written as follows

$$\operatorname{Re} F_{\ell\pm}(s) = F^L_{\ell\pm}(s) + \frac{1}{\pi} \int_{s_{th}}^{\infty} \frac{\operatorname{Im} F_{\ell\pm}(s')}{s'-s} \, ds' . \tag{A.7.4}$$

In analogy to the treatment of potential scattering, the first term is usually interpreted as describing the "forces" between pion and nucleon. It is sometimes called the "generalized potential". The second term describes "rescattering", because $\operatorname{Im} F_{\ell\pm}(s)$ is determined by the unitarity relation, where the final state is reached via intermediate states.

One uses the name "long range forces" for the contributions from the nearby parts of the cuts, for instance from the short nucleon cut, the front of the circle (ϱ- and "ε"-exchange) and the Δ(1232)-exchange part of the crossed cut. These terms have an appreciable energy dependence in the physical region.

The "short range forces" belong to the distant parts of the cuts. It is assumed that the sum of all these terms is slowly varying in the physical region. This conjecture is somewhat vague. Within the framework of the partial wave dispersion relations, it is not possible to make theoretical predictions for the short range forces. For $\ell>0$, the short range effects are suppressed by the denominator of the reduced partial waves $F_{\ell\pm}(s)$.

Other methods for the determination of $F^L_{\ell\pm}(s)$ will be mentioned in Sect. A.7.2.

7.1.3 Unitary sum rules

Since $\mathrm{Re}F_{\ell\pm}(s)$ is bounded by unitarity, it must go to zero at least as fast as $s^{-(\ell+1/2)}$, i.e. faster than $1/s$ as $s\to\infty$ if $\ell>0$. Therefore, it can be neglected in the limit $s\to\infty$, if we assume that the terms on the right hand side of Eq. (A.7.4) go to zero as $1/s$. It follows that

$$\lim_{s\to\infty} sF^L_{\ell\pm}(s)-\frac{1}{\pi}\int_{s_{\mathrm{th}}}^{\infty}\mathrm{Im}F_{\ell\pm}(s)\mathrm{d}s=0\,;\qquad \ell>0. \tag{A.7.5}$$

Higher sum rules can be obtained in a similar way. In the sum rule for $\ell>1$ one has to add a factor s in front of $F^L_{\ell\pm}$ and of $\mathrm{Im}F_{\ell\pm}$.

The assumptions are fulfilled in simple models, but they cannot be proved and therefore, the unitary sum rules have the status of a conjecture. They can be applied in order to estimate the magnitude of the unknown contribution of the sum over all distant left hand cuts.

7.1.4 Behaviour of $f_{\ell\pm}(s)$ at $s=0$

$f_{\ell\pm}(s)$ is defined by the projection integral Eq. (A.3.7), (A.3.8) over the invariant amplitudes, which goes over the range $-1<\cos\theta<+1$ or $-4q^2<t<0$. According to Eq. (A.1.12), the approximation $4q^2\approx(m^2-\mu^2)^2/s$ is valid for small s, i.e. $q^2\to\infty$ as $s\to 0$.

A possible singularity of $f_{\ell\pm}(s)$ at $s\to\pm 0$ can only arise from the fact that the lower limit of the projection integral goes to $t=\mp\infty$. Its nature can be derived, if one knows the invariant amplitude in this limit.

Figure A.6.10a shows that the invariant amplitudes at $s=0$ and $t\to-\infty$ belong to u-channel backward scattering at very high energies, whereas the amplitudes at $s=0$, $t\to+\infty$ belong to $\pi\pi N\bar{N}$ forward scattering at very high energies. Because of crossing symmetry, the amplitudes for $t\to-\infty$ agree with those for s-channel backward scattering and the amplitudes for $t\to+\infty$ agree with those for t-channel backward scattering (in both cases a sign could be different). Furthermore, one can conclude from the Phragmén-Lindeloef theorem that the invariant amplitudes have the same asymptotic behaviour at $s=0$ and $t\to\pm\infty$ (cf. Sects. A.6.4.4 and A.6.6).

Assuming that backward scattering can be described by the Regge model (Sect. A.9.2), one finds that the partial waves behave as

$$f_{\ell\pm}(s)\to c_{\ell\pm}s^{-\alpha_0-1/2}\,;\qquad s\to\pm 0. \tag{A.7.6}$$

Taking $\Delta(1232)$-exchange, one has to insert $\alpha_0\approx 0$ and the partial waves diverge as $s^{-0.5}$. A reggeized nucleon exchange contribution follows with $\alpha_0=-0.4$ (see Sect. 2.6.7.4 for new values of α_0).

The s^{-1}-dependence in Eq. (A.7.2b) belongs to the exchange of an elementary nucleon ($\alpha_0=1/2$). It follows from the decrease of the experimental backward amplitude that the partial wave amplitude has a weaker singularity than the s^{-1}-behaviour of the nucleon pole term projection in Eq. (A.7.2a, b). The pole term in Eq. (A.7.2c) is finite as $s=0$.

The continuation of the partial waves in the neighborhood of $s=0$ is given by

$$\mathrm{Re}f_{\ell\pm}(0-)=\mathrm{Re}f_{\ell\pm}(0+)\,;\qquad \mathrm{Im}f_{\ell\pm}(0-)=-\mathrm{Im}f_{\ell\pm}(0+). \tag{A.7.7}$$

7.1.5 Partial wave dispersion relations in the W-plane

Several authors preferred to write the dispersion relations in the $W=\sqrt{s}$-plane. One of the reasons is to avoid the kinematic cut $s<0$. The half-plane $\mathrm{Re}W>0$ corresponds to the whole of the physical sheet of the s-plane.

The cut structure in the W-plane is shown in Fig. A.7.3. The cut along the imaginary axis separates the left and right half-planes. However, a unique relation is given by the MacDowell reflection symmetry

$$f_{\ell+}(W)=-f_{(\ell+1)-}(-W), \tag{A.7.8}$$

which follows from

$$f_1(W,z)=-f_2(-W,z). \tag{A.7.9}$$

Equation (A.7.9) is a consequence of the fact that the invariant amplitudes in Eq. (A.2.31) depend only on s,t,u.

In the W-plane, a modified version of the reduced partial waves of Eq. (A.7.3) is preferable

$$\bar{F}_{\ell+}(W)=\frac{W}{E+m}\,\frac{f_{\ell+}(W)}{q^{2\ell}}\,;\quad \bar{F}_{(\ell+1)-}(W)=\frac{W}{E-m}\,\frac{f_{(\ell+1)-}(W)}{q^{2\ell}}\,;\quad \frac{2W^2}{(W\pm m)^2-\mu^2}=\frac{W}{E\pm m}. \tag{A.7.10}$$

The MacDowell symmetry relation has the same form as in Eq. (A.7.8)

$$\bar{F}_{\ell+}(W) = -\bar{F}_{(\ell+1)-}(-W) \tag{A.7.8a}$$

and the partial wave dispersion relation reads

$$\mathrm{Re}\,\bar{F}_{\ell\pm}(W) = \bar{V}_{\ell\pm}(W) + \frac{1}{\pi} \int_{m+\mu}^{\infty} \frac{dW'}{W'-W} \mathrm{Im}\,\bar{F}_{\ell\pm}(W'). \tag{A.7.4a}$$

The generalized potential $\bar{V}_{\ell\pm}$ includes all left hand cut contributions and the nucleon pole.

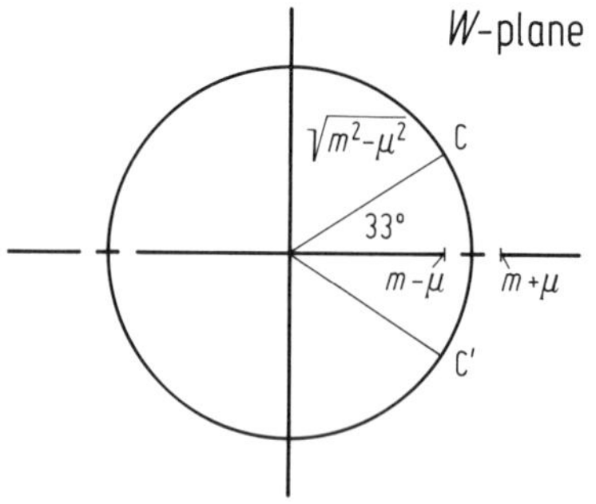

Fig. A.7.3. Cuts of the s-channel partial wave amplitudes in the W-plane. The t-channel partial wave expansion converges only between the points C and C′ on the circle. The two short cuts are nucleon cuts. The values of $f_{\ell+}(W)$ in the right half plane (Re $W>0$) are connected with the values of $f_{(\ell+1)-}$ in the left half plane by Eq. (A.7.8).

7.2 s-channel partial wave relations

7.2.1 Partial wave relations derived from fixed-t dispersion relations

The projection formulas Eqs. (A.3.7), (A.3.8) express the s-channel partial waves in terms of invariant amplitudes A and B. One can insert the fixed-t dispersion relations (A.6.30) for A and B and replace Im A and Im B under the dispersion integrals by the partial wave expansion Eq. (A.3.6). The result can be written as follows (omitting the isospin index)

$$f_{J\varepsilon}(s) = f_{J\varepsilon,\mathrm{N}}(s) + \frac{1}{\pi} \int_{s_{\mathrm{th}}}^{\infty} \sum_{J',\varepsilon'} K_{J\varepsilon,J'\varepsilon'}(s,s')\,\mathrm{Im}\,f_{J'\varepsilon'}(s')\,ds', \tag{A.7.11}$$

where the notation $J\varepsilon$ is used instead of $\ell\pm$ [see Eq. (A.3.15)].

The kernels $K_{J\varepsilon,J'\varepsilon'}(s,s')$ are very complicated known functions of the kinematic variables. They become simpler, if the variables W, W' are used instead of s, s' and if the modified reduced partial waves (A.7.10) are introduced

$$\begin{aligned}
\bar{F}_{\ell+}(W) &= \bar{F}_{\ell+,\mathrm{N}}(W) + \frac{1}{\pi} \int_{m+\mu}^{\infty} dW' \sum_{\ell'=0}^{\infty} \{K_{\ell\ell'}(W,W')\,\mathrm{Im}\,\bar{F}_{\ell'+}(W') + K_{\ell\ell'}(W,-W')\,\mathrm{Im}\,\bar{F}_{(\ell'+1)-}(W')\};\\
\bar{F}_{(\ell+1)-}(W) &= \bar{F}_{(\ell+1)-,\mathrm{N}}(W) - \frac{1}{\pi} \int_{m+\mu}^{\infty} dW' \sum_{\ell'=0}^{\infty} \{K_{\ell\ell'}(-W,W')\,\mathrm{Im}\,\bar{F}_{\ell'+}(W') + K_{\ell\ell'}(-W,-W')\,\mathrm{Im}\,\bar{F}_{(\ell'+1)-}(W')\}.
\end{aligned} \tag{A.7.12}$$

The integrals are principal value integrals only in that term of the sum, which has the same quantum number as the partial wave on the left hand side.

The projections of the nucleon pole terms are given by

$$f^I_{\ell\pm,\mathrm{N}}(W) = \frac{m^2 f^2}{W}(W-m)(E+m)Z^I_\ell(W) + (W+m)(E-m)Z^I_{\ell\pm1}(W); \tag{A.7.13}$$

with

$$Z^I_\ell(W) = \frac{2\delta_{\ell 0}}{m^2-W^2} + \frac{\varepsilon^I}{q^2} Q_\ell(y); \quad y = 1 - \frac{W^2-m^2-2\mu^2}{2q^2}; \quad Q_\ell(y) = \frac{1}{2}\int_{-1}^{+1} dx \frac{P_\ell(x)}{y-x}. \tag{A.7.14}$$

$\varepsilon^I = +1$ and -1 for the isospin even and odd combination respectively. The s-channel nucleon exchange term only contributes to the $0+$ and $1-$ partial waves and vanishes for $\pi^+ p$ scattering.

Since the partial wave expansion (A.3.6) is used for Im A and Im B, the fixed-t dispersion relation is valid only in the strip $4\mu^2 > t > -26\mu^2$ [see Eq. (A.6.40)]. This strip covers the whole range $-1 < \cos\theta < +1$ in the s-interval $20\mu^2 < s < 98\mu^2$ (Fig. A.6.5) which includes s-channel scattering up to a pion laboratory momentum of 520 MeV/c, the interval between the u- and s-channel thresholds $(m-\mu)^2 = 33\mu^2 < s < 60\mu^2 = (m+\mu)^2$ and part of the u-channel range $20\mu^2 < s < 33\mu^2$ (Fig. A.7.4).

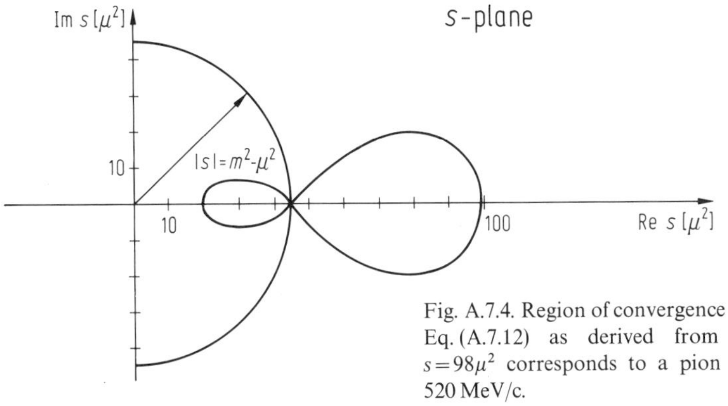

Fig. A.7.4. Region of convergence of the partial wave relation Eq. (A.7.12) as derived from Mandelstam analyticity. $s = 98\mu^2$ corresponds to a pion laboratory momentum of 520 MeV/c.

The generalized potential

In terms of the amplitudes of Eq. (A.7.10), the partial wave relations can be written (after separating the principal value part of the kernel)

$$\operatorname{Re}\bar{F}_{\ell\pm}(W) = \bar{V}_{\ell\pm}(W) + \frac{1}{\pi} \mathop{⨍}_{m+\mu}^{\infty} \frac{\mathrm{d}W'}{W'-W} \operatorname{Im}\bar{F}_{\ell\pm}(W'). \tag{A.7.15}$$

This relation is analogous to Eq. (A.7.4a) insofar as the real part is decomposed into a sum of "left hand cut contributions", which is called the "generalized potential", and a rescattering term. It is an advantage, that the potential can now be calculated with reasonable accuracy from a convergent expansion in terms of πN partial waves, whereas the potential in Eq. (A.7.4a) has distant cuts which cannot be calculated from partial waves.

One should note that $\bar{V}_{\ell\pm}$ in Eq. (A.7.4a) has u- and t-channel contributions, but the same function is derived from u-channel contributions alone in the partial wave relation approach (A.7.15).

We conclude that there are several equivalent possibilities for the description of the "forces" between pion and nucleon. See also Sect. A.7.2.2.

Partial wave crossing relations

The real parts of the s-channel partial waves $f_{\ell\pm}(s)$ in the unphysical region $s < (m+\mu)^2$ can be calculated from the partial wave relation Eq. (A.7.11). In part of this region, $0 < s < (m-\mu)^2$, one can calculate Re $f_{\ell\pm}(s)$ in a second way, which should lead exactly to the same result.

Let $W = W_c$ be an energy in the range $0 < W < m-\mu$. The projection integral Eq. (A.3.8) is taken over the range $-1 < z_c < +1$ along a line $s = s_c = \text{const}$ in the u-channel physical region (Fig. A.7.5).

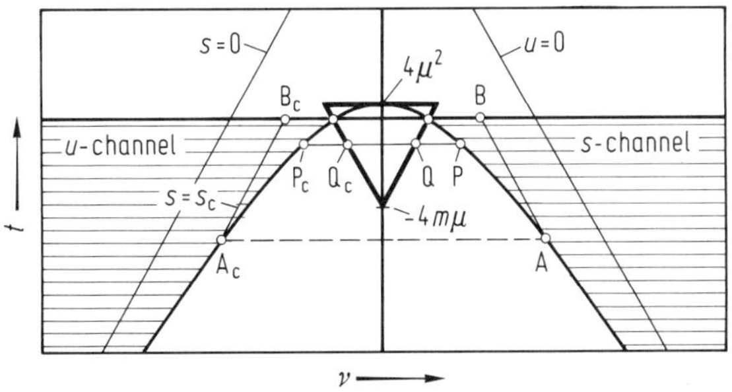

Fig. A.7.5. Projections in partial wave crossing relations. s-channel partial waves at the unphysical energy s_c are obtained by calculating the projection integral over the range $A_c - B_c$. Due to crossing symmetry Eq. (A.6.28) the invariant amplitudes are the same (possibly up to a factor) as in the range A – B of the physical region. — This figure also shows the integration range in the projection formula for t-channel partial waves, Eq. (A.3.50), which is the line $P_c - P$. The invariant amplitudes are real within the triangle, i.e. in the range $Q_c - Q$. The limits of the integrals in Eqs. (A.7.33) and (A.7.36) are $\pm\nu_{th}$ belonging to Q and Q_c, and $\pm\nu_b$ belonging to P and P_c.

In order to determine the invariant amplitudes A and B along this line, crossing symmetry is used, Eq. (A.6.28), which tells us that A and B are the same (possibly up to a factor ± 1) along a line u=const in the s-channel physical region, where these amplitudes can be expressed in terms of phase shifts, Eq. (A.3.6).

The result for the projection integral reads

$$\operatorname{Re} f^{\pm}_{\ell+}(W_c) = \pm \int_{-1}^{+1} dz_c \sum_{\ell'=0}^{\infty} [C_{\ell\ell'}(W_c, z_c; W, z) \operatorname{Re} f^{\pm}_{\ell'+}(W) - C_{\ell\ell'}(W_c, z_c; -W, z) \operatorname{Re} f^{\pm}_{(\ell'+1)-}(W)]. \tag{A.7.16}$$

This relation and similar relations for $\operatorname{Re} f^{\pm}_{(\ell+1)-}(W)$ and for the imaginary parts of these partial waves are called *partial wave crossing relations*.

Notation: from the given value of $s_c = W_c^2$ and from the integration variable z_c one calculates q_c^2, using Eq. (A.1.12), and $t = -2q_c^2(1-z_c)$. Then $s = W^2$ and q^2 follow from $s + s_c + t = 2m^2 + 2\mu^2$ and from Eq. (A.1.12). Finally $z = 1 + t/2q^2$.

7.2.2 Partial wave relations derived from hyperbolic dispersion relations

In Sect. A.6.7 we have mainly treated the "interior dispersion relations", which are a one-parameter subset of the dispersion relations along hyperbolas, Eq. (A.6.83), belonging to a fixed scattering angle θ_π in the laboratory system. The parameter a is determined by θ_π and $b = \lambda(a, m^2, \mu^2)$, since all curves must pass the s-channel threshold. Now we consider another subset: a is kept fixed and b is determined such that the curve passes a point (s_1, t_1) or (s_1, z_1) in the Mandelstam plane

$$b(s_1, t_1, a) = (s-a)(u-a) = (s_1 - a)(\Sigma - s_1 - t_1 - a).$$

For given values of a and s_1, the members of the subset are characterized by the value of z_1 or of t_1.

Since we intend to perform a partial wave projection, we need a set of hyperbolas which cover the integration range $-1 < z < +1$ between threshold and an energy s. Furthermore, we demand that both the s- and t-channel partial wave expansions converge along the hyperbolas. It turns out that these conditions can be fulfilled for $a = -23\mu^2$ as long as $s < 97\mu^2$ corresponding to the pion laboratory momentum $k < 510$ MeV/c.

The s-channel partial wave projection of the hyperbolic dispersion relation reads

$$\begin{aligned} f_{\ell+}(W) = N_{\ell+}(W) &+ \frac{1}{\pi} \int_{m+\mu}^{\infty} dW' \sum_{\ell'=0}^{\infty} \{K_{\ell\ell'}(W, W') \operatorname{Im} f_{\ell'+}(W') + K_{\ell\ell'}(W, -W') \operatorname{Im} f_{(\ell'+1)-}(W')\} \\ &+ \frac{1}{\pi} \int_{4\mu^2}^{\infty} dt' \sum_J \{G_{\ell J}(W, t') \operatorname{Im} f^J_+(t') + H_{\ell J}(W, t') \operatorname{Im} f^J_-(t')\} \end{aligned} \tag{A.7.18}$$

and a similar equation for $f_{(\ell+1)-}(W)$. The fixed value of the parameter a is not listed in the kernels. $N_{\ell\pm}$ denotes the nucleon pole term projection.

In analogy to Eqs. (A.7.4a) and (A.7.15), one can define a generalized potential after having decomposed the kernel into a principal value part and the rest

$$\operatorname{Re} f_{\ell\pm}(W) = V_{\ell\pm}(W) + \frac{1}{\pi} \int_{m+\mu}^{\infty} \frac{\operatorname{Im} f_{\ell\pm}(W')}{W' - W} dW'. \tag{A.7.19}$$

In the present version, the "potential" or "driving force" has contributions from u-channel exchanges as well as from t-channel exchanges. It is an advantage in comparison with the usual partial wave dispersion relation that the potential is now expressed by a convergent expansion. There are no distant left hand cut contributions which are unknown in principle.

7.3 Analytic continuation of s-channel partial wave amplitudes into the second sheet

For the discussion of the analytic continuation we use the partial wave amplitude $f_{\ell\pm}(s)$, Eq. (A.3.2), whose analytic properties are treated in Sect. 7.1. $f_{\ell\pm}(s)$ is a real analytic function

$$f_{\ell\pm}(s^*) = f^*_{\ell\pm}(s) \tag{A.7.20}$$

and has a cut for $s > s_{th} = (m+\mu)^2$. In the s-range below the inelastic threshold s_{in}, the unitarity relation (A.5.20) reads

$$\begin{aligned} &\operatorname{Im} f_{\ell\pm}(s+i0) = q(s+i0) f_{\ell\pm}(s+i0) f^*_{\ell\pm}(s+i0); \quad s_{th} < s < s_{in}; \\ &\operatorname{Im} f^{-1}_{\ell\pm}(s+i0) = -q(s+i0). \end{aligned} \tag{A.7.21}$$

Both sides of the first equation are equal to $\sin^2\delta_{\ell\pm}/q$. The second equation is equivalent to the first one. It shows that the cut of $q(s)$ must be chosen at $-\infty < s < (m-\mu)^2$ and $(m+\mu)^2 < s < \infty$ [see Eq. (A.1.12) for the location of the branch points]. Therefore, the relation

$$q(s^*) = -q^*(s) \qquad \text{(A.7.22)}$$

is valid for $q(s)$.

Using this equation and the real analyticity of $f_{\ell\pm}(s)$, one can rewrite the unitarity relation as follows

$$f_{\ell\pm}(s+\mathrm{i}0) = \frac{f_{\ell\pm}(s-\mathrm{i}0)}{1+2\mathrm{i}q(s-\mathrm{i}0)f_{\ell\pm}(s-\mathrm{i}0)} \equiv f_{\ell\pm}^{\mathrm{II}}(s-\mathrm{i}0)\,. \qquad \text{(A.7.23)}$$

The right hand side is the boundary value of an analytic function

$$f_{\ell\pm}^{\mathrm{II}}(s) = \frac{f_{\ell\pm}(s)}{1+2\mathrm{i}q(s)f_{\ell\pm}(s)}, \qquad \text{(A.7.24)}$$

which is the continuation of $f_{\ell\pm}(s)$ into the second sheet. The connection between $f_{\ell\pm}(s)$ below the cut and $f_{\ell\pm}^{\mathrm{II}}(s)$ is given by

$$f_{\ell\pm}(s-\mathrm{i}0) = f_{\ell\pm}^{\mathrm{II}}(s+\mathrm{i}0)\,. \qquad \text{(A.7.25)}$$

$f_{\ell\pm}(s)$ and $f_{\ell\pm}^{\mathrm{II}}(s)$ can be represented by two analytic functions $d_{\ell\pm}(s)$ and $a_{\ell\pm}(s)$, which have no cut at $(m+\mu)^2 < s < s_{\mathrm{in}}$ and are real in this range

$$f_{\ell\pm}(s) = d_{\ell\pm}(s) + \mathrm{i}q(s)a_{\ell\pm}(s); \quad f_{\ell\pm}^{\mathrm{II}}(s) = d_{\ell\pm}(s) - \mathrm{i}q(s)a_{\ell\pm}(s)\,. \qquad \text{(A.7.26)}$$

These functions are related to the partial waves as follows

$$\begin{aligned} &\mathrm{Re}f_{\ell\pm}(s) = d_{\ell\pm}(s); \quad \mathrm{Im}f_{\ell\pm}(s) = q(s)a_{\ell\pm}(s); \quad (m+\mu)^2 < s < s_{\mathrm{in}}\,; \\ &d_{\ell\pm}(s) = f_{\ell\pm}^{\mathrm{II}}(s)\{1+\mathrm{i}q(s)f_{\ell\pm}(s)\}\,; \quad a_{\ell\pm}(s) = f_{\ell\pm}(s)f_{\ell\pm}^{\mathrm{II}}(s)\,. \end{aligned} \qquad \text{(A.7.27)}$$

The phase shifts and partial waves can be expressed in terms of the diagonal elements of the S- and K-matrices. The formulas read in the elastic region [$2T_{\ell\pm}$ is the matrix element of T in Eq. (A.5.1)]

$$\begin{aligned} &S_{\ell\pm}(s) = \exp\{2\mathrm{i}\delta_{\ell\pm}(s)\} = 1 + 2\mathrm{i}q(s)f_{\ell\pm}(s) = 1 + 2\mathrm{i}T_{\ell\pm}(s) = \frac{f_{\ell\pm}(s)}{f_{\ell\pm}^{\mathrm{II}}(s)}; \\ &K_{\ell\pm}(s) = \frac{1}{\mathrm{i}}\frac{S_{\ell\pm}-1}{S_{\ell\pm}+1} = \tan\delta_{\ell\pm}(s) = q(s)\frac{a_{\ell\pm}(s)}{d_{\ell\pm}(s)} = \frac{q(s)f_{\ell\pm}(s)}{1+\mathrm{i}q(s)f_{\ell\pm}(s)}\,. \end{aligned} \qquad \text{(A.7.28)}$$

$S_{\ell\pm}(s)$ is real analytic and fulfills the extended unitarity relation

$$S_{\ell\pm}(s^*) = S_{\ell\pm}^*(s); \quad S_{\ell\pm}(s)S_{\ell\pm}^{\mathrm{II}}(s) = 1\,. \qquad \text{(A.7.29)}$$

$f_{\ell\pm}(s)$ is regular in the first sheet of the s-plane, except for the cuts shown in Fig. A.7.1. However, $f_{\ell\pm}^{\mathrm{II}}(s)$, which is equal to $f_{\ell\pm}(s)/S_{\ell\pm}(s)$, has poles at the location of the zeros of $S_{\ell\pm}(s)$. These poles belong to the nucleon resonances, i.e. to the excited states of the nucleon, if they lie near the cut $s \geqq (m+\mu)^2$.

It follows from the unitarity relation for $S_{\ell\pm}(s)$ and from Eq. (A.7.27) that $S_{\ell\pm}^{\mathrm{II}}(s)$, $d_{\ell\pm}(s)$, and $a_{\ell\pm}(s)$ have poles at the same positions. The resonance poles of $d_{\ell\pm}(s)$ and $a_{\ell\pm}(s)$ cancel in $K_{\ell\pm}(s)$, Eq. (A.7.28) but, if the width of the resonance is not too large, $K_{\ell\pm}(s)$ has a pole in the neighborhood on the real axis, $\delta_{\ell\pm}$ assuming the value $\pi/2$. This can be seen from the formula

$$f_{\ell\pm}^{\mathrm{II}}(s) = \frac{1}{q(s)\{K_{\ell\pm}^{-1}+\mathrm{i}\}}, \qquad \text{(A.7.30)}$$

if $K_{\ell\pm}^{-1}(s)$ is approximated by a linear term in the neighborhood of its zero.

Since $d_{\ell\pm}(s)$, $a_{\ell\pm}(s)$ and also $s(q^2)$ are regular functions near threshold, one can express $d_{\ell\pm}(s)$ and $a_{\ell\pm}(s)$ by power series in q^2. If the threshold behaviour of the phase shifts is taken into account, this leads to the expansions for $K_{\ell\pm}(s)$, $K_{\ell\pm}^{-1}(s)$, and $f_{\ell\pm}(s)$ in Sect. A.3.5.2.

At threshold, the partial waves $f_{\ell\pm}(s)$ show a cusp phenomenon, because $qa_{\ell\pm}(s)$ contributes to the imaginary part for $s > s_{\mathrm{th}}$ and to the real part for $s < s_{\mathrm{th}}$, since $q = \mathrm{i}\,{}_{+}\sqrt{\mu^2-\omega_q^2}$ in this region.

7.4 The Froissart-Gribov representation for $\pi\pi N\bar{N}$ wave amplitudes

7.4.1 The projection formulas

The Froissart-Gribov representation is of interest for a study of the analytic properties of the $\pi\pi N\bar{N}$ partial waves $f_\pm^J(t)$ and for a determination of these partial waves at $t<0$ from the πN partial waves $f_{\ell\pm}(s)$.

This representation is obtained by inserting the fixed-t dispersion relations (A.6.30), (A.6.34), (A.6.35) into the projection formula (A.3.50) for the t-channel partial waves. Then, the order of the two integrations is interchanged and the projection integrals are calculated using the formulas

$$Q_J(Z')=\frac{1}{2}\int_{-1}^{+1}\frac{dZ}{Z'-Z}P_J(Z);\quad (-1)^J Q_J(Z')=\frac{1}{2}\int_{-1}^{+1}\frac{dZ}{Z'+Z}P_J(Z). \tag{A.7.31}$$

It is important to take into account the $\pm i0$-prescription as described in Sect. A.3.4.

We show an intermediate step of the calculation of $f_+^J(t)$ from $C^-(\nu,t)$

$$\begin{aligned} f_+^J(t+i0) &= f_{+N}^J+\frac{m}{4\pi^2}\frac{p_-^2}{(p_-q_-)^{J+1}}\int_{\nu_{th}}^{\infty} d\nu' \operatorname{Im} C^-(\nu'+i0,t)Q_J(Z'+i0)\{1-(-1)^J\} \\ &= f_{+N}^J+\frac{m}{2\pi^2}\frac{p_-^2}{(p_-q_-)^{J+1}}\int_{\nu_{th}}^{\infty} d\nu' \operatorname{Im} C^-(\nu'+i0,t)Q_J(Z'+i0), \quad \text{if } J \text{ is odd}; \\ &= 0, \quad \text{if } J \text{ is even}. \end{aligned} \tag{A.7.32a}$$

The isospin even amplitude C^+ contributes only, if J is even

$$\begin{aligned} f_+^J(t+i0) &= f_{+N}^J+\frac{m}{2\pi^2}\frac{p_-^2}{(p_-q_-)^{J+1}}\int_{\nu_{th}}^{\infty} d\nu' \operatorname{Im} C^+(\nu'+i0,t)\{Q_J(Z'+i0)-\delta_{J0}/Z'\} \\ &\quad +\frac{p_-^2}{4\pi}C^+(0,t)\delta_{J0}; \quad J \text{ even}. \end{aligned} \tag{A.7.32b}$$

The representation for the partial waves $f_-^J(t)$ is written in a more compact form. $\operatorname{Im} B^+$ and $\operatorname{Im} B^-$ have to be used for even and odd values of J, respectively

$$f_-^J(t+i0)=f_{-N}^J+\frac{m}{2\pi^2}\frac{\sqrt{J(J+1)}}{(2J+1)(p_-q_-)^J}\int_{\nu_{th}}^{\infty} d\nu' \operatorname{Im} B^\pm(\nu'+i0,t)\{Q_{J-1}(Z'+i0)-Q_{J+1}(Z'+i0)\}. \tag{A.7.33}$$

The following notation is used in the above formulas

$$Z=\frac{m\nu}{p_-q_-}=-\cos\theta_t;\quad Z'=\frac{m\nu'}{p_-q_-};\quad p_-=\sqrt{m^2-t/4};\quad q_-=\sqrt{\mu^2-t/4}. \tag{A.7.34}$$

The Legendre functions of the second kind, $Q_J(Z)$, have a cut along $-1<Z<+1$. The boundary values from above and below follow from Eqs. (A.6.16) and (A.7.31)

$$Q_J(Z\pm i0)=Q_J^{pv}(Z)\mp i\frac{\pi}{2}P_J(Z);\quad -1<Z<+1. \tag{A.7.35}$$

The real-valued function $Q_J^{pv}(Z)$ is obtained by taking the principal value of the integral in Eq. (A.7.31).

If this formula is inserted into the Froissart-Gribov representations (A.7.32), (A.7.33), the evaluation of the integrals gives $\operatorname{Re} f_\pm^J(t)$ and $\operatorname{Im} f_\pm^J(t+i0)$. The expression for $\operatorname{Re} f_\pm^J(t)$ has a singularity at

$$\nu'=\nu_b=\{s^2-(m^2-\mu^2)^2\}/4ms,$$

which belongs to $Z'=-\cos\theta_t'=1$ in Eq. (A.7.31) and also to backward scattering ($t=-4q^2$) in the πN channel. Q_J^{pv} from the left hand side and Q_J from the right hand side go both to infinity as $\ln|1-Z'|$, if ν' approaches ν_b. However, the integrals exist and there is no serious problem with the numerical evaluation.

The integral in the expression for $\operatorname{Im} f_\pm^J$ goes only from ν_{th} to ν_b, because Q_J is real for larger values of ν' (see Fig. A.7.5)

$$\begin{aligned} \operatorname{Im} f_+^J(t+i0) &= \operatorname{Im} f_{+N}^J(t+i0)-\frac{m}{4\pi}\frac{p_-^2}{(p_-q_-)^{J+1}}\int_{\nu_{th}}^{\nu_b} d\nu' \operatorname{Im} C^\pm(\nu'+i0,t)P_J(Z'); \\ \operatorname{Im} f_-^J(t+i0) &= \operatorname{Im} f_{-N}^J(t+i0)-\frac{m}{4\pi}\frac{\sqrt{J(J+1)}}{(2J+1)(p_-q_-)^J}\int_{\nu_{th}}^{\nu_b} d\nu' \operatorname{Im} B^\pm(\nu'+i0,t)\{P_{J-1}(Z')-P_{J+1}(Z')\}. \end{aligned} \tag{A.7.36}$$

7.4.2 The nucleon pole term contributions

In several applications one needs explicit formulas for the nucleon pole term projections, which can be obtained either by inserting the nucleon pole terms of the fixed-t dispersion relations (A.6.30), (A.6.34), (A.6.35) into the projection formulas (A.3.50) or by inserting Eq. (A.6.31) into the Froissart-Gribov representation instead of adding the pole terms.

$$f^J_{+\mathrm{N}}(t+\mathrm{i}0)=\frac{g^2}{4\pi}\frac{m}{(p_-q_-)^J}\{Z_\mathrm{N}Q_J(Z_\mathrm{N}+\mathrm{i}0)-\delta_{J0}\};$$
$$f^J_{-\mathrm{N}}(t+\mathrm{i}0)=\frac{g^2}{4\pi}\frac{\sqrt{J(J+1)}}{(2J+1)(p_-q_-)^J}\{Q_{J-1}(Z_\mathrm{N}+\mathrm{i}0)-Q_{J+1}(Z_\mathrm{N}+\mathrm{i}0)\},\qquad \text{(A.7.37)}$$

where $Z_\mathrm{N}=m\nu_B/p_-q_-=(t-2\mu^2)/4p_-q_-$ and $g^2/4\pi=f^2(2m/\mu)^2=14.28$, if $f^2=0.079$.

Using Eq. (A.7.35), one finds for the imaginary parts at $t<4\mu^2-\mu^4/m^2=3.98\mu^2$

$$\mathrm{Im}f^J_{+\mathrm{N}}(t+\mathrm{i}0)=-g^2\frac{mZ_\mathrm{N}}{8(p_-q_-)^J}P_J(Z_\mathrm{N});$$
$$\mathrm{Im}f^J_{-\mathrm{N}}(t+\mathrm{i}0)=g^2\frac{\sqrt{J(J+1)}}{8(2J+1)(p_-q_-)^J}\{P_{J+1}(Z_\mathrm{N})-P_{J-1}(Z_\mathrm{N})\}.\qquad \text{(A.7.38)}$$

The imaginary parts vanish for $t>4\mu^2-\mu^4/m^2$.

In particular, the imaginary parts of the lowest partial waves are

$$\mathrm{Im}f^0_{+\mathrm{N}}(t+\mathrm{i}0)=-g^2\frac{m}{32p_-q_-}(t-2\mu^2);\quad \mathrm{Im}f^1_{+\mathrm{N}}(t+\mathrm{i}0)=-g^2\frac{m}{8p_-q_-}Z^2_\mathrm{N};$$
$$\mathrm{Im}f^1_{-\mathrm{N}}(t+\mathrm{i}0)=-g^2\frac{Z^2_\mathrm{N}-1}{8\sqrt{2}p_-q_-}.\qquad \text{(A.7.39)}$$

Table A.7.2 is useful for discussions of the rapid variation of the $f^J_{\pm\mathrm{N}}$.

Table A.7.2. The rapid variation of $Z_\mathrm{N}(t)$.

t/μ^2	$-\infty$	-8	0	2	$4-\mu^2/m^2$	4	5	6	8	16
Z_N	-1	-0.21	$-\mu/2m$	0	$+1$	∞	0.23i	0.21i	0.23i	0.32i

The following formulas are needed for the calculation of the real parts

$$Q^{\mathrm{pv}}_0(Z)=\frac{1}{2}\ln\frac{1+Z}{1-Z}=Z+\frac{Z^3}{3}+\dots$$
$$Q^{\mathrm{pv}}_1(Z)=\frac{Z}{2}\ln\frac{1+Z}{1-Z}-1=-1+Z^2+\frac{Z^4}{3}+\dots\qquad t<4\mu^2-\mu^4/m^2;\quad -1<Z_\mathrm{N}<+1;\qquad \text{(A.7.40)}$$
$$Q^{\mathrm{pv}}_2(Z)=\frac{1}{4}(3Z^2-1)\ln\frac{1+Z}{1-Z}-\frac{3}{2}Z=-2Z+\frac{4}{3}Z^3+\dots,$$

$$Q_0(Z)=\frac{1}{2}\ln\frac{Z+1}{Z-1}=\frac{1}{Z}+\frac{1}{3Z^3}+\dots$$
$$Q_1(Z)=\frac{Z}{2}\ln\frac{Z+1}{Z-1}-1=\frac{1}{3Z^2}+\frac{1}{5Z^4}+\dots\qquad 4\mu^2-\mu^4/m^2<t<4\mu^2;\quad 1<Z_\mathrm{N}<\infty\qquad \text{(A.7.41)}$$
$$Q_2(Z)=\frac{1}{4}(3Z^2-1)\ln\frac{Z+1}{Z-1}-\frac{3}{2}Z=\frac{2}{15Z^3}+\frac{4}{35Z^5}+\dots.$$

In the range $4\mu^2<t$ the value of Z_N is purely imaginary and Q_0 can be expressed as follows ($q_-\to-\mathrm{i}q_t$)

$$\mathrm{i}Q_0(Z_\mathrm{N})=-\arctan|Z_\mathrm{N}|+\pi/2.\qquad \text{(A.7.42)}$$

The pole term projections read

$$f^0_{+\mathrm{N}}(t)=\frac{g^2}{4\pi}m\{Z_\mathrm{N}Q_0(Z_\mathrm{N})-1\};\quad f^1_{+\mathrm{N}}(t)=\frac{g^2}{4\pi}\frac{m}{p_-q_t}\{Z_\mathrm{N}Q_0(Z_\mathrm{N})-1\}\,Z_\mathrm{N}\mathrm{i};$$
$$f^1_{-\mathrm{N}}(t)=\frac{g^2}{4\pi}\frac{1}{\sqrt{2}p_-q_t}\{\mathrm{i}Q_0(Z_\mathrm{N})(1-Z_\mathrm{N}^2)+\mathrm{i}Z_\mathrm{N}\};\tag{A.7.43}$$

7.4.3 The Froissart-Gribov projection at $t=4\mu^2$

If $t\to 4\mu^2$, the variable Z goes to infinity and the Q_J-functions can be replaced by the leading term of the asymptotic expansion

$$Q_J(Z)=\frac{J!}{(2J+1)!!}\frac{1}{Z^{J+1}}+\ldots;\quad (2J+1)!!=1\cdot 3\cdot 5\ldots(2J+1).\tag{A.7.44}$$

We obtain for the real partial waves

$$f^0_+(4\mu^2)=\frac{m^2-\mu^2}{4\pi}C^+(0,4\mu^2);\quad f^0_{+\mathrm{N}}(4\mu^2)=0;$$
$$f^J_+(4\mu^2)=\frac{g^2}{4\pi}m\frac{J!(\mu^2/2)^{-J}}{(2J+1)!!}+\frac{J!}{(2J+1)!!}\frac{m^2-\mu^2}{2\pi^2m^J}\int_{\nu_\mathrm{th}}^{\infty}\mathrm{d}\nu\frac{\operatorname{Im}C^\pm(\nu,4\mu^2)}{\nu^{J+1}};\quad J>0;\tag{A.7.45}$$
$$f^J_-(4\mu^2)=\frac{g^2}{4\pi}\frac{(J-1)!}{(2J+1)!!}\sqrt{J(J+1)}\left\{\frac{\mu^2}{2}\right\}^{-J}+\frac{\sqrt{J(J+1)}\,(J-1)!}{2\pi^2(2J+1)!!m^{J-1}}\int_{\nu_\mathrm{th}}^{\infty}\mathrm{d}\nu\frac{\operatorname{Im}B^\pm(\nu,4\mu^2)}{\nu^J};\quad J>0.$$

In these formulas, it is understood that C^+, B^+ have to be inserted for even J, and C^-, B^- for odd J.

7.5 t-channel partial wave dispersion relations

The analytic properties of the t-channel partial waves $f^J_\pm(t)$ follow from those of the invariant amplitudes and from the Froissart-Gribov projection formulas (Sect. A.7.4). The partial wave amplitudes have a right hand cut from $t=4\mu^2$ to ∞ and a left hand cut from $-\infty$ to $t=t_\mathrm{nc}=4\mu^2-\mu^4/m^2=3.98\mu^2$. Part of the left hand cut $(0<t<t_\mathrm{nc})$ belongs only to the projection of the nucleon pole term $f^J_{\pm\mathrm{N}}(t)$. It is not present in the amplitudes

$$\tilde f^J_\pm(t)\equiv f^J_\pm(t)-f^J_{\pm\mathrm{N}}(t).\tag{A.7.46}$$

The dispersion relation reads

$$f^J_\pm(t)=\frac{1}{\pi}\int_{-\infty}^{t_\mathrm{nc}}\frac{\operatorname{Im}f^J_\pm(t'+\mathrm{i}0)}{t'-t}\mathrm{d}t'+\frac{1}{\pi}\int_{4\mu^2}^{\infty}\frac{\operatorname{Im}f^J_\pm(t'+\mathrm{i}0)}{t'-t}\mathrm{d}t'.\tag{A.7.47}$$

For real values of t, one obtains $\operatorname{Re}f^J_\pm(t)$, by taking the principal value integrals.

The asymptotic behaviour of $f^J_\pm(t)$ follows from Eqs. (A.3.56) and (A.3.57), using the fact that $S^J_\pm(t)$ is bounded by the unitarity condition. It turns out that for $J>0$ the decrease is at least as fast as

$$f^J_+(t)\propto t^{-J+1/2};\quad f^J_-(t)\propto t^{-J}\quad\text{as}\quad t\to\infty.\tag{A.7.48}$$

The combinations $\Gamma_1(t)$, $\Gamma_2(t)$ of Eq. (A.3.52) go to zero at least as

$$\Gamma_1(t)\propto t^{-1};\quad \Gamma_2(t)\propto t^{-3/2}\quad\text{as}\quad t\to\infty.\tag{A.7.49}$$

In the remaining case of $f^0_+(t)$ a subtraction is required, but it is not necessary to introduce a subtraction constant. The dispersion relation is written for $f^0_+(t)/p_t^2$, which remains finite at $t=4m^2$, where the denominator vanishes [see Eq. (A.3.75)].

The behaviour of the partial waves near $t=4\mu^2$ and $t=4m^2$ is discussed in Sects. A.3.5.3 and A.3.5.4, respectively.

In applications of the partial wave dispersion relations, the amplitude is frequently represented as a ratio ("N/D-method")

$$f_{\pm}^{J}(t) \equiv \frac{N_{\pm}^{J}(t)}{D^{J}(t)}, \tag{A.7.50}$$

where $D^J(t)$ has only the right hand cut of $f_{\pm}^J(t)$ and $N_{\pm}^J(t)$ has only the left hand cut. However, one should note that, in contrast to the case of pion-pion scattering, it is not justified to assume that the function $N_{\pm}^J(t)$ is slowly variable on the nearby part of the right hand cut. The reason is that the branch point of the left hand cut ($t=t_{nc}=3.98\mu^2$) lies now very near to that of the right hand cut ($t=4\mu^2$).

Instead of the exact D-function, one usually considers a truncation of its phase representation which can be evaluated, using the experimental pion-pion phase shifts $\delta_J^I(t)$

$$D^J(t) = \exp\left\{-\frac{1}{\pi}\int_{4\mu^2}^{t_{\max}} \frac{dt'}{t'-t}\,\delta_J^I(t')\right\}, \tag{A.7.51a}$$

where $I=0$ and 1 for even and odd values of J, respectively. The integral is taken over the t-range in which the contribution of the 4π intermediate states to the unitarity relation (Sect. A.5.2) is either zero or negligible.

Since we are free to multiply $N_{\pm}^J(t)$ and $D^J(t)$ by a factor without affecting $f_{\pm}^J(t)$, we can perform a subtraction in the dispersion integral which leads to the normalization $D^J(0)=1$ and to a suppression of the contributions from the range $t>t_{\max}$

$$D^J(t) = \exp\left\{-\frac{t}{\pi}\int_{4\mu^2}^{t_{\max}} \frac{dt'}{t'(t'-t)}\,\delta_J^I(t')\right\}. \tag{A.7.51b}$$

In the special case $J=1$, one can take $D^1(t) \propto F_\pi^{-1}(t)$, where F_π is the pion form factor [see Eqs. (A.11.32) and (A.11.27)].

7.6 The Chew-Mandelstam model for $\pi\pi$-partial wave amplitudes

In this subsection we mention some results for $\pi\pi$-partial waves, which are needed for the discussion of $\pi\pi N\bar{N}$-partial waves and of the pion form factor (Sects. 2.4.6, 2.5.2, and A.11.2).

The partial wave expansion of the invariant $\pi\pi$-scattering amplitude is usually written

$$F^I(s,\cos\theta) = \sum_{\ell=0}^{\infty} (2\ell+1) A_\ell^I(s) P_\ell(\cos\theta); \quad A_\ell^I(s) \equiv \frac{\sqrt{s}}{2q_{\pi\pi}} T_\ell^I(s);$$
$$q_{\pi\pi}^2 = \frac{1}{4}(s-4\mu^2); \quad T_\ell^I(s) = \frac{1}{2i}\{\eta_\ell^I \exp(2i\delta_\ell^I)-1\}; \quad a_\ell^I = \lim_{s\to 4\mu^2} (\delta_\ell^I / q_{\pi\pi}^{2\ell+1}); \tag{A.7.52}$$

where I denotes the isospin, $\sqrt{s}=M_{\pi\pi}$ the total energy and θ the scattering angle, both in the c.m. system. The normalization of the amplitude follows from the expression for the elastic differential cross section

$$\frac{d\sigma}{d\Omega_{c.m.}}(\pi\pi\to\pi\pi) = \frac{16}{s}|F|^2. \tag{A.7.53}$$

The analytic properties of the partial wave amplitude are derived from the Mandelstam representation. $A_\ell^I(s)$ is regular in the physical sheet of the s-plane, except for cuts at $s<0$ and $s>4\mu^2$.

In the following we replace s by t, because the formulas will be applied to t-channel exchanges ($q_{\pi\pi}\to q_t$).

For several applications, it is useful to have a parametrization of the partial wave which describes correctly the analytic properties above and below threshold ($t=4\mu^2$) and has a resonance at a larger t-value. We consider only the p-wave amplitude in the elastic range ($\eta_1^1=1$)

$$h_1^1(t) \equiv \frac{2}{q_t^2} A_1^1(t) = \frac{\sqrt{t}}{q_t^3} e^{i\delta_1^1(t)} \sin\delta_1^1(t), \tag{A.7.54}$$

where $q_t = \sqrt{t/4-\mu^2}$ is the c.m. momentum.

The CHEW-MANDELSTAM parametrization reads

$$h_1^1(t)=\frac{N(t)}{D(t)}=\frac{\gamma}{A-Bt+\gamma h(t)}; \quad h_1^1(0)=\frac{\gamma}{t_\varrho}, \qquad \text{(A.7.55)}$$

where $N(t)$ is chosen to be a constant. $h(t)$ describes the analytic structure near $t=4\mu^2$. For $t<4\mu^2$ we use $q_-=\sqrt{\mu^2-t/4}=-\mathrm{i}q_t$

$$h(t)=\frac{1}{\pi}\left[-\mu^2+\frac{t}{3}-t^2\int_{4\mu^2}^{\infty}\frac{\left(\frac{t'}{4}-\mu^2\right)^{3/2}\mathrm{d}t'}{t'^{5/2}(t'-t)}\right]=\begin{cases}\frac{2}{\pi}\frac{q_t^3}{\sqrt{t}}\ln\left(\frac{q_t}{\mu}+\frac{\sqrt{t}}{2\mu}\right)-\mathrm{i}\frac{q_t^3}{\sqrt{t}}; & t>4\mu^2;\\ -\frac{2}{\pi}\frac{q_-^3}{\sqrt{t}}\arctan\frac{\sqrt{t}}{2q_-}; & 0<t<4\mu^2;\\ -\frac{2}{\pi}\frac{q_-^3}{\sqrt{-t}}\ln\left(\frac{q_-}{\mu}+\frac{\sqrt{-t}}{2\mu}\right); & t<0.\end{cases} \qquad \text{(A.7.56)}$$

The parametrization can be written as a generalized effective range formula

$$\mathrm{Re}\frac{1}{h_1^1(t)}=\frac{q_t^3}{\sqrt{t}}\cot\delta_1^1=\frac{1}{\gamma}\{A-Bt\}+\mathrm{Re}\,h(t)=N^{-1}\,\mathrm{Re}\,D. \qquad \text{(A.7.57)}$$

Taking this result at $t=4\mu^2$ and using $h(4\mu^2)=0$, one obtains an expression for the scattering length

$$a_1^1=\frac{\gamma}{2\mu(A-4\mu^2B)}=\frac{N}{2\mu\,\mathrm{Re}\,D(4\mu^2)}. \qquad \text{(A.7.58)}$$

The resonance position follows from the zero of $\mathrm{Re}(1/h_1^1)$. The two parameters A/γ and B/γ can be fixed for instance by the values of the scattering length and of the resonance position.

Since the D-function has a zero at a very large negative t-value ($t_0\approx-10^7m_\varrho^2$), it should have been divided by $(1-t/t_0)$, but this factor can be ignored, because t_0 is so large.

Table A.7.3. Parameters of the CHEW-MANDELSTAM model for $h_1^1(t)$. m_ϱ and Γ_ϱ almost agree with an average over all reactions [82 R ROOS: $m_\varrho=769\pm3$ MeV, $\Gamma_\varrho=154\pm5$ MeV]. The prediction from the linearity assumption of soft pion theory and current algebra is sometimes called the "WEINBERG value".

Input	$m_\varrho=775$ MeV	$\Gamma_\varrho=149$ MeV from fit to F_π, Sect. A.11.2
Parameters	$A/\gamma=0.352\ \mathrm{GeV}^2$	$B/\gamma=0.695$;
Prediction	$a_1^1=0.033\,\mu^{-3}$	$a_1^1=1/(12\pi\mu f_\pi^2)=0.030\,\mu^{-3}$ from soft pion theory

If a_1^1, Eq. (A.7.58), is equated to the result derived from soft pion theory, (Table A.7.3), one obtains a generalized "KSFR-relation"

$$f_\varrho^2=\frac{m_\varrho^2}{f_\pi^2}\left(1-\frac{4\mu^2}{m_\varrho^2}\right)\left(1-\frac{2q_\varrho^4}{3\pi f_\pi^2}\frac{\mathrm{d}}{\mathrm{d}t}\frac{\mathrm{Re}\,h}{q_t^2}\Big|_{t=t_\varrho}\right)^{-1}\left(1+\frac{\Gamma_\varrho}{m_\varrho}d\right)^{-1}, \qquad \text{(A.7.59)}$$

where q_ϱ is the value of q_t at $t=t_\varrho$. f_ϱ^2 is introduced by assuming complete ϱ-dominance and the formalism of 67L KROLL-LEE-ZUMINO

$$\frac{f_\varrho^2}{4\pi}=\frac{f_{\varrho\pi\pi}^2}{4\pi}=\frac{3}{2}\gamma. \qquad \text{(A.7.60)}$$

d is defined in Eq. (A.11.36). Its numerical value is $d=0.48$.

8 Low energy models

8.1 Nucleon and resonance exchanges

In models for πN amplitudes at low energies or in the unphysical region below threshold, the most important contributions come from the exchange of nucleons and of Δ(1232) and ϱ-resonances. In the following we list the usual expressions for the exchange contributions (Fig. A.8.1) and describe the theoretical constraints which show, how these terms can be added without double counting.

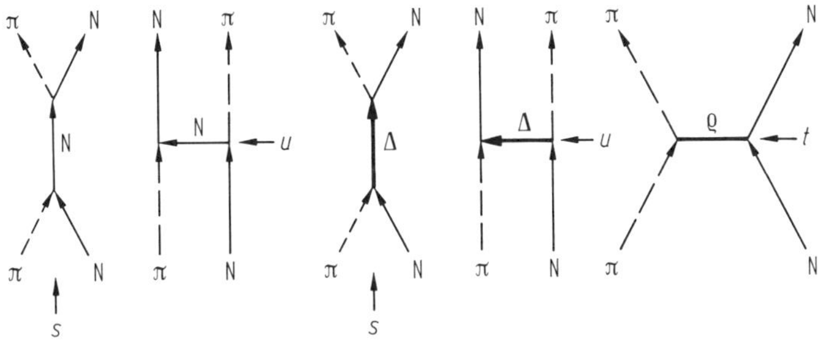

Fig. A.8.1. Feynman diagrams for the exchange of nucleons, isobars and rho-mesons.

8.1.1 Nucleon exchange

Nucleon exchange is described by the formulas obtained from second order Feynman graphs, inserting the experimental value of the coupling constant. If *pseudoscalar* coupling is used, the result agrees with the pole terms of the fixed-t dispersion relations for the amplitudes $A^{\pm}$ and $B^{\pm}$, which are denoted by $B_N^{\pm}(\nu, t)$ and listed in Eq. (A.6.30). There is no contribution to the amplitudes $A^{\pm}$.

The nucleon Born terms resulting from 2nd order Feynman graphs and *pseudovector* coupling are

$$A^+_{\mathrm{N,pv}} = \frac{g^2}{m}; \quad B^+_{\mathrm{N,pv}} = B^+_{\mathrm{N}} = \frac{g^2}{m}\frac{\nu}{\nu_B^2-\nu^2}; \quad \nu_B = \frac{t-2\mu^2}{4m};$$
$$C^+_{\mathrm{N,pv}} = \frac{g^2}{4m^3}\frac{(\mu^2-t/2)^2(1-t/4m^2)+\nu^2 t}{(\nu_B^2-\nu^2)(1-t/4m^2)}; \quad D^+_{\mathrm{N,pv}} = \frac{g^2}{m}\frac{\nu_B^2}{\nu_B^2-\nu^2}; \tag{A.8.1}$$

$$A^-_{\mathrm{N,pv}} = 0; \quad B^-_{\mathrm{N,pv}} = B^-_{\mathrm{N}} - \frac{g^2}{2m^2} = \frac{g^2}{m}\left\{\frac{\nu_B}{\nu_B^2-\nu^2} - \frac{1}{2m}\right\};$$
$$C^-_{\mathrm{N,pv}} = \frac{g^2}{2m^2}\frac{\nu^2-\mu^2+\frac{t}{2}-\nu_B^2}{(\nu_B^2-\nu^2)(1-t/4m^2)}\nu; \quad D^-_{\mathrm{N,pv}} = \frac{g^2}{2m^2}\frac{\nu^2-\mu^2+\frac{t}{2}-\nu_B^2}{\nu_B^2-\nu^2}\nu. \tag{A.8.2}$$

In contrast to the pseudoscalar Born terms, the pseudovector Born terms $C^{\pm}_{\mathrm{N,pv}}$ and $D^{\pm}_{\mathrm{N,pv}}$ are very small at threshold.

In general, the effect of the neutron-proton mass difference can be neglected, as it was done in the above formulas. However, it can be appreciable in special cases and therefore, we add the exact expressions.

We consider only $\pi^{\pm}$p elastic scattering. Therefore, all kinematic variables have to be calculated with the proton mass. However, the pole position is now given by $s = m_{\mathrm{n}}^2$ and $u = m_{\mathrm{n}}^2$. This is taken into account by a new definition for ν_B

$$\nu_B(t) \equiv \frac{1}{2m_{\mathrm{p}}}(m_{\mathrm{n}}^2 - m_{\mathrm{p}}^2 - \mu^2 + t/2). \tag{A.8.3}$$

The nucleon Born terms for the isospin even and odd combinations and pseudoscalar coupling read

$$A_N^+(\nu,t) = -\frac{g^2}{m_p}\frac{\nu_B}{\nu_B^2-\nu^2}\Delta m; \qquad A_N^-(\nu,t) = -\frac{g^2}{m_p}\frac{\nu}{\nu_B^2-\nu^2}\Delta m; \tag{A.8.4}$$

$$B_N^+(\nu,t) = \frac{g^2}{m_p}\frac{\nu}{\nu_B^2-\nu^2}; \qquad B_N^-(\nu,t) = \frac{g^2}{m_p}\frac{\nu_B}{\nu_B^2-\nu^2}; \qquad A_N^\pm = -\Delta m B_N^\mp; \tag{A.8.5}$$

$$D_N^+(\nu,t) = \frac{g^2}{m_p}\frac{\nu^2-\nu_B\Delta m}{\nu_B^2-\nu^2}; \qquad D_N^-(\nu,t) = \frac{g^2}{m_p}\frac{\nu_B-\Delta m}{\nu_B^2-\nu^2}\nu, \tag{A.8.6}$$

where $\Delta m = m_n - m_p = 1.293$ MeV.

It is remarkable that the tiny mass difference of 0.14% leads to a large change for the location of the point on the t-axis in the Mandelstam plane, where the two pole lines $\nu = \pm\nu_B$ intersect each other: $t = 1.75\mu^2$, instead of $t = 2\mu^2$, if the mass difference is neglected. The pseudovector Born terms read

$$A_{N,pv}^+ = \frac{g^2}{m_p}\left\{\frac{2m_p}{m_n+m_p} - \frac{\nu_B\Delta m}{\nu_B^2-\nu^2}\right\}; \qquad A_{N,pv}^- = A_N^-; \tag{A.8.7}$$

$$B_{N,pv}^+ = \frac{g^2}{m_p}\frac{\nu}{\nu_B^2-\nu^2} = B_N^+; \qquad B_{N,pv}^- = B_N^- - \frac{2g^2}{(m_n+m_p)^2}. \tag{A.8.8}$$

8.1.2 Isobar exchange

i) Isobar exchange contributions to s-channel amplitudes

The exchange of the isobar $\Delta(1232)$ in the s-and u-channels is described by the 2nd order Feynman graph. To evaluate this graph one needs to know the propagator and the interaction.

The most general spin 3/2 propagator reads

$$P_{\mu\nu} = \frac{1}{-i\gamma\cdot p + M}\left[\delta_{\mu\nu} - \frac{1}{3}\gamma_\mu\gamma_\nu + \frac{i}{3M}(\gamma_\mu p_\nu - \gamma_\nu p_\mu) + \frac{2}{3M^2}p_\mu p_\nu\right]$$
$$+ \frac{i}{6M^2}\left\{\frac{2(A^*+1)}{2A^*+1}\gamma_\mu p_\nu + \frac{2(A+1)}{2A+1}\gamma_\nu p_\mu + \left|\frac{A+1}{2A+1}\right|^2\gamma_\mu(\gamma\cdot p)\gamma_\nu - iM\frac{2AA^*+A+A^*}{|2A+1|^2}\gamma_\mu\gamma_\nu\right\}. \tag{A.8.9}$$

M is the isobar mass and A is a parameter, which determines the spin 1/2 admixture to the spin 3/2-field. The A-dependent terms do not contribute on the isobar mass shell.

The interaction term in the Lagrangian is chosen such that the result for the exchange amplitude does not depend on A, but it depends on another parameter Z

$$L_{int} = g_\Delta \bar{\Delta}^\mu \Theta_{\mu\nu} N \partial^\nu \pi + \text{h.c.}; \qquad \Theta_{\mu\nu} = \delta_{\mu\nu} + \{\tfrac{1}{2}(1+4Z)A + Z\}\gamma_\mu\gamma_\nu. \tag{A.8.10}$$

Δ^μ is the spin 3/2-field, N the nucleon field and π the pion field. The isospin dependence is suppressed.

The Δ-exchange amplitude consists of a pole term, which is unique, and a non-pole term which depends on Z. The pole term reads

$$A_{\Delta,p}^\pm(\nu,t) = \begin{pmatrix}2\\-1\end{pmatrix}\frac{g_\Delta^2}{18m}(\alpha_1+\alpha_2 t)\left(\frac{1}{\nu_\Delta-\nu} \pm \frac{1}{\nu_\Delta+\nu}\right);$$
$$B_{\Delta,p}^\pm(\nu,t) = \begin{pmatrix}2\\-1\end{pmatrix}\frac{g_\Delta^2}{18m}(\beta_1+\beta_2 t)\left(\frac{1}{\nu_\Delta-\nu} \mp \frac{1}{\nu_\Delta+\nu}\right), \tag{A.8.11}$$

where the upper and lower coefficients in the brackets belong to the isospin index $\pm$ and

$$\alpha_1 = 3(M+m)q_\Delta^2 + (M-m)(E_\Delta+m)^2; \qquad \alpha_2 = \tfrac{3}{2}(M+m);$$
$$\beta_1 = 3q_\Delta^2 - (E_\Delta+m)^2; \qquad \beta_2 = \tfrac{3}{2}; \tag{A.8.12}$$
$$2m\nu_\Delta = M^2 - m^2 - \mu^2 + t/2 = 2m\omega_\Delta + t/2.$$

An index Δ indicates that the variable has to be taken at $W = M$, except for g_Δ which denotes the $\pi N\Delta$ coupling constant.

Isobar pole terms in dispersion theory

The isobar pole term has the same kinematical structure in a Feynman graph calculation and in an approximation to the *unsubtracted* fixed-t dispersion relations for the amplitudes $A^\pm$, $B^\pm$, in which only the P33 contribution to the imaginary part is taken into account and a narrow resonance approximation is used for $\mathrm{Im}\,T_{33} \equiv \mathrm{Im}\,T_{1+}^{3/2}$

$$\frac{\mathrm{Im}\,T_{33}}{q^3} \to \frac{g_\Delta^2}{4\pi}\frac{E+m}{2m}\frac{\pi}{3}\delta(\omega-\omega_\Delta).$$

$$\mathrm{Im}\,A_\Delta^\pm(v,t) \to \begin{pmatrix}2\\-1\end{pmatrix} g_\Delta^2 \frac{\alpha_1+\alpha_2 t}{18m}\pi\delta(\omega-\omega_\Delta); \qquad \mathrm{Im}\,B_\Delta^\pm(v,t) \to \begin{pmatrix}2\\-1\end{pmatrix} g_\Delta^2 \frac{\beta_1+\beta_2 t}{18m}\pi\delta(\omega-\omega_\Delta). \tag{A.8.13}$$

A comparison of Eq. (A.8.11) with the pole term approximation to the dispersion relation shows that g_Δ^2 is determined by

$$\frac{g_\Delta^2}{4\pi} = \frac{3}{\pi}\int_\mu^{\omega_{\max}} \frac{2m}{E+m}\frac{\mathrm{Im}\,T_{33}}{q^3}\,\mathrm{d}\omega. \tag{A.8.14}$$

The cut-off is needed, because there exist further P33 resonances at higher energies.

One can ask for an "effective Δ-pole term" approximation, in which the parameters g_Δ^2 and the isobar mass M are adjusted in such a way that the Δ-pole terms Eq. (A.8.11) agree as well as possible with the exact evaluation of the P33 contribution to the fixed-t dispersion relation (A.6.30) on the left wing of the resonance structure ($0<\omega<\mu$, see Fig. A.8.2).

The result is somewhat different for the A- and B-amplitudes and it depends slightly on t. An overall average for small $|t|$ is

$$\frac{g_\Delta^2}{4\pi} = 13.5\,\mathrm{GeV}^{-2} = 0.26\mu^{-2} \pm 5\%; \qquad M = 1219\,\mathrm{MeV} \pm 1.5\%. \tag{A.8.15}$$

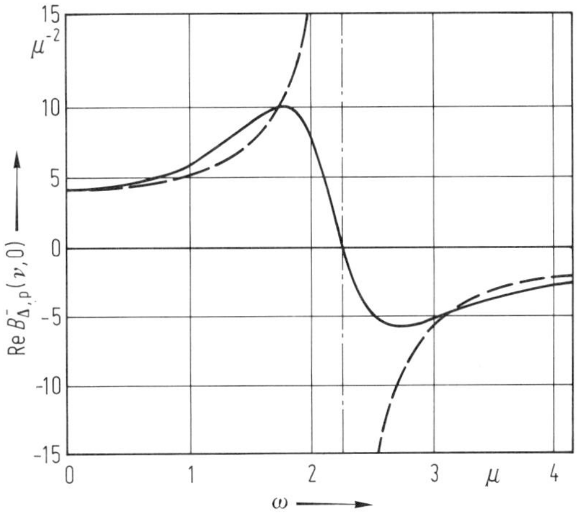

Fig. A.8.2. Effective pole approximation for Δ-exchange. The figure shows $\mathrm{Re}\,B_\Delta^-(\omega,t=0)$ as calculated from the P33-contribution to $\mathrm{Im}\,B_\Delta^-$ in the dispersion relation Eq. (A.6.30) without the nucleon pole term. The dashed line is the pole approximation of Eq. (A.8.11). The parameters, Eq. (A.8.16), belong to an average over fits for the amplitudes A and B at small $|t|$ in the range $0<\omega<\mu$.

It is interesting to note that the real part of the complex Δ-pole location lies even farther below the conventional mass value 1232 MeV (see Table 2.4.1.2). The value for g_Δ^2 agrees with that obtained from Eq. (A.8.14). The parameters of Eq. (A.8.12) for this effective pole term approximation are (Fig. A.8.2)

$$\alpha_1 = 486\mu^3; \quad \beta_1 = -178\mu^2; \quad \alpha_2 = 23.2\mu; \quad \beta_2 = 1.5;$$
$$M = 8.74\mu; \quad E_\Delta + m = 13.62\mu; \quad q_\Delta^2 = 2.40\mu^2; \quad \omega_\Delta = 2.25\mu. \tag{A.8.16}$$

If the value of the coupling constant is derived from the Δ-width, using the interaction term of the Lagrangian in 1st order, one obtains with $\Gamma = 115$ MeV

$$\frac{g_\Delta^2}{4\pi} = \Gamma(M)\frac{3M}{q_\Delta^3(E_\Delta+m)} = 19\,\mathrm{GeV}^{-2}. \tag{A.8.17}$$

This result is higher by 40% than the earlier one. It should not be used in applications of pole term formulas near or below threshold.

Some authors define a dimensionless coupling constant

$$\frac{G^2}{4\pi} = \frac{g_\Delta^2}{4\pi}\frac{M}{3}(E_\Delta+m) = \frac{g_\Delta^2}{4\pi}\frac{1}{6}\{(M+m)^2-\mu^2\}, \tag{A.8.18}$$

The factor of $\frac{g_\Delta^2}{4\pi}$ on the right hand side is 0.78 GeV².

The Δ-pole term contributions to the coefficients of the expansion (A.3.59) are

$$a^+_{00}=\frac{2g^2_\Delta}{9m}\frac{\alpha_1}{\omega_\Delta};\quad a^+_{10}=\frac{a^+_{00}}{\omega^2_\Delta};\quad a^+_{01}=\left(\frac{\alpha_2}{\alpha_1}-\frac{1}{4m\omega_\Delta}\right)a^+_{00};\quad a^+_{11}=\left(\frac{\alpha_2}{\alpha_1}-\frac{3}{4m\omega_\Delta}\right)\frac{a^+_{00}}{\omega^2_\Delta};$$
$$b^+_{00}=\frac{2g^2_\Delta}{9m}\frac{\beta_1}{\omega^2_\Delta};\quad b^+_{10}=\frac{b^+_{00}}{\omega^2_\Delta};\quad b^+_{01}=\left(\frac{\beta_2}{\beta_1}-\frac{1}{2m\omega_\Delta}\right)b^+_{00};\quad b^+_{11}=\left(\frac{\beta_2}{\beta_1}-\frac{1}{m\omega_\Delta}\right)\frac{b^+_{00}}{\omega^2_\Delta}. \tag{A.8.19}$$

In order to obtain the coefficients of a^-_{mn}, b^-_{mn}, one has to multiply by a factor $-1/2$.

Non-pole contributions

Next, we list the expressions for the non-pole contributions to the Δ-exchange amplitude. They are non-resonant and at most linear in t, i.e. they contribute only to S- and P-waves

$$A^+_{\Delta,\mathrm{np}}=\frac{-4g^2_\Delta}{9M}\left\{(E_\Delta+m)(2M-m)+\left(2+\frac{m}{2M}\right)\mu^2-(2\mu^2-t)Y\right\};$$
$$B^+_{\Delta,\mathrm{np}}/\nu=-\frac{16g^2_\Delta}{9M}\frac{m}{M}Z^2;\qquad A^-_{\Delta,\mathrm{np}}/\nu=-\frac{8g^2_\Delta}{9}\frac{m}{M}Y; \tag{A.8.20}$$
$$B^-_{\Delta,\mathrm{np}}=\frac{g^2_\Delta}{9}\left\{\left(1+\frac{m}{M}\right)^2+\frac{8m}{M}Y+\frac{4}{M^2}\left[\left(\mu^2-\frac{t}{2}\right)Z^2-\mu^2 Z\right]\right\};$$

where

$$Y=\left(2+\frac{m}{M}\right)Z^2+\left(1+\frac{m}{M}\right)Z. \tag{A.8.20a}$$

Our formulas are written for real Z, although this parameter could be complex-valued.

There is a tendency for a cancellation between the pole terms and non-pole terms of the amplitudes $C^\pm$ in the low energy region. For instance, the total Δ-exchange contribution to $C^+(0,2\mu^2)=A^+(0,2\mu^2)$ is very small and independent of Z

$$A^+_{\Delta,\mathrm{p}}(0,2\mu^2)+A^+_{\Delta,\mathrm{np}}(0,2\mu^2)=\frac{2}{9}g^2_\Delta\frac{2W+m}{s(s-m^2)}=0.0075\mu^{-1}=3\cdot10^{-4}A^+_{\Delta,\mathrm{p}}(0,2\mu^2). \tag{A.8.21}$$

Some authors used Δ-exchange formulas derived from another treatment, in which the spin-1/2 admixture is described by a parameter c. The results agree with the above formulas, if the relation

$$Z=-\frac{1}{2}\frac{c}{2c+1} \tag{A.8.22}$$

exists between the parameters.

$c=-1/3$ corresponding to $Z=1/2$ gives the exchange amplitudes obtained from the Rarita-Schwinger formalism. In this case, the pole and non-pole contributions to the $C^\pm$-amplitudes cancel exactly at threshold

$$C^+_{\Delta,\mathrm{p}}(\mu,0)=-C^+_{\Delta,\mathrm{np}}(\mu,0)=\frac{8}{9}g^2_\Delta m\frac{E_\Delta+m}{2m}\left(2-\frac{m}{M}\right);$$
$$C^-_{\Delta,\mathrm{p}}(\mu,0)=-C^-_{\Delta,\mathrm{np}}(\mu,0)=-\frac{4}{9}g^2_\Delta m\frac{E_\Delta+m}{2m}\frac{1}{M}. \tag{A.8.23}$$

Another popular choice is $c=-1$, corresponding to $Z=-1/2$. Theoretical arguments have been presented in favour of $Z=\pm1/2$, but a completely satisfactory prescription for the calculation of the Δ-exchange amplitudes does not yet exist.

ii) Isobar exchange contributions to *t*-channel amplitudes

In applications of $\pi\pi N\bar{N}$ partial wave amplitudes (Sect. 2.5), it is of interest to know the Δ-exchange contributions in the pole term approximation. We start from the narrow resonance approximation for $\mathrm{Im}\,B^\pm$, Eq. (A.8.13), and for $\mathrm{Im}\,C^\pm$

$$\mathrm{Im}\,C^\pm(\nu,t)=\begin{pmatrix}2\\-1\end{pmatrix}g^2_\Delta\frac{W(E+m)}{18(m^2-t/4)}\left\{2q^2+t\left(2-\frac{E}{m}\right)-\frac{3t^2}{8m(E+m)}\right\}\pi\delta(\nu-\nu_B), \tag{A.8.24}$$

and insert it into the FROISSART-GRIBOV representation (A.7.32), (A.7.33). The result for the $J=1$ partial waves reads

$$\mathrm{Re} f^1_{+\Delta}(t) = \frac{g_\Delta^2}{4\pi}\Phi_+(t) Z_\Delta^2 Q_1(Z_\Delta)\mu\,; \qquad Z_\Delta = \frac{m\nu_\Delta}{p_- q_-} = \frac{m}{p_- q_-}(\omega_\Delta + t/4m)\,;$$
$$\mathrm{Re} f^1_{-\Delta}(t) = \frac{g_\Delta^2}{4\pi}\Phi_-(t) Z_\Delta \{Q_0(Z_\Delta) - Q_2(Z_\Delta)\}\,; \tag{A.8.25}$$

where

$$\Phi_+(t) = -\frac{M(E_\Delta+m)}{9m\mu\omega_\Delta^2}(1+t/4m\omega_\Delta)^{-2}\left\{2q_\Delta^2 + t\left(2-\frac{E_\Delta}{m}\right) - \frac{3t^2}{8m(E_\Delta+m)}\right\};$$
$$\Phi_-(t) = -\frac{\sqrt{2}}{27}\frac{\beta_1+\beta_2 t}{m\omega_\Delta}\left(1-\frac{t}{4m\omega_\Delta}\right)^{-1}. \tag{A.8.26}$$

The numerical values of the parameters are listed in Eqs. (A.8.15) and (A.8.16).

If the unsubtracted dispersion integral for $A^+(\nu,t)$ exists, $\mathrm{Re}\tilde{f}^0_+$ can be obtained from

$$\mathrm{Re}\tilde{f}^0_+(t) = \frac{p_-^2}{2\pi^2}\int_{\nu_{th}}^{\infty}\frac{d\nu'}{\nu'}\{\mathrm{Im}\,A^+(\nu',t) + \mathrm{Im}\,C^+(\nu',t)Q_1(Z')\}\,. \tag{A.8.27}$$

The isobar contribution to $f^0_+(t)$ follows by inserting Eqs. (A.8.13) and (A.8.24).

Equation (A.8.27) differs from Eq. (A.7.32b) only insofar as the subtraction function $C^+(0,t) \equiv A^+(0,t)$ is replaced by the unsubtracted dispersion relation (A.6.30) for $A^+(0,t)$.

8.1.3 Rho-exchange

The couplings of the ϱ-meson to the nucleon and to two pions are described by the following interaction terms in a phenomenological Lagrangian

$$L_{\varrho NN} = \bar{N}\left\{G^V_{NN\varrho}\gamma_\mu\varrho^\mu - \frac{1}{4m}G^T_{NN\varrho}\sigma_{\mu\nu}(\partial^\mu\varrho^\nu - \partial^\nu\varrho^\mu)\right\}\frac{1}{2}\boldsymbol{\tau}N\,;$$
$$L_{\varrho\pi\pi} = G_{\varrho\pi\pi}(\boldsymbol{\pi}\times\partial^\mu\boldsymbol{\pi})\boldsymbol{\varrho}_\mu\,. \tag{A.8.28}$$

From $L_{\varrho\pi\pi}$ one obtains in lowest order perturbation theory for the ϱ-decay width

$$\Gamma(\varrho\to\pi^+\pi^-) = \frac{g^2_{\varrho\pi\pi}}{4\pi}\frac{2}{3}\frac{q_\varrho^3}{m_\varrho^2}\,; \qquad q_\varrho^2 = \frac{m_\varrho^2}{4} - \mu^2\,. \tag{A.8.29}$$

The ϱ-exchange contribution to the pion-nucleon scattering amplitudes in 2nd order perturbation theory is given in Table A.8.1. It is zero for the isospin even combination.

Table A.8.1. ϱ-exchange contributions to πN scattering amplitudes. $m_\varrho = 0.769$ GeV, $t_\varrho = 0.591$ GeV$^2 = 30.3\ \mu^2$. $G^2_{\varrho\pi\pi}/4\pi \approx 2.6$, $(G^V_{NN\varrho})^2/4\pi \approx 2.2$, $G^T_{NN\varrho}/G^V_{NN\varrho} \approx 6.6$ (see Sect. 2.4.6).

Invariant amplitude	First term of t-channel p. w. expansion	vector coupling	tensor coupling
A^-/ν	$-12\pi\Gamma_2(t)$	0	$\frac{-G_{\varrho\pi\pi}G^T_{NN\varrho}}{t_\varrho - t}$
B^-	$\frac{12\pi}{\sqrt{2}}f^1_-(t)$	$\frac{G_{\varrho\pi\pi}G^V_{NN\varrho}}{t_\varrho - t}$	$\frac{G_{\varrho\pi\pi}G^T_{NN\varrho}}{t_\varrho - t}$
C^-/ν	$\frac{12\pi m}{m^2 - t/4}f^1_+(t)$	$\frac{G_{\varrho\pi\pi}G^V_{NN\varrho}}{t_\varrho - t}\frac{1}{1-t/4m^2}$	$\frac{G_{\varrho\pi\pi}G^T_{NN\varrho}}{t_\varrho - t}\frac{t}{4m^2 - t}$
$D^-/\nu = A^-/\nu + B^-$	$12\pi\Gamma_1(t)$	$\frac{G_{\varrho\pi\pi}G^V_{NN\varrho}}{t_\varrho - t}$	0

8.2 The current algebra model

It is convenient to work with the amplitudes $A(\nu, t)$, $B(\nu, t)$, and $D(\nu, t) \equiv A + \nu B$. Unless stated otherwise, we consider "on-shell" amplitudes, both pions having their physical mass μ.

A decomposition of the πN amplitudes is derived from the Ward identity, using current algebra and the PCAC relation

$$\pi^{\mathrm{i}}(\mathbf{x}) \equiv \frac{\sqrt{2}}{f_\pi \mu^2} \partial^\mu A_\mu^{\mathrm{i}}(\mathbf{x}); \quad \mathrm{i} = 1, 2, 3, \tag{A.8.30}$$

which defines the interpolating pion field $\pi^{\mathrm{i}}(\mathbf{x})$ by means of the divergence of the weak axial-vector current

$$\begin{aligned}
A^+(\nu, t) &= A^+_{\mathrm{N,pv}}(\nu, t) + A^+_{\mathrm{ax}}(\nu, t) + \frac{2}{f_\pi^2}\sigma_{\mathrm{NN}}(t); \\
B^+(\nu, t) &= B^+_{\mathrm{N,pv}}(\nu, t) + B^+_{\mathrm{ax}}(\nu, t); \\
D^+(\nu, t) &= D^+_{\mathrm{N,pv}}(\nu, t) + D^+_{\mathrm{ax}}(\nu, t) + \frac{2}{f_\pi^2}\sigma_{\mathrm{NN}}(t); \\
A^-(\nu, t) &= A^-_{\mathrm{N,pv}}(\nu, t) + A^-_{\mathrm{ax}}(\nu, t) - \frac{2}{f_\pi^2}F_{2\mathrm{v}}(t)\nu; \\
B^-(\nu, t) &= B^-_{\mathrm{N,pv}}(\nu, t) + B^-_{\mathrm{ax}}(\nu, t) + \frac{2}{f_\pi^2}G_{\mathrm{Mv}}(t); \\
D^-(\nu, t) &= D^-_{\mathrm{N,pv}}(\nu, t) + D^-_{\mathrm{ax}}(\nu, t) + \frac{2}{f_\pi^2}F_{1\mathrm{v}}(t)\nu.
\end{aligned} \tag{A.8.31}$$

$f_\pi = 132$ MeV $= 0.945\mu$ is the $\pi^+ \to \mu^+ + \nu$ decay constant.

The first two terms on the right hand side are derived from the axial-vector nucleon scattering amplitudes: $q^\lambda A^+_{\lambda\mu} q'^\mu = A^+_{\mathrm{N,pv}} + A^+_{\mathrm{ax}}$, etc.

The first term belongs to nucleon exchange. In order to express the coupling constant g_{A} of the axial-vector-NN vertex by the πNN coupling constant g, one uses the Goldberger-Treiman *relation*

$$g_{\mathrm{A}}(0) = \frac{f_\pi}{\sqrt{2}m} K(0) g. \tag{A.8.32}$$

$K(\mathbf{q}^2)$ is the form factor of the πNN vertex and $\mathbf{q}$ the 4-momentum of the pion. If we insert the values $g_{\mathrm{A}}(0) = 1.2546 \pm 0.0063$, $f_\pi = (0.945 \pm 0.001)\mu$, $g = 13.4 \pm 0.15$, we obtain $K(0) = 0.942 \pm 0.015$.

The main contribution to the second term belongs to the exchange of $\Delta(1232)$. It is frequently calculated as described in Sect. A.8.1.2.

The terms in the last column follow from the evaluation of the SU(2)×SU(2) charge-current equal-time commutators and are therefore sometimes denoted by a subscript "ETC". In the isospin odd case, they are known from the analysis of electron-nucleon scattering (see Sects. A.11.1 and 2.5.2).

Since it is difficult to calculate the contributions with the subscript "ax" in Eq. (A.8.31), it is important to know that they vanish in two cases, which belong to unphysical values of the pion mass squared q^2 and q'^2. We shall use the notation $A^+(\nu, \nu_B; q^2, q'^2)$; $\nu_B = (t - q^2 - q'^2)/4m$.

The Adler-Weisberger *relation* is a statement on the amplitude $\tilde{D}^- \equiv D^- - D^-_{\mathrm{N}}$:

$$\lim_{\nu \to 0} \tilde{D}^-(\nu, 0; 0, 0)/\nu = \frac{1 - g_{\mathrm{A}}^2(0)}{f_\pi^2} = -0.64\mu^{-2}. \tag{A.8.33}$$

Adlers *consistency condition* reads

$$A^+(0, 0; 0, \mu^2) = A^+(0, 0; \mu^2, 0) = \frac{g^2}{m} K(0). \tag{A.8.34}$$

The πN *sigma term* is defined by

$$A^+(0, 0; 0, 0) - \frac{g^2}{m} K^2(0) = -\frac{2}{f_\pi^2}\sigma_{\mathrm{NN}}(0). \tag{A.8.35}$$

In order to determine $\sigma_{\rm NN}$ from the on-shell amplitude in the unphysical region, one can assume a linear dependence of

$$A^+(0,0;q^2,q'^2)-\frac{g^2}{m}K(q^2)K(q'^2) \tag{A.8.36}$$

on q^2 and q'^2. Equations (A.8.34) and (A.8.35) then lead to

$$A^+(0,2\mu^2)-\frac{g^2}{m}\equiv A^+(0,0;\mu^2,\mu^2)-\frac{g^2}{m}=\frac{2}{f_\pi^2}\sigma_{\rm NN}(0)+O(\mu^4). \tag{A.8.37}$$

The first term on the left hand side is written in the earlier notation $A^+(v,t)$. The point $v=0$, $t=2\mu^2$ is sometimes called the "CHENG-DASHEN point".

It is interesting to compare Eq. (A.8.37) with the first Eq. (A.8.31), taken at $v=0$, $t=2\mu^2$, $v_B=0$. Current algebra and chiral invariance impose the following constraint on the on-shell amplitude $A^+_{\rm ax}$

$$A^+_{\rm ax}(v,t)=\frac{g^2}{m}\{(\mu/m)^4a+(2v_B/m)b+(v/m)^2c\}, \tag{A.8.38}$$

where a, b, c are non-singular functions of v and t or v_B (up to logarithms). The smallness of $A^+_{\rm ax}$ at $v=v_B=0$ is in agreement with the small value found in Eq. (A.8.21) for the isobar contribution.

The two expressions differ insofar as Eq. (A.8.31) contains in the present case $\sigma_{\rm NN}(2\mu^2)$, whereas Eq. (A.8.37) gives $\sigma_{\rm NN}(0)$ and this is the quantity which can be predicted from quark models. According to 71L PAGELS the difference is fairly large

$$\sigma_{\rm NN}(2\mu^2)-\sigma_{\rm NN}(0)=\frac{3}{16\pi}\left[\frac{g_{\rm A}}{f_\pi/\mu}\right]^2\mu+O(\mu^2\ln\mu^2)=15\,{\rm MeV}, \tag{A.8.39}$$

but this result is not generally accepted (Sect. 2.5.1.3).

In old papers, the isovector nucleon *form* factors are usually described in the *vector dominance model*, which leads to

$$F_{1v}(t)=\frac{0.5}{t_\varrho-t}\frac{t_\varrho}{f_\varrho}G^{\rm V}_{\rm NN\varrho};\quad F_{2v}(t)=\frac{0.5}{t_\varrho-t}\frac{t_\varrho}{f_\varrho}G^{\rm T}_{\rm NN\varrho};\quad G^{\rm T}_{\rm NN\varrho}/G^{\rm V}_{\rm NN\varrho}=3.7, \tag{A.8.40}$$

where t_ϱ/f_ϱ is the coupling constant for the $\gamma-\varrho$ transition (Fig. A.8.3). It is defined by the matrix element of the isovector electromagnetic current (see p. 586 for an improved version of the model)

$$\langle 0|j^{\rm v}_\mu(0)|\varrho\rangle=\frac{t_\varrho}{f_\varrho}\varepsilon_\mu;\quad \Gamma(\varrho\to{\rm e}^+{\rm e}^-)\approx\frac{\alpha^2}{3}m_\varrho\frac{4\pi}{f_\varrho^2};\quad \frac{f_\varrho^2}{4\pi}\approx 2.3. \tag{A.8.41}$$

ε_μ is the polarization vector of the ϱ. Assuming ϱ-dominance also for the pion form factor, one finds $f_\varrho=G_{\varrho\pi\pi}=G^{\rm V}_{\rm NN\varrho}$.

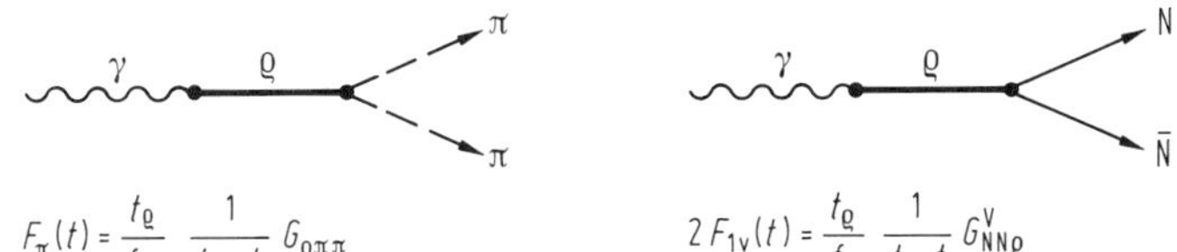

Fig. A.8.3. Pion and nucleon form factors in the ϱ-dominance approximation. t_ϱ/f_ϱ is the $\gamma-\varrho$ coupling constant. The normalization $F_\pi(0)=2F_{1v}(0)=1$ leads to $f_\varrho=G_{\varrho\pi\pi}=G^{\rm V}_{\rm NN\varrho}$.

The last term in Eq. (A.8.31) for D^- agrees with the direct evaluation of the ϱ-exchange contribution to πN scattering, Table A.8.1, if $f_\varrho=G_{\varrho\pi\pi}$ is inserted and the *KSFR relation*

$$f_\varrho^2 f_\pi^2=m_\varrho^2 \tag{A.8.42}$$

is fulfilled.

Furthermore, t-channel exchanges in the isospin even amplitudes were frequently described by "ε-exchange" or "σ-exchange". Nowadays, it is clear that simple pole or Breit-Wigner formulas for ϱ, ε or σ-exchange are not adequate for a quantitative treatment (Sects. 2.4.6 and 2.5.3).

8.3 The dispersion model

The aim is to rewrite the fixed-t dispersion relations (A.6.30), (A.6.34), (A.6.35) in such a way that the low energy πN amplitudes appear as sums of s, t, and u-channel exchanges plus a remainder. Since nucleon exchange is already taken into account by the nucleon pole terms, the problem is to find a suitable decomposition of the dispersion integrals.

As a first step, the unitarity relation is used in order to write $\mathrm{Im}\,C$ and $\mathrm{Im}\,B$ as sums of an "elastic contribution", which belongs to πN intermediate states (including charge-exchange), and the rest which is called the "inelastic contribution" in analogy to Eq. (A.5.19)

$$\mathrm{Im}\,C = \mathrm{Im}\,C_{\mathrm{el}} + \mathrm{Im}\,C_{\mathrm{in}}; \qquad \mathrm{Im}\,B = \mathrm{Im}\,B_{\mathrm{el}} + \mathrm{Im}\,C_{\mathrm{in}}. \tag{A.8.43}$$

In practice, this can be done by calculating the imaginary parts from partial wave expansions and summing over the first and the second terms of Eq. (A.5.20) separately. Then, $\mathrm{Re}\,C_{\mathrm{el}}$ and $\mathrm{Re}\,B_{\mathrm{el}}$ are defined by the dispersion integrals over $\mathrm{Im}\,C_{\mathrm{el}}$, $\mathrm{Im}\,B_{\mathrm{el}}$, respectively, omitting the nucleon pole terms. The elastic contributions to the real and imaginary parts are dominated by the P33 contributions, i.e. by $\Delta(1232)$-exchange in the s- and u-channels. In the narrow resonance approximation, one obtains the pole term formulas of Sect. A.8.1.2. The next step is different for the isospin odd and even amplitudes.

8.3.1 Isospin odd amplitudes

The decomposition of $\mathrm{Im}\,C^-$ and $\mathrm{Im}\,B^-$ is inserted into the Froissart-Gribov representations (A.7.32) and (A.7.33). We define

$$\begin{aligned}
\mathrm{Re}\,f^{1}_{+\,\mathrm{in}}(t) &= \frac{m}{2\pi^2 q_-^2}\int_{\nu_{\mathrm{th}}}^{\infty} d\nu'\,\mathrm{Im}\,C^-_{\mathrm{in}}(\nu',t)\,Q_1(Z'); \\
\mathrm{Re}\,f^{1}_{-\,\mathrm{in}}(t) &= \frac{m}{3\sqrt{2\pi^2}\,p_- q_-}\int_{\nu_{\mathrm{th}}}^{\infty} d\nu'\,\mathrm{Im}\,B^-_{\mathrm{in}}(\nu',t)\,\{Q_0(Z') - Q_2(Z')\},
\end{aligned} \tag{A.8.44}$$

multiply Eq. (A.8.44) by the same factor which occurs in the partial wave expansion, Table A.3.3, and subtract it from the fixed-t dispersion relation. The result is

$$\begin{aligned}
\mathrm{Re}\,C^-(\nu,t) &= C^-_{\mathrm{N}}(\nu,t) + \mathrm{Re}\,C^-_{\mathrm{el}}(\nu,t) + \frac{12\pi m\nu}{p_-^2}\,\mathrm{Re}\,f^{1}_{+\,\mathrm{in}}(t) \\
&\quad + \frac{2\nu}{\pi}\int_{\nu_{\mathrm{th}}}^{\infty} d\nu'\,\mathrm{Im}\,C^-_{\mathrm{in}}(\nu',t)\left\{\frac{1}{\nu'^2-\nu^2} - \frac{3m^2}{p_-^2 q_-^2}\,Q_1(Z')\right\};
\end{aligned} \tag{A.8.45}$$

$$\begin{aligned}
\mathrm{Re}\,B^-(\nu,t) &= B^-_{\mathrm{N}}(\nu,t) + \mathrm{Re}\,B^-_{\mathrm{el}}(\nu,t) + 6\sqrt{2\pi}\,\mathrm{Re}\,f^{1}_{-\,\mathrm{in}}(t) \\
&\quad + \frac{2}{\pi}\int_{\nu_{\mathrm{th}}}^{\infty} d\nu'\,\mathrm{Im}\,B^-_{\mathrm{in}}(\nu',t)\left\{\frac{\nu'}{\nu'^2-\nu^2} - \frac{m}{p_- q_-}\{Q_0(Z') - Q_2(Z')\}\right\}.
\end{aligned} \tag{A.8.46}$$

If one inserts the asymptotic expansion of the Legendre functions

$$Q_1(Z) = \frac{1}{3Z^2} + \frac{1}{5Z^4} + \ldots; \qquad Q_0(Z) - Q_2(Z) = \frac{1}{Z} + \frac{1}{5Z^3} + \ldots, \tag{A.8.47}$$

it is seen that the leading term just cancels the leading term of the expansions of the first terms in the curly brackets, i.e. the Q-functions act like a strong cut-off. In Eqs. (A.8.45) and (A.8.46), the high energy part of $\mathrm{Im}\,C^-_{\mathrm{in}}$ and $\mathrm{Im}\,B^-_{\mathrm{in}}$, which comes mainly from reggeized ϱ-exchange, is replaced by the terms $\propto \mathrm{Re}\,f^{1}_{\pm\,\mathrm{in}}$.

These terms contain the effect of ϱ-exchange, if ϱ is treated as a spin-one particle. This follows from the fact that, due to G-parity, $\mathrm{Im}\,C^-_{\mathrm{el}}$ and $\mathrm{Im}\,B^-_{\mathrm{el}}$ have at least 4 pions in the intermediate states, if one considers the $\pi\pi\mathrm{N}\bar{\mathrm{N}}$ amplitudes (Fig. A.5.1). ϱ is coupled only to two-pion states and therefore, ϱ-exchange belongs to $\mathrm{Im}\,C^-_{\mathrm{in}}$ and $\mathrm{Im}\,B^-_{\mathrm{in}}$.

Near threshold, the integral in the second line of Eq. (A.8.45) is negligible (2% of $\mathrm{Re}\,C^-$ at $t=0$). In the first line of this equation, we have in addition to the nucleon exchange term the amplitude $\mathrm{Re}\,C^-_{\mathrm{el}}$, which is dominated by isobar exchange. The last term in this line is an improved version of the ϱ-exchange contribution. It differs appreciably from the result of 2nd order perturbation theory (Table A.8.1).

9.2 Exchange of reggeized nucleons and nucleon resonances

The treatment of fermion Regge pole exchanges introduces several new problems, which were not encountered in our discussion of boson exchanges in the t-channel.

9.2.1 Analytic continuation of πN partial waves in J

The basic idea is the same as in Sect. A.9.1.1: one derives a FROISSART-GRIBOV representation by inserting the fixed-s dispersion relation (A.6.9) into the projection formula (A.3.8) and carries out the projection integral.

For this purpose, it is necessary to write the fixed-s relation in such a way that the dependence on $\cos\theta$ is exhibited. It reads for the amplitude A (ignoring subtractions)

$$A(s,t,u)=\frac{1}{\pi}\int_{4\mu^2}^{\infty}\frac{A_t(s,x,\tilde{u})\mathrm{d}x}{x+2q^2(1-\cos\theta)}+\frac{1}{\pi}\int_{x_0}^{\infty}\frac{A_u(s,\tilde{t},x+u_b)\mathrm{d}x}{x+2q^2(1+\cos\theta)}. \tag{A.9.24}$$

The variable x agrees with t' in the first integral and with $u'-u_b$ in the second integral. $u_b=(m^2-\mu^2)^2/s$ is the value of u in the backward direction and $x_0=(m+\mu)^2-u_b$.

If one inserts the dispersion relation into the projection formula (A.3.8) and does the integral over $\cos\theta$, it is seen that the condition for Carlson's theorem is violated due to a factor $(-1)^{\ell}$ which occurs in the u-channel contribution.

Therefore, one introduces two different continuations in ℓ, which are called partial waves of *even* $(\tau=+1)$ and *odd* $(\tau=-1)$ *signature* τ

$$A^{(\tau)}(\ell,s)\equiv\frac{1}{\pi q^2}\int_{4\mu^2}^{\infty}\mathrm{d}x\{A_t(s,x,u)+\tau A_u(s,t,x+u_b)\}Q_{\ell}(1+x/2q^2), \tag{A.9.25}$$

where

$$A^{(+)}(\ell,s)=A_{\ell}(s)\ \text{for even}\ \ell; \qquad A^{(-)}(\ell,s)=A_{\ell}(s)\ \text{for odd}\ \ell. \tag{A.9.26}$$

$B^{(\tau)}$ is defined in a similar way.

One can also introduce a *signatured invariant amplitude* $A^{(\tau)}$

$$A^{(\tau)}(s,z)\equiv A_{\mathrm{R}}(s,z)+\tau A_{\mathrm{L}}(s,-z), \tag{A.9.27}$$

where "R" and "L" refer to the right hand (t-channel) and left hand (u-channel) contribution to the fixed-s dispersion relation and $z=\cos\theta$ is written instead of t in the argument of A.

Both amplitudes $A^{(\pm)}$ have right hand cut contributions only

$$A^{(\tau)}(s,z)=\frac{1}{\pi}\int_{4\mu^2}^{\infty}\frac{A_t(s,x,\tilde{u})+\tau A_u(s,\tilde{t},x+u_b)}{x+2q^2(1-z)}\mathrm{d}x. \tag{A.9.28}$$

Therefore, their FROISSART-GRIBOV representation leads directly to the unique continuation in ℓ. The original amplitude can be reconstructed as follows

$$A(s,z)=\tfrac{1}{2}\{A^{(+)}(s,z)+A^{(+)}(s,-z)+A^{(-)}(s,z)-A^{(-)}(s,-z)\}. \tag{A.9.29}$$

The $A^{(\pm)}(\ell,s)$ are the partial wave projections of the $A^{(\pm)}(s,z)$, i.e.

$$A^{(\pm)}(s,z)=\sum_{\ell=0}^{\infty}(\ell+\tfrac{1}{2})A^{(\pm)}(\ell,s). \tag{A.9.30}$$

The continuation of the partial wave amplitude $f_{\ell\pm}(s)$ follows, if one inserts $A(\ell,s)$ into Eq. (A.3.7). One obtains four continued partial waves, which can be classified for instance according to the values of the signature τ and the naturality η, or τ and the parity P. We denote the continued partial wave by $f(J,s;\tau,\eta)$. It agrees with the physical partial wave $f_{\ell\pm}(s)$ in the following cases [see Eq. (A.3.14) for η]

$$f(J,s;\tau,\eta)\equiv f_{\ell\pm}(s),\quad \text{if}\quad \ell=J+\eta/2\quad\text{and}\quad \tau=(-1)^{J-1/2}. \tag{A.9.31}$$

9.2.2 Exchange of Regge poles in the s-channel

The partial wave expansion Eq. (A.3.4) for f_1 and f_2 can be re-expressed by a Sommerfeld-Watson transform, using the relation $P'_\ell(-z)=(-1)^{\ell+1}P'_\ell(z)$ in order to construct signatured amplitudes $f_i^{(\tau)}$

$$f_1^{(\pm)}(s,z)=\frac{i}{4}\int_{C_1}\frac{dJ}{\cos\pi J}\left\{f(J,s;\pm,-)\{P'_{J+1/2}(-z)\pm P'_{J+1/2}(z)\}-f(J,s;\pm,+)\{P'_{J-1/2}(-z)\mp P'_{J-1/2}(z)\}\right\}, \tag{A.9.32}$$

$$f_2^{(\pm)}(s,z)=-\frac{i}{4}\int_{C_1}\frac{dJ}{\cos\pi J}\left\{f(J,s;\pm,-)\{P'_{J-1/2}(-z)\mp P'_{J-1/2}(z)\}-f(J,s;\pm,+)\{P'_{J+1/2}(-z)\pm P'_{J+1/2}(z)\}\right\}. \tag{A.9.33}$$

Before displacing the contour of the integral, we substitute instead of $f(J,W;\tau,\eta)$ new amplitudes, which have the advantage of being free of kinematic singularities in the W-plane

$$h(J,W;\tau,\eta)\equiv\frac{32\pi W^2}{(W-\eta m)^2-\mu^2}\,\frac{f(J,W;\tau,\eta)}{(2q^2)^{J-1/2}}. \tag{A.9.34}$$

Now, the contour of the Sommerfeld-Watson integral is displaced as shown in Fig. A.9.1. Assuming for instance that a Regge pole at $J=\alpha$ has the quantum numbers $\tau=\eta=-1$, we find for its contribution to the amplitudes (written for the variables W and z)

$$\begin{aligned} f_{1R}(W,z)&=-\frac{E+m}{32W}\,\frac{(2q^2)^{\alpha-1/2}b(W)}{\cos\pi\alpha}\{P'_{\alpha+1/2}(-z)-P'_{\alpha+1/2}(z)\};\\ f_{2R}(W,z)&=\frac{E+m}{32W}\,\frac{(2q^2)^{\alpha-1/2}b(W)}{\cos\pi\alpha}\{P'_{\alpha-1/2}(-z)+P'_{\alpha-1/2}(z)\}, \end{aligned} \tag{A.9.35}$$

where $b(W)=\lim_{J\to\alpha}(J-\alpha)h(J,W;--)$ is the residue of the pole of $h(J,W;--)$ at $J=\alpha(W)$. [The notation $b(W)$ and $\alpha(W)$ omits the dependence on τ and η.]

If $\alpha(W)$ is increasing, $\cos\pi\alpha$ has zeros at $\alpha(W)=1/2, 3/2, 5/2, \ldots$, of which those at $1/2, 5/2, \ldots$ are cancelled by zeros of the numerator. The poles at $\alpha=3/2, 7/2, \ldots$ belong to a family of s-channel resonances (Regge recurrences of the $J=3/2$ resonance).

In the vicinity of the poles, the amplitudes f_{1R}, f_{2R} can be approximated by Breit-Wigner terms

$$\begin{aligned} G(s,z)&=f_1+\cos\theta f_2=\frac{1}{q}\frac{x_e(J+1/2)}{\varepsilon-i}P_\ell(z);\\ f_2(s,z)&=\frac{1}{q}\frac{x_e}{\varepsilon-i}(-1)^{J-\ell+1/2}P'_\ell(z). \end{aligned} \tag{A.9.36}$$

x_e represents the elasticity parameter and $\varepsilon=(M^2-s)/M\Gamma$, where M is the mass and Γ the full width of the resonance. The mass and width are determined by the trajectory $\alpha(W)$: the mass follows from $\mathrm{Re}\,\alpha(M)=J$ and the width from

$$\Gamma=\left.\frac{2\,\mathrm{Im}\,\alpha(M)}{d\mathrm{Re}\,\alpha/dW}\right|_{W=M}. \tag{A.9.37}$$

The amplitudes $h(J,W;\tau,\eta)$ fulfill the generalized MacDowell symmetry relation [see Eq. (A.7.8)]

$$h(J,W;\tau,\eta)=-h(J,-W;\tau,-\eta). \tag{A.9.38}$$

If $h(J,W;\tau,\eta)$ has a Regge pole at $J=\alpha(W)$, then $h(J,W;\tau,-\eta)$ has a Regge pole at $J=\alpha(-W)$. The trajectories and the residue functions are related by

$$\alpha(W;\tau,\eta)=\alpha(-W;\tau,-\eta);\qquad b(W;\tau,\eta)=-b(-W;\tau,-\eta). \tag{A.9.39}$$

$\alpha(W;\tau,\eta)$ and $b(W;\tau,\eta)$ are presumed to be real analytic in the cut W-plane, with cuts from $-\infty$ to $-(m+\mu)$ and from $(m+\mu)$ to $+\infty$.

The existence of the MacDowell symmetry was ignored in Eq. (A.9.35). One can take into account the MacDowell partner of the Regge pole, which is a pole of the amplitude $h(J,W;-,+)$ at $J=\alpha(-W)$ and leads to a second term in f_{1R}, Eq. (A.9.35). Then, f_2 can be obtained from $f_2(W,z)=-f_1(-W,z)$.

The Regge model does not predict the shape of $\alpha(W)$. If $\mathrm{Re}\,\alpha(W; -, +) = \mathrm{Re}\,\alpha(-W; -, -)$ passes $J = 3/2$, $7/2, \ldots$ for positive values of W, the MacDowell pair describes two families of resonances, which have the same J but opposite parity (parity doublets).

If we restrict ourselves to the leading trajectories, there are 8 different Regge poles, which can be distinguished by the quantum numbers isospin $I = 1/2$ or $3/2$, signature $\tau = \pm 1$ and naturality $\eta = P\tau = \pm 1$ (or parity $P = \pm 1$). Some of their properties are listed in Table A.9.1.

Table A.9.1. Classification of nucleon resonances and Regge poles. $\tau = (-1)^{J-1/2}$, $P = (-1)^{\ell+1}$, $\eta = 2(\ell - J) = P\tau$. If the isospin is included, the usual notation is $\ell(2I)\,(2J)$. For instance D13 for the $I = 1/2$, $J = 3/2$, $\ell = 2$ state. The states in the two upper lines and those in the two lower lines are MacDowell partners.

$I=\frac{1}{2}$	$I=\frac{3}{2}$	τ	η	P	$\ell\pm$	$\ell(2J)$
N_α	Δ_α	+	+	+	$\ell-$	P1, F5, H9, ...
N_β	Δ_β	+	−	−	$\ell+$	S1, D5, G9, ...
N_γ	Δ_γ	−	+	−	$\ell-$	D3, G7, I11, ...
N_δ	Δ_δ	−	−	+	$\ell+$	P3, F7, H11, ...

9.2.3 Exchange of Regge poles in the u-channel

As a first step, the $s-u$ crossing operation is applied to the expression for a Regge pole exchange in the s-channel, taking into account one trajectory and its MacDowell reflection.

Then, the result is simplified by using an approximation for the Legendre functions. Next, one adds the contributions of several Regge poles and calculates the s-channel amplitudes from the u-channel amplitudes. The final step is the calculation of the s-channel differential cross section and polarization.

An approximation for the Legendre function

We denote the c.m. momentum and the scattering angle in the u-channel by q_u and θ_u and find by applying the crossing operation (Sect. A.1.5)

$$4q_u^2 = u - 2(m^2 + \mu^2) + (m^2 - \mu^2)^2/u\,;$$
$$z_u = \cos\theta_u = 1 + \frac{t}{2q_u^2} = -\frac{1}{2q_u^2}\left[s - m^2 - \mu^2 + \frac{u}{2} - \frac{(m^2-\mu^2)^2}{2u}\right] = -\frac{1}{4q_u^2}\left[s - t - \frac{(m^2-\mu^2)^2}{u}\right]. \tag{A.9.40}$$

In s-channel scattering, the largest effect of u-channel Regge pole exchanges is expected near the backward direction at high energies. Since $|z_u| < 1$ in the backward cone $[u_b = u(180°) = (m^2 - \mu^2)^2/s > u > 0]$, it seems that the Legendre function $P'_{\alpha+1/2}(-z_u)$ cannot be approximated by the leading term of its asymptotic expansion, similar to the procedure in Sect. A.9.1.3, Eq. (A.9.13).

However, the behaviour of the Regge pole amplitudes near $u = 0$ can only be determined by a detailed investigation, because q_u and z_u are singular at $u = 0$. It turns out that the leading term of the u-channel Regge pole contribution is obtained by omitting the singular part of z_u, corresponding to the prescription

$$P'_{\alpha+1/2}(-z_u) \text{ has to be replaced by } \frac{2\Gamma(\alpha+1)}{\sqrt{\pi}\,\Gamma(\alpha+1/2)}\left(\frac{s-t}{2q_u^2}\right)^{\alpha-1/2}. \tag{A.9.41}$$

Furthermore, the non-leading terms have singularities at $u = 0$ which must be cancelled, because the full amplitude is regular at this point. This cancellation is usually performed by assuming suitable lower lying Regge trajectories ("daughter trajectories").

Exchange of one pair of trajectories

We start from the formula for the exchange of one Regge trajectory [example: Eq. (A.9.35)] and add a second term for the MacDowell partner. The $s-u$ crossing operation leads to u-channel amplitudes, which are written as functions of W_u and s: $f_1(W_u, s) = -f_2(-W_u, s)$. The result is simplified by inserting the approximation

Eq. (A.9.41) and by retaining only the highest power of $(s-t)$. From $(s-t)=2(s-m^2-\mu^2+u/2)$ one can see that in our applications $(s-t)$ is a large positive quantity. The result for the contribution of the exchange of a MacDowell pair of Regge poles to the u-channel amplitude is

$$f_1(W_u,s)=-\frac{E_u+m}{16\sqrt{\pi}W_u}\frac{\Gamma(\alpha+1)}{\Gamma(\alpha+1/2)}\frac{b}{\sin\pi(\alpha-1/2)}\{1+\tau e^{-i\pi(\alpha-1/2)}\}(s-t)^{\alpha-1/2}. \quad (A.9.42)$$

The trajectory and the residue functions are given by

$$\alpha(\pm W_u); \quad b(\pm W_u) \quad \text{for} \quad \eta=P\tau=\mp 1. \quad (A.9.43)$$

s-channel amplitudes can be derived either from Eq. (A.2.36), which leads to $f_1(W,u)$, $f_2(W,u)$, or from Eq. (A.2.25), which leads to the invariant amplitudes.

We introduce amplitudes $F^{(\eta)}(W_u,s)$, because they were used in practical calculations and are closely related to Eq. (A.2.25)

$$\begin{aligned} F^{(+)}(W_u,s)&=-A(s,t,u)-(W_u+m)B(s,t,u)=\frac{8\pi W_u}{E_u-m}f_2(W_u,s);\\ F^{(-)}(W_u,s)&=A(s,t,u)-(W_u-m)B(s,t,u)=\frac{8\pi W_u}{E_u+m}f_1(W_u,s).\end{aligned} \quad (A.9.44)$$

These amplitudes have the generalized MacDowell symmetry

$$F^{(-)}(W_u,s)=-F^{(+)}(-W_u,s). \quad (A.9.45)$$

Now, we can write the final result for the u-channel exchange of a Regge pole and its MacDowell reflection. It is equivalent to Eq. (A.9.42)

$$\begin{aligned} &F^{(+)}(W_u,s)=\gamma^{(+)}(W_u)R[\alpha^{(+)}(W_u),s];\\ &F^{(-)}(W_u,s)=\gamma^{(-)}(W_u)R[\alpha^{(-)}(W_u),s];\\ &R[\alpha(W_u),s]=\frac{1}{\Gamma(\alpha+1/2)}\frac{1+\tau e^{-i\pi(\alpha-1/2)}}{\sin\pi(\alpha-1/2)}\left(\frac{s-t}{2s_0}\right)^{\alpha-1/2};\\ &\alpha^{(+)}(W_u)=\alpha^{(-)}(-W_u); \quad \gamma^{(+)}(W_u)=-\gamma^{(-)}(-W_u); \quad \text{if } u>0;\\ &\alpha^{(+)}(W_u)=[\alpha^{(-)}(W_u)]^*; \quad \gamma^{(+)}(W_u)=-[\gamma^{(-)}(W_u)]^*; \quad W_u=i\sqrt{|u|} \quad \text{if } u<0.\end{aligned} \quad (A.9.46)$$

Exchanges of several pairs of trajectories

The amplitudes for Regge pole exchanges in the u-channel are denoted by a superscript "N" and "Δ", if the isospin is $I=1/2$ and $I=3/2$ respectively. The most important cases are the exchanges of the trajectories N_α, which starts with the nucleon, and Δ_δ, which starts with Δ(1232).

The contributions of these u-channel exchange amplitudes to $\pi^\pm p$ elastic and $\pi^- p$ charge-exchange amplitudes follow from the $s-u$ isospin crossing relations of Table A.2.1

$$A_+=\frac{1}{3}(2A^N+A^\Delta); \quad A_-=A^\Delta; \quad A_0=\frac{\sqrt{2}}{3}(A^N-A^\Delta). \quad (A.9.47)$$

Differential cross section and polarization

We list the exact expression for the differential cross section in terms of the combinations F_1 and F_2, which are simply related to the invariant amplitudes A and B

$$F_1=\frac{1}{2}(F^{(-)}-F^{(+)})=A+mB; \quad F_2=\frac{1}{2W_u}(F^{(-)}+F^{(+)})=-B; \quad (A.9.48)$$

$$\frac{d\sigma}{du}=\frac{1}{64\pi m^2k^2}\left\{|F_1|^2(s+u+2m^2-2\mu^2)+|F_2|^2[u(-s+4m^2)+(m^2-\mu^2)^2]+4m\,\mathrm{Re}(F_1^*F_2)(u+m^2-\mu^2)\right\}. \quad (A.9.49)$$

For backward scattering the formula reads

$$\left.\frac{d\sigma}{du}\right|_{180^\circ} = \frac{E^2}{16\pi m^2 k^2}\left|F_1 + \frac{m}{EW}(m^2-\mu^2)F_2\right|^2 = \frac{E^2}{16\pi m^2 k^2}|C|^2 . \qquad \text{(A.9.50)}$$

Finally, we mention the expression for the polarization parameter in terms of F_1 and F_2

$$P\frac{d\sigma}{du} = -\frac{\sin\theta}{16\pi W}\,\mathrm{Im}(F_1^* F_2). \qquad \text{(A.9.51)}$$

9.2.4 Line reversal

Starting from a reaction

(I) $a+b \to c+d$,

one defines *line-reversed reactions* (see also Sect. A.6.1)

(II) $\bar{c}+b \to \bar{a}+d$; (III) $\bar{d}+b \to c+\bar{a}$.

If (I) is dominated by the exchange of a t-channel Regge trajectory, this should also be true for (II). The amplitudes for (II) differ from those for (I) by changing $z_t \to -z_t$ in the t-channel partial wave expansion and in the Regge pole formula, because of the interchange of the initial particles in the t-channel.

If (I) is dominated by the exchange of a u-channel Regge trajectory, this should also be true for (III). Now, the amplitudes are related by $z_u \to -z_u$.

In $\pi^\pm p$ scattering, the only interesting case is $\pi^\pm p$ near-backward scattering, the line-reversed reaction being $\bar{p}p \to \pi^+\pi^-$ (Fig. A.9.2). The situation is most favourable for near-backward $\pi^- p$ scattering and $\bar{p}p \to \pi^+\pi^-$, where the angle between the $\bar{p}$ and π^- momenta is $\approx 180°$, since N-exchange is forbidden and Δ-exchange should be dominant. In near-backward $\pi^+ p$ scattering, N-exchange is dominant but Δ-exchange is not negligible.

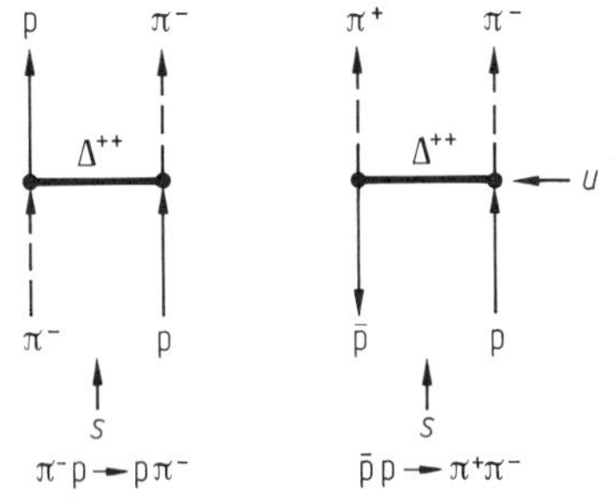

Fig. A.9.2. Line-reversal. $\pi^- p$ backward scattering, which is sometimes denoted by $\pi^- p \to p\pi^-$, is compared with the line-reversed reaction $\bar{p}p \to \pi^+\pi^-$, where the angle between the incoming $\bar{p}$ and the outgoing π^- is 180°. The incoming and outgoing particles are exchanged at the left-hand vertex, whereas the right-hand vertex remains unchanged. (The arrow for $\bar{p}$ should be reversed.)

Using the notation of Eq. (A.9.48), the line-reversed amplitudes for the u-channel exchange of a fermion trajectory pair of signature τ are related by

$$\bar{F}_1(u,s) = \tau F_1(u,s); \quad \bar{F}_2(u,s) = \tau F_2(u,s), \qquad \text{(A.9.52)}$$

where $\bar{F}_1$ and $\bar{F}_2$ belong to the amplitudes for the reaction $\bar{p}p \to \pi\pi$.

In the general case, the cross section for $\pi p \to \pi p$ is given by Eq. (A.9.49), whereas it reads for $\bar{p}p \to \pi\pi$

$$\frac{d\bar{\sigma}}{du} = \frac{1}{128\pi q_{\bar{p}p}^2 s}\{|\bar{F}_1|^2(s-4m^2) + |\bar{F}_2|^2(-us-m^2+\mu^2-u^2) - 4m\,\mathrm{Re}(\bar{F}_1^*\bar{F}_2)(u+m^2-\mu^2)\}.$$

The factor 2 in the denominator comes from the average over the initial spin states.

In both reactions, the c.m. energy squared is denoted by s (although the second reaction is a "t-channel reaction"). q and $q_{\bar{p}p}$ are the appropriate c.m. momenta of the incoming particles. u is in both reactions the 4-momentum transfer squared between the target nucleon and the near-backward pion.

As $s \to \infty$ at fixed u, the cross sections approach the asymptotic values

$$\frac{d\sigma}{du} = \frac{1}{16\pi s}\{|F_1|^2 - u|F_2|^2\}; \quad \frac{d\bar{\sigma}}{du} = \frac{1}{32\pi s}\{|\bar{F}_1|^2 - u|\bar{F}_2|^2\}. \qquad \text{(A.9.53)}$$

Assuming the dominance of a single trajectory pair, one finds in the asymptotic region

$$\frac{d\bar{\sigma}}{du} = \frac{1}{2}\frac{d\sigma}{du} \quad \text{for} \quad s \to \infty . \tag{A.9.54}$$

10 Isospin bounds

The consequences of isospin invariance for measurable quantities follow from the relation

$$F_+(\pm) - F_-(\pm) - \sqrt{2} F_0(\pm) = 0 , \tag{A.10.1}$$

where $F(\pm)$ are the transversity amplitudes (Sect. A.2.1.2) and lower indices $\pm, 0$ refer to the reactions $\pi^\pm p \to \pi^\pm p$ and $\pi^- p \to \pi^0 n$. Equation (A.10.1) is represented graphically by the isospin triangle of Fig. A.10.1. Using the notation

$$\Sigma(\pm) \equiv \frac{d\sigma}{d\Omega}(1 \pm P) = |F(\pm)|^2 , \tag{A.10.2}$$

we define

$$\Lambda(\pm) \equiv \lambda(\Sigma_+(\pm), \Sigma_-(\pm), 2\Sigma_0(\pm)) , \tag{A.10.3}$$

where the lower indices of Σ have the same meaning as in Eq. (A.10.1). $\lambda(x, y, z)$ is the "triangle function"

$$\lambda(x, y, z) \equiv (x + y - z)^2 - 4xy \equiv x^2 + y^2 + z^2 - 2xy - 2xz - 2yz = -16A^2 , \tag{A.10.4}$$

which was used already in Eqs. (A.1.12) and (A.1.25) in another context. A is the area of a triangle whose sides have the lengths $\sqrt{x}, \sqrt{y}, \sqrt{z}$.

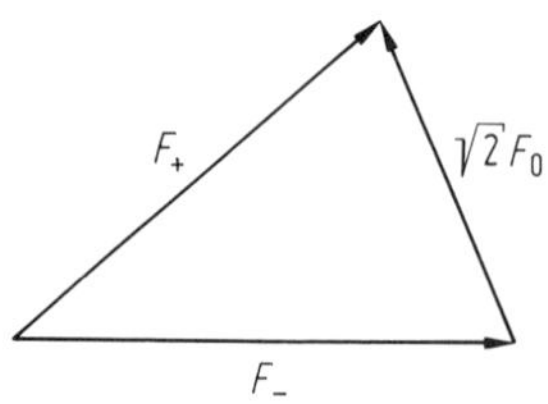

Fig. A.10.1. The isospin triangle for transversity amplitudes.

10.1 Equalities and bounds for a complete set of data

If all measurable quantities are known for a pair of variables (s, t), one can derive from Eq. (A.10.1) that the following relations are fulfilled

$$\frac{d\sigma_+}{d\Omega}\frac{d\sigma_-}{d\Omega}(1 - \xi_+ \cdot \xi_-) = 2\frac{d\sigma_-}{d\Omega}\frac{d\sigma_0}{d\Omega}(1 - \xi_- \cdot \xi_0) = 2\frac{d\sigma_0}{d\Omega}\frac{d\sigma_+}{d\Omega}(1 - \xi_0 \cdot \xi_+) \equiv \frac{H}{2} \geq 0 . \tag{A.10.5}$$

$\xi = (A, P, R)$ is the unit spin-rotation vector. It follows that a necessary condition for the compatibility with isospin invariance is

$$-\lambda\left(\frac{d\sigma_+}{d\Omega}, \frac{d\sigma_-}{d\Omega}, 2\frac{d\sigma_0}{d\Omega}\right) \geq H . \tag{A.10.6}$$

10.2 Bounds for differential cross sections and polarizations

The bounds are given by the triangular inequalities for $|F(\pm)|$

$$\sqrt{\Sigma_+(\pm)}+\sqrt{\Sigma_-(\pm)}\geqq\sqrt{2\Sigma_0(\pm)}, \text{ etc.} \tag{A.10.7}$$

or, equivalently, by [writing Σ instead of $\Sigma(\pm)$]

$$\Lambda(\pm)\equiv(\Sigma_--\Sigma_+)^2-4\Sigma_0(\Sigma_-+\Sigma_+-\Sigma_0)\leqq 0, \tag{A.10.8}$$

which is the condition that the area of the triangle must be positive or zero. A third equivalent version reads

$$(\sqrt{\Sigma_+}-\sqrt{\Sigma_-})^2\leqq 2\Sigma_0\leqq(\sqrt{\Sigma_+}+\sqrt{\Sigma_-})^2. \tag{A.10.9}$$

It can be used if one wants for instance to get a bound for P_0 in the case that the differential cross sections and P_+, P_- are known.

Starting from the above bounds, one can derive others in which one of the polarizations does not occur, for instance

$$(P_+-P_-)^2\leqq\left[8\frac{d\sigma_0}{d\Omega}\frac{d\sigma^+}{d\Omega}-\left(\frac{d\sigma_-}{d\Omega}-\frac{d\sigma_+}{d\Omega}\right)^2\right]\left[\frac{d\sigma_+}{d\Omega}\frac{d\sigma_-}{d\Omega}\right]^{-1}=4\sin^2\chi, \tag{A.10.10}$$

where

$$\frac{d\sigma^+}{d\Omega}=\frac{1}{2}\left[\frac{d\sigma_+}{d\Omega}+\frac{d\sigma_-}{d\Omega}-\frac{d\sigma_0}{d\Omega}\right]; \quad 2\sqrt{\frac{d\sigma_-}{d\Omega}\frac{d\sigma_+}{d\Omega}}\cos\chi=\frac{d\sigma_+}{d\Omega}+\frac{d\sigma_-}{d\Omega}-2\frac{d\sigma_0}{d\Omega}. \tag{A.10.11}$$

χ is an angle in an isospin triangle, whose sides have lengths equal to the square roots of the differential cross sections (Fig. A.10.2). If the bound Eq. (A.10.10) is saturated, P_+ is equal to $-P_-$.

Continuing the inequality Eq. (A.10.10) to the left by $0\leqq(P_+-P_-)^2$, one obtains an inequality for the differential cross sections

$$\lambda\left(\frac{d\sigma_+}{d\Omega},\frac{d\sigma_-}{d\Omega},2\frac{d\sigma_0}{d\Omega}\right)\leqq 0, \tag{A.10.12}$$

i.e. the above inequalities for the $\Sigma(\pm)$ are also valid for $d\sigma/d\Omega$. The bound is saturated, if the triangle in Fig. A.10.2 is degenerated, the three vectors being parallel or antiparallel.

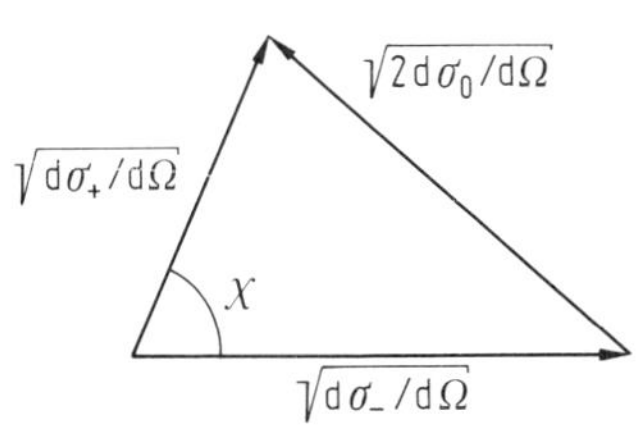

Fig. A.10.2. Definition of the angle χ in Eq. (A.10.11). From this figure one can also derive the triangle inequalities for the differential cross sections.

Since it is difficult to measure the small difference between the elastic cross sections at high energies, a bound for this quantity is of interest. Instead of starting with Eq. (A.10.10), we prefer a stronger version derived from Eq. (A.10.9) using the approximation

$$\Sigma_-(-)-\Sigma_+(-)\approx\left\{\frac{d\sigma_-}{d\Omega}-\frac{d\sigma_+}{d\Omega}\right\}+\frac{d\sigma^+}{d\Omega}(P_+-P_-). \tag{A.10.13}$$

One obtains

$$\frac{d\sigma_-}{d\Omega}-\frac{d\sigma_+}{d\Omega}\leqq\left[8\frac{d\sigma^+}{d\Omega}(1-P^+)\frac{d\sigma_0}{d\Omega}(1-P_0)\right]^{1/2}-\frac{d\sigma^+}{d\Omega}(P_+-P_-). \tag{A.10.14}$$

$d\sigma^+/d\Omega$ and P^+ follow from Eqs. (A.4.13), if one inserts the isospin even amplitudes. See also Eq. (A.10.11). Finally, we mention a bound for integrated differential cross sections σ^{el}

$$(\sigma_-^{el}-\sigma_+^{el})^2\leqq 4\sigma_0^{el}(\sigma_-^{el}+\sigma_+^{el}-\sigma_0^{el}). \tag{A.10.15}$$

10.3 Saturation of isospin bounds

If one of the isospin bounds Eq. (A.10.9) is saturated, the isospin triangle (Fig. A.10.1) is degenerated. The complex vectors of all amplitudes are parallel or antiparallel to each other and one of them can be zero. In particular, the difference of the phases of the isospin 1/2 and 3/2 transversity amplitudes $F^{1/2}(\pm)$ and $F^{3/2}(\pm)$, $\Delta\phi(\pm)=\phi_1(\pm)-\phi_3(\pm)$, is either zero or π.

The ratio $F^{1/2}/F^{3/2}$ can be expressed in terms of $\Sigma_+, \Sigma_-, \Sigma_0$

$$\frac{F^{1/2}}{F^{3/2}}=\varrho\exp(\mathrm{i}\Delta\phi);$$

$$\varrho=\frac{3(\Sigma_0+\Sigma_-)-\Sigma_+}{2\Sigma_+}; \qquad \cos\Delta\phi=\frac{\Sigma_++3\Sigma_--6\Sigma_0}{|8\Sigma_+(3\Sigma_0+3\Sigma_--\Sigma_+)|^{1/2}}. \tag{A.10.16}$$

The saturation occurs along certain lines in the (s,t) plane, which can be calculated from cross section and polarization data, using Eq. (A.10.9) (Fig. 4.4.2). Zeros of the transversity amplitudes for physical s,t-values can be determined by studying the variation of ϱ along these lines.

Table A.10.1. Condition for zeros of transversity amplitudes at physical s, t-values. All zeros lie on saturation lines $\Delta\phi(s,t)=0$ or π at points where ϱ passes the values listed in the table. These points are the intersections of zero trajectories (Sect. 2.4.3) with the physical s, t-plane.

Zero of	$F_+=F^{3/2}$	F^+	F_-	$F^{1/2}$	F_0	F_+
$\varrho=$	∞	2	1/2	0	1	∞
$\Delta\phi=$	singular	π	π	singular	0	singular
Eq. (A.10.9)	upper bound saturated			lower bound saturated		

11 Electromagnetic nucleon and pion form factors

11.1 The nucleon form factors

The nucleon form factors are defined by the expressions for the matrix element of the electromagnetic current of the nucleon

$$\begin{aligned}\langle \boldsymbol{p}'|j_\mu(0)|\boldsymbol{p}\rangle &= e\bar{u}(\boldsymbol{p}')\left\{\gamma_\mu F_1(t)+\frac{\mathrm{i}}{2m}\sigma_{\mu\nu}(p'^\nu-p^\nu)F_2(t)\right\}u(\boldsymbol{p})\\
&=\frac{e}{m}\bar{u}(\boldsymbol{p}')\left\{P_\mu F_1(t)+\frac{\mathrm{i}}{2}\sigma_{\mu\nu}(p'^\nu-p^\nu)G_M(t)\right\}u(\boldsymbol{p})\\
&=e\bar{u}(\boldsymbol{p}')\left\{-\frac{P_\mu}{m}F_2(t)+\gamma_\mu G_M(t)\right\}u(\boldsymbol{p})\\
&=\frac{e/m}{1-t/4m^2}\bar{u}(\boldsymbol{p}')\left\{P_\mu G_E(t)+\frac{1}{2m}(\gamma_\mu \boldsymbol{P}\!\!\!/\,\varkappa\!\!\!/-\varkappa\!\!\!/\,\boldsymbol{P}\!\!\!/\,\gamma_\mu)G_M(t)\right\}u(\boldsymbol{p});\end{aligned} \tag{A.11.1}$$

the notation is explained in Eqs. (A.1.4) and (A.2.12)

$$t=(\mathbf{p}'-\mathbf{p})^2;\quad \mathbf{P}=\frac{1}{2}(\mathbf{p}+\mathbf{p}');\quad \boldsymbol{\varkappa}=\frac{1}{2}(\mathbf{p}'-\mathbf{p});\quad \sigma_{\mu\nu}=\frac{\mathrm{i}}{2}(\gamma_\mu\gamma_\nu-\gamma_\nu\gamma_\mu). \tag{A.11.2}$$

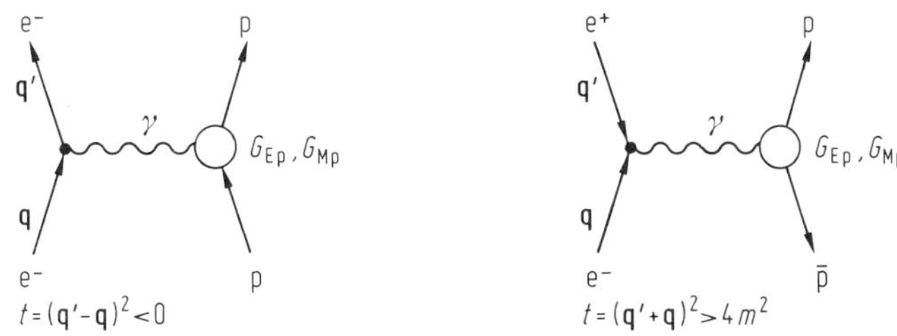

Fig. A.11.1. Electromagnetic nucleon form factors. The same form factors describe the one-gamma exchange contributions to the reactions $e^-p \to e^-p$ and $\bar{p}p \to e^+e^-$. t is "spacelike" for the first reaction and "timelike" for the second one.

The relation between the Dirac and Pauli form factors $F_1(t)$, $F_2(t)$ and the Sachs form factors $G_E(t)$ and $G_M(t)$ is obtained by inserting Gordon's decomposition of the current, Eq. (A.2.14)

$$G_E = F_1 + \frac{t}{4m^2} F_2; \qquad G_M = F_1 + F_2; \tag{A.11.3}$$

$$F_1 = \frac{G_E - \frac{t}{4m^2} G_M}{1 - t/4m^2}; \qquad F_2 = \frac{G_M - G_E}{1 - t/4m^2}. \tag{A.11.4}$$

The proton and neutron form factors are denoted by subscripts p and n and the isoscalar and isovector combinations by subscripts s and v respectively. For instance

$$G_{Es} = \tfrac{1}{2}(G_{Ep} + G_{En}); \qquad G_{Ev} = \tfrac{1}{2}(G_{Ep} - G_{En}). \tag{A.11.5}$$

The slope of the form factors at $t=0$ is usually expressed in terms of a "nucleon radius" $\sqrt{\langle r^2 \rangle}$

$$F(t) = F(0)\,(1 + \tfrac{1}{6}\langle r^2 \rangle t + \ldots). \tag{A.11.6}$$

This notion has its origin in the nonrelativistic description of a scattering process, in which a point charge particle interacts with a given charge distribution $\varrho(r)$. In the lowest order, the scattering amplitude differs only by a form factor $F(Q^2)$ from that belonging to scattering against a point charge target, where

$$F(Q^2) = \iiint \varrho(r) e^{i\boldsymbol{Q}\boldsymbol{r}} dV = \frac{4\pi}{Q} \int_0^\infty \varrho(r) \sin(Qr) r dr, \tag{A.11.7}$$

$\boldsymbol{Q}$ being the 3-momentum transfer. The mean square radius of the charge distribution is given by

$$\langle r^2 \rangle = \int_0^\infty r^2 \varrho(r) 4\pi r^2 dr = -6 \left. \frac{dF/dQ^2}{F(Q^2)} \right|_{Q^2=0}. \tag{A.11.8}$$

It agrees with Eq. (A.11.6) in the nonrelativistic limit ($Q^2 = -t$).

Equation (A.11.6) can be used for all form factors except G_{En} and F_{1n}, which vanish at $t=0$. In the literature, the slope of G_{En} at $t=0$ is expressed in several different ways (Foldy term: $V_0 = -4083$ eV)

$$\left. \frac{dG_{En}}{dt} \right|_{t=0} = \left. \frac{dF_{1n}}{dt} \right|_{t=0} + \frac{F_{2n}(0)}{4m^2} = \frac{V_0}{3e^2/\hbar c} \frac{r_e^3}{\hbar c} = 5.18 \cdot 10^{-6}\, V_0 [\mathrm{fm}^2]; \quad V_0 \text{ in eV}; \quad (V_0 \approx -3760\,\mathrm{eV});$$

$$= b_{ne} \frac{\lambda\!\!\!^{-}{}_e^2}{2r_e} \frac{m_e}{m} = 14.4 b_{ne} [\mathrm{fm}^2]; \quad b_{ne} \text{ in fm}; \quad (b_{ne} \approx -1.35 \cdot 10^{-2}\,\mathrm{fm}). \tag{A.11.9}$$

In the first equation, the *Foldy term* $F_{2n}(0)/4m^2$ is strongly dominant. The 2nd equation belongs to a description of the interaction by a square well potential of radius $r_e = e^2/m_e c^2 = 2.818$ fm, where m_e is the electron mass. The 3rd equation is written for the neutron-electron scattering length b_{ne} ($\lambda\!\!\!^{-}{}_e = \hbar/m_e c$).

Table A.11.1 gives the normalization of the nucleon form factors, their slopes at $t=0$ and the nucleon radii.

Table A.11.1. Normalization, slopes and radii of nucleon form factors. $G(t)=G(0)+G'(0)t+\cdots$. The radii $r=\sqrt{\langle r\rangle^2}$ follow from $G'(0)=G(0)r^2/6$. The determination of the slopes is treated in Sect. 2.5.2. $G_D(t)$ is the dipole formula, Eq. (A.11.20). $G_\varrho(t)$ is the form factor following from ϱ-dominance in the narrow resonance approximation: $F_\varrho(t)=t_\varrho/(t_\varrho-t)$, where $t_\varrho=m_\varrho^2 (m_\varrho=769$ MeV).

	$G(0)$	$G'(0)$		r		$F(0)$	$F'(0)$		r
		fm^2	GeV^{-2}	fm			fm^2	GeV^{-2}	fm
G_{Ep}	1	0.117	3.00	0.836	F_{1p}	1	0.097	2.48	0.761
G_{Mp}	2.793	0.331	8.50	0.843	F_{2p}	1.793	0.234	6.01	0.883
G_{En}	0	−0.0195	−0.50	—	F_{1n}	0	0.0017	0.04	—
G_{Es}	0.5	0.049	1.26	0.77	F_{1s}	0.5	0.049	1.26	0.77
G_{Ev}	0.5	0.068	1.75	0.91	F_{1v}	0.5	0.048	1.23	0.76
G_{Mn}	−1.913	−0.243	−6.24	0.87	F_{2n}	−1.913	−0.245	−6.29	0.88
G_{Ms}	0.440	0.044	1.13	0.77	F_{2s}	−0.060	−0.007	−0.18	—
G_{Mv}	2.353	0.287	7.37	0.86	F_{2v}	1.853	0.238	6.11	0.88
G_D	1	0.109	2.81	0.81	G_ϱ	1	0.066	1.69	0.63

Behaviour at $t=4m^2$

At the $e^-e^+\to N\bar{N}$ threshold $t=4m^2$, the form factors are restricted by the condition

$$G_E(4m^2)=G_M(4m^2), \tag{A.11.10}$$

which is similar to the constraint Eq. (A.3.75), since it ensures that near threshold the $N\bar{N}$ D-wave contribution to the cross section is negligible in comparison with that of the S-wave [Eq. (A.3.81)]. Furthermore, this constraint cancels the poles of F_1 and F_2 at $t=4m^2$ in Eq. (A.11.4).

Dispersion relation

The form factors are real at $t<4\mu^2$ and complex-valued at $t>4\mu^2$. It is generally assumed that they fulfill unsubtracted dispersion relations of the type

$$\mathrm{Re}\,F(t)=\frac{1}{\pi}\int_{4\mu^2}^{\infty}\frac{\mathrm{Im}\,F(t')}{t'-t}\,dt', \tag{A.11.11}$$

which can be written for F_1 and F_2 and as well for G_E and G_M. The lower limit is $9\mu^2$ in the case of the isoscalar combinations. If $tF(t)\to 0$ for $t\to-\infty$ as suggested by the data, one can write a superconvergence sum rule

$$\int_{4\mu^2}^{\infty}\mathrm{Im}\,F(t)\,dt=0. \tag{A.11.12}$$

Extended unitarity leads to expressions for the spectral functions $\mathrm{Im}\,F_{iv}$, $\mathrm{Im}\,G_{Ev}$, $\mathrm{Im}\,G_{Mv}$ in terms of the pion form factor and the $J=1$ $\pi\pi N\bar{N}$ partial waves (Fig. A.11.2)

$$\begin{aligned}
\mathrm{Im}\,F_{iv}(t)&=\frac{q_t^3}{\sqrt{t}}F_\pi(t)\Gamma_i^*(t)=\frac{q_t^3}{\sqrt{t}}|F_\pi(t)|^2J_i(t);\\
\mathrm{Im}\,G_{Ev}(t)&=\frac{1}{m}\frac{q_t^3}{\sqrt{t}}F_\pi(t)f_+^{1*}(t)=\frac{1}{m}\frac{q_t^3}{\sqrt{t}}|F_\pi(t)|^2J_+(t); \qquad 4\mu^2<t<16\mu^2;\\
\mathrm{Im}\,G_{Mv}(t)&=\frac{1}{\sqrt{2}}\frac{q_t^3}{\sqrt{t}}F_\pi(t)f_-^{1*}(t)=\frac{1}{\sqrt{2}}\frac{q_t^3}{\sqrt{t}}|F_\pi(t)|^2J_-(t).
\end{aligned} \tag{A.11.13}$$

The expressions on the right hand side follow from an N/D-type ansatz for the $\pi\pi N\bar{N}$ partial waves

$$f_\pm^1(t)=\frac{J_\pm(t)}{D(t)}=J_\pm(t)F_\pi(t);\quad \Gamma_i(t)=\frac{J_i(t)}{D(t)}=J_i(t)F_\pi(t). \tag{A.11.14}$$

They are exactly valid up to $16\mu^2$, but it is generally assumed that the 4 pion contribution can be neglected up to the $\omega\pi$ threshold, because the inelasticity in $\pi\pi$-scattering is small.

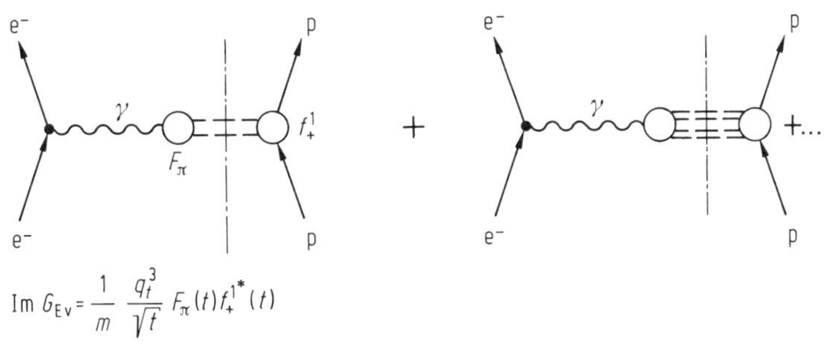

Fig. A.11.2. t-channel intermediate states in e^-p scattering. The imaginary part of the isovector nucleon form factor is decomposed into 2-pion, 4-pion ... contributions. The 2-pion contribution can be predicted from the pion form factor and the $J=1$ $\pi\pi N\bar{N}$ partial waves. It contains an accurate description of the ϱ-exchange.

Cross sections

In the lowest order (one gamma exchange), the differential cross section for $e^-p \to e^-p$ scattering is given by the *Rosenbluth formula*

$$\frac{d\sigma}{d\Omega_{lab}}(e^-p \to e^-p) = \frac{d\sigma}{d\Omega_{NS}} \left\{ \frac{G_E^2 - \frac{t}{4m^2} G_M^2}{1 - t/4m^2} - \frac{t}{2m^2} G_M^2 \tan^2(\Theta/2) \right\};$$

$$\frac{d\sigma}{d\Omega_{NS}} = \frac{\alpha^2}{4E^2} \frac{\cos^2(\Theta/2)}{\sin^4(\Theta/2)} \left\{ 1 + \frac{2E}{m} \sin^2(\Theta/2) \right\}^{-1}. \tag{A.11.15}$$

The "no structure" cross section in the second line is the Born approximation for scattering between a Dirac electron and a spinless point charge of proton mass m. $\alpha = e^2/\hbar c \approx 1/137$ is the fine structure constant. In terms of the total energies of the incoming and outgoing electron E and E' and of the scattering angle Θ in the laboratory system, t is given by (extreme relativistic approximation)

$$t = -4E^2 \sin^2(\Theta/2) \left\{ 1 + \frac{2E}{m} \sin^2(\Theta/2) \right\}^{-1} = 2m(E' - E);$$

$$E/E' = 1 + \frac{2E}{m} \sin^2(\Theta/2). \tag{A.11.16}$$

The cross section for the reaction $e^+e^- \to \bar{N}N$ in the center of mass frame reads

$$\frac{d\sigma}{d\Omega}(e^+e^- \to \bar{N}N) = \frac{\alpha^2\beta}{4t} \left\{ (1 + \cos^2\theta)|G_M|^2 + \frac{4m^2}{t} \sin^2\theta |G_E|^2 \right\}, \tag{A.11.17}$$

where $\beta = \sqrt{1 - 4m^2/t}$ is the velocity of the nucleon and θ its angle of emission.

Since one-photon exchange is assumed, the final state must have the quantum numbers $J^P = 1^-$, which can be fulfilled only by ${}^3S_1$ and ${}^3D_1$ states [see Eq. (A.3.73)]. In the limit $t \to 4m^2$ the cross section becomes isotropic, corresponding to S-wave dominance, if we demand $G_E(4m^2) = G_M(4m^2)$ as mentioned above in Eq. (A.11.10).

The cross section for the inverse reaction differs from Eq. (A.11.17) only by a phase space factor

$$\frac{d\sigma}{d\Omega}(\bar{N}N \to e^+e^-) = \beta^{-2} \frac{d\sigma}{d\Omega}(e^+e^- \to \bar{N}N). \tag{A.11.18}$$

For the reaction on the left hand side, a "$1/v$-law" is valid in analogy to Eq. (A.3.81).

Parametrization of the nucleon form factors

According to the *scaling law*, three of the form factors are approximately equal, if normalized to unity at $t=0$. The common decrease can be described by the *dipole law*, which leads to the value 0.81 fm for the "radius"

$$G_{Ep}(t) \approx G_{Mp}(t)/g_p \approx G_{Mn}(t)/g_n \approx G_D(t); \tag{A.11.19}$$

$$G_D(t) = (1-t/t_D)^{-2}; \quad t_D = 0.71\,\mathrm{GeV}^2 \hat{=} 18.2\,\mathrm{fm}^{-2} \hat{=} 36.4\mu^2. \tag{A.11.20}$$

Since $F_{1n}(t)$ turns out to be small, the form factor $G_{En}(t)$ is approximately given by the "Foldy-term" [see Eq. (A.11.9)] and the above approximation gives for small t

$$G_{En}(t) \approx \frac{t}{4m^2} G_D(t). \tag{A.11.21}$$

The "iso-scaling law" states that

$$\frac{G_{Es}(t)}{G_{Es}(0)} \approx \frac{G_{Ms}(t)}{G_{Ms}(0)}; \quad \frac{G_{Ev}(t)}{G_{Ev}(0)} \approx \frac{G_{Mv}(t)}{G_{Mv}(0)}. \tag{A.11.22}$$

For quantitative calculations it is necessary to check whether the accuracy of these empirical relations is sufficient (Sect. 2.5.2).

The relation $2F_{1v}(t) \approx F_\pi(t)$ between nucleon and pion form factors should not be used, because the spectral functions $2\,\mathrm{Im}\,F_{1v}$ and $\mathrm{Im}\,F_\pi$ differ strongly between $t=4\mu^2$ and the ϱ-peak (Fig. 5.2.1).

11.2 The pion form factor

The pion form factor is defined by matrix elements of the isovector electromagnetic current, taken between charged pion states in the space-like region ($t<0$) and between a $\pi^+\pi^-$-state and the vacuum in the time-like region

$$\begin{aligned} \langle \pi^\pm(q')|j_\mu^v(0)|\pi^\pm(q)\rangle &= (q'+q)_\mu F_\pi(t); \quad t=(q'-q)^2<0; \\ \langle \pi^+(q')\pi^-(q)|j_\mu^v(0)|0\rangle &= (q'-q)_\mu F_\pi(t); \quad t=(q'+q)^2>4\mu^2. \end{aligned} \tag{A.11.23}$$

It follows from time-reversal invariance that $F_\pi(t)$ is real in the space-like region. It becomes complex-valued for $t>4\mu^2$.

The normalization is $F_\pi(0)=1$. The definition of the "pion radius" is analogous to that of the nucleon radius in Eq. (A.11.6)

$$F_\pi(t) = 1 + \tfrac{1}{6}\langle r_\pi^2\rangle t + \dots. \tag{A.11.24}$$

The dispersion relation for $F_\pi(t)$ reads

$$F_\pi(t) = \frac{1}{\pi}\int_{4\mu^2}^{\infty} \frac{\mathrm{Im}\,F_\pi(t')}{t'-t}\,dt', \tag{A.11.25}$$

where $\mathrm{Im}\,F_\pi(t)$ is restricted by a sum rule which ensures that $F_\pi(0)=1$.

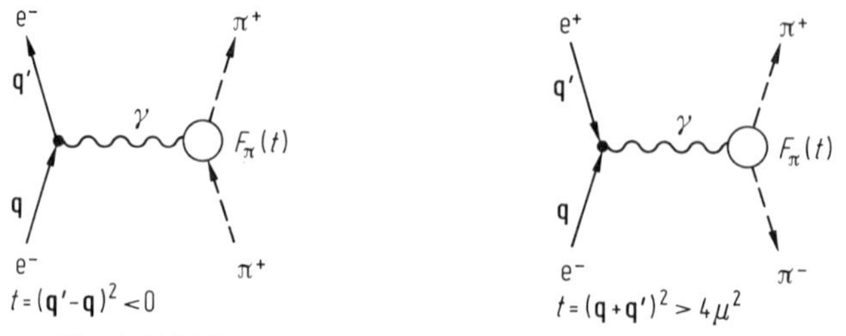

Fig. A.11.3. Electromagnetic pion form factor. The same form factor describes the one-gamma exchange contributions to the reactions $e^-\pi^+ \to e^-\pi^+$ and $e^-e^+ \to \pi^-\pi^+$.

In the range where 4-pion intermediate states in the extended unitarity relation can be neglected, the phase of $F_\pi(t)$ is given by the $I=J=1$ $\pi\pi$-scattering phase shift $\delta_1^1(t)$. $\operatorname{Im} F_\pi(t)$ can be expressed in terms of the $\pi\pi$-partial wave $h_1^1(t)=N/D=NF_\pi$ (see Sects. A.7.6 and 2.4.6.3)

$$\operatorname{Im} F_\pi(t)=\frac{q_t^3}{\sqrt{t}}F_\pi(t)h_1^{1*}(t)=\frac{q_t^3}{\sqrt{t}}N(t)|F_\pi(t)|^2\,; \qquad 4\mu^2<t<16\mu^2\,. \tag{A.11.26}$$

If $F_\pi(t)$ is free of zeros, one can write a phase representation

$$F_\pi(t)=\exp\left[\frac{t}{\pi}\int_{4\mu^2}^{\infty}\frac{\delta(t')}{t'(t'-t)}\mathrm{d}t'\right]; \qquad F_\pi\propto t^{-\delta(\infty)/\pi} \quad \text{as} \quad t\to\infty\,. \tag{A.11.27}$$

Otherwise, this representation is valid for $F_\pi(t)$ divided by a polynomial, which cancels the zeros of F_π.

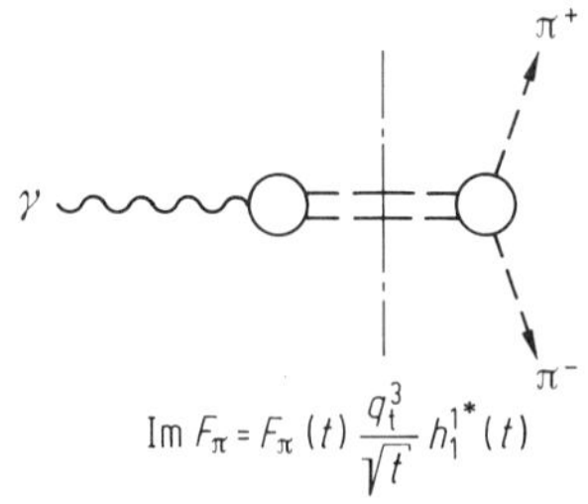

Fig. A.11.4. Two-pion intermediate states of the $\gamma\pi^+\pi^-$ vertex.

Assuming that $F_\pi(t)$ has no zeros and that its modulus is known for all $t>4\mu^2$, one can write a dispersion relation for the function $\ln F_\pi(t)/(4\mu^2-t)^{1/2}$ and arrive at results for the phase at $t>4\mu^2$ and for $F_\pi(t)$ at $t<0$

$$\begin{aligned}\ln F_\pi(t)&=(-t)\sqrt{4\mu^2-t}\,\frac{1}{\pi}\int_{4\mu^2}^{\infty}\frac{\ln|F_\pi(t')|\mathrm{d}t'}{t'(t'-t)\sqrt{t'-4\mu^2}}; \qquad t<0;\\ \delta(t)&=t\sqrt{t-4\mu^2}\,\frac{1}{\pi}\int_{4\mu^2}^{\infty}\frac{\ln|F_\pi(t')|\mathrm{d}t'}{t'(t'-t)\sqrt{t'-4\mu^2}}; \qquad t>4\mu^2\,.\end{aligned} \tag{A.11.28}$$

By letting $t\to\infty$ one obtains the sum rule

$$\int_{4\mu^2}^{\infty}\mathrm{d}t\,\frac{\ln|F_\pi(t)|}{t\sqrt{t-4\mu^2}}=0\,. \tag{A.11.29}$$

At $t>4\mu^2$ the modulus of the pion form factor is measurable, because the reaction $e^+e^-\to\pi^+\pi^-$ is strongly dominated by the one-photon exchange graph. The expression for the differential cross section reads

$$\frac{\mathrm{d}\sigma}{\mathrm{d}\Omega}(e^+e^-\to\pi^+\pi^-)=\frac{\alpha^2}{8t}\left(1-\frac{4\mu^2}{t}\right)^{3/2}\sin^2\theta|F_\pi(t)|^2\,. \tag{A.11.30}$$

It is necessary to consider corrections from $\pi^+\pi^-$ production via ω-exchange, which is forbidden by G-parity conservation in strong interactions, but allowed to order α, if electromagnetic interactions are included.

Models for the pion form factor

It turns out that ϱ-dominance gives a reasonable first approximation. In the space-like region one obtains, ignoring finite width effects

$$F_\pi(t)=\frac{1}{1-t/m_\varrho^2}; \qquad r_\pi^2=6/m_\varrho^2\,. \tag{A.11.31}$$

Fits to the most recent data in this region (78L Bebek, see Table 2.5.2.2) suggest an effective mass value $m_\varrho^2=0.46\ \mathrm{GeV}^2$, which is smaller than the rho-mass squared ($0.59\ \mathrm{GeV}^2$), the corresponding values of the radius being 0.71 and 0.63 fm, respectively.

In the time-like region one needs at least two parameters: the location and the width of the rho resonance. A simple Breit-Wigner parametrization would be applicable only near the resonance peak. Therefore, one prefers a Chew-Mandelstam-type effective range parametrization, where two parameters are sufficient in order to describe in addition the behaviour in the neighborhood of threshold.

The ansatz for $F_\pi(t)$ follows immediately, if one notes that the inverse of the D-function for $\pi\pi$-scattering in the $I=J=1$ state, Eq. (A.7.55), has the correct analytic properties and fulfills also the unitarity condition (A.11.26), aside from a normalization factor.

$$F_\pi(t)=\frac{D(0)}{D(t)}\equiv\frac{\Delta_\varrho(0)}{\Delta_\varrho(t)}. \qquad \text{(A.11.32)}$$

The second equation expresses the fact that the D-function is proportional to the $\delta_{\mu\nu}$-part of the ϱ-propagator

$$\Delta_\varrho(t)=A-Bt+\gamma h(t);\quad \Delta_\varrho(0)=t_\varrho \quad \text{[67 L Kroll; Gourdin (74 R) uses a different convention]}. \qquad \text{(A.11.33)}$$

In the neighbourhood of the resonance peak, $F_\pi(t)$ can be approximated in such a way that one obtains a Breit-Wigner type formula

$$F_\pi(t)\approx\frac{t_\varrho\{1+(\Gamma_\varrho/m_\varrho)d\}}{(t_\varrho-t)-\mathrm{i}m_\varrho\Gamma_\varrho(q_t/q_\varrho)^3(m_\varrho/\sqrt{t})};\quad t \text{ near } t_\varrho. \qquad \text{(A.11.34)}$$

Γ_ϱ is approximately the full width at half height, i.e. at $\sqrt{t}=m_\varrho\pm\Gamma_\varrho/2$ the value of $|F_\pi|^2$ is about half of its maximum value. At $t=t_\varrho$ we have

$$F_\pi(t_\varrho)=\mathrm{i}(d+m_\varrho/\Gamma_\varrho)=\mathrm{i}m_\varrho/(R\Gamma_\varrho). \qquad \text{(A.11.35)}$$

Since the parametrization (A.11.32) occurs in most of the recent discussions of the pion form factor, it is of interest to list the formulas which relate the different parameters with m_ϱ and Γ_ϱ.

$$t_\varrho=m_\varrho^2;\quad q_\varrho=\sqrt{t_\varrho/4-\mu^2};\quad d=\frac{3\mu^2}{\pi q_\varrho^2}\ln\frac{m_\varrho+2q_\varrho}{2\mu}+\frac{m_\varrho}{2\pi q_\varrho}\left(1-\frac{2\mu^2}{q_\varrho^2}\right);$$

$$R=-\frac{\mathrm{d}}{\mathrm{d}t}\operatorname{Re}\Delta_\varrho(t)\Big|_{t=t_\varrho}=\left(1+\frac{\Gamma_\varrho}{m_\varrho}d\right)^{-1};\quad \gamma=\frac{\Gamma_\varrho t_\varrho}{q_\varrho^3}R;\quad c_1=\gamma\operatorname{Re}h(t_\varrho)/q_\varrho^2;$$

$$c_2=\gamma q_\varrho^2\frac{\mathrm{d}}{\mathrm{d}t}\operatorname{Re}h(t)/q_t^2\Big|_{t=t_\varrho}=\gamma q_\varrho^2\left[\frac{1}{2\pi t_\varrho}+\frac{\mu^2}{\pi q_\varrho m_\varrho^3}\ln\frac{m_\varrho+2q_\varrho}{2\mu}\right]. \qquad \text{(A.11.36)}$$

$$A=t_\varrho(R+c_2)+c_1\mu^2;\quad B=R+c_2+c_1/4;\quad \langle r_\pi^2\rangle=\frac{6}{t_\varrho}\left(B-\frac{\gamma}{3\pi}\right).$$

Using complete ϱ-dominance and the formalism of 67 L Kroll-Lee-Zumino, which is based on the "current-field identity" $j_\mu^\nu(\mathbf{x})=m_\varrho^2\varrho_\mu(\mathbf{x})/f_\varrho$ (61 L Gell-Mann, 69 R Sakurai, p. 54), one obtains from the above parametrization

$$\frac{f_\varrho^2}{4\pi}=\frac{G_{\varrho\pi\pi}^2}{4\pi}=\frac{3}{2}\gamma=\frac{3}{2}t_\varrho N;$$

$$\Gamma(\varrho\to\pi^+\pi^-)\equiv\Gamma_\varrho=\frac{f_\varrho^2}{4\pi}\frac{2q_\varrho^3}{3t_\varrho}\frac{1}{R};\quad \Gamma(\varrho\to e^+e^-)=\frac{\alpha^2}{3}\frac{4\pi}{f_\varrho^2}\frac{m_\varrho}{R}. \qquad \text{(A.11.37)}$$

The N-function is constant and the expressions for the decay rates differ from the result of lowest order perturbation theory, Eq. (A.8.29) by a finite width correction.

Table A.11.2 lists the numerical values of the parameters which occur in Eqs. (A.11.36) and (A.11.37), (A.7.55).

Table A.11.2. Parameters of a "Gounaris-Sakurai fit" to the pion form factor, Eq. (A.11.32).

Input: $m_\varrho=0.775$ GeV, $\Gamma_\varrho=0.149$ GeV

$t_\varrho=0.601\ \mathrm{GeV}^2$	$A=0.611\ \mathrm{GeV}^2$	$B=1.206$	$q_\varrho=0.361$ GeV
$c_1=0.865$	$c_2=0.074$	$\gamma=1.735$	$R=0.916$
$d=0.48$	$r_\pi=0.63$ fm	$\operatorname{Im}F_\pi(t_\varrho)=5.68$	$f_\varrho^2/4\pi=2.60$
$a_1^1=0.033\mu^{-3}$	$N=0.056\mu^{-2}$	$h(0)=-\mu^2/\pi$	$h(4\mu^2)=0$

It is interesting to note that r_π differs only by 1% from the value $\sqrt{6/t_\varrho}$ of Eq. (A.11.31). A differs only slightly from t_ϱ. a_1^1 is near the "Weinberg value" $a_1^1=1/(12\pi f_\pi^2\mu)=0.030\mu^{-3}$.

11.3 Analytic continuation of electromagnetic form factors into the second sheet

The continuation is performed with the same method as in Sect. A.7.3. We also include the continuation of the partial wave $f_{\pm}^{J}(t)$, because the result is of interest for Sect. A.3.5.3.

The unitarity relations Eqs. (A.5.23) and (A.11.26) are written as follows

$$\operatorname{Im} F_{\pi}(t) = \varrho_{\pi}(t) F_{\pi}(t) A_1^{1*}(t); \quad \operatorname{Im} f_{\pm}^{J}(t) = \varrho_{\pi}(t) f_{\pm}^{J}(t) A_J^{I*}(t); \quad 4\mu^2 < t < 16\mu^2, \tag{A.11.38}$$

where $\varrho_{\pi}(t) = 2q_t/\sqrt{t}$ and the cuts are chosen such that $\varrho_{\pi}(t^*) = -\varrho_{\pi}^*(t)$. $A_J^I(t) = T_J^I(t)\sqrt{t}/2q_t$ is defined in Eq. (A.7.52) and $I=0$ or 1 for even and for odd J, respectively.

One obtains the following expressions for the pion form factor and the $\pi\pi N\bar{N}$-amplitude in the second sheet

$$F_{\pi}^{\mathrm{II}}(t) = \frac{F_{\pi}(t)}{1 + 2\mathrm{i}\varrho_{\pi}(t) A_1^1(t)}; \quad f_{\pm}^{J,\mathrm{II}}(t) = \frac{f_{\pm}^{J}(t)}{1 + 2\mathrm{i}\varrho_{\pi}(t) A_J^I(t)}. \tag{A.11.39}$$

The situation is different for the nucleon isovector form factors, because these form factors do not occur on the right hand side of the unitarity relation, Eq. (A.11.13), which is now written

$$\operatorname{Im} G_{\mathrm{Ev}}(t) = \varrho_{\mathrm{E}}(t) f_{+}^{1}(t) F_{\pi}^*(t); \quad \operatorname{Im} G_{\mathrm{Mv}}(t) = \varrho_{\mathrm{M}}(t) f_{-}^{1}(t) F_{\pi}^*(t), \tag{A.11.40}$$

where

$$\varrho_{\mathrm{E}}(t) = q_t^3/m\sqrt{t}; \quad \varrho_{\mathrm{M}}(t) = q_t^3/\sqrt{2t}. \tag{A.11.41}$$

Again, the cuts are chosen such that $\varrho_{\mathrm{E}}(t^*) = -\varrho_{\mathrm{E}}^*(t)$ and $\varrho_{\mathrm{M}}(t^*) = -\varrho_{\mathrm{M}}^*(t)$. The form factors in the second sheet are obtained by substituting one of the expressions of Eq. (A.11.39) into Eq. (A.11.40)

$$\begin{aligned} G_{\mathrm{Ev}}^{\mathrm{II}}(t) &= G_{\mathrm{Ev}}(t) - 2\mathrm{i}\varrho_{\mathrm{E}}(t) \frac{f_{+}^{1}(t) F_{\pi}(t)}{1 + 2\mathrm{i}\varrho_{\pi}(t) A_1^1(t)}; \\ G_{\mathrm{Mv}}^{\mathrm{II}}(t) &= G_{\mathrm{Mv}}(t) - 2\mathrm{i}\varrho_{\mathrm{M}}(t) \frac{f_{-}^{1}(t) F_{\pi}(t)}{1 + 2\mathrm{i}\varrho_{\pi}(t) A_1^1(t)}. \end{aligned} \tag{A.11.42}$$

The functions $G_{\mathrm{Ev}}^{\mathrm{II}}(t)$ and $G_{\mathrm{Mv}}^{\mathrm{II}}(t)$ have the correct analytic properties. In order to show that they also have the correct boundary values for real $t > 4\mu^2$, we start from Eq. (A.11.40)

$$\begin{aligned} 2\mathrm{i} \operatorname{Im} G_{\mathrm{Ev}}(t) &= G_{\mathrm{Ev}}(t+\mathrm{i}0) - G_{\mathrm{Ev}}(t-\mathrm{i}0) = 2\mathrm{i}\varrho_{\mathrm{E}}(t+\mathrm{i}0) f_{+}^{1}(t+\mathrm{i}0) F_{\pi}(t-\mathrm{i}0) \\ &= 2\mathrm{i}\varrho_{\mathrm{E}}(t+\mathrm{i}0) f_{+}^{1}(t+\mathrm{i}0) F_{\pi}^{\mathrm{II}}(t+\mathrm{i}0). \end{aligned} \tag{A.11.43}$$

Eq. (A.11.42) is obtained, if we solve Eq. (A.11.43) for $G_{\mathrm{Ev}}(t-\mathrm{i}0) = G_{\mathrm{Ev}}^{\mathrm{II}}(t+\mathrm{i}0)$ and substitute Eq. (A.11.39) for $F_{\pi}^{\mathrm{II}}(t+\mathrm{i}0)$.

It is seen that $f_{\pm}^{J,\mathrm{II}}$, $G_{\mathrm{Ev}}^{\mathrm{II}}$, and $G_{\mathrm{Mv}}^{\mathrm{II}}$ inherit the branch point singularity of $f_{\pm}^{J}(t)$ at $t = 4\mu^2 - \mu^4/m^2$.

12 References

In this section the reader can find references of books, reviews and original papers, in which the formulas of the Appendix are derived and discussed. In some cases I have also mentioned publications on closely related topics not treated in this volume.

Notation: "80 R" refers to a book or review published in 1980, whose full reference is given in Sect. 3.1. The original papers and preprints are denoted by the letter "L" (80 L etc.) and listed in Sect. 3.2.

The basic formulas are treated in a fairly large number of books and lecture notes, for instance in 61 R CINI-CORINALDESI, 61 R CINI-VITALE, 64 R KALLEN, 67 R GASIOROWICZ, 67 R PILKUHN, 68 R MUIRHEAD, 69 R CENCE, 69 R MOORHOUSE. The books 70 R MARTIN-SPEARMAN, 73 R BRANSDEN-MOORHOUSE, and 74 R PERL cover the theory of pion-nucleon scattering in greater detail, but they can also be used as an introduction.

Other sources of formulas belonging to many sections of the Appendix are: BARUTs book (67 R) and 60 R GOLDBERGER, 64 R GOLDBERGER-WATSON, 67 R HAMILTON-1 and 2, 69 R BARGER-CLINE, 74 R HAMILTON-PETERSEN.

Finally, we mention some books in which pion-nucleon scattering is treated within the framework of quantum field theory: 59 R HAMILTON, 61 R SCHWEBER, 63 R and 69 R NISHIJIMA, 64 R HAGEDORN-2, 65 R BJORKEN-DRELL, 69 R ROMAN, 76 R BOGOLJUBOV-SHIRKOV, 80 R ITZYGSON-ZUBER.

In the following we shall give references belonging to the different sections of the Appendix, omitting frequently the above mentioned books, reviews and lecture notes.

1. Kinematics

60 R GOLDBERGER, 64 R HAGEDORN-1, 70 R MARTIN-SPEARMAN. Fixed lab. scattering angle: 77 L BORIE. Continuation to the physical t-channel region: 64 L TRUEMAN-WICK, 67 R PILKUHN, 70 R MARTIN-SPEARMAN.

2. Scattering amplitudes

60 R GOLDBERGER, 70 R MARTIN-SPEARMAN, 73 R BRANSDEN-MOORHOUSE. Transversity amplitudes: 69 L GERSTEN, 72 L and 76 L BARRELET. s-channel lab. frame amplitudes: 69 L MICHAEL, 75 L ZOVKO. t-channel amplitudes: 60 L FRAZER-FULCO-1, 70 R MARTIN-SPEARMAN. u-channel amplitudes: 69 R BARGER-CLINE. Helicity amplitudes: 59 L and 61 R JACOB, 64 R JACOB-CHEW, 64 L TRUEMAN-WICK, 70 R MARTIN-SPEARMAN. Isospin structure: 60 R ROMAN, 66 R WERLE, 68 R MUIRHEAD, 71 R CARRUTHERS. Electromagnetic corrections: see the references in Sects. 2.1.17 and 2.6.2.

3. Expansions of scattering amplitudes

60 R GOLDBERGER, 70 R MARTIN-SPEARMAN, 73 R BRANSDEN-MOORHOUSE. Our choice for the d-functions agrees with that in 82 R ROOS (Particle Data Group) and in the following references: 59 L JACOB-WICK, 70 R MARTIN-SPEARMAN, 73 R BRANSDEN-MOORHOUSE, 74 R PERL, 79 R PILKUHN, 79 R NAGELS. Tables of Clebsch-Gordan coefficients can be found in 82 R ROOS, p. 24. Another convention was used in 77 R COLLINS.

Parity-conserving helicity amplitudes and naturality: 64 L GELL-MANN, 69 R BARGER-CLINE (p. 59), 70 R MARTIN-SPEARMAN (p. 438), 77 R COLLINS (p. 127). Expansion of transversity amplitudes: 72 L BARRELET.

Impact parameter representations: 62 L BLANKENBECLER-GOLDBERGER, 65 L COTTINGHAM-PEIERLS, 68 R and 71 R ISLAM, 71 R ZACHARIASEN. Review of the earlier work on eikonal approximations: 59 R GLAUBER. Two-dimensional Fourier transform: 68 L CHOU-YANG. Extensions and discussions of the formalism: 66 L ADACHI, 66 L LUMING, 66 L PREDAZZI, 67 L ARNOLD, 67 L HENZI, 73 L HÖHLER-1, 73 L and 74 L KUPSCH. 76 L ISLAM. Use of longitudinal and transverse momentum variables: 75 L ELVEKJAER-PETERSEN.

t-channel partial wave expansion: 60 L FRAZER-FULCO-1. s-channel threshold and subthreshold expansion: 72 L HÖHLER-1. Behaviour near $t=4\mu^2$: 60 L FRAZER-FULCO-1, 75 L HÖHLER-4, 79 R HÖHLER-KAISER-KOCH-PIETARINEN. Expansion at the t-channel threshold $t=4m^2$: 60 L FRAZER-FULCO-1, 66 R WERLE, 68 L COHEN-TANNOUDJI, 68 L JACKSON, 70 R MARTIN-SPEARMAN, 79 R PILKUHN (p. 157).

4. Measurable quantities

Description of spin states: 70 R MARTIN-SPEARMAN. Polarization and spin-rotation measurements: 67 R HAMILTON-2 (Course B), 72 L DE LESQUEN, 74 R HAMILTON-PETERSEN, 74 L KELLY, 80 R BOURRELY-LEADER-SOFFER.

Legendre coefficients: Tables of the coefficients in terms of partial wave amplitudes can be found in 76 L AYED and 79 L VON SCHLIPPE, and formulas for the general case in 78 L OYANAGI and 79 L HÖHLER-1.

$\bar{N}N \leftrightarrow \pi\pi$ cross sections: 60 L FRAZER-FULCO-1, 66 R WERLE, 67 R BARUT, 80 R PENNINGTON.

Electromagnetic contributions to cross sections and polarization parameters: 78 L BORIE-2, 80 R BOURRELY-LEADER-SOFFER and the references given in Sects. 2.1.17 and 2.6.2.

5. Unitarity

5.1 Unitarity for πN scattering amplitudes

Books and reviews: 64 R GOLDBERGER-WATSON, 66 R EDEN-LANDSHOFF-OLIVE-POLKINGHORNE, 67 R HAMILTON-2 (Course B), 67 R BARUT, 67 R and 73 R MARTIN, 70 R MARTIN-CHEUNG, 70 R MARTIN-SPEARMAN, 79 R PILKUHN.

Elastic unitarity: 61 R CINI-VITALE, 67 R BARUT, 68 R COLLINS-SQUIRES, 59 L MANDELSTAM-2.

Unitarity for impact parameter transforms: 62 L BLANKENBECLER, 65 L COTTINGHAM, 73 L HÖHLER-1.

5.2 Unitarity for $\pi\pi N\bar{N}$-amplitudes

60 L FRAZER-FULCO-1.

5.3 Unitarity bounds

Books and reviews: 67 R EDEN-1 and 2, 71 R and 73 R EDEN, 70 R MARTIN-CHEUNG, 70 R SOMMER, 71 R ATKINSON, 71 R SINGH, 72 R ROY, 72 R YNDURAIN, 72 R VAN HIEU, 81 R FISCHER.

Positivity: 67 R and 73 R MARTIN, 70 R MARTIN-CHEUNG, 70 R SOMMER. Inequalities for Legendre polynomials: 72 R ROY (Appendix).

Table A.5.1: 61 L FROISSART, 63 L MARTIN, 70 R SOMMER, 72 R ROY.

Slope of the diffraction peak: 64 L MAC DOWELL-MARTIN, 71 L SINGH, 75 L CORNILLE, 76 L MAHOUX, 76 L CHUNG, 77 L AUBERSON, 79 L GANGAL.

Shape of the diffraction peak: 70 L SINGH-ROY, 79 L ROY [Eq. (A.5.38)], 74 L UCHIYAMA, 77 L RAINA. Additional constraints: 70 L JACOBS, 71 R EINHORN, 71 L SAVIT, 72 L HAHN.

Scaling: 76 L CORNILLE-1 and 2, 76 R MARTIN.

Helicity flip amplitude: 71 L MUETTER.

Constraint on the phases: 74 L CORNILLE.

Bounds at finite energies: 71 L COMMON, 72 R YNDURAIN, 73 L EDEN, 74 L GRASSBERGER, 76 L COMMON.

Partial wave amplitudes at large ℓ: 70 R MARTIN-SPEARMAN (p. 386), 67 R EDEN-1 (p. 170), 79 R HAAN.

5.4 POMERANCHUK-type theorems

58 L POMERANCHUK, 63 L LOGUNOV, 63 L VAN HOVE, 67 R EDEN-1 and 2, 71 R EDEN, 72 R ROY, 78 L FISCHER-SALY, 81 R FISCHER.

6. Symmetries and analytic properties of invariant amplitudes

6.1 Behaviour of invariant amplitudes under symmetry operations

See for instance: 60 R CHEW, 66 R WERLE, 67 R BARUT, 68 R HAMILTON-2 (Course B), 69 R BURKHARDT, 70 R MARTIN-SPEARMAN, 71 R CARRUTHERS.

6.2 The MANDELSTAM representation

Fundamental papers: 58 L, 59 L MANDELSTAM-1 and 2.

Books and reviews: 60 R, 61 R, 65 R, and 66 R CHEW, 61 R JACKSON, 61 R CINI-VITALE, 62 R AMATI-FUBINI, 62 R BERESTETSKI, 62 R MANDELSTAM, 64 R BARUT, 64 R JACOB-CHEW, 65 R FROISSART-OMNES, 66 R EDEN-LANDSHOFF-OLIVE-POLKINGHORNE, 67 R BARUT, 67 R EDEN-1, 67 R HAMILTON-2, 68 R COLLINS-SQUIRES, 69 R BURKHARDT, 69 R SHIRKOV-SEREBRYAKOV-MESHCHERYAKOV, 70 R OMNES, 74 R QUEEN-VIOLINI, 77 R COLLINS.

6.2.4 Convergence of the partial wave expansions

s-channel: 58 L LEHMANN, 63 R HAMILTON-WOOLCOCK, 66 L MARTIN, 67 L SOMMER, 70 R MARTIN-CHEUNG, 70 R SOMMER.

t-channel: 60 L FRAUTSCHI-WALECKA, 60 L FRAZER-FULCO-2. Along a hyperbolic path: 77 L BORIE.

6.3 Fixed-*t* dispersion relations

Proofs of the dispersion relation (only books and reviews): 58 R BOGOLJUBOV-MEDVEDEV-POLIVANOV, 59 R LEHMANN, 60 R GOLDBERGER, 60 R OMNES, 64 R HAGEDORN-2, 70 R SOMMER, 76 R BOGOLJUBOV-SHIRKOV.

Applications: 57 L CHEW ("CGLN"), 60 L HÖHLER, 61 R CINI-VITALE, 62 R AMATI-FUBINI, 63 R HAMILTON-WOOLCOCK, 65 R LYTH, 71 L HÖHLER-2, 73 R HÖHLER, 74 R HAMILTON-PETERSEN. See also the references in Sects. 2.1.7 and 2.1.16.

Subtractions: 58 L MANDELSTAM, 70 R SOMMER, 70 L HÖHLER-1.

Evaluation of the dispersion relations: see the references in Sects. 2.1.7 and 2.1.16.

The expansion method: PIETARINEN (72 L-2 and 3, 73 L, 76 L-1, 78 L-1). Other expansion methods: CIULLI (62 L, 69 L, 73 L, 73 R), CIULLI-POMPONIU-SABBA-STEFANESCU (75 R), CUTKOSKY (68 L, 69 L, 71 R, 73 L, 79 L-1), HAMILTON-PETERSEN (74 R), LANGBEIN (75 L-1, 79 L), LICHARD (76 L-1 and 2, 77 L), PISUT (69 L), PRESNAJDER (71 L, 75 L).

Finite contour dispersion relations: 70 R PHILLIPS, 71 L NIELSEN, 70 L BARGER-2, 73 L BAACKE, 74 R HAMILTON-PETERSEN. See also 63 L IGI. This method has the disadvantage that the constraints from experimental data at high energies are partially or completely ignored.

6.4 Forward dispersion relations

55 L GOLDBERGER-1 and 2, 55 L and 56 L OEHME, 60 R GOLDBERGER.

Connection with causality: 56 L TOLL, 59 R LEHMANN, 60 R BOHR, 60 R HAMILTON, 60 R HILGEVOORD, 70 L CREUTZ, 70 L OEHME-1, 70 R SOMMER, 72 R NUSSENZVEIG, 76 R BOGOLJUBOV-SHIRKOV, 76 R IAGOLNITZER.

Subtractions: 61 L SUGAWARA, 64 L JIN-2, 65 L KHURI, 70 L HÖHLER-1. Additional real term in the dispersion relation for C^-: 75 L JOYNSON. See also 80 R HÖHLER-KAISER (Sect. 8.2).

Phase-modulus correlation: 63 L SUGAWARA, 65 L JIN, 67 R EDEN-1, 68 L ODORICO, 72 L MCCLURE, 76 L and 81 R FISCHER.

Constraints for the asymptotic behaviour of C^+: 65 L KHURI.

Modified dispersion relations: 57 L GILBERT, 65 L ADLER, 67 L KHALFIN, 68 L LIU-OKUBO, 70 L CIULLI, 71 L ROSS, 72 L and 74 L SHEPARD, 73 R CIULLI, 73 L FISCHER, 74 L LOMSADZE, 74 R QUEEN-VIOLINI, 80 R HÖHLER-KAISER.

Analytic parametrization of the high energy amplitude: 67 R EDEN-1, 70 L BARGER-2, 72 L BARGER-5, 73 L BOURRELY, 75 L HÖHLER-1.

Derivative analyticity relations: 74 L BRONZAN, 74 L EICHMANN, 74 R JACKSON, 75 R HÖHLER, 75 L SIDHU, 75 L SUKHATME, 78 L and 81 R FISCHER, 80 R HÖHLER-KAISER.

Continuation into the second sheet: 61 R OEHME, 67 R BARUT. Quasipole: 65 L BARUT, 67 L SAWADA.

6.5 The dispersion relation for the amplitude $\partial C^+/\partial t$ at $t=0$

71 L HÖHLER-1, 81 L KOCH.

6.6 Fixed-*u* and fixed-*s* dispersion relations

65 R LYTH, 66 L OADES, 70 L SCHLAILE, 78 L and 82 L KOCH-2.

6.7 Hyperbolic dispersion relations

71 L GREENBERG, 73 L HITE-1 and 2, 77 L HEDEGAARD-JENSEN, 82 L KOCH.

Interior dispersion relation: HITE (74 L, 75 L, 76 L, 78 L), 75 L JACOB, 76 L MOIR, 76 L and 78 L KAISER, 77 L BORIE-KAISER. (Printing error in Sect. 3.2: 73 HITE-2 appeared in 1972.)

6.8 Fixed center-of-mass angle dispersion relations

61 L EFREMOV, 69 R SHIRKOV-SEREBRYAKOV-MESHCHERYAKOV, 77 L HÖHLER-1.

6.9 Backward dispersion relations

62 L Atkinson, 66 L and 69 R Lovelace, 68 L Odorico, 70 L Engels-1 and 2, 70 L Nielsen, 77 L Borie, 78 L Pietarinen-2, 79 L Cutkosky-1. See the references in Sect. 2.6.7 for methods used for the evaluation of the dispersion relation.

6.10 Fixed-ν dispersion relations

72 L Engels, 74 L Nielsen-1, 77 L Höhler-2, 79 R Höhler-Kaiser-Koch-Pietarinen.

7. Analytic properties of partial wave amplitudes

7.1 s-channel partial wave dispersion relations

Fundamental papers: 59 L Mac Dowell, 60 L Frautschi-Walecka, 60 L Frazer-Fulco-2, 62 L Hamilton-2.

Books, reviews and lecture notes: Hamilton (67 R-1 and 2, 68 R, 71 R-1 and 2), 67 R Barut, 67 R Donnachie, 67 R Pilkuhn, 69 R Burkhardt, 70 R Martin-Spearman, 72 R Hamilton-Tromborg, 76 R Kaschluhn.

S-waves: 61 L Hamilton-1 and 2, 62 L Hamilton-2, 70 L Nielsen, 71 L Hamilton.

P-waves: 62 L Hamilton-2, 64 L Donnachie-1, 67 L Kaschluhn, 72 R Hamilton-Tromborg, 72 L Lea, 73 L Brandt, 76 L Hamilton, 81 L Hutt, 81 L Sawada-1 and 2.

Higher partial waves and resonances: 64 L Donnachie-2, 65 L Donnachie-1 and 2, 70 L, 71 L and 72 L Gustafson, 72 R Gustafson-Hamilton.

Constraints from analyticity and elastic unitarity in the t-channel: 79 R Haan.

Some other investigations on partial wave dispersion relations: 68 L Frye-Warnock (N/D method), 69 L Petersen, 71 L and 72 L Brandt, 72 L Choudhary, 74 L and 80 L Nenciu.

The most recent evaluation: 80 L Hutt-Koch.

Generalized potential: 61 R Chew (Sects. 7 and 8), 70 L Engels-3, 70 R Steiner.

Unitary sum rule: 65 L Donnachie-1, 67 R Donnachie, 67 R Hamilton-2 (Course B).

Behaviour at $s=0$: 69 L Jakob-Steiner, 70 L Nielsen, 70 L Petersen, 71 R Hamilton, 72 L Elvekjaer, 75 L Diu.

Dispersion relation in the W-plane: 60 L Frautschi-Walecka.

7.2 s-channel partial wave relations

Approximation: 57 L Chew ("CGLN"), 63 R Hamilton-Woolcock; exact version for the S-wave: 60 L Höhler-Dietz; general case: 69 L Baacke-1, 70 R Baacke-Steiner, 71 R Steiner, 74 L Hamilton-Petersen. New evaluation: R. Koch (work in progress), see Fig. 2.2.22.

Generalized potential: 70 L Engels-3, 70 R Steiner. Partial wave crossing relations: 71 L Jakob-Steiner, 71 R Steiner.

Partial wave relations derived from hyperbolic dispersion relations: 73 L Hite-Steiner-1. Evaluation: J. Stahov (work in progress).

7.3 Analytic continuation of s-channel partial waves into the second sheet

61 L and 61 R Oehme, 65 R Barton, 65 L and 67 R Barut.

7.4 The Froissart-Gribov representation for $\pi\pi$-$N\bar{N}$ partial wave amplitudes

Froissart (unpubl.), 62 L Gribov, 67 R Eden-1, 68 R Collins-Squires, 70 R Martin-Spearman, 71 R Collins, 77 R Collins.

An application of the projection formula at $t=4\mu^2$: 71 L Common.

7.5 t-channel partial wave dispersion relations

60 L Frazer-Fulco-1. Recent evaluations: see Sect. 2.4.6.

7.6 The Chew-Mandelstam model for $\pi\pi$-partial waves

60 L Chew-Mandelstam, 76 R Martin-Morgan-Shaw (Sect. 9.2). See also 68 L and 69 L Gounaris, 74 R Gourdin.

KSFR relation: 66 L Kawarabayashi, 66 L Riazuddin.

8. Low energy models

8.1 Nucleon and resonance exchanges

Reviews on "polology": 59 R Chew, 62 R Amati-Fubini.

Nucleon exchange Born terms (second order perturbation theory): 50 L Ashkin [Amati-Fubini (62 R, p. 387) mention that the ps and pv *pole* contributions are the same, but the *Born terms* are different; the equivalence theorem is valid for instance in pion photoproduction. See 56 R Bethe-De Hoffmann-Schweber, Vol. I, Sect. 26]. Effect of the neutron-proton mass difference: 77 L Gensini, 77 L Höhler-2, 79 R Pilkuhn (p. 280).

$\Delta(1232)$ isobar exchange: 56 R Umezawa, 58 L Fronsdal, 67 L Raman, 67 L Schnitzer, 68 L Peccei, 70 L Engels-1, 71 L Hagen, 71 L Nath, 72 L Höhler-1 [Fig. A.8.2 and Eq. (A.8.15)], 73 L, 75 L, and 77 L Olsson, 75 L Jenkins.

ϱ-exchange: 60 L and 69 R Sakurai, 61 L Gell-Mann, 67 R Hamilton-1.

8.2 The current algebra model

PCAC relation: 65 L Adler-1. Eq. (A.8.31): 71 L Brown, 71 L Höhler-2, 73 L and 75 L Olsson, 75 R Pagels.

Simplest vector dominance model for the isovector nucleon form factor: 67 R Hamilton-1. KSFR relation: 66 L Kawarabayashi, 66 L Riazuddin. Further references are given in Sect. 2.5.1.5.

Books and reviews: 68 R Adler-Dashen, 68 R Renner, 69 R Schnitzer, 70 R Weinberg, 71 R Weinstein, 72 R Dashen, 72 R Lee, 73 R Alfaro-Fubini-Furlan-Rosetti, 74 R Reya, 75 R Pagels, 79 R Scadron, 81 R Scadron, 82 R Dominguez.

8.3 The dispersion model

Basic ideas: 57 L Chew (fixed-t dispersion relation, pole approximation for Δ-exchange), 59 L Mandelstam (subtraction of the lowest t-channel partial wave), 60 L Cini-Fubini (decomposition into elastic and inelastic contributions).

Improved versions: 60 L Höhler, 60 L Bowcock, 60 L Frazer-3, 62 L Carrassi, 62 L and 63 L Dietz, 68 L Höhler-1, 71 L Höhler-2.

Investigations and modifications of the static model of Chew and Low (56 L): 55 R Wick, 61 R Thirring, 62 R Henley-Thirring, 65 R Serebryakov-Shirkov, 69 R Shirkov-Serebryakov-Meshcheryakov, 74 R Meshcheryakov, 78 L Ernst, 80 L Ernst, 81 L McLeod, 81 L Alvarez-Estrada.

9. Regge models

The number of papers on Regge pole theory and its applications is very large. We quote some of the books and reviews: 63 R Frautschi, 63 R Omnes-Froissart, 63 R Squires, 64 R Oehme-1 and 2, 65 R and 66 R Chew, 66 R Leader, 66 R Omnes, 67 R Bertocchi, 67 R Eden-1, 67 R van Hove, 68 R Collins-Squires, 69 R Barger-Cline, 69 R Fox, 69 R Hite, 70 R Drechsler, 70 R Phillips, 71 R Collins-2, 71 R Oehme-1, 2, and 3, 71 R Squires, 72 R Chiu, 72 R Hara, 72 R Tavkhelidze, 73 R Fox-Quigg, 75 R Michael, 77 R Collins, 77 R Irving-Worden and two chapters in 82 R Collins-Martin.

There exists a bibliography of "Regge Phenomenology" (75 R Collins-Gault). The authors collected those papers whose main purpose is the quantitative comparison of high-energy two-body hadronic scattering data with Regge pole and cut models. It covers the period 1962–1974.

Regge formalism for relativistic particles with spin: 63 L Mandelstam-1; for pion-nucleon scattering: 63 L Singh, 66 L Desai.

Regge cuts and absorption: 63 L Mandelstam-2, 68 R Collins-Squires, 71 R Collins-1, 72 R Schrempp, 76 R Kane-Seidl, 77 R Collins, 77 R Irving-Worden.

Exchange of nucleons and of nucleon resonances: 67 R Eden-1, 69 R Barger-Cline, 71 R Collins, 77 R Irving-Worden.

Line-reversal: 67 L Barger, 69 R Barger-Cline.

An attempt to determine J-plane singularities of $\pi\pi N\bar{N}$ amplitudes: 72 L Pietarinen-1.

See also the references in Sects. 2.6.4.9 and 2.6.5.

10. Isospin bounds

A review: 72 R Roy. Bounds for a complete set of data: 72 L Doncel. Bounds for polarization parameters: 61 L Michel, 71 L Dass. Saturation of isospin bounds: 72 L Höhler-2 and 3.

11. Electromagnetic nucleon and pion form factors

Books and reviews: 61 R DRELL-ZACHARIASEN, 61 R FRAZER, 72 R BARTOLI-FELICETTI-SILVESTRINI, 74 R GOURDIN, 74 R RASZILLIER.

11.1 Nucleon form factors

Further books and reviews: 63 R HOFSTADTER, 66 R GOURDIN, 67 R WILSON, 68 R BERNSTEIN, 76 R HÖHLER.

Scattering of polarized electrons: 69 R DOMBEY.

Fundamental papers on the dispersion approach: 58 L CHEW, 58 L FEDERBUSH, 60 L FRAZER-FULCO-1, 62 L SACHS. The most recent evaluation: 76 L HÖHLER. Nucleon radii: 75 L HÖHLER-3 and 76 L HÖHLER, 80 L SIMON-2. FOLDY term: 58 R FOLDY. Review of vector dominance models: 75 R ZOVKO.

Sidewise dispersion relations: 60 L BINCER, 65 L DRELL, 68 L DRELL, 69 L LOVE, 73 L BLUVSTEIN, 74 L DEO, 74 L LEE. The predictive power is weaker than in the case of dispersion relations in t because of uncertainties of the input for some important intermediate states.

Determination of the spectral function by analytic continuation: 64 L LEVINGER-PEIERLS and other papers listed in 71 R PFISTER. The authors ignored a nearby singularity in the second sheet (see Sect. 2.5.2).

$e^+e^- \leftrightarrow N\bar{N}$: 61 L CABIBBO-GATTO.

11.2 The pion form factor

Further books and reviews: 79 R AMALDI-FUBINI-FURLAN, 79 R WIIK-WOLF. GOUNARIS-SAKURAI-fit: 60 L CHEW-MANDELSTAM, 60 L FRAZER-FULCO-1, 68 L and 69 L GOUNARIS, 74 R GOURDIN. A recent dispersion fit: 81 L HEYN-LANG.

Bounds: 74 L BONNEAU. Causality: 69 L ALVAREZ-ESTRADA.

11.3 Analytic continuation into the second sheet

61 L and 61 R OEHME, 65 R BARTON, 74 R RASZILLIER.

Subject index

Tables (T) and figures (F) of πN amplitudes

Tables (T) and figures (F) of πN cross sections and polarization parameters

Tables (T) and figures (F) of parameters of the πN and ππ systems

Coupling constants and decay widths

Reactions other than $\pi N \to \pi N$

$\pi\pi \leftrightarrow N\bar{N}$

$\pi\pi \leftrightarrow \pi\pi$

NN → NN

eN → eN

$e^+e^- \leftrightarrow N\bar{N}$

$e^+e^- \to \pi^+\pi^-$

$\gamma p \to \gamma p$ and $\gamma p \to VN$

$\gamma p \to N\pi$

$e^-p \to e^-n\pi^+$, $\pi^-p \to e^+e^-n$ and $e^-\pi \to e^-\pi$

Table of the kinematical variables

The formulas and the values used for the nucleon and pion mass are given in Table A.1.2

k MeV/c	T_π MeV	W GeV	s GeV^2	s/μ^2	$2q^2$ GeV^2	ω/μ	E GeV
20.	1.4	1.079	1.164	59.8	0.001	1.010	0.938
40.	5.6	1.083	1.172	60.2	0.002	1.040	0.939
60.	12.4	1.089	1.185	60.8	0.005	1.088	0.940
80.	21.3	1.096	1.202	61.7	0.009	1.153	0.941
100.	32.1	1.105	1.222	62.7	0.014	1.230	0.942
120.	44.5	1.116	1.245	63.9	0.020	1.319	0.944
140.	58.1	1.127	1.271	65.2	0.027	1.416	0.945
160.	72.8	1.139	1.298	66.6	0.035	1.521	0.947
180.	88.2	1.152	1.327	68.1	0.043	1.632	0.950
200.	104.3	1.165	1.358	69.7	0.052	1.747	0.952
220.	121.0	1.178	1.389	71.3	0.061	1.867	0.954
240.	138.1	1.192	1.421	72.9	0.071	1.989	0.957
260.	155.5	1.206	1.454	74.6	0.082	2.114	0.960
280.	173.3	1.219	1.487	76.3	0.093	2.242	0.963
300.	191.3	1.233	1.521	78.1	0.104	2.371	0.966
320.	209.5	1.247	1.555	79.8	0.116	2.501	0.969
340.	228.0	1.261	1.590	81.6	0.128	2.633	0.972
360.	246.5	1.275	1.624	83.4	0.140	2.766	0.975
380.	265.3	1.288	1.660	85.2	0.153	2.900	0.978
400.	284.1	1.302	1.695	87.0	0.166	3.035	0.982
420.	303.0	1.315	1.730	88.8	0.179	3.171	0.985
440.	322.0	1.329	1.766	90.7	0.193	3.307	0.988
460.	341.1	1.342	1.802	92.5	0.207	3.444	0.992
480.	360.3	1.356	1.838	94.3	0.221	3.582	0.995
500.	379.5	1.369	1.874	96.2	0.235	3.719	0.999
520.	398.8	1.382	1.910	98.1	0.249	3.858	1.002
540.	418.2	1.395	1.946	99.9	0.264	3.996	1.006
560.	437.6	1.408	1.983	101.8	0.278	4.135	1.010
580.	457.0	1.421	2.019	103.7	0.293	4.274	1.013
600.	476.4	1.434	2.056	105.5	0.308	4.414	1.017
620.	495.9	1.447	2.092	107.4	0.323	4.553	1.021
640.	515.5	1.459	2.129	109.3	0.339	4.693	1.025
660.	535.0	1.472	2.166	111.2	0.354	4.833	1.028
680.	554.6	1.484	2.203	113.1	0.370	4.974	1.032
700.	574.2	1.496	2.239	115.0	0.385	5.114	1.036
720.	593.8	1.509	2.276	116.8	0.401	5.255	1.040
740.	613.5	1.521	2.313	118.7	0.417	5.395	1.043

k MeV/c	T_π MeV	W GeV	s GeV^2	s/μ^2	$2q^2$ GeV^2	ω/μ	E GeV
760.	633.1	1.533	2.350	120.6	0.433	5.536	1.047
780.	652.8	1.545	2.387	122.5	0.449	5.677	1.051
800.	672.5	1.557	2.424	124.4	0.465	5.818	1.055
820.	692.2	1.569	2.461	126.3	0.481	5.960	1.059
840.	711.9	1.580	2.498	128.2	0.497	6.101	1.063
860.	731.7	1.592	2.535	130.1	0.514	6.242	1.066
880.	751.4	1.604	2.572	132.0	0.530	6.384	1.070
900.	771.2	1.615	2.609	133.9	0.547	6.525	1.074
920.	791.0	1.627	2.646	135.8	0.563	6.667	1.078
940.	810.7	1.638	2.683	137.7	0.580	6.809	1.082
960.	830.5	1.649	2.720	139.6	0.597	6.951	1.086
980.	850.3	1.661	2.757	141.6	0.613	7.092	1.089
1000.	870.1	1.672	2.795	143.5	0.630	7.234	1.093
1020.	889.9	1.683	2.832	145.4	0.647	7.376	1.097
1040.	909.8	1.694	2.869	147.3	0.664	7.518	1.101
1060.	929.6	1.705	2.906	149.2	0.681	7.660	1.105
1080.	949.4	1.716	2.943	151.1	0.698	7.802	1.109
1100.	969.2	1.726	2.981	153.0	0.715	7.945	1.113
1120.	989.1	1.737	3.018	154.9	0.732	8.087	1.116
1140.	1008.9	1.748	3.055	156.8	0.749	8.229	1.120
1160.	1028.8	1.759	3.092	158.7	0.766	8.371	1.124
1180.	1048.7	1.769	3.130	160.7	0.783	8.513	1.128
1200.	1068.5	1.780	3.167	162.6	0.801	8.656	1.132
1250.	1118.2	1.806	3.260	167.4	0.844	9.012	1.141
1300.	1167.9	1.831	3.353	172.1	0.887	9.368	1.151
1350.	1217.6	1.857	3.447	176.9	0.931	9.724	1.160
1400.	1267.4	1.882	3.540	181.7	0.975	10.081	1.170
1450.	1317.1	1.906	3.633	186.5	1.019	10.437	1.179
1500.	1366.9	1.931	3.727	191.3	1.063	10.794	1.188
1550.	1416.7	1.955	3.820	196.1	1.107	11.150	1.198
1600.	1466.5	1.978	3.914	200.9	1.152	11.507	1.207
1650.	1516.3	2.002	4.007	205.7	1.196	11.864	1.216
1700.	1566.1	2.025	4.101	210.5	1.241	12.221	1.225
1750.	1616.0	2.048	4.194	215.3	1.286	12.578	1.234
1800.	1665.8	2.071	4.288	220.1	1.330	12.935	1.243
1850.	1715.7	2.093	4.381	224.9	1.375	13.293	1.252
1900.	1765.5	2.115	4.475	229.7	1.420	13.650	1.261